Jan William S…

CALIFORNIA POLYTECHN… …RSITY

Solutions Manual

L. G. WADE, JR.

WHITMAN COLLEGE

ORGANIC CHEMISTRY

SECOND EDITION

PRENTICE HALL, ENGLEWOOD CLIFFS, NEW JERSEY 07632

Production Editor: John A. Nestor
Manufacturing Buyers: Paula Massenaro and Lori Bulwin
Supplements Acquisitions Editor: Alison Munoz
Acquisitions Editor: Dan Joraanstad

Printed in the United States of America

10 9 8 7

ISBN 0-13-643362-6

Prentice-Hall International (UK) Limited, *London*
Prentice-Hall of Australia Pty. Limited, *Sydney*
Prentice-Hall Canada Inc., *Toronto*
Prentice-Hall Hispanoamericana, S.A., *Mexico*
Prentice-Hall of India Private Limited, *New Delhi*
Prentice-Hall of Japan, Inc., *Tokyo*
Simon & Schuster Asia Pte. Ltd., *Singapore*
Editora Prentice-Hall do Brasil, Ltda., *Rio de Janeiro*

CONTENTS

PREFACE

Including "Hints for Passing Organic Chemistry"

Do you want to pass your course in organic chemistry? Here is my best advice, based on twenty years of observing students learning organic chemistry:

Hint #1: *Do the problems*. Seems straightforward, but humans, including students, try to take the easy way out until they discover there is no short-cut. Unless you have a measured IQ above 200 and comfortably cruise in the top 1% of your class, *do the problems*. Usually your teacher (professor or teaching assistant) will recommend certain ones; try to do all those recommended. If you do half of them, you will be half-prepared come test time. (Do you want your surgeon coming to your appendectomy having practiced only *half* the procedure?) And when you do the problems, keep this Solutions Manual CLOSED. Avoid looking at *my* answer before you write *your* answer—your trying and struggling with the problem is the most valuable part of the problem. Remember that the primary goal of doing these problems is *not* just getting the right answer, but understanding the material well enough to get right answers to the questions you haven't seen yet.

Hint #2: *Keep up*. Getting behind in your work in a course that moves as quickly as this one is the Kiss of Death. For most students, organic chemistry is the most rigorous intellectual challenge they have faced so far in their studies. Some are taken by surprise at the diligence it requires. Don't think that you can study all of the material the couple of days before the exam—well, you can, but you won't pass. Study organic chemistry like a foreign language: try to do some every day so that the freshly-trained neurons stay sharp.

Hint #3: *Get help when you need it*. Use your teacher's office hours when you have difficulty. Many schools have tutoring centers (in which organic chemistry is a popular offering). Here's a secret: absolutely the best way to cement this material in your brain is to get together with a few of your fellow students and make up problems for each other, then correct and discuss them. When *you* write the problems, you will gain great insight into what this is all about.

So what is the point of this Solutions Manual? First, I can't do your studying for you. Second, since I am not leaning over your shoulder as you write your answers, I can't give you direct feedback on what you write and think—the print medium is limited in its usefulness. What I *can* do for you is: 1) provide correct answers; the publishers, Professor Wade, Professor Mattern (my reviewer), and I have gone to great lengths to assure that what I have written is correct, for we all understand how it can shake a student's confidence to discover that the answer book flubbed up; 2) provide a considerable degree of rigor; beyond the fundamental requirement of correctness, I have tried to flesh out these answers, being complete but succinct; 3) provide insight into how to solve a problem and into where the sticky intellectual points are. Insight is the toughest to accomplish, but over the years, I have come to understand where students have trouble, so I have tried to anticipate your questions and to add enough detail so that the concept, as well as the answer, is clear.

It is difficult for students to understand or acknowledge that their teachers are human (some are more human than others). Since I am human (mostly), I can and do make mistakes. If there are mistakes in this book, they are my sole responsibility, and I am sorry. If you find one, PLEASE let me know so that it can be corrected in future printings. Nip it in the bud.

Acknowledgements

No project of this scope is ever done alone. These are team efforts, and there are several people who have assisted and facilitated in one fashion or another who deserve my thanks.

Professor L. G. Wade is a remarkable person. He has gone to extraordinary lengths to make the textbook as clear, organized, informative and insightful as possible. He has solicited and followed my suggestions on his text, and his comments on my solutions have been perceptive and valuable. We agreed early on that our primary goal is to help the students learn a fascinating and challenging subject, and all of our efforts have been directed toward that goal. I have appreciated our collaboration.

Professor Daniell Mattern has reviewed every problem and every solution. His care, diligence, precision and clarity have contributed beyond measure to the quality of this book. Once again, I have relied on Dan to set me straight. As a co-worker and long-time friend, he is the best.

The people at Prentice-Hall have made this project possible. Good books would not happen without their dedication, professionalism, and experience. Among the many people who contributed are: Lee Englander, who connected me with this project; Dan Joraanstad, Chemistry Editor; Alison Munoz, Science and Math Supplements Editor; Deborah Wechsler, Production Editor; John Nestor, production; and assistants Lynne Breitfeller and Lisa Tarabokjia.

Finally, I appreciate my friends who supported me throughout this project: Dave, Big John, Little John, Barb, Cheryl, Michael, Norma, Tom, Ashley, Sara, Melissa, Kendra, Dane and Cathy. The students are too numerous to list, but it is for them that all this happens.

Jan William Simek
Chemistry Department
Cal Poly State University
San Luis Obispo, CA 93407

DEDICATION

To my inspirational chemistry teachers:

Joe Plaskas, who made the batter;

Kurt Kaufman, who baked the cake;

Carl Djerassi, who put on the icing;

and to my parents:

Ervin J. and Imilda B. Simek,

who had the original concept.

SYMBOLS AND ABBREVIATIONS

Below is a list of symbols and abbreviations used in this Solutions Manual, consistent with those used in the textbook by Wade. (Do not expect all of these to make sense to you now. You will learn them throughout your study of organic chemistry.)

BONDS

—	a single bond
=	a double bond
≡	a triple bond
	a bond in three dimensions, coming out of the paper toward the reader
	a bond in three dimensions, going behind the paper away from the reader
------	a stretched bond, in the process of forming or breaking

ARROWS

⟶	in a reaction, shows direction from reactants to products
⇌	signifies equilibrium (not to be confused with resonance)
⟷	signifies resonance (not to be confused with equilibrium)
	shows direction of electron movement: the arrowhead with one barb shows movement of one electron; the arrowhead with two barbs shows movement of a pair of electrons
+⟶	shows polarity of a bond or molecule, the arrowhead signifying the more negative end of the dipole

SUBSTITUENT GROUPS

Me	a methyl group, CH_3
Et	an ethyl group, CH_2CH_3
Pr	a propyl group, a three carbon group (two possible arrangements)
Bu	a butyl group, a four carbon group (four possible arrangements)
R	the general abbreviation for an alkyl group (or any substituent group not under scrutiny)

CONTINUED ON NEXT PAGE

SUBSTITUENT GROUPS, continued

Ph a phenyl group, the name of a benzene ring as a substituent, represented:

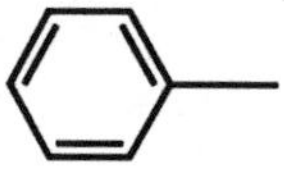

or

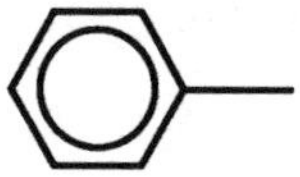

Ar the general name for an aromatic group

Ac an acetyl group:

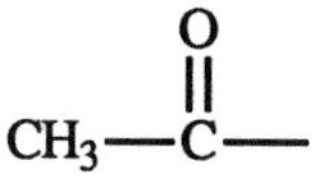

Cy a cyclohexyl group:

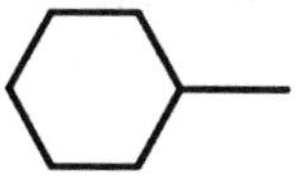

Ts tosyl, or *p*-toluenesulfonyl group:

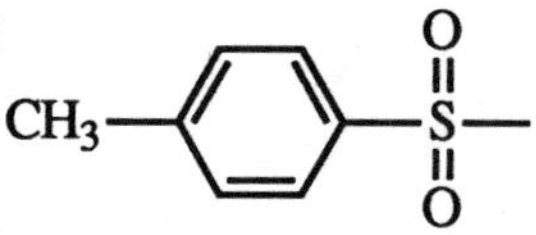

Boc a *t*-butoxycarbonyl group (amino acid and peptide chemistry):

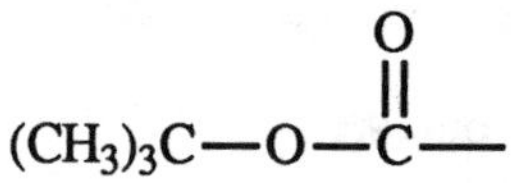

Z, or Cbz a carbobenzoxy (benzyloxycarbonyl) group (amino acid and peptide chemistry):

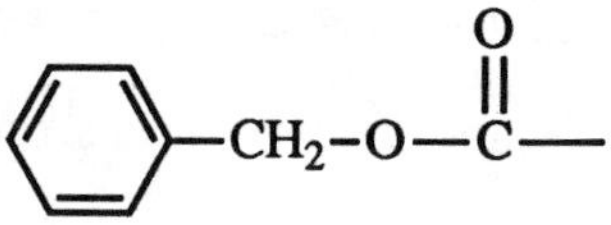

REAGENTS AND SOLVENTS

DCC **dicyclohexylcarbodiimide**

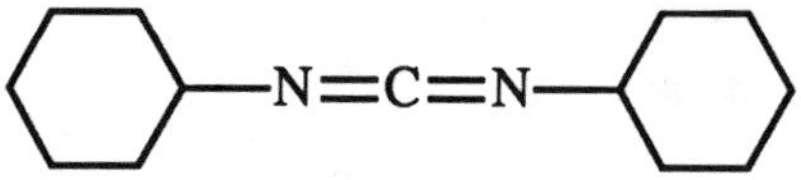

DMSO **dimethylsulfoxide**

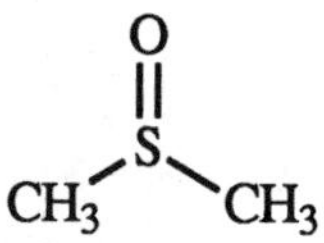

CONTINUED ON NEXT PAGE

Symbols and Abbreviations, continued

REAGENTS AND SOLVENTS, continued

ether	diethyl ether, $CH_3CH_2OCH_2CH_3$
MCPBA	*meta*-chloroperoxybenzoic acid 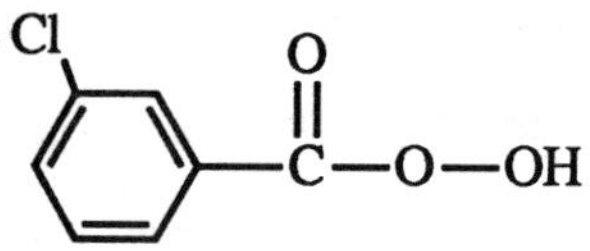
MVK	methyl vinyl ketone 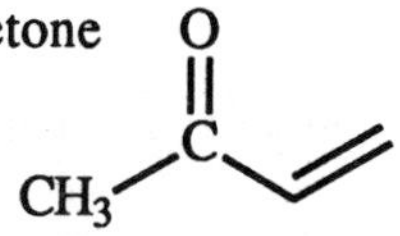
NBS	*N*-bromosuccinimide 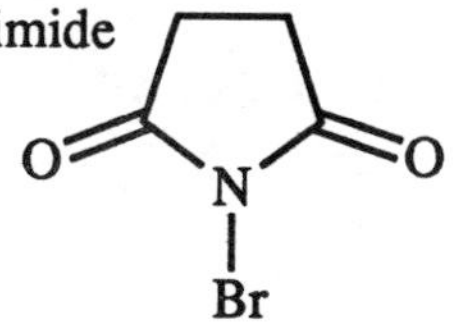
PCC	pyridinium chlorochromate, $CrO_3 \cdot HCl \cdot$ 
Sia_2BH	disiamylborane 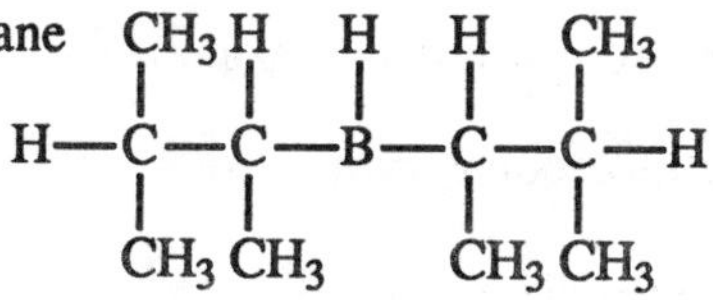
THF	tetrahydrofuran 

SPECTROSCOPY

IR	infrared spectroscopy
NMR	nuclear magnetic resonance spectroscopy
MS	mass spectrometry
UV	ultraviolet spectroscopy
ppm	parts per million, a unit used in NMR

CONTINUED ON NEXT PAGE

Symbols and Abbreviations, continued

SPECTROSCOPY, continued

Hz	hertz, cycles per second, a unit of frequency
MHz	megahertz, millions of cycles per second
TMS	**t**etra**m**ethyl**s**ilane, $(CH_3)_4Si$, the reference compound in NMR
s, d, t, q	**s**inglet, **d**oublet, **t**riplet, **q**uartet, referring to the number of peaks an NMR absorption gives
nm	nanometers, 10^{-9} meters (usually used as a unit of wavelength)
m/z	mass-to-charge ratio, in mass spectrometry
δ	in NMR, chemical shift value, measured in ppm
λ	wavelength
ν	frequency

OTHER

a, ax	axial (in chair forms of cyclohexane)
e, eq	equatorial (in chair forms of cyclohexane)
HOMO	**h**ighest **o**ccupied **m**olecular **o**rbital
LUMO	**l**owest **u**noccupied **m**olecular **o**rbital
NR	no reaction
o, m, p	*ortho, meta, para* (positions on an aromatic ring)
Δ	when written over an arrow: "heat" when written before a letter: "change in"
δ^+, δ^-	partial positive charge, partial negative charge
hν	energy from electromagnetic radiation (light)
$[\alpha]_D$	specific rotation at the D line of sodium (589 nm)

1-1 Na $1s^22s^22p^63s^1$

Mg $1s^22s^22p^63s^2$

Al $1s^22s^22p^63s^23p_x^{\,1}$

Si $1s^22s^22p^63s^23p_x^{\,1}3p_y^{\,1}$

P $1s^22s^22p^63s^23p_x^{\,1}3p_y^{\,1}3p_z^{\,1}$

S $1s^22s^22p^63s^23p_x^{\,2}3p_y^{\,1}3p_z^{\,1}$

Cl $1s^22s^22p^63s^23p_x^{\,2}3p_y^{\,2}3p_z^{\,1}$

Ar $1s^22s^22p^63s^23p_x^{\,2}3p_y^{\,2}3p_z^{\,2}$

1-2 In this book, lines between atomic symbols represent covalent bonds between those atoms. Nonbonding electrons are indicated with heavy dots.

(a) 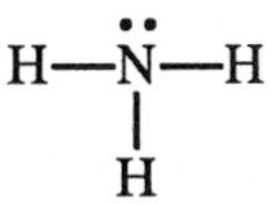

(b) 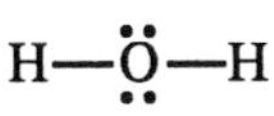

(c) 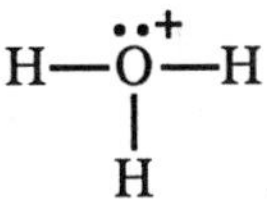

(d) 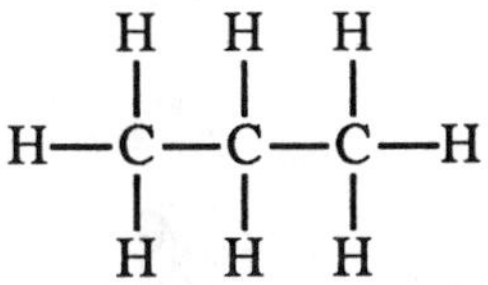

(e) 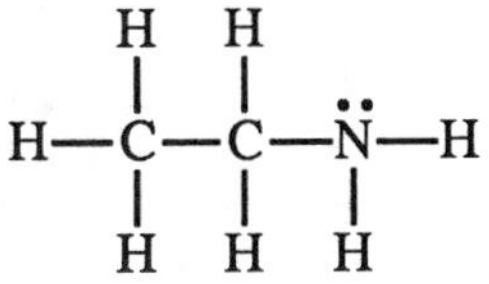

(f) 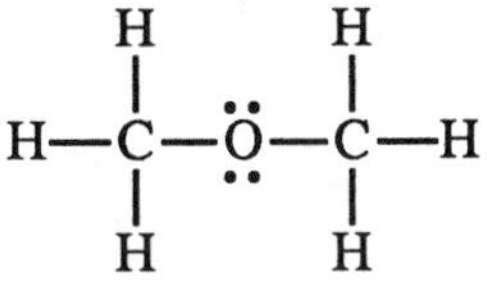

(g) 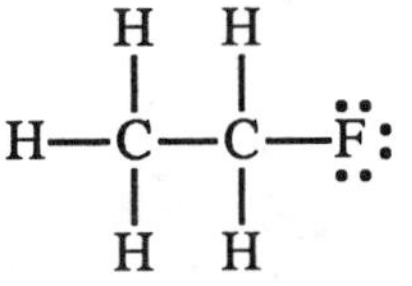

(h) 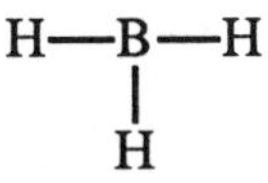

(i) 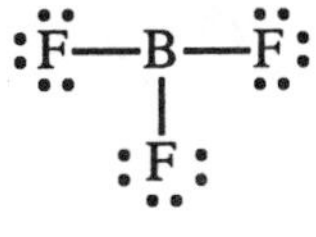

The compounds in (h) and (i) are unusual in that boron does not have an octet of electrons.

1-3 (a)  (b)  (c) 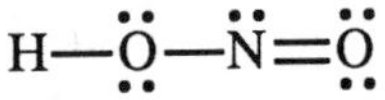

(d) 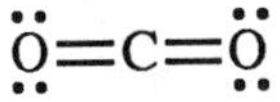 (e) 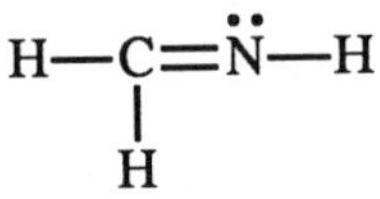 (f) 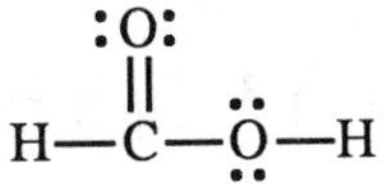

(g) 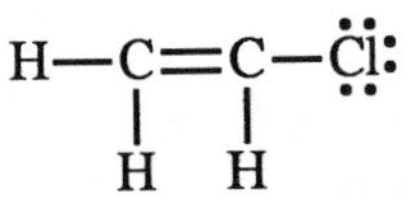 (h) 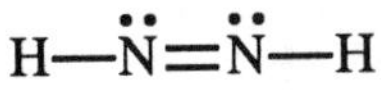 (i) 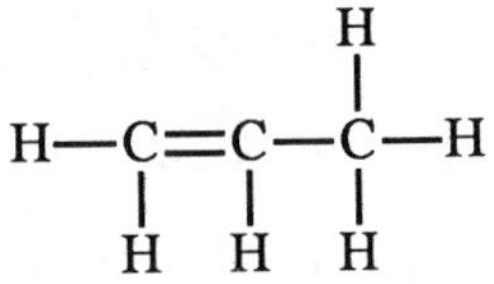

(j) 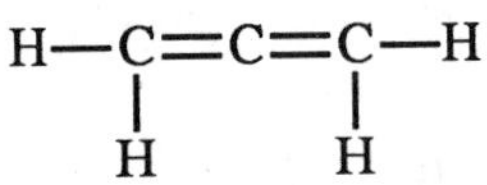 (k) 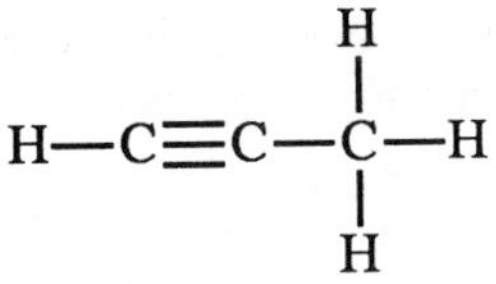

1-4 (a) 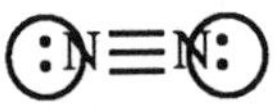 (b) 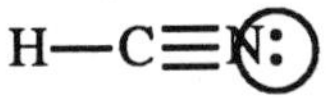 (c) 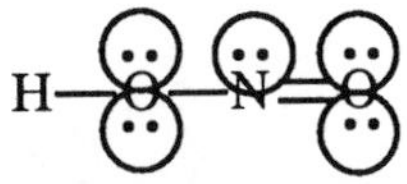

(d) 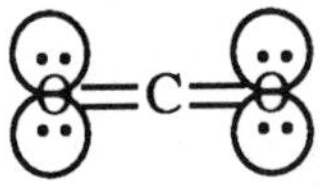 (e) 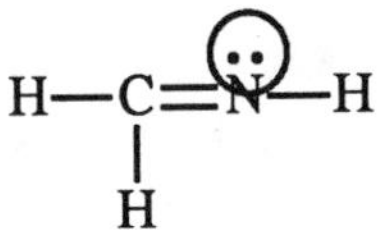 (f) 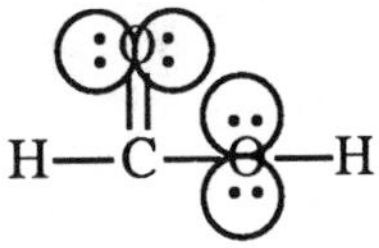

(g) 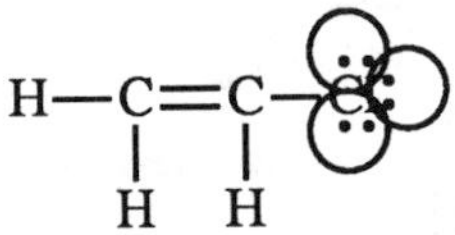 (h) 

1-5 The symbols " δ^+ " and " δ^- " indicate bond polarity by showing partial charge. (In the arrow symbolism, the arrow should point to the partial negative charge.)

(a) $\overset{\delta^+}{C}-\overset{\delta^-}{Cl}$ (b) $\overset{\delta^+}{C}-\overset{\delta^-}{O}$ (c) $\overset{\delta^+}{C}-\overset{\delta^-}{N}$ (d) $\overset{\delta^+}{C}-\overset{\delta^-}{S}$

(e) $\overset{\delta^-}{C}-\overset{\delta^+}{B}$ (f) $\overset{\delta^+}{N}-\overset{\delta^-}{Cl}$ (g) $\overset{\delta^+}{N}-\overset{\delta^-}{O}$ (h) $\overset{\delta^-}{N}-\overset{\delta^+}{S}$

(i) $\overset{\delta^-}{N}-\overset{\delta^+}{B}$ (j) $\overset{\delta^+}{B}-\overset{\delta^-}{Cl}$

1-6 Non-zero formal charges are shown beside the atoms.

(a) 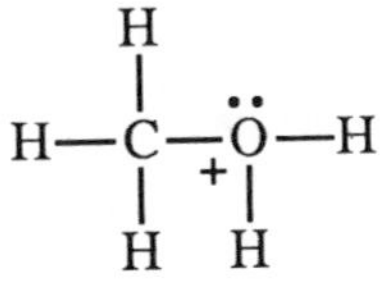

(b) 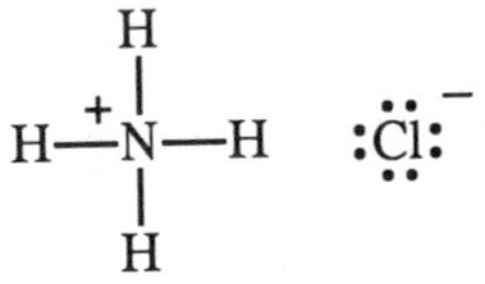

(c) 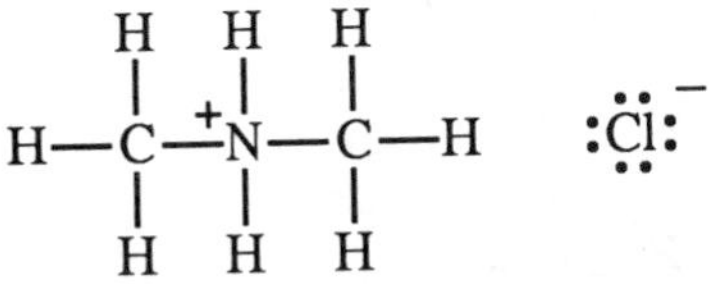

(d) 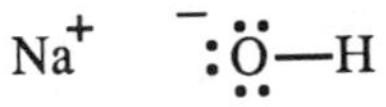

(e) 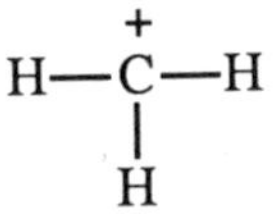

(f) 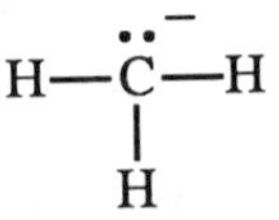

(g) 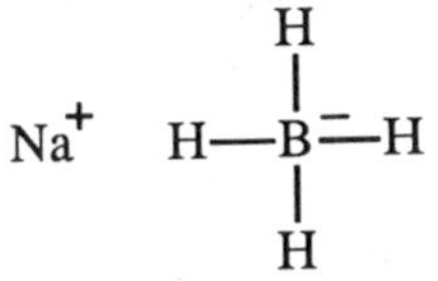

(h) 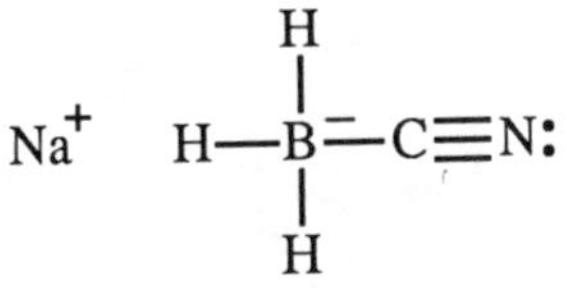

(i) 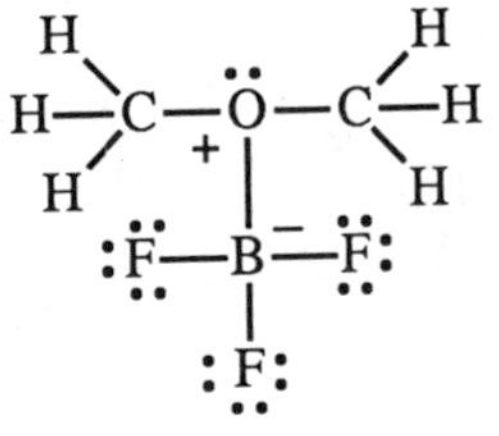

(j) 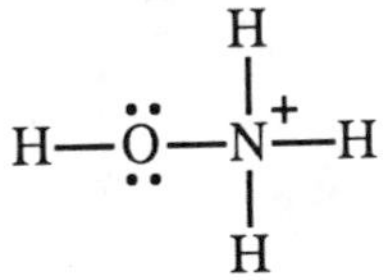

(k) 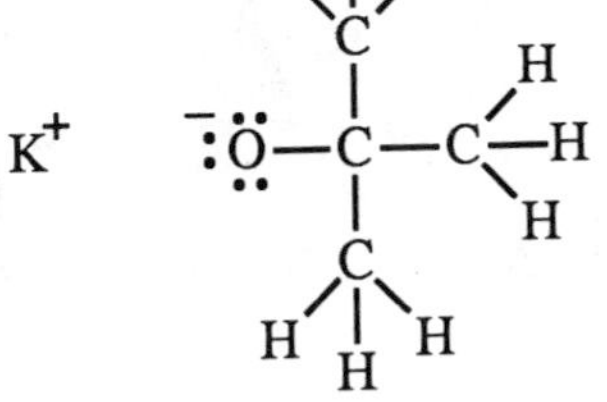

(l) 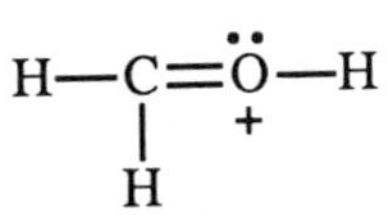

1-7 Resonance structures in which all atoms have full octets are the most significant contributors. Remember that, in resonance forms, ATOMS KEEP THEIR POSITIONS—ONLY ELECTRONS ARE SHOWN IN DIFFERENT POSITIONS.

(a) 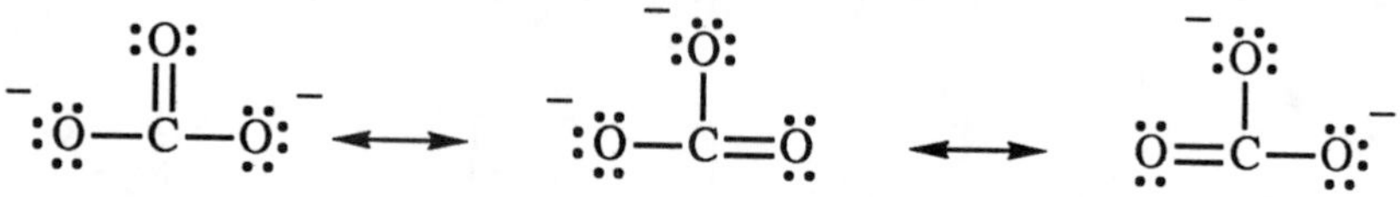

(b) 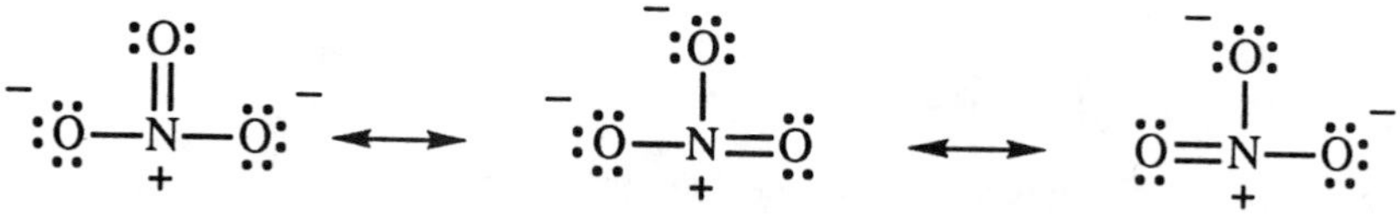

(c) 

(d) 

(e) 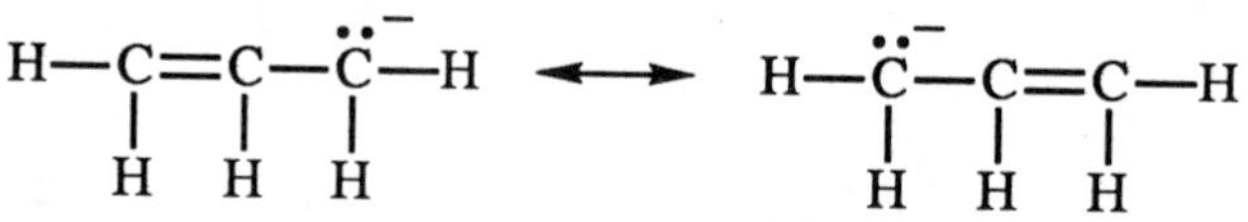

(f) Sulfur can have up to 12 electrons around it.

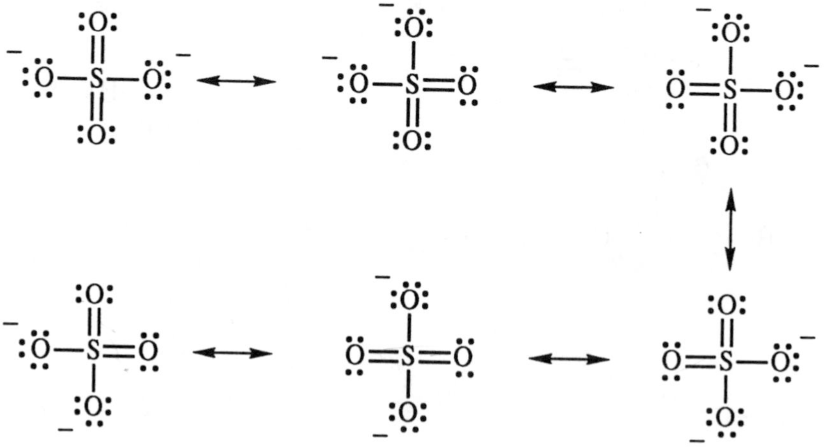

1-8 Major resonance contributors will have the lowest energy.

(a)

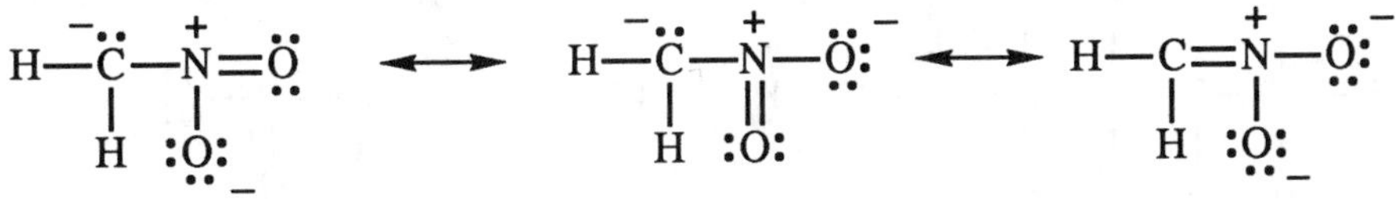

minor — minor — major (negative charge on electronegative atoms)

(b)

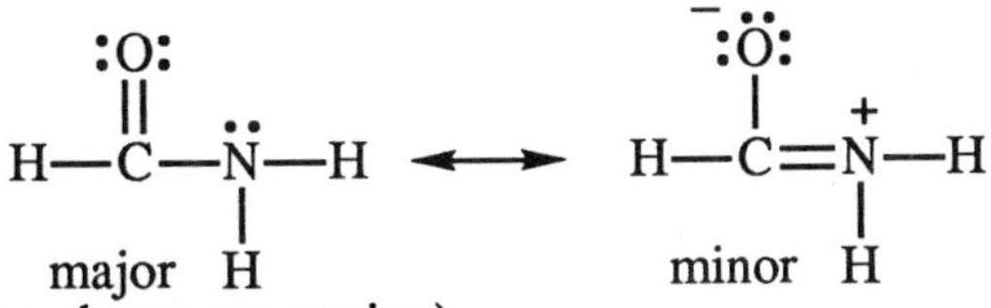

major (no charge separation) — minor

(c)

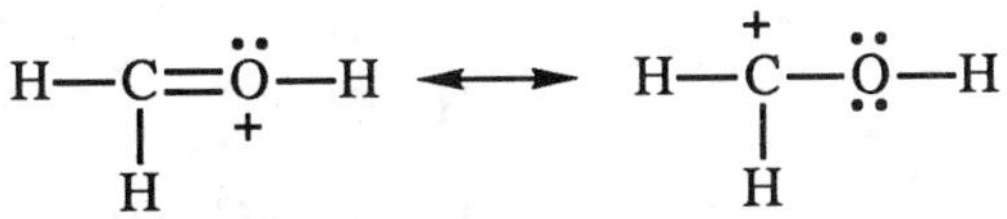

major (octets) — minor

(d)

major (negative charge on electronegative atom) — minor

(e)

minor — major (negative charge on electronegative atom)

(f)

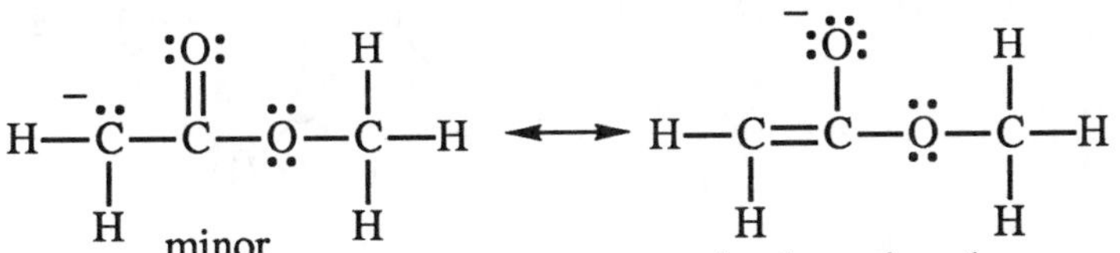

minor — major (negative charge on electronegative atom)

(g)

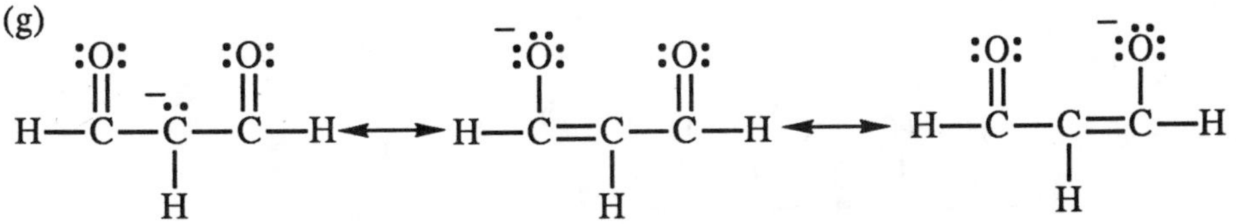

minor — major — major

these two have equivalent energy

1-9

(a) 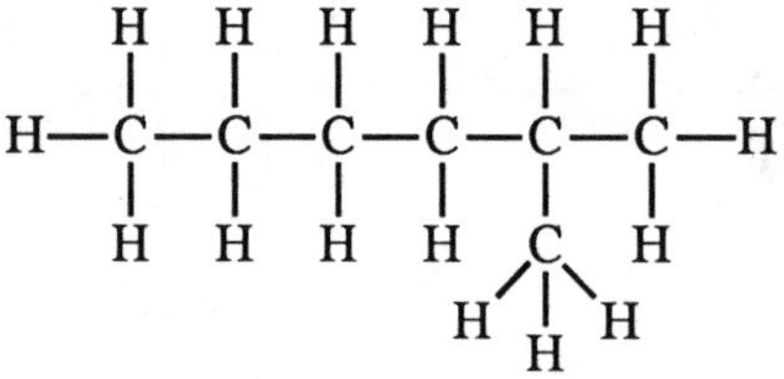

(b) 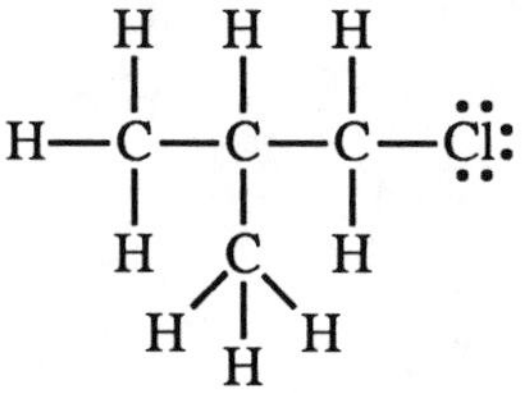

(c) 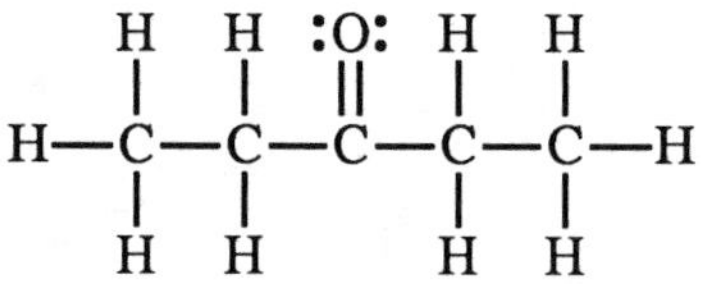

(d) 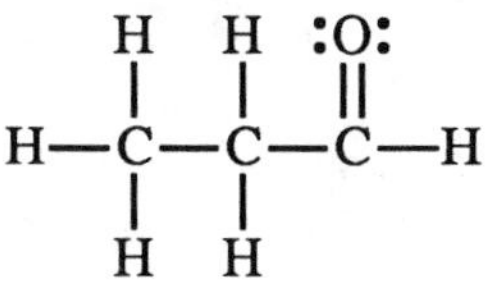

(e) 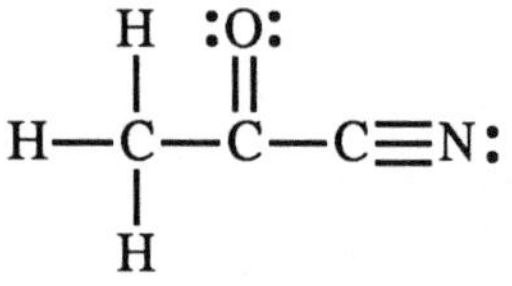

(f) 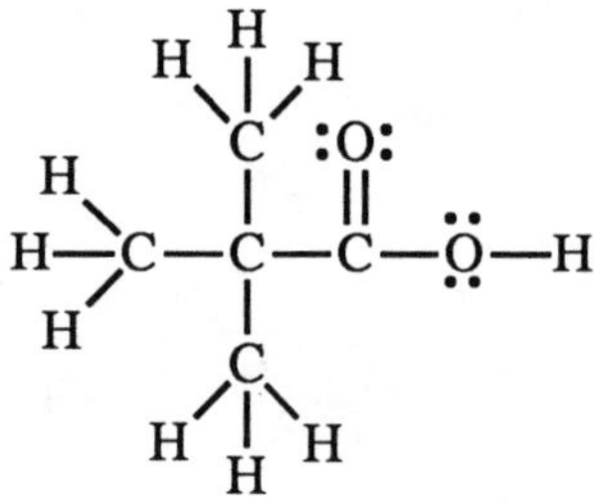

(g) (same as (c))

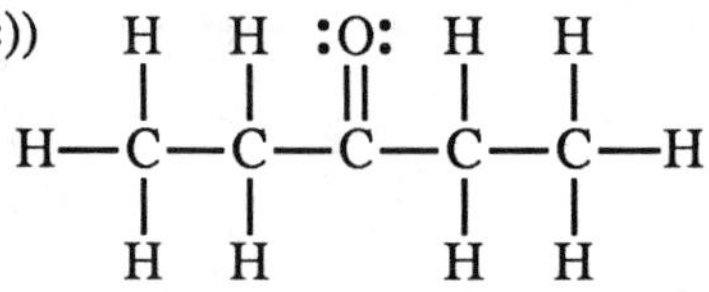

1-10

(a)

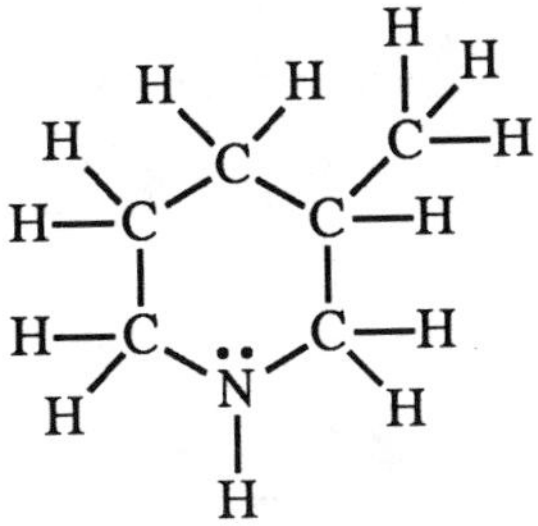

(b)

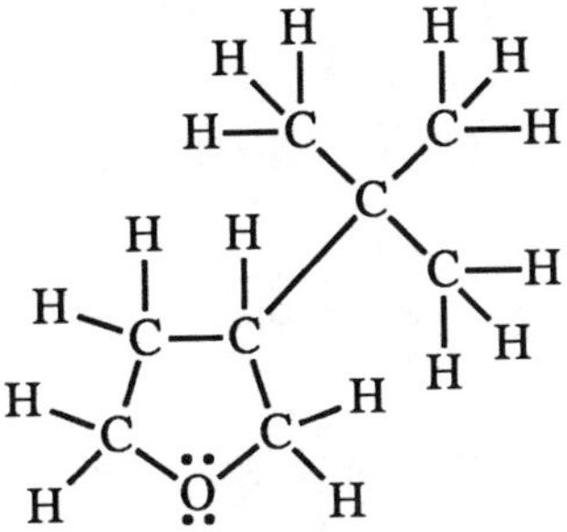

(c)

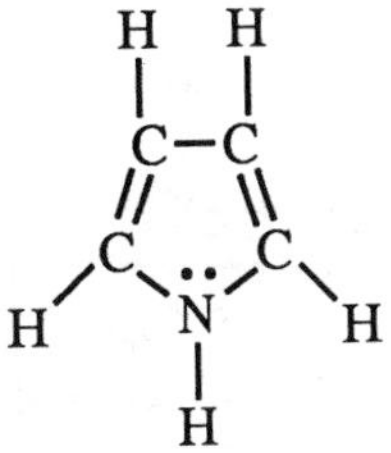

(d)

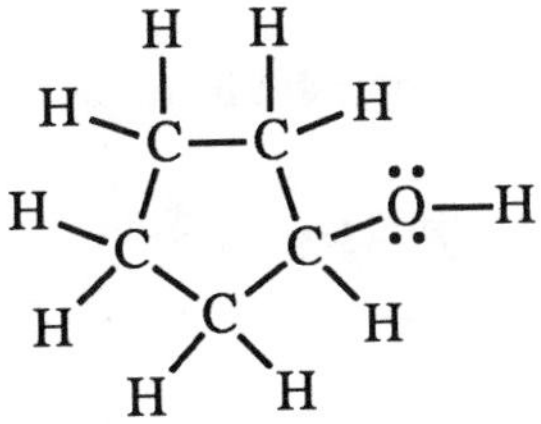

(e)

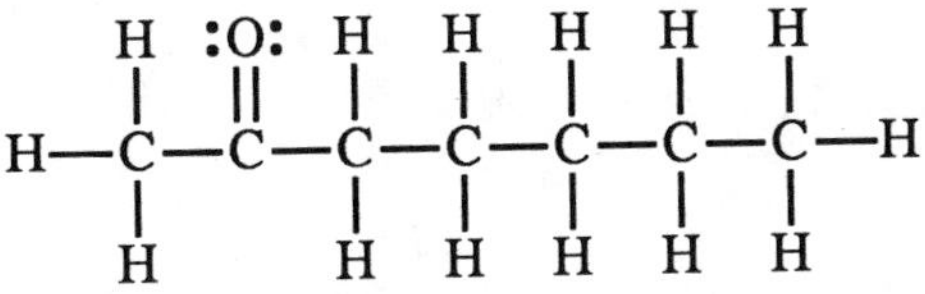

(f)

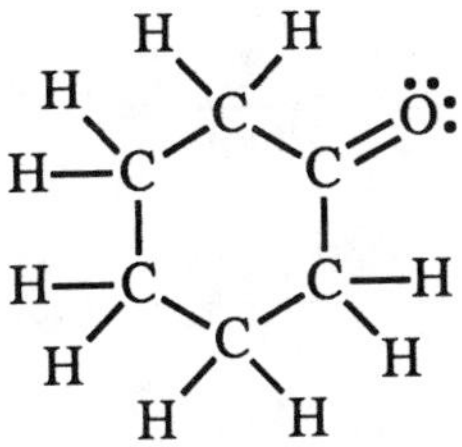

(g)

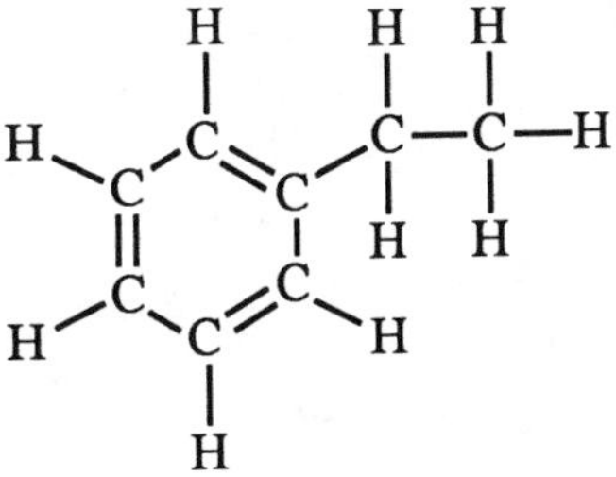

1-11 If the percent values do not sum to 100%, the remainder must be oxygen. Assume 100 g of sample; percents then translate directly to grams of each element.

There are usually many possible structures for a molecular formula. Yours may be different from the examples shown here.

(a) $$\frac{40.0 \text{ g C}}{12.0 \text{ g/mole}} = 3.33 \text{ moles C} \div 3.33 \text{ moles} = 1 \text{ C}$$

$$\frac{6.67 \text{ g H}}{1.01 \text{ g/mole}} = 6.60 \text{ moles H} \div 3.33 \text{ moles} = 1.98 \approx 2 \text{ H}$$

$$\frac{53.33 \text{ g O}}{16.0 \text{ g/mole}} = 3.33 \text{ moles O} \div 3.33 \text{ moles} = 1 \text{ O}$$

empirical formula = CH_2O ⇨ empirical weight = 30

molecular weight = 90, three times the empirical weight ⇨

three times the empirical formula = molecular formula = $C_3H_6O_3$

possible structure:

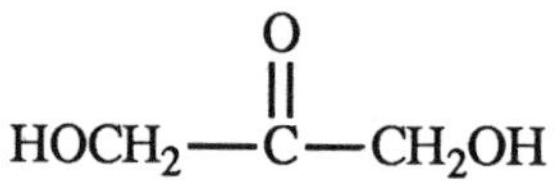

(b) $$\frac{32.0 \text{ g C}}{12.0 \text{ g/mole}} = 2.67 \text{ moles C} \div 1.34 \text{ moles} = 1.99 \approx 2 \text{ C}$$

$$\frac{6.67 \text{ g H}}{1.01 \text{ g/mole}} = 6.60 \text{ moles H} \div 1.34 \text{ moles} = 4.93 \approx 5 \text{ H}$$

$$\frac{18.7 \text{ g N}}{14.0 \text{ g/mole}} = 1.34 \text{ moles N} \div 1.34 \text{ moles} = 1 \text{ N}$$

$$\frac{42.6 \text{ g O}}{16.0 \text{ g/mole}} = 2.66 \text{ moles O} \div 1.34 \text{ moles} = 1.99 \approx 2 \text{ O}$$

empirical formula = $C_2H_5NO_2$ ⇨ empirical weight = 75

molecular weight = 75, same as the empirical weight ⇨

empirical formula = molecular formula = $C_2H_5NO_2$

possible structure:

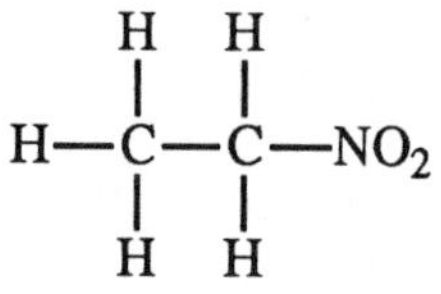

1-11 (c) $\frac{37.2\text{ g C}}{12.0\text{ g/mole}} = 3.10\text{ moles C} \div 1.55\text{ moles} = 2\text{ C}$

$\frac{7.75\text{ g H}}{1.01\text{ g/mole}} = 7.67\text{ moles H} \div 1.55\text{ moles} = 4.95 \approx 5\text{ H}$

$\frac{55.0\text{ g Cl}}{35.45\text{ g/mole}} = 1.55\text{ moles Cl} \div 1.55\text{ moles} = 1\text{ Cl}$

empirical formula = **C_2H_5Cl** ⇨ empirical weight = 64.46

molecular weight = 64, same as the empirical weight ⇨

empirical formula = molecular formula = **C_2H_5Cl**

only structure:

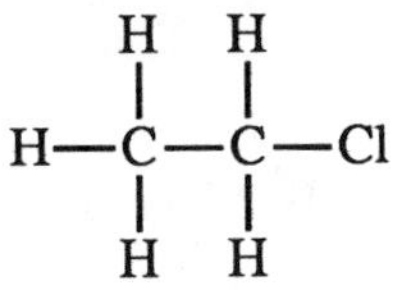

(d) $\frac{38.4\text{ g C}}{12.0\text{ g/mole}} = 3.20\text{ moles C} \div 1.60\text{ moles} = 2\text{ C}$

$\frac{4.80\text{ g H}}{1.01\text{ g/mole}} = 4.75\text{ moles H} \div 1.60\text{ moles} = 2.97 \approx 3\text{ H}$

$\frac{56.8\text{ g Cl}}{35.45\text{ g/mole}} = 1.60\text{ moles Cl} \div 1.60\text{ moles} = 1\text{ Cl}$

empirical formula = **C_2H_3Cl** ⇨ empirical weight = 62.45

molecular weight = 125, twice the empirical weight ⇨

twice the empirical formula = molecular formula = **$C_4H_6Cl_2$**

possible structure:

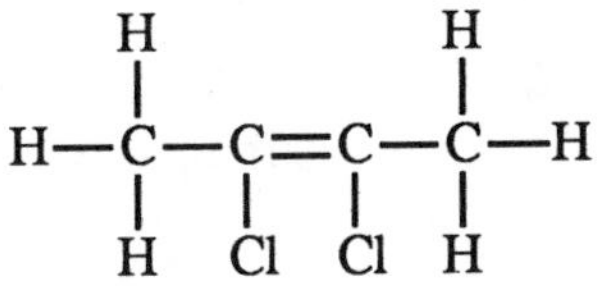

1-12 (a)

$$5.00 \text{ g HBr} \times \frac{1 \text{ mole HBr}}{80.9 \text{ g HBr}} = 0.0618 \text{ moles HBr}$$

$$0.0618 \text{ moles HBr} \Longrightarrow 0.0618 \text{ moles } H_3O^+ \text{ (100\% dissociated)}$$

$$\frac{0.0618 \text{ moles } H_3O^+}{100 \text{ mL}} \times \frac{1000 \text{ mL}}{1 \text{ L}} = \frac{0.618 \text{ moles } H_3O^+}{1 \text{ L solution}}$$

$$pH = -\log_{10}[H_3O^+] = -\log_{10}(0.618) = \boxed{0.209}$$

(b)

$$2.00 \text{ g NaOH} \times \frac{1 \text{ mole NaOH}}{40.0 \text{ g NaOH}} = 0.0500 \text{ moles NaOH}$$

$$0.0500 \text{ moles NaOH} \Longrightarrow 0.0500 \text{ moles } ^-OH \text{ (100\% dissociated)}$$

$$\frac{0.0500 \text{ moles } ^-OH}{50. \text{ mL}} \times \frac{1000 \text{ mL}}{1 \text{ L}} = \frac{1.0 \text{ moles } ^-OH}{1 \text{ L solution}}$$

$$[H_3O^+] = \frac{10^{-14}}{[^-OH]} = \frac{10^{-14}}{1.0} = 1.0 \times 10^{-14}$$

$$pH = -\log_{10}[H_3O^+] = -\log_{10}(1.0 \times 10^{-14}) = \boxed{14.0}$$

1-13

(a) $HCOOH + {}^-CN \rightleftharpoons HCOO^- + HCN$ FAVORS PRODUCTS

HCOOH		^-CN		$HCOO^-$		HCN	
stronger acid, pK_a 3.76	+	stronger base	⇌	weaker base	+	weaker acid, pK_a 9.22	FAVORS PRODUCTS

(b) $CH_3COO^- + CH_3OH \rightleftharpoons CH_3COOH + CH_3O^-$ FAVORS REACTANTS

CH_3COO^-		CH_3OH		CH_3COOH		CH_3O^-	
weaker base	+	weaker acid, pK_a 15.5	⇌	stronger acid, pK_a 4.74	+	stronger base	FAVORS REACTANTS

(c) $CH_3OH + NaNH_2 \rightleftharpoons CH_3O^- \, Na^+ + NH_3$ FAVORS PRODUCTS

CH_3OH		$NaNH_2$		$CH_3O^- \, Na^+$		NH_3	
stronger acid, pK_a 15.5	+	stronger base	⇌	weaker base	+	weaker acid, pK_a 33	FAVORS PRODUCTS

1-13 continued

(d) $Na^+ \ ^-OCH_3$ + HCN ⇌ $HOCH_3$ + NaCN — FAVORS PRODUCTS

$Na^+ \ ^-OCH_3$	+ HCN	⇌ $HOCH_3$	+ NaCN	FAVORS PRODUCTS
stronger base	stronger acid pK_a 9.22	weaker acid pK_a 15.5	weaker base	

(e) HCl + H_2O ⇌ H_3O^+ + Cl^- — FAVORS PRODUCTS

HCl	+ H_2O	⇌ H_3O^+	+ Cl^-	FAVORS PRODUCTS
stronger acid	stronger base	weaker acid	weaker base	

(f) H_3O^+ + CH_3O^- ⇌ H_2O + CH_3OH — FAVORS PRODUCTS

H_3O^+	+ CH_3O^-	⇌ H_2O	+ CH_3OH	FAVORS PRODUCTS
stronger acid	stronger base	weaker base	weaker acid	

1-14

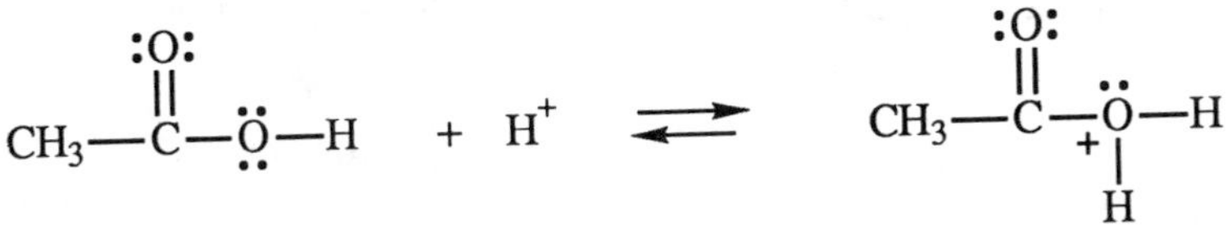

Protonation of the double-bonded oxygen gives three resonance forms (as shown in Solved Problem 1-5); protonation of the single-bonded oxygen gives only one. In general, the more resonance forms a species has, the more stable it is, so the proton would bond to the oxygen that gives a more stable species, that is, the double-bonded oxygen.

1-15 In Solved Problem 1-4, the structures of ethanol and methylamine are shown to be similar to methanol and ammonia, respectively. We must infer that their acid-base properties are also be similar.

(a) This problem can be viewed in two ways. 1) Quantitatively, the pK_a values determine the order of acidity. 2) Qualitatively, the stabilities of the conjugate bases determine the order of acidity (see Solved Problem 1-4 for structures): the conjugate base of acetic acid, acetate ion, is resonance-stabilized, so acetic acid is the most acidic; the conjugate base of ethanol has a negative charge on a very electronegative oxygen atom; the conjugate base of methylamine has a negative charge on a mildly electronegative nitrogen atom, so methylamine is the least acidic.

acetic acid	>	ethanol	>	methylamine
pK_a 4.74		$pK_a \approx 15.5$		$pK_a \approx 33$
strongest acid				weakest acid

(b) Ethoxide ion is the conjugate base of ethanol, so it must be a stronger base than ethanol; Solved Problem 1-4 indicates ethoxide is analogous to hydroxide in base strength. Methylamine has pK_b 3.36. The basicity of methylamine is between the basicity of ethoxide ion and ethanol.

ethoxide ion	>	methylamine	>	ethanol
strongest base				weakest base

1-16

(a) 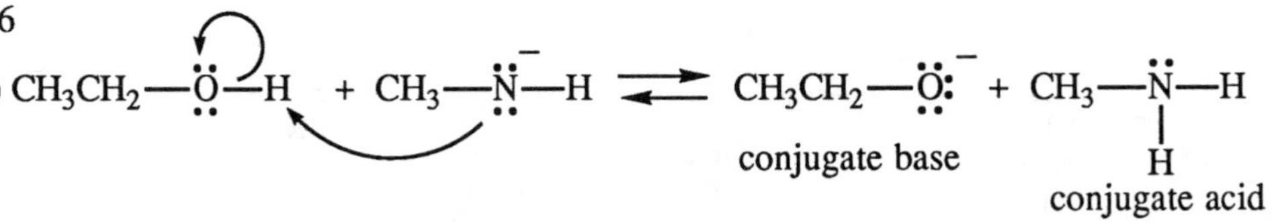

(b) 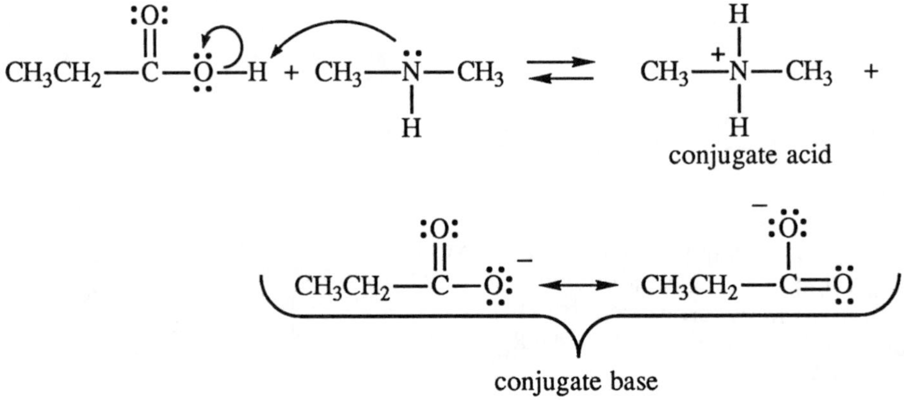

(c) 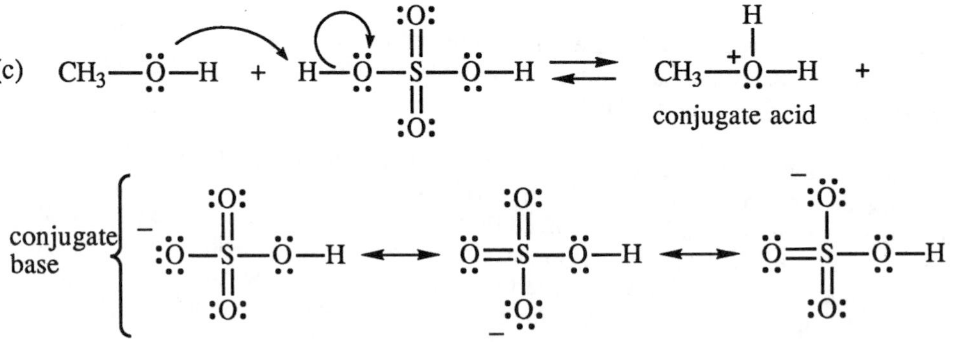

1-16 continued

(d)

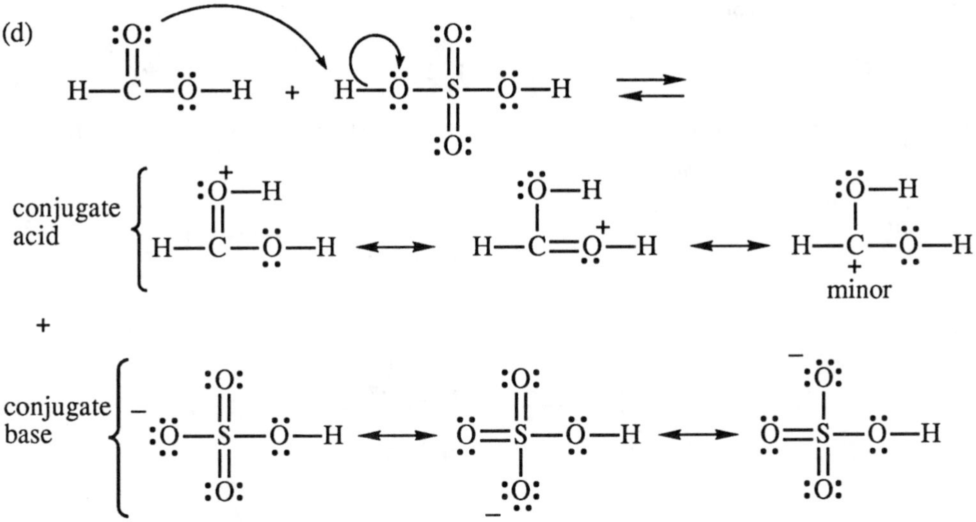

conjugate acid
minor
+
conjugate base

(e)

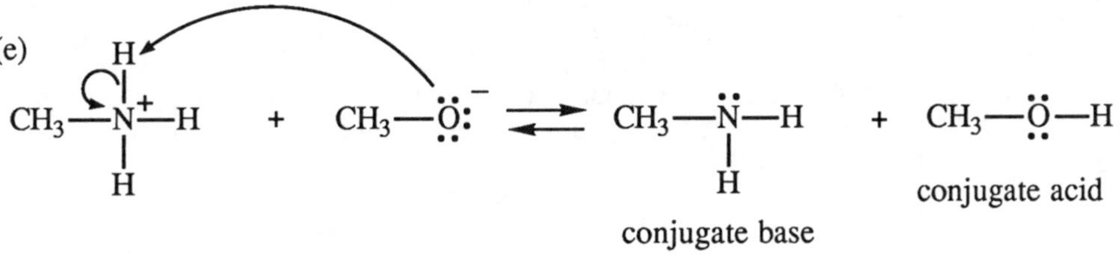

conjugate base
conjugate acid

1-17 (a) and (b) are presented in the Solved Problem.

(c)

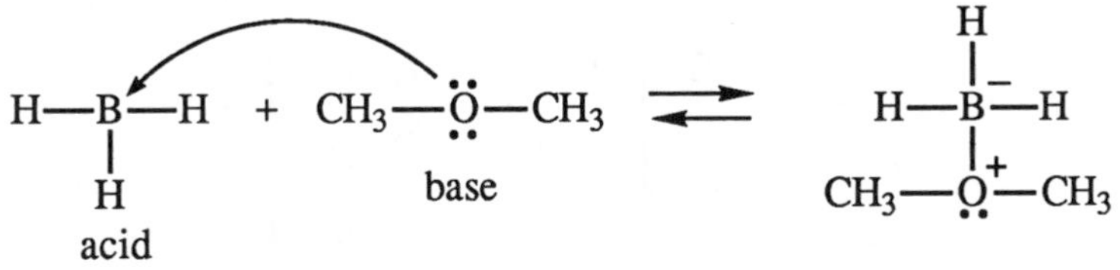

acid
base

(d)

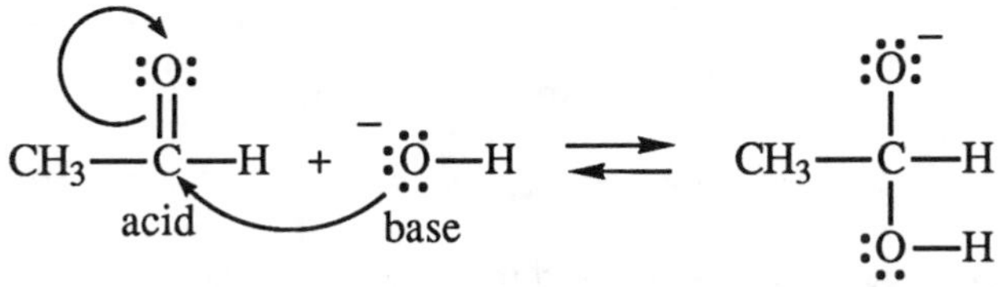

acid
base

1-17 continued

(e) Bronsted-Lowry—proton transfer

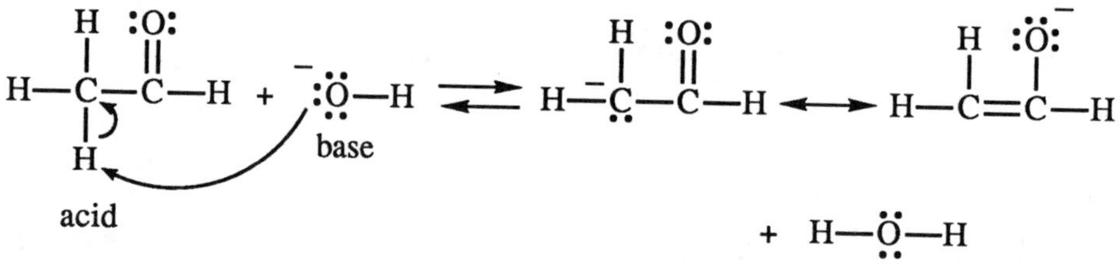

(f)

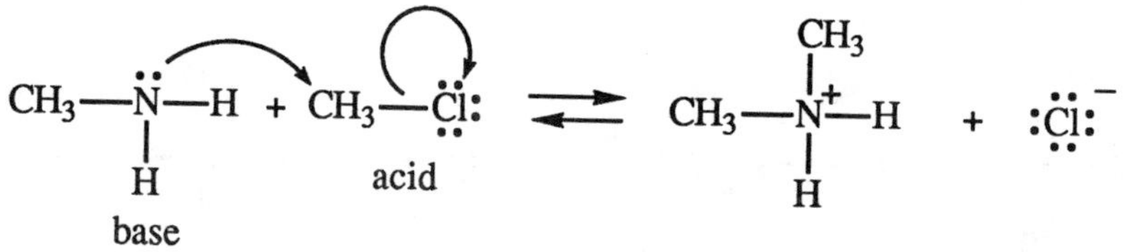

1-18 Learning organic chemistry is similar to learning a foreign language: new vocabulary, new grammar (reactions), some new concepts, and even a new alphabet (the symbolism of chemistry). This type of definition question is intended to help you review the vocabulary and concepts in each chapter. All of the definitions and examples are presented in the Glossary and in the chapter so this Solutions Manual will not repeat them. Use these questions to evaluate your comprehension and to guide your review of the important concepts in the chapter.

1-19 (a) carbon (b) oxygen (c) phosphorus (d) chlorine

1-20

valence e^- →	1	2	3	4	5	6	7	8
	H							He($2e^-$)
	Li	Be	B	C	N	O	F	Ne
					P	S	Cl	
							Br	
							I	

1-21 (a) ionic (b) covalent (H—O^-) and ionic (Na^+ ^-OH)

(c) covalent, but the C—Li bond is strongly polarized

(d) covalent (e) covalent (CH_3—O^-) and ionic (Na^+ $^-OCH_3$)

(f) covalent (HCO_2^-) and ionic (HCO_2^- Na^+) (g) covalent

1-22

(a) 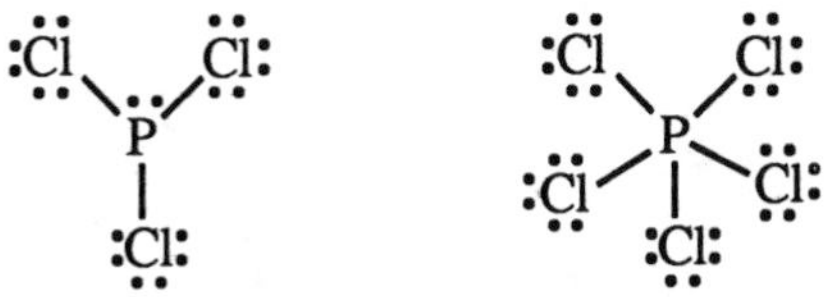

(b) 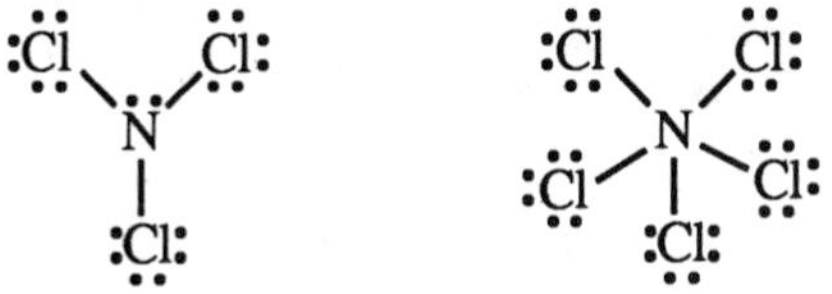 does not exist

NCl_5 violates the octet rule; nitrogen can have no more than eight electrons (or four atoms) around it. Phosphorus, a third-row element, can have more than eight electrons, so PCl_5 is a stable, isolable compound.

1-23

(a) 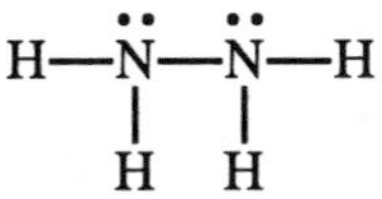

(b) 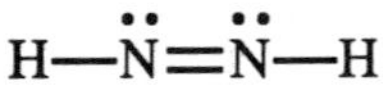

(c) 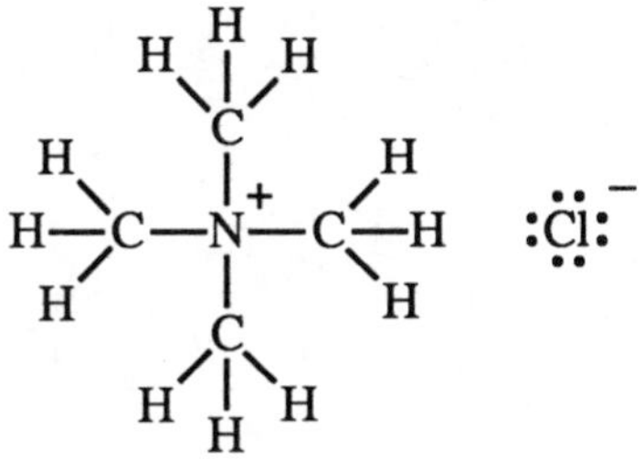

(d) 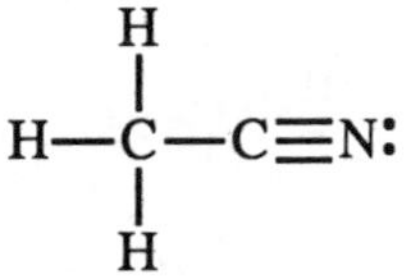

(e) 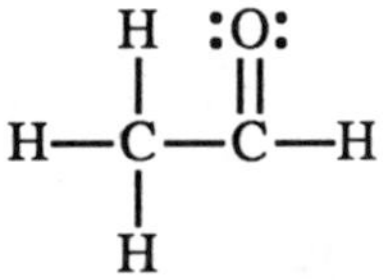

(f) 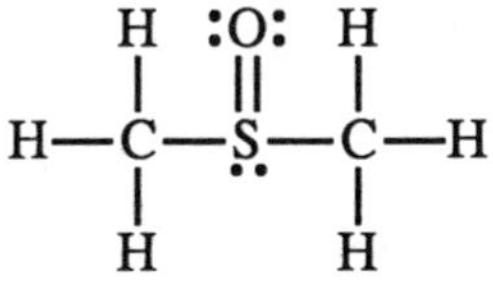

(g) 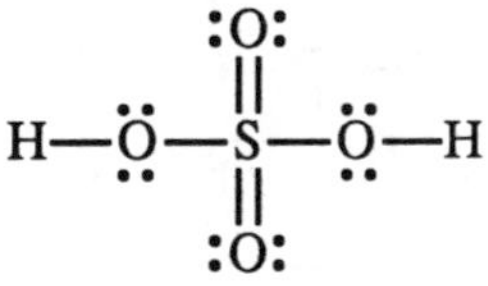

(h) 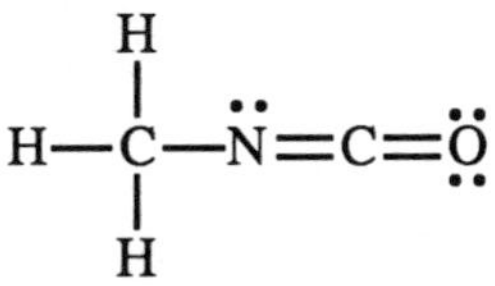

(i) 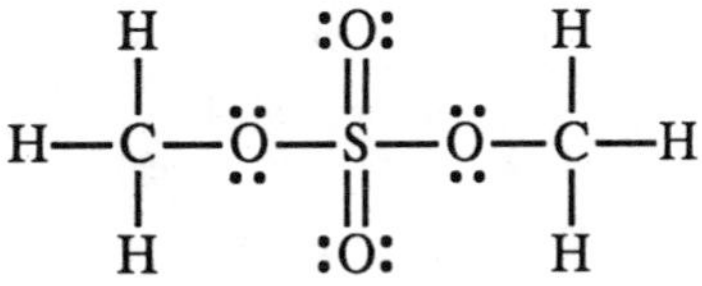

1-23 continued

(j) 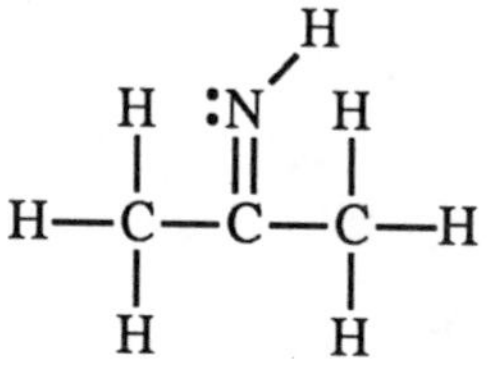

(k) 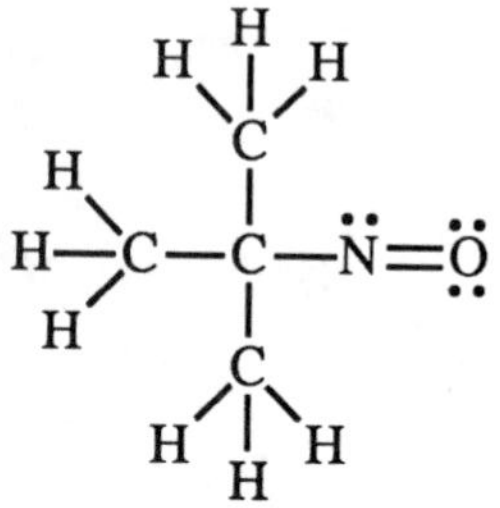

1-24

(a) 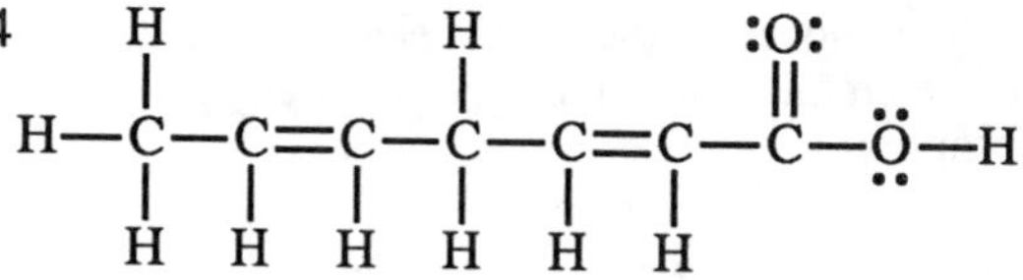

(b) 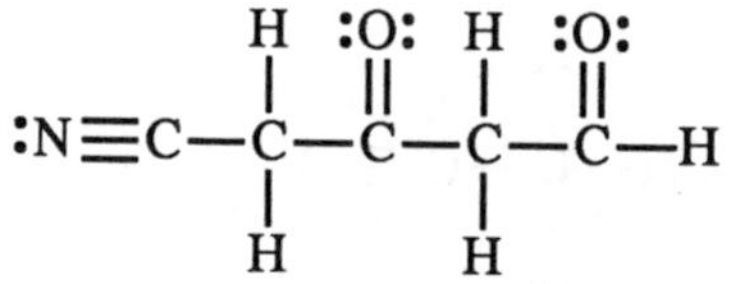

(d) 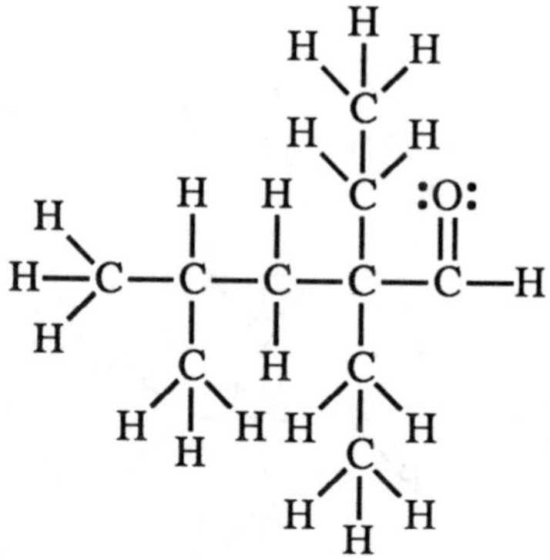

(c) 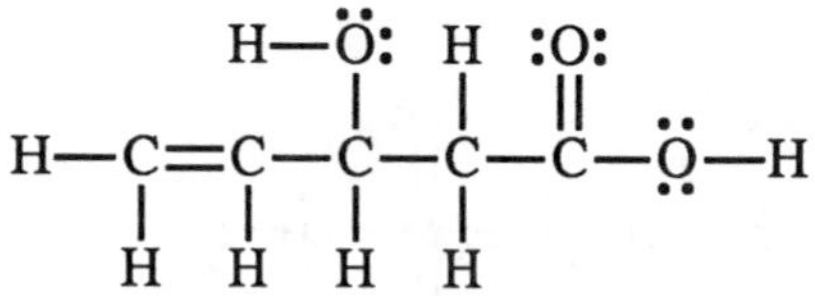

1-25 (a) 

OR 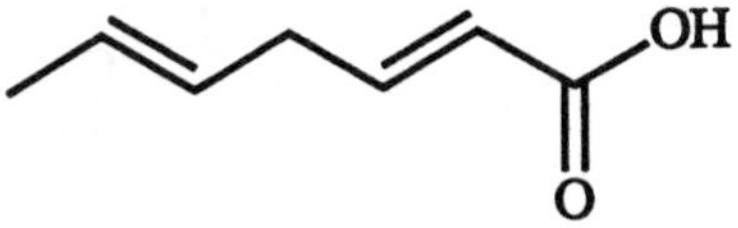

(b) 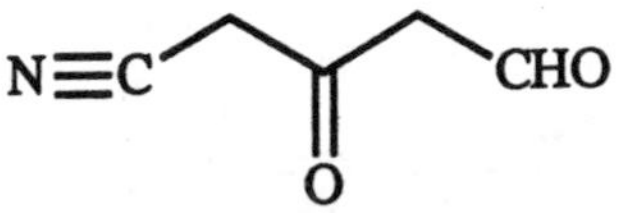

OR 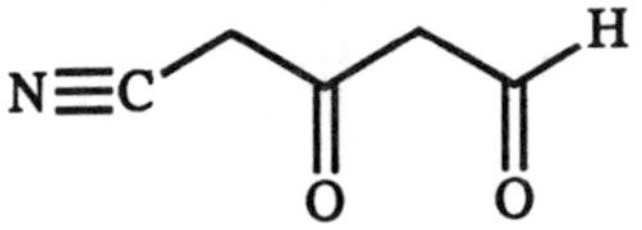

(c) 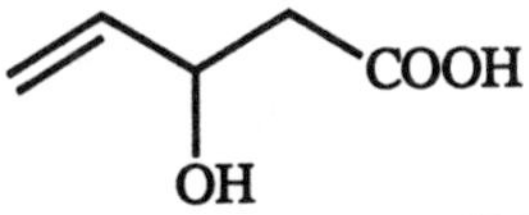

OR 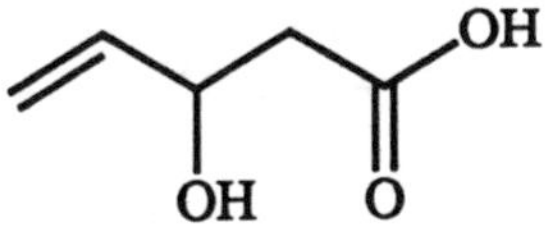

(d) 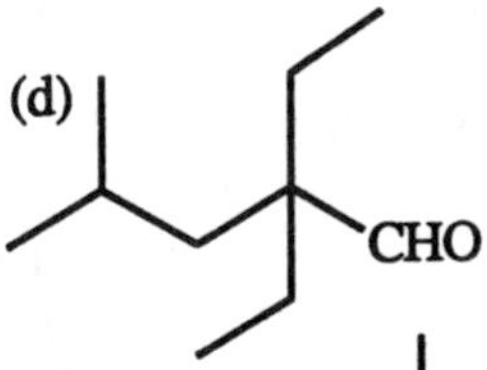

OR 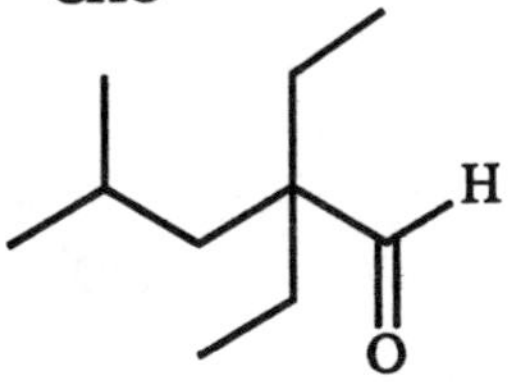

1-26

(a) 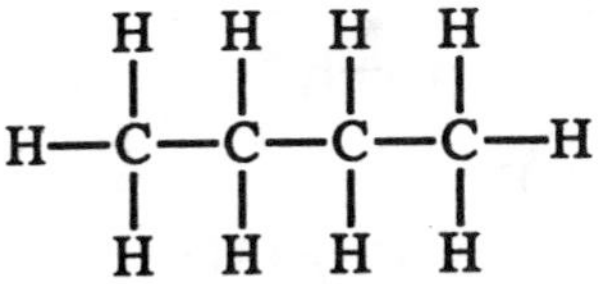 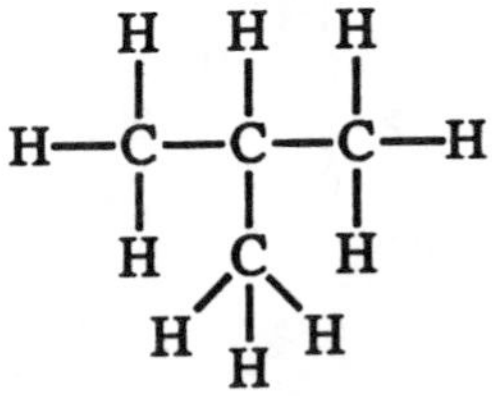

(b) 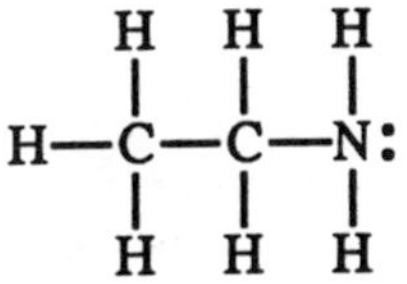 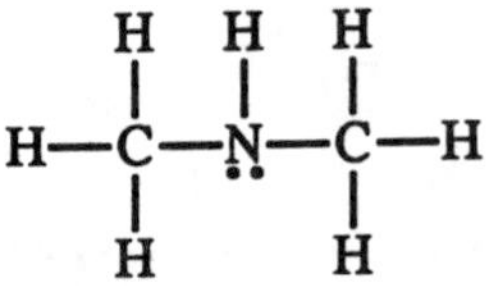

(c) There are several other possibilities as well.

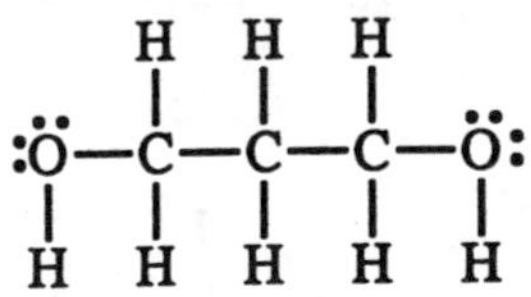 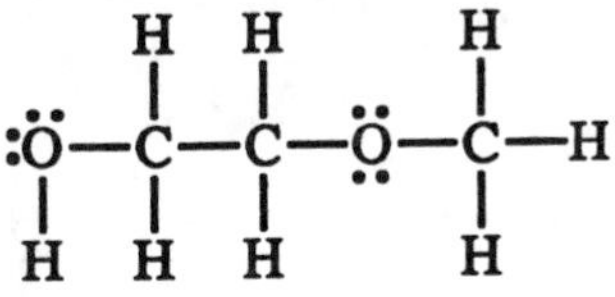 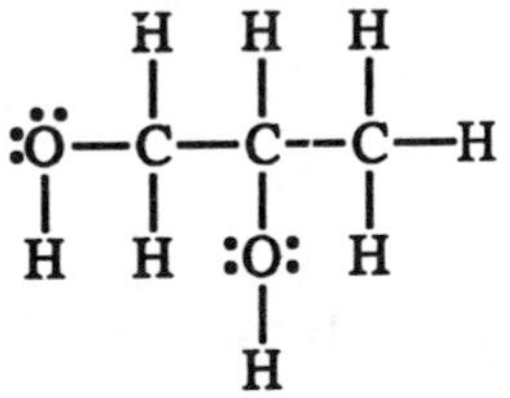

(d) 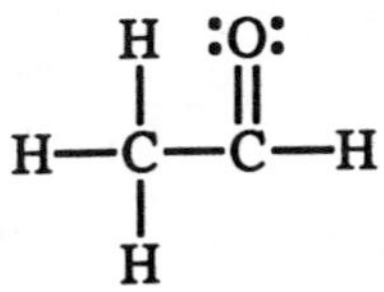 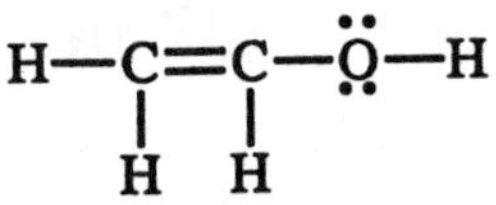 also 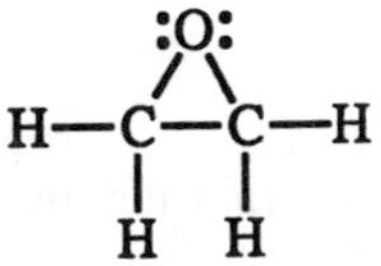

1-27

(a)

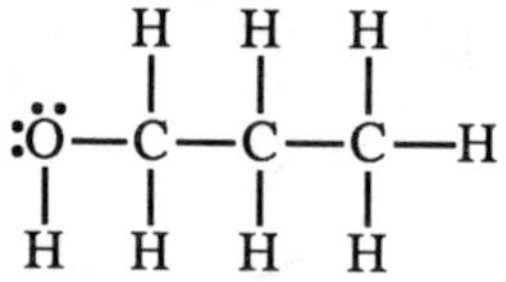

$HOCH_2CH_2CH_3$

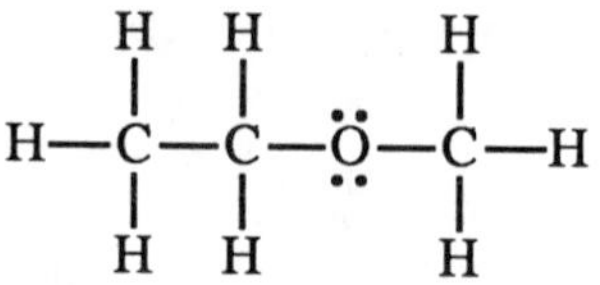

$CH_3CH_2OCH_3$

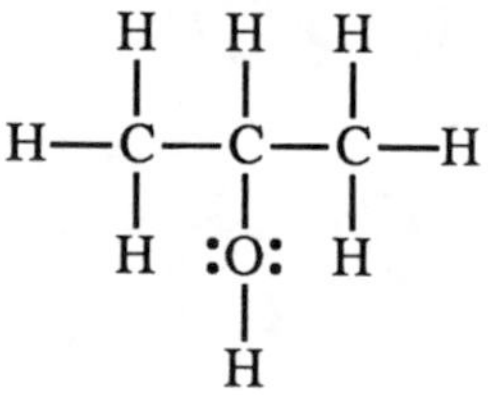

$CH_3CH(OH)CH_3$

(b) There are several other possibilities as well.

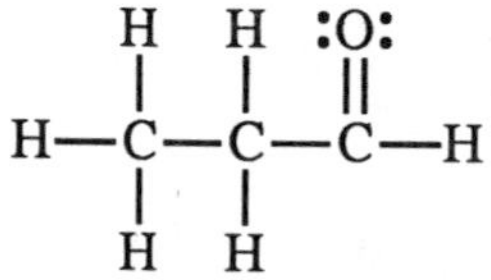

CH_3CH_2CHO

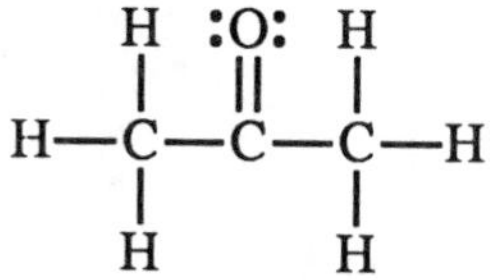

CH_3COCH_3

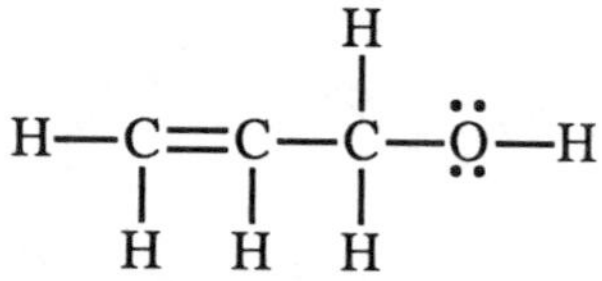

$CH_2{=}CHCH_2OH$

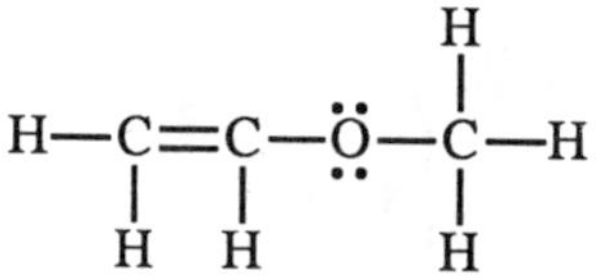

$CH_2{=}CHOCH_3$

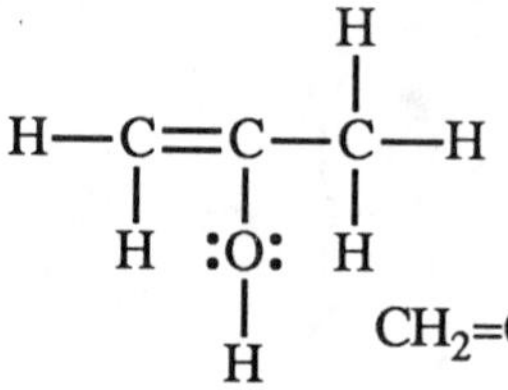

$CH_2{=}CH(OH)CH_3$

1-28

one carbon:

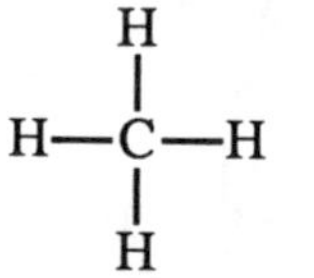

CH_4

two carbons:

C_2H_2

C_2H_4

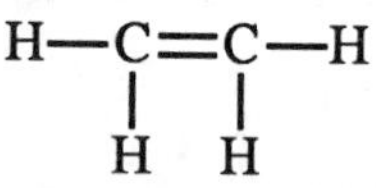

C_2H_6

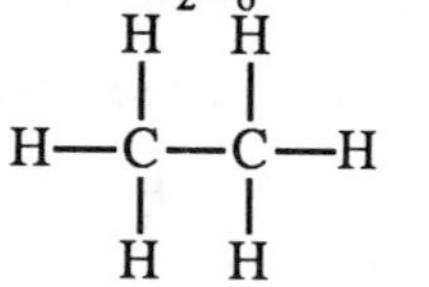

three carbons:

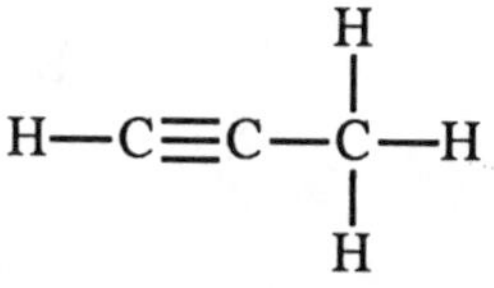

C_3H_4

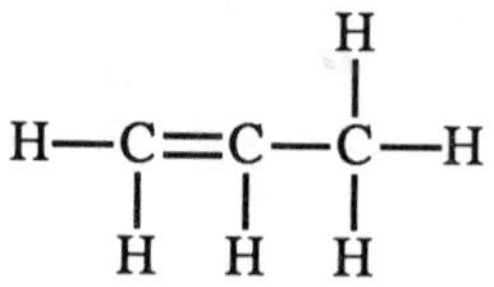

C_3H_6

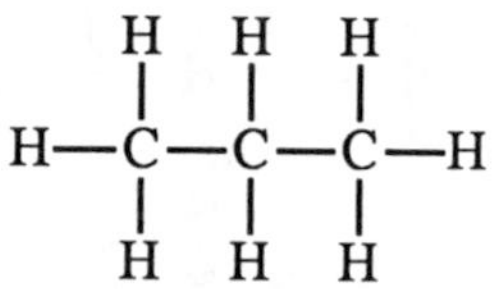

C_3H_8

General rule: molecular formulas of stable hydrocarbons must have an even number of hydrogens.

1-29 (a) 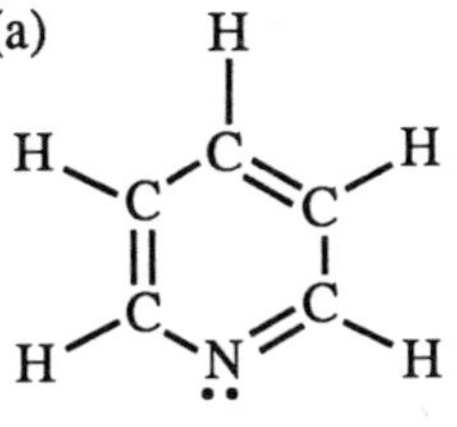 (b) 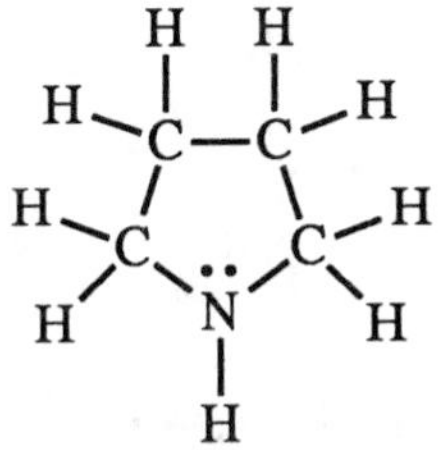 (c) 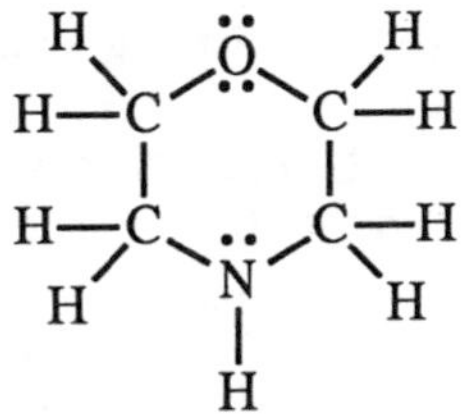

(d) 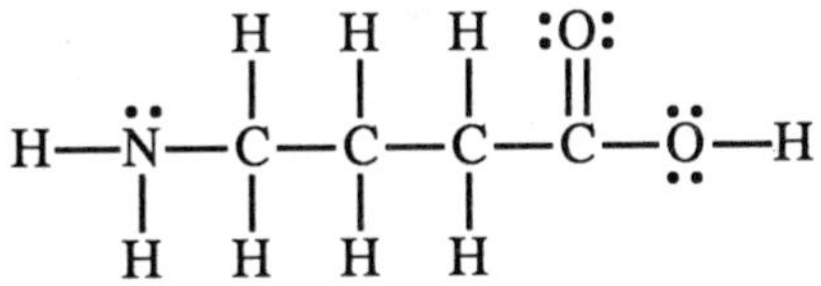

1-30 (a) C_5H_5N (b) C_4H_9N (c) C_4H_9NO (d) $C_4H_9NO_2$

1-31 (a) 100% − 62.1% C − 10.3% H = 27.6% oxygen

$$\frac{62.0 \text{ g C}}{12.0 \text{ g/mole}} = 5.17 \text{ moles C} \div 1.73 \text{ moles} = 2.99 \approx 3 \text{ C}$$

$$\frac{10.4 \text{ g H}}{1.01 \text{ g/mole}} = 10.3 \text{ moles H} \div 1.73 \text{ moles} = 5.95 \approx 6 \text{ H}$$

$$\frac{27.6 \text{ g O}}{16.0 \text{ g/mole}} = 1.73 \text{ moles O} \div 1.73 \text{ moles} = 1 \text{ O}$$

(b) empirical formula = **C_3H_6O** ⇨ empirical weight = 58

molecular weight = 117, about double the empirical weight ⇨

double the empirical formula = molecular formula = **$C_6H_{12}O_2$**

(c) There are many possible structures.

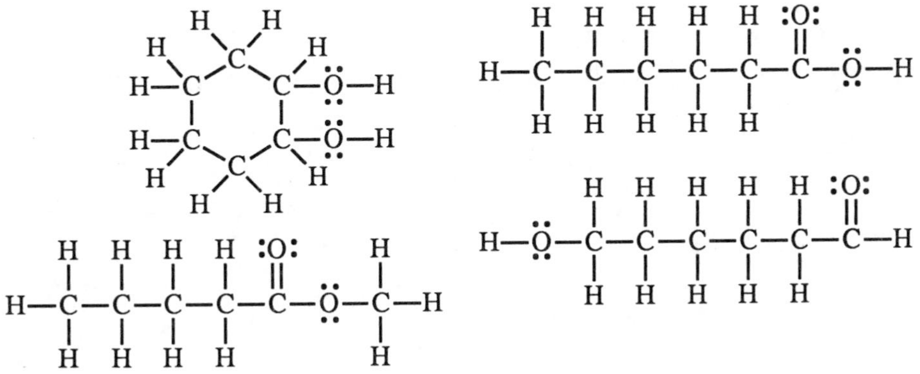

1-32 Non-zero formal charges are shown by the atoms.

(a) 

(b) 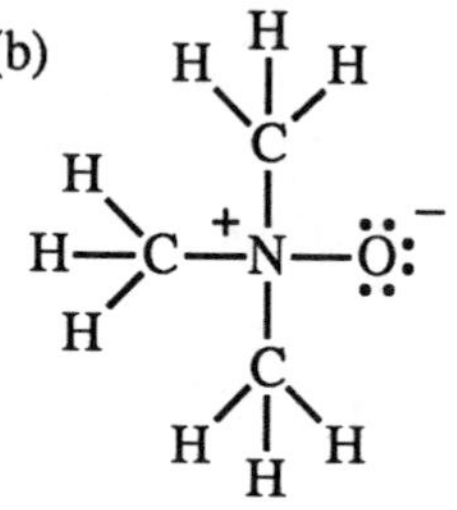

(c) 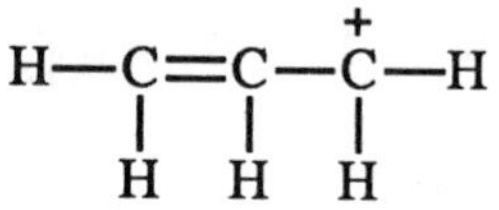

(d) 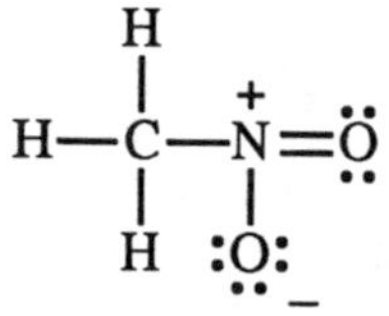

(e) 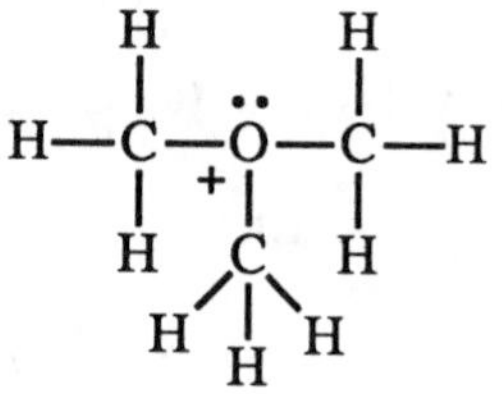

1-33 The symbols " δ^+ " and " δ^- " indicate bond polarity by showing partial charge. Electronegativity differences greater than 0.5 are considered large.

(a) $\overset{\delta^+}{\mathrm{C}}\text{—}\overset{\delta^-}{\mathrm{Cl}}$ large

(b) $\overset{\delta^-}{\mathrm{C}}\text{—}\overset{\delta^+}{\mathrm{H}}$ small

(c) $\overset{\delta^-}{\mathrm{C}}\text{—}\overset{\delta^+}{\mathrm{Li}}$ large

(d) $\overset{\delta^+}{\mathrm{C}}\text{—}\overset{\delta^-}{\mathrm{N}}$ small

(e) $\overset{\delta^+}{\mathrm{C}}\text{—}\overset{\delta^-}{\mathrm{O}}$ large

(f) $\overset{\delta^-}{\mathrm{C}}\text{—}\overset{\delta^+}{\mathrm{B}}$ large

(g) $\overset{\delta^-}{\mathrm{C}}\text{—}\overset{\delta^+}{\mathrm{Mg}}$ large

(h) $\overset{\delta^-}{\mathrm{N}}\text{—}\overset{\delta^+}{\mathrm{H}}$ large

(i) $\overset{\delta^-}{\mathrm{O}}\text{—}\overset{\delta^+}{\mathrm{H}}$ large

(j) $\overset{\delta^+}{\mathrm{C}}\text{—}\overset{\delta^-}{\mathrm{Br}}$ small

1-34 (a) different compounds—a hydrogen atom has changed position

(b) resonance forms—only the position of electrons is different

(c) resonance forms—only the position of electrons is different

(d) resonance forms—only the position of electrons is different

(e) different compounds—a hydrogen atom has changed position

(f) resonance forms—only the position of electrons is different

(g) resonance forms—only the position of electrons is different

(h) different compounds—a hydrogen atom has changed position

(i) resonance forms—only the position of electrons is different

(j) resonance forms—only the position of electrons is different

1-35 (a)

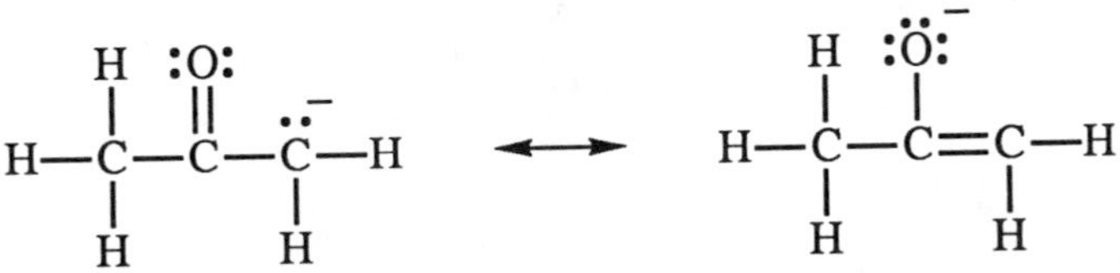

(b)

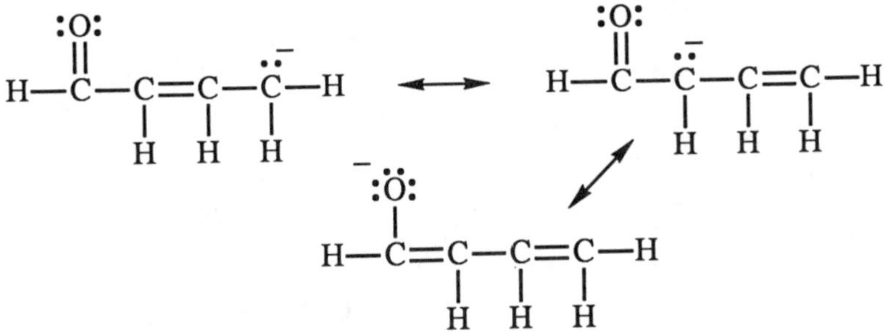

(c)

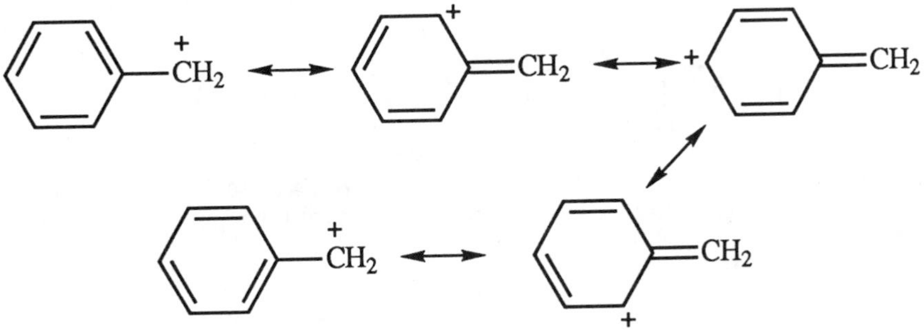

1-35 continued

(d)

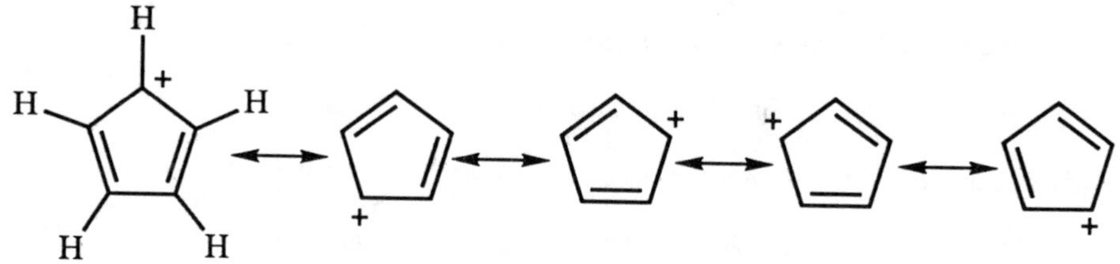

(e)

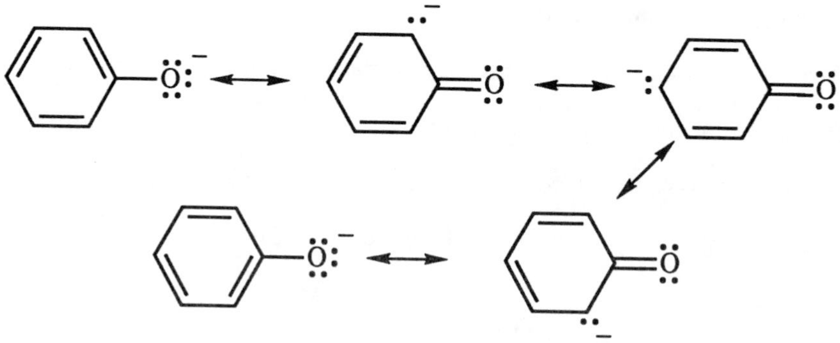

(f)

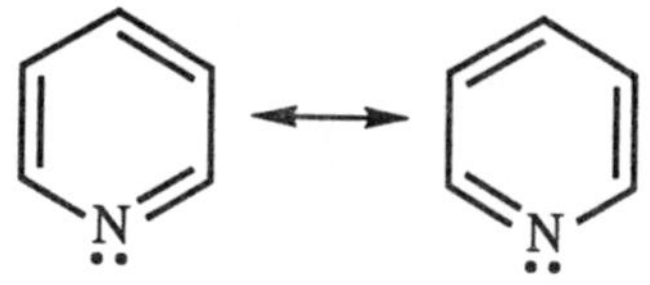

(g)

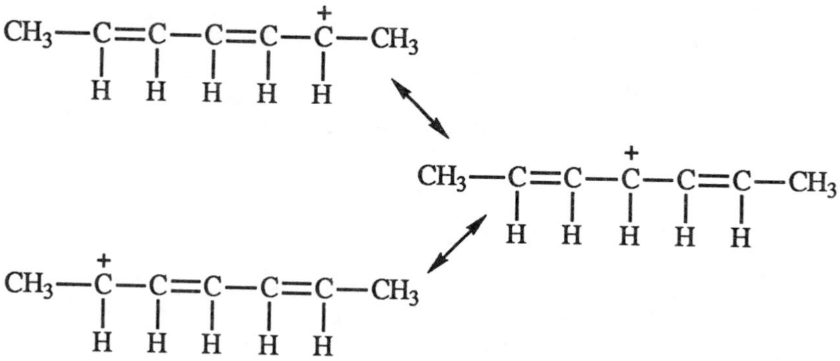

(h) no resonance forms—the charge must be on an atom next to a double or triple bond, or next to a non-bonded pair of electrons, in order for resonance to delocalize the charge

1-36

(a) 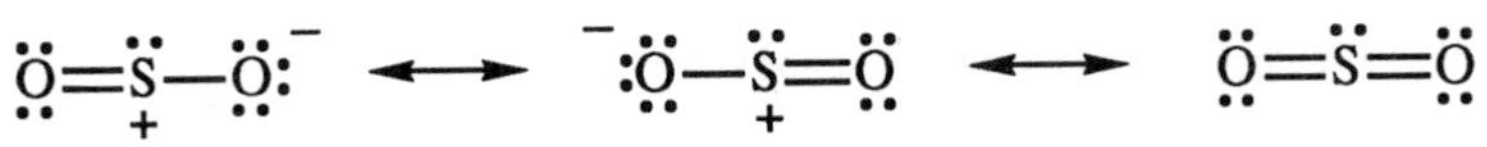

(b) 

(c) The last resonance form of SO_2 has no equivalent form in O_3. Sulfur, a row three element, can have more than eight electrons around it, whereas oxygen, a row two element, must adhere strictly to the octet rule.

1-37

(a)

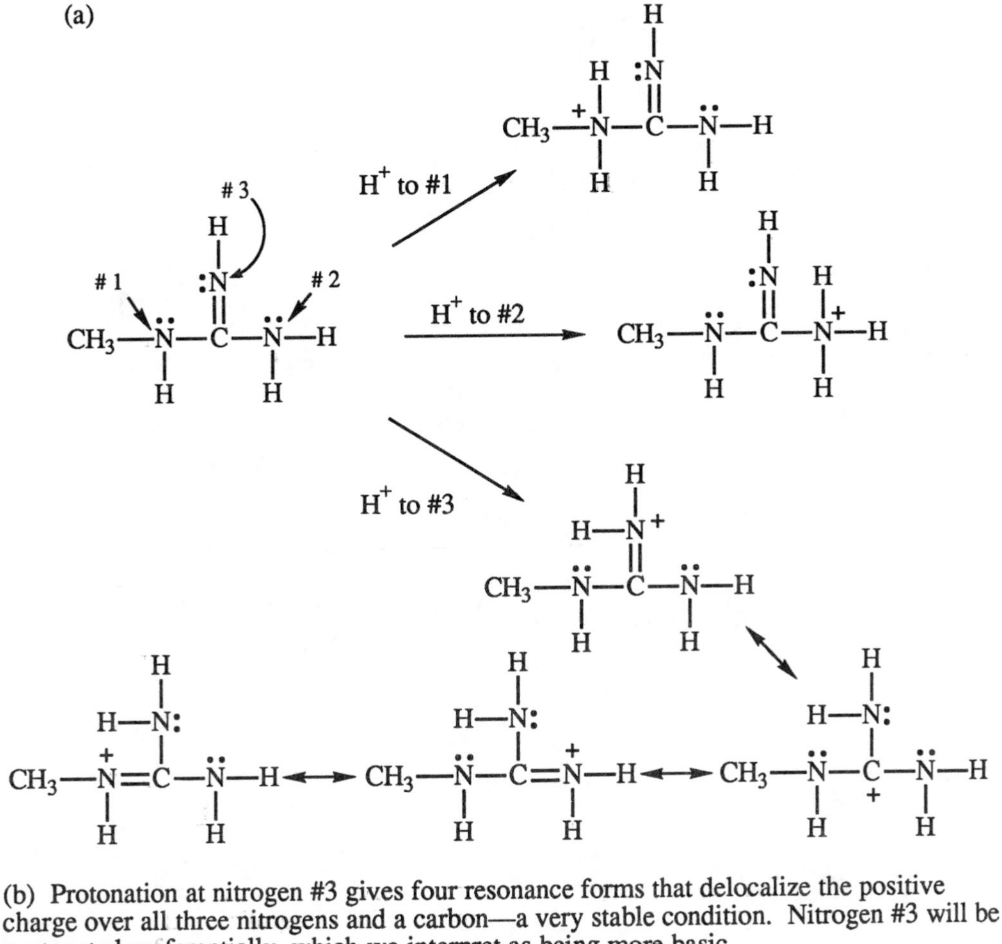

(b) Protonation at nitrogen #3 gives four resonance forms that delocalize the positive charge over all three nitrogens and a carbon—a very stable condition. Nitrogen #3 will be protonated preferentially, which we interpret as being more basic.

1-38

(a)

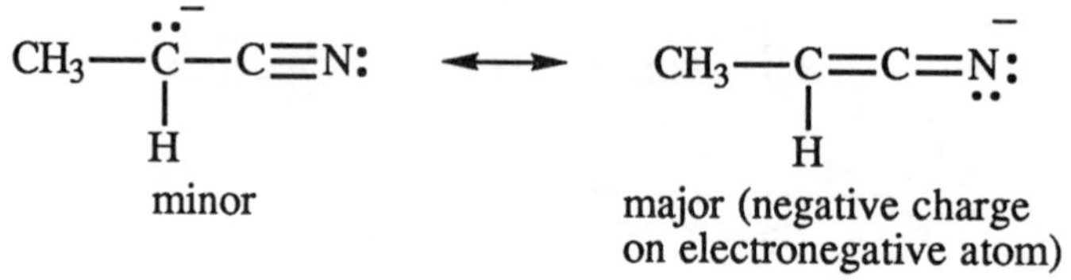
minor
major (negative charge
on electronegative atom)

(b)

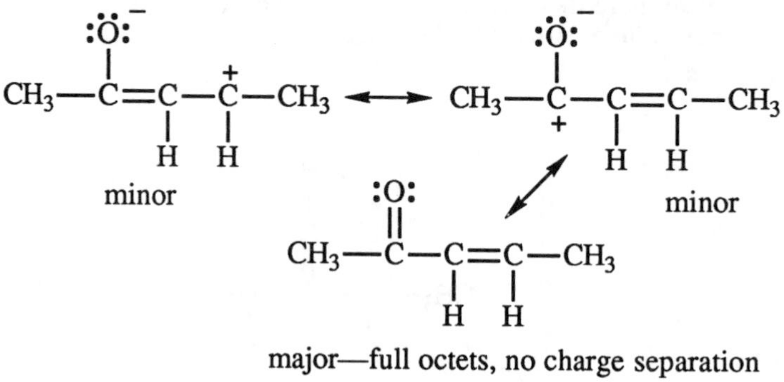
minor
minor
major—full octets, no charge separation

(c)

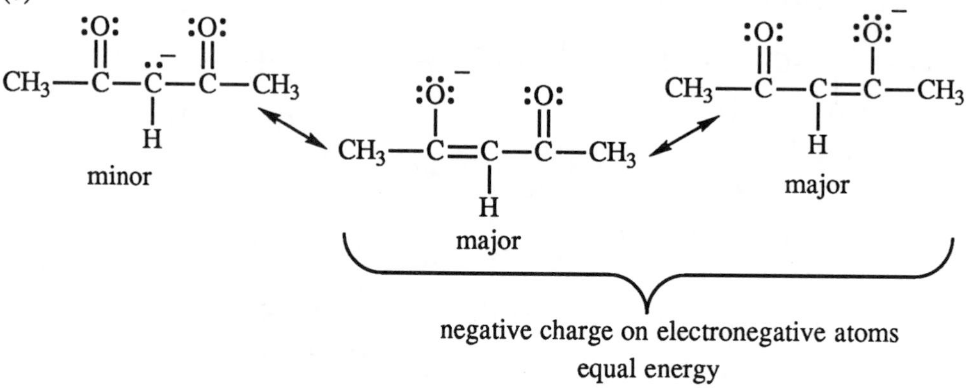
minor
major
major
negative charge on electronegative atoms
equal energy

1-38 continued

(d)

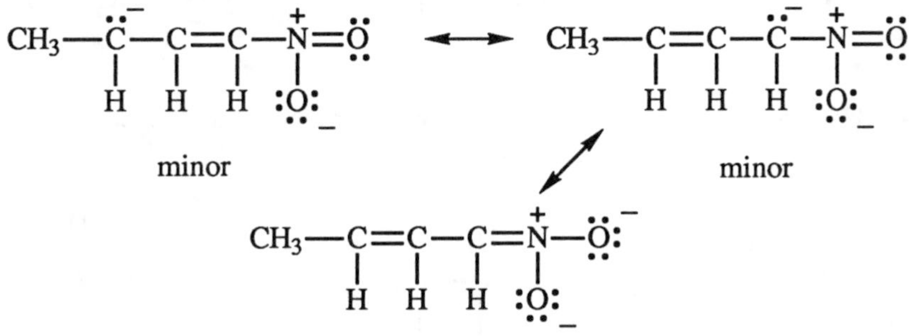

minor ... minor

major (negative charge on electronegative atoms)

(Note: the two structures below are resonance forms also, with the double bonds in the NO_2 in different positions from the first two structures in part (d). Usually, chemists omit drawing these forms *with the understanding that their presence is implied!* The importance of understanding resonance forms cannot be overestimated.

)

(e)

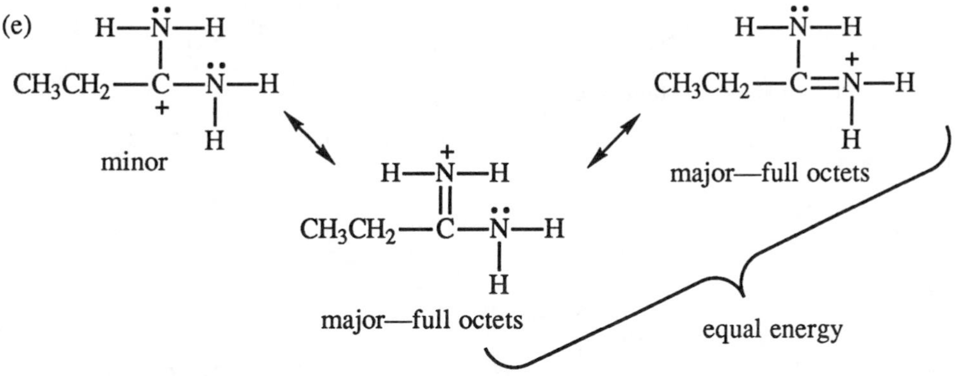

1-39

(a) 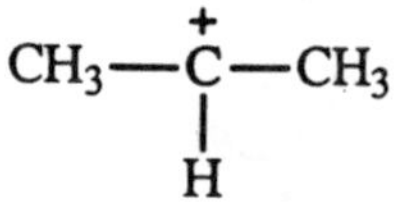

no resonance forms

more stable—resonance stabilized

(b) 

more stable—resonance stabilized

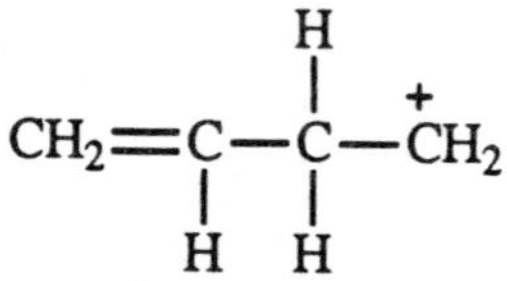

no resonance forms

(c) 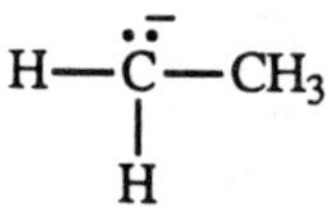

no resonance forms

more stable—resonance stabilized

(d) 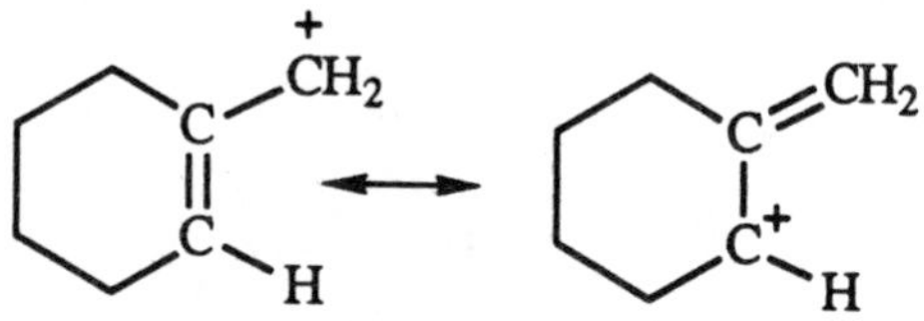

more stable—resonance stabilized

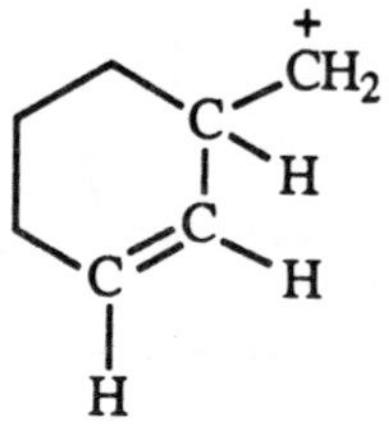

no resonance forms

(e) 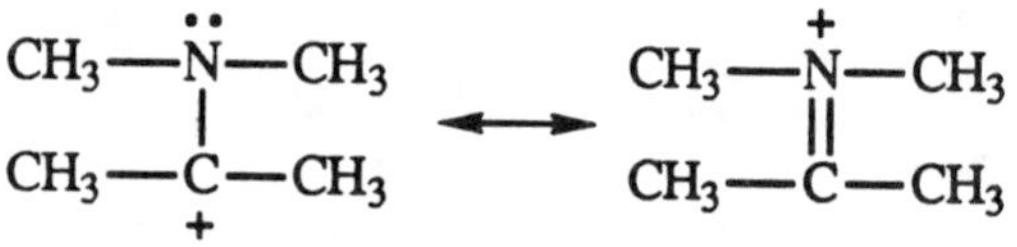

more stable—resonance stabilized

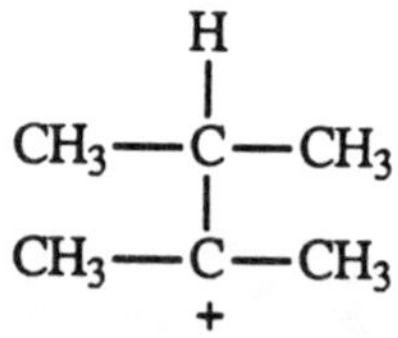

no resonance forms

1-40 These pK_a values from the text, Table 1-6, provide the answers.

least acidic						most acidic
NH_3	<	CH_3OH	<	CH_3COOH	<	H_2SO_4
33		15.5		4.74		< 0

1-41 Conjugate bases of the weakest acids will be the strongest bases. The pK_a's of the conjugate acids are listed here. (The relative order of the first two cannot be determined without the pK_a value of protonated acetic acid.)

least basic								most basic
HSO_4^- , CH_3COOH	<	CH_3COO^-	<	CH_3O^-	<	NaOH	<	$^-NH_2$
< 0 ?		from 4.74		from 15.5		from 15.7		from 33

1-42

(a) $pK_a = -\log_{10} K_a = -\log_{10}(5.2 \times 10^{-5}) =$ **4.3** for phenylacetic acid

for propionic acid, pK_a 4.87: $K_a = 10^{-4.87} = \mathbf{1.35 \times 10^{-5}}$

(b) phenylacetic acid is 3.8 times stronger than propionic acid

$$\frac{5.2 \times 10^{-5}}{1.35 \times 10^{-5}} = 3.8$$

1-43

(a)

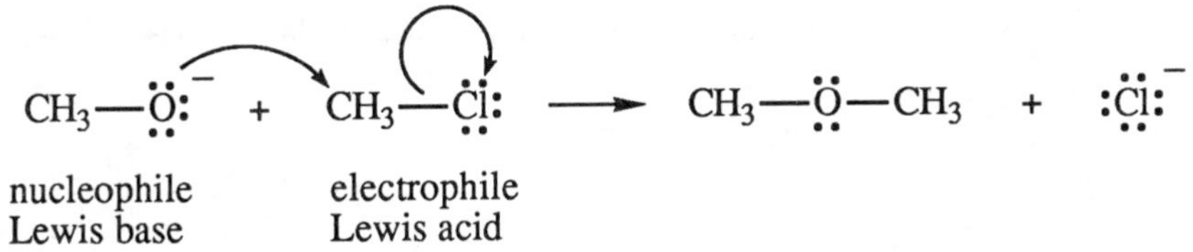

nucleophile
Lewis base
electrophile
Lewis acid

(b)

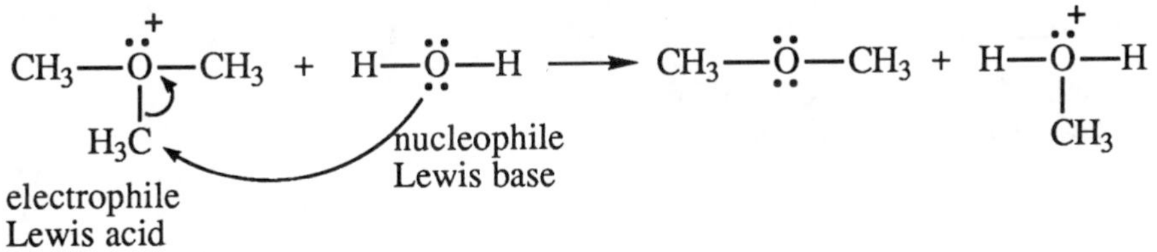

electrophile
Lewis acid
nucleophile
Lewis base

(c)

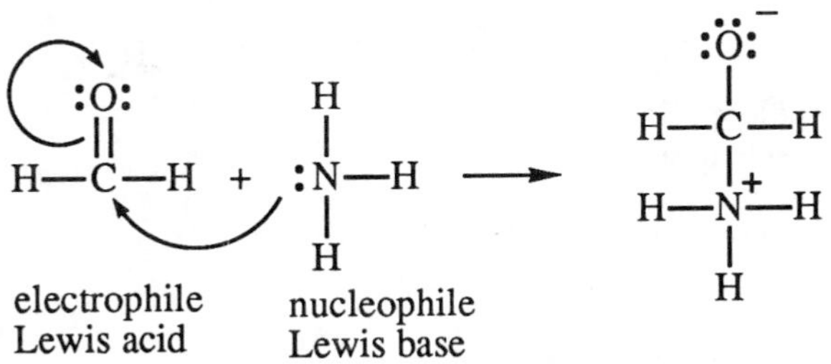

electrophile
Lewis acid
nucleophile
Lewis base

1-43 continued

(d) 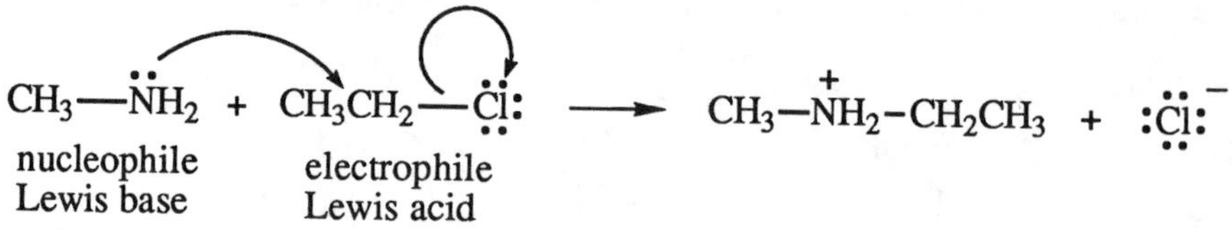

nucleophile
Lewis base
electrophile
Lewis acid

(e) 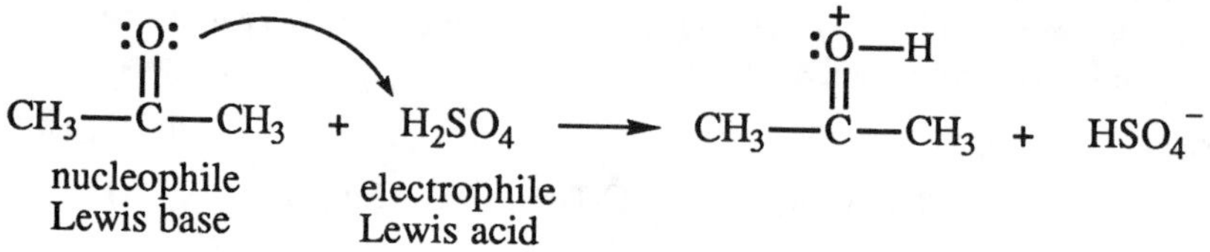

nucleophile
Lewis base
electrophile
Lewis acid

(f) 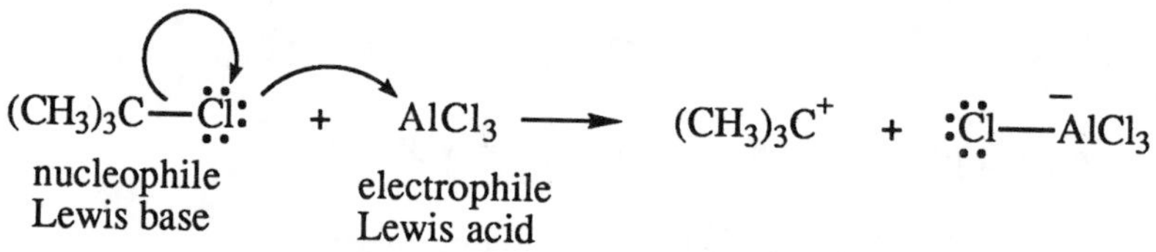

nucleophile
Lewis base
electrophile
Lewis acid

(g) 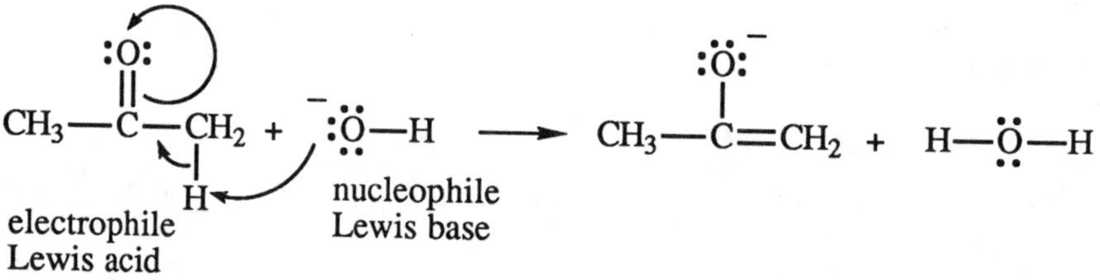

electrophile
Lewis acid
nucleophile
Lewis base

(h) 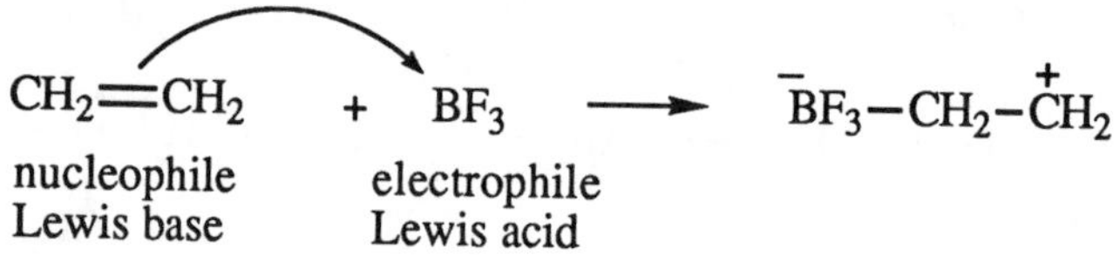

nucleophile
Lewis base
electrophile
Lewis acid

(i) 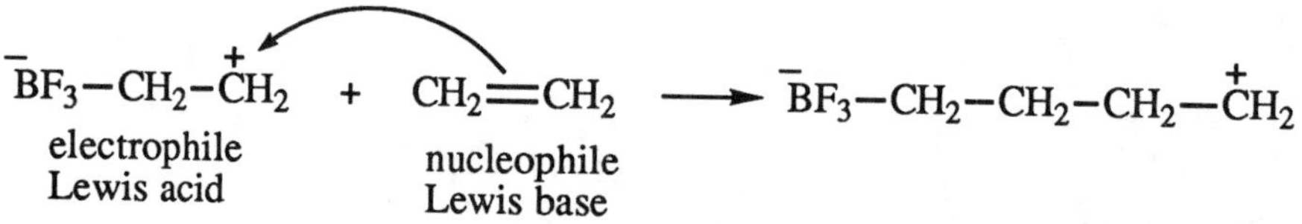

electrophile
Lewis acid
nucleophile
Lewis base

1-44

(a) $H_2SO_4 + CH_3COO^- \rightleftharpoons HSO_4^- + CH_3COOH$

(b) $CH_3COOH + (CH_3)_3N\!: \rightleftharpoons CH_3COO^- + (CH_3)_3N^+{-}H$

(c) 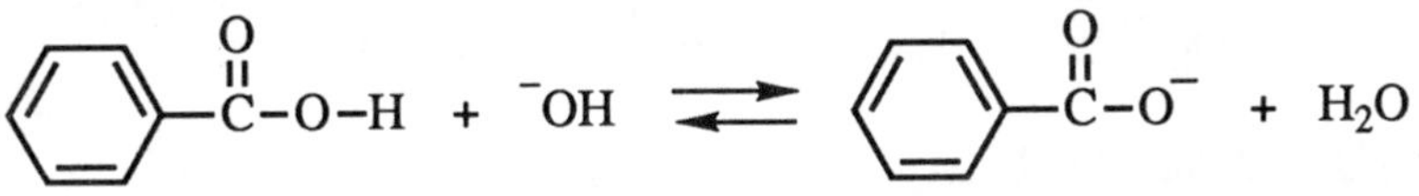

(d) $(CH_3)_3N^+{-}H + {}^-OH \rightleftharpoons (CH_3)_3N\!: + H_2O$

(e) $HO{-}\overset{O}{\overset{\|}{C}}{-}OH + 2\ {}^-OH \rightleftharpoons {}^-O{-}\overset{O}{\overset{\|}{C}}{-}O^- + 2\,H_2O$

(f) $H_2O + NH_3 \rightleftharpoons HO^- + {}^+NH_4$

(g) $HCOOH + CH_3O^- \rightleftharpoons HCOO^- + CH_3OH$

1-45 (a) $CH_3CH_2{-}O{-}H + CH_3{-}Li \longrightarrow CH_3CH_2{-}O^-\ Li^+ + CH_4$

(b) The conjugate acid of CH_3Li is CH_4. Table 1-6 gives the pK_a of CH_4 as > 40, one of the weakest acids known. The conjugate base of one of the weakest acids known must be one of the strongest bases known.

1-46 From the amounts of CO_2 and H_2O generated, the milligrams of C and H in the original sample can be determined, thus giving by difference the amount of oxygen in the 5.00 mg sample. From these values, the empirical formula and empirical weight can be calculated.

(a) how much carbon in 14.54 mg CO_2

$$14.54 \text{ mg } CO_2 \times \frac{1 \text{ mmole } CO_2}{44.01 \text{ mg } CO_2} \times \frac{1 \text{ mmole C}}{1 \text{ mmole } CO_2} \times \frac{12.01 \text{ mg C}}{1 \text{ mmole C}} = 3.968 \text{ mg C}$$

how much hydrogen in 3.97 mg H_2O

$$3.97 \text{ mg } H_2O \times \frac{1 \text{ mmole } H_2O}{18.016 \text{ mg } H_2O} \times \frac{2 \text{ mmoles H}}{1 \text{ mmole } H_2O} \times \frac{1.008 \text{ mg H}}{1 \text{ mmole H}} = 0.444 \text{ mg H}$$

how much oxygen in 5.00 mg estradiol

5.00 mg estradiol - 3.968 mg C - 0.444 mg H = 0.59 mg O

calculate empirical formula

$$\frac{3.968 \text{ mg C}}{12.01 \text{ mg/mole}} = 0.3304 \text{ mmoles C} \div 0.037 \text{ mmoles} = 8.93 \approx 9 \text{ C}$$

$$\frac{0.444 \text{ mg H}}{1.008 \text{ mg/mole}} = 0.440 \text{ mmoles H} \div 0.037 \text{ mmoles} = 11.9 \approx 12 \text{ H}$$

$$\frac{0.59 \text{ mg O}}{16.00 \text{ mg/mole}} = 0.037 \text{ mmoles O} \div 0.037 \text{ mmoles} = 1 \text{ O}$$

empirical formula = $C_9H_{12}O$ ⟹ empirical weight = 136

(b) molecular weight = 272, exactly twice the empirical weight

twice the empirical formula = molecular formula = $C_{18}H_{24}O_2$

CHAPTER 2—STRUCTURE AND PROPERTIES OF ORGANIC MOLECULES

2-1 The fundamental principle of organic chemistry is that a molecule's chemical and physical properties depend on the molecule's structure—the structure-function or structure-reactivity correlation. It is essential that you understand the three-dimensional nature of organic molecules, and there is no better device to assist you than a molecular model set. You are strongly encouraged to use models regularly when reading the text and working the problems.

(a) requires use of models

(b)

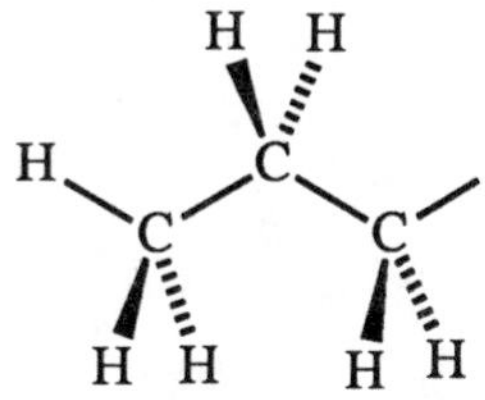

2-2 The hybridization of oxygen is sp^3 since it has two sigma bonds and two pairs of nonbonding electrons. The reason that the bond angle of 104.5° is less than the perfect tetrahedral angle of 109.5° is that the lone pairs in the two sp^3 orbitals are repelling each other more strongly than the electron pairs in the sigma bonds, thereby compressing the bond angle.

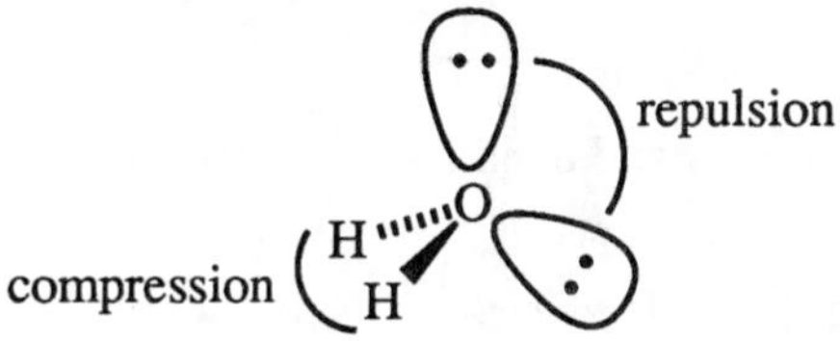

2-3 Each double-bonded carbon is sp^2 hybridized with bond angles about 120°; geometry around the sp^2 carbons is trigonal planar, that is, all four carbons, the two hydrogens on the sp^2 carbons, and one hydrogen on each of the sp^3 carbons are all in one plane. Each carbon on the end is sp^3 hybridized with tetrahedral geometry and bond angles about 109°.

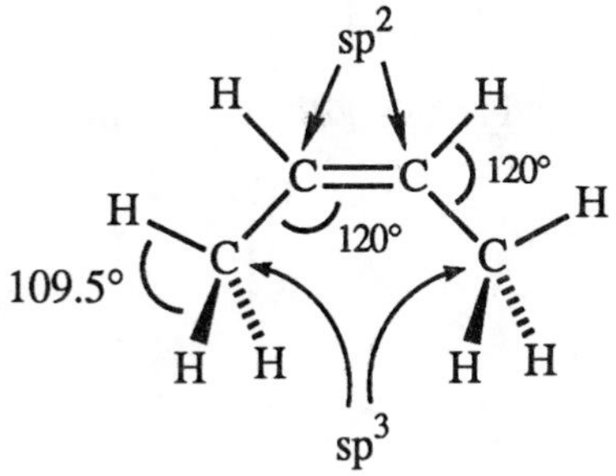

2-4 The hybridization of the nitrogen and the triple-bonded carbon are sp, giving linear geometry (C—C—N are linear) and a bond angle around the triple-bonded carbon of 180°. The CH_3 carbon is sp^3 hybridized, tetrahedral, with bond angles about 109°.

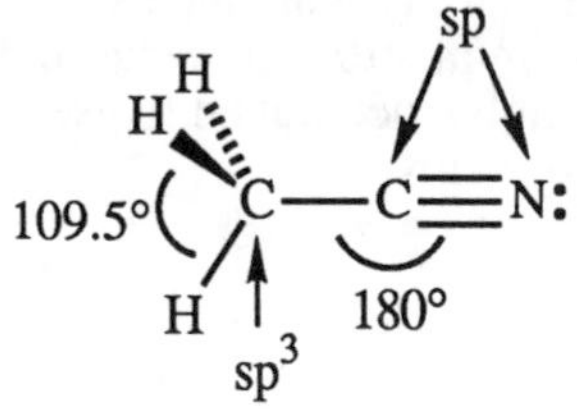
sp
109.5°
180°
sp3

2-5 (a) linear, bond angle 180°

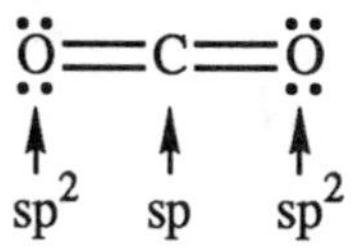
sp2
sp
sp2

(b) all atoms are sp^3; tetrahedral geometry and bond angles of 109° around each atom

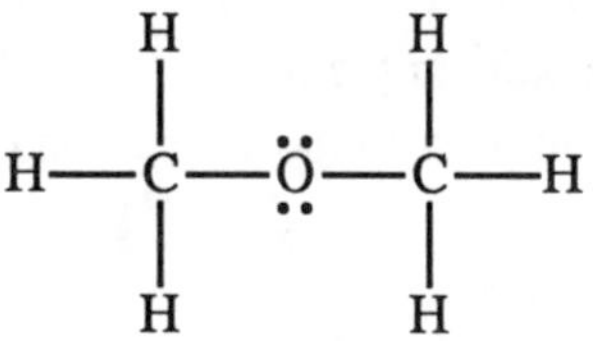

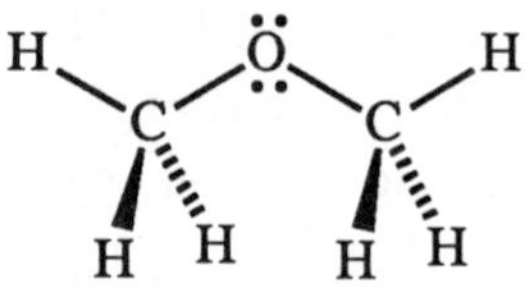

(c) all atoms are sp^3; tetrahedral geometry and bond angles of 109° around each atom

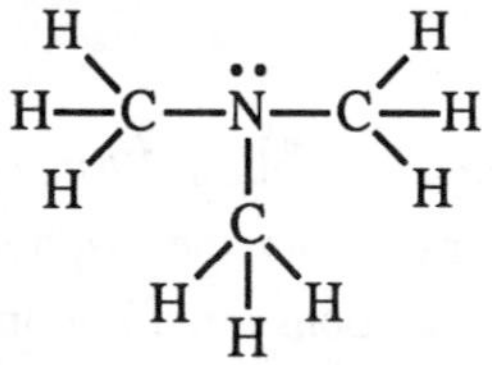

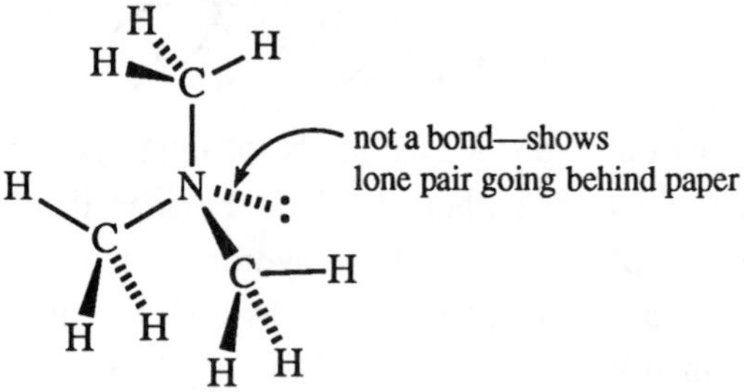
not a bond—shows
lone pair going behind paper

(d) trigonal planar around the carbon, bond angles 120°; tetrahedral around the single-bonded oxygen, bond angle 109°

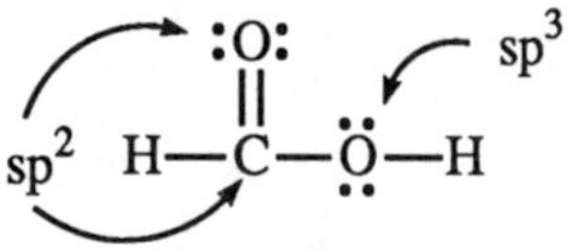
sp3
sp2

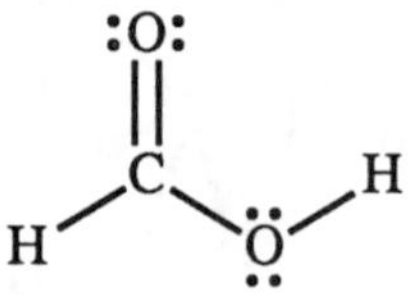

2-5 continued

(e) carbon and nitrogen both sp, linear, bond angle 180°

H—C≡N:

(f) trigonal planar around the sp^2 carbons, bond angles 120°; around the sp^3 carbon, tetrahedral geometry and 109° angles

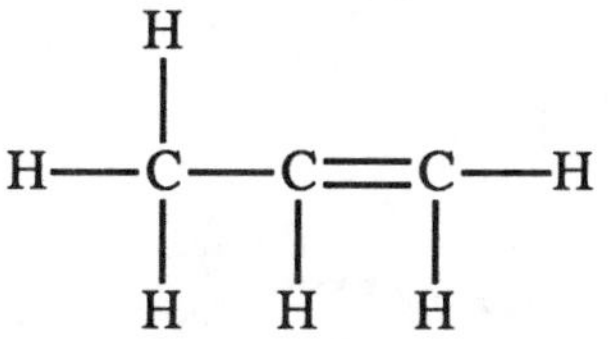

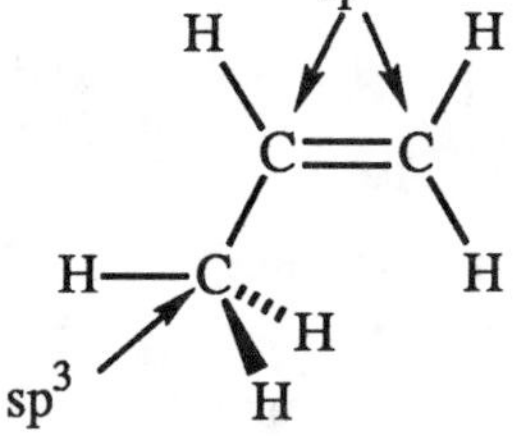

(g) trigonal planar, bond angle about 120°

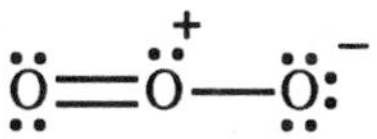

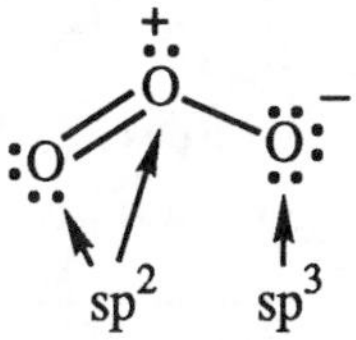

(the other resonance form of ozone shows that this atom must be sp^2)

2-6 Carbon-2 is sp hybridized. If the p orbitals making the pi bond between C-1 and C-2 are in the plane of the paper (putting the hydrogens in front of and behind the paper), then the other p orbital on C-2 must be perpendicular to the plane of the paper, making the pi bond between C-2 and C-3 perpendicular to the paper. This necessarily places the hydrogens on C-3 in the plane of the paper. (Models will surely help.)

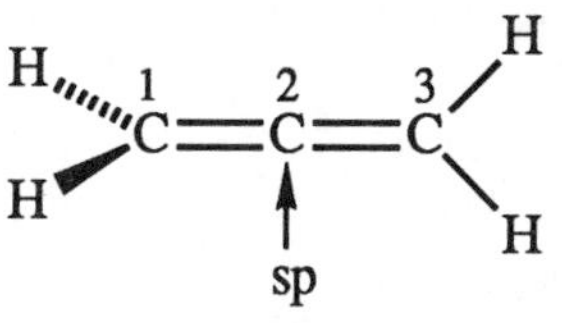

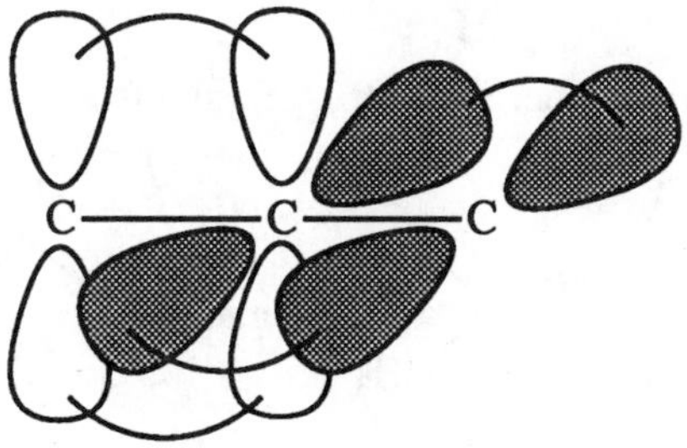

2-7

(a)

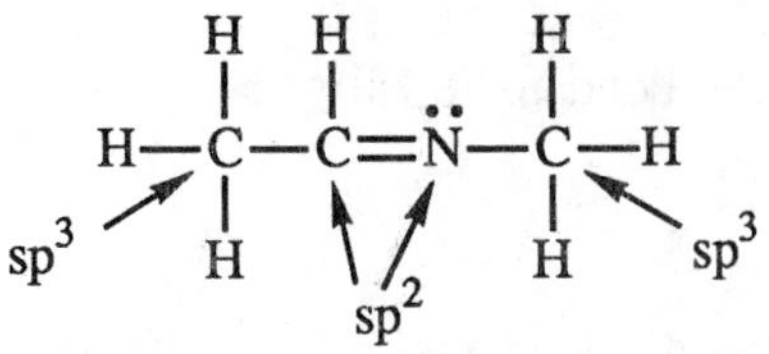

(b)

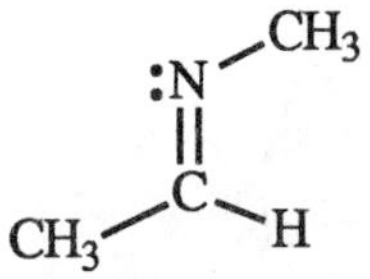

NOT INTER-
CONVERTIBLE

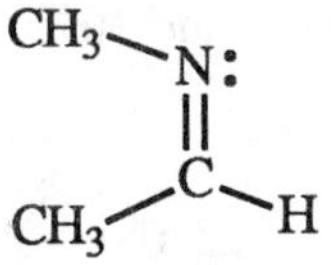

two CH_3's on opposite sides of the C=N

two CH_3's on the same side of the C=N

(c) the CH_3 on the N is on the same side as another CH_3 no matter how it is drawn—only one possible structure

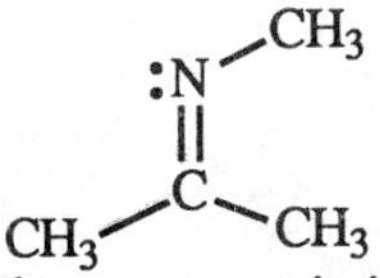

2-8 2-Methyl-2-butene does not have geometric isomers. In order to have geometric isomerism, the two groups on each double-bonded atom must be different. In this case, one sp^2 carbon has two groups the same, two CH_3's.

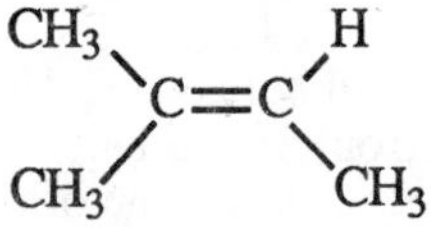

same as

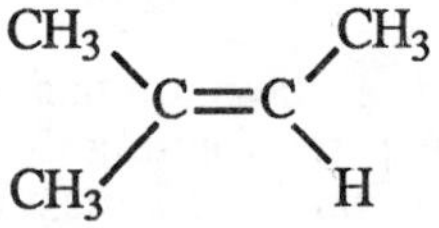

2-9

(a)

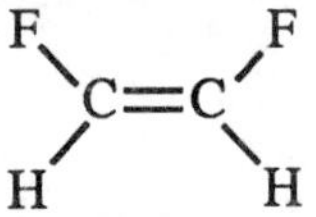

and

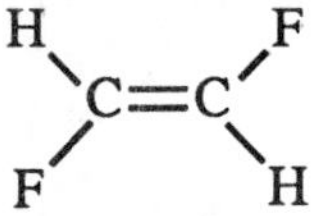

no geometric isomerism in (b), (c), and (d)

(e)

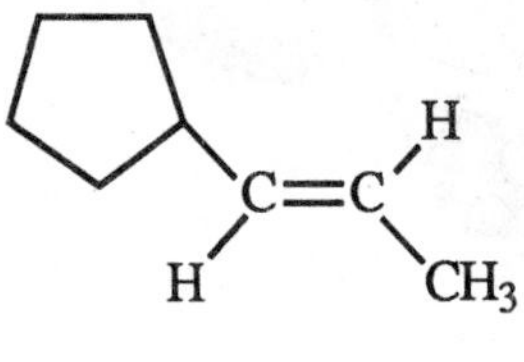

and

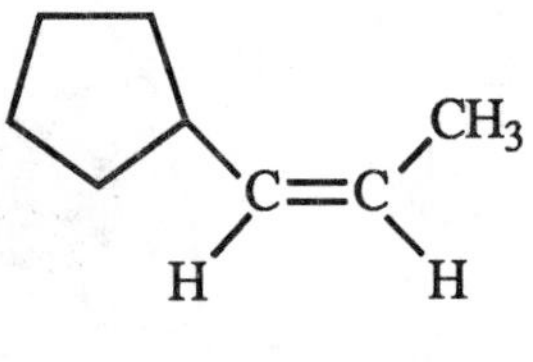

(f)

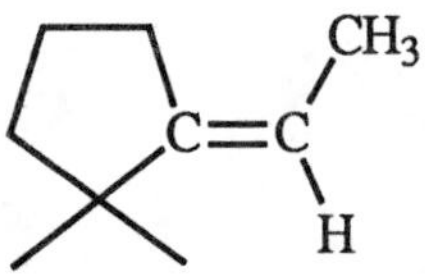

and

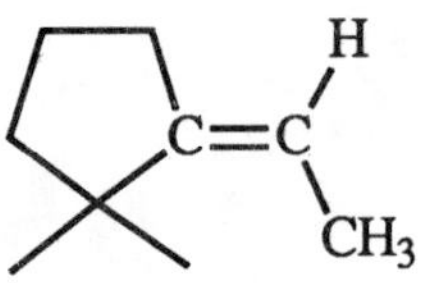

2-10 Models will be helpful here.

(a) structural isomers
(b) geometric isomers
(c) structural isomers
(d) same compound
(e) same compound
(f) same compound
(g) not isomers
(h) structural isomers
(i) same compound
(j) structural siomers
(k) structural isomers

2-11 (a) $2.4\ D = 4.8 \times \delta \times 1.21\ Å$

$\delta = 0.41$, or 41% of a positive charge on carbon and 41% of a negative charge on oxygen

(b)

A B

Resonance form A must be the major contributor. If B were the major contributor, the value of the charge separation would be between 0.5 and 1.0. Even though B is "minor," is it quite significant, explaining in part the high polarity of the C=O.

2-12 Both NH_3 and NF_3 have a pair of nonbonding electrons on the nitrogen. In NH_3, the *direction* of polarization of the N—H bonds is *toward* the nitrogen; thus, all three bond polarities and the lone pair polarity reinforce each other. In NF_3, on the other hand, the direction of polarization of the N—F bonds is *away* from the nitrogen; the three bond polarities cancel the lone pair polarity, so the net result is a very small *molecular* dipole moment.

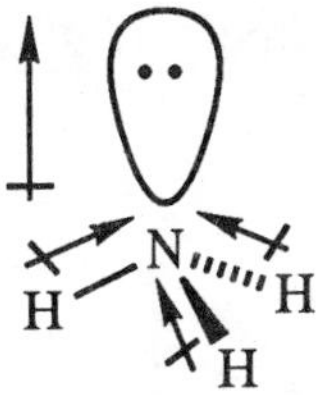

polarities reinforce;
large dipole moment

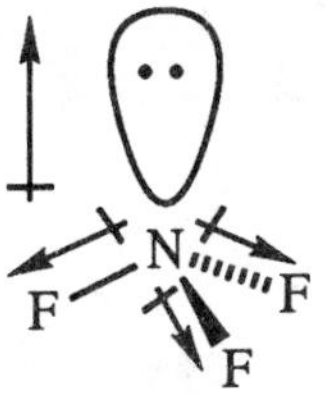

polarities oppose;
small dipole moment

2-13 Some magnitudes of dipole moments are difficult to predict; however, the direction of the dipole should be straightforward, in most cases. Actual values of molecular dipole moments are given in parentheses. (Each halogen atom has three nonbonded pairs, not shown below.)

The C—H is usually considered non-polar.

(a)

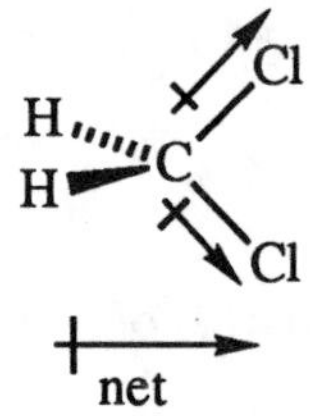

large dipole (1.54)

(b)

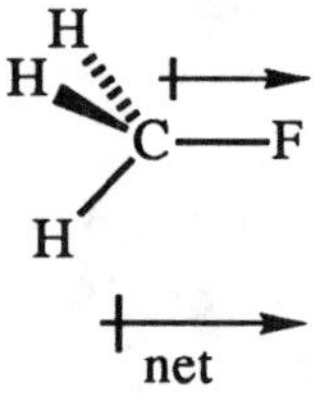

large dipole (1.81)

(c)

net dipole = 0

(d)

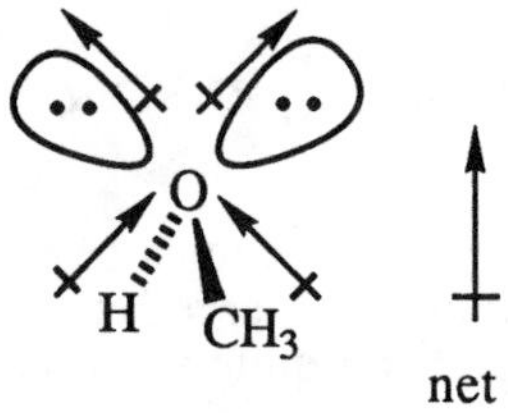

large dipole (1.70)

(e)

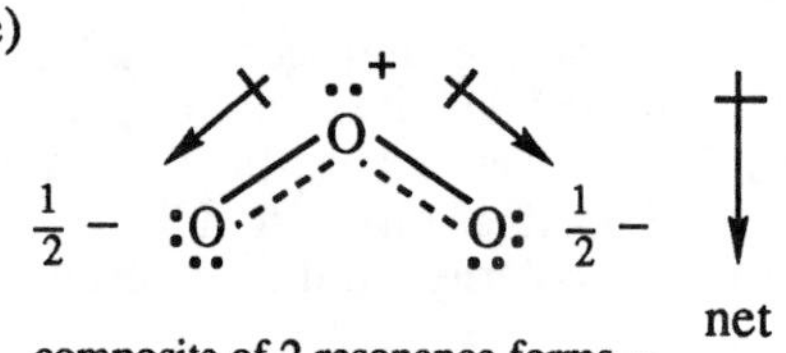

composite of 2 resonance forms—
see solution to 1-36(b)

small dipole (0.52)

(f)

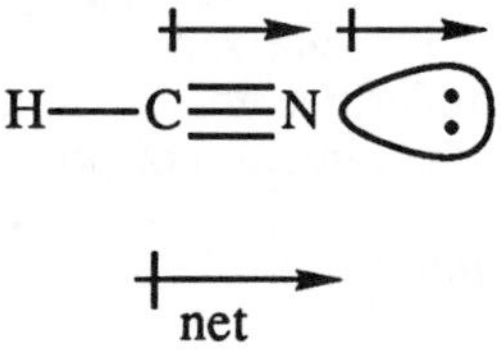

large dipole (2.95)

(g)

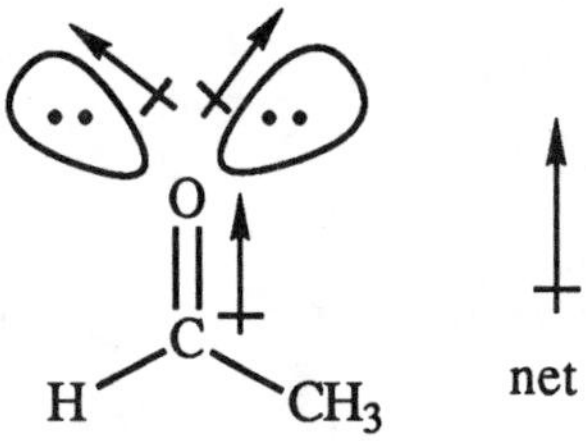

large dipole (2.72)

(h)

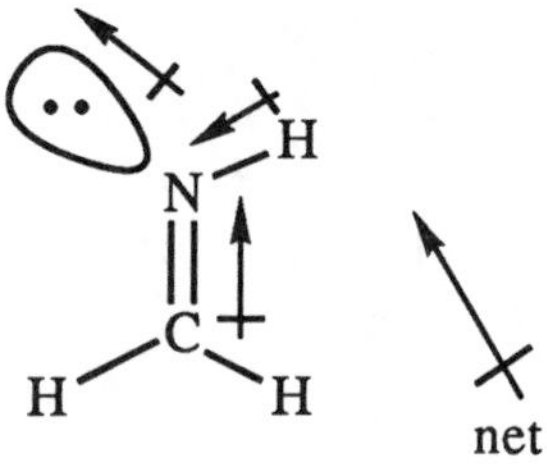

large dipole

2-13 continued

(i)

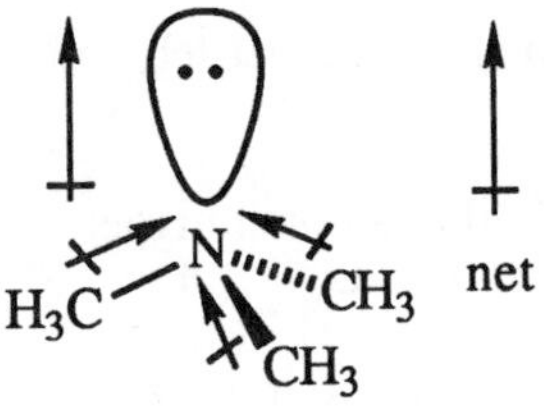

small dipole (0.67)

(j)

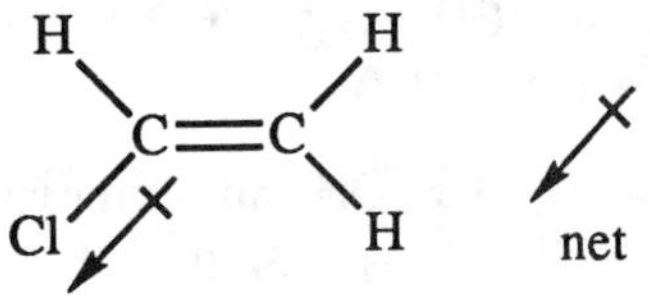

large dipole (1.45)

(k)

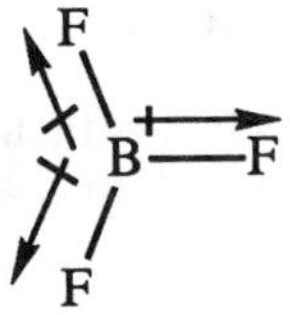

net dipole = 0

(l)

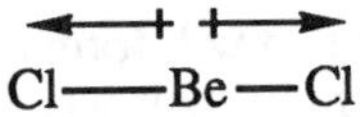

net dipole = 0

(m)

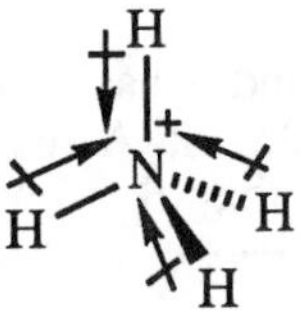

net dipole = 0

2-14 With chlorines on the same side of the double bond, the bond dipole moments reinforce each other, resulting in a large net dipole. With chlorines on opposite sides of the double bonds, the bond dipole moments exactly cancel each other, resulting in a zero net dipole.

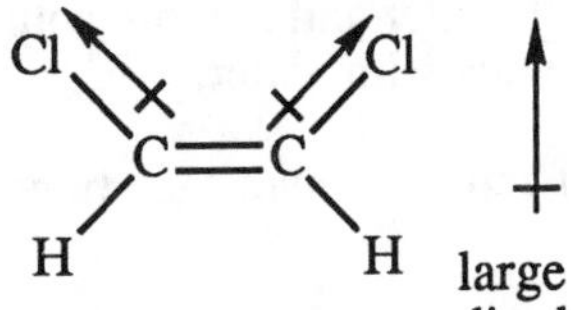

large net dipole

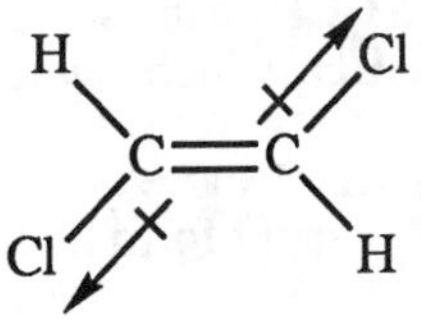

net dipole = 0

2-15

(a)

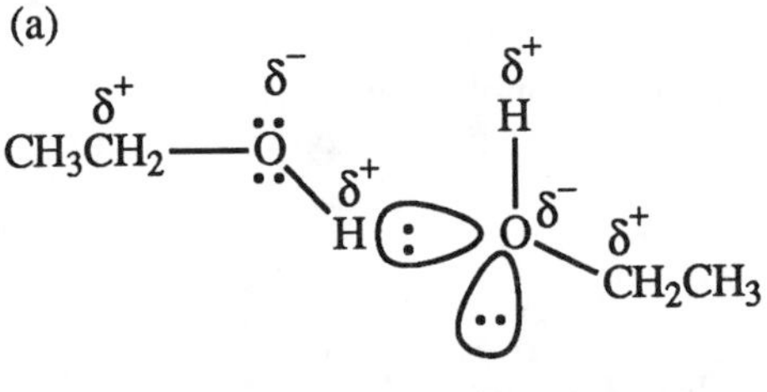

(b)

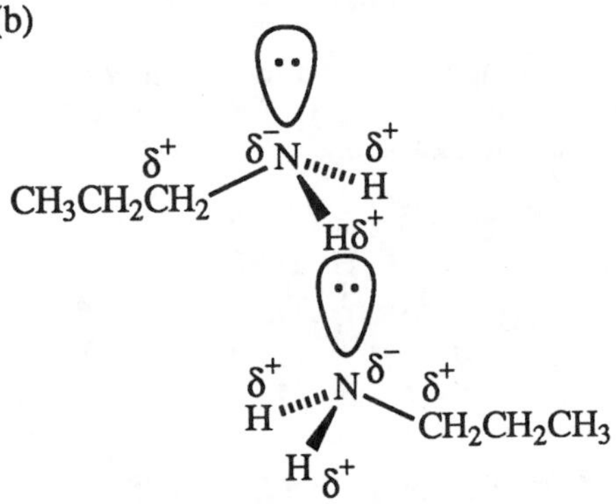

2-16

(a) $(CH_3)_2CHCH_2CH_2CH(CH_3)_2$ has less branching and boils at a higher temperature than $(CH_3)_3CC(CH_3)_3$.

(b) $CH_3(CH_2)_5CH_2OH$ can form hydrogen bonds and will boil at a much higher temperature than $CH_3(CH_2)_6CH_3$ which cannot form hydrogen bonds.

(c) $HOCH_2(CH_2)_4CH_2OH$ can form hydrogen bonds at both ends and has no branching; it will boil at a much higher temperature than $(CH_3)_3CCH(OH)CH_3$.

(d) $(CH_3CH_2CH_2)_2NH$ has an N—H bond and can form hydrogen bonds; it will boil at a higher temperature than $(CH_3CH_2)_3N$ which cannot form hydrogen bonds.

(e) Compound A has two N—H bonds and will form more hydrogen bonds than compound B will; A also has a higher molecular weight than B; A will boil at a higher temperature than B.

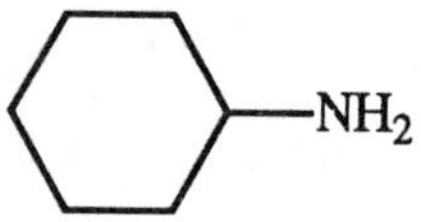

A

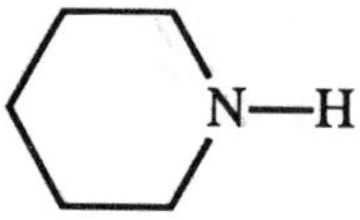

B

2-17

(a) $CH_3CH_2OCH_2CH_3$ can form hydrogen bonds with water and is more soluble than $CH_3CH_2CH_2CH_2CH_3$ which cannot form hydrogen bonds with water.

(b) $CH_3CH_2NHCH_3$ is more water soluble because it can form hydrogen bonds; $CH_3CH_2CH_2CH_3$ cannot form hydrogen bonds.

(c) CH_3CH_2OH is more soluble in water. The polar O—H group forms hydrogen bonds with water, overcoming the resistance of the non-polar CH_3CH_2 group toward entering the water. In $CH_3CH_2CH_2CH_2OH$, however, the hydrogen bonding from only one OH group cannot carry a four-carbon chain into the water; this substance is only slightly soluble in water.

(d) Both compounds form hydrogen bonds with water at the double-bonded oxygen, but only the smaller molecule (CH_3COCH_3) dissolves. The cyclic compound has too many non-polar CH_2 groups to dissolve.

2-18

(a)

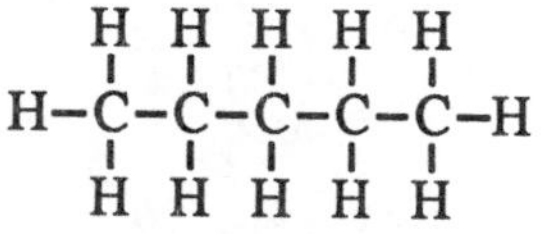

alkane

(Usually, we use the term "alkane" only when no other groups are present.)

(b)

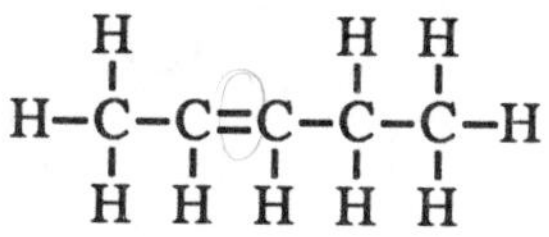

alkene

2-18 continued

(c)

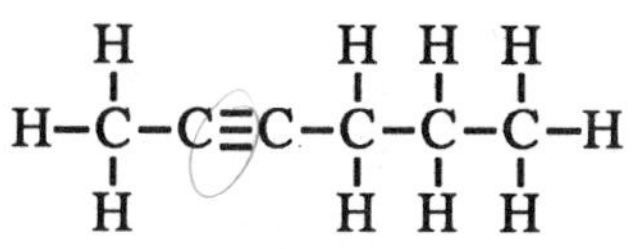

alkyne

(d)

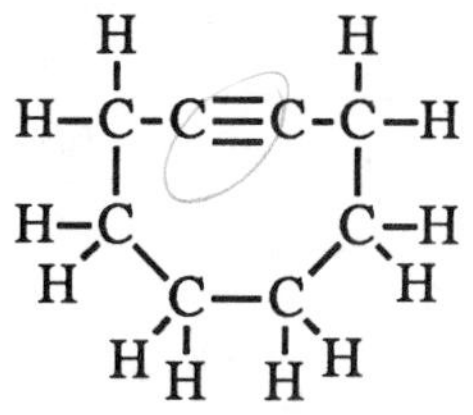

cycloalkyne

(e)

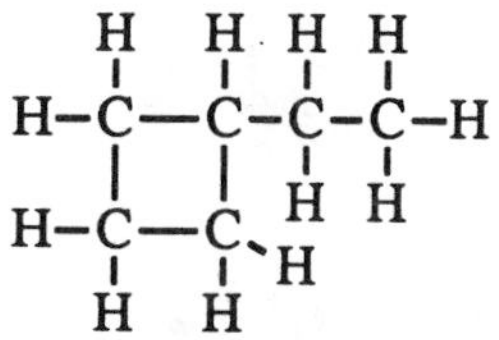

cycloalkane

(f)

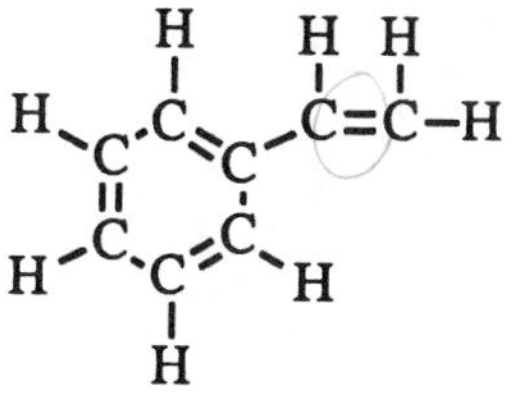

aromatic hydrocarbon
and alkene

(g)

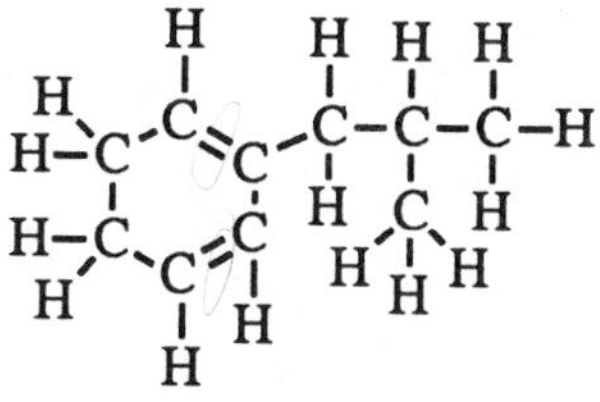

cycloalkene

(h)

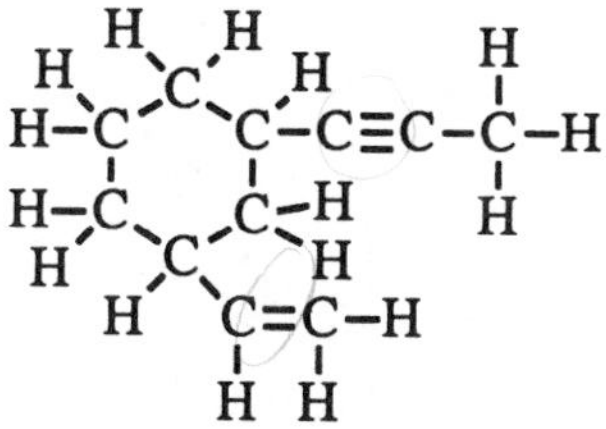

alkyne, alkene

(i)

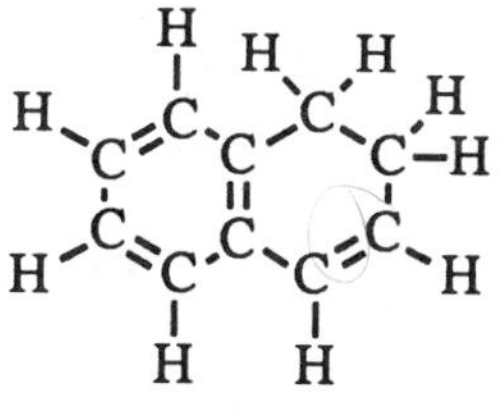

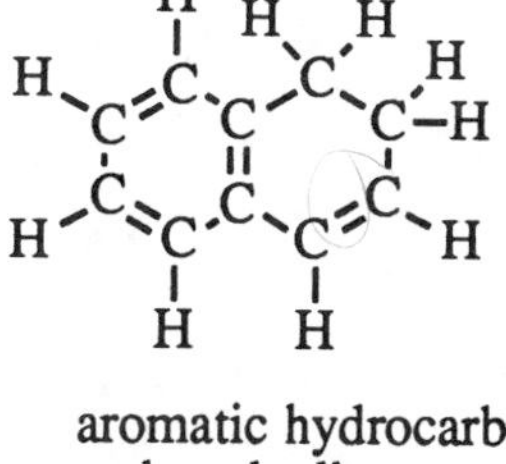

aromatic hydrocarbon
and cycloalkene

2-19

(a)

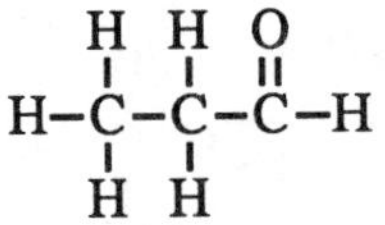

aldehyde

(b)

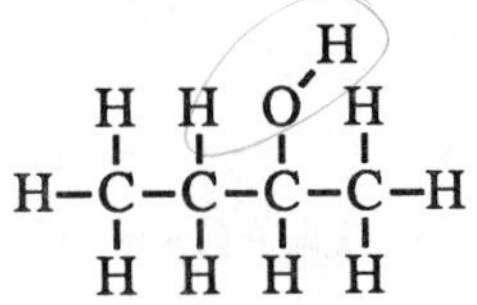

alcohol

(c)

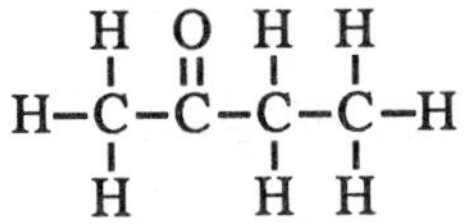

ketone

(d)

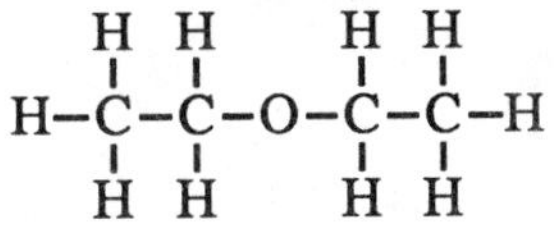

ether

(e)

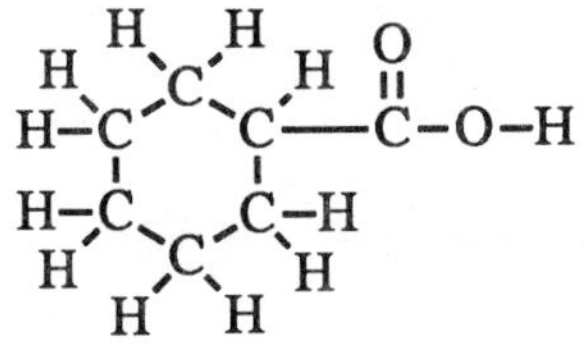

carboxylic acid

(f)

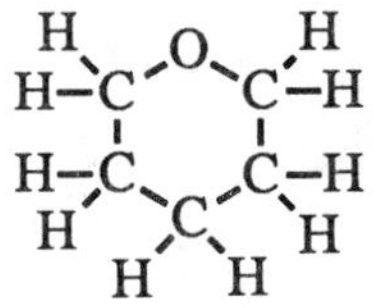

ether

(g)

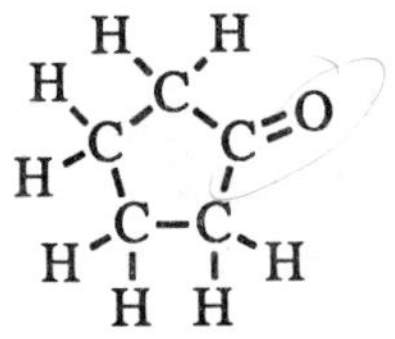

ketone

(h)

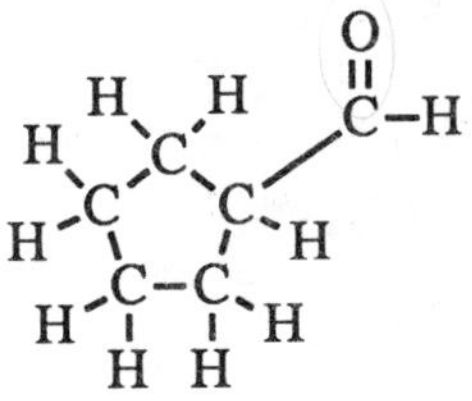

aldehyde

(i)

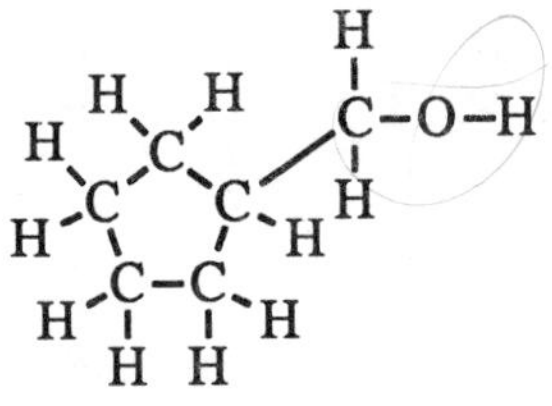

alcohol

2-20

(a) 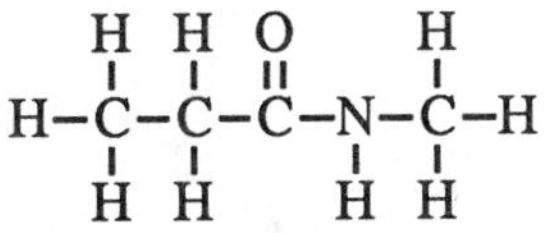

amide

(b) 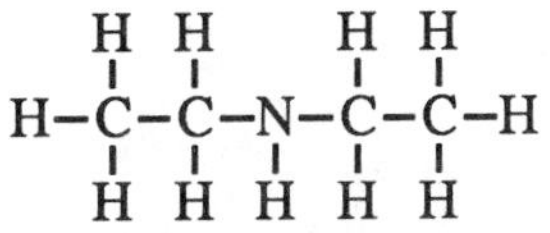

amine

(c) 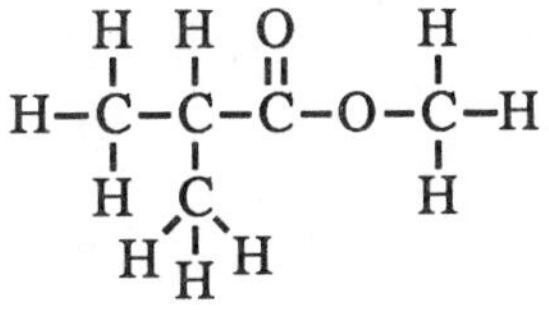

ester

(d) 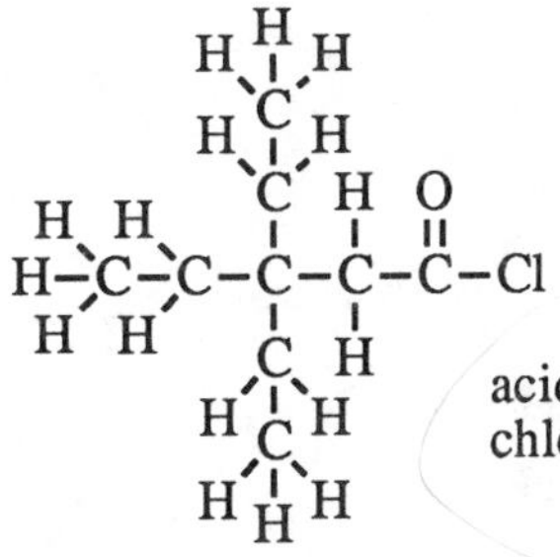

acid chloride

(e) 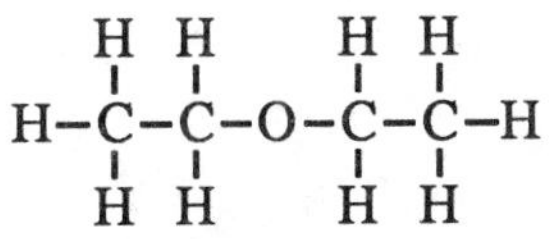

ether

(f) 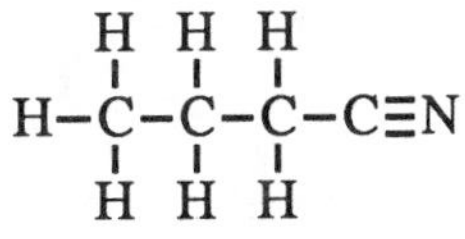

nitrile

(g) 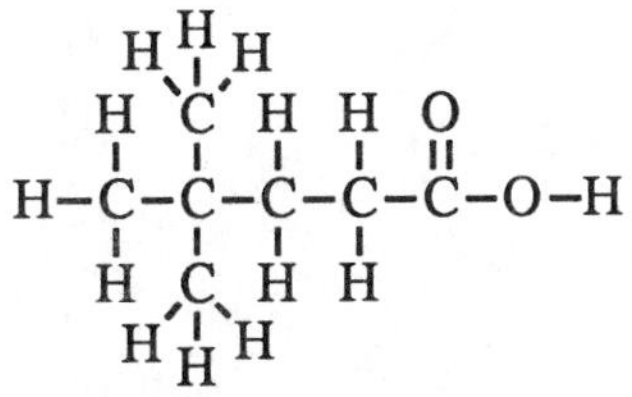

carboxylic acid

(h) 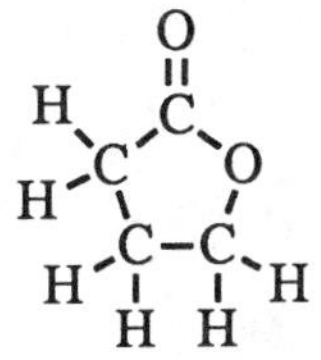

cyclic ester

(i) 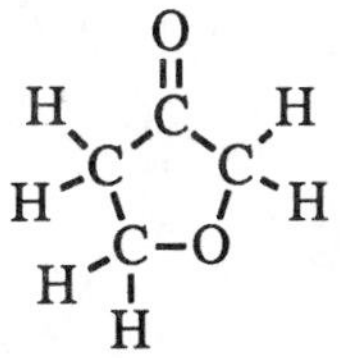

ketone and ether

(j) 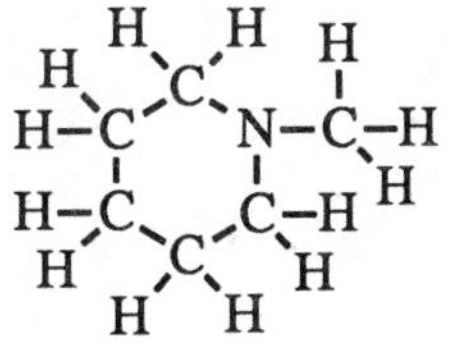

amine

2-20 continued

(k)

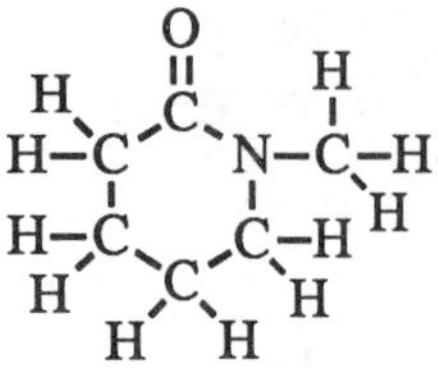

cyclic amide

(l)

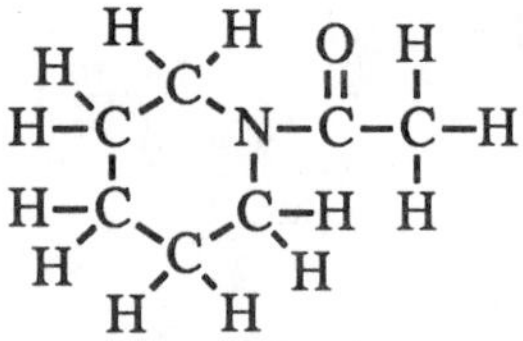

amide

(m)

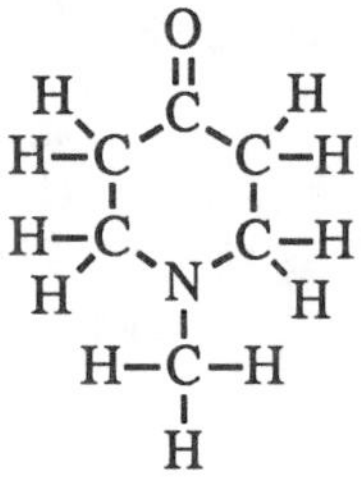

ketone and amine

(n)

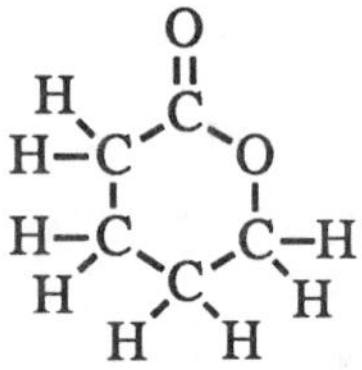

cyclic ester

(o)

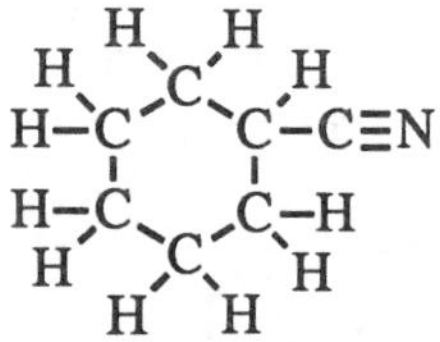

nitrile

(p)

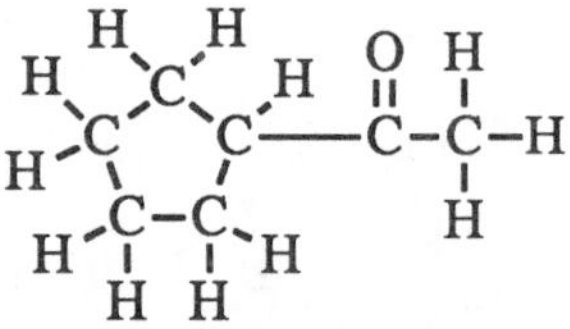

ketone

2-21 When the identity of a functional group depends on several atoms, all of those atoms should be circled. For example, an ether is an oxygen between two carbons, so the oxygen and both carbons should be circled. A ketone is a carbonyl group between two other carbons, so all those atoms should be circled.

(a) $CH_2{=}CHCH_2CH_3$

alkene

(b) $CH_3{-}O{-}CH_3$

ether

2-21 continued

(c)

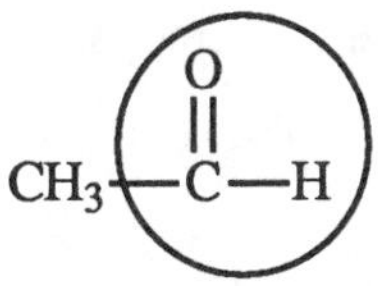

O
CH3—C—H

aldehyde

(d)

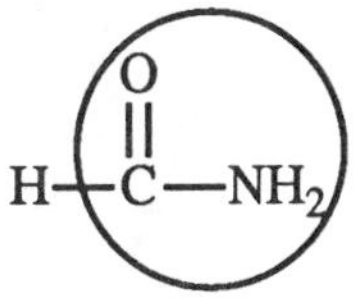

O
H—C—NH2

amide

(e)

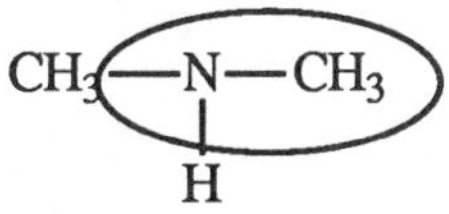

CH3—N—CH3
H

amine

(f)

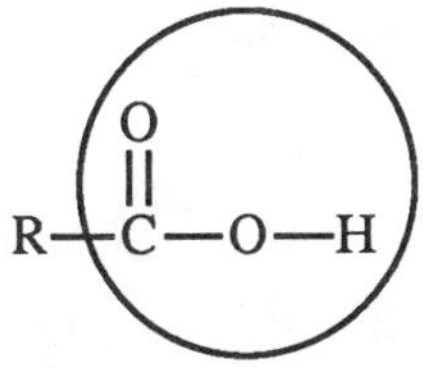

O
R—C—O—H

carboxylic acid
("R" is the general abbreviation
for an alkyl group)

(g)

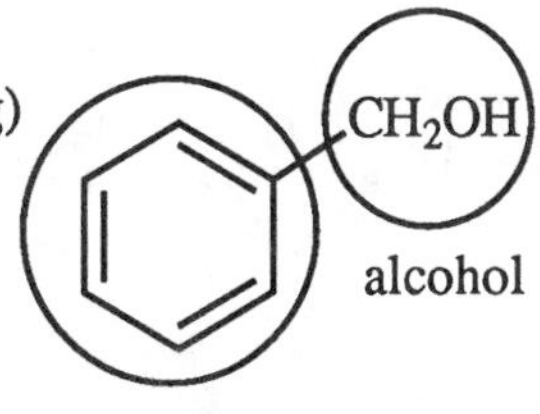

CH2OH
alcohol

aromatic

(h)

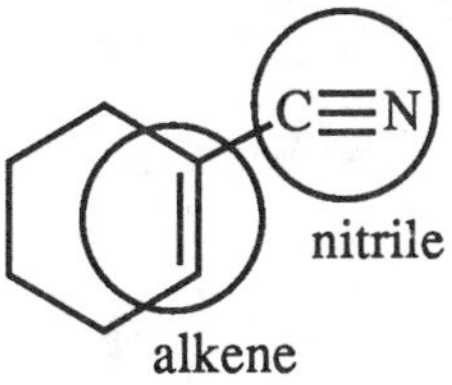

C≡N
nitrile
alkene

(i)

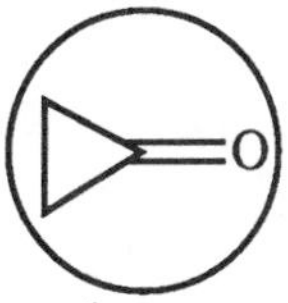

O

ketone

(j)

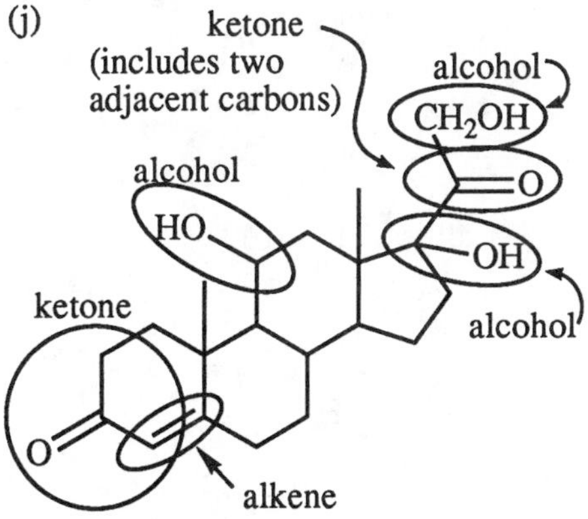

ketone
(includes two
adjacent carbons)
alcohol
CH2OH
O
alcohol
HO
OH
alcohol
ketone
O
alkene

(k)

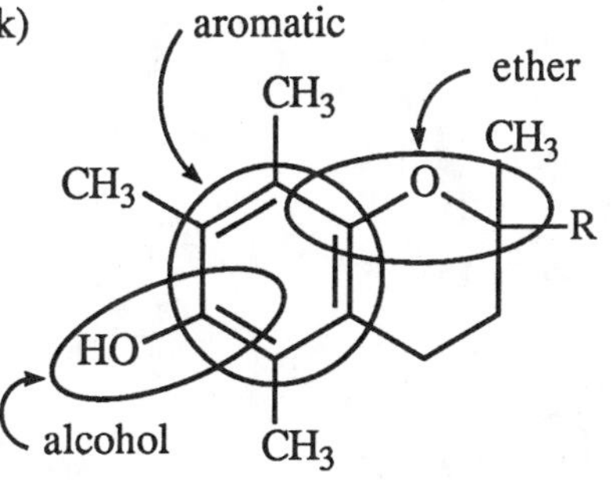

aromatic
CH3
ether
CH3
CH3
O
R
HO
alcohol
CH3

$R = (CH_2CH_2CH_2CH(CH_3))_3CH_3$

2-22 Refer to the Glossary and the text for definitions and examples.

2-23 Models show that the tetrahedral geometry of CH_2Cl_2 precludes stereoisomers.

2-24

(a)

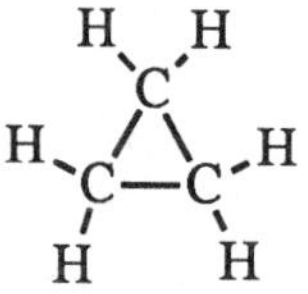

(b) Cyclopropane must have 60° bond angles compared with the usual sp^3 bond angle of 109.5° in an acyclic molecule.

(c) Like a bent spring, bonds that deviate from their normal angles or positions are highly strained. Cyclopropane is reactive because breaking the ring relieves the strain.

2-25

(a)

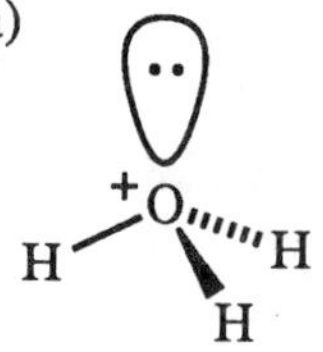

sp^3, ≈ 109°

(b)

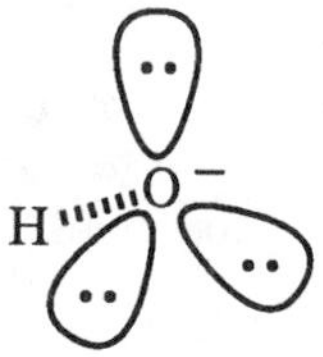

sp^3 ?, no bond angle

(c)

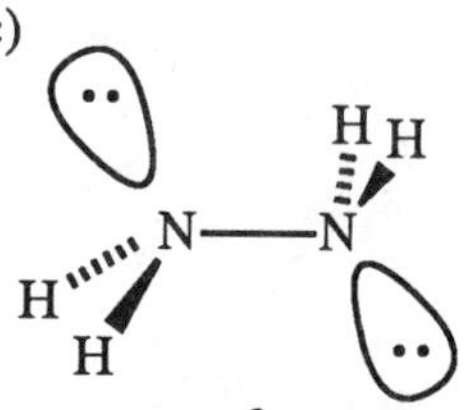

both sp^3, all ≈ 109°

(d)

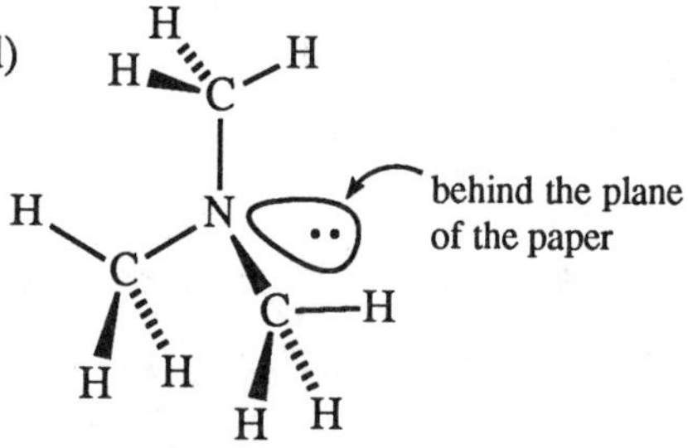
behind the plane
of the paper

all sp^3, all ≈ 109°

(e)

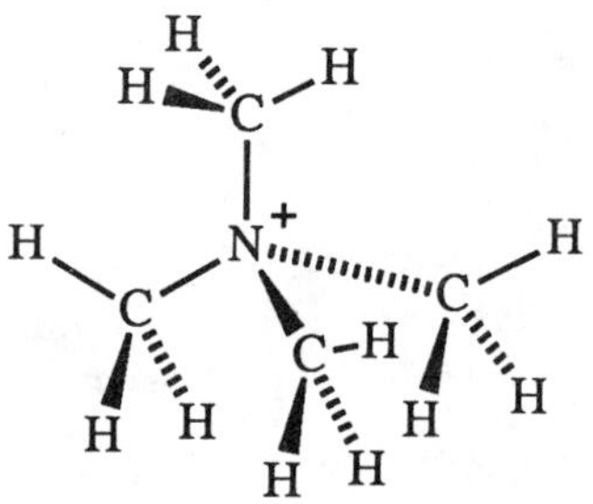

all sp^3, all ≈ 109°

(f)

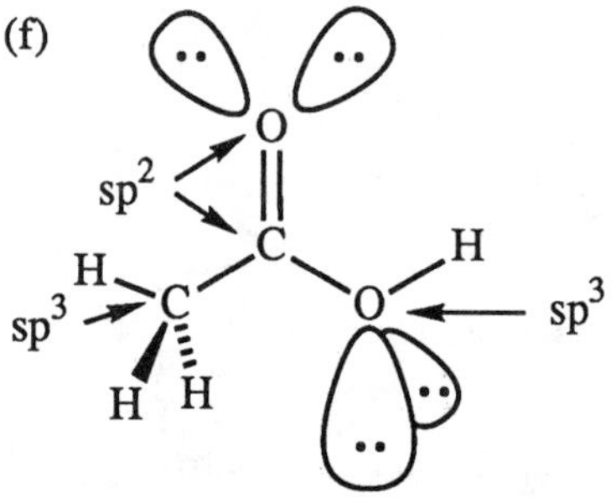
sp2
sp3
sp3

angles around sp^3 atoms ≈ 109°
angles around sp^2 carbon ≈ 120°

(g)

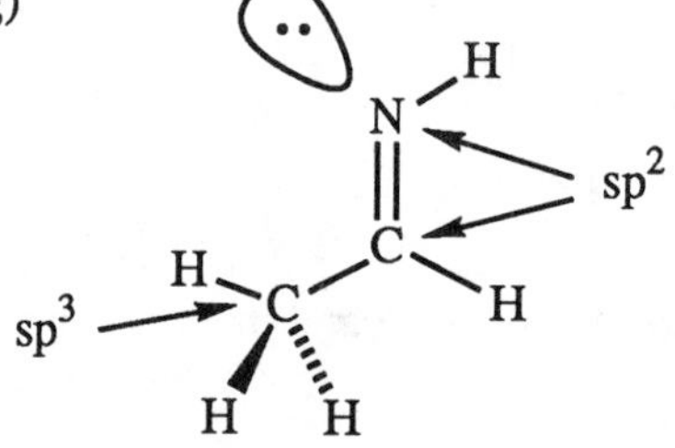
sp2
sp3

angles around sp^3 atom ≈ 109°
angles around sp^2 atoms ≈ 120°

2-25 continued

(h)

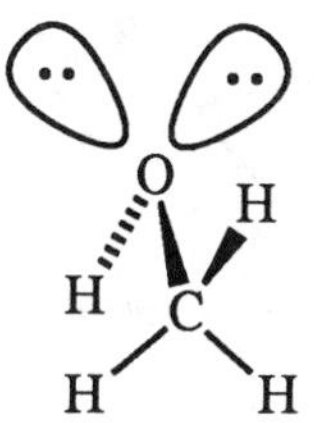
O
H
H
C
H
H

both sp^3, all ≈ 109°

(i)

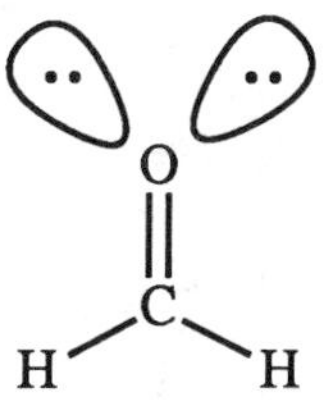
O
C
H
H

both sp^2, all ≈ 120°

2-26 For simplicity in these pictures, hydrogens bonded to sp^3 atoms are not labeled, although their bonds are shown. These bonds are s-sp^3 overlap.

(a)

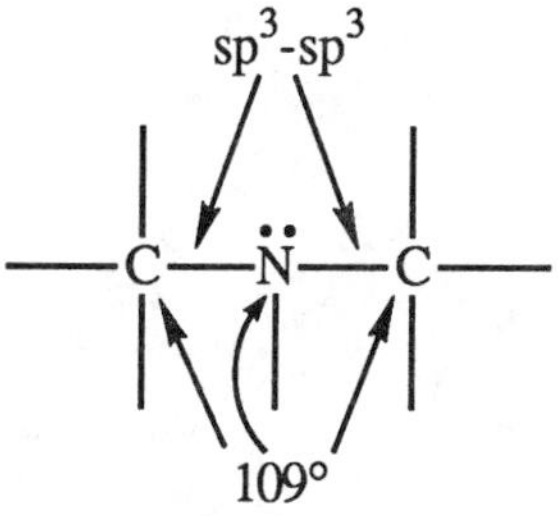
sp3-sp3
C
N
C
109°

(b)

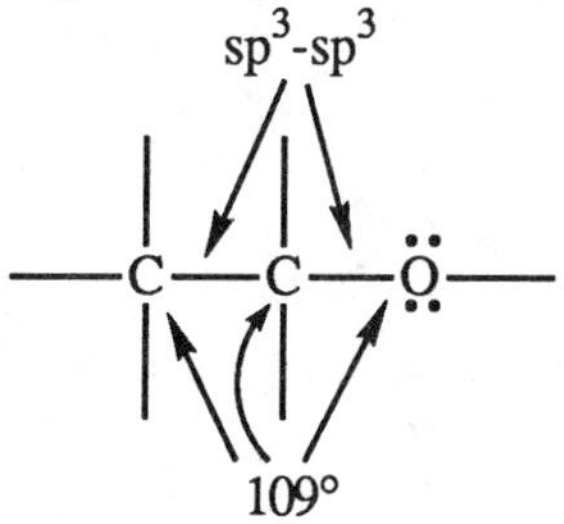
sp3-sp3
C
C
O
109°

(c)

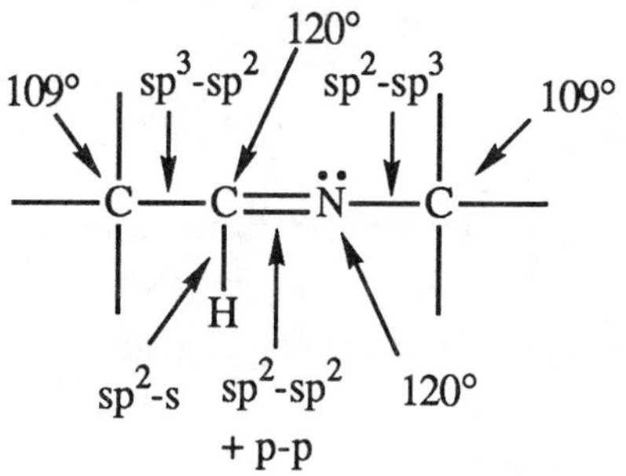
120°
109°
sp3-sp2
sp2-sp3
109°
C
C
N
C
H
sp2-s
sp2-sp2
+ p-p
120°

(d)

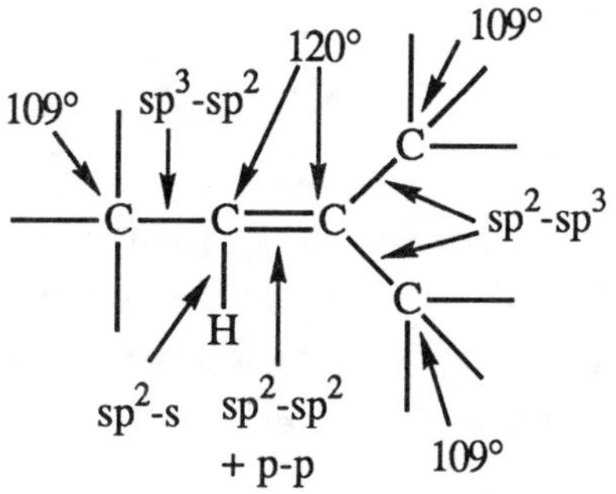
109°
120°
109°
sp3-sp2
C
C
C
C
C
sp2-sp3
H
sp2-s
sp2-sp2
+ p-p
109°

(e)

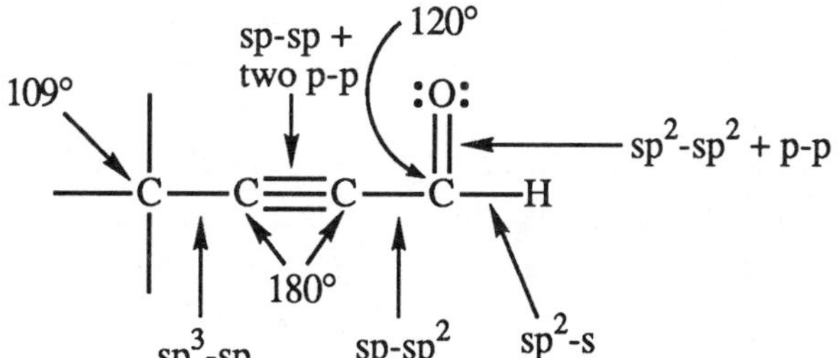
120°
sp-sp +
two p-p
109°
:O:
sp2-sp2 + p-p
C
C
C
C
H
180°
sp3-sp
sp-sp2
sp2-s

2-26 continued

(f)

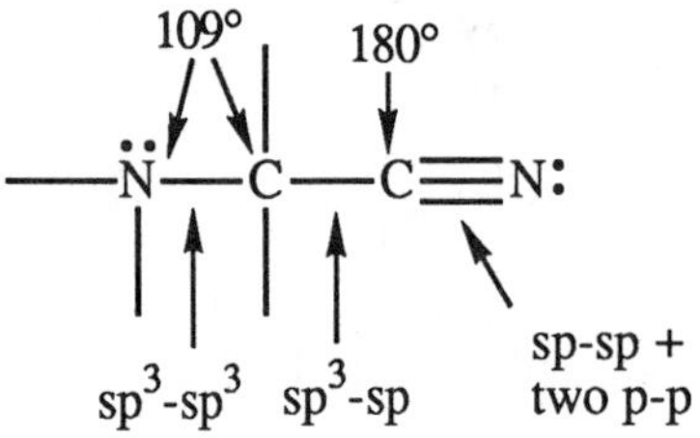
109°
180°
sp³-sp³
sp³-sp
sp-sp +
two p-p

(g)

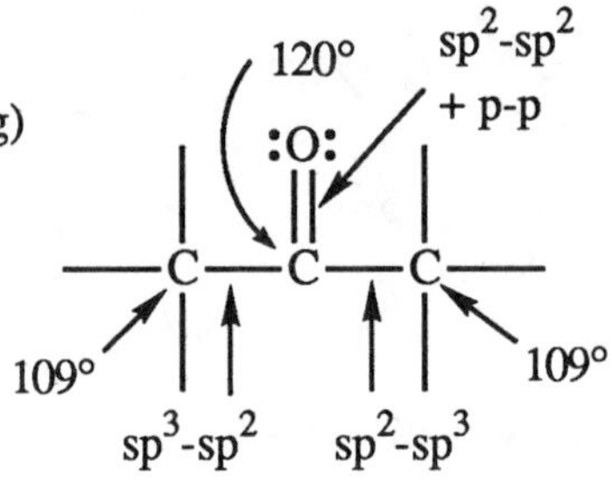
120°
sp²-sp²
+ p-p
109°
sp³-sp²
sp²-sp³
109°

(h)

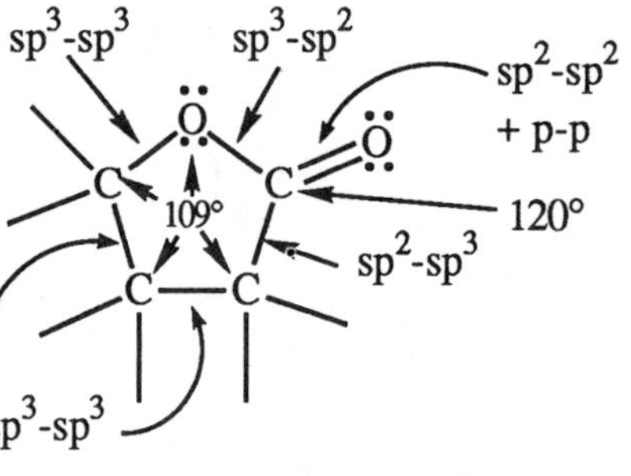
sp³-sp³
sp³-sp²
sp²-sp²
+ p-p
109°
120°
sp²-sp³
sp³-sp³

(i)

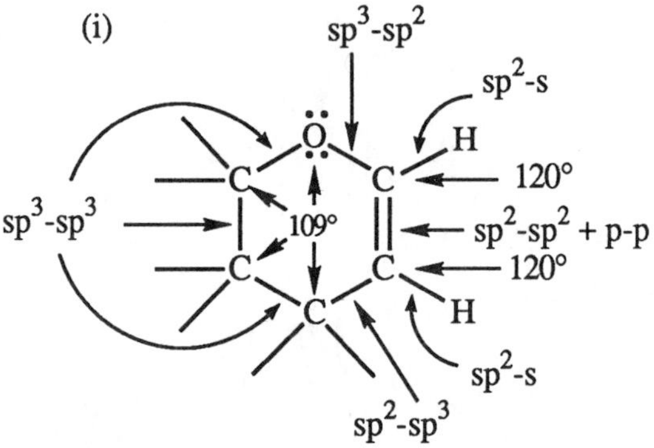
sp³-sp²
sp²-s
H
120°
sp³-sp³
109°
sp²-sp² + p-p
120°
H
sp²-s
sp²-sp³

2-27 The second resonance form of formamide is a minor but significant resonance contributor. It shows that the nitrogen-carbon bond has some double bond character, requiring that the nitrogen be sp^2 hybridized with bond angles approaching 120°.

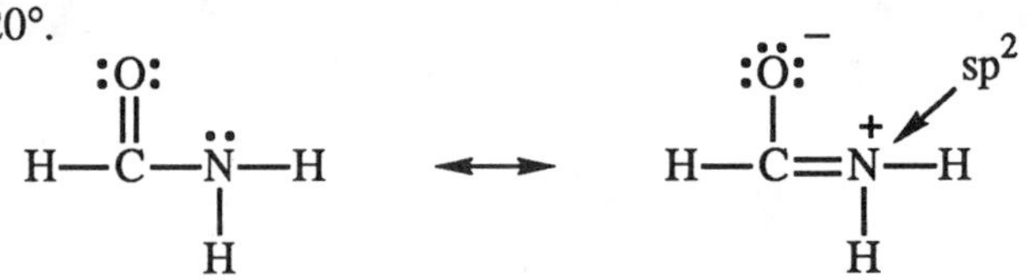
sp²

2-28 The major resonance contributor shows a carbon-carbon double bond, suggesting that both carbons are sp^2 hybridized with trigonal planar geometry. The CH_3 carbon is sp^3 hybridized with tetrahedral geometry.

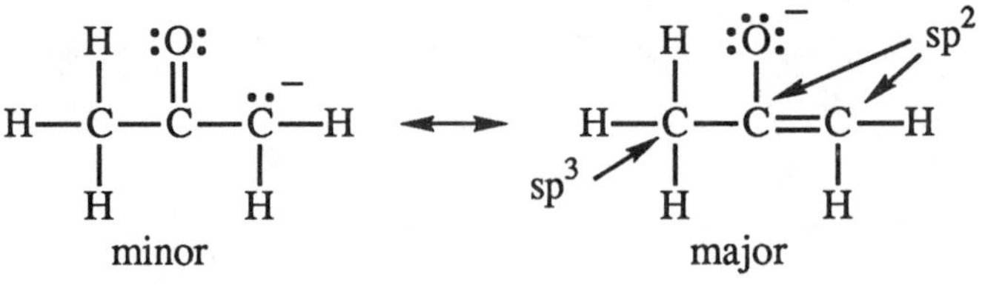
sp²
sp³
minor
major

2-29 (a)

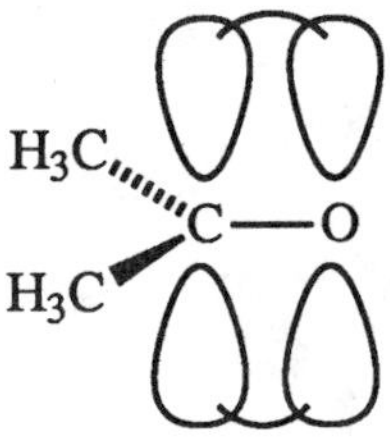

(b)

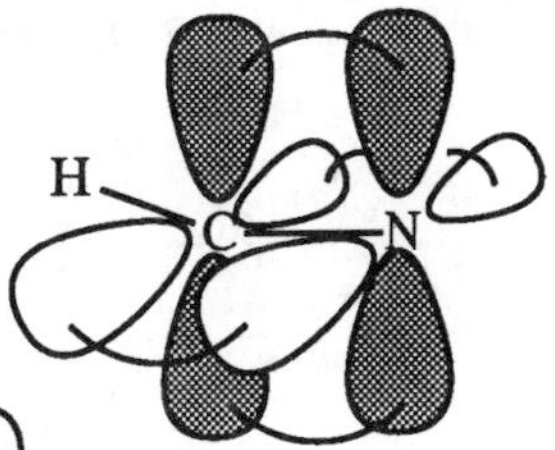

(c)

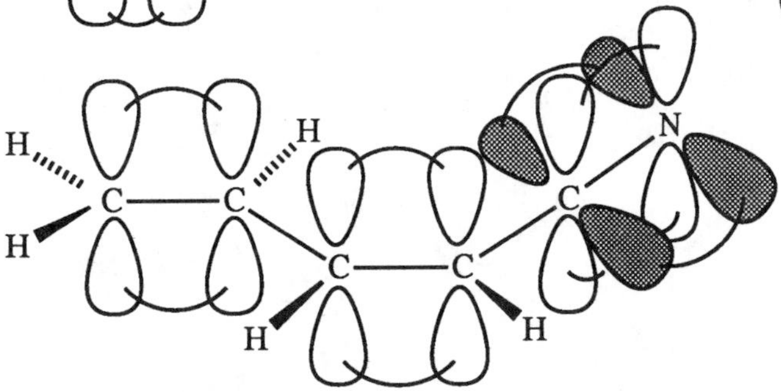

(d)

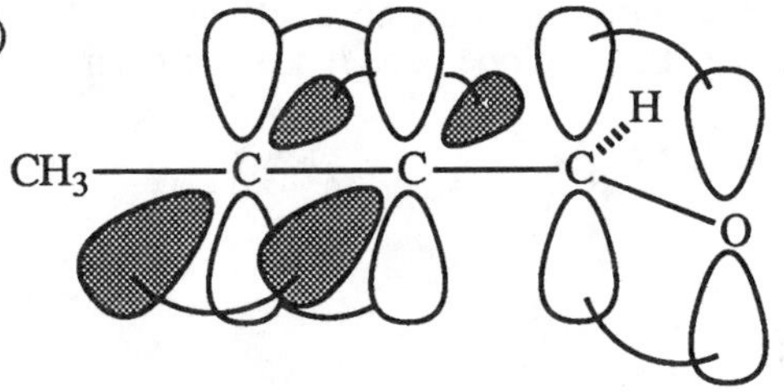

2-30
(a)

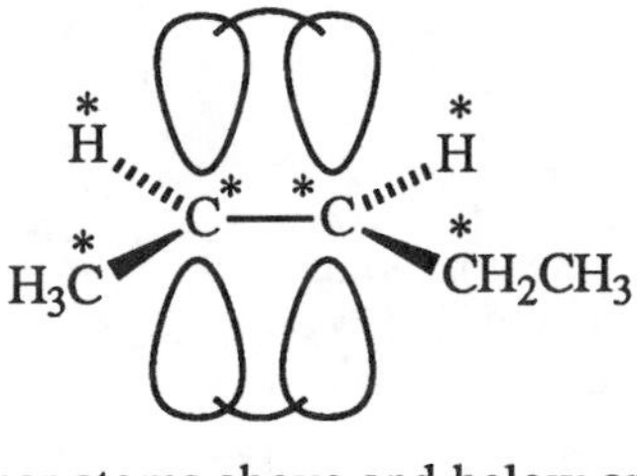

(b) The coplanar atoms above and below are marked with asterisks.

(c)

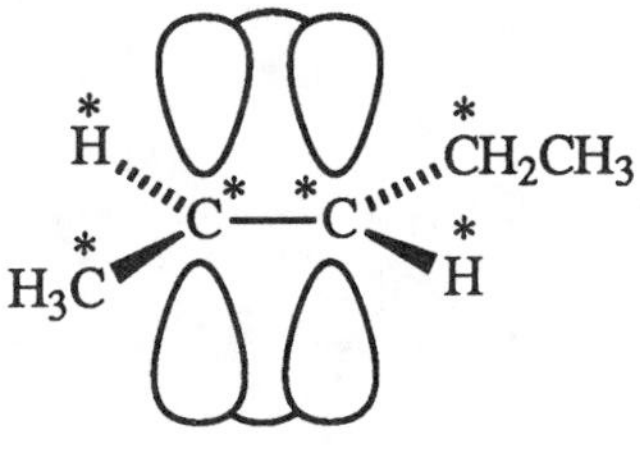

(d)

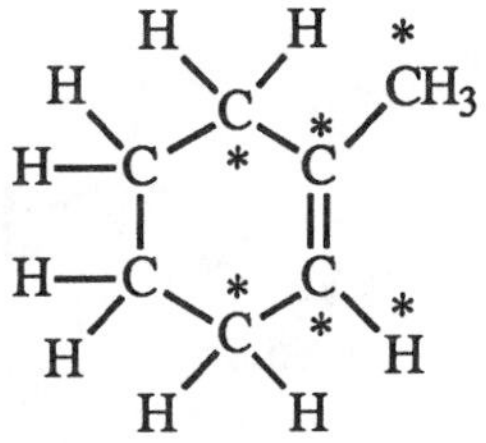

2-31 Collinear atoms are marked with asterisks.

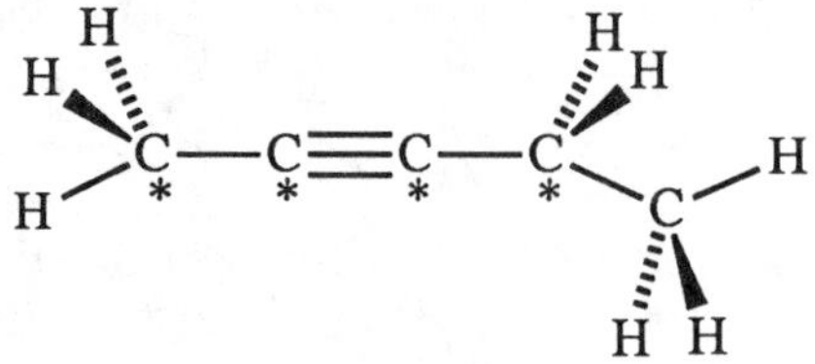

2-32 (a) no geometric isomerism

(b)

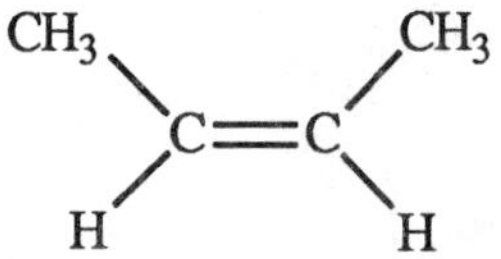

cis

and

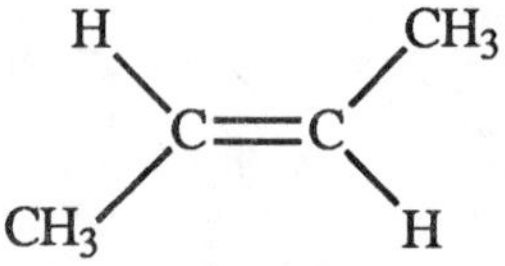

trans

(c) no geometric isomerism

(d) Theoretically, cyclopentene could show geometric isomerism. In reality, the *trans* form is too unstable to exist because of the necessity of stretched bonds and deformed bond angles. *trans*-Cyclopentene has never been detected.

cis

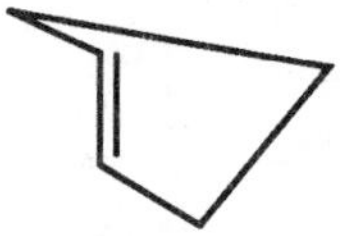

"*trans*"

(e)

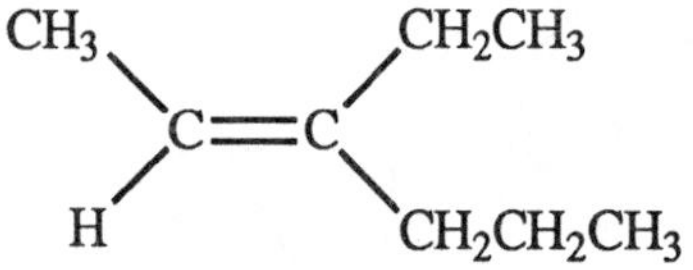

and

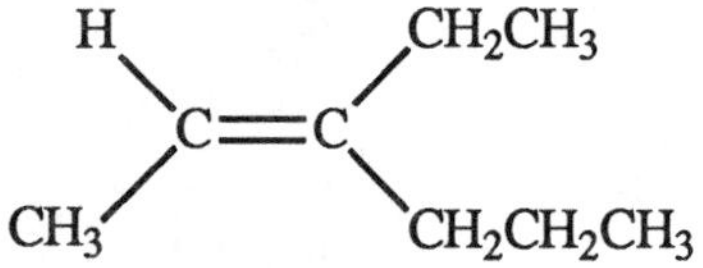

2-33 (a) structural isomers
(b) structural isomers
(c) geometric isomers
(d) structural isomers
(e) geometric isomers
(f) same compound
(g) geometric isomers

2-34 CO_2 is linear; its bond dipoles cancel, so it has no net dipole. SO_2 is bent, so its bond dipoles do not cancel.

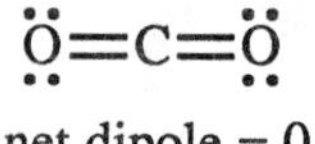

net dipole = 0

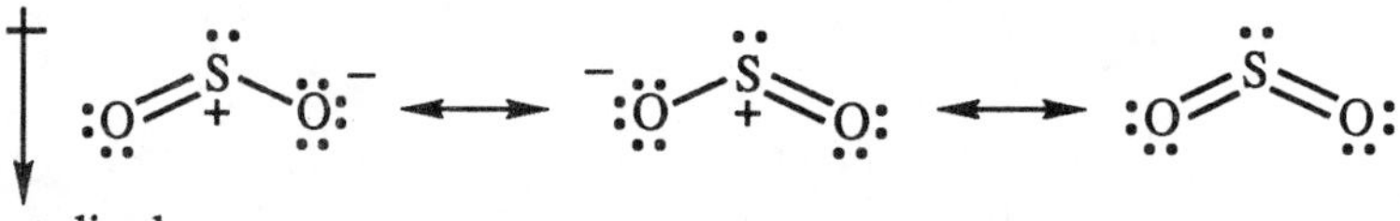

net dipole

2-35 Some magnitudes of dipole moments are difficult to predict; however, the direction of the dipole should be straightforward in most cases. Actual values of molecular dipole moments are given in parentheses. (The C—H bond is usually considered non-polar.)

(a)

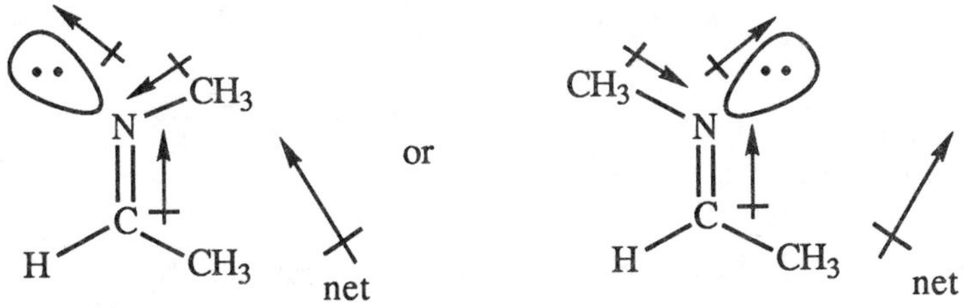

large dipole

large dipole

(b)

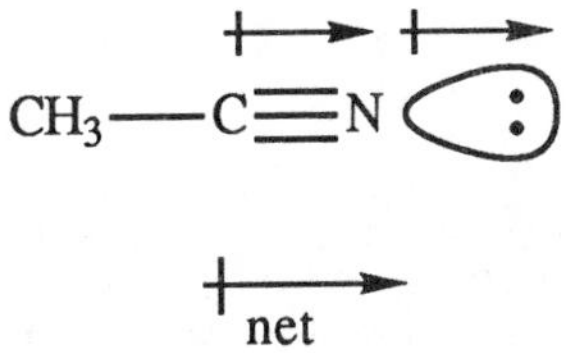

large dipole (3.96)

(c)

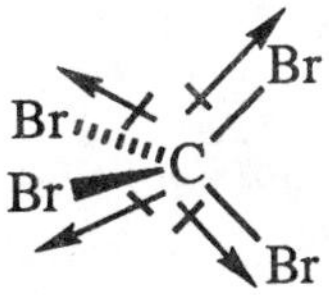

net dipole = 0

(d)

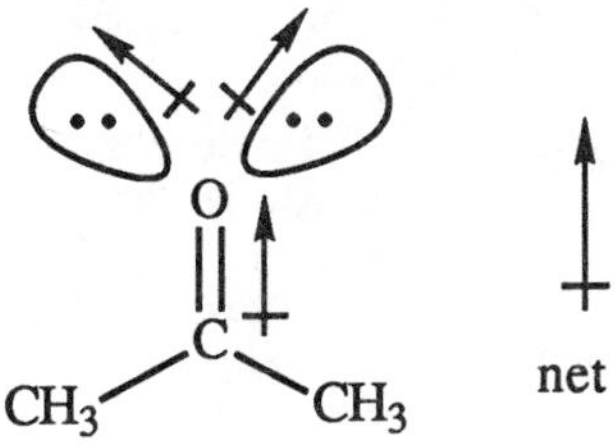

large dipole (2.89)

(f)

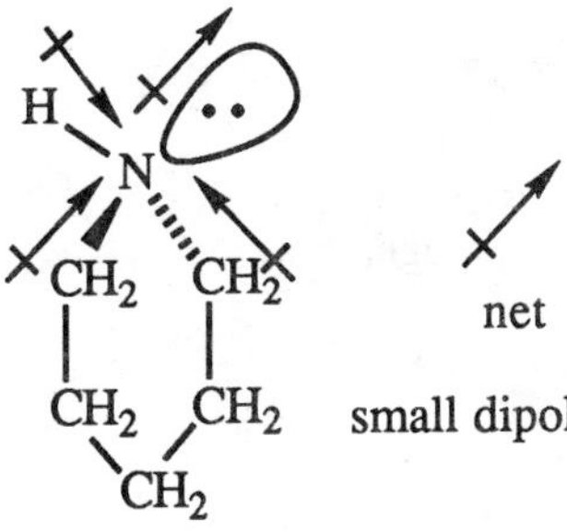

small dipole

2-35 continued

(e)

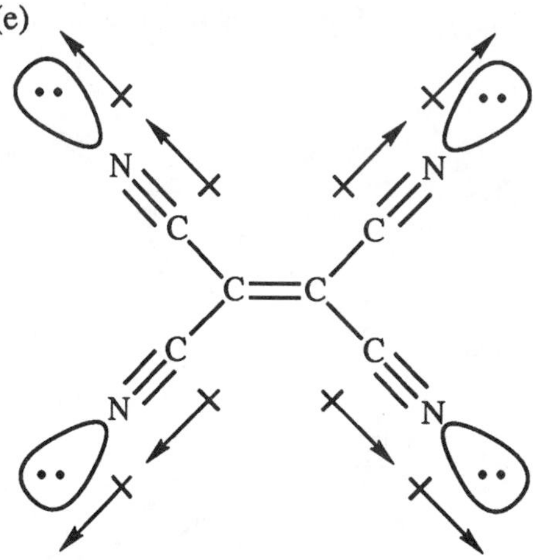

net dipole = 0

(g)

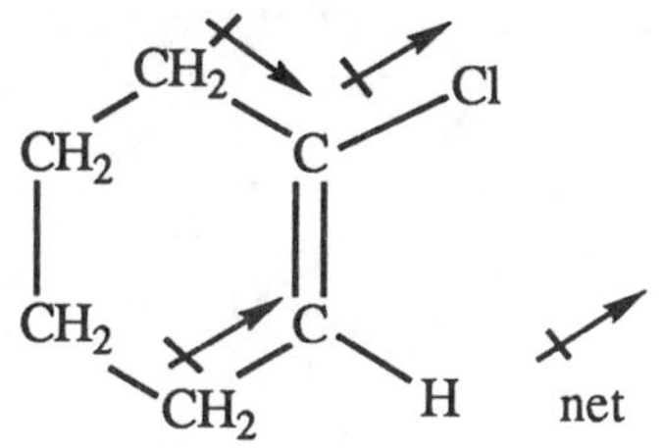

large dipole

2-36 Both diethyl ether and 1-butanol have one oxygen, so they can both hydrogen bond with water (water supplies the H for hydrogen bonding with diethyl ether); their water solubilities should be similar. The boiling point of 1-butanol is much higher because these molecules can hydrogen bond with each other, thus requiring more energy to separate one molecule from another. Diethyl ether molecules cannot hydrogen bond with each other, so it is relatively easy to separate them.

CH_3CH_2—O—CH_2CH_3

diethyl ether

can hydrogen bond with water
cannot hydrogen bond with itself

$CH_3CH_2CH_2CH_2$—OH

1-butanol

can hydrogen bond with water
can hydrogen bond with itself

2-37

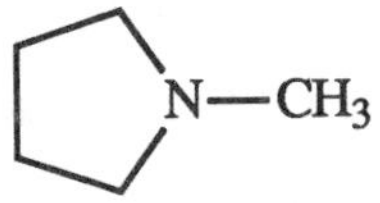

N-methylpyrrolidine
b.p. 81°C

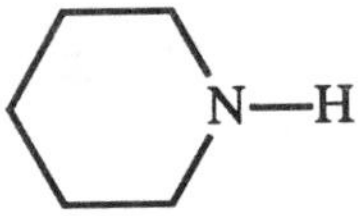

piperidine
b.p. 106°C

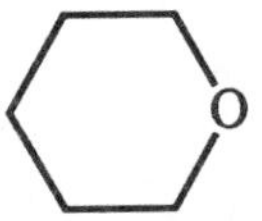

tetrahydropyran
b.p. 88°C

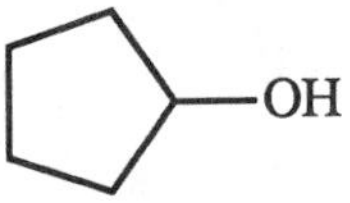

cyclopentanol
b.p. 141°C

(a) Piperidine has an N—H bond, so it can hydrogen bond with other molecules of itself. N-Methylpyrrolidine has no N—H, so it cannot hydrogen bond and will require less energy (lower boiling point) to separate one molecule from another.

(b) Two effects need to be explained: 1) Why does cyclopentanol have a higher boiling point than tetrahydropyran? and 2) Why do the oxygen compounds have a greater difference in boiling points than the analogous nitrogen compounds?

The answer to the first question is the same as in (a): cyclopentanol can hydrogen bond with its neighbors while tetrahydropyran cannot.

The answer to the second question lies in the text, Table 2-1, that shows the bond dipole moments for C—O and H—O are much greater than C—N and H—N; bonds to oxygen are more polarized, with greater charge separation than bonds to nitrogen.

How is this reflected in the data? The boiling points of tetrahydropyran (88°C) and N-methylpyrrolidine (81°C) are close; tetrahydropyran molecules would have a slightly stronger dipole-dipole attraction, and tetrahydropyran is a little less "branched" than N-methylpyrrolidine, so it is reasonable that tetrahydropyran boils at a slightly higher temperature. The large difference comes when comparing the boiling points of cyclopentanol (141°C) and piperidine (106°C). The greater polarity of O—H versus N—H is reflected in a more negative oxygen (more electronegative than nitrogen) and a more positive hydrogen, resulting in a much stronger intermolecular attraction. The conclusion is that hydrogen bonding due to O—H is much stronger than that due to N—H.

2-38 (a) can hydrogen bond with itself and with water
(b) can hydrogen bond only with water
(c) can hydrogen bond with itself and with water
(d) can hydrogen bond only with water
(e) cannot hydrogen bond
(f) cannot hydrogen bond
(g) can hydrogen bond only with water
(h) can hydrogen bond with itself and with water
(i) can hydrogen bond only with water
(j) can hydrogen bond only with water
(k) can hydrogen bond only with water
(l) can hydrogen bond with itself and with water

2-39 Higher-boiling compounds are listed.

(a) $CH_3CH(OH)CH_3$ —hydrogen bonding

(b) $CH_3CH_2CH_2CH_2CH_3$ —higher molecular weight

(c) $CH_3CH_2CH_2CH_2CH_3$ —less branching

(d) $CH_3CH_2CH_2CH_2CH_2Cl$ —higher molecular weight, dipole-dipole interaction

2-40 (a)

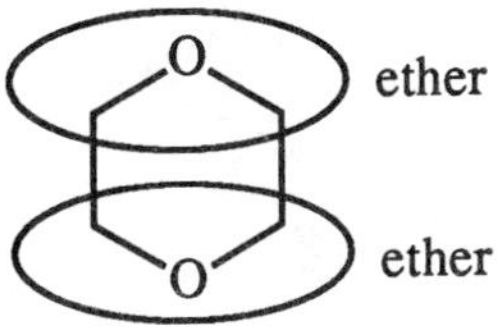
ether
ether

(b)

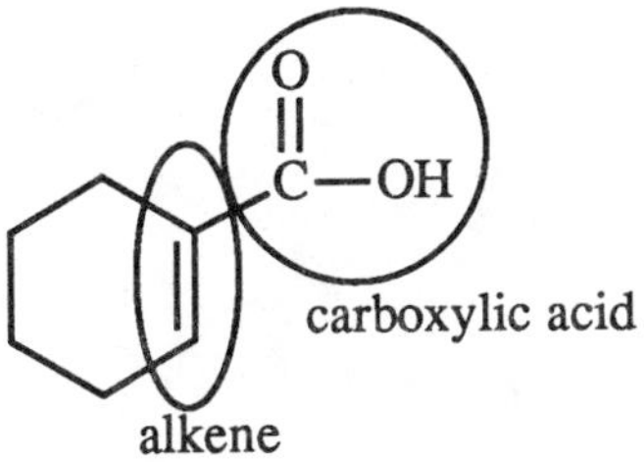
carboxylic acid
alkene

(c)

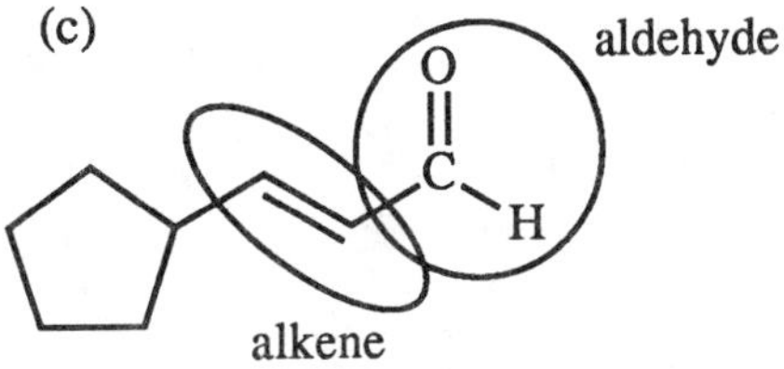
aldehyde
alkene

(d)

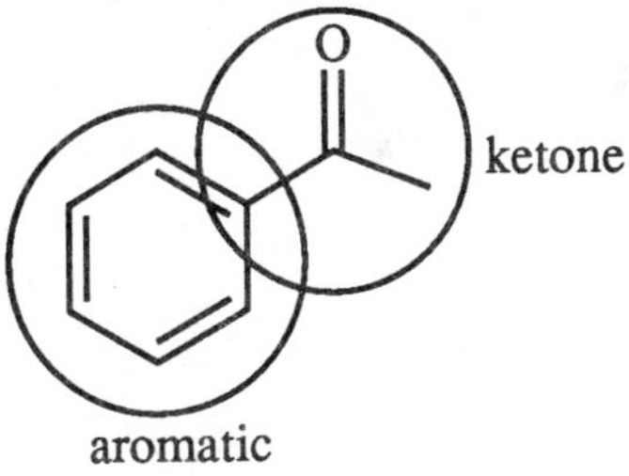
ketone
aromatic

(e)

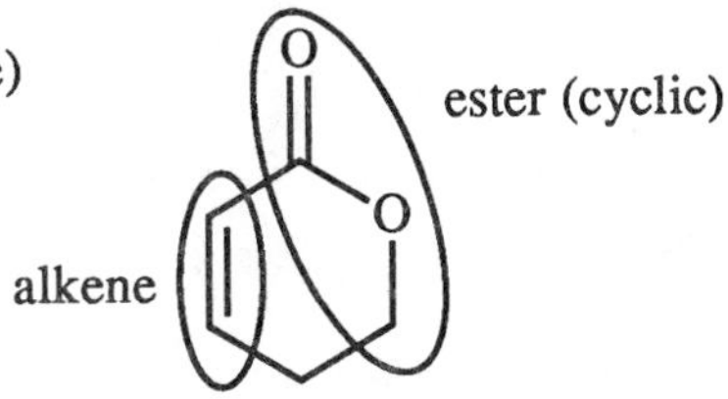
ester (cyclic)
alkene

(f)

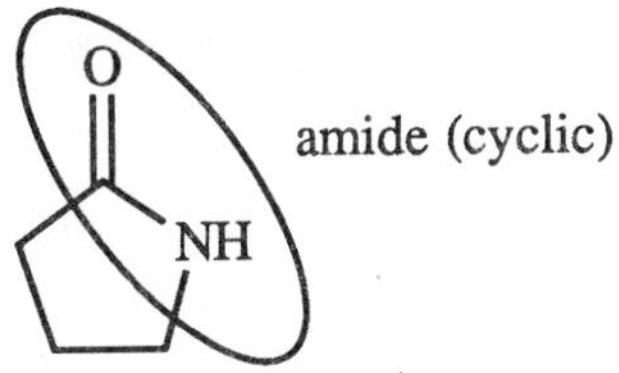
amide (cyclic)

(g)

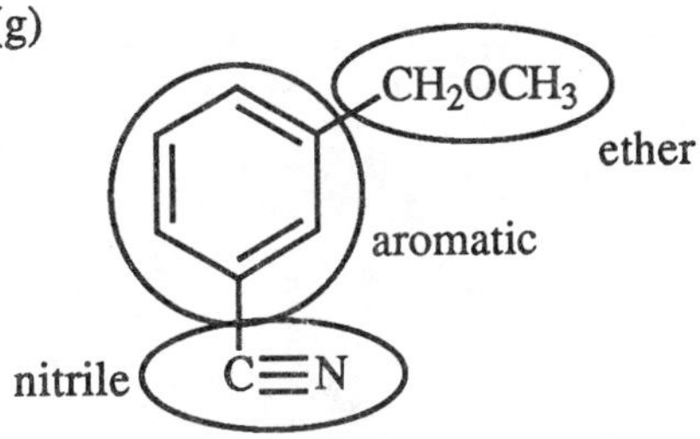
ether
aromatic
nitrile

(h)

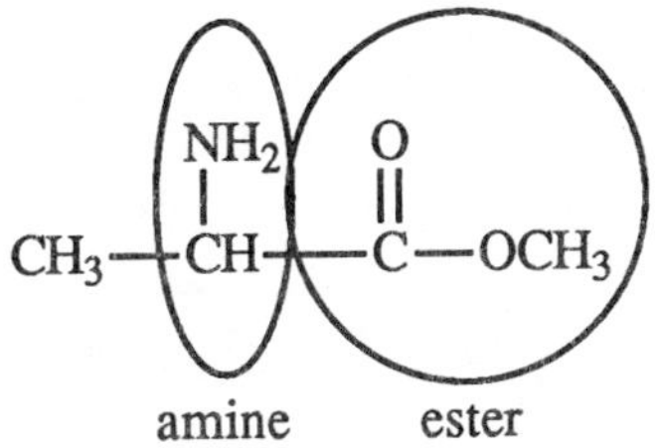
amine
ester

2-41

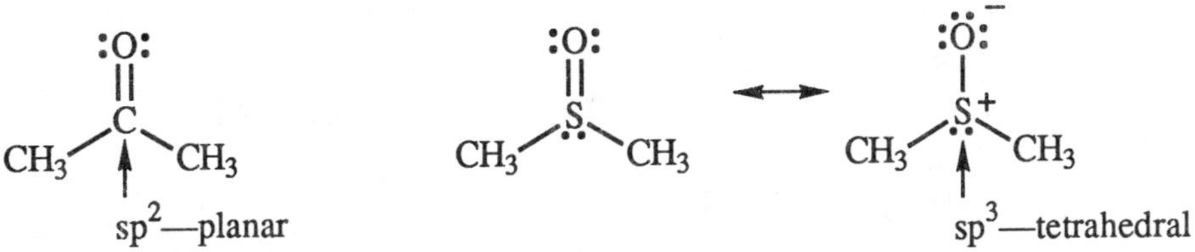
sp²—planar
sp³—tetrahedral

The key to this problem is understanding that sulfur has a *lone pair of electrons*. The second resonance form shows four pairs of electrons around the sulfur, an electronic configuration requiring sp^3 hybridization. Sulfur in DMSO cannot be sp^2 like carbon in acetone, so we would expect sulfur's geometry to be pyramidal (tetrahedral).

CHAPTER 3: STRUCTURE AND STEREOCHEMISTRY OF ALKANES

3-1 (a) C_nH_{2n+2} where n = 30 gives $C_{30}H_{62}$

(b) C_nH_{2n+2} where n = 44 gives $C_{44}H_{90}$

3-2 (a) (lowest b.p.) octane < nonane < decane (highest b.p.)

(b) (lowest b.p.) $(CH_3)_3C—C(CH_3)_3$ < $CH_3CH_2C(CH_3)_2CH_2CH_2CH_3$ < octane (highest b.p.)

3-3 (a) (lowest m.p.) octane < nonane < decane (highest m.p.)

(b) (lowest m.p.) octane < $CH_3CH_2C(CH_3)_2CH_2CH_2CH_3$ < $(CH_3)_3C—C(CH_3)_3$ (highest m.p.)

3-4 (a) 3-methylpentane
(b) 4-isopropyl-2-methyldecane

3-5 (a)

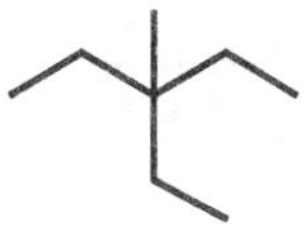

(b)

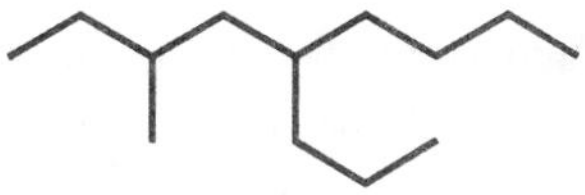

(c)

(d)

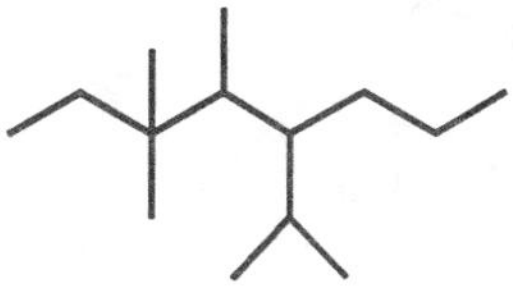

3-6 Separate numbers from letters with hyphens. Separate numbers from numbers with commas.

(a) 2-methylbutane
(b) 2,2-dimethylpropane
(c) 3-ethyl-2-methylhexane
(d) 2,4-dimethylhexane
(e) 3-ethyl-2,2,4,5-tetramethylhexane
(f) 4-*t*-butyl-3-methylheptane

3-7 (Hints: *systematize* your approach to these problems. For the isomers of a six carbon formula, for example, start with the isomer containing all six carbons in a straight chain, then the isomers containing a five-carbon chain, then a four-carbon chain, *etc*. Carefully check your answers to AVOID DUPLICATE STRUCTURES.)

(a)

n-hexane

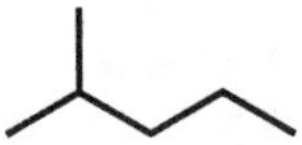

2-methylpentane

3-methylpentane

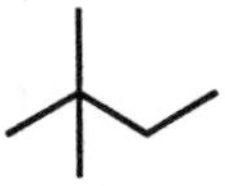

2,2-dimethylbutane

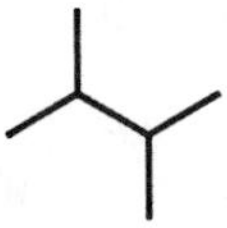

2,3-dimethylbutane

(b)

n-heptane

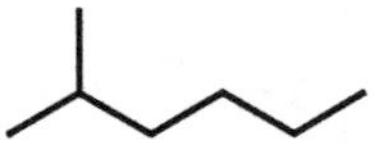

2-methylhexane

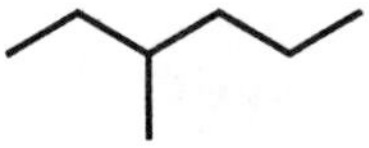

3-methylhexane

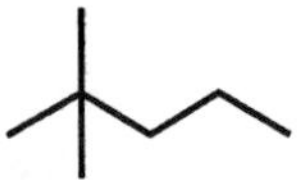

2,2-dimethylpentane

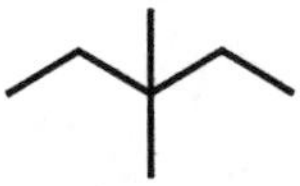

3,3-dimethylpentane

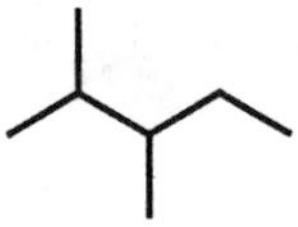

2,3-dimethylpentane

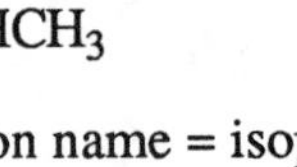

2,4-dimethylpentane

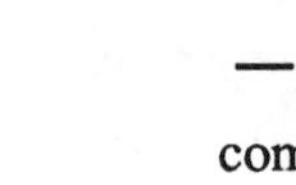

3-ethylpentane

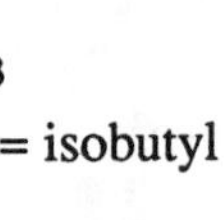

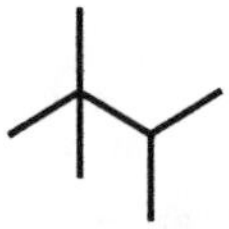

2,2,3-trimethylbutane

3-8 (a)

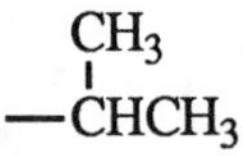
CH3
—CHCH3

common name = isopropyl

(b)

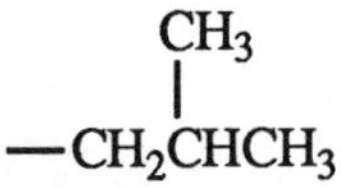
CH3
—CH2CHCH3

common name = isobutyl

(c)

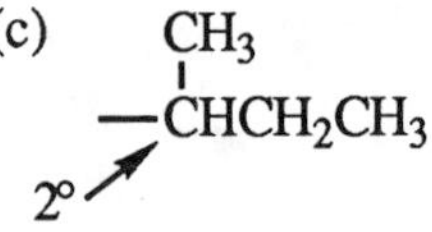
CH3
—CHCH2CH3
2°

common name = *sec*-butyl

(d)

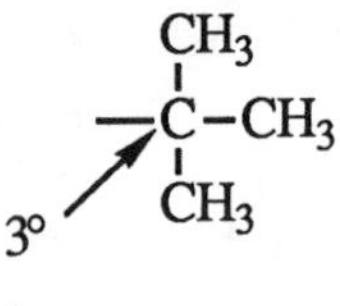
CH3
—C—CH3
CH3
3°

common name = *t*-butyl or *tert*-butyl

3-9 (a)

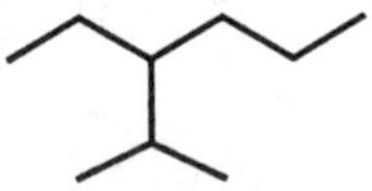

(b)

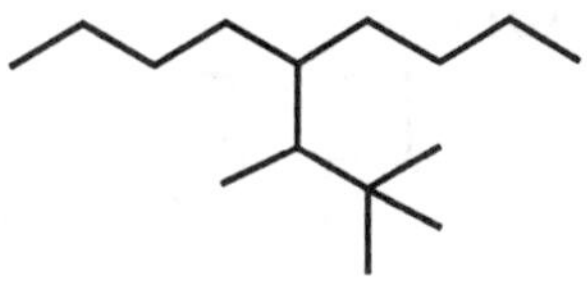

3-10 (a) C_9H_{20} (b) $C_{15}H_{32}$

3-11

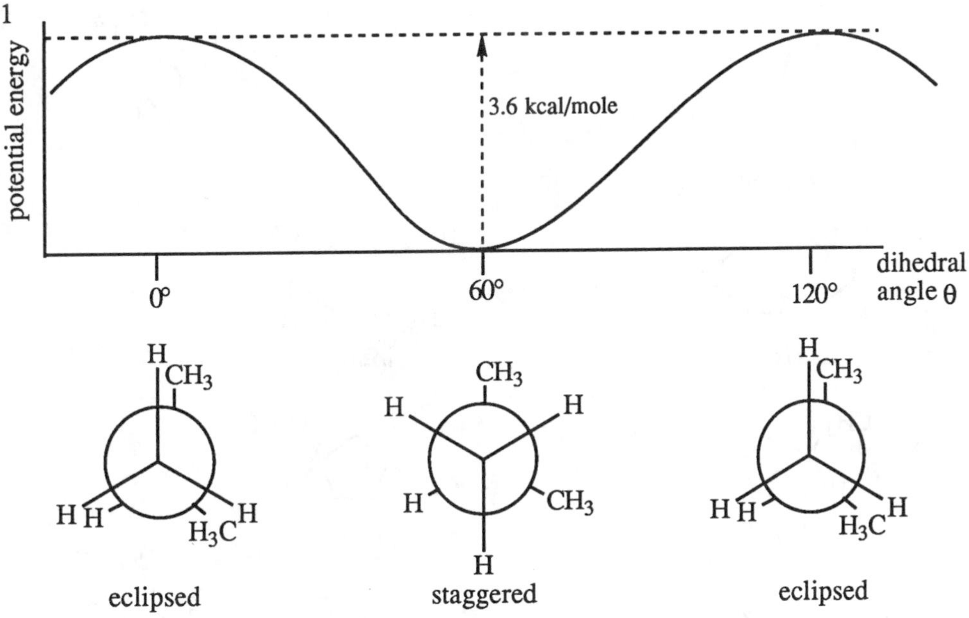

3-12

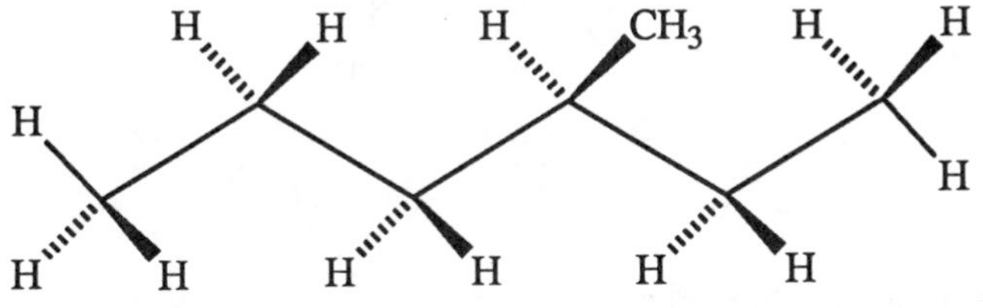

all bonds are staggered

3-13 (a) 3-*sec*-butyl-1,1-dimethylcyclopentane
(b) 3-cyclopropyl-1,1-dimethylcyclohexane
(c) 4-cyclobutylnonane

3-14 (a) $C_{12}H_{24}$

(b) C_9H_{18}

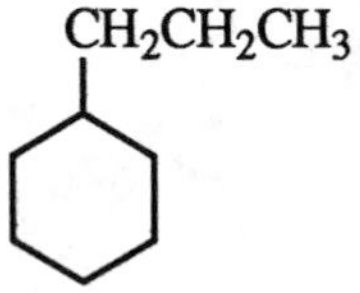

(c) C_8H_{14}

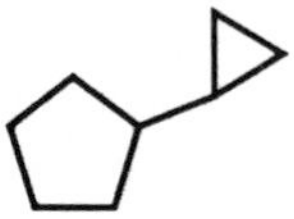

(d) $C_{10}H_{20}$

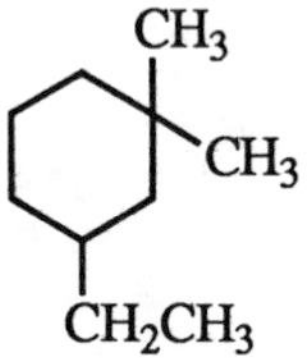

3-15 (a) no geometric isomerism possible

(b)

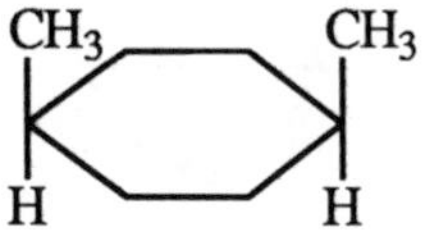

cis

and

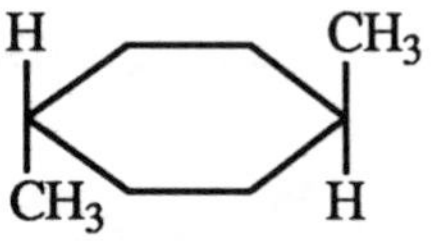

trans

(c)

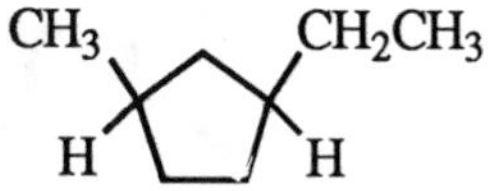

cis

and

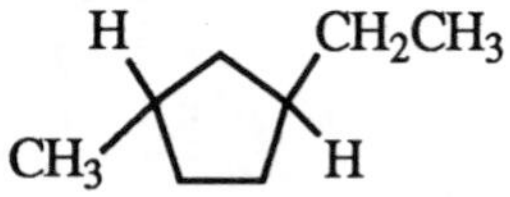

trans

(d)

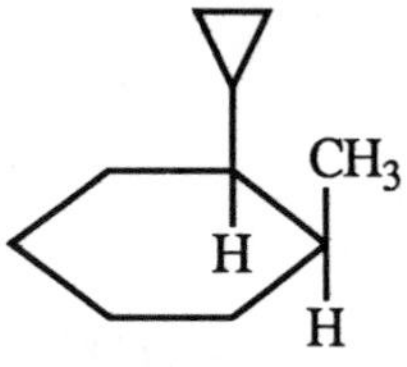

cis

and

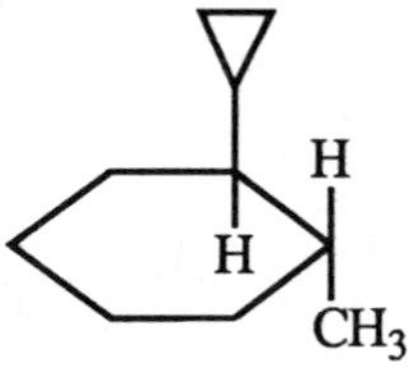

trans

3-16 (a) *cis*-1-methyl-3-*n*-propylcyclobutane
(b) *trans*-1-*t*-butyl-3-ethylcyclohexane
(c) *trans*-1,2-dimethylcyclopropane

3-17 The Newman projection of *cis*-1,2-dimethylcyclopropane shows the two methyls are eclipsed with each other; in the *trans* isomer, the methyls are still eclipsed, but with hydrogens, not each other—a lower energy.

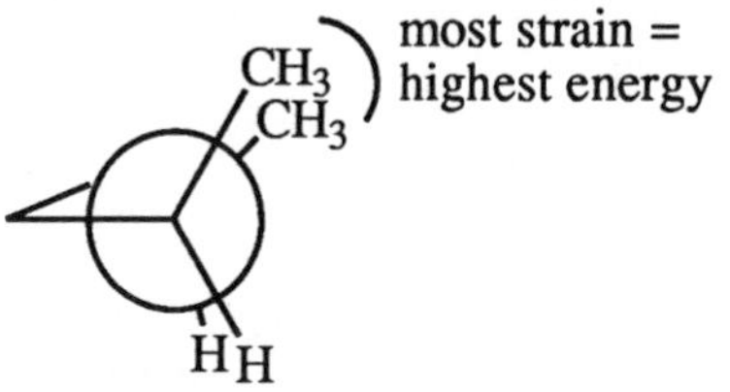

cis

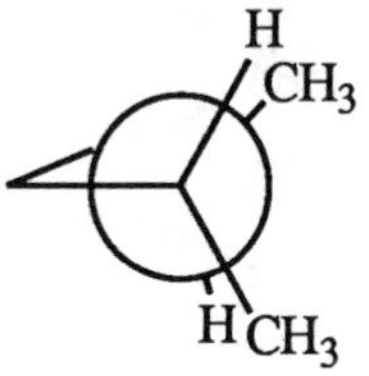

trans

3-18 *trans*-1,2-Dimethylcyclobutane is more stable than *cis* because the two methyls can be farther apart in the *trans*, as shown in the Newman projections.

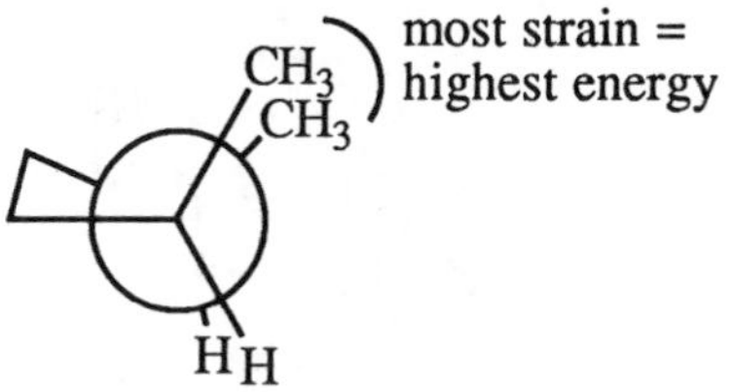

cis

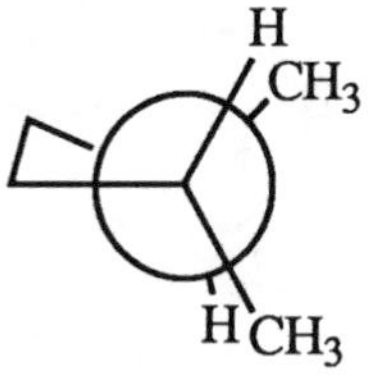

trans

In the 1,3-dimethylcyclobutanes, however, the *cis* allows the methyls to be farther from other atoms and therefore more stable than the *trans*.

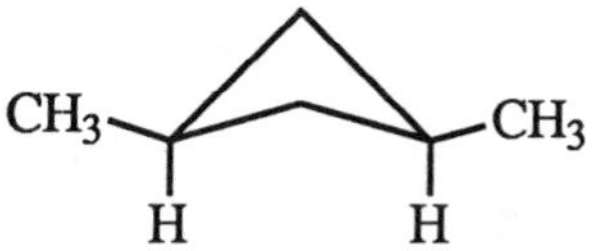

cis

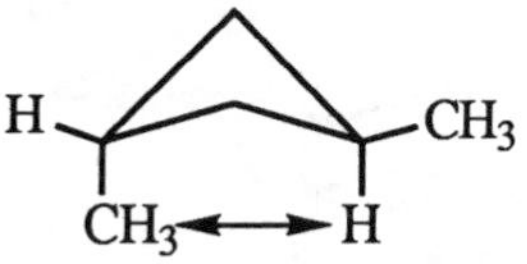

trans

3-19 (a)

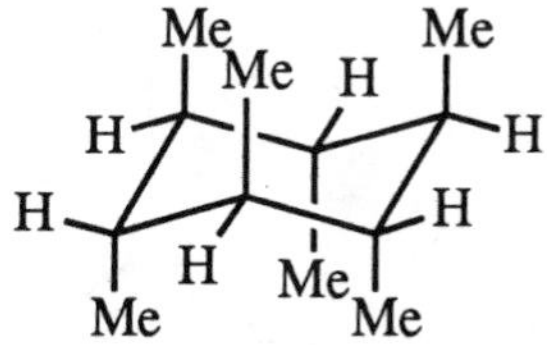

all methyls axial
(all H's equatorial)

(b)

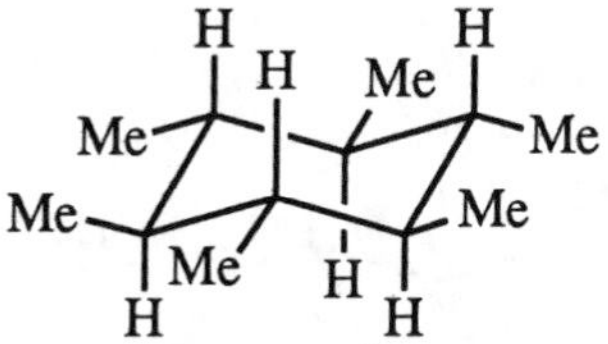

all methyls equatorial
(all H's axial)

3-20

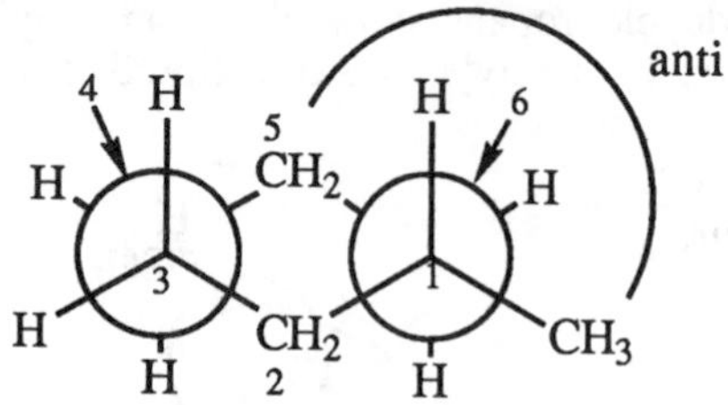

3-21 The isopropyl group can rotate so that its hydrogen is near the axial hydrogens on carbons 3 and 5, similar to a methyl group's hydrogen, and therefore similar to a methyl group in energy. The *t*-butyl group, however, must point a methyl group toward the hydrogens on carbons 3 and 5, giving severe diaxial interactions, causing the energy of this conformer to jump dramatically.

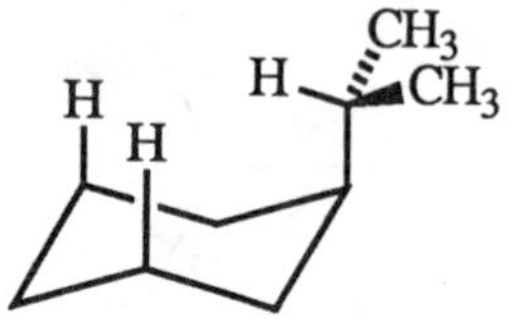

isopropylcyclohexane

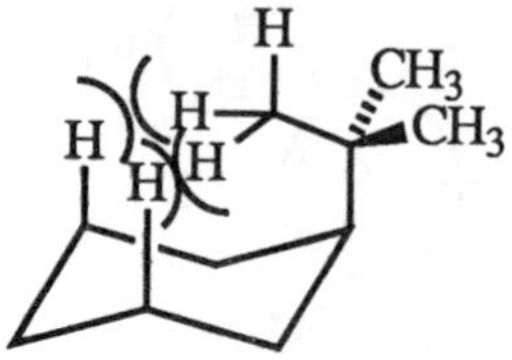

t-butylcyclohexane

3-22 The most stable conformers have the substituents equatorial.

(a)

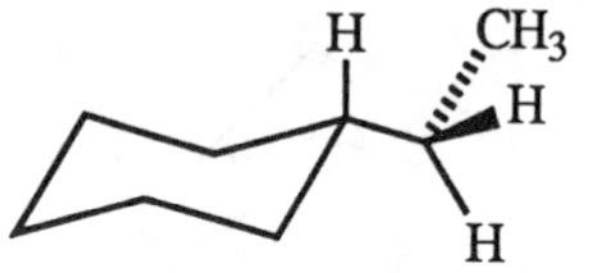

(b)

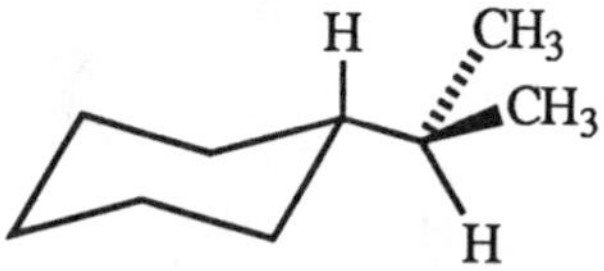

(c)

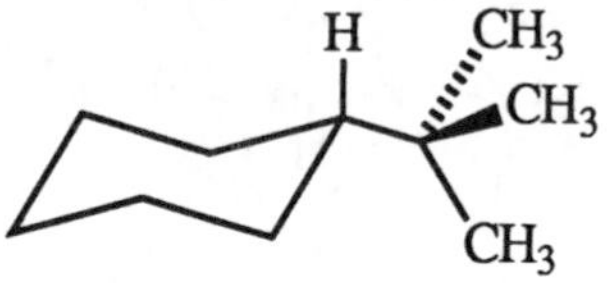

3-23 (a) *cis*

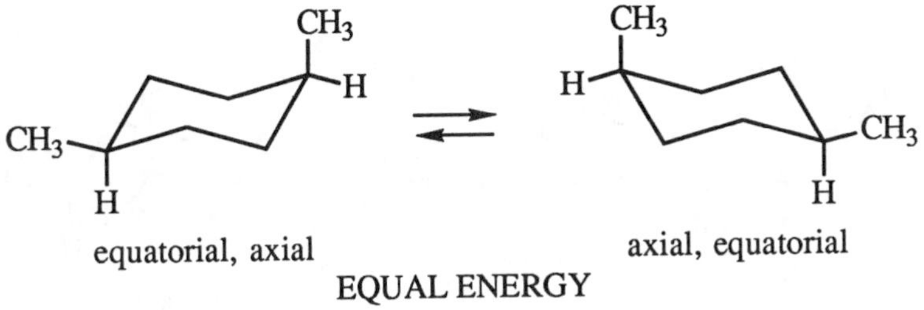

CH3
H
CH3
H
CH3
H
H
CH3
equatorial, axial
axial, equatorial
EQUAL ENERGY

(b) *trans*

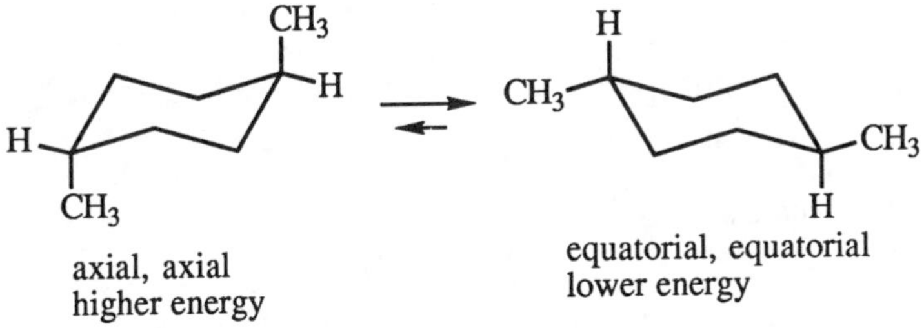

CH3
H
H
CH3
H
CH3
CH3
H
axial, axial
higher energy
equatorial, equatorial
lower energy

(c) The *trans* isomer is more stable because both substituents can be in the preferred equatorial positions.

3-24

Positions	*cis*	*trans*
1,2	(e,a) or (a,e)	(e,e) or (a,a)
1,3	(e,e) or (a,a)	(e,a) or (a,e)
1,4	(e,a) or (a,e)	(e,e) or (a,a)

3-25 (a)

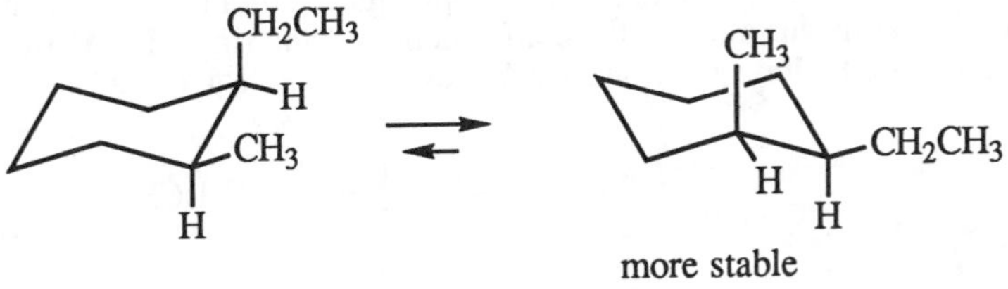

CH2CH3
H
CH3
H
CH3
CH2CH3
H
H
more stable

3-25 continued

(b)

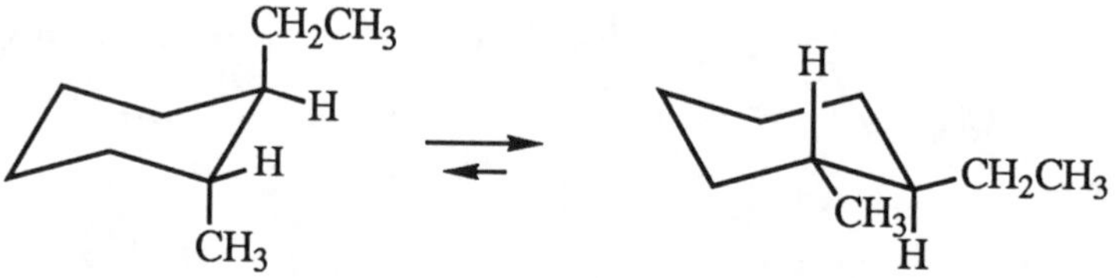

CH2CH3
H
H
CH3
H
CH2CH3
CH3
H

more stable

(c)

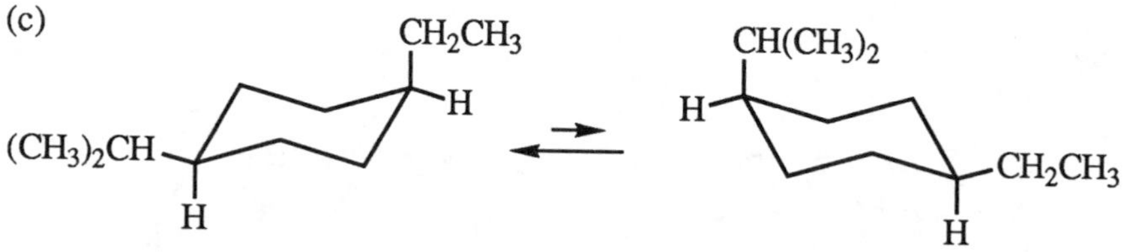

CH2CH3
H
(CH3)2CH
H
CH(CH3)2
H
CH2CH3
H

more stable

(d)

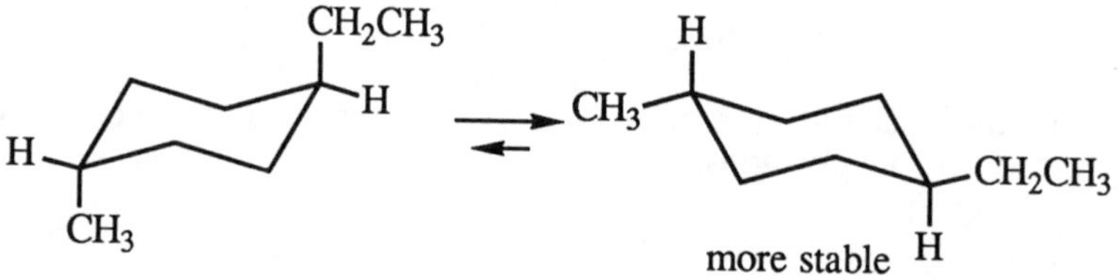

CH2CH3
H
H
CH3
H
CH3
CH2CH3
H

more stable

3-26 (a) *cis*-1,3-dimethylcyclohexane
(b) *cis*-1,4-dimethylcyclohexane
(c) *trans*-1,2-dimethylcyclohexane
(d) *cis*-1,3-dimethylcyclohexane
(e) *cis*-1,3-dimethylcyclohexane
(f) *trans*-1,4-dimethylcyclohexane

3-27 (a)

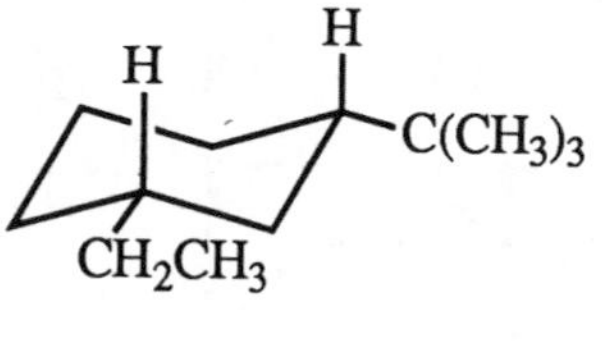

H
H
C(CH3)3
CH2CH3

(b)

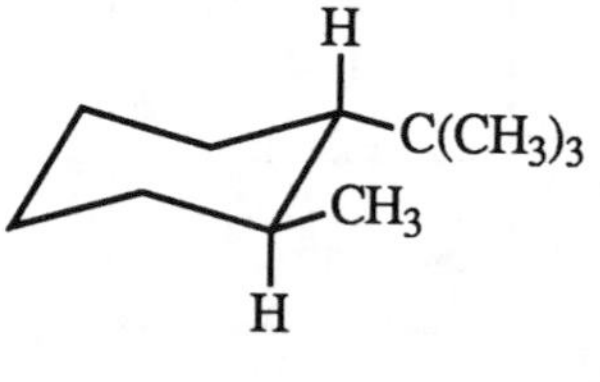

H
C(CH3)3
CH3
H

(c) Bulky substituents like *t*-butyl adopt equatorial rather than axial positions, even if that means altering the conformation of the ring. The twist boat conformation allows both bulky substituents to be "equatorial".

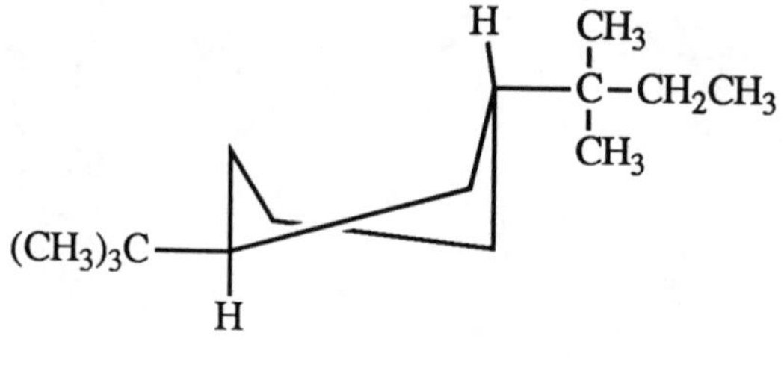

H
CH3
C–CH2CH3
CH3
(CH3)3C
H

3-28 (a) bicyclo[3.2.0]heptane
(b) bicyclo[3.2.1]octane
(c) bicyclo[2.2.2]octane
(d) bicyclo[3.1.1]heptane

3-29 Using models is essential for this problem.

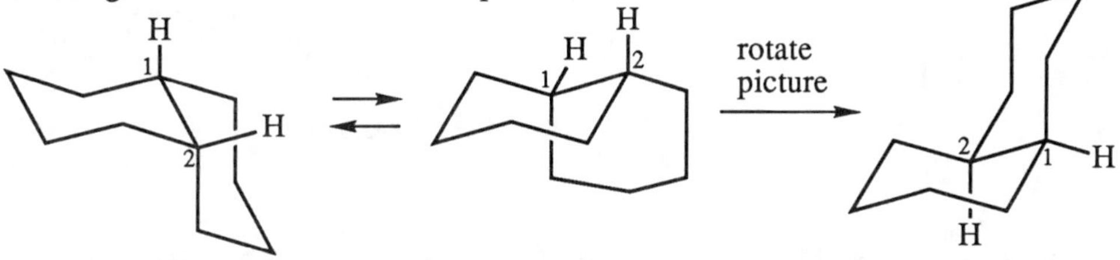

from Figure 3-31

3-30 Refer to the Glossary and the text for definitions and examples.

3-31 (a) The four structures on the left are all *n*-butane. The structure at the top-right is 2-methylpropane (isobutane).

(b) The two structures at the top-left and bottom-left are both *cis*-2-butene. The two structures at the top-center and bottom-center are both 1-butene. The unique structure at the upper right is *trans*-2-butene. The unique structure at the lower right is 2-methylpropene.

(c) The two structures at the top-left and top-center are both *cis*-1,2-dimethylcyclopentane. The two structures at the top-right and bottom-left are both *trans*-1,2-dimethylcyclopentane. The structure at the bottom-right is different from all the others, *cis*-1,3-dimethylcyclopentane.

3-32 (a)

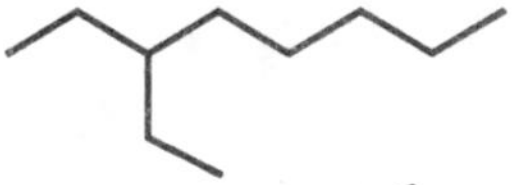

(b)

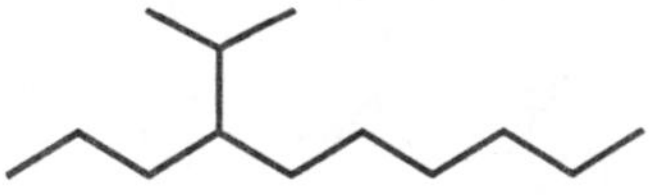

(c)

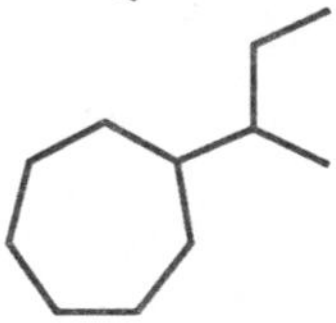

(d)

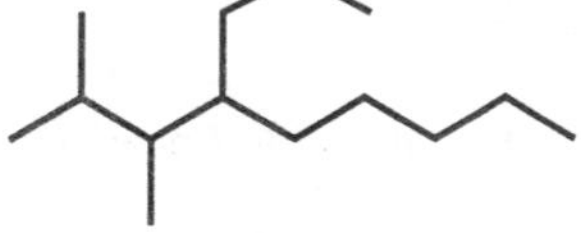

(e)

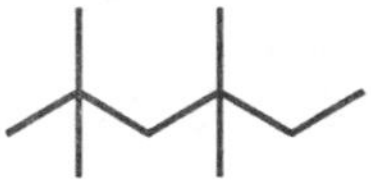

(f)

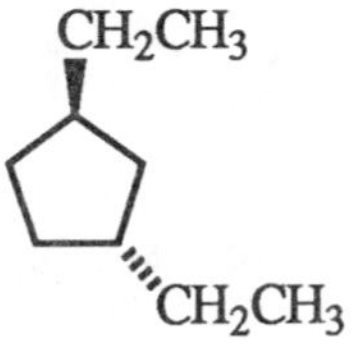

3-32 continued

(g) 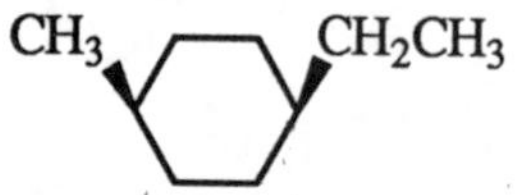
CH3
CH2CH3

(h) 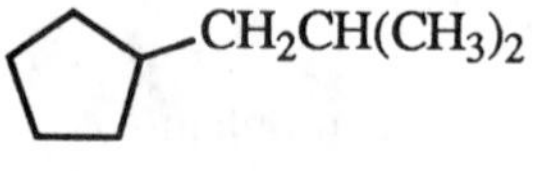
CH2CH(CH3)2

(i) 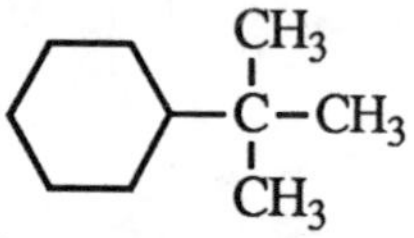
CH3
C–CH3
CH3

(j) 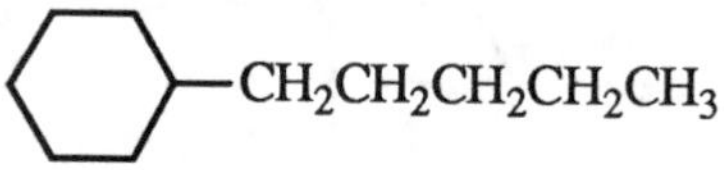
CH2CH2CH2CH2CH3

(k) 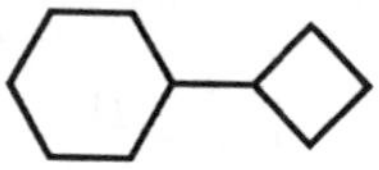

3-33 There are many possible answers to each of these problems. The ones shown here are examples of correct answers. Your answers may be different AND correct.

(a) 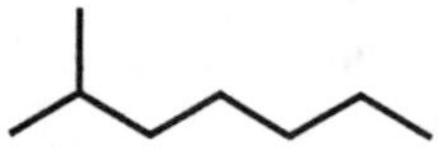

2-methylheptane

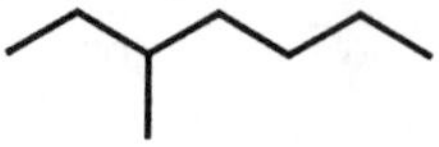

3-methylheptane

(b) 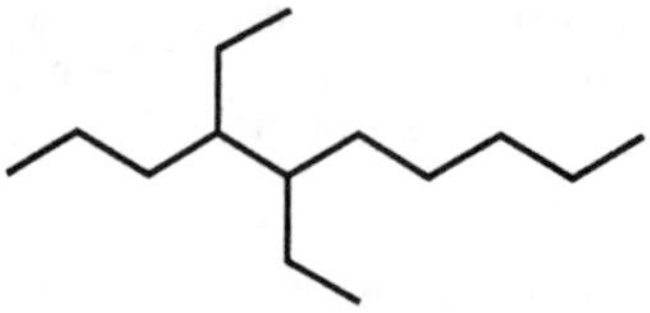

4,5-diethyldecane

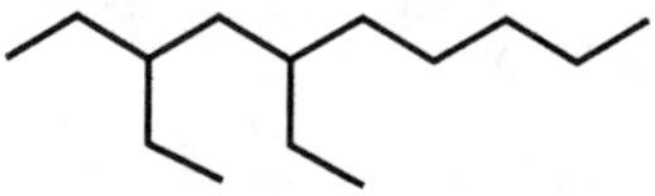

3,5-diethyldecane

(c) 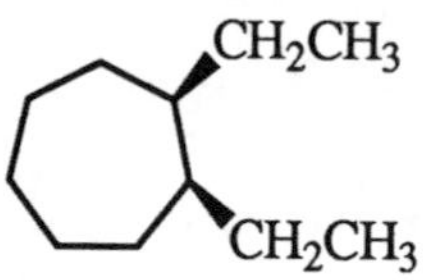
CH2CH3
CH2CH3

cis-1,2-diethylcycloheptane

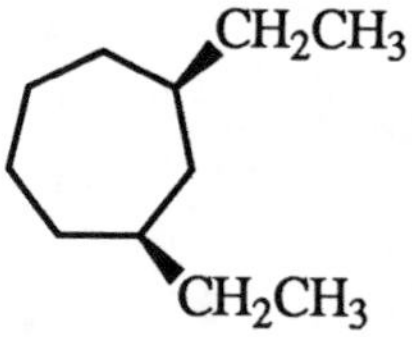
CH2CH3
CH2CH3

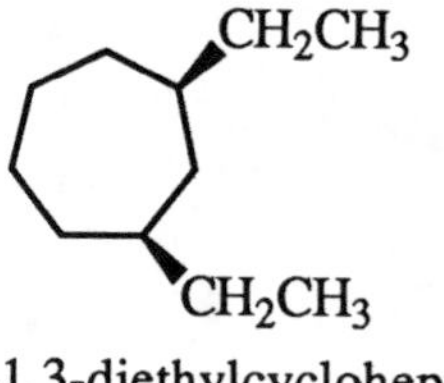

cis-1,3-diethylcycloheptane

(d) only two possible answers

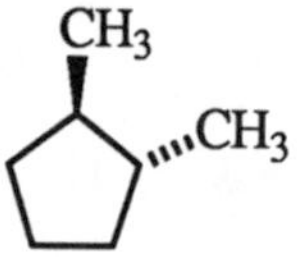
CH3
CH3

trans-1,2-dimethylcyclopentane

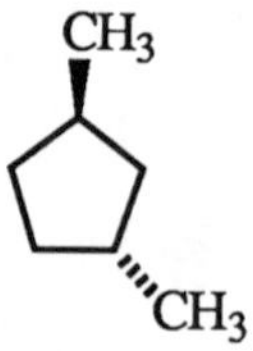
CH3
CH3

trans-1,3-dimethylcyclopentane

3-33 continued

(e)

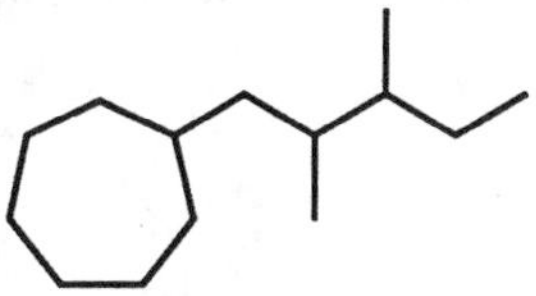

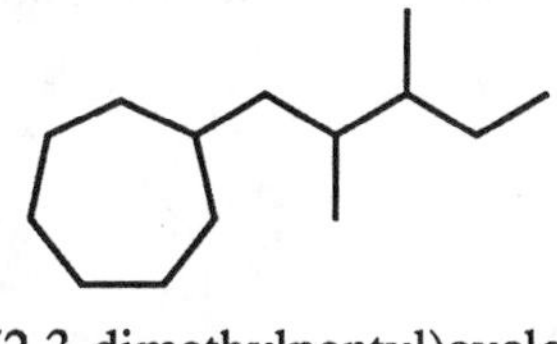

(2,3-dimethylpentyl)cycloheptane

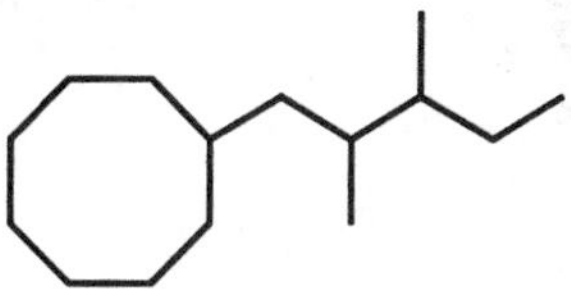

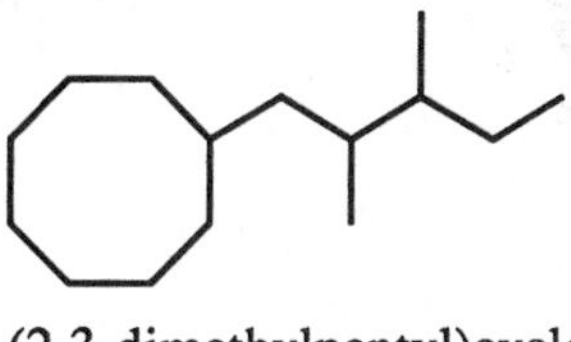

(2,3-dimethylpentyl)cyclooctane

3-34

CH_3-OH

CH_3CH_2-OH

$CH_3CH_2CH_2-OH$

$CH_3CH_2CH_2CH_2-OH$

$CH_3CH_2CH_2CH_2CH_2-OH$

$CH_3CH_2CH_2CH_2CH_2CH_2-OH$

3-35 (a) 3-ethyl-2,2,6-trimethylheptane
(b) 3-ethyl-2,6,7-trimethyloctane
(c) 3,7-diethyl-2,2,8-trimethyldecane
(d) 2-ethyl-1,1-dimethylcyclobutane
(e) bicyclo[4.1.0]heptane
(f) *cis*-1-ethyl-3-*n*-propylcyclopentane
(g) (1,1-diethylpropyl)cyclohexane
(h) *cis*-1-ethyl-4-isopropylcyclodecane

3-36 There are eighteen isomers of C_8H_{18}. Here are eight of them. Yours may be different.

n-octane

2-methylheptane

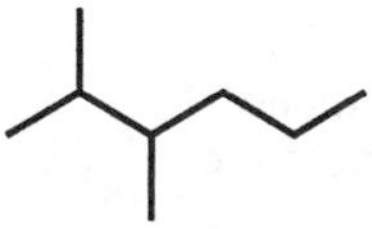

2,3-dimethylhexane

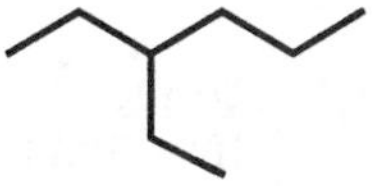

3-ethylhexane

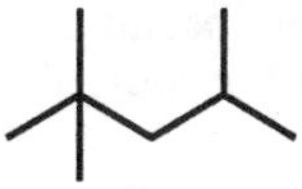

2,2,4-trimethylpentane

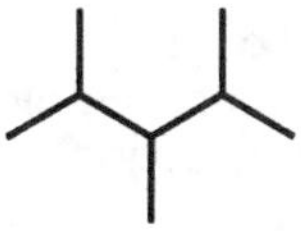

2,3,4-trimethylpentane

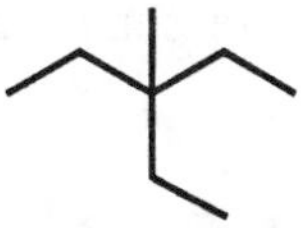

3-ethyl-3-methylpentane

2,2,3,3-tetramethylbutane

3-37 (a)

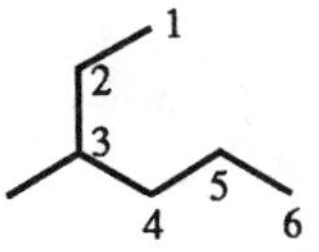

correct name: 3-methylhexane (longer chain)

(b)

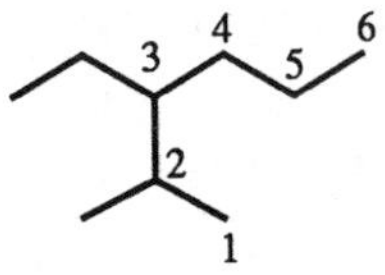

correct name: 3-ethyl-2-methylhexane
(more branching with this numbering)

(c)

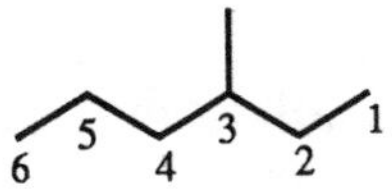

correct name: 3-methylhexane
(begin numbering at end closest to substituent)

(d)

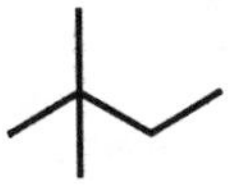

correct name: 2,2-dimethylbutane (include a position number for each substituent, regardless of redundancies)

(e)

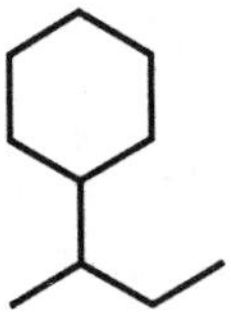

correct name: *sec*-butylcyclohexane or
(1-methylpropyl)cyclohexane
(the longer chain or ring is the base name)

(f)

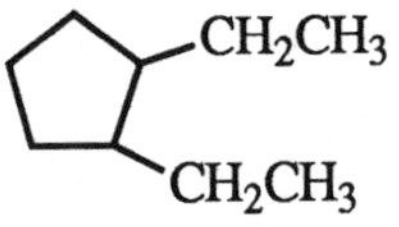

correct name: 1,2-diethylcyclopentane
(position numbers are the lowest possible)

3-38 (a) *n*-Octane has a higher boiling point than 2,2,3-trimethylpentane because linear molecules boil higher than branched molecules of the same molecular weight (increased van der Waals interaction).

(b) 2-Methylnonane has a higher boiling point that *n*-heptane because it has a significantly higher molecular weight than *n*-heptane.

(c) *n*-Nonane boils higher than 2,2,5-trimethylhexane for the same reason as in (a).

3-39

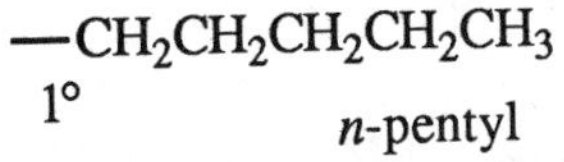

n-pentyl

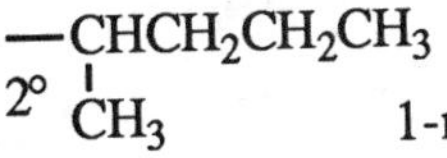

1-methylbutyl

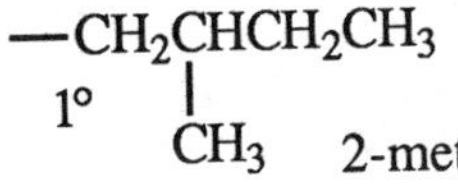

2-methylbutyl

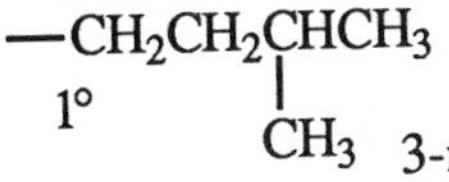

3-methylbutyl (isopentyl)

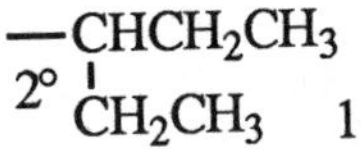

1-ethylpropyl

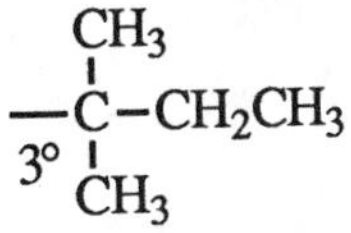

1,1-dimethylpropyl (*t*-pentyl)

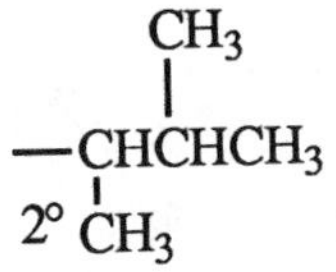

1,2-dimethylpropyl

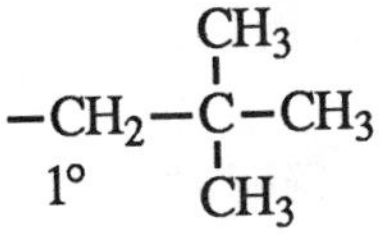

2,2-dimethylpropyl (*neo*-pentyl)

3-40 (a)

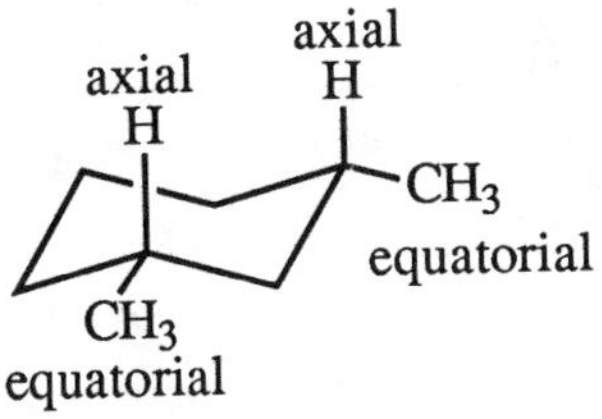

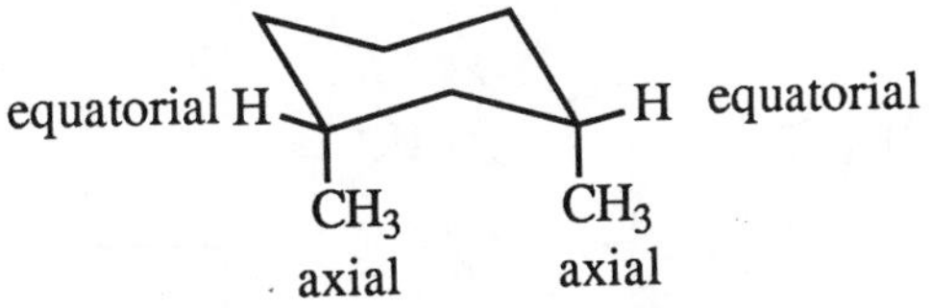

(b) more stable (lower energy) less stable (higher energy)

(c) From Wade Section 3-13, each gauche interaction raises the energy 0.9 kcal/mole (3.8 kJ/mole), and each axial methyl has two gauche interactions, so the energy is
2 methyls x 2 interactions per methyl x 0.9 kcal/mole per interaction =
3.6 kcal/mole (15.2 kcal/mole)

(d) The steric strain from the 1,3-diaxial interaction of the methyls must be the difference between the total energy and the energy due to gauche interactions:

5.4 kcal/mole - 3.6 kcal/mole = 1.8 kcal/mole
(23 kJ/mole - 15.2 kJ/mole = 7.8 kJ/mole)

3-41 (a)

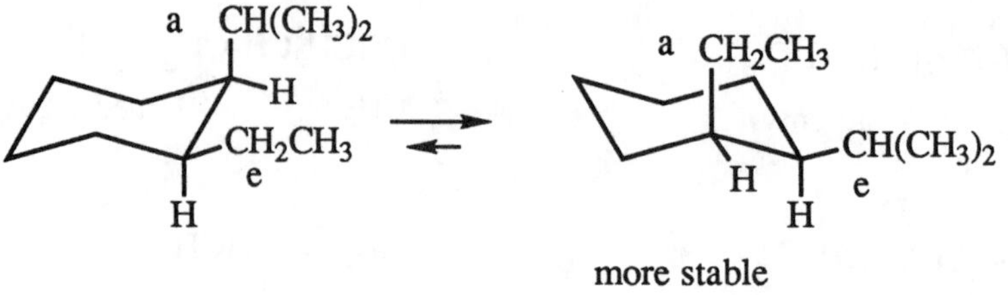
a CH(CH3)2
H
CH2CH3
e
H
a CH2CH3
CH(CH3)2
H
e
H
more stable

(b)

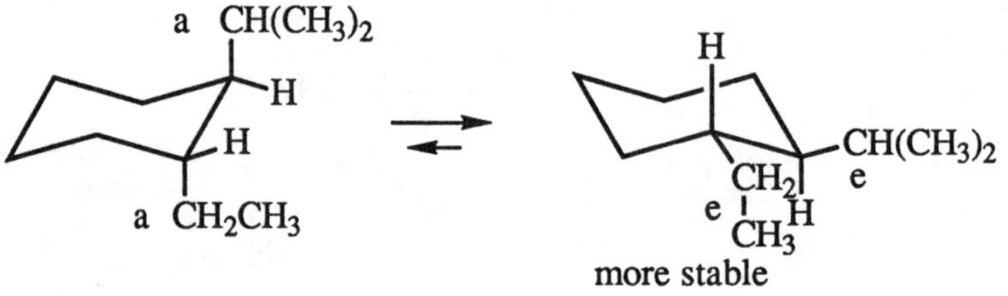
a CH(CH3)2
H
H
a CH2CH3
H
CH(CH3)2
CH2
e
e
H
CH3
more stable

(c)

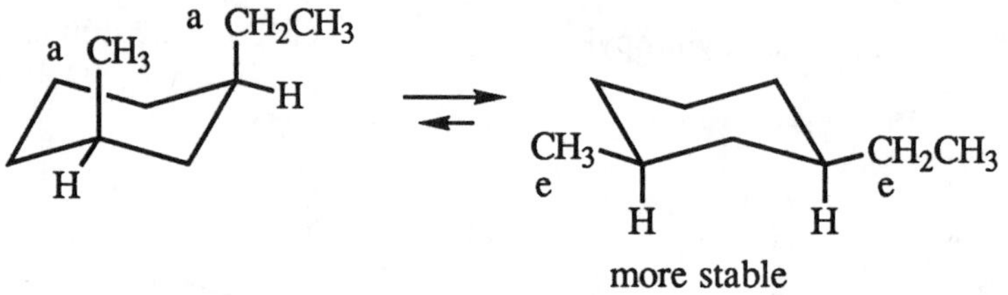
a CH2CH3
a CH3
H
H
CH3
e
CH2CH3
e
H
H
more stable

(d)

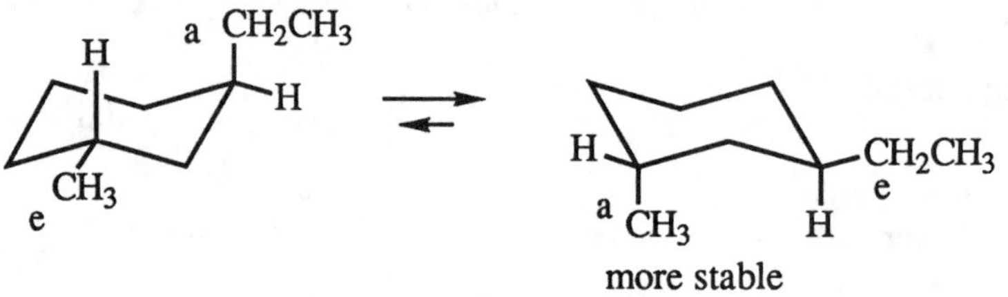
a CH2CH3
H
H
CH3
e
H
CH2CH3
e
a CH3
H
more stable

(e)

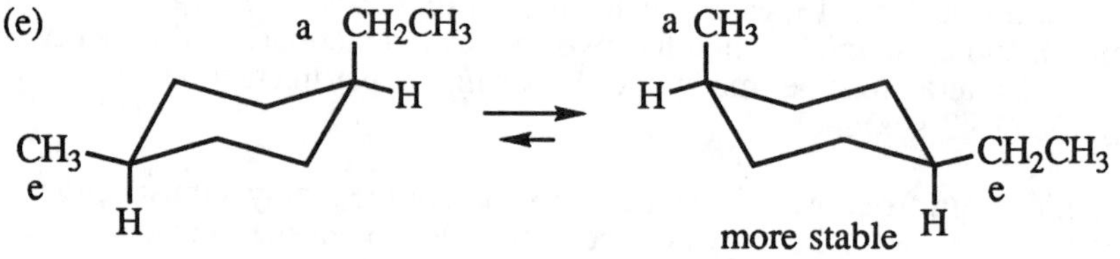
a CH2CH3
H
CH3
e
H
a CH3
H
CH2CH3
e
more stable
H

3-42 (Using models is essential to this problem.)

In both *cis*- and *trans*-decalin, the cyclohexane rings can be in chair conformations. The relative energies will depend on the number of axial substituents.

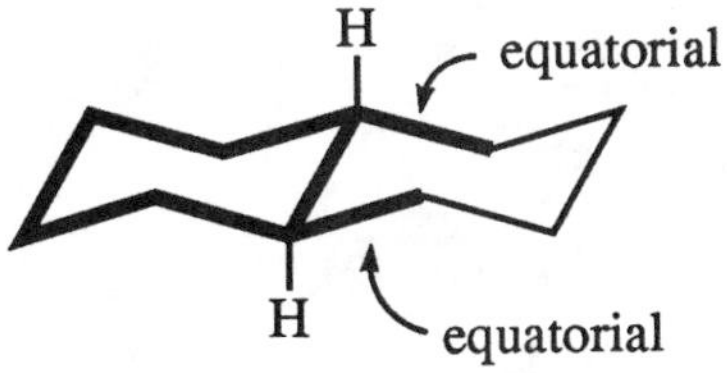

trans
no axial substituents
MORE STABLE

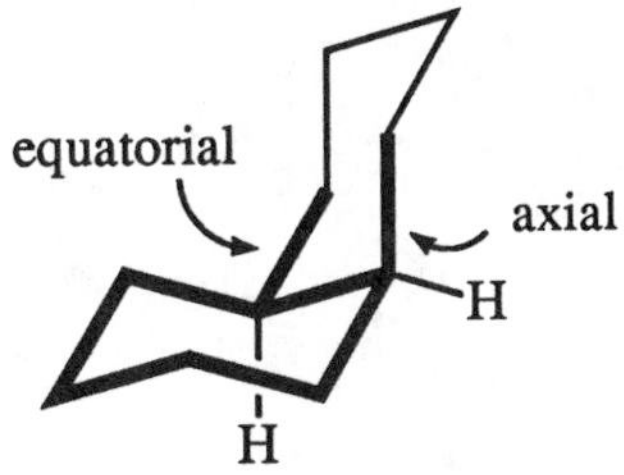

cis
one axial substituent

3-43 chair form of glucose—all substituents equatorial

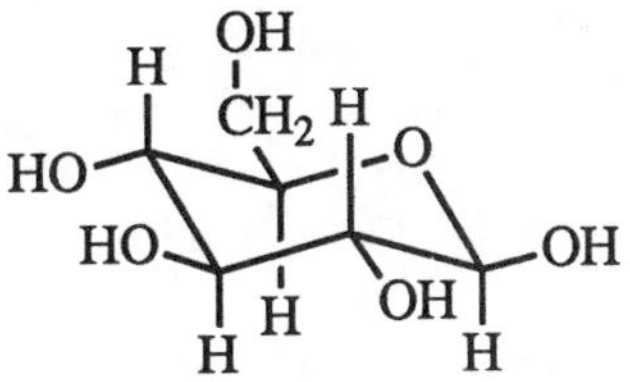

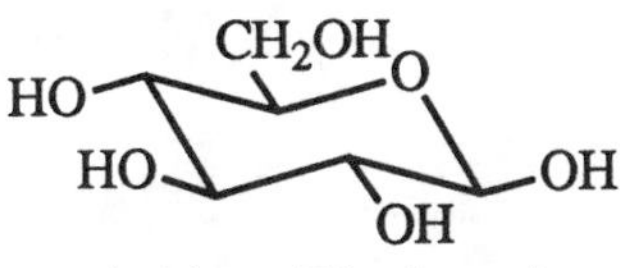

(without H's shown)

CHAPTER 4: THE STUDY OF CHEMICAL REACTIONS

4-1 (a) 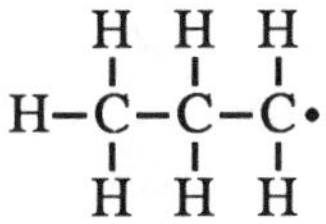 (b) 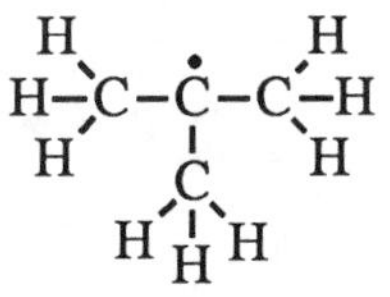 (c) 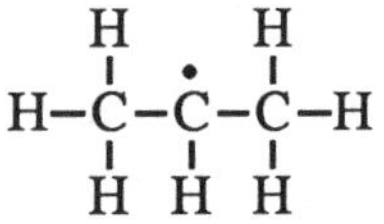 (d) 

4-2 (a)

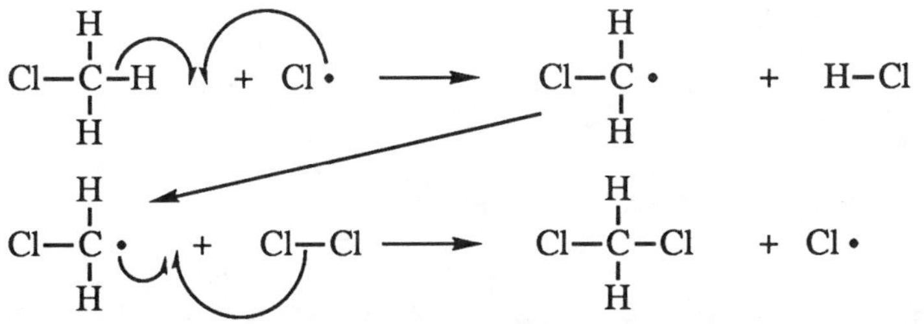

(b) Free-radical halogenation substitutes a halogen atom for a hydrogen. Even if a molecule has only one type of hydrogen, substitution of the first of these hydrogens forms a new compound. Any remaining hydrogens in this product can compete with the initial reactant for the available halogen. Thus, chlorination of methane, CH_4, produces all possible substitution products: CH_3Cl, CH_2Cl_2, $CHCl_3$, and CCl_4.

If a molecule has different types of hydrogens, the reaction can generate a mixture of the possible substitution products.

4-3 (a) This mechanism requires that one photon of light be added for each CH_3Cl generated, a quantum yield of 1. The actual quantum yield is several hundred or thousand. The high quantum yield suggests a chain reaction, but this mechanism is not a chain; it has no propagation steps.

(b) This mechanism conflicts with at least two experimental observations. First, the energy of light required to break a H-CH_3 bond is 104 kcal/mole (435 kJ/mole, from Table 4-2); the energy of light determined by experiment to initiate the reaction is only 60 kcal/mole of photons (251 kJ/mole of photons), much less than the energy needed to break this H-C bond. Second, as in (a), each CH_3Cl produced would require one photon of light, a quantum yield of 1, instead of the actual number of several hundred or thousand. As in (a), there is no provision for a chain process, since all the radicals generated are also consumed in the mechanism.

4-4 (a) The twelve hydrogens of cyclohexane are all on equivalent 2° carbons. Replacement of any one of the twelve will lead to the same product, chlorocyclohexane. *n*-Hexane, however, has hydrogens in three different positions: on carbon-1 (equivalent to carbon-6), carbon-2 (equivalent to carbon-5), and carbon-3 (equivalent to carbon-4). Monochlorination of *n*-hexane will produce a mixture of all three possible isomers: 1-, 2-, and 3-chlorohexane.

(b) The best conversion of cyclohexane to chlorocyclohexane would require the ratio of cyclohexane/chlorine to be a large number. If the ratio were small, as the concentration of chlorocyclohexane increased during the reaction, chlorine would begin to substitute for a second hydrogen of chlorocyclohexane, generating unwanted products. The goal is to have chlorine attack a molecule of cyclohexane before it ever encounters a molecule of chloro-cyclohexane, so the concentration of cyclohexane should be kept high.

4-5 (a)

$$K_{eq} = e^{-\Delta G°/RT} \quad \text{(convert to cal)}$$

$$= e^{-(-500\ \text{cal/mole})/((1.987\ \text{cal/kelvin-mole}) \cdot (298\ \text{kelvin}))}$$

$$= e^{500/592} = e^{0.845} = \boxed{2.3}$$

(b)

$$K_{eq} = 2.3 = \frac{[CH_3SH]\,[HBr]}{[CH_3Br]\,[H_2S]}$$

	$[CH_3SH]$	$[H_2S]$	$[CH_3SH]$	$[HBr]$
initial concentrations:	1	1	0	0
final concentrations	1 – x	1 – x	x	x

$$K_{eq} = 2.3 = \frac{x \cdot x}{(1-x)(1-x)} = \frac{x^2}{1 - 2x + x^2} \Longrightarrow x^2 = 2.3\,x^2 - 4.6\,x + 2.3$$

$$0 = 1.3\,x^2 - 4.6\,x + 2.3 \Longrightarrow x = 0.60\ ,\quad 1 - x = 0.40$$

(using quadratic equation)

$$\boxed{[CH_3SH] = [HBr] = 0.60\ M;\quad [CH_3Br] = [H_2S] = 0.40\ M}$$

4-6 2 acetone $\rightleftharpoons$ diacetone

Assume that the initial concentration of acetone is 1 molar, and 5% of the acetone is converted to diacetone. NOTE THE MOLE RATIO.

	[acetone]	[diacetone]
initial concentrations:	1 M	0
final concentrations:	0.95 M	0.025 M

$$K_{eq} = \frac{[\text{diacetone}]}{[\text{acetone}]^2} = \frac{0.025}{(0.95)^2} = \boxed{0.028}$$

$$\Delta G° = -2.303\,RT \log_{10} K_{eq} = -2.303\,(0.592\ \text{kcal/mole})\,(-1.56)$$

$$= \boxed{+2.1\ \text{kcal/mole} \quad (+8.9\ \text{kJ/mole})}$$

4-7 $\Delta S°$ will be negative since two molecules are combined into one, a loss of freedom of motion. Since $\Delta S°$ is negative, $-T\Delta S°$ is positive; but $\Delta G°$ is a large negative number since the reaction goes to completion. Therefore, $\Delta H°$ must also be a large negative number, necessarily larger in value than $\Delta G°$. We can explain this by formation of two strong C-H bonds (98 kcal/mole each) after breaking a strong H-H bond (104 kcal/mole) and a WEAKER C=C pi bond.

4-8 (a) $\Delta S°$ is positive—one molecule became two smaller molecules with greater freedom of motion
(b) $\Delta S°$ is negative—two smaller molecules combined into one larger molecule with less freedom of motion
(c) $\Delta S°$ cannot be predicted since the number of molecules in reactants and products is the same

4-9 (a)

initiation (1) $Cl{-}Cl \xrightarrow{h\nu} 2\ Cl\cdot$

propagation
(2) $Cl\cdot + H{-}CH_2CH_3 \longrightarrow H{-}Cl + \cdot CH_2CH_3$
(3) $Cl{-}Cl + \cdot CH_2CH_3 \longrightarrow Cl{-}CH_2CH_3 + Cl\cdot$

termination
(4) $Cl\cdot + \cdot Cl \longrightarrow Cl{-}Cl$
(5) $Cl\cdot + \cdot CH_2CH_3 \longrightarrow Cl{-}CH_2CH_3$
(6) $CH_3CH_2\cdot + \cdot CH_2CH_3 \longrightarrow CH_3CH_2CH_2CH_3$

(b)

step (1)	break Cl-Cl	$\Delta H°$ = + 58 kcal/mole (+ 242 kJ/mole)
step (2)	break H-CH_2CH_3	$\Delta H°$ = + 98 kcal/mole (+ 410 kJ/mole)
	make H-Cl	$\Delta H°$ = - 103 kcal/mole (- 431 kJ/mole)
	step (2)	**$\Delta H°$ = −5 kcal/mole (−21 kJ/mole)**
step (3)	break Cl-Cl	$\Delta H°$ = + 58 kcal/mole (+242 kJ/mole)
	make Cl-CH_2CH_3	$\Delta H°$ = - 81 kcal/mole (- 339 kJ/mole)
	step (3)	**$\Delta H°$ = −23 kcal/mole (−97 kJ/mole)**

(c) $\Delta H°$ for the reaction is the sum of the $\Delta H°$ values of the individual propagation steps:

– 5 kcal/mole + – 23 kcal/mole = **– 28 kcal/mole**
(– 21 kJ/mole + – 97 kJ/mole = **– 118 kJ/mole**)

4-10 (a)

initiation (1) $Br{-}Br \xrightarrow{h\nu} 2\ Br\cdot$

propagation
(2) $Br\cdot + H{-}CH_3 \longrightarrow H{-}Br + \cdot CH_3$
(3) $Br{-}Br + \cdot CH_3 \longrightarrow Br{-}CH_3 + Br\cdot$

step (1)	break Br-Br	$\Delta H°$ = + 46 kcal/mole (+ 192 kJ/mole)
step (2)	break H-CH_3	$\Delta H°$ = + 104 kcal/mole (+ 435 kJ/mole)
	make H-Br	$\Delta H°$ = - 88 kcal/mole (- 368 kJ/mole)
	step (2)	**$\Delta H°$ = +16 kcal/mole (+67 kJ/mole)**
step (3)	break Br-Br	$\Delta H°$ = + 46 kcal/mole (+ 192 kJ/mole)
	make Br-CH_3	$\Delta H°$ = - 70 kcal/mole (- 293 kJ/mole)
	step (3)	**$\Delta H°$ = −24 kcal/mole (−101 kJ/mole)**

(b) $\Delta H°$ for the reaction is the sum of the $\Delta H°$ values of the individual propagation steps:

+ 16 kcal/mole + – 24 kcal/mole = **– 8 kcal/mole**
(+ 67 kJ/mole + – 101 kJ/mole = **– 34 kJ/mole**)

4-11 (a) first order: the exponent of [$(CH_3)_3CCl$] in the rate law = 1
(b) zeroth order: [CH_3OH] does not appear in the rate law (its exponent is zero)
(c) first order: the sum of the exponents in the rate law = 1

4-12 (a) there is a linear relationship between the rate and [CH_3Cl], so the exponent = 1 and the rate equation is first order with respect to chloromethane
(b) there is a linear relationship between the rate and [^-CN], so the exponent = 1 and the rate equation is first order with respect to the cyanide ion
(c) rate = k_r [CH_3Cl] [^-CN]
(d) overall, second order: the sum of the exponents in the rate law = 2

4-13 (a) the reaction rate depends on neither [ethylene] nor [hydrogen], so it is zeroth order in both species. The overall reaction must be zeroth order.
(b) rate = k_r
(c) The rate law does not depend on the concentration of the reactants. It must depend, therefore, on the only other chemical present, the catalyst. Apparently, whatever is happening on the surface of the catalyst determines the rate, regardless of the concentrations of the two gases. Increasing the surface area of the catalyst, or simply adding more catalyst, would speed the reaction.

4-14 (a)

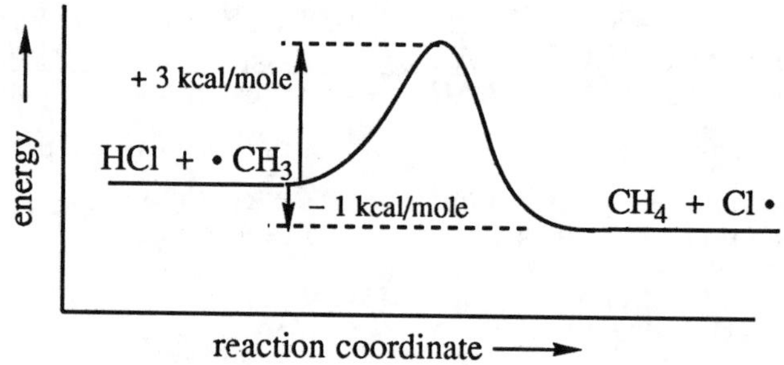

(b) E_a = + 3 kcal/mole (+ 13 kJ/mole)
(c) $\Delta H°$ = − 1 kcal/mole (− 4 kJ/mole)

4-15 (a)

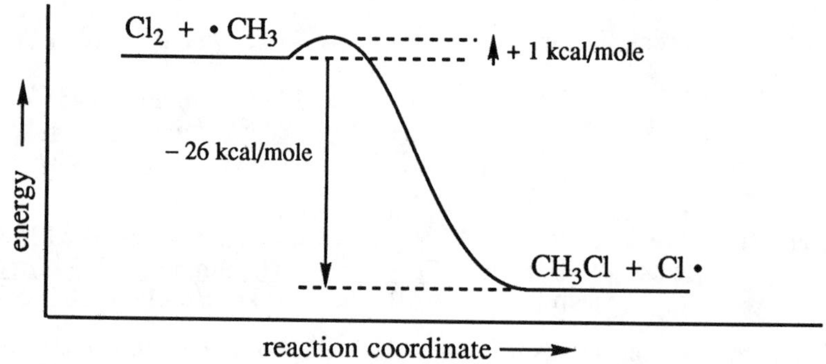

4-15 continued

(b) reverse: $CH_3Cl + Cl\bullet \longrightarrow Cl_2 + \bullet CH_3$

(c) reverse: E_a = + 26 kcal/mole + + 1 kcal/mole = **+ 27 kcal/mole**
(+ 109 kJ/mole + + 4 kJ/mole = **+ 113 kJ/mole**)

4-16 (a)

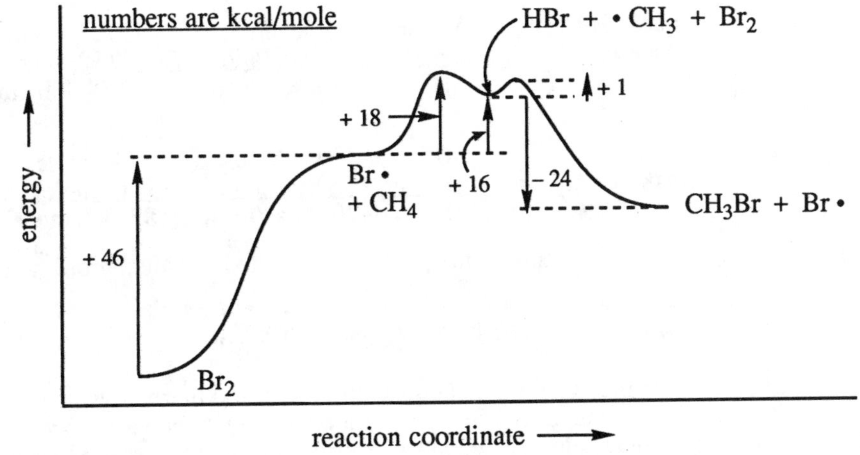

(b) The step leading to the highest energy transition state is rate-determining. In this mechanism, the first propagation step is rate-determining:

$$Br\bullet + CH_4 \longrightarrow HBr + \bullet CH_3$$

(c) (1) δ• Br·······Br δ• ("δ •" means partial radical character on the atom)

(2)

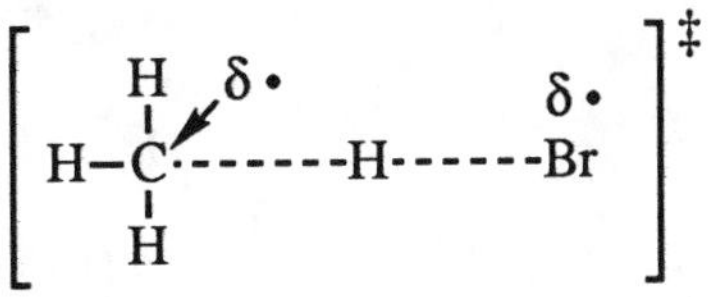

(3)

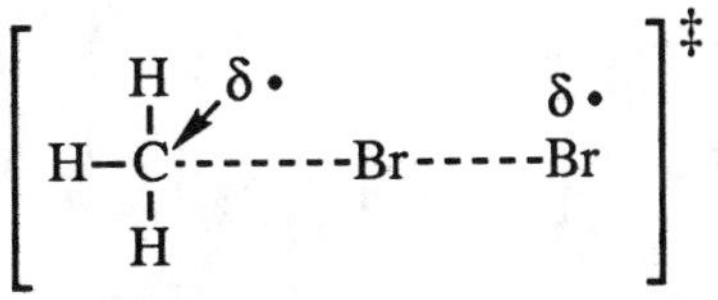

(d) ΔH° for the reaction is the sum of the ΔH° values of the individual propagation steps:

+ 16 kcal/mole + – 24 kcal/mole = **– 8 kcal/mole**
(+ 67 kJ/mole + – 101 kJ/mole = **– 34 kJ/mole**)

4-17 (a) initiation (1) $I{-}I \xrightarrow{h\nu} 2\ I\bullet$

propagation (2) $I\bullet + H{-}CH_3 \longrightarrow H{-}I + \bullet CH_3$

propagation (3) $I{-}I + \bullet CH_3 \longrightarrow I{-}CH_3 + I\bullet$

step (1)	break I-I	$\Delta H°$ = + 36 kcal/mole (+ 151 kJ/mole)
step (2)	break H-CH_3	$\Delta H°$ = + 104 kcal/mole (+ 435 kJ/mole)
	make H-I	$\Delta H°$ = - 71 kcal/mole (- 297 kJ/mole)
	step (2)	**$\Delta H°$ = +33 kcal/mole (+138 kJ/mole)**
step (3)	break I-I	$\Delta H°$ = + 36 kcal/mole (+ 151 kJ/mole)
	make I-CH_3	$\Delta H°$ = - 56 kcal/mole (- 234 kJ/mole)
	step (3)	**$\Delta H°$ = –20 kcal/mole (–83 kJ/mole)**

(b) $\Delta H°$ for the reaction is the sum of the $\Delta H°$ values of the individual propagation steps:

+ 33 kcal/mole + – 20 kcal/mole = **+ 13 kcal/mole**
(+ 138 kJ/mole + – 83 kJ/mole = **+ 55 kJ/mole)**

(c) Iodination of methane is unfavorable for both kinetic and thermodynamic reasons. Kinetically, the rate of the first propagation step must be very slow because it is very endothermic; the activation energy must be at least + 33 kcal/mole. Thermodynamically, the overall reaction is endothermic, so an equilibrium would favor reactants, not products; there is no energy decrease to drive the reaction to products.

4-18 Propane has six primary hydrogens and two secondary hydrogens, a ratio of 3 : 1. If primary and secondary hydrogens were replaced by chlorine at equal rates, the chloropropane isomers would reflect the same 3 : 1 ratio, that is, 75% 1-chloropropane and 25% 2-chloropropane.

4-19 (a)

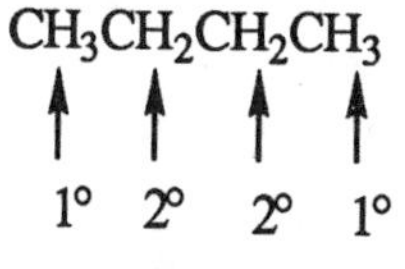
CH3CH2CH2CH3
1° 2° 2° 1°

(b)

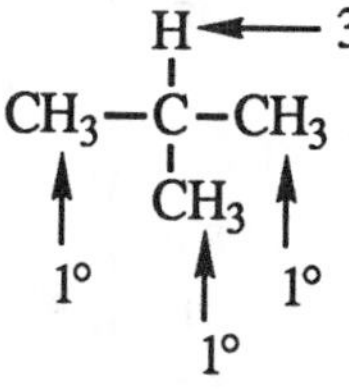
H 3°
CH3–C–CH3
CH3
1° 1° 1°

(c)

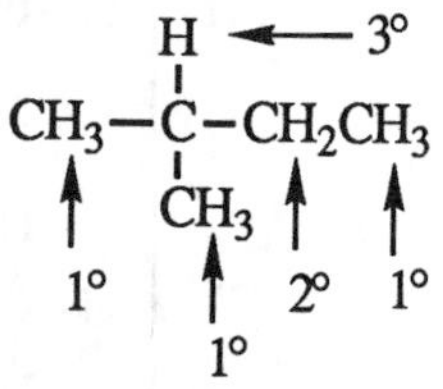
H 3°
CH3–C–CH2CH3
CH3
1° 1° 2° 1°

(d)

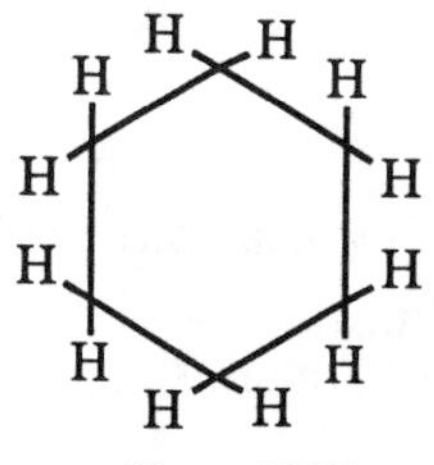

all are 2° H

(e)

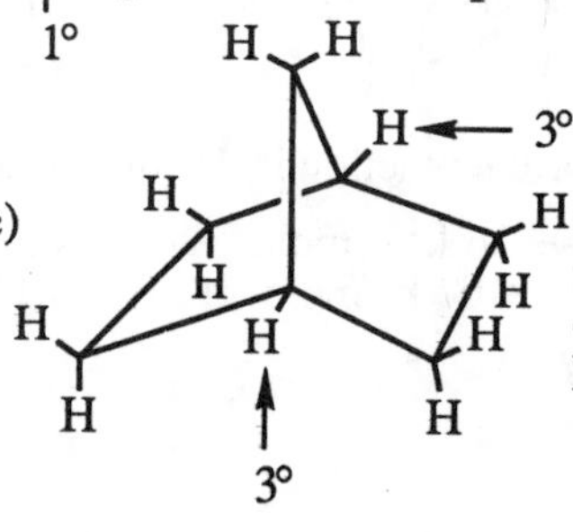
H 3°
3°

all are 2° H except bridgehead H's (labeled)

4-20 3° H abstraction

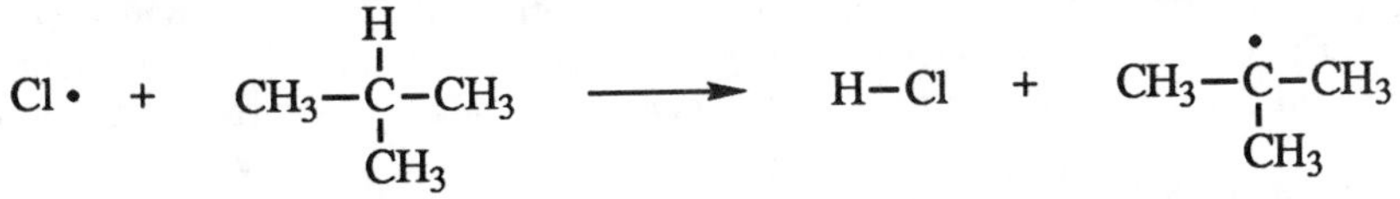

break 3° H-$C(CH_3)_3$ — $\Delta H°$ = + 91 kcal/mole (+ 381 kJ/mole)
make H-Cl — $\Delta H°$ = - 103 kcal/mole (- 431 kJ/mole)
overall 3° H abstraction — **$\Delta H°$ = −12 kcal/mole (−50 kJ/mole)**

1° H abstraction

break 1° H-$CH_2CH(CH_3)_2$ — $\Delta H°$ = + 98 kcal/mole (+ 410 kJ/mole)
make H-Cl — $\Delta H°$ = - 103 kcal/mole (- 431 kJ/mole)
overall 1° H abstraction — **$\Delta H°$ = −5 kcal/mole (−21 kJ/mole)**

(Note: $\Delta H°$ for abstraction of a 1° H from both ethane and propane are + 98 kcal/mole (+ 410 kJ/mole). It is reasonable to use this same value for abstraction of the 1° H in isobutane.)

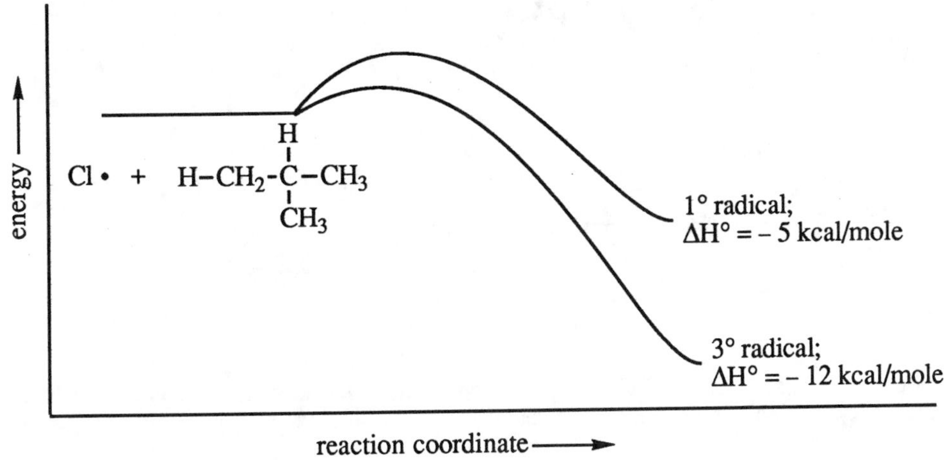

Since $\Delta H°$ for forming the 3° radical is more negative than $\Delta H°$ for forming the 1° radical, it is reasonable to infer that the activation energy leading to the 3° radical is lower than the activation energy leading to the 1° radical.

4-21 2-Methylbutane can produce four mono-chloro isomers. To calculate the relative amount of each in the product mixture, multiply the numbers of hydrogens which could lead to that product times the reactivity for that type of hydrogen. Each relative amount divided by the sum of all the amounts will provide the percent of each in the product mixture.

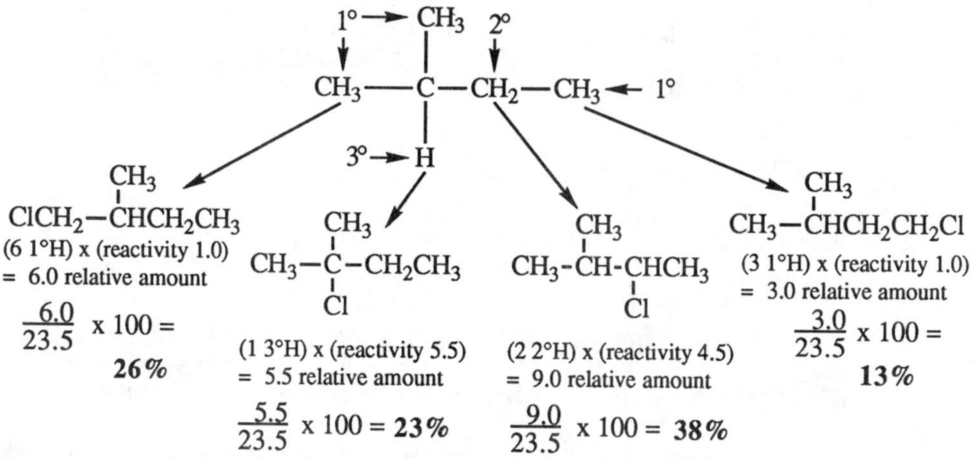

4-22 (a) When *n*-heptane is burned, only 1° and 2° radicals can be formed (from either C-H or C-C bond cleavage). These are high energy, unstable radicals which rapidly form other products. When isooctane (2,2,4-trimethylpentane, below) is burned, 3° radicals can be formed from either C-H or C-C bond cleavage. The 3° radicals are lower in energy than 1° or 2°, relatively stable, with lowered reactivity. Slower combustion translates to less "knocking."

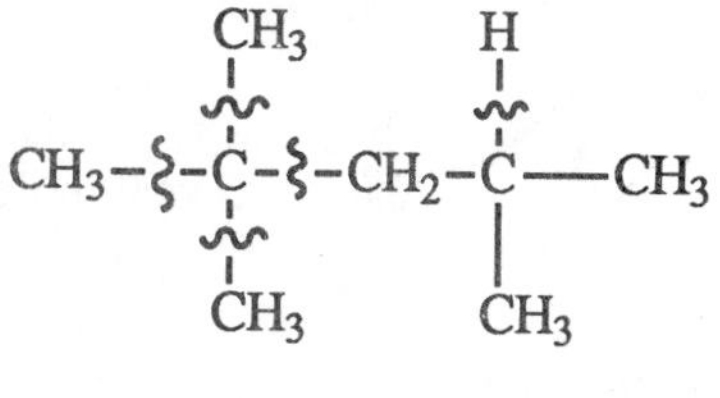

isooctane (2,2,4-trimethylpentane)

Any indicated bond cleavage will produce a 3° radical.

(b) When the alcohol hydrogen is abstracted from *t*-butyl alcohol, a relatively stable *t*-butoxy radical ($(CH_3)_3C\text{-}O\bullet$) is produced. This low energy radical is slower to react than alkyl radicals, moderating the reaction and producing less "knocking."

4-23 (a) 1° H abstraction

$$F\bullet \; + \; CH_3CH_2CH_3 \longrightarrow H{-}F \; + \; \bullet CH_2CH_2CH_3$$

break 1° $H\text{-}CH_2CH_2CH_3$	$\Delta H°$ = + 98 kcal/mole (+ 410 kJ/mole)
make H-F	$\Delta H°$ = - 136 kcal/mole (- 569 kJ/mole)
overall 1° H abstraction	**$\Delta H°$ = −38 kcal/mole (−159 kJ/mole)**

2° H abstraction

$$F\bullet \; + \; CH_3CH_2CH_3 \longrightarrow H{-}F \; + \; CH_3\dot{C}HCH_3$$

break 2° $H\text{-}CH(CH_3)_2$	$\Delta H°$ = + 95 kcal/mole (+ 397 kJ/mole)
make H-F	$\Delta H°$ = - 136 kcal/mole (- 569 kJ/mole)
overall 2° H abstraction	**$\Delta H°$ = −41 kcal/mole (−172 kJ/mole)**

(b) Free radical fluorination is extremely exothermic. In exothermic reactions, the transition states resemble the starting materials more than the products, so while the 1° and 2° radicals differ by about 3 kcal/mole (13 kJ/mole), the transition states will differ by only a tiny amount. For fluorination, then, the rate of abstraction for 1° and 2° hydrogens will be virtually identical. Product ratios will depend on statistical factors only.

Fluorination is difficult to control, but if propane were mono-fluorinated, the product mixture would reflect the ratio of the types of hydrogens: six 1° H to two 2° H, or 3 : 1 ratio, giving 75% 1-fluoropropane and 25% 2-fluoropropane.

4-24 (a) $$C_2H_5{-}D \; + \; Cl_2 \longrightarrow \underbrace{C_2H_5{-}Cl \; + \; DCl}_{7\%} \; + \; \underbrace{C_2H_4DCl \; + \; HCl}_{93\%}$$

D replacement: 7% ÷ 1 D = 7 (reactivity factor)
H replacement: 93% ÷ 5 H = 18.6 (reactivity factor)
relative reactivity of H : D abstraction = 18.6 ÷ 7 = 2.7

Each hydrogen is abstracted 2.7 times faster than deuterium.

(b) In both reactions of chlorine with either methane or ethane, the first propagation step is rate-determining. The reaction of chlorine atom with methane is *endothermic* by 1 kcal/mole (4 kJ/mole), while for ethane this step is *exothermic* by 5 kcal/mole (21 kJ/mole). By the Hammond Postulate, differences in activation energy are most pronounced in *endothermic* reactions where the transition states most resemble the products. Therefore, a change in the methane molecule causes a greater change in its transition state energy than the same change in the ethane molecule causes in its transition state energy. Deuterium will be abstracted more slowly in both methane and ethane, but the rate effect will be more pronounced in methane than in ethane.

4-25

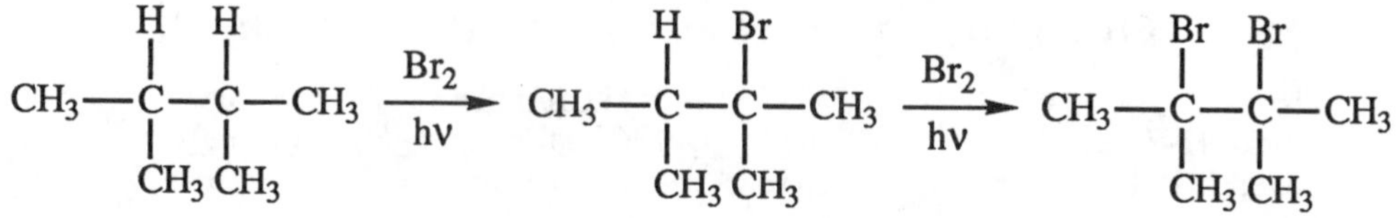

Mechanism

initiation 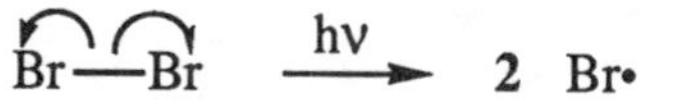

propagation 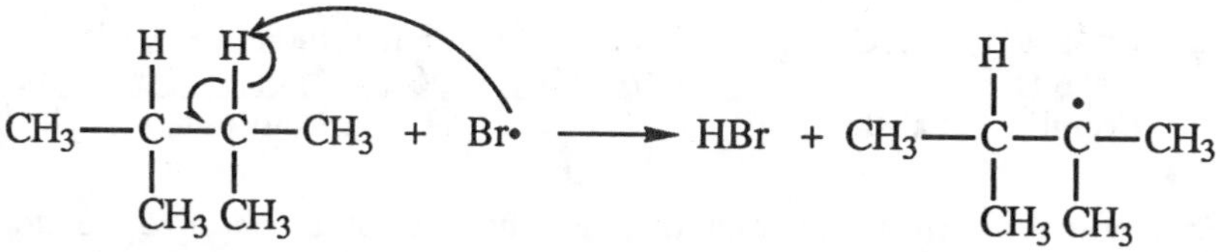

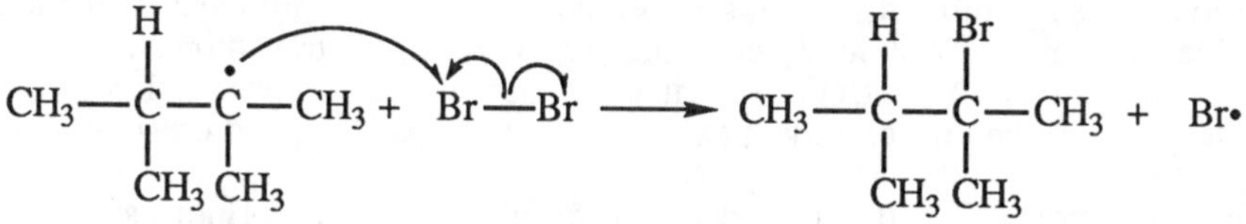

4-26 The triphenylmethyl cation is so stable because of the delocalization of the charge. The more resonance forms a species has, the more stable it will be.

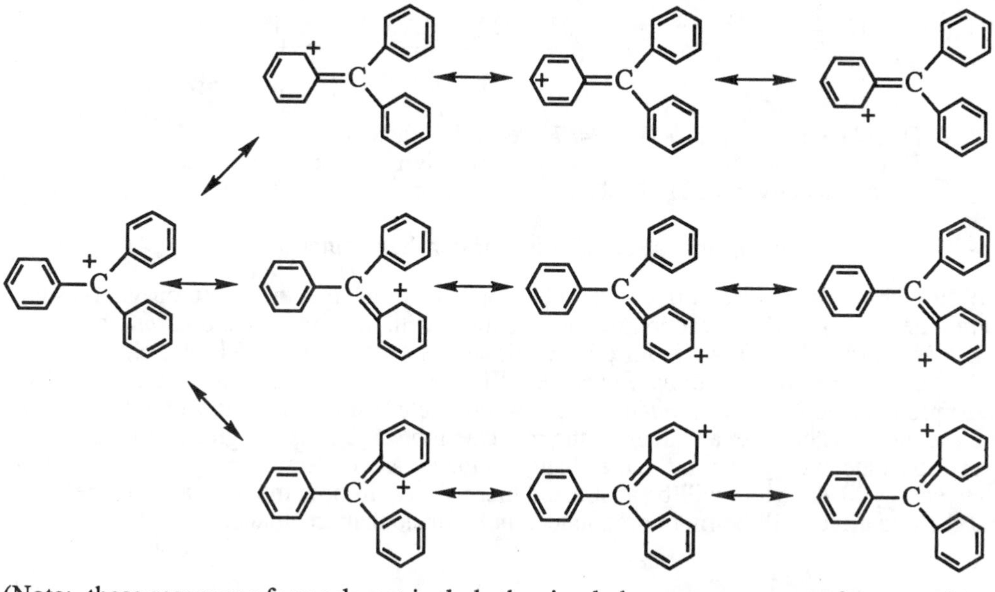

(Note: these resonance forms do not include the simple benzene resonance forms as shown below; they are significant, but repetitive. Refer to the note in the solution to 1-38(d).)

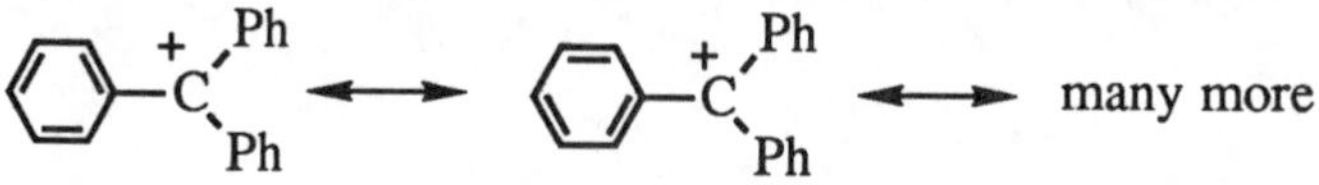

4-27 most stable (c) > (b) > (a) least stable

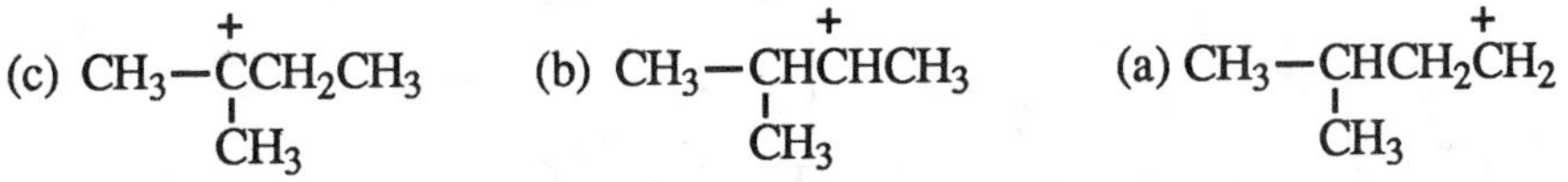

3° 2° 1°

4-28 most stable (c) > (b) > (a) least stable

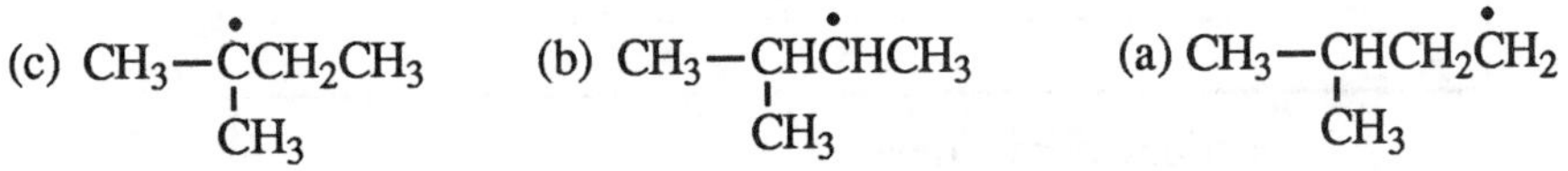

3° 2° 1°

4-29

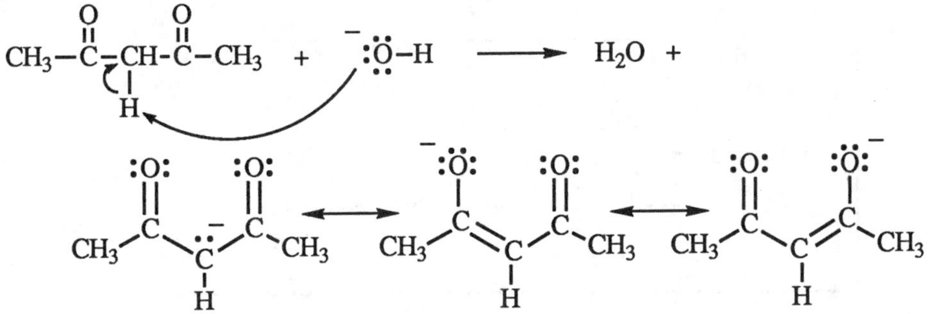

4-30

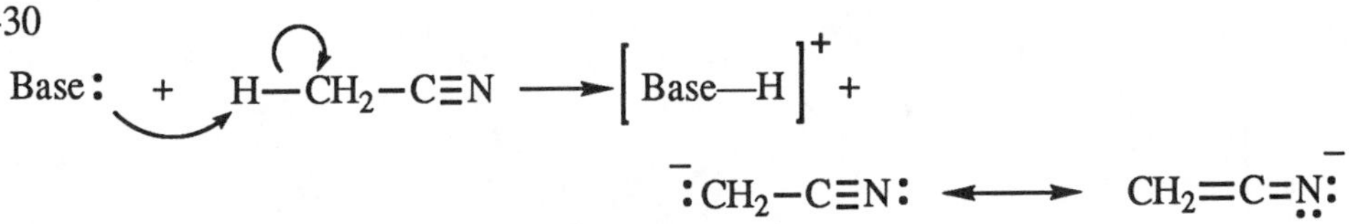

4-31

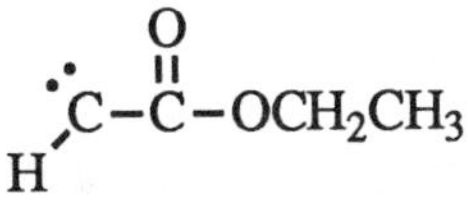

4-32 Refer to the Glossary and the text for definitions and examples.

4-33

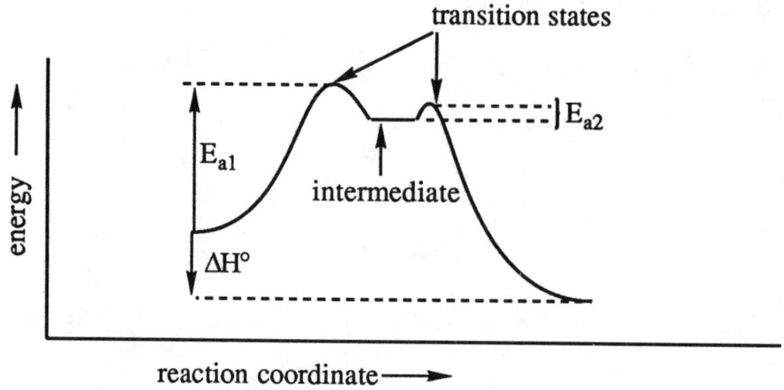

(b) ΔH° is negative (decreases), so the reaction is exothermic
(d) The first transition state determines the rate since it is the highest energy point.

4-34

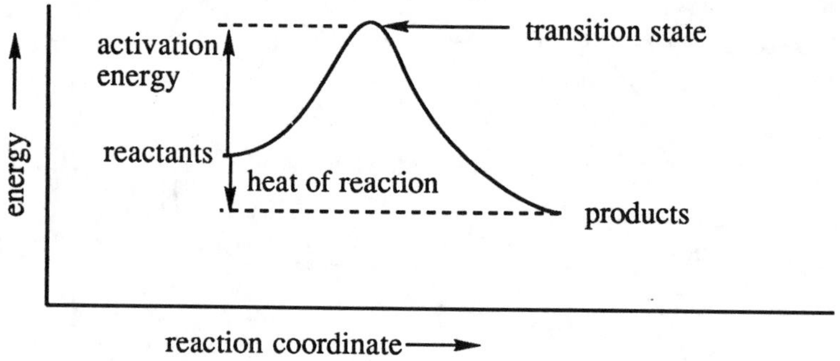

4-35

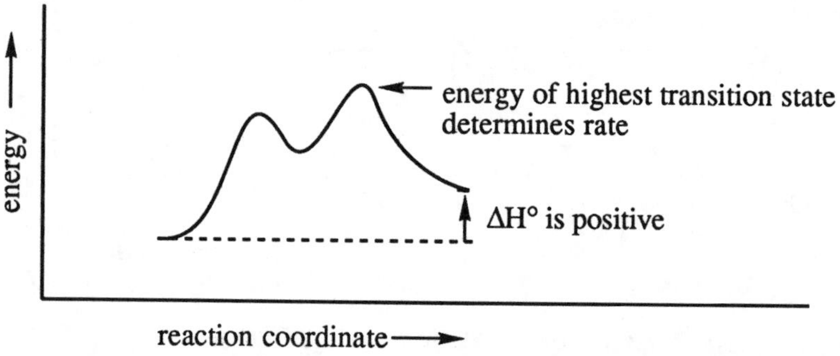

4-36 The rate law is first order with respect to the concentrations of hydrogen ion and of *t*-butyl alcohol, zeroth order with respect to the concentration of chloride ion, second order overall.

rate = k_r $[(CH_3)_3COH]\,[H^+]$

4-37 (a) $CH_3CH_2CH(CH_3)_2$

1° 2° 3° 1°

(b) $(CH_3)_3CCH_2C(CH_3)_3$

1° 2° 1°

(c)

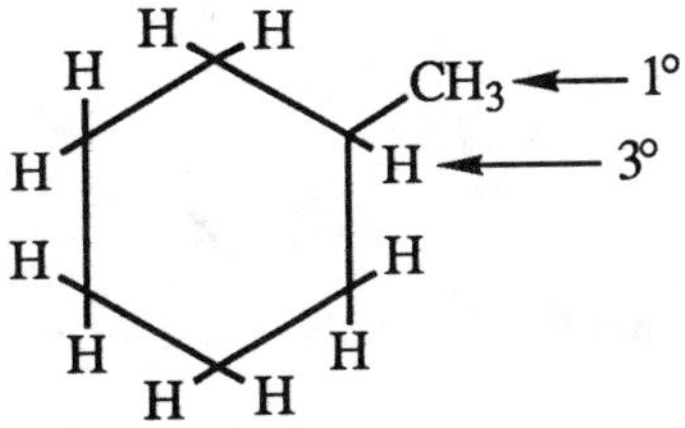

all are 2° H except as labeled

(d)

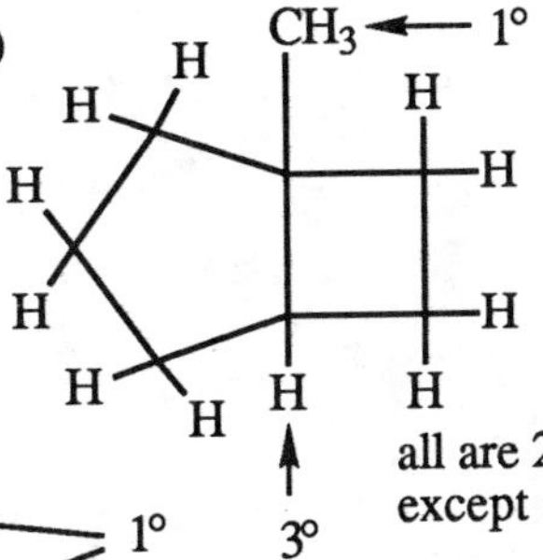

all are 2° H except as labeled

(e)

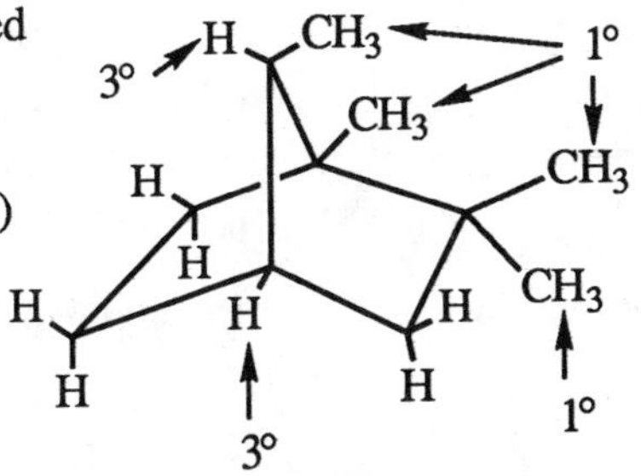

all are 2° H except as labeled

4-38 (a) break $H—CH_2CH_3$ and I—I, make $I—CH_2CH_3$ and H—I
kcal/mole: (+ 98 + + 36) + (− 53 + − 71) = **+ 10 kcal/mole**
kJ/mole: (+ 410 + + 151) + (− 222 + − 297) = **+ 42 kJ/mole**

(b) break $CH_3CH_2—Cl$ and H—I, make $CH_3CH_2—I$ and H—Cl
kcal/mole: (+ 81 + + 71) + (− 53 + − 103) = **− 4 kcal/mole**
kJ/mole: (+ 339 + + 297) + (− 222 + − 431) = **− 17 kJ/mole**

(c) break $(CH_3)_3C—OH$ and H—Cl, make $(CH_3)_3C—Cl$ and H—OH
kcal/mole: (+ 91 + + 103) + (− 79 + − 119) = **− 4 kcal/mole**
kJ/mole: (+ 381 + + 431) + (− 331 + − 498) = **− 17 kJ/mole**

(d) break $CH_3CH_2—CH_3$ and H—H, make $CH_3CH_2—H$ and $H—CH_3$
kcal/mole: (+ 85 + + 104) + (− 98 + − 104) = **− 13 kcal/mole**
kJ/mole: (+ 356 + + 435) + (− 410 + − 435) = **− 54 kJ/mole**

(e) break $CH_3CH_2—OH$ and H—Br, make $CH_3CH_2—Br$ and H—OH
kcal/mole: (+ 91 + + 88) + (− 68 + − 119) = **− 8 kcal/mole**
kJ/mole: (+ 381 + + 368) + (− 285 + − 498) = **− 34 kJ/mole**

4-39

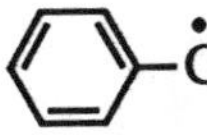

$\dot{C}H_2$ > $CH_2{=}CH\dot{C}H_2$ > $(CH_3)_3\dot{C}$ > $(CH_3)_2\dot{C}H$ > $CH_3\dot{C}H_2$ > $\dot{C}H_3$

85	87	91	95	98	104
(356)	(364)	(381)	(397)	(410)	(435)
most stable					least stable

Numbers are bond dissociation energies, kcal/mole (kJ/mole).

4-40

(a)

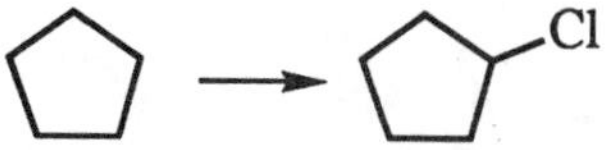

Only one product; chlorination would work.

(b)

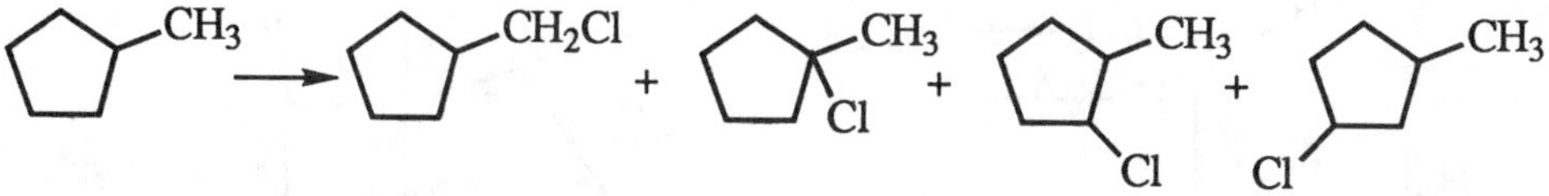

Chlorination would produce four positional isomers and would not be a good method to make only one of these.

(c)

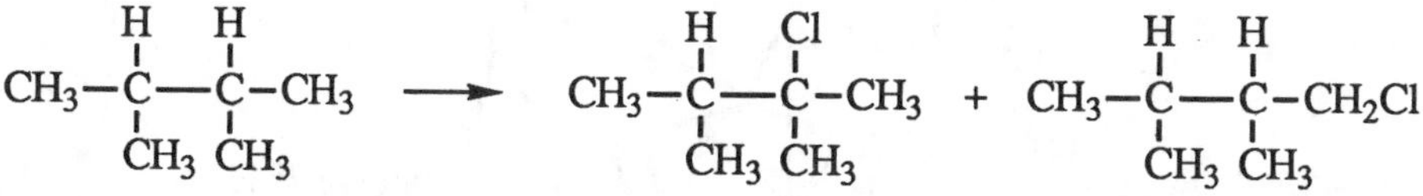

Chlorination would produce two positional isomers and would not be a good method to make only one of these.

(d)

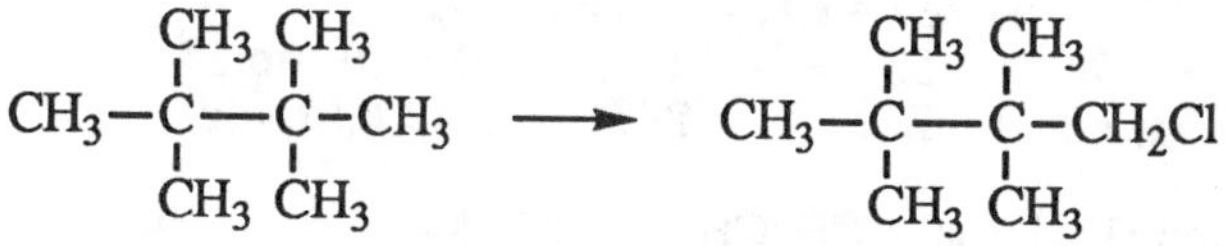

Only one product; chlorination would work.

4-41

initiation

propagation

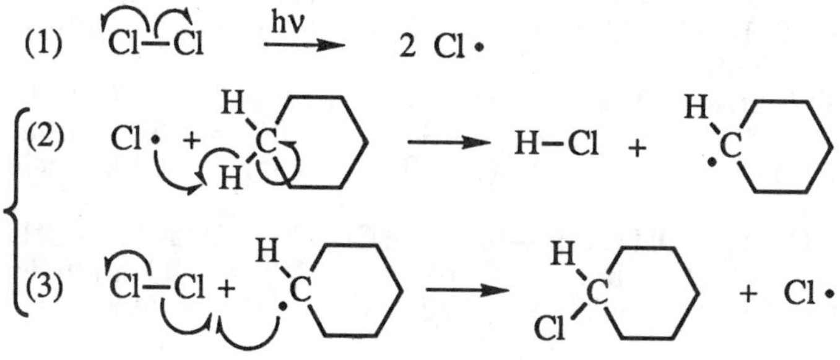

Termination steps are any two radicals combining.

4-42

(a) $\dot{C}H_2$ resonance: $CH_2{=}CH{-}\dot{C}H_2 \longleftrightarrow \dot{C}H_2{-}CH{=}CH_2$

(b)

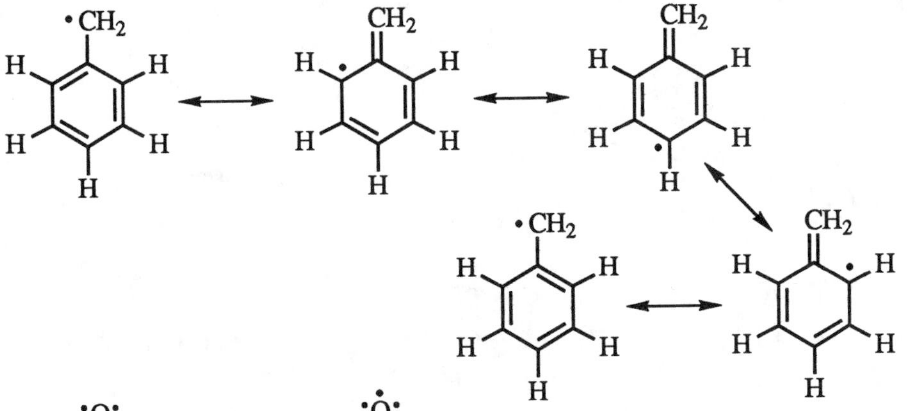

(c)

(d)

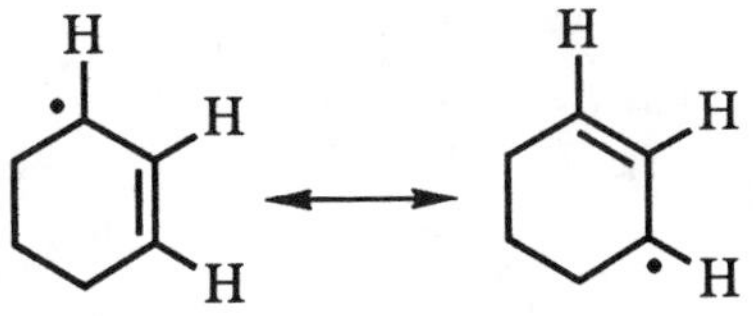

(e)

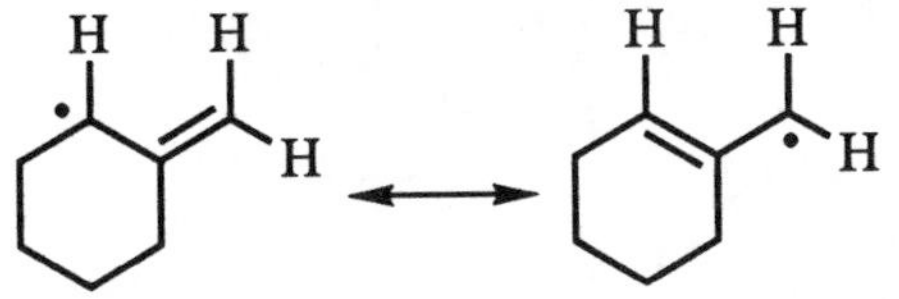

4-43 Cyclohexene has three types of hydrogens which could be abstracted by bromine atom. They are shown below, labeled with their bond dissociation energy in kcal/mole (kJ/mole).

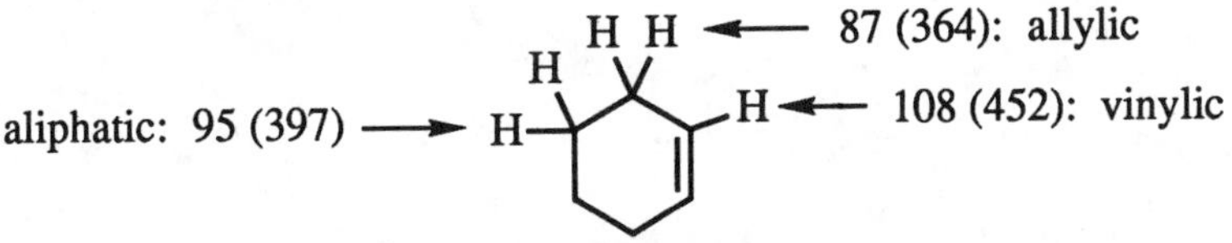
87 (364): allylic
108 (452): vinylic
aliphatic: 95 (397)

Bromine atom will abstract the hydrogen with the lowest bond dissociation energy at the fastest rate, in this case, the allylic hydrogen. The allylic hydrogen is most easily abstracted because the radical produced is stabilized by resonance.

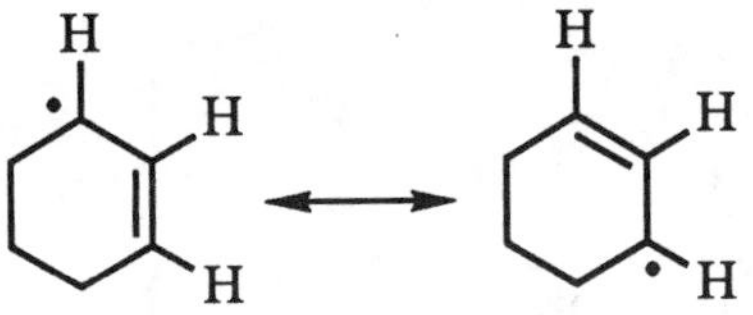

4-44 Where mixtures are possible, only the major product is shown.

(a) 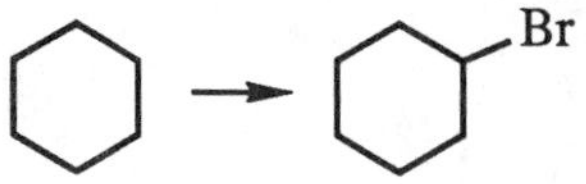
Br

only one product possible

(b) 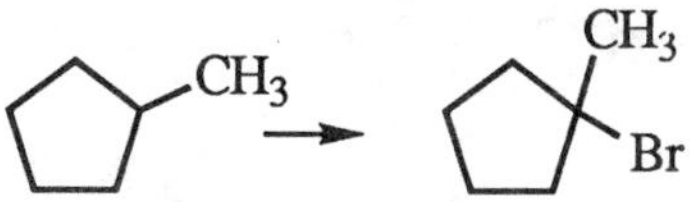
CH3
CH3
Br

3° hydrogen abstracted fastest

(c) 
CH3 H
CH3—C—C—CH3
CH3 CH3
CH3 Br
CH3—C—C—CH3
CH3 CH3

3° hydrogen abstracted fastest

(d) 
Br

decalin

3° hydrogen abstracted fastest

(e) 
Br

3° hydrogen abstracted fastest

(f) 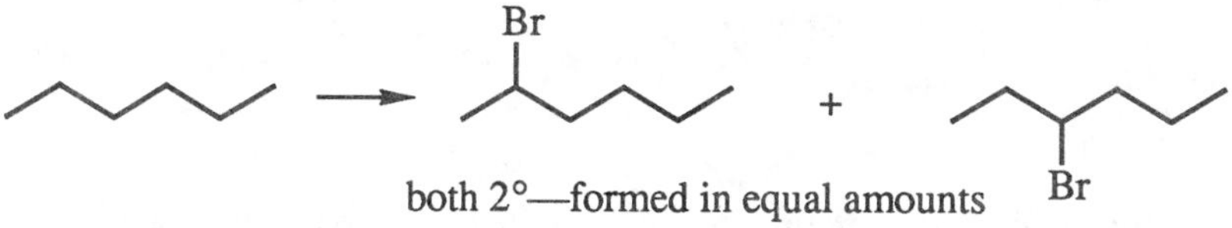
Br
+
Br

both 2°—formed in equal amounts

(g) 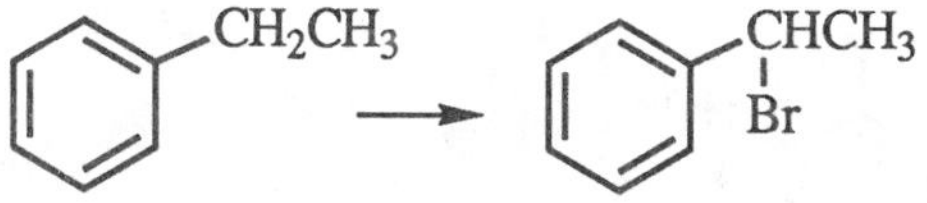
CH2CH3
CHCH3
Br

from resonance-stabilized radical

(h) 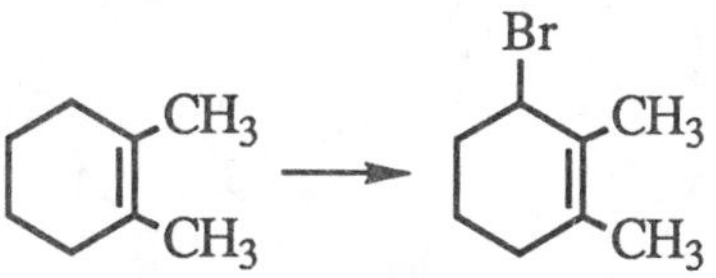
CH3
CH3
Br
CH3
CH3

from resonance-stabilized radical

4-45

(a) As CH_3Cl is produced, it can compete with CH_4 for available Cl •, generating CH_2Cl_2. This can generate $CHCl_3$, *etc.*

propagation steps

$$CH_4 + Cl\bullet \longrightarrow HCl + \bullet CH_3$$
$$\bullet CH_3 + Cl_2 \longrightarrow ClCH_3 + Cl\bullet$$
$$ClCH_3 + Cl\bullet \longrightarrow HCl + \bullet CH_2Cl$$
$$\bullet CH_2Cl + Cl_2 \longrightarrow CH_2Cl_2 + Cl\bullet$$
$$CH_2Cl_2 + Cl\bullet \longrightarrow HCl + \bullet CHCl_2$$
$$\bullet CHCl_2 + Cl_2 \longrightarrow CHCl_3 + Cl\bullet$$
$$CHCl_3 + Cl\bullet \longrightarrow HCl + \bullet CCl_3$$
$$\bullet CCl_3 + Cl_2 \longrightarrow CCl_4 + Cl\bullet$$

(b) To maximize CH_3Cl and minimize formation of polychloromethanes, the ratio of methane to chlorine must be kept high (see problem 4-2).

To guarantee that all hydrogens are replaced with chlorine to produce CCl_4, the ratio of chlorine to methane must be kept high.

4-46

n-Pentane can produce three mono-chloro isomers. To calculate the relative amount of each in the product mixture, multiply the numbers of hydrogens which could lead to that product times the reactivity for that type of hydrogen. Each relative amount divided by the sum of all the amounts will provide the percent of each in the product mixture.

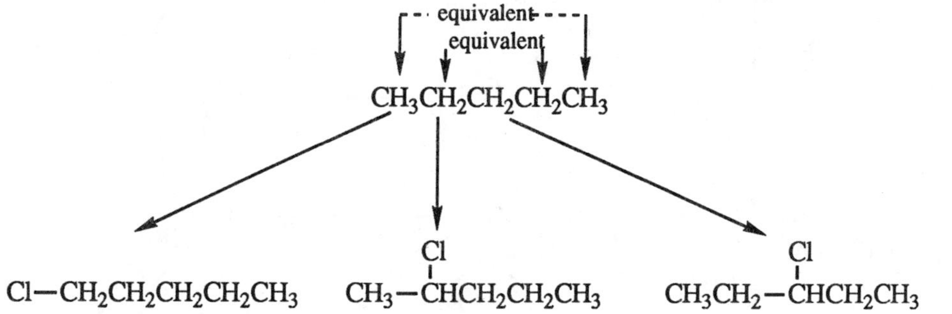

(6 1° H) x (reactivity 1.0) = 6.0 relative amount

(4 2° H) x (reactivity 4.5) = 18.0 relative amount

(2 2° H) x (reactivity 4.5) = 9.0 relative amount

total amount = 33.0

$\frac{6.0}{33.0}$ x 100 = **18%** $\frac{18.0}{33.0}$ x 100 = **55%** $\frac{9.0}{33.0}$ x 100 = **27%**

4-47

(a) The second propagation step in the chlorination of methane is highly exothermic ($\Delta H° = -26$ kcal/mole (-109 kJ/mole)). The transition state resembles the reactants, that is, the Cl-Cl bond will be slightly stretched and the Cl-CH_3 bond will just be starting to form.

$$\overset{\delta\bullet}{Cl} \text{-----} Cl \text{------------} \overset{\delta\bullet}{CH_3}$$

stronger weaker

(b) The second propagation step in the bromination of methane is highly exothermic ($\Delta H° = -24$ kcal/mole (-101 kJ/mole)). The transition state resembles the reactants, that is, the Br-Br bond will be slightly stretched and the Br-CH_3 bond will just be starting to form.

$$\overset{\delta\bullet}{Br} \text{-----} Br \text{------------} \overset{\delta\bullet}{CH_3}$$

stronger weaker

4-48 Two mechanisms are possible depending on whether HO • reacts with chlorine or cyclopentane.

Mechanism 1

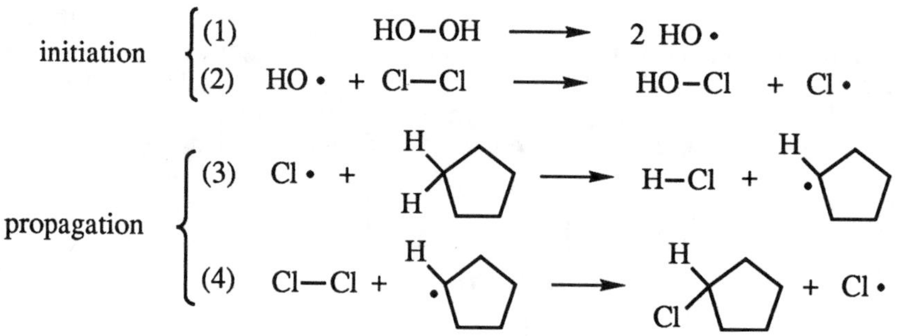
initiation
(1) HO−OH ⟶ 2 HO•
(2) HO• + Cl—Cl ⟶ HO−Cl + Cl•
propagation
(3) Cl• +
H−Cl +
(4) Cl—Cl +
+ Cl•
H
Cl

Mechanism 2

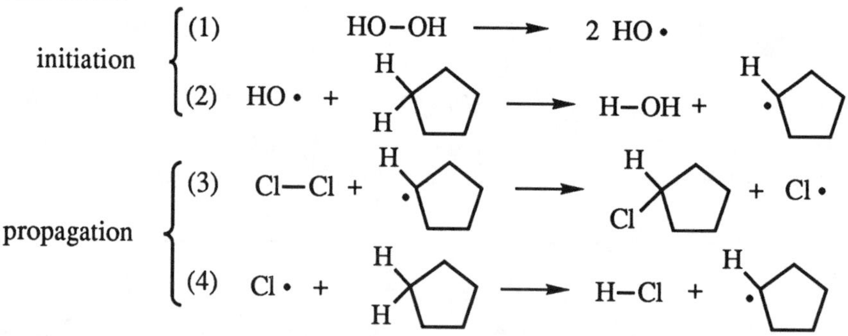
initiation
(1) HO−OH ⟶ 2 HO•
(2) HO• +
H−OH +
propagation
(3) Cl—Cl +
+ Cl•
(4) Cl• +
H−Cl +
H
Cl

The energies of the propagation steps determine which mechanism is followed. The bond dissociation energy of HO—Cl is about 50 kcal/mole, making initiation step (2) in mechanism 1 *endothermic*. In mechanism 2, initiation step (2) is *exothermic* by about 24 kcal/mole; mechanism 2 is preferred.

4-49

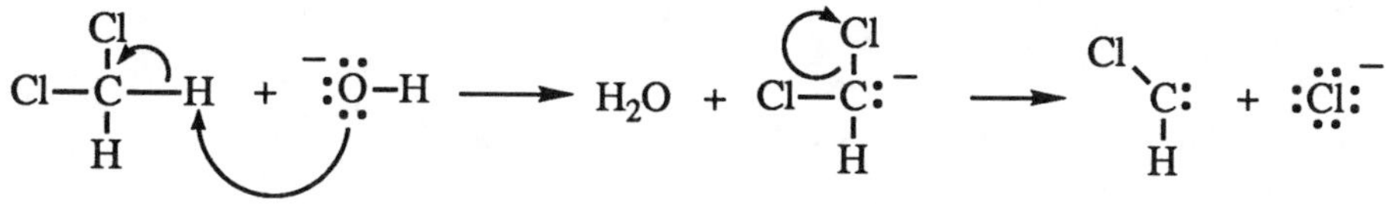

4-50 The following critical equation is the key to this problem:

$$\Delta G = \Delta H - T\Delta S$$

At 1400 K, the equilibrium constant is 1; therefore:

$$K_{eq} = 1 \quad \Longrightarrow \quad \Delta G = 0 \quad \Longrightarrow \quad \Delta H = T\Delta S$$

Assuming ΔH is about the same at 1400 K as it is at calorimeter temperature:

$$\Delta S = \frac{\Delta H}{T} = \frac{-32.7 \text{ kcal/mole}}{1400 \text{ K}} = \frac{-32{,}700}{1400} \text{ cal/kelvin-mole}$$

$$= -23 \text{ cal/kelvin-mole}$$

$$= -23 \text{ e.u. ("entropy units")}$$

This is a large *decrease* in entropy, consistent with two molecules combining into one.

5-1 (a) vinyl halide
(b) alkyl halide
(c) aryl halide
(d) alkyl halide
(e) vinyl halide
(f) aryl halide

5-2 (a)

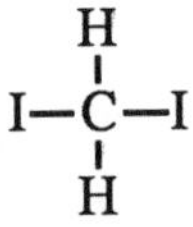

(b)

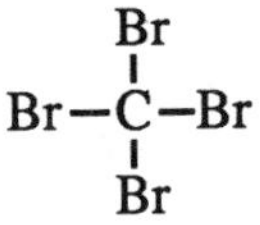

(c)

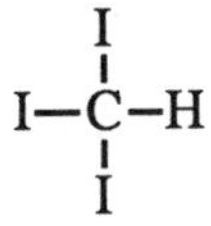

(d)

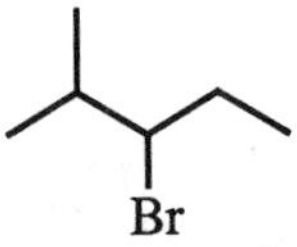

(e)

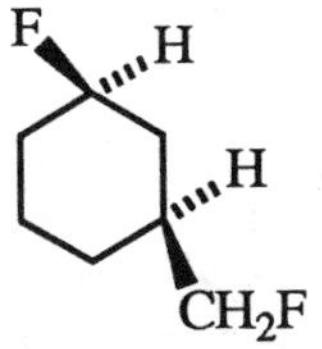

5-3 IUPAC name; common name; degree of halogen-bearing carbon

(a) 1-chloro-2-methylpropane; isobutyl chloride; 1°
(b) 3-bromo-4-methylhexane; no common name; 2°
(c) 1,1-dichloroethane; no common name; 1°
(d) 2-bromo-1,1,1-trichloroethane; no common name; all 1°
(e) trichloromethane; chloroform; methyl
(f) 2-bromo-2-methylpropane; *t*-butyl bromide; 3°
(g) 2-bromobutane; *sec*-butyl bromide; 2°
(h) 1-chloro-2-methylbutane; no common name; 1°
(i) diiodomethane; methylene iodide; methyl
(j) 4-fluoro-1,1-dimethylcyclohexane; no common name; 2°
(k) *cis*-1-bromo-2-chlorocyclobutane; no common name; both 2°
(l) *trans*-1,3-dichlorocyclopentane; no common name; all 2°

5-4

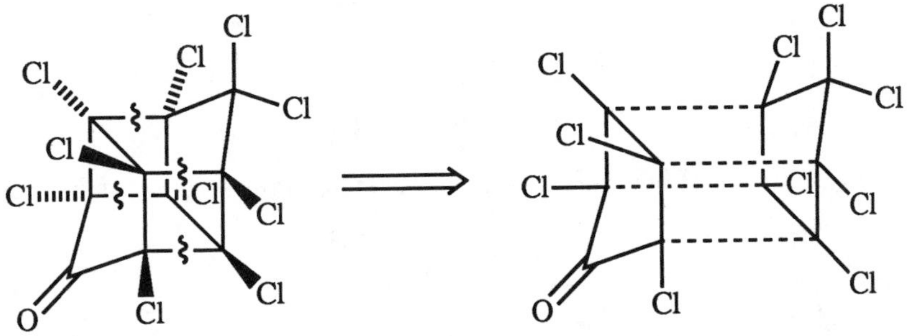

kepone

5-4 continued

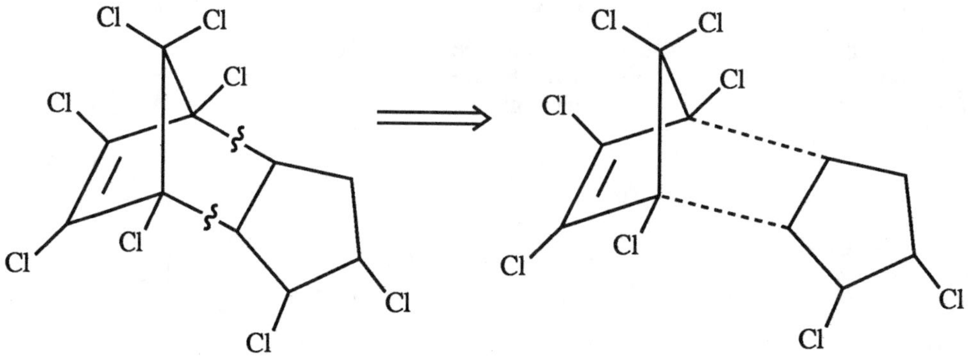
Cl
Cl
Cl
Cl
Cl
Cl
Cl
Cl

chlordane

5-5 From the text, Section 2-11, the bond dipole moment depends not only on bond length but also on charge separation, which in turn depends on the difference in electronegativities of the two atoms connected by the bond. Because chlorine's electronegativity (3.2) is significantly higher than iodine's (2.7), the C—Cl bond dipole is greater than that of C—I, despite C—I being a longer bond.

5-6 (a) *n*-Butyl bromide has a higher molecular weight and less branching, and boils at a higher temperature than isopropyl bromide.

(b) *t*-Butyl bromide has a higher molecular weight and a larger halogen, and despite its greater branching, boils at a higher temperature than isopropyl chloride.

(c) *n*-Butyl bromide has a higher molecular weight and a larger halogen, and boils at a higher temperature than *n*-butyl chloride.

5-7 From Table 3-2, the density of hexane is 0.66; it will float on the water layer (d 1.00). From Table 5-3, the density of chloroform is 1.50; water will float on the chloroform.

5-8

(a) 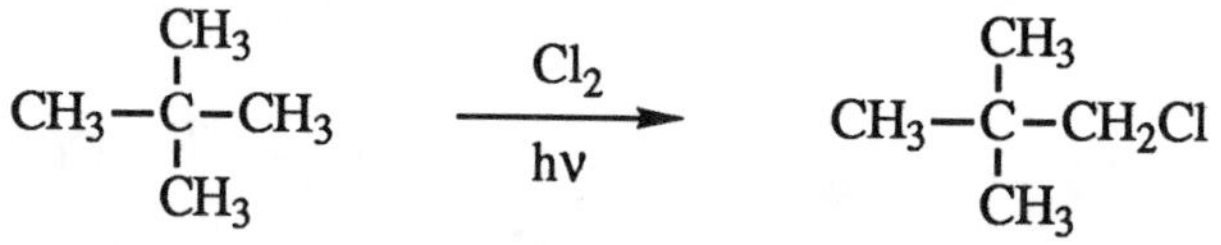
CH3
CH3—C—CH3
CH3
Cl2
hν
CH3
CH3—C—CH2Cl
CH3

has only one type of hydrogen—only one monochlorine isomer can be produced

(b) 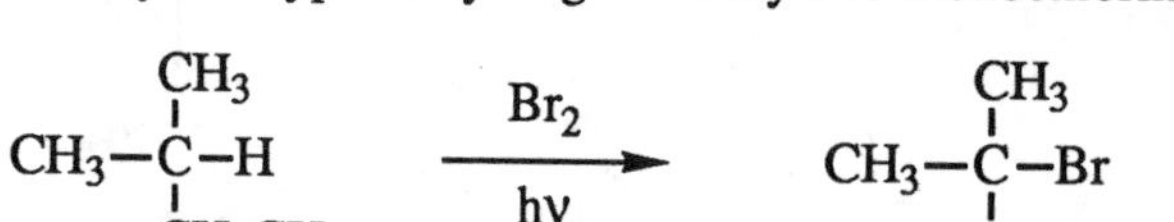
CH3
CH3—C—H
CH2CH3
Br2
hν
CH3
CH3—C—Br
CH2CH3

bromination has a strong preference for abstracting hydrogens (like 3°) that give stable radical intermediates

5-8 continued

(c) 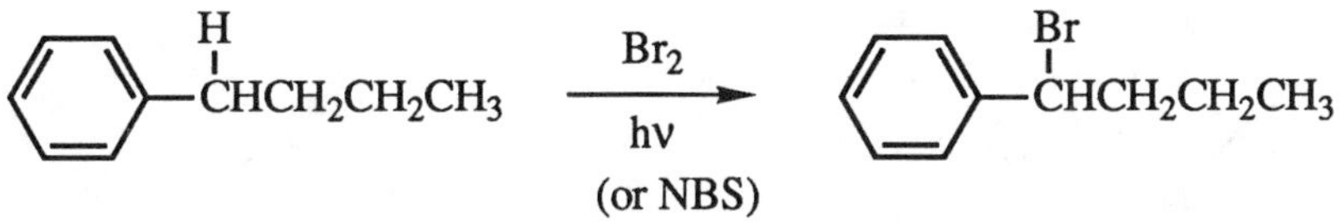

bromine atom will abstract the hydrogen giving the most stable radical; in this case, the radical intermediate will be stabilized by resonance with the benzene ring

5-9 propagation steps

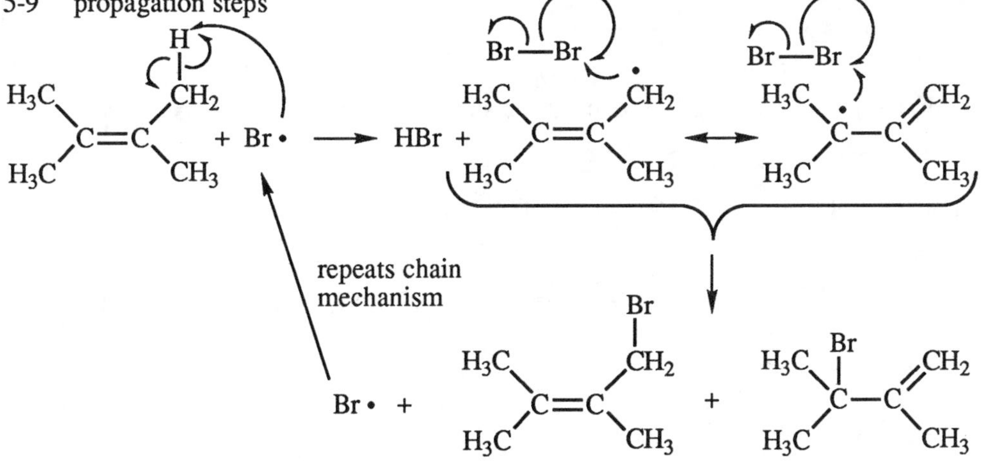

The resonance-stabilized allylic radical intermediate has radical character on both the 1° and 3° carbons, so bromine can bond to either of these carbons producing two isomeric products.

5-10 (a) substitution—Br is replaced
(b) elimination—H and OH are lost
(c) elimination—both Br atoms are lost

5-11 (a) $CH_3(CH_2)_4CH_2—OCH_2CH_3$ (b) $CH_3(CH_2)_4CH_2—CN$

(c) $CH_3(CH_2)_4CH_2—OH$

5-12 The rate law is first order in both CH_3I and hydroxide. If the concentration of CH_3I is lowered to one-fifth the original value, the rate must decrease to one-fifth; if the concentration of hydroxide is doubled, the rate must also double. Thus, the rate must decrease to 0.02 mole/L per second:

$$\text{rate} = \underbrace{(0.05 \text{ mole/L per second})}_{\text{original rate}} \times \underbrace{\frac{(0.1\text{ M})}{(0.5\text{ M})}}_{\text{change in } CH_3I} \times \underbrace{\frac{(2.0\text{ M})}{(1.0\text{ M})}}_{\text{change in NaOH}} = \underbrace{0.02 \text{ mole/L per second}}_{\text{new rate}}$$

5-13

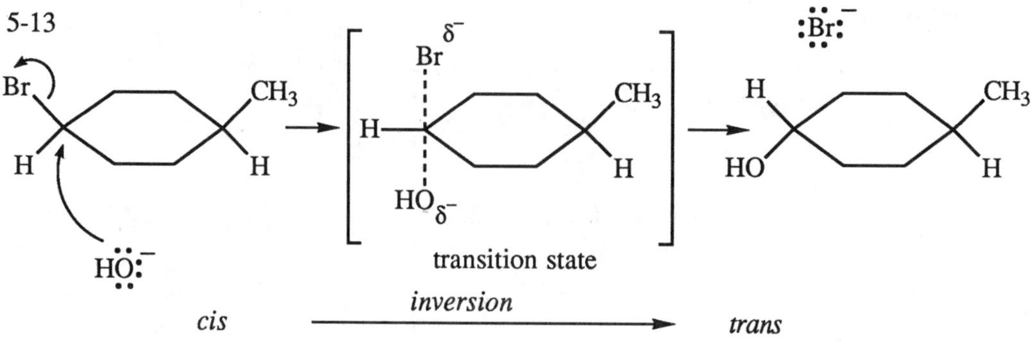
Br
H
CH3
H
HO:⁻
δ⁻
Br
H
CH3
H
HO δ⁻
transition state
:Br:⁻
H
HO
CH3
H
cis
inversion
trans

5-14 Both organic and inorganic products are shown here for completeness.

(a) $(CH_3)_3C{-}O{-}CH_2CH_3$ + KBr

(b) $CH_3CH_2CH_2CH_2{-}C{\equiv}CH$ + NaCl

(c) $(CH_3)_2CHCH_2{-}NH_3^+\ Br^-$ $\xrightarrow{NH_3}$ $(CH_3)_2CHCH_2{-}NH_2$ + $NH_4^+\ Br^-$

(d) $CH_3CH_2{-}CN$ + NaI

(e)  + NaCl
I

(f)  + Hg_2ClF
F

5-15 All reactions in this problem follow the same pattern; the only difference is the nucleophile ($^-$:Nuc). Only the nucleophile is listed below. (Cations like Na^+ or K^+ accompany the nucleophile but are simply spectator ions and do not take part in the reaction; they are not shown here.)

Cl
+ ⁻:Nuc
Nuc
+ Cl⁻

1-chlorobutane

(a) HO^- (b) F^- from Hg_2F_2 (c) I^- (d) ^-CN

(e) $HC{\equiv}C^-$ (f) $^-OCH_2CH_3$ (g) excess NH_3 (or $^-NH_2$)

5-16 In all cases, the less hindered structure is the better S_N2 substrate.

(a) 2-methyl-1-iodopropane (1° versus 3°)

(b) cyclohexyl bromide (2° versus 3°)

(c) isopropyl bromide (no substituent on neighboring carbon)

(d) 2-chlorobutane (even though this is a 2° halide, it is easier to attack than the neopentyl type in 2,2-dimethyl-1-chlorobutane—see below)

(e) isopropyl iodide (same reason as in (d))

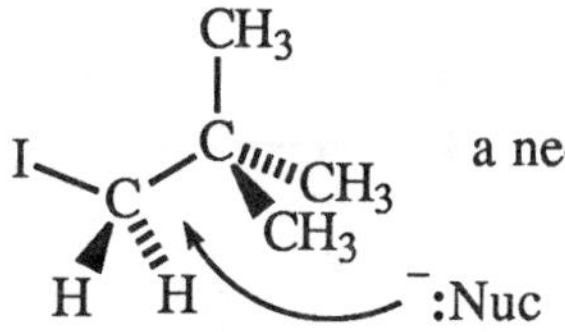

CH3
I
C
C
CH3
CH3
H
H
:Nuc

a neopentyl halide—hindered to backside attack

5-17

(a) $(CH_3CH_2)_2NH$ is a better nucleophile—less hindered

(b) $(CH_3)_2S$ is a better nucleophile—S is larger, more polarizable than O

(c) PH_3 is a better nucleophile—P is larger, more polarizable than N

(d) CH_3S^- is a better nucleophile—anions are better than neutral atoms of the same element

(e) $(CH_3)_3N$ is a better nucleophile—less electronegative than oxygen, better able to donate an electron pair

(f) $CH_3CH_2CH_2O^-$ is a better nucleophile—less branching, less steric hindrance

(g) I^- is a better nucleophile—larger, more polarizable

5-18

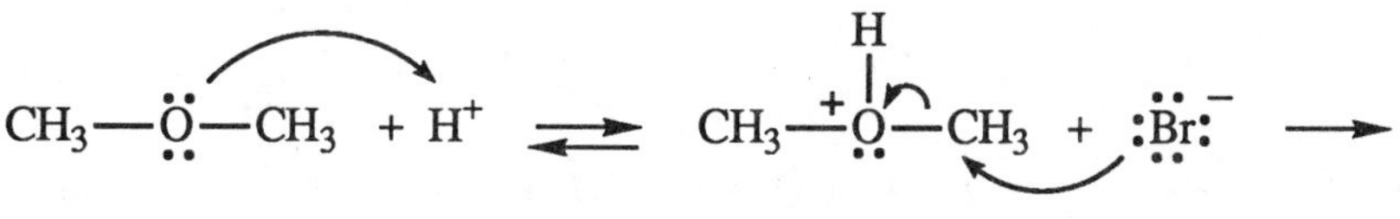

CH3—O—CH3 + H+
H
CH3—O—CH3 + :Br:

Protonation converts OCH_3 to a good leaving group.

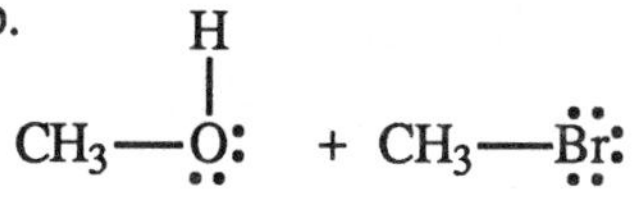

H
CH3—O:
+ CH3—Br:

5-19 methyl iodide > methyl chloride > ethyl chloride > isopropyl bromide >> neopentyl bromide , *t*-butyl iodide

Establishing the relative order of neopentyl bromide and *t*-butyl iodide would be difficult because both would be extremely slow.

5-20 Ionization is the rate-determining step in S_N1. Anything that stabilizes the intermediate will speed the reaction. Both of these compounds form resonance-stabilized intermediates.

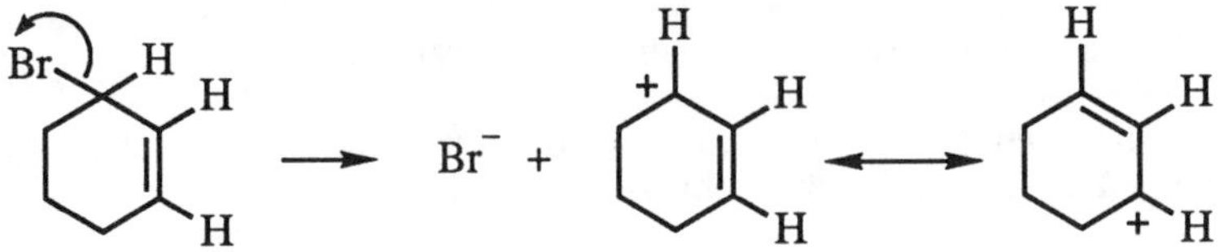

allylic

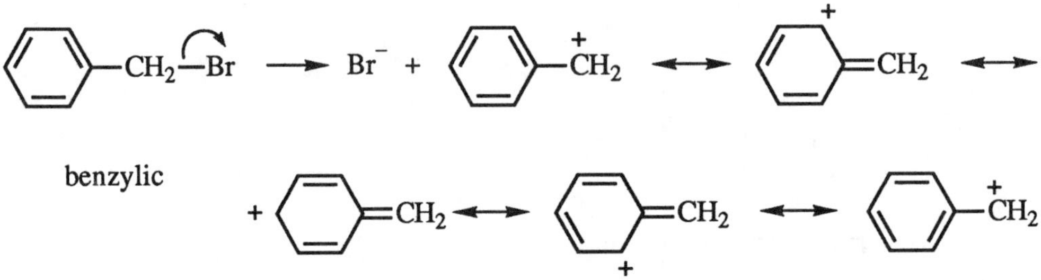

benzylic

5-21

(a) 2-iodo-2-methylbutane (3°) is faster than ethyl iodide (1°)

(b) 2-iodo-2-methylbutane is faster than *t*-butyl chloride (iodide is a better leaving group than chloride)

(c) 3-bromocyclohexene (2°) is faster than *n*-propyl bromide (1°) (3-bromocyclohexene can form a resonance-stabilized intermediate as shown in Problem 5-20)

(d) cyclohexyl bromide (2°) is faster than methyl iodide

5-22

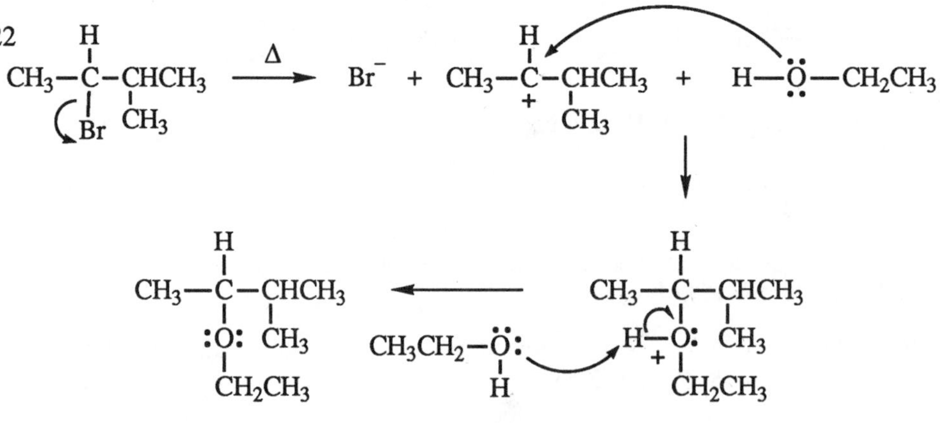

5-23 It is important to analyze the structure of carbocations to consider if migration of any groups from adjacent carbons will lead to a more stable carbocation. As a general rule, if rearrangement would lead to a more stable carbocation, a carbocation will rearrange. (Beginning with this problem, only those unshared electons pairs involved in a particular step will be shown.)

(a)

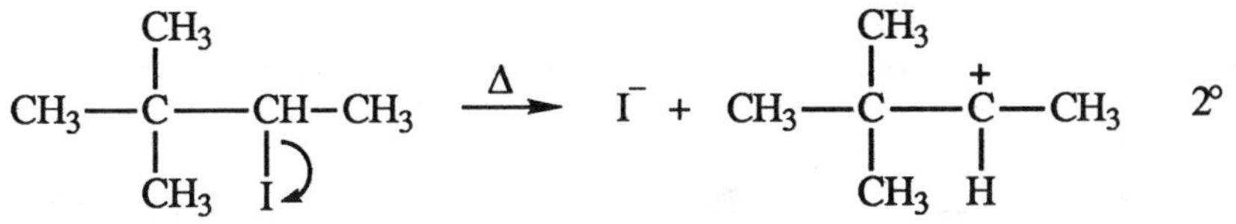

nucleophilic attack on unrearranged carbocation

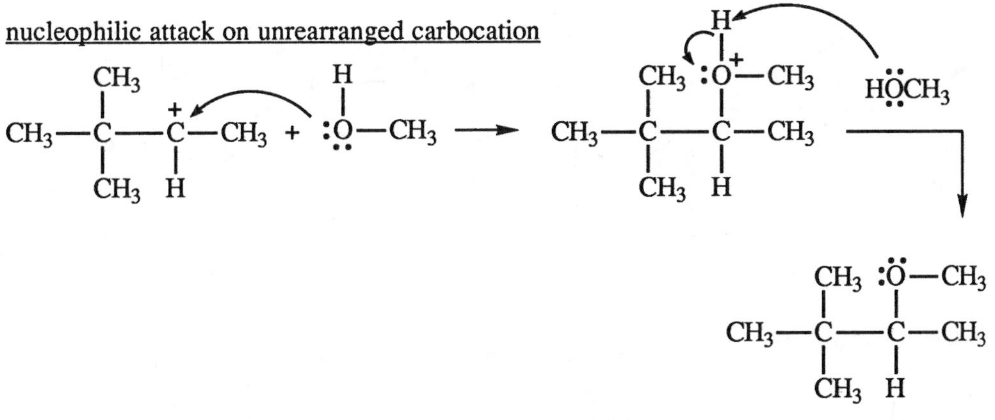

unrearranged product

nucleophilic attack after carbocation rearrangement

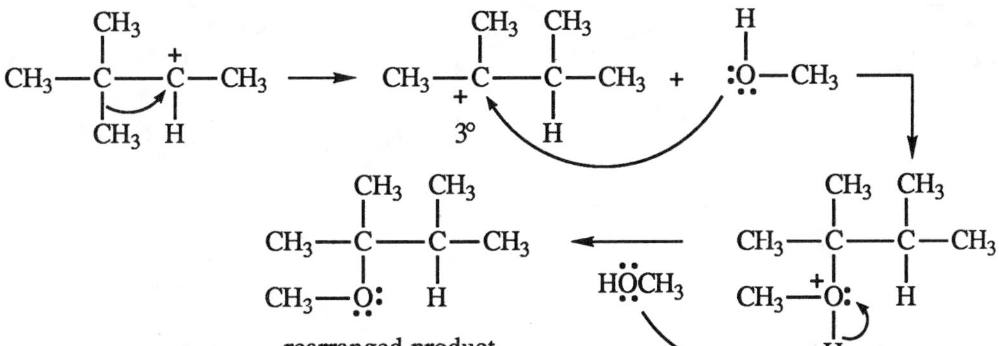

rearranged product

Methyl shift to the 2° carbocation forms a more stable 3° carbocation.

5-23 continued

(b)

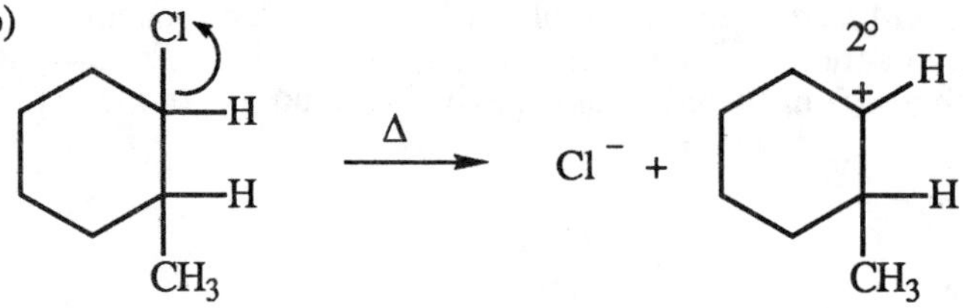

nucleophilic attack on unrearranged carbocation

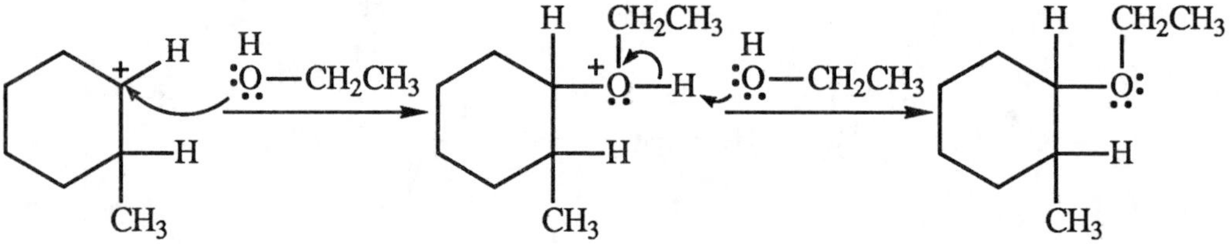

unrearranged product

nucleophilic attack after carbocation rearrangement

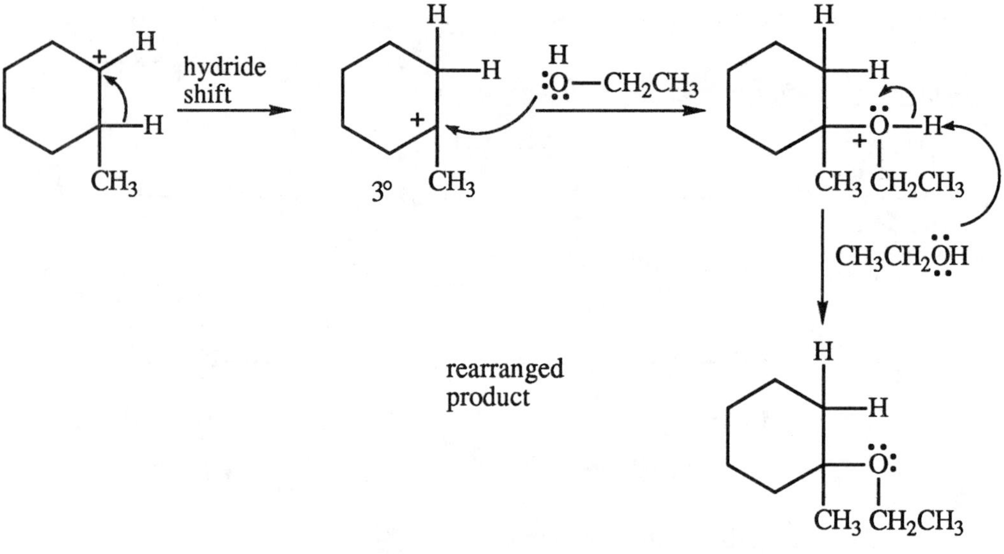

rearranged product

Hydride shift to the 2° carbocation forms a more stable 3° carbocation.

5-23 continued

(c)

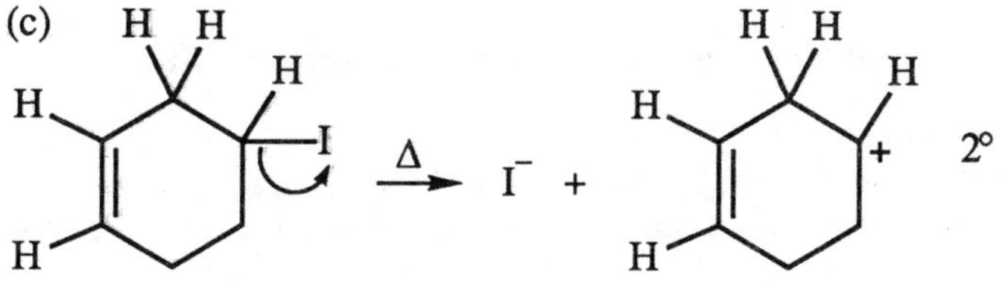

Δ
2°

<u>nucleophilic attack on unrearranged carbocation</u>

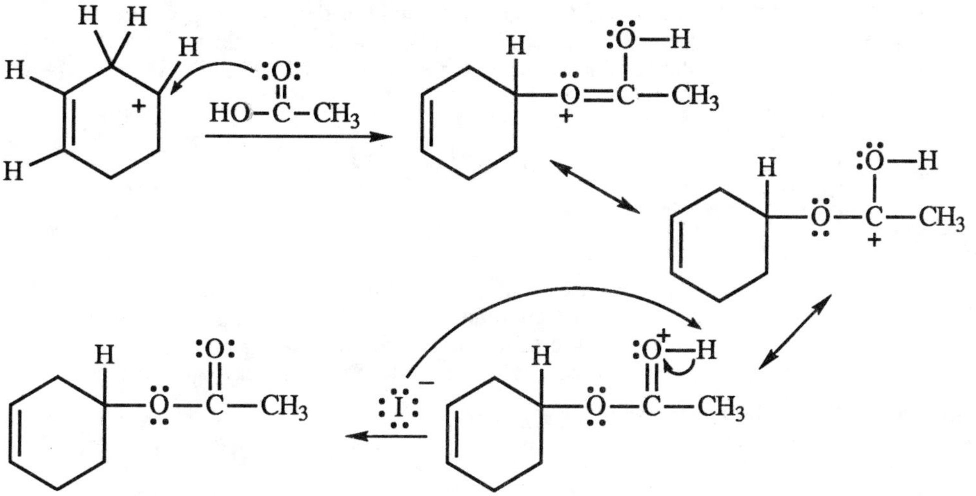

unrearranged product

<u>nucleophilic attack after carbocation rearrangement</u>

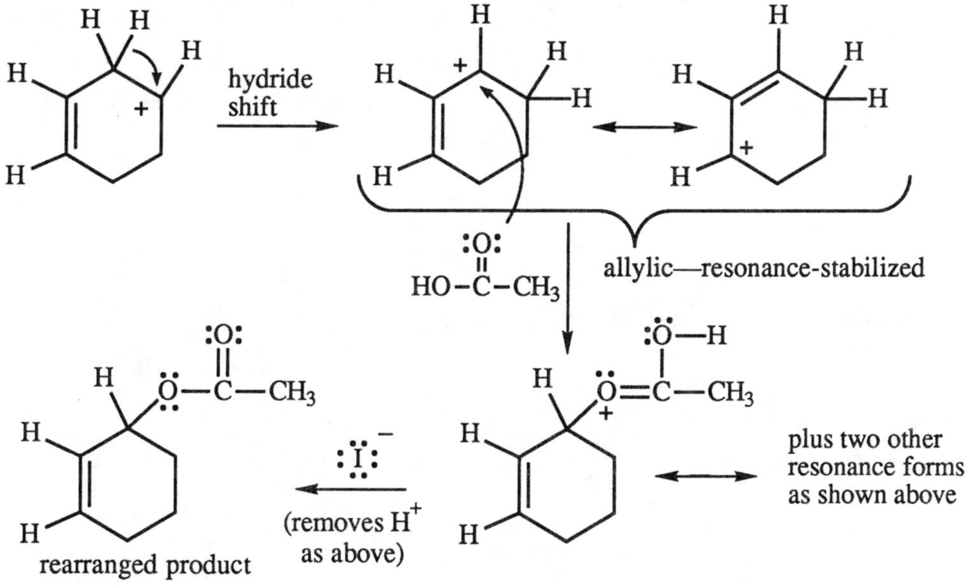

hydride shift
allylic—resonance-stabilized
plus two other resonance forms as shown above
(removes H+ as above)
rearranged product

5-23 continued

Comments on 5-23(c)

(1) The hydride shift to a 2° carbocation generates an allylic, resonance-stabilized 2° carbocation.

(2) The double-bonded oxygen of acetic acid is more nucleophilic because of the resonance forms it can have after attack. (See Solved Problem 1-5 and Problem 1-14 in the text.)

(3) Attack on only one carbon of the allylic carbocation is shown. In reality, both positive carbons would be attacked in equal amounts, but they would give the identical product *in this case*. In other compounds, however, attack on the different carbons might give different products. ALWAYS CONSIDER ALL POSSIBILITIES.

(d)

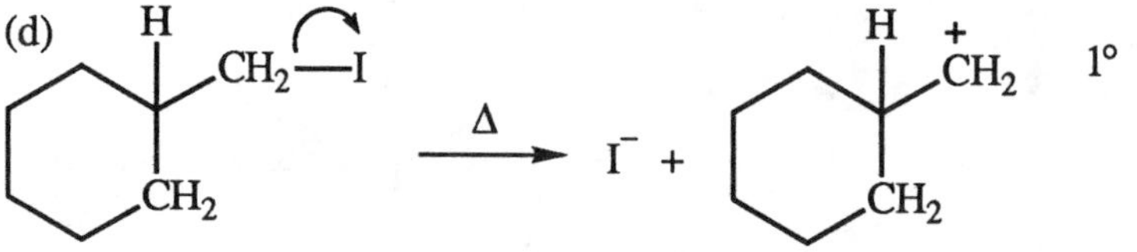

H
CH_2—I
CH_2
Δ
I^- +
H +
CH_2
1°
CH_2

The 1° carbocation initially formed is very unstable; some chemists believe that rearrangement occurs at the same time as the leaving group leaves. At most, the 1° carbocation has a very short lifetime.

hydride shift followed by nucleophilic attack

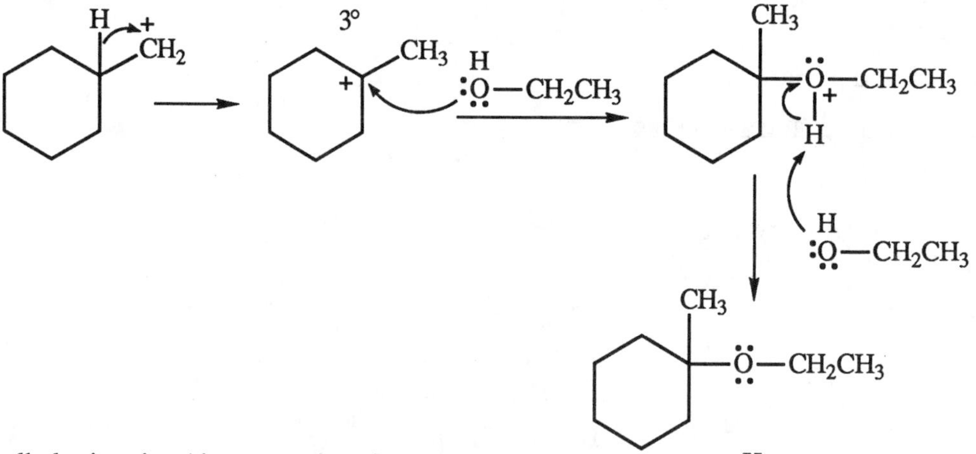

H +
CH_2
3°
CH_3
H
:O—CH_2CH_3
CH_3
O—CH_2CH_3
H
H
:O—CH_2CH_3
CH_3
O—CH_2CH_3

alkyl migration (ring expansion) followed by nucleophilic attack

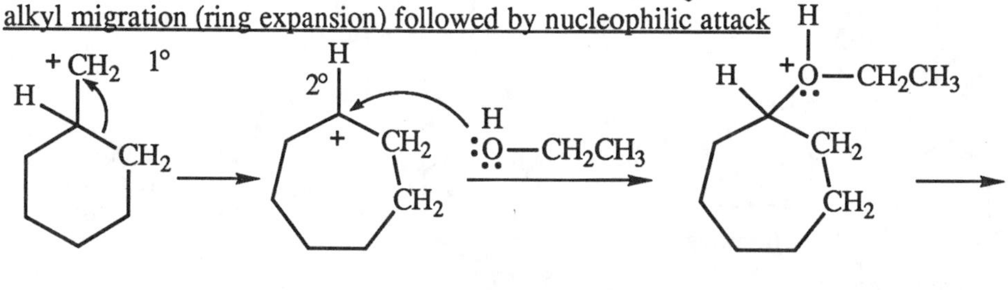

+ CH_2 1°
H
CH_2
H
2°
CH_2
CH_2
H
:O—CH_2CH_3
H
H
+O—CH_2CH_3
CH_2
CH_2

5-23 continued

(d) continued

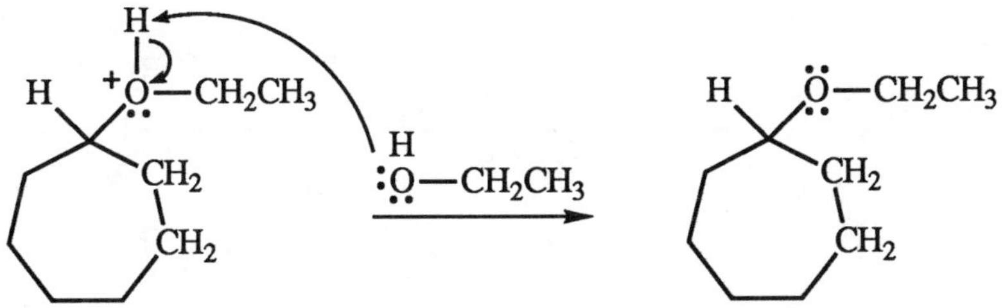

H
H
+O
CH2CH3
CH2
CH2
H
:O—CH2CH3
H
O—CH2CH3
CH2
CH2

5-24

(a)

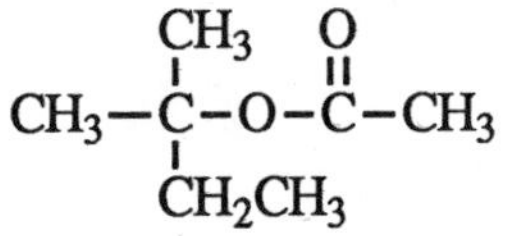

S_N1, 3°

(b)

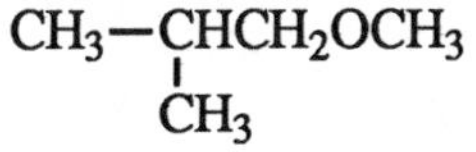

S_N2, 1°

(c)

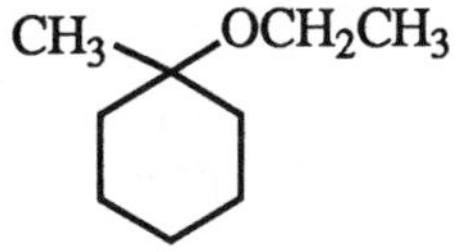

S_N1, 3°

(d)

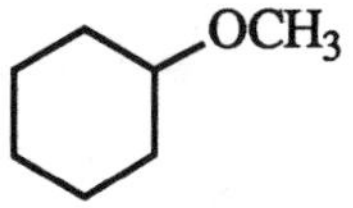

S_N1, weak nucleophile

(e)

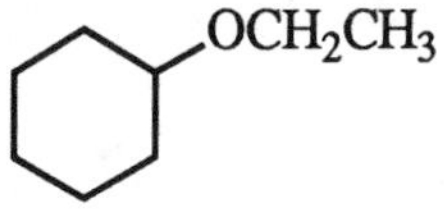

S_N2, strong nucleophile

5-25

(a) Methyl shift may occur simultaneously with ionization.

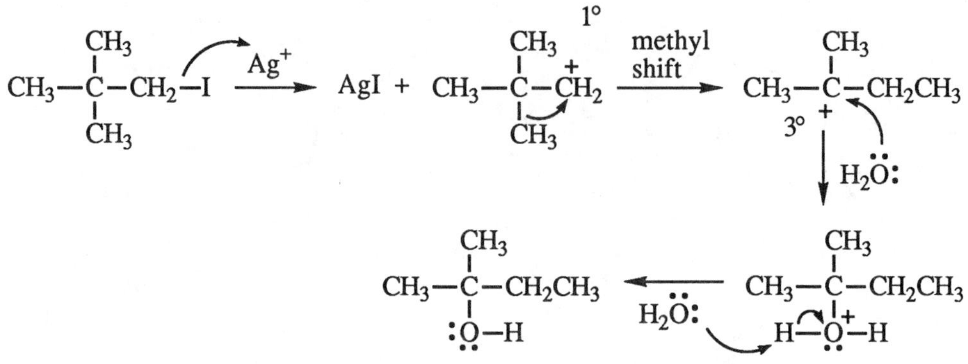

CH3
CH3—C—CH2—I
CH3
Ag+
AgI +
1°
CH3—C—CH2
methyl shift
CH3—C—CH2CH3
3°
H2O:
CH3—C—CH2CH3
:O—H
H—O—H

(b) Alkyl shift may occur simultaneously with ionization.

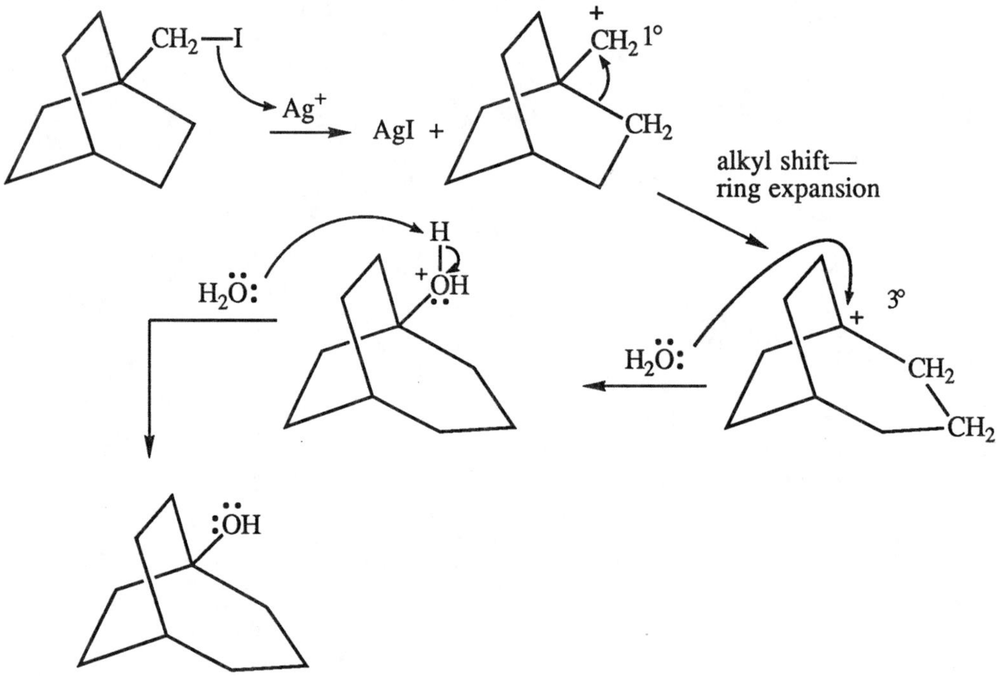

CH2—I
Ag+
AgI +
CH2 1°
CH2
alkyl shift—
ring expansion
3°
CH2
CH2
H2O:
H
+OH
H2O:
:OH

5-26

(a) 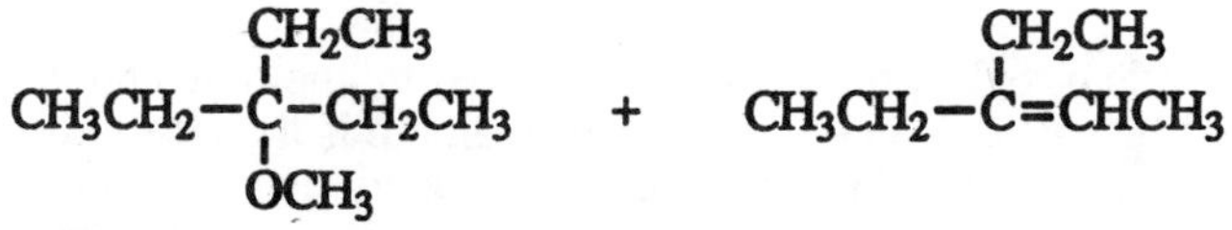

(b) 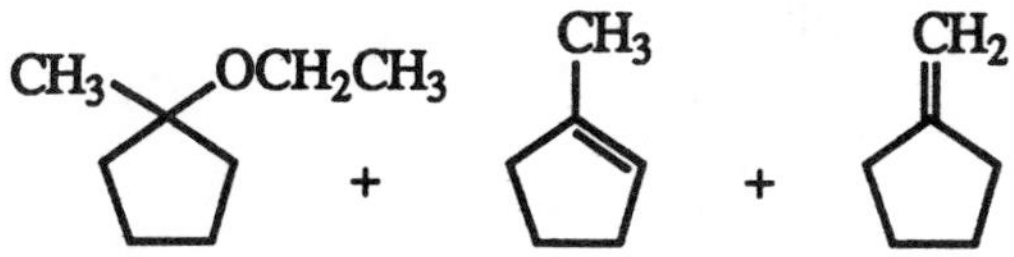

(c) <u>without rearrangement</u>

<u>with rearrangement</u>

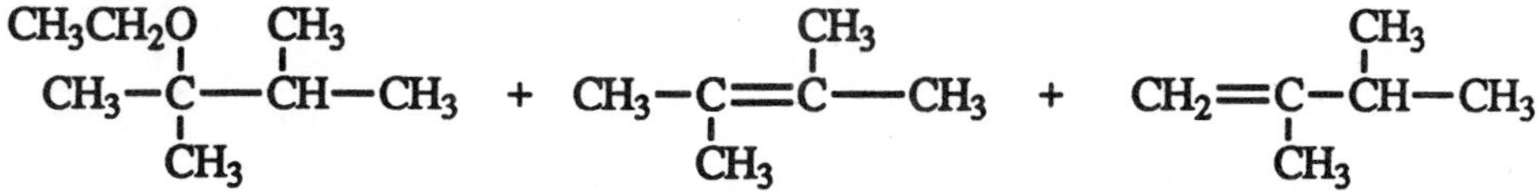

(d) 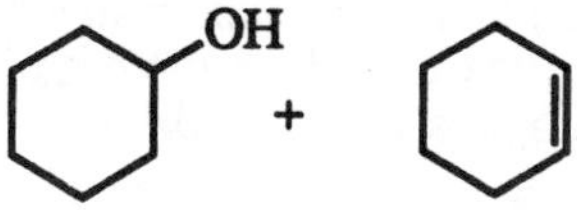

5-27

(a) $CH_3CH{=}CHCH_3$ + $CH_3CH_2CH{=}CH_2$

cis and *trans*

(b) 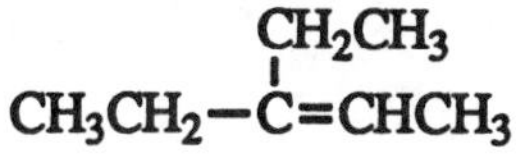

(c) 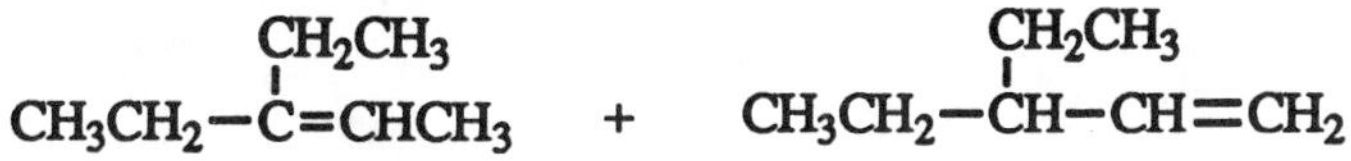

(d) 

5-28

In systems where free rotation is possible, the H to be abstracted by the base and the leaving group (Br here) must be anti-coplanar. The E2 mechanism is a concerted, one-step mechanism, so the arrangement of the other groups around the carbons in the starting material is retained in the product; there is no intermediate which would allow time for rotation of groups. (Models will help.)

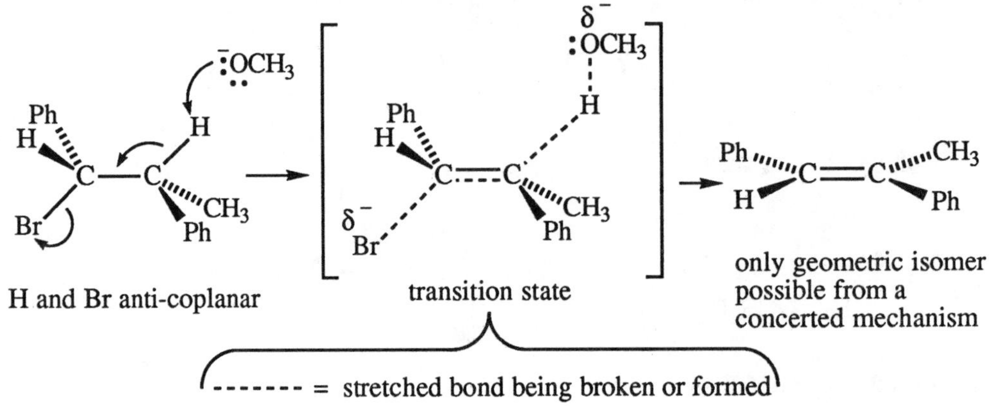

:ÖCH3
Ph
H
C
C
H
CH3
Br
Ph
H and Br anti-coplanar
δ−
:OCH3
H
transition state
δ−
Br
CH3
Ph
Ph
CH3
C=C
H
Ph
only geometric isomer possible from a concerted mechanism
------- = stretched bond being broken or formed

5-29

E2 elimination requires that the H and the leaving group be anti-coplanar; in a chair cyclohexane, this requires that the two groups be *trans* diaxial. When the bromine atom is in an axial position, however, there are no hydrogens in axial positions on adjacent carbons, so no elimination can occur.

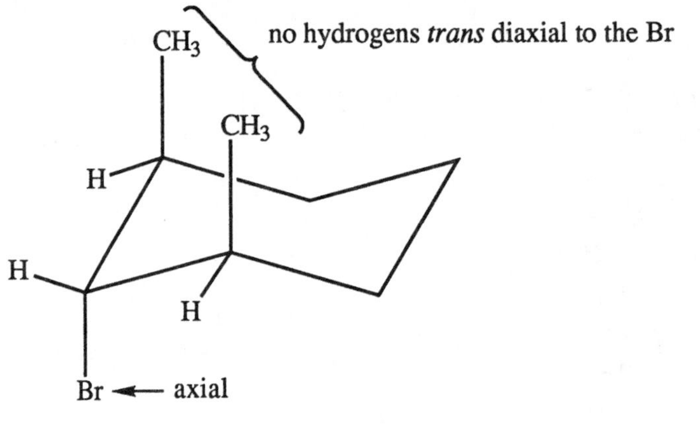

CH3
no hydrogens *trans* diaxial to the Br
CH3
H
H
H
Br
axial

5-30 Models are a big help for this problem.

(a)

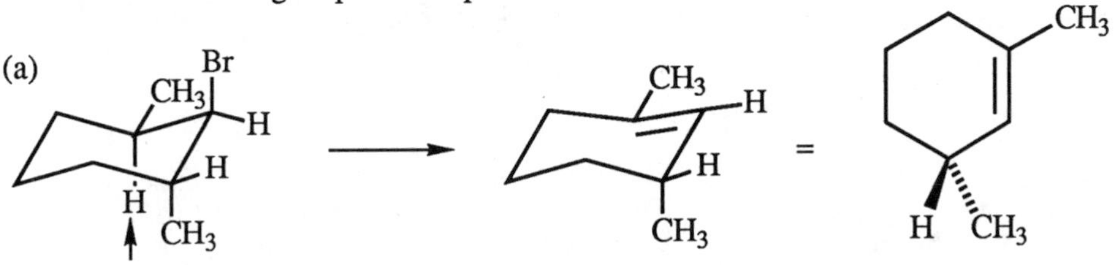

the only H *trans* diaxial

the only elimination product

(b)

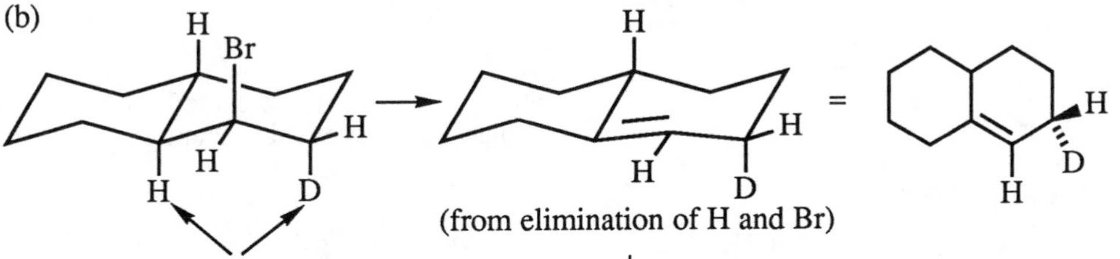

(from elimination of H and Br)

both *trans* diaxial— gives two products

+

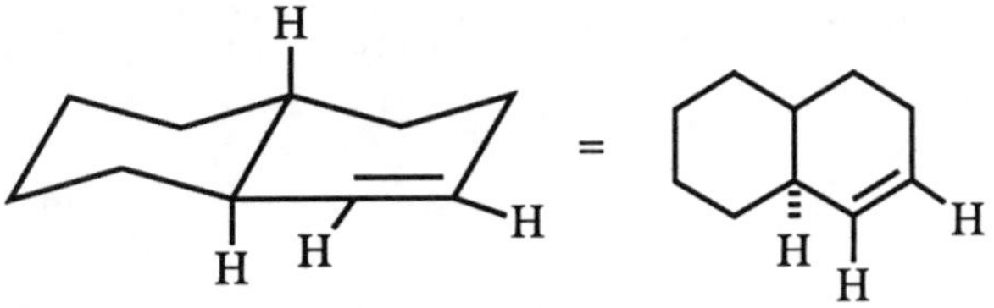

(from elimination of D and Br)

5-31 without rearrangement

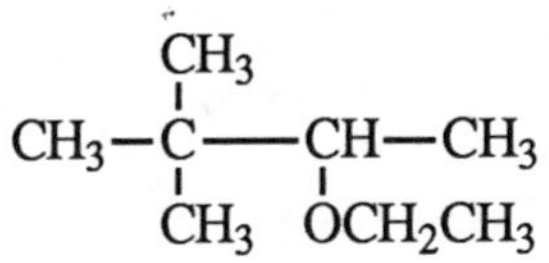

with rearrangement

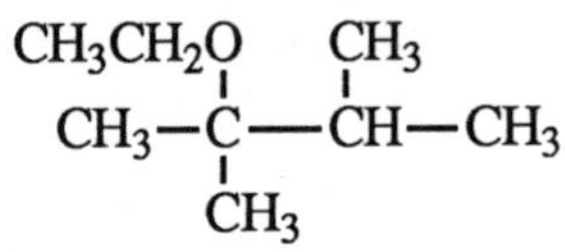

5-32

S_N2

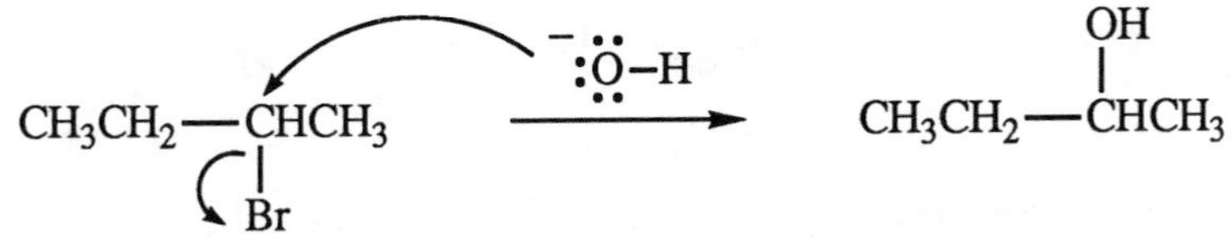

5-33

(a) E2 $CH_3CH{=}CHCH_2CH_3$ + $CH_2{=}CHCH_2CH_2CH_3$
cis and *trans*

(b) E1

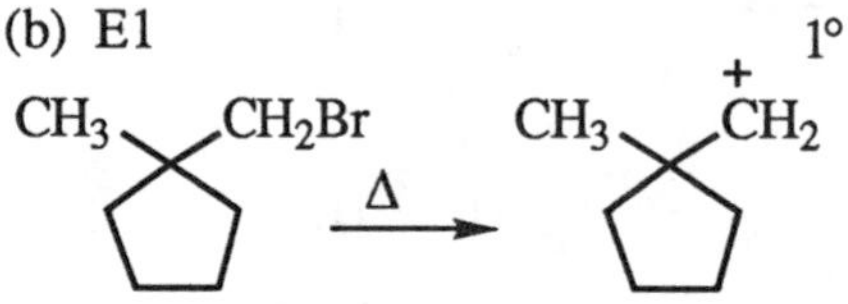

This short-lived 1° carbocation will undergo either a methyl shift or an alkyl shift (ring expansion), perhaps simultaneously with ionization.

<u>methyl shift</u>

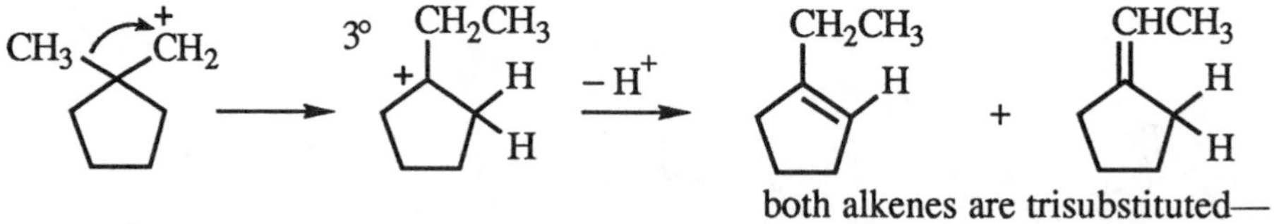

both alkenes are trisubstituted—formed in about equal amounts

<u>alkyl shift (ring expansion)</u>

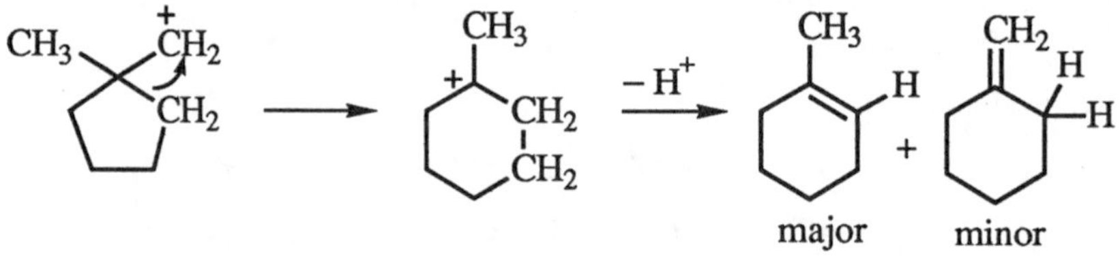
major
minor

(c) E1

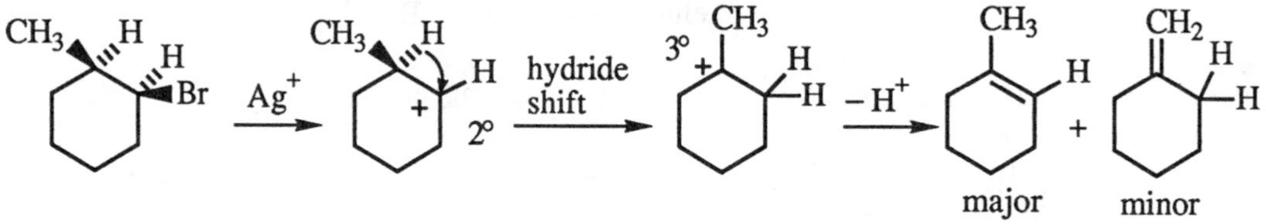
hydride shift
major
minor

(d) E2

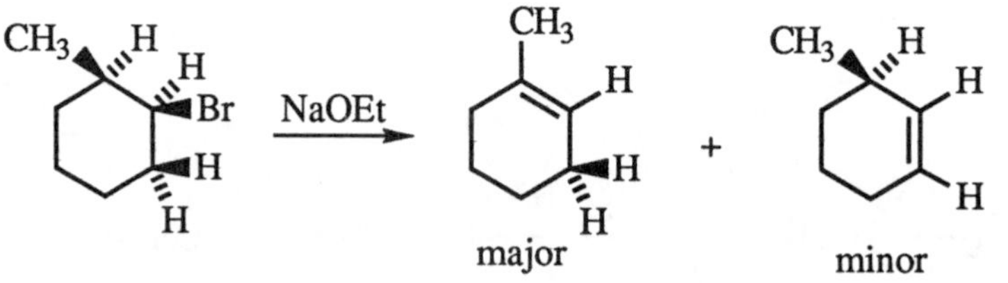
NaOEt
major
minor

(Note: if the problem had specified *trans*-1-bromo-2-methylchclohexane, only the second alkene of this pair would be correct.)

5-33 continued

(e) Methyl shift may occur simultaneously with ionization.

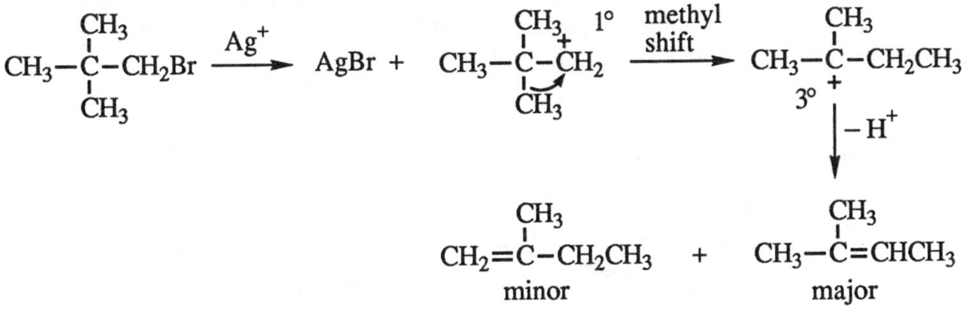

Ag^+
AgBr +
1°
methyl shift
3°
$-H^+$
minor
major

(f) no reaction if under E2 conditions; heating in a polar solvent with $NaOCH_3$ will produce the same product mixture as in 5-33(e)

5-34

(a) Ethoxide is a strong base/nucleophile—second-order conditions. The 1° bromide favors substitution over elimination, so S_N2 will predominate over E2.

$CH_3CH_2OCH_2CH_3$	$CH_2{=}CH_2$
substitution—major	elimination—minor

(b) Ethoxide is a strong base/nucleophile—second-order conditions. The 3° bromide is hindered and cannot undergo S_N2 by backside attack. E2 is the only route possible.

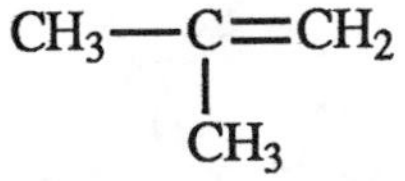

(c) Ethoxide is a strong base/nucleophile—second-order conditions. The 2° bromide can undergo S_N2 or E2 reaction, the relative amounts depending on solvent and temperature.

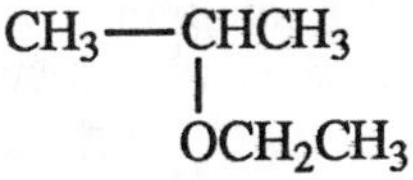

substitution

$CH_3{-}CH{=}CH_2$

elimination

(d) Hydroxide is a strong base/nucleophile—second-order conditions. The 1° bromide is more likely to undergo S_N2 than E2, but both products will be observed.

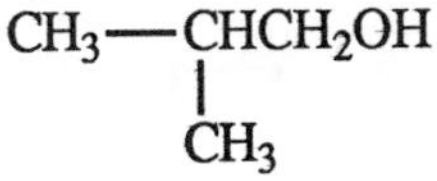

substitution—major

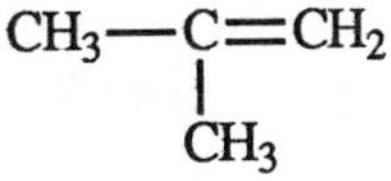

elimination—minor

5-34 continued

(e) Silver nitrate in aqueous ethanol ionizes alkyl halides and there is no strong base present—first-order conditions. The isobutyl cation will rearrange to the *t*-butyl cation which in turn can be attacked by either water or ethanol as nucleophile. S_N1 will predominate over E1 since there is no indication of high temperature.

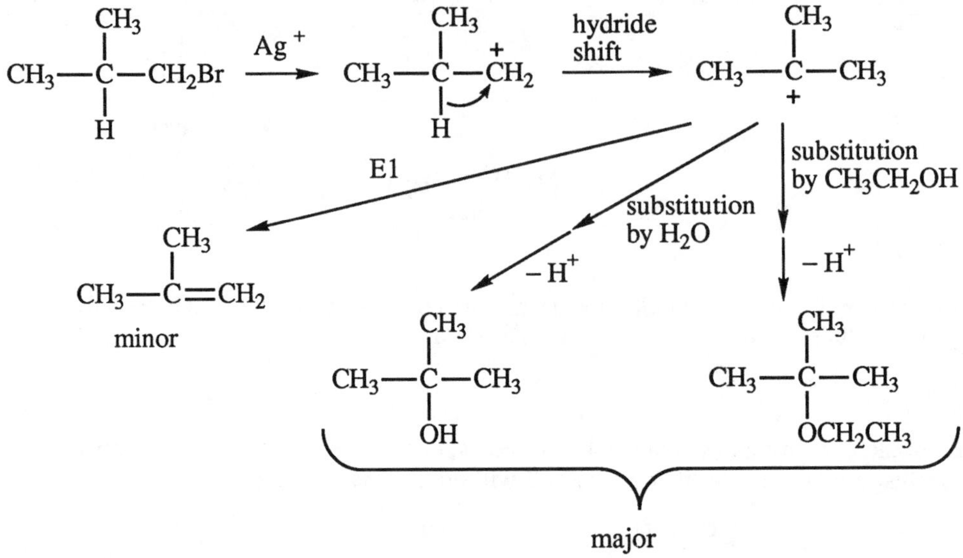
Ag+
hydride shift
E1
substitution by H2O
substitution by CH3CH2OH
- H+
- H+
minor
major

(f) Heating a 3° halide in methanol is quintessential first-order conditions, either E1 or S_N1 (solvolysis).

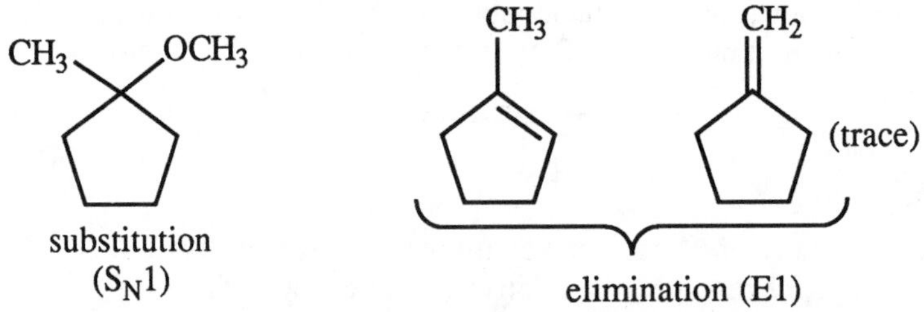
(trace)
substitution
(SN1)
elimination (E1)

(g) Silver nitrate in methanol ionizes alkyl halides—first-order conditions. Simultaneous rearrangement by either hydride shift or by alkyl shift (ring expansion) produces carbocations which are likely to undergo S_N1 with methanol at room temperature; higher temperature would favor E1.

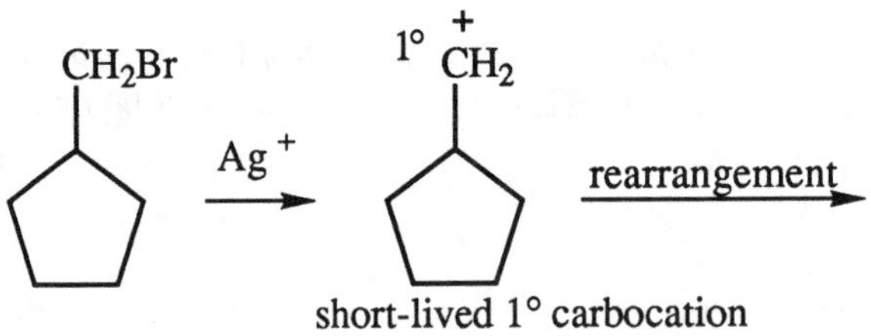
1°
Ag+
rearrangement
short-lived 1° carbocation

5-34 continued

(g)

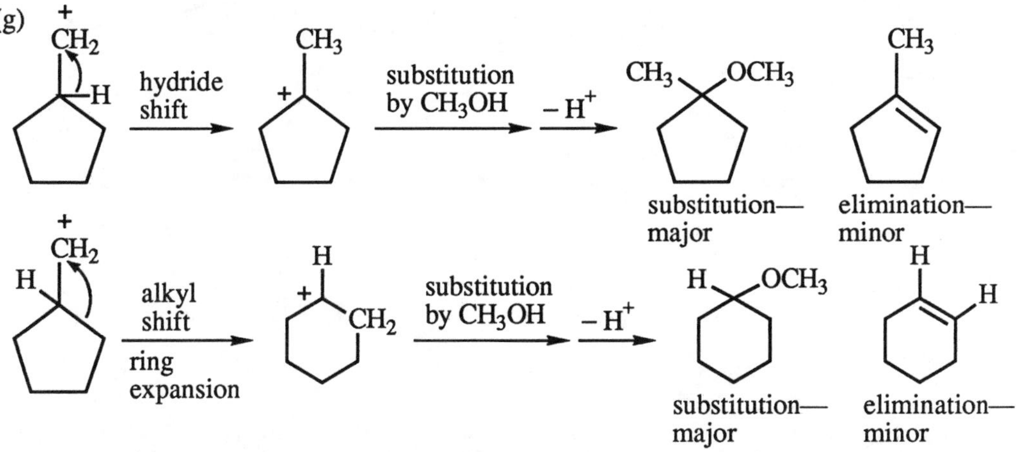

+
CH_2
H
hydride
shift
CH_3
+
substitution
by CH_3OH
$- H^+$
CH_3
OCH_3
CH_3
substitution—
major
elimination—
minor
+
CH_2
H
alkyl
shift
ring
expansion
H
+
CH_2
substitution
by CH_3OH
$- H^+$
H
OCH_3
H
H
substitution—
major
elimination—
minor

5-35 The stereochemical requirement of E2 elimination is anti-coplanar; in cyclohexanes, this translates to *trans* diaxial. Both dibromides are *trans*, but because the *t*-butyl group must be in an equatorial position, only the left molecule can have the bromines diaxial. The one on the right has both bromines locked into equatorial positions, from which they cannot undergo E2 elimination.

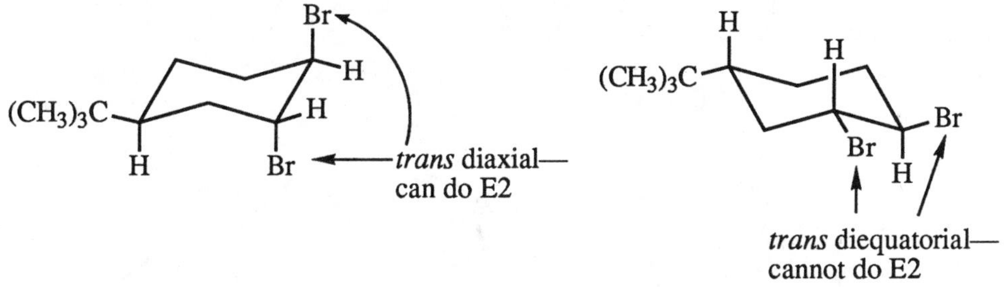

Br
H
$(CH_3)_3C$
H
H
Br
trans diaxial—
can do E2
H
H
$(CH_3)_3C$
Br
Br
H
trans diequatorial—
cannot do E2

5-36

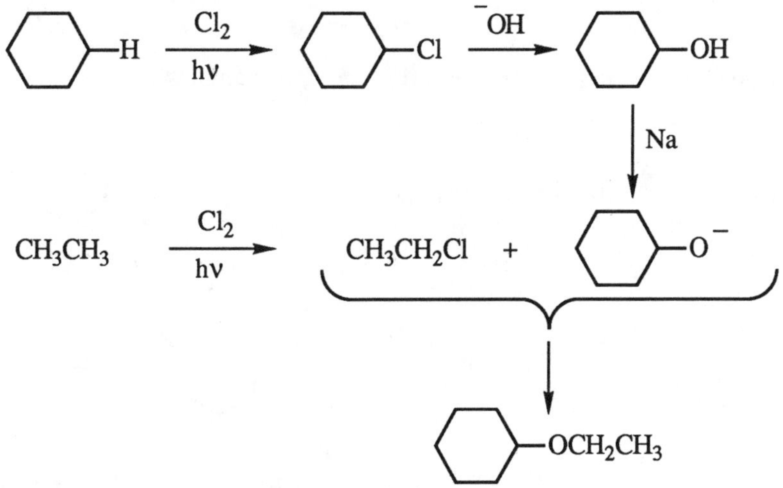

H
Cl_2
hν
Cl
^-OH
OH
Na
CH_3CH_3
Cl_2
hν
CH_3CH_2Cl +
O^-
OCH_2CH_3

5-37

(a) 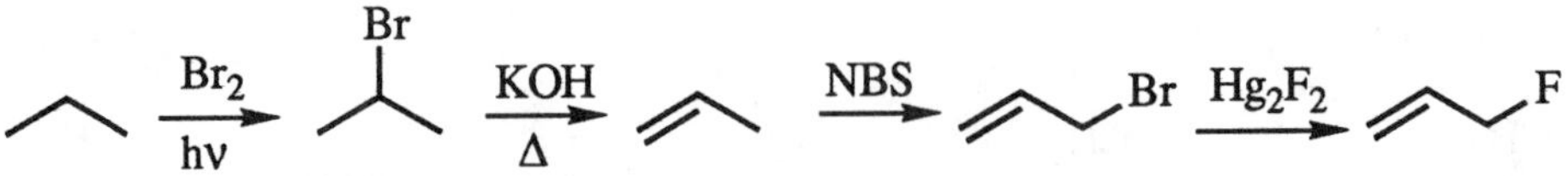

(b) 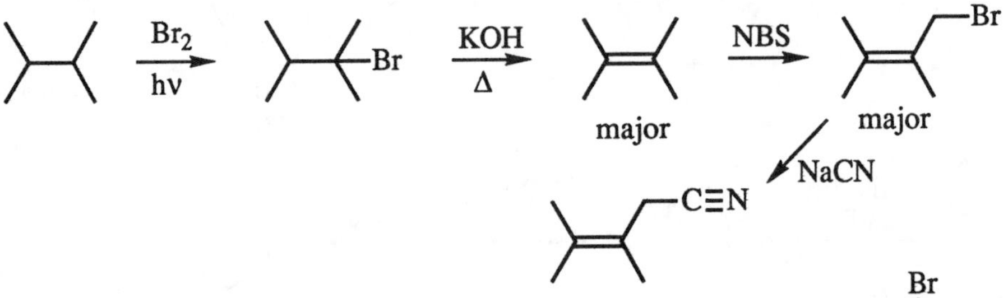

(c) 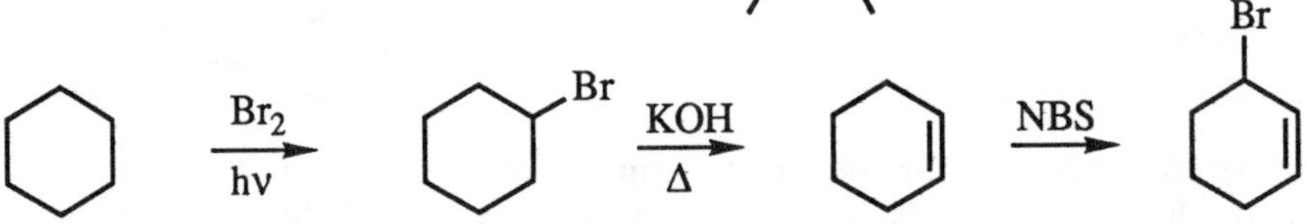

(d) 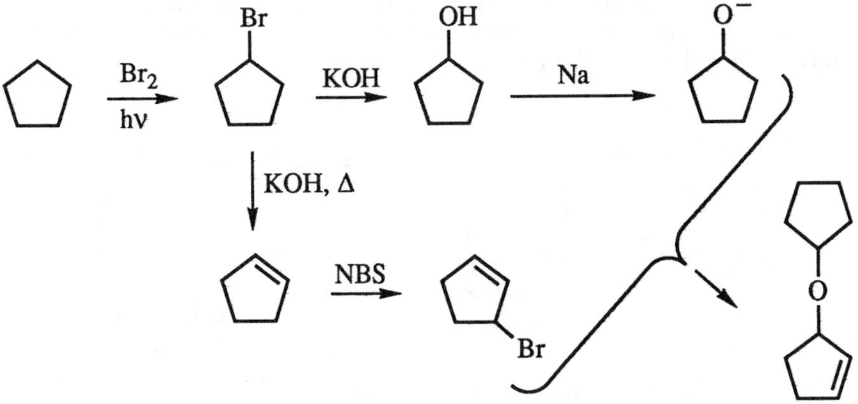

5-38 Refer to the Glossary and the text for definitions and examples.

5-39

(a) 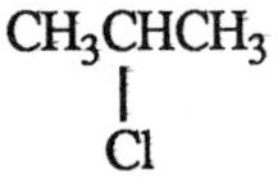

(b) 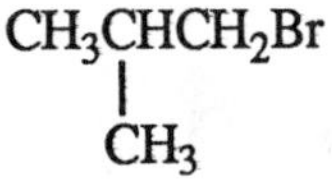

(c) 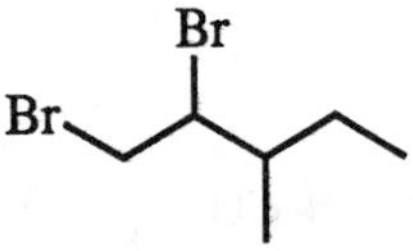

(d) 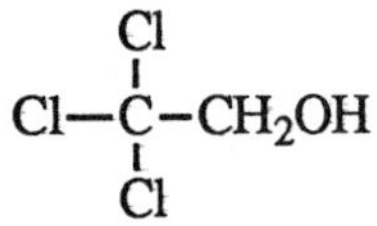

(e) 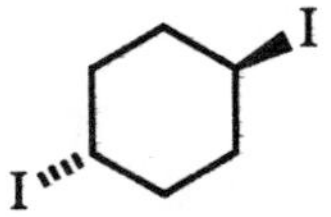

5-40 (a) 2-bromo-2-methylpentane
(b) 1-chloro-1-methylcyclohexane
(c) 1,1-dichloro-3-fluorocycloheptane
(d) 4-(2-bromoethyl)-3-(fluoromethyl)-2-methylheptane
(e) 4,4-dichloro-5-cyclopropyl-1-iodoheptane
(f) *cis*-1,2-dichloro-1-methylcyclohexane

5-41 Ease of backside attack (less steric hindrance) decides which undergoes S_N2 faster in all these examples except (b).

(a)  faster than 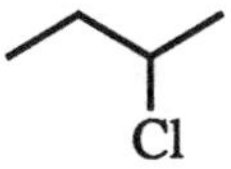

(b)  faster than  (leaving group ability)

(c) 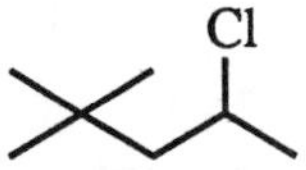 faster than 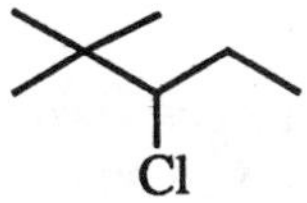

(d)  faster than 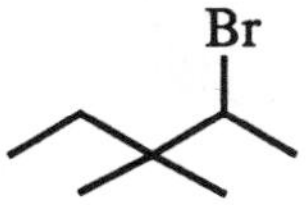

(e) 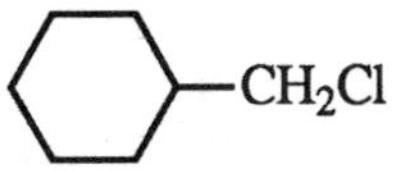 faster than 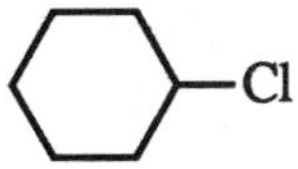

5-42 Formation of the more stable carbocation decides which undergoes S_N1 faster in all these examples except (d).

(a) 3°  faster than 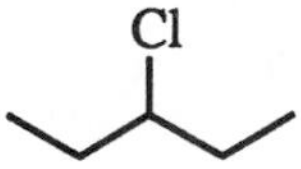 2°

(b) 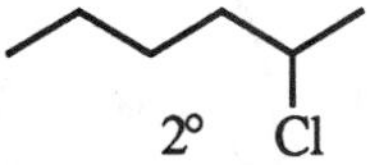 2° faster than  1°

(c) 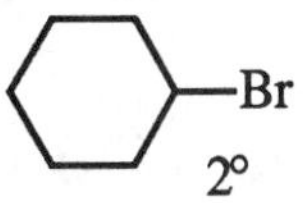 2° faster than 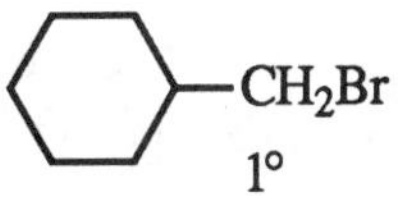 1°

5-42 continued

(d) 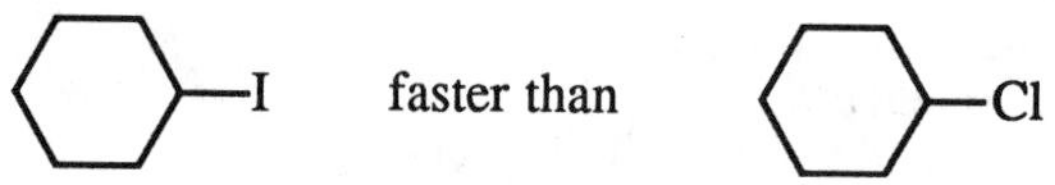
I
faster than
Cl

(leaving group ability)

(e) 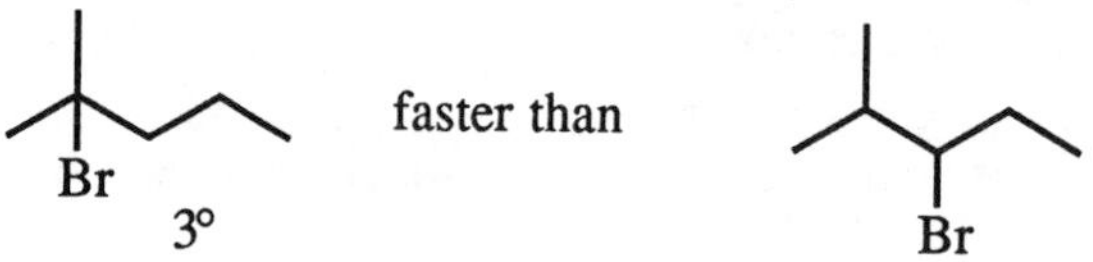
Br
3°
faster than
Br

(f) 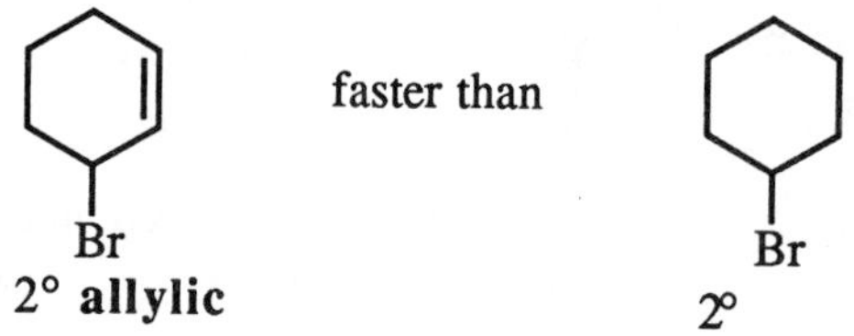
Br
2° allylic
faster than
Br
2°

5-43 For S_N2, reactions should be designed such that the nucleophile attacks the least highly substituted alkyl halide. ("X" stands for a halide: Cl, Br, or I.)

(a) 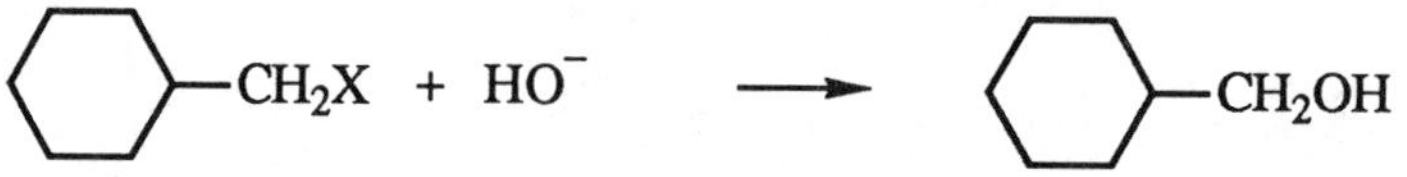
CH2X + HO−
CH2OH

(b) 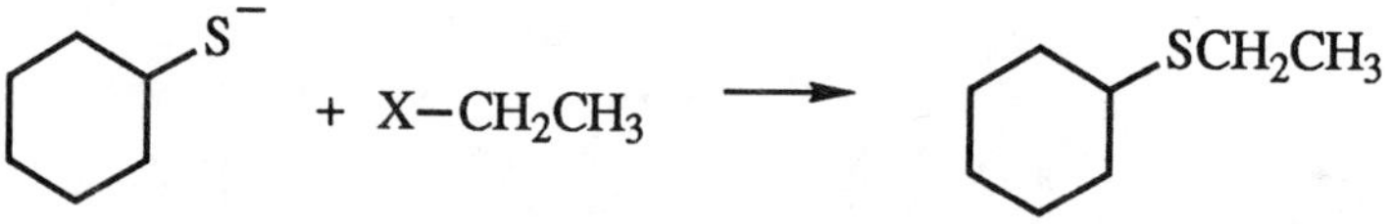
S−
+ X−CH2CH3
SCH2CH3

(c) $(CH_3)_3C{-}O^- \; + \; X{-}CH_3 \longrightarrow (CH_3)_3C{-}O{-}CH_3$

(d) 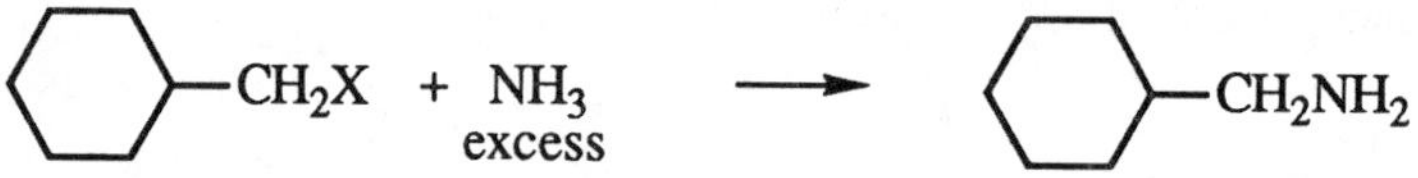
CH2X + NH3
excess
CH2NH2

(e) $H_2C{=}CHCH_2X \; + \; {}^{-}CN \longrightarrow H_2C{=}CHCH_2CN$

(f) 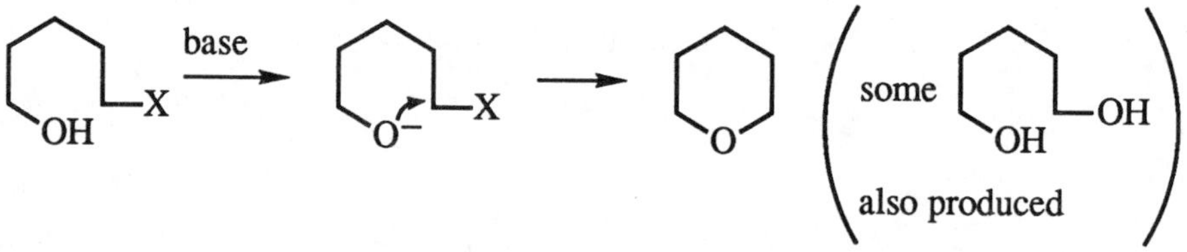
X
OH
base
X
O−
O
some
OH
OH
also produced

(g) $HC{\equiv}C^- \; + \; X{-}CH_2CH_2CH_3 \longrightarrow HC{\equiv}C{-}CH_2CH_2CH_3$

5-44 (a) (1)

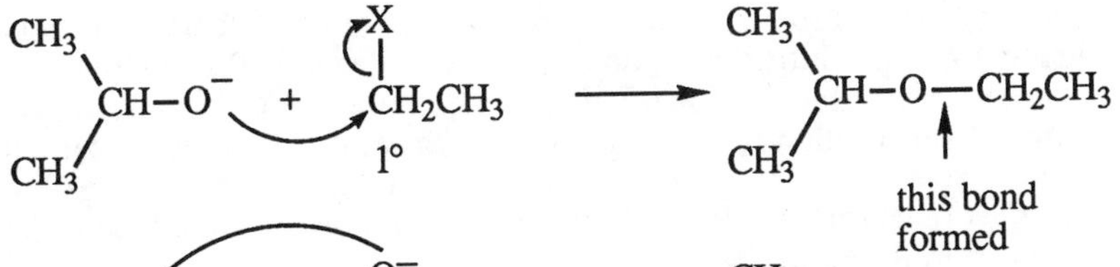
1°
this bond formed

(2)

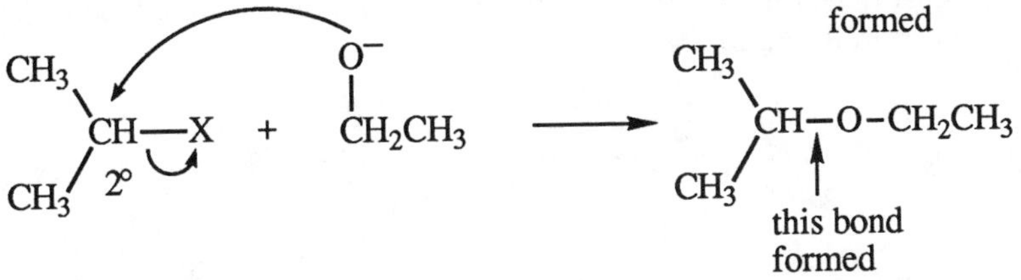
2°
this bond formed

Synthesis (1) would give a better yield of the desired ether product. (1) uses S_N2 attack of a nucleophile on a 1° carbon, while (2) requires attack on a more hindered 2° carbon. Reaction (2) would give a lower yield of substitution, with more elimination.

(b) CANNOT DO S_N2 ON A 3° CARBON!

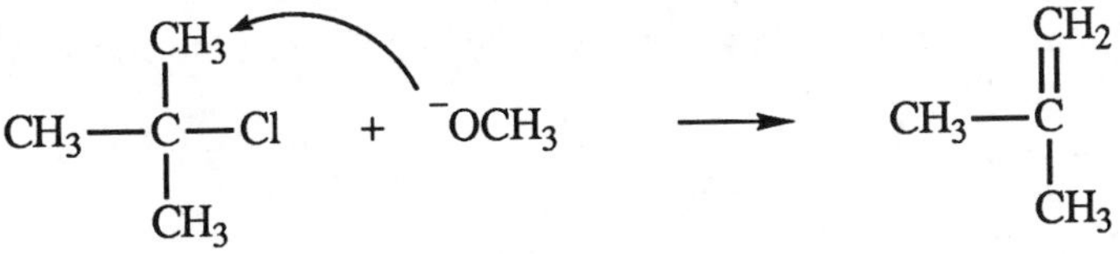

elimination (E2) competes

Better to do S_N2 on a methyl carbon:

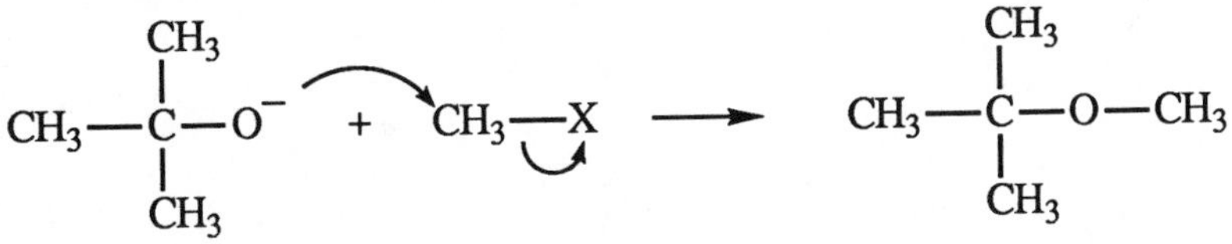

5-45 (a) S_N2—second order: reaction rate doubles

(b) S_N2—second order: reaction rate increases six times

(c) Virtually all reaction rates, including this one, increase with a temperature increase.

5-46 The key to this problem is that iodide ion is both an excellent nucleophile AND leaving group. Substitution on chlorocyclohexane is faster with iodide than with cyanide (see Table 5-6 for relative nucleophilicities). Once iodocyclohexane is formed, substitution by cyanide is much faster on iodocyclohexane than on chlorocyclohexane because iodide is a better leaving group than chloride. So two fast reactions involving iodide replace a slower single reaction, resulting in an overall rate increase.

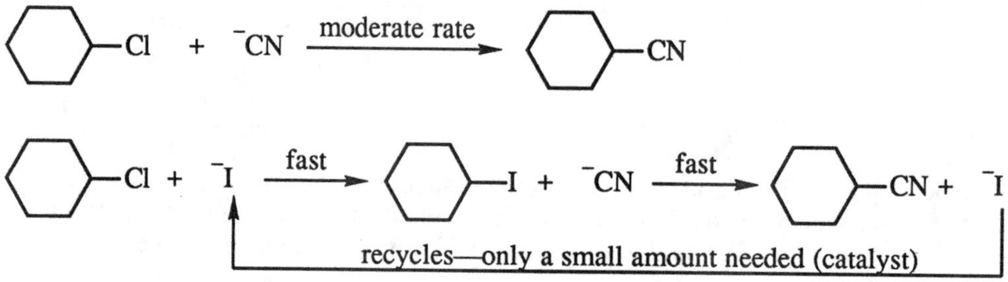

5-47 This is an S_N1 reaction; the rate law depends only on the substrate concentration, not on the nucleophile concentration.

(a) no change in rate

(b) the rate triples, dependent only on [*t*-butyl bromide]

(c) Virtually all reaction rates, including this one, increase with a temperature increase.

5-48

(a) 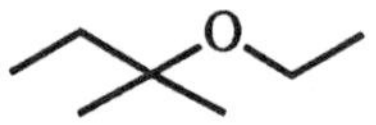

(b) 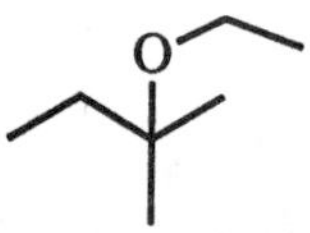

rearrangement

(c) 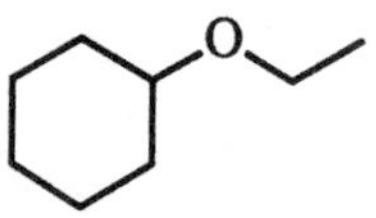

(d) 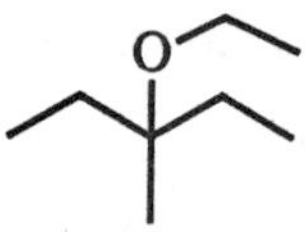

rearrangement

5-49

(a)

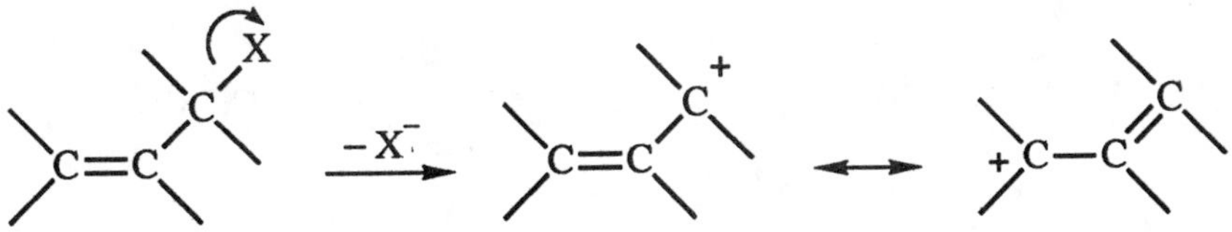

(b) (i)

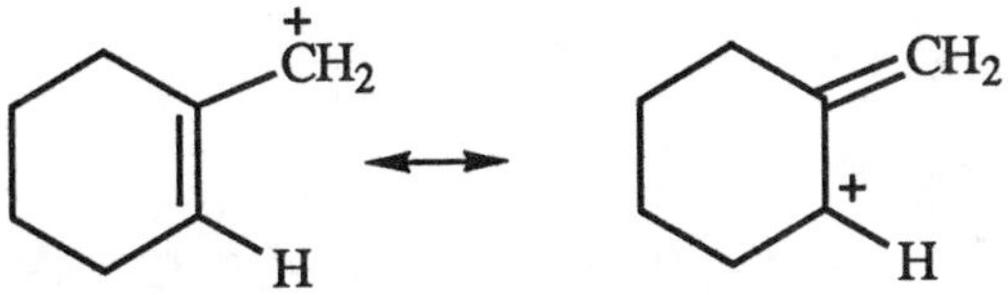

(ii)

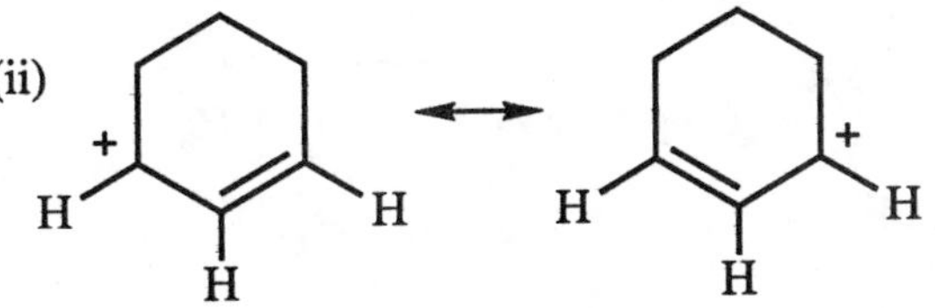

(iii)

5-50

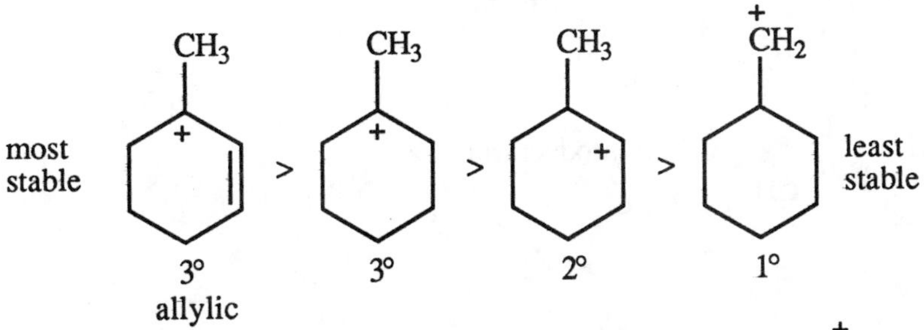

most stable
least stable
3° allylic
3°
2°
1°

5-51

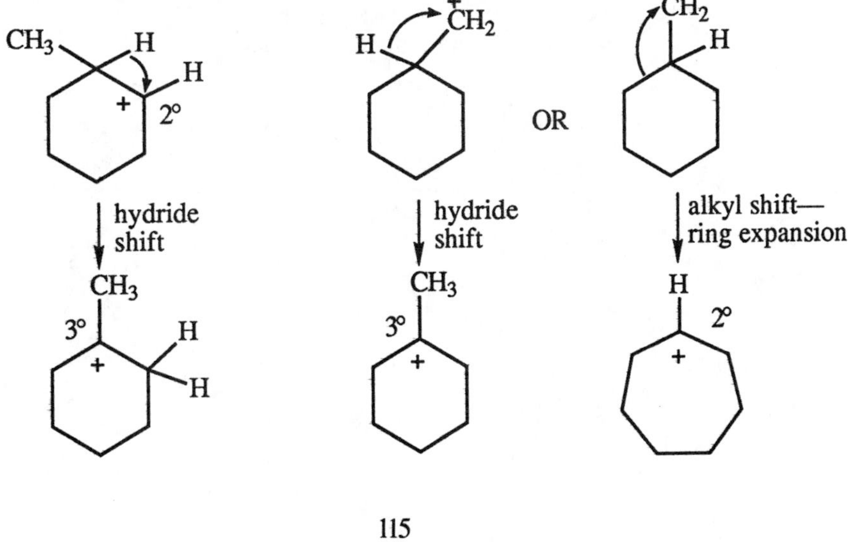

hydride shift
hydride shift
OR
alkyl shift—ring expansion
3°
3°
2°

5-52

(a) $CH_3CH_2OCH_2CH_3$

(b) 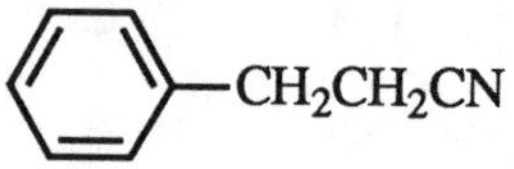

(c) 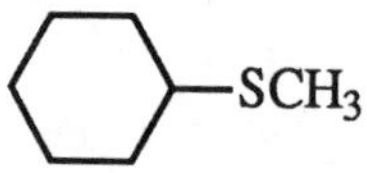

(d) $CH_3(CH_2)_8CH_2I$

(e) 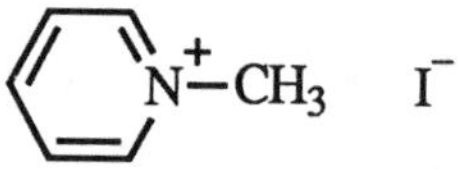

(f) $(CH_3)_3C-CH_2CH_2NH_2$

(g) 

(h) 

5-53

(a) 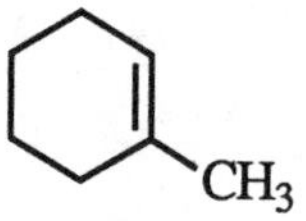 + 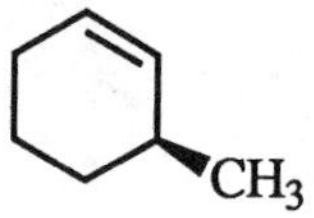

major minor

(b) 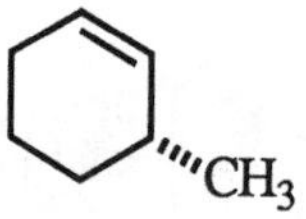 only product from E2

(c)  only E2 product possible (mixture of *E* and *Z*)

5-54

(a)

major minor

(b)

major minor

(c)

major minor

(d)

major minor

5-55 (a)

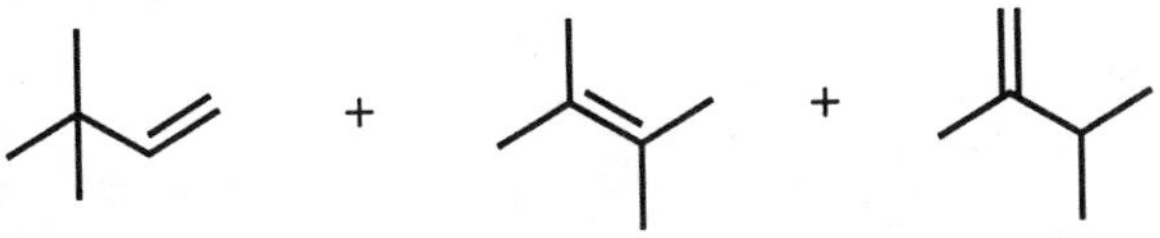

minor major minor

rearrangement

(b)

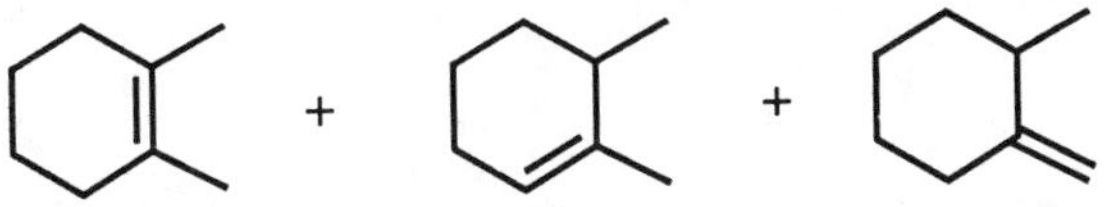

major minor trace amount

(c)

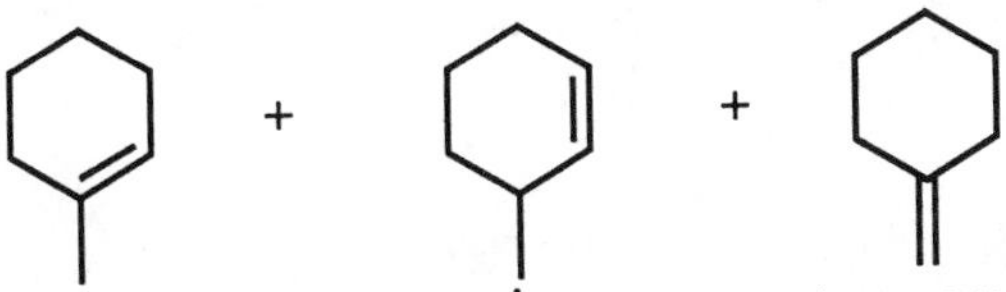

major minor trace amount (rearrangement)

5-56 The allylic carbocation has two resonance forms showing that two carbons share the positive charge. The ethanol nucleophile can attack either of these carbons, giving the S_N1 products; or loss of an adjacent H will give the E1 product.

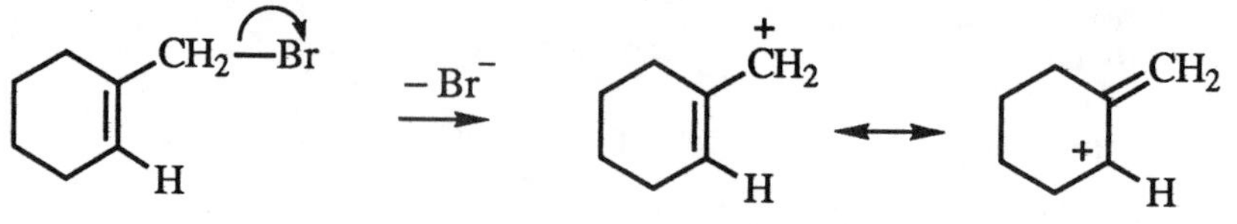

<u>S_N1</u>

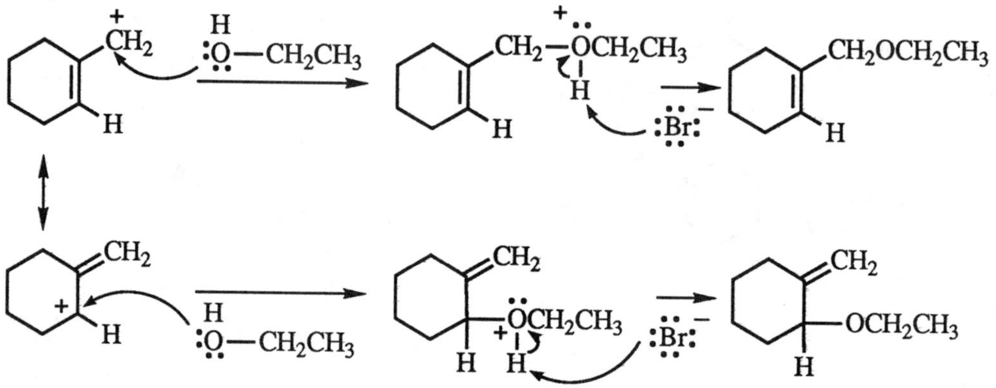

<u>E1</u>

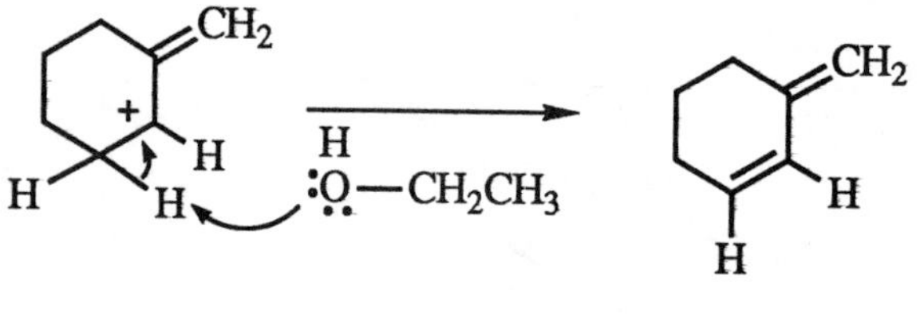

5-57

(a) 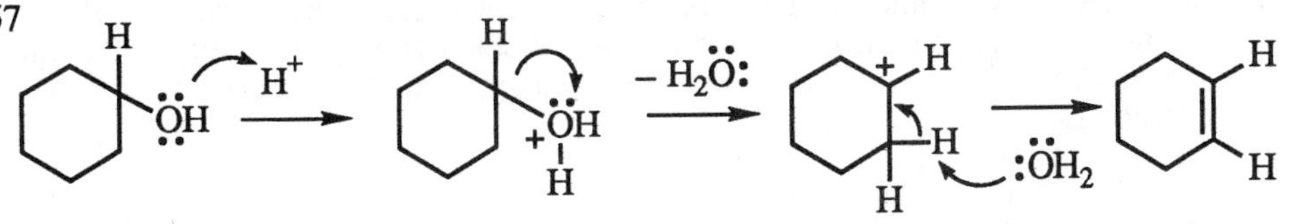

(b) S_N1 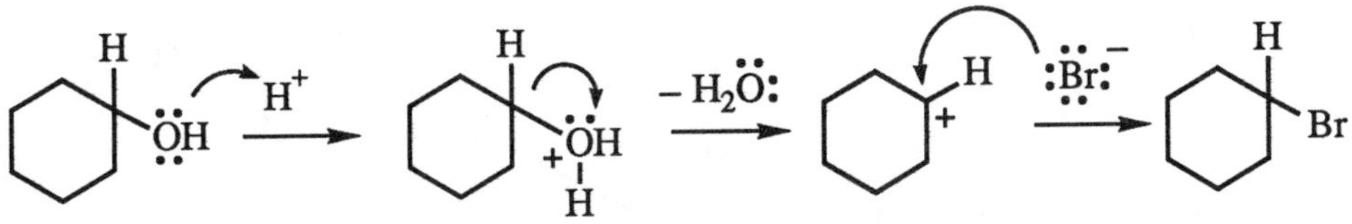

S_N2 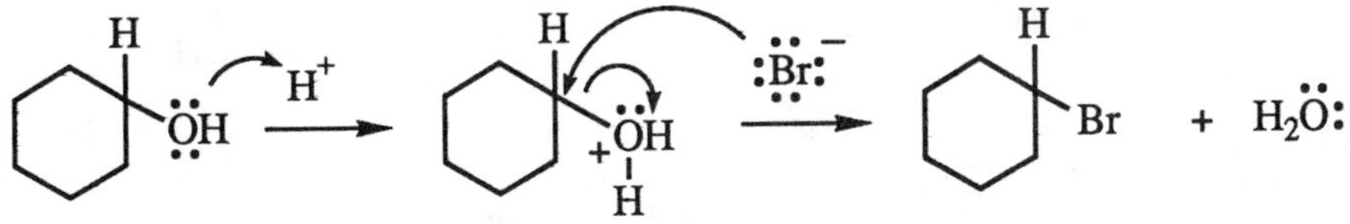

5-58 (a) $CH_3CH_2CH_2{=}CH_2$

(b) No reaction—the two bromines are not *trans* and therefore cannot be *trans*-diaxial.

(c) 

(d) 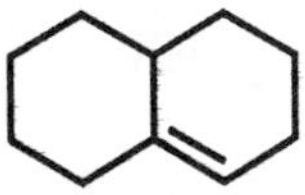

(e) No reaction—the bromines are *trans*, but they are diequatorial because of the locked conformation of the *trans*-decalin system. E2 can occur only when the bromines are *trans* diaxial.

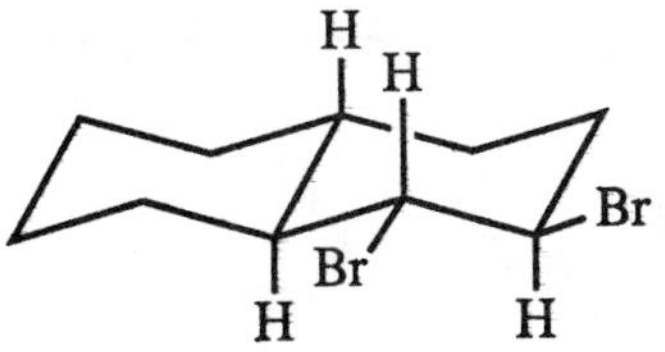

bromines are diequatorial—
cannot undergo E2

5-59 NBS generates bromine which produces bromine radical. Bromine radical abstracts an allylic hydrogen, resulting in a resonance-stabilized allylic radical. The allylic radical can bond to bromine at either of the two carbons with radical character.

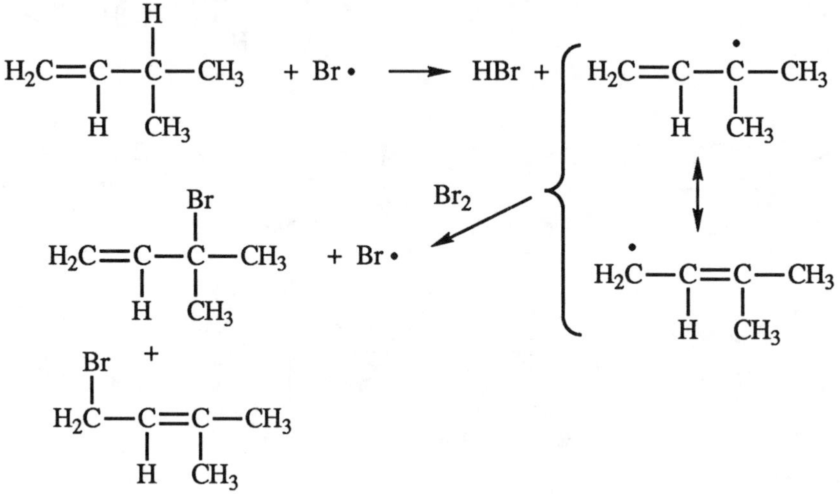

5-60 The bromine radical from NBS will abstract whichever hydrogen produces the most stable intermediate; in this structure, that is a benzylic hydrogen, giving the resonance-stabilized benzylic radical.

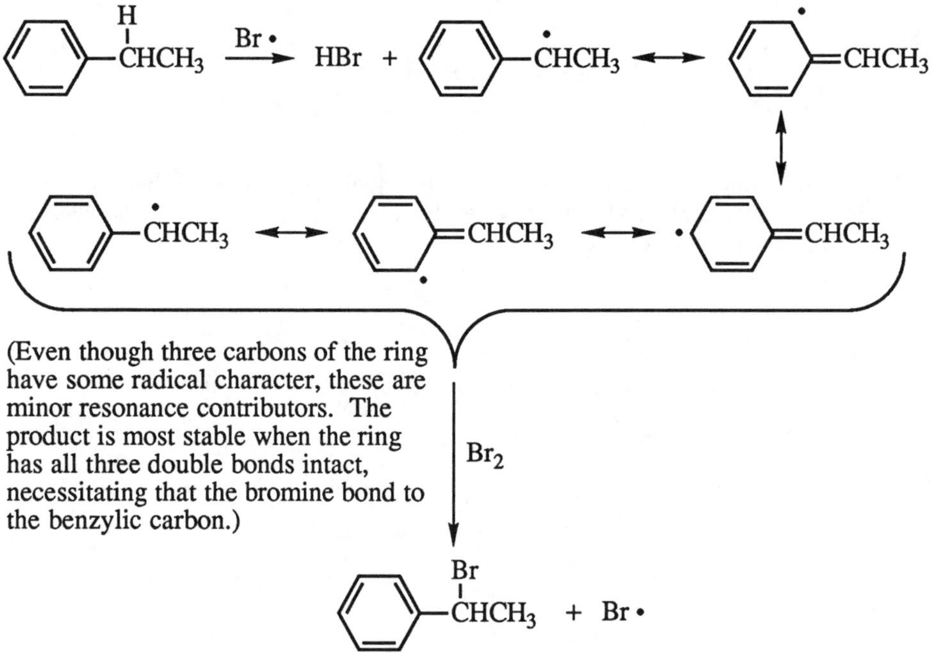

(Even though three carbons of the ring have some radical character, these are minor resonance contributors. The product is most stable when the ring has all three double bonds intact, necessitating that the bromine bond to the benzylic carbon.)

5-61 All five products come from rearranged carbocations. Rearrangement, which may occur simultaneously with ionization, can occur by hydride shift to the 3° methylcyclopentyl cation, or by ring expansion to the cyclohexyl cation.

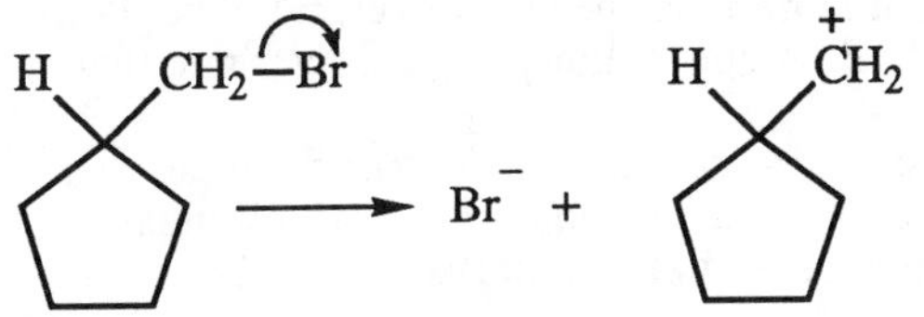

<u>hydride shift</u>

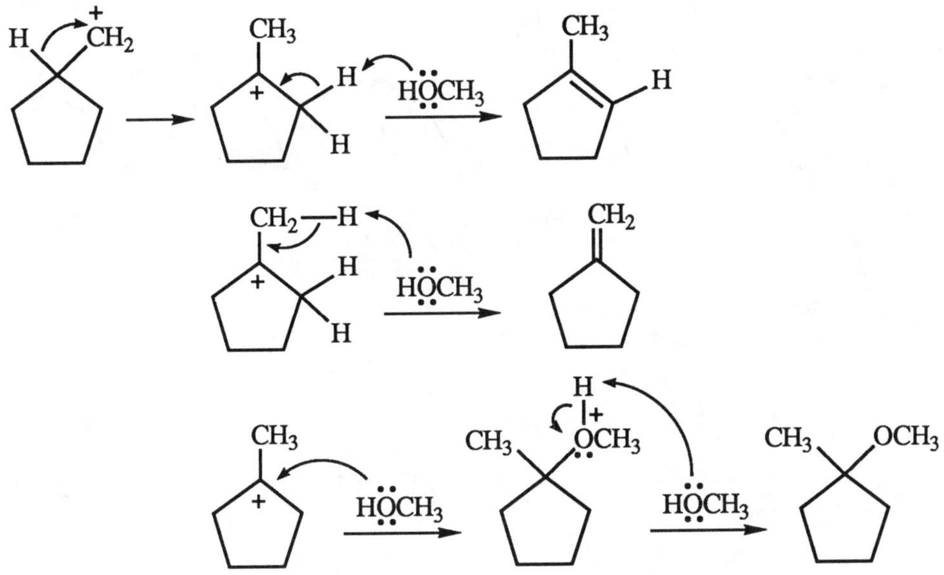

<u>alkyl shift (ring expansion)</u>

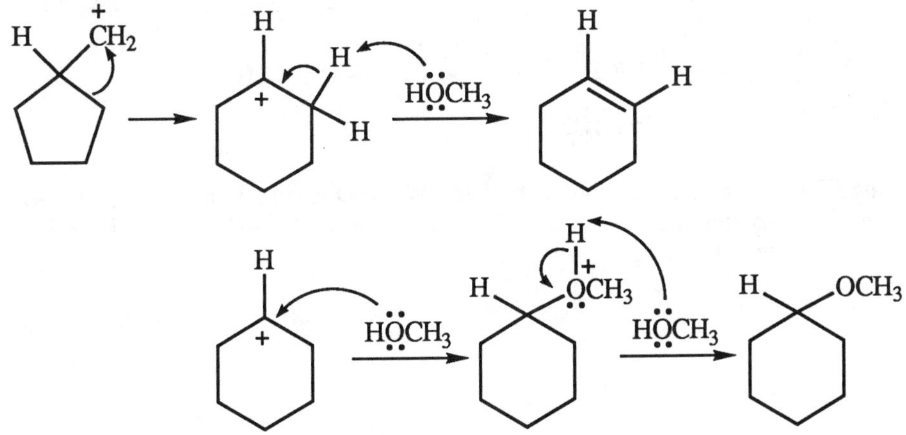

5-62

In E2, the two groups to be eliminated must be coplanar. In conformationally mobile systems like acyclic molecules, or in cyclohexanes, anti-coplanar is the preferred orientation. In rigid systems like norbornanes, however, SYN-coplanar is the only possible orientation and E2 will occur, although at a slower rate than anti-coplanar.

The structure having the H and the Cl syn-coplanar is the *trans*, which undergoes the E2 elimination. (It is possible that the *other* H and Cl eliminate from the *trans* isomer; the results from this reaction cannot distinguish between these two possibilities.)

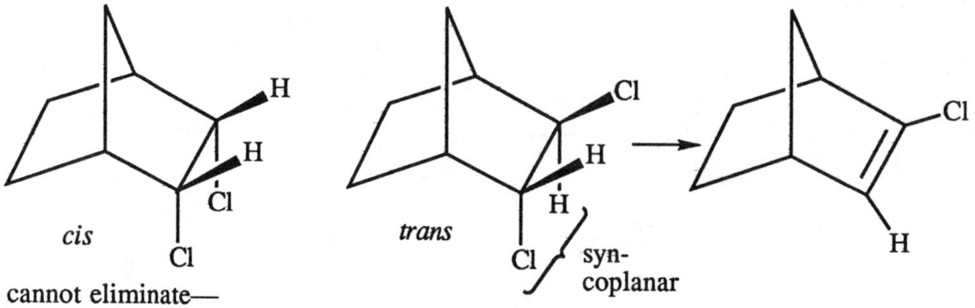
H
H
Cl
cis
Cl
Cl
H
H
trans
Cl
syn-
coplanar
Cl
H

cannot eliminate—
H and Cl not coplanar

5-63

(a) E2—
one step

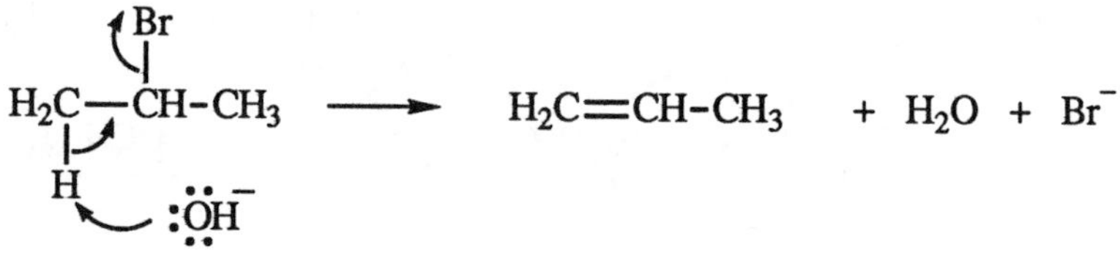
Br
H₂C—CH-CH₃
H
:OH⁻
H₂C=CH-CH₃
+ H₂O + Br⁻

S_N2—
one step

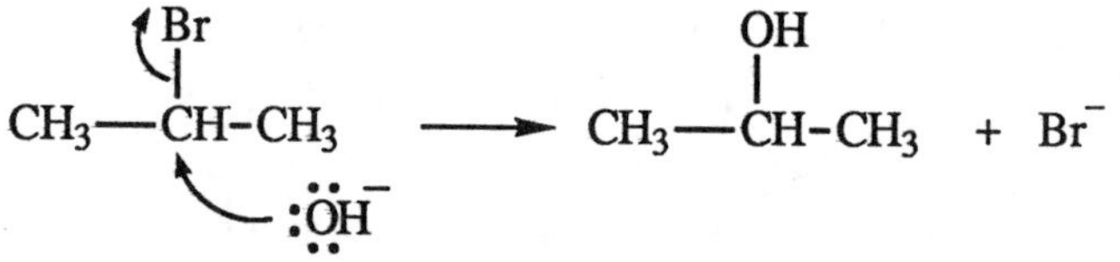
Br
CH₃—CH-CH₃
:OH⁻
OH
CH₃—CH-CH₃
+ Br⁻

(b) In the E2 reaction, a C—H bond is broken. When D is substituted for H, a C—D bond is broken, slowing the reaction. In the S_N2 reaction, no C—H (C—D) bond is broken, so the rate is unchanged.

5-63 continued

(c) These are first-order reactions. The slow, rate-determining step is the first step in each mechanism.

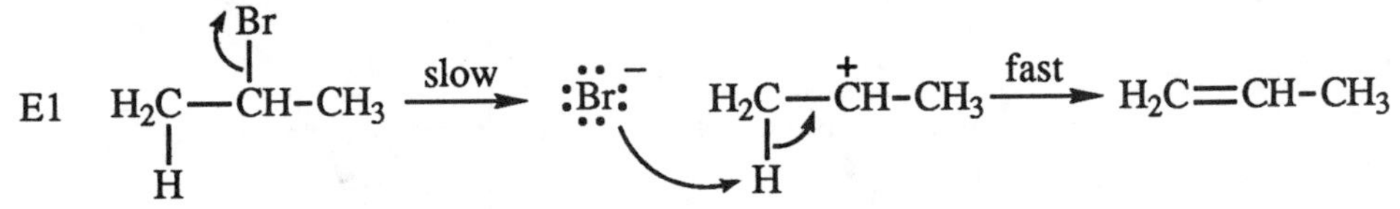

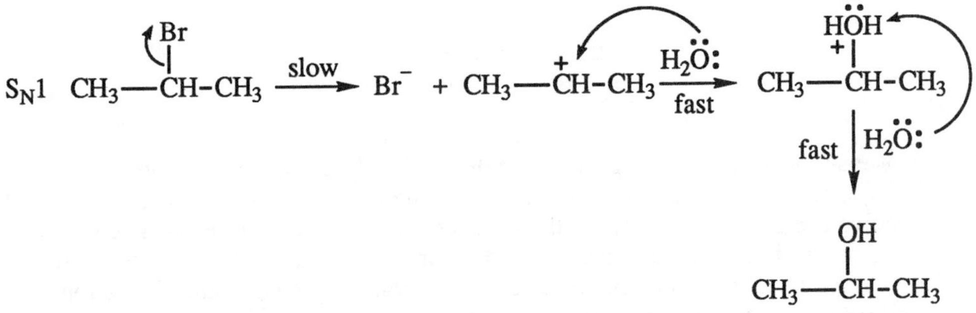

The only mechanism of these two involving C—H bond cleavage is the E1, but the C—H cleavage does NOT occur in the slow, rate-determining step. Kinetic isotope effects are observed only when C—H (C—D) bond cleavage occurs in the rate-determining step. Thus, we would expect to observe *no change in rate* for the deuterium-substituted molecules in the E1 or S_N1 mechanisms. (In fact, this technique of measuring isotope effects is one of the most useful tools chemists have for determining what mechanism a reaction follows.)

5-64 Both products are formed through E2 reactions. The difference is whether a D or an H is removed by the base. As explained in Problem 5-63, C—D cleavage can be up to 7 times slower than C—H cleavage, so the product from C—H cleavage should be formed about 7 times as fast. This rate preference is reflected in the 7 : 1 product mixture. ("Ph" is the abbreviation for a benzene ring.)

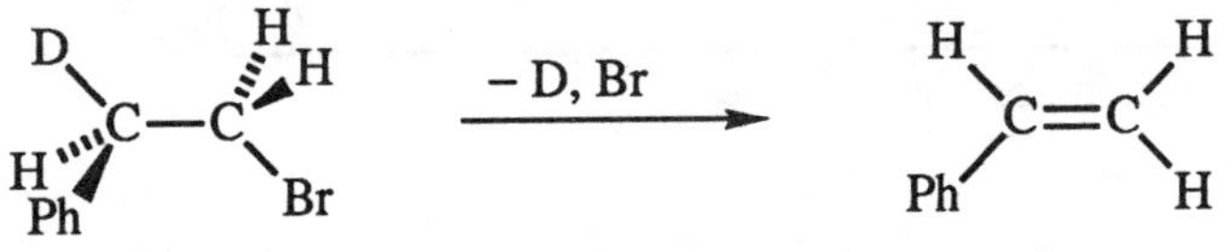

requires C—D bond cleavage; slow; minor product

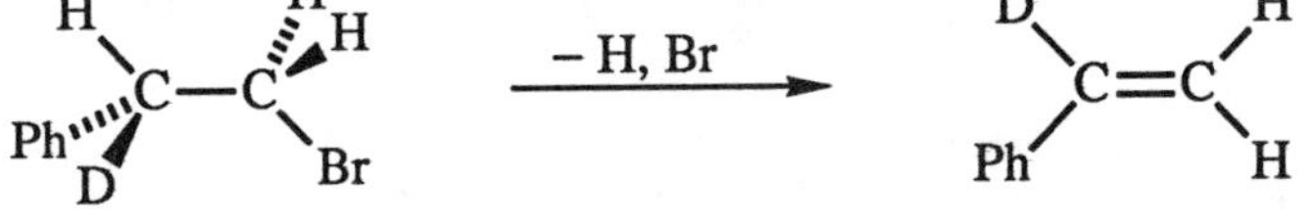

requires C—H bond cleavage; 7 times faster; major product

5-65 The energy, and therefore the structure, of the transition state determines the rate of a reaction. Any factor which lowers the energy of the transition state will speed the reaction.

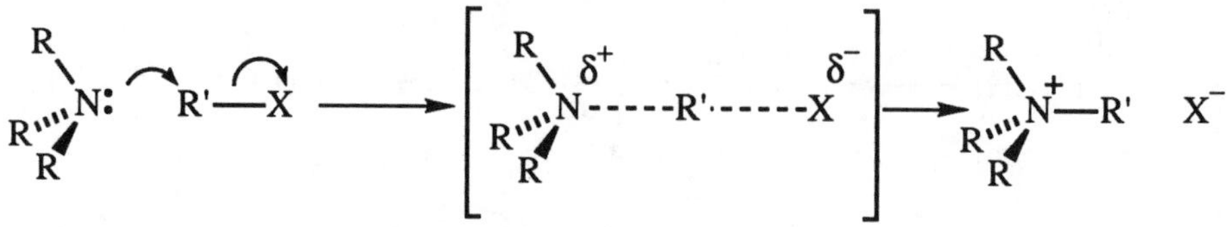

transition state—
developing charge

This example of S_N2 is unusual in that the nucleophile is a neutral molecule—it is not negatively charged. The transition state is beginning to show the positive and negative charges of the products (ions), so the transition state is more charged than the reactants. The polar transition state will be stabilized in a more polar solvent through dipole-dipole interactions, so the rate of reaction will be enhanced in a polar solvent.

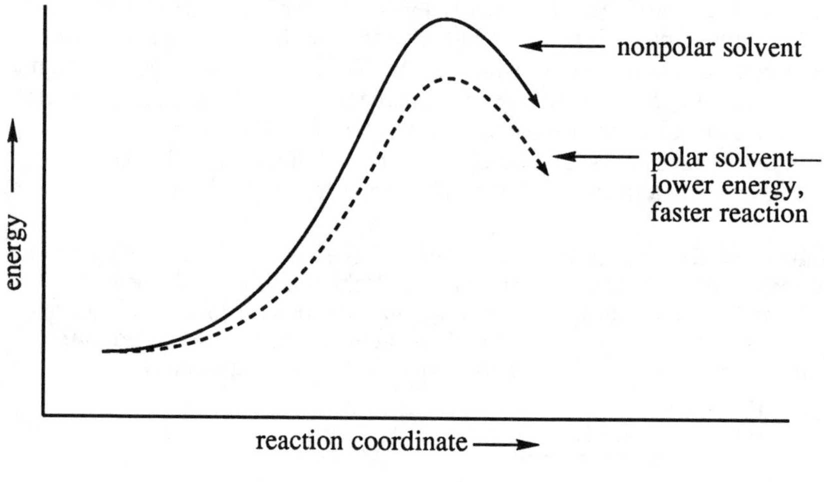
nonpolar solvent
polar solvent—
lower energy,
faster reaction
energy
reaction coordinate

5-66 The problem is how to explain this reaction:

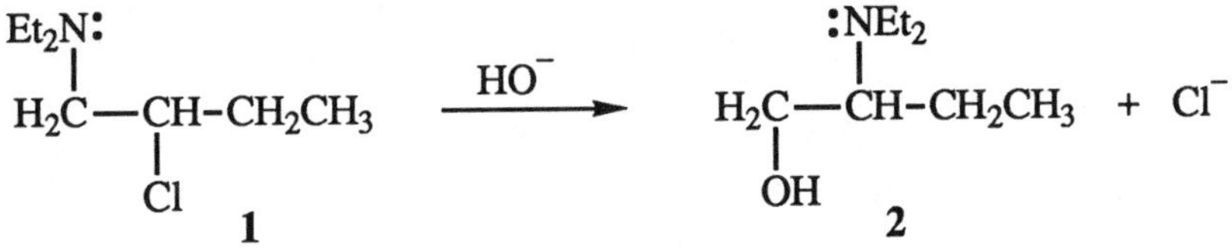

1 2

facts: 1) second order, but several thousand times faster than similar second order reactions without the NEt_2 group

2) NEt_2 group migrates

Solution

Clearly, the NEt_2 group is involved. The nitrogen is a nucleophile and can do an internal nucleophilic substitution (S_Ni), a very fast reaction for entropy reasons because two different molecules do not have to come together.

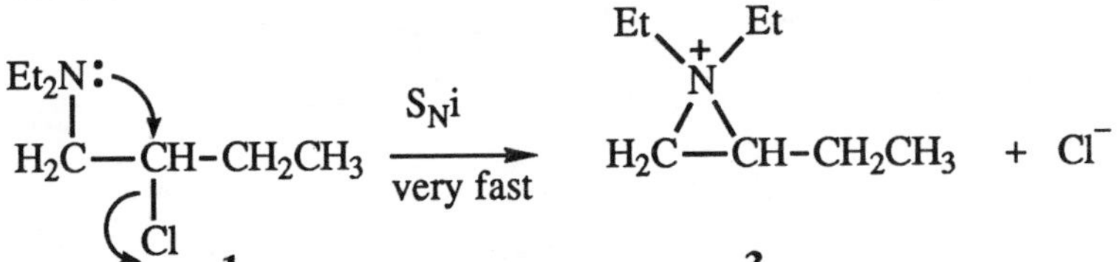

1 3

The slower step is attack of HO^- on intermediate **3**; the N is a good leaving group because it is positively charged. Where will HO^- attack **3**? On the less substituted carbon, in typical S_N2 fashion.

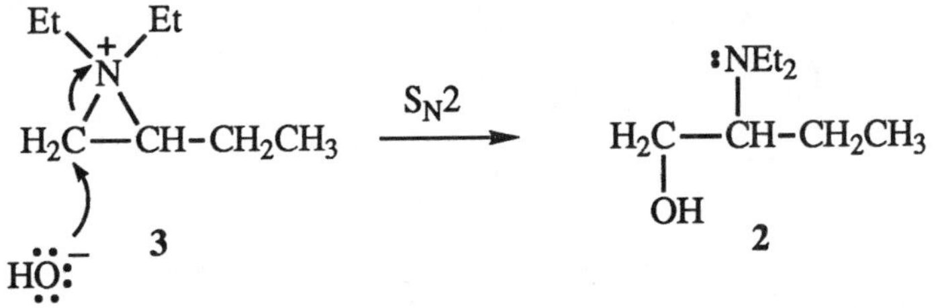

3 2

This overall reaction is fast because of the *neighboring group assistance* in forming **3**. It is second order because the HO^- group and **3** collide in the slow step (not the *only* step, however). And the NEt_2 group "migrates", although in two steps.

CHAPTER 6—STEREOCHEMISTRY

6-1 The chiral objects are the spring, the writing desk, the screw-cap bottle, the rifle and the knotted rope; the spring, the bottle top, and the rope each have a twist in one direction, and the rifle and desk are clearly made for right-handed users. All the other objects are achiral and would feel equivalent to right- or left-handed users.

6-2 (a)

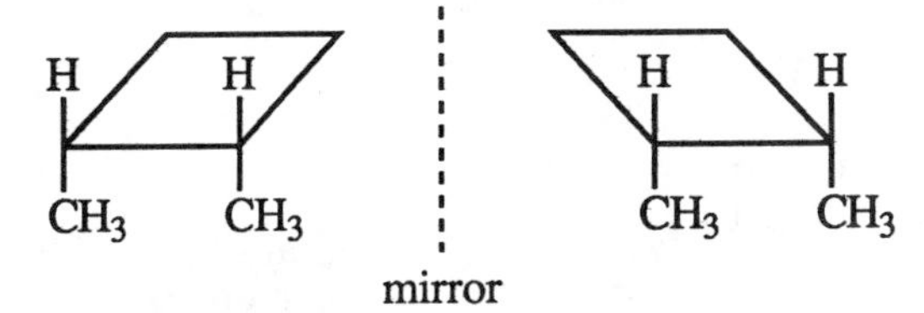

achiral—
identical mirror images

(b)

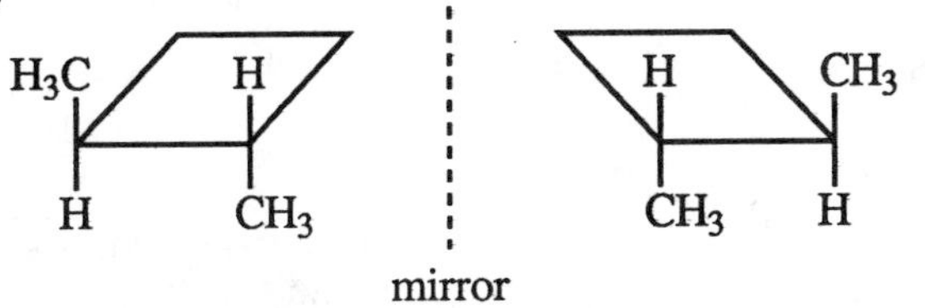

chiral—
enantiomers

(c)

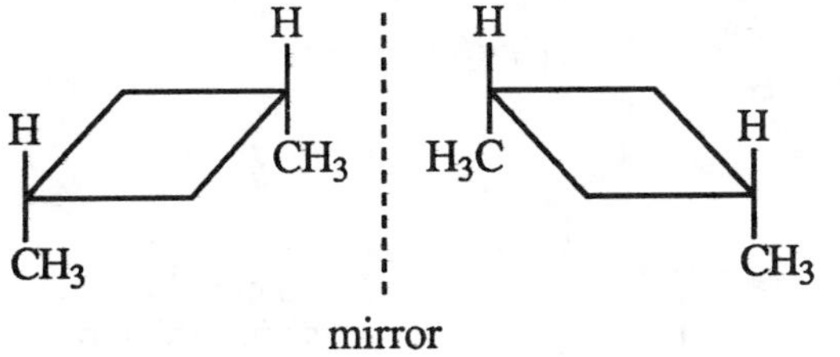

achiral—
identical mirror images

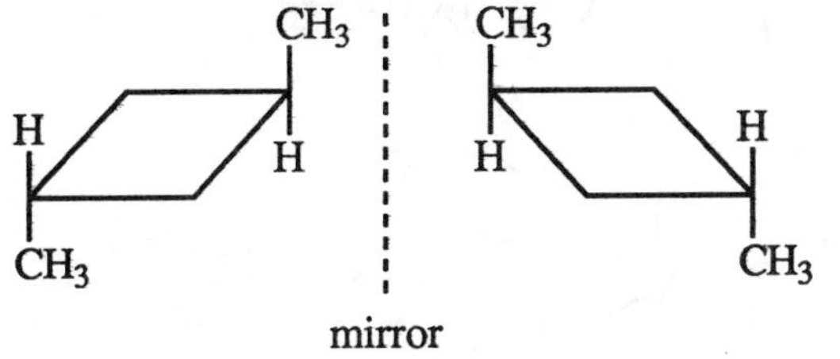

achiral—
identical mirror images

(d)

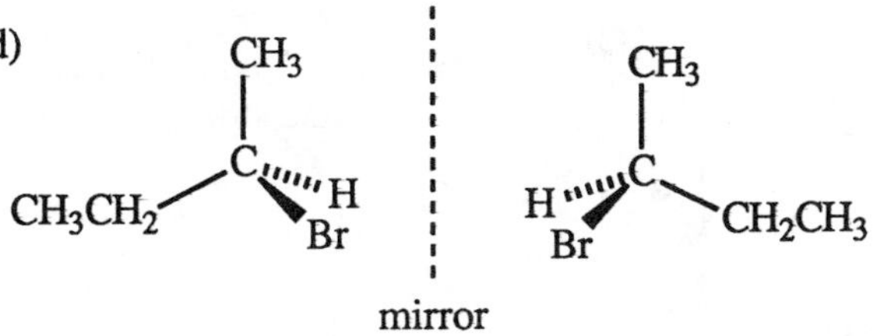

chiral—
enantiomers

6-2 continued

(e)

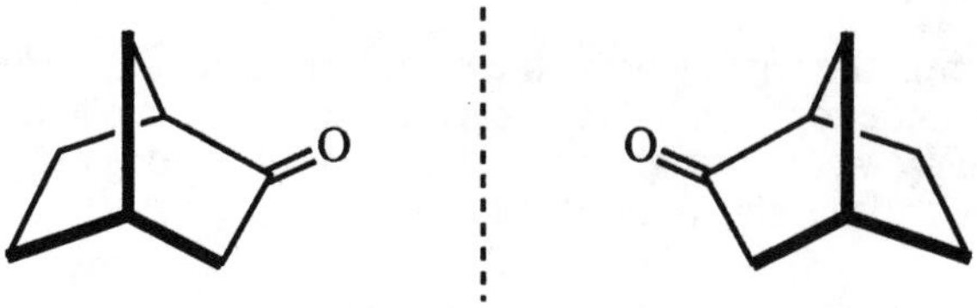

chiral—
enantiomers

mirror

(f)

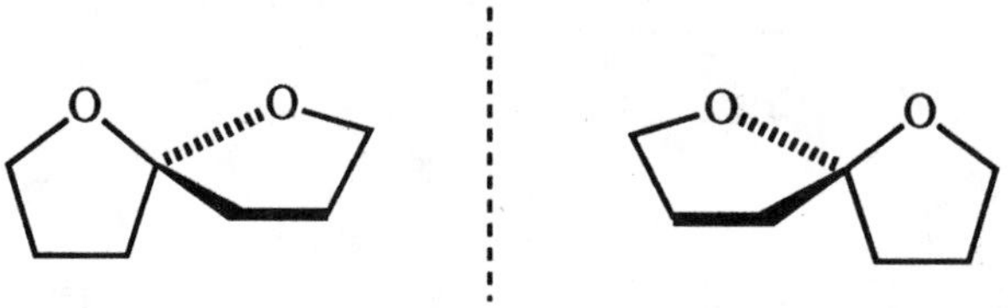

chiral—
enantiomers

mirror

6-3

(a)

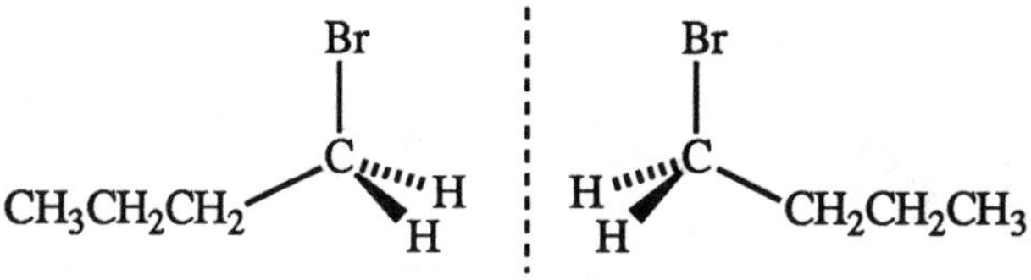

no chiral carbons—
same structure

(b)

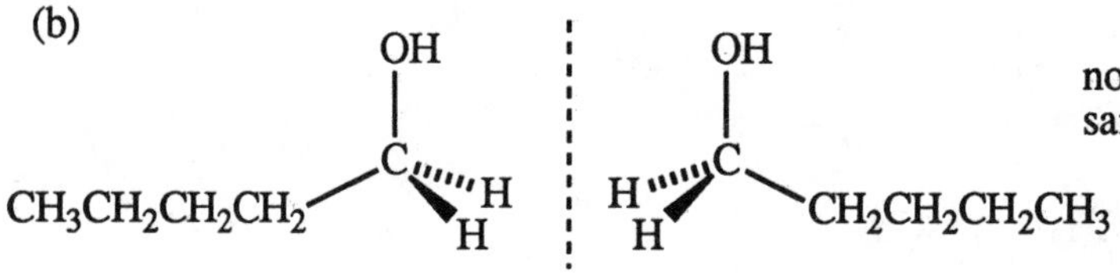

no chiral carbons—
same structure

(c)

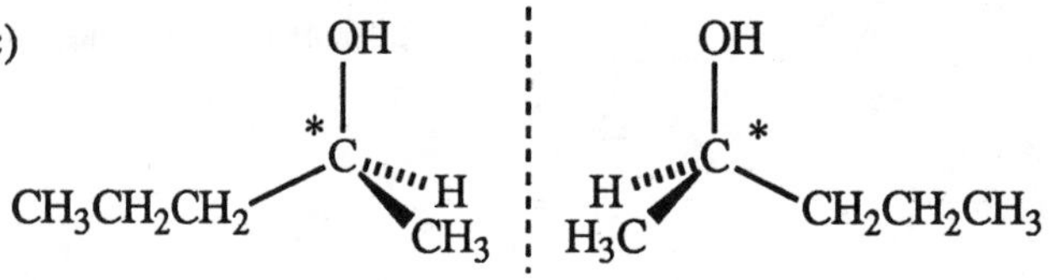

enantiomers

(d)

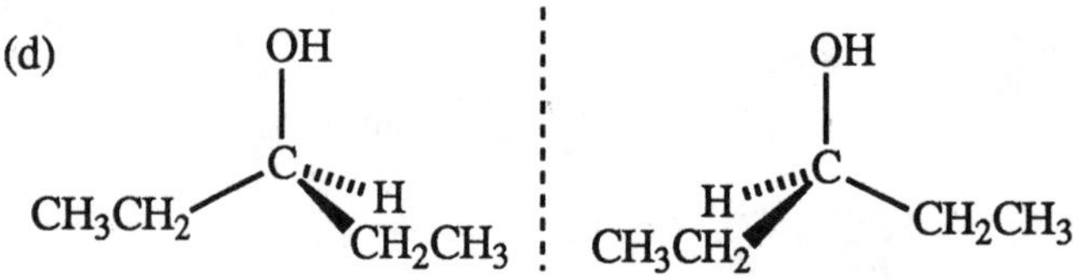

no chiral carbons—
same structure

6-3 continued

(e)

no chiral carbon—
same structure

(f)

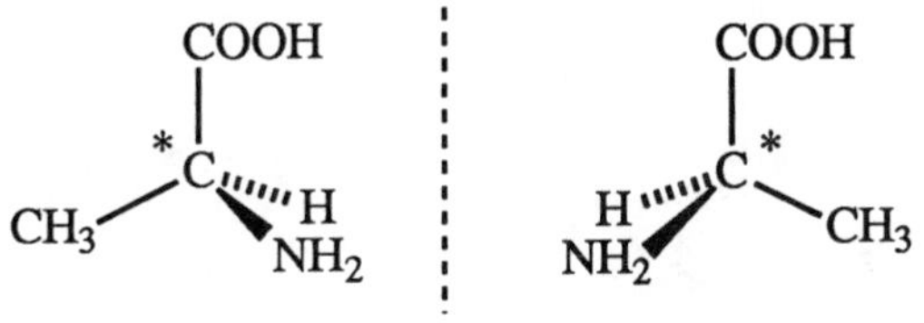

enantiomers

(g)

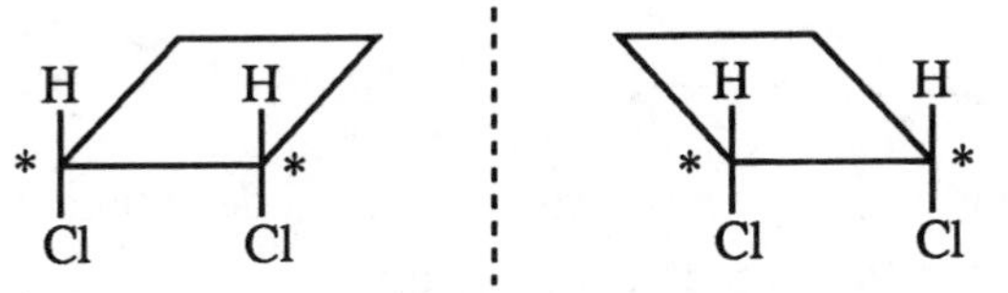

same structure

(h)

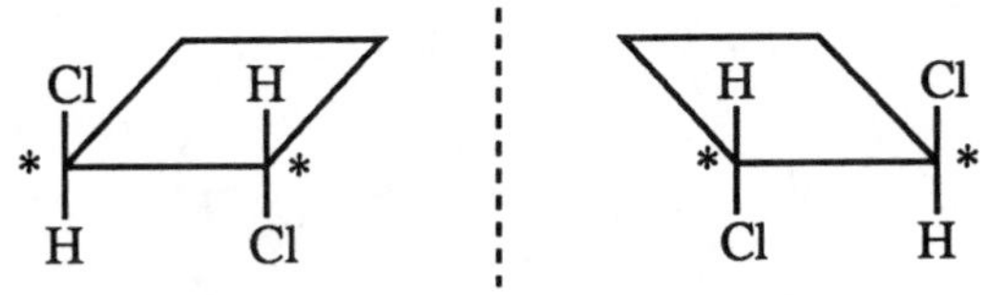

enantiomers

(i)

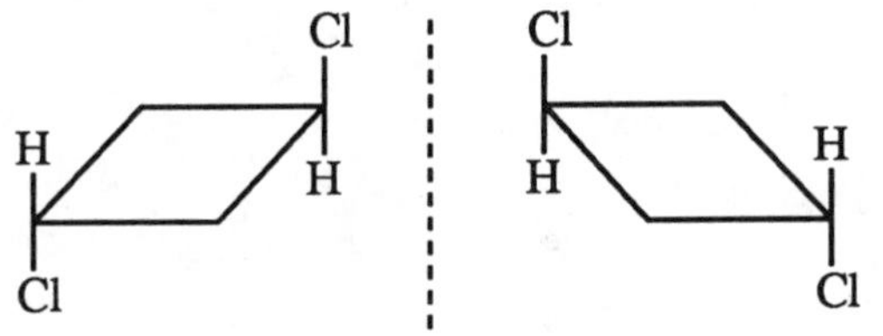

no chiral carbons—
same structure

(j)

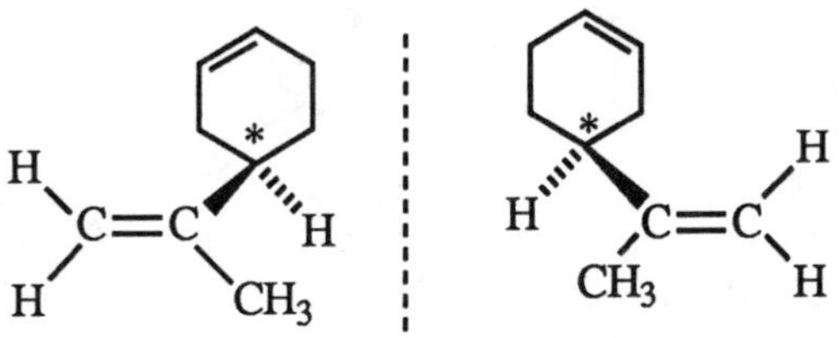

enantiomers

(k)

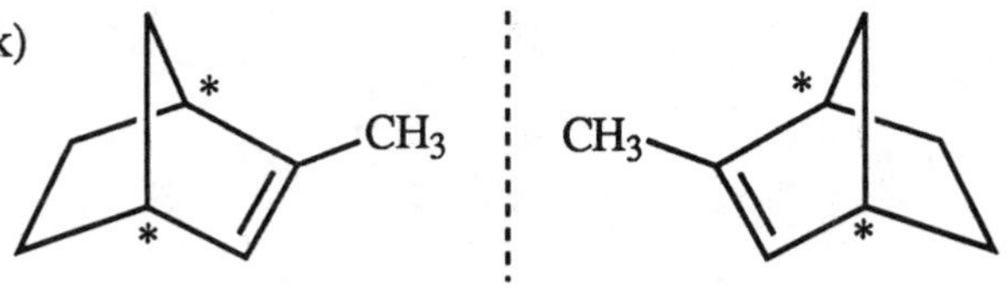

enantiomers

6-4

(a)

(b)

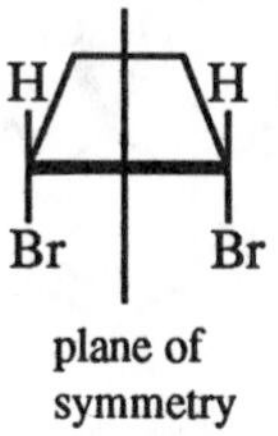

(c) chiral—no plane of symmetry

(d) chiral—no plane of symmetry

(e) chiral—no plane of symmetry

(f)

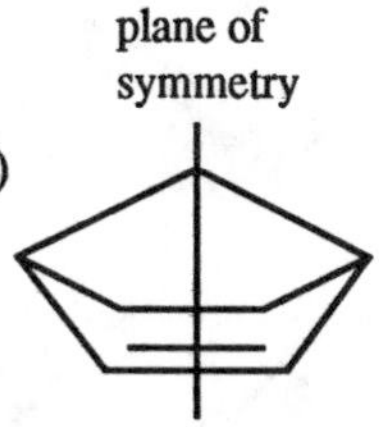

(g)

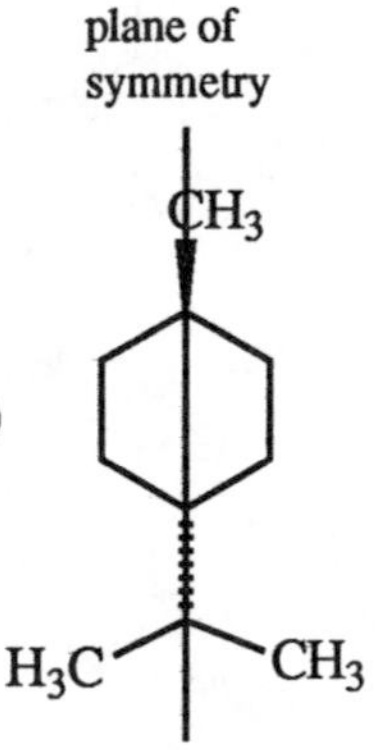

6-5 ALWAYS place the 4th priority group away from you. Then determine if the sequence 1→2→3 is clockwise (*R*) or counter-clockwise (*S*).

(a)

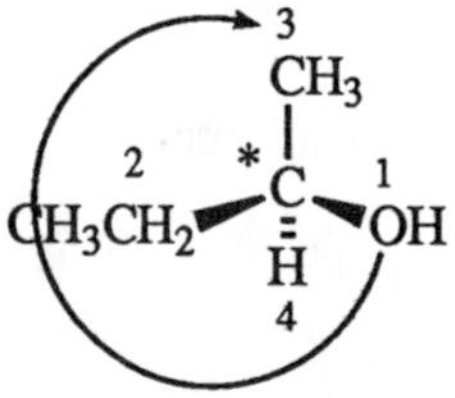

R

(b)

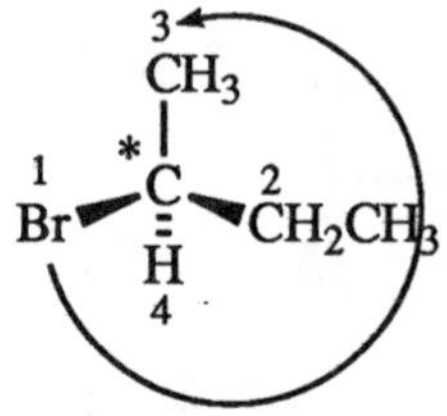

S

(c)

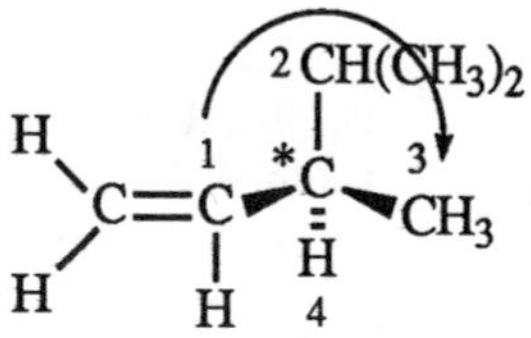

R

(d)

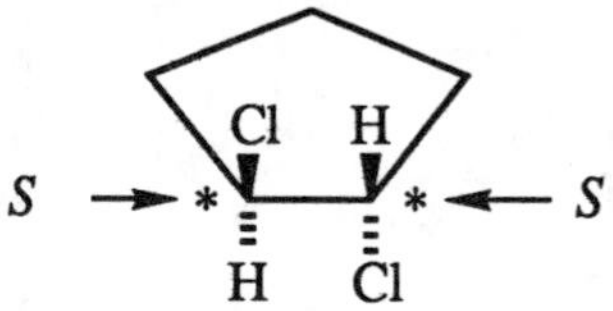

(e)

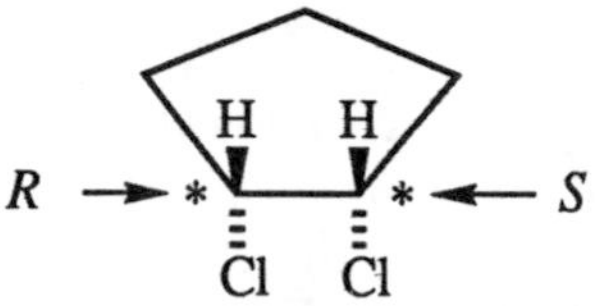

(f)

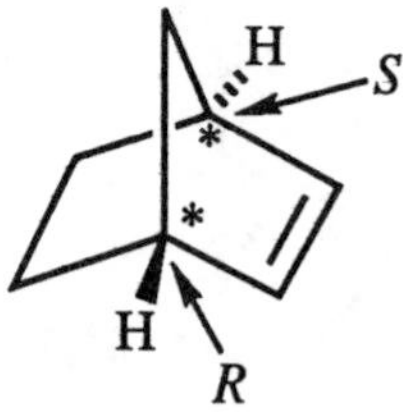

(g)

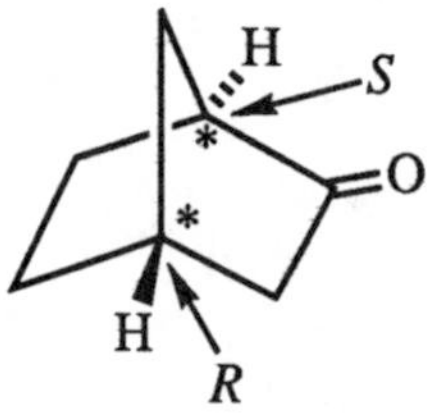

(h)

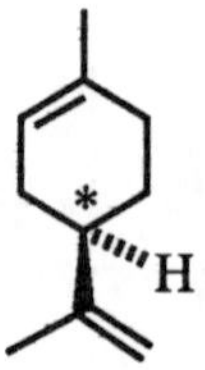

R

6-6 There are no chiral carbons in 6-3 (a), (b), (d), (e), or (i).

(c)

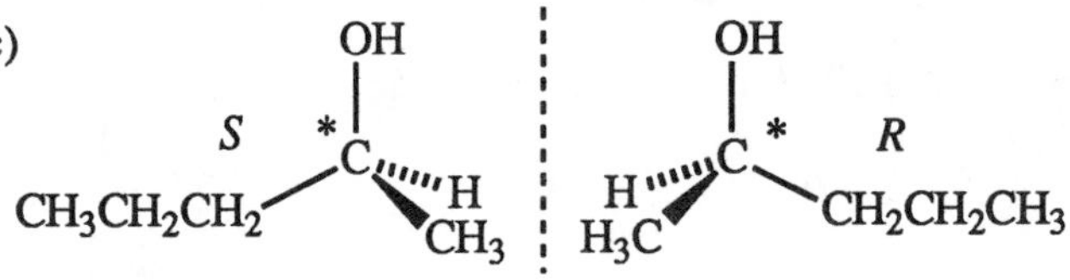

(f)

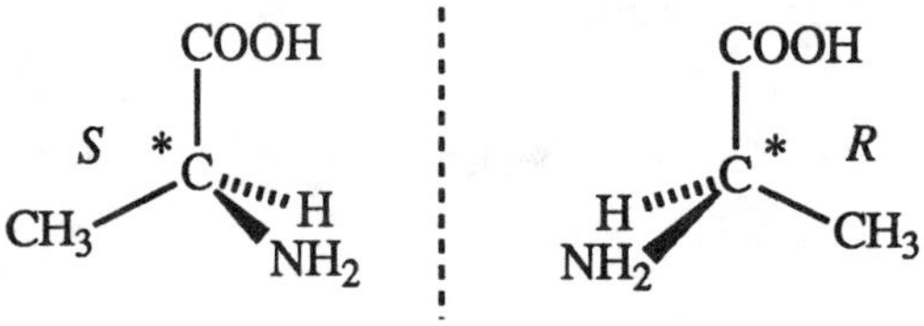

(g)

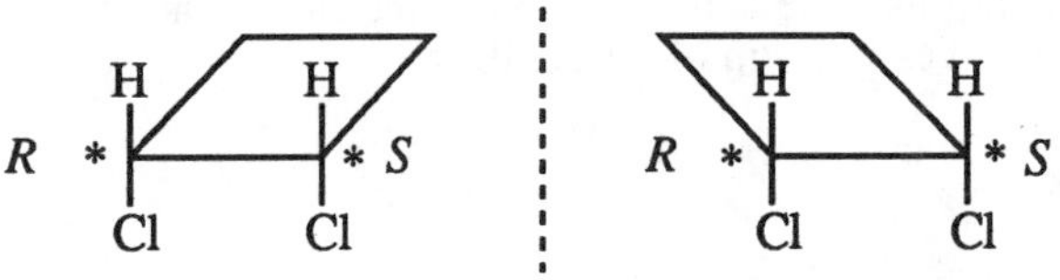

(h)

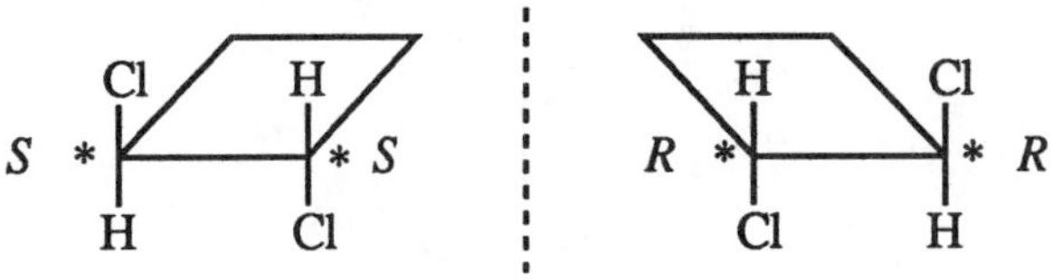

(j)

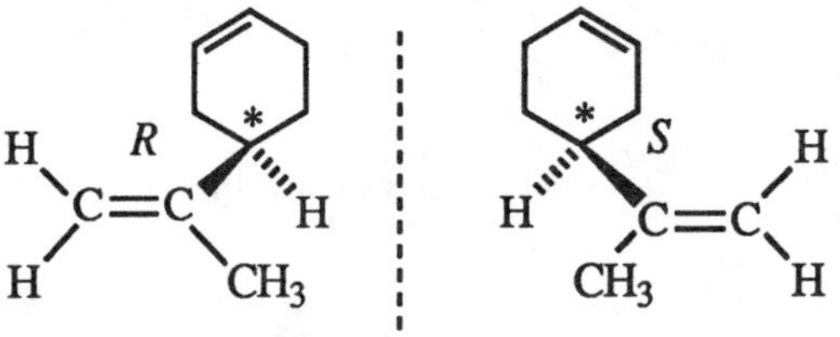

(k)

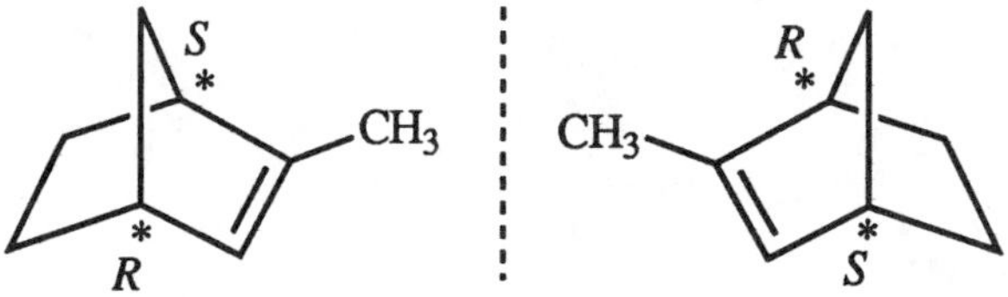

6-7 2.0 g / 10 mL = 0.20 g/mL ; 100 mm = 1 dm

$$[\alpha]_D^{25} = \frac{+1.74°}{(0.20)\ (1)} = +8.7° \text{ for (+)-glyceraldehyde}$$

6-8 0.5 g / 10 mL = 0.05 g/mL ; 20 cm = 2 dm

$$[\alpha]_D^{25} = \frac{-5.0°}{(0.05)\ (2)} = -50° \text{ for (–)-epinephrine}$$

6-9 Measure using a solution of about half the concentration of the first. The value will be either + 90° or – 90°, which gives the sign of the rotation.

6-10

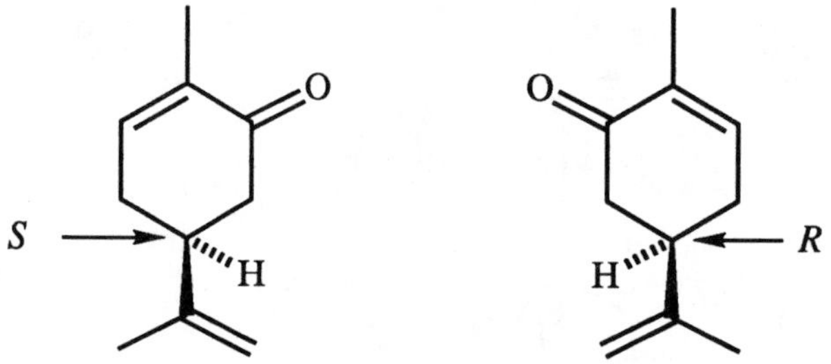

(+)-carvone (caraway seed) (–)-carvone (spearmint)

6-11

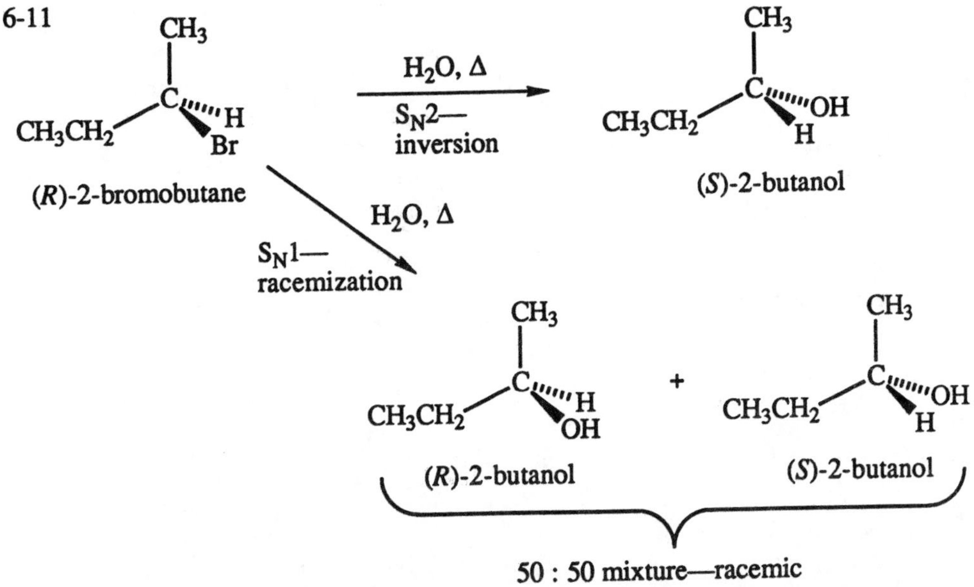

H_2O, Δ
S_N2— inversion
(R)-2-bromobutane
(S)-2-butanol
H_2O, Δ
S_N1— racemization
(R)-2-butanol
(S)-2-butanol
50 : 50 mixture—racemic

If S_N1, which gives racemization, occurs exactly twice as fast as S_N2, which gives inversion, then the racemic mixture (50 : 50 *R* + *S*) is 66.7% of the mixture and the rest, 33.3%, is the *S* enantiomer from S_N2. Therefore, the excess of one enantiomer over the racemic mixture must be 33.3%, the enantiomeric excess. (In the racemic mixture, the *R* and *S* "cancel" each other algebraically as well as in optical rotation.)

The optical rotation of pure (*S*)-2-butanol is + 13.5°. The optical rotation of this mixture is:

$$33.3\% \times +13.5^\circ = +4.5^\circ$$

6-12 The rotation of pure (+)-2-butanol is + 13.5° .

$$\frac{\text{observed rotation}}{\text{rotation of pure enantiomer}} = \frac{+0.45^\circ}{+13.5^\circ} \times 100\% = 3.3\% \text{ optical purity}$$
$$= 3.3\% \text{ e.e.} = \text{excess of (+) over (–)}$$

To calculate percentages of (+) and (–): (two equations in two unknowns)

$$(+) + (-) = 100\% \Longrightarrow (-) = 100\% - (+)$$

$$(+) - (-) = 3.3\% \Longrightarrow (+) - (100\% - (+)) = 3.3\%$$

$$2\,(+) = 103.3\%$$

(+) = 51.6% (rounded)
(–) = 48.4%

6-13

(a)

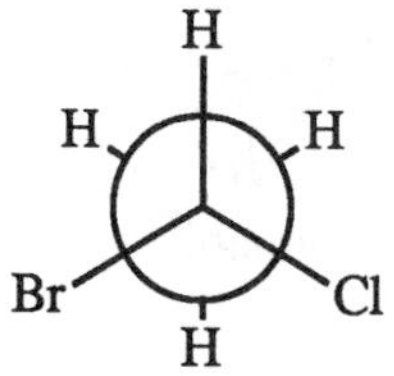

chiral—
optically active

(b)

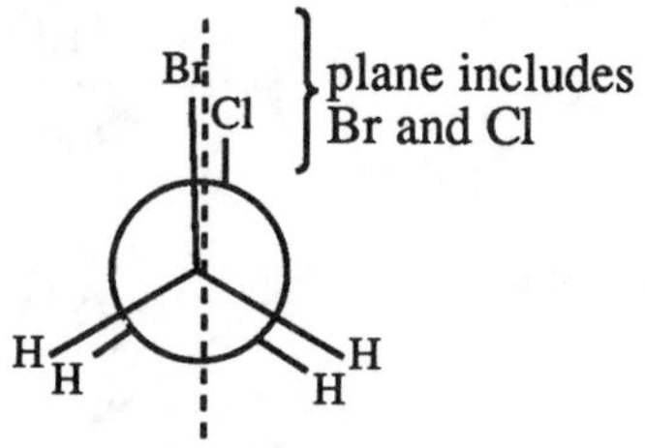

plane of symmetry containing
Br—C—C—Cl; not optically active

(c)

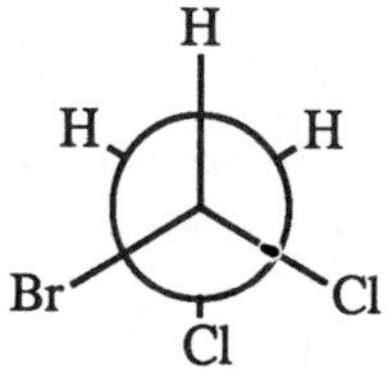

chiral—
optically active

(d)

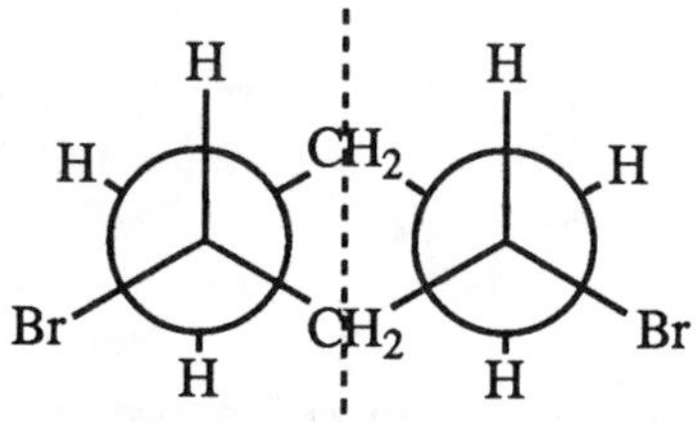

plane of symmetry—not optically active despite the presence of two chiral carbons

(e)

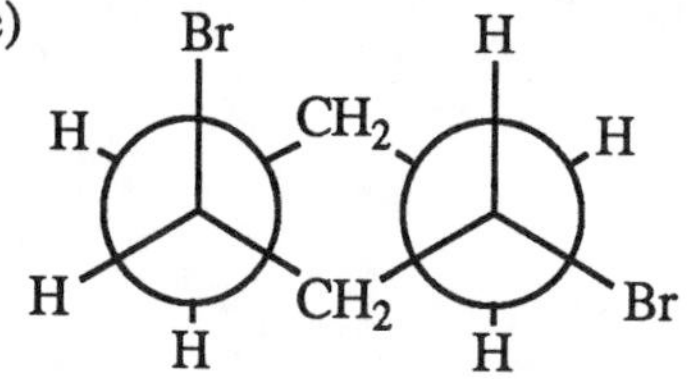

no plane of symmetry—
optically active
(other chair form is equivalent—no plane of symmetry)

(f)

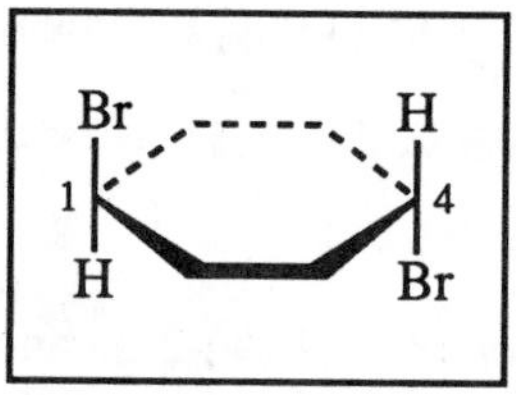

plane of symmetry through C-1 and C-4—not optically active

6-14

(a)

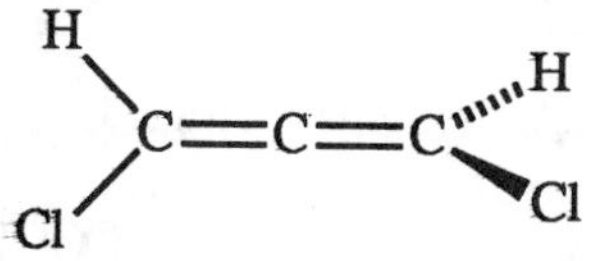

no chiral carbons, but the molecule is chiral (an allene)

(b)

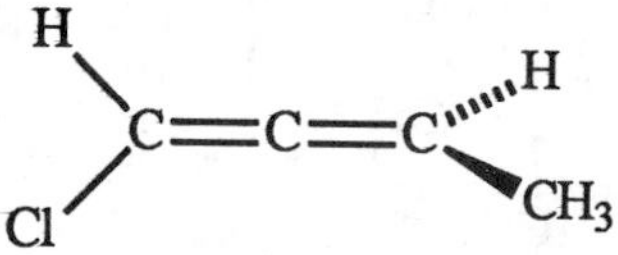

no chiral carbons, but the molecule is chiral (an allene)

(c)

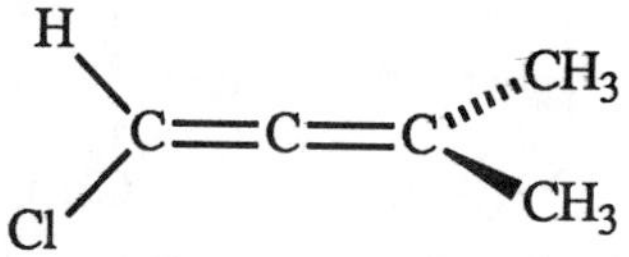

no chiral carbons; this allene has a plane of symmetry between the two methyls (the plane of the paper), including all the other atoms; not a chiral molecule

(d)

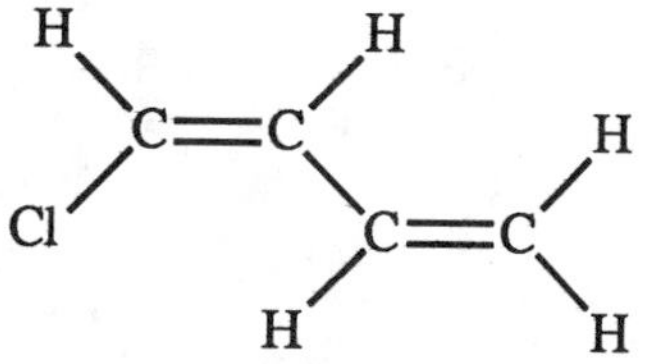

planar molecule—not chiral

(e)

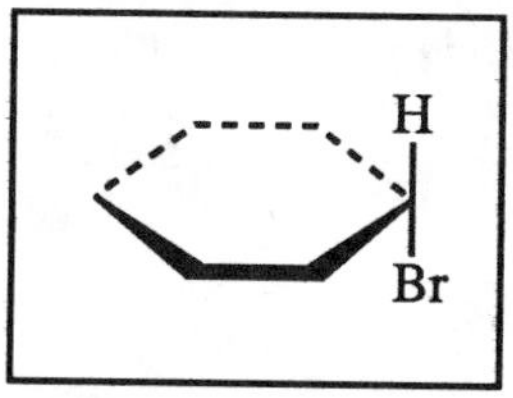

plane of symmetry bisecting the molecule; not chiral

(f)

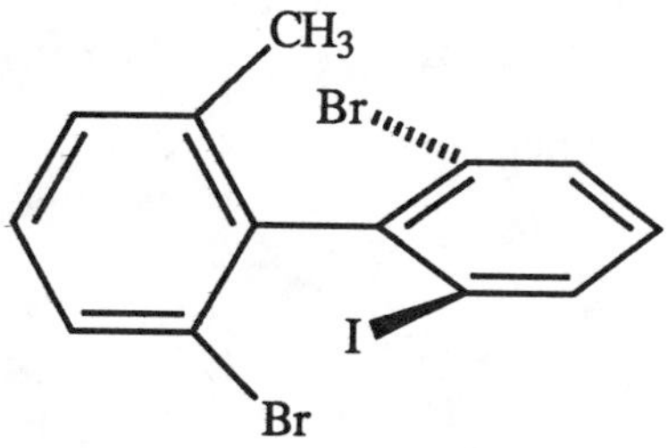

no chiral carbons, but the molecule is chiral due to restricted rotation

(g)

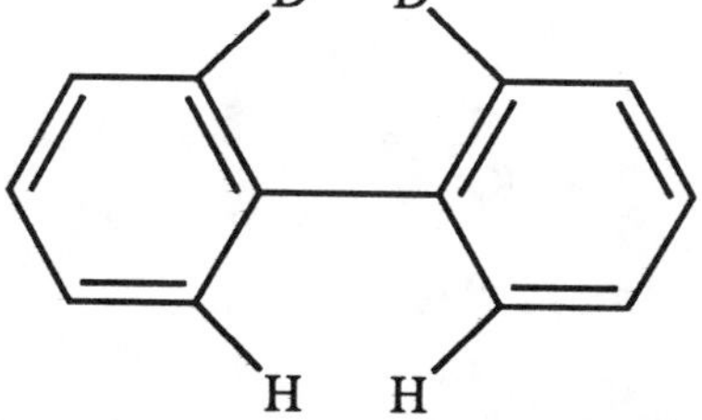

no chiral carbons, and the groups are not large enough to restrict rotation; not a chiral compound

(h)

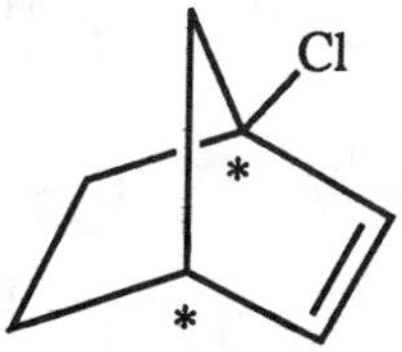

two chiral carbons; a chiral compound

6-15

(a)

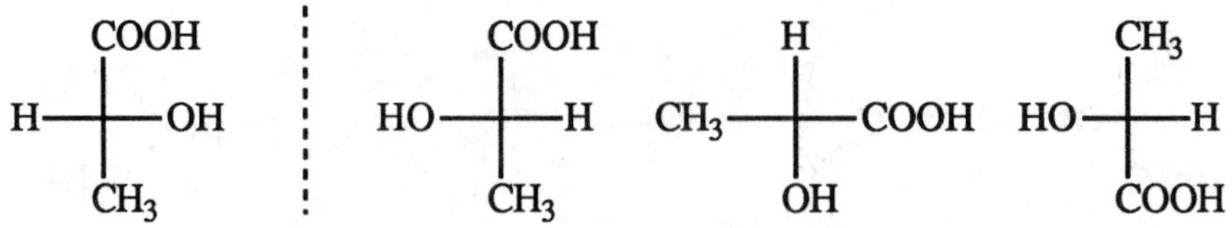

enantiomer | enantiomer | same

(b)

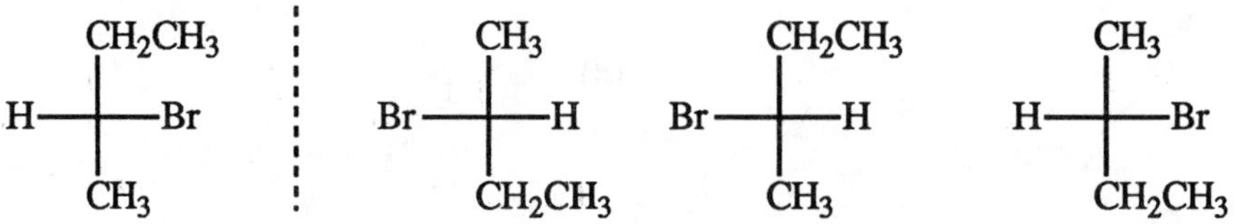

same | enantiomer | enantiomer

(c)

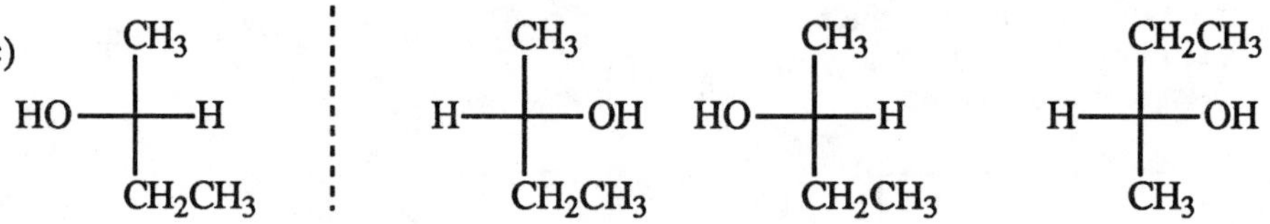

(*R*)-2-butanol | enantiomer | same | same

6-16

(a)

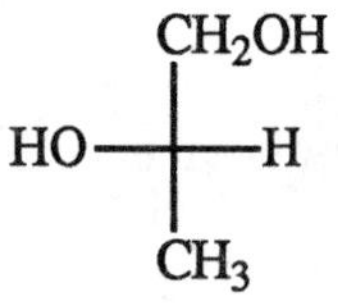

(b)

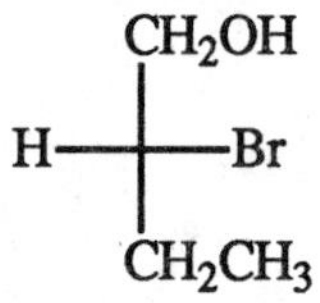

(c)

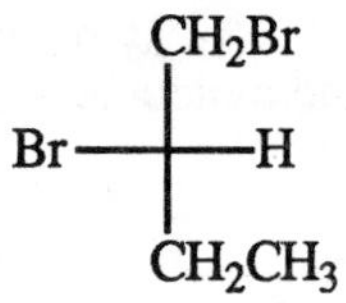

(d)

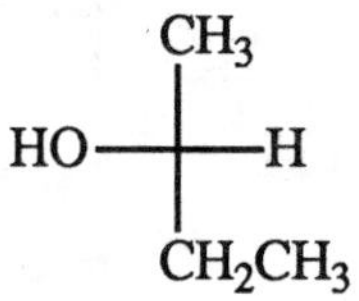

(e)

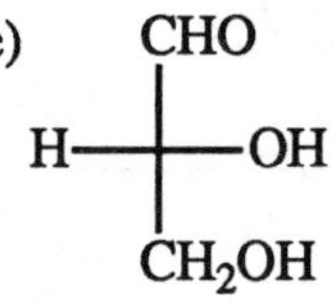

6-17

(a) 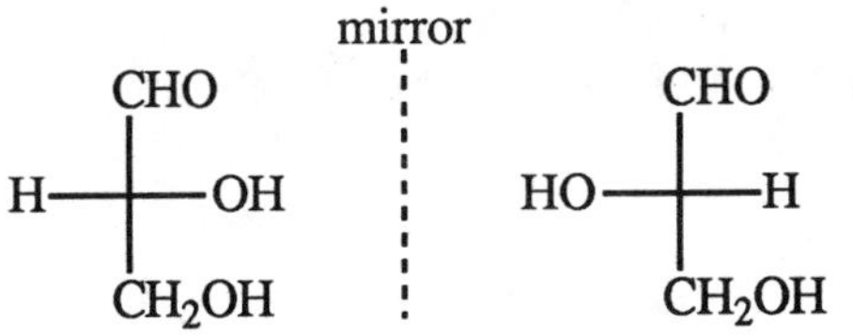

180 ° rotation gives different structures; no plane of symmetry: **enantiomers**

(b) 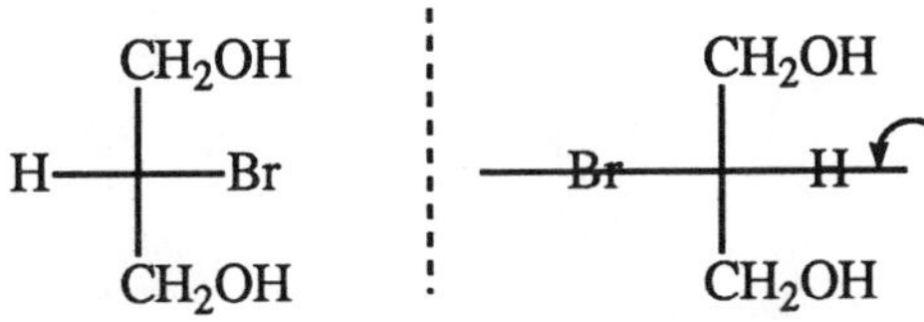

180° rotation gives same structures; plane of symmetry: **same structure**

(c) 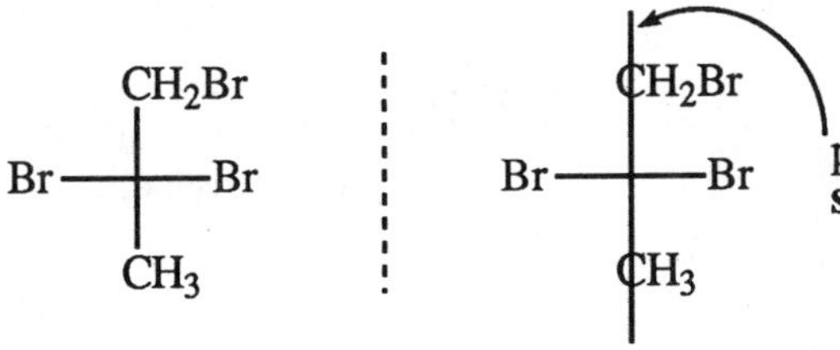

plane of symmetry: **same structure**

(d) 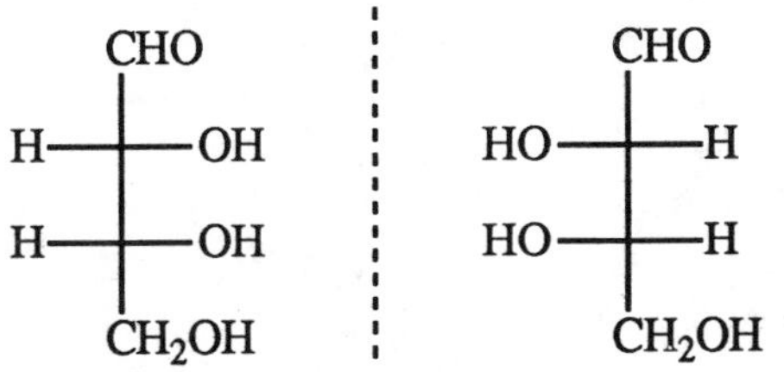

180 ° rotation gives different structures; no plane of symmetry: **enantiomers**

(e) 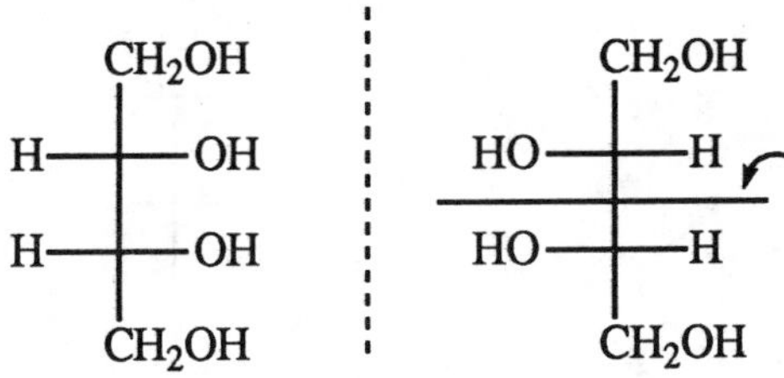

180° rotation gives same structures; plane of symmetry: **same structure**

(f) 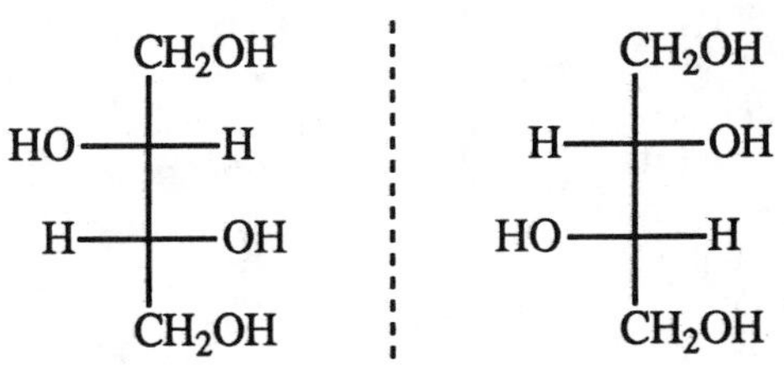

180 ° rotation gives different structures; no plane of symmetry: **enantiomers**

6-18 (a) from problem 6-17: (b) and (c) have no chiral centers; (a) *R* ; (d) 2*R*,3*R* ; (e) 2*S*,3*R* (numbering down); (f) 2*R*, 3*R*

(b) *R* (c) *S* (d) *S*

6-19 (a) enantiomers
(b) diastereomers
(c) diastereomers
(d) structural isomers
(e) enantiomers
(f) diastereomers
(g) enantiomers
(h) same compound

6-20

(a)

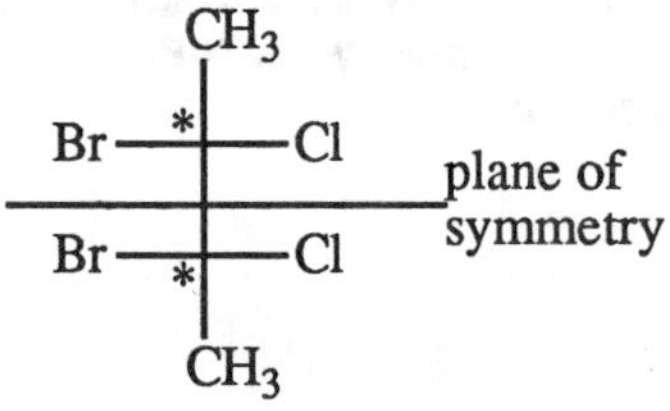

meso structure
not optically active

(b)

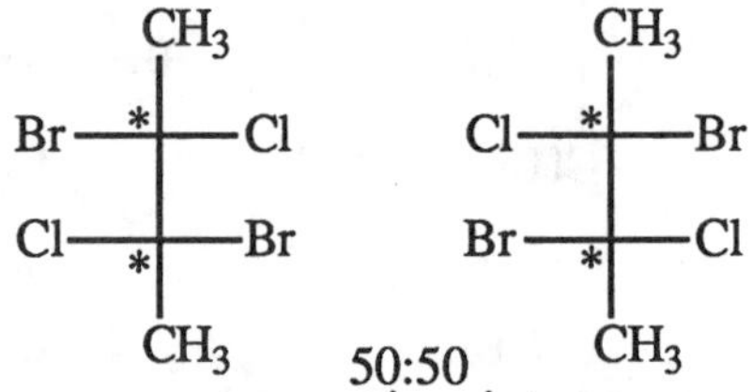

50:50
racemic mixture
not optically active

(c)

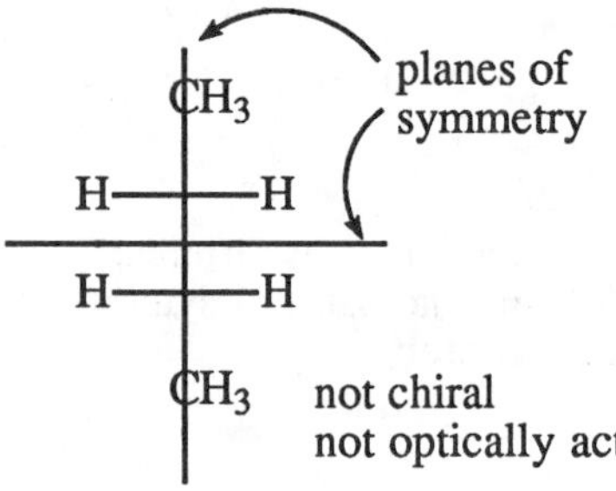

not chiral
not optically active

(d)

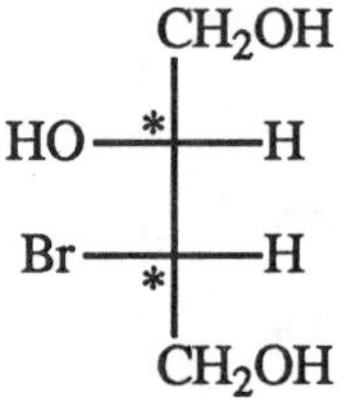

optically active

(e)

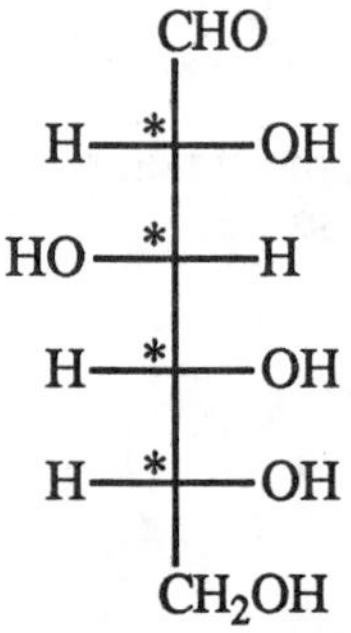

optically active

(f)

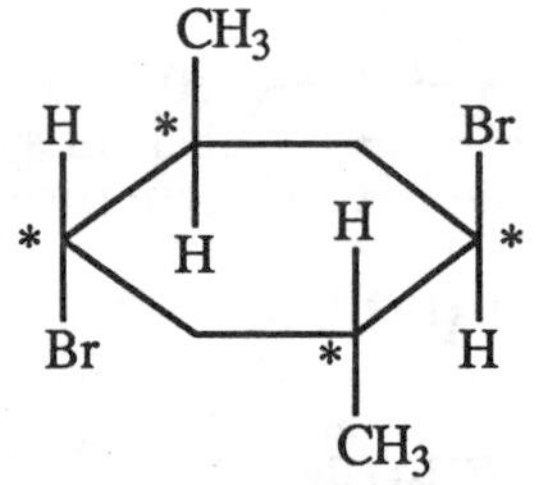

not optically active—
superimposable on its
mirror image—
technically, this is
a meso structure
(may require models!)

(g)

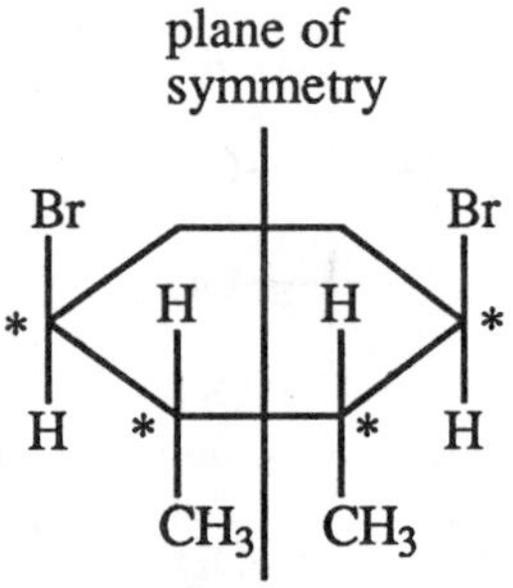

meso—
not optically active

6-21

(a)

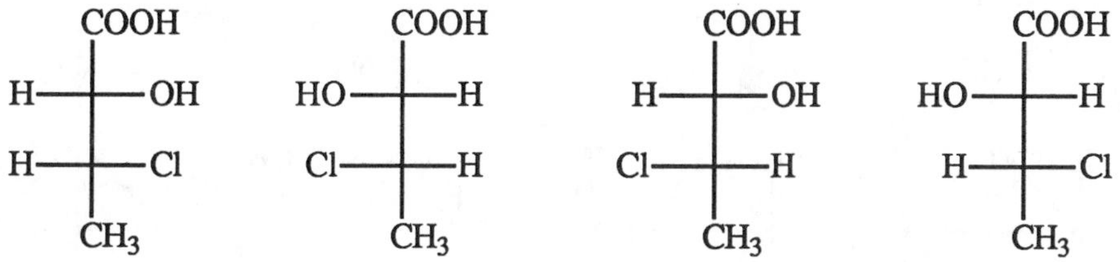

A B C D

enantiomers: A and B; C and D
diastereomers: A and C, A and D, B and C, B and D

(b)

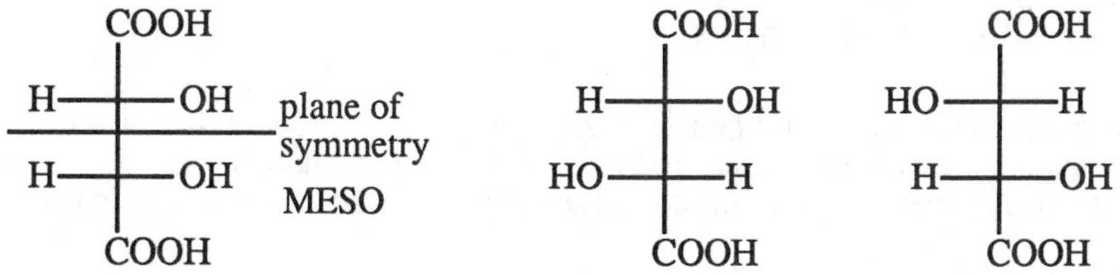

E F G

enantiomers: F and G
diastereomers: E and F, E and G

(c)

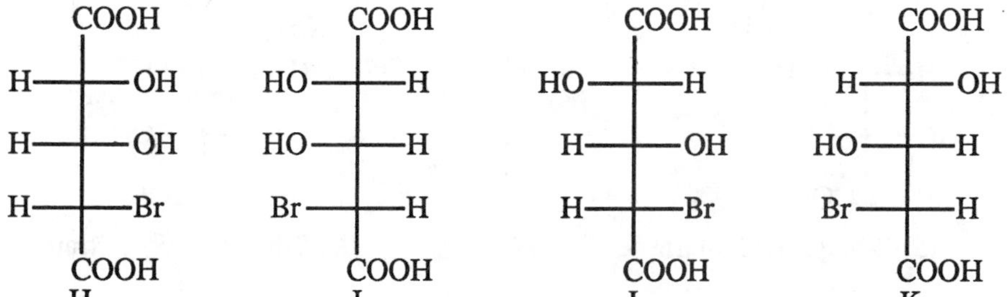

H I J K

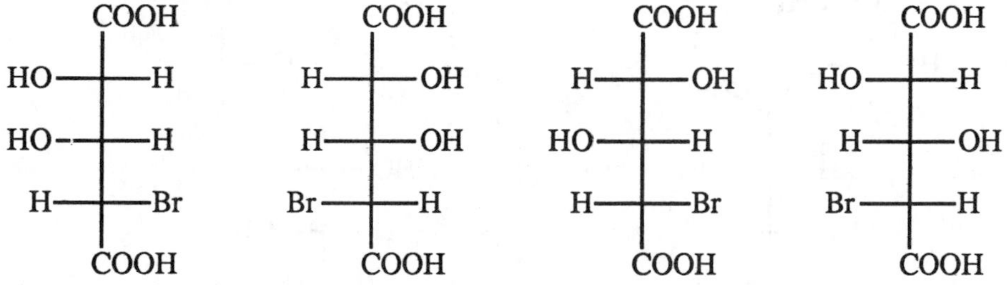

L M N O

enantiomers: H and I, J and K, L and M, N and O
diastereomers: any pair which is not enantiomeric

6-21 continued

(d)

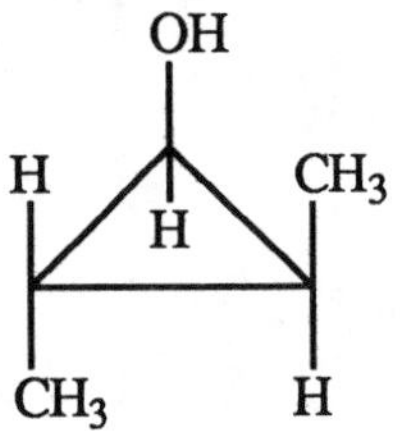

P

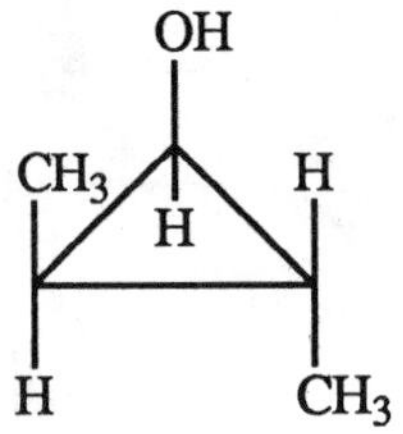

Q

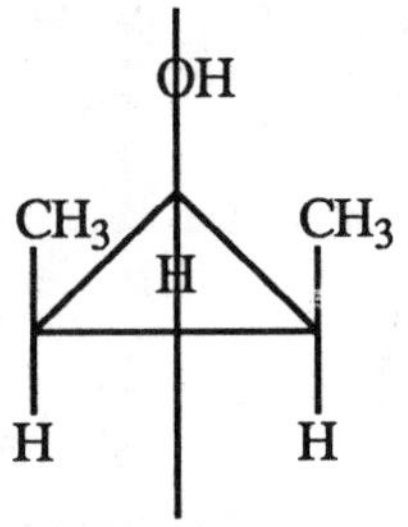

plane of symmetry
MESO

R

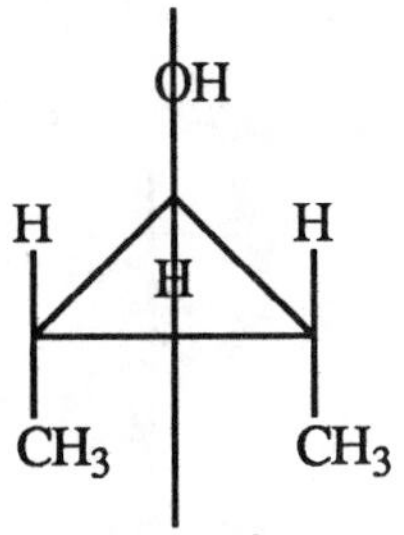

plane of symmetry
MESO

S

enantiomers: P and Q
diastereomers: P and R, P and S, Q and R, Q and S, R and S

6-22 Any diastereomeric pair could be separated by a physical process like distillation or crystallization. Diastereomers are found in parts (a), (b), and (d). The structures in (c) are enantiomers; they could not be separated by normal physical means.

6-23

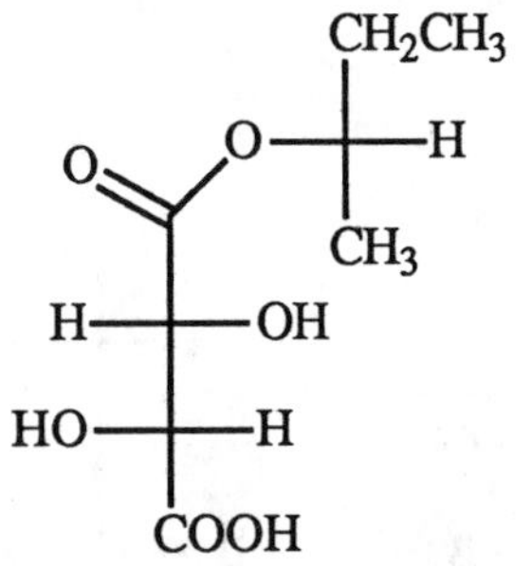

(*S*)-2-butyl (*R*,*R*)-tartrate

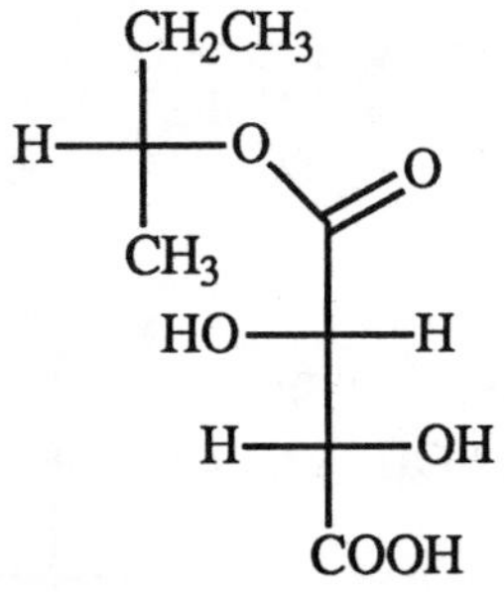

(*R*)-2-butyl (*S*,*S*)-tartrate

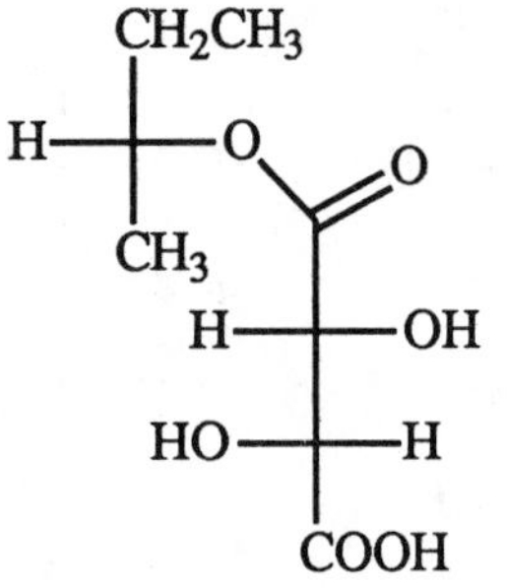

(*R*)-2-butyl (*R*,*R*)-tartrate

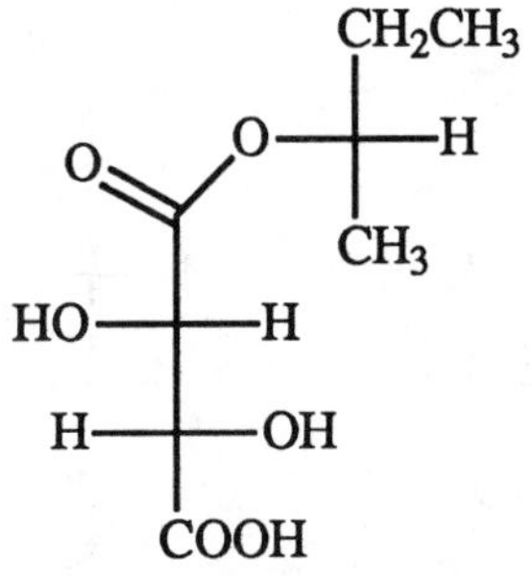

(*S*)-2-butyl (*S*,*S*)-tartrate

6-24 Reactions in (a) and (c) would also give some elimination products; only the substitution products are shown here.

(a) S_N2 gives inversion:
only product:

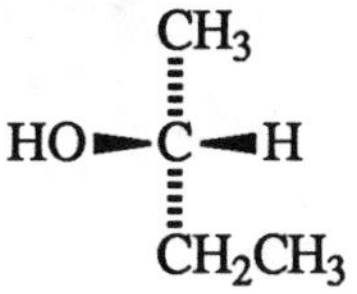

(b) S_N2 only

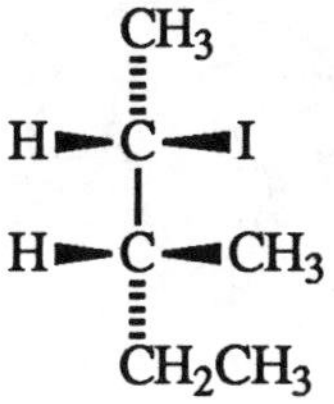

(c) solvolysis, S_N1, racemization

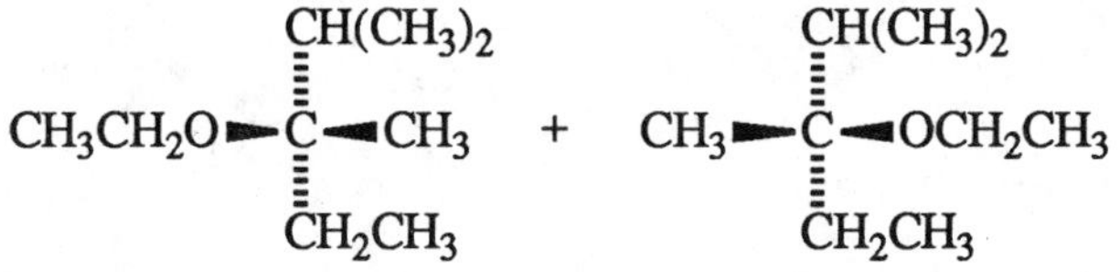

6-25

(a)

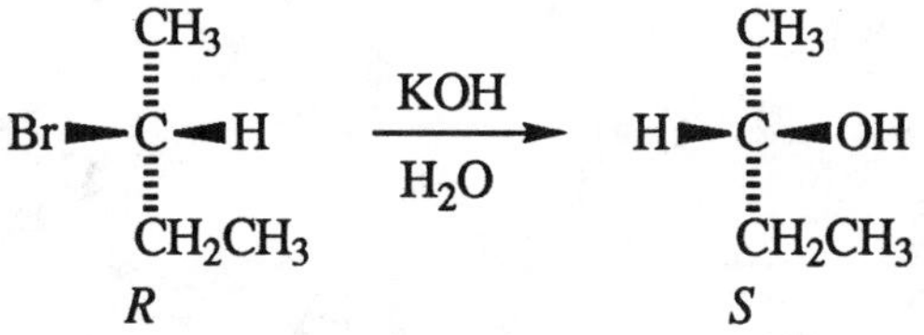

S_N2—inversion

(b)

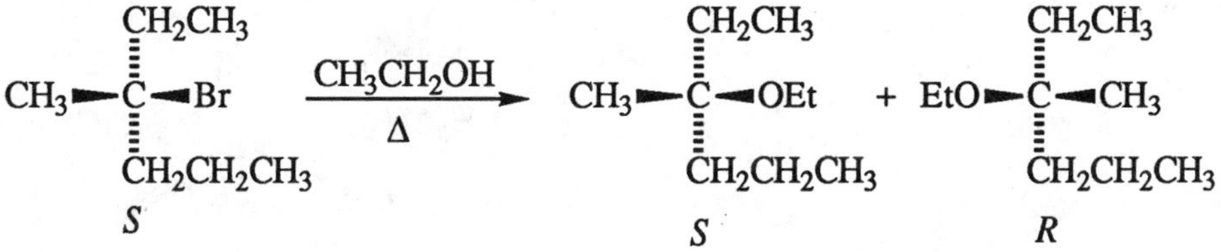

S_N1—racemization

(c)

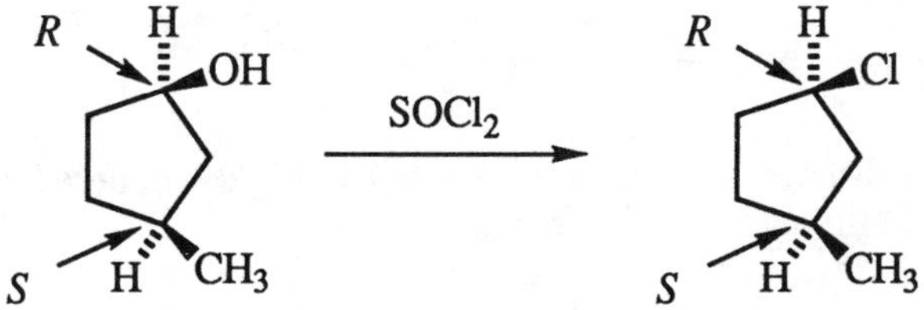

retention

6-26

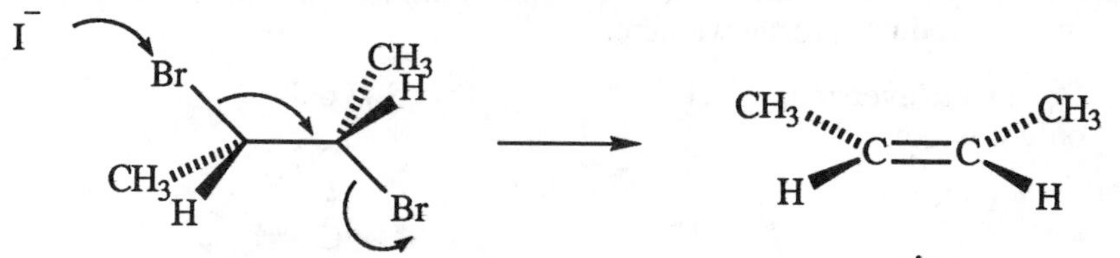

S,S *cis*

6-27

(a)

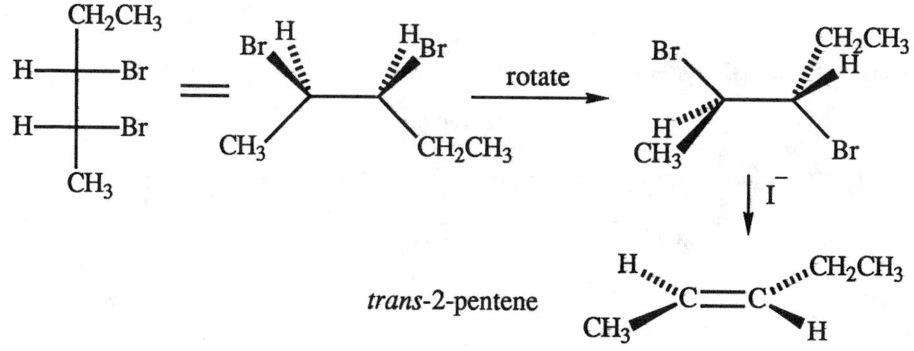
rotate
trans-2-pentene

(b)

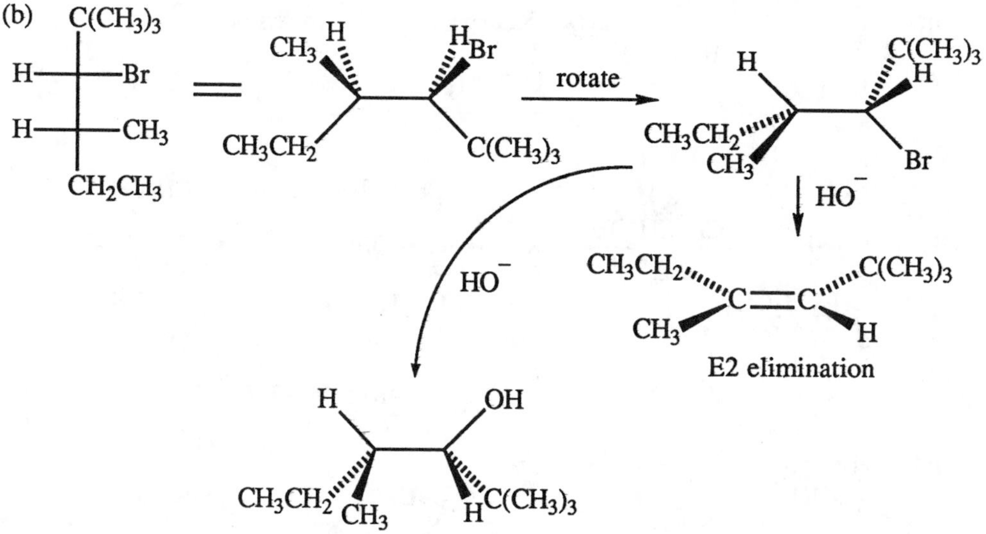
rotate
E2 elimination

substitution product—since the substrate is a neopentyl halide and highly hindered, the S_N2 substitution is slow, and elimination is favored.

6-27 (c)

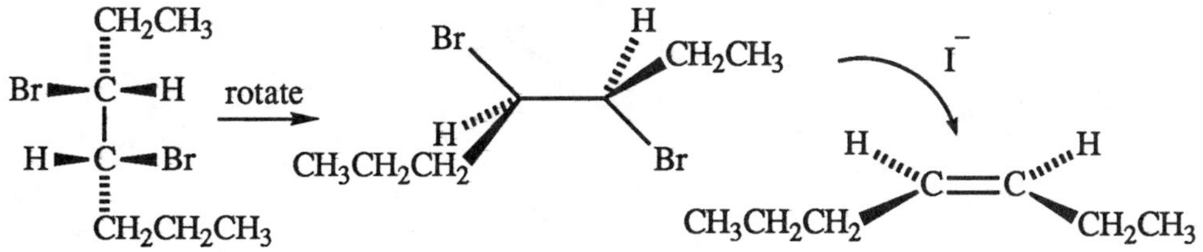

6-28 Refer to the Glossary and the text for definitions and examples.

6-29

(a)

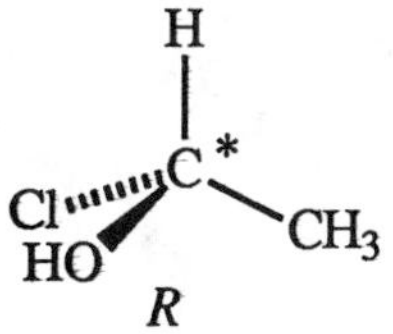

chiral

(b)

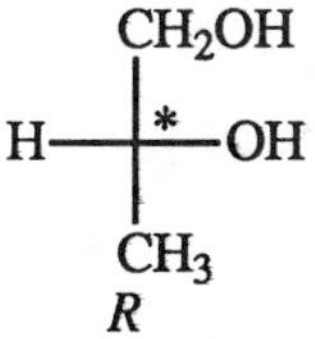

chiral

(c)

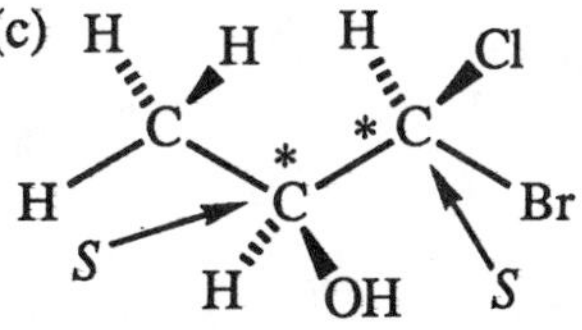

chiral

(d)

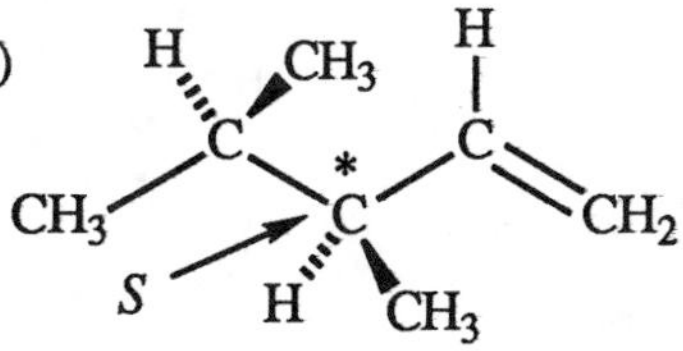

chiral

(e)

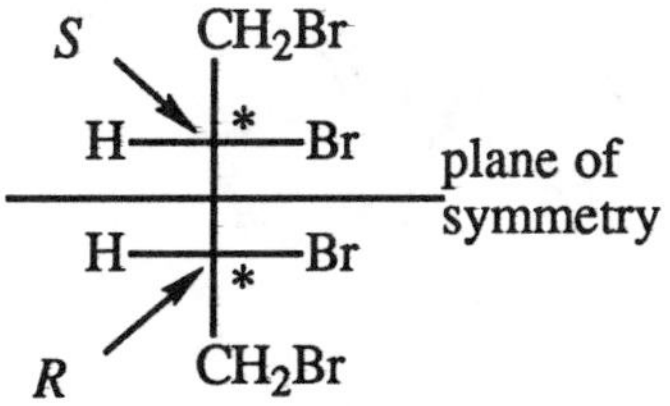

meso; achiral

(f)

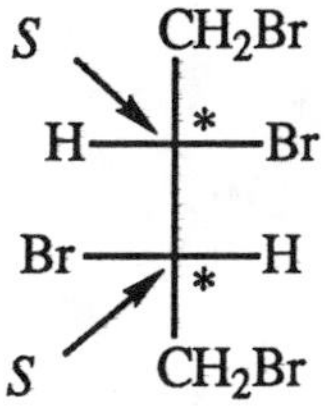

chiral

(g)

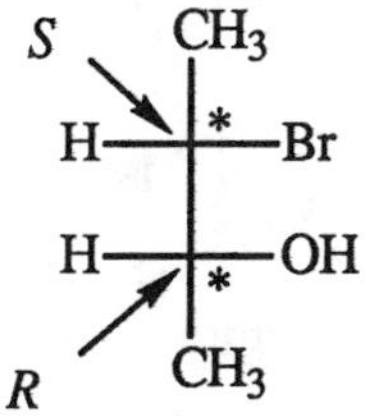

chiral

(h)

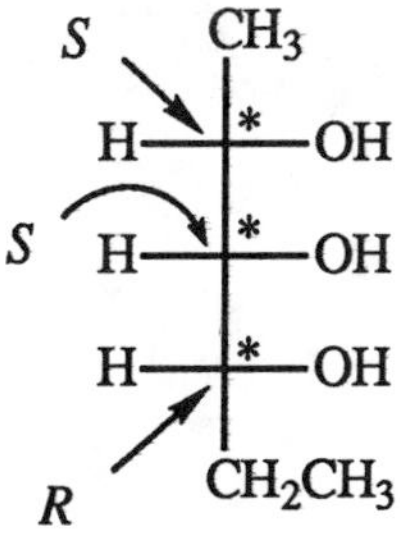

chiral

6-29 continued

(i)

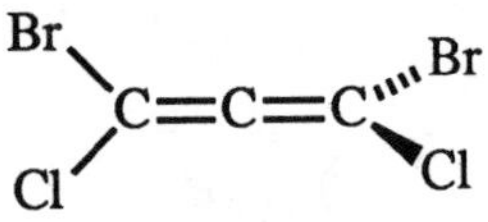

Br
Br
C=C=C
Cl
Cl

chiral molecule, but
no chiral centers

(j)

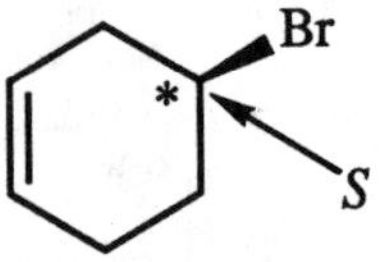

Br
*
S

chiral

(k)

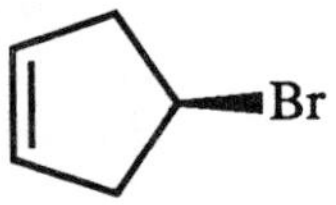

Br

achiral

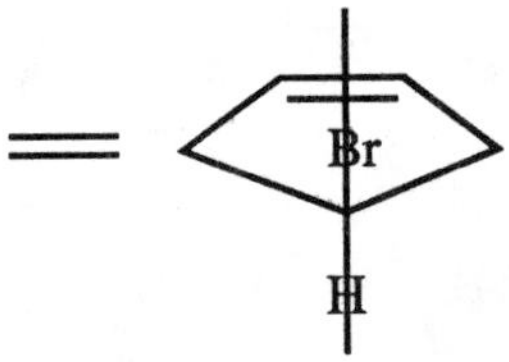

Br
H

plane of symmetry

(l)

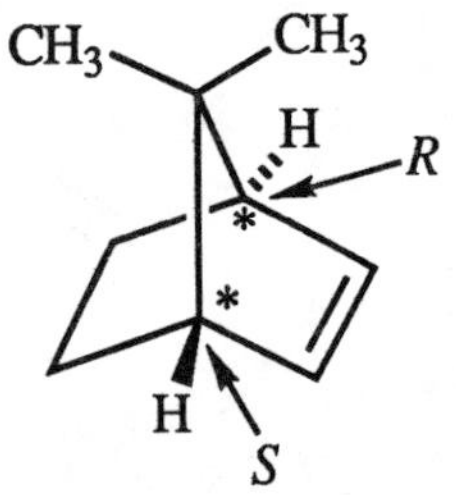

CH3
CH3
H
R
*
*
H
S

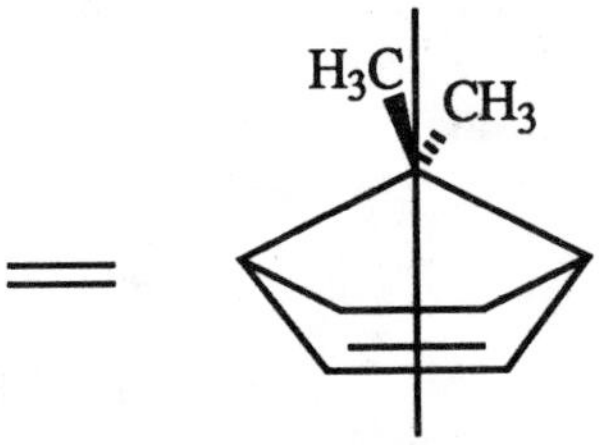

H3C
CH3

plane of symmetry
meso; achiral

(m)

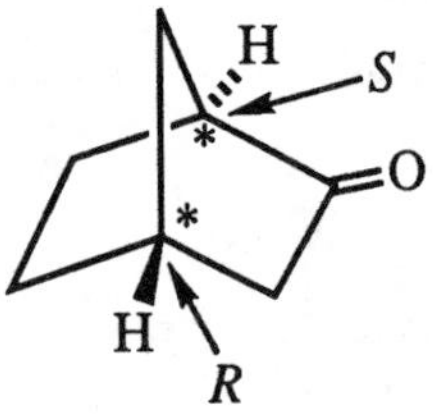

H
S
*
*
O
H
R

chiral

(n)

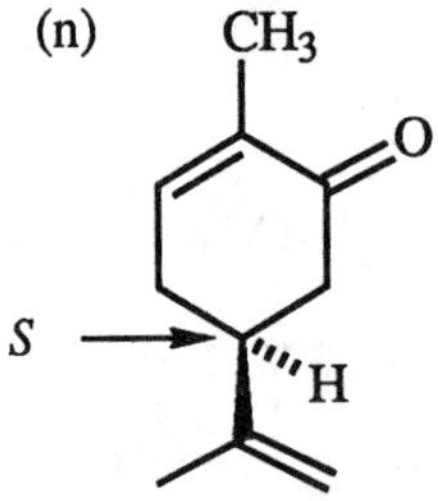

CH3
O
S
H

chiral

6-30

(a)
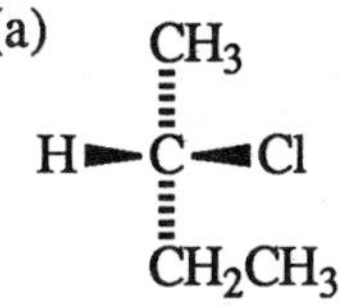

(b)
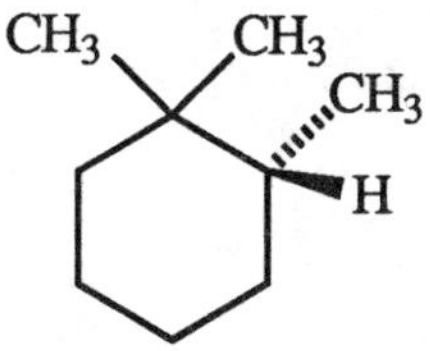

(c)
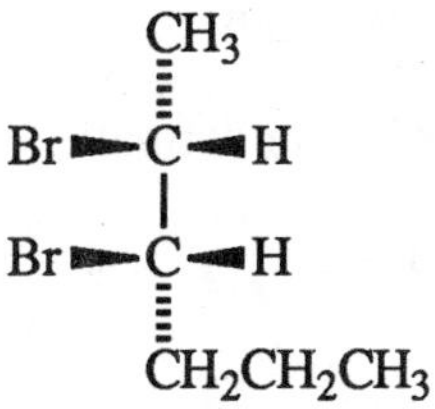

(d)
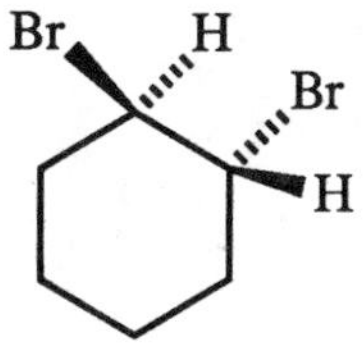

(e)
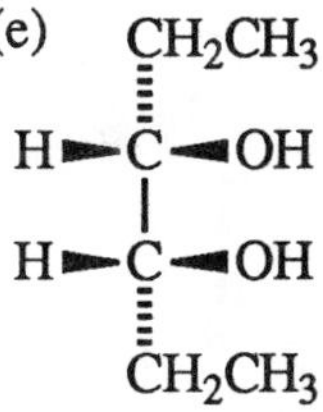

(f)
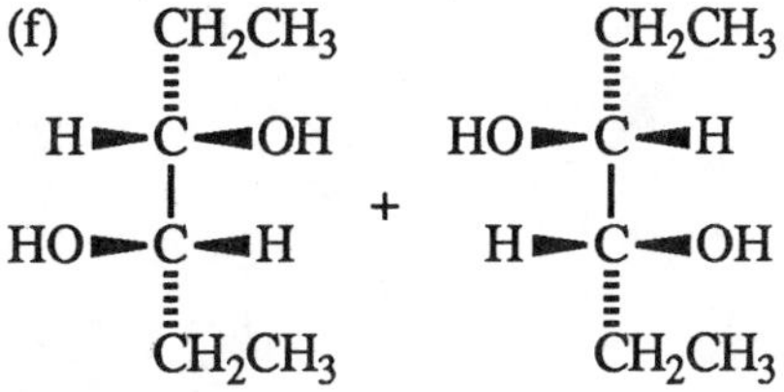

6-31

(a)
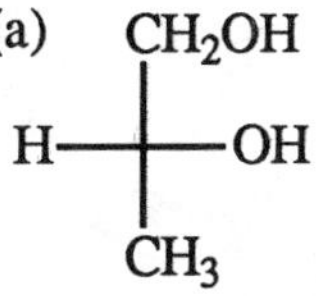

(b)
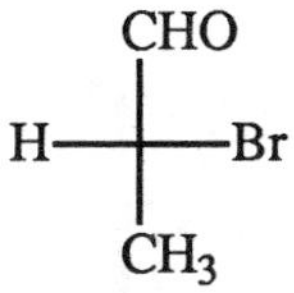

(c)
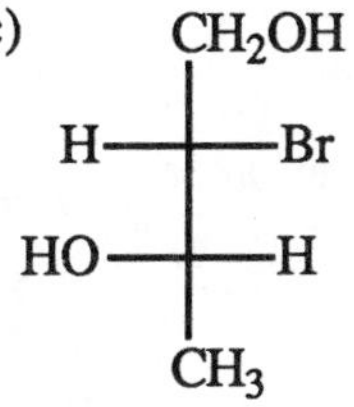

(d)
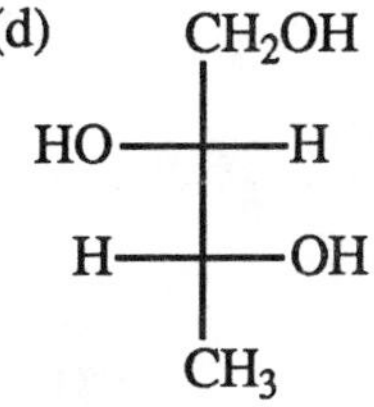

6-32

(a)
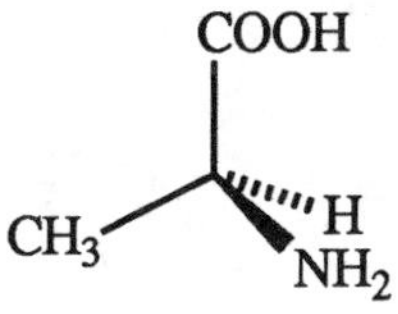

(b)
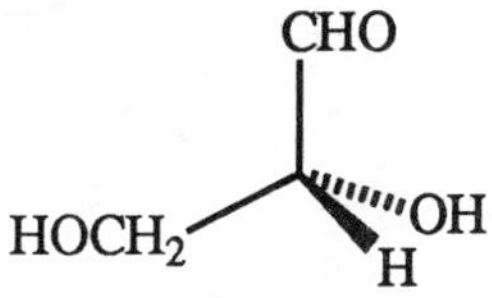

(c)
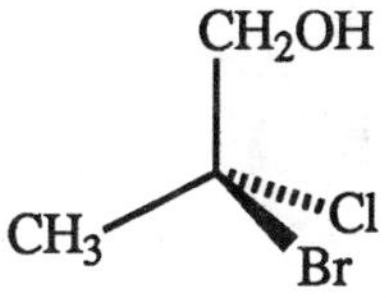

(d)
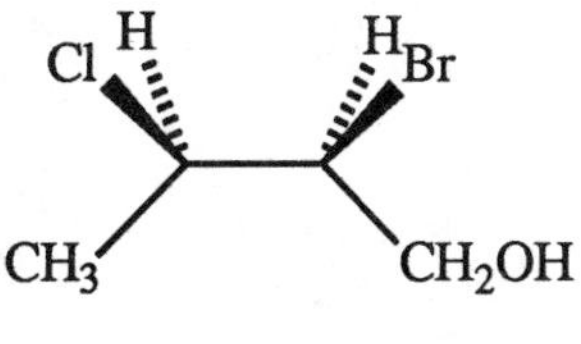

6-33 (a) same (meso)
(b) enantiomers
(c) enantiomers
(d) enantiomers
(e) enantiomers
(f) diastereomers
(g) enantiomers
(h) same compound

6-34

(a)

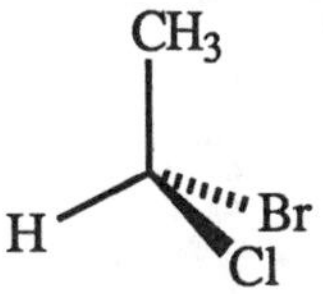

CH_3
H
Br
Cl

(b)

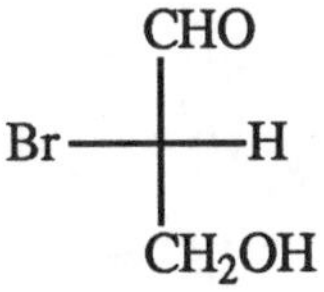

CHO
Br
H
CH_2OH

(c)

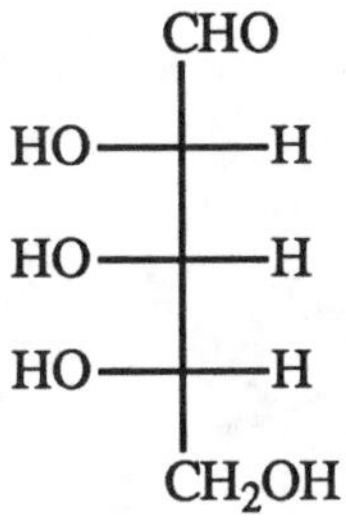

CHO
HO
H
HO
H
HO
H
CH_2OH

(d)

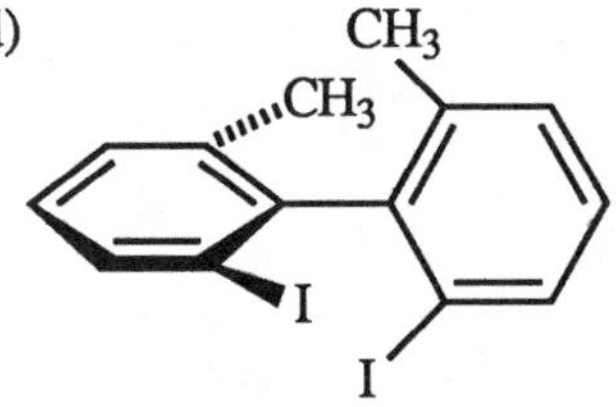

CH_3
CH_3
I
I

(e)

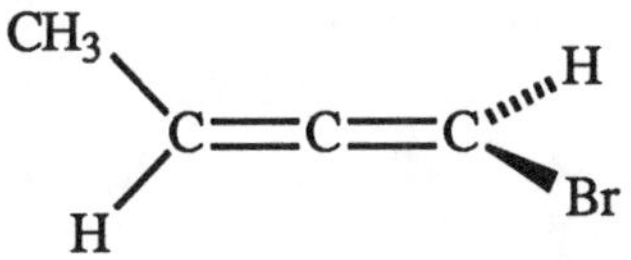

CH_3
H
C=C=C
H
Br

(f) plane of symmetry—no enantiomer

(g)

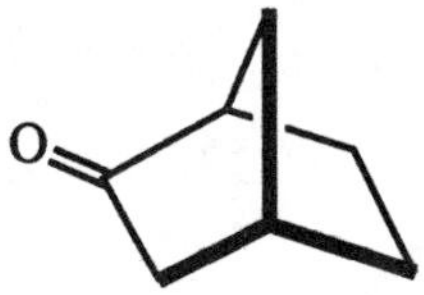

O

(h)

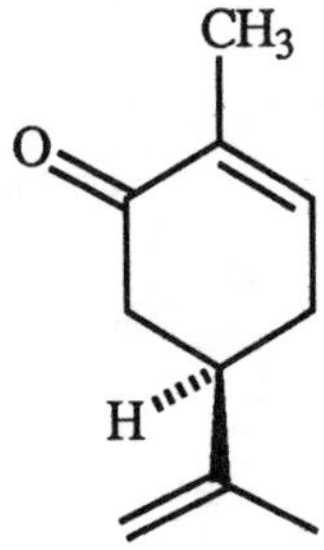

CH_3
O
H

(i)

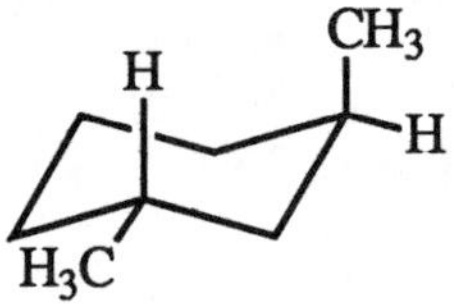

H
CH_3
H
H_3C

6-35 substitution

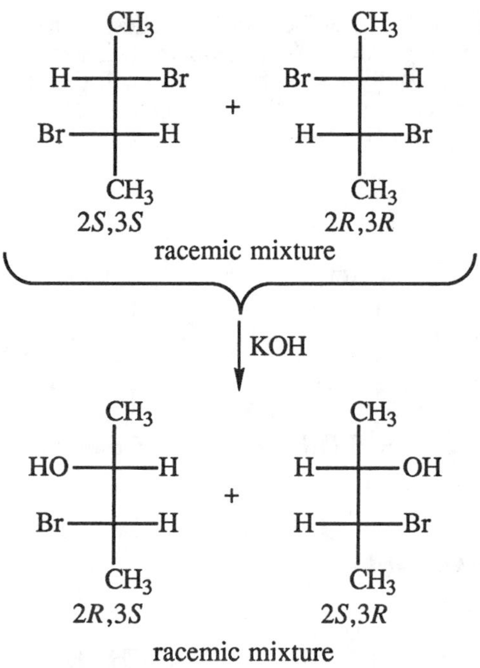

Regardless of which bromine is substituted on each molecule, the same mixture of products results.

Each of the substitution products has one chiral center inverted from the starting material. The mechanism that accounts for inversion is S_N2. If an S_N1 process were occurring, the product mixture would also contain 2*R*,3*R* and 2*S*,3*S* diastereomers. Their absence argues against an S_N1 process occurring here.

elimination

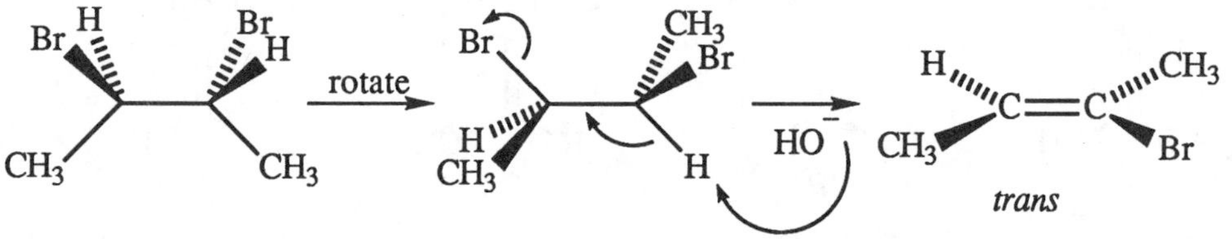

The other enantiomer gives the same product (you should prove this to yourself).

The absence of *cis* product is evidence that only the E2 elimination is occurring, not E1.

6-36

(a) 1.00 g / 20.0 mL = 0.050 g/mL ; 20.0 cm = 2.00 dm

$$[\alpha]_D^{25} = \frac{-1.25°}{(0.0500)\ (2.00)} = -12.5°$$

(b) 0.050 g / 2.0 mL = 0.025 g/mL ; 2.0 cm = 0.20 dm

$$[\alpha]_D^{25} = \frac{+0.043°}{(0.025)\ (0.20)} = +8.6°$$

6-37

The 40% of the mixture that is (–)-tartaric acid will cancel the optical rotation of the 40% of the mixture that is (+)-tartaric acid, leaving only 20% of the mixture as excess (+)-tartaric acid to give measurable optical rotation. The specific rotation will therefore be only 20% of the rotation of pure (+)-tartaric acid: $+12.0° \times 20\% = +2.4°$

6-38 all structures in this problem are chiral

(a)

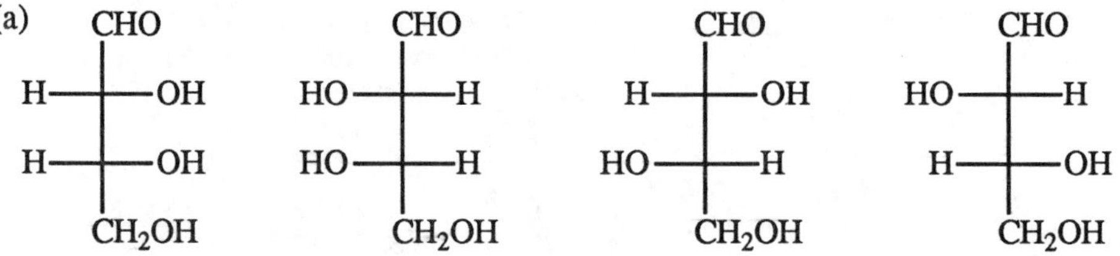

E F G H

enantiomers: E and F, G and H
diastereomers: E and G, E and H, F and G, F and H

(b)

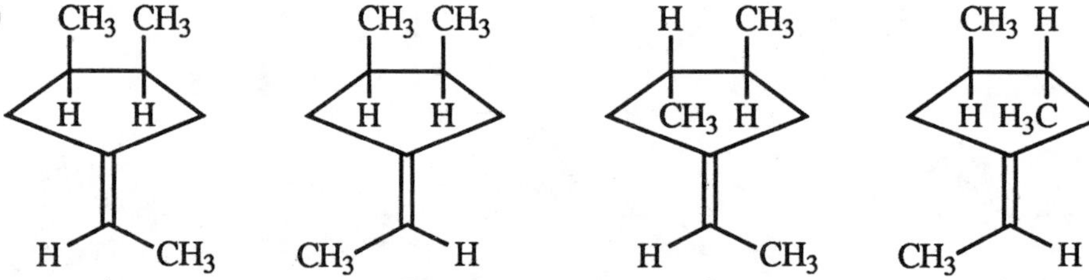

A B C D

enantiomers: A and B, C and D
diastereomers: A and C, A and D, B and C, B and D

6-39

(a)

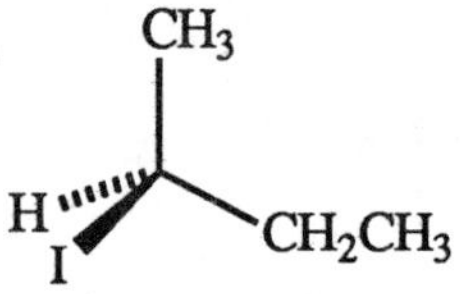

(b) Rotation of the enantiomer will be equal in magnitude, opposite in sign: – 15.90° .

(c) The rotation – 7.95° is what percent of – 15.90° ?

$$\frac{-7.95°}{-15.90°} \times 100\% = 50\% \text{ e.e.}$$

There is 50% excess of (*R*)-2-iodobutane over the racemic mixture; that is, another 25% must be *R* and 25% must be *S*. The total composition is 75% (*R*)-(–)-2-iodobutane and 25% (*S*)-(+)-2-iodobutane.

6-40

(a) $\dfrac{+15.58°}{+15.90°}$ x 100% = 98% of original optical activity = 98% e.e.

Thus, 98% of the *S* enantiomer and 2% racemic mixture gives an overall composition of 99% *S* and 1% *R*.

(b) The 1% of radioactive iodide has produced exactly 1% of the *R* enantiomer. Each substitution must occur with inversion, a classic S_N2 mechanism.

6-41

(a) An S_N2 mechanism with inversion will convert *R* to its enantiomer, *S*. An accumulation of excess *S* does not occur because it can also react with bromide, regenerating *R*. The system approaches a racemic mixture at equilibrium.

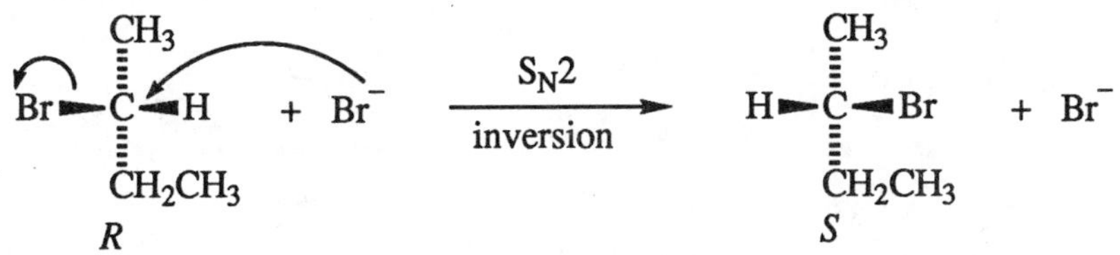

(b) In order to undergo substitution and therefore inversion, HO^- would have to be the leaving group, but HO^- is never a leaving group in S_N2. No reaction can occur.

(c) Once the —OH is protonated, it can leave as H_2O. Racemization occurs in the S_N1 mechanism because of the planar, achiral carbocation intermediate which "erases" all stereochemistry of the starting material. Racemization occurs in the S_N2 mechanism by establishing an equilibrium of *R* and *S* enantiomers, as explained in 6-41(a).

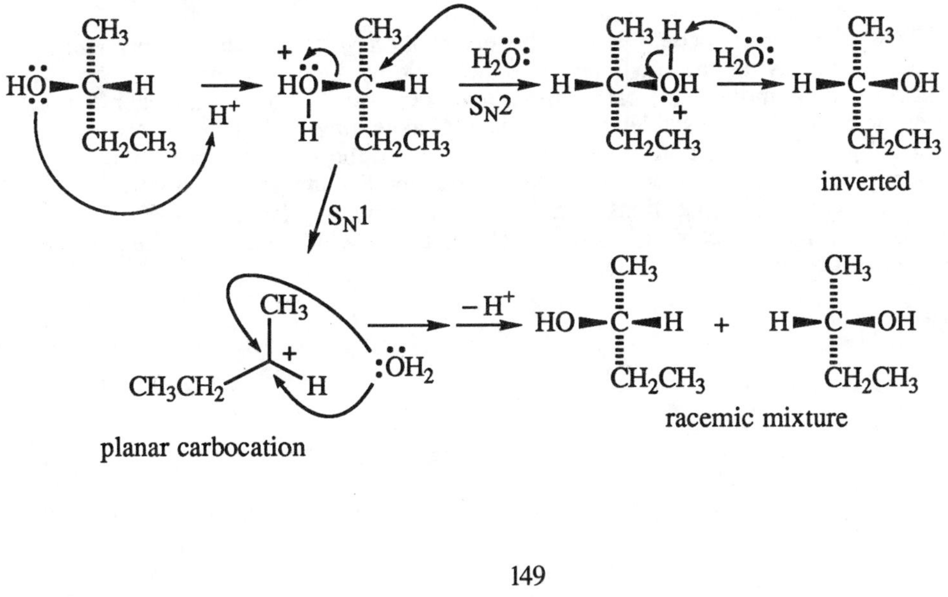

6-42 "Ph" = phenyl ; "OTs" = tosylate

(a)

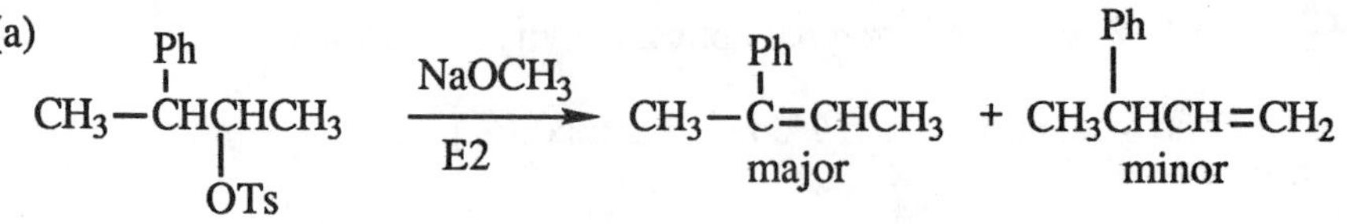
Ph
CH3—CHCHCH3
OTs
NaOCH3
E2
Ph
CH3—C=CHCH3
major
+
Ph
CH3CHCH=CH2
minor

(b) H and OTs must be anti-coplanar in the transition state

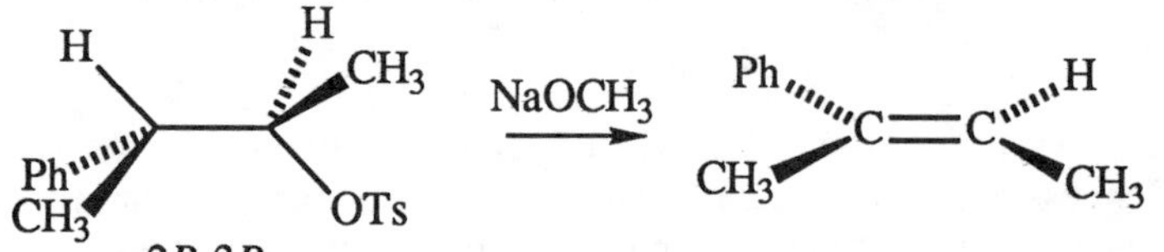
H
H
CH3
Ph
CH3
OTs
NaOCH3
Ph
H
C=C
CH3
CH3

2*R*,3*R*

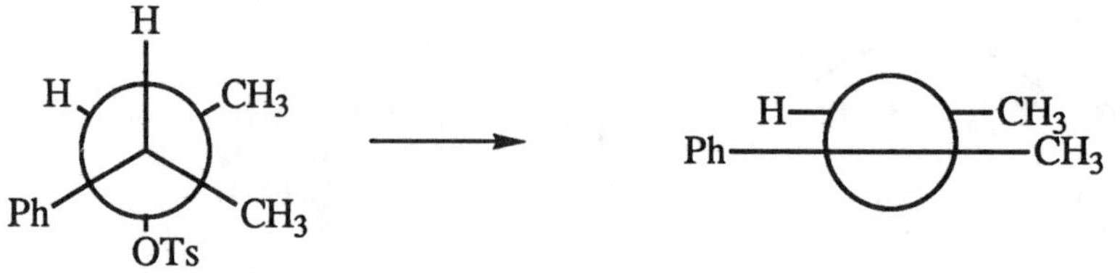
H
H
CH3
Ph
CH3
OTs
H
Ph
CH3
CH3

(c)

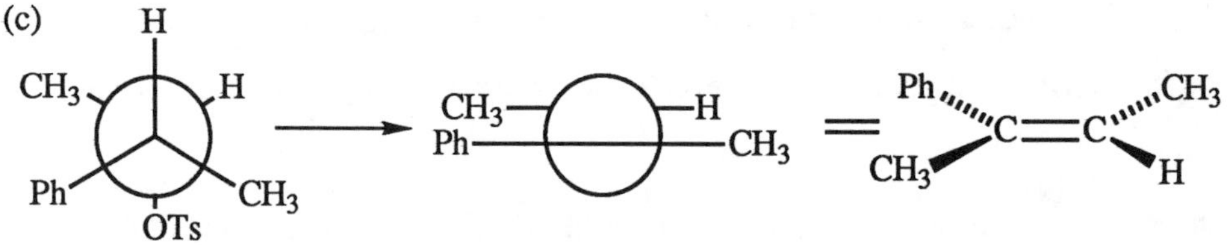
H
CH3
H
Ph
CH3
OTs
CH3
Ph
H
CH3
Ph
CH3
C=C
CH3
H

2*S*,3*R*

(d) The 2*S*,3*S* is the mirror image of 2*R*,3*R*; it would give the mirror image of the alkene that 2*R*,3*R* produced (with two methyl groups *cis*). The alkene product is planar, not chiral, so its mirror image is the same: the 2*S*,3*S* and the 2*R*,3*R* give the same alkene.

6-43 Two related factors could explain this observation. First, as carbocation stability increases, the leaving group will be less tightly held by the carbocation for stabilization; the more stable carbocations are more "free" in solution, meaning more exposed. Second, more stable carbocations will have longer lifetimes, allowing the leaving group to drift off in the solvent, leading to more possibility for the incoming nucleophile to attack from the side that the leaving group just left.

The less stable carbocations hold tightly to their leaving groups, preventing nucleophiles from attacking this side. Backside attack with inversion is the preferred stereochemical route in this case.

6-44 Let us assume that Kenyon and Phillips started with optically pure 1-phenyl-2-propanol. Under treatment with potassium and bromoethane, no bonds to the chiral center are broken, so the optical purity is retained. We can say with confidence that the specific rotation of optically pure 2-ethoxy-1-phenylpropane is + 23.5° .

Formation of the tosylate breaks no bonds to the chiral center, but subsequent replacement of the tosylate by ethoxy does. If this substitution occurred with 100% inversion, the optical rotation of the product would be – 23.5°, equal in magnitude but opposite in sign from the above. The actual value, – 19.9°, has the correct sign showing the inversion, but not all molecules went by inversion since the magnitude is not 23.5°. Some of the nucleophile must have attacked from the same side that the tosylate just left.

How can we explain this? An S_N1 reaction in which some carbocation is generated would give some retention of configuration as part of the racemization. Is this chemically reasonable? Yes. We have seen many times that a 2° carbocation, when heated in a weakly nucleophilic solvent, will give S_N1.

Furthermore, we can calculate how much S_N2 versus S_N1 occurred in this reaction.

$$\frac{-19.9^\circ}{-23.5^\circ} \times 100\% = 85\% \text{ of the highest possible optical purity}$$

Thus, 85% of the molecules must have undergone inversion (S_N2) while 15% underwent racemization (S_N1).

6-45 The symmetry of this molecule is crucial.

(a)

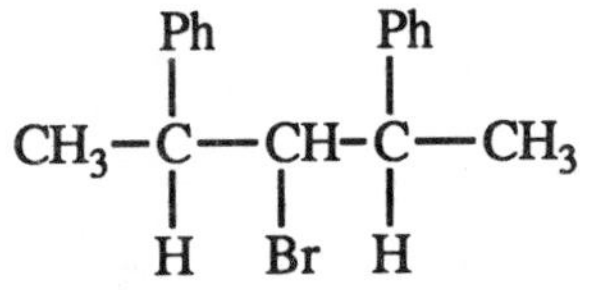

Regardless of which adjacent H is removed by *t*-butoxide, the product will be 2,4-diphenyl-2-pentene.

(b) Here are a three-dimensional representation and a Fischer projection of the required diastereomer. On both carbons 2 and 4, the H has to be anti-coplanar with the bromine while leaving the other groups to give the same product. Not coincidentally, the correct diastereomer is a meso structure.

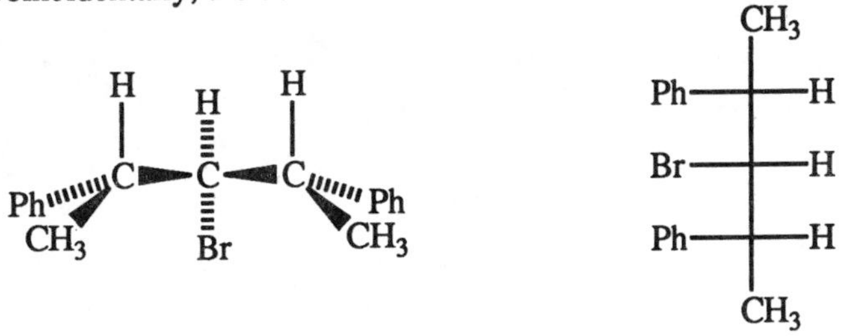

6-46

(a)

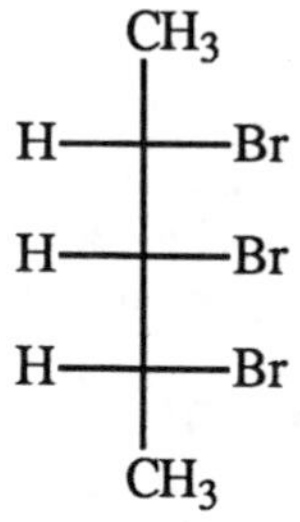

A
meso

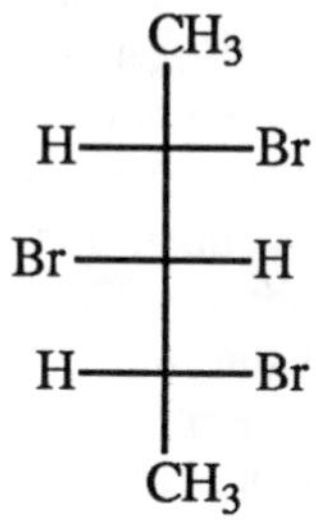

B
meso

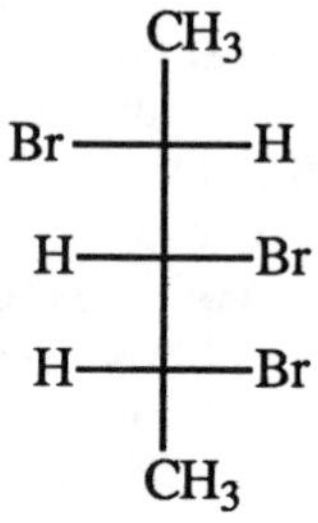

C
chiral

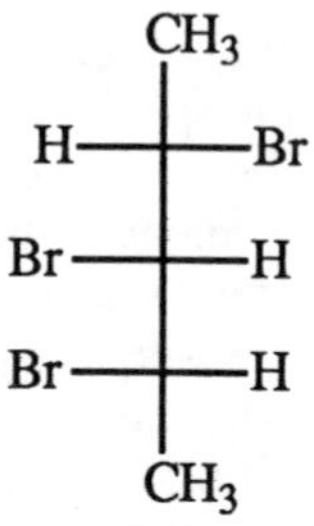

D
chiral

(b) A: 2*S*,4*R* equivalent to 2*R*,4*S*; B: 2*S*,4*R* equivalent to 2*R*,4*S*; C: 2*R*,4*R*; D: 2*S*,4*S*

(c) This is one of the challenging types of problems in stereochemistry. Each C-3 in structures A and B is chiral, because the substituent groups are different (*R* versus *S*), yet structures A and B have planes of symmetry passing through C-3. In structures C and D, C-3 is not chiral because both substituent groups are either *R* or *S*, yet the molecules are chiral. Chemists classify these carbons as "pseudochiral" because they contradict our notions of chiral carbons and chiral molecules.

CHAPTER 7—ALKENES: STRUCTURE AND SYNTHESIS

7-1 The number of elements of unsaturation in a hydrocarbon formula is given by:

$$\frac{2(\#C) + 2 - (\#H)}{2}$$

(a) $C_6H_{12} \Rightarrow \dfrac{2(6) + 2 - (12)}{2}$ = 1 element of unsaturation

(b) Many examples are possible. Yours may not match these, but all must have either a double bond or a ring, that is, one element of unsaturation.

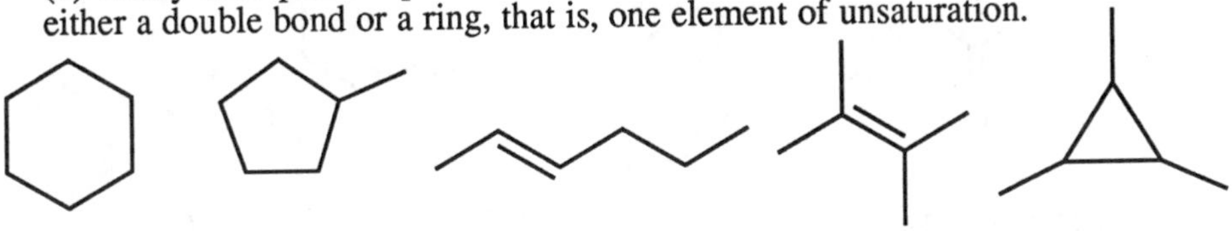

7-2 $C_4H_6 \Rightarrow \dfrac{2(4) + 2 - (6)}{2}$ = 2 elements of unsaturation

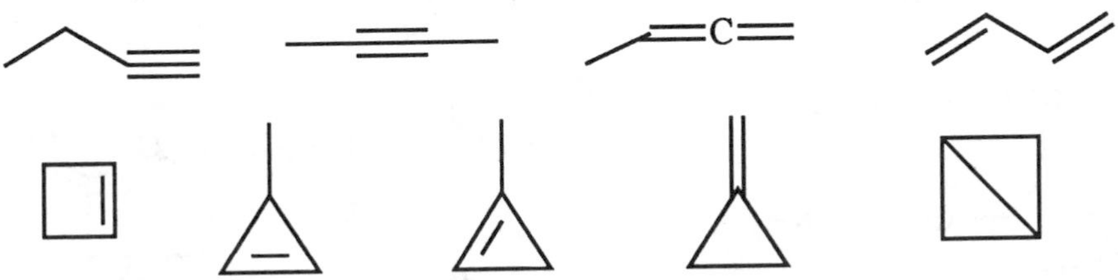

7-3 Many examples of C_4H_6NOCl are possible. Yours may not match these, but all must contain two elements of unsaturation.

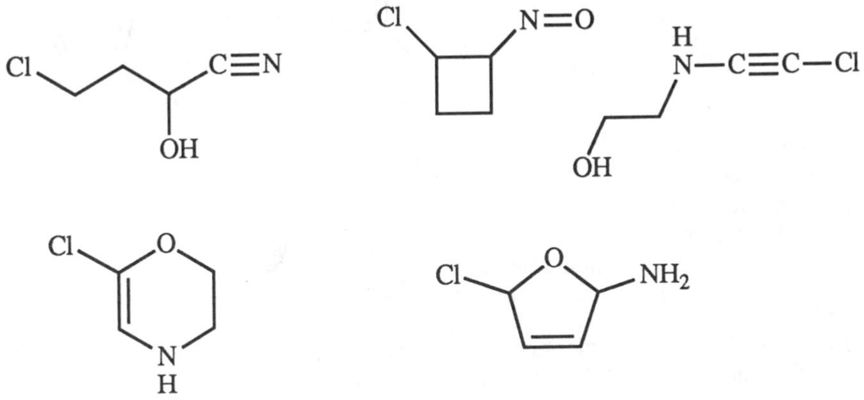

7-4 Many examples of these formulas are possible. Yours may not match these, but they must have the correct number of elements of unsaturation.

(a) $C_3H_4Cl_2 \Rightarrow C_3H_6$ = **1**

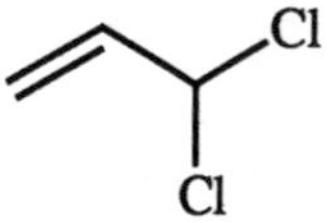
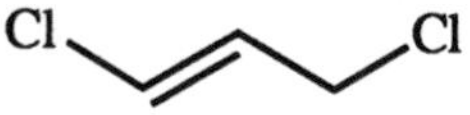

(b) $C_4H_8O \Rightarrow C_4H_8$ = **1**

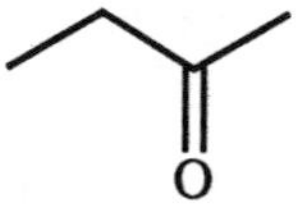

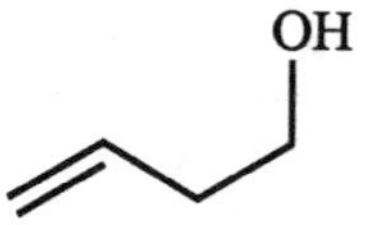

(c) $C_4H_4O_2 \Rightarrow C_4H_4$ = **3**

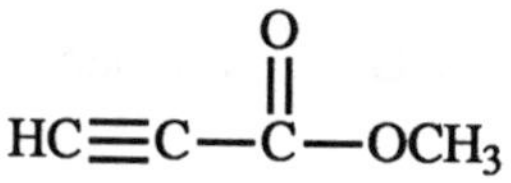
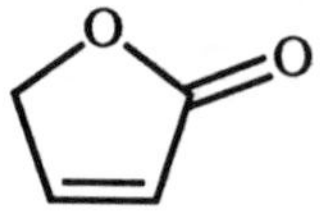
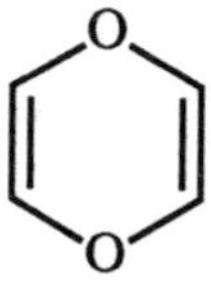

(d) $C_5H_5NO_2 \Rightarrow C_{5.5}H_5$ = **4**

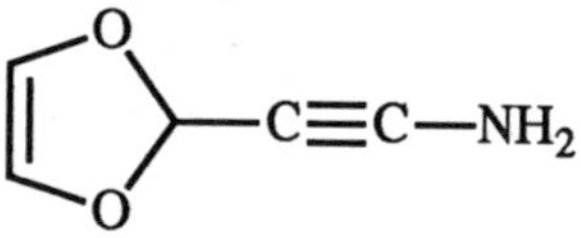
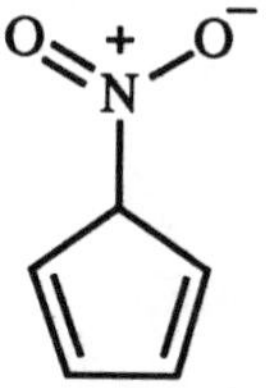
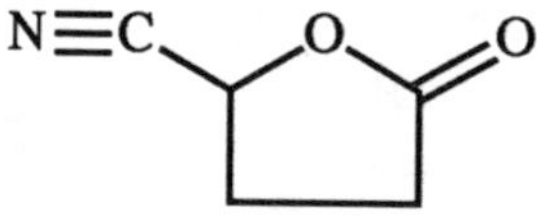

(e) $C_6H_3NClBr \Rightarrow C_{6.5}H_5$ = **5**

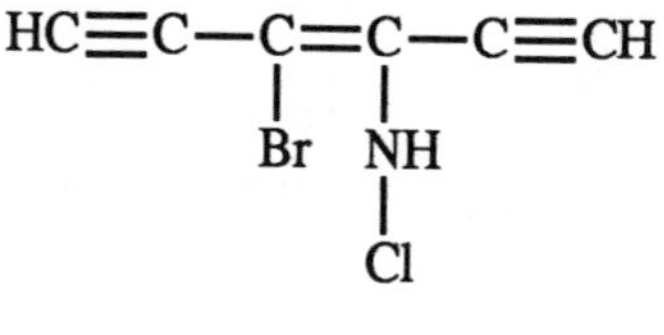
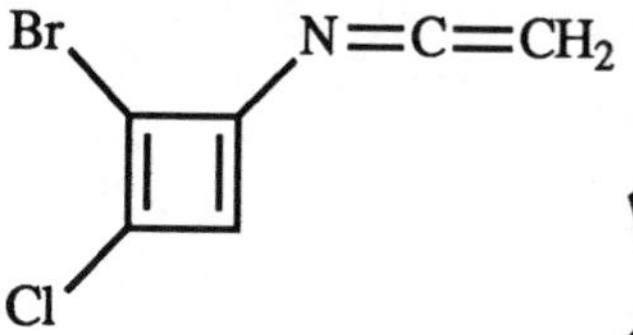
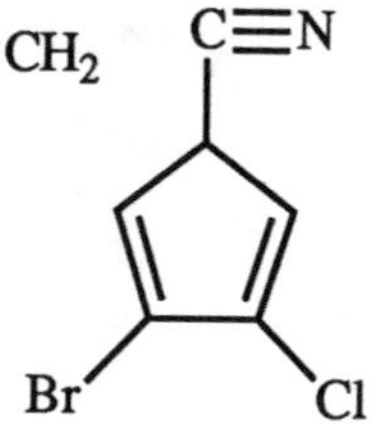

7-5 (a) 4-methyl-1-pentene
(b) 2-ethyl-1-hexene
(c) 1,4-pentadiene
(d) 1,2,4-pentatriene
(e) 2,5-dimethyl-1,3-cyclopentadiene
(f) 4-vinyl-1-cyclohexene ("1" is optional)
(g) 3-phenyl-1-propene ("1" is optional)
(h) *trans*-3,4-dimethyl-1-cyclopentene ("1" is optional)
(i) 7-methylene-1,3,5-cycloheptatriene

7-6 (b), (e), and (f) do not show *cis,trans* isomerism

(a)

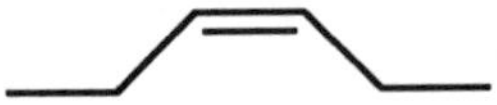

(Z)-3-hexene
(*cis*-3-hexene)

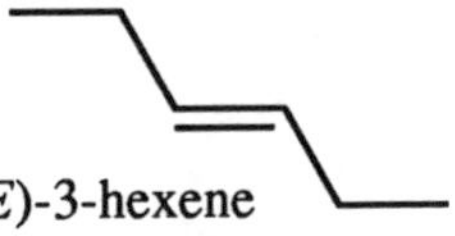

(*E*)-3-hexene
(*trans*-3-hexene)

(c)

(2Z,4Z)-2,4-hexadiene

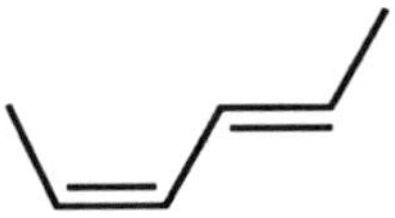

(2Z,4*E*)-2,4-hexadiene

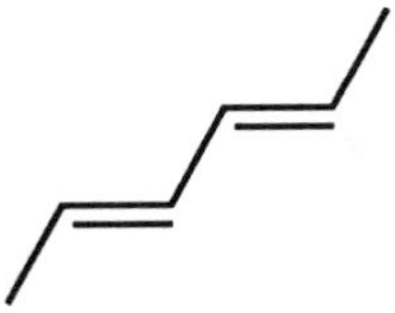

(2*E*,4*E*)-2,4-hexadiene

(d)

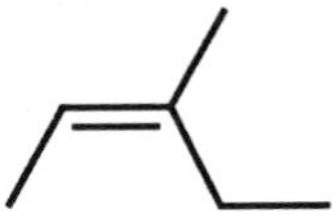

(Z)-3-methyl-2-pentene

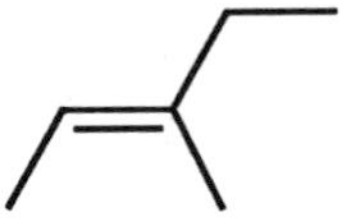

(*E*)-3-methyl-2-pentene

"*Cis*" and "*trans*" are not clear for this example; "*E*" and "Z" are unambiguous.

7-7

(a)

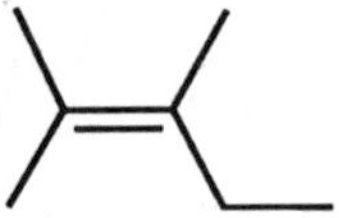

2,3-dimethyl-2-pentene
(neither *cis* nor *trans*)

(b)

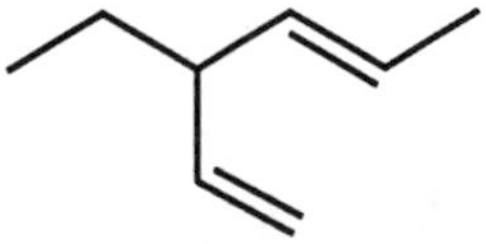

3-ethyl-1,4-hexadiene
(*cis* or *trans* not specified; the vinyl group is part of the main chain)

(c)

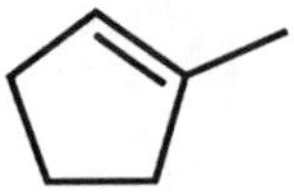

1-methylcyclopentene

(d)

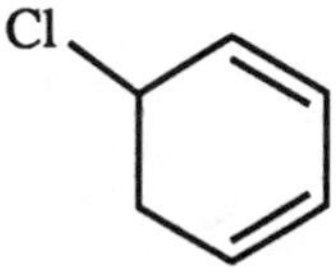

5-chloro-1,3-cyclohexadiene
(positions of double bonds need to be specified)

(e)

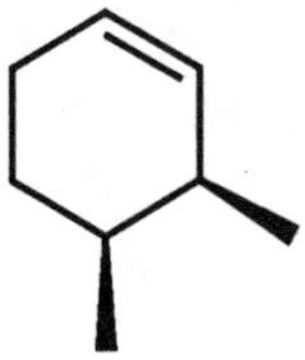

cis-3,4-dimethylcyclohexene
(could also have drawn *trans*)

(f)

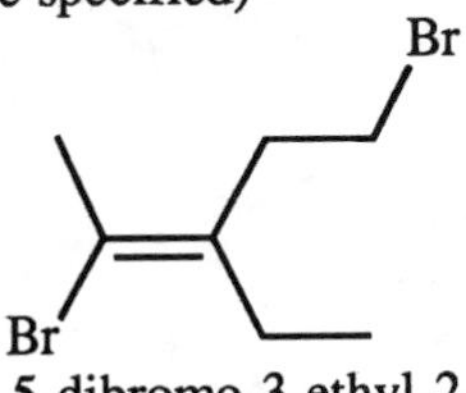

(*E*)-2,5-dibromo-3-ethyl-2-pentene
(*cis* does not apply)

7-8

(a)

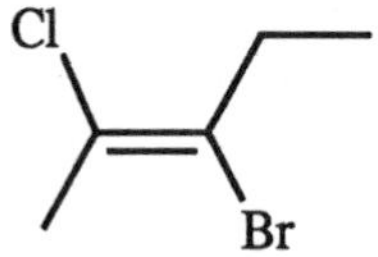

(*E*)-3-bromo-2-chloro-2-pentene

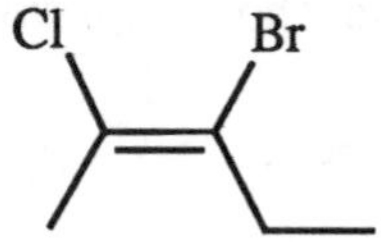

(*Z*)-3-bromo-2-chloro-2-pentene

(b)

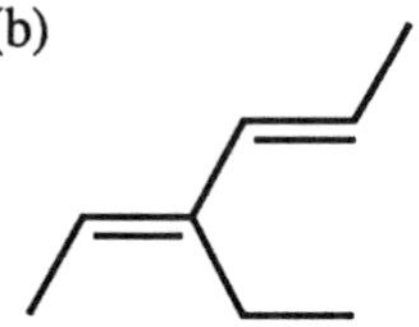

(2*E*,4*E*)-3-ethyl-2,4-hexadiene

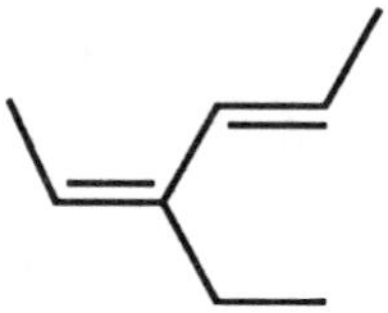

(2*Z*,4*E*)-

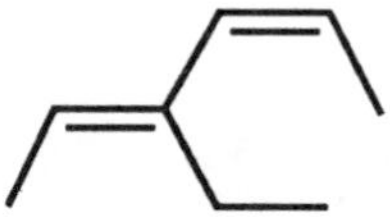

(2*E*,4*Z*)-

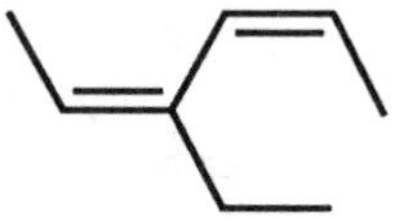

(2*Z*,4*Z*)-

(c) no geometric isomers

(d)

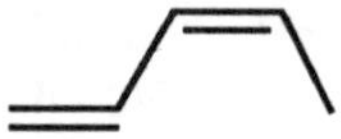

(*Z*)-1,3-pentadiene

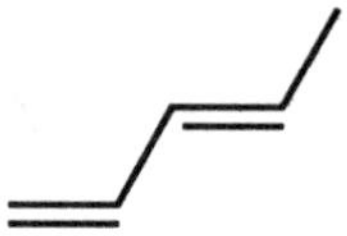

(*E*)-1,3-pentadiene

(e)

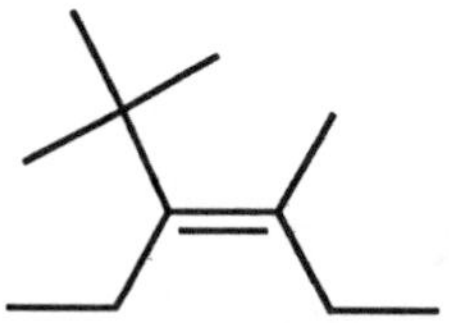

(*E*)-3-*t*-butyl-4-methyl-3-hexene
(more correct name: (*E*)-3-ethyl-2,2,4-trimethyl-3-hexene)

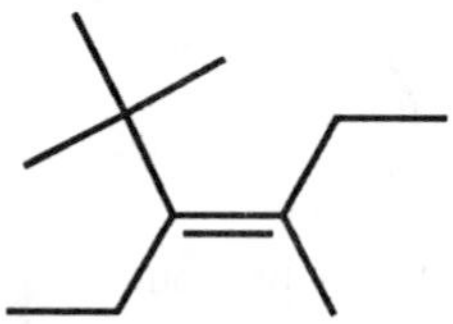

(*Z*)-3-*t*-butyl-4-methyl-3-hexene

(f)

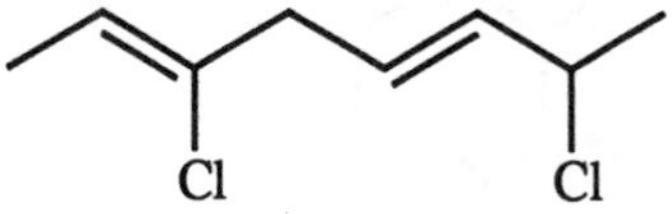

(2*Z*,5*E*)-3,7-dichloro-2,5-octadiene

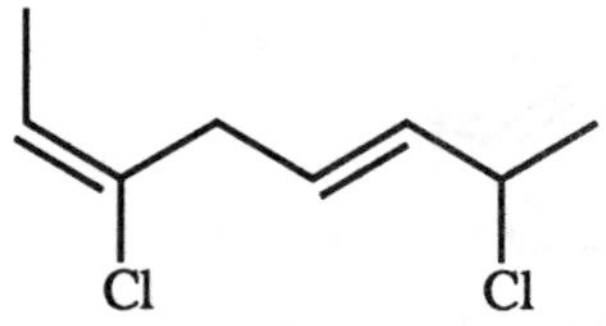

(2*E*,5*E*)-3,7-dichloro-2,5-octadiene

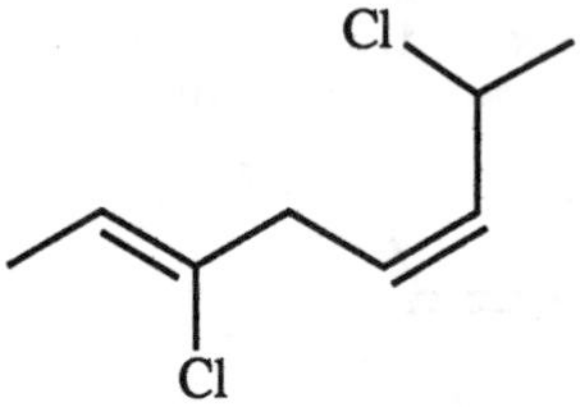

(2*Z*,5*Z*)-3,7-dichloro-2,5-octadiene

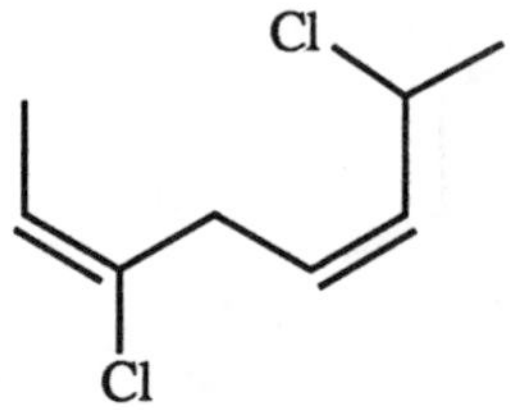

(2*E*,5*Z*)-3,7-dichloro-2,5-octadiene

7-8 continued

(g) no geometric isomers (an *E* double bond would be too highly strained)

(h)

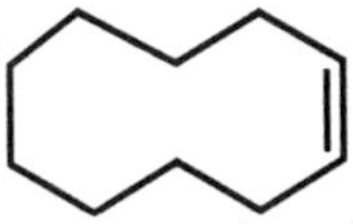

(*Z*)-cyclodecene

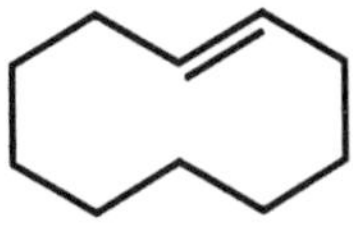

(*E*)-cyclodecene

(i)

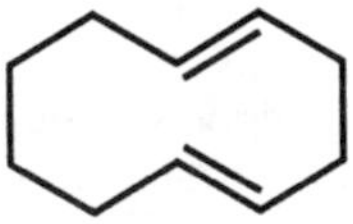

(1*E*,5*E*)-1,5-cyclodecadiene

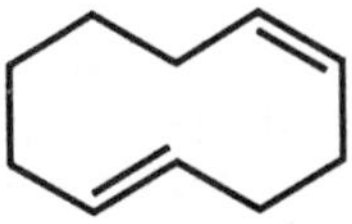

(1*Z*,5*E*)-1,5-cyclodecadiene

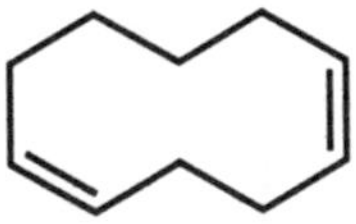

(1*Z*,5*Z*)-1,5-cyclodecadiene

7-9 From Table 7-1, approximate heats of hydrogenation can be determined for similarly substituted alkenes. The energy difference is approximately 1.4 kcal/mole (6 kJ/mole), the more highly substituted alkene being more stable.

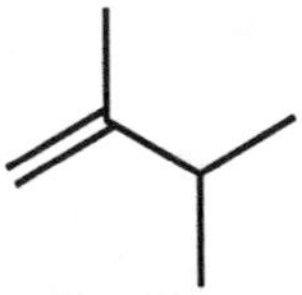

gem-disubstituted
28.0 kcal/mole
(117 kJ/mole)

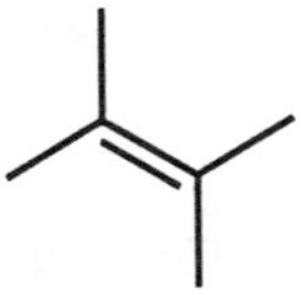

tetrasubstituted
26.6 kcal/mole
(111 kJ/mole)

7-10 Use the relative values in Figure 7-6.

(a) 2 x (*trans*-disubstituted − *cis*-disubstituted) =
2 x (5.2 − 4.2) = 2 kcal/mole more stable for *trans,trans*
[2 x (22 − 18) = 8 kJ/mole]

(b) *gem*-disubstituted − monosubstituted = 4.8 − 2.7 = 2.1 kcal/mole
(20 − 11 = 9 kJ/mole)

2-methyl-1-butene is more stable

(c) trisubstituted − *gem*-disubstituted = 5.9 − 4.8 = 1.1 kcal/mole
(25 − 20 = 5 kJ/mole)

2-methyl-2-butene is more stable

(d) tetrasubstituted − *gem*-disubstituted = 6.2 − 4.8 = 1.4 kcal/mole
(26 − 20 = 6 kJ/mole)

2,3-dimethyl-2-butene is more stable

7-11

(a) 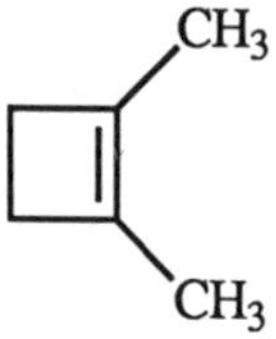
CH3
CH3

strained but stable

(b) could not exist—ring size must be 8 atoms or greater to include *trans* double bond

(c) 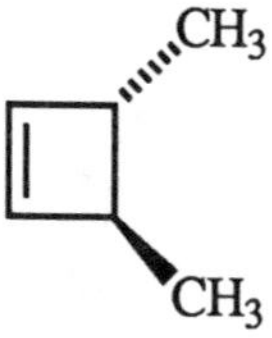
CH3
CH3

Despite the ambiguity of this name, we must assume that the *trans* refers to the two methyls since this compound can exist, rather than the *trans* referring to the alkene, a molecule which could not exist.

(d) 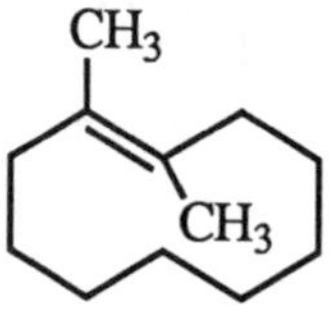
CH3
CH3

stable—*trans* in 10-membered ring

(e) unstable at room temperature—cannot have *trans* alkene in 7-membered ring (possibly isolable at very low temperature—this type of experiment is one of the challenges chemists attack with gusto)

(f) stable—alkene not at bridgehead

(g) unstable—violation of Bredt's Rule (alkene at bridgehead in 6-membered ring)

(h) stable—alkene at bridgehead in 8-membered ring

7-12 (a) The dibromo compound should boil at a higher temperature because of its much larger molecular weight.

(b) The *cis* should boil at a higher temperature than the *trans* as the *trans* has a zero dipole moment and therefore no dipole-dipole interactions.

(c) 1,2-Dichlorocyclohexene should boil at a higher temperature because of its much larger molecular weight.

7-13

(a)

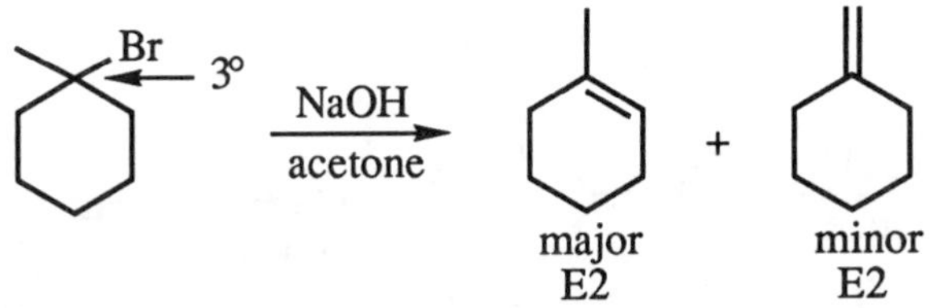

Br
3°
NaOH
acetone
+
major
E2
minor
E2

(b)

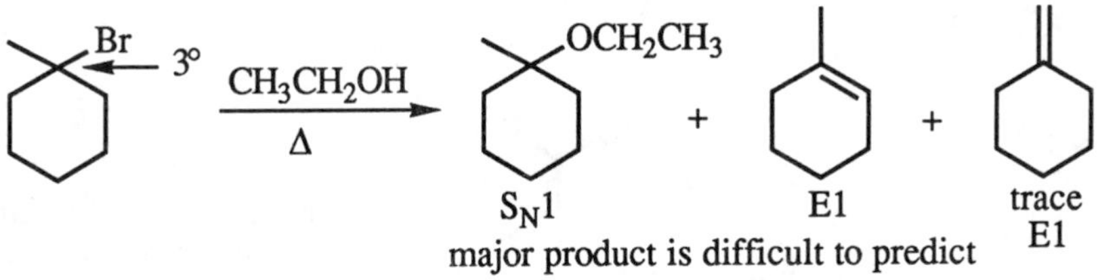

Br
3°
CH_3CH_2OH
Δ
OCH_2CH_3
+
+
S_N1
E1
trace
E1
major product is difficult to predict

(c)

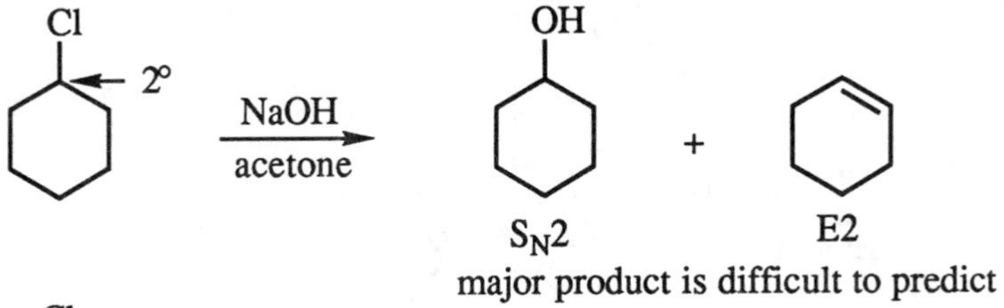

Cl
2°
NaOH
acetone
OH
+
S_N2
E2
major product is difficult to predict

(d)

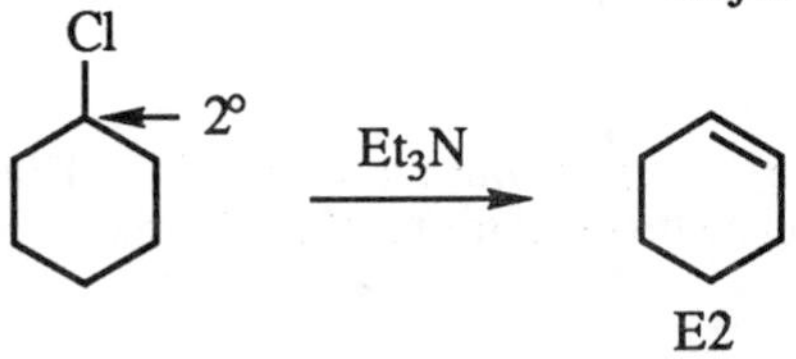

Cl
2°
Et_3N
E2

Et_3N is a bulky base and a poor nucleophile, minimizing S_N2

(e)

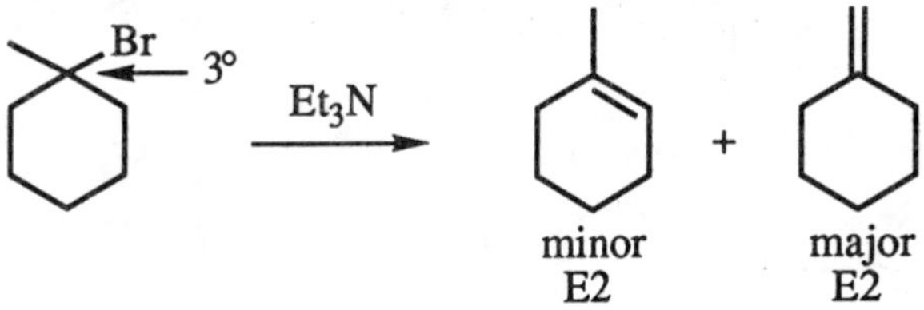

Br
3°
Et_3N
+
minor
E2
major
E2

hindered base gives Hofmann product as major isomer

(f)

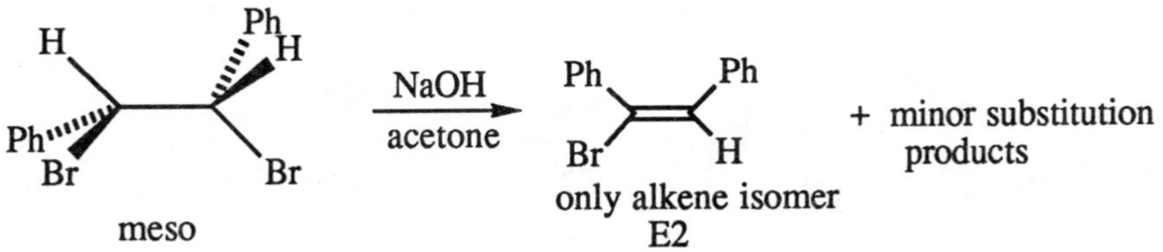

H
Ph
H
Ph
Br
Br
meso
NaOH
acetone
Ph
Ph
Br
H
only alkene isomer
E2
+ minor substitution products

(g)

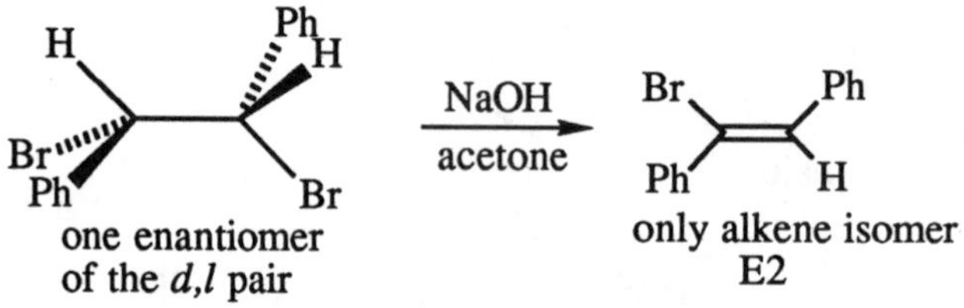

H
Ph
H
Br
Ph
Br
one enantiomer
of the *d,l* pair
NaOH
acetone
Br
Ph
Ph
H
only alkene isomer
E2

+ minor substitution products

(h)

H
Cl
NaOH
acetone
HO
H

S_N2 is the only mechanism possible; elimination would violate Bredt's Rule

(i)

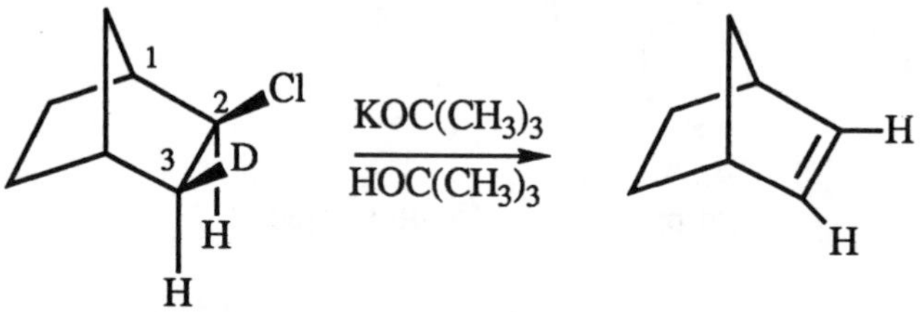

1
2
3
Cl
D
H
H
KOC(CH3)3
HOC(CH3)3
H
H

Models show that the H on C-3 cannot be anti-coplanar with the Cl on C-2. Thus, this E2 elimination must occur with a *syn*-coplanar orientation. (See Problem 5-62.)

7-14

(a)

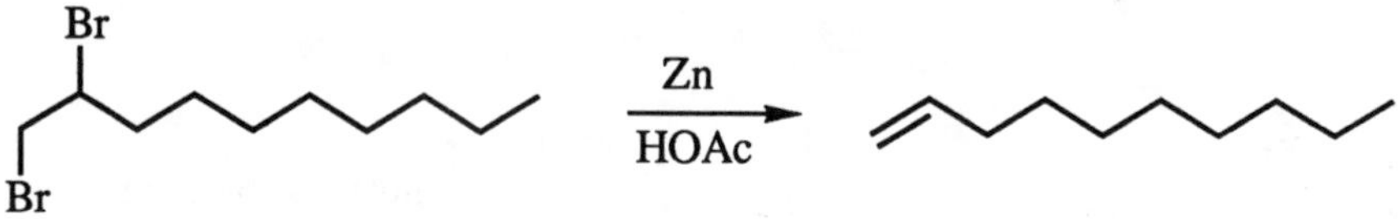

Br
Br
Zn
HOAc

(b)

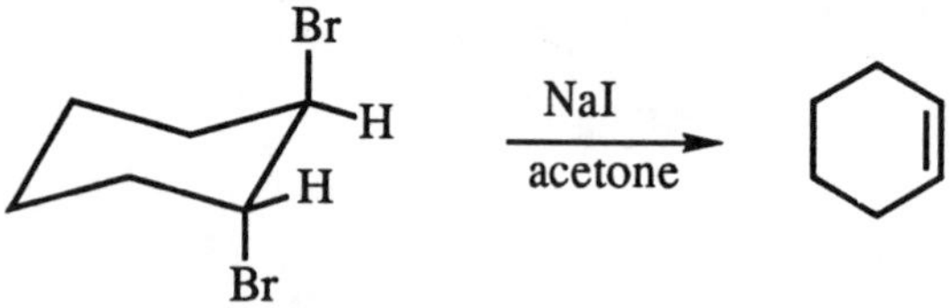

Br
H
H
Br
NaI
acetone

(c)

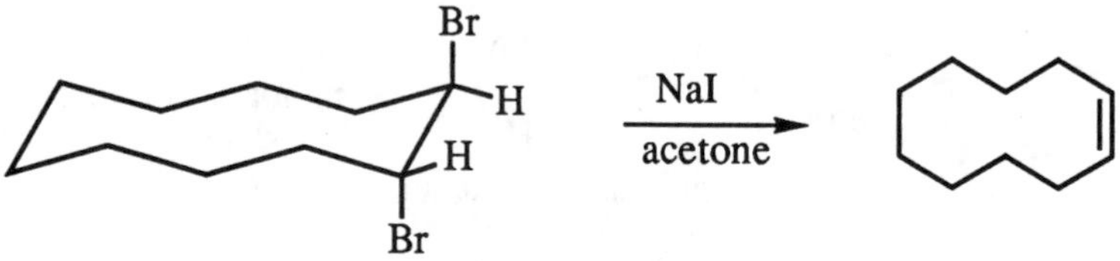

Br
H
H
Br
NaI
acetone

7-14 continued

(d)

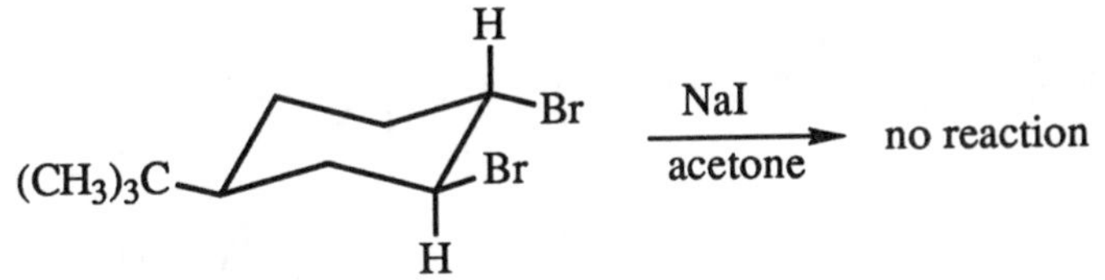

The correct answer is "no reaction" because the two bromines are not coplanar (neither the preferred anti nor the possible syn orientations) and are locked in this conformation by the *t*-butyl group. Given time, however, the excellent nucleophile iodide can displace bromide by backside attack (S_N2), eventually generating anti-coplanar *trans*-diiodo, which could then eliminate.

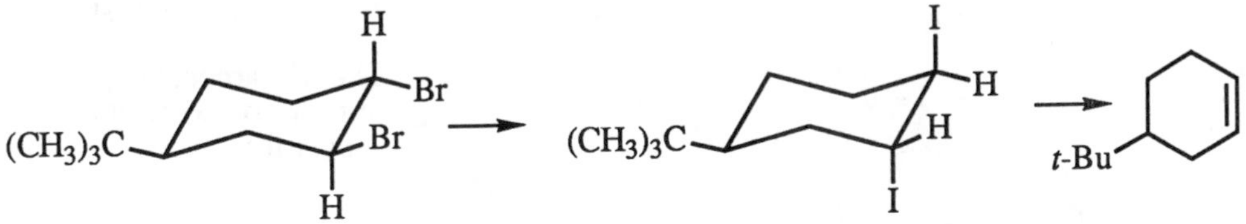

(e)

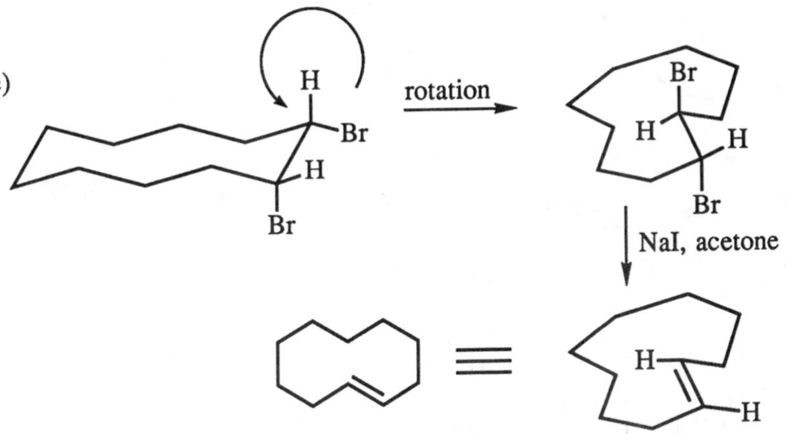

In the first conformation shown, the Br's are not coplanar and cannot eliminate. Rotation in this large ring can place the two bromines anti-coplanar, generating a *trans* alkene in the ring. (Use models!)

7-15

(a) a basic and nucleophilic mechanism: $Ba(OH)_2$ is a strong base

(b) an acidic and electrophilic mechanism: the catalyst is H^+

(c) a free radical chain reaction: the catalyst is a peroxide which initiates free radical reactions

(d) an acidic and electrophilic mechanism: the catalyst BF_3 is a strong Lewis acid

7-16

(a)

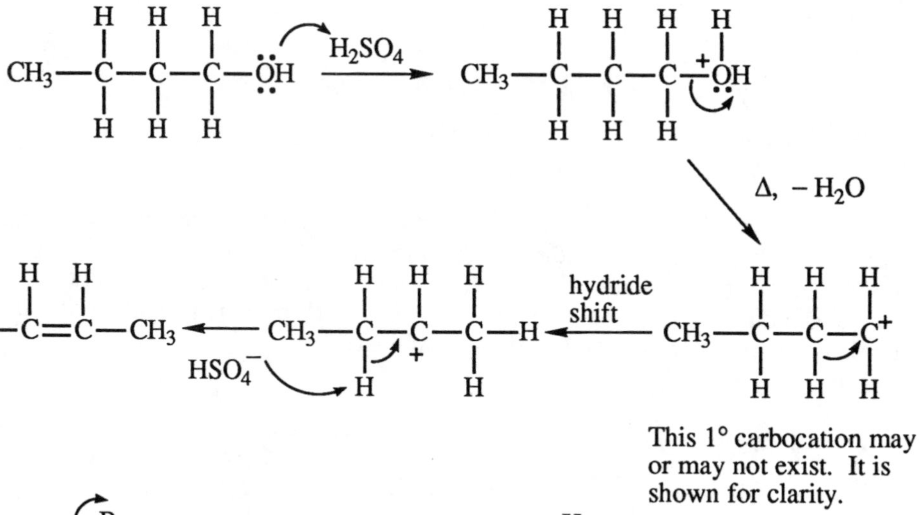

H₂SO₄
Δ, − H₂O
hydride shift
This 1° carbocation may or may not exist. It is shown for clarity.

(b)

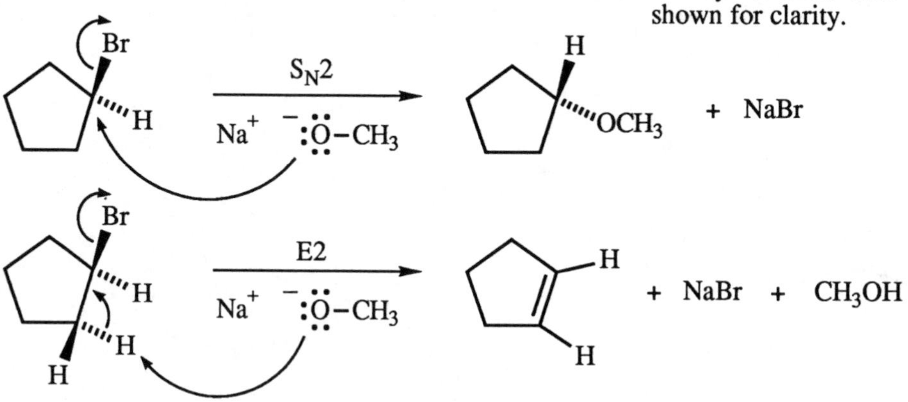

S$_N$2
E2
+ NaBr
+ NaBr + CH₃OH

7-17

(a)

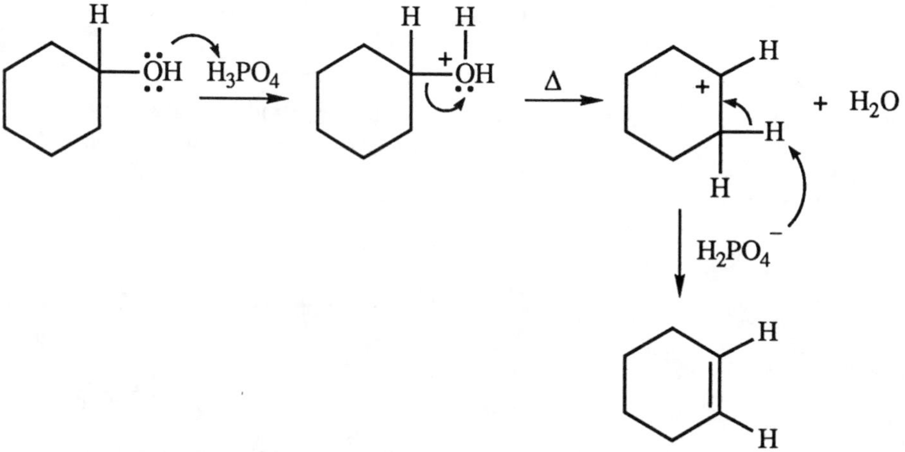

H₃PO₄
Δ
+ H₂O

7-17 continued

(b)

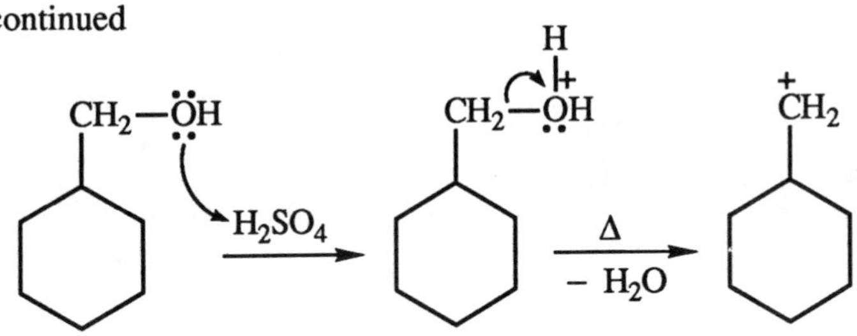

This 1° carbocation may or may not exist. It is shown for clarity.

Two possible rearrangements

1. Hydride shift

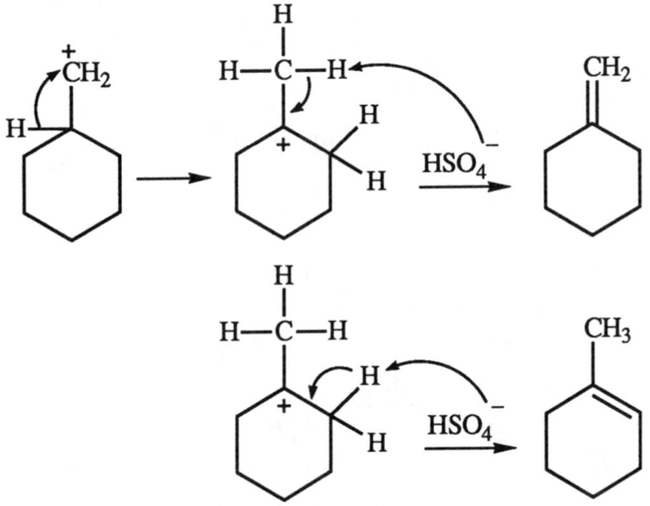

2. Alkyl shift—ring expansion

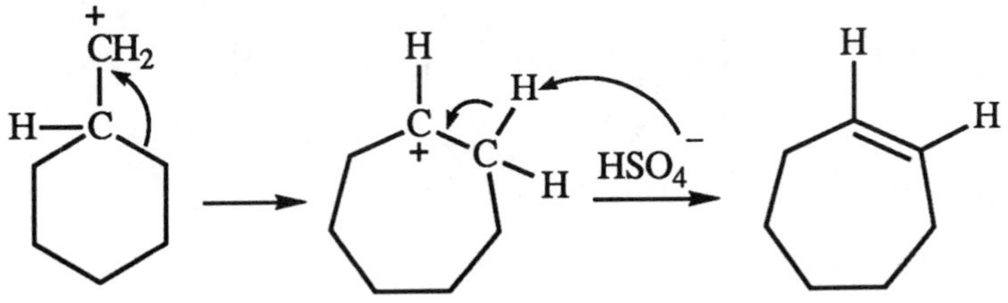

7-17 continued

(c)

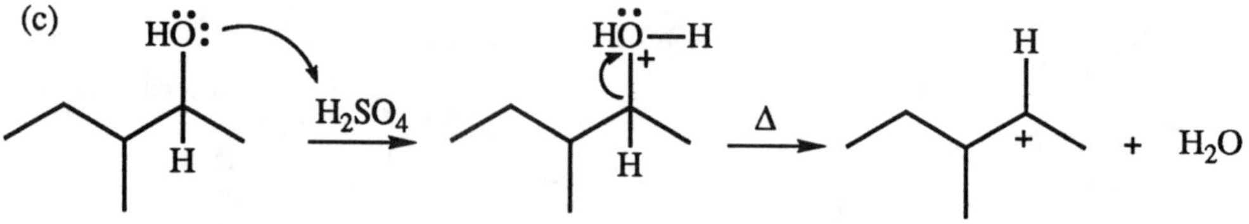

without rearrangement

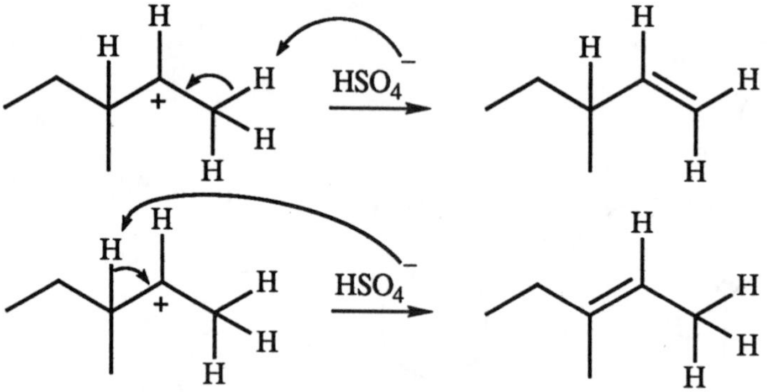

with hydride shift

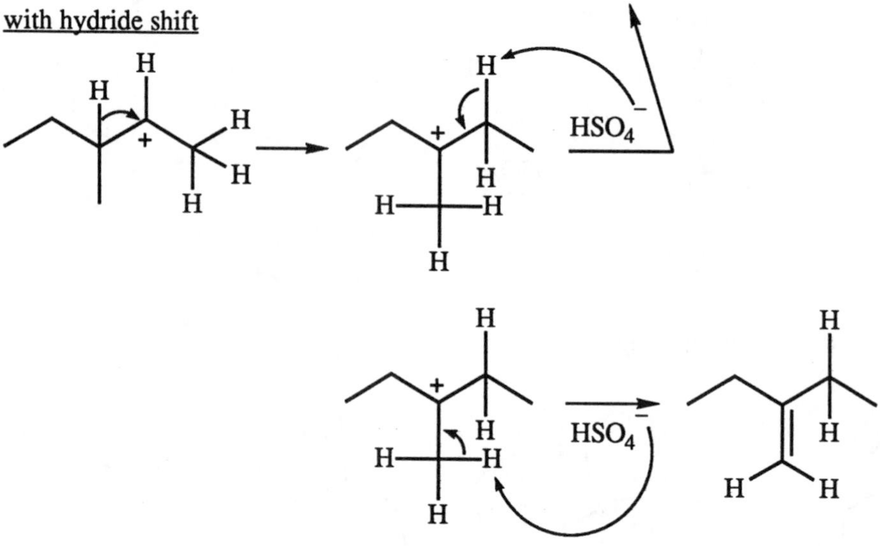

7-18 (a) $\Delta G° = \Delta H° - T\Delta S°$

$= +27{,}600 \text{ cal} - 298 \text{ K } (28.0 \text{ cal/K})$

$= +19{,}300 \text{ cal} = +19.3 \text{ kcal}$

$\Delta G°$ is positive, the reaction is **disfavored**

(b) $\Delta G_{1000} = +27{,}600 \text{ cal} - 1273 \text{ K } (28.0 \text{ cal/K})$

$= -8000 \text{ cal} = -8.0 \text{ kcal}$

ΔG is negative, the reaction is **favored**

7-19 Refer to the Glossary and the text for definitions and examples.

7-20

(a) 

(b) 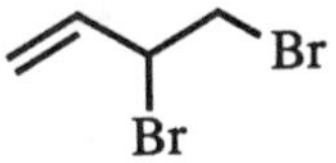

(c) 

(d) 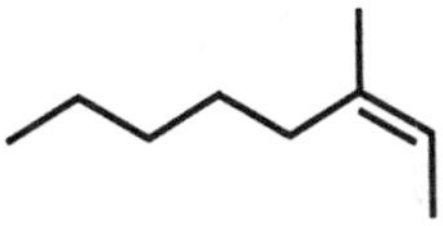

(e) 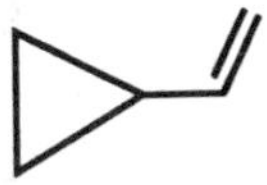

(f) 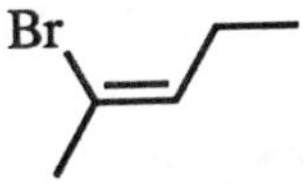

(g) 

7-21 (a) 2-ethyl-1-pentene
(b) 3-ethyl-2-pentene
(c) (3*E*,6*E*)-1,3,6-octatriene
(d) (*E*)-4-ethyl-3-heptene
(e) 1-cyclohexyl-1,3-cyclohexadiene

7-22 (a) *E* (b) neither (c) *Z* (d) *Z*

7-23

(a) 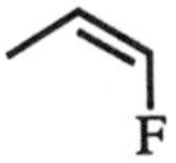

(*Z*)-1-fluoro-1-propene

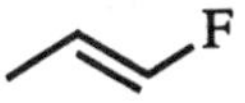

(*E*)-1-fluoro-1-propene

2-fluoro-1-propene

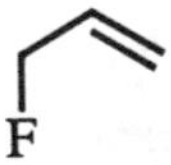

3-fluoro-1-propene

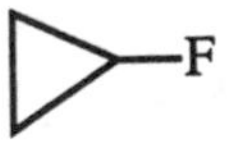

fluorocyclopropane

(b) Cholesterol, $C_{27}H_{46}O$, has five elements of unsaturation. If only one of those is a pi bond, the other four must be rings.

7-24

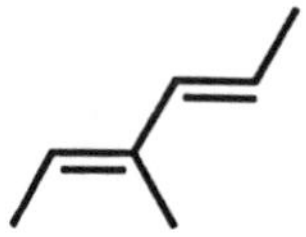 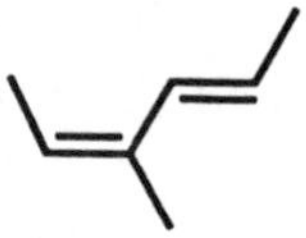 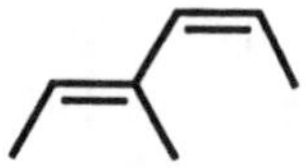 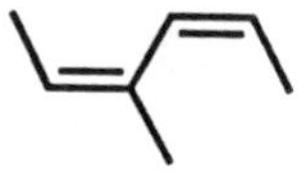

(a)	*trans,trans*	*cis,trans*	*trans,cis*	*cis,cis*
(b)	2*E*,4*E*	2Z,4*E*	2*E*,4Z	2Z,4Z

This problem is intended to show the difficulty of using *cis-trans* nomenclature with any but the simplest alkenes. The *E-Z* nomenclature is preferred for all four of these isomers.

7-25 (a), (d), and (e) have no geometric isomers

(b)

trans-2-pentene
(*E*)-2-pentene

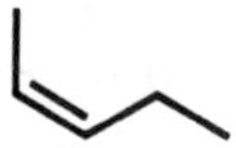

cis-2-pentene
(Z)-2-pentene

(c)

trans-3-hexene
(*E*)-3-hexene

cis-3-hexene
(Z)-3-hexene

(f)

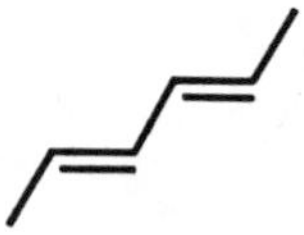

trans,trans-2,4-hexadiene
(2*E*,4*E*)-2,4-hexadiene

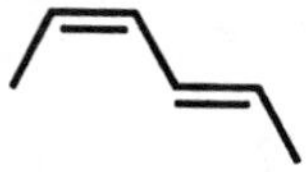

cis,trans-2,4-hexadiene
(2Z,4*E*)-2,4-hexadiene

cis,cis-2,4-hexadiene
(2Z,4Z)-2,4-hexadiene

7-26

(a)

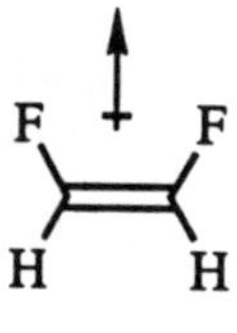

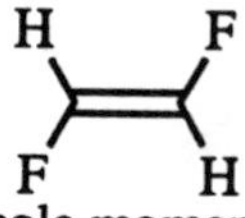

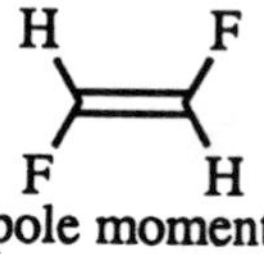

dipole moment = 0

(b)

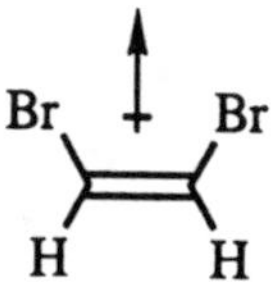

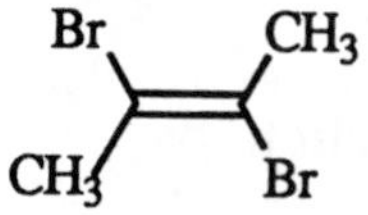

dipole moment = 0

7-26 continued

(c)

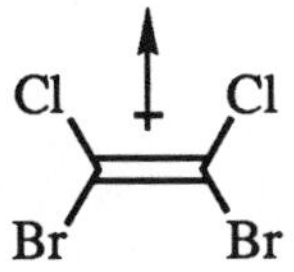

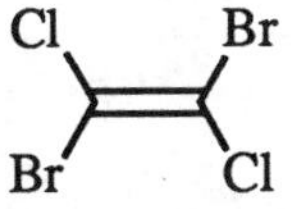

dipole moment = 0

(d)

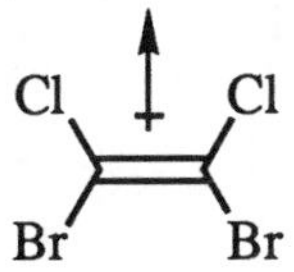

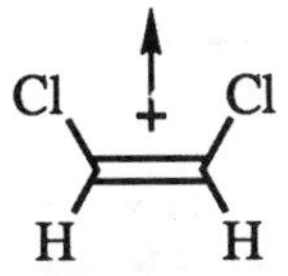

larger dipole moment
(no bromines opposing
the dipole of the chlorines)

7-27

(a)

(b)

major minor

(c)

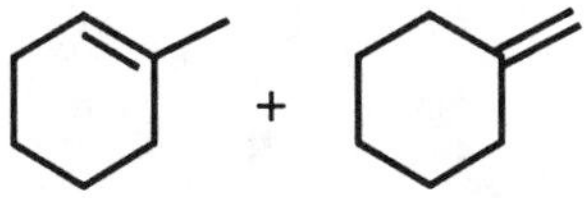

major minor

(d)

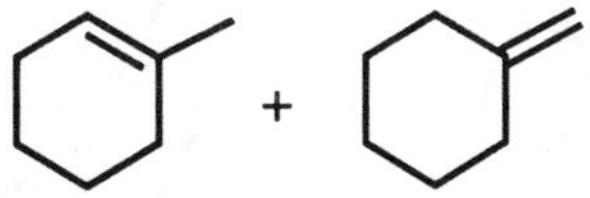

minor major

7-28

(a)

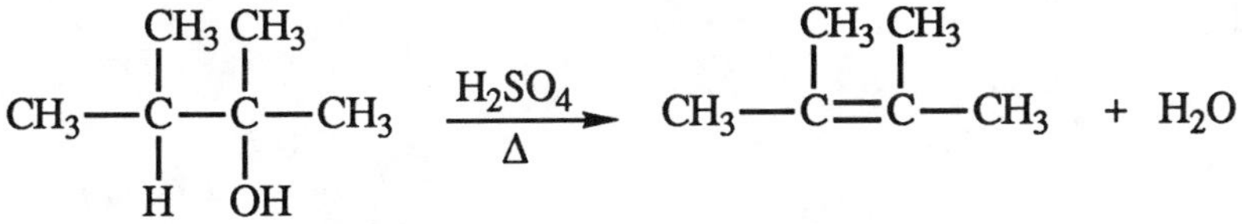
CH3 CH3
CH3—C—C—CH3
H OH
H2SO4
Δ
CH3 CH3
CH3—C=C—CH3
+ H2O

(b)

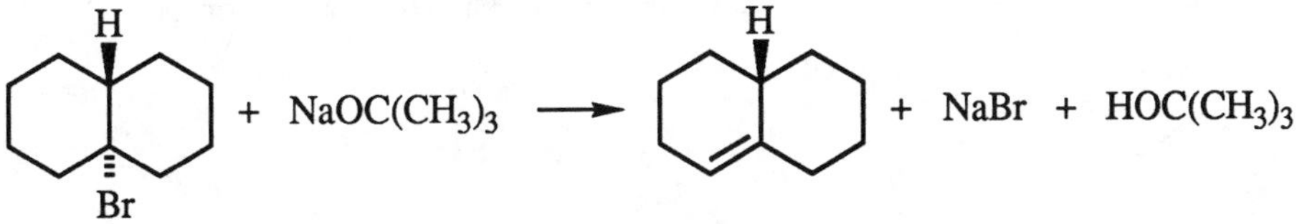
H
Br
+ NaOC(CH3)3
H
+ NaBr + HOC(CH3)3

(c)

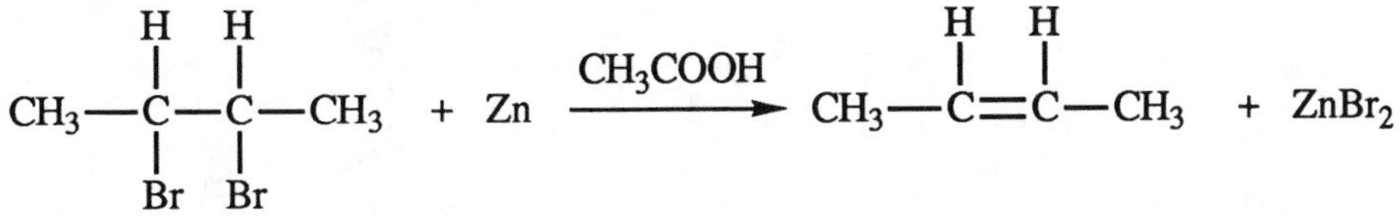
H H
CH3—C—C—CH3
Br Br
+ Zn
CH3COOH
H H
CH3—C=C—CH3
+ ZnBr2

7-29

(a)

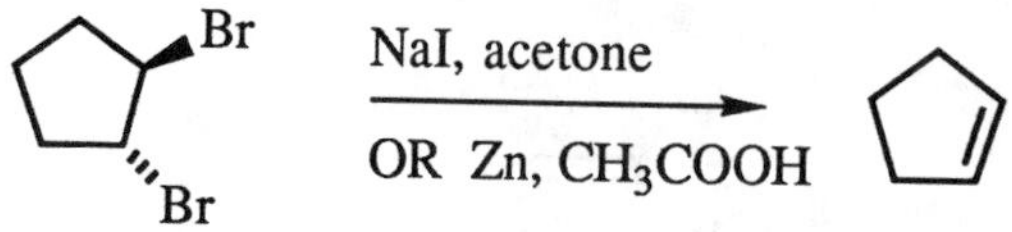
Br
Br
NaI, acetone
OR Zn, CH3COOH

(b)

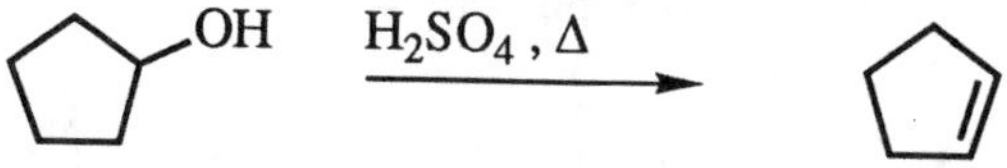
OH
H2SO4 , Δ

(c)

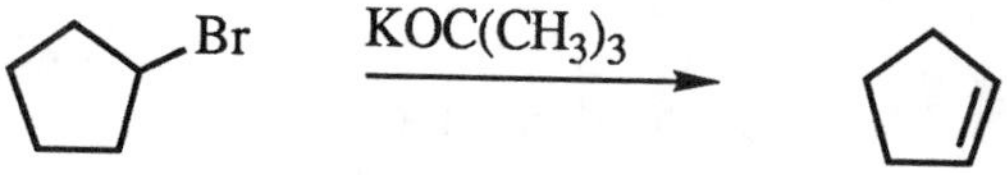
Br
KOC(CH3)3

(d)

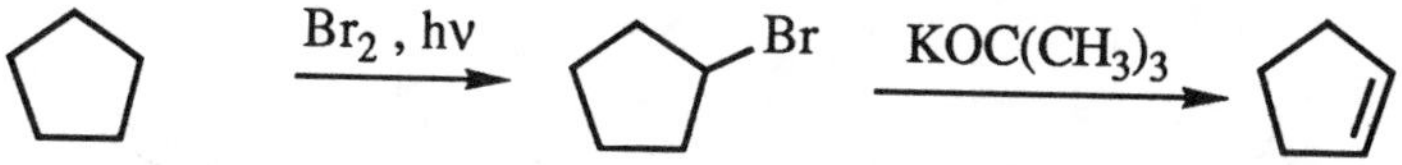
Br2 , hν
Br
KOC(CH3)3

7-30

(a)

only product

(b)

major

minor

(c)

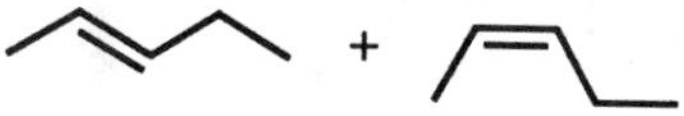

(d)

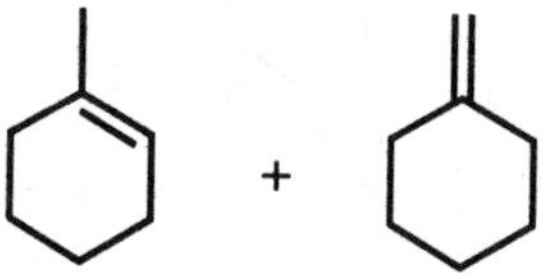

major

minor

(e)

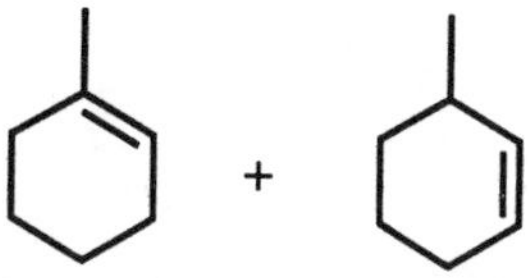

major

minor

7-31 The bromides are shown here. Chlorides or iodides would also work.

(a)

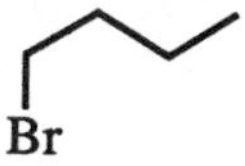
Br

(b)

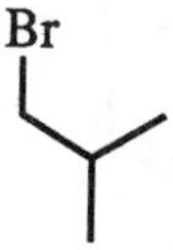
Br

or

Br

(c)

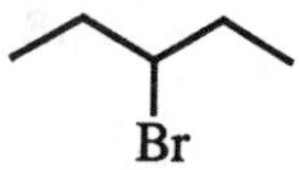
Br

(d)

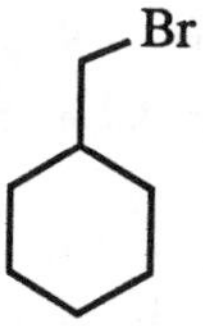
Br

(e)

Br

7-32

(a)  (b) 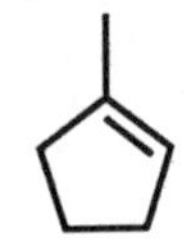 (c) 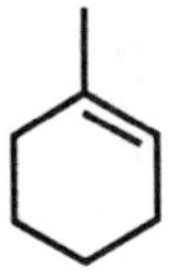 (d) 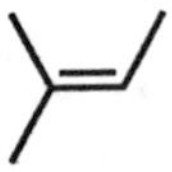

rearrangement

7-33

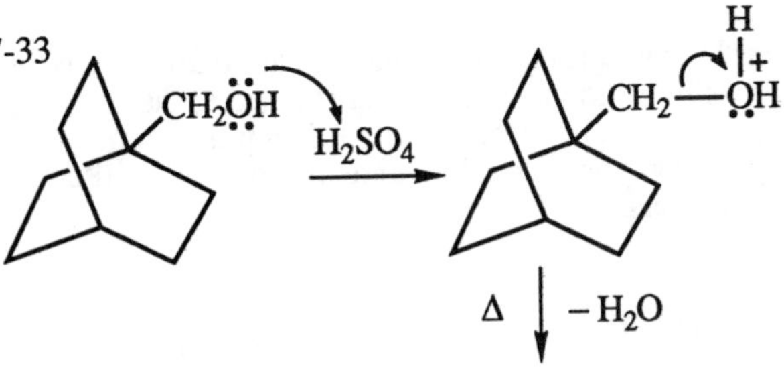

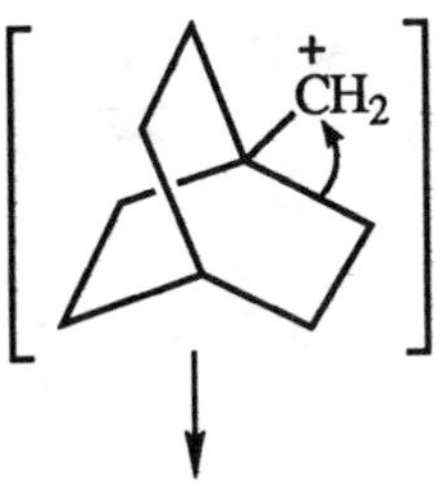

1° carbocation—terrible!—will rearrange; can't do hydride shift, must do alkyl shift = ring expansion

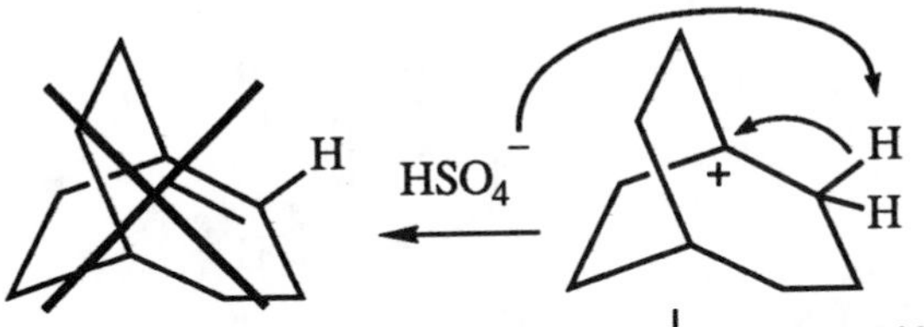

this is an unstable carbocation even though it is 3°; bridgehead carbons cannot be sp^2 (planar—try to make a model), so this carbocation does a hydride shift to a 2°, *more stable* carbocation

abstraction of adjacent H gives bridgehead alkene—violates Bredt's Rule

hydride shift

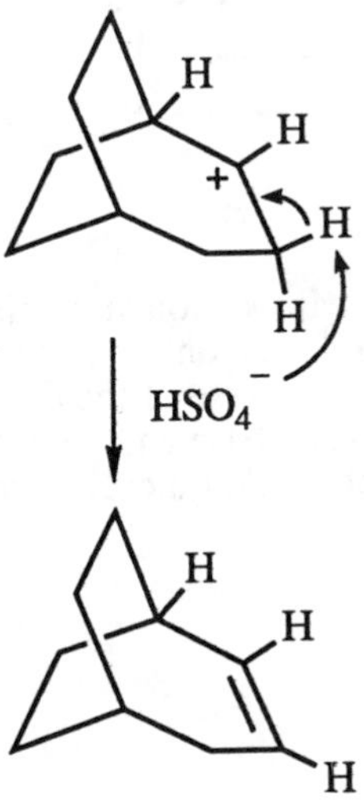

this 2° carbocation loses an adjacent H to form an alkene; can't form at bridgehead (Bredt's Rule)—only one other choice

7-34 (a) There are two reasons why alcohols do not dehydrate with strong base. The potential leaving group, hydroxide, is itself a strong base and therefore a terrible leaving group. Second, the strong base deprotonates the —OH faster than any other reaction can occur, consuming the base and making the leaving group anionic and therefore even worse.

$$ROH + {}^{-}OC(CH_3)_3 \rightleftharpoons RO^{-} + HOC(CH_3)_3$$

(b) A halide is already a decent leaving group. Since halides are extremely weak bases, the halogen atom is not easily protonated, and even if it were, the leaving group ability is not significantly enhanced. The hard step is to remove the adjacent H, something only a strong base can do—and strong bases will not be present under strong acid conditions.

7-35

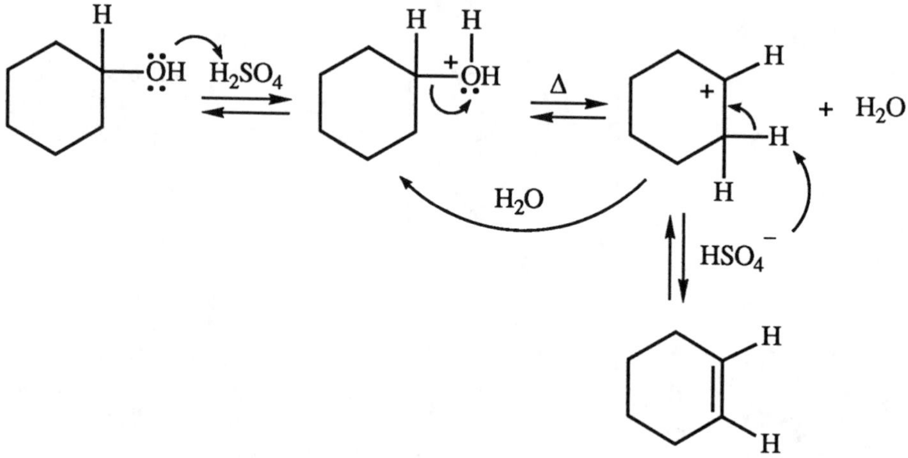

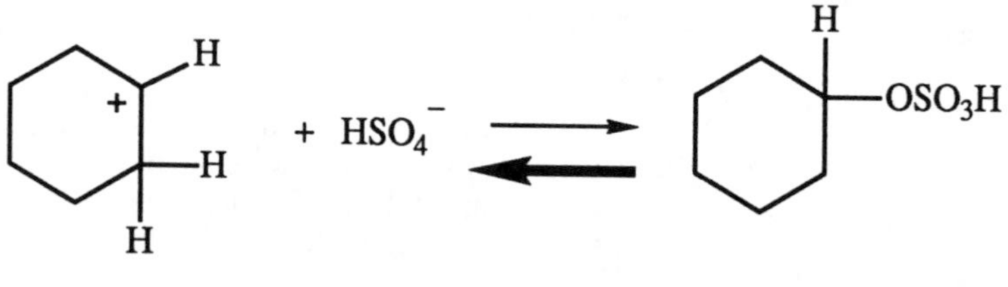

E1 works well because only one carbocation and only one alkene are possible. Substitution is not a problem here. The only nucleophiles are water, which would simply form starting material by a reverse of the dehydration, and bisulfate anion. Bisulfate anion is an extremely weak base and poor nucleophile; if it did attack the carbocation, the unstable product would quickly re-ionize, with no net change, back to the carbocation.

7-36

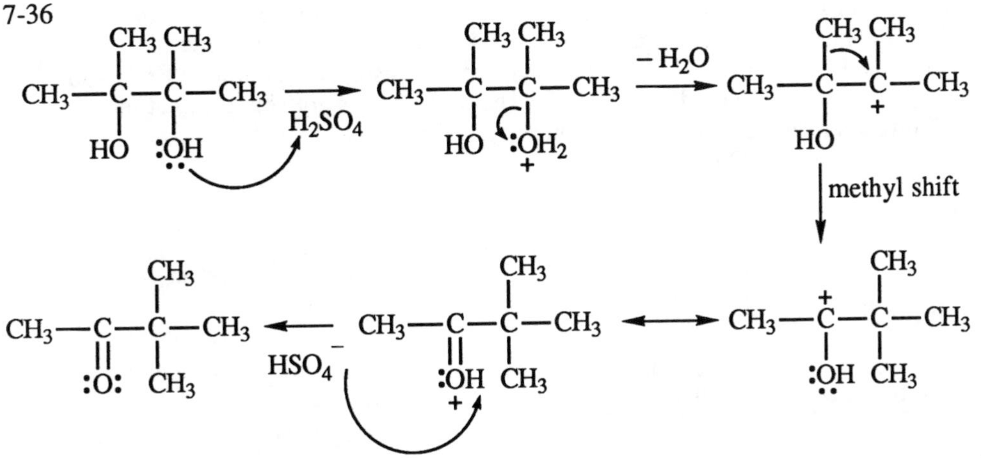

The driving force for this rearrangement is the great stability of the resonance-stabilized, protonated carbonyl group.

7-37

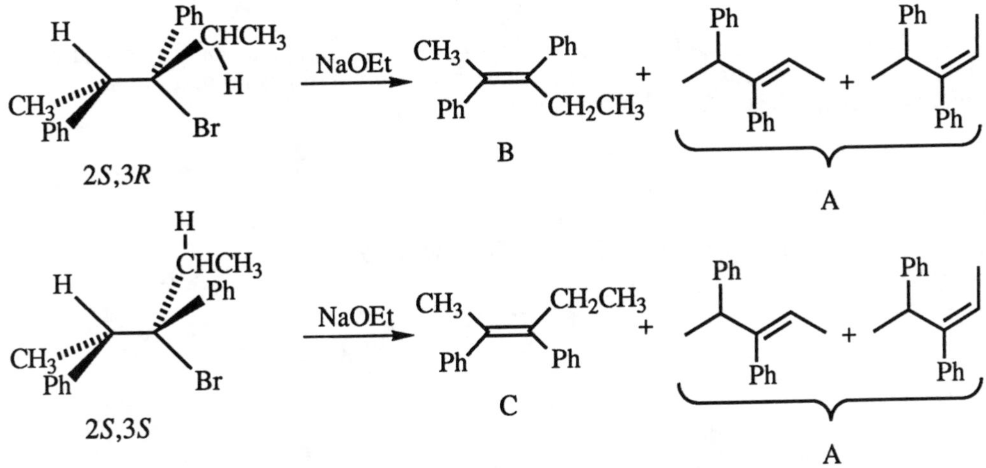

E2 dehydrohalogenation requires anti-coplanar arrangement of H and Br, so specific geometric isomers (B or C) are generated depending on the stereochemistry of the starting material. Removing a hydrogen from C-4 (achiral) will give about the same mixture of *cis* and *trans* (A) from either diastereomer.

7-38 Breaking a C—D bond requires more energy and is therefore slower than breaking a C—H bond. Since breaking a C—H bond is required in the rate-determining step of the E2 mechanism but not in the S_N2 mechanism, substituting D for H will slow the E2 but will not affect the S_N2. The S_N2 rate has not changed, but more S_N2 product is generated since the competing E2 reaction is so much slower (roughly 7 times slower with D than with H).

7-39

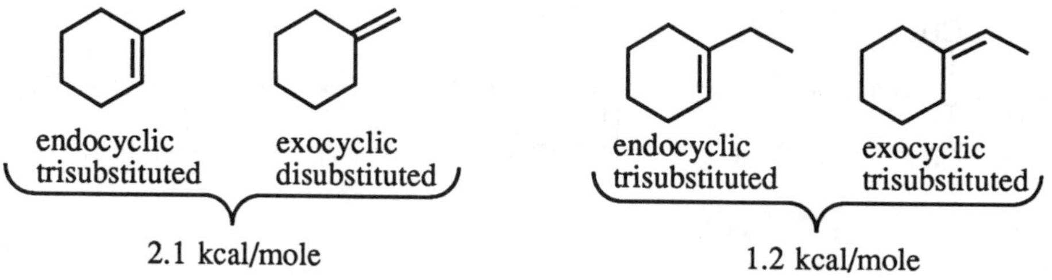

A standard principle of science is to compare experiments which differ by only one variable. Changing more than one variable clouds the interpretation, possibly to the point of invalidating the experiment.

The first set of structures compares endo and exocyclic double bonds, but the degree of substitution on the alkene is also different, so this comparison is not valid—we are not isolating simply the exo or endocyclic effect.

The second pair is a much better measure of endo versus exocyclic stability because both alkenes are trisubstituted, so the degree of substitution plays no part in the energy values. Thus, 1.2 kcal/mole is a better value.

7-40

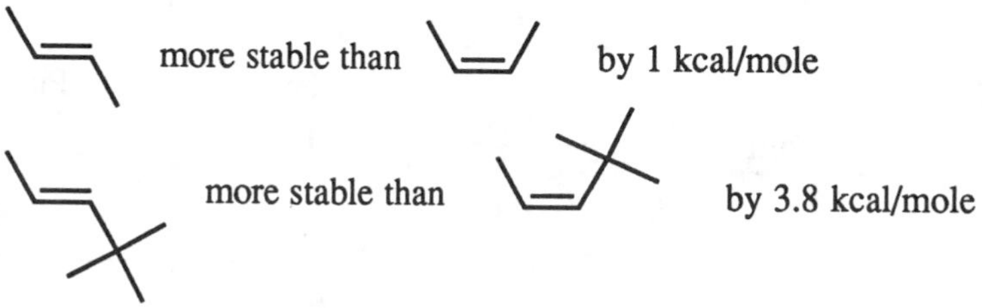

Steric crowding by the *t*-butyl group is responsible for the energy difference. In *cis*-2-butene, the two methyl groups have only slight interaction. In the 4,4-dimethyl-2-pentenes, however, the larger size of the *t*-butyl group crowds the methyl group in the *cis* isomer, increasing its strain and therefore its energy.

CHAPTER 8—REACTIONS OF ALKENES

8-1

(a) 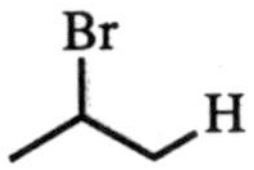

(b) 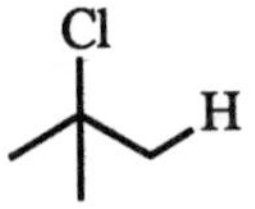

(c) 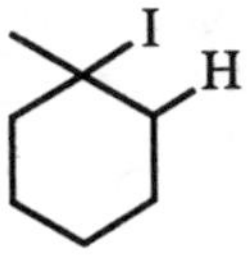

(d) 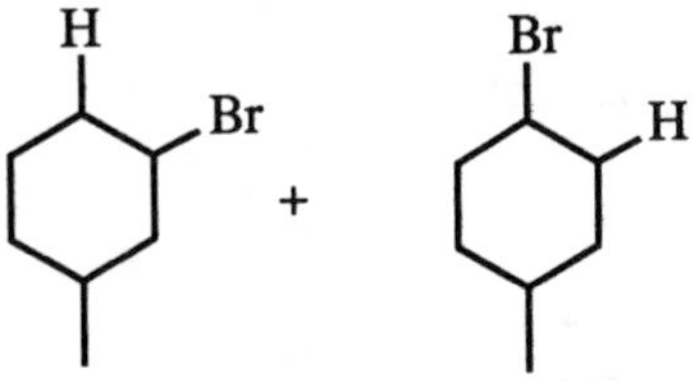

produced in about equal amounts (mixture of *cis* and *trans*)

8-2

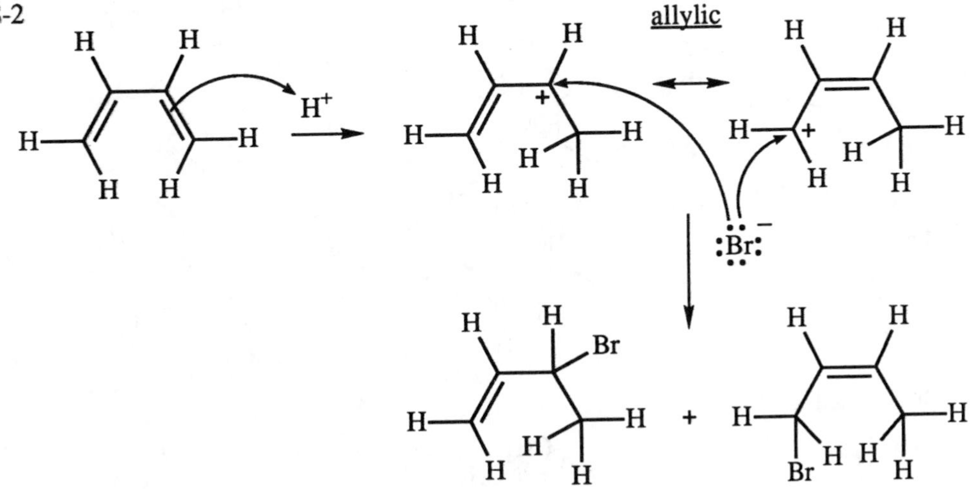

3-bromo-1-butene   1-bromo-2-butene

Because the allylic carbocation has partial positive charge at two carbons, the bromide nucleophile can bond at either electrophilic carbon, giving two products.

8-3

(a) 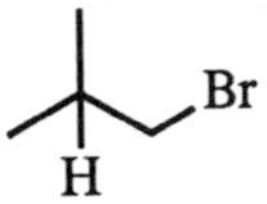

(b) 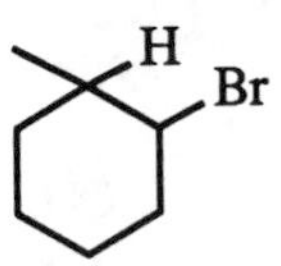

(c) 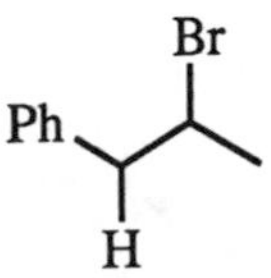

8-4 (a)

(b)

(c)

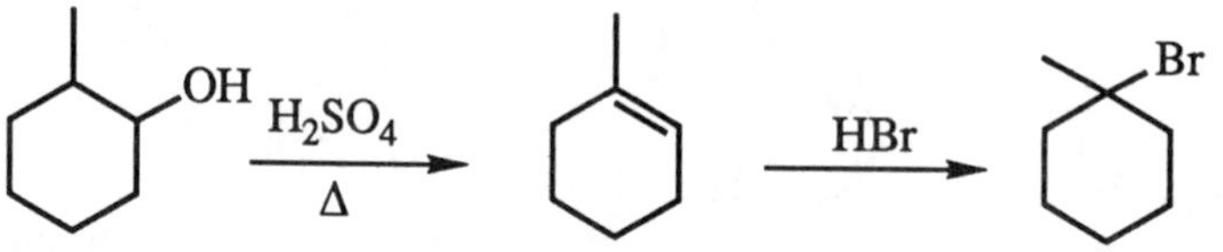

(d)

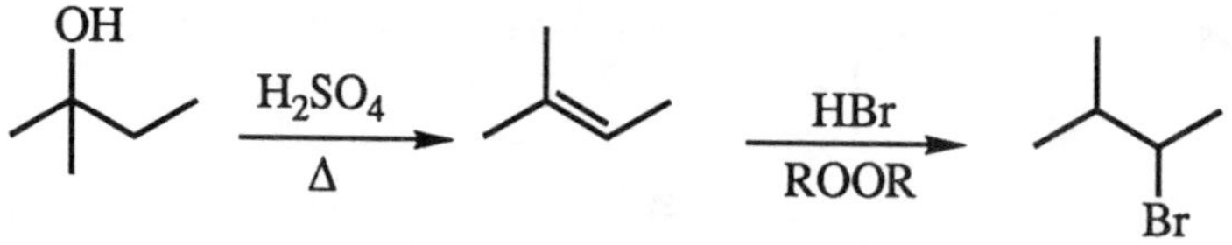

8-5 (a)

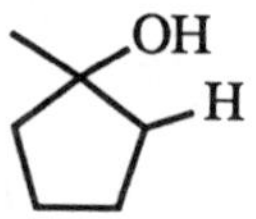

(b)

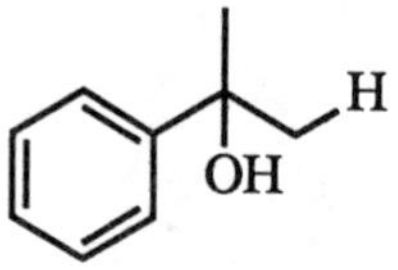

(c)

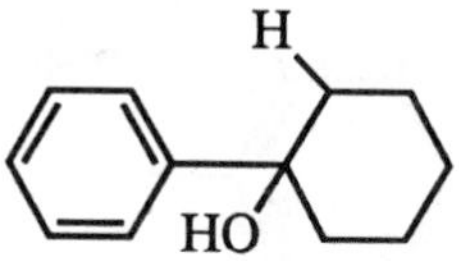

8-6

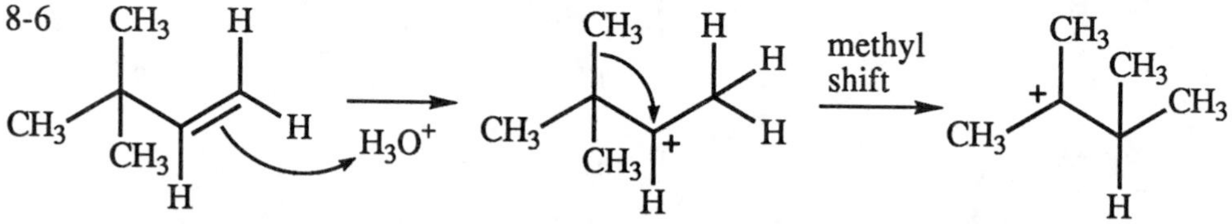

nucleophilic attack by water

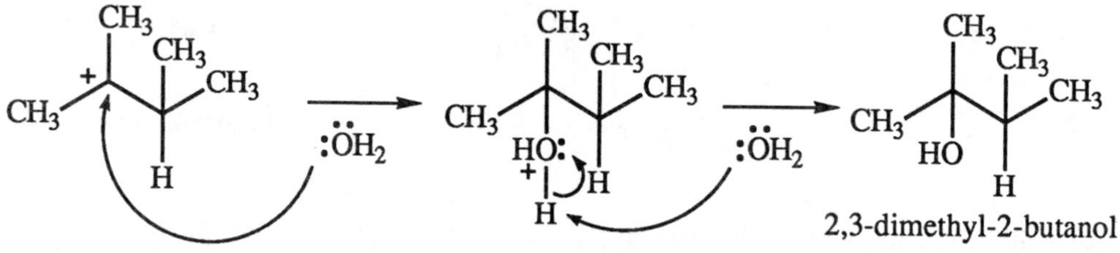

2,3-dimethyl-2-butanol

+ H_3O^+

proton removal

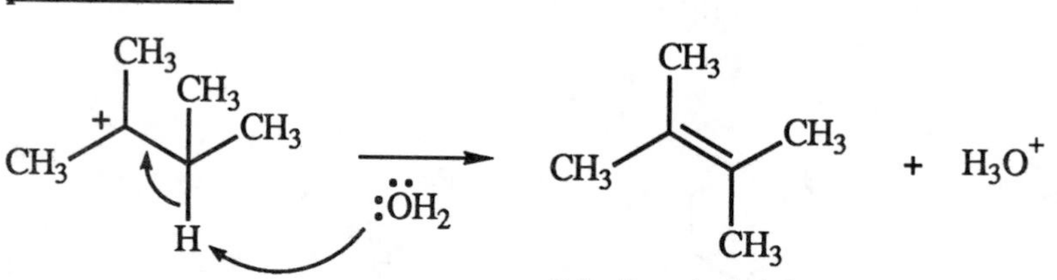

2,3-dimethyl-2-butene

8-7

(a)

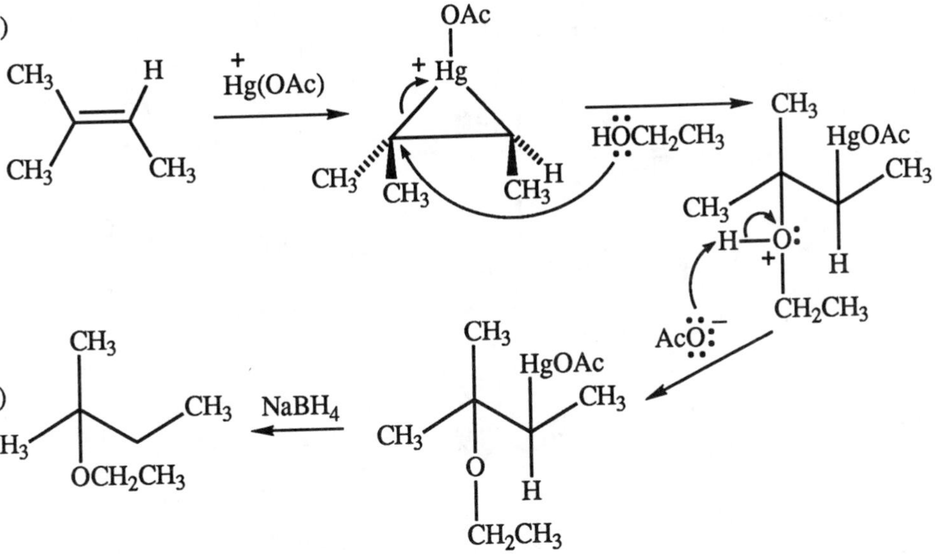

(b)

8-8

(a)

(b)

(c)

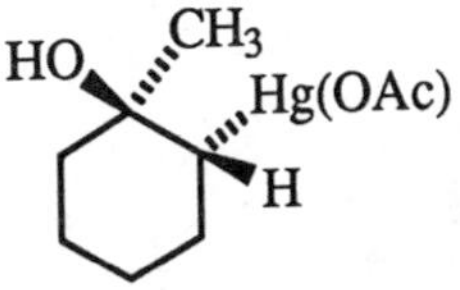

(d)

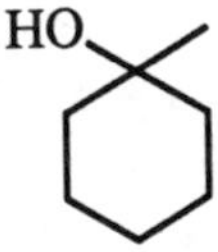

(e)

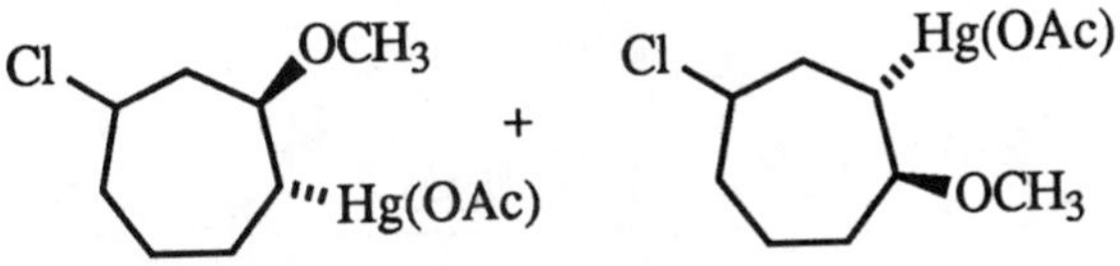

(f)

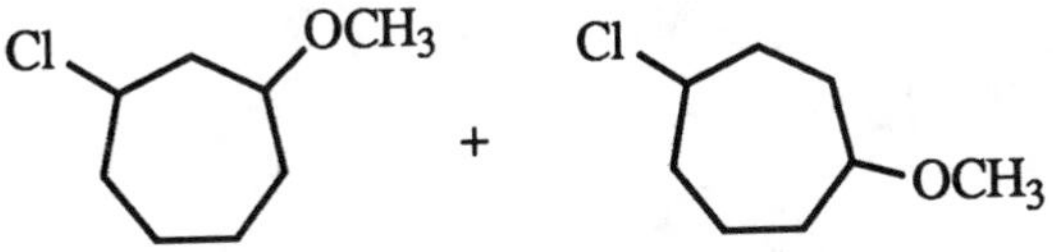

8-9

(a) 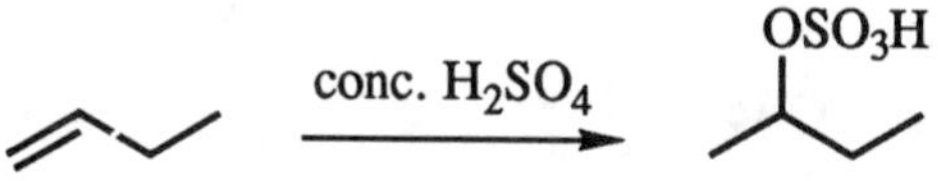
conc. H_2SO_4
OSO_3H

(b) 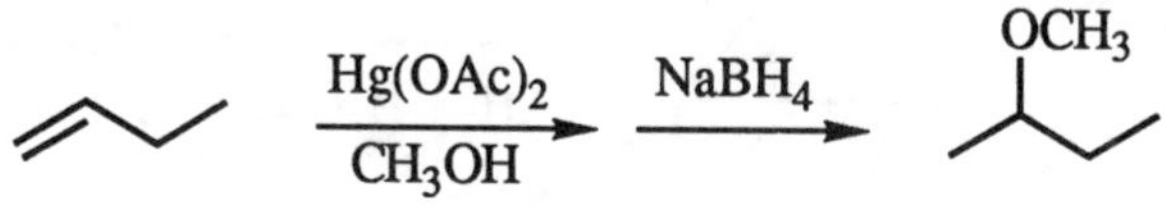
$Hg(OAc)_2$
CH_3OH
$NaBH_4$
OCH_3

(c) 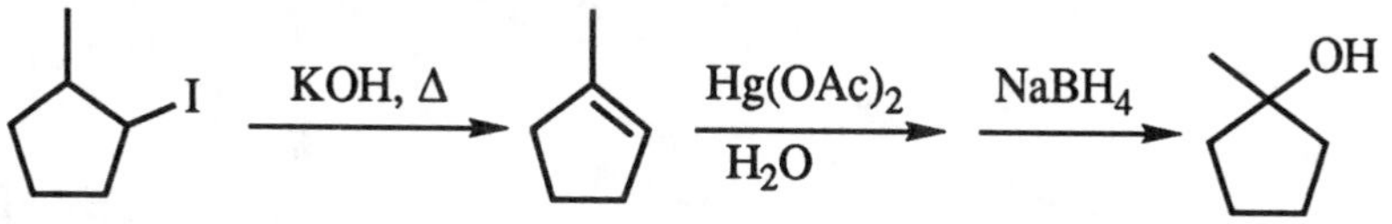
I
KOH, Δ
$Hg(OAc)_2$
H_2O
$NaBH_4$
OH

8-10

(a), (b) 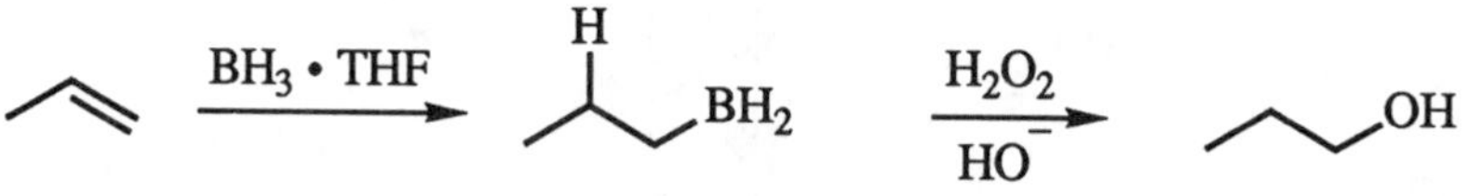
$BH_3 \cdot THF$
H
BH_2
H_2O_2
HO^-
OH

(c), (d) 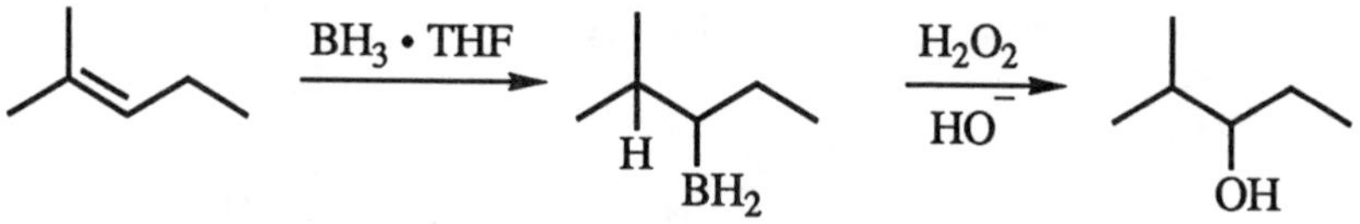
$BH_3 \cdot THF$
H
BH_2
H_2O_2
HO^-
OH

(e), (f) 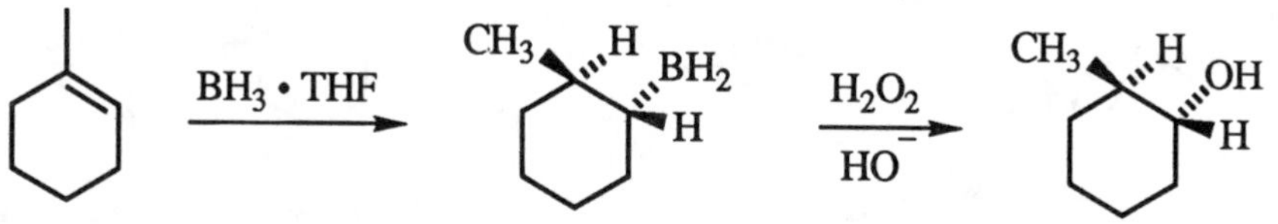
$BH_3 \cdot THF$
CH_3
H
BH_2
H
H_2O_2
HO^-
CH_3
H
OH
H

8-11

(a) 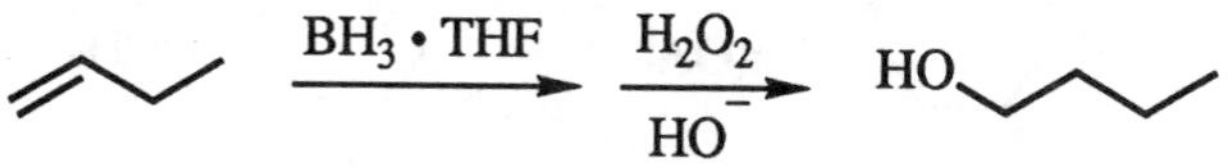
$BH_3 \cdot THF$
H_2O_2
HO^-
HO

(b) 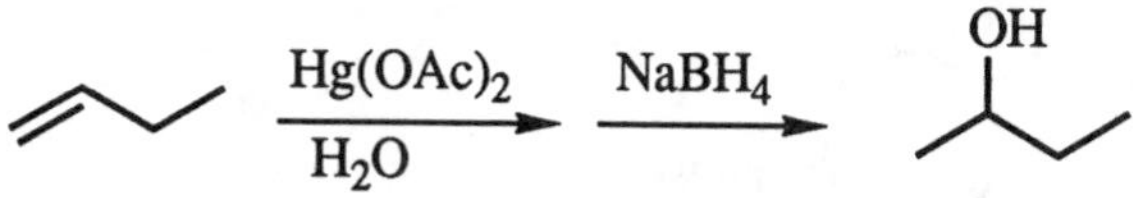
$Hg(OAc)_2$
H_2O
$NaBH_4$
OH

(c) 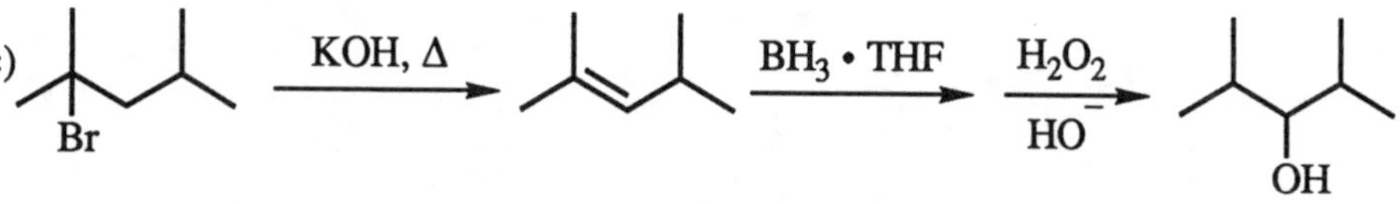
Br
KOH, Δ
$BH_3 \cdot THF$
H_2O_2
HO^-
OH

8-12

Instead of borane attacking the bottom face of 1-methylcyclopentene, it is equally likely to attack the top face, leading to the enantiomer.

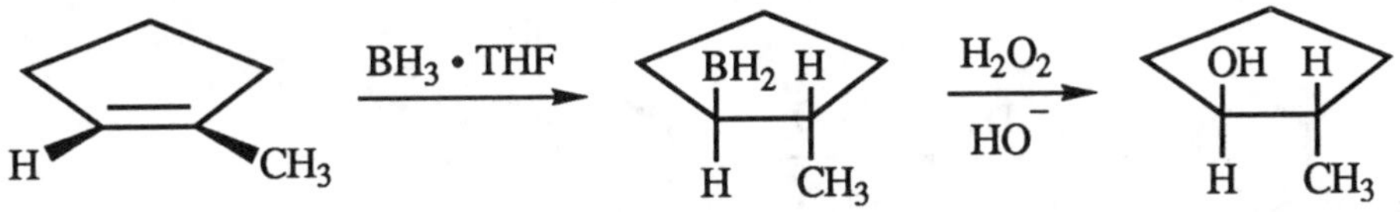

8-13

(a)

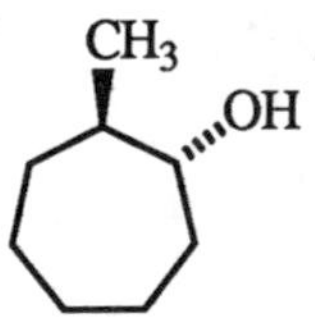

(b)

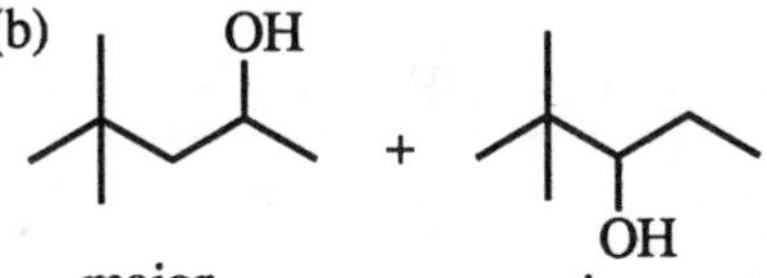

major

minor—steric hindrance to attack of BH_3

(c)

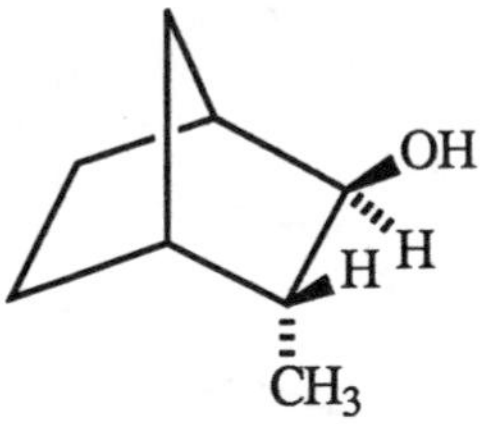

8-14

(a)

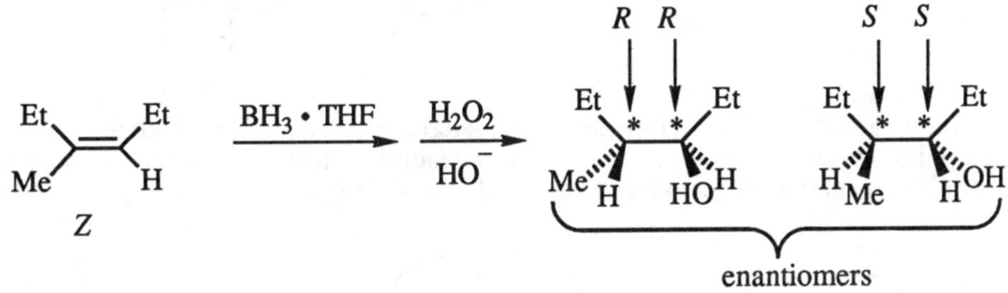

(b)

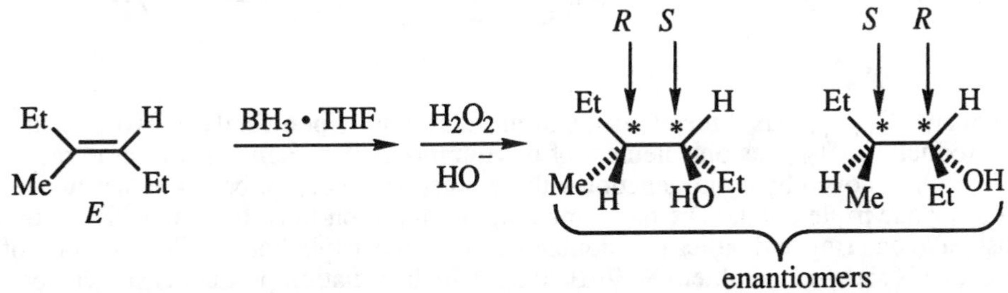

The enantiomeric pair produced from the Z-alkene is diastereomeric with the other enantiomeric pair produced from the *E*-alkene. These answers underscore that hydroboration-oxidation is stereospecific, that is, each alkene gives a specific set of stereoisomers, not a random mixture.

8-15

(a) 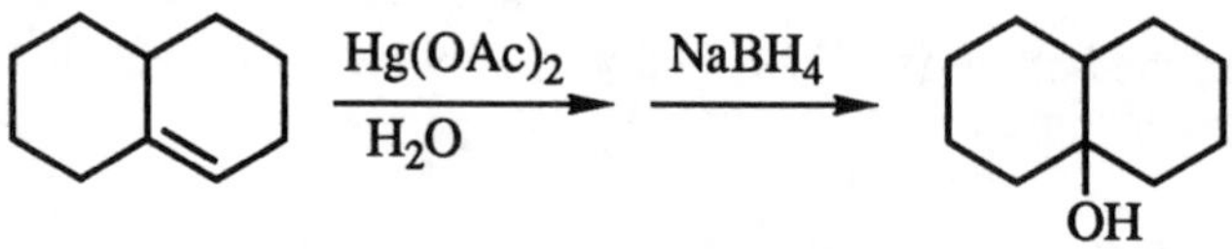

(b) 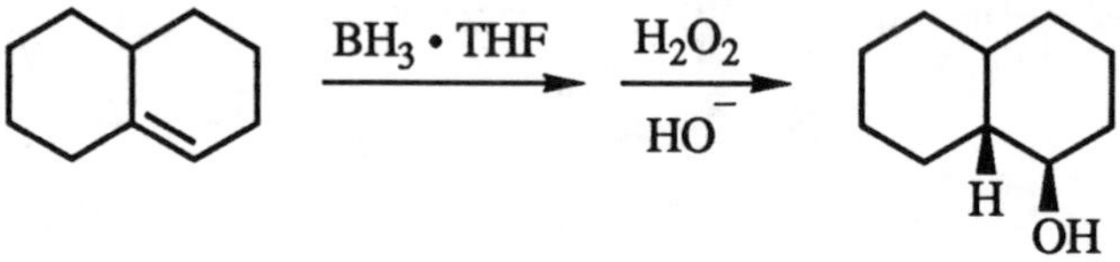

(c) 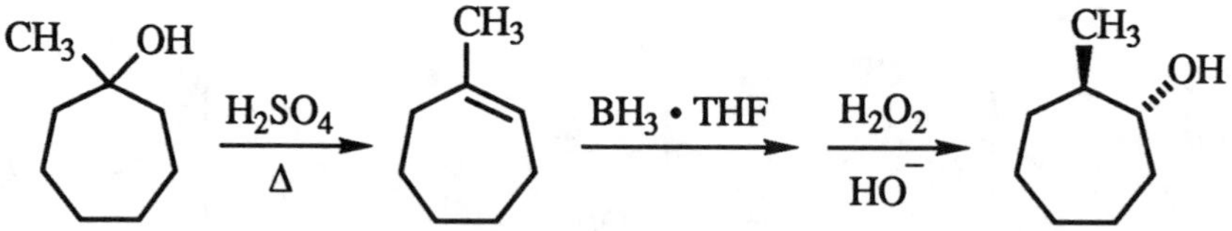

8-16 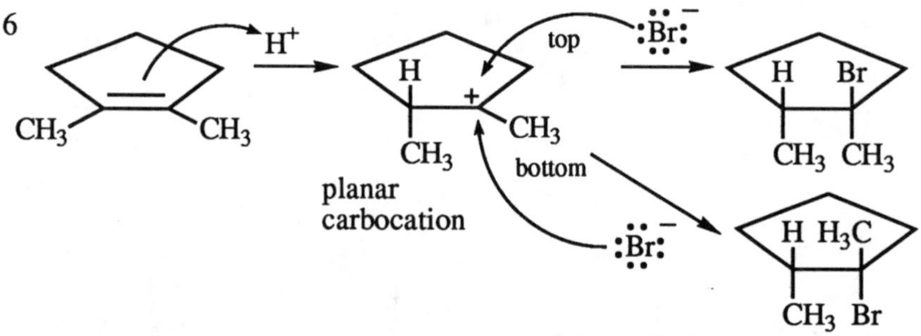

The *planar* carbocation is responsible for non-stereoselectivity. The bromide nucleophile can attack from the top or bottom, leading to a mixture of stereoisomers. The addition is therefore a mixture of syn and anti addition.

8-17

(a)  (b)  (c) 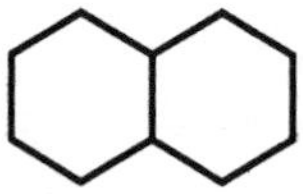 (d) 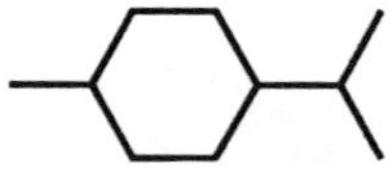

8-18

Limonene, $C_{10}H_{16}$, has three elements of unsaturation. Upon catalytic hydrogenation, the product, $C_{10}H_{20}$, has one element of unsaturation. Two elements of unsaturation have been removed by hydrogenation—these must have been pi bonds, either two double bonds or one triple bond. The one remaining unsaturation must be a ring. Thus, limonene must have one ring and either two double bonds or one triple bond. (The structure of limonene is shown in Problem 8-17(d), and the hydrogenation product is shown above in the solution to 8-17(d).)

8-19 Methylene inserts into the circled bonds.

(a) 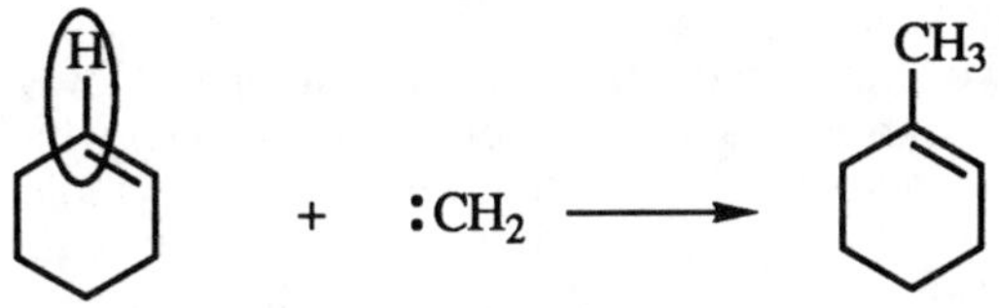

H
+
:CH_2
CH_3

(b) 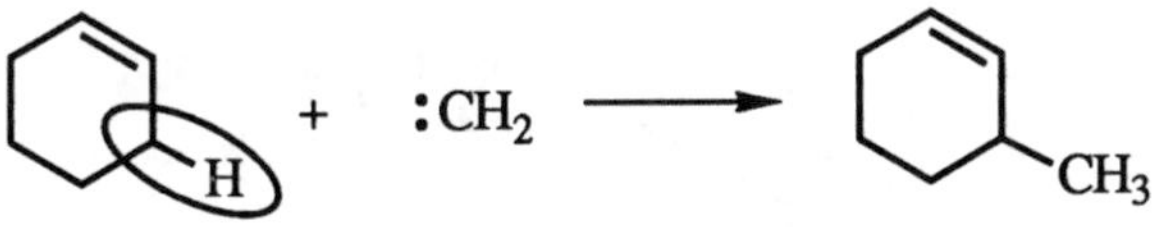

H
+
:CH_2
CH_3

(c) 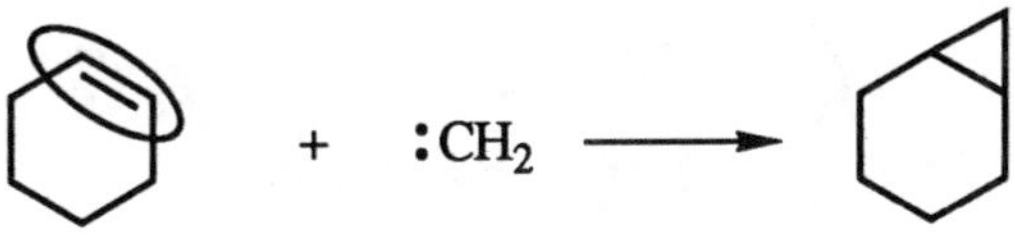

+
:CH_2

8-20

(a) 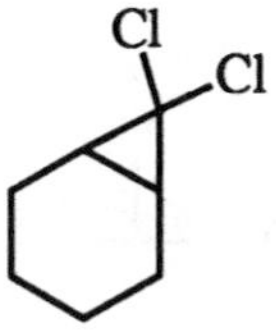

Cl
Cl

(b) 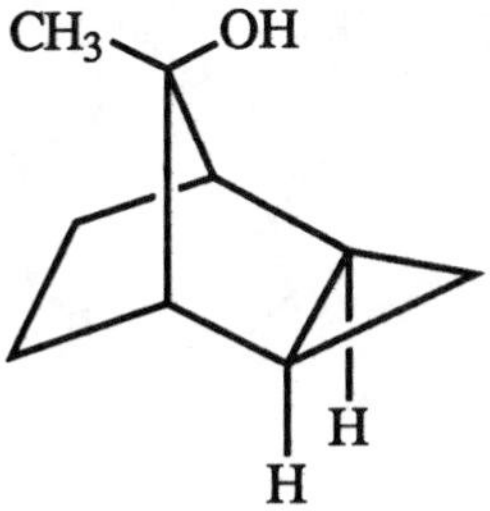

CH_3
OH
H
H

(c) 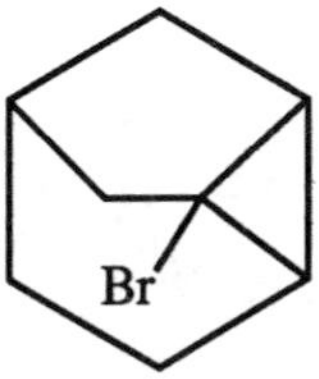

Br

8-21

(a) 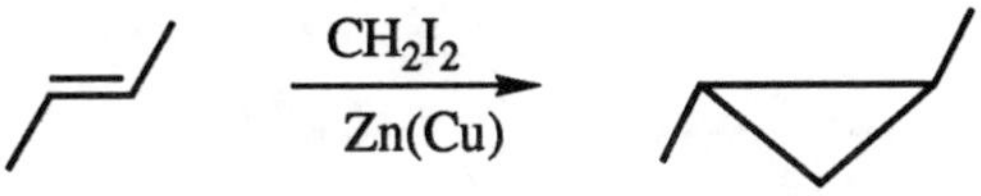

CH_2I_2
Zn(Cu)

(b) 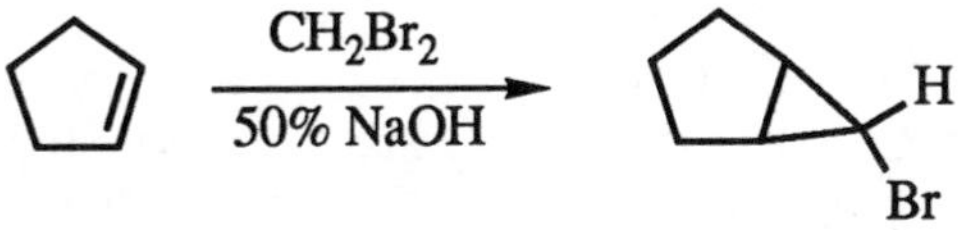

CH_2Br_2
50% NaOH
H
Br

(c) 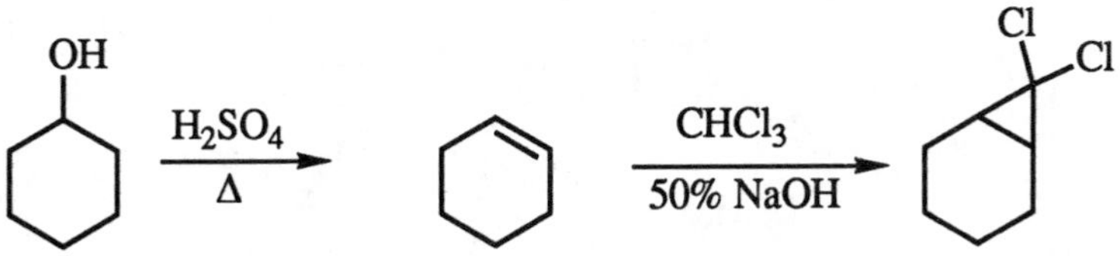

OH
H_2SO_4
Δ
$CHCl_3$
50% NaOH
Cl
Cl

8-22 During bromine addition to either the *cis*- or *trans*-alkene, two new chiral centers are being formed. Neither alkene (nor bromine) is optically active, so the product cannot be optically active.

The *cis*-2-butene gives two chiral products, a racemic mixture. However, *trans*-2-butene, because of it symmetry, gives only one *meso* product which can never be chiral. The "optical *in*activity" is built into this symmetric molecule.

This can be seen by following what happens to the configuration of the chiral centers from the intermediates to products, below. (The key lies in the symmetry of the intermediate and *inversion* of configuration when bromide attacks.)

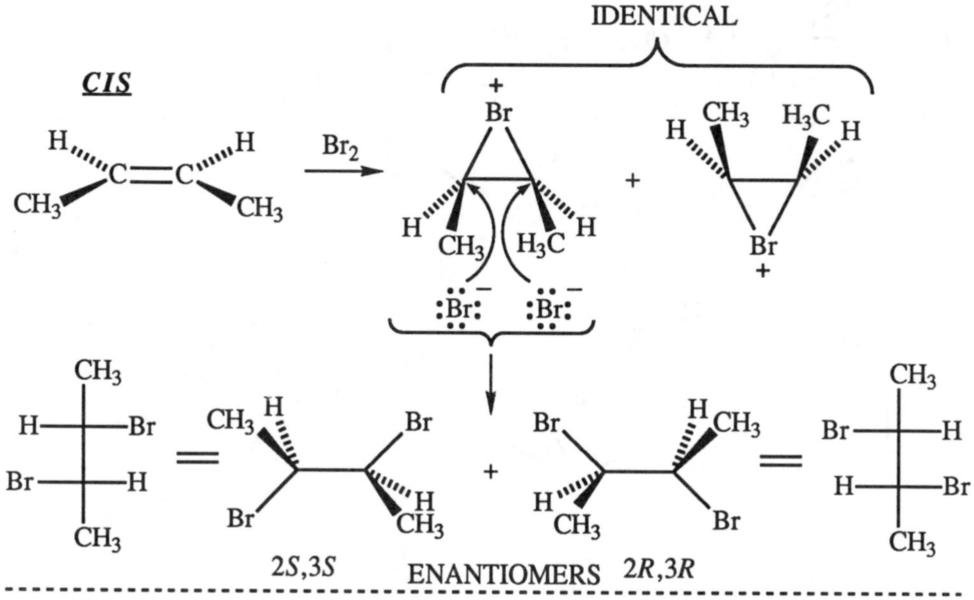
IDENTICAL
CIS
Br₂
2S,3S
ENANTIOMERS
2R,3R

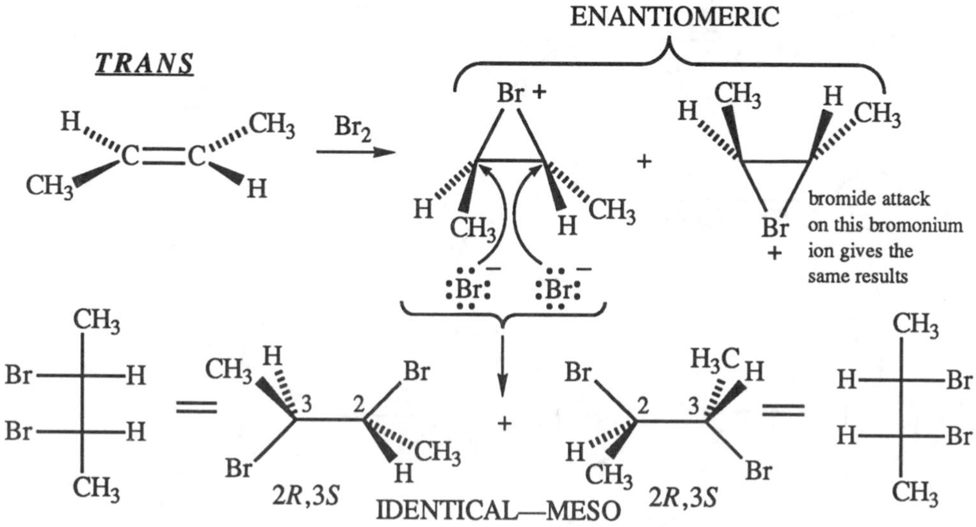
ENANTIOMERIC
TRANS
Br₂
bromide attack on this bromonium ion gives the same results
2R,3S
IDENTICAL—MESO
2R,3S

8-22 continued

CONCLUSION: anti addition of a symmetric reagent to a symmetric *cis*-alkene gives racemic product, while anti addition to a *trans*-alkene gives meso product. (We will see shortly that syn addition to a *cis*-alkene gives meso product, and syn addition to a *trans*-alkene gives racemic product. Stay tuned.)

8-23 Enantiomers of chiral products are also produced but not shown.

(a)

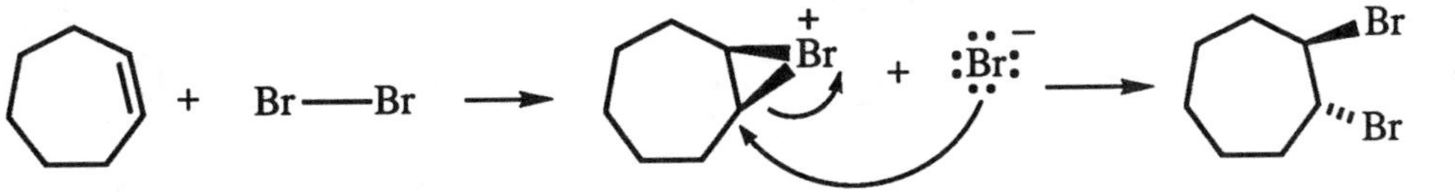

(b)

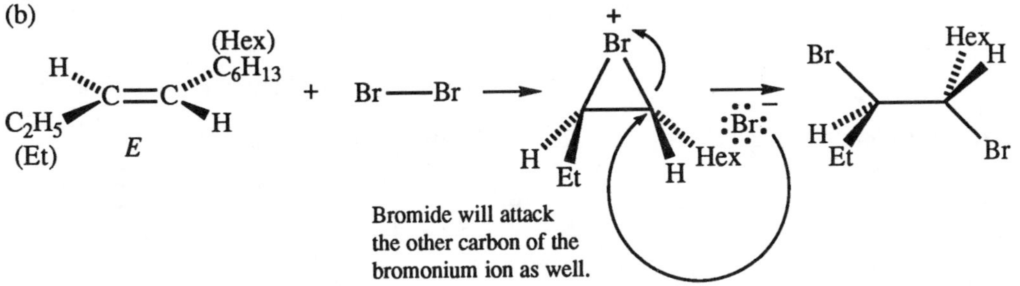
(Hex)
C_6H_{13}
C_2H_5
(Et)
E
Bromide will attack
the other carbon of the
bromonium ion as well.

(c)

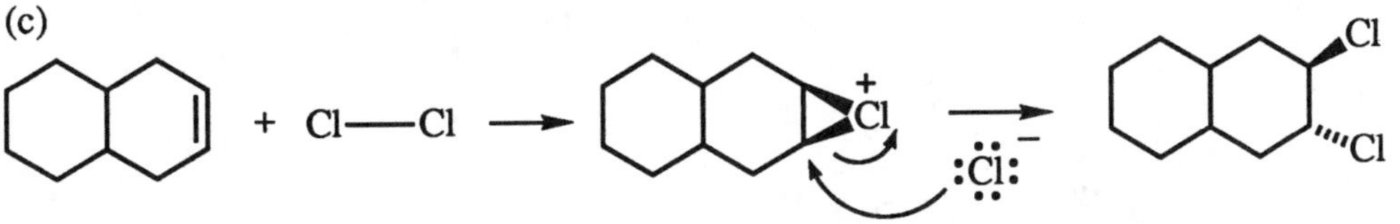

(d) Three new chiral centers are produced in this reaction. All stereoisomers will be produced with the restriction that the two adjacent chlorines on the ring must be *trans*.

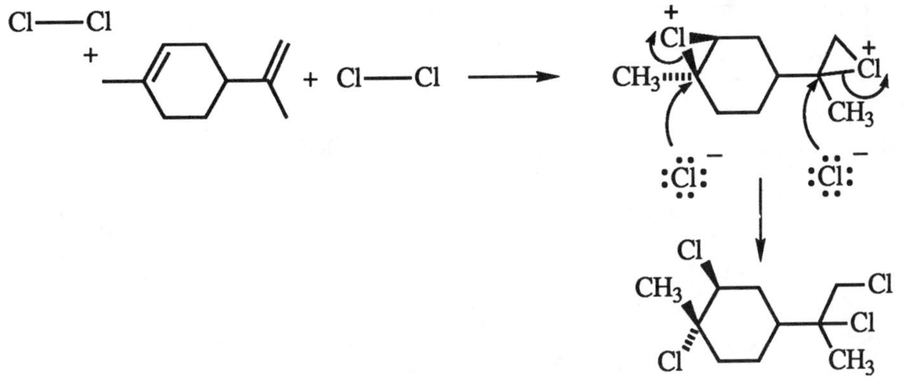

8-24 The *trans* product results from water attacking the bromonium ion from the face opposite the bromine. Equal amounts of the two enantiomers result from the equal probability that water will attack either C-1 or C-2.

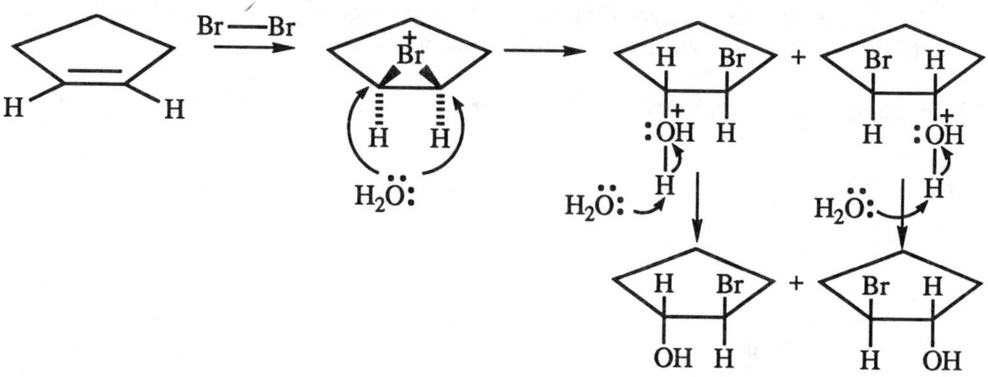

equal amounts of enantiomers

8-25 The chiral products shown here will actually be racemic mixtures.

(a)

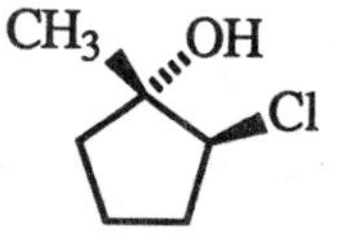

(b)

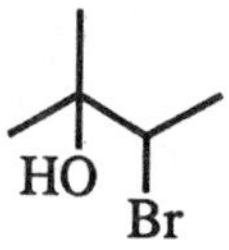

(c)

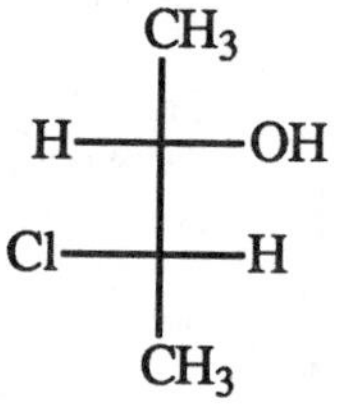

(d)

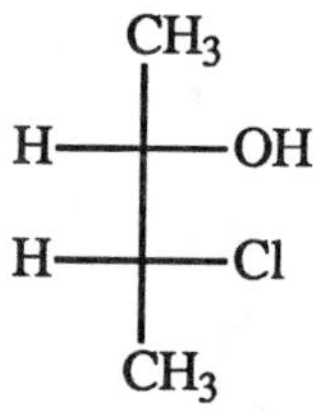

(e)

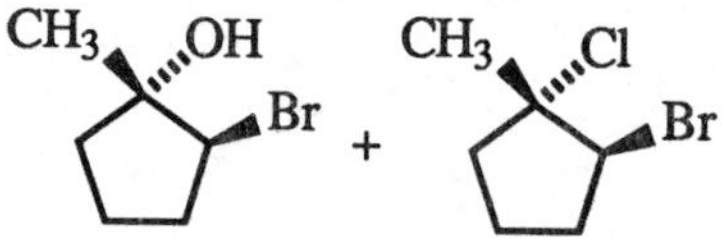

8-26

(a)

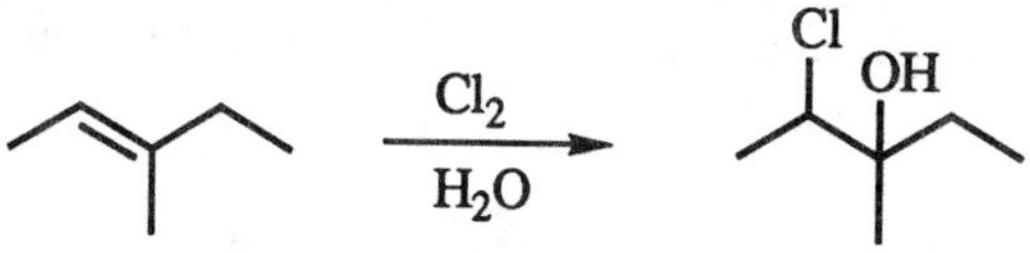

(b)

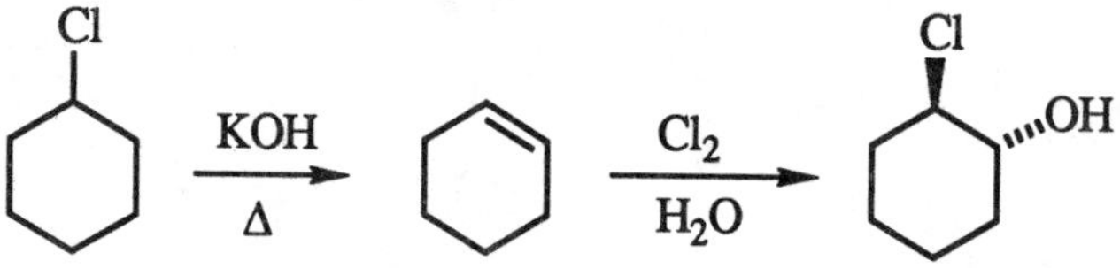

(c)

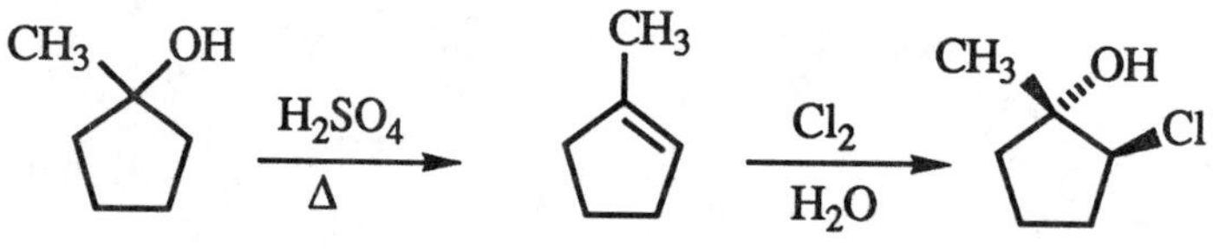

8-27

(a)

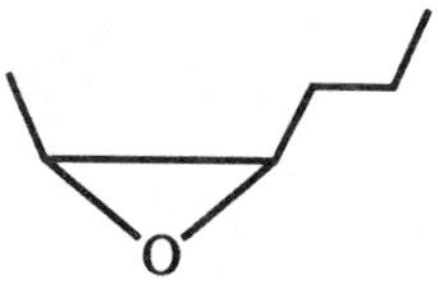

(b)

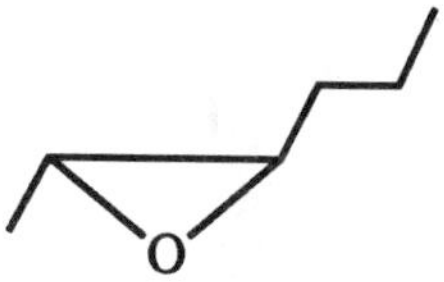

(c)

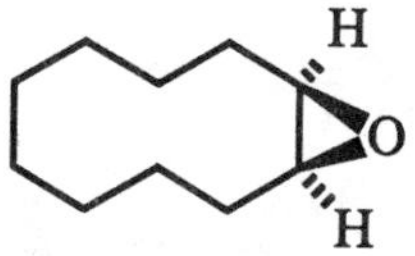

(d)

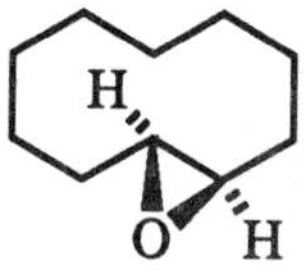

8-28

(a)

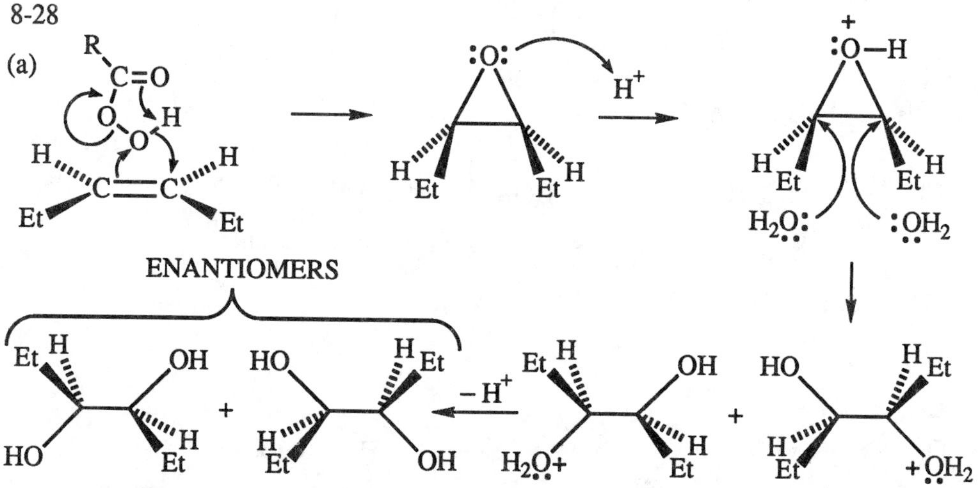
ENANTIOMERS

(b)

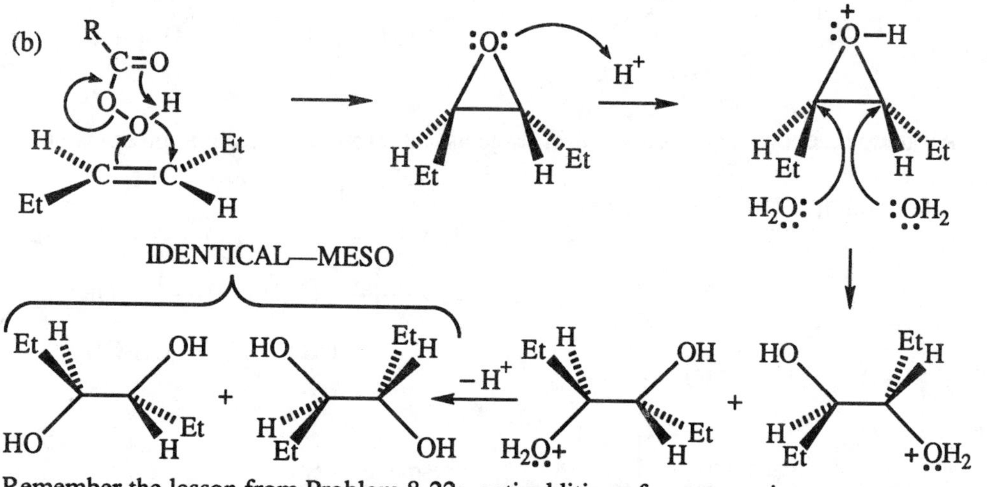
IDENTICAL—MESO

Remember the lesson from Problem 8-22: anti addition of a symmetric reagent to a symmetric *cis*-alkene gives racemic product, while anti addition to a *trans*-alkene gives meso product.

8-29

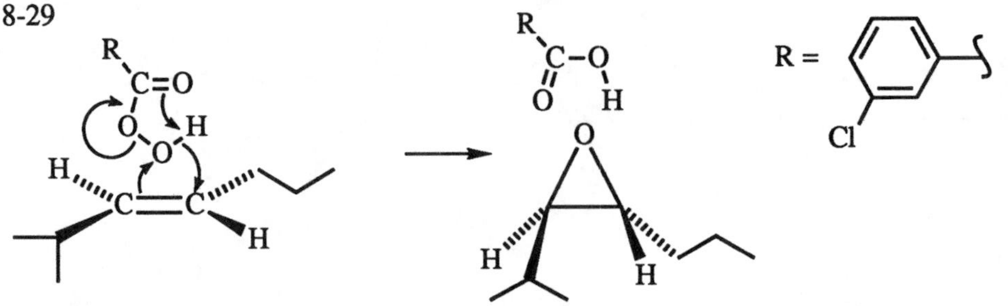

R
C=O
O
O
H
H
C=C
H
R
C-O
O
H
O
H
H
R =
Cl

8-30

(a)

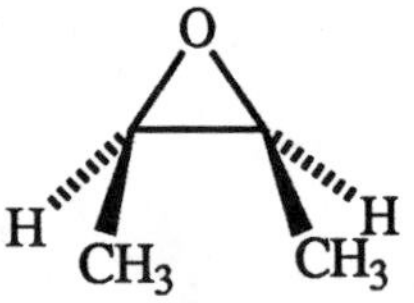

O
H
CH3
H
CH3

(b)

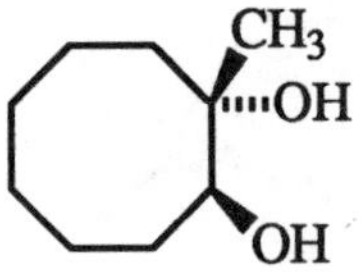

CH3
OH
OH

(c)

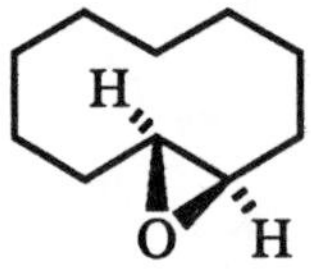

H
O
H

(d)

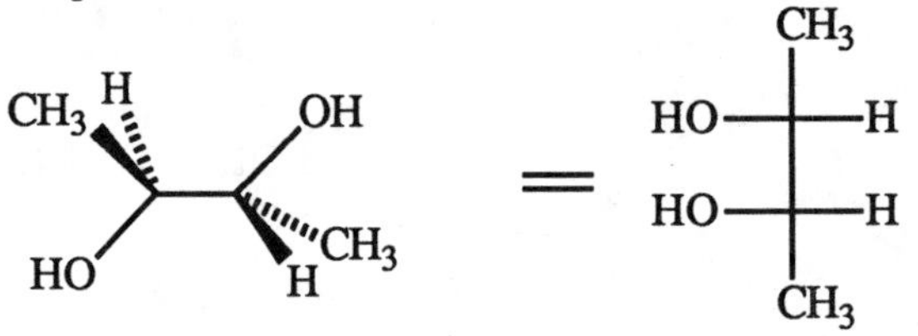

CH3
H
OH
HO
CH3
H
CH3
HO
H
HO
H
CH3

anti addition to a *trans*-alkene gives meso product

8-31

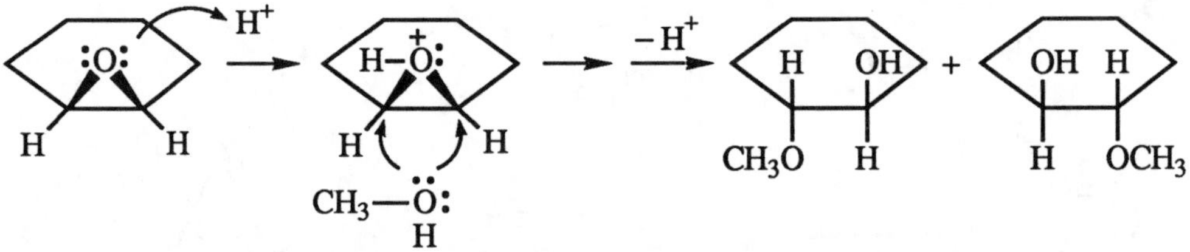

:O:
H
H
H+
H-O:
+
H
H
CH3—O:
H
– H+
H
OH
CH3O
H
+
OH
H
H
OCH3

8-32 All these reactions begin with achiral reagents; therefore, all the chiral products are racemic

(a)

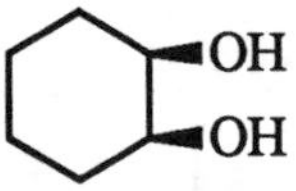

OH
OH

(b)

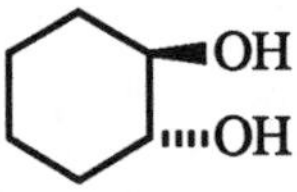

OH
OH

(c)

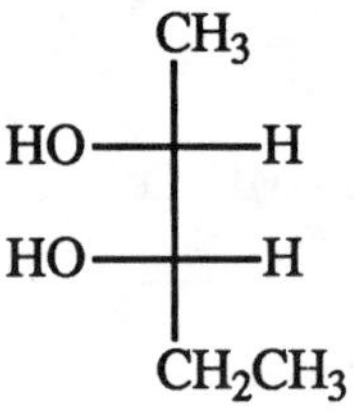

CH3
HO
H
HO
H
CH2CH3

(d)

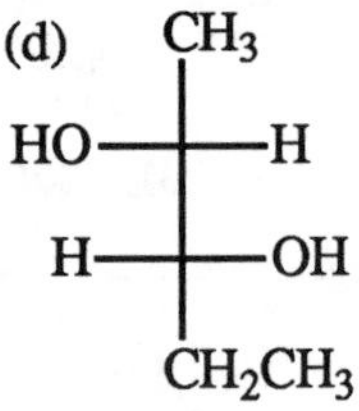

CH3
HO
H
H
OH
CH2CH3

(e)

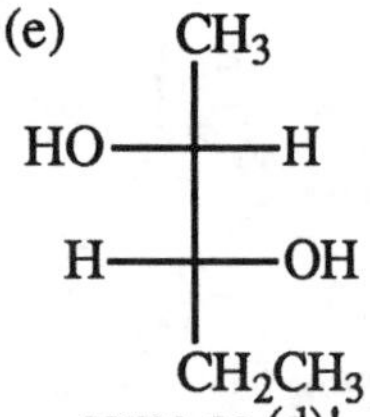

CH3
HO
H
H
OH
CH2CH3

same as (d)!

(f)

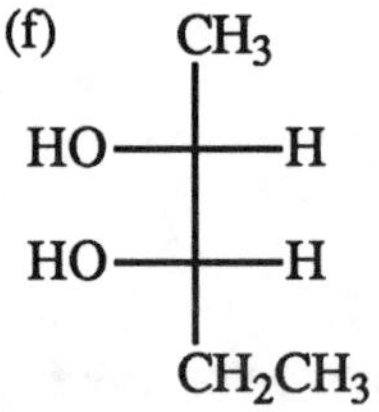

CH3
HO
H
HO
H
CH2CH3

same as (c)!

See the answer to 8-33.

8-33

(a) 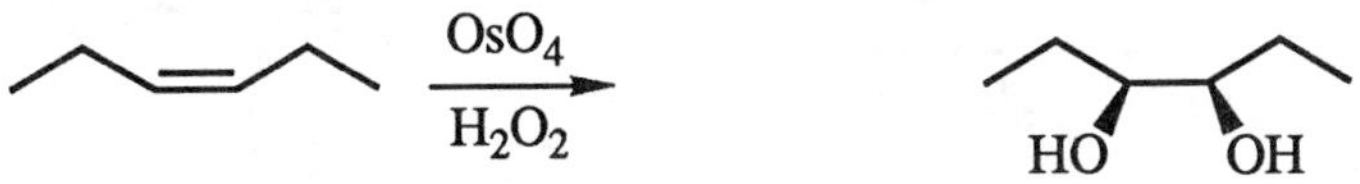  meso

(b) 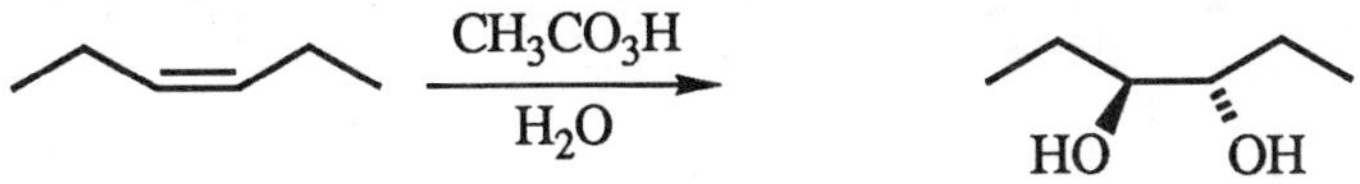 racemic (*d,l*)

(c) 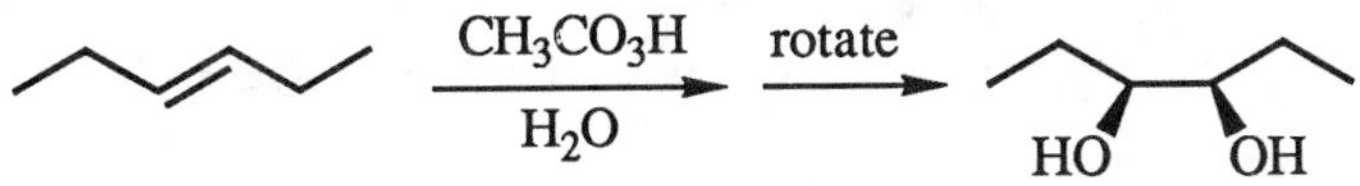 meso

(d) 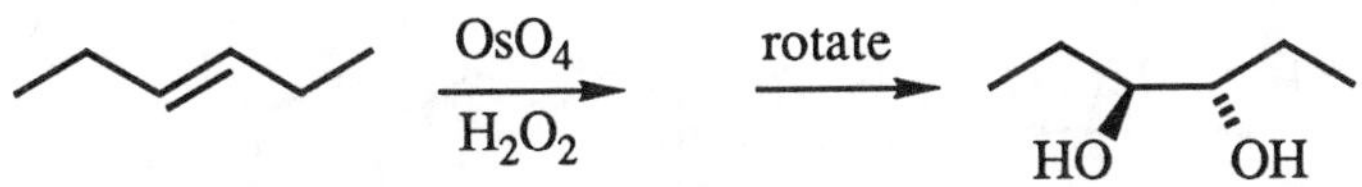 racemic (*d,l*)

Have you noticed yet? For symmetric alkenes and symmetric reagents (addition of two identical X groups):

cis-alkene + **syn** addition → meso

cis-alkene + **anti** addition → racemic

trans-alkene + **syn** addition → racemic

trans-alkene + **anti** addition → meso

<u>Just like math!!</u>

+1 x +1 = +1

+1 x −1 = −1

−1 x +1 = −1

−1 x −1 = +1

8-34

Solve these ozonolysis problems by working backwards, that is, by "reattaching" the two carbons of the new carbonyl groups into alkenes. Here's a hint. When you cut a circular piece of string, you still have only one piece. When you cut a linear piece of string, you have two pieces. Same with molecules. If ozonolysis forms only one product with two carbonyls, the alkene had to have been in a ring. If ozonolysis gives two molecules, the alkene had to have been in a chain.

(a) 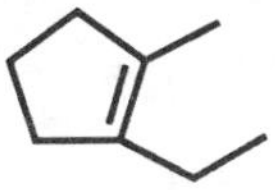

(b) 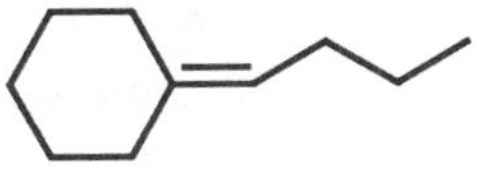

(c) 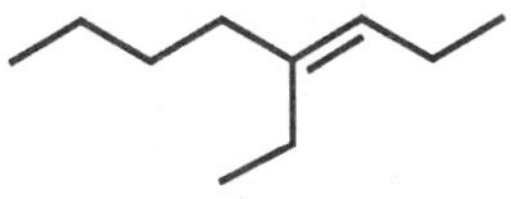

E or *Z*

8-35

(a) 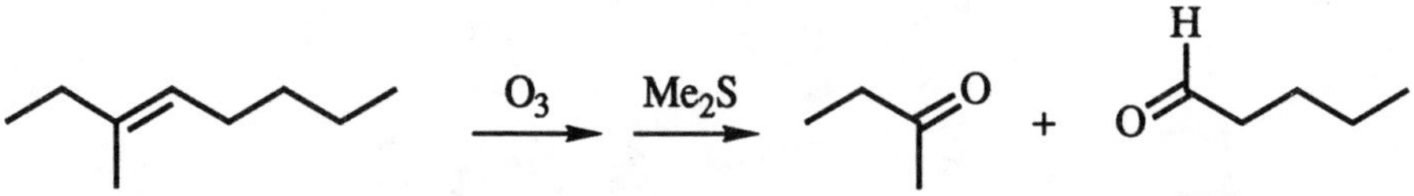

(b) 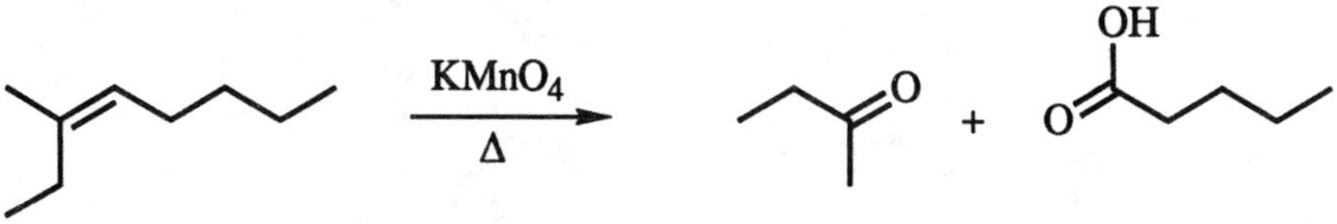

(c) 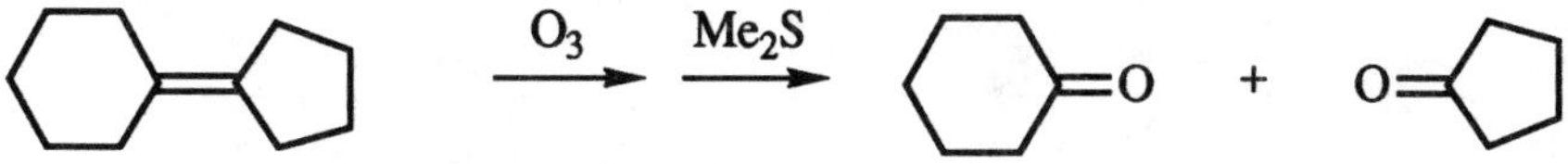

(d) 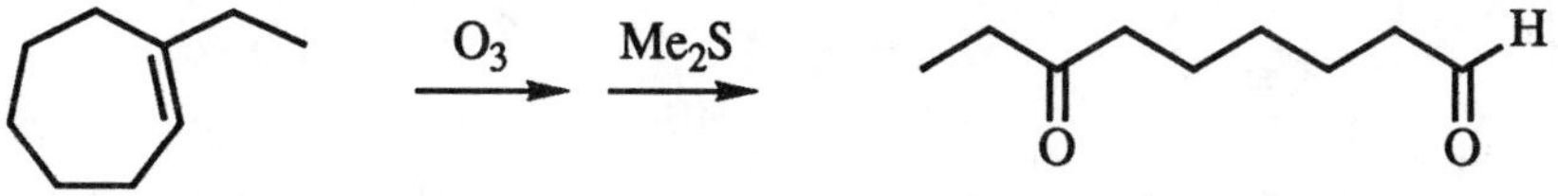

(e) 

(f) 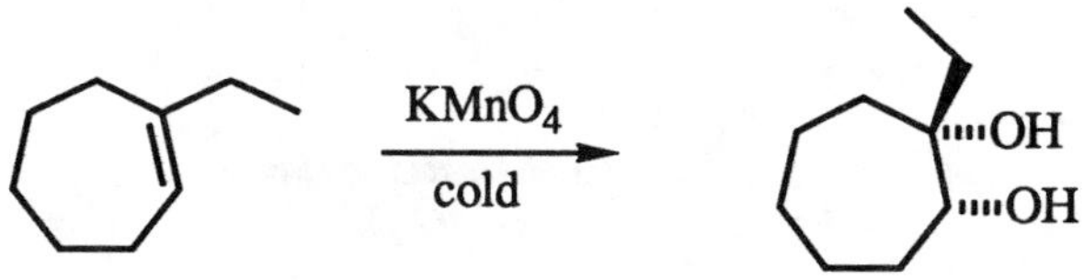

8-36 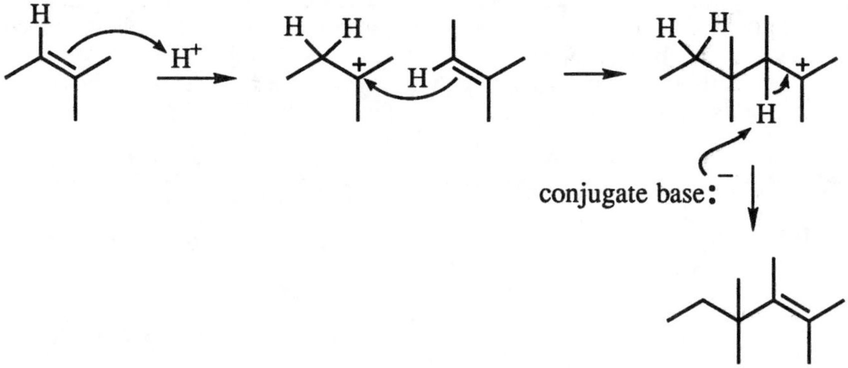

8-37

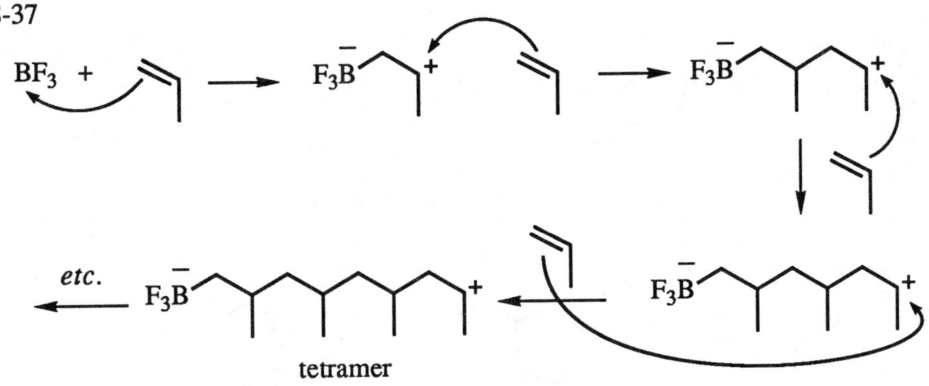
BF3 +
F3B
F3B
F3B
etc.
F3B
tetramer

8-38

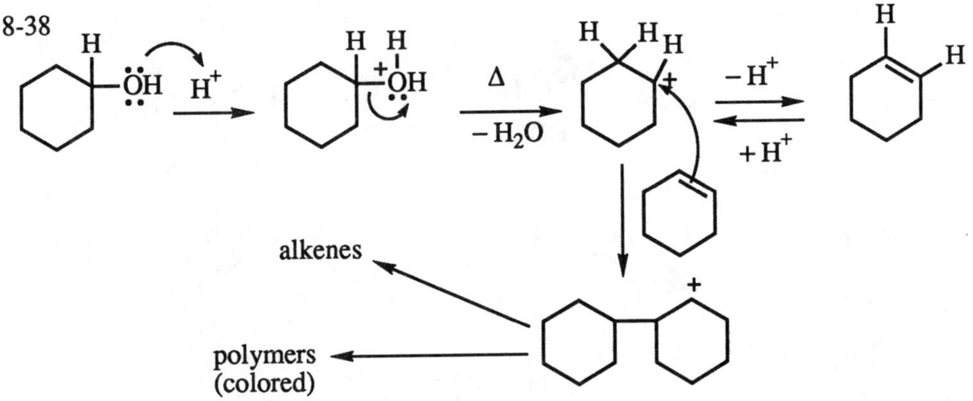
H
ÖH
H+
H H
+ÖH
Δ
− H2O
H
H H
− H+
+ H+
H
H
alkenes
polymers
(colored)

8-39 Refer to the Glossary and the text for definitions and examples.

8-40

(a) 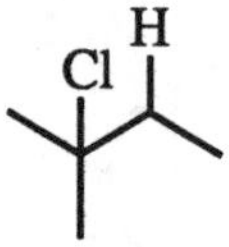
H
Cl

(b) 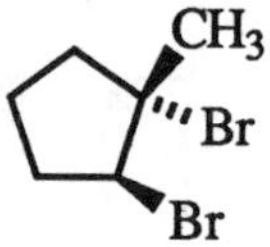
CH3
Br
Br

(c) 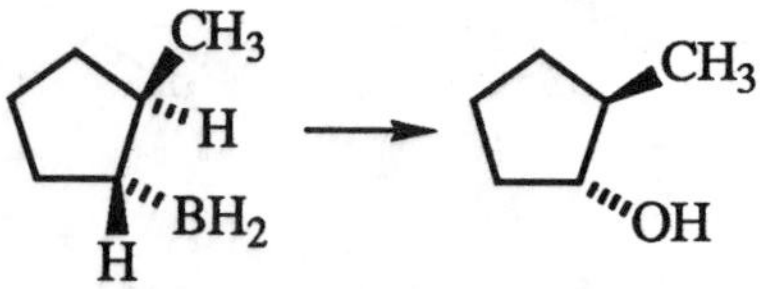
CH3
H
BH2
H
CH3
OH

intermediate

(d) 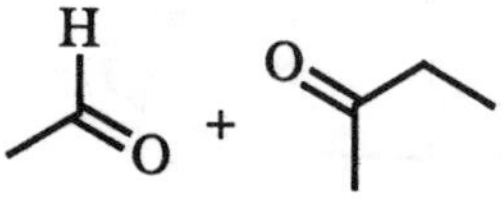
H
O
+
O

(e) 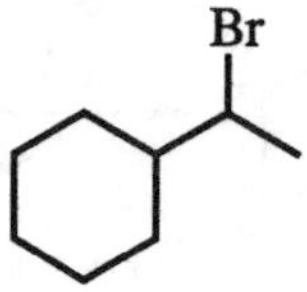
Br

(f) 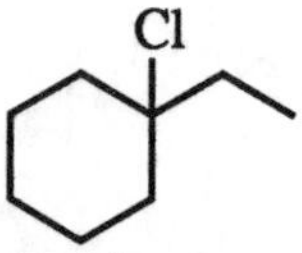
Cl

peroxides do not affect HCl addition

(g) 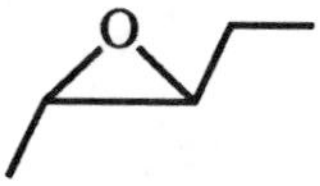
O

(h) 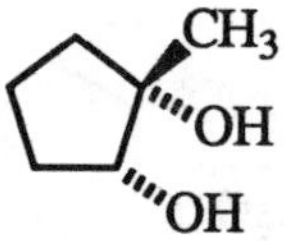
CH3
OH
OH

(i) 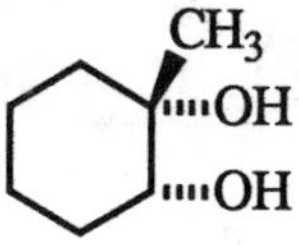
CH3
OH
OH

(j) 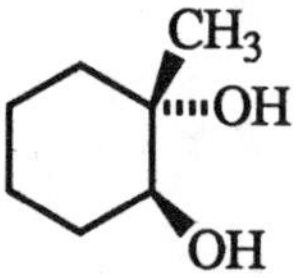
CH3
OH
OH

(k) 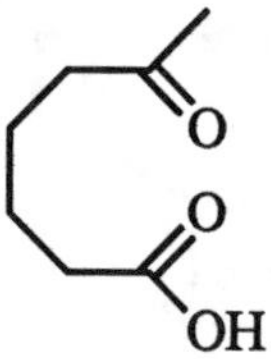
O
O
OH

(l) 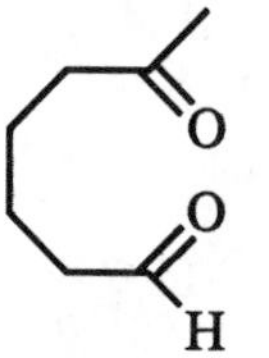
O
O
H

(m) 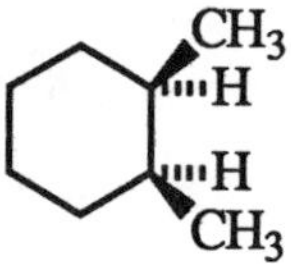
CH3
H
H
CH3

(n) 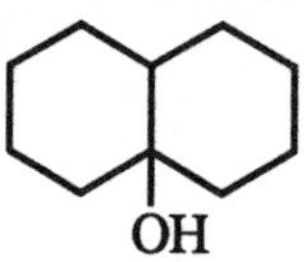
OH

(o) 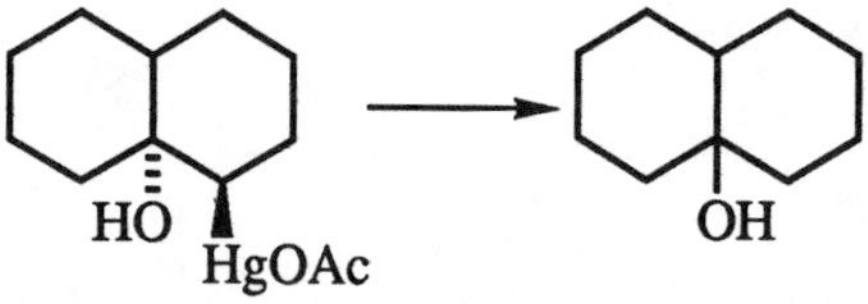
HO
HgOAc
OH

intermediate

(p) 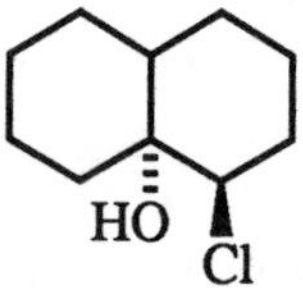
HO
Cl

8-41

(a)

<u>initiation</u>

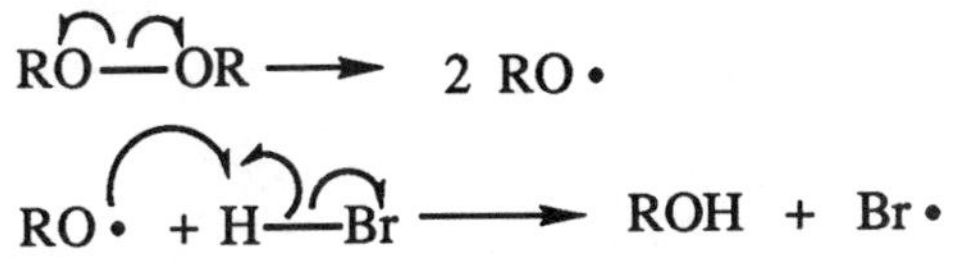

<u>propagation</u>

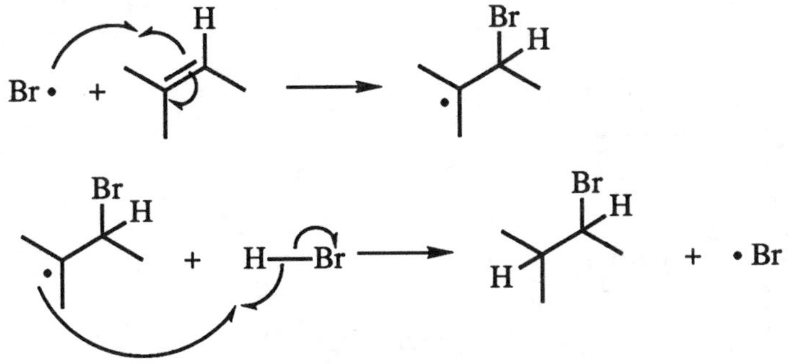

(b)

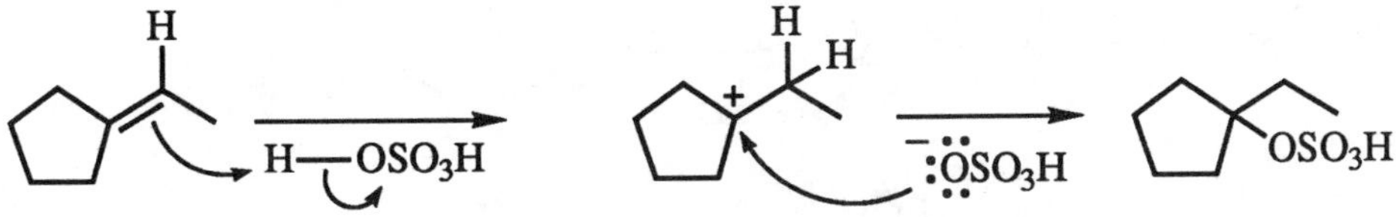

(c)

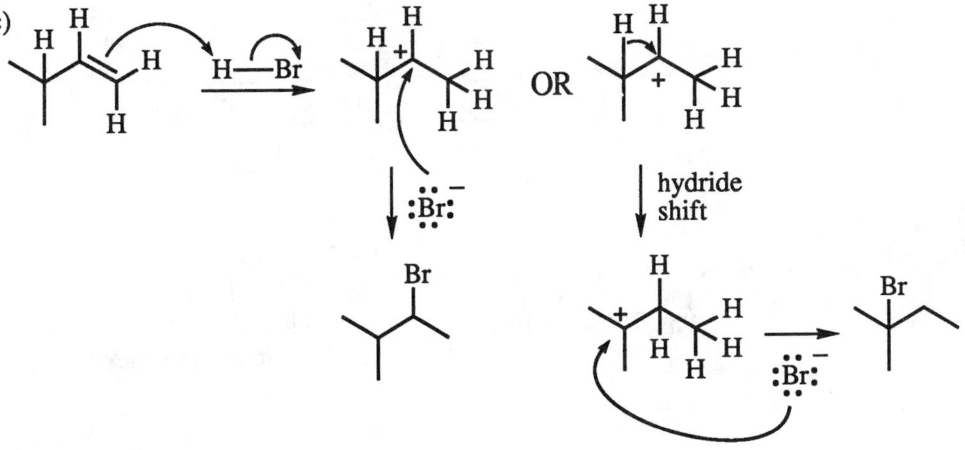
OR
hydride shift

8-41 continued

(d)

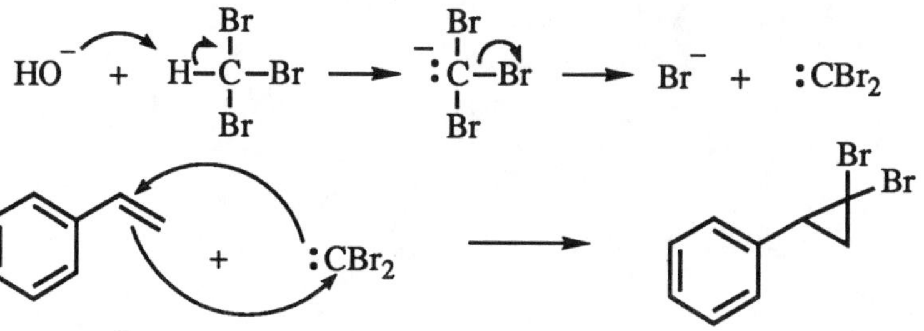

(e)

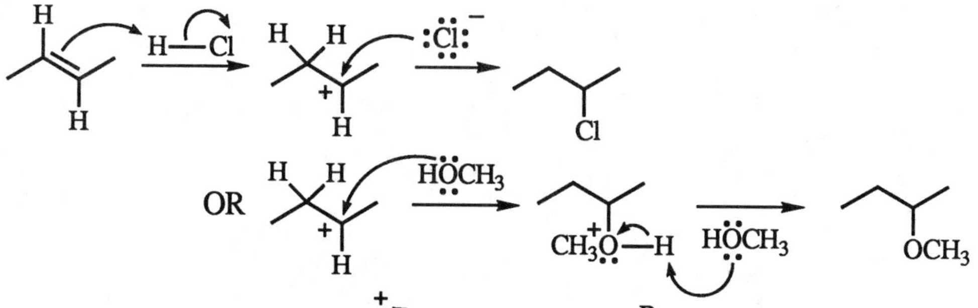

(f)

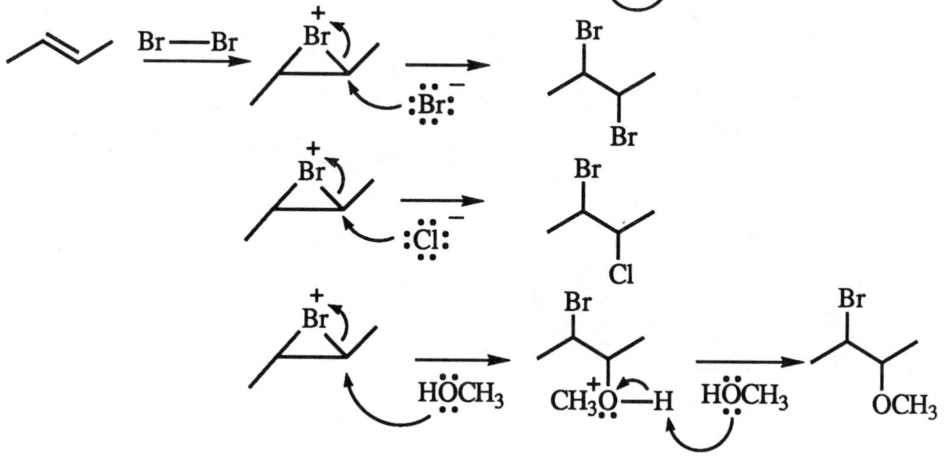

(g)

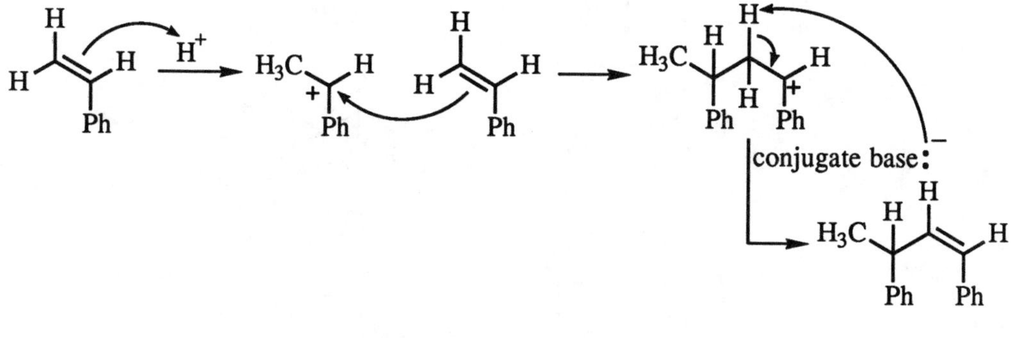

8-42

(a) 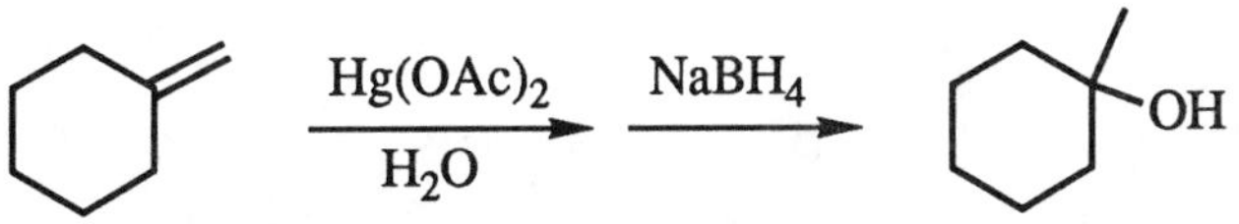
Hg(OAc)$_2$
H_2O
$NaBH_4$
OH

(b) 
HBr
ROOR
Br

(c) 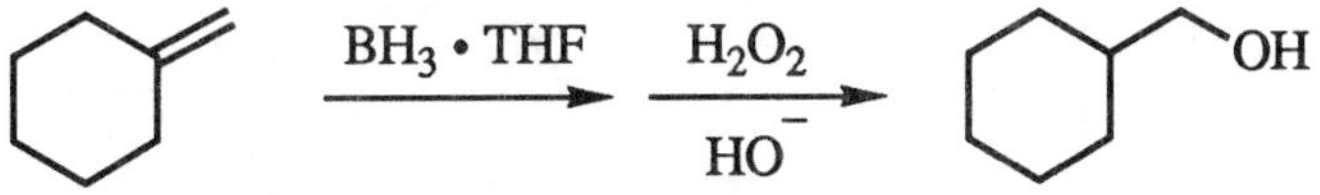
BH_3 • THF
H_2O_2
HO^-
OH

(d) 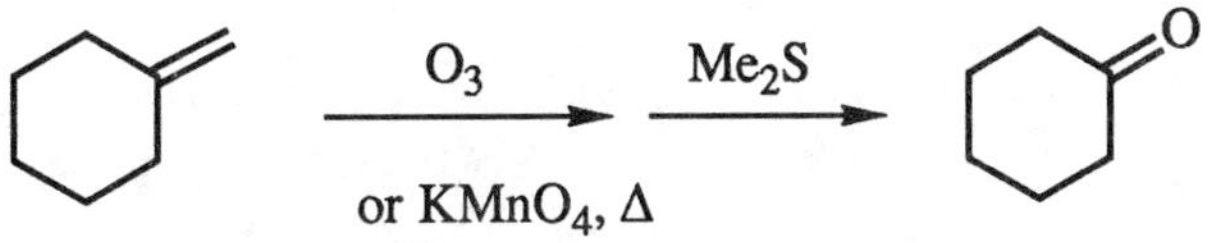
O_3
Me_2S
or $KMnO_4$, Δ
O

(e) 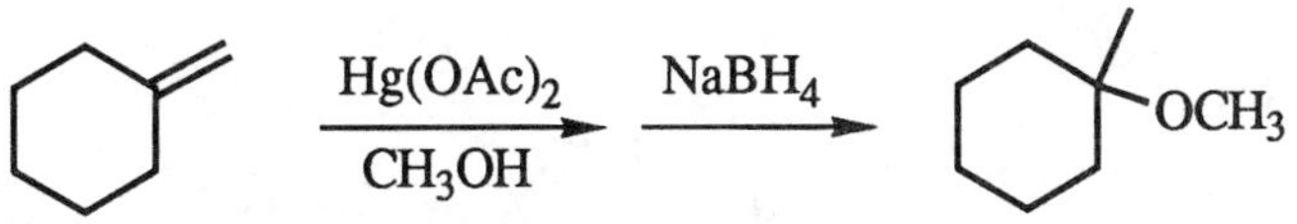
Hg(OAc)$_2$
CH_3OH
$NaBH_4$
OCH_3

(f) 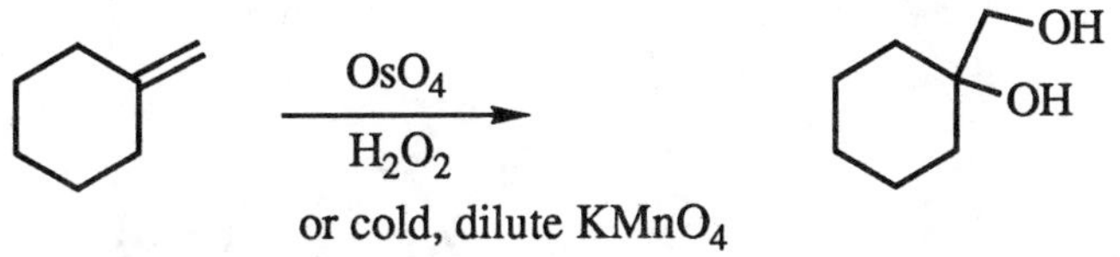
OsO_4
H_2O_2
or cold, dilute $KMnO_4$
OH
OH

(g) 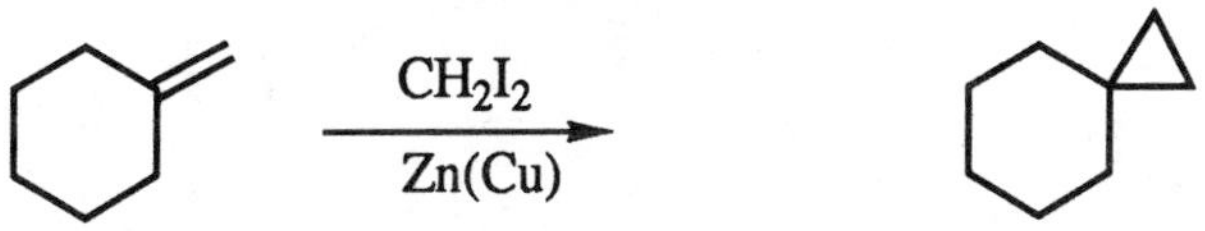
CH_2I_2
Zn(Cu)

(h) 
Cl_2
H_2O
Cl
OH

(i) 
$CHBr_3$
50% NaOH
Br
Br

8-43

(a) 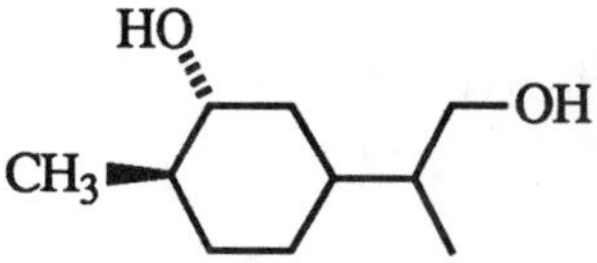

(b) 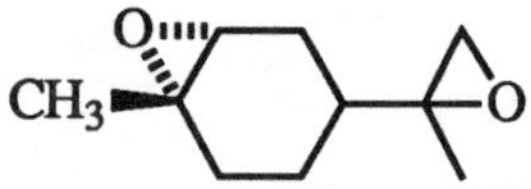

(c) 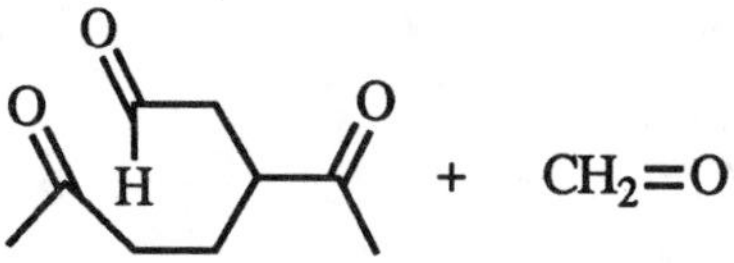

(d) 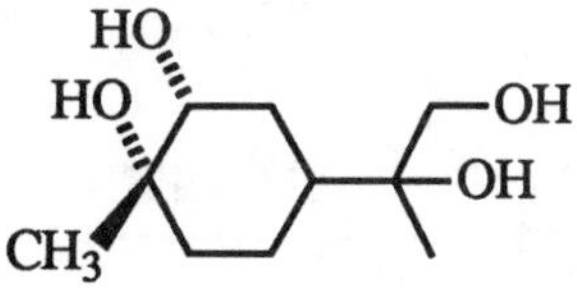

(e) 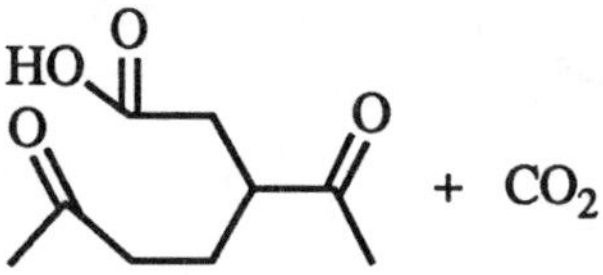

(f) 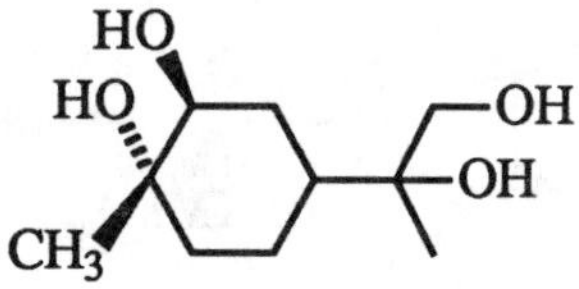

(g) 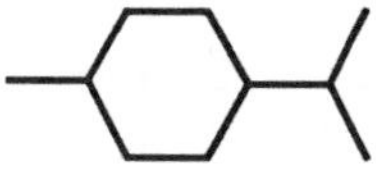

(h) 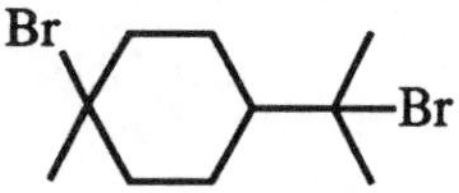

(i) 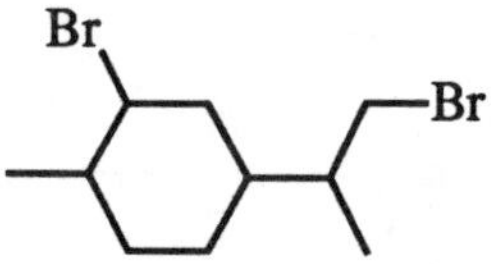

(j) 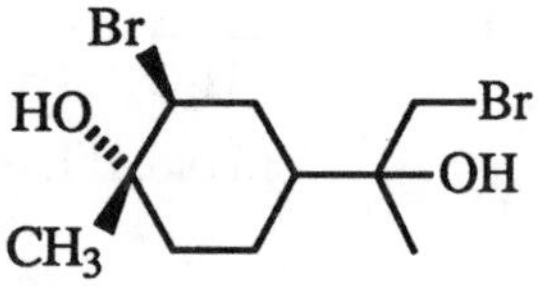

(k) 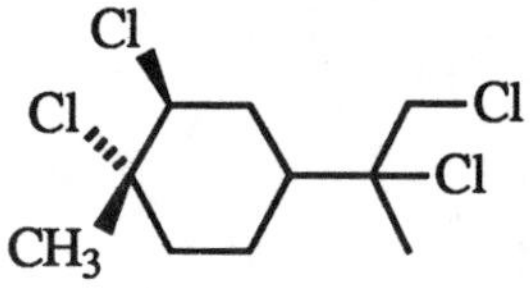

(l) 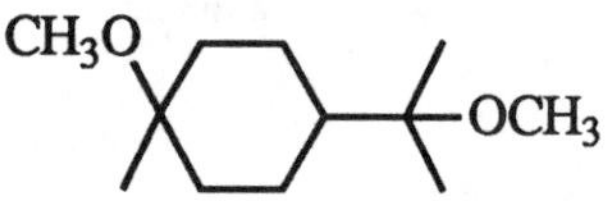

(m) 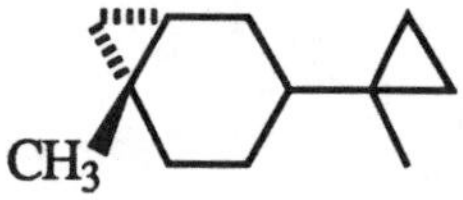

8-44

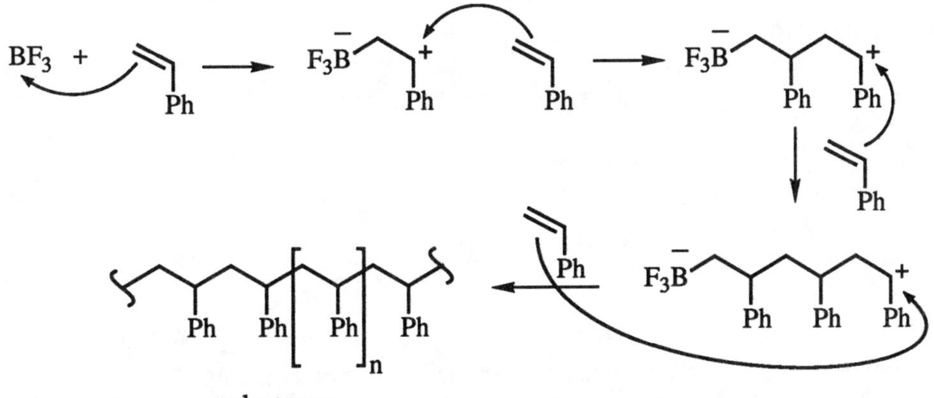

8-45

$$CH_2{=}CH\overset{\overset{\displaystyle O}{\|}}{C}OCH_2CH_3$$ ethyl acrylate

8-46 In each case, the compound (boxed) which produces the more stable carbocation is more reactive.

(a)

(b)

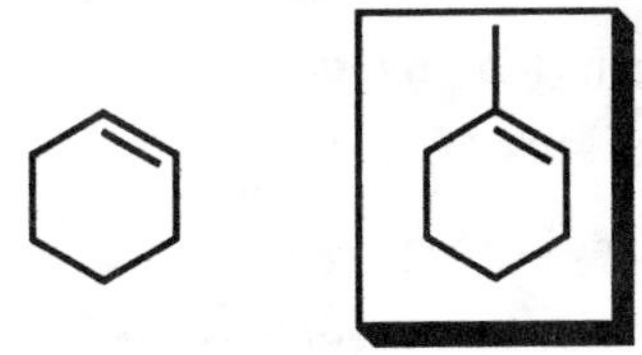

(c)

8-47 Once the bromonium ion is formed, it can be attacked by either nucleophile, bromide or chloride, leading to the mixture of products.

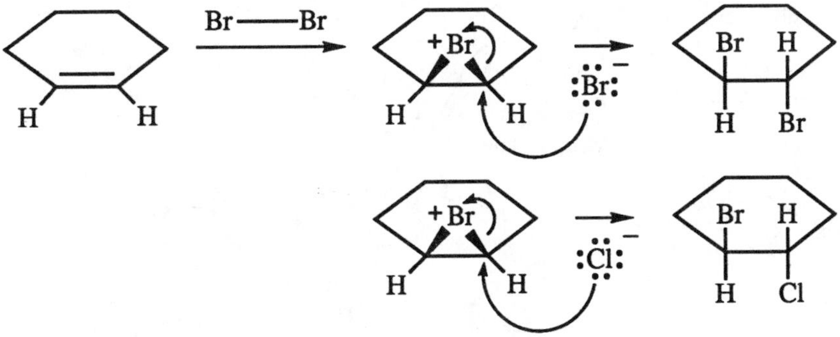

8-48 Two possible orientations of attack of bromine radical are possible:

(A) anti-Markovnikov

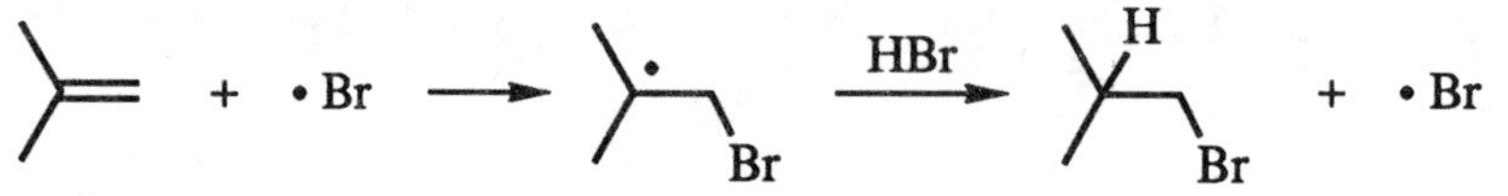

3° radical

(B) Markovnikov

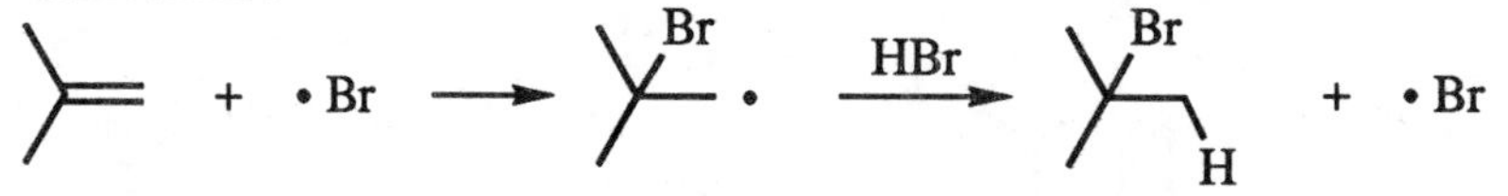

1° radical

The first step in the mechanism is endothermic and rate determining. The 3° radical produced in anti-Markovnikov attack (A) of bromine radical is several kcal/mole more stable than the 1° radical generated by Markovnikov attack (B). The Hammond Postulate tells us that it is reasonable to assume that the activation energy for anti-Markovnikov addition is lower than for Markovnikov addition. This defines the first half of the energy diagram.

The relative stabilities of the final products are somewhat difficult to predict. (Remember that stability of final products does not necessarily reflect relative stabilities of intermediates; this is why a thermodynamic product can be different from a kinetic product.) From bond dissociation energies (kcal/mole) in Table 4-2:

anti-Markovnikov		Markovnikov	
H to 3° C	91	H to 1° C	98
Br to 1° C	68	Br to 3° C	65
	159 kcal/mole		163 kcal/mole

If it takes more energy to break bonds in the Markovnikov product, it must be lower in energy, therefore, more stable—OPPOSITE OF STABILITY OF THE INTERMEDIATES! Thus, it is the anti-Markovnikov product that is the kinetic product, not the thermodynamic product; the anti-Markovnikov product is obtained since its rate-determining step has the lower activation energy.

Now we are ready to construct the energy diagram.

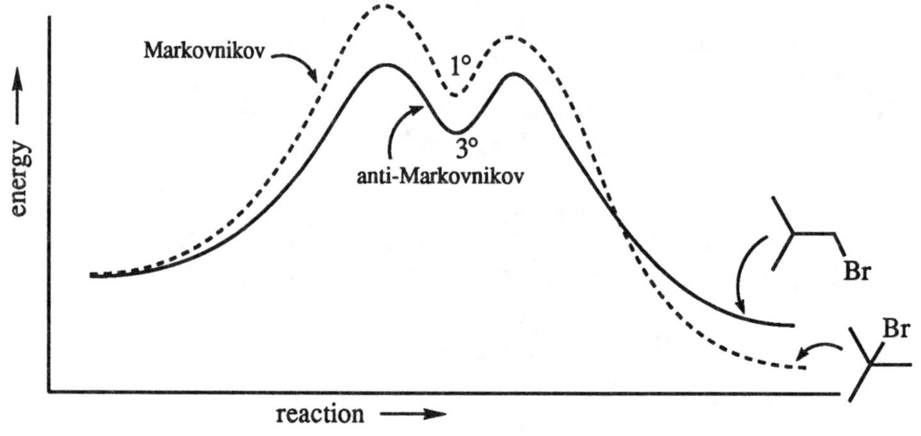

8-49

(a)

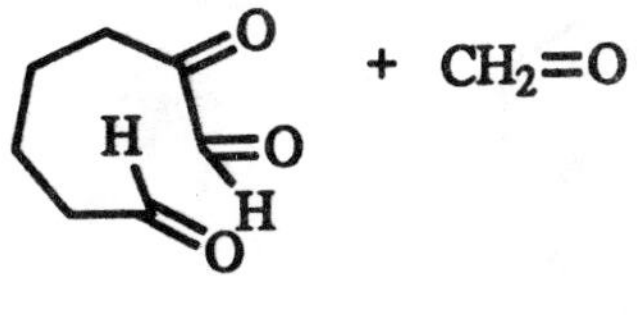
H
O
O
H
O
+ CH2=O

(b)

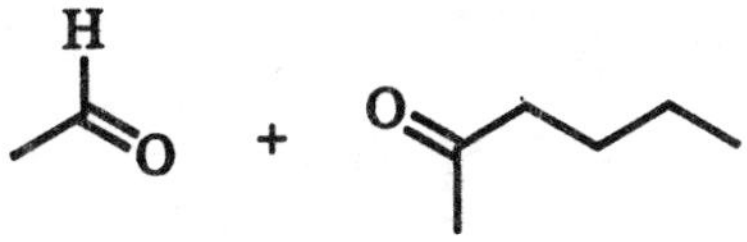
H
O
+
O

(c)

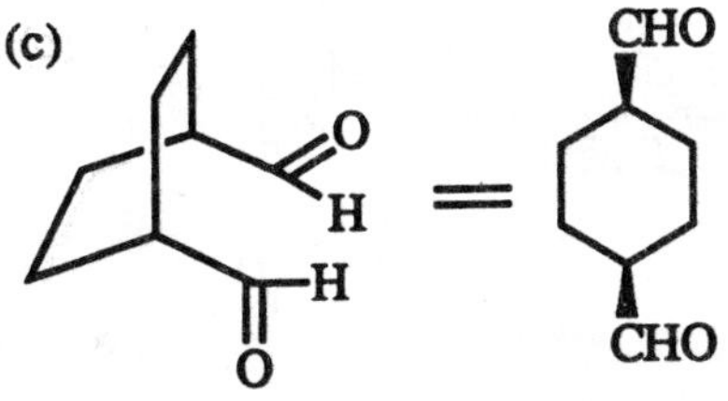
O
H
H
O
CHO
CHO

(d)

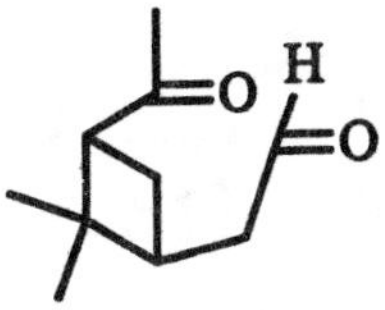
O
H
O

8-50

(a)

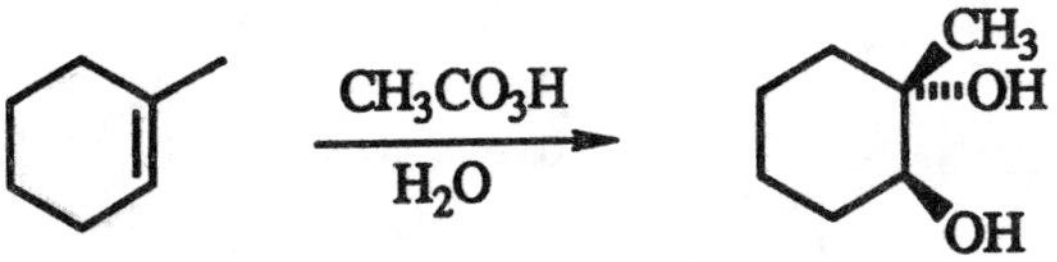
CH3CO3H
H2O
CH3
OH
OH

(b)

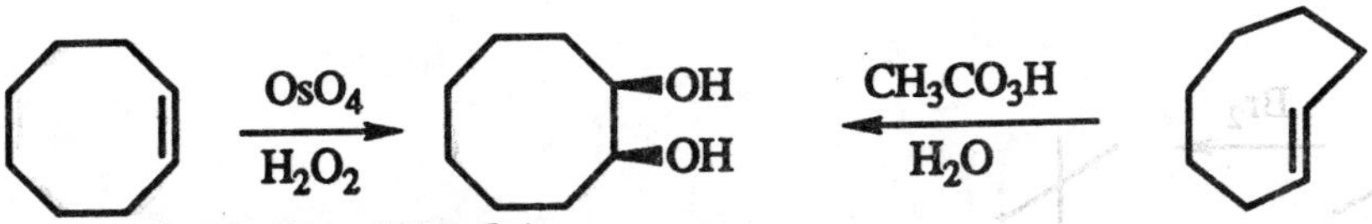
OsO4
H2O2
OH
OH
CH3CO3H
H2O

(or cold, dilute $KMnO_4$)

cis + **syn** *trans* + **anti**

(c)

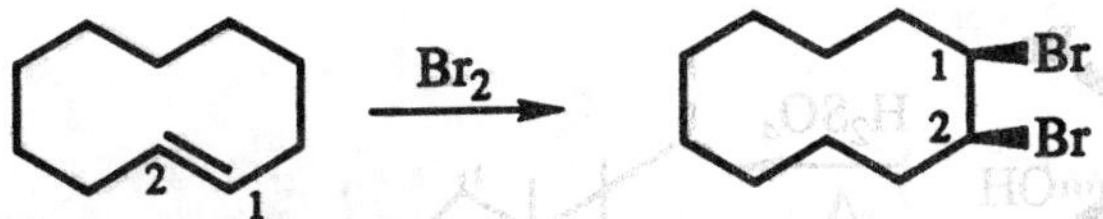
2
1
Br2
1
2
Br
Br

anti addition of Br_2 <u>requires</u> *trans* alkene to give meso product

This structure shows a *trans* alkene in a 10-membered ring. It is a more convenient depiction than the structure to the right.

rotate around C-2

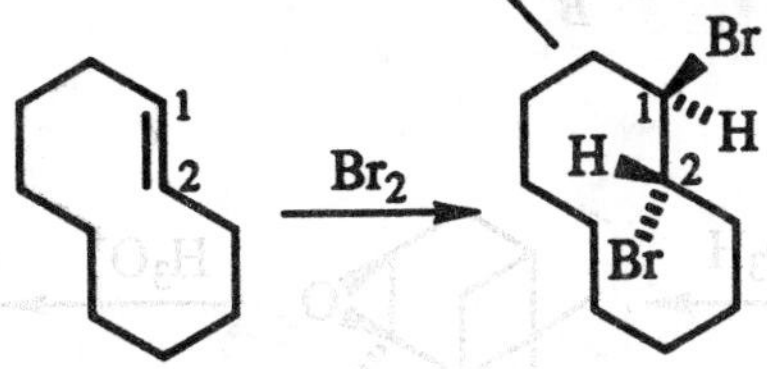
1
2
Br2
Br
1
H
H
2
Br

trans-cyclodecene

(d)

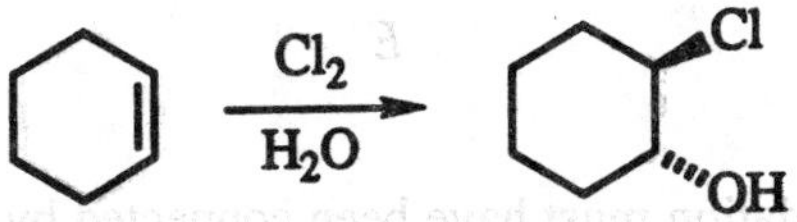
Cl2
H2O
Cl
OH

8-50 continued

(e) 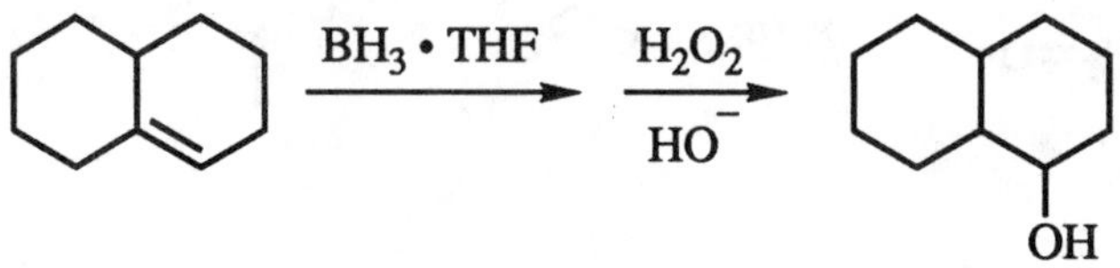

(f) 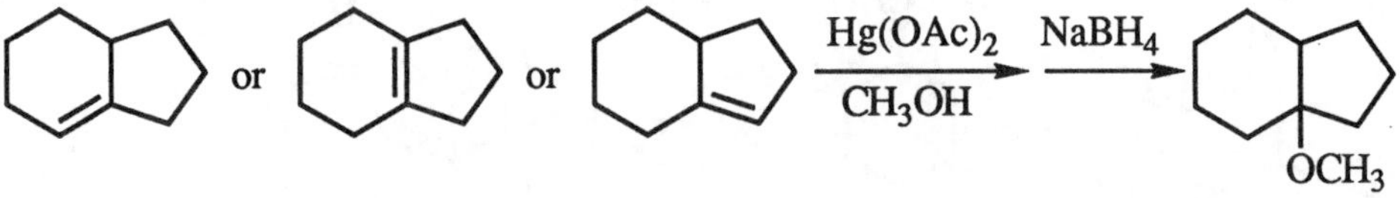

8-51 The clue to the structure of α-pinene is the ozonolysis. Working backwards shows us the alkene position.

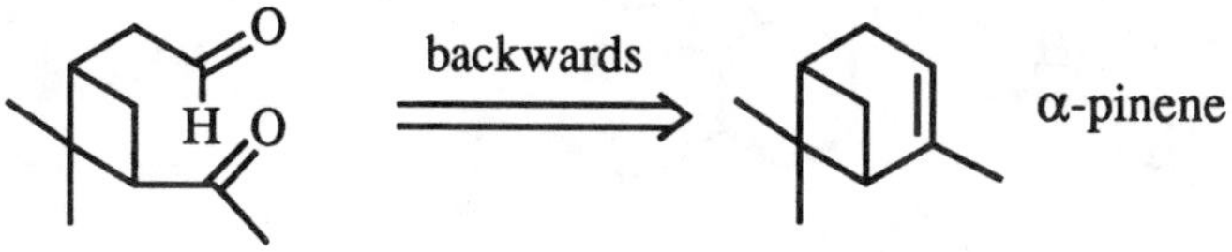

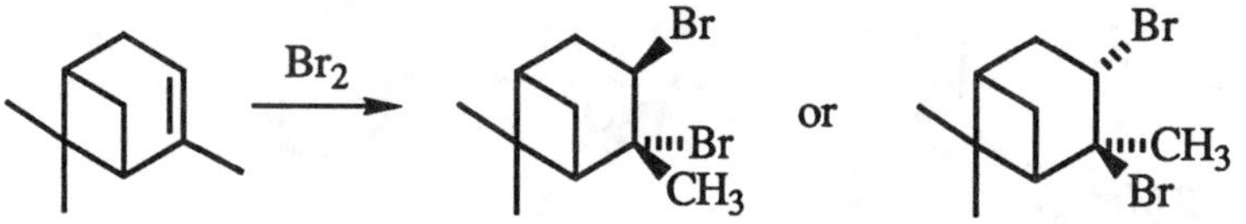

A *A*

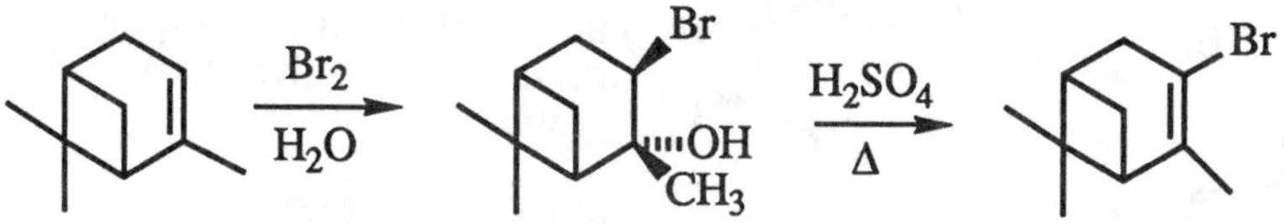

Saytzeff product

B *C*

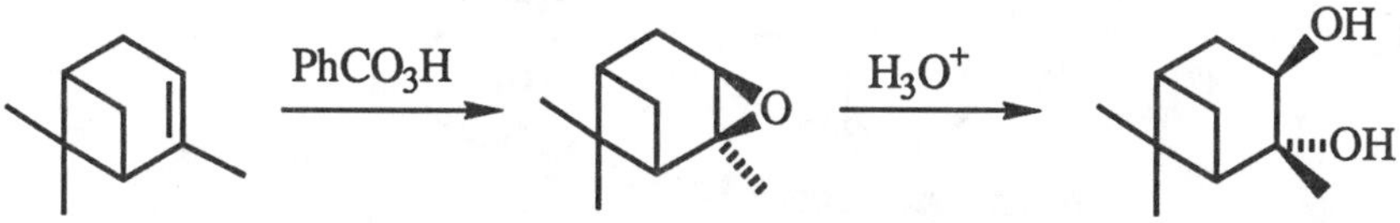

D *E*

8-52 The two products from permanganate oxidation must have been connected by a double bond at the carbonyl carbons. Whether the alkene was *E* or *Z* cannot be determined by this experiment.

$$CH_3(CH_2)_{12}CH{=}CH(CH_2)_7CH_3$$

8-53

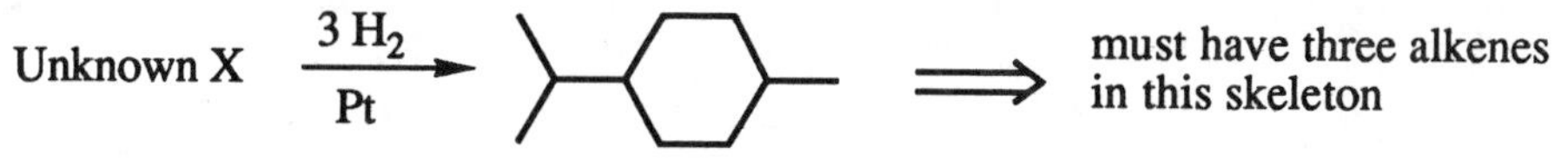
Unknown X
3 H2
Pt
must have three alkenes in this skeleton

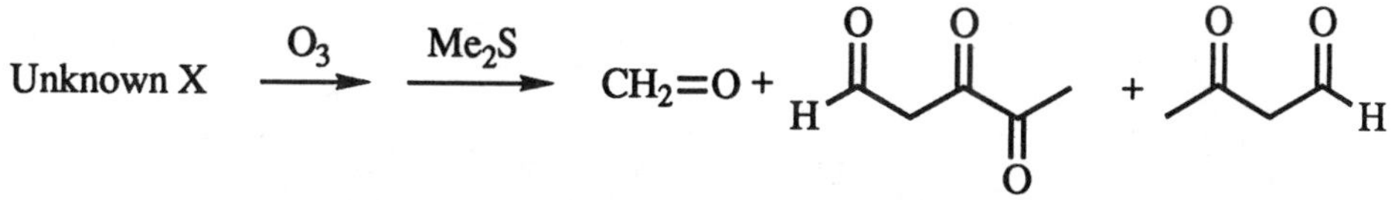
Unknown X
O3
Me2S
CH2=O +
O
O
H
O
+
O
O
H

There are several ways to attack a problem like this. One is the trial-and-error method, that is, put double bonds in all possible positions until the ozonolysis products match. There are times when the trial-and-error method is useful (as in simple problems where the number of possibilities is few), but this is not one of them.

Let's try logic. Analyze the ozonolysis products carefully—what do you see? There are only two methyl groups, so one of the three terminal carbons in the skeleton (C-8, C-9, or C-10) has to be a $=CH_2$. Do we know which terminal carbon has the double bond? Yes, we can deduce that. If C-10 were double-bonded to C-4, then after ozonolysis, C-8 and C-9 must still be attached to C-7. However, in the ozonolysis products, there is no branched chain, that is, no combination of C-8 + C-9 + C-7 + C-1. What if C-7 had a double bond to C-1? Then we would have acetone, CH_3COCH_3, as an ozonolysis product—we don't. Thus, we can't have a double bond from C-4 to C-10. One of the other terminal carbons (C-8) must have a double bond to C-7.

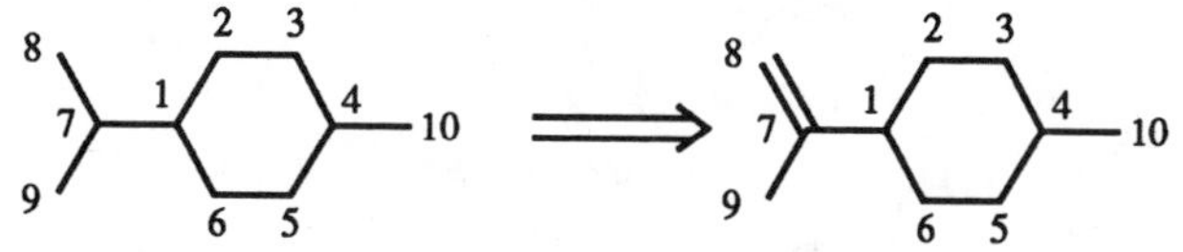
8
2 3
1
4
7
10
9
6 5
8
2 3
1
4
7
10
9
6 5

The other two double bonds have to be in the ring, but where? The products do not have branched chains, so double bonds must appear at both C-1 and C-4. There are only two possibilities for this requirement.

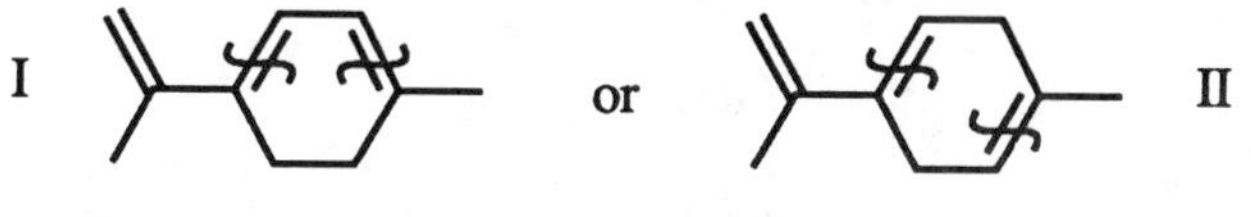
I
or
II

Ozonolysis of I would give fragments containing one carbon, two carbons, and seven carbons. Ozonolysis of II would give fragments containing one carbon, four carbons, and five carbons. Aha! Our mystery structure must be II.

(Editorial: Science is more than a collection of facts. The application of observation and logic to solve problems by *deduction* and *inference* are critical scientific skills, ones that distinguish humans from algae.)

8-54 In this type of problem, begin by determining which bonds are broken and which are formed. These will always give clues as to what is happening.

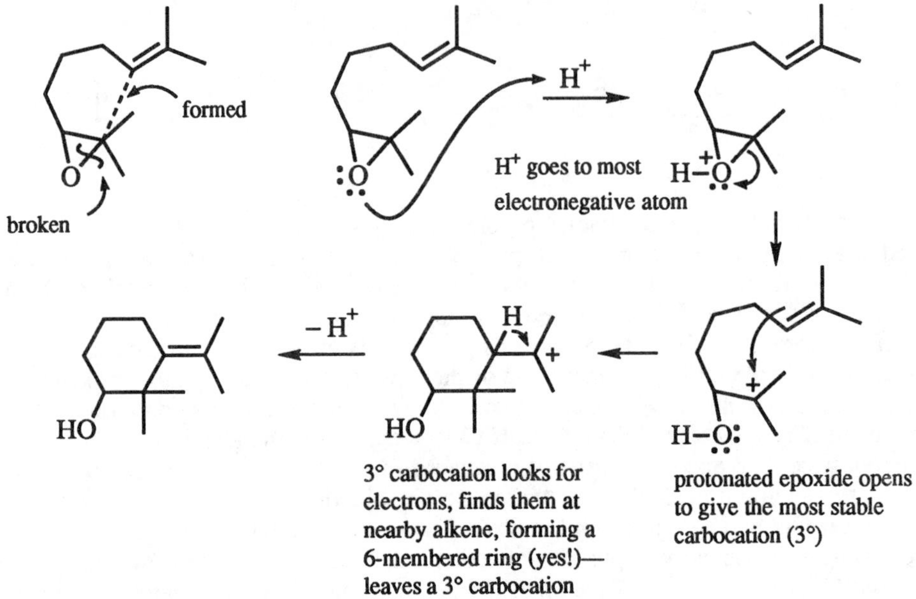

formed
broken
H⁺
H⁺ goes to most electronegative atom
H–O
– H⁺
H
HO
HO
H–O:
3° carbocation looks for electrons, finds them at nearby alkene, forming a 6-membered ring (yes!)—leaves a 3° carbocation
protonated epoxide opens to give the most stable carbocation (3°)

8-55 See answer to Problem 8-33.

(a)

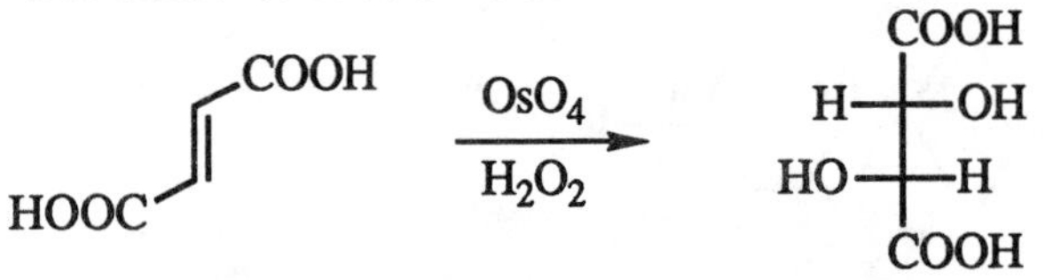

COOH
HOOC
OsO4
H2O2
COOH
H—OH
HO—H
COOH

trans + **syn** → racemic

(b)

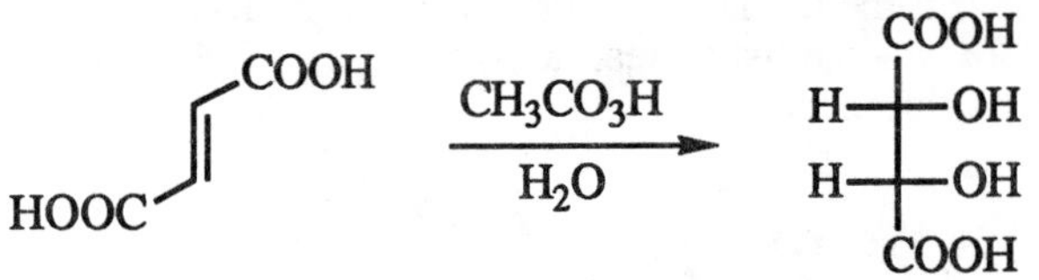

COOH
HOOC
CH3CO3H
H2O
COOH
H—OH
H—OH
COOH

trans + **anti** → meso

(c)

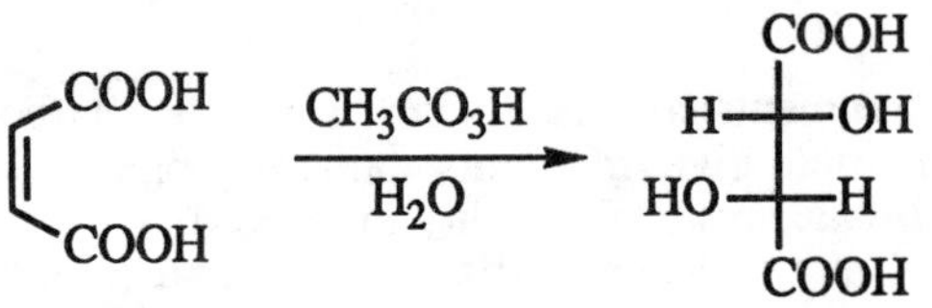

COOH
COOH
CH3CO3H
H2O
COOH
H—OH
HO—H
COOH

cis + **anti** → racemic

(d)

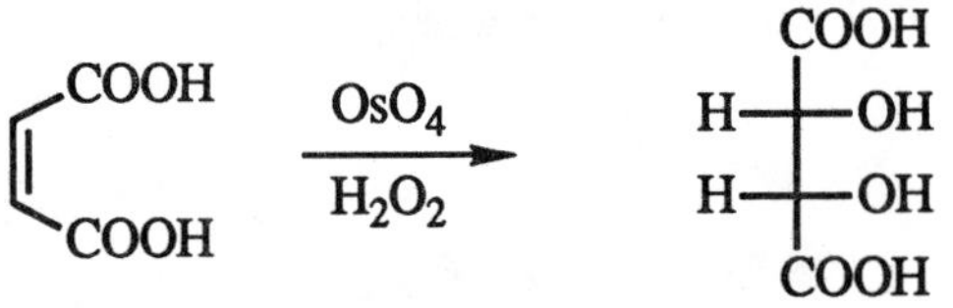

COOH
COOH
OsO4
H2O2
COOH
H—OH
H—OH
COOH

cis + **syn** → meso

8-56

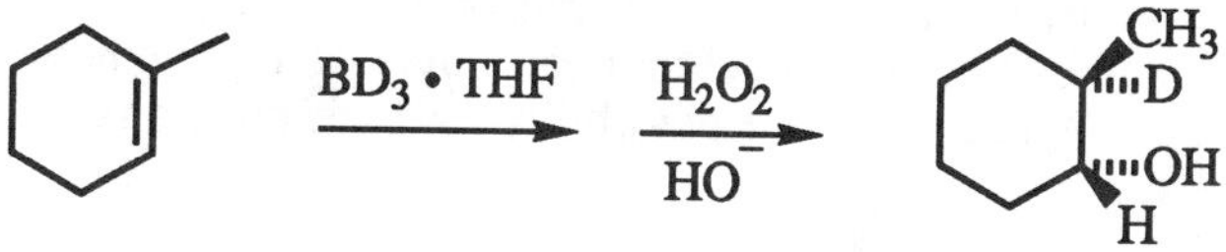

8-57 (a) Fumarase catalyzes the addition of H and OH, a hydration reaction.
(b) Fumaric acid is planar and cannot be chiral. Malic acid does have a chiral center and is chiral. The enzyme-catalyzed reaction produces only the *S* enantiomer, so the product must be optically active.
(c) One of the fundamental rules of stereochemistry is that optically inactive starting materials produce optically inactive products. Sulfuric-acid-catalyzed hydration would produce a racemic mixture of malic acid, that is, equal amounts of *R* and *S*.
(d) If the product is optically active, then either the starting materials or the catalyst were chiral. We know that water and fumaric acid are not chiral, so we must infer that fumarase is chiral.
(e) The D and the OD are on the "same side" of the Fischer projection (sometimes called the "erythro" stereoisomer). These are produced from either: (1) syn addition to *cis* alkenes, or (2) anti addition to *trans* alkenes. We know that fumaric acid is *trans*, so the addition of D and OD must necessarily be anti.
(f) Hydroboration is a syn addition.

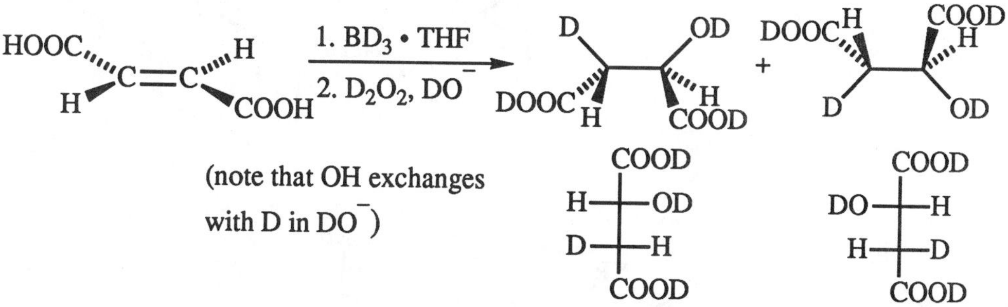

(note that OH exchanges with D in DO^-)

As expected, *trans* alkene plus syn addition puts the two groups on the "opposite" side of the Fischer projection (sometimes called "threo").

8-58 Each alkene will produce two carbonyls upon ozonolysis or permanganate oxidation. Oxidation of the unknown generated four carbonyls, so the unknown must have had two alkenes. There is only one possibility for their positions.

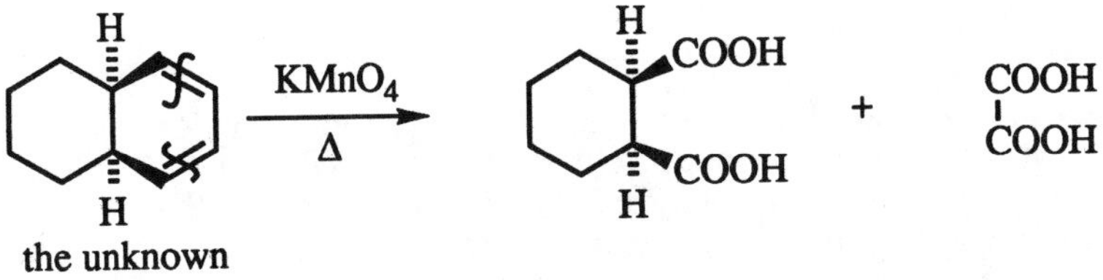

8-59

(a)

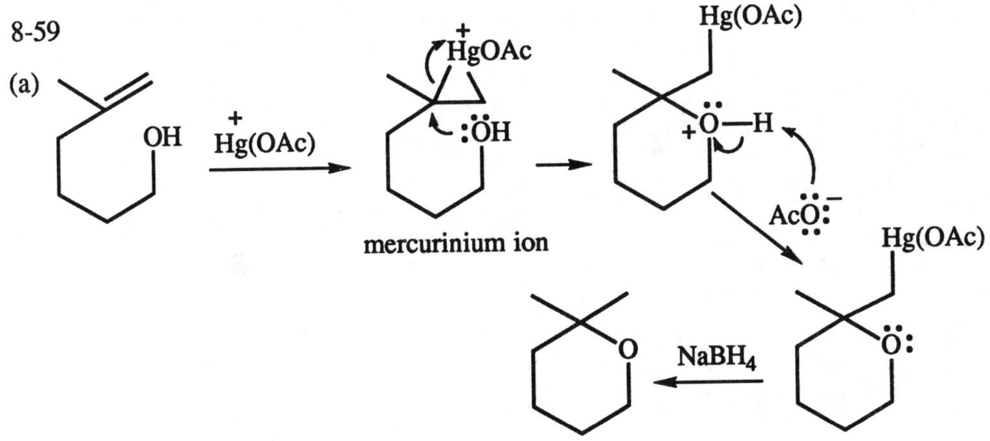

(b)

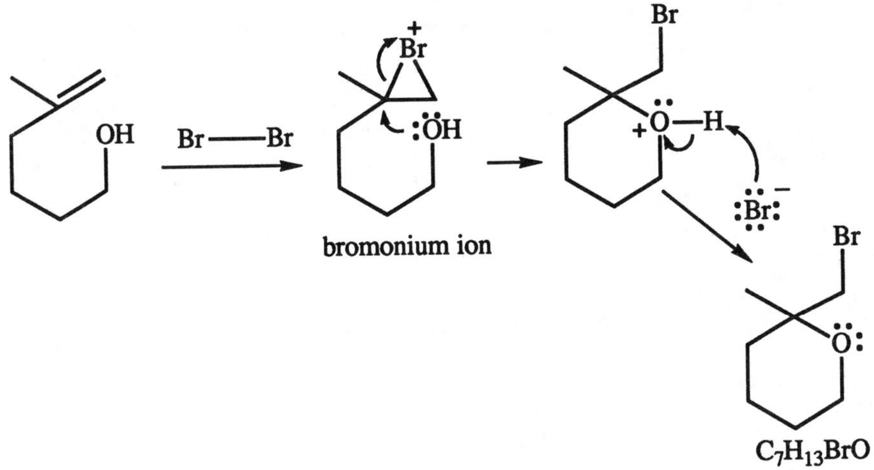

8-60 The addition of BH_3 to an alkene is reversible. Given heat and time, the borane will eventually "walk" its way to the end of the chain through a series of addition-elimination cycles. The most stable alkylborane has the boron on the end carbon; eventually, the series of equilibria lead to that product which is oxidized to the primary alcohol.

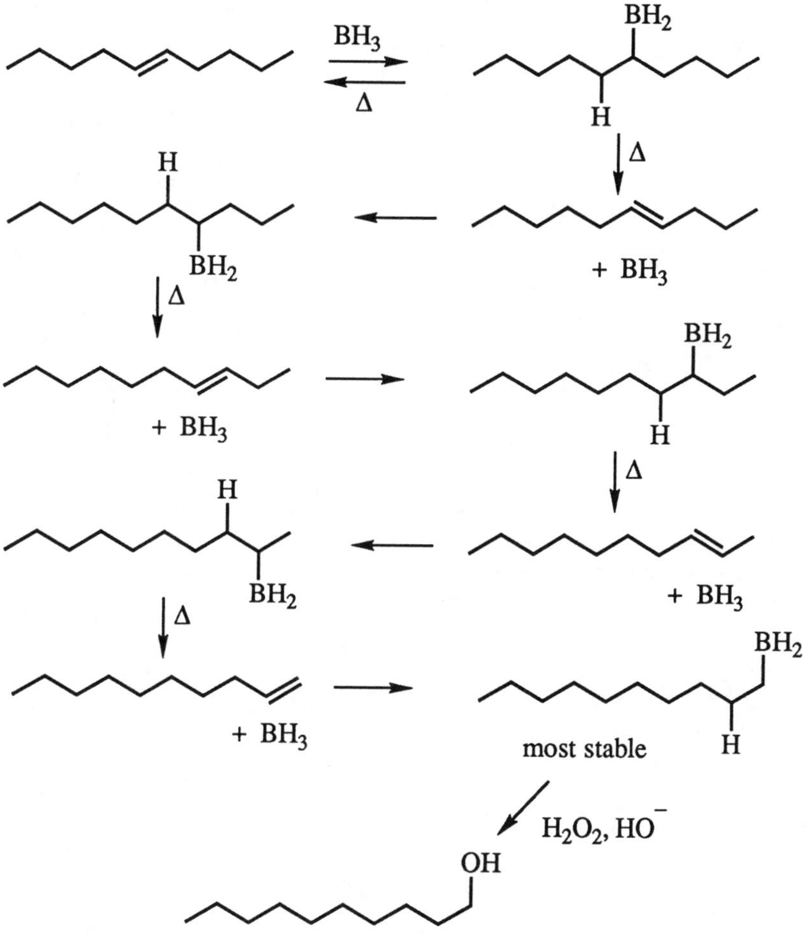

8-61 First, we explain *how* the mixture of stereoisomers results, then *why*.

We have seen many times that the bridged halonium ion permits attack of the nucleophile only from the opposite side.

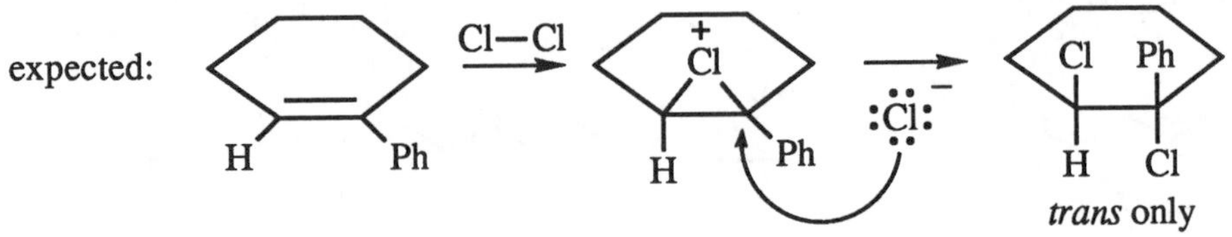

A mixture of *cis* and *trans* could result only if attack of chloride were possible from both top and bottom, something possible only if a *carbocation* existed at this carbon.

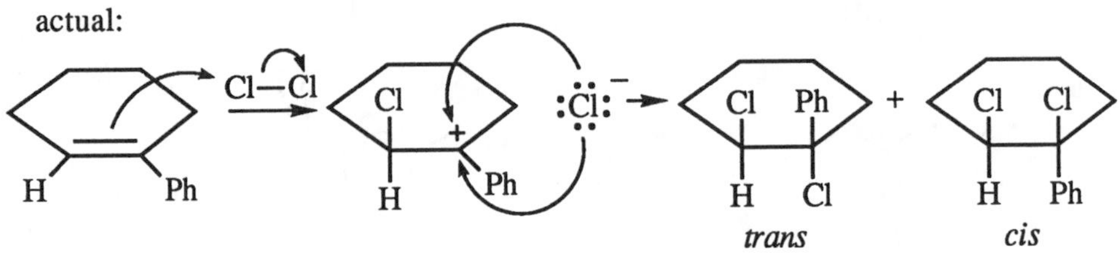

Why does a carbocation exist here? Not only is it 3°, *it is also next to a benzene ring (benzylic) and therefore resonance-stabilized.* This resonance stabilization would be forfeited in a halonium ion intermediate.

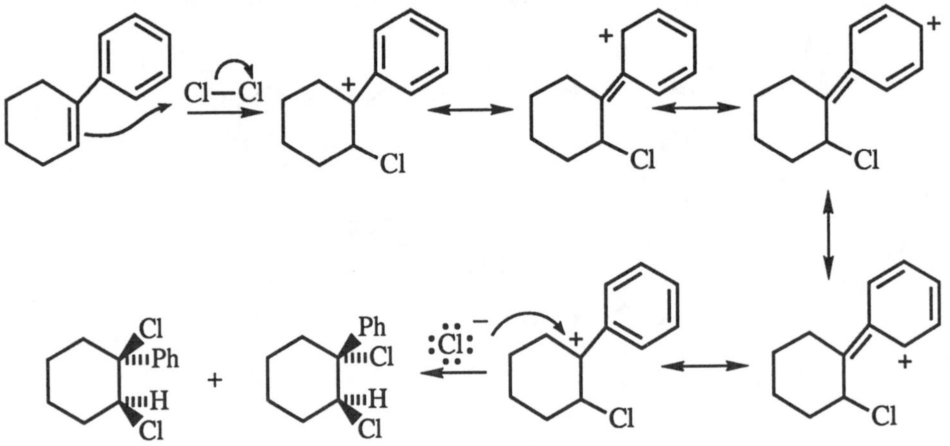

8-62

Unknown X, C_5H_9Br, has one element of unsaturation. X reacts with neither bromine nor $KMnO_4$, so the unsaturation in X cannot be an alkene; it must be a ring.

Upon treatment with strong base, X loses H and Br to give Y, C_5H_8, which does react with bromine and $KMnO_4$; it must have an alkene and a ring. Only one isomer is formed.

Ozonolysis of Y gives Z, $C_5H_8O_2$, which contains all the original carbons, so the alkene cleaved in the ozonolysis had to be in the ring.

There are several possible answers:

1) 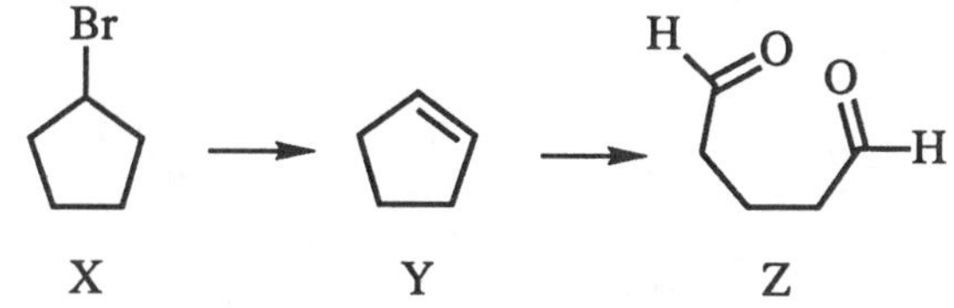

2) 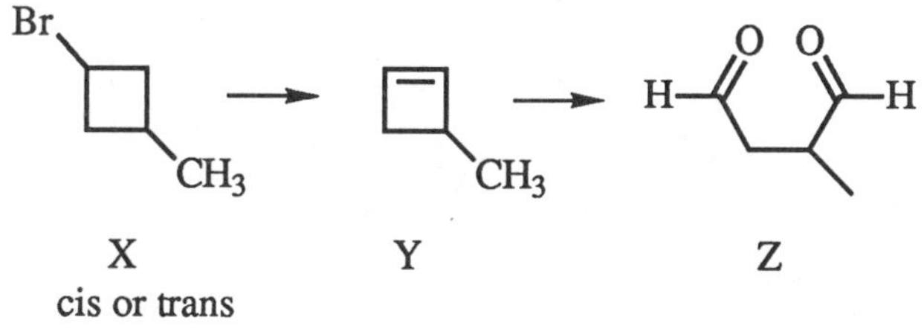

3) 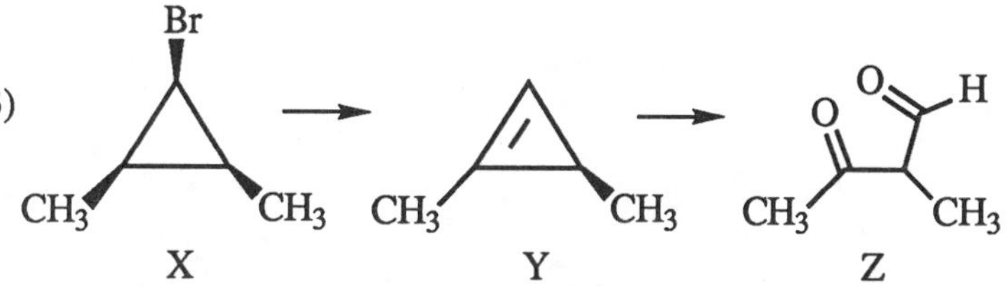

(On paper, example 3 looks fine. In reality, cyclopropane rings are so strained that they react with bromine or $KMnO_4$.)

Examples of structures which fit the molecular formula for X but not the chemistry

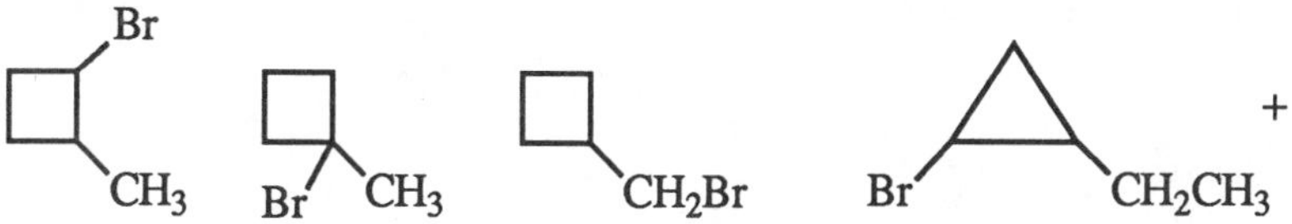

+ several more

CHAPTER 9—STRUCTURE AND SYNTHESIS OF ALCOHOLS

9-1 (a) 2-phenyl-2-propanol
(b) 5-bromo-2-heptanol
(c) 4-methyl-3-cyclohexen-1-ol ("1" is optional)
(d) *trans*-2-methyl-1-cyclohexanol ("1" is optional)
(e) (*E*)-2-chloro-3-methyl-2-penten-1-ol
(f) (2*R*,3*S*)-2-bromo-3-hexanol

9-2 IUPAC name first, then common name.

(a) cyclopropanol; cyclopropyl alcohol
(b) 2-methyl-2-propanol; *t*-butyl alcohol
(c) 1-cyclobutyl-2-propanol; no common name
(d) 3-methyl-1-butanol; isopentyl alcohol (also isoamyl alcohol)

9-3

(a)

1-propanol

2-propanol

(b)

1-butanol

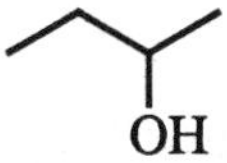

2-butanol

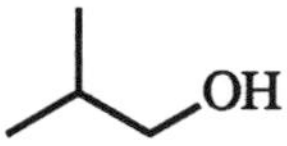

2-methyl-1-propanol

2-methyl-2-propanol

(c) C_4H_8O has one element of unsaturation, either a double bond or a ring.

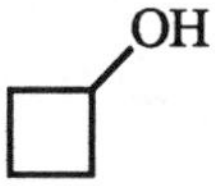

cyclobutanol

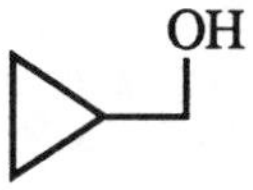

cyclopropylmethanol

1-methylcyclopropanol

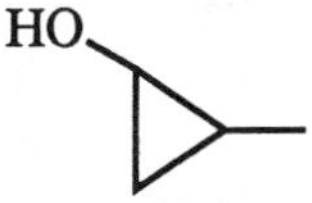

2-methylcyclopropanol
cis or *trans*

3-buten-1-ol

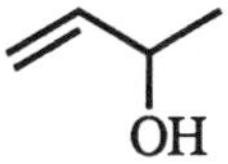

3-buten-2-ol

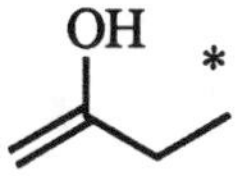

1-buten-2-ol

1-buten-1-ol (*E* or *Z*)

2-buten-1-ol
(*E* or *Z*)

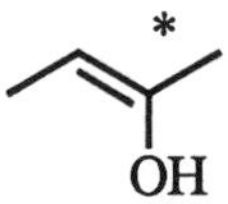

2-buten-2-ol
(*E* or *Z*)

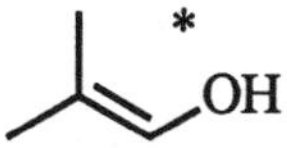

2-methyl-1-
propen-1-ol

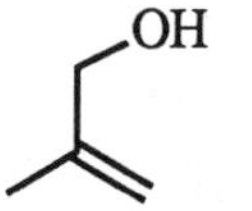

2-methyl-2-
propen-1-ol

*These compounds with the OH bonded directly to the alkene are called "enols" or "vinyl alcohols." You will learn later that these are unstable, although the structures are legitimate.

9-4 (a) 8,8-dimethyl-2,7-nonanediol
(b) 1,8-octanediol
(c) *cis*-2-cyclohexene-1,4-diol
(d) 3-cyclopentyl-2,4-heptanediol

9-5 (a) Cyclohexanol is more soluble because its alkyl group is more compact than in 1-hexanol.
(b) 4-Methylphenol is more soluble because its hydrocarbon portion is more compact than in 1-heptanol, and phenols form particularly strong hydrogen bonds with water.
(c) 3-Ethyl-3-hexanol is more soluble because its alkyl protion is more spherical than in 2-octanol.
(d) Cyclooctane-1,4-diol is more soluble because it has two OH groups which can hydrogen bond with water, whereas 2-hexanol has only one OH group. (The ratio of carbons to OH is 4 to 1 in the former compound and 6 to 1 in the latter; the smaller this ratio, the more soluble.)
(e) These are enantiomers and will have identical solubility.

9-6 Dimethylamine molecules can hydrogen bond among themselves so it takes more energy (higher temperature) to separate them from each other. Trimethylamine has no N-H and cannot hydrogen bond, so it takes less energy to separate these molecules from each other.

9-7 (a) Methanol is more acidic than *t*-butyl alcohol. The greater the substitution, the lower the acidity.
(b) 1-Chloroethanol is more acidic because the electron-withdrawing chlorine atom is closer to the OH group than in 2-chloroethanol.
(c) 2,2-Dichloroethanol is more acidic because two electron-withdrawing chlorine atoms increase acidity more than just the one chlorine in 2-chloroethanol.

9-8 Resonance forms of phenoxide anion show the negative charge delocalized onto the ring only at carbons 2, 4, and 6:

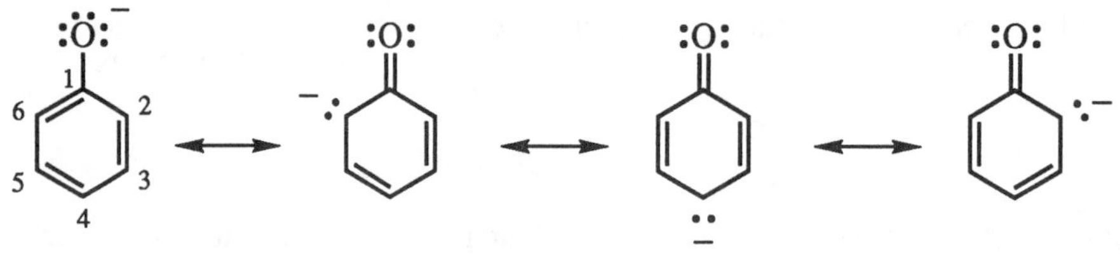

Nitro group at position 2

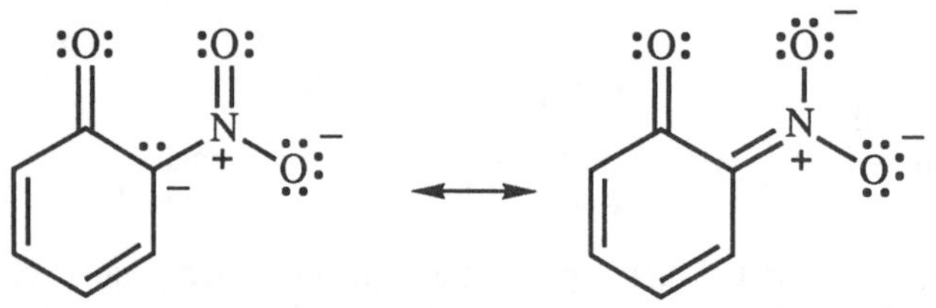

Nitro at position 2 delocalizes negative charge.

9-8 **continued**

Nitro group at position 3

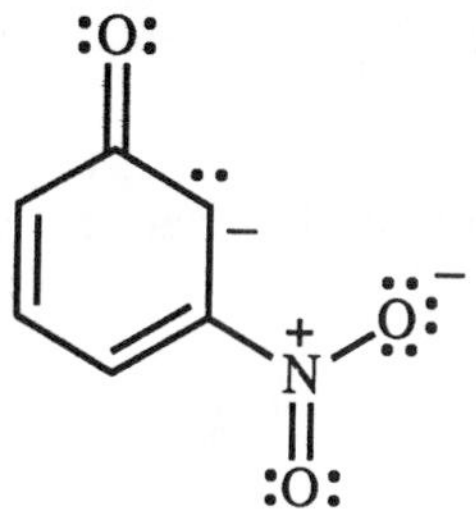

Nitro at position 3 cannot delocalize negative charge at position 2 or 4—no resonance stabilization.

Nitro group at position 4

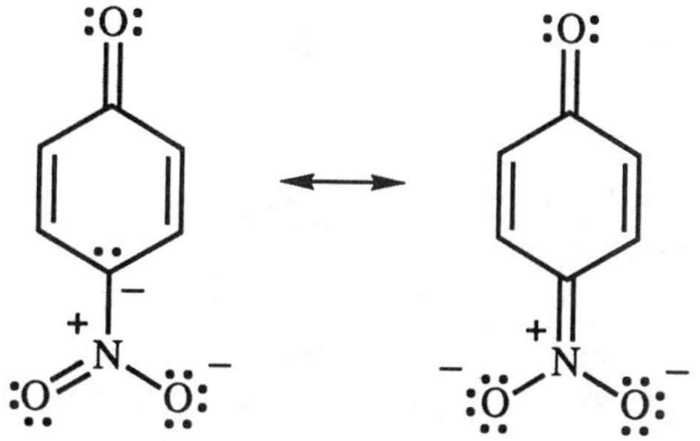

Nitro at position 4 delocalizes negative charge.

Only when the nitro group is at one of the negative carbons will the nitro have a stabilizing effect. Thus, 2-nitrophenol and 4-nitrophenol are substantially more acidic than phenol itself, but 3-nitrophenol is only slightly more acidic than phenol.

9-9 (a) The structure on the left is a phenol because the OH is bonded to a benzene ring. As a phenol, it will be acidic enough to react with sodium hydroxide to generate a phenoxide ion which will be fairly soluble in water.
(b) Both of these organic compounds will be soluble in an organic solvent like dichloromethane. Shaking this organic solution with aqueous sodium hydroxide will ionize the phenol, making it more polar and water soluble; it will be extracted from the organic layer into the water layer, while the alcohol will remain in the organic solvent. Separating these immiscible solvents will separate the original compounds. The alcohol can be retrieved by evaporating the organic solvent. The phenol can be isolated by acidifying the basic aqueous solution and filtering if the phenol is a solid, or separating the layers if the phenol is a liquid.

9-10 The Grignard reaction needs a solvent containing an ether functional group: (b), (f), (g), and (h) are possible solvents. Dimethyl ether, (b), is a gas at room temperature, however, so it would have to be liquified at low temperature for it to be a useful solvent.

9-11 (a) CH_3CH_2MgBr

(b) 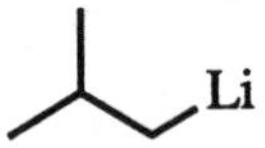 + LiI

(c) 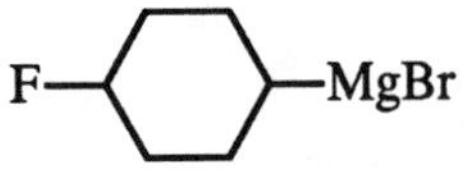

(d) 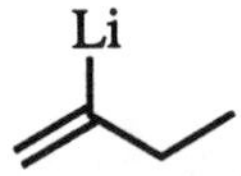 + LiCl

9-12 Any of three halides—chloride, bromide, iodide, but not fluoride—can be used.

(a) 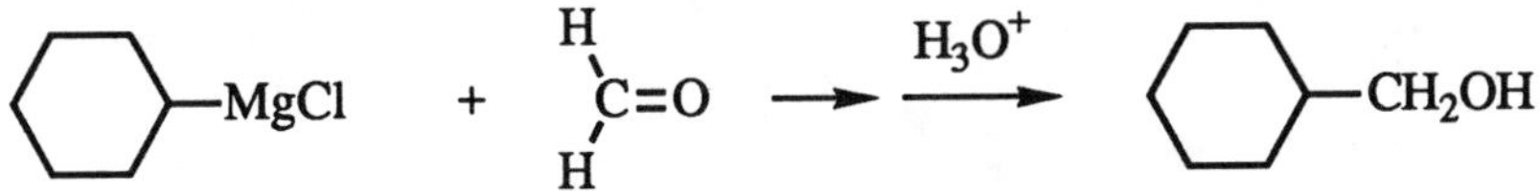

(b) 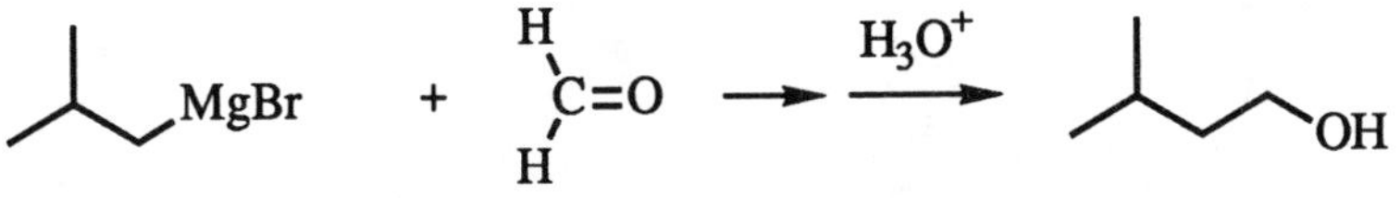

(c) 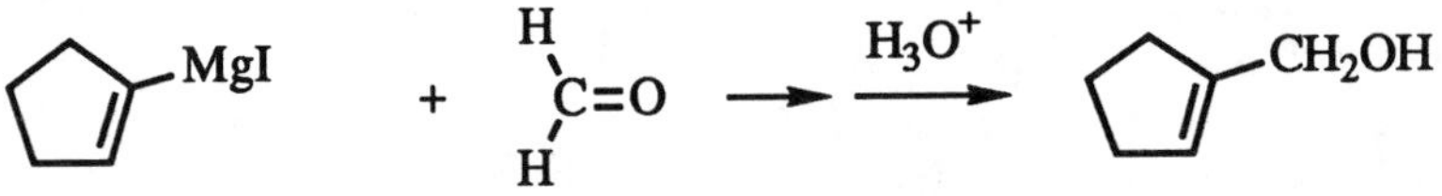

9-13 Any of three halides—chloride, bromide, iodide, but not fluoride—can be used.

(a) two methods

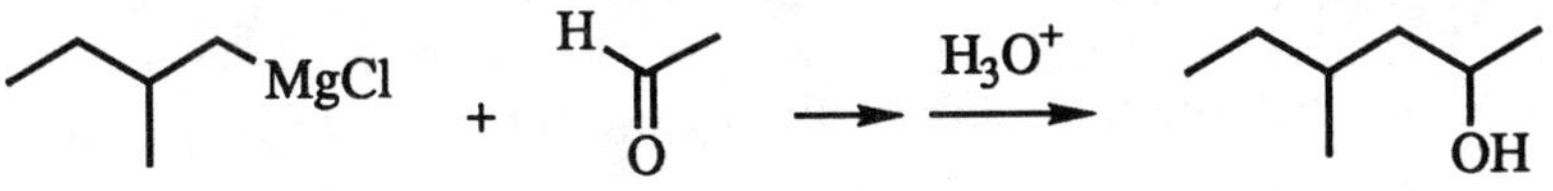

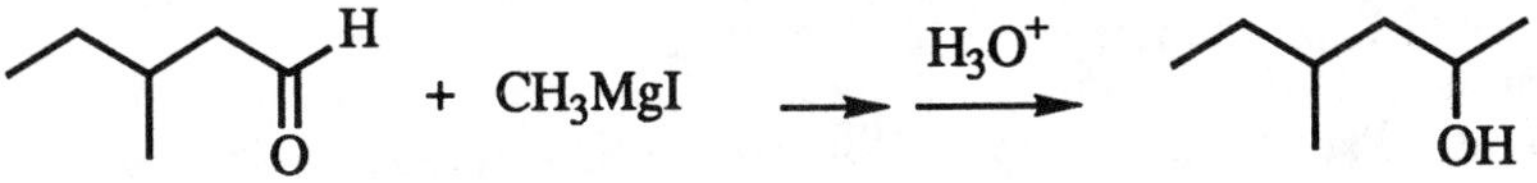

9-13 continued

(b) two methods

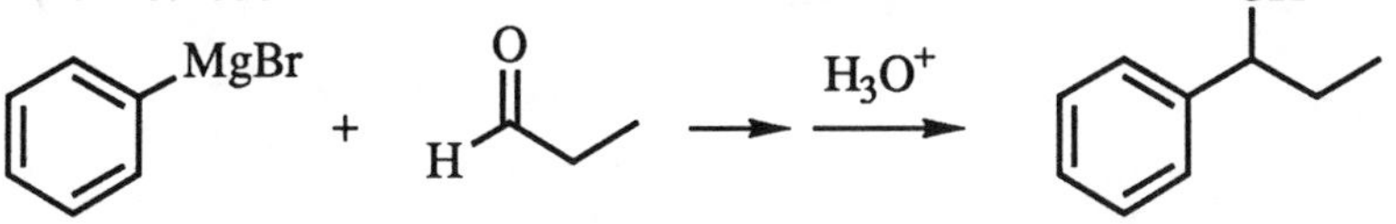

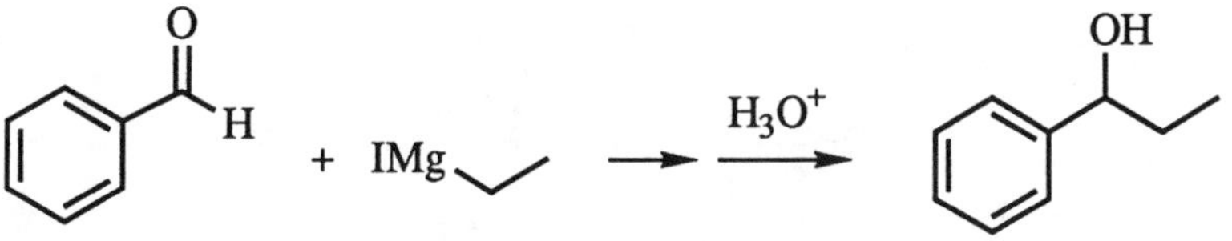

(c)

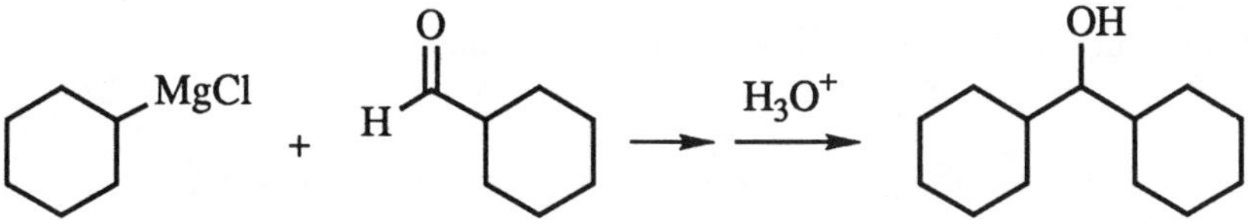

9-14

(a)

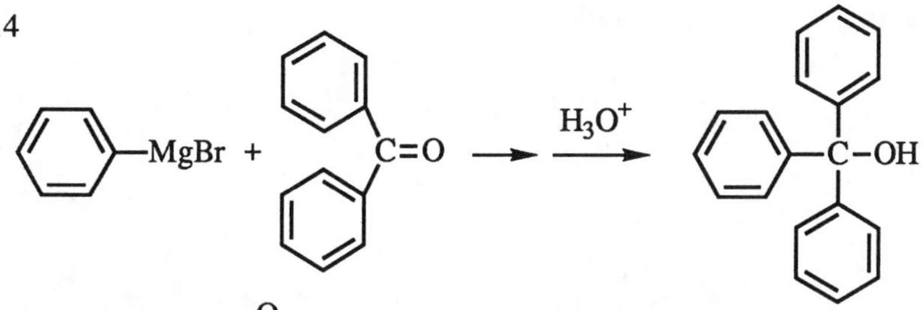

(b)

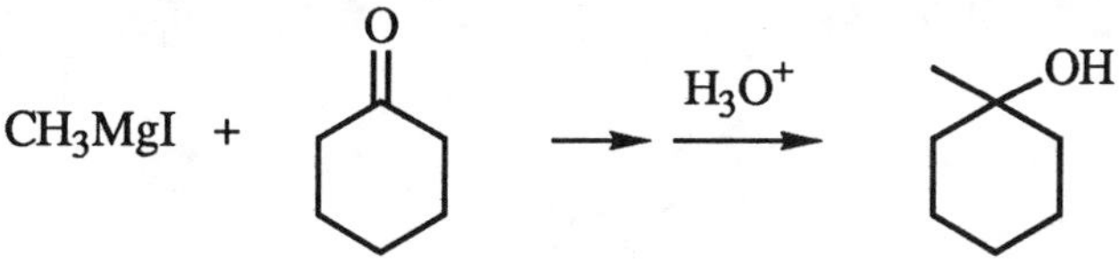

(c) two methods

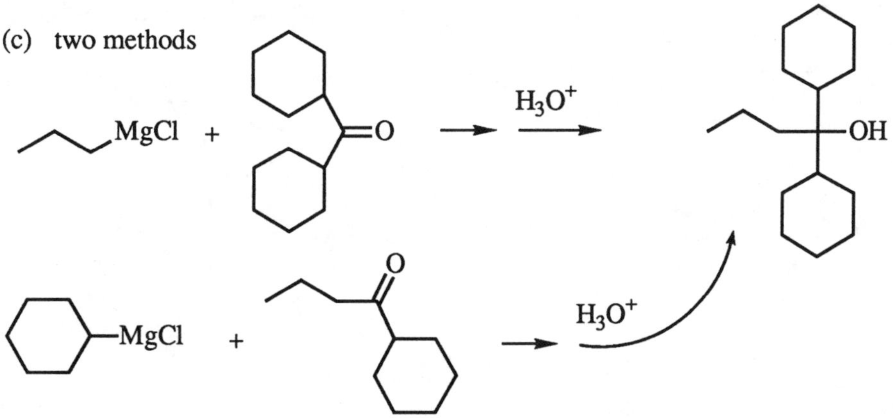

9-15

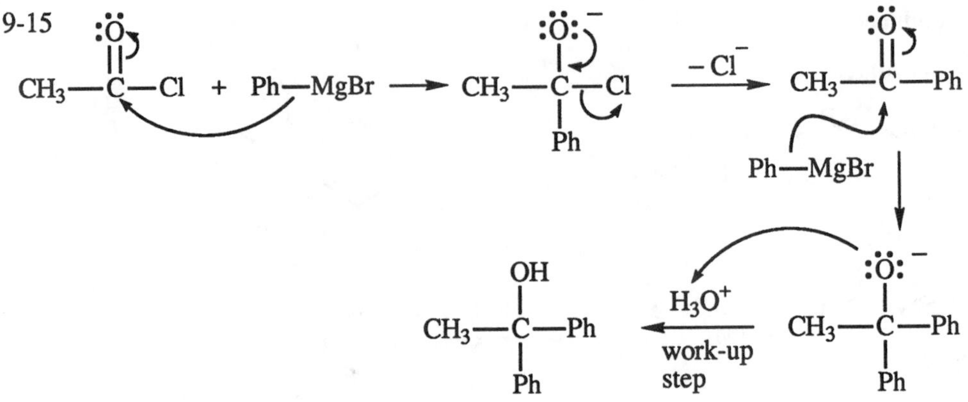
- Cl⁻
Ph—MgBr
H₃O⁺
work-up
step

9-16 Either acid chlorides or esters will work in these problems.

(a)

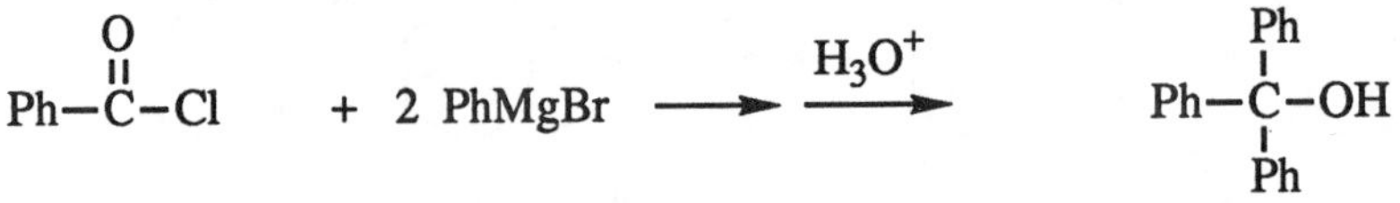
+ 2 PhMgBr
H₃O⁺

(b)

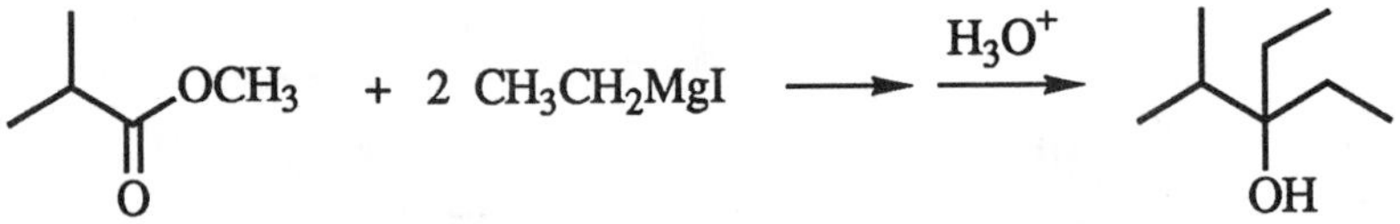
+ 2 CH₃CH₂MgI
H₃O⁺

(c)

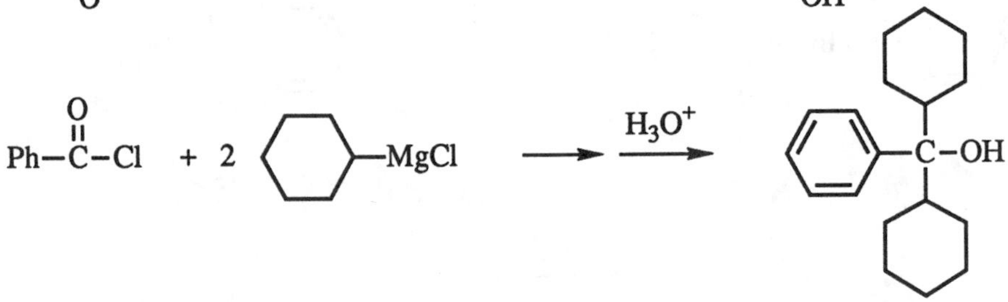
+ 2
H₃O⁺

9-17

(a)

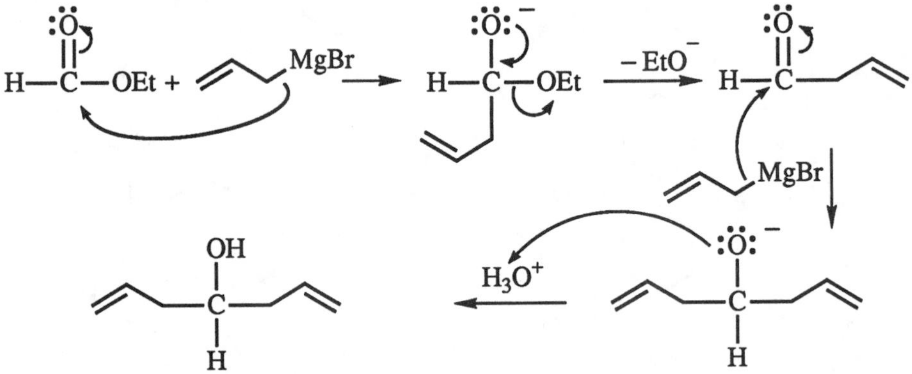
- EtO⁻
H₃O⁺

9-17 continued

(b) (i)
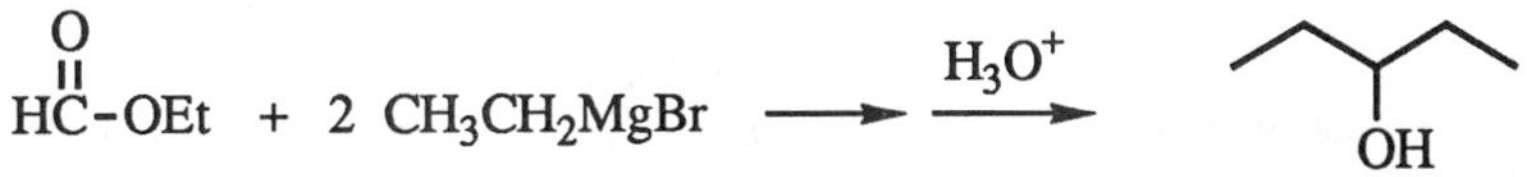
HC-OEt
O
+ 2 CH3CH2MgBr
H3O+
OH

(ii)
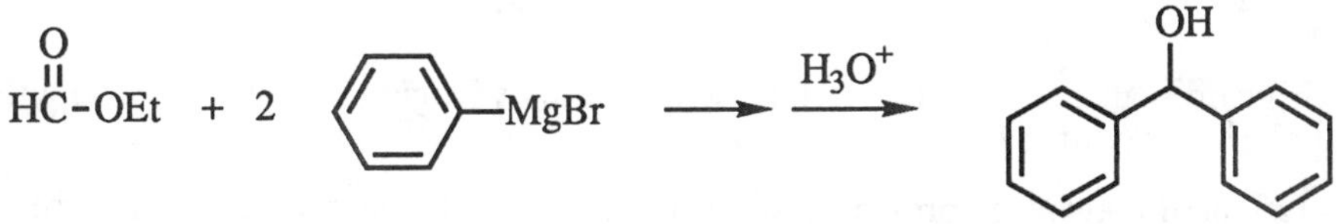
HC-OEt
O
+ 2
MgBr
H3O+
OH

(iii)
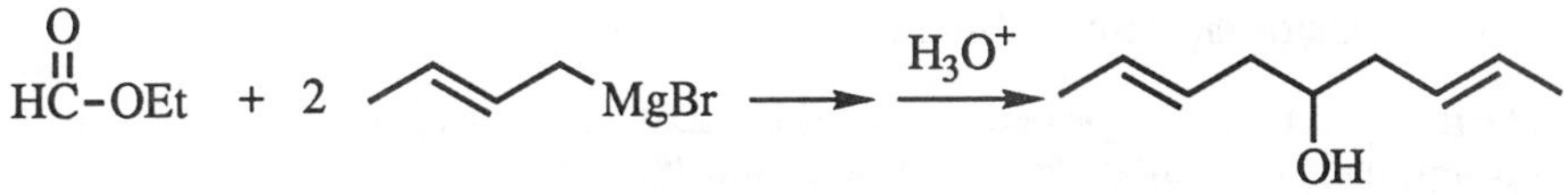
HC-OEt
O
+ 2
MgBr
H3O+
OH

9-18

(a)

MgBr +
O
H3O+
OH

(b)

MgCl +
O
H3O+
OH

(c)
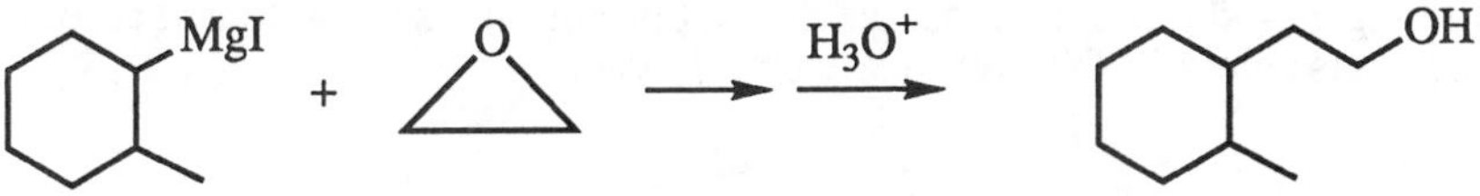
MgI
+
O
H3O+
OH

9-19 There are several possible synthetic routes to each structure, of which these are representative. Your answers may be different and still be correct.

(a)
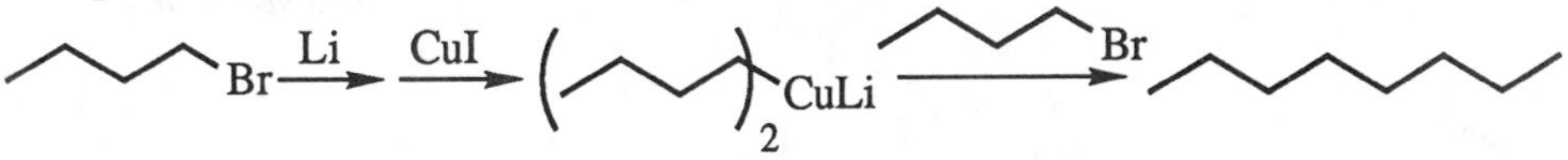
Br
Li
CuI
CuLi
2
Br

(b)
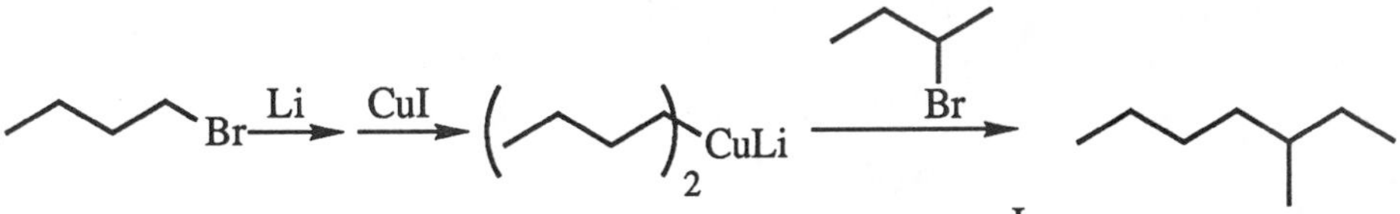
Br
Li
CuI
CuLi
2
Br

(c)
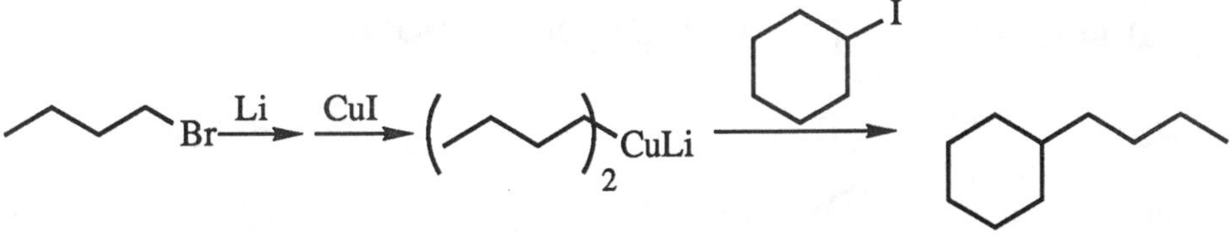
Br
Li
CuI
CuLi
2
I

(d)

Br
Li
CuI
CuLi
2
I

9-20 (a) CH_3D + Mg(OD)I (b) $CH_3CH_2CH_2CH_3$ + $LiOCH_2CH_3$

(c)  + $CH_3C(=O)$-OLi (d) 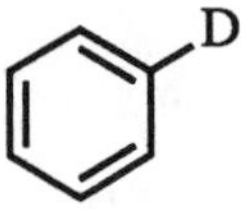 + Mg(OD)Br

9-21 Grignard reagents are incompatible with acidic hydrogens and with electrophilic, polarized multiple bonds.

(a) As Grignard reagent is formed, it would instantaneously react with N—H present in other molecules of the same substance.

(b) As Grignard reagent is formed, it would immediately react with the ester functional group present in other molecules of the same substance.

(c) Care must be taken in how reagents are written above and below arrows. If reagents are numbered "1. ... 2. ... *etc.*", it means they are added sequentially, the same as writing reagents over separate arrows. If reagents written around an arrow are not numbered, it means they are added all at once in the same mixture. In this problem, the ketone is added in the presence of aqueous acid. The acid will immediately protonate and destroy the Grignard reagent before reaction with the ketone can occur.

(d) The ethyl Grignard reagent will be immediately protonated and consumed by the OH. This reaction *could* be made to work, however, by adding two equivalents of ethyl Grignard reagent—the first to consume the OH proton, the second to add across the ketone. Aqueous acid will then protonate both oxygens.

9-22

(a) $CH_3(CH_2)_8CH_2OH$ (b) no reaction (c) no reaction ($PhCOO^-$ before acid work-up)

(d) 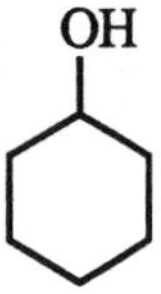

(e) 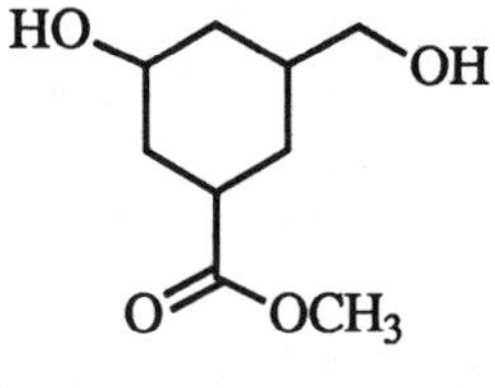

9-23

(a) $CH_3(CH_2)_8CH_2OH$ (b) $CH_3CH_2CH_2OH$ + $HOCH_3$

(c) $PhCH_2OH$ (d) 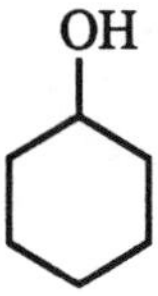

(e) 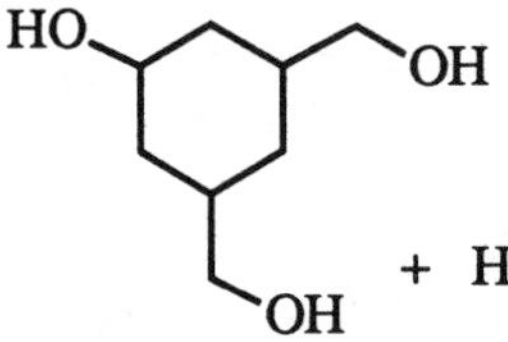

+ $HOCH_3$

9-24

(a) 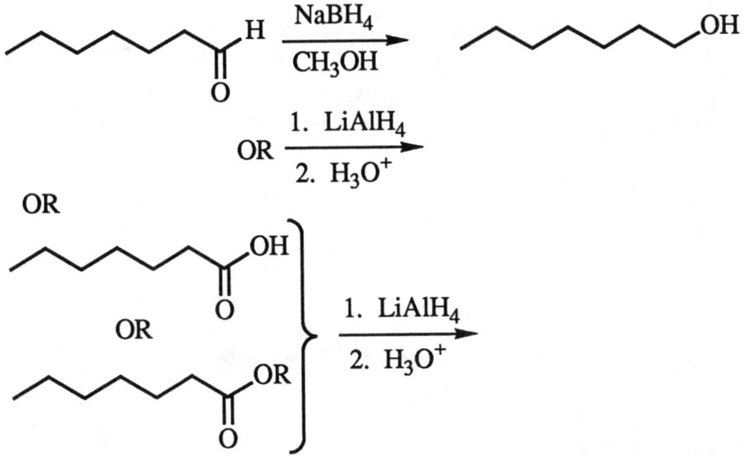
H
NaBH4
CH3OH
OH
O
OR
1. LiAlH4
2. H3O+
OR
OH
O
OR
1. LiAlH4
2. H3O+
OR
O

(b) 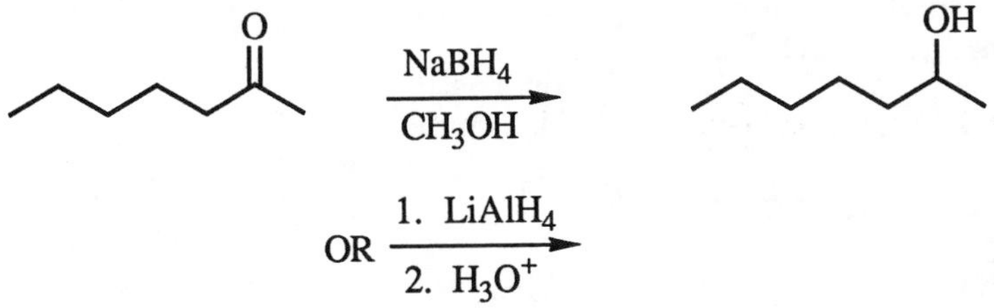
O
NaBH4
CH3OH
OH
OR
1. LiAlH4
2. H3O+

(c) 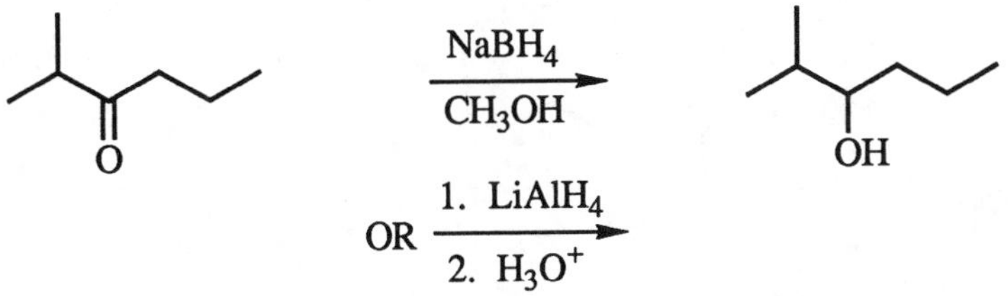
NaBH4
CH3OH
O
OH
OR
1. LiAlH4
2. H3O+

(d) 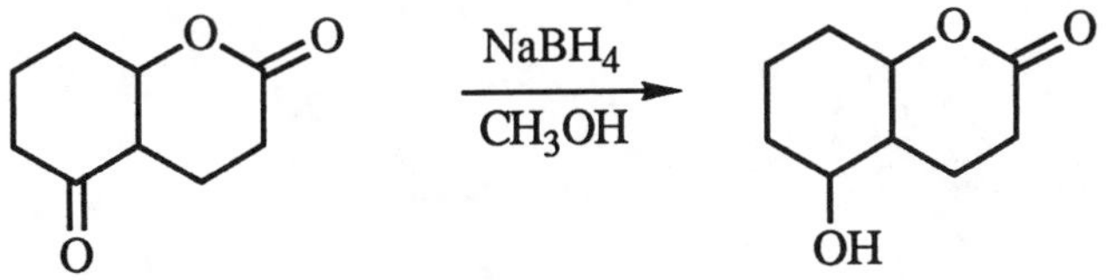
O
O
NaBH4
CH3OH
O
O
O
OH

9-25 (a) 4-methyl-2-pentanethiol
(b) (Z)-2,3-dimethyl-2-pentene-1-thiol
(c) 2-cyclohexene-1-thiol

9-26

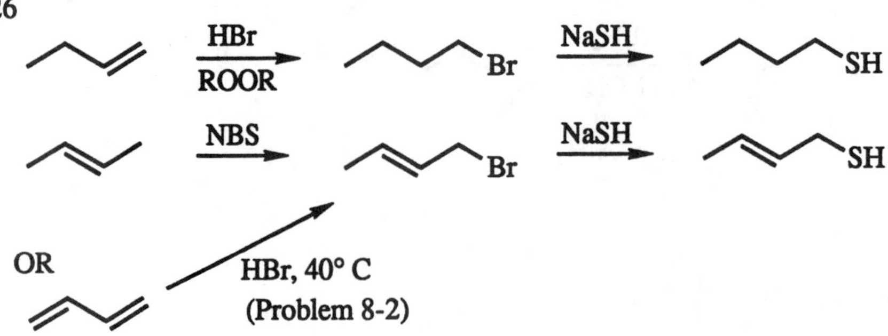

9-27 Refer to the Glossary and the text for definitions and examples.

9-28 (a) 3-*n*-propyl-2-heptanol; 2°
(b) 4-(1-bromoethyl)-3-heptanol; 2°
(c) 6-chloro-3-phenyl-3-octanol; 3°
(d) (*E*)-4,5-dimethyl-3-hexen-1-ol; 1°
(e) 3-bromo-3-cyclohexen-1-ol; 2° ("1" is optional)
(f) *cis*-4-chloro-2-cyclohexen-1-ol; 2° ("1" is optional)

9-29 (a) 4-chloro-1-phenyl-1,5-hexanediol
(b) *trans*-1,2-cyclohexanediol
(c) 3-nitrophenol
(d) 4-bromo-2-chlorophenol

9-30

(a)
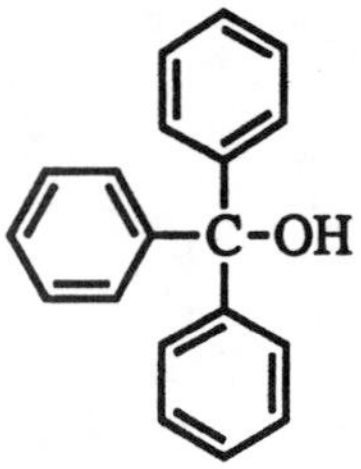

(b)
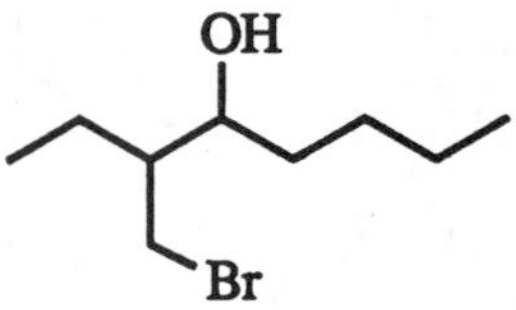

(c)
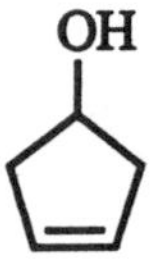

(d)
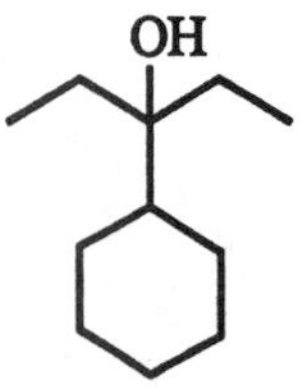

(e)
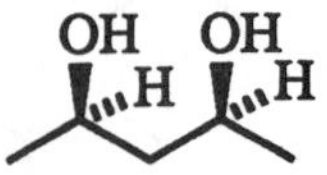

(f)
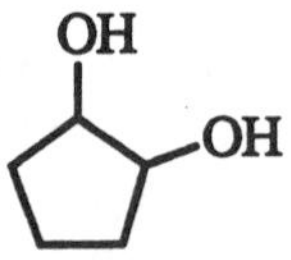

(g)

(h)
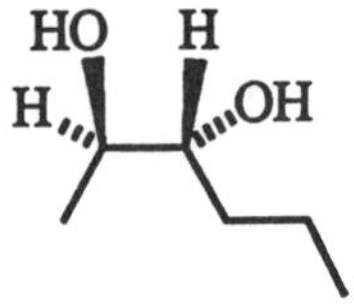

(i)

(j) $CH_3S—SCH_3$

9-31 (a) 1-Hexanol will boil at a higher temperature as it is less branched than 3,3-dimethyl-1-butanol.
(b) 2-Hexanol will boil at a higher temperature because its molecules hydrogen bond with each other, whereas molecules of 2-hexanone have no intermolecular hydrogen bonding.
(c) 1,5-Hexanediol will boil at a higher temperature as it has two OH groups for hydrogen bonding. 2-Hexanol has only one group for hydrogen bonding.

9-32 (a) 3-Chlorophenol is more acidic than cyclopentanol. In general, phenols are many orders of magnitude more acidic than alcohols.
(b) 2-Chlorocyclohexanol is slightly more acidic than cyclohexanol; the proximity of the electronegative chlorine to the OH increases its acidity.
(c) Cyclohexanecarboxylic acid is more acidic than cyclohexanol. In general, carboxylic acids are many orders of magnitude more acidic than alcohols.
(d) 1-Butanol is slightly more acidic than 2,2-dimethyl-1-butanol, for two reasons. In the latter compound, the two methyl groups on carbon-2 are slightly electron-donating and will increase the electron density on the oxygen, destabilizing the alkoxide anion. Second, in solution, the two methyl groups prevent stabilization of the alkoxide by solvation with molecules of solvent, lowering the stability of the anion, making the equilibrium less favorable.

9-33 (a) 2-Methyl-2-propanol is the most soluble as it is the most highly branched and therefore the lease hydrophobic among these alcohols of identical molecular weight.
(b) 1,2-Cyclohexanediol is the most soluble as it has two OH groups for hydrogen bonding. Cyclohexanol has only one OH group; chlorocyclohexane cannot hydrogen bond and is the least soluble.
(c) Cyclohexanol is the most soluble as it can hydrogen bond. Chlorocyclohexane cannot hydrogen bond, and 4-methylcyclohexanol has the added hydrophobic methyl group, decreasing its water solubility.

9-34

(a) 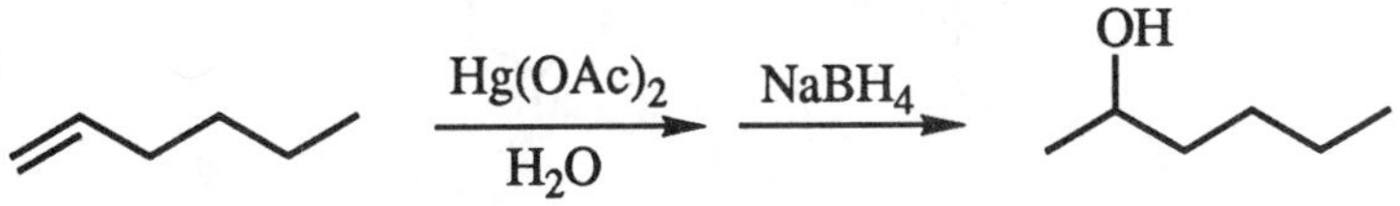

(b) 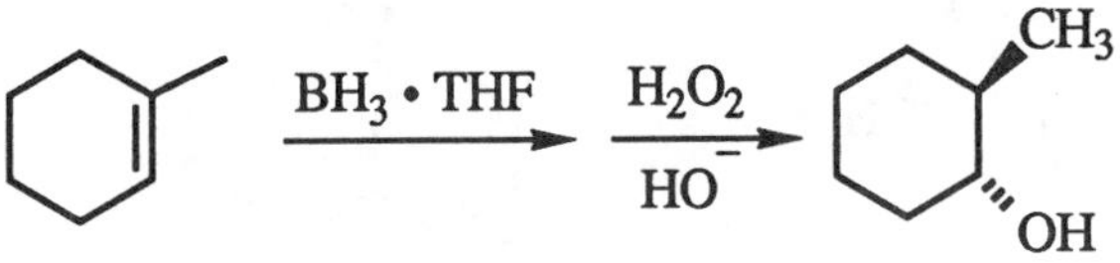

(c) 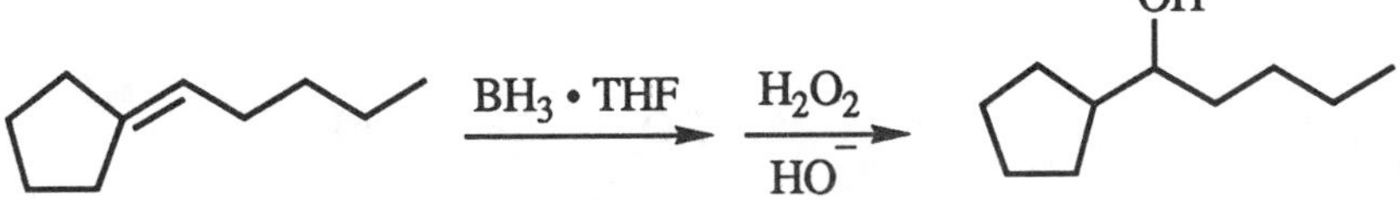

(d) 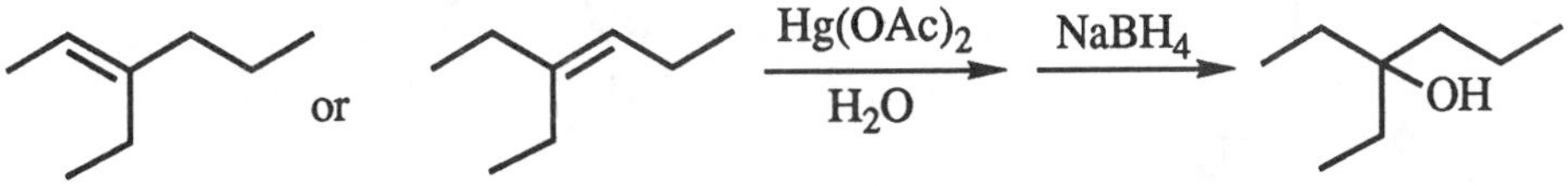

9-34 continued

(e)

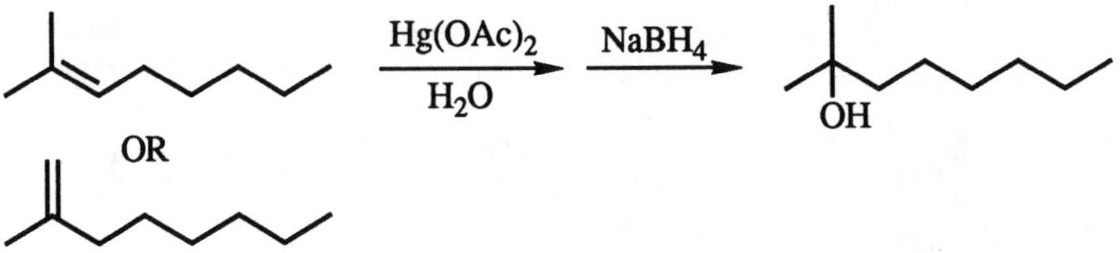

(f)

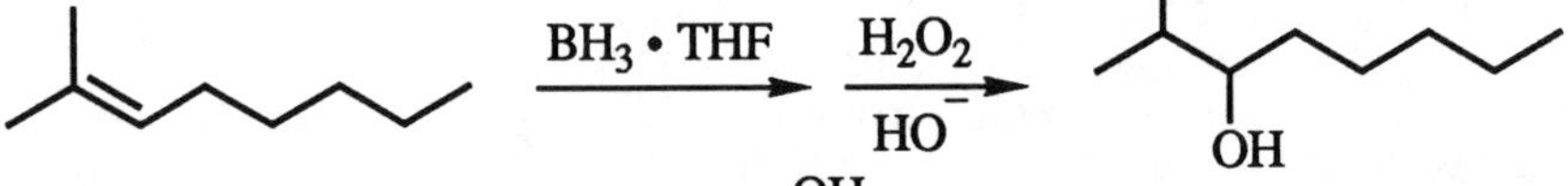

9-35

(a)

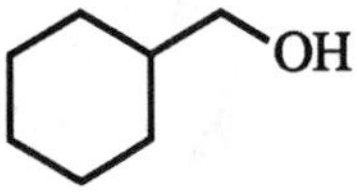

(b)

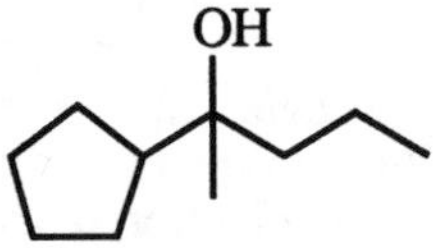

(c)

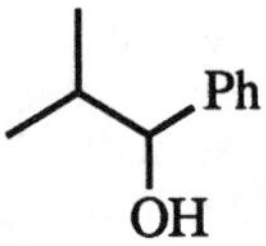

(d)

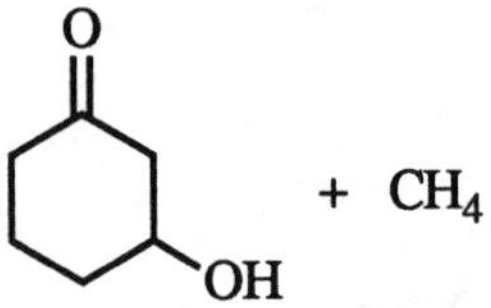

(assume only one equivalent of Grignard reagent added)

(e)

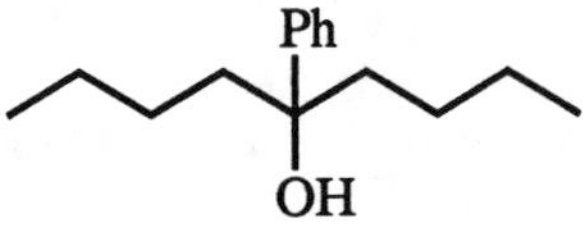

(f)

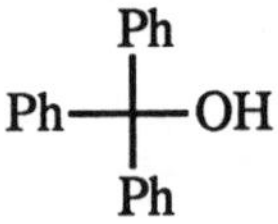

(g)

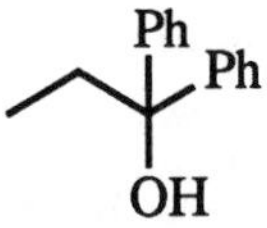

(h)

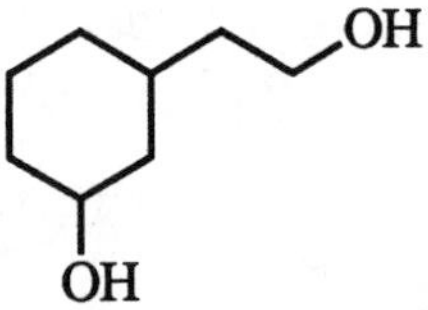

(i)

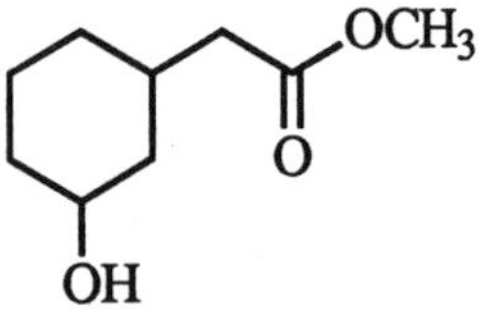

(j)

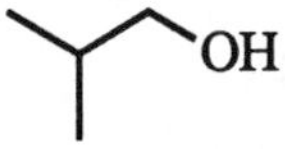

(k)

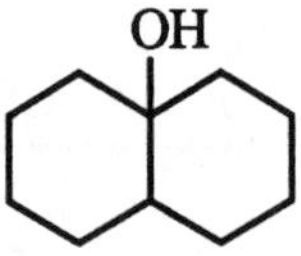

(l)

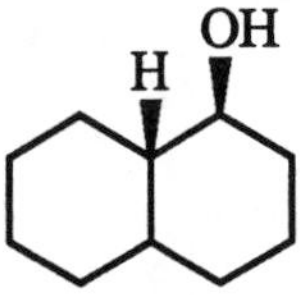

(m)

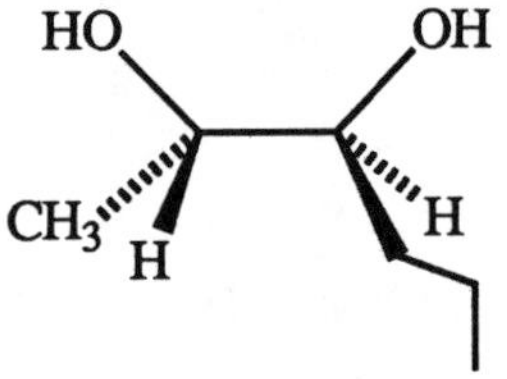

(n)

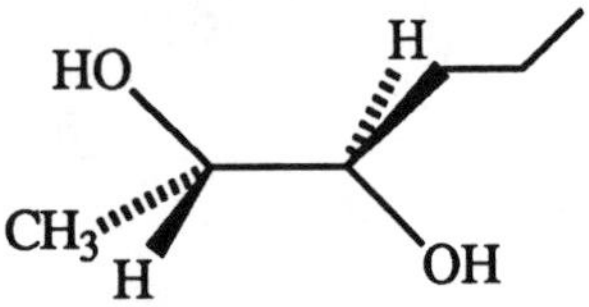

(o)

9-36

(a) 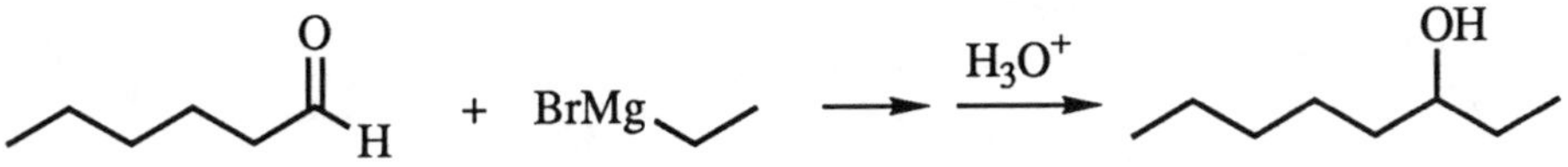

O
H
+ BrMg
H3O+
OH

(b) 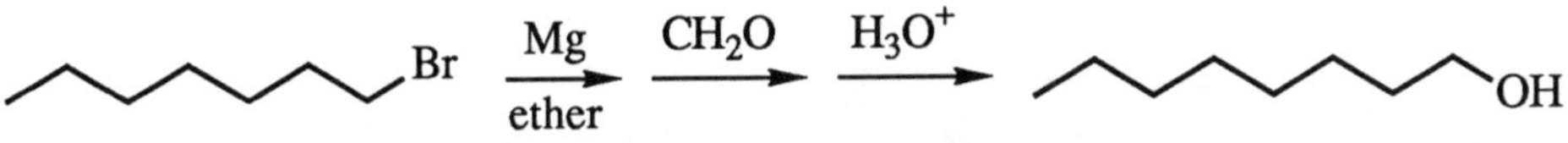

Br
Mg
ether
CH2O
H3O+
OH

(c) 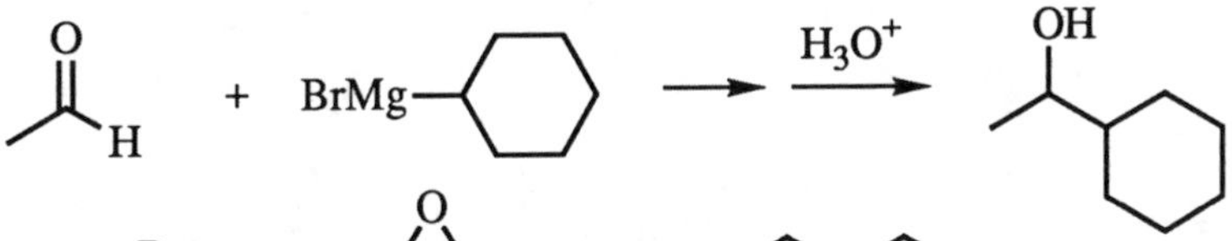

O
H
+ BrMg
H3O+
OH

(d) 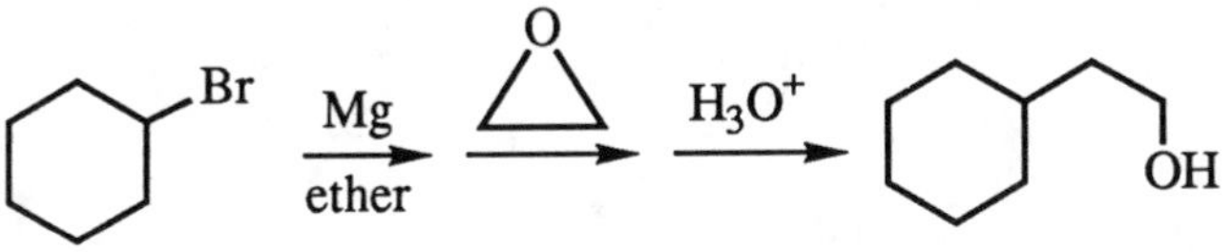

Br
Mg
ether
O
H3O+
OH

(e) 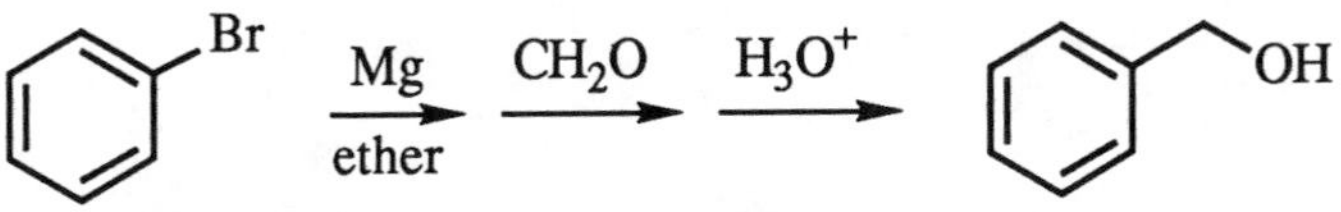

Br
Mg
ether
CH2O
H3O+
OH

(f) 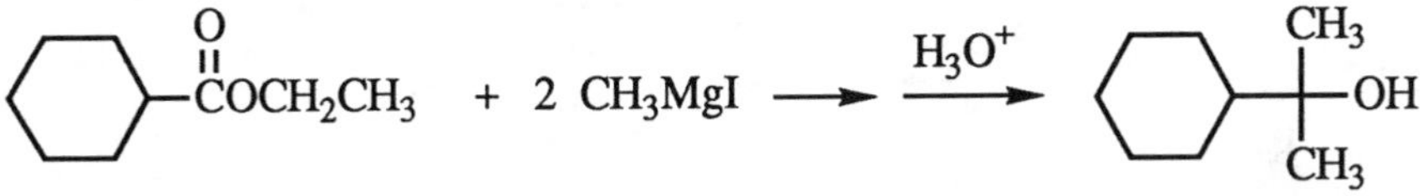

O
COCH2CH3
+ 2 CH3MgI
H3O+
CH3
OH
CH3

(g) 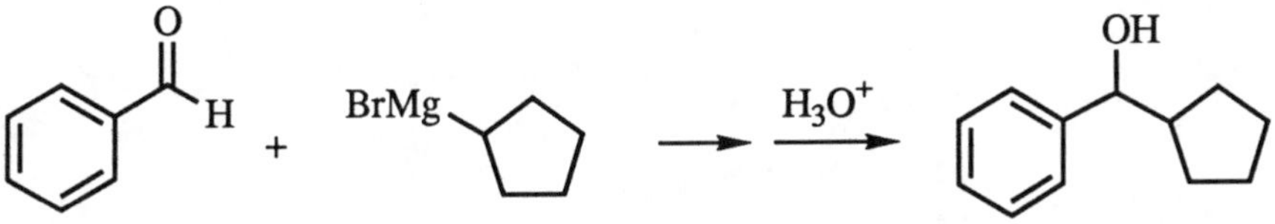

O
H
+
BrMg
H3O+
OH

9-37

(a) 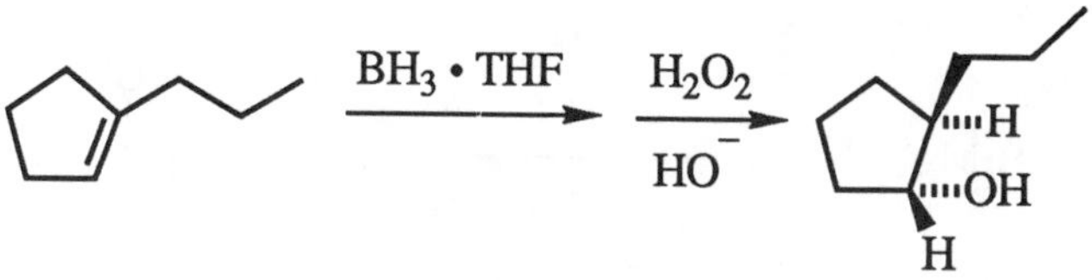

(b) 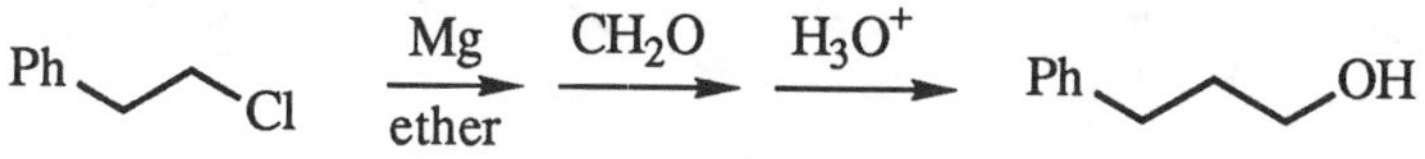

(c) 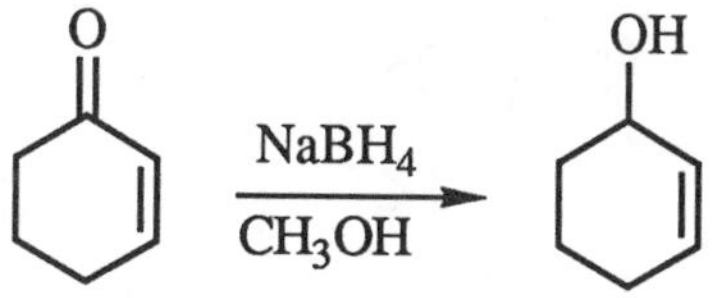

(d) 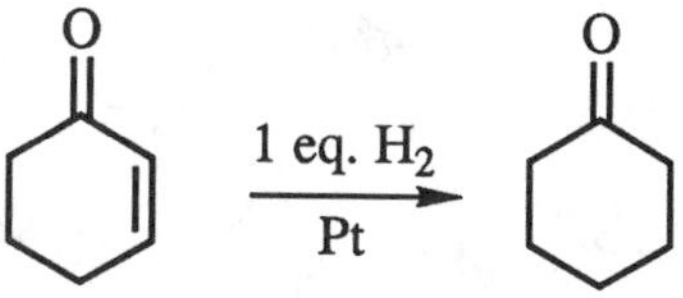

(e) 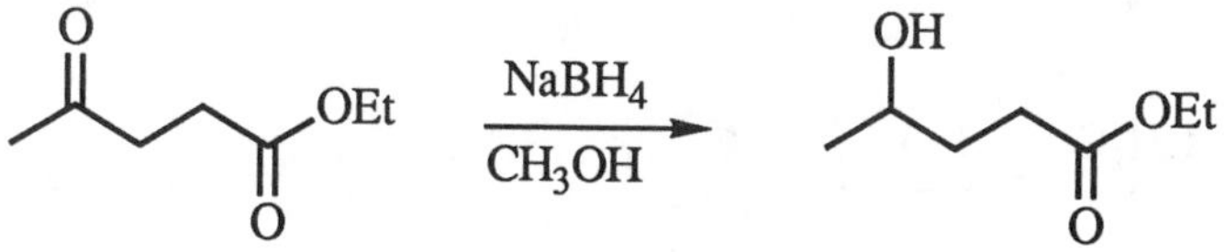

(f) 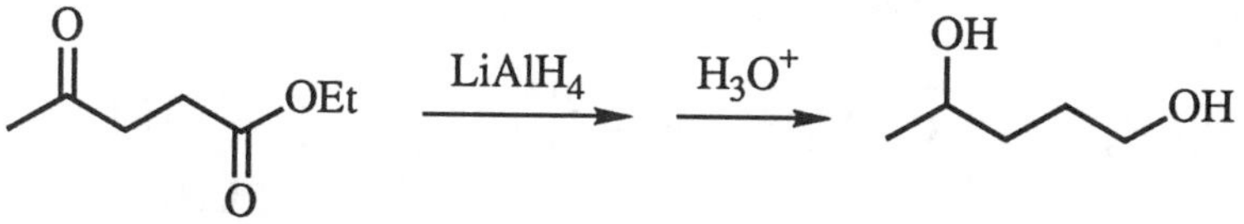

9-38

(a) 

(b) 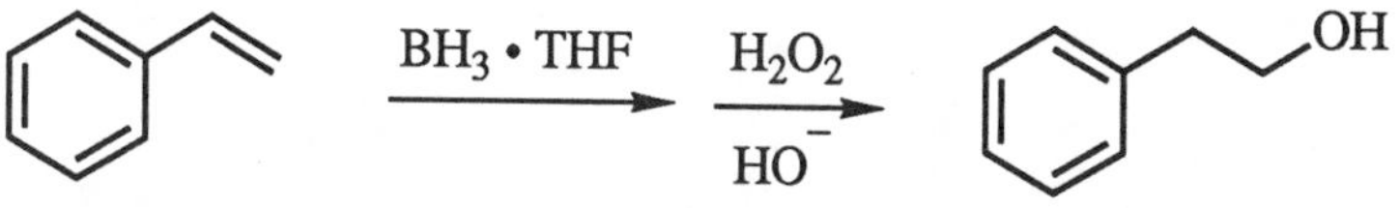

(c) 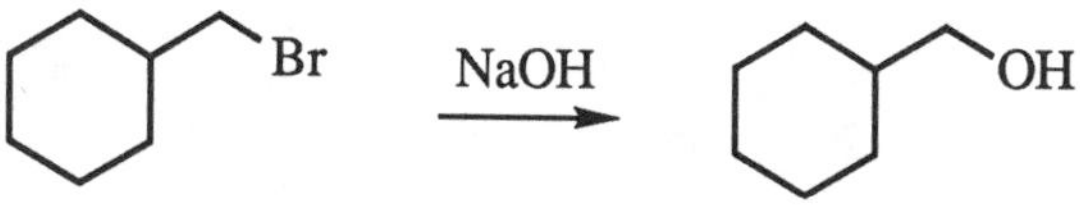

(d) 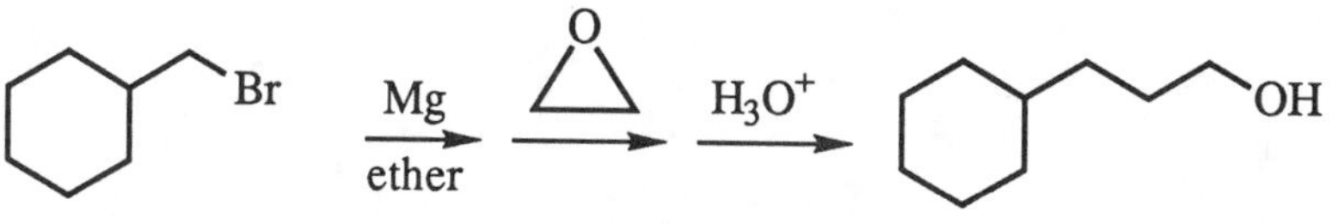

9-38 continued

(e)

Br
+ NaSH
SH

(f)

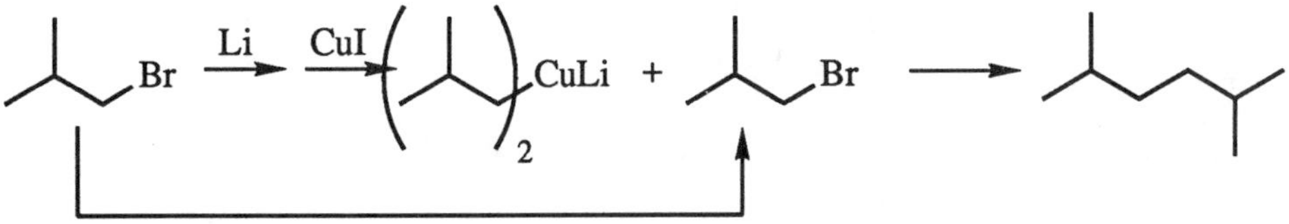

Br
Li
CuI
CuLi
2
+
Br

9-39

(a)

CH3CH2O
+
OH
CH3CH2OH +
O

products favored

(b) $KOH + CH_3CH_2OH \rightleftharpoons H_2O + CH_3CH_2OK$

reactants favored

(c) $(CH_3)_3CO^- + CH_3CH_2OH \rightleftharpoons (CH_3)_3COH + CH_3CH_2O^-$

products favored

(d)

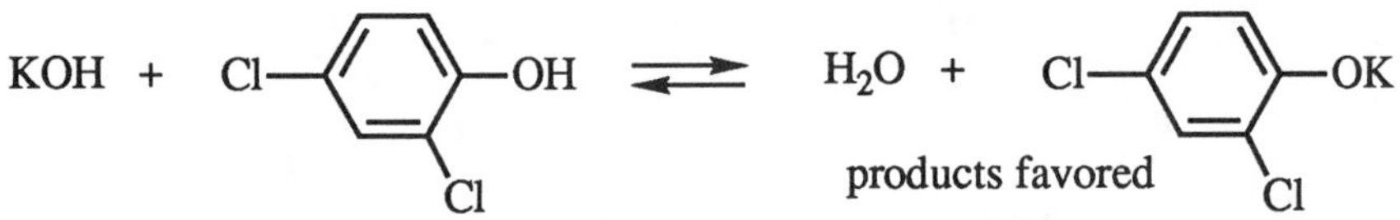

KOH +
Cl
OH
Cl
H2O +
Cl
OK
Cl

products favored

(e)

OH
+ CH3O
O
+ CH3OH

products favored

(f) $(CH_3)_3CO^- + H_2O \rightleftharpoons (CH_3)_3COH + HO^-$

products favored

9-40

(a)

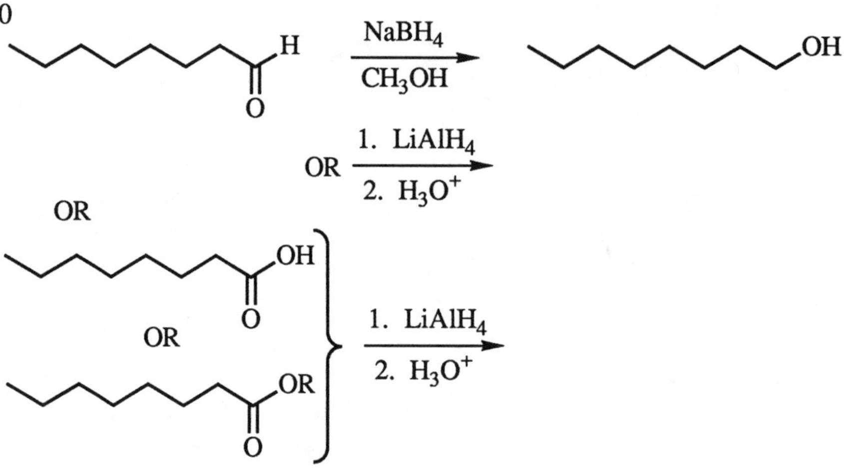

H
O
NaBH4
CH3OH
OH
OR
1. LiAlH4
2. H3O+
OR
OH
O
OR
1. LiAlH4
2. H3O+
OR
O

(b)

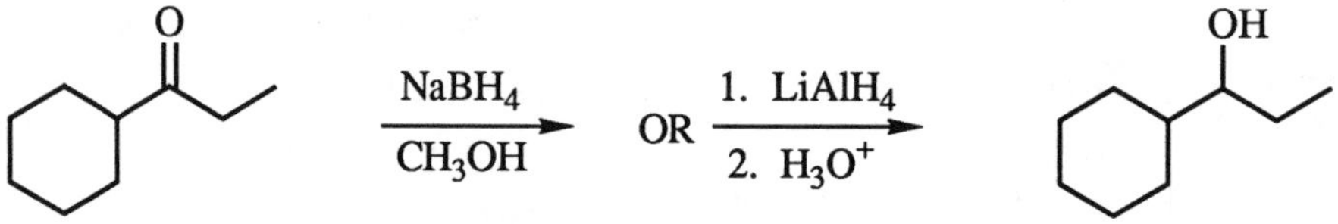

O
NaBH4
CH3OH
OR
1. LiAlH4
2. H3O+
OH

(c)

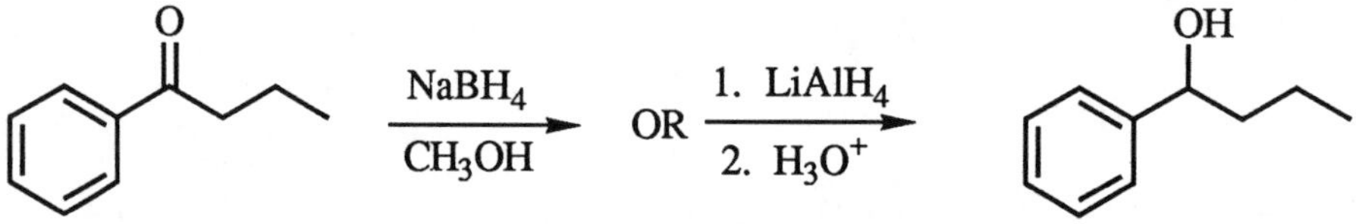

O
NaBH4
CH3OH
OR
1. LiAlH4
2. H3O+
OH

(d)

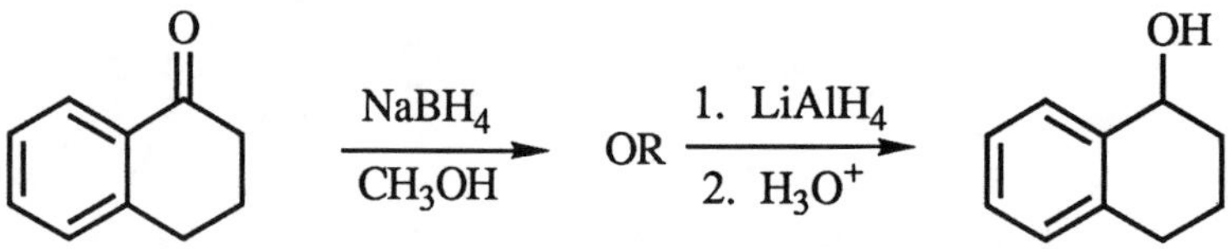

O
NaBH4
CH3OH
OR
1. LiAlH4
2. H3O+
OH

(e)

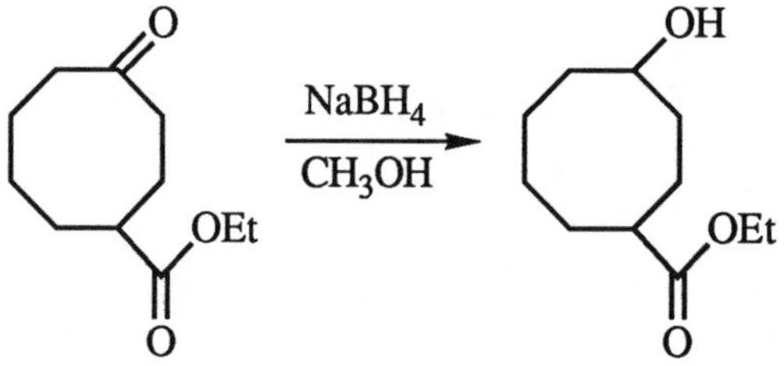

O
NaBH4
CH3OH
OEt
O
OH
OEt
O

(f)

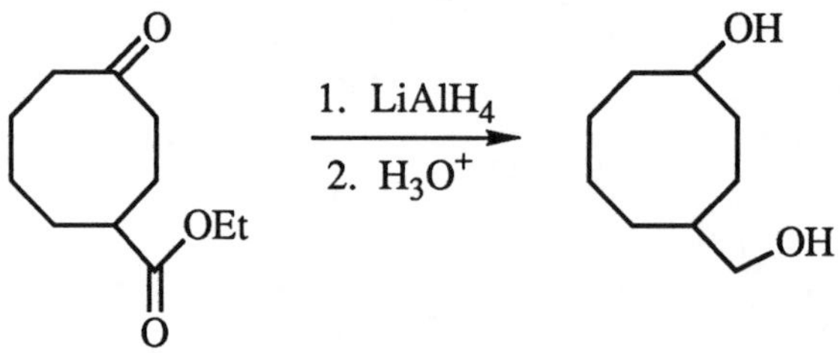

O
1. LiAlH4
2. H3O+
OEt
O
OH
OH

9-41 All steps are reversible.

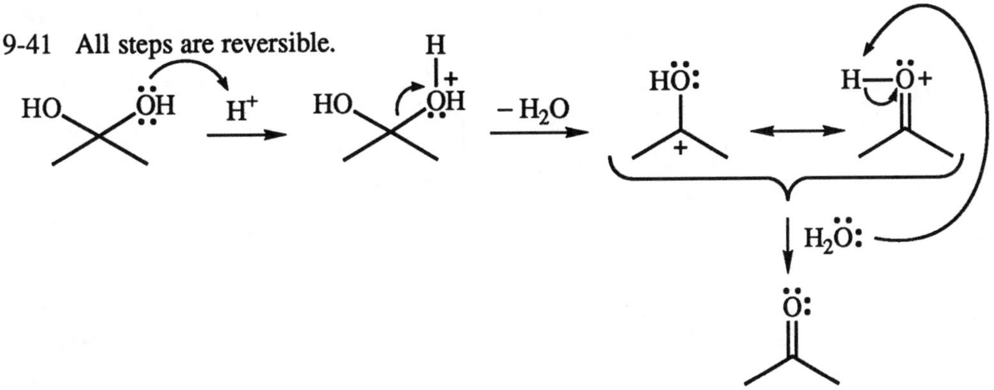

9-42

(a)

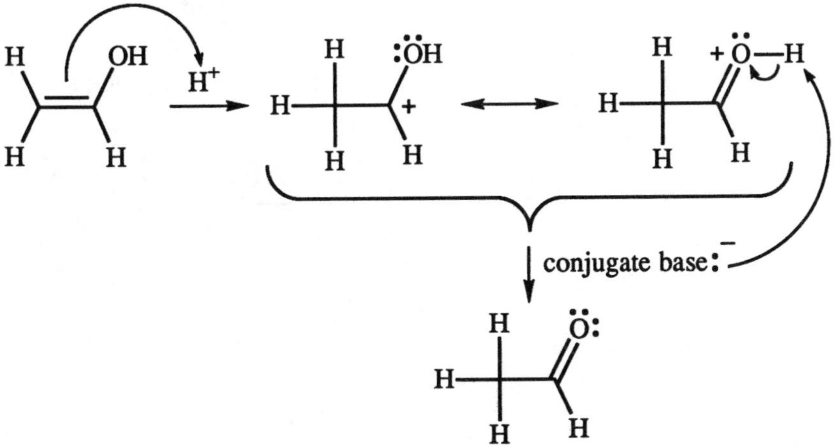

(b)

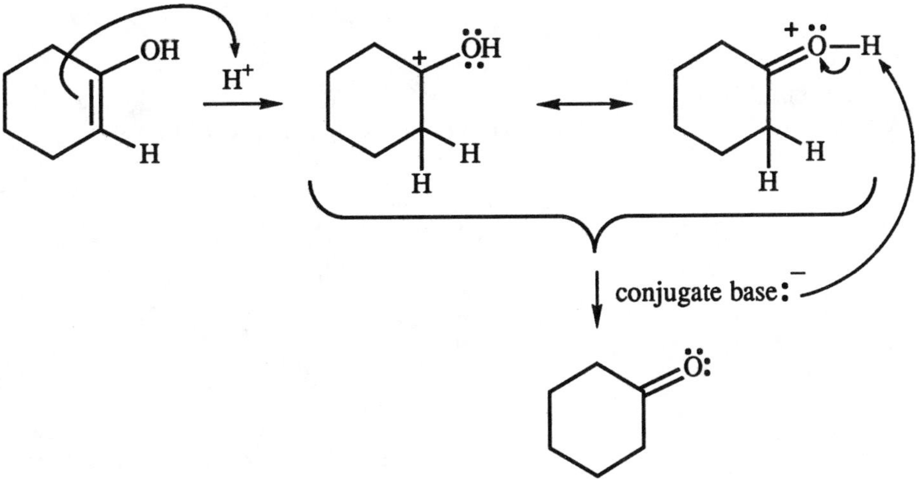

9-43

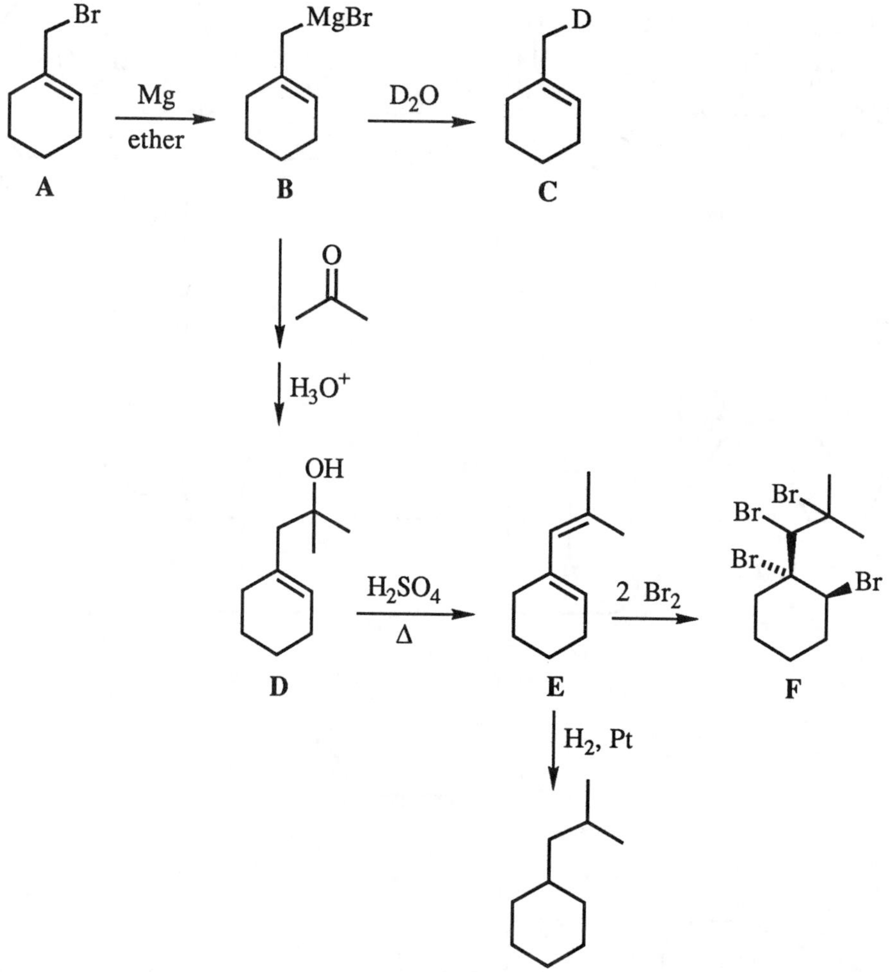

Br
Mg
ether
MgBr
D₂O
D
A
B
C
O
H₃O⁺
OH
H₂SO₄
Δ
2 Br₂
Br
Br
Br
Br
D
E
F
H₂, Pt

9-44 This mechanism is similar to cleavage of the epoxide in ethylene oxide by Grignard reagents. The driving force for the reaction is relief of ring strain in the 4-membered cyclic ether.

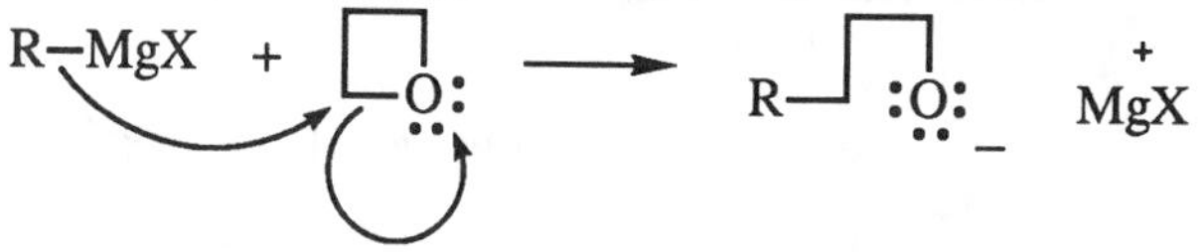

R—MgX +
O
R
O
MgX

10-1 (a) both are oxidations
(b) oxidation, oxidation, reduction, oxidation
(c) one carbon is oxidized and one carbon is reduced—no net change
(d) reduction
(e) neither oxidation nor reduction

10-2

(a) 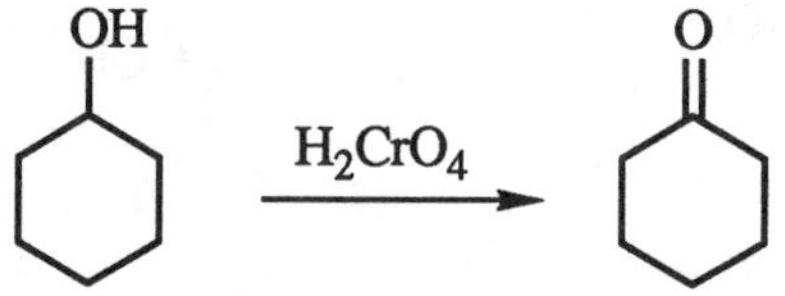

(b) 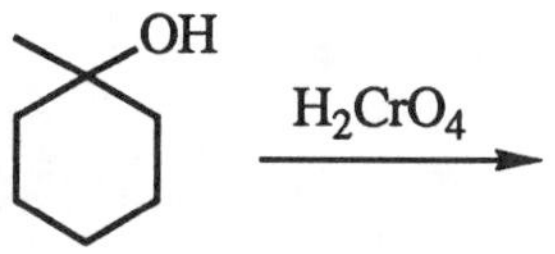 no reaction

(c) 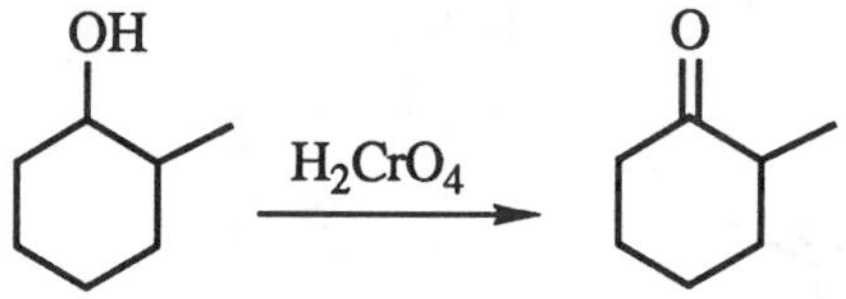

(d) 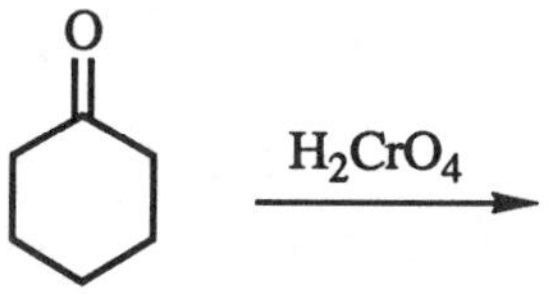 no reaction

(e) 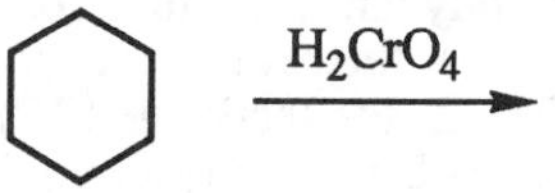 no reaction

(f) $CH_3C(=O){-}OH \xrightarrow{H_2CrO_4}$ no reaction

(g) $CH_3CH_2OH \xrightarrow{H_2CrO_4} CH_3C(=O){-}OH$

(h) $CH_3C(=O){-}H \xrightarrow{H_2CrO_4} CH_3C(=O){-}OH$

10-3

(a) 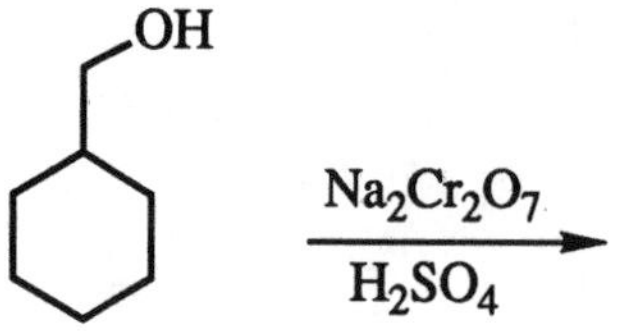
OH
$Na_2Cr_2O_7$
H_2SO_4
COOH

(b) 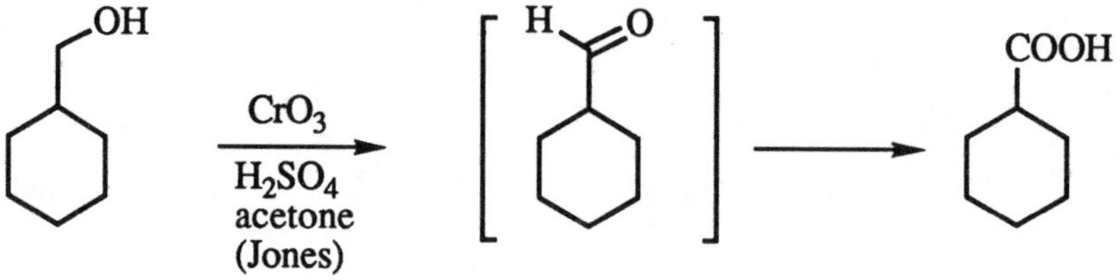
OH
CrO_3
H_2SO_4
acetone
(Jones)
H
O
COOH

(c) 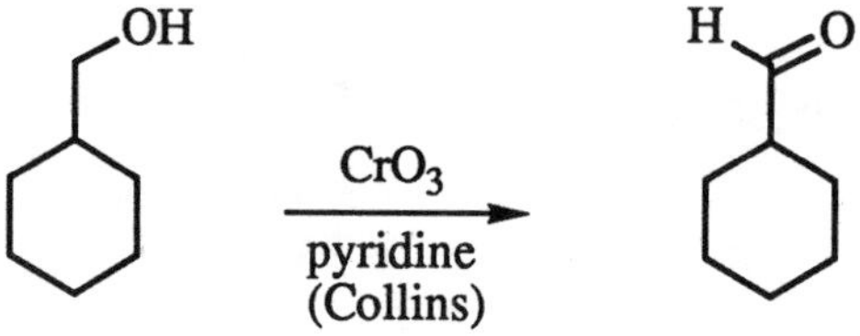
OH
CrO_3
pyridine
(Collins)
H
O

(d) 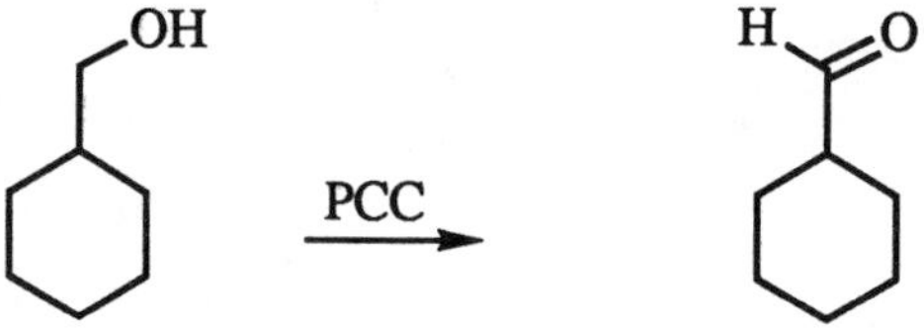
OH
PCC
H
O

10-4

(a) Dehydrogenation does not occur at 25° C—either: 1) there is a high kinetic barrier (a high activation energy) for this reaction, or 2) it is thermodynamically unfavorable, with $\Delta G > 0$. The latter possibility is supported by the fact that the reverse reaction (catalytic hydrogenation of a carbonyl) *is* spontaneous at 25° C (see section 9-12C) and therefore has $\Delta G < 0$. This makes the question of kinetics academic—a reaction that cannot proceed must be uselessly slow.

(b) and (c) Kinetics will improve with increasing temperature for virtually all reactions, so both the hydrogenation and dehydrogenation reactions will go faster. In this case, however, the question is how to favor the dehydrogenation reaction. The answer is that thermodynamics will favor this reaction as the temperature is raised. The key is the fundamental thermodynamic equation $\Delta G = \Delta H - T\Delta S$. We can estimate that $\Delta H > 0$ since the product ketone plus hydrogen is less stable than the starting alcohol. Also, $\Delta S > 0$ since one molecule is converted to two: therefore, $-T\Delta S < 0$. At low temperature (25° C), ΔH dominates because T is so small, so $\Delta G > 0$. At a high enough temperature, the $-T\Delta S$ term will begin to overwhelm ΔH, and ΔG will become negative. For the reaction in question, this must be the case at 300° C.

10-5

(a)

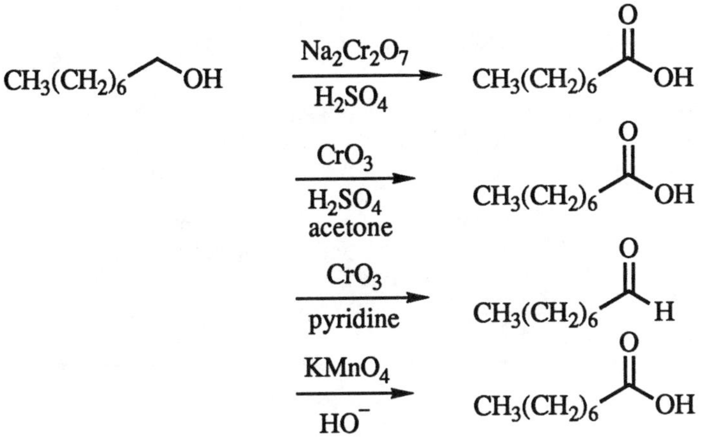

CH3(CH2)6 OH
Na2Cr2O7
H2SO4
CH3(CH2)6 OH
O
CrO3
H2SO4
acetone
CH3(CH2)6 OH
O
CrO3
pyridine
CH3(CH2)6 H
O
KMnO4
HO⁻
CH3(CH2)6 OH
O

(b) all four reagents give the same ketone product with a secondary alcohol

CH3(CH2)4
OH
CH3(CH2)4
O

(c)

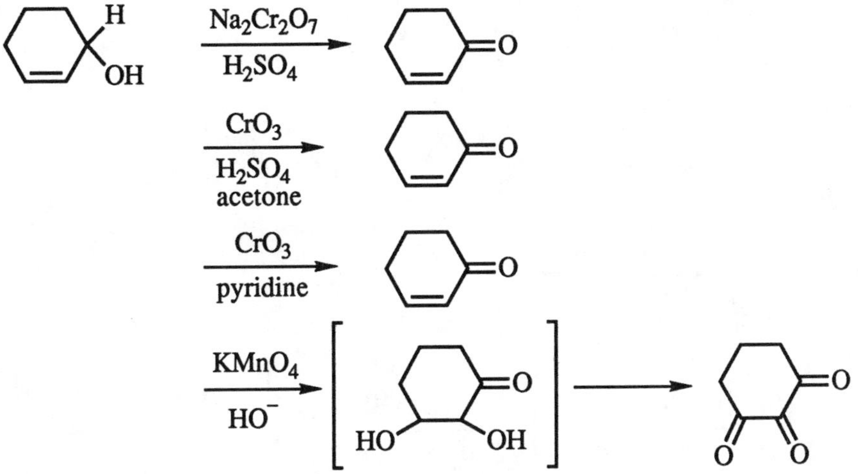

H
OH
Na2Cr2O7
H2SO4
=O
CrO3
H2SO4
acetone
=O
CrO3
pyridine
=O
KMnO4
HO⁻
=O
HO
OH
=O
O
O

(d) all four reagents give no reaction with a tertiary alcohol

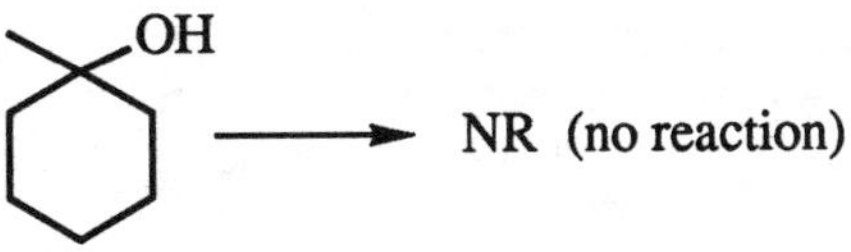

OH
NR (no reaction)

Note to the student: For simplicity, this book will use these standard laboratory methods of oxidation:

—PCC (pyridinium chlorochromate) to oxidize 1° alcohols to aldehydes;
—H_2CrO_4 (chromic acid) to oxidize 1° alcohols to carboxylic acids;
—CrO_3, H_2SO_4, acetone (Jones reagent) to oxidize 2° alcohols to ketones.

Understand that other choices are legitimate; for example, Collins reagent (CrO_3/pyridine) works about as well as PCC in the preparation of aldehydes, and Collins reagent or PCC will oxidize a 2° alcohol to a ketone as well as chromic acid. If you have a question about the appropriateness of a reagent you choose, consult the table in the text before Problem 10-2.

10-6

(a) 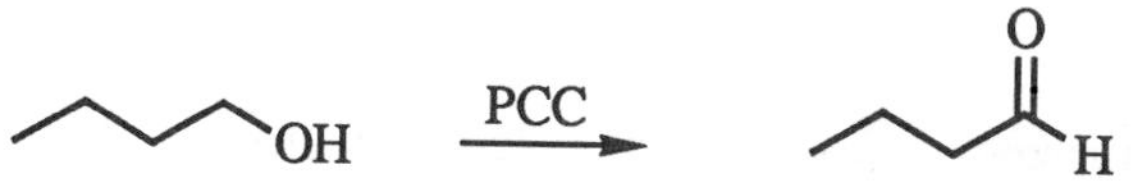

(b) 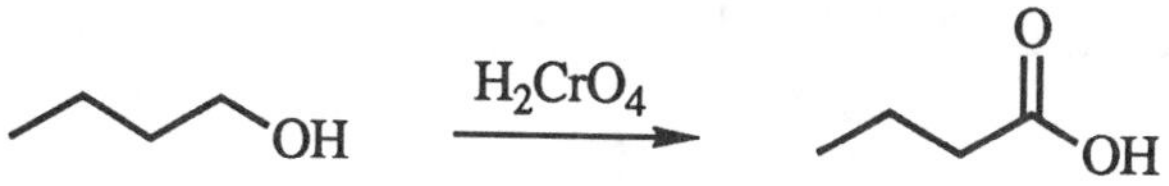

(c) 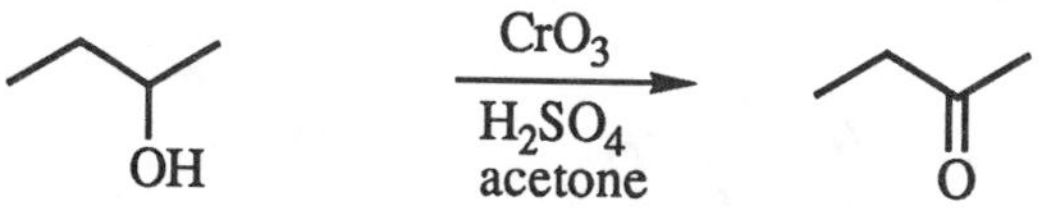

(d) 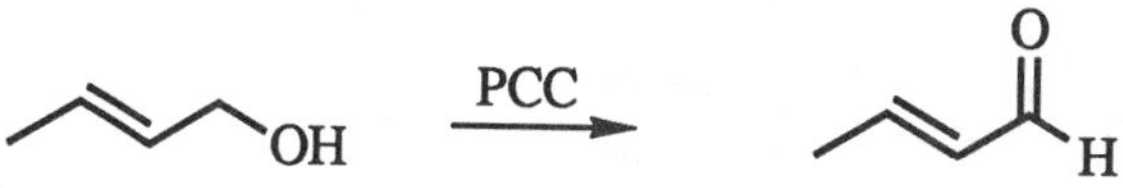

(e) 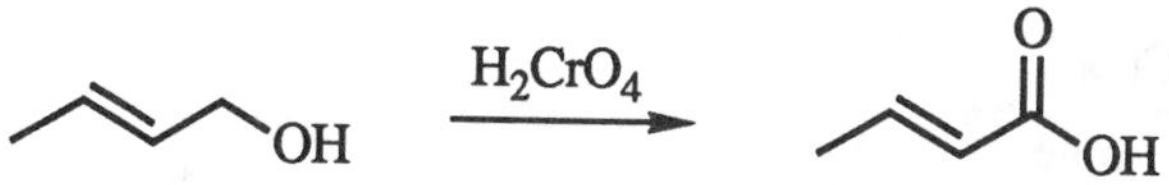

(f) 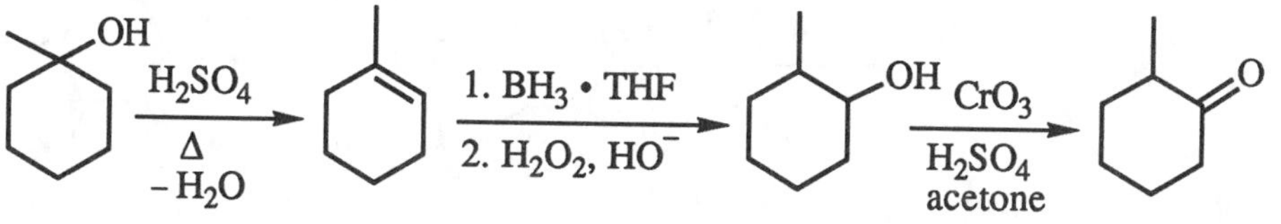

10-7 A chronic alcoholic has induced more ADH enzyme to be present to handle large amounts of imbibed ethanol, so requires more ethanol "antidote" molecules to act as a competitive inhibitor to "tie up" the extra enzyme molecules.

10-8

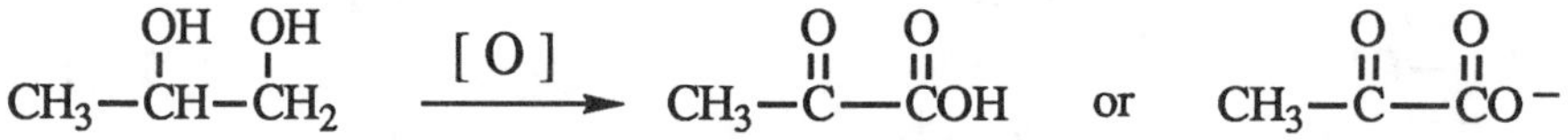

[O]
or

pyruvic acid — pyruvate

a normal metabolite in the breakdown of glucose ("blood sugar")

10-9 From this problem on, "Ts" will refer to the "tosyl" or "*p*-toluenesulfonyl" group:

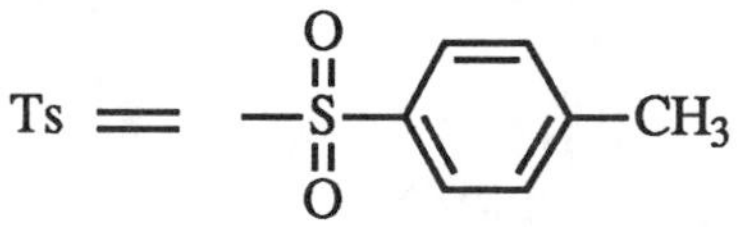

(a)

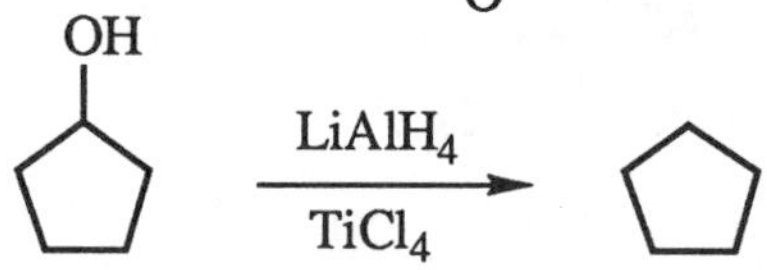

LiAlH4
TiCl4

(b)

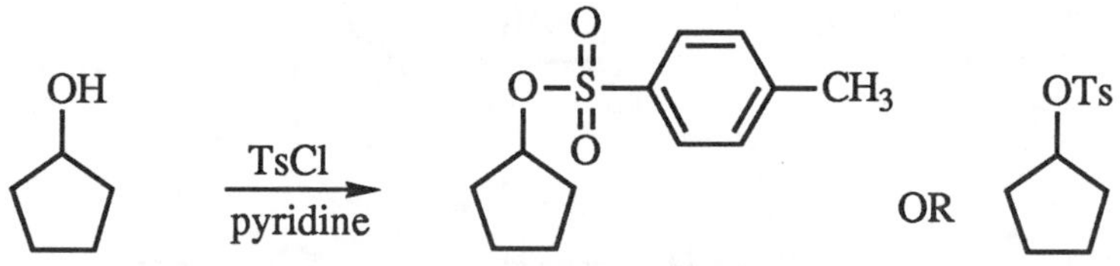

TsCl
pyridine
OR

(c)

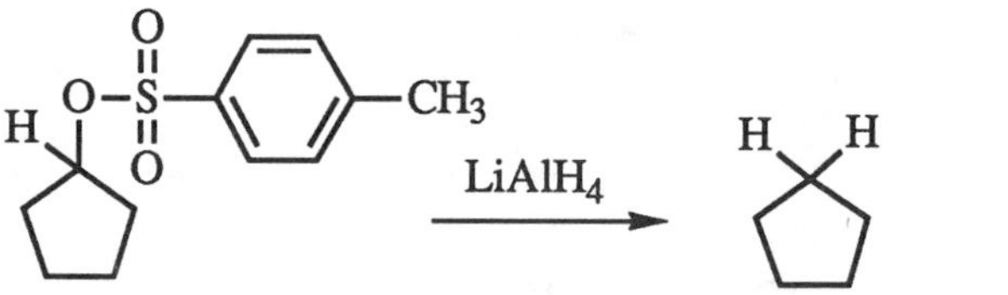

LiAlH4

(d)

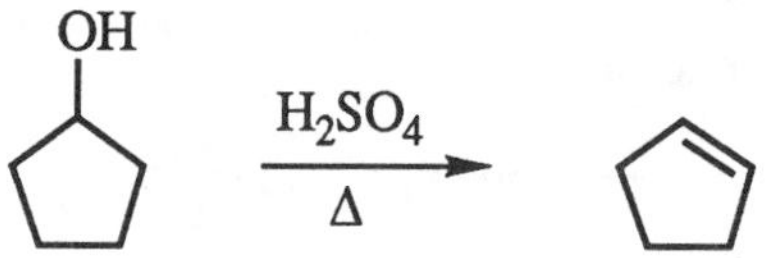

H2SO4
Δ

(e)

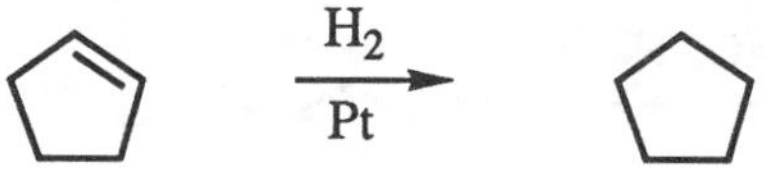

H2
Pt

10-10

(a) $CH_3CH_2{-}OTs \; + \; KO{-}C(CH_3)_2{-}CH_3 \longrightarrow CH_3CH_2O{-}C(CH_3)_2{-}CH_3 \; + \; KOTs$

(E2 is also possible with this hindered base; the product would be ethylene, $CH_2{=}CH_2$.)

(b) 
OTs
+ NaI
I
+ NaOTs

(c) 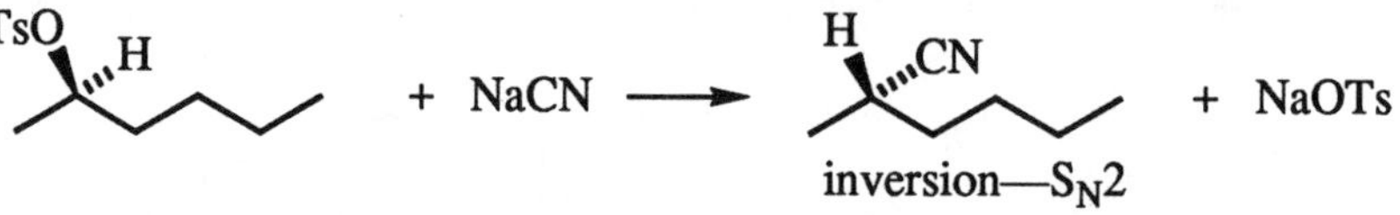
TsO
H
+ NaCN
H
CN
+ NaOTs
inversion—SN2

(d) 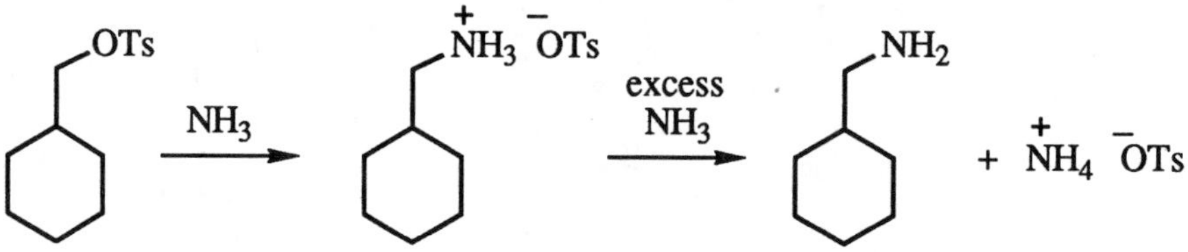
OTs
NH3
NH3 OTs
excess
NH3
NH2
+ NH4 OTs

(e) 
OTs
+ Na+ :C≡CH
C≡CH
+ NaOTs

10-11

(a) 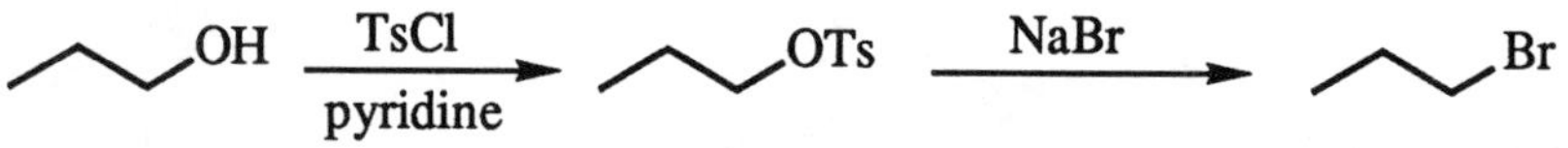
OH
TsCl
pyridine
OTs
NaBr
Br

(b) 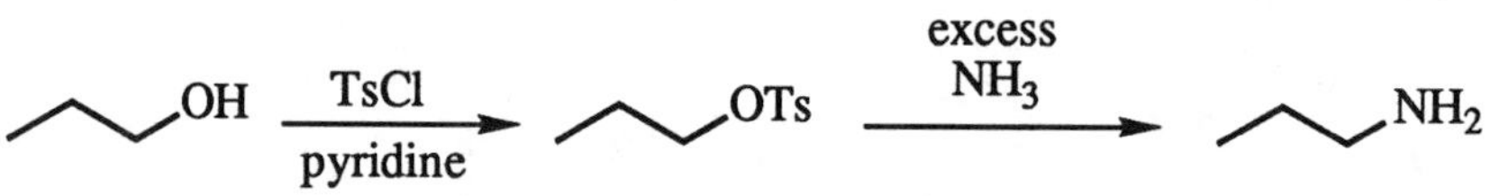
OH
TsCl
pyridine
OTs
excess
NH3
NH2

(c) 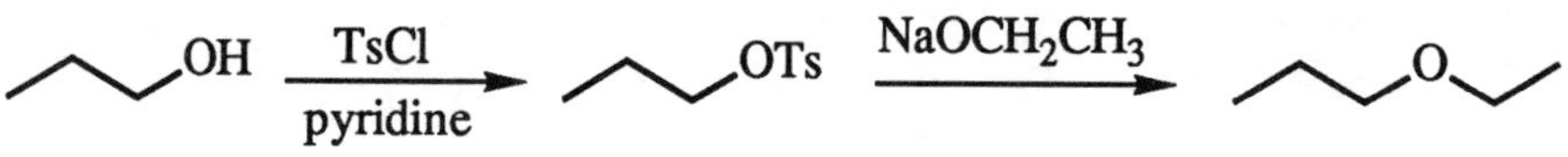
OH
TsCl
pyridine
OTs
NaOCH2CH3
O

(d) 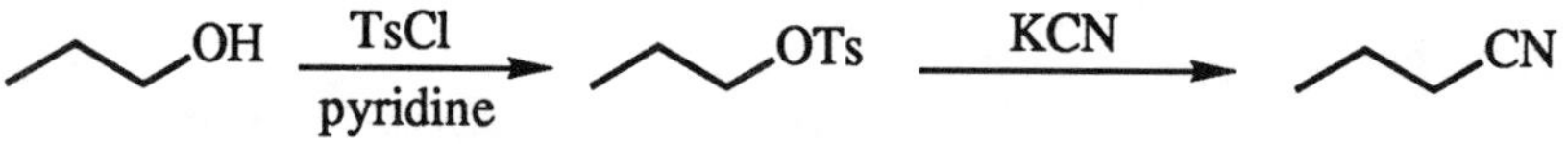
OH
TsCl
pyridine
OTs
KCN
CN

10-12

(a) either S_N1 or S_N2 on 2° alcohols

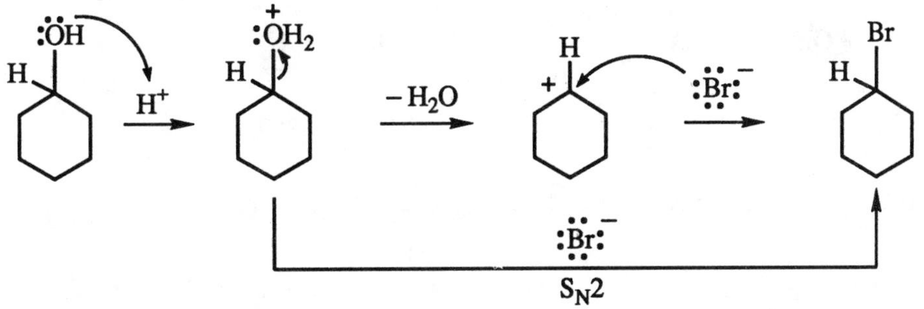

(b) S_N2 on 1° alcohols

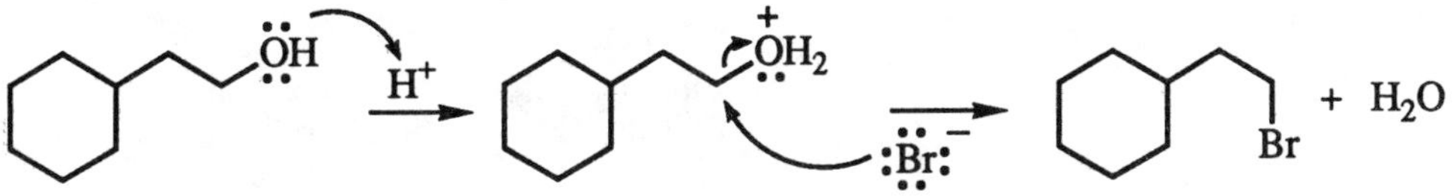

10-13

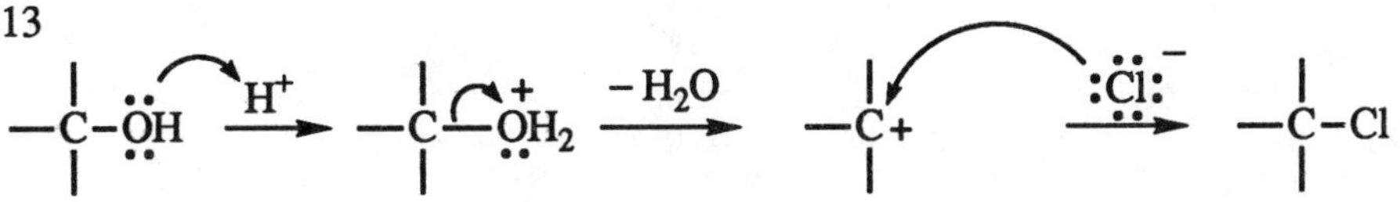

10-14 The two standard qualitative tests are:

1) chromic acid—distinguishes 3° alcohol from either 1° or 2°

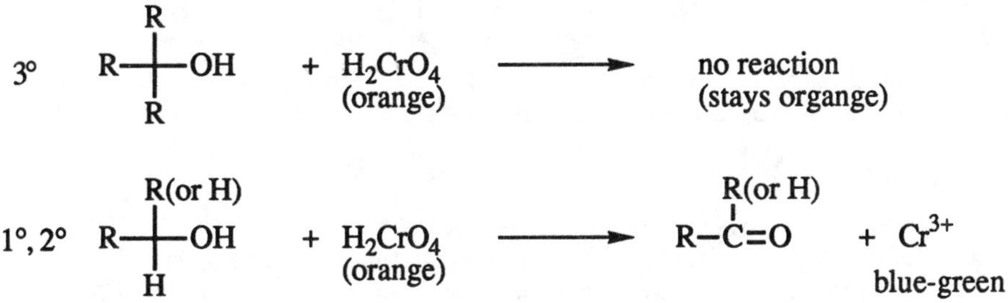

10-14 continued

2) <u>Lucas test</u>—distinguishes 1° from 2° from 3° alcohol by the rate of reaction

3° 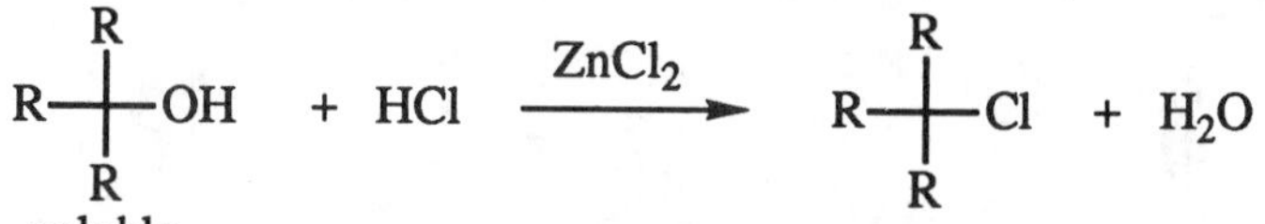

soluble — insoluble—"cloudy" in < 1 minute

2° 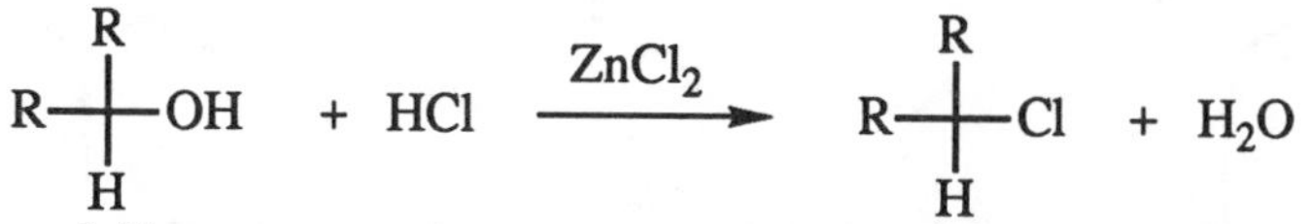

soluble — insoluble—"cloudy" in 1-5 minutes

1° 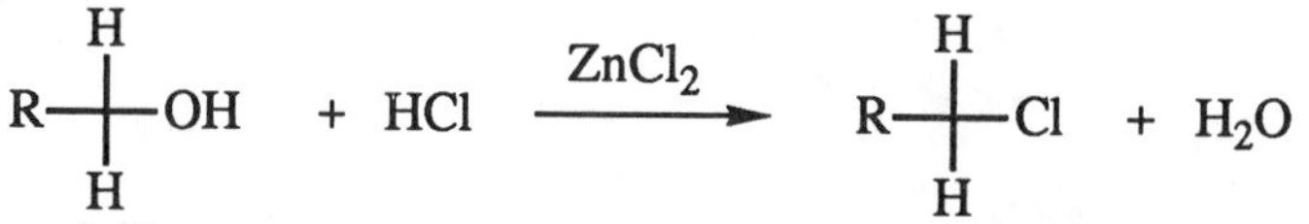

soluble — insoluble—"cloudy" in > 6 minutes (no observable reaction at room temp.)

(a)

 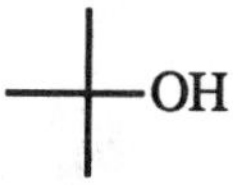

Lucas: cloudy in 1-5 min. — cloudy in < 1 min.

H_2CrO_4: immediate blue-green — no reaction—stays orange

(b)

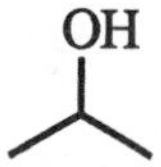 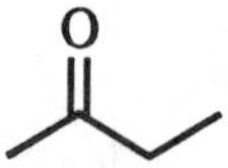

Lucas: cloudy in 1-5 min. — no reaction

H_2CrO_4: immediate blue-green — no reaction—stays orange

(c)

 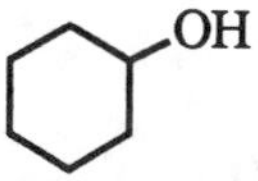

Lucas: no reaction — cloudy in 1-5 min.

H_2CrO_4: DOES NOT DISTINGUISH—immediate blue-green for both

(d)

Lucas: cloudy in < 1 min. ** — no reaction

H_2CrO_4: DOES NOT DISTINGUISH—immediate blue-green for both

(**Remember that allylic cations are resonance-stabilized and are about as stable as 3° cations. Thus, they will react as fast as 3° in the Lucas test, even though they may be 1°. Be careful to notice the subtle but important structural features!)

(e)

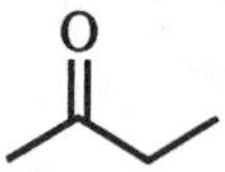 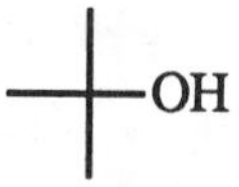

Lucas: no reaction — cloudy in < 1 min.

H_2CrO_4: DOES NOT DISTINGUISH—stays orange for both

10-15

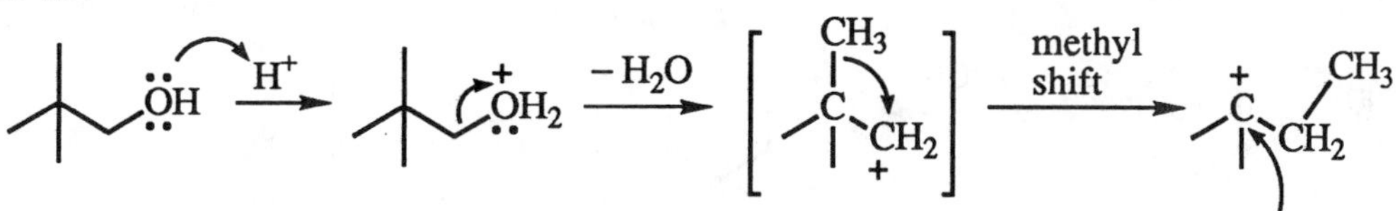
H+
−H2O
methyl
shift

Even though 1°, the neopentyl carbon is hindered to backside attack, so S_N2 cannot occur easily. Instead, an S_N1 mechanism occurs, with rearrangement.

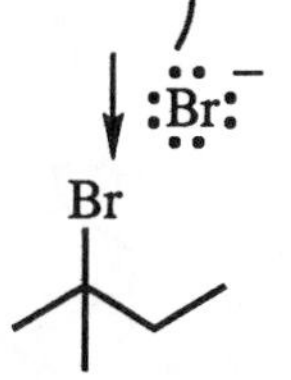
Br

10-16

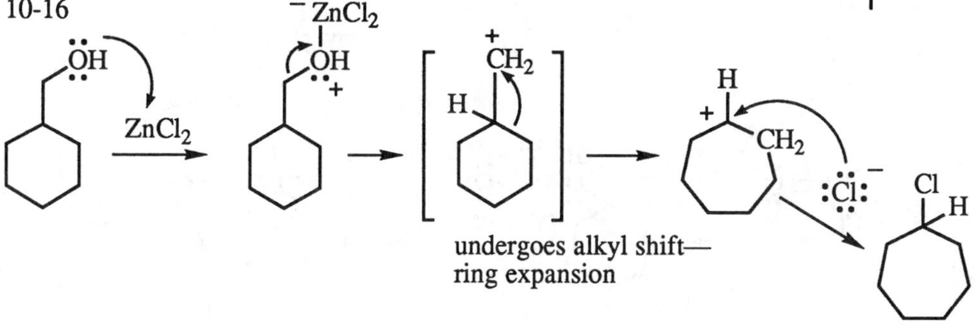
ZnCl2
undergoes alkyl shift—
ring expansion

10-17

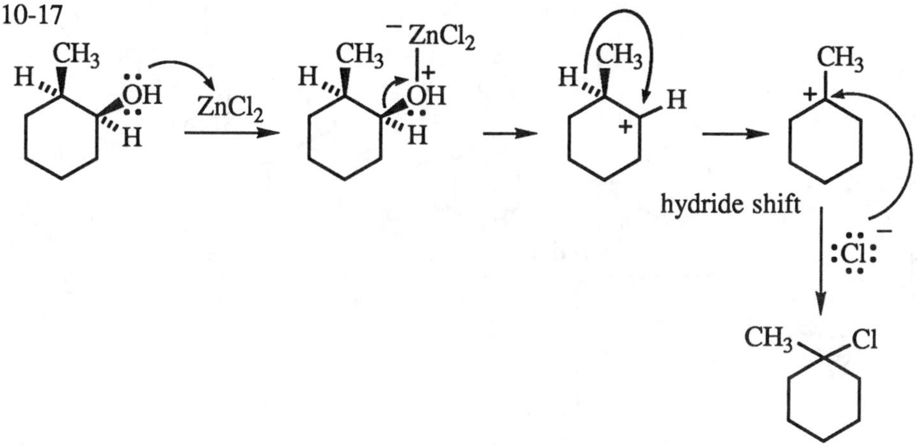
ZnCl2
hydride shift

10-18

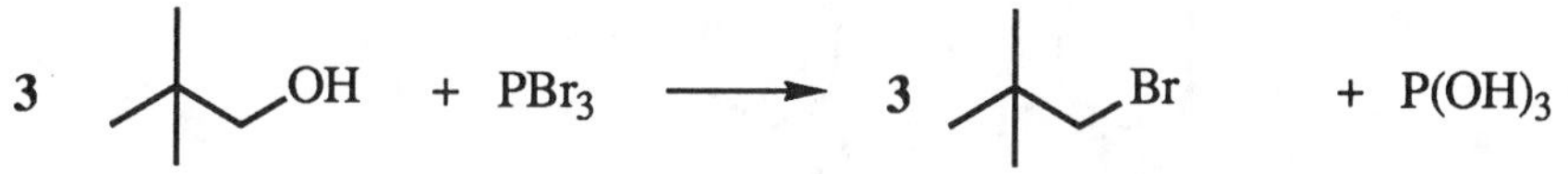
3
+ PBr3
3
+ P(OH)3

$$6\ CH_3(CH_2)_{14}CH_2OH + 2\ P + 3\ I_2 \longrightarrow 6\ CH_3(CH_2)_{14}CH_2I + 2\ P(OH)_3$$

$$6\ CH_3OH + 2\ P + 3\ I_2 \longrightarrow 6\ CH_3I + 2\ P(OH)_3$$

10-19

(a)

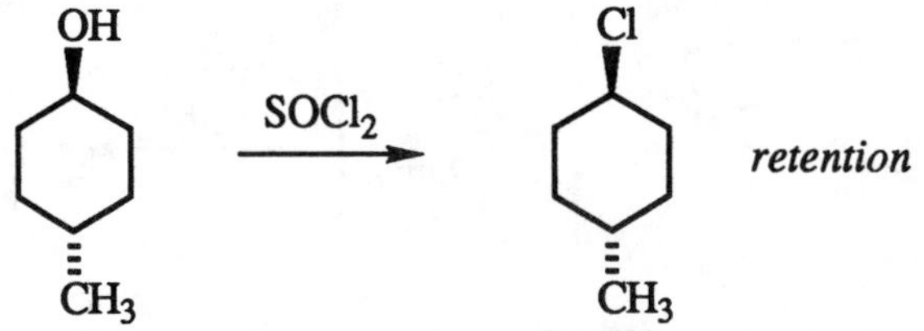

OH
SOCl2
Cl
retention
CH3
CH3

(b)

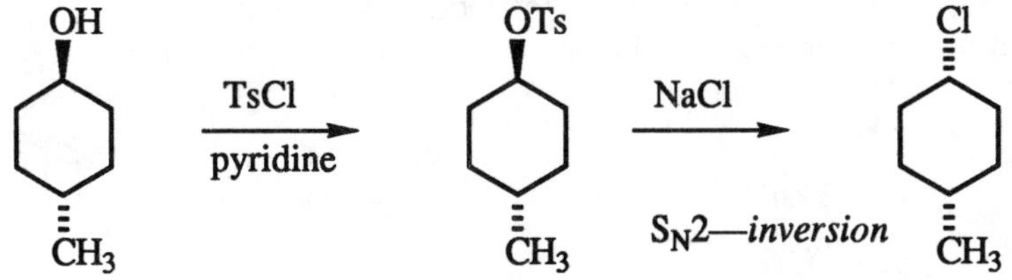

OH
TsCl
pyridine
OTs
NaCl
SN2—inversion
Cl
CH3
CH3
CH3

10-20

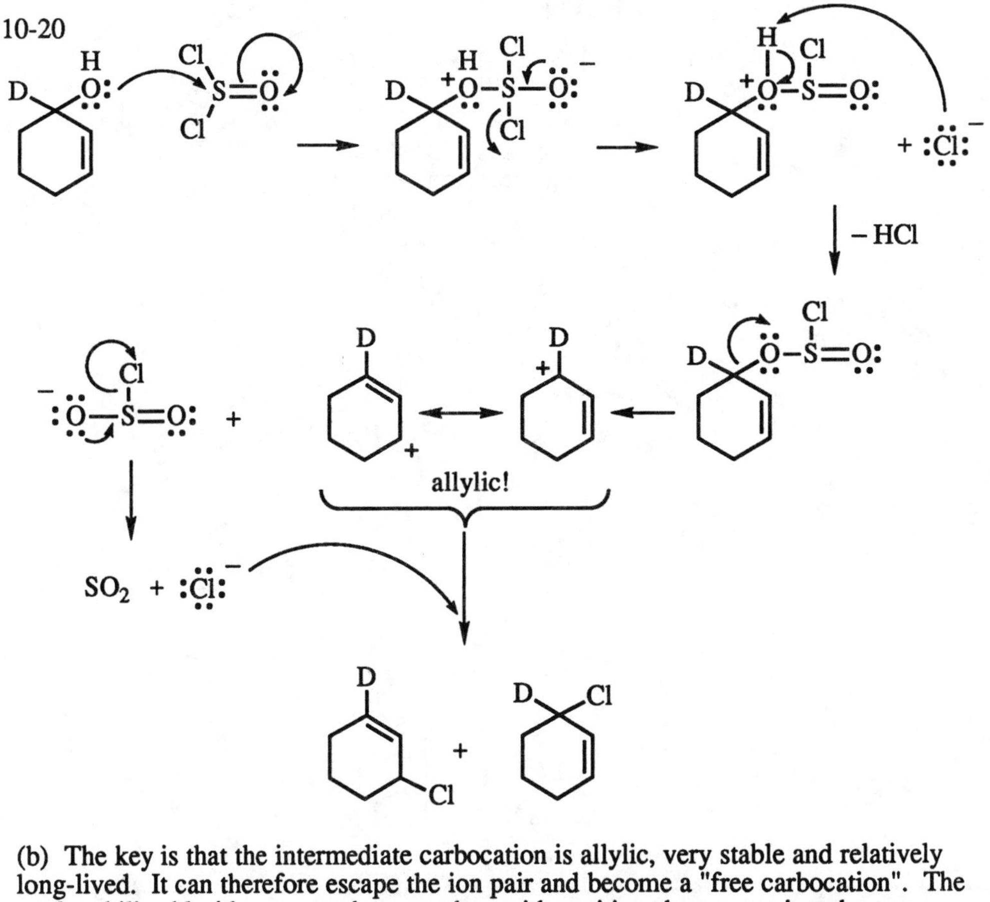

- HCl
allylic!

(b) The key is that the intermediate carbocation is allylic, very stable and relatively long-lived. It can therefore escape the ion pair and become a "free carbocation". The nucleophilic chloride can attack any carbon with positive charge, not just the one closest. Since two carbons have partial positive charge, two products result.

10-21

(a)

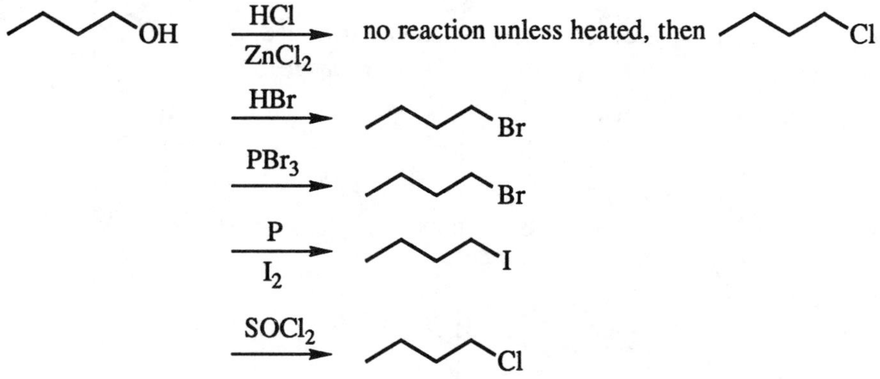
OH
HCl
ZnCl2
no reaction unless heated, then
Cl
HBr
Br
PBr3
Br
P
I2
I
SOCl2
Cl

(b)

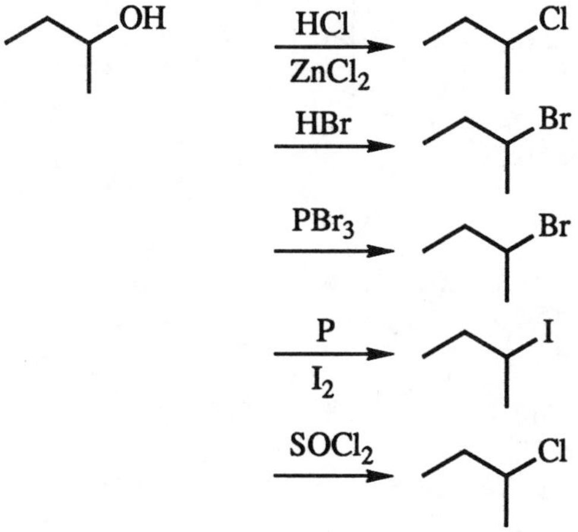
OH
HCl
ZnCl2
Cl
HBr
Br
PBr3
Br
P
I2
I
SOCl2
Cl

(c)

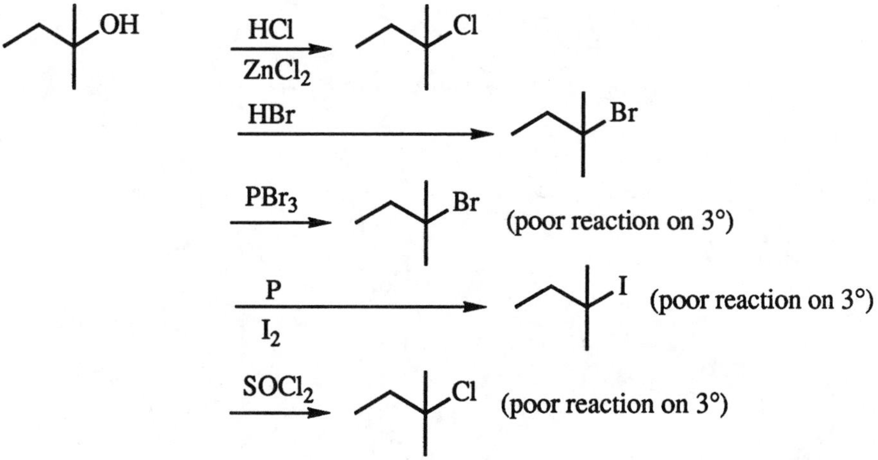
OH
HCl
ZnCl2
Cl
HBr
Br
PBr3
Br
(poor reaction on 3°)
P
I2
I
(poor reaction on 3°)
SOCl2
Cl
(poor reaction on 3°)

10-21 continued

(d)

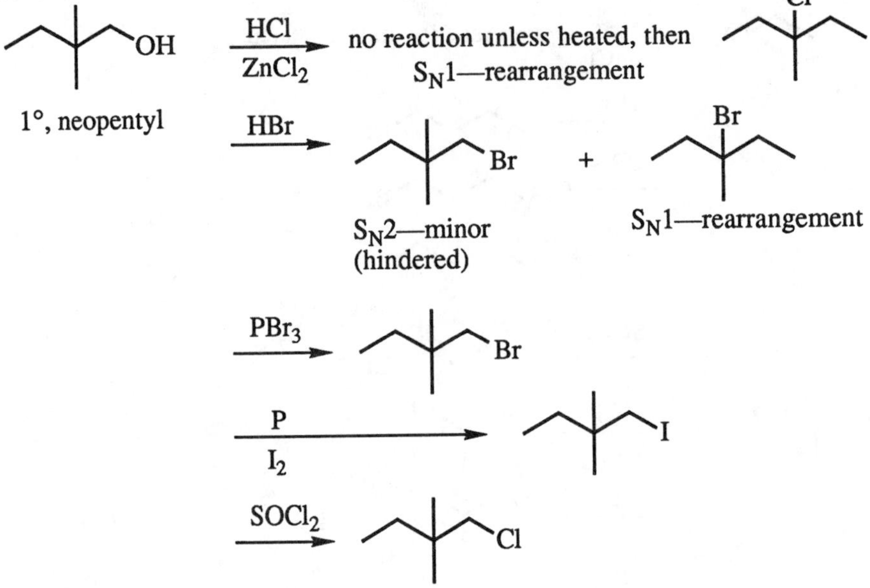

OH
HCl
$ZnCl_2$
no reaction unless heated, then
S_N1—rearrangement
Cl
1°, neopentyl
HBr
Br
+
Br
S_N2—minor
(hindered)
S_N1—rearrangement
PBr_3
Br
P
I_2
I
$SOCl_2$
Cl

10-22

(a)

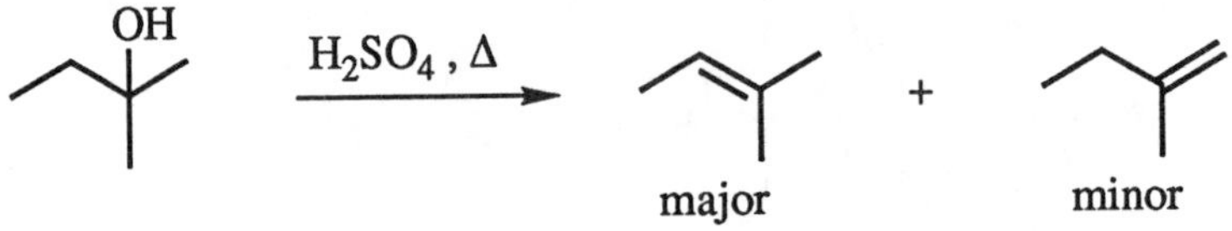

OH
H_2SO_4 , Δ
+
major
minor

(b)

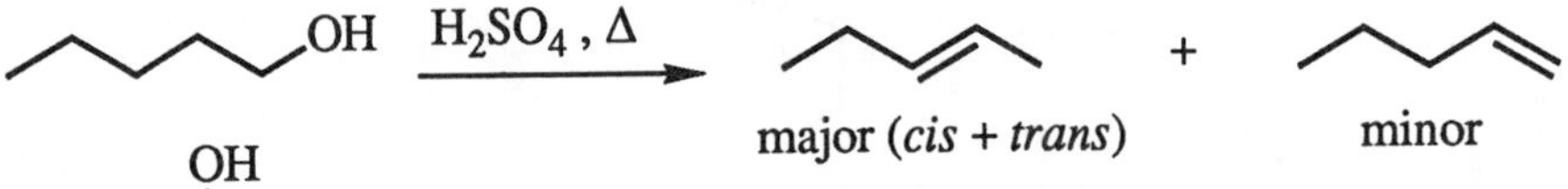

OH
H_2SO_4 , Δ
+
major (*cis* + *trans*)
minor

(c)

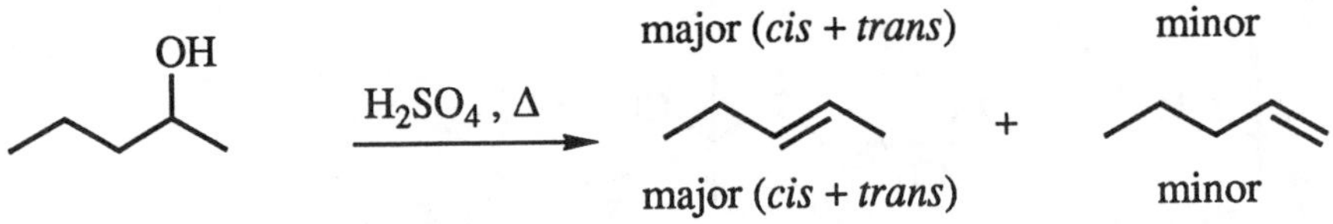

OH
H_2SO_4 , Δ
+
major (*cis* + *trans*)
minor

(d)

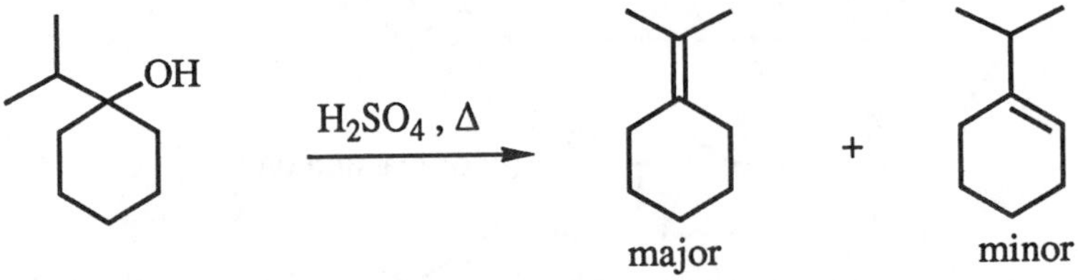

OH
H_2SO_4 , Δ
+
major
minor

(e)

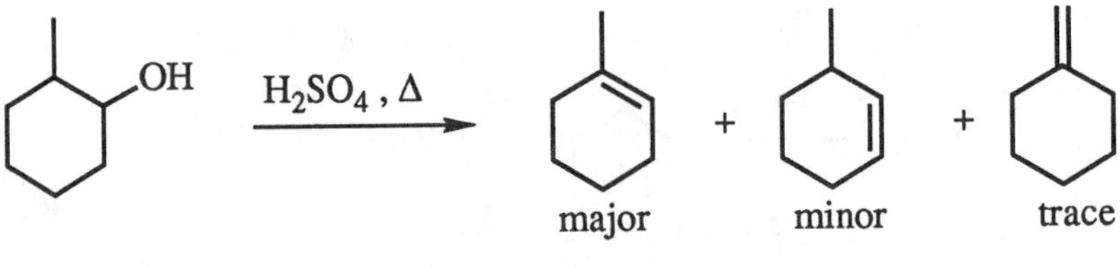

OH
H_2SO_4 , Δ
+
+
major
minor
trace

10-23

Both mechanisms begin with protonation of the oxygen.

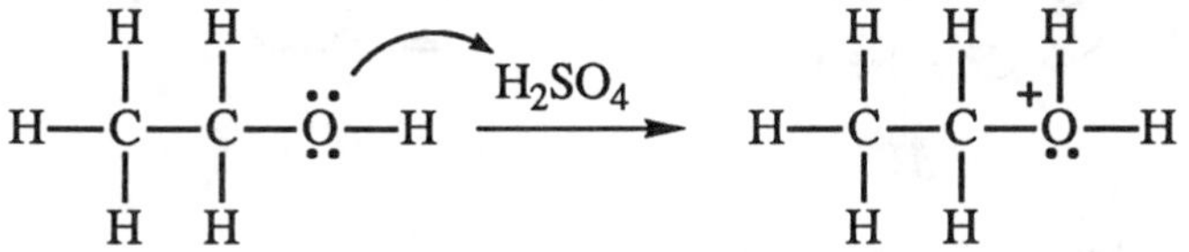

One mechanism involves another molecule of ethanol acting as a base, giving elimination.

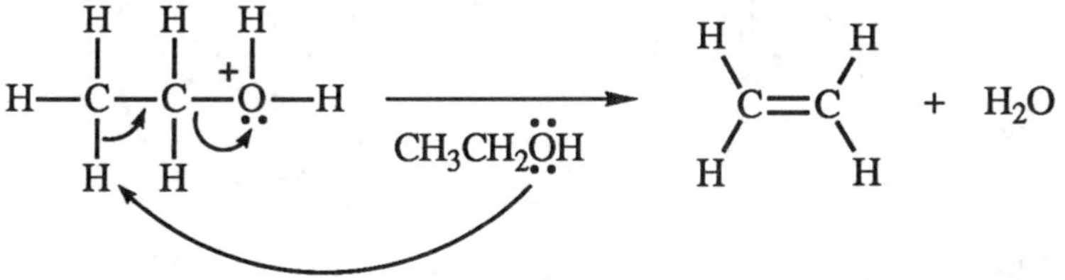

The other mechanism involves another molecule of ethanol acting as a nucleophile, giving substitution.

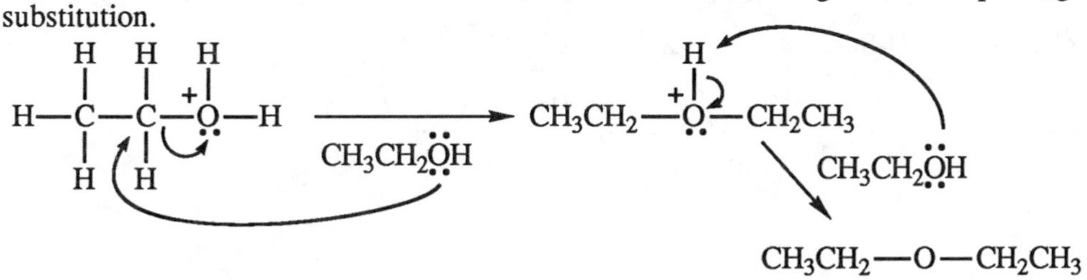

10-24 An equimolar mixture of methanol and ethanol would produce all three possible ethers. The difficulty in separating these compounds would preclude this method from being a practical route to any one of them. This method is practical only for symmetric ethers, that is, where both alkyl groups are identical.

$$CH_3CH_2OH + HOCH_3 \xrightarrow[\Delta]{H^+} H_2O + CH_3CH_2OCH_3 + CH_3OCH_3 + CH_3CH_2OCH_2CH_3$$

10-25

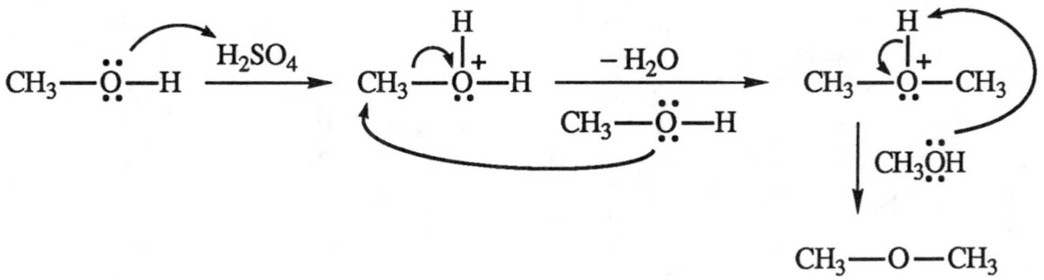

10-26

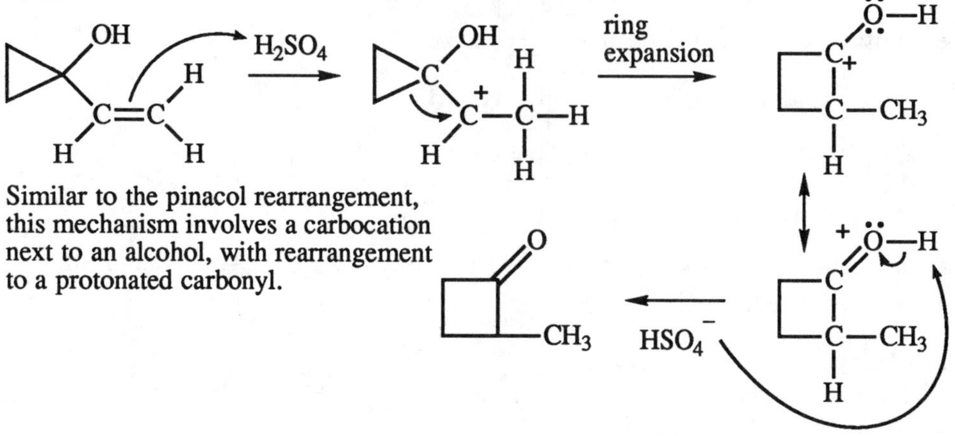

Similar to the pinacol rearrangement, this mechanism involves a carbocation next to an alcohol, with rearrangement to a protonated carbonyl.

10-27

(a)

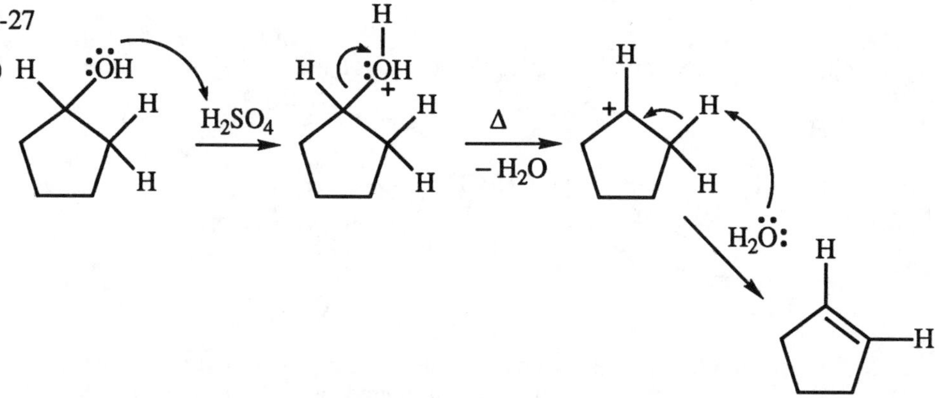

(b)

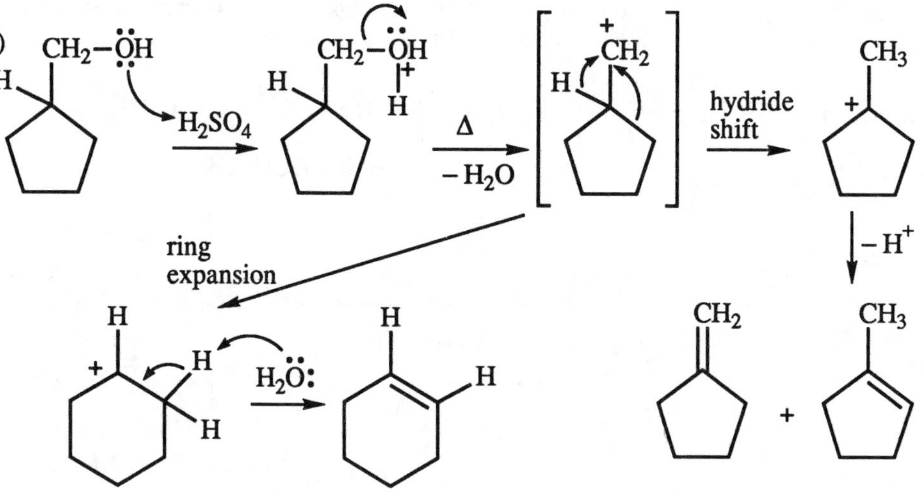

10-27 continued

(c)

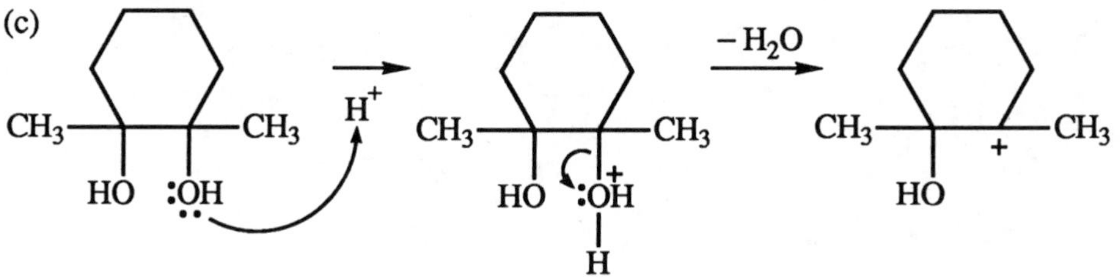

Methyl shift

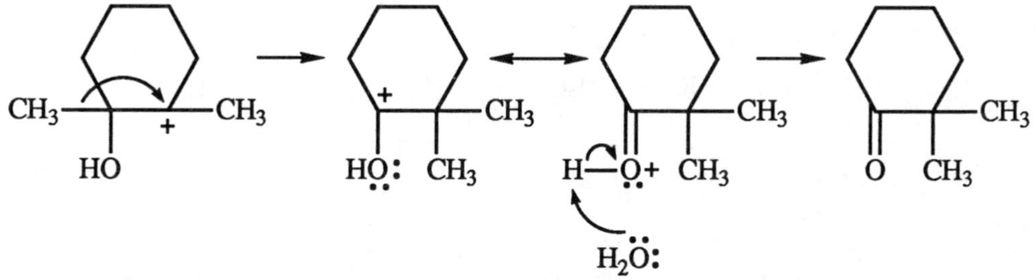

Alkyl shift—ring *contraction*

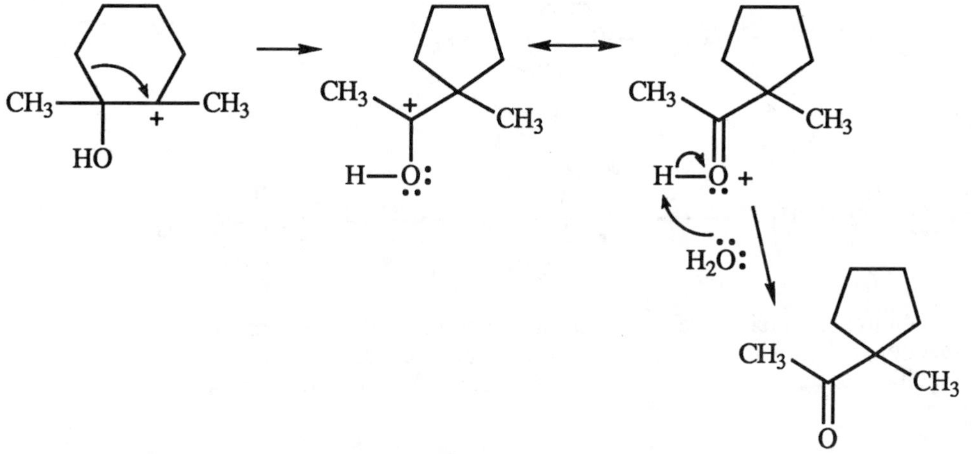

10-28

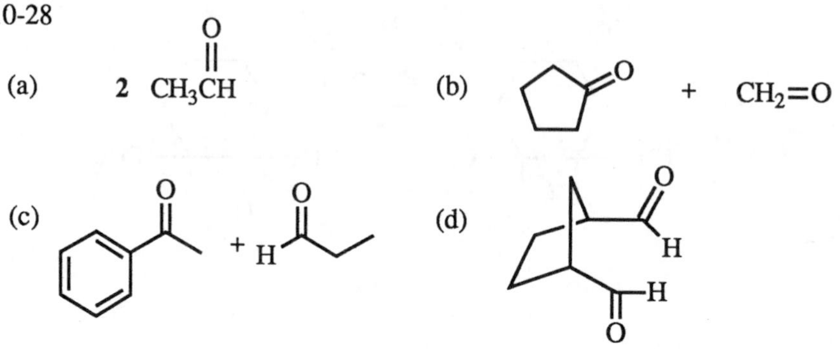

(a) $2\ CH_3\overset{O}{\overset{\|}{C}}H$

(b) 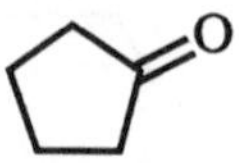 + $CH_2{=}O$

(c) 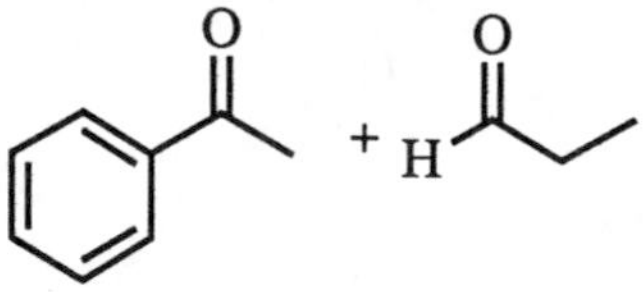

(d) 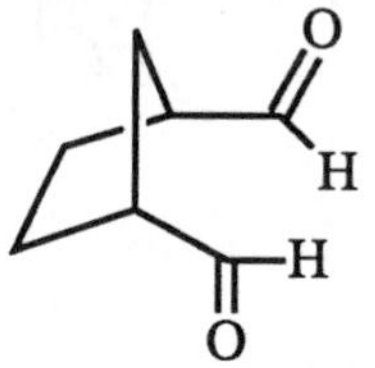

10-29

(a) $CH_3CH_2CH_2\overset{O}{\overset{\|}{C}}{-}Cl$ + $HOCH_2CH_2CH_3$

(b) $CH_3(CH_2)_3OH$ + $Cl{-}\overset{O}{\overset{\|}{C}}CH_2CH_3$

(c) 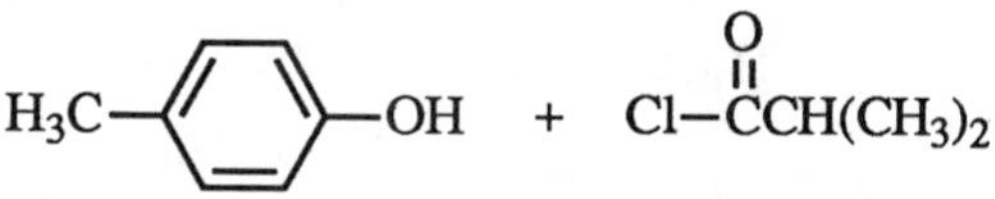

(d) 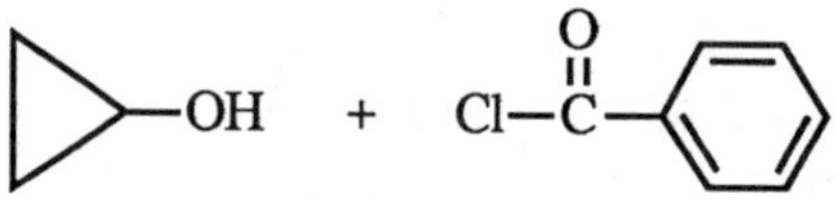

10-30

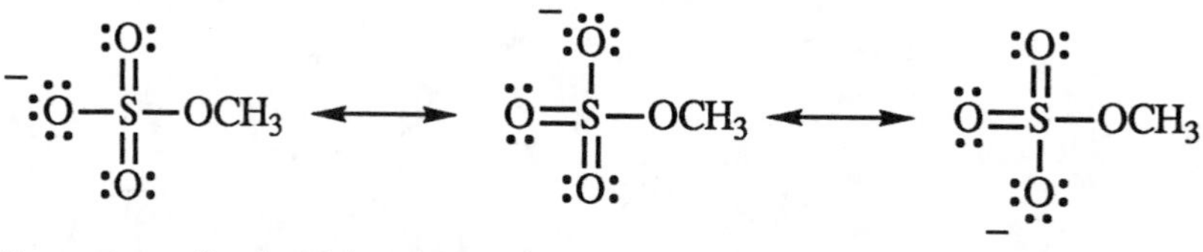

10-31 Proton transfer (acid-base) reactions are much faster than almost any other reaction. Methoxide will act as a base and remove a proton from the oxygen much faster than methoxide will act as a nucleophile and displace water.

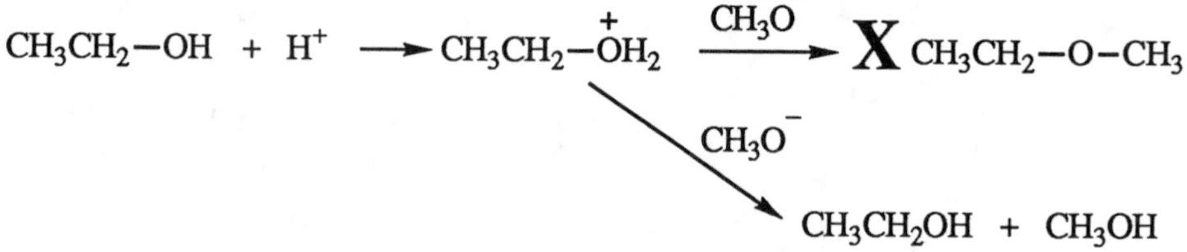
CH3CH2—OH + H+
CH3CH2—OH2
CH3O−
X CH3CH2—O—CH3
CH3O−
CH3CH2OH + CH3OH

10-32 Williamson ether synthesis is an S_N2 reaction, so it works best when the alkoxide attacks the *least substituted* carbon (methyl better than 1° better than 2°).

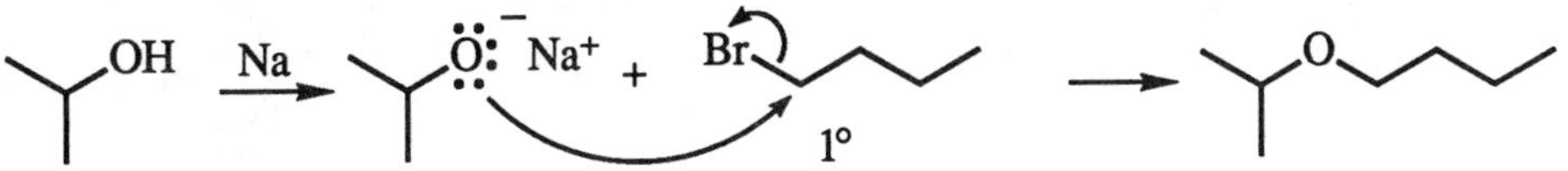

OH
Na
O: Na+
+
Br
1°
O

10-33

(a) 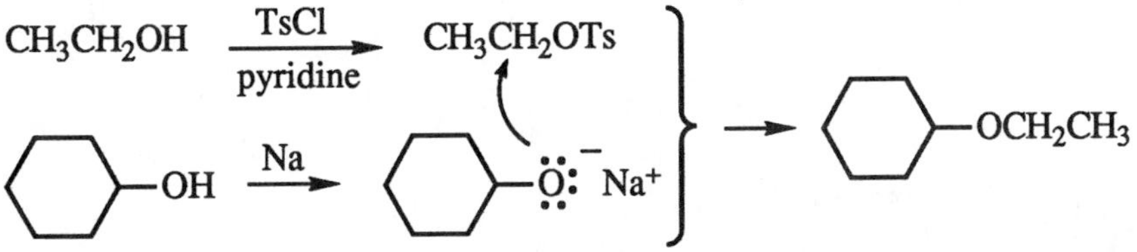

CH3CH2OH
TsCl
pyridine
CH3CH2OTs
OH
Na
O: Na+
OCH2CH3

(b) There are two problems with this attempted bimolecular dehydration. First, all three possible ether combinations of cyclohexanol and ethanol would be produced. Second, hot sulfuric acid are the conditions for dehydrating secondary alcohols like cyclohexanol, so elimination would compete with substitution.

10-34

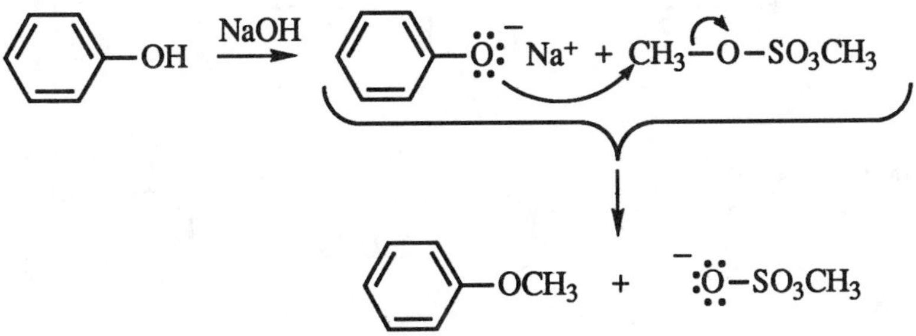

OH
NaOH
O: Na+
+
CH3-O-SO3CH3
OCH3
+
:O-SO3CH3

10-35

(a) 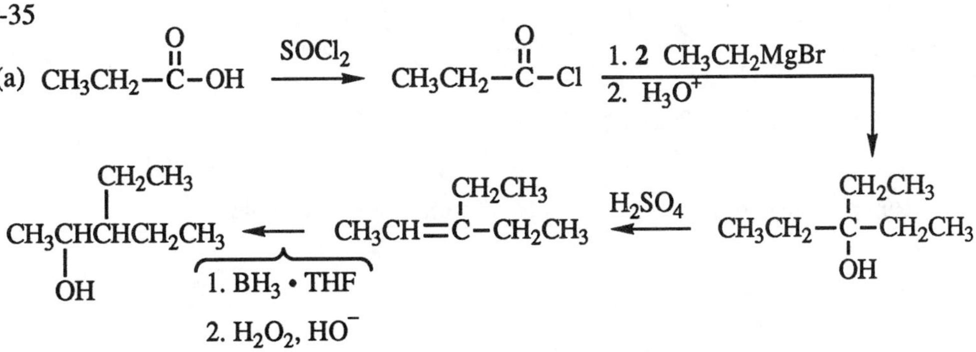

CH3CH2-C-OH
SOCl2
CH3CH2-C-Cl
1. 2 CH3CH2MgBr
2. H3O+
CH2CH3
CH3CH2-C-CH2CH3
OH
H2SO4
CH3CH=C-CH2CH3
CH3CHCHCH2CH3
1. BH3 • THF
2. H2O2, HO−

10-35 continued

(b)

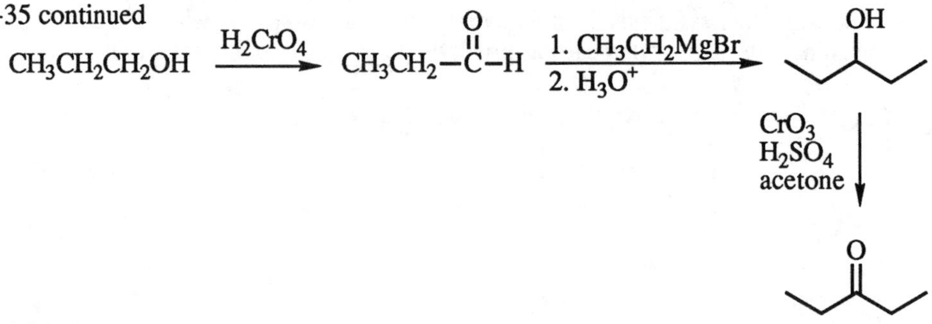

10-36

For parts (a) and (b):

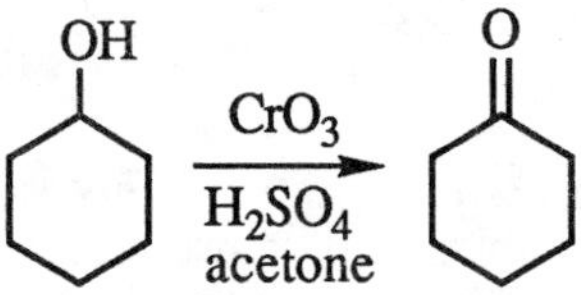

(a)

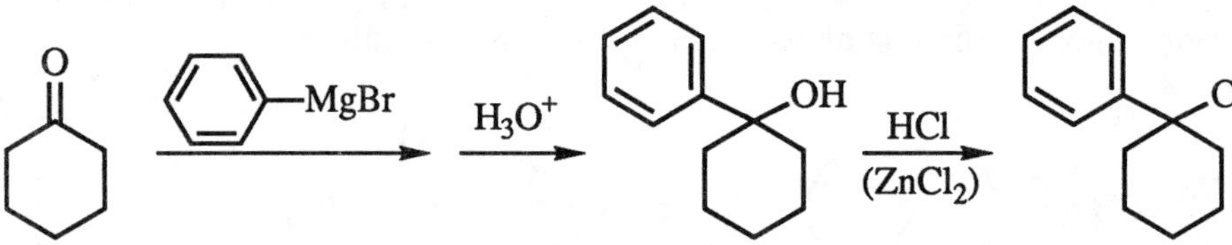

(b)

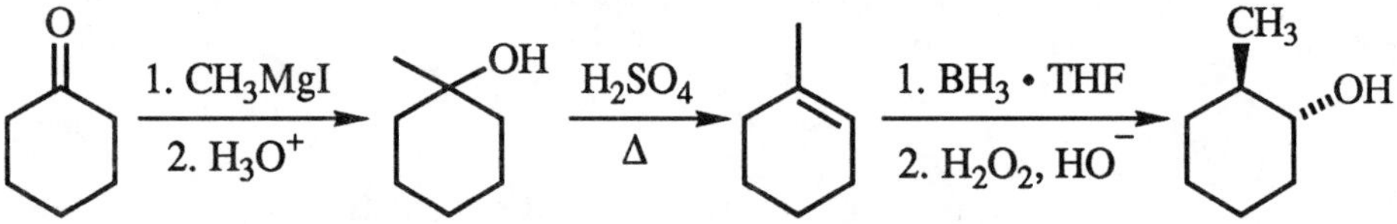

(c)

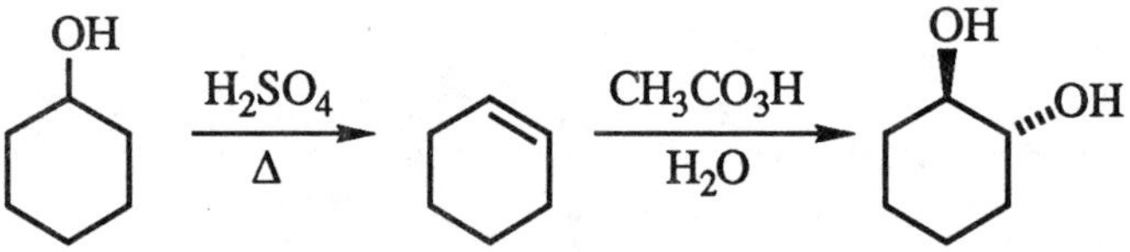

10-37 Refer to the Glossary and the text for definitions and examples.

10-38

(a)

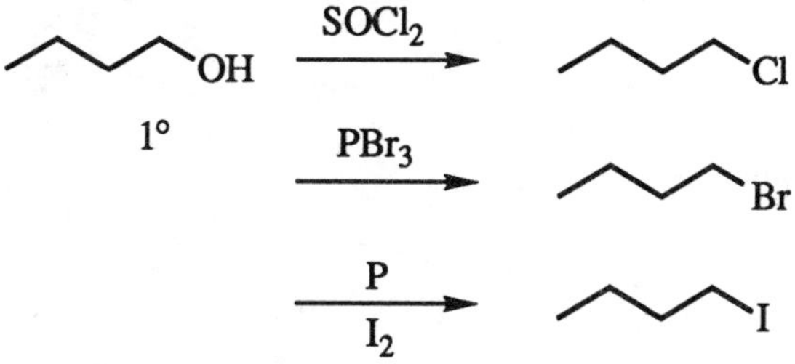
OH
1°
SOCl2
PBr3
P
I2
Cl
Br
I

(b)

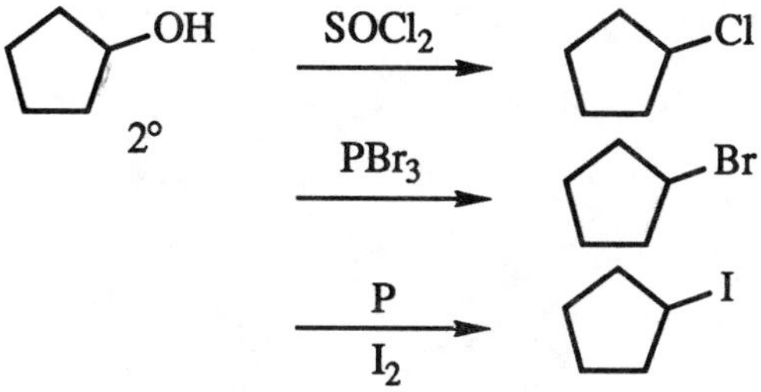
OH
2°
SOCl2
PBr3
P
I2
Cl
Br
I

(c)

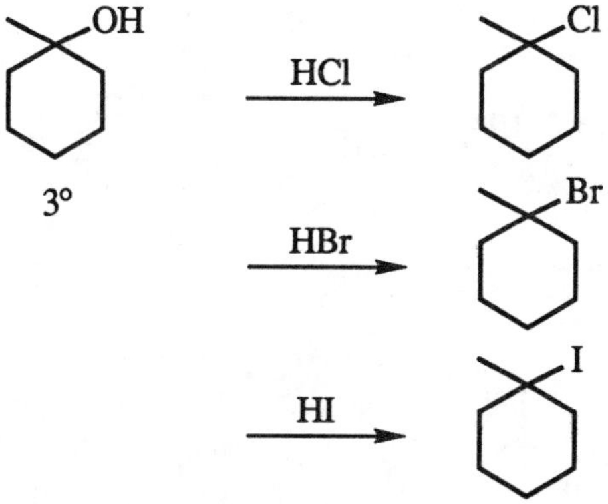
OH
3°
HCl
HBr
HI
Cl
Br
I

(d)

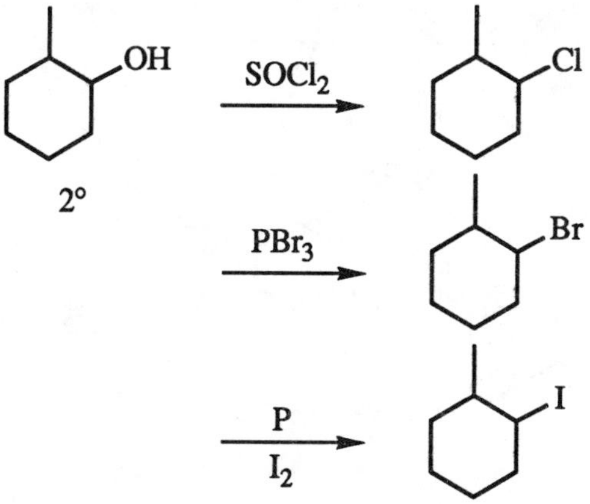
OH
2°
SOCl2
PBr3
P
I2
Cl
Br
I

10-39

(a) 

R

(b) 

R
(from inversion)

(c) 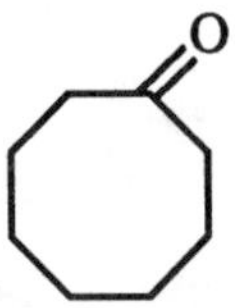

(d) 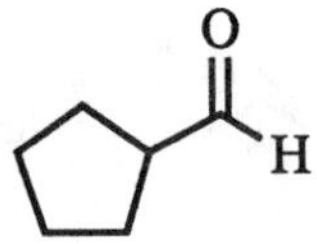

(e) 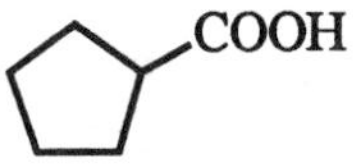

(f) 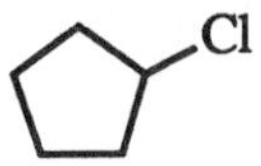

(g) 

(h) 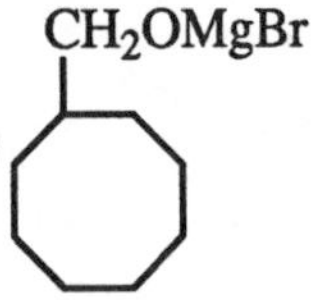

+ CH_3CH_3

(i) 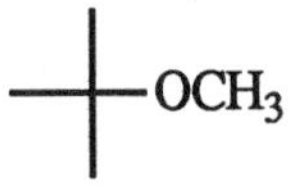

(j)  + CH_3OH

(k) 

(l) 

(m) 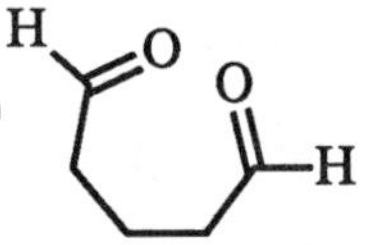

(n) 

(o) 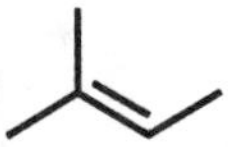 + 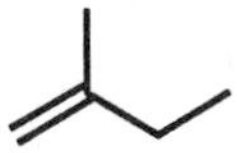

major + minor

10-40

(a) 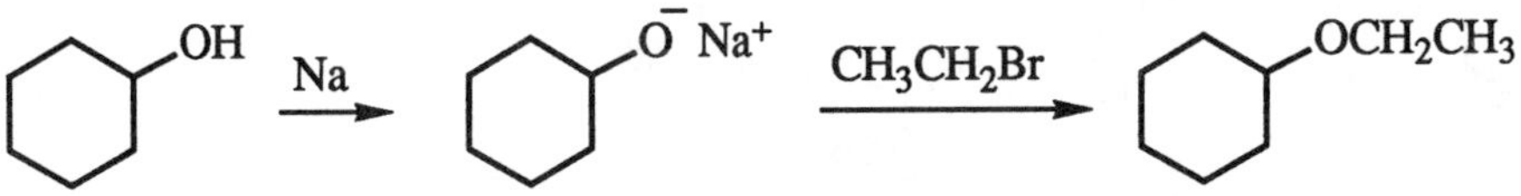

(b) 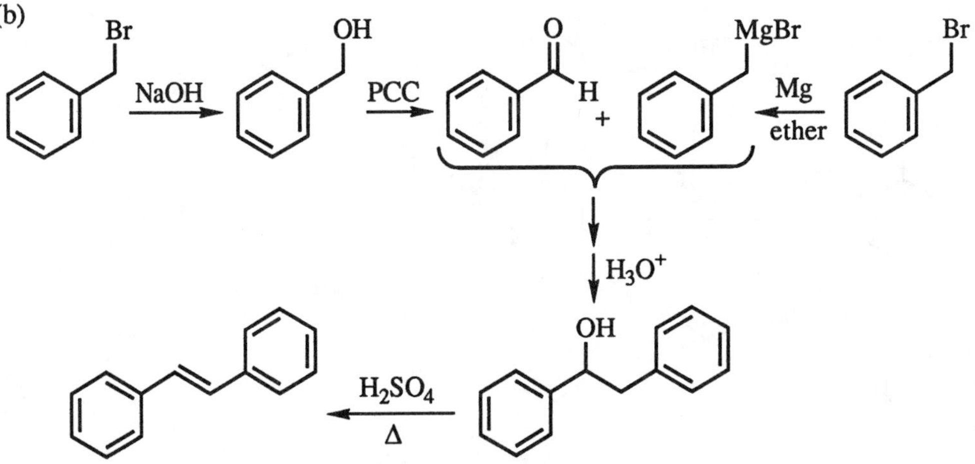

10-40 continued

(c)

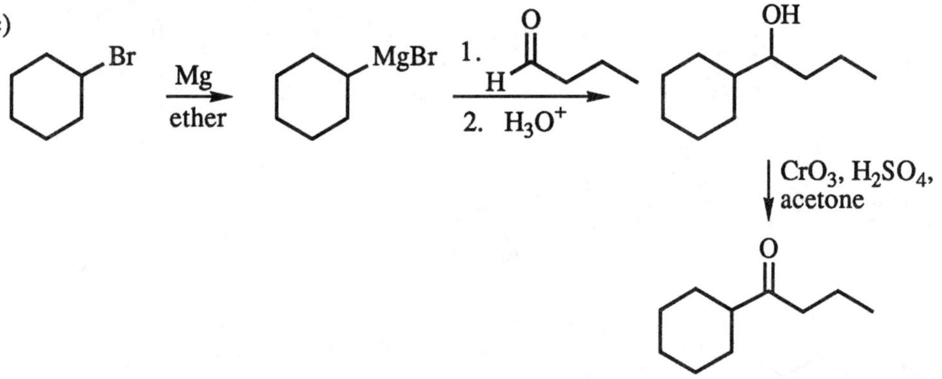

(d)

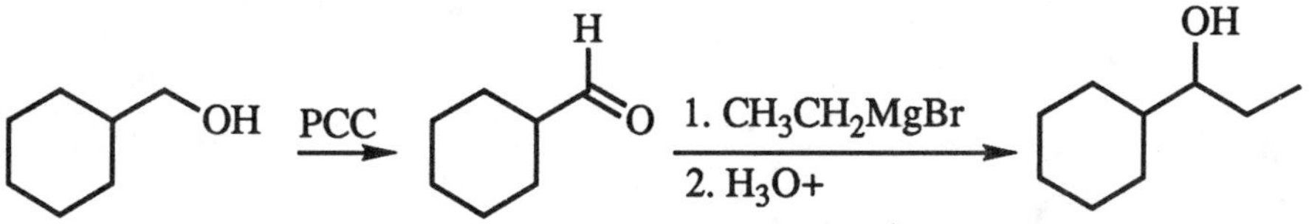

10-41 Major product for each reaction is shown.

(a)  (b)  (c) 

(d)  (e)  (f) 

10-42

(a) $CH_3CH_2CH_2COOCH_3$ (b) 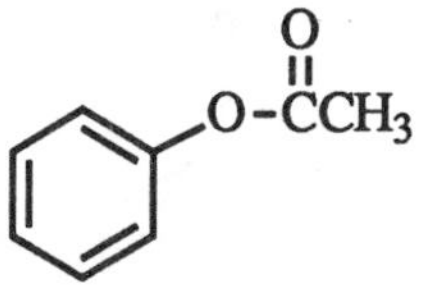 (c) 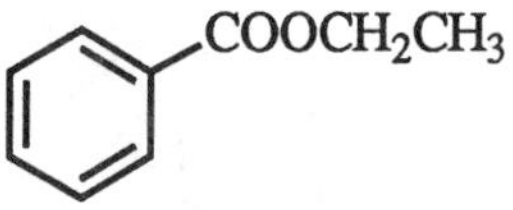

(d) $CH_3CH_2O{-}\overset{O}{\overset{\|}{P}}{-}OH$ with OCH_2CH_3 on P (e) CH_3ONO_2

10-43

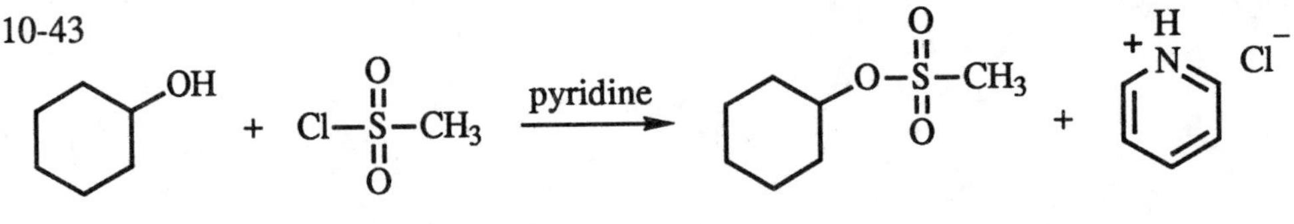

10-44

(a)

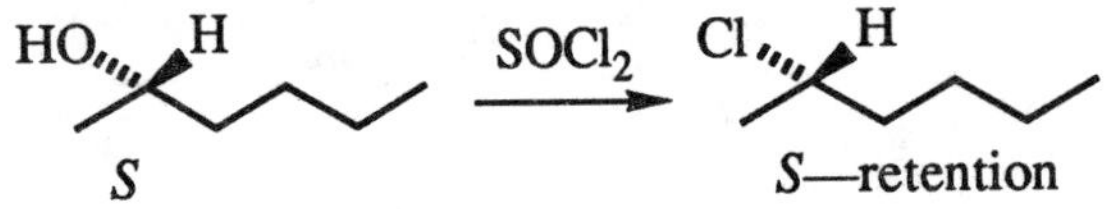

(b)

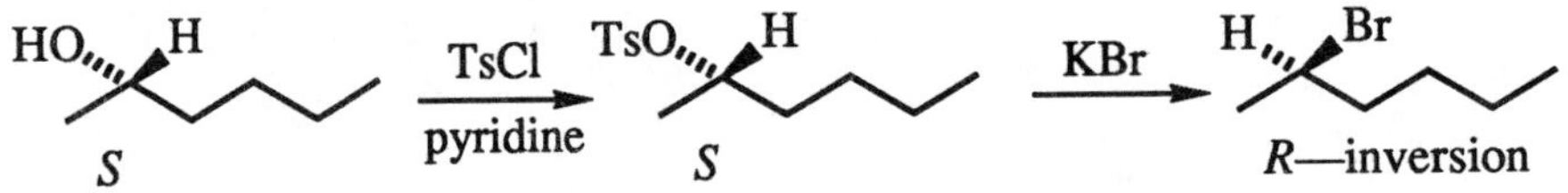

10-45

(a)

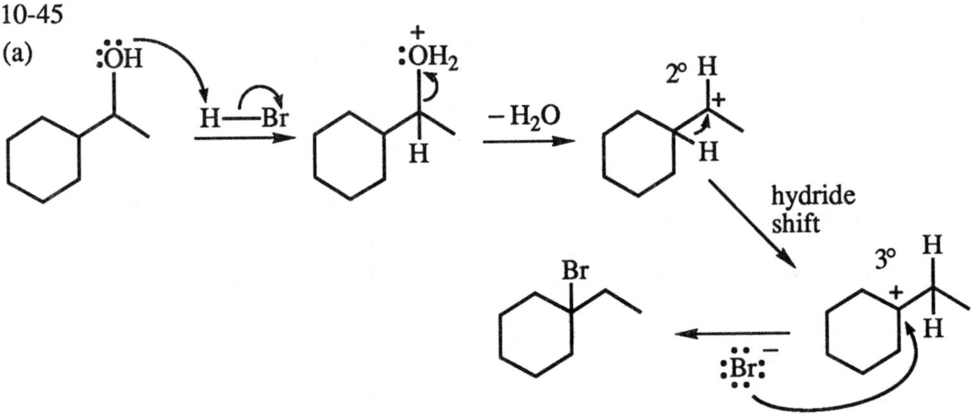

(b) PBr_3 converts alcohols to bromides without rearrangement because no carbocation intermediate is produced.

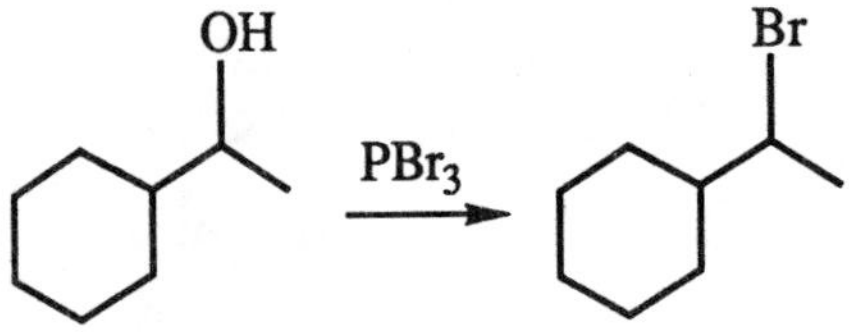

10-46

(a) 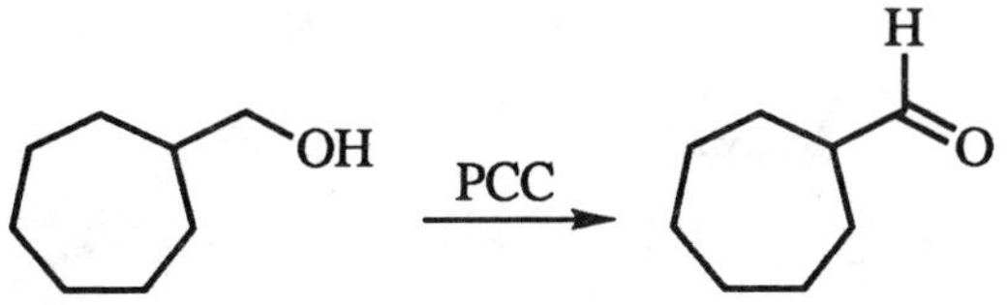

(b) 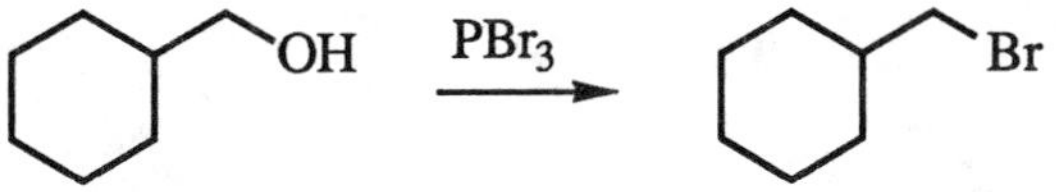

(c) 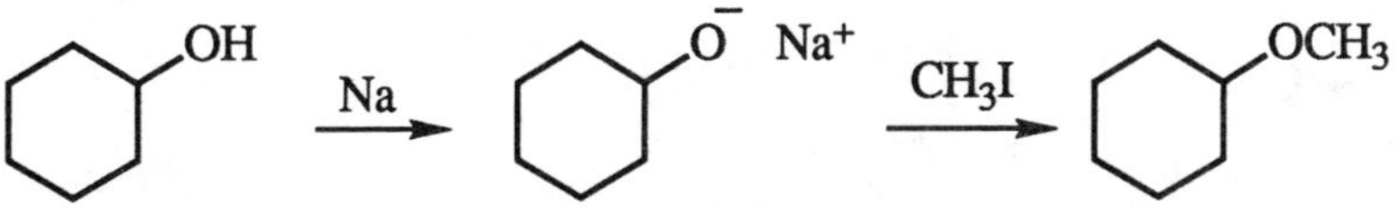

(d) 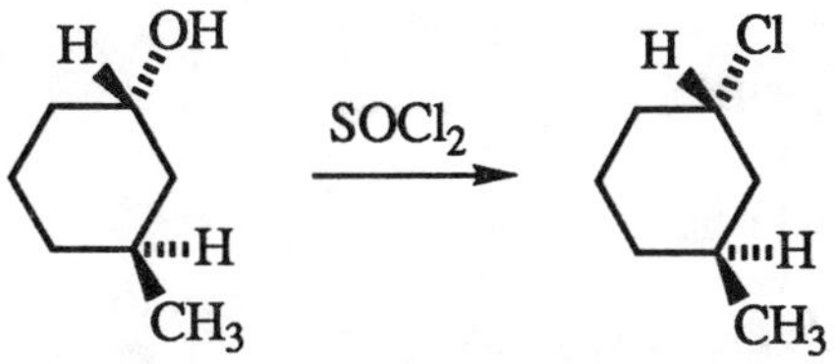

(Alternatively, the *cis* isomer could be made into the tosylate, followed by substitution by chloride with inversion of configuration.)

(e) 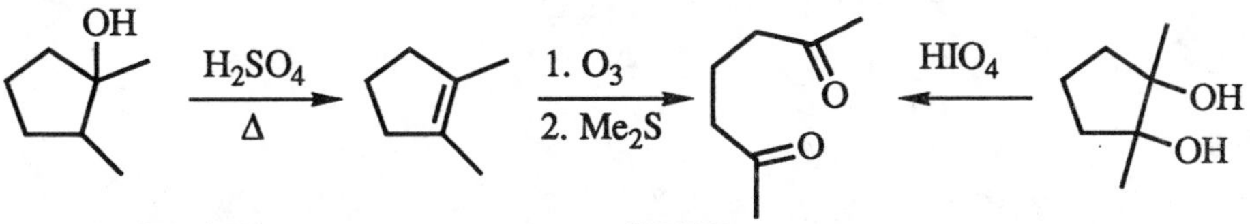

(f) 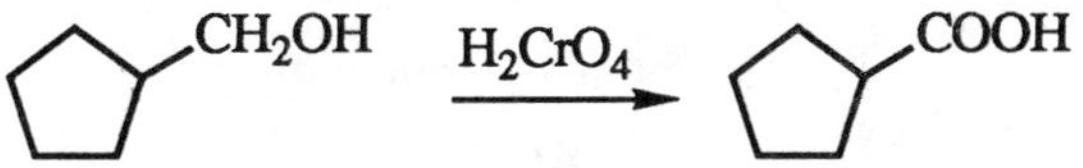

(g) 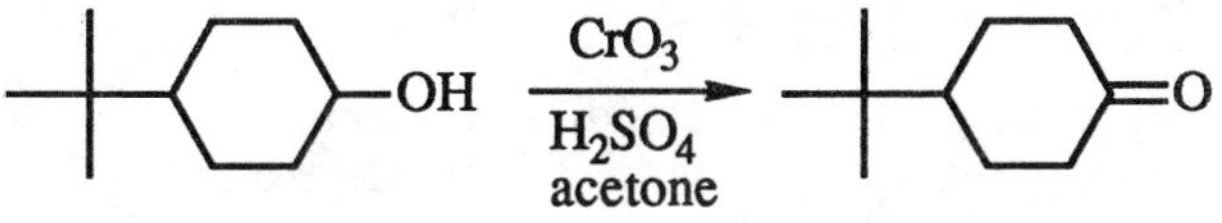

(h) 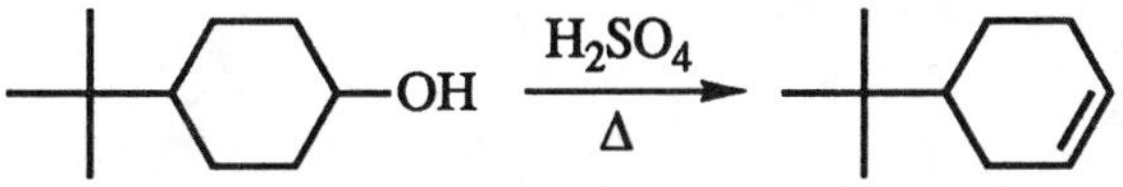

(i) 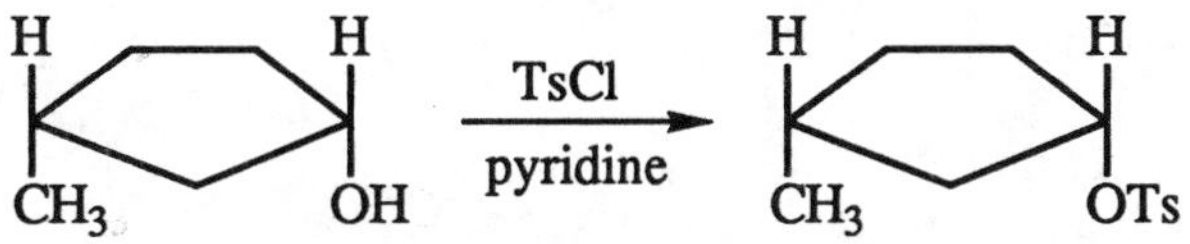

(j) 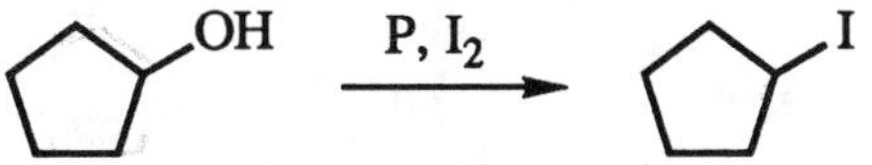

10-47

(a) 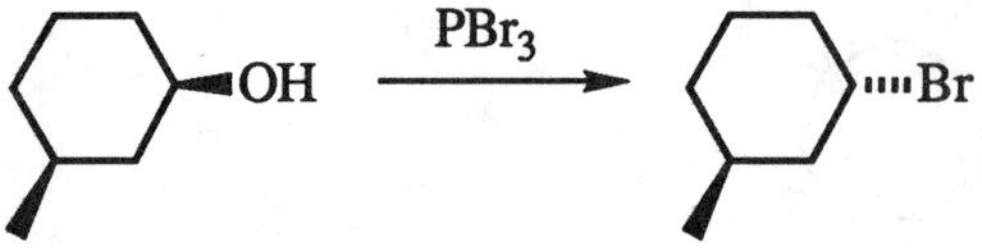
OH
PBr₃
Br

(b) 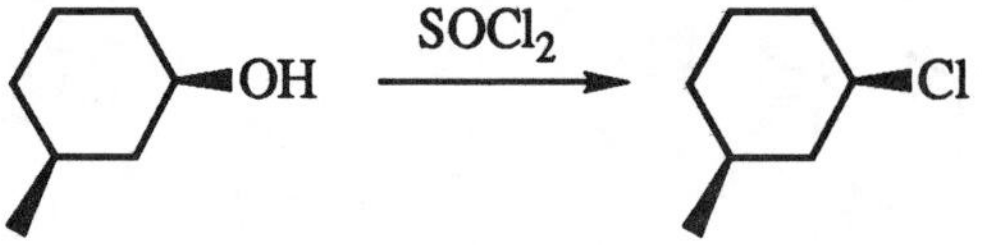
OH
SOCl₂
Cl

(c) 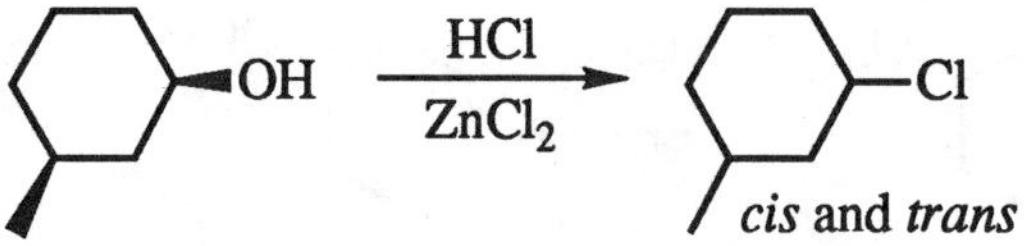
OH
HCl
ZnCl₂
Cl
cis and trans

(d) 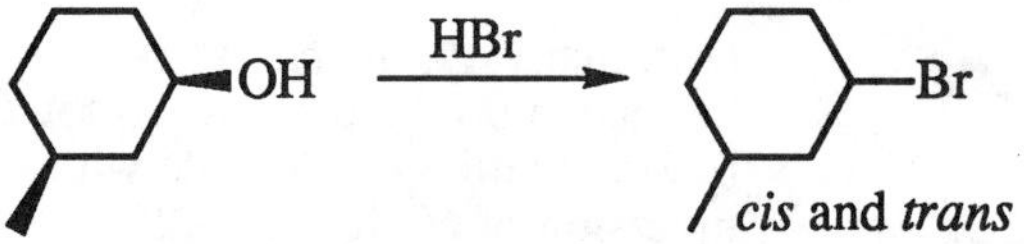
OH
HBr
Br
cis and trans

(e) 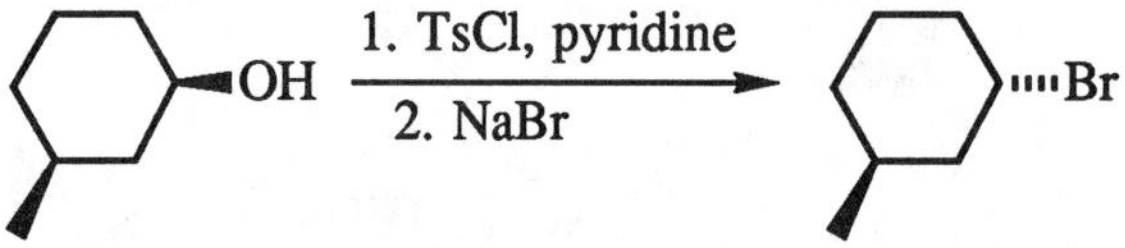
OH
1. TsCl, pyridine
2. NaBr
Br

10-48

(a)  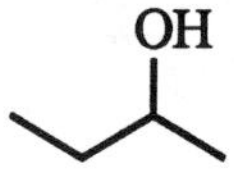

Lucas:	no reaction	cloudy in 1-5 min.

(b) 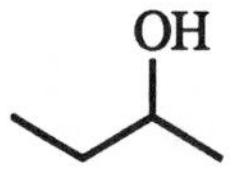 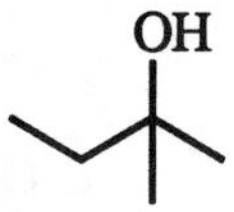

Lucas:	cloudy in 1-5 min.	cloudy in < 1 min.
H_2CrO_4:	immediate blue-green	no reaction—stays orange

(c) 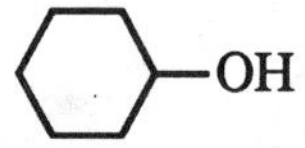 

Lucas:	cloudy in 1-5 min.	no reaction
H_2CrO_4:	immediate blue green	no reaction—stays orange

(d) 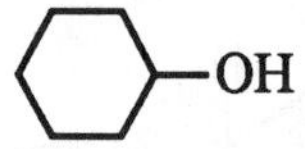 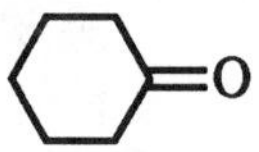

Lucas:	cloudy in 1-5 min.	no reaction
H_2CrO_4:	immediate blue green	no reaction—stays orange

(e) 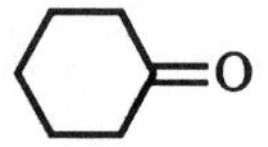 

Lucas:	no reaction	cloudy in < 1 min.

10-49

(a)

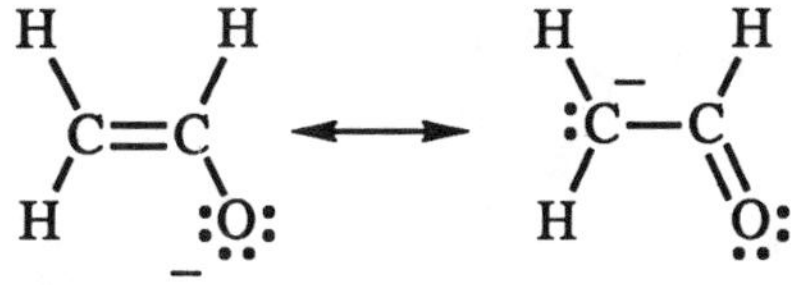

(b)

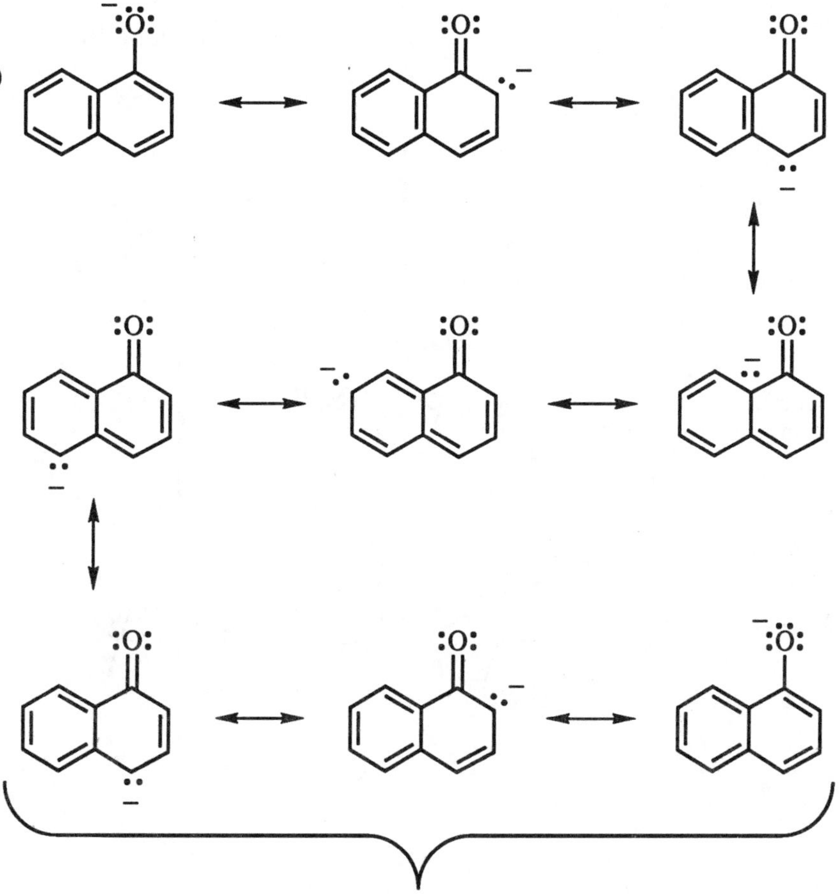

These last three resonance forms are similar to the first three but with the positions of the electrons in the left ring changed. To be rigorously correct, these three should be included, but most chemists would not write them since they do not reveal extra charge delocalization, with the understanding that they would still be significant.

(c)

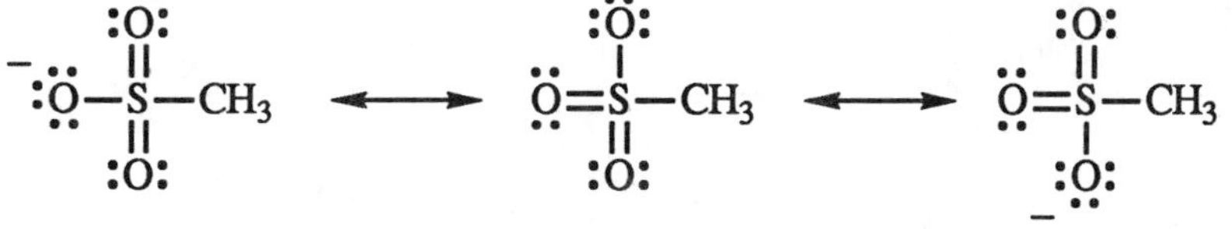

10-50

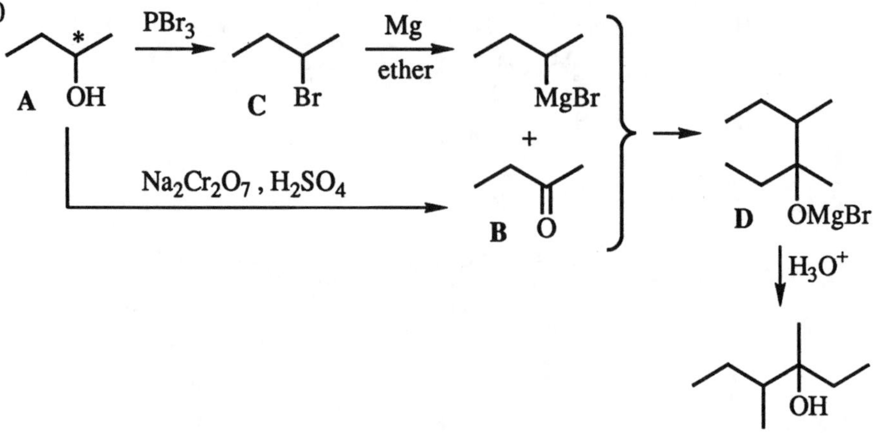
PBr3
Mg
ether
A
OH
C
Br
MgBr
+
Na2Cr2O7 , H2SO4
B
O
D
OMgBr
H3O+
OH

10-51

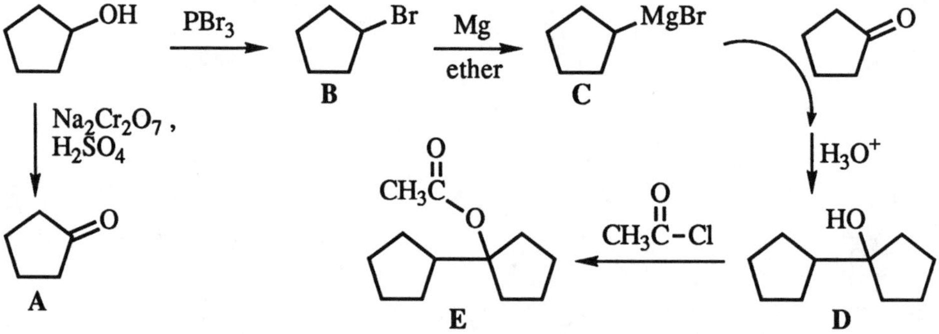
OH
PBr3
Br
Mg
ether
MgBr
O
B
C
Na2Cr2O7 ,
H2SO4
H3O+
O
CH3C
O
O
CH3C-Cl
HO
O
A
E
D

10-52

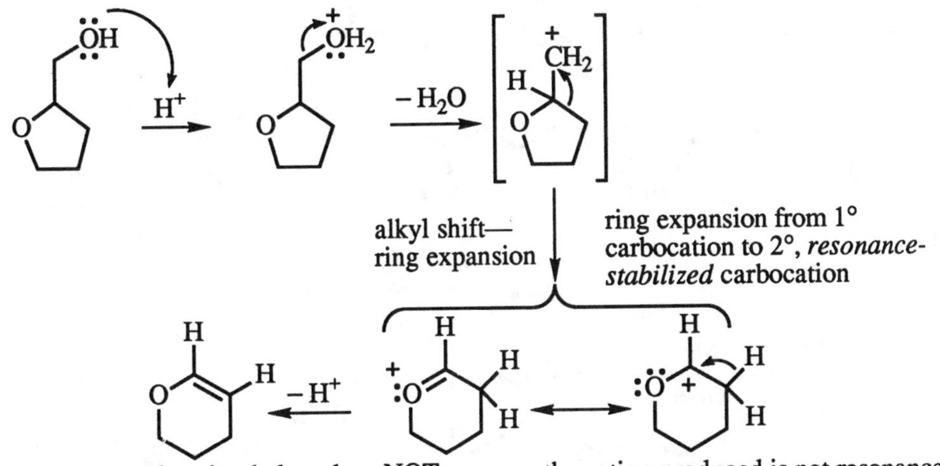

H^+
$-H_2O$
alkyl shift—
ring expansion
ring expansion from 1° carbocation to 2°, *resonance-stabilized* carbocation
$-H^+$

Note that the migration below does NOT occur as the cation produced is not resonance-stabilized.

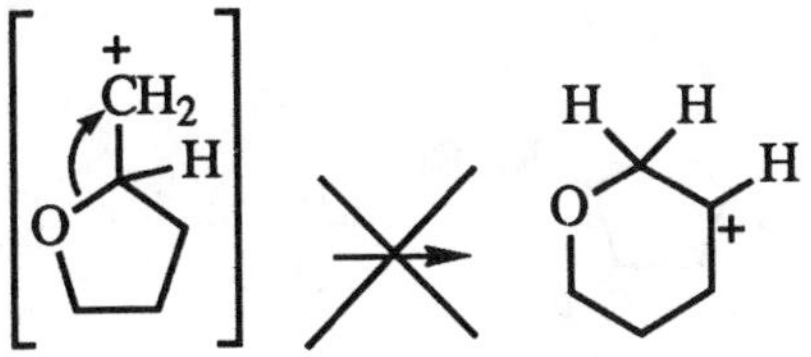

10-53

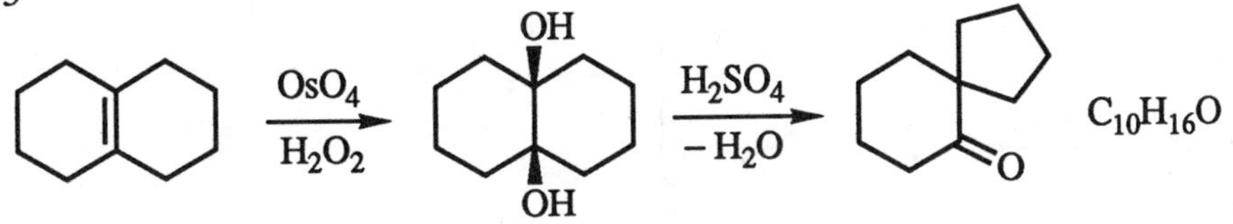

OsO_4
H_2O_2
H_2SO_4
$-H_2O$
$C_{10}H_{16}O$

10-54

(a)

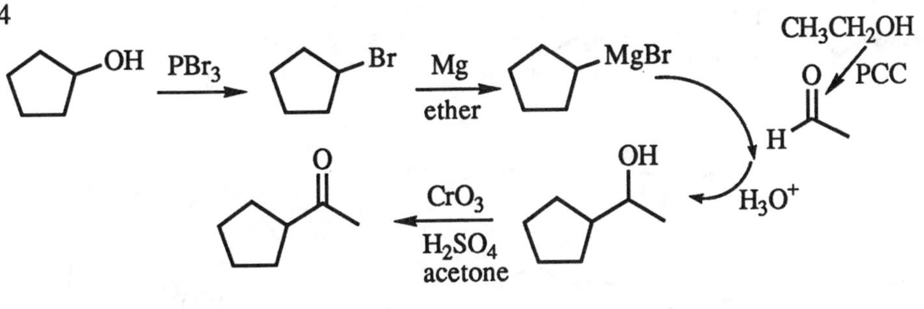

PBr_3
Mg
ether
CH_3CH_2OH
PCC
H_3O^+
CrO_3
H_2SO_4
acetone

(b)

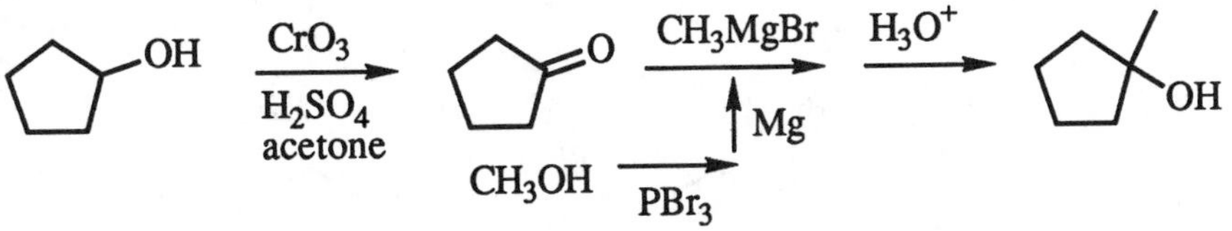

CrO_3
H_2SO_4
acetone
CH_3MgBr
H_3O^+
Mg
CH_3OH
PBr_3

10-54 continued

(c)

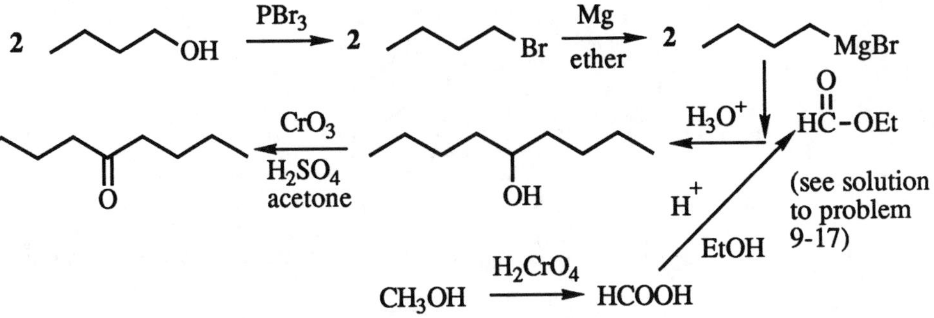
2
OH
PBr3
2
Br
Mg
ether
2
MgBr
O
HC-OEt
H3O+
CrO3
H2SO4
acetone
OH
O
(see solution
to problem
9-17)
H+
EtOH
H2CrO4
CH3OH
HCOOH

(d)

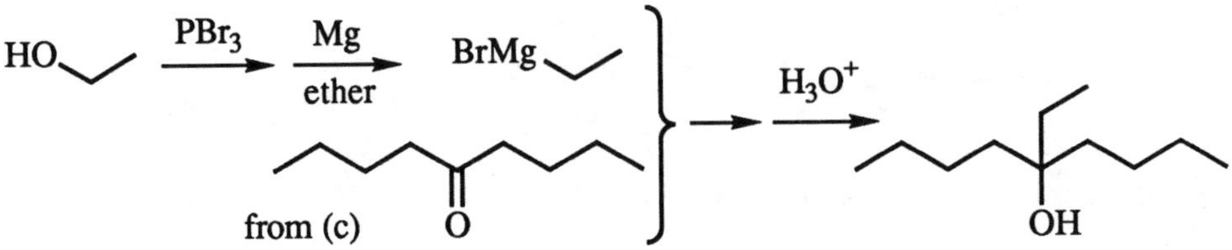
HO
PBr3
Mg
ether
BrMg
H3O+
from (c)
O
OH

(e)

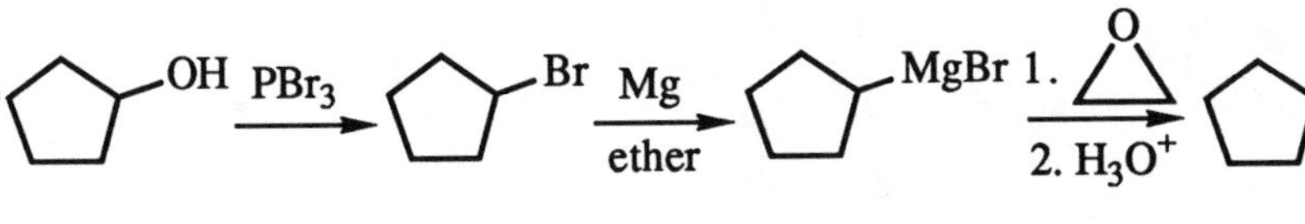
OH
PBr3
Br
Mg
ether
MgBr
1.
O
2. H3O+
OH

(f)

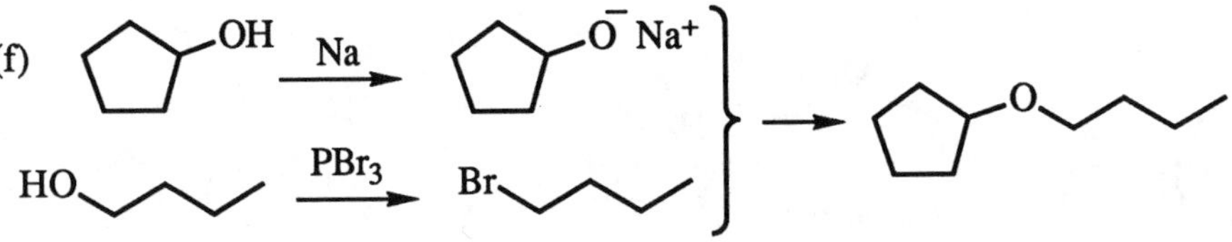
OH
Na
O− Na+
O
HO
PBr3
Br

(g)

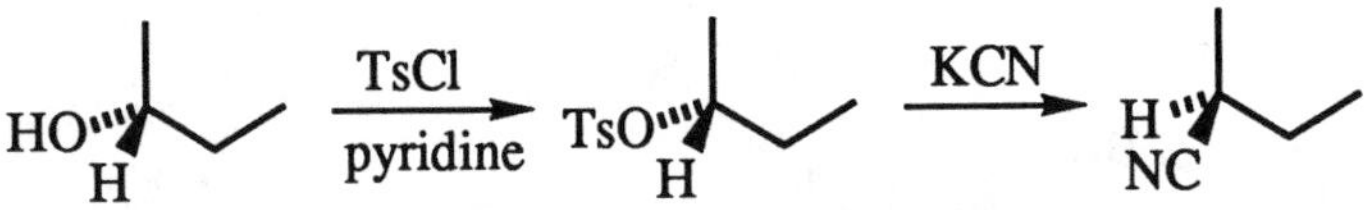
HO
H
TsCl
pyridine
TsO
H
KCN
H
NC

(h)

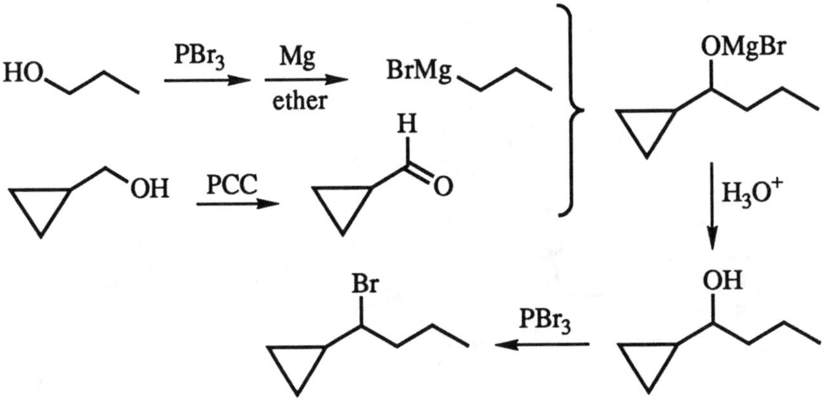
HO
PBr3
Mg
ether
BrMg
H
OH
PCC
O
OMgBr
H3O+
OH
PBr3
Br

10-55

(a)

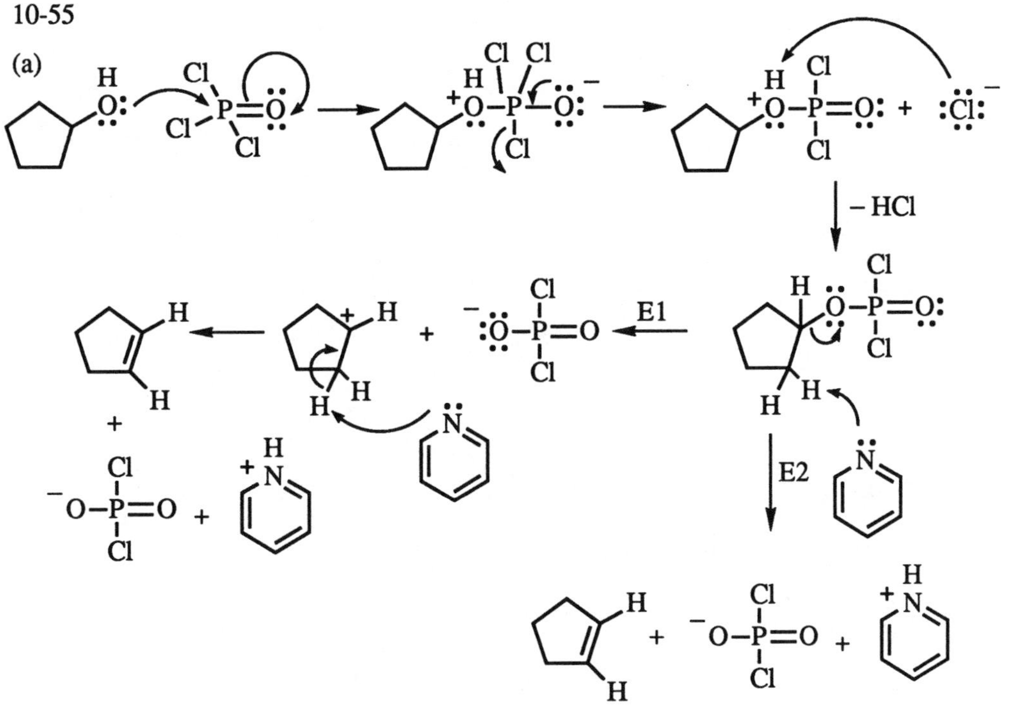

(b)

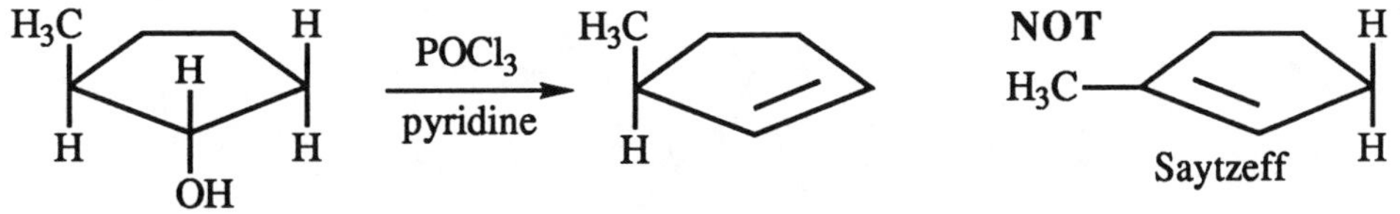

There must be a stereochemical requirement in this elimination. If the Saytzeff alkene is not produced because the methyl group is *trans* to the leaving group, then the H and the leaving group must be *trans* and the elimination must be anti—the characteristic stereochemistry of E2 elimination. This evidence differentiates between the two possibilities in part (a).

10-56

Compound X :  —must be a 1° or 2° alcohol with an alkene; no reaction with Lucas leads to a 1° alcohol; can't be allylic as this would give a positive Lucas test

Compound Y :  —must be a cyclic ether, not an alcohol and not an alkene; other isomers of cyclic ethers possible

CHAPTER 11—INFRARED SPECTROSCOPY AND MASS SPECTROMETRY

11-1 The table is completed by recognizing that: $(\overline{\nu})(\lambda) = 10{,}000$

$\overline{\nu}$ (cm^{-1})	1600	1640	1700	**2198**	2500	**3003**	**3300**
λ (μm)	**6.25**	**6.10**	**5.88**	4.55	4.00	3.33	3.03

11-2

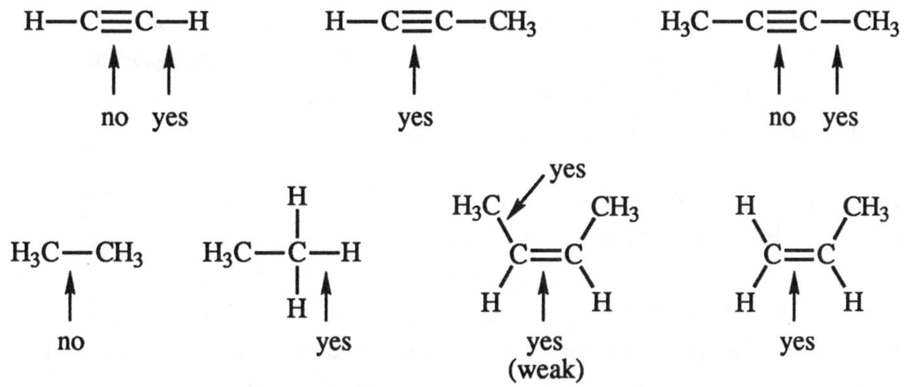

11-3 (a) alkene: C=C at 1640 cm^{-1}, =C—H at 3020 cm^{-1}
(b) alkane: no peaks indicating sp or sp^2 carbons present
(c) terminal alkyne: C≡C at 2150 cm^{-1}, ≡C—H at 3300 cm^{-1}; =C—H at 3050 cm^{-1}, C=C at 1640 cm^{-1} (perhaps benzene)

11-4 (a) 1° amine, R—NH_2: the two peaks at 3300-3360 cm^{-1} indicate two N—H bonds. This spectrum also shows a C=C at 1640 cm^{-1}.
(b) carboxylic acid: the extremely broad absorption in the 2500-3500 cm^{-1} range, with a "shoulder" around 2500-2700 cm^{-1}, and a C=O at 1710 cm^{-1}, are compelling evidence for a carboxylic acid.
(c) alcohol: strong, broad O—H at 3350 cm^{-1}

11-5 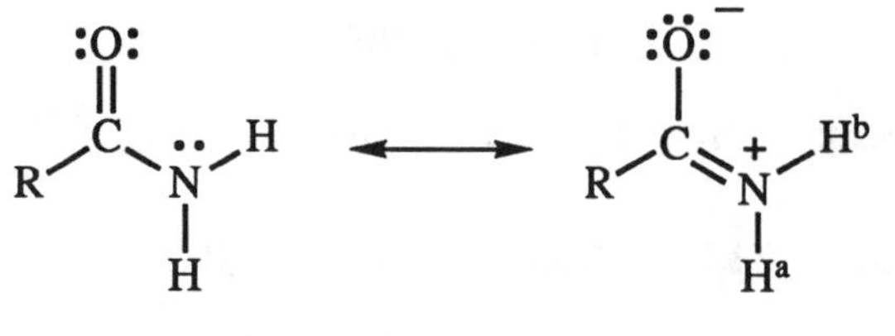

The structure on the right is a significant resonance contributor because all atoms still have octets and the negative charge is on the more electronegative atom. The C—N bond must therefore have some double bond character, suggesting that rotation around the C—N bond is not "free", but rather is restricted. The two hydrogens H^a and H^b bear a *trans* and *cis* relationship, respectively, to the oxygen; these hydrogens are not equivalent and could be expected to show different absorption frequencies. We observe this phenomenon as a double peak in the N—H region of the IR.

11-6 (a) conjugated ketone: the small peak at 3100 cm^{-1} suggests =C—H, and the strong peak at 1675 cm^{-1} is consistent with a ketone conjugated with the alkene. (Note that no C=C is apparent in this spectrum; when the intensity of the conjugated C=O is so strong, the weaker C=C can be hidden by the strong C=O peak.)
(b) ester: the C=O absorption at 1725 cm^{-1} (higher than the ketone's 1710 cm^{-1}), in conjunction with the strong C—O at 1240 cm^{-1} , points to an ester.
(c) amide: the two peaks at 3200-3400 cm^{-1} are likely to be an NH_2 group; the strong peak at 1640 cm^{-1} is too strong for an alkene, so it must be a different type of C=X, in this case a C=O, so low because it is part of an amide.

11-7 (a) The small peak at 1650 cm^{-1} indicates a C=C, consistent with the =C—H at 3080 cm^{-1} . This appears to be a simple alkene.
(b) The strong absorption at 1720 cm^{-1} is unmistakably a C=O. The sharp peak at 2710 cm^{-1} and the small spike at 2800 cm^{-1} (look carefully!) confirm that this is an aldehyde.
(c) The strong peak at 1640 cm^{-1} is C=C, probably conjugated as it is unusually strong. The 1700 cm^{-1} peak appears to be a conjugated C=O, undeniably carboxylic acid because of the strong, broad O—H absorption from 2500-3500 cm^{-1} .
(d) The C=O absorption at 1735 cm^{-1} coupled with C—O at 1250 cm^{-1} suggests an ester.

11-8 (a) The M and M+2 peaks of equal intensity identify the presence of bromine. The mass of M (156) minus the weight of the lighter isotope of bromine (79) gives the mass of the rest of the molecule: 156 – 79 = 77. The C_6H_5 (phenyl) group weighs 77; this compound is bromobenzene, C_6H_5Br .

(b) The m/z 127 peak shows that iodine is present. The molecular ion minus iodine gives the remainder of the molecule: 156 – 127 = 29 . The C_2H_5 (ethyl) group weighs 29; this compound is iodoethane, C_2H_5I .

(c) The M and M+2 peaks have relative intensities of about 3:1, a sure sign of chlorine. The mass of M minus the mass of the lighter isotope of chlorine gives the mass of the remainder of the molecule: 90 – 35 = 55. A fragment of mass 55 is not one of the common alkyl groups (15, 29, 43, 57, *etc.*, increasing in increments of 14 mass units (CH_2)), so the presence of an atom like oxygen must be considered. In addition to the chlorine atom, mass 55 could be C_4H_7 or C_3H_3O .

(d) The odd-mass molecular ion indicates the presence of an odd number of nitrogen atoms (always begin by assuming *one* nitrogen). The rest of the molecule must be: 115 – 14 = 101; this is most likely C_7H_{17} . The formula $C_7H_{17}N$ is the correct formula of a molecule with no elements of unsaturation.

11-9

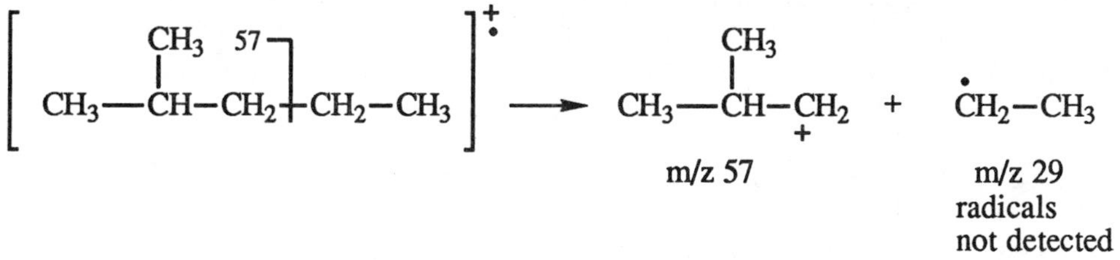

11-10

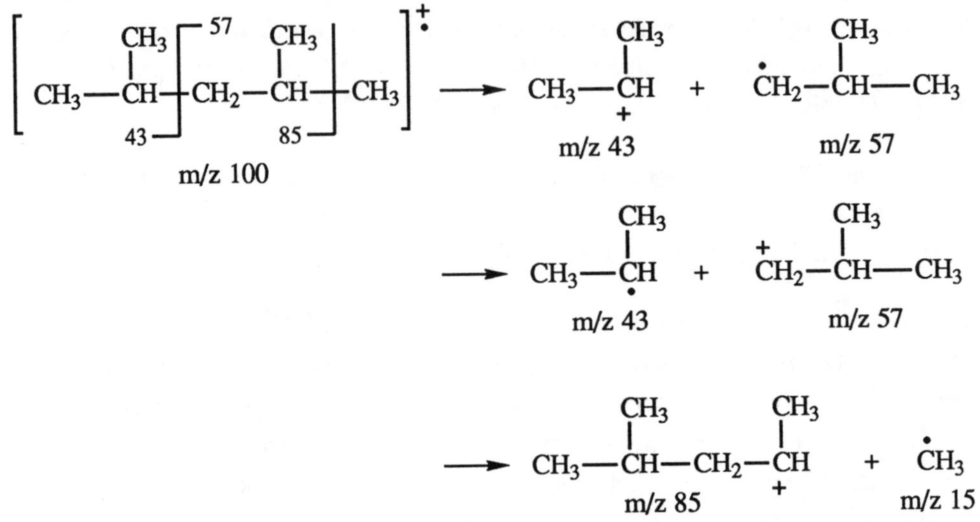

11-11 2,6-Dimethyl-4-heptanol, $C_9H_{20}O$, has molecular weight 144. The highest mass peak at 126 is *not* the molecular ion, but rather is the loss of water (18) from the molecular ion.

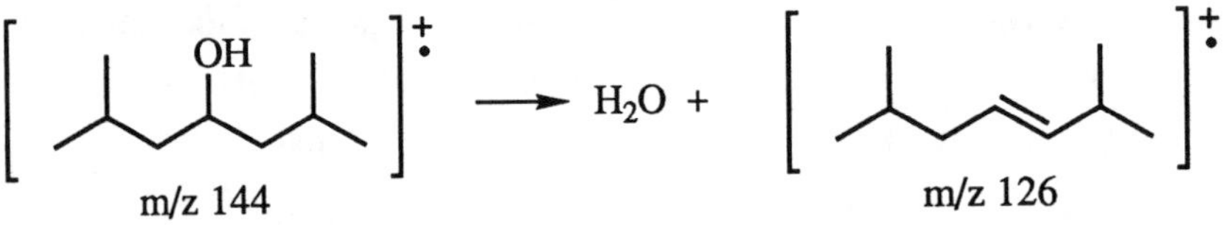

The peak at m/z 111 is loss of another 15 (CH_3) from the fragment of m/z 126. This is called allylic cleavage; it generates a 2°, allylic, resonance-stabilized carbocation.

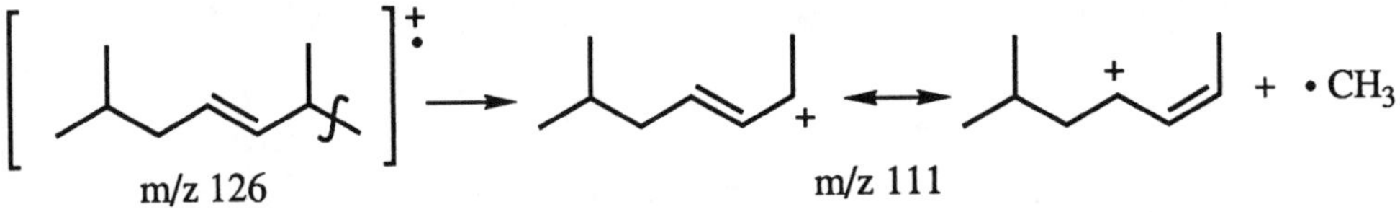

The peak at m/z 87 results from fragmentation on one side of the alcohol:

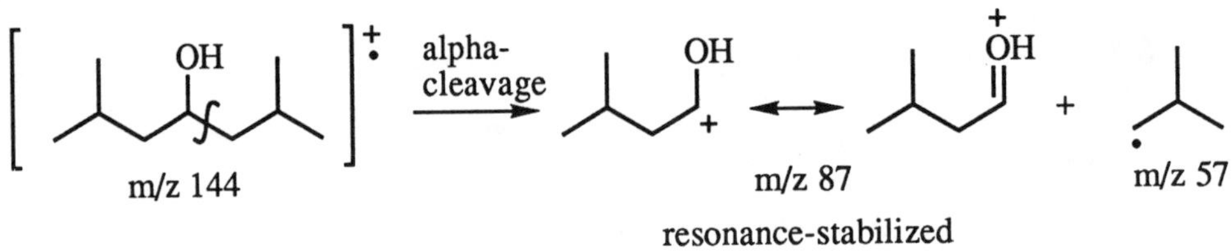

11-12

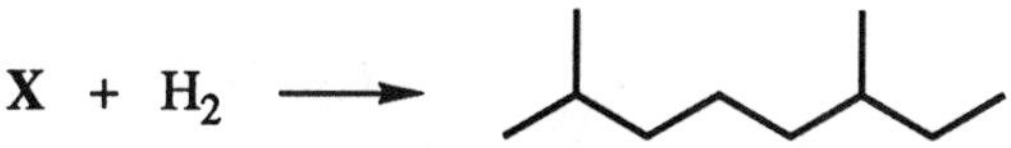

molecular weight 142

X has molecular weight 140 (from the mass spec), so **X** has one double bond—where?

A mass spec fragment of 57 is C_4H_9, a butyl group with no unsaturation; a mass spec fragment of 83 is C_6H_{11}, a six-carbon group with one unsaturation. If we presume that the cleavage is most likely allylic, then the alkene must be in one of only two possible positions. **X** must be one of these:

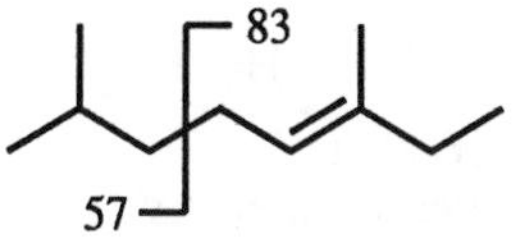

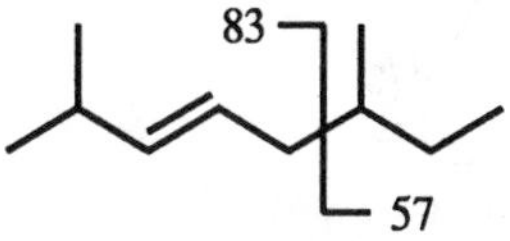

11-13 Refer to the Glossary and the text for definitions and examples.

11-14 (a) 1603 cm^{-1} (b) 2959 cm^{-1} (c) 1709 cm^{-1}

(d) 1739 cm^{-1} (e) 2212 cm^{-1} (f) 3300 cm^{-1}

11-15

(a)

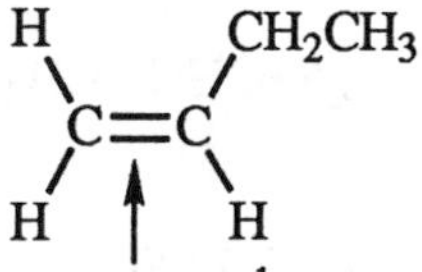

1660 cm^{-1}

and

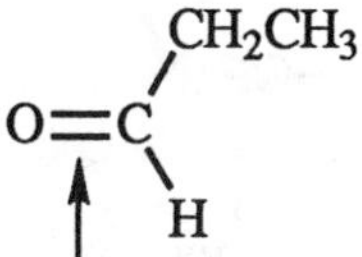

1710 cm^{-1}
stronger—larger dipole

(b)

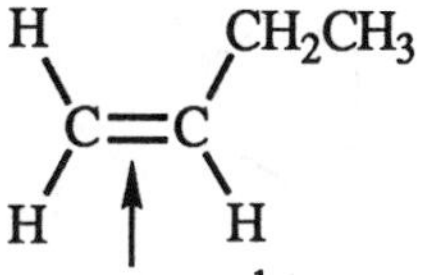

1660 cm^{-1}

and

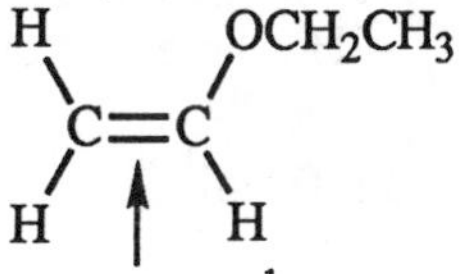

1640 cm^{-1}
stronger—larger dipole

(c)

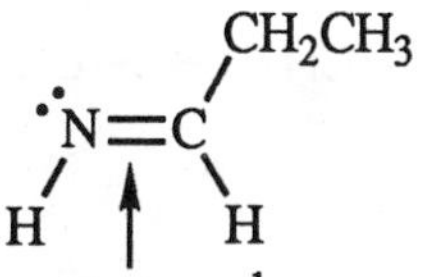

1660 cm^{-1}
stronger—larger dipole

and

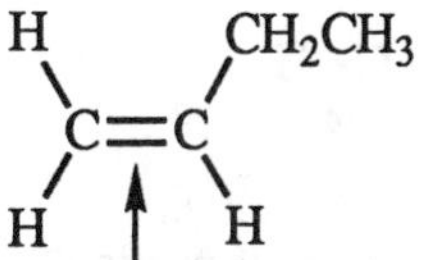

1660 cm^{-1}

11-15 continued

(d) 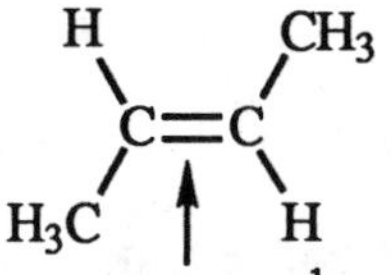
1660 cm^{-1}
(no dipole moment)

and

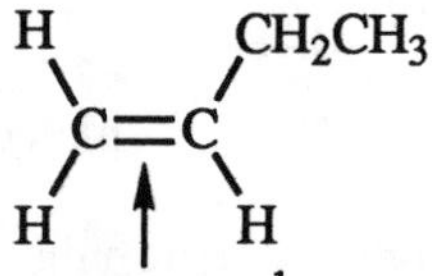
1660 cm^{-1}
stronger—larger dipole

11-16

(a) 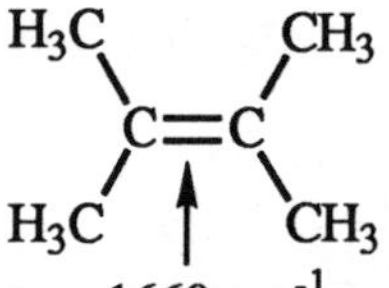
1660 cm^{-1}
weak or non-existent

and

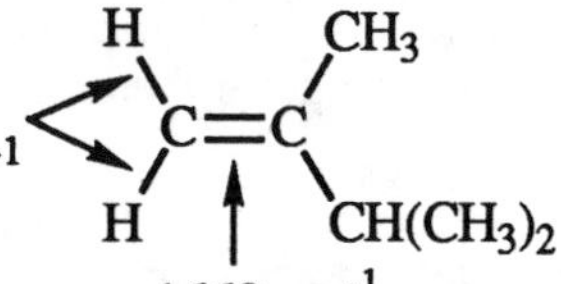
3000-
3100 cm-1
1660 cm^{-1}
moderate intensity

(b) 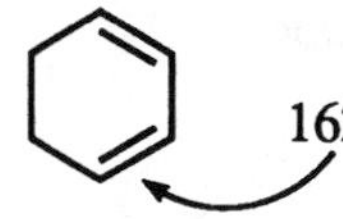
1620 cm-1
conjugated

and

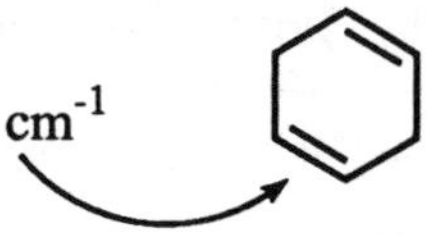
1645 cm-1
not conjugated

(c) $CH_3(CH_2)_5-C\equiv C-H$ — 3300 cm^{-1} (C–H); C≡C: 2100-2200 cm^{-1}, moderate intensity

and

$CH_3(CH_2)_3-C\equiv C-CH_2CH_3$ — C≡C: 2100-2200 cm^{-1}, weak or non-existent

(d) 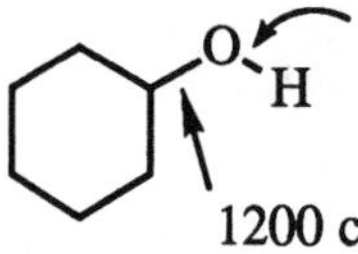
3300 cm-1
broad, strong
1200 cm-1

and

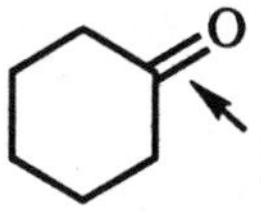
1710 cm-1
strong

(e) both carbonyls show strong absorptions around 1710 cm^{-1}

$CH_3(CH_2)_3-C(=O)-H$ — aldehyde C–H: 2700-2800 cm^{-1}, two small peaks

and

$CH_3(CH_2)_2-C(=O)-CH_3$

11-16 continued

(f) 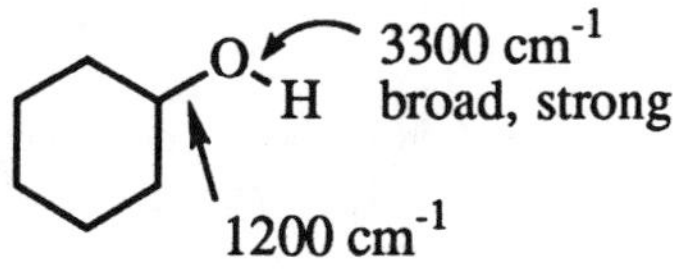

3300 cm-1
broad, strong
1200 cm-1

and 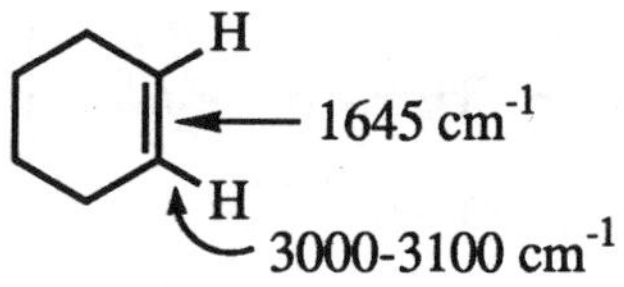

1645 cm-1
3000-3100 cm-1

(g) both carbonyls show strong absorptions around 1710 cm^{-1}

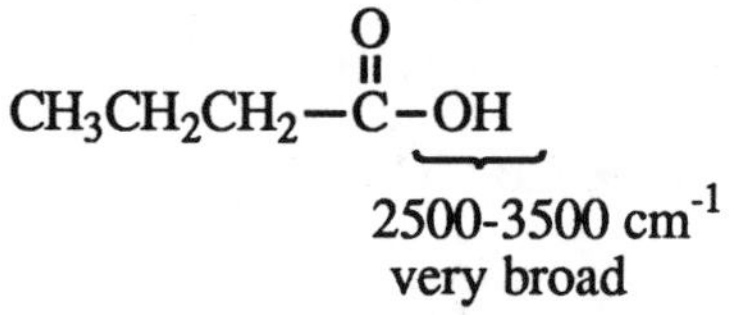

CH3CH2CH2-C-OH
2500-3500 cm-1
very broad

and 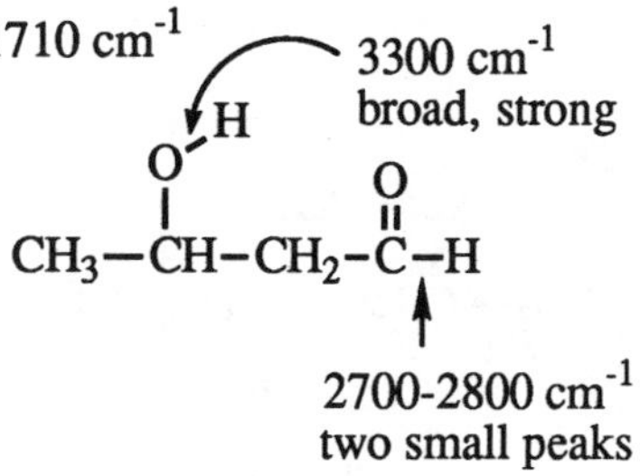

3300 cm-1
broad, strong
CH3-CH-CH2-C-H
2700-2800 cm-1
two small peaks

(h) 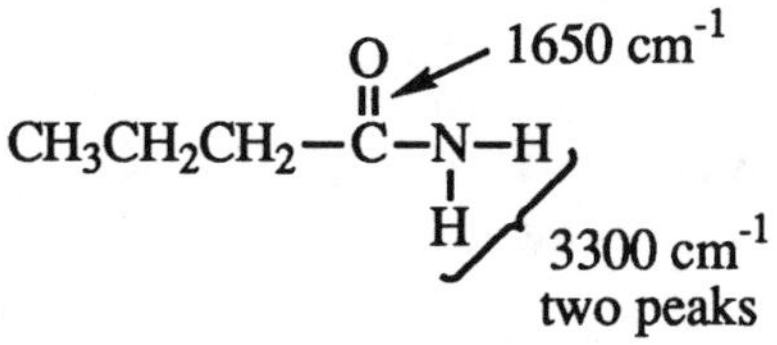

1650 cm-1
CH3CH2CH2-C-N-H
3300 cm-1
two peaks

and 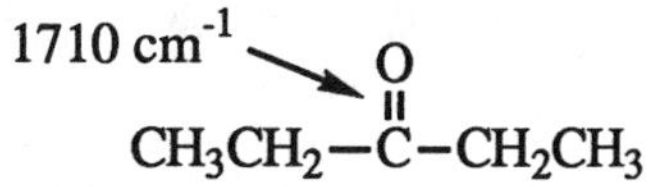

1710 cm-1
CH3CH2-C-CH2CH3

11-17

(a) 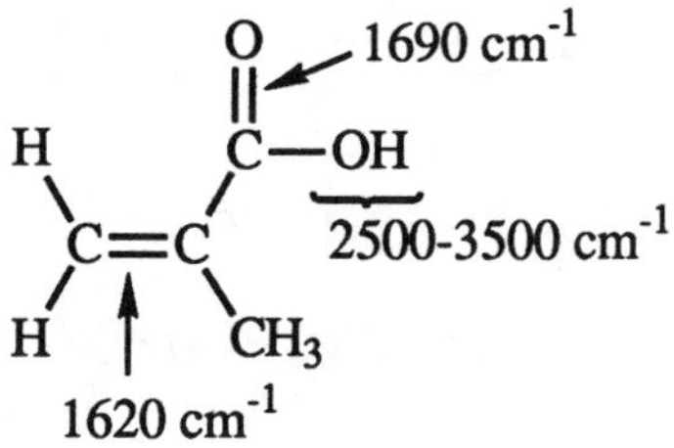

1690 cm-1
C-OH
2500-3500 cm-1
1620 cm-1

(b) 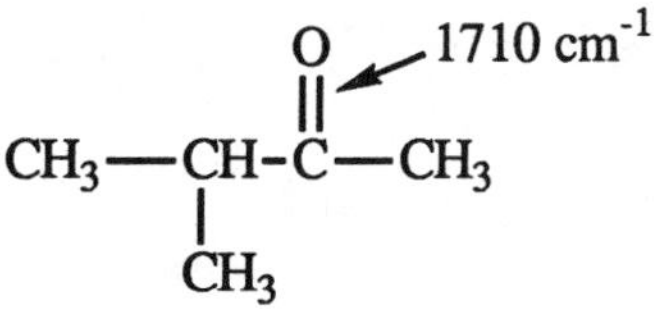

1710 cm-1
CH3-CH-C-CH3
CH3

(c) 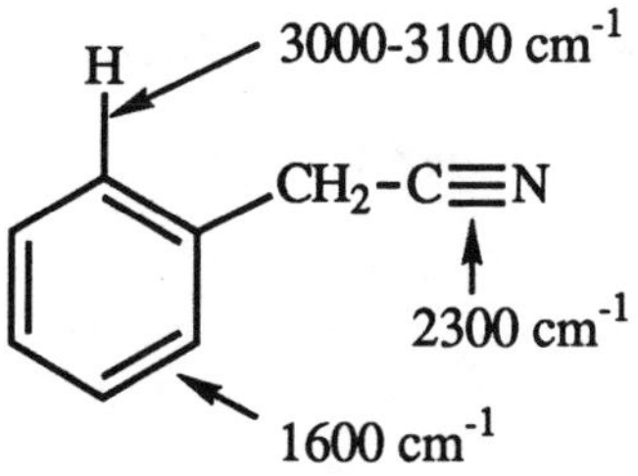

3000-3100 cm-1
CH2-C≡N
2300 cm-1
1600 cm-1

(d) 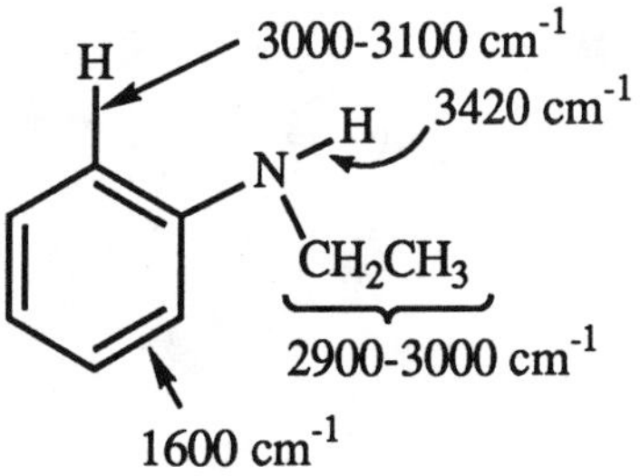

3000-3100 cm-1
3420 cm-1
CH2CH3
2900-3000 cm-1
1600 cm-1

11-18

(a)

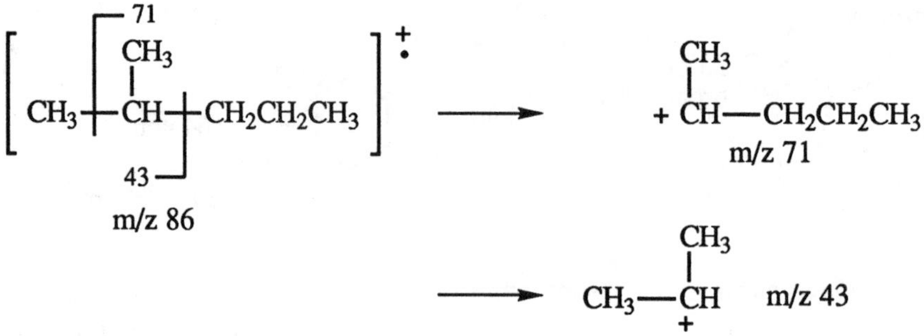

m/z 86
m/z 71
m/z 43

(b)

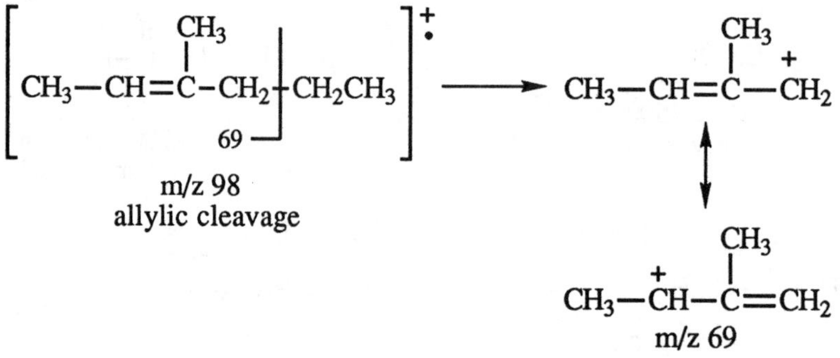

m/z 98
allylic cleavage
m/z 69

(c)

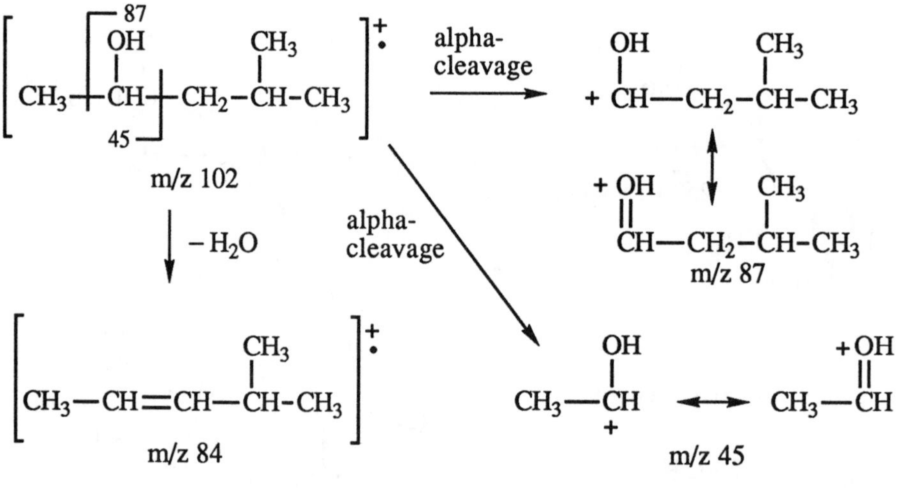

alpha-cleavage
m/z 102
− H_2O
alpha-cleavage
m/z 87
m/z 84
m/z 45

11-19

(a)

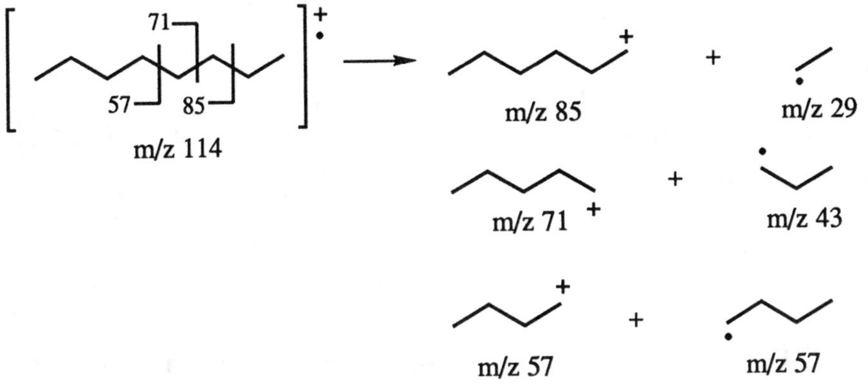
71
57
85
m/z 114
m/z 85
m/z 29
m/z 71
m/z 43
m/z 57
m/z 57

(b)

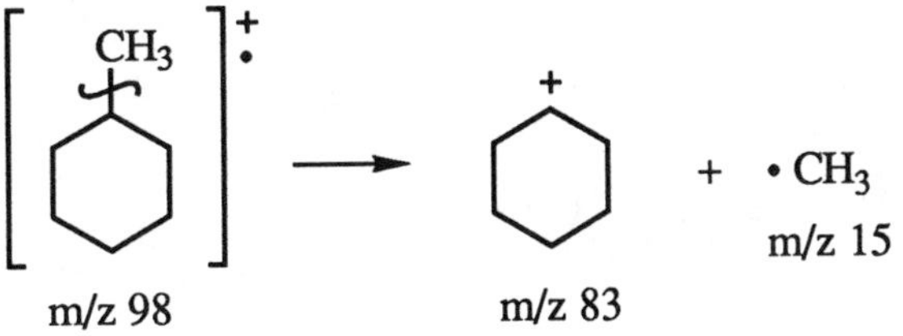
CH3
m/z 98
m/z 83
• CH3
m/z 15

(c)

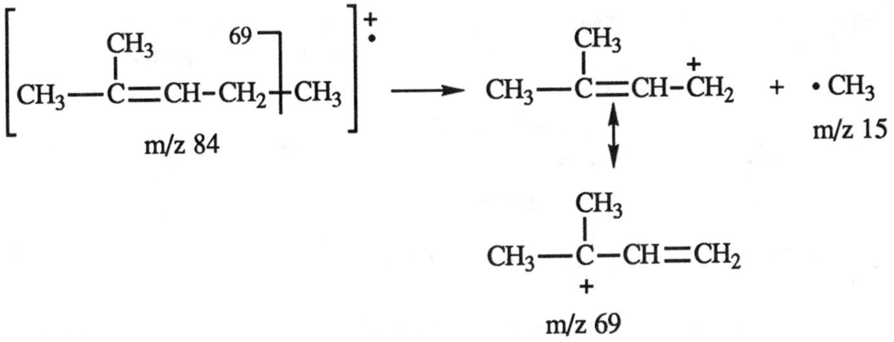
CH3
69
CH3—C=CH-CH2-CH3
m/z 84
CH3—C=CH-CH2
• CH3
m/z 15
CH3—C-CH=CH2
m/z 69

11-19 continued

(d)

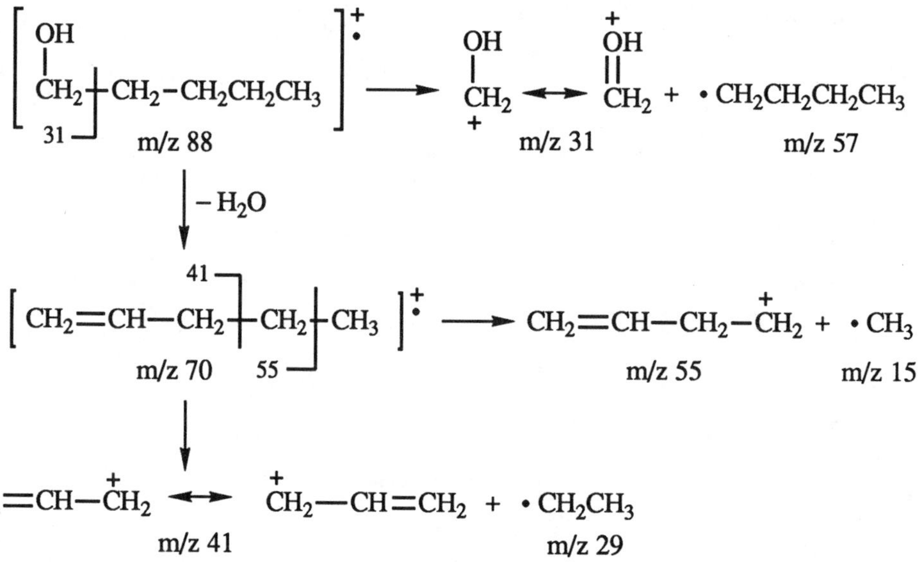
OH
CH2
CH2-CH2CH2CH3
31
m/z 88
OH
CH2
OH
CH2
• CH2CH2CH2CH3
m/z 31
m/z 57
- H2O
41
CH2=CH—CH2
CH2
CH3
m/z 70
55
CH2=CH—CH2—CH2
• CH3
m/z 55
m/z 15
CH2=CH—CH2
CH2—CH=CH2
• CH2CH3
m/z 41
m/z 29

11-20

(a) Deuterium has twice the mass of hydrogen, but similar spring constant, k. Compare the frequency of C—D vibration to C—H vibration by setting up a ratio, changing only the mass (substitute 2m for m).

$$\frac{\nu_D}{\nu_H} = \frac{\sqrt{k/2m}}{\sqrt{k/m}} = \frac{\sqrt{1/2}\ \sqrt{k/m}}{\sqrt{k/m}} = \sqrt{1/2} = 0.707$$

$\nu_D = 0.707\ \nu_H = 0.707\ (3000\ \text{cm}^{-1}) \approx$ **2100 cm^{-1}**

(b) The functional group most likely to be confused with a C—D stretch is the alkyne, which appears in the same region and is often very weak.

11-21

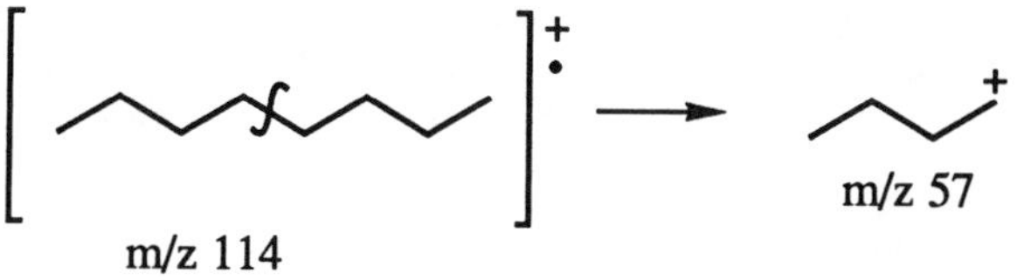

a 1° carbocation—
not favorable

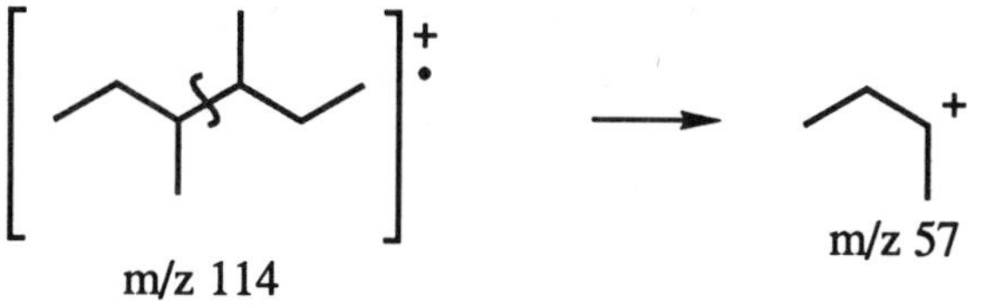

a 2° carbocation—
reasonable

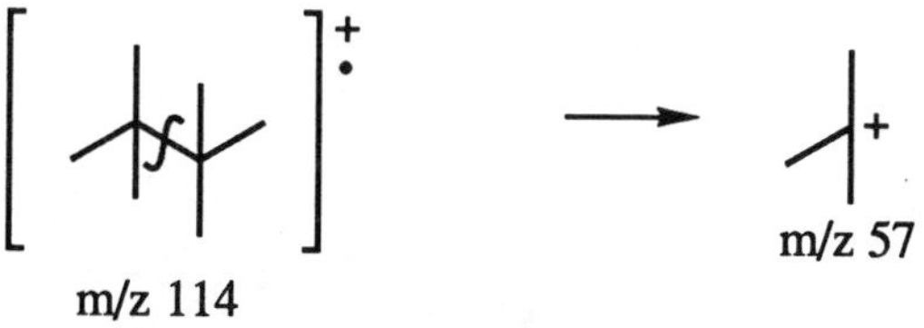

a 3° carbocation—
the best

The most likely fragmentation of 2,2,3,3-tetramethylbutane will give a 3° carbocation, the most stable of the common alkyl cations. The molecular ion should be small or non-existent while m/z 57 is likely to be the base peak.

CHAPTER 12—NUCLEAR MAGNETIC RESONANCE SPECTROSCOPY

Note to the student: A benzene ring can be written with three alternating double bonds or with a circle in the ring. All of the carbons and hydrogens in an unsubstituted benzene ring are equivalent, regardless of which symbol is used.

 equivalent to 

12-1

(a) $\dfrac{130\text{ Hz}}{60 \times 10^{6}\text{ Hz}} = 2.17 \times 10^{-6} = 2.17$ ppm downfield from TMS

(b) difference in magnetic field $= 14{,}092$ gauss $\times\ (2.17 \times 10^{-6}) = 0.031$ gauss

(c) The chemical shift does not change with field strength: δ 2.17 at both 60 MHz and 100 MHz.

(d) $(2.17\text{ ppm}) \times (100\text{ MHz}) = (2.17 \times 10^{-6}) \times (100 \times 10^{6}\text{ Hz}) = 217\text{ Hz}$

12-2 Numbers are chemical shift values, in ppm, derived from Table 12-3 and the Appendix. *Your predictions should be in the given range, or within 0.5 ppm of the given value.*

(a)

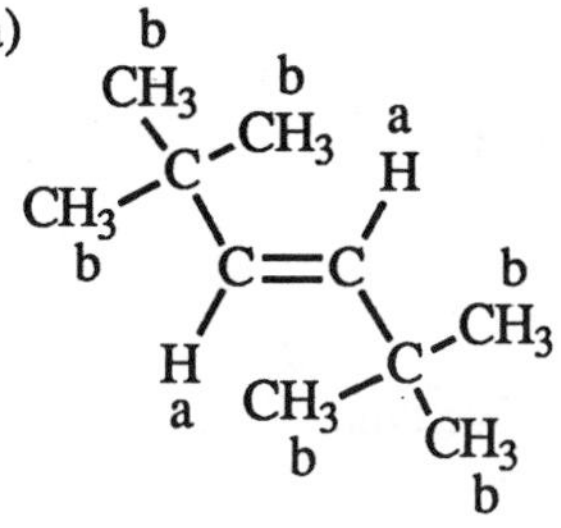

a = δ 5-6
b = δ 0.9

(b)

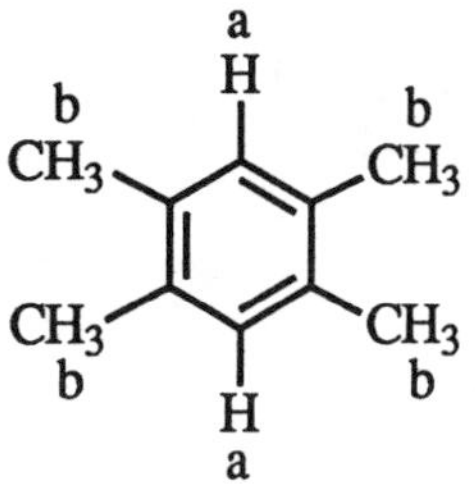

a = δ 7.2
b = δ 2.3

(c)

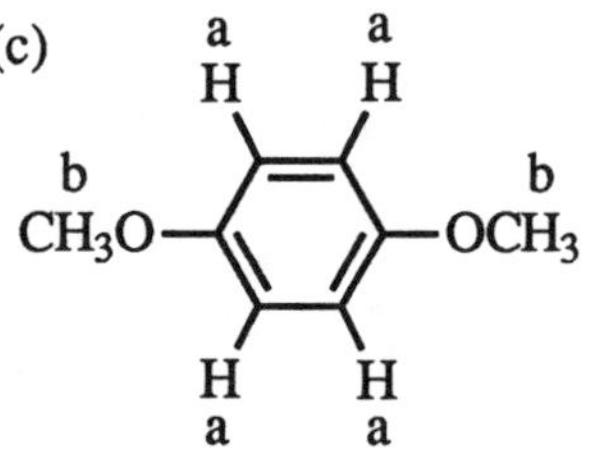

a = δ 7.2
b = δ 3.6

(d)

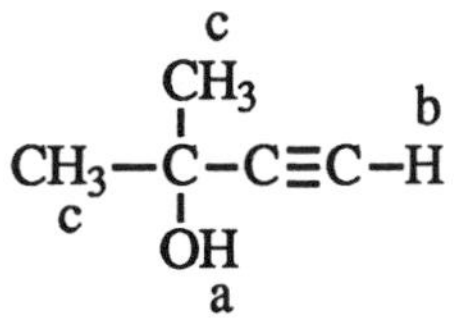

a = δ 2-5
b = δ 2.5
c = δ 1-2

12-2 continued

(e)

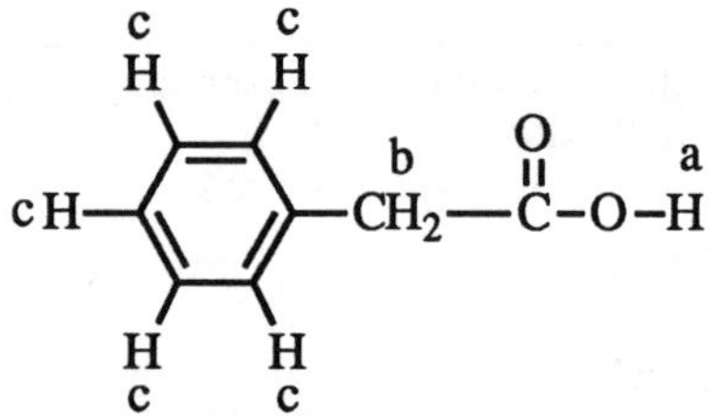

a = δ 10-12
b ≈ δ 3 (between two deshielding groups)
c = δ 7.2
(Note that the hydrogens labeled "c" are not equivalent. They appear at roughly the same chemical shift because the substituent is neither strongly electron-donating nor withdrawing.)

(f)

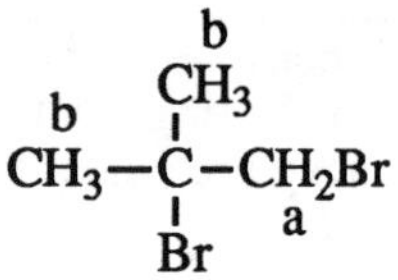

a = δ 3-4
b = δ 1-2

12-3

(a) c b a

$CH_3CH_2CH_2Cl$

three types of H

(b)

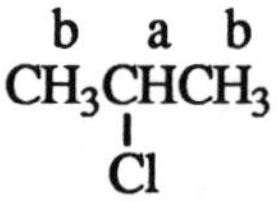

two types of H

(c)

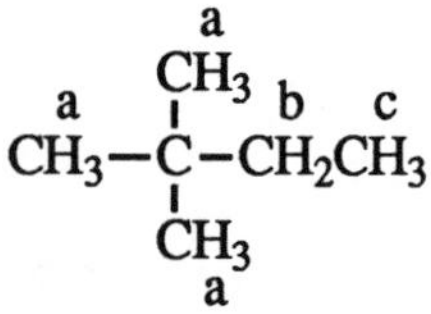

three types of H

(d)

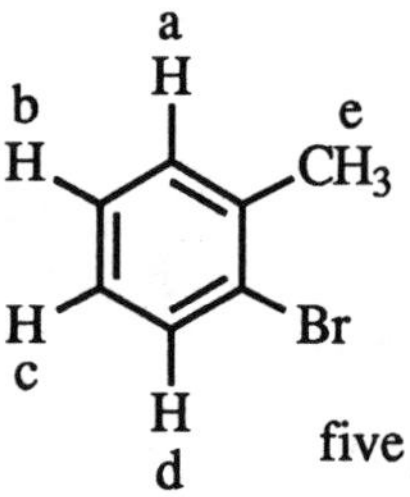

five types of H

12-4

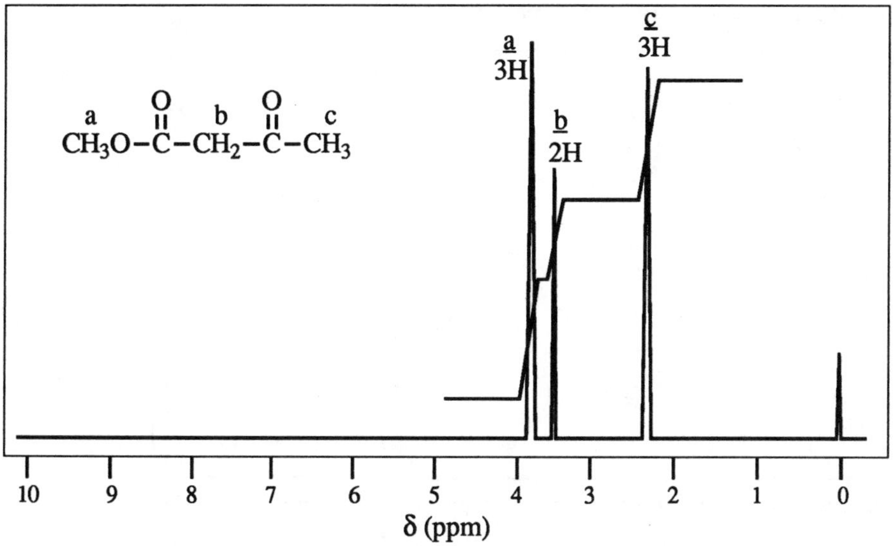

12-5 The three spectra are identified with their structures. Data are given as chemical shift values, with the integration ratios of each peak given in parentheses.

Spectrum (a)

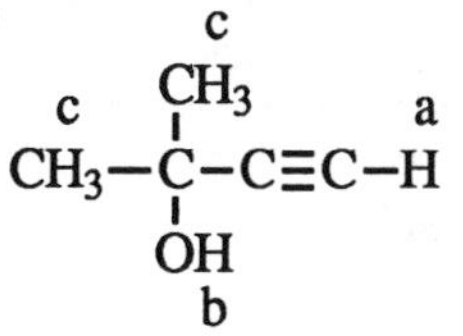

a = δ 2.5 (1) (1H)
b = δ 2.3 (1) (1H)
c = δ 1.5 (6) (6H)

Spectrum (b)

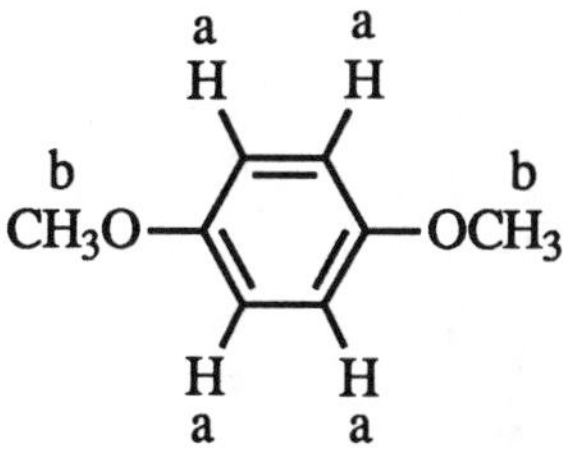

a = δ 6.7 (2) (4H)
b = δ 3.7 (3) (6H)

Spectrum (c)

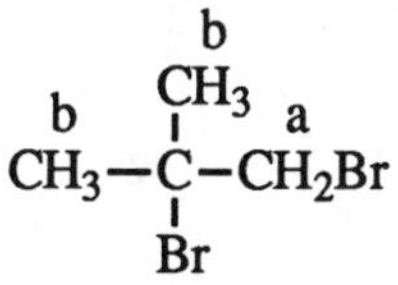

a = δ 3.9 (1) (2H)
b = δ 1.8 (3) (6H)

(This is the compound in Problem 12-2(f). You may wish to check your answer to that question against the spectrum.)

12-6 Chemical shift values are approximate and may vary from yours. The splitting and integration values should match exactly, however.

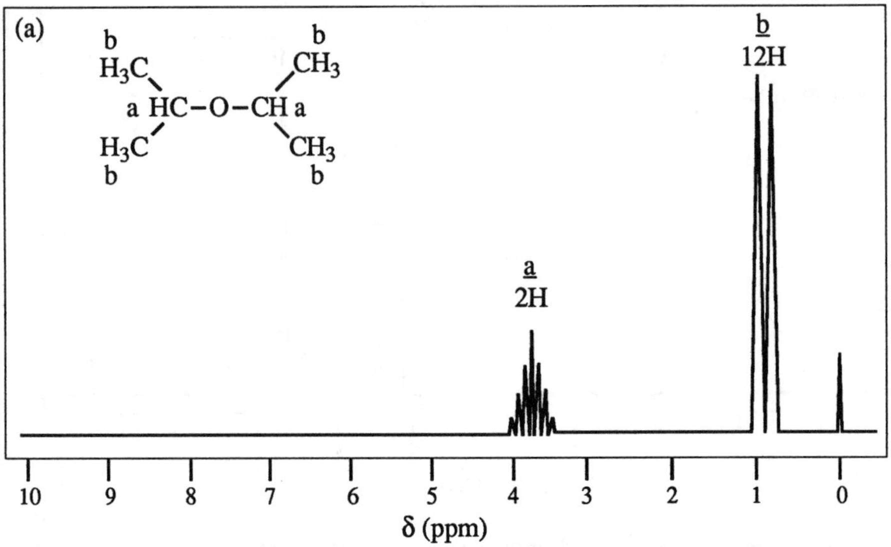

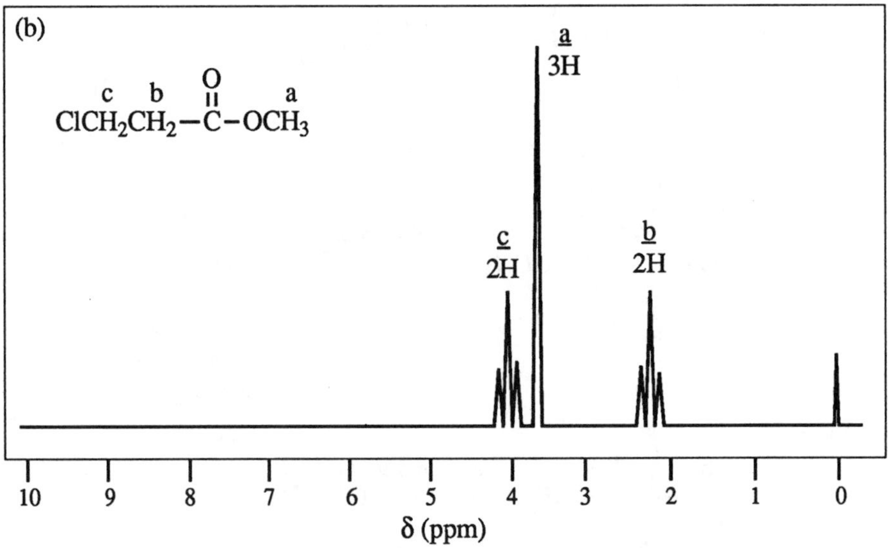

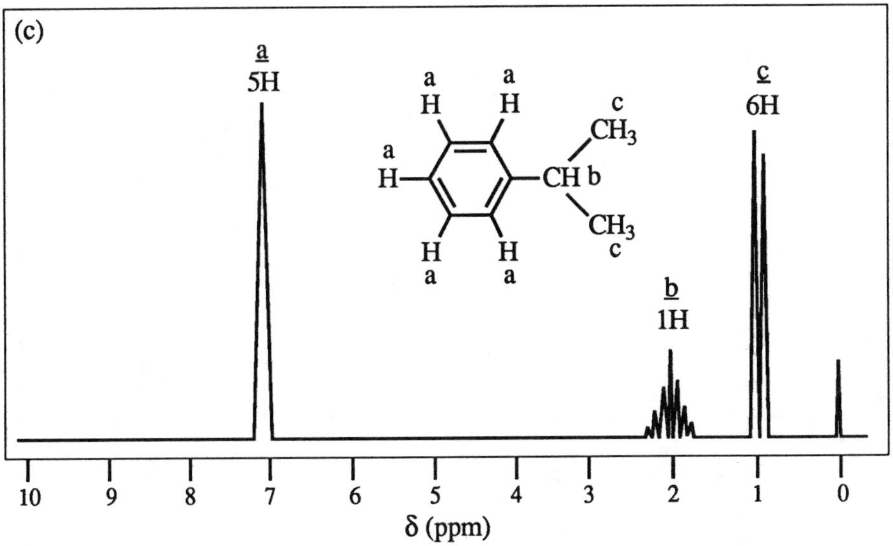

Note: the three types of aromatic protons above are *accidentally equivalent* because the isopropyl group has little effect on their chemical shifts.

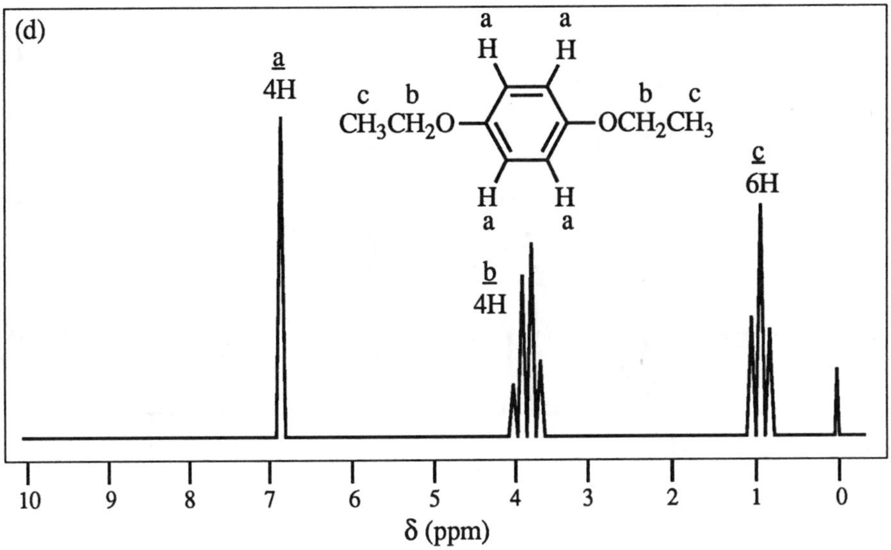

12-6 continued

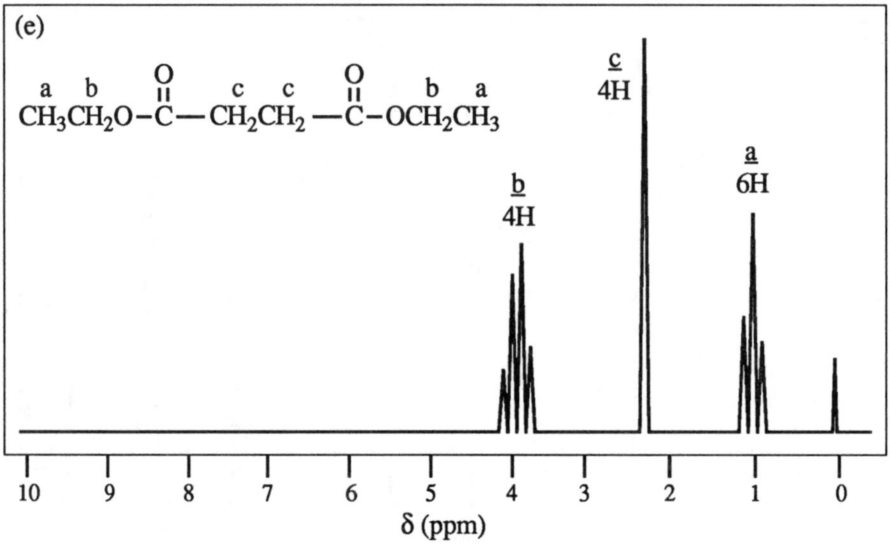
(e)
a b c c b a
$CH_3CH_2O-\overset{O}{\overset{\|}{C}}-CH_2CH_2-\overset{O}{\overset{\|}{C}}-OCH_2CH_3$
c
4H
b
4H
a
6H
10
9
8
7
6
5
4
3
2
1
0
δ (ppm)

Note: no splitting is observed when *chemically equivalent* hydrogens are adjacent to each other, as with hydrogens "c" above.

12-7

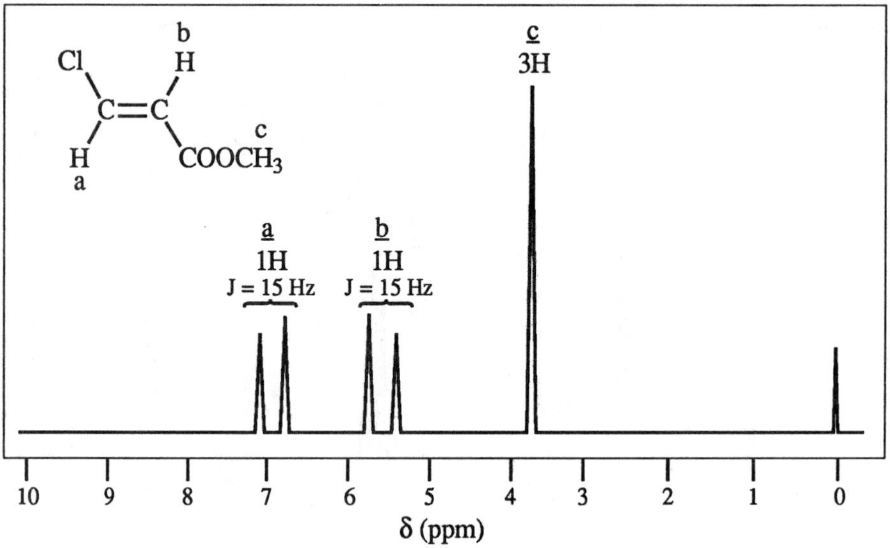
Cl
b
H
C=C
H
a
c
$COOCH_3$
c
3H
a
1H
J = 15 Hz
b
1H
J = 15 Hz
10
9
8
7
6
5
4
3
2
1
0
δ (ppm)

12-8

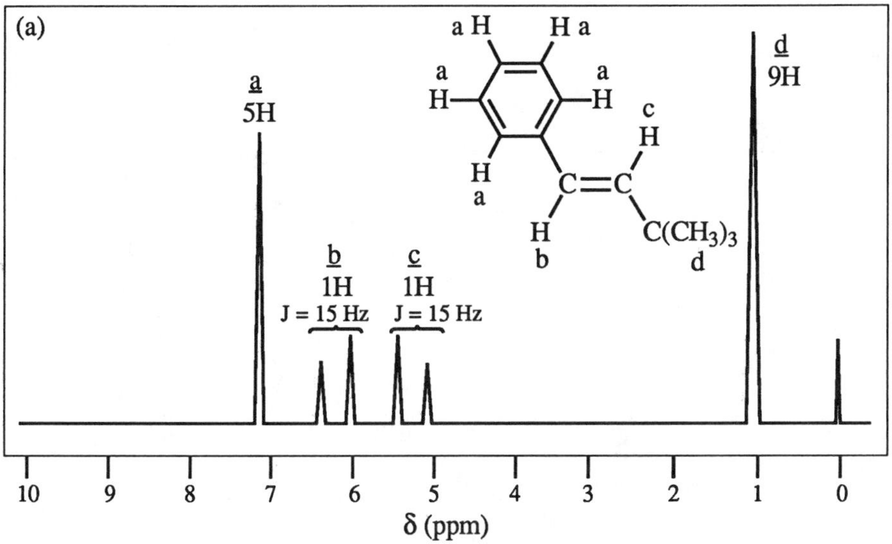

Note: the three types of aromatic protons above are *accidentally equivalent.*

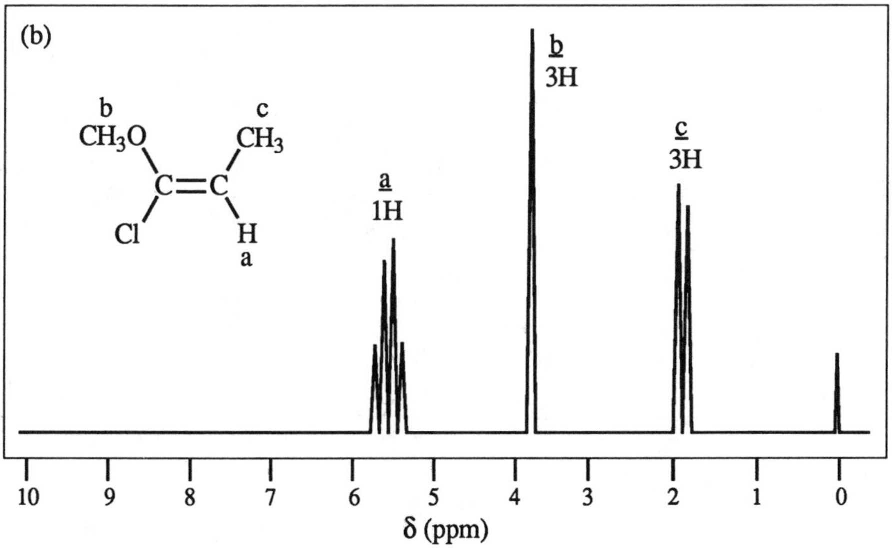

12-8 continued

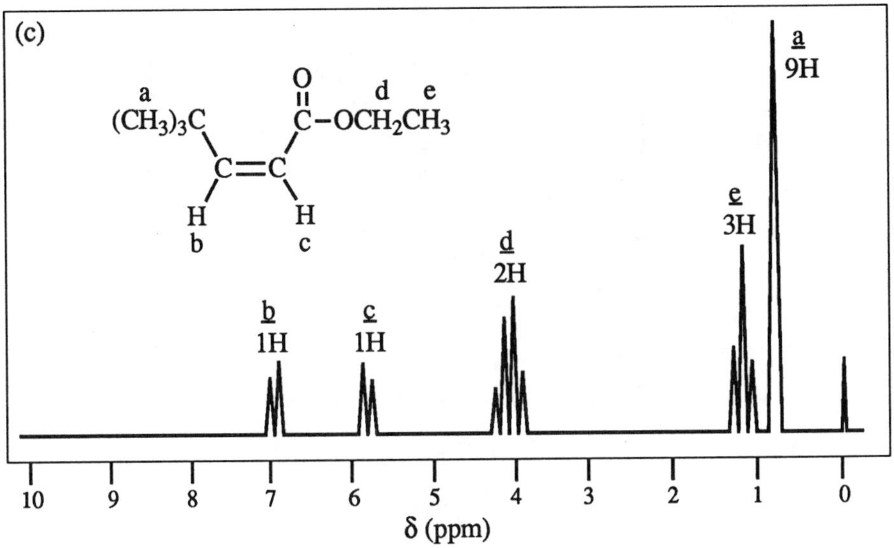

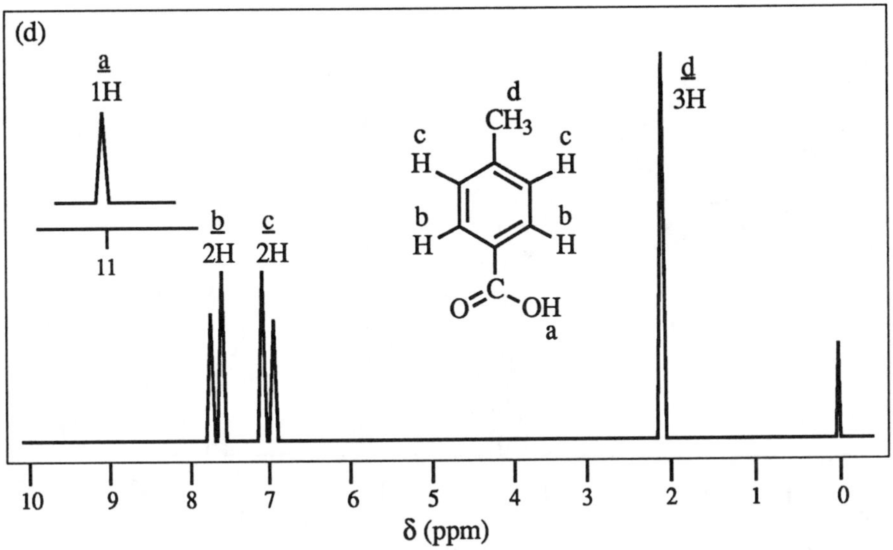

Note: the electron-withdrawing carbonyl group deshields the protons adjacent to it, moving them downfield from their usual position at δ 7.2.

12-9 The formula C_3H_2NCl has three elements of unsaturation. The IR peak at 1650 cm^{-1} indicates an alkene, while the absorption at 2200 cm^{-1} must be from a nitrile (not enough carbons left for an alkyne). These two groups account for the three elements of unsaturation. So far, we have:

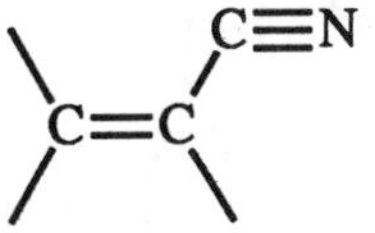

+ 2H + Cl

The NMR gives the coupling constant for the two protons as 14 Hz. This large J value shows the two protons as *trans* (*cis*, J = 10 Hz; geminal, J = 2 Hz). The structure must be:

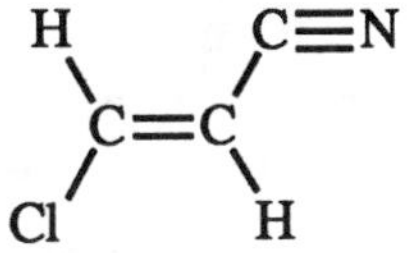

12-10

(a) C_3H_7Cl—no elements of unsaturation; 3 types of protons in the ratio of 2 : 2 : 3 .

a b c
Cl—$CH_2CH_2CH_3$

a = δ 3.5 (triplet, 2H); b = δ 1.8 (multiplet, 2H);
c = δ 1.1 (triplet, 3H)

(b) $C_9H_{10}O_2$—5 elements of unsaturation; four protons in the aromatic region of the NMR indicate a disubstituted benzene; the pair of doublets with J = 8 Hz indicate the substituents are on opposite sides of the ring (*para*).

+ 3 C + 6 H + 2 O + 1 element of unsaturation

The other NMR signals are two 3H singlets, two CH_3 groups. One at δ 3.9 must be a CH_3O group. The other at δ 2.4 is most likely a CH_3 group on the benzene ring.

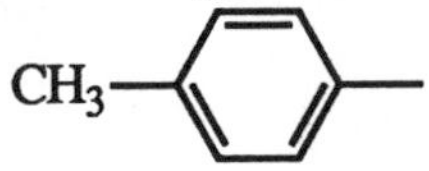

+ OCH_3 + C + O 1 element of unsaturation

One way to assemble these pieces consistent with the NMR is:

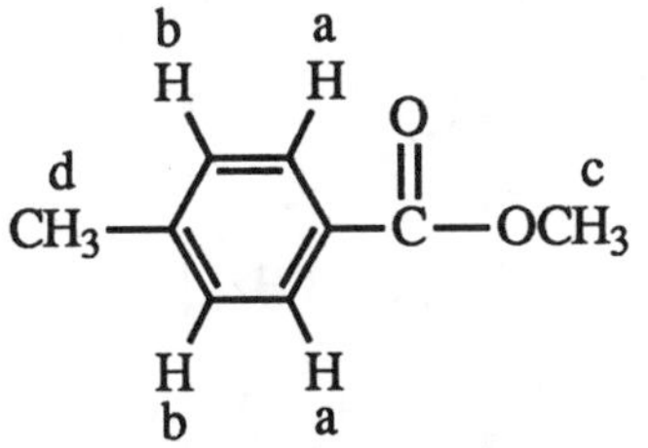

a = δ 7.9 (doublet, 2H)
b = δ 7.2 (doublet, 2H)
c = δ 3.9 (singlet, 3H)
d = δ 2.4 (singlet, 3H)

(Another plausible structure is to have the methoxy group directly on the ring and to put the carbonyl between the ring and the methyl. This does not fit the chemical shift values quite as well as the above structure, as the methyl would appear around δ 2.1 or 2.2 instead of 2.4 .)

12-11 H_c, δ 5.1

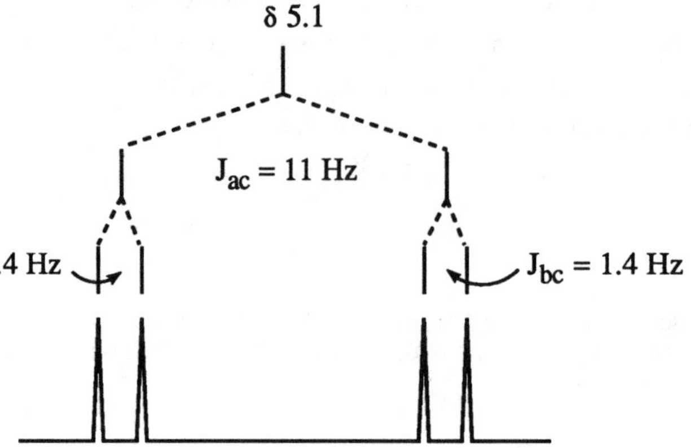

12-12

(a)

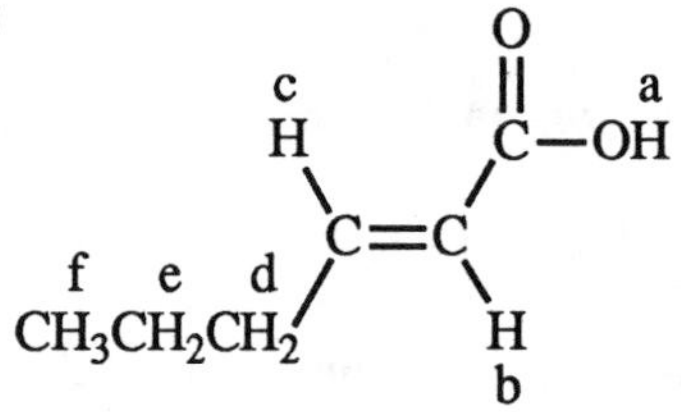

a = δ 12.1 (singlet, 1H)
b = δ 5.7 (doublet, 1H)
c = δ 7.0 (multiplet, 1H)
d = δ 2.25 (multiplet, 2H)
e = δ 1.5 (multiplet, 2H)
f = δ 0.9 (triplet, 3H)

(b) The vinyl proton at δ 7.0 is H_c; it is coupled with H_b and H_d, with two different coupling constants, J_{bc} and J_{cd}, respectively. The value of J_{bc} can be measured most precisely from the signal for H_b at δ 5.7; the two peaks at δ 5.55 and δ 5.85 are separated by 0.3 ppm, or 18 Hz in a 60 MHz spectrum. The value of J_{cd} appears to be about the standard value 8 Hz, judging from the signal at δ 7.0. The splitting tree would thus appear:

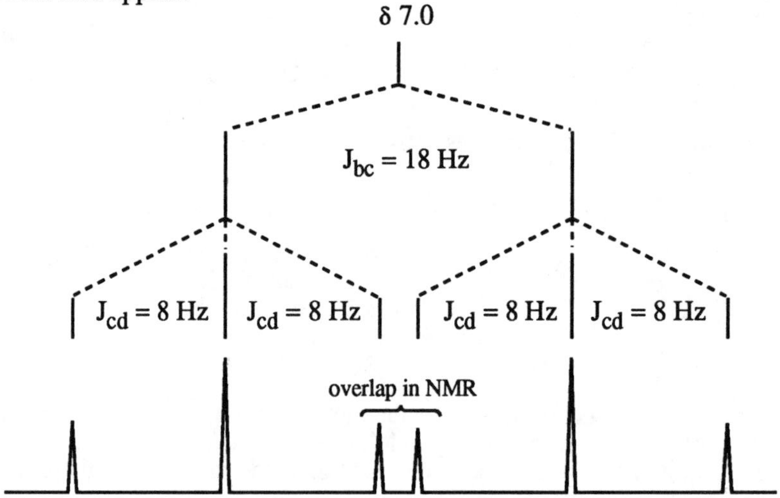

12-13 (a)

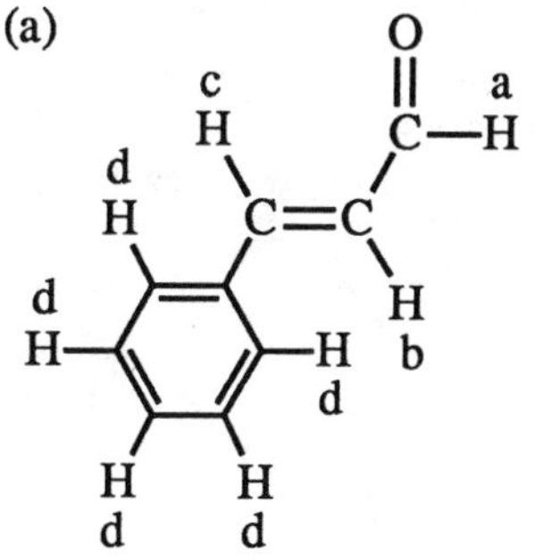

O
a
c
H
C—H
d
H
C=C
d
H
H—
—H
b
d
H
H
d
d

a = δ 9.7 (doublet, 1H)
b = δ 6.6 (multiplet, 1H)
c = δ 7.4 (doublet, 1H)
d = δ 7.4 (singlet, 5H)

Note 1: the three types of aromatic protons labeled "d" are accidentally equivalent.

Note 2: the doublet for H_c at δ 7.4 OVERLAPS the 5H singlet of H_d.

(b) J_{ab} can be determined most accurately from H_a at δ 9.7: $J_{ab} \approx 8$ Hz.

J_{bc} can be measured from H_b at δ 6.6, as the distance between either the first and third peaks or the second and fourth peaks (see diagram below): $J_{bc} \approx 18$ Hz.

(c)

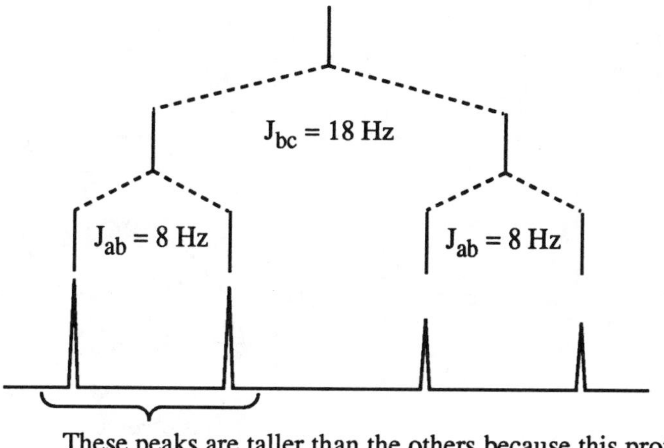

δ 6.6
J_{bc} = 18 Hz
J_{ab} = 8 Hz
J_{ab} = 8 Hz

These peaks are taller than the others because this proton is coupled to another proton downfield, at δ 7.4; the peaks "point" to signals of coupled protons as shown in Figure 12-25 in the text.

12-14

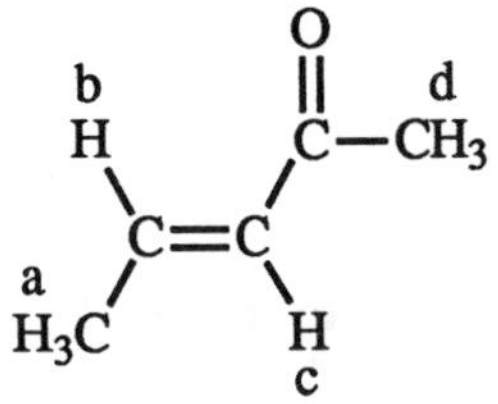

O
b
H
d
C—CH₃
C=C
a
H₃C
H
c

(a) a = δ 1.7
b ≈ δ 6.8
c = δ 5-6
d = δ 2.1

(b) a = doublet
b = multiplet (two overlapping quartets—see (c))
c = doublet
d = singlet

(c) using J_{bc} = 15 Hz and J_{ab} = 7 Hz:

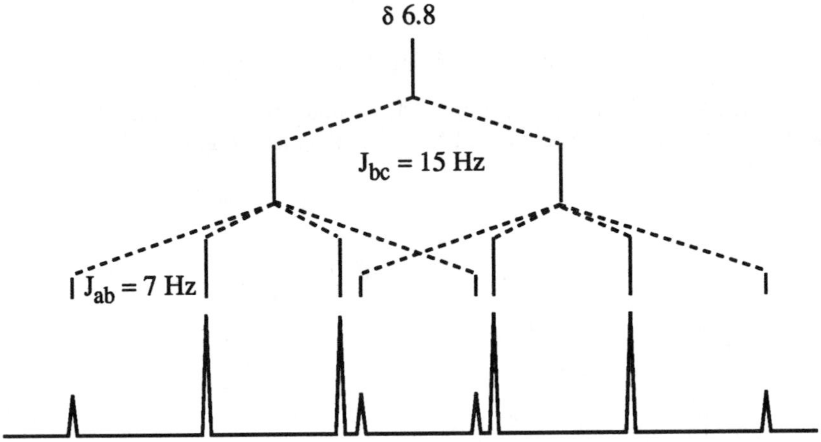

δ 6.8
J_bc = 15 Hz
J_ab = 7 Hz

12-15

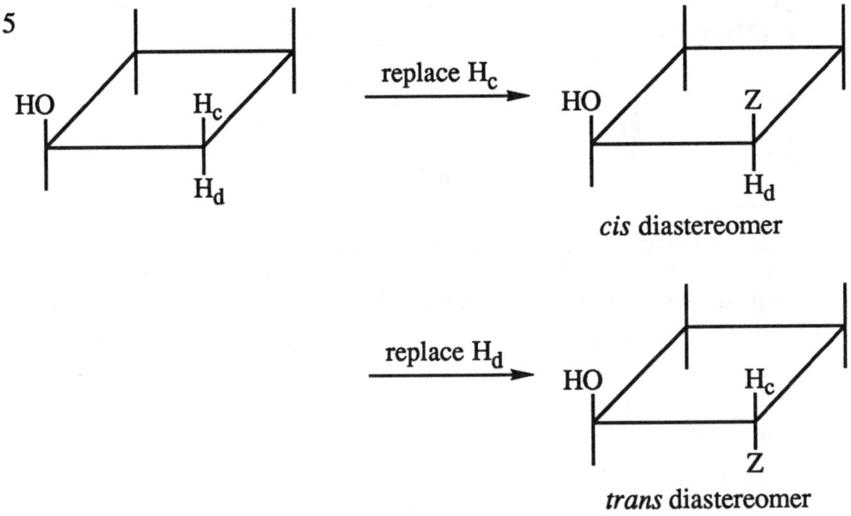

HO
H_c
H_d
replace H_c
HO
Z
H_d
cis diastereomer
replace H_d
HO
H_c
Z
trans diastereomer

12-16

(a)

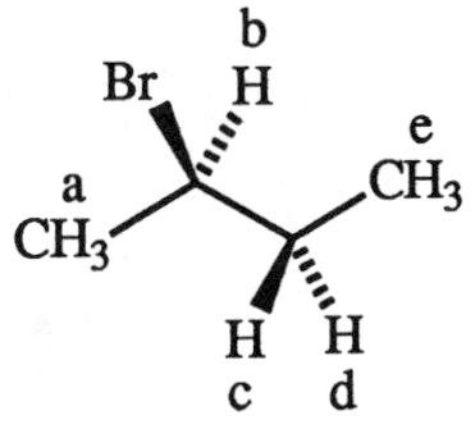

This compound has five types of protons.
H_c and H_d are diastereotopic.

a = δ 1.5
b = δ 3.6
c,d = δ 1.7
e = δ 1.0

(b)

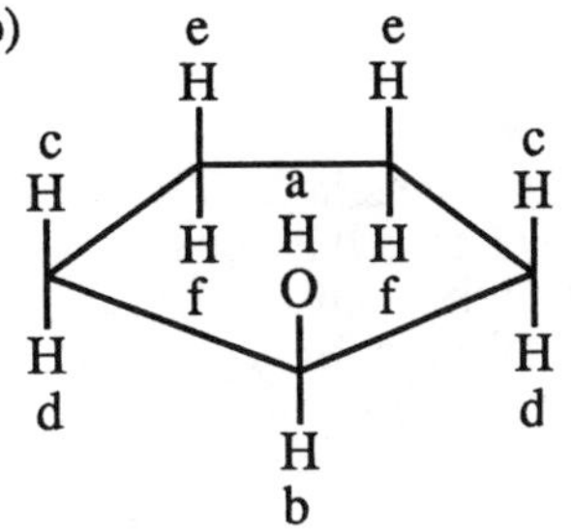

This compound has six types of protons.
H_c and H_d are diastereotopic, as are H_e and H_f.

a = δ 2-5
b = δ 3.9
c,d = δ 1.6
e,f = δ 1.3

(c)

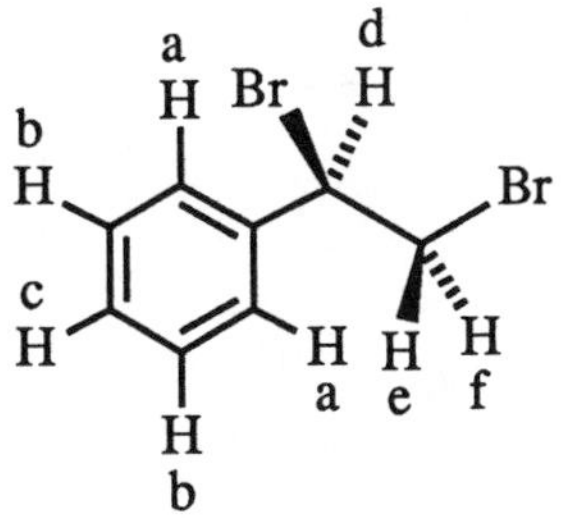

This compound has six types of protons.
H_e and H_f are diastereotopic.

a,b,c = δ 7.2
d ≈ δ 5.0
e,f = δ 3.6

(d)

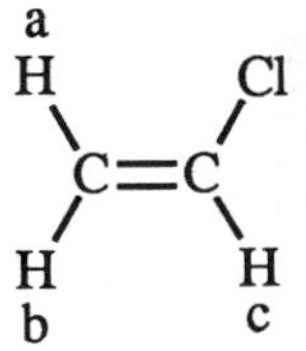

This compound has three types of protons.
H_a and H_b are diastereotopic.

a,b = δ 5-6
c = δ 7-8

12-17

(a)

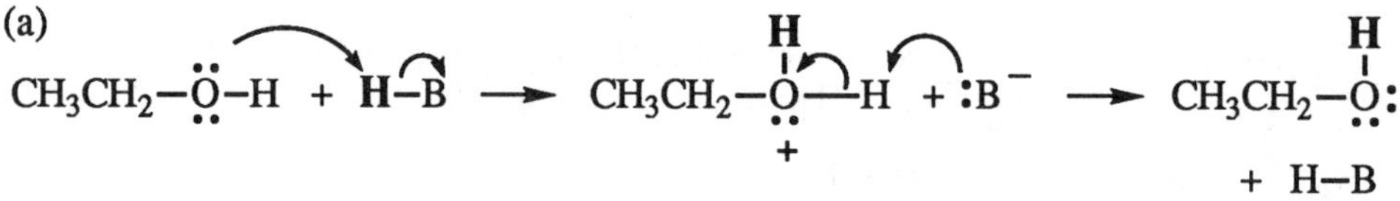

(b)

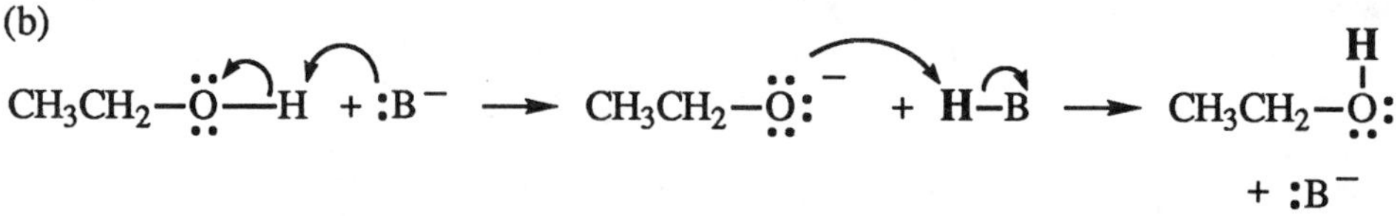

12-18 The protons from the OH in ethanol exchange with the deuteriums in D_2O. Thus, the OH in ethanol is replaced with OD which does not absorb in the NMR. What happens to the H? It becomes HOD which can usually be seen as a broad singlet around δ 5.25. (If the solvent is $CDCl_3$, the HOD will float on top of the solvent, out of the spectrometer beam, and its signal will be missing.)

$$ROH + D_2O \longrightarrow ROD + HOD$$

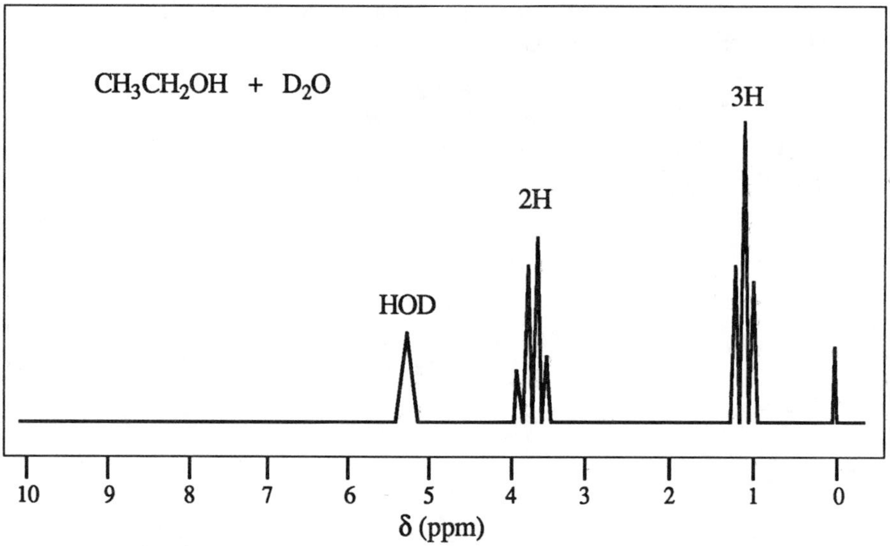

12-19

(a) The formula $C_4H_{10}O_2$ has no elements of unsaturation, so the oxygens must either be alcohol or ether functional groups. The doublet at δ 1.1 represents 3H and must be a CH_3 next to a CH. The peaks centered at δ 1.6 integrating to 2H appear to be an uneven quartet and signify a CH_2 between two sets of non-equivalent protons. The broad, 2H singlet at δ 3.3 has the earmark of OH protons, so apparently the molecule is a di-alcohol. The 3H multiplet centered around δ 3.8 is very complex; while the splitting is hard to interpret, the chemical shift suggests that these 3 hydrogens are on carbons bonded to oxygen. To put the pieces together:

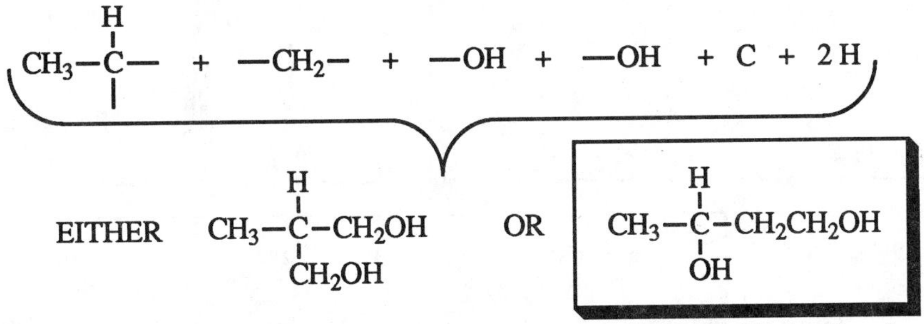

CH3—C— + —CH2— + —OH + —OH + C + 2 H
EITHER
CH3—C—CH2OH
CH2OH
OR
CH3—C—CH2CH2OH
OH

The structure on the left does not fit the NMR; it would show a 4H doublet about δ 3.8. The structure on the right must be correct. The multiplet at δ 3.8 are the overlapping signals of the CH and the CH_2 bonded to the OH's.

(b) The formula C_2H_7NO has no elements of unsaturation. The N must be an amine and the O must be an alcohol or ether. Two triplets, each with 2H, are certain to be —CH_2CH_2—. Since there are no carbons left, the N and O, with enough hydrogens to fill their valences, must go on the ends of this chain. The rapidly exchanging OH and NH_2 protons appear as a broad, 3H singlet at δ 2.4.

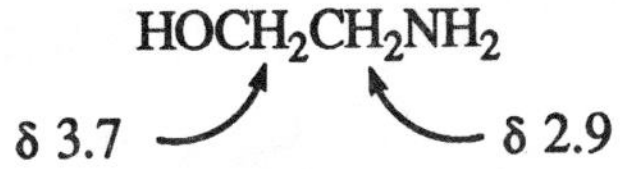

HOCH2CH2NH2
δ 3.7
δ 2.9

12-20

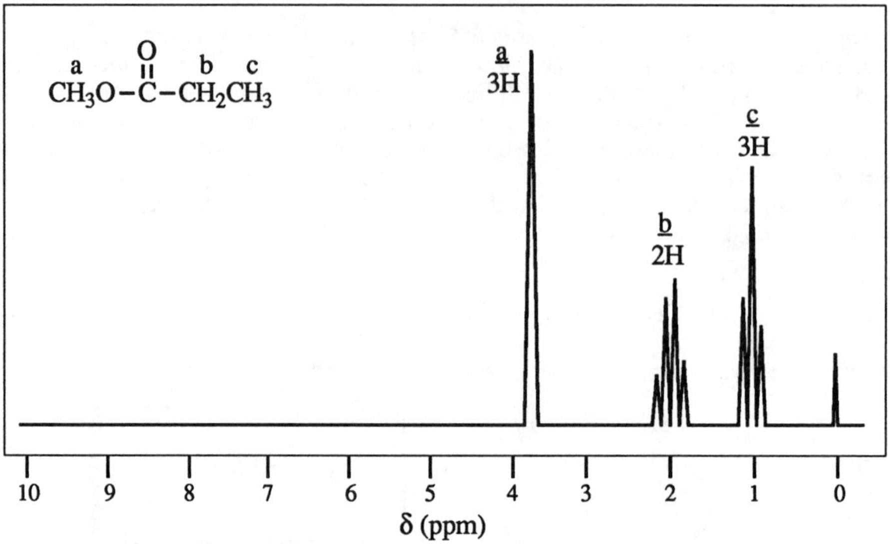

a O b c
CH3O–C–CH2CH3
a
3H
b
2H
c
3H
10 9 8 7 6 5 4 3 2 1 0
δ (ppm)

The key is what protons are adjacent to the oxygen. The CH_3O in methyl propionate, above, absorbs at δ 3.9, whereas the CH_2 next to the carbonyl absorbs at δ 2.2. This is in contrast to ethyl acetate, below.

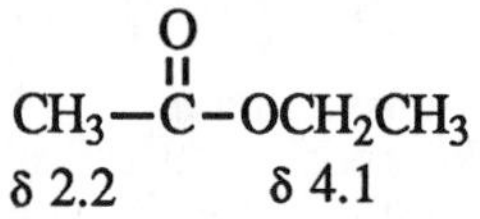

O
CH3—C–OCH2CH3
δ 2.2
δ 4.1

12-21

(a) H_a = δ 4.0 (singlet, 1H)
H_b = δ 3.4 (doublet, 2H)
H_c = δ 1.8 (multiplet, 1H)
H_d = δ 0.9 (doublet, 6H)

(b) The methine (CH) absorption is split into 7 peaks by 6 equivalent protons on the adjacent methyl groups. The signal is also split into triplets by the adjacent methylene (CH_2). Theoretically, each of seven peaks is split into a triplet, for a total of 21 peaks. You would look ragged too!

12-22

(a) The formula $C_4H_8O_2$ has one element of unsaturation. The 1H singlet at δ 12.2 indicates carboxylic acid. The 1H multiplet and the 6H doublet scream isopropyl group.

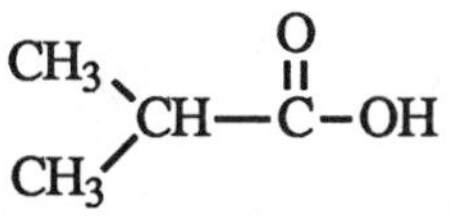

(b) The formula C_8H_8O has five elements of unsaturation. The broad 5H singlet at δ 7.2 indicates monosubstituted benzene. The peak at δ 9.7 is unmistakably an aldehyde, trying to be a triplet because it is weakly coupled to an adjacent CH_2, whose own signal at δ 3.7 appears to be a weakly coupled doublet.

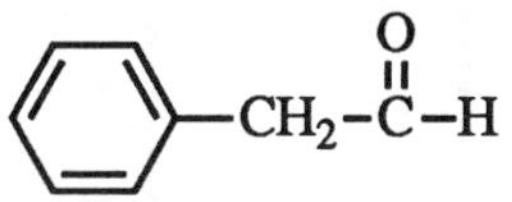

(c) The formula $C_5H_8O_2$ has two elements of unsaturation. A 3H singlet at δ 2.3 is probably a CH_3 next to carbonyl. The 2H quartet and the 3H triplet are certain to be ethyl; with the CH_2 at δ 2.7, this also appears to be next to carbonyl.

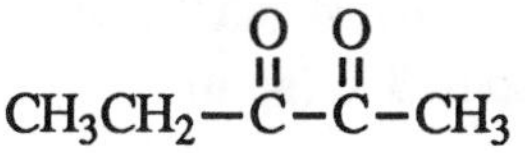

(d) The formula C_4H_8O has one element of unsaturation, and the signals from δ 4.7-6.5 indicate a vinyl pattern (CH_2=CH—). The complex quartet for 1H at δ 4.1 is a CH bonded to an alcohol, next to CH_3. The OH appears as a 1H singlet at δ 2.4, and the CH_3 next to CH is a doublet at δ 1.0. Put together:

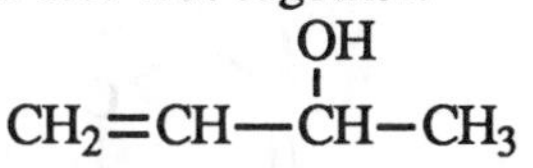

(e) The formula $C_6H_{10}O$ has two elements of unsaturation. No splitting means that none of the CH_x groups have neighboring hydrogens. (Sometimes this simplifies the problem, sometimes it makes it harder.) The 6H singlet at δ 1.4 must be two CH_3 groups in identical environments; a reasonable first guess is that they are on the same carbon: $(CH_3)_2C$. The 3H singlet at δ 3.3 is probably a CH_3O, although it is slightly upfield; it must be in a shielded environment. So far:

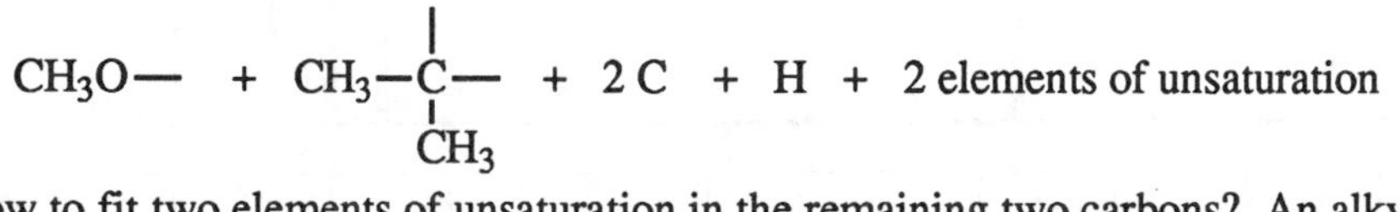

How to fit two elements of unsaturation in the remaining two carbons? An alkyne! The lone H must be on the end of the alkyne, at δ 2.4 in the NMR. The puzzle solved:

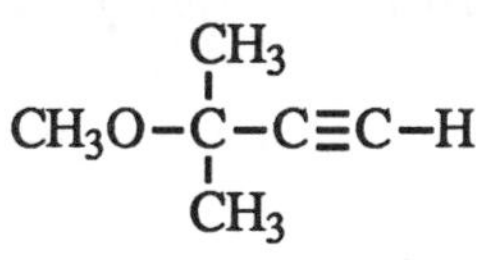

12-23 Chemical shift values are estimates from Figure 12-41, except in (d), where the values are exact.

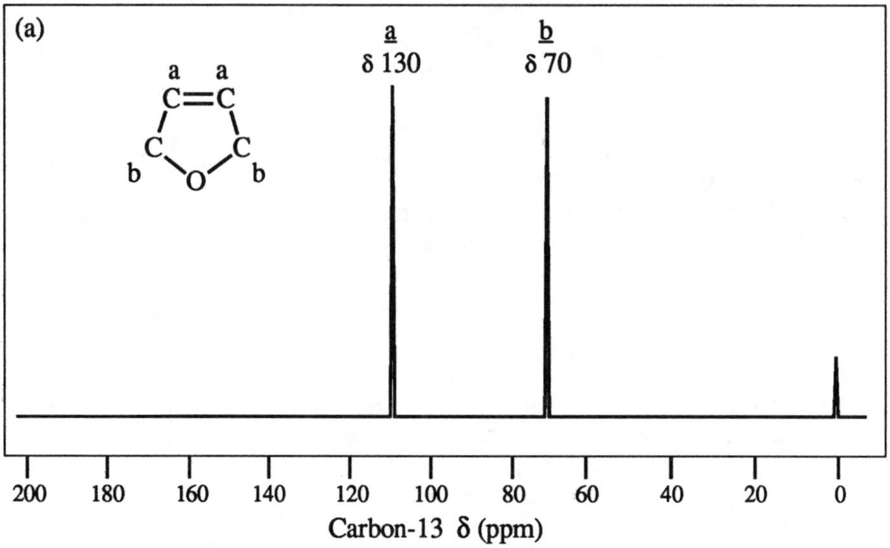

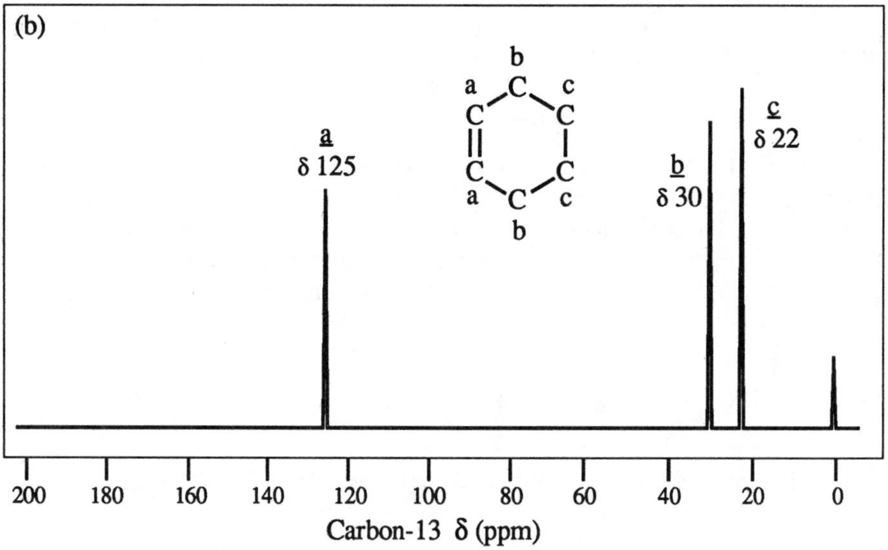

12-23 continued

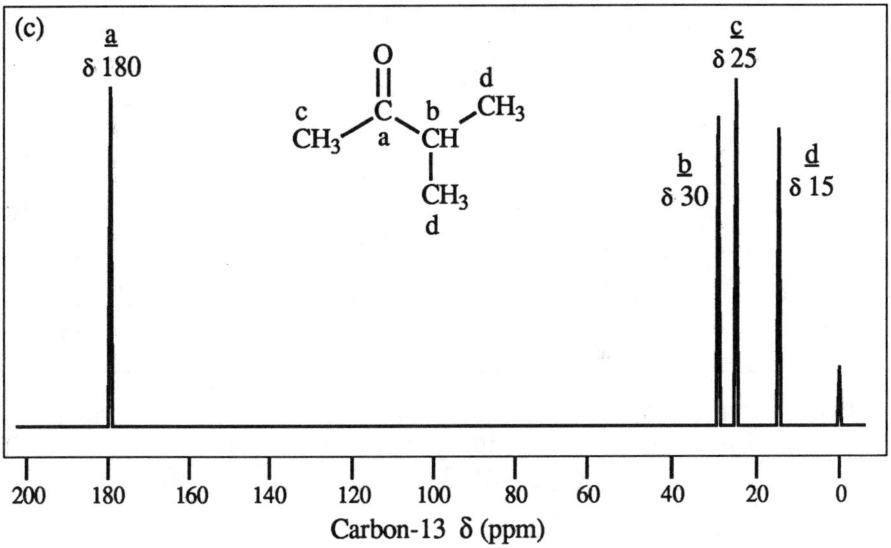

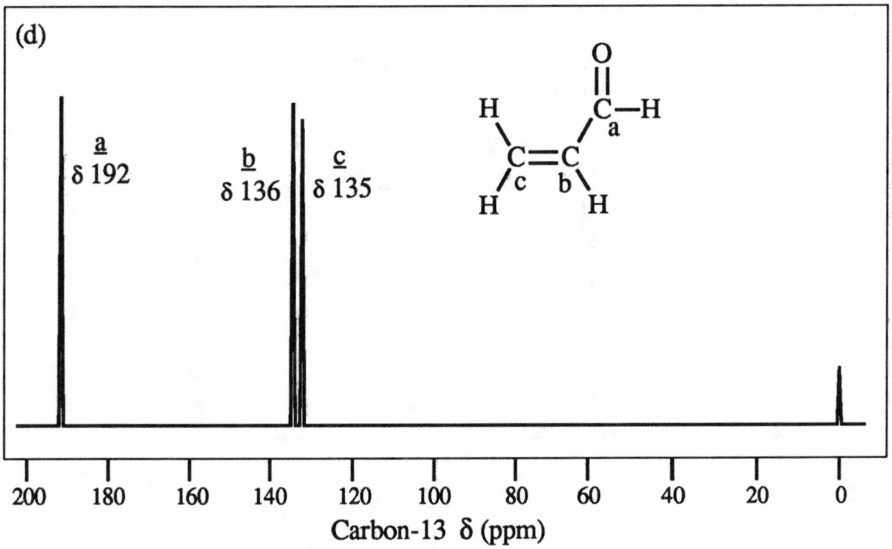

12-24

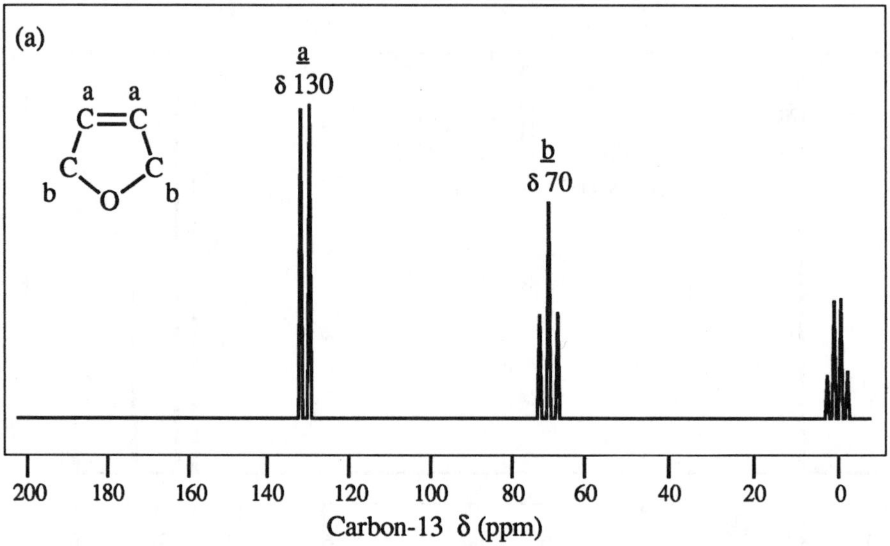
(a)
a a
C=C
C C
b O b
a
δ 130
b
δ 70
200 180 160 140 120 100 80 60 40 20 0
Carbon-13 δ (ppm)

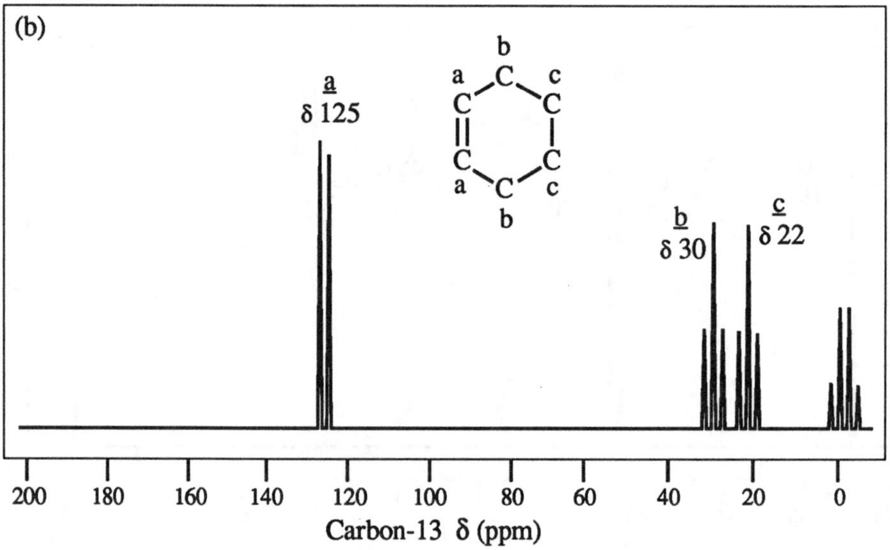
(b)
a
δ 125
b
a C c
C C
C C
a C c
b
b
δ 30
c
δ 22
200 180 160 140 120 100 80 60 40 20 0
Carbon-13 δ (ppm)

12-24 continued

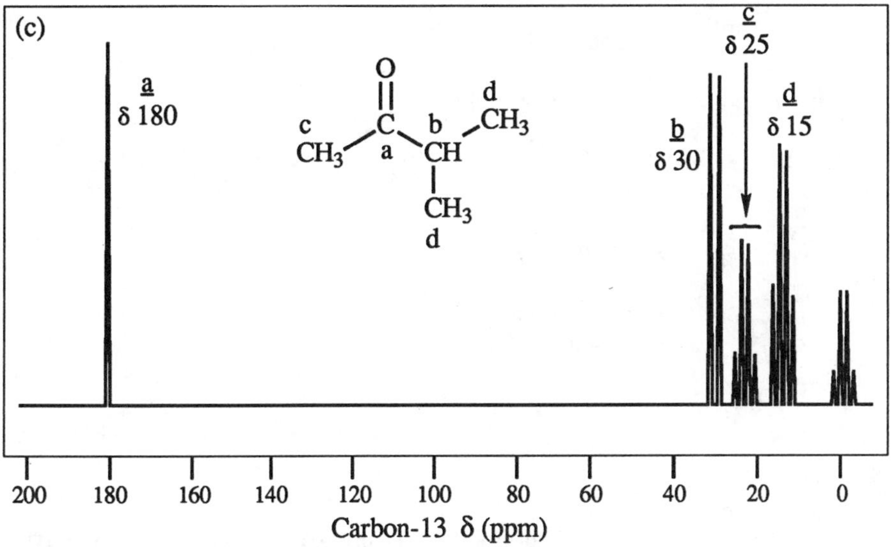

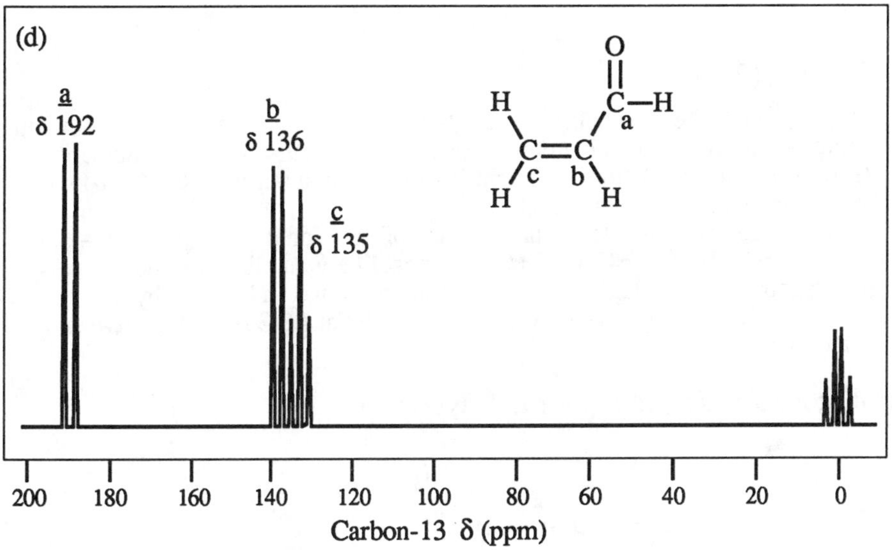

12-25

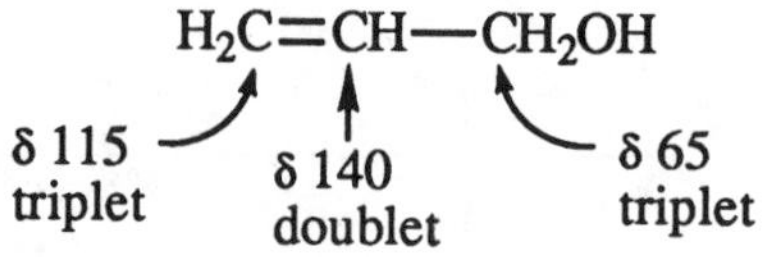

$H_2C=CH-CH_2OH$
δ 115
triplet
δ 140
doublet
δ 65
triplet

Allyl bromide is easily hydrolyzed with water.

$$H_2C=CH-CH_2Br \xrightarrow{H_2O} H_2C=CH-CH_2OH + HBr$$

12-26

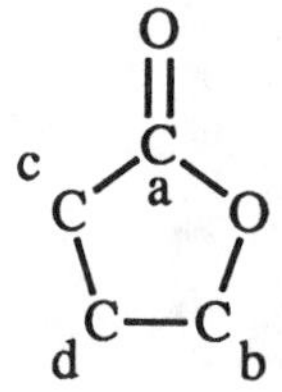

O
C
a
c
C
O
C
C
d
b

a = δ 180, singlet
b = δ 70, triplet
c = δ 28, triplet
d = δ 22, triplet

12-27

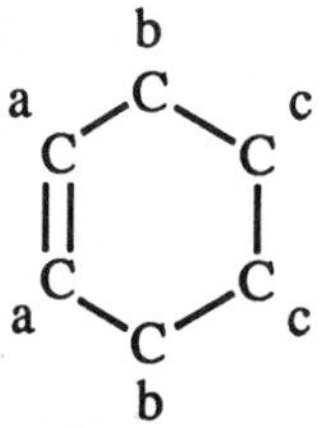

b
a
c
a
c
b

a = δ 125, doublet
b = δ 28, triplet
c = δ 22, triplet

Using PBr_3 instead of H_2SO_4/NaBr would give a higher yield of bromocyclohexane.

12-28 <u>Compound 2</u>

<u>Mass spectrum</u>: the molecular ion at m/z 136 shows a peak at 138 of about equal height, indicating a bromine atom is present: 136 – 79 = 57. The fragment at m/z 57 is the base peak; this fragment is most likely a butyl group, C_4H_9, so a likely molecular formula is C_4H_9Br.
<u>Infrared spectrum</u>: Notable for the absence of functional groups: no O—H, no N—H, no =C—H, no C=C, no C=O ⇒ most likely an alkyl bromide.
<u>NMR spectrum</u>: The 6H doublet at δ 1.0 suggests two CH_3's split by an adjacent H—an isopropyl group. The 2H doublet at δ 3.3 is a CH_2 between a CH and the Br.

Putting the pieces together gives isobutyl bromide.

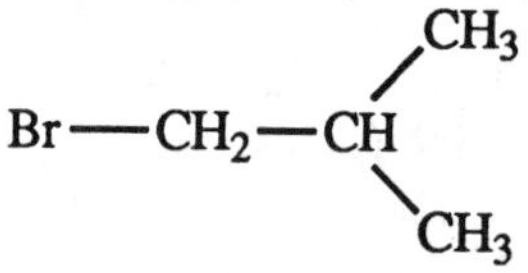

CH_3
Br—CH_2—CH
CH_3

12-29

The formula $C_9H_{11}Br$ indicates four elements of unsaturation, just enough for a benzene ring.

Here is the most accurate method for determining the number of protons per signal from integration values *when the total number of protons is known.* Add the integration heights: 4.4 cm + 13.0 cm + 6.7 cm = 24.1 cm. Divide by the total number of hydrogens: 24.1 cm ÷ 11H = 2.2 cm/H. Each 2.2 cm of integration height = 1H, so the ratio of hydrogens is 2 : 6 : 3.

The 2H singlet at δ 7.1 means that only two hydrogens remain on the benzene ring, that is, it has 4 substituents. The 6H singlet at δ 2.3 must be two CH_3's on the benzene ring in identical environments. The 3H singlet at δ 2.2 is another CH_3 in a slightly different environment from the first two. Substitution of the three CH_3's and the Br in the most symmetric way leads to:

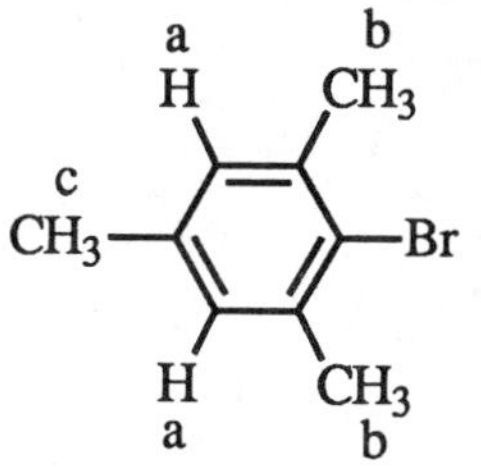

a = δ 7.1 (singlet, 2H)
b = δ 2.3 (singlet, 6H)
c = δ 2.2 (singlet, 3H)

12-30 The numbers in italics indicate the number of peaks in each signal.

(a) $CH_3—CH_2—CCl_2—CH_3$
3 (CH₃), *4* (CH₂), *1* (CH₃)

(b) $CH_3—CH(CH_3)—OH$
CH *7*; OH *1*; CH_3 *2*
(assume OH exchanging rapidly—no splitting)

(c) $CH_3—CH(CH_3)—CH_3$
CH *10*; CH_3 *2*

(d)

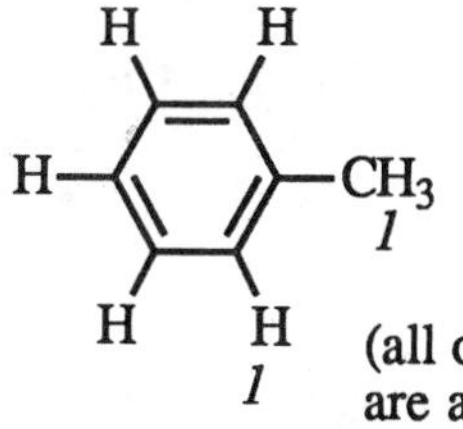

(all of these benzene H's are accidentally equivalent and do not split each other)

(e)

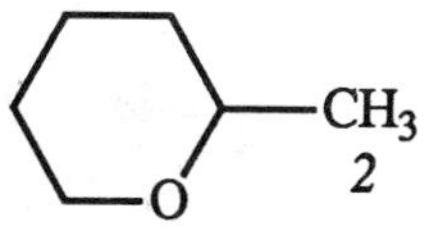

12-31 Consult Appendix 1 for chemical shift values. *Your predictions should be in the given range, or within 0.5 ppm of the given value.*

(a)

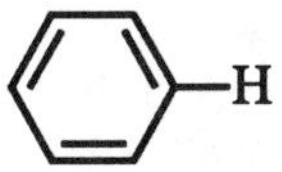

all at δ 7.2

(b)

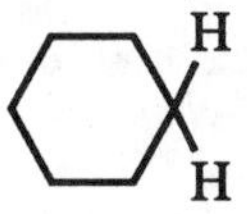

all at δ 1.3

(c)

$CH_3-O-CH_2-CH_2-CHCl_2$

δ 3.4 (CH₃), δ 3.8 (OCH₂), δ 1.6 (CH₂), δ 5.5 (CHCl₂)

(d)

$CH_3-CH_2-C\equiv C-H$

δ 1.5, δ 2.2, δ 2.5

(e)

$CH_3-CH_2-C(=O)-CH_3$

δ 1.0, δ 2.3, δ 2.1

(f)

$(CH_3)_2CH-CH_2-CH_2-OH$

CH₃ δ 0.9; CH δ 1.4; CH₂ δ 1.5; CH₂ δ 3.8; OH δ 2-5

(g)

$CH_3-CH_2-C(=O)-H$

δ 1.1, δ 2.4, δ 9-10

(h)

$CH_3-CH=CH-C(=O)-H$

δ 1.7 (CH₃), δ 6-7 (CH=), δ 5-6 (=CH), δ 9-10 (CHO)

(i)

$HOOC-CH_2-CH_2-C(=O)-O-CH(CH_3)_2$

δ 10-12, δ 2.3, δ 2.3, δ 4.0 (CH), δ 1.1 (CH₃)

(j)

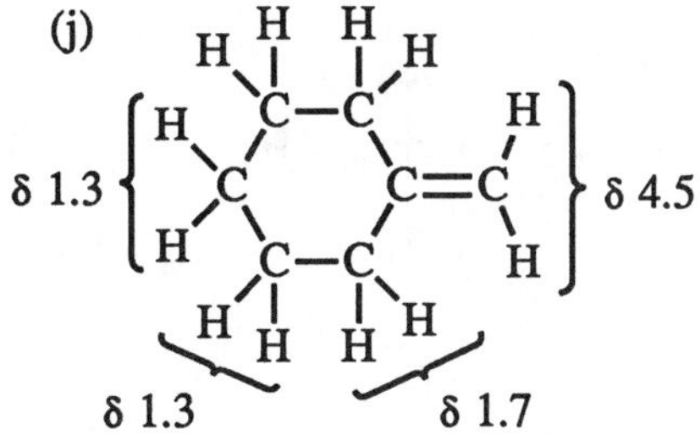

δ 1.3
δ 4.5
δ 1.3
δ 1.7

(k)

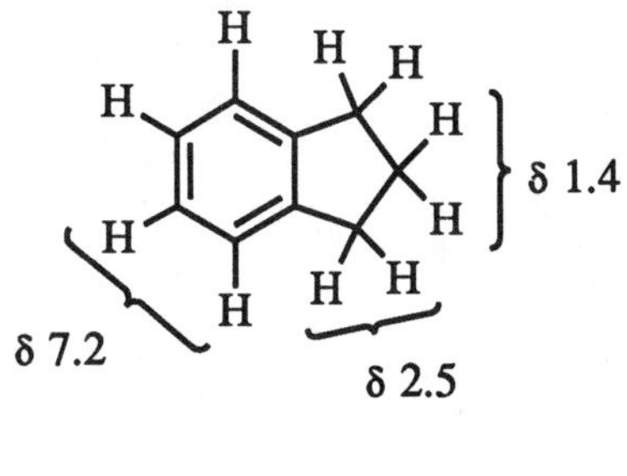

δ 1.4
δ 7.2
δ 2.5

(l)

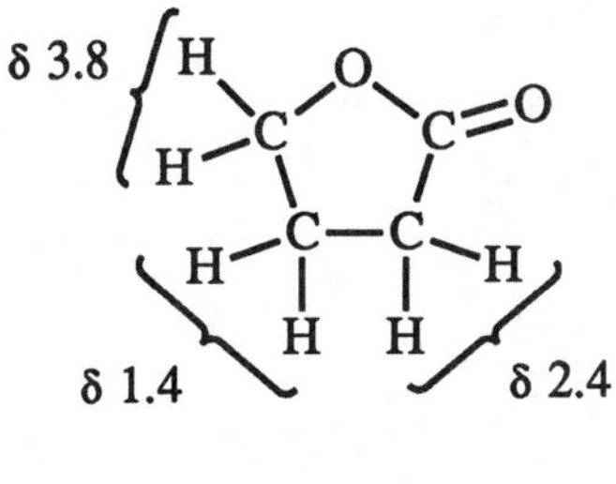

δ 3.8
δ 1.4
δ 2.4

12-32

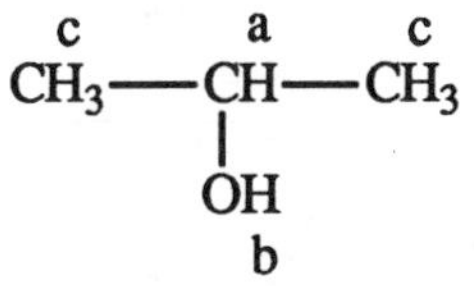

a = δ 4.0 (septet, 1H)
b = δ 2.2 (singlet, 1H)
c = δ 1.2 (doublet, 6H)

12-33

(a) δ 4.00 = 4.00 ppm = $(4.00 \times 10^{-6}) \times (60 \times 10^{6}\ \text{Hz}) = 240\ \text{Hz}$

The signal is 240 Hz downfield from TMS.

(b) The chemical shift *in ppm* would not change: δ 4.00. At 100 MHz:

$(4.00 \times 10^{-6}) \times (100 \times 10^{6}\ \text{Hz}) = 400\ \text{Hz}$

The signal is 400 Hz downfield from TMS.

(c) Coupling constants do not change with field strength: J = 7 Hz, regardless of field strength.

12-34

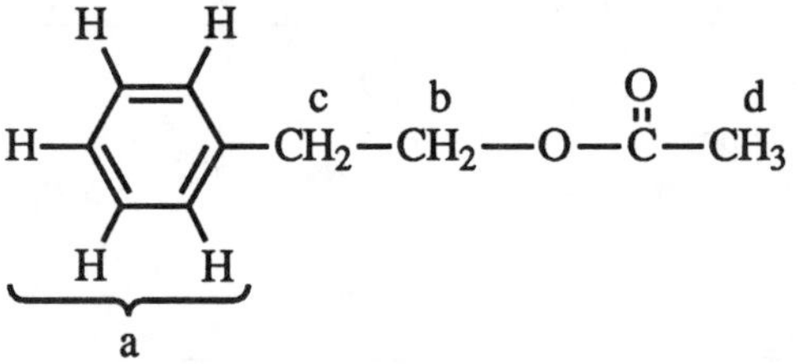

a = δ 7.3 (singlet, 5H)
b = δ 4.3 (triplet, 2H)
c = δ 2.9 (triplet, 2H)
d = δ 2.0 (singlet, 3H)

12-35

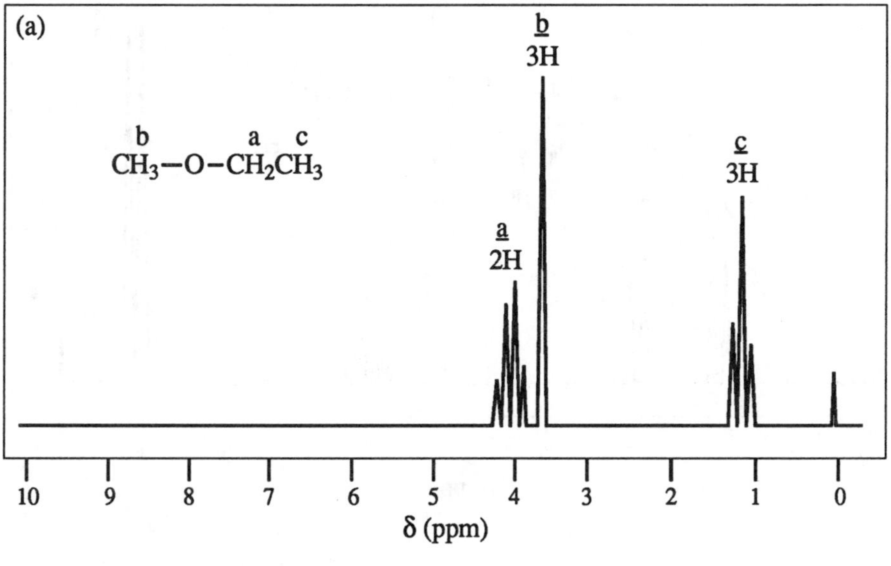

12-35 continued

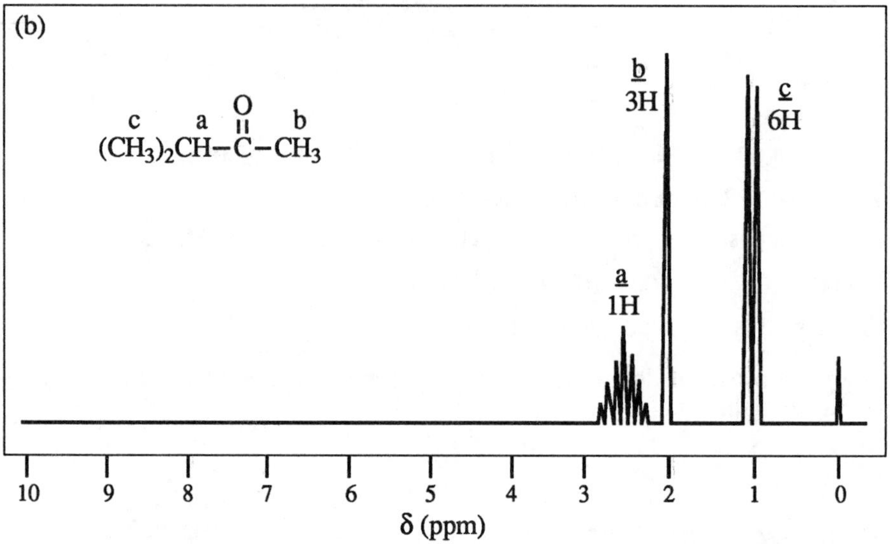

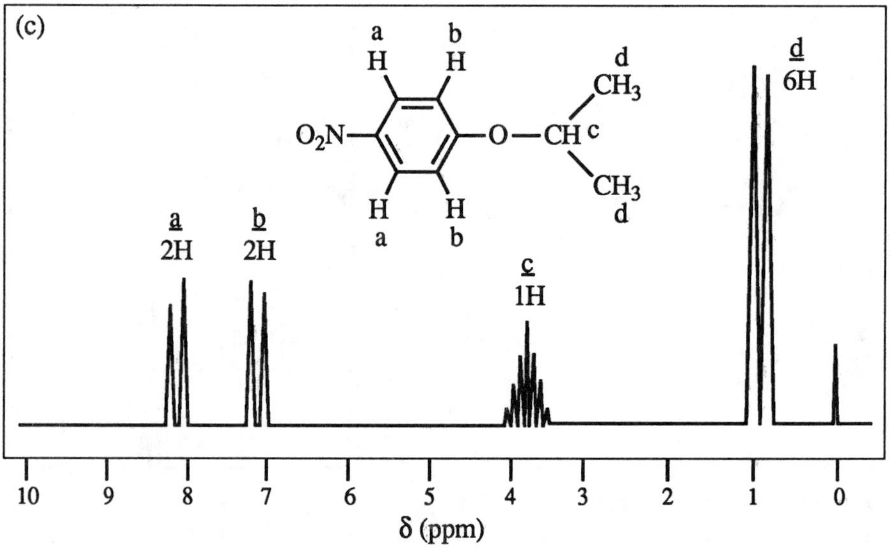

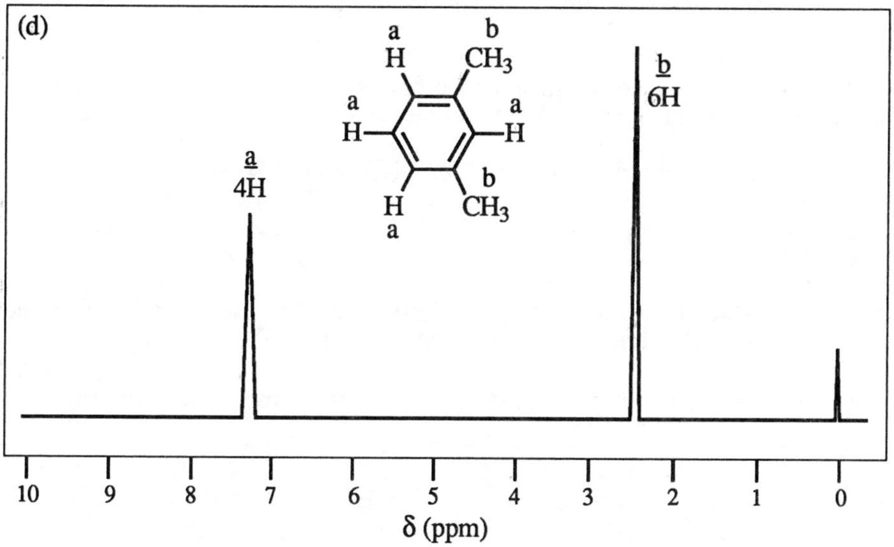

Note: all three types of aromatic protons above are accidentally equivalent.

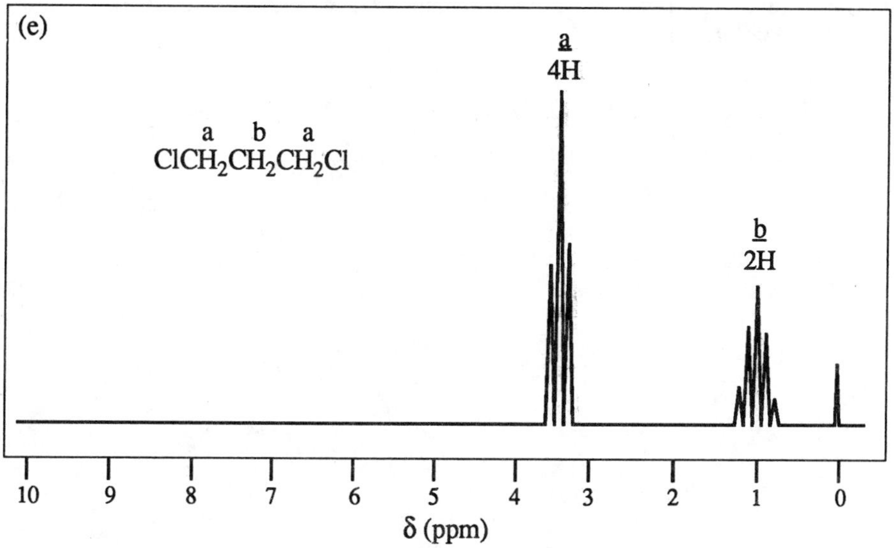

12-36

(a) The NMR of 1-bromopropane would have three sets of signals, whereas the NMR of 2-bromopropane would have only two sets (a septet and a doublet, the typical isopropyl pattern).

(b) The NMR of the aldehyde would show a triplet at δ 9-10 and an ethyl pattern, whereas the ketone would exhibit only one singlet at δ 2.0.

(c) The most obvious difference is the chemical shift of the CH_3 singlet. In the compound on the left, the CH_3 singlet would appear at δ 2.1, while the compound on the right would show the CH_3 singlet at δ 3.8. Refer to 12-20 for the spectrum of the second compound.

(d) The NMR of 1-butyne would show three signals: an ethyl pattern and a 1H singlet at δ 2.5. The NMR of 2-butyne would have only one singlet about δ 2.4.

12-37 The numbers below are chemical shift values in the carbon NMR spectra.

(a)

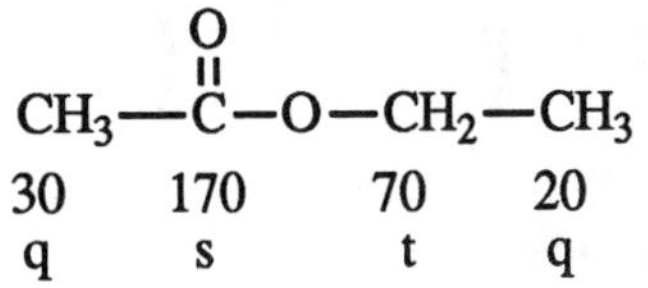
O
CH3—C—O—CH2—CH3
30
q
170
s
70
t
20
q

(b)

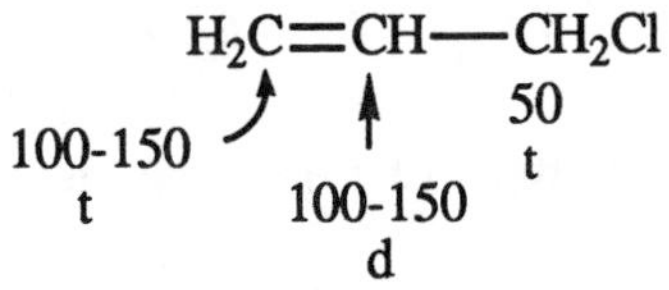
H2C=CH—CH2Cl
100-150
t
100-150
d
50
t

12-38 The multiplicity of the peaks in this off-resonance decoupled spectrum show two different CH's and a CH_3. There is only one way to assemble these pieces with three chlorines.

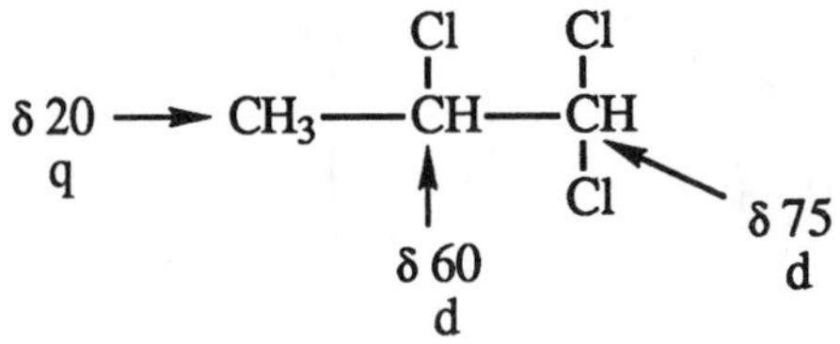
Cl
Cl
δ 20
q
CH3—CH—CH
Cl
δ 60
d
δ 75
d

12-39

(a) The formula $C_4H_8Br_2$ has no elements of unsaturation. The 3H triplet at δ 1.7 and the 1H signal at δ 4.3 indicates the fragment, CH_3CHBr. The rest of the molecule must be CH_2CH_2Br to give the remaining sets of 2H peaks.

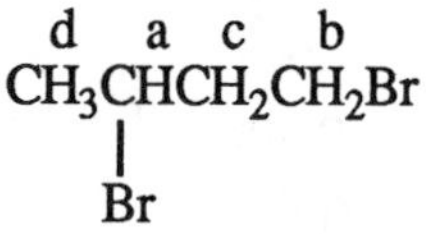

a = δ 4.3 (sextet, 1H)
b = δ 3.5 (triplet, 2H)
c = δ 2.2 (apparent quartet, 2H)
d = δ 1.7 (doublet, 3H)

(b) The formula C_3H_8O has no elements of unsaturation. Four sets of signals could only be 1-propanol.

b a c d
$HOCH_2CH_2CH_3$

a = δ 3.6 (triplet, 2H)
b = δ 3.1 (singlet, 1H)
c = δ 1.5 (6 peaks, 2H)
d = δ 0.9 (triplet, 3H)

12-40

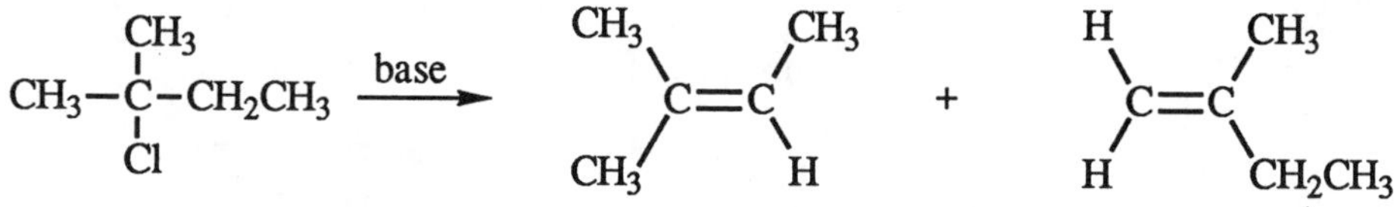

Isomer A Isomer B

(a)

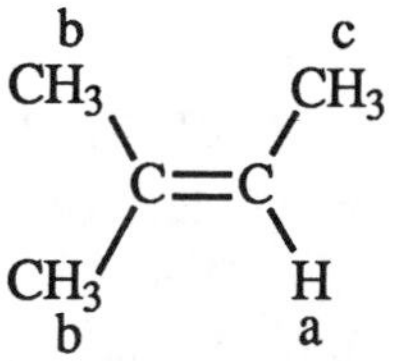

Isomer A

a = δ 5.1 (quartet, 1H)
b = δ 1.6 (singlet, 6H)
c = δ 1.5 (doublet, 3H)

Note that H_c overlaps with H_b.

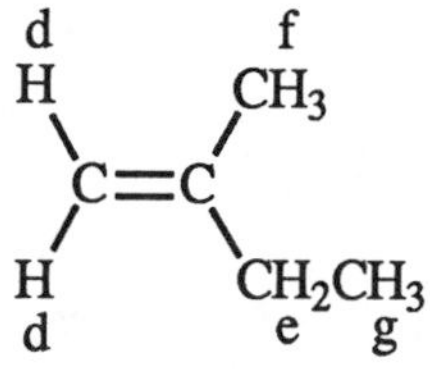

Isomer B

d = δ 4.6 (singlet, 2H)
e = δ 2.0 (quartet, 2H)
f = δ 1.7 (singlet, 3H)
g = δ 1.0 (triplet, 3H)

(b) With NaOH as base, the more highly substituted alkene, Isomer A, would be expected to predominate—the Saytzeff Rule. With KO-*t*-Bu as a hindered, bulky base, the less substituted alkene, Isomer B, would predominate (the Hofmann product).

12-41

Compound 1

Mass spectrum: The molecular ion of m/z 117 suggests the presence of an odd number of nitrogens.

Infrared spectrum: No NH or OH appears. Hydrogens bonded to both sp^2 and sp^3 carbon are indicated around 3000 cm^{-1} . The characteristic C≡N peak appears at 2200 cm^{-1} and aromatic C=C is suggested by the peak at 1600 cm^{-1}.

NMR spectrum: Five aromatic protons are shown in the NMR as a singlet at δ 7.3. A CH_2 singlet appears at δ 3.7.

Assemble the pieces:

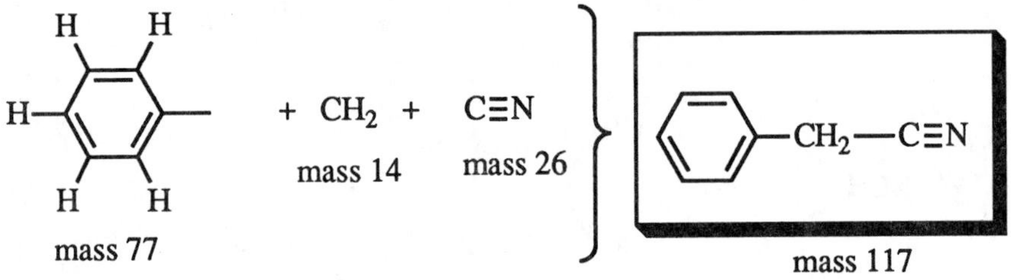

Compound 2

Mass spectrum: The molecular ion at 86 suggests no Cl, no Br, and no N. The loss of 15 to m/z 71 indicates a CH_3 . The peak at m/z 43 is usually either a $C_3H_7^+$ or $CH_3C{=}O^+$.

Infrared spectrum: The dominant functional group peak is at 1700 cm^{-1}, a carbonyl.

NMR spectrum: The two dominant features are a 3H singlet at δ 2.0 (CH_3 next to carbonyl) and an isopropyl pattern, $(CH_3)_2CH$, most likely on the other side of the carbonyl.

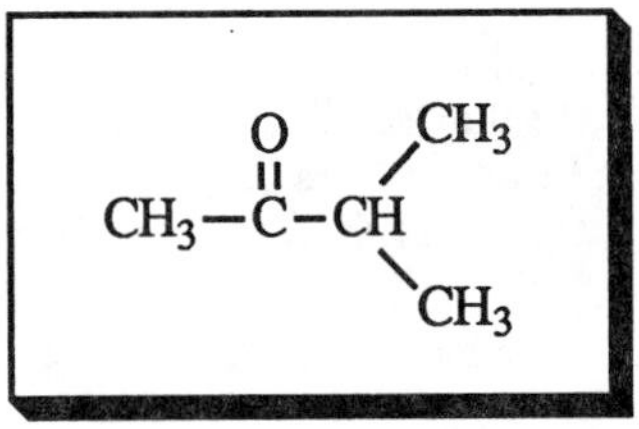

$C_5H_{10}O$

mass 86

CHAPTER 13—ETHERS AND EPOXIDES

13-1 The four polar solvents decrease in polarity in this order: water, ethanol, ethyl ether, and dichloromethane. The three solutes decrease in polarity in this order: sodium acetate, 2-naphthol, and naphthalene. The guiding principle in determining solubility is, "Like dissolves like." Compounds of similar polarity will dissolve (in) each other. Thus, sodium acetate will dissolve in water, will dissolve only slightly in ethanol, and will be virtually insoluble in ethyl ether and dichloromethane. 2-Naphthol will be insoluble in water, somewhat soluble in ethanol, and soluble in ether and dichloromethane. Naphthalene will be insoluble in water, partially soluble in ethanol, and soluble in ethyl ether and dichloromethane. (Actual solubilities are difficult to predict, but you should be able to predict *trends*.)

13-2

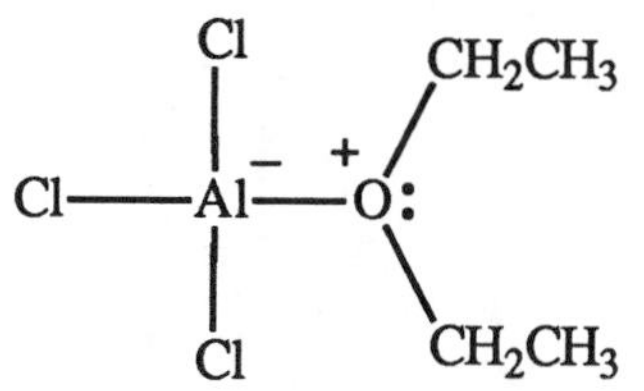

13-3 IUPAC name first; then common name.

(a) methoxycyclopropane; cyclopropyl methyl ether
(b) 2-ethoxypropane; ethyl isopropyl ether
(c) 1-chloro-2-methoxyethane; 2-chloroethyl methyl ether
(d) 2-*t*-butoxybutane; *sec*-butyl *t*-butyl ether
(e) *trans*-2-methoxy-1-cyclohexanol; no common name

13-4

(a)

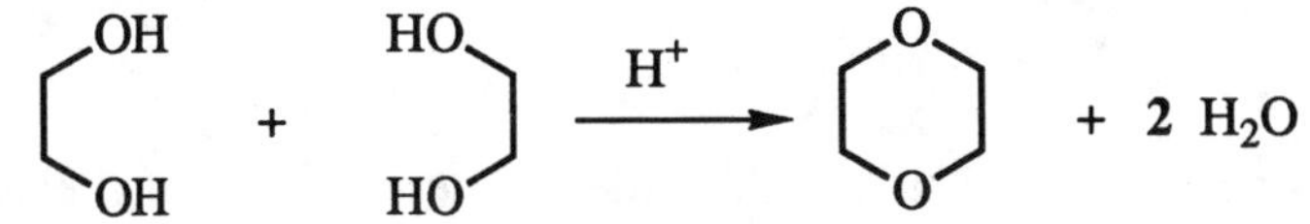

(b)

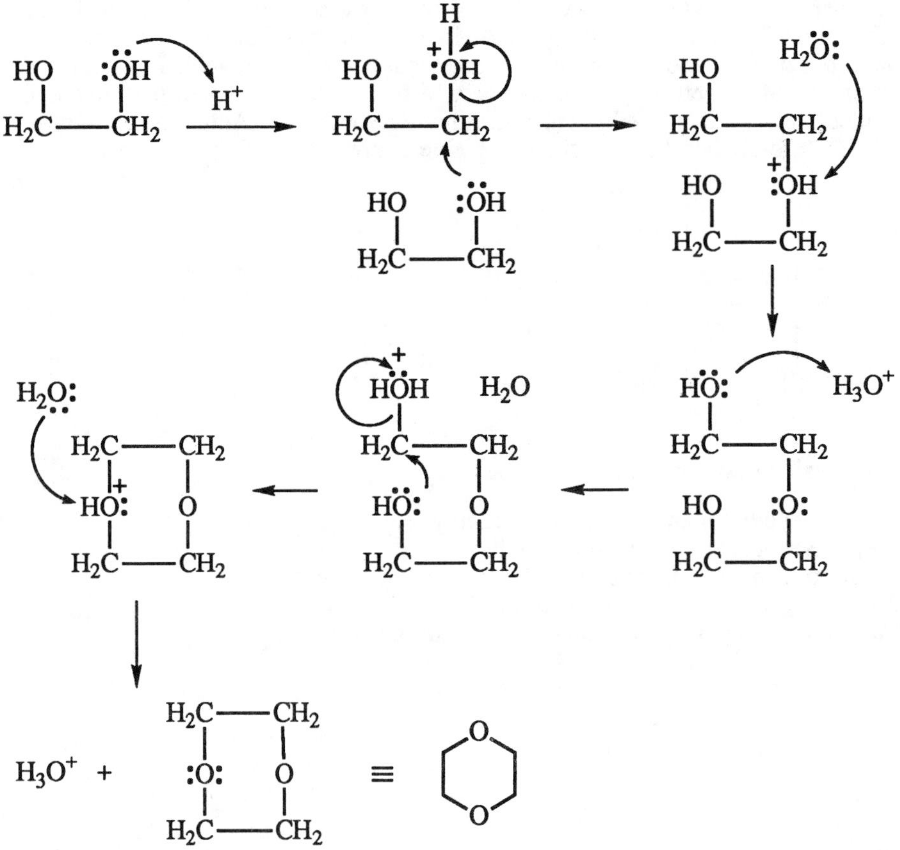

13-5 (a) dihydropyran
(b) 2-chloro-1,4-dioxane
(c) 3-isopropylpyran
(d) *trans*-2,3-diethyloxirane; *trans*-3,4-epoxyhexane; *trans*-3-hexene oxide
(e) 3-bromo-2-ethoxyfuran
(f) 3-bromo-2,2-dimethyloxetane

13-6

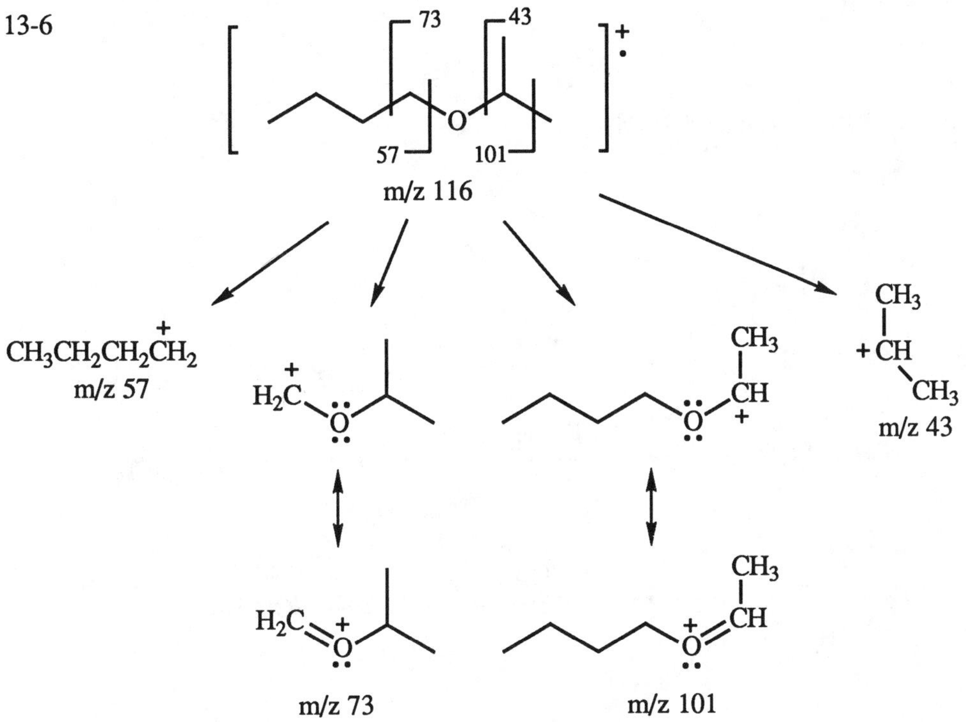
73
43
57
101
m/z 116
$CH_3CH_2CH_2\overset{+}{C}H_2$
m/z 57
$H_2\overset{+}{C}$
CH_3
CH
+
CH_3
+CH
CH_3
m/z 43
H_2C
+
m/z 73
CH_3
CH
+
m/z 101

13-7 S_N2 reactions, including the Williamson ether synthesis, work best when the nucleophile attacks a 1° or methyl carbon. Instead of attempting to form the bond from oxygen to the 3° carbon, as shown in the Solved Problem, form the bond from oxygen to the 1° carbon.

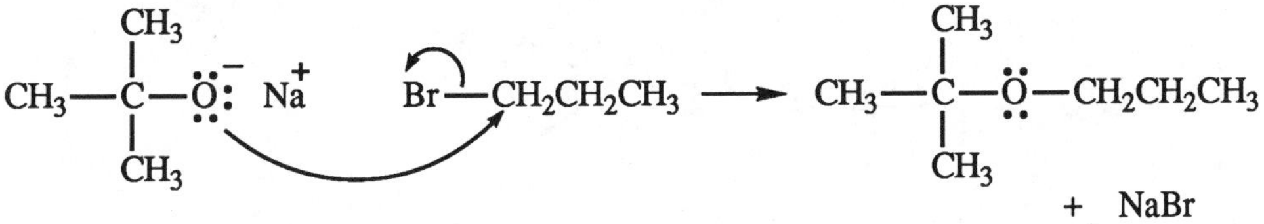
CH_3
CH_3—C—O: $\overset{+}{Na}$
CH_3
Br—$CH_2CH_2CH_3$
CH_3
CH_3—C—O—$CH_2CH_2CH_3$
CH_3
+ NaBr

13-8

(a)

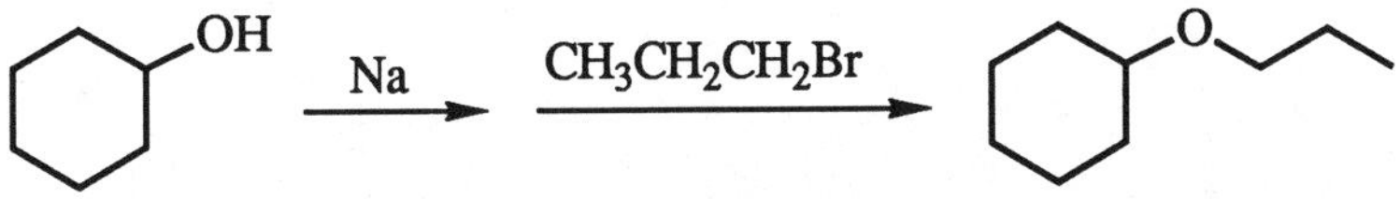
OH
Na
$CH_3CH_2CH_2Br$
O

(b)

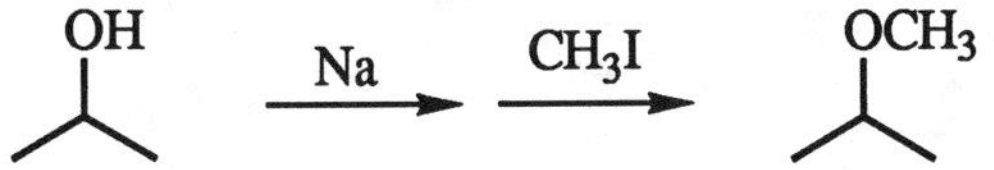
OH
Na
CH_3I
OCH_3

13-8 continued

(c)

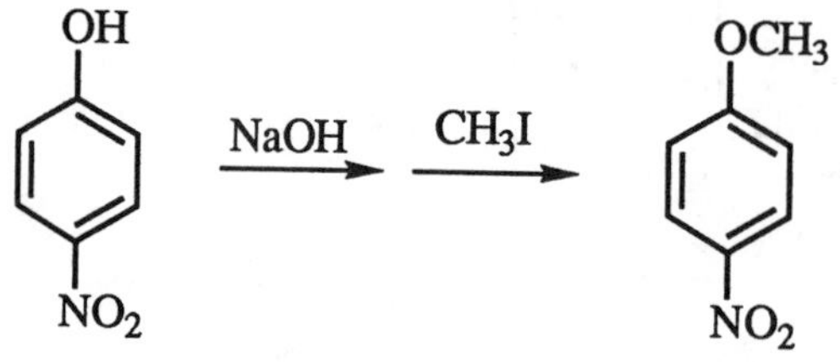
OH
NO2
NaOH
CH3I
OCH3
NO2

(d) $CH_3CH_2CH_2OH \xrightarrow{Na} \xrightarrow{CH_3CH_2Br} CH_3CH_2CH_2OCH_2CH_3$

$CH_3CH_2OH \xrightarrow{Na} \xrightarrow{CH_3CH_2CH_2Br} CH_3CH_2CH_2OCH_2CH_3$

(e)

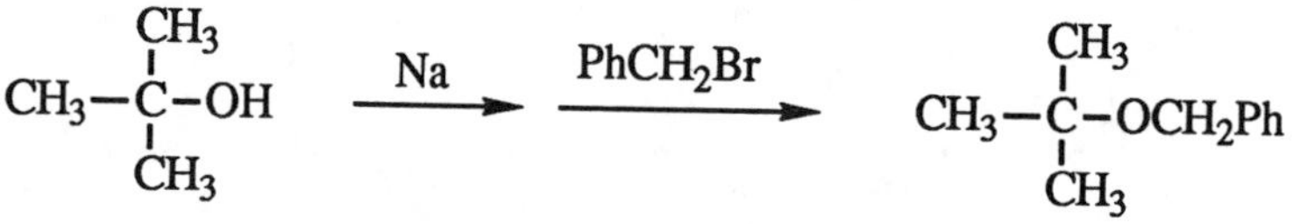
CH3
CH3—C—OH
CH3
Na
PhCH2Br
CH3
CH3—C—OCH2Ph
CH3

13-9

(a) (1)

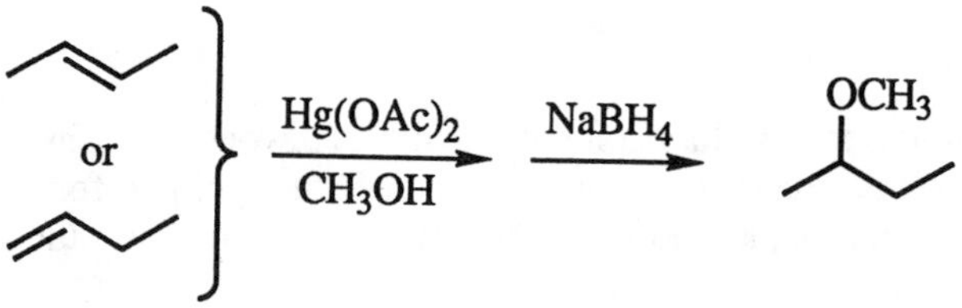
or
Hg(OAc)2
CH3OH
NaBH4
OCH3

(2)

OH
Na
CH3I
OCH3

(b) (1)

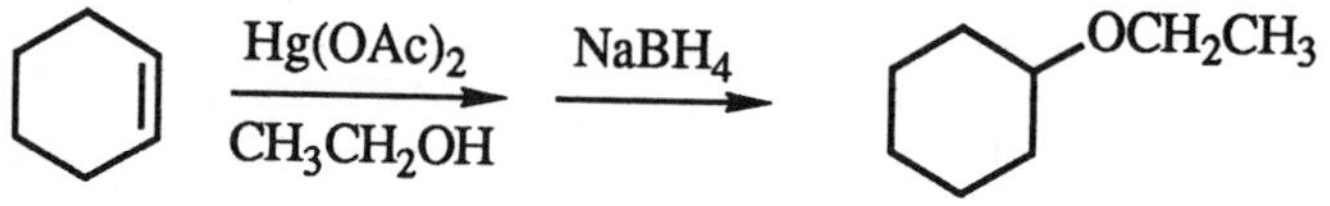
Hg(OAc)2
CH3CH2OH
NaBH4
OCH2CH3

(2)

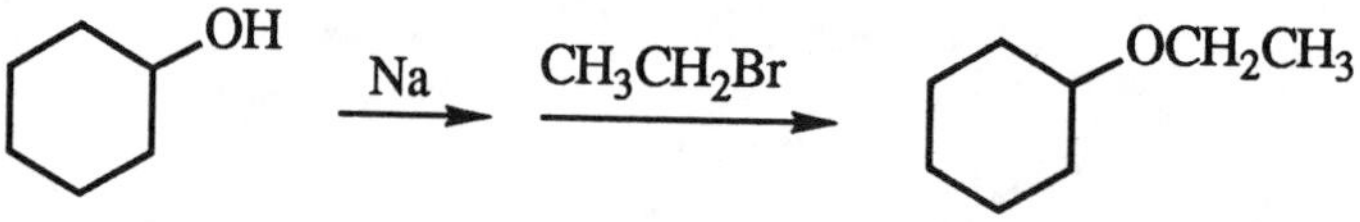
OH
Na
CH3CH2Br
OCH2CH3

13-9 continued

(c) (1) Alkoxymercuration is not practical here; the product does not have Markovnikov orientation.

(2)

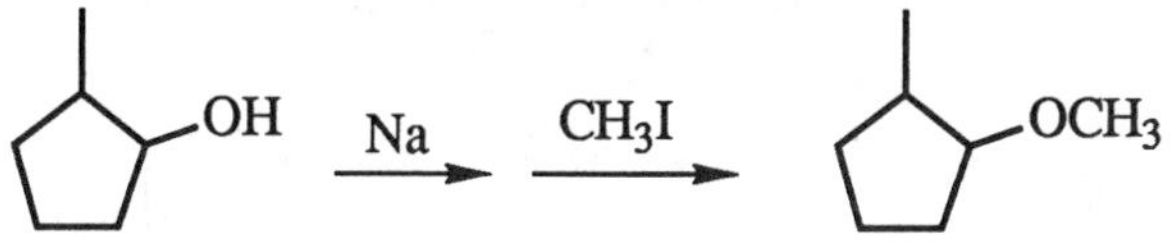
OH
Na
CH_3I
OCH_3

(d) (1)

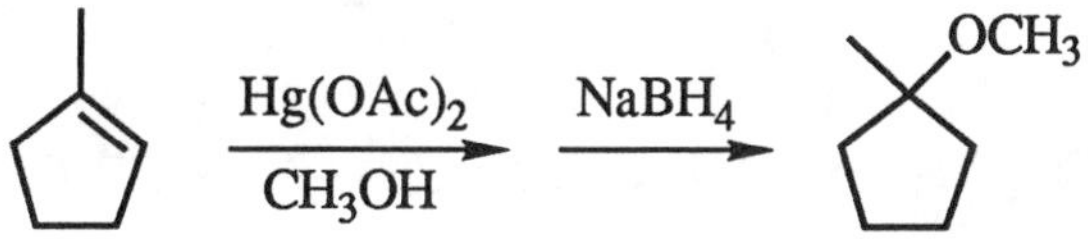
$Hg(OAc)_2$
CH_3OH
$NaBH_4$
OCH_3

(2)

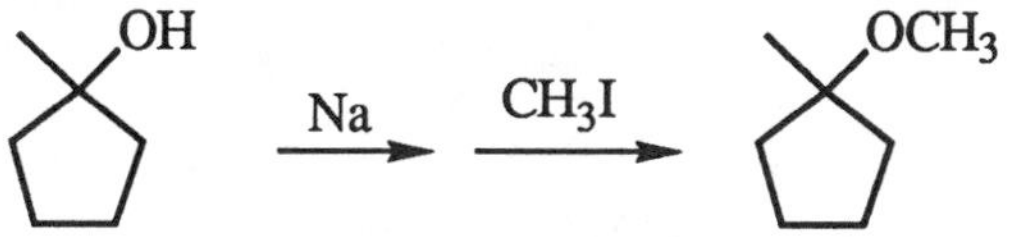
OH
Na
CH_3I
OCH_3

(e) (1)

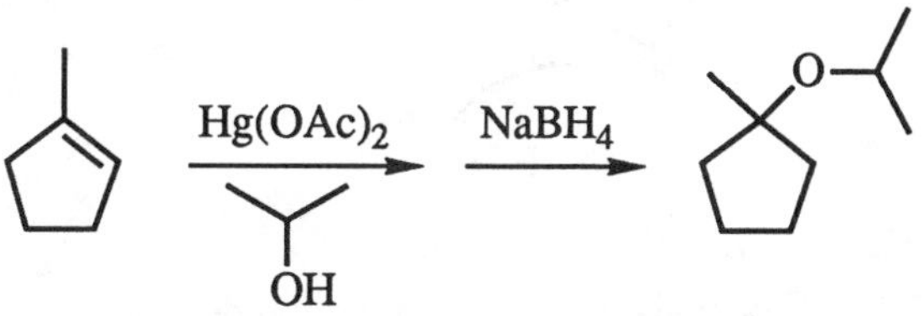
$Hg(OAc)_2$
OH
$NaBH_4$
O

(2) Williamson ether synthesis would give a poor yield of product as the halide is on a 2° carbon.

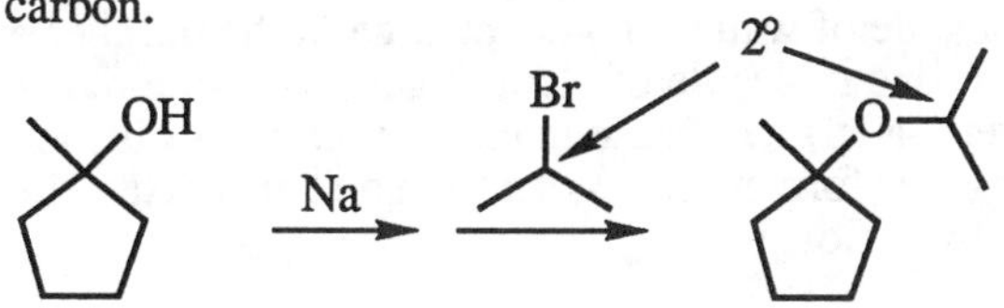
OH
Na
Br
2°
O

(f) (1)

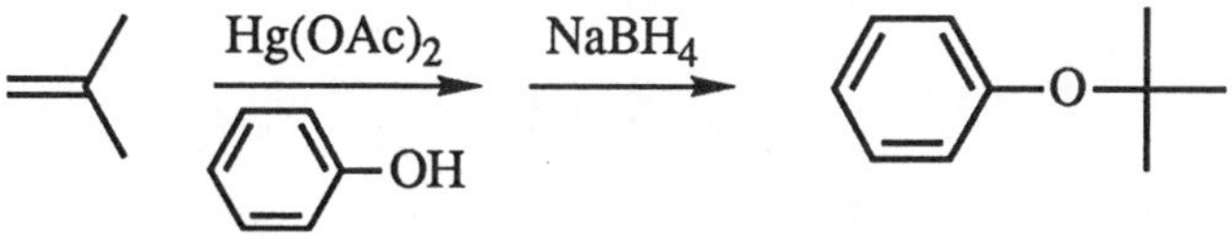
$Hg(OAc)_2$
OH
$NaBH_4$
O

(2) Williamson ether synthesis is not feasible here. S_N2 does not work on either a benzene or a 3° halide.

13-10 Bimolecular dehydration is a random process. Heating a mixture of ethanol and methanol with acid will produce all possible combinations: dimethyl ether, ethyl methyl ether, and diethyl ether.

13-11

Ether formation

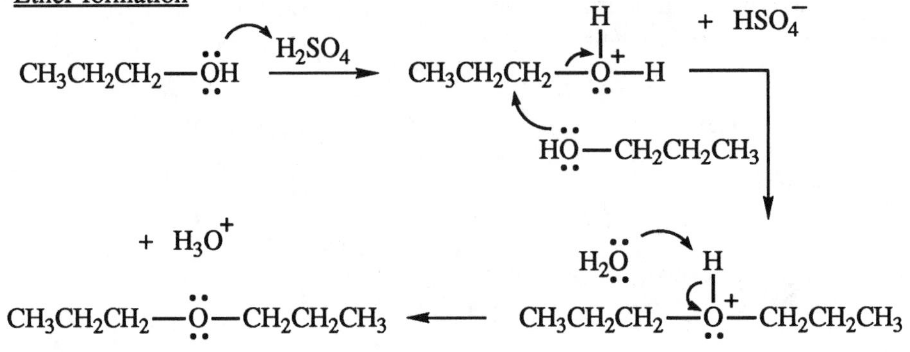

Dehydration

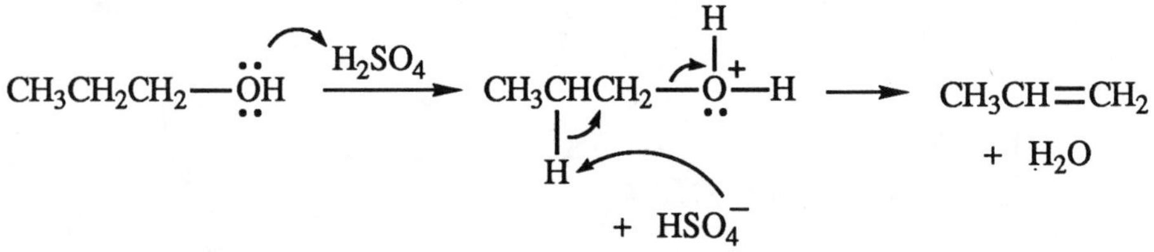

Remember $\Delta G = \Delta H - T\Delta S$? Thermodynamics of a reaction depend on the sign and magnitude of ΔG. As temperature increases, the entropy term grows in importance. In ether formation, the ΔS is small because two molecules of alcohol give one molecule of ether plus one molecule of water—no net change in the number of molecules. In dehydration, however, one molecule of alcohol generates one molecule of alkene plus one molecule of water—a large increase in entropy. So $T\Delta S$ is more important for dehydration than for ether formation. As temperature increases, the competition will shift toward more dehydration.

13-12

(a) This symmetrical ether at 1° carbons could be produced in good yield by bimolecular dehydration.

(b) This unsymmetrical ether could not be produced in high yield by bimolecular dehydration. Williamson synthesis would be preferred.

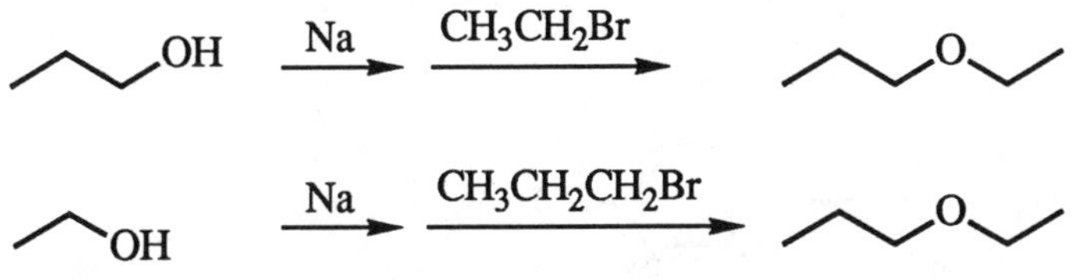

(c) Even though this ether is symmetrical, both carbons are 2°, so bimolecular dehydration would give low yields. Unimolecular dehydration to give alkenes would be the dominant pathway. Alkoxymercuration-demercuration is the preferred route.

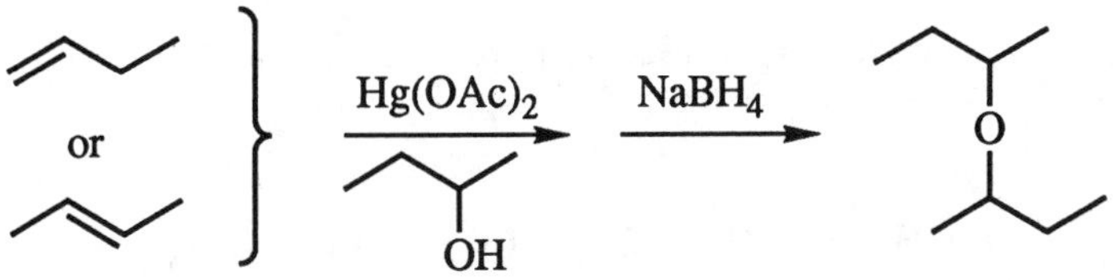

13-13

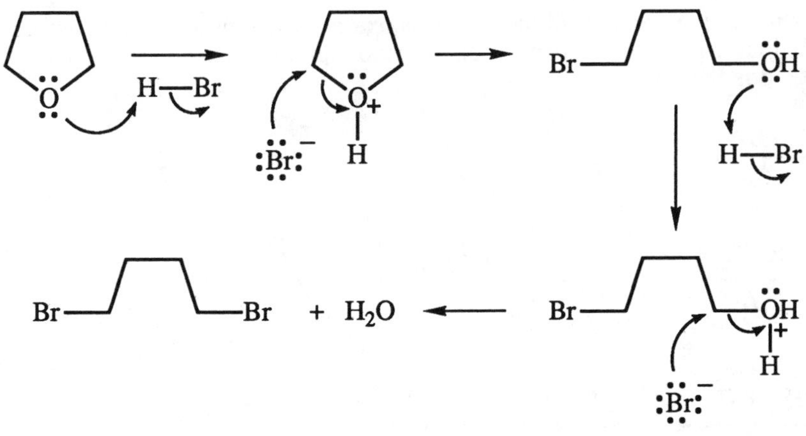

13-14

(a) 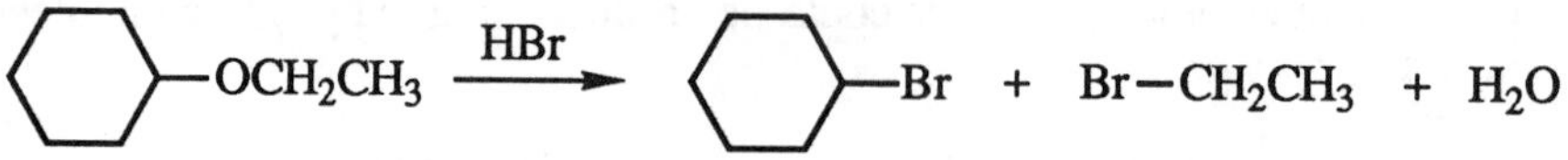
OCH2CH3
HBr
Br + Br—CH2CH3 + H2O

(b) 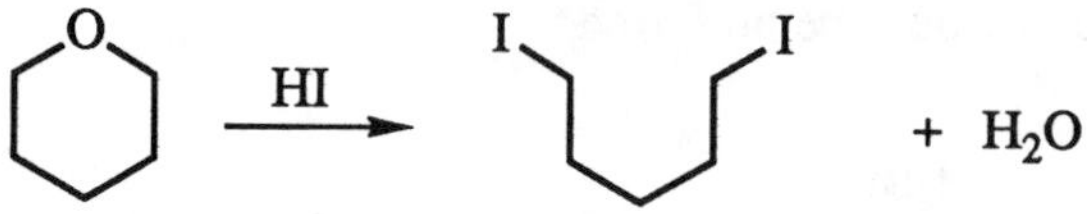
O
HI
I
I
+ H2O

(c) 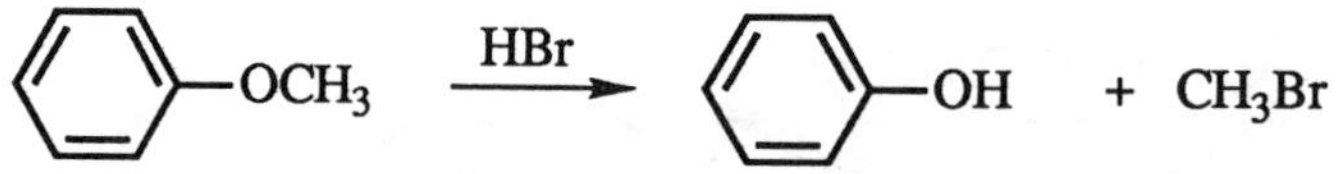
OCH3
HBr
OH + CH3Br

(d) 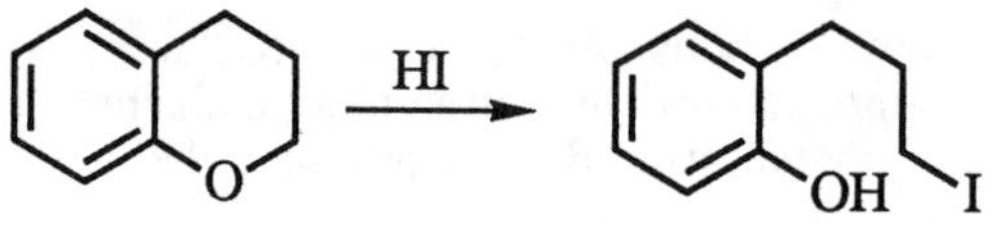
O
HI
OH
I

(e) 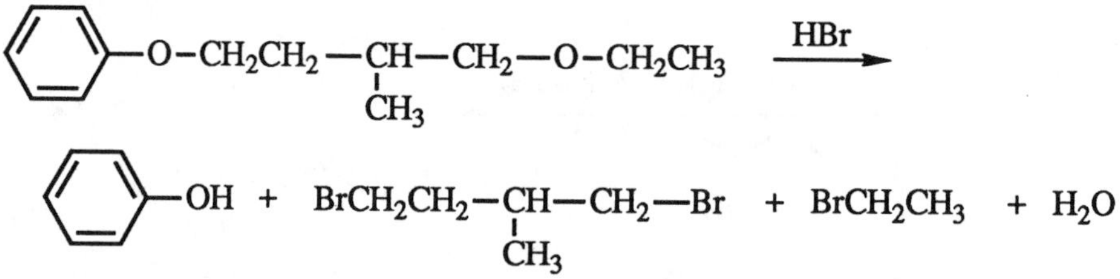
O–CH2CH2—CH—CH2—O–CH2CH3
CH3
HBr
OH + BrCH2CH2–CH—CH2—Br + BrCH2CH3 + H2O
CH3

13-15

Generally, chemists prefer the peroxyacid method of epoxide formation to the halohydrin method. These solutions will show the peroxyacid method, but this does not imply that the halohydrin method is incorrect.

(a) 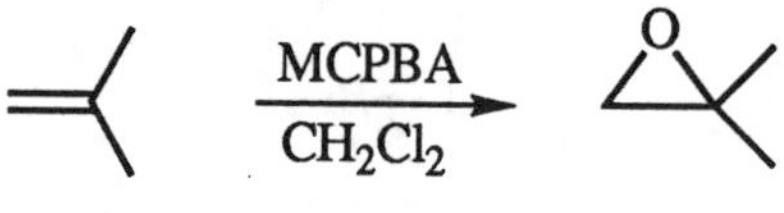
MCPBA
CH2Cl2
O

(b) 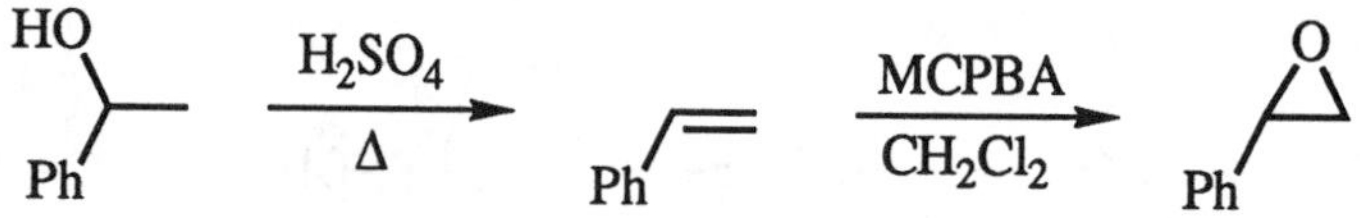
HO
Ph
H2SO4
Δ
Ph
MCPBA
CH2Cl2
O
Ph

(c) 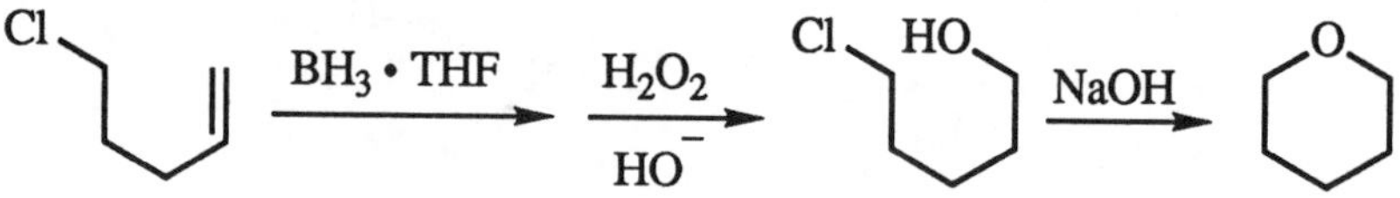
Cl
BH3 • THF
H2O2
HO−
Cl
HO
NaOH
O

13-15 continued

(d)

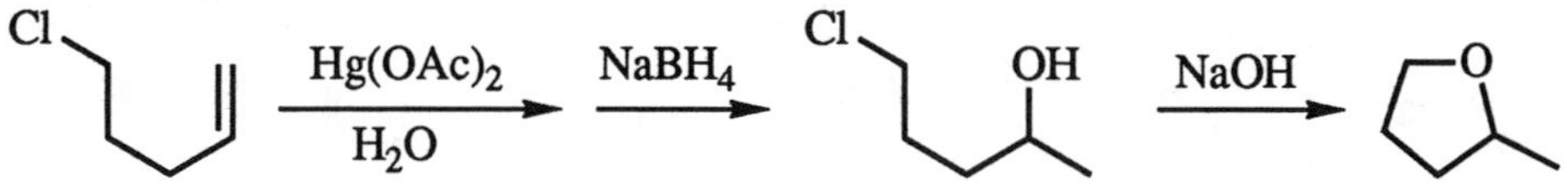

(e)

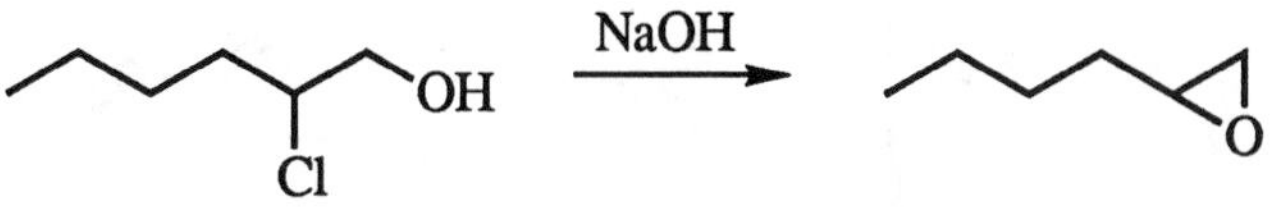

13-16

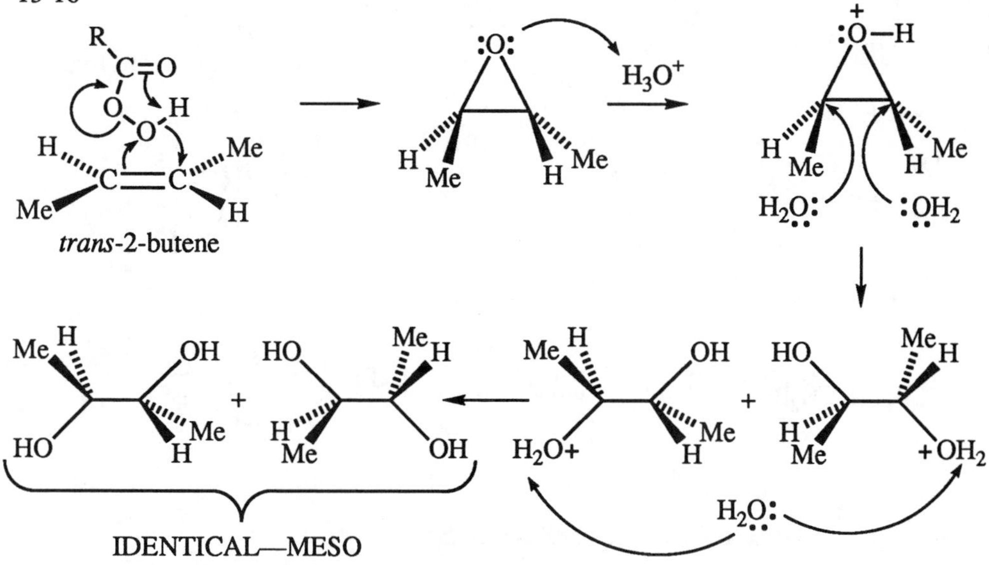

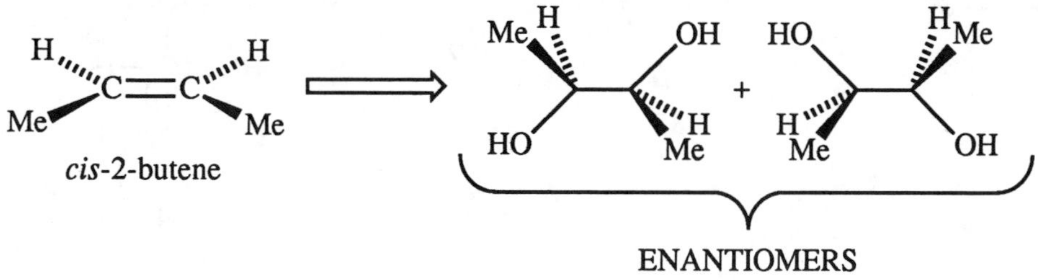

13-17

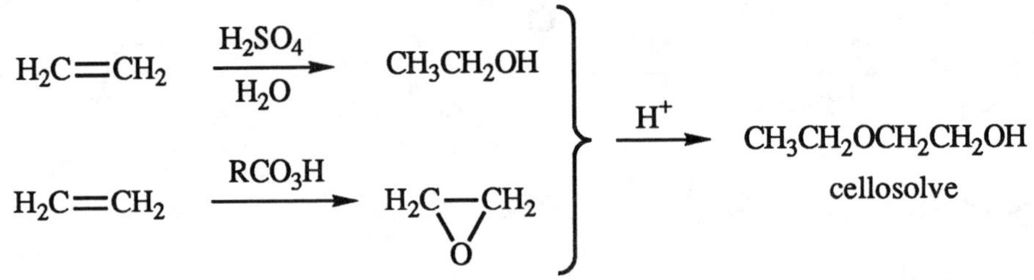
H2C=CH2
H2SO4
H2O
CH3CH2OH
H2C=CH2
RCO3H
H2C—CH2
O
H+
CH3CH2OCH2CH2OH
cellosolve

13-18

anhydrous HBr

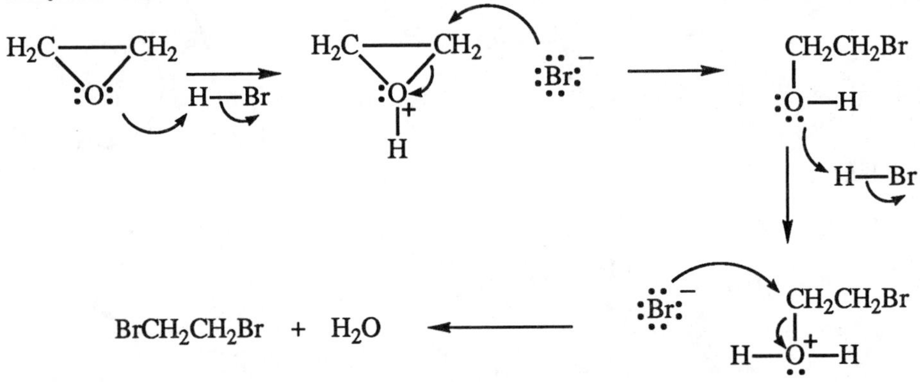
H2C—CH2
:O:
H—Br
H2C—CH2
:O+
H
:Br:−
CH2CH2Br
:O—H
H—Br
:Br:−
CH2CH2Br
H—O+—H
BrCH2CH2Br + H2O

aqueous HBr—many more H_2O nucleophiles than Br^- nucleophiles

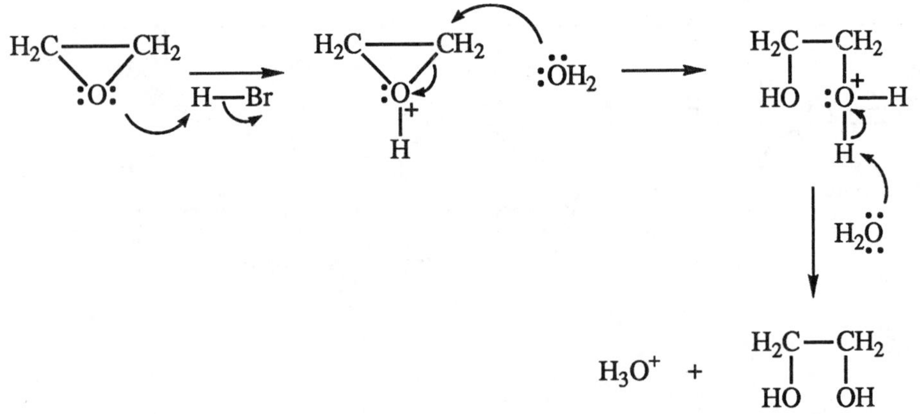
H2C—CH2
:O:
H—Br
H2C—CH2
:O+
H
:OH2
H2C—CH2
HO :O+—H
H
H2O
H3O+ +
H2C—CH2
HO OH

13-19

The cyclization of squalene via the epoxide is an excellent (and extraordinary) example of how Nature uses organic chemistry to its advantage. In one enzymatic step, Nature forms four rings and eight chiral centers! Out of 256 possible stereoisomers, only one is formed!

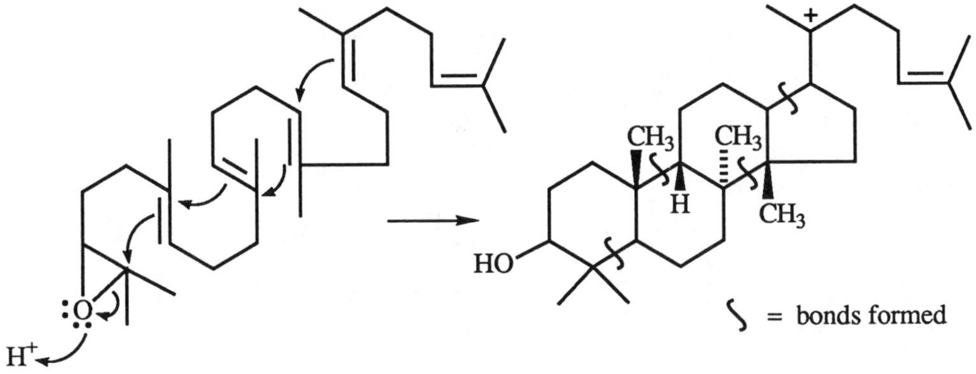

13-20

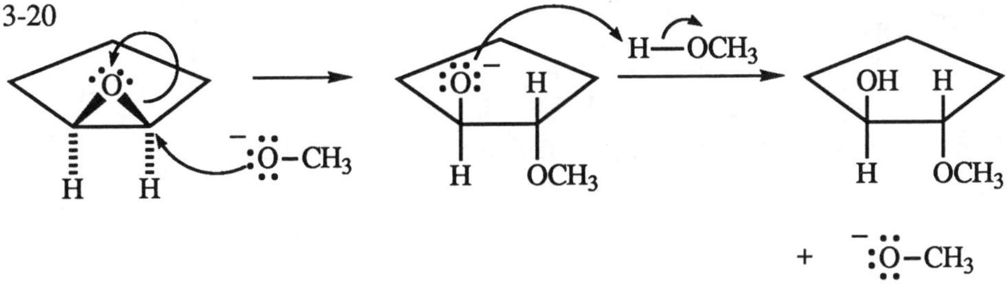

13-21

(a) 

(b) 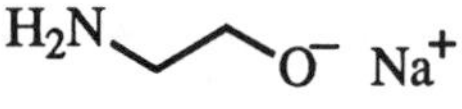

(c) 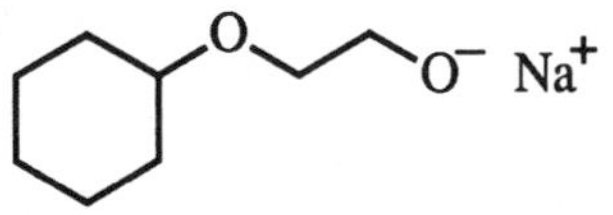

13-22

(a) 

(b) 

(c) 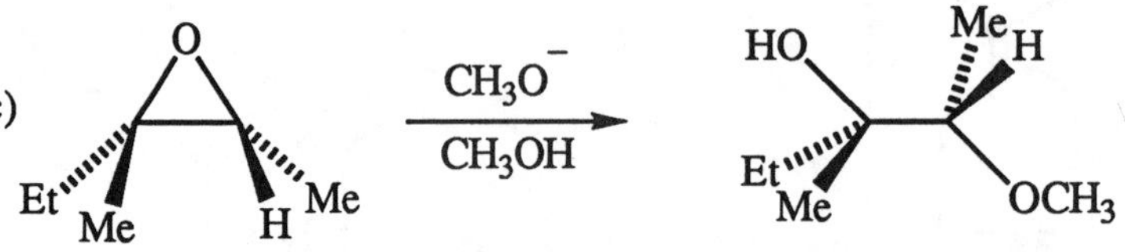

(d) 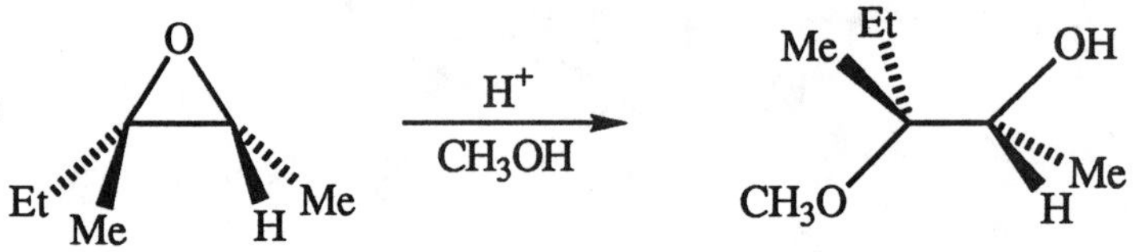

13-23

(a) 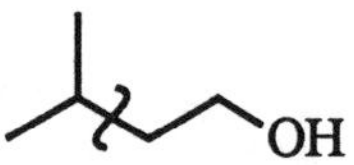

(b) 

(c) 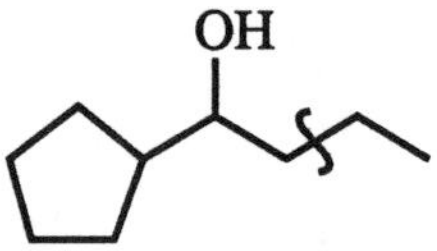

 = bond formed

13-24 Refer to the Glossary and the text for definitions and examples.

13-25

(a) 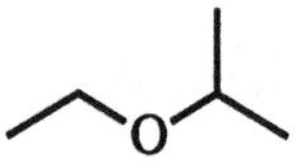

(b) 

(c) 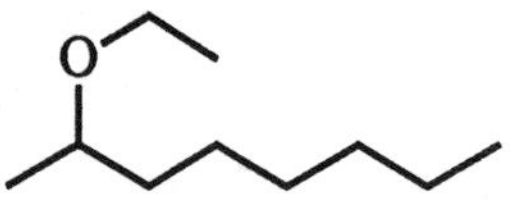

(d) 

(e) 

(f) 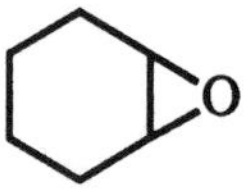

(g) 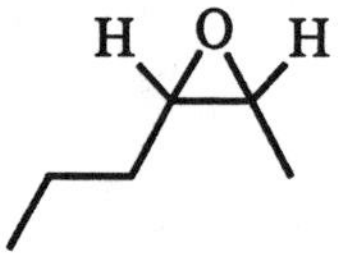

(h) 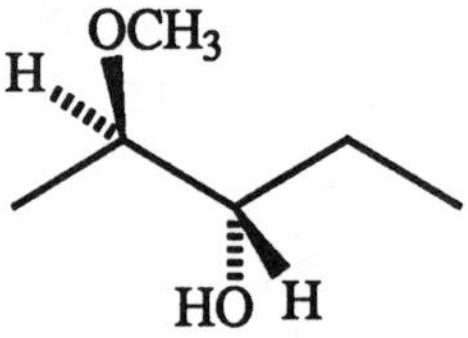

13-26

(a) *sec*-butyl isopropyl ether
(b) isobutyl *t*-butyl ether
(c) ethyl phenyl ether
(d) chloromethyl *n*-propyl ether
(e) *trans*-cyclohexene glycol
(f) cyclopentyl methyl ether
(g) propylene oxide
(h) cyclopentene oxide

13-27

(a) 2-methoxy-1-propanol
(b) ethoxybenzene or phenoxyethane
(c) methoxycyclopentane
(d) 2,2-dimethoxy-1-cyclopentanol
(e) *trans*-1-methoxy-2-methylcyclohexane
(f) *trans*-3-chloro-1,2-epoxycycloheptane
(g) *trans*-1-methoxy-1,2-epoxybutane; or, *trans*-2-ethyl-3-methoxyoxirane
(h) 3-bromooxetane
(i) 1,3-dioxane

13-28

(a) 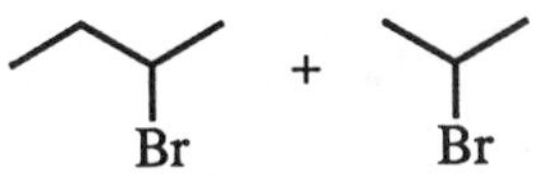 + H_2O

(b) 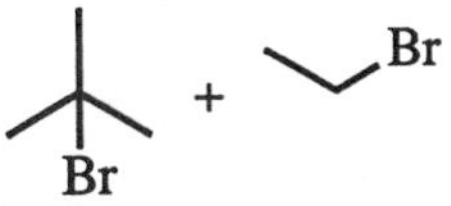 + H_2O

(c) no reaction

(d) no reaction

(e) 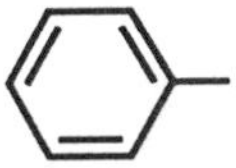 —OH + CH_3CH_2I

(f) 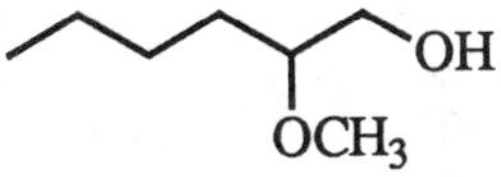

(g) 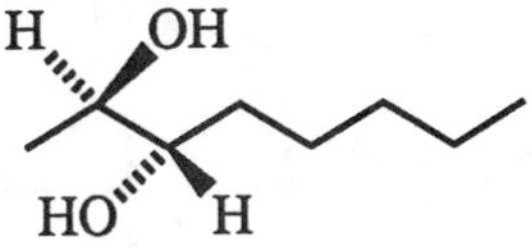

(h) 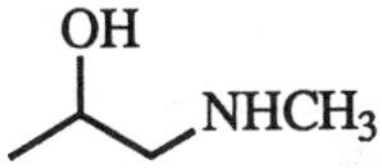

(i) 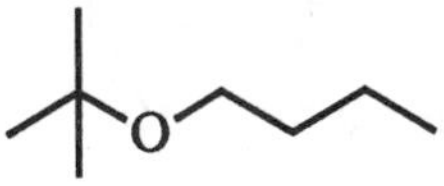

(j) 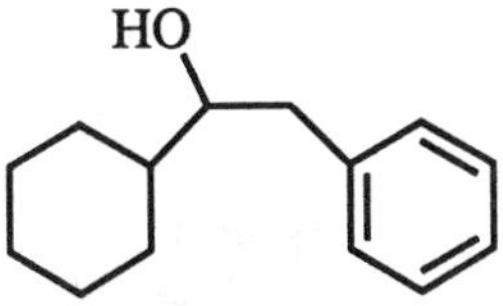

(k) 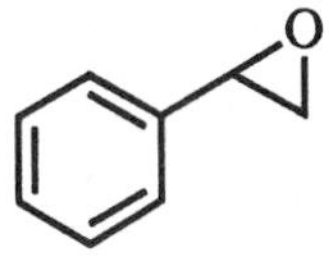

(l) 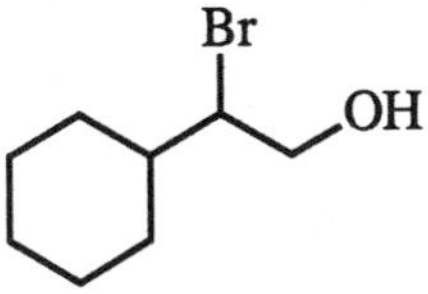

(m) 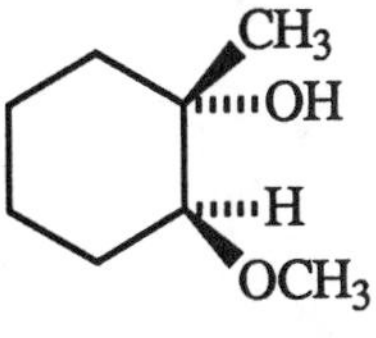

(n) 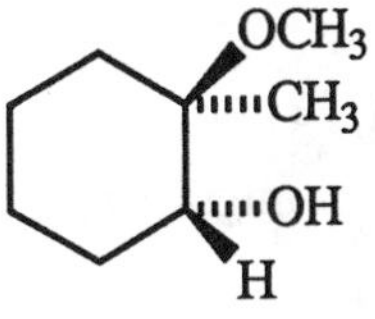

13-29

(a) On long-term exposure to air, ethers form peroxides. Peroxides are explosive when concentrated or heated. (For exactly this reason, ethers should *never* be distilled to dryness.)

(b) Peroxide formation can be prevented by excluding oxygen. Ethers can be checked for the presence of peroxides, and peroxides can be destroyed safely by treatment with reducing agents.

13-30

(a)

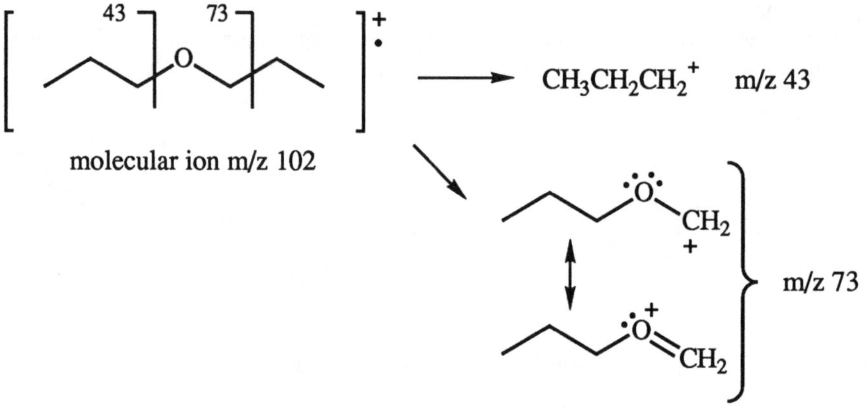

(b)

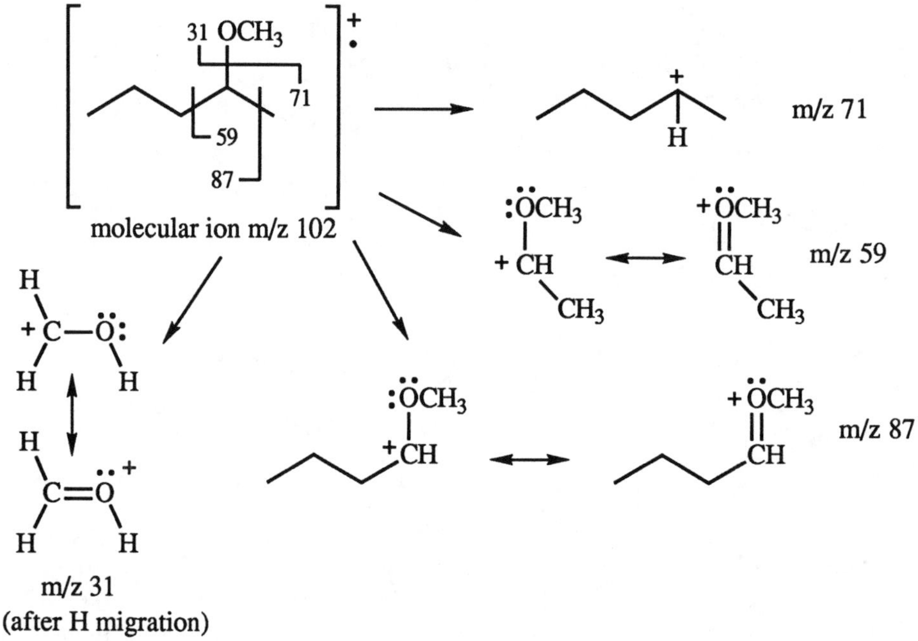

13-31

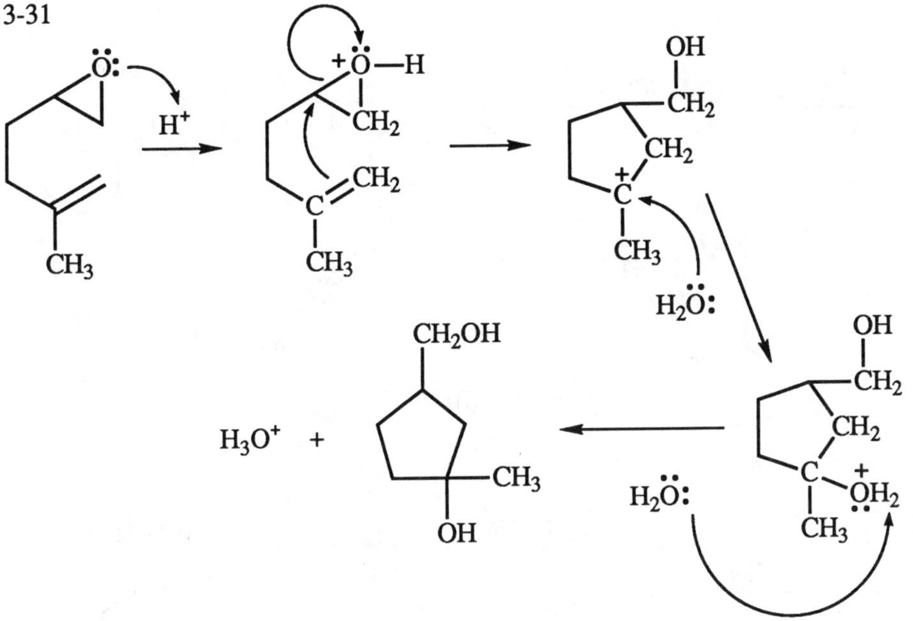
Ö:
H⁺
+Ö—H
CH₂
CH₂
C
CH₃
CH₃
OH
CH₂
CH₂
C
CH₃
H₂Ö:
OH
CH₂
CH₂
C
ÖH₂
CH₃
H₂Ö:
CH₂OH
CH₃
OH
H₃O⁺ +

13-32

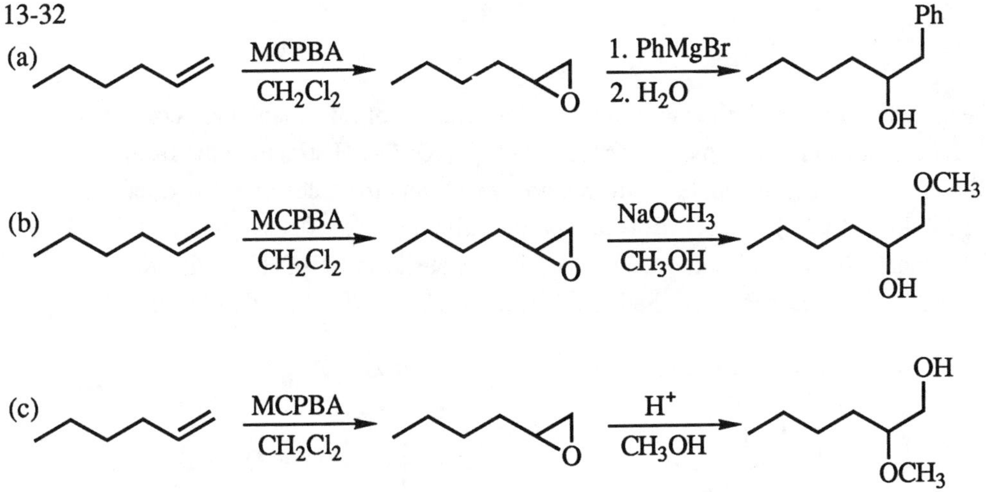
(a)
MCPBA
CH₂Cl₂
O
1. PhMgBr
2. H₂O
Ph
OH
(b)
MCPBA
CH₂Cl₂
O
NaOCH₃
CH₃OH
OCH₃
OH
(c)
MCPBA
CH₂Cl₂
O
H⁺
CH₃OH
OH
OCH₃

13-33

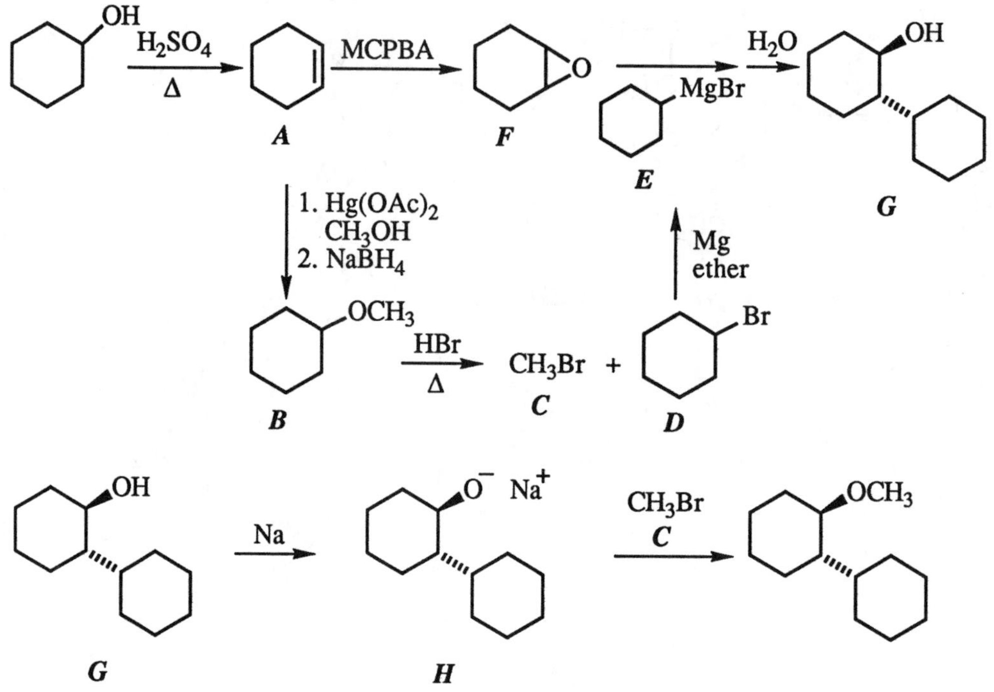

13-34

The student turned in the wrong product! Three pieces of information are consistent with the desired product: molecular formula $C_4H_{10}O$; O—H stretch in the IR at 3300 cm^{-1} (although it should be strong, not weak); and mass spectrum fragment at m/z 59 (loss of CH_3). The NMR of the product should have a 9H singlet at δ 1.0 and a 1H singlet between δ 2 and δ 5. Instead, the NMR shows CH_3CH_2 bonded to oxygen. The student isolated diethyl ether, ***the typical solvent used in Grignard reactions***.

Predicted product	Isolated product
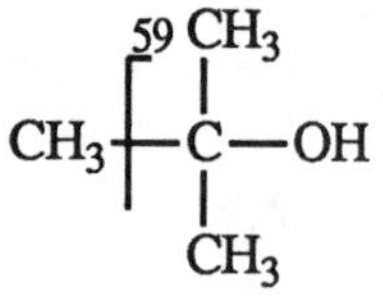	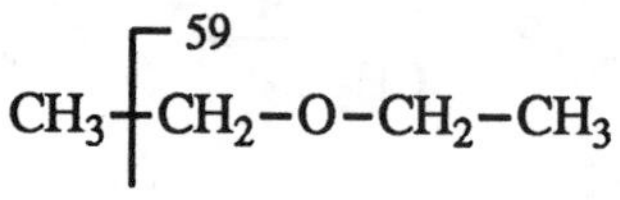
$C_4H_{10}O$	$C_4H_{10}O$
O—H at 3300 cm^{-1}	O—H at 3300 cm^{-1} due to water contamination

13-35

(a)

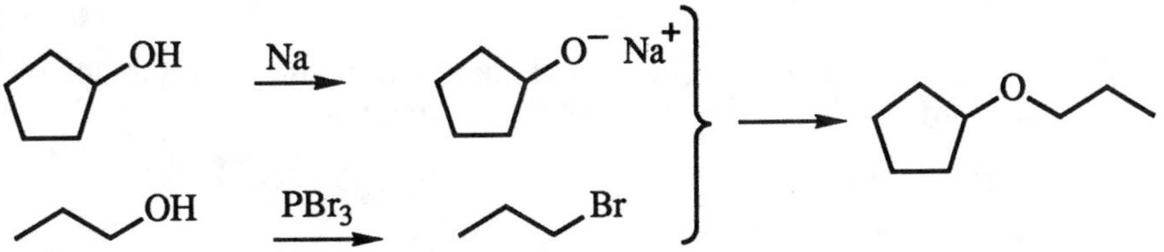
OH
Na
O⁻ Na⁺
OH
PBr3
Br
O

(b)

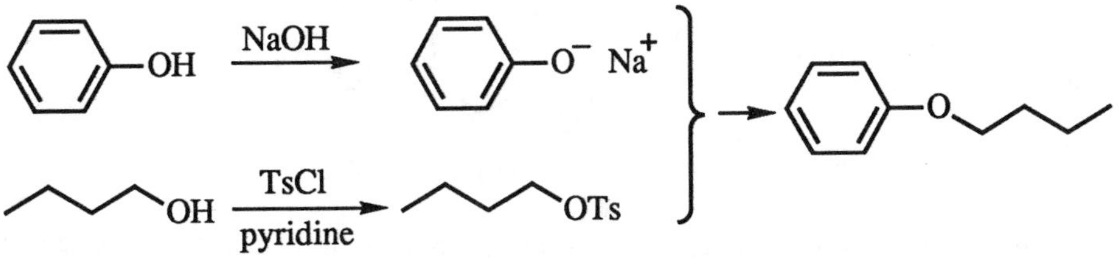
OH
NaOH
O⁻ Na⁺
OH
TsCl
pyridine
OTs
O

(c)

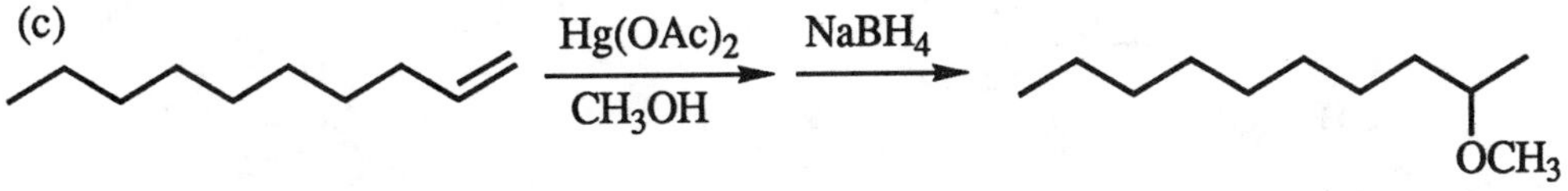
Hg(OAc)2
CH3OH
NaBH4
OCH3

(d)

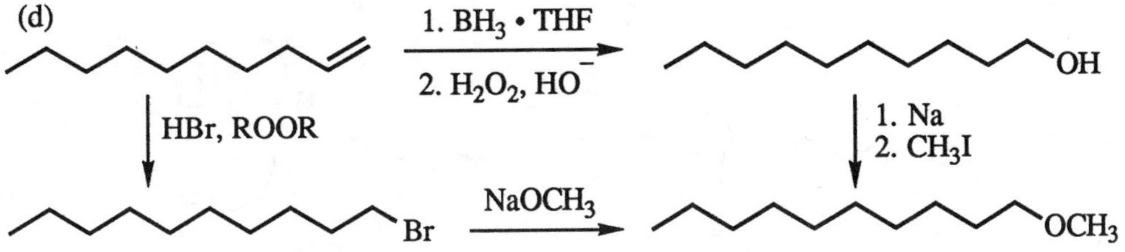
1. BH3 • THF
2. H2O2, HO⁻
OH
HBr, ROOR
1. Na
2. CH3I
Br
NaOCH3
OCH3

(e)

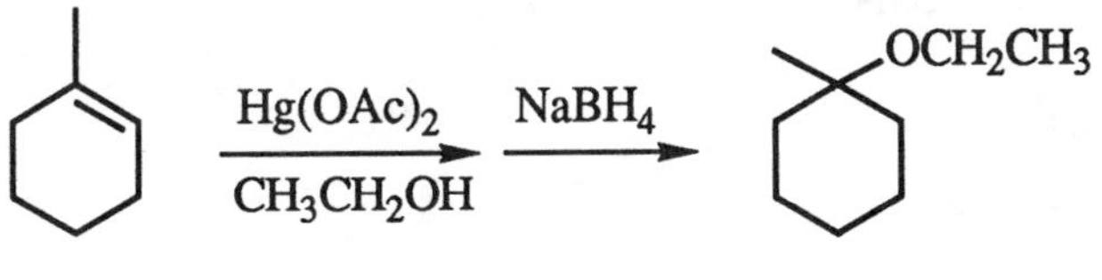
Hg(OAc)2
CH3CH2OH
NaBH4
OCH2CH3

(f)

MCPBA
CH2Cl2
O

13-36

In the first sequence, no bond is broken to the chiral center, so the configuration of the product is the same as the configuration of the starting material.

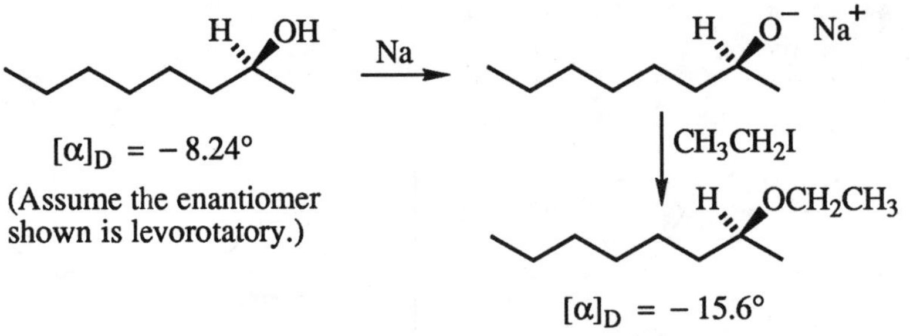
H OH
Na
H O⁻ Na⁺
[α]D = − 8.24°
(Assume the enantiomer shown is levorotatory.)
CH3CH2I
H OCH2CH3
[α]D = − 15.6°

In the second reaction sequence, however, bonds to the chiral carbon are broken twice, so the stereochemistry of each process must be considered.

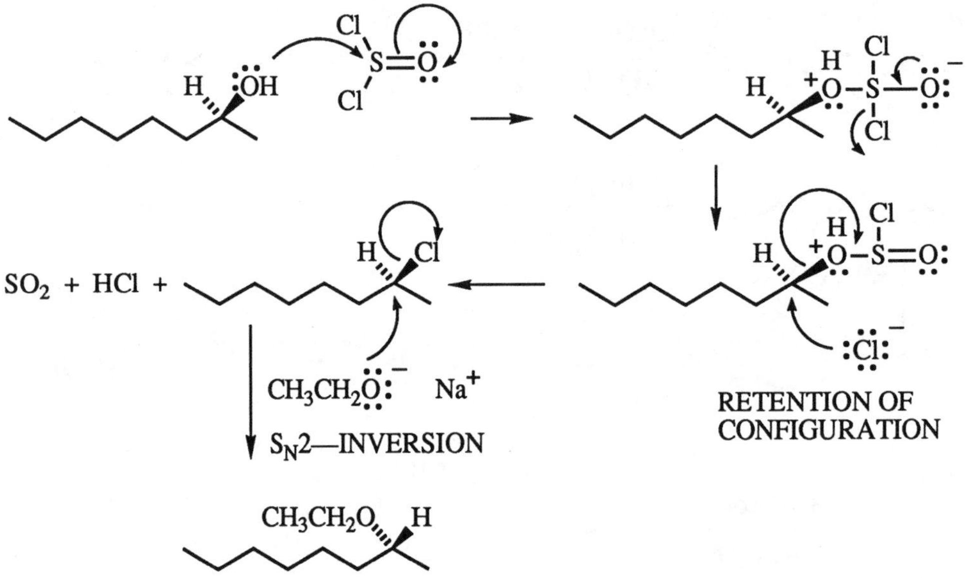
Cl
H OH
S=O
Cl
H Cl
H +O-S-O:
Cl
H Cl
H +O-S=O:
:Cl:
RETENTION OF CONFIGURATION
SO2 + HCl +
H Cl
CH3CH2O: Na+
SN2—INVERSION
CH3CH2O H

The second sequence involves *retention* followed by *inversion*, thereby producing the *enantiomer* of the 2-ethoxyoctane generated by the first sequence. The optical rotation of the final product will have equal magnitude but opposite sign, $[\alpha]_D = +15.6°$.

13-37

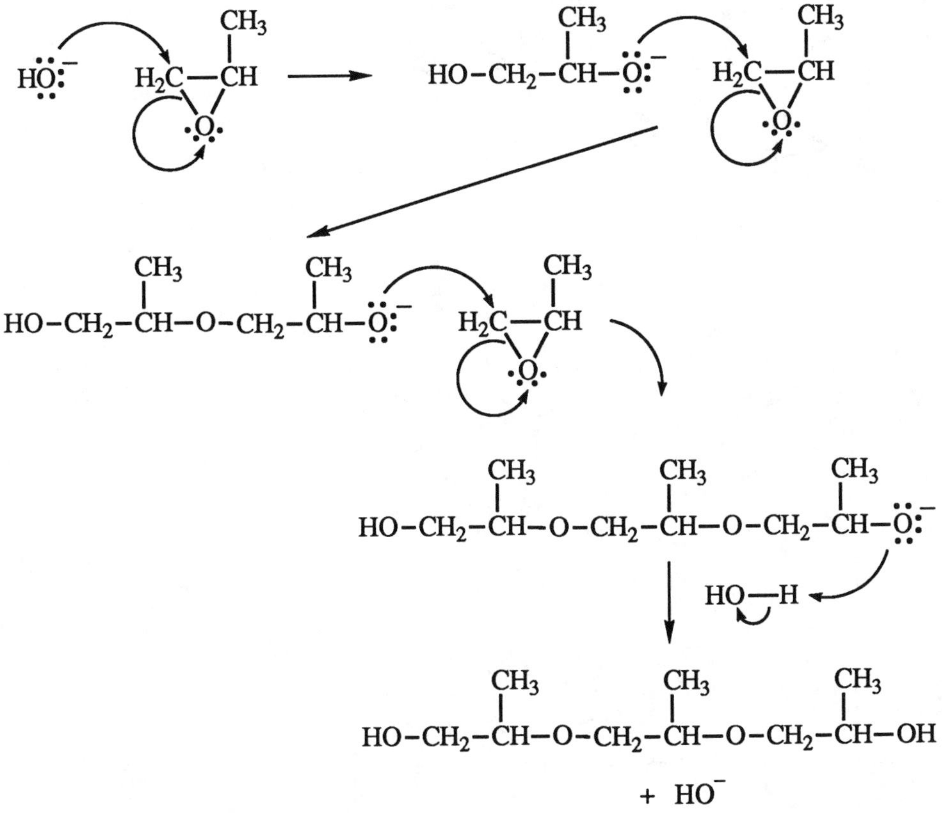

13-38

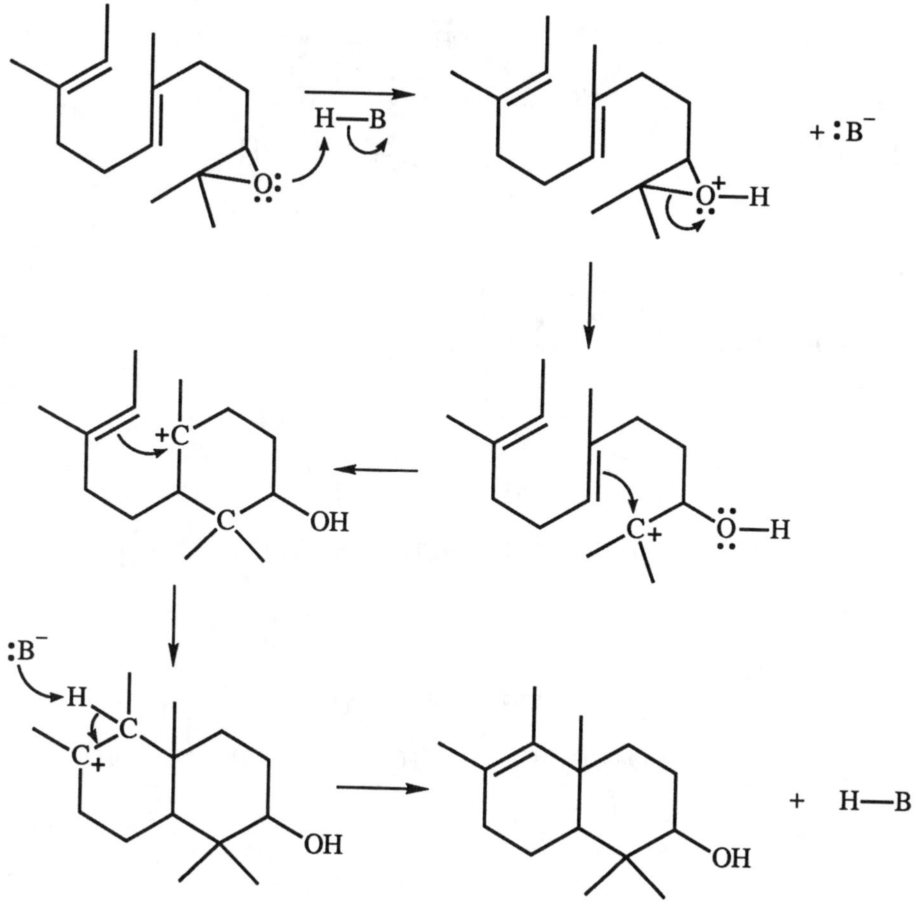

This process resembles the cyclization of squalene oxide to lanosterol. (See the solution to problem 13-19.) In fact, pharmaceutical synthesis of steroids uses the same type of reaction called a "biomimetic cyclization".

13-39

(a)

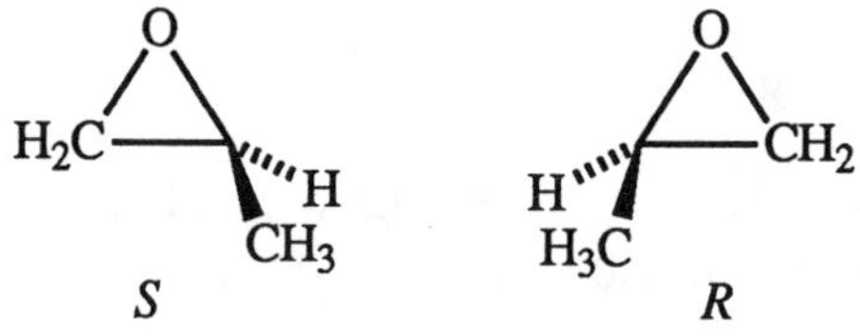

(b)

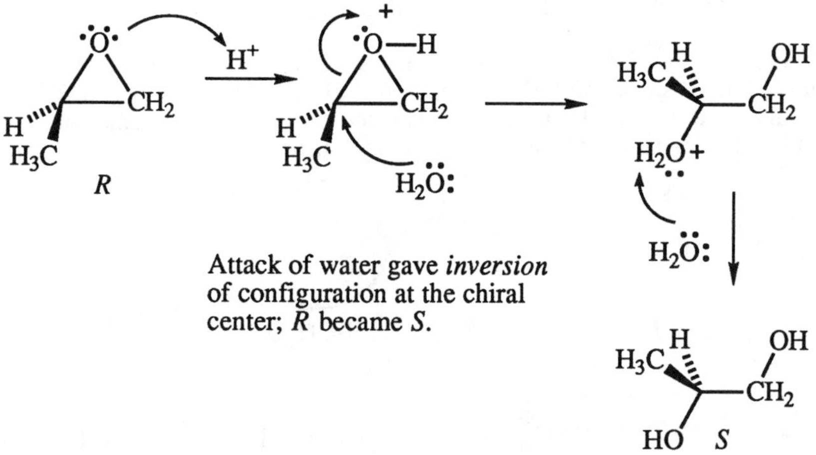

Attack of water gave *inversion* of configuration at the chiral center; *R* became *S*.

(c)

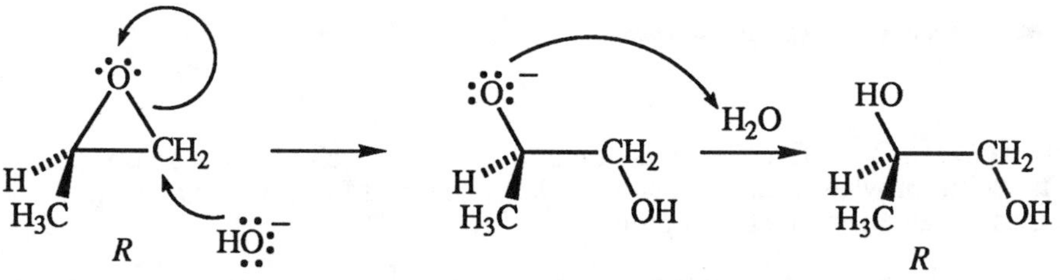

No bond to the chiral center was broken. Configuration is retained; *R* stays as *R*.

(d) The difference in these mechanisms lies in where the nucleophile attacks. Attack at the chiral carbon gives inversion; attack at the achiral carbon retains the configuration at the chiral carbon. These products are enantiomers and must necessarily have optical rotations of opposite sign.

13-40

methyl cellosolve $CH_3OCH_2CH_2OH$

To begin, what can be said about methyl cellosolve? Its molecular weight is 76; its IR would show C—O in the 1000-1200 cm^{-1} region and a strong O—H around 3300 cm^{-1}; and its NMR would show three sets of signals in the ratio of 3 : 4 : 1. (The two CH_2 groups are not identical but their environments are roughly the same, so that their NMR signals can be expected to be accidentally equivalent.)

The unknown has molecular weight 134; this is double the weight of methyl cellosolve, minus 18 (water). The IR shows no OH, only ether C—O. The NMR shows no OH, only H—C—O in the ratio of 3 : 4. Apparently, two molecules of methyl cellosolve have combined in an acid-catalyzed, bimolecular dehydration.

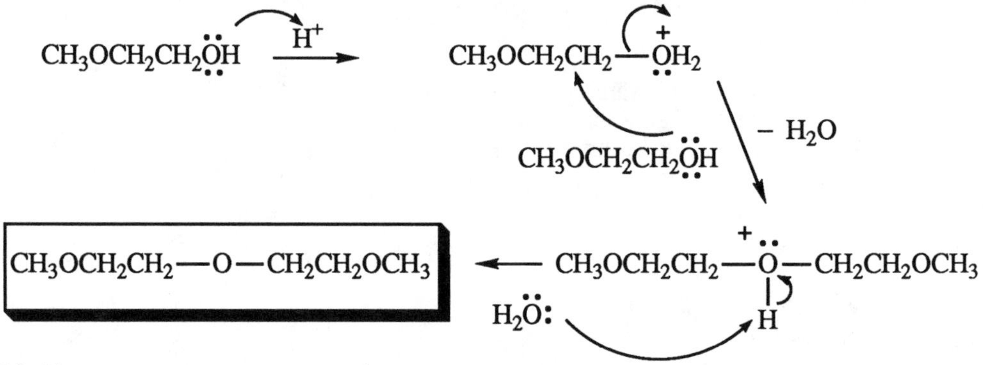

13-41

The formula $C_8H_{10}O$ has five elements of unsaturation (enough (4) for a benzene ring). The IR is useful for what is does *not* show. There is neither OH nor C=O, so the oxygen must be an ether functional group.

The NMR shows a 5H singlet at δ 7.2, a monosubstituted benzene. No peaks in the δ 4.5-6.0 range indicate the absence of an alkene, so the remaining element of unsaturation must be a ring. The three protons are non-equivalent, with complex splitting.

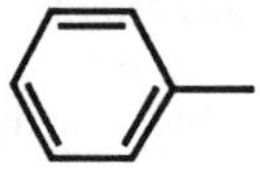 + 2 C + O + 3 H + ring

ether

These pieces can be assembled in only one manner consistent with the data.

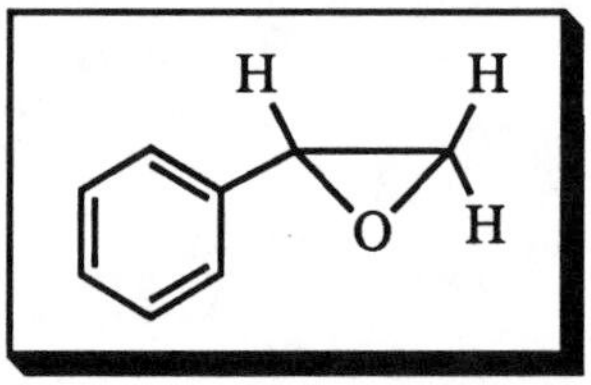

(Note that the CH_2 hydrogens are not equivalent (one is *cis* and one is *trans* to the phenyl) and therefore have distinct chemical shifts.)

CHAPTER 14—ALKYNES

14-1 Other structures are possible in each case.

(a) C_6H_{10} $CH_3CH_2C{\equiv}CCH_2CH_3$ $CH_3CH_2CH_2CH_2C{\equiv}CH$

(b) C_8H_{12} 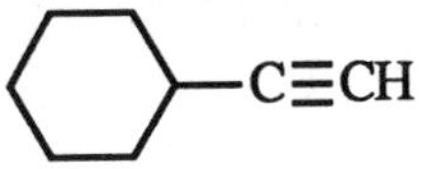 $HC{\equiv}C{-}CH_2CH{=}CHCH_2CH_2CH_3$

(c) C_7H_{10} 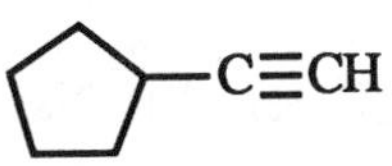 $CH_2{=}CH{-}C{\equiv}C{-}CH_2CH_2CH_3$

14-2 The asterisk (*) denotes acetylenic hydrogens of terminal alkynes.

(a) $CH_3CH_2{-}C{\equiv}C{-}\overset{*}{H}$ (1-butyne) $CH_3{-}C{\equiv}C{-}CH_3$ (2-butyne)

(b) $CH_3CH_2CH_2{-}C{\equiv}C{-}\overset{*}{H}$ (1-pentyne) $CH_3CH_2{-}C{\equiv}C{-}CH_3$ (2-pentyne) $CH_3\underset{}{\overset{CH_3}{\overset{|}{C}}}H{-}C{\equiv}C{-}\overset{*}{H}$ (3-methyl-1-butyne)

14-3 This reaction is exothermic ($\Delta H°$ = – 56 kcal/mole) as well as having an increase in entropy. Thermodynamically, at 1500° C, an increase is entropy will have a large effect on ΔG (remember $\Delta G = \Delta H - T\Delta S$?). Kinetically, almost any activation energy barrier will be overcome at 1500° C. Acetylene would likely decompose into its elements:

$$C_2H_2 \xrightarrow{1500°\ C} 2\ C + H_2$$

14-4 The formula C_6H_{10} has two elements of unsaturation, just enough for one alkyne. The presence of the alkyne is shown by the IR peaks around 2140 cm^{-1} for the C≡C and 3300 cm^{-1} for the ≡C—H. The NMR shows a 1H singlet at δ 2.0 (terminal alkyne C—H) and another singlet at δ 1.2 integrating to 9H, certainly a *t*-butyl group. The compound is *t*-butylacetylene, or 3,3-dimethyl-1-butyne.

$$CH_3{-}\underset{CH_3}{\underset{|}{\overset{CH_3}{\overset{|}{C}}}}{-}C{\equiv}C{-}H$$

14-5 Adding sodium amide to the mixture will produce the sodium salt of 1-hexyne, leaving 1-hexene untouched.

$$\left.\begin{array}{r}CH_2{=}CHCH_2CH_2CH_2CH_3\\ H{-}C{\equiv}C{-}CH_2CH_2CH_2CH_3\end{array}\right\} \xrightarrow{NaNH_2} \left\{\begin{array}{l}CH_2{=}CHCH_2CH_2CH_2CH_3\\ Na^+\ {}^-{:}C{\equiv}C{-}CH_2CH_2CH_2CH_3\ \text{(non-volatile salt)}\end{array}\right.$$

Distillation will remove the 1-hexene, leaving the non-volatile salt behind.

14-6 The key to this problem is to understand that *a proton donor will react only with the conjugate base of a* ***weaker*** *acid.*

(a) $H{-}C{\equiv}C{-}H \;+\; NaNH_2 \longrightarrow H{-}C{\equiv}C{:}^{-}\;Na^{+} \;+\; NH_3$

(b) $H{-}C{\equiv}C{-}H \;+\; CH_3Li \longrightarrow H{-}C{\equiv}C{:}^{-}\;Li^{+} \;+\; CH_4$

(c) no reaction: $NaOCH_3$ is not a strong enough base

(d) no reaction: NaOH is not a strong enough base

(e) $H{-}C{\equiv}C{:}^{-}\;Na^{+} \;+\; CH_3OH \longrightarrow H{-}C{\equiv}C{-}H \;+\; NaOCH_3$
(opposite of (c))

(f) $H{-}C{\equiv}C{:}^{-}\;Na^{+} \;+\; H_2O \longrightarrow H{-}C{\equiv}C{-}H \;+\; NaOH$
(opposite of (d))

(g) no reaction : $H{-}C{\equiv}C{:}^{-}\;Na^{+}$ is not a strong enough base

(h) no reaction: $NaNH_2$ is not a strong enough base

(i) $CH_3OH \;+\; NaNH_2 \longrightarrow NaOCH_3 \;+\; NH_3$

14-7

$$H{-}C{\equiv}C{-}H \xrightarrow{NaNH_2} H{-}C{\equiv}C{:}^{-}\;Na^{+} \xrightarrow{CH_3CH_2Br} H{-}C{\equiv}C{-}CH_2CH_3$$

$$H{-}C{\equiv}C{-}CH_2CH_3 \xrightarrow{NaNH_2} Na^{+}\;{}^{-}{:}C{\equiv}C{-}CH_2CH_3 \xrightarrow{CH_3(CH_2)_5Br} CH_3(CH_2)_5{-}C{\equiv}C{-}CH_2CH_3$$

14-8

(a) $H-C\equiv C-H \xrightarrow[\text{2. } CH_3CH_2CH_2CH_2Br]{\text{1. } NaNH_2} H-C\equiv C-CH_2CH_2CH_2CH_3$

(b) $H-C\equiv C-H \xrightarrow[\text{2. } CH_3CH_2CH_2Br]{\text{1. } NaNH_2} H-C\equiv C-CH_2CH_2CH_3 \xrightarrow[\text{2. } CH_3I]{\text{1. } NaNH_2} CH_3-C\equiv C-CH_2CH_2CH_3$

(c) $H-C\equiv C-H \xrightarrow[\text{2. } CH_3CH_2Br]{\text{1. } NaNH_2} H-C\equiv C-CH_2CH_3 \xrightarrow[\text{2. } CH_3CH_2Br]{\text{1. } NaNH_2} CH_3CH_2-C\equiv C-CH_2CH_3$

(d) cannot be synthesized by an S_N2 reaction—would require attack on a 2° alkyl halide

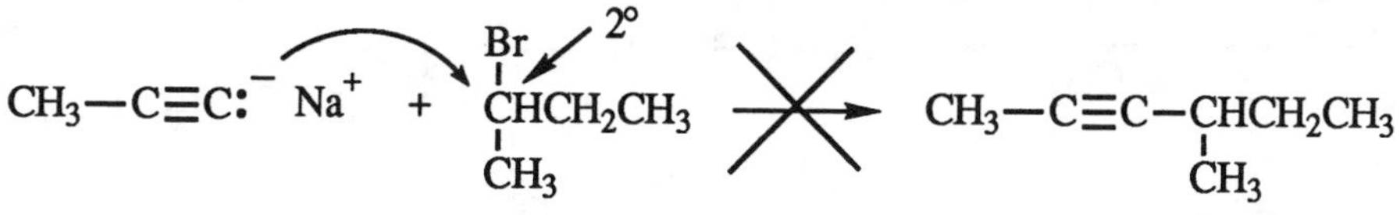

not produced

(e) $H-C\equiv C-H \xrightarrow[\text{2. } CH_3I]{\text{1. } NaNH_2} CH_3-C\equiv C-H \xrightarrow[\text{2. } BrCH_2CH(CH_3)CH_3]{\text{1. } NaNH_2} CH_3-C\equiv C-CH_2CH(CH_3)CH_3$

(f)

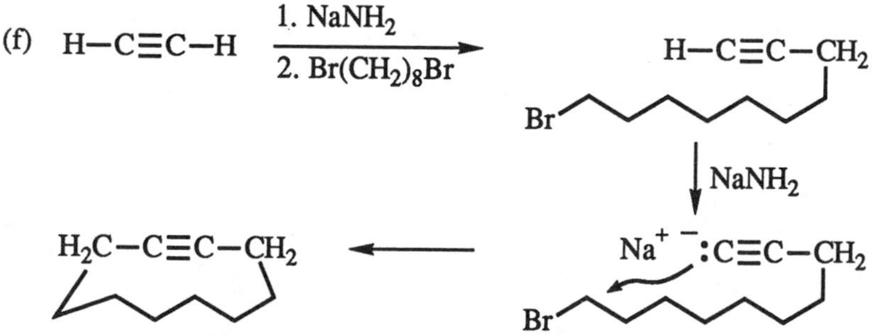

Intramolecular cyclization of large rings must be carried out in dilute solution so the last S_N2 displacement will be *intra*molecular and not *inter*molecular.

14-9

(a) 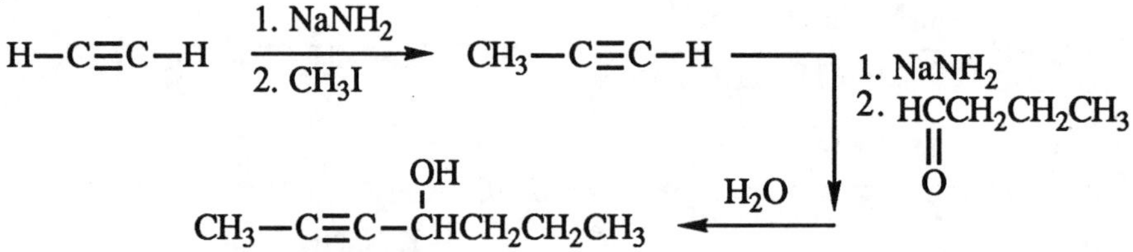
H−C≡C−H
1. NaNH2
2. CH3I
CH3−C≡C−H
1. NaNH2
2. HCCH2CH2CH3
O
H2O
OH
CH3−C≡C−CHCH2CH2CH3

(b) 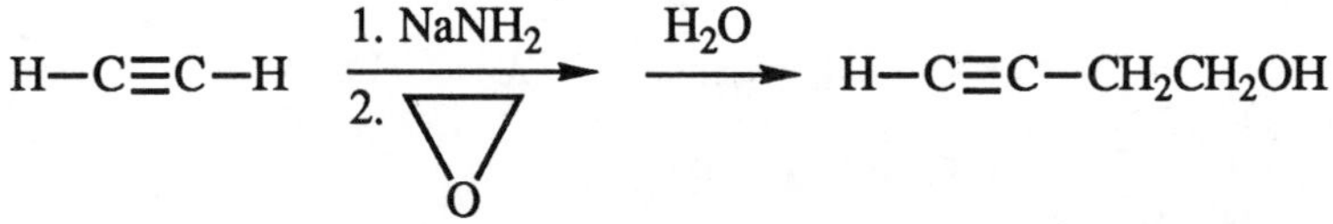
H−C≡C−H
1. NaNH2
2.
O
H2O
H−C≡C−CH2CH2OH

(c) 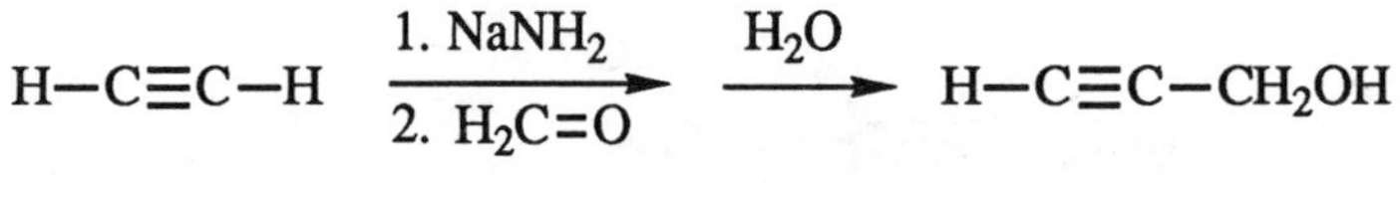
H−C≡C−H
1. NaNH2
2. H2C=O
H2O
H−C≡C−CH2OH

(d) 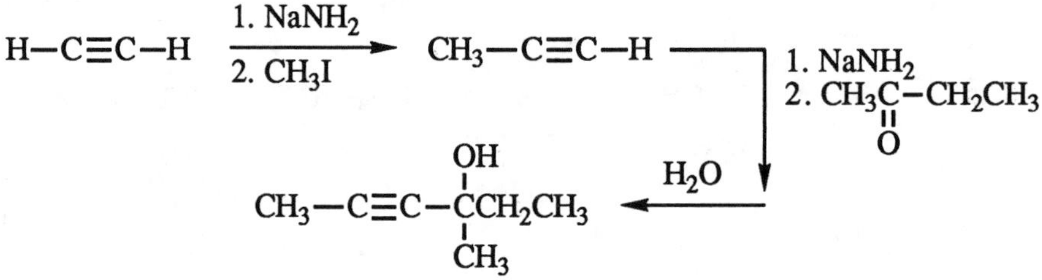
H−C≡C−H
1. NaNH2
2. CH3I
CH3−C≡C−H
1. NaNH2
2. CH3C−CH2CH3
O
H2O
OH
CH3−C≡C−CCH2CH3
CH3

14-10

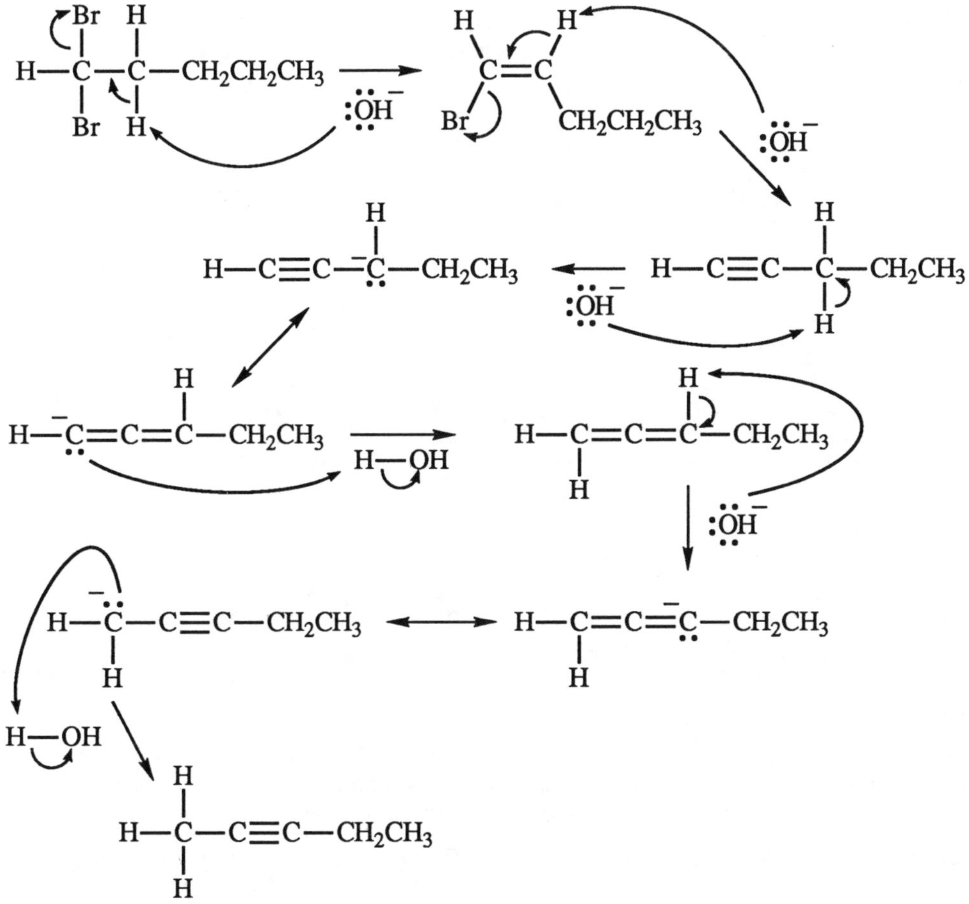

14-11 To determine the equilibrium constant in the reaction:

terminal alkyne $\rightleftharpoons$ internal alkyne $\Delta G = -4.0$ kcal/mole (− 17.0 kJ/mole)

$$\Delta G = -RT \ln K_{eq}$$

$$K_{eq} = e^{\left(-\frac{\Delta G}{RT}\right)} = e^{\left(\frac{4.0}{(0.00198)(473)}\right)} = e^{4.27} = 72$$

$$\frac{[\text{internal}]}{[\text{terminal}]} = \frac{72}{1} = \frac{98.6\%\ \text{internal}}{1.4\%\ \text{terminal}}$$

14-12

(a)

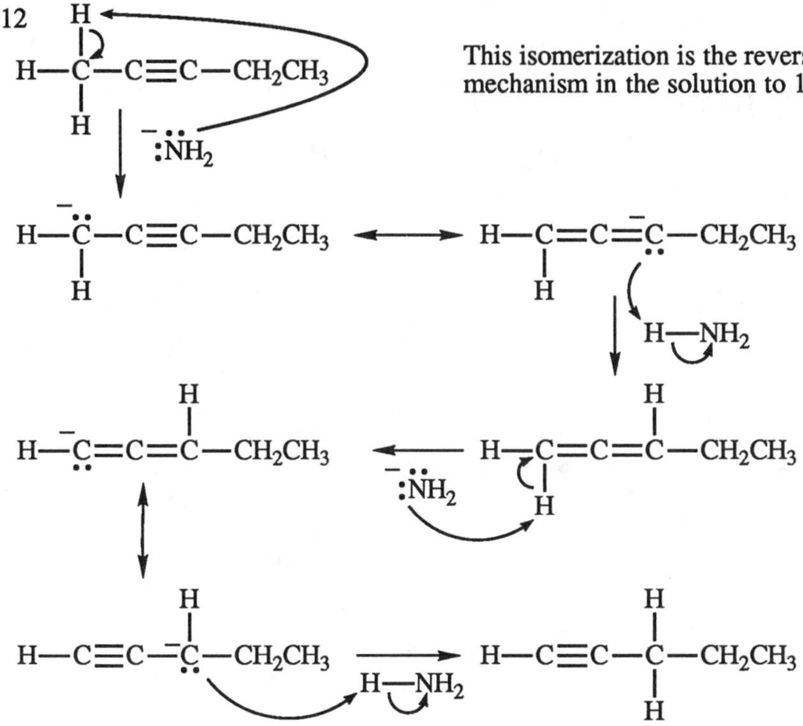

This isomerization is the reverse of the mechanism in the solution to 14-10.

(b) All steps in part (a) are reversible. With a weaker base like KOH, an equilibrium mixture of 1-pentyne and 2-pentyne would result. With the strong base $NaNH_2$, however, the final terminal alkyne is deprotonated to give the acetylide ion:

$$H{-}C{\equiv}C{-}CH_2CH_2CH_3 \; + \; NaNH_2 \longrightarrow Na^{+} \; {}^{-}{:}C{\equiv}C{-}CH_2CH_2CH_3 \; + \; NH_3$$

Because 1-pentyne is about 10 pK units more acidic than ammonia, this deprotonation is *not reversible*. The acetylide ion is produced and can't go back. Le Châtelier's Principle tells us that the reaction will try to replace the 1-pentyne that is being removed from the reaction mixture, so eventually all of the 2-pentyne will be drawn into the 1-pentyne anion "sink".

(c) Using the weaker base KOH at 200°C will restore the equilibrium between the two alkyne isomers with 2-pentyne predominating.

14-13

(a)

(1) $CH_3-CHBr-CHBr-CH_3 \xrightarrow[\Delta]{KOH} CH_3-C\equiv C-CH_3$

$\uparrow$ Br_2, CCl_4

(2) $CH_3CH=CHCH_3$

(b)

(1) $H-CHBr-CHBr-(CH_2)_5CH_3 \xrightarrow[150°]{NaNH_2} H-C\equiv C-(CH_2)_5CH_3$

$\uparrow$ Br_2, CCl_4

(2) $CH_2=CH(CH_2)_5CH_3$

(c)

(1) $CH_3-CHBr-CHBr-(CH_2)_4CH_3 \xrightarrow[\Delta]{KOH} CH_3-C\equiv C-(CH_2)_4CH_3$

$\uparrow$ Br_2, CCl_4

(2) $CH_3CH=CH(CH_2)_4CH_3$

(c)

(1)

(2)

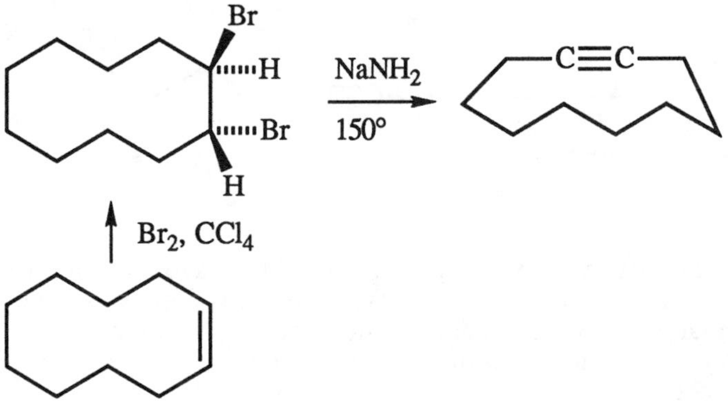

14-14

(a) 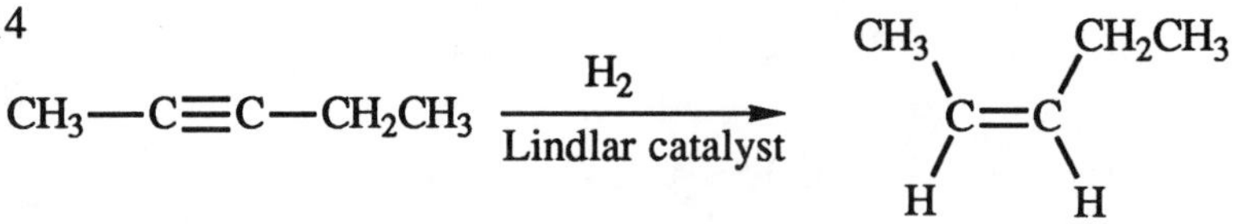
CH3—C≡C—CH2CH3
H2
Lindlar catalyst
CH3
CH2CH3
C=C
H
H

(b) 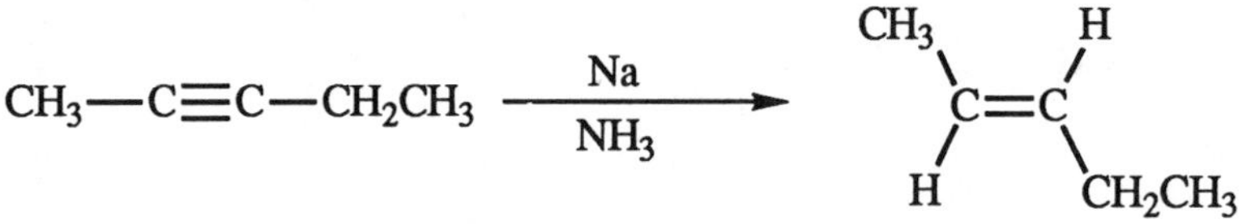
CH3—C≡C—CH2CH3
Na
NH3
CH3
H
C=C
H
CH2CH3

(c)

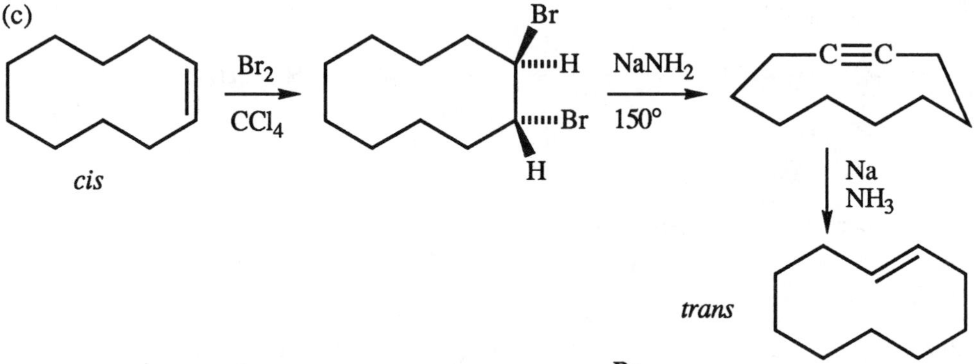
cis
Br2
CCl4
Br
H
Br
H
NaNH2
150°
C≡C
Na
NH3
trans

(d)

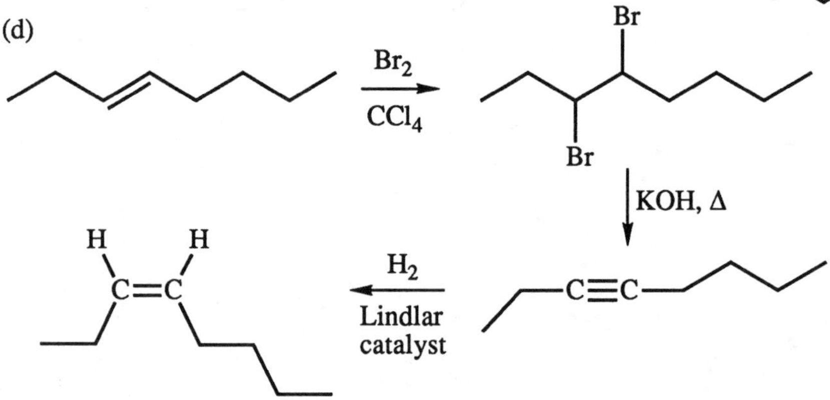
Br2
CCl4
Br
Br
KOH, Δ
C≡C
H2
Lindlar
catalyst
H
H
C=C

14-15 The goal is to add only one equivalent of bromine, always avoiding an excess of bromine. If the alkyne is added to the bromine, the first drops of alkyne will encounter a large excess of bromine. Instead, adding bromine to the alkyne will always ensure an excess of alkyne and should give a good yield of dibromo product.

14-16

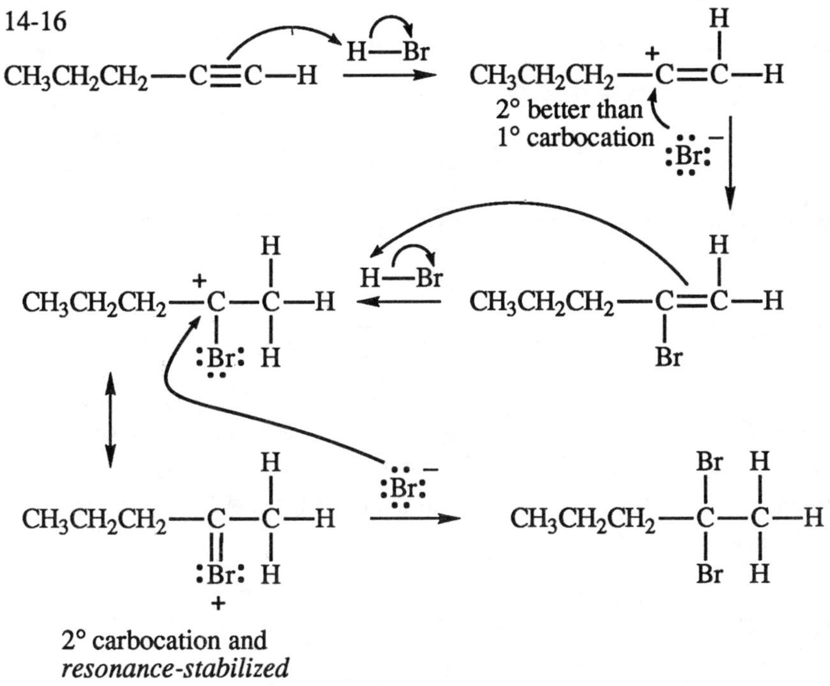

2° better than
1° carbocation
2° carbocation and
resonance-stabilized

14-17

(a)

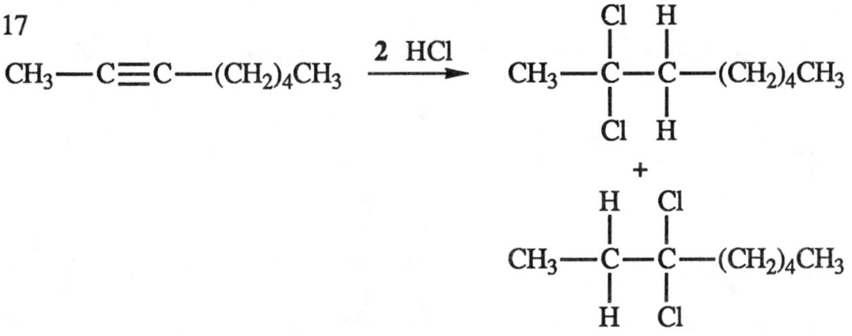

2 HCl

(b) Addition occurs to make the carbocation intermediate at the carbon with the halogen because of *resonance stabilization.*

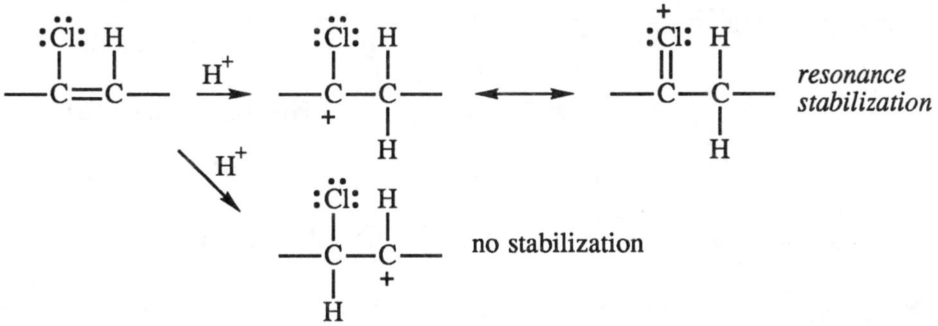

resonance
stabilization
no stabilization

14-18

<u>initiation</u>:

$$RO{-}OR \xrightarrow{h\nu} 2\ RO\bullet$$

$$RO\bullet + H{-}Br \longrightarrow RO{-}H + Br\bullet$$

<u>propagation</u>:

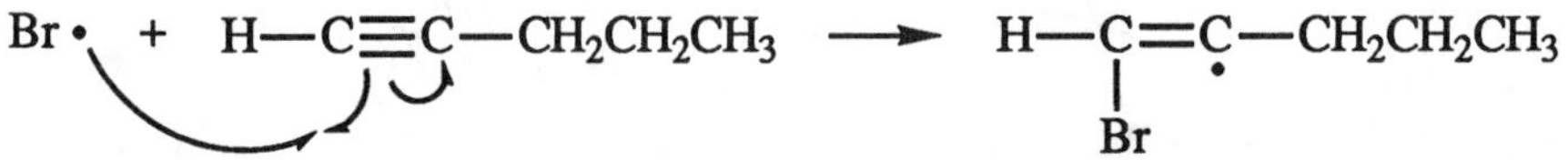

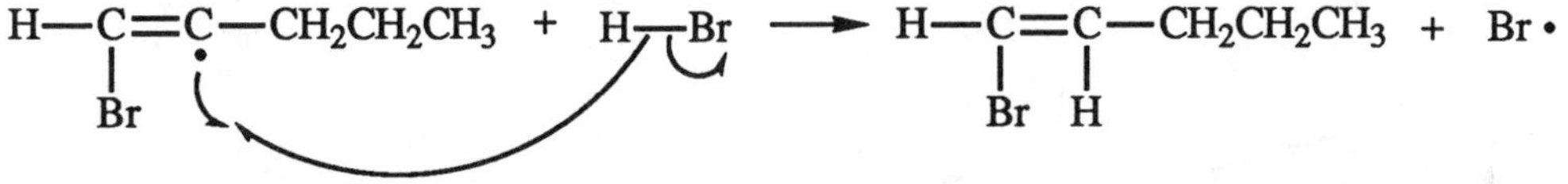

The 2° radical is more stable than 1°. The anti-Markovnikov orientation occurs because the bromine radical attacks first, rather than the H^+ attacking first as in electrophilic (ionic) addition.

14-19

(a) $H{-}C{\equiv}C{-}(CH_2)_3CH_3 \xrightarrow{\text{1 eq. } Cl_2} H{-}C(Cl){=}C(Cl){-}(CH_2)_3CH_3$ *E* + *Z*

(b) $H{-}C{\equiv}C{-}(CH_2)_3CH_3 \xrightarrow[\text{ROOR}]{\text{HBr}} H{-}C(Br){=}C(H){-}(CH_2)_3CH_3$ *E* + *Z*

(c) $H{-}C{\equiv}C{-}(CH_2)_3CH_3 \xrightarrow{\text{HBr}} H{-}C(H){=}C(Br){-}(CH_2)_3CH_3$

(d) $H{-}C{\equiv}C{-}(CH_2)_3CH_3 \xrightarrow[CCl_4]{2\ Br_2} H{-}CBr_2{-}CBr_2{-}(CH_2)_3CH_3$

(e) $H{-}C{\equiv}C{-}(CH_2)_3CH_3 \xrightarrow[\text{Lindlar catalyst (or Na, } NH_3)]{H_2} H{-}C(H){=}C(H){-}(CH_2)_3CH_3 \xrightarrow{\text{HBr}} CH_3{-}CH(Br){-}(CH_2)_3CH_3$

(f) $H{-}C{\equiv}C{-}(CH_2)_3CH_3 \xrightarrow{\text{2 HBr}} H{-}CH_2{-}CBr_2{-}(CH_2)_3CH_3$

14-20

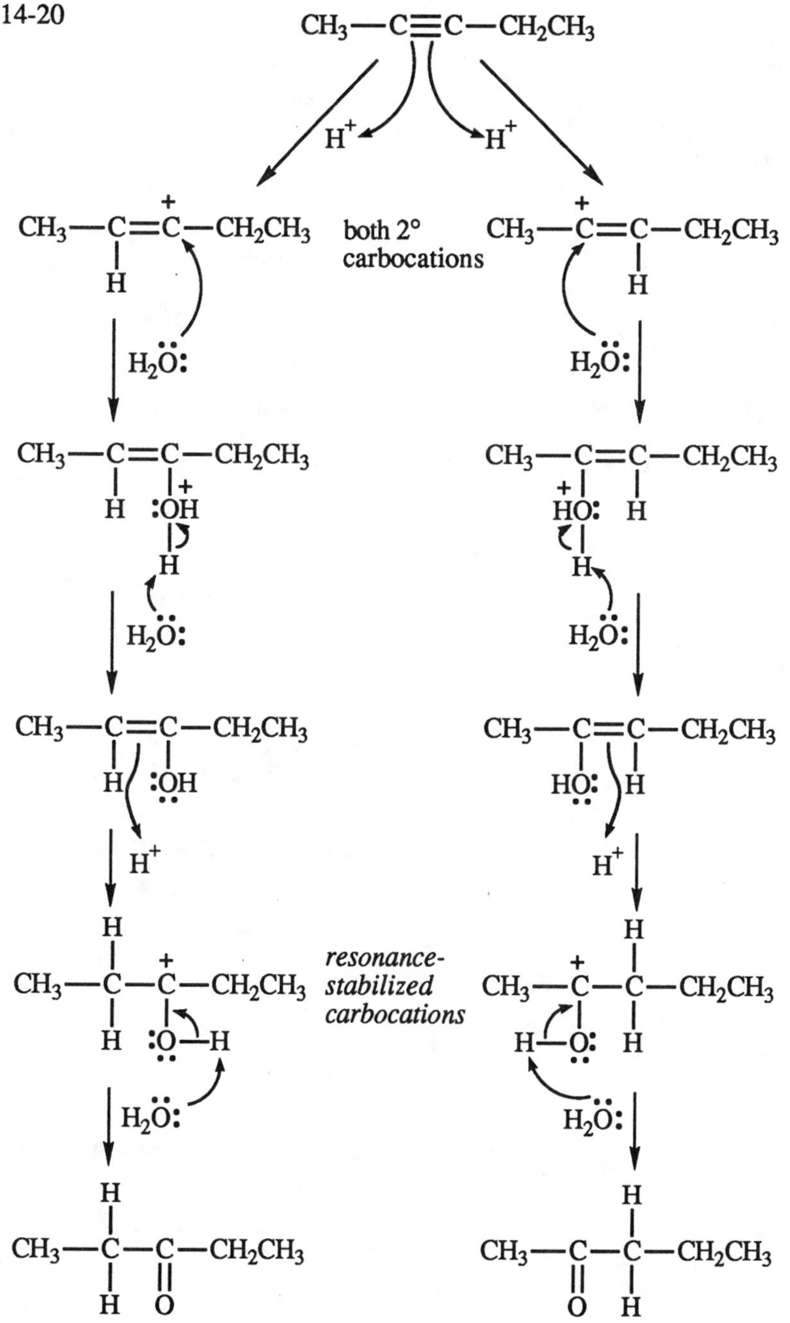
both 2° carbocations
resonance-stabilized carbocations

3-pentanone 2-pentanone

The role of the mercury catalyst is not shown in this mechanism. As a Lewis acid, it may act like the proton in the first step, helping to form vinyl cations.

14-21

(a) 2-Butyne is symmetric. Either orientation produces the same product.

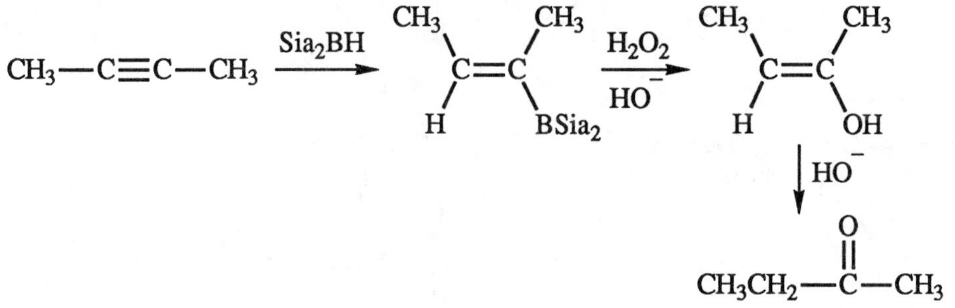

(b) 2-Pentyne is not symmetric. Different orientations of attack will lead to different products.

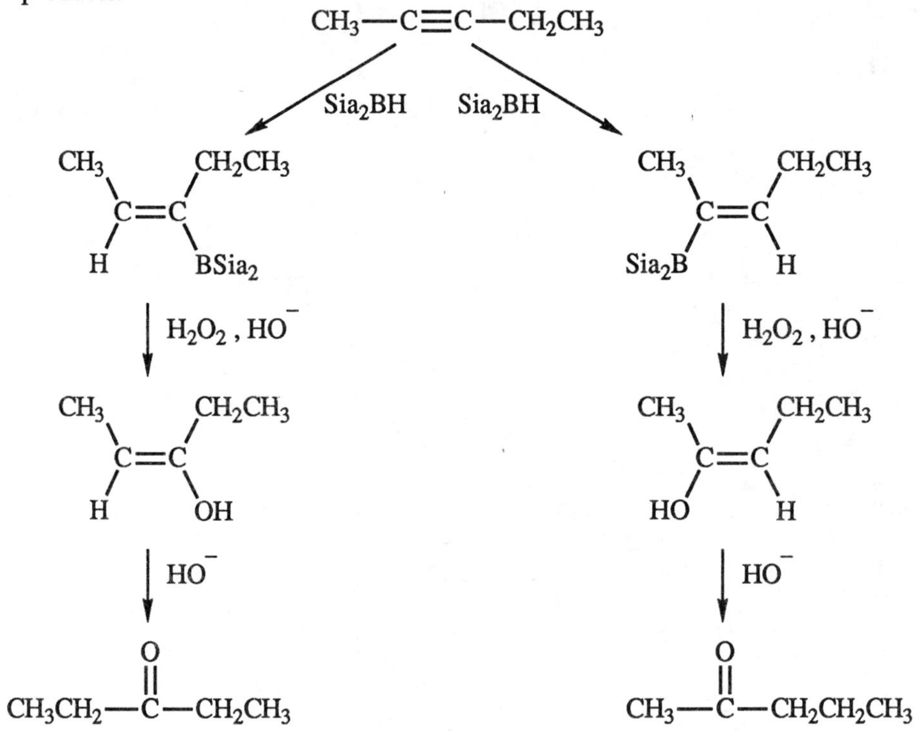

14-22

(a) (1) $CH_3\overset{O}{\overset{\parallel}{C}}CH_2CH_2CH_2CH_3$ (2) $H\overset{O}{\overset{\parallel}{C}}CH_2CH_2CH_2CH_2CH_3$

(b) (1) $CH_3\overset{O}{\overset{\parallel}{C}}CH_2CH_2CH_2CH_3$ + $CH_3CH_2\overset{O}{\overset{\parallel}{C}}CH_2CH_2CH_3$

(2) same mixture as in (b) (1)

(c) (1) $CH_3CH_2\overset{O}{\overset{\parallel}{C}}CH_2CH_2CH_3$ (2) $CH_3CH_2\overset{O}{\overset{\parallel}{C}}CH_2CH_2CH_3$

(d) (1) 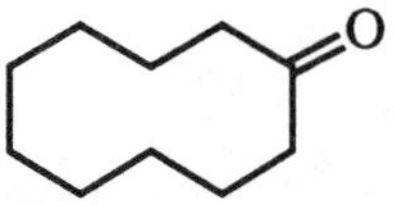

(2) 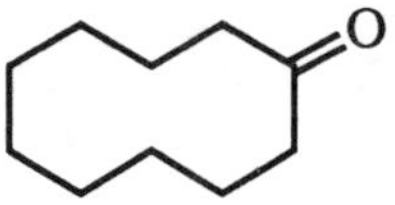

14-23

(a) 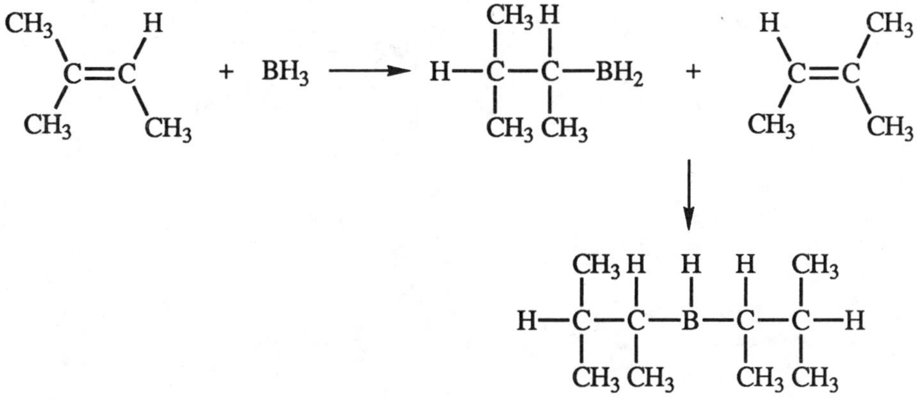

disiamylborane, Sia_2BH

(b) There is too much steric hindrance in Sia_2BH for the third B—H to add across another alkene. The reagent can add to alkynes because alkynes are linear and attack is not hindered by bulky substituents.

14-24

(a) (1) $H-C(=O)-C(=O)-(CH_2)_3CH_3$ (2) $CO_2 + HO-C(=O)-(CH_2)_3CH_3$

(b) (1) $CH_3-C(=O)-C(=O)-CH_2CH_2CH_3$ (2) $CH_3-C(=O)-OH + HO-C(=O)-CH_2CH_2CH_3$

(c) (1) $CH_3CH_2-C(=O)-C(=O)-CH_2CH_3$ (2) $CH_3CH_2-C(=O)-OH$

(d) (1) $CH_3CH(CH_3)-C(=O)-C(=O)-CH_2CH_3$ (2) $CH_3CH(CH_3)-C(=O)-OH + HO-C(=O)-CH_2CH_3$

(e) (1)

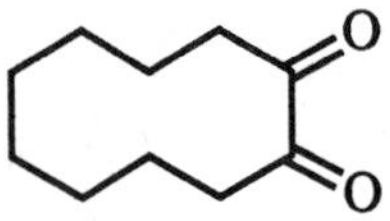

(2)

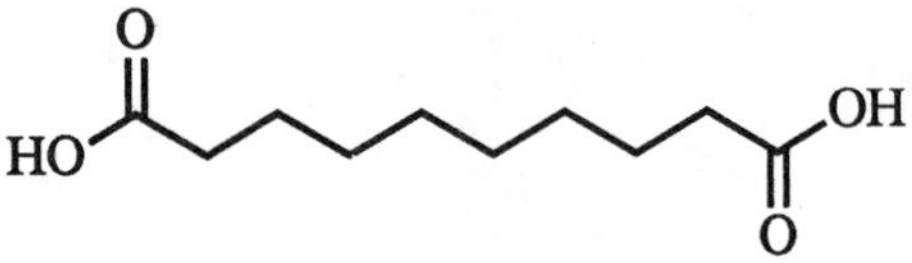

14-25

(a) $CH_3-C\equiv C-(CH_2)_4-C\equiv C-CH_3$ (b)

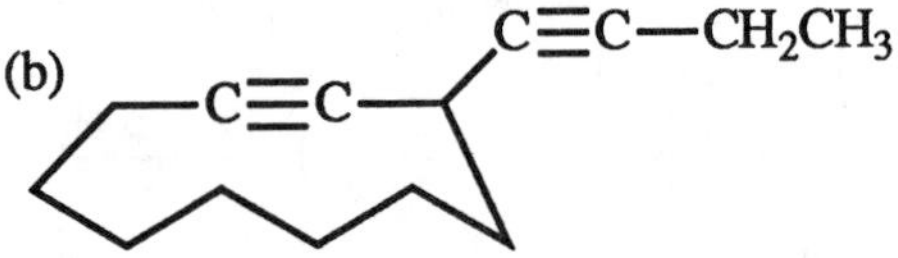

14-26

(a)

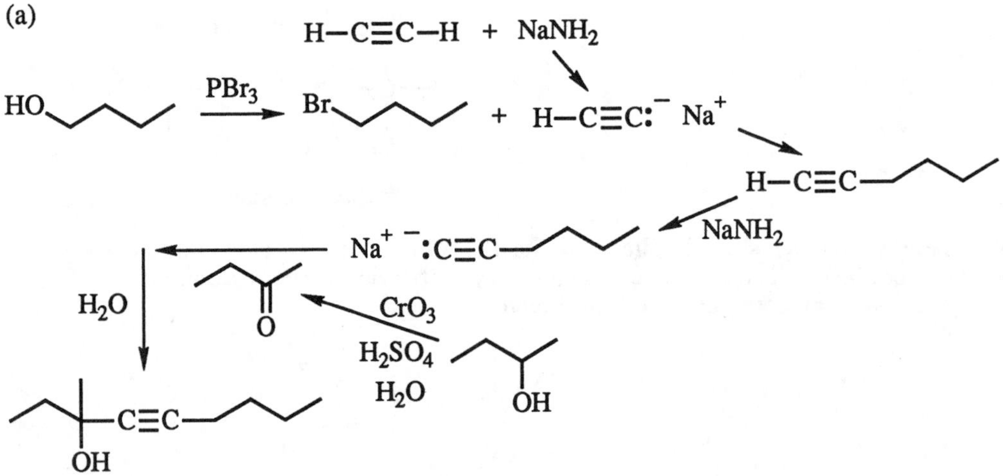

14-26 continued

(b)

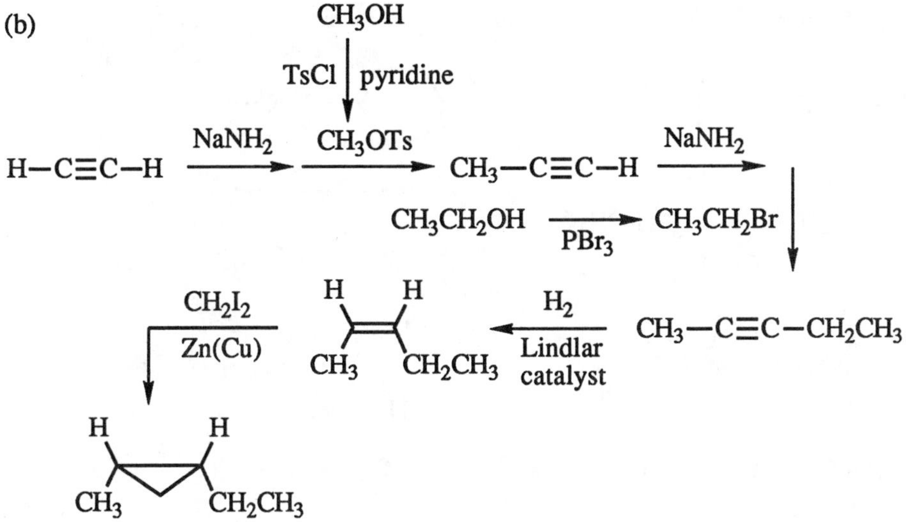

(c)

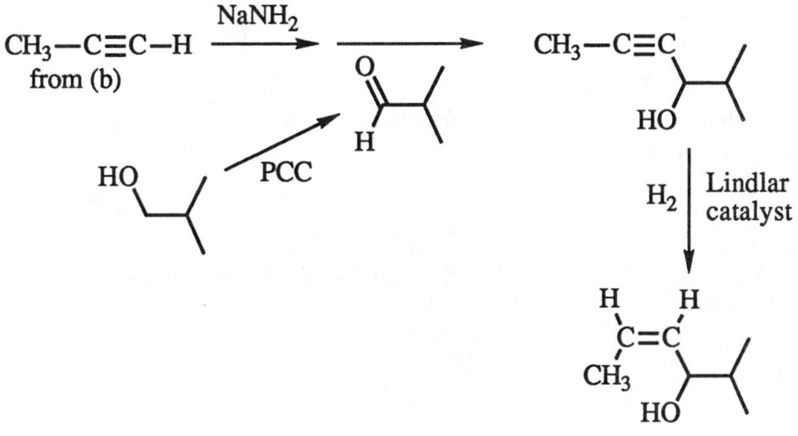

14-27 Refer to the Glossary and the text for definitions and examples.

14-28

(a) $CH_3CH_2-C\equiv C-(CH_2)_4CH_3$

(b) $CH_3-C\equiv C-(CH_2)_4CH_3$

(c) 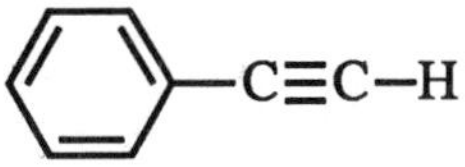

(d) 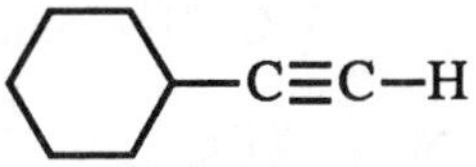

(e) $CH_3CH_2-C\equiv C-CH(CH_3)CH_2CH_2CH_3$

(f) 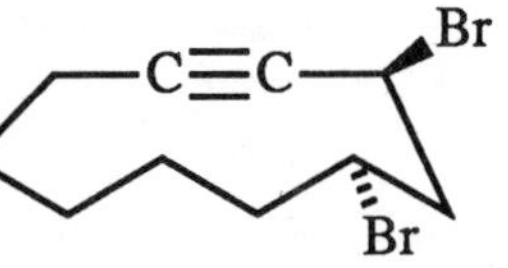

(g) $CH_3CH(OH)-C\equiv C-(CH_2)_3CH_3$

(h) 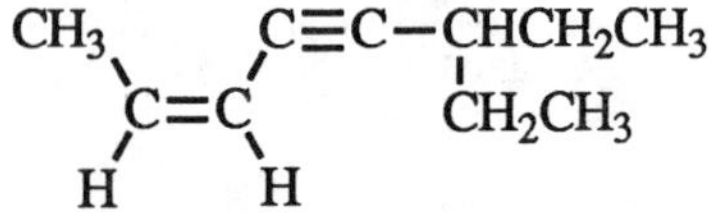

(i) $H-C\equiv C-CH_2-C\equiv C-CH_2CH_3$

(j) $H-C\equiv C-CH=CH_2$

(k) 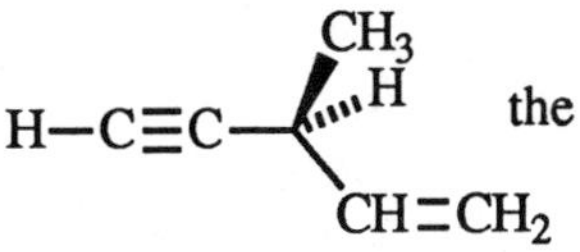 the correct IUPAC name is (*S*)-3-methyl-1-penten-4-yne

14-29

(a) ethylmethylacetylene

(b) phenylacetylene

(c) *sec*-butyl-*n*-propylacetylene

(d) *sec*-butyl-*t*-butylacetylene

14-30

(a) 4-phenyl-2-pentyne

(b) 4,4-dibromo-2-pentyne

(c) 2,6,6-trimethyl-3-heptyne

(d) (*E*)-3-methyl-2-hepten-4-yne

(e) 3-methyl-4-hexyn-3-ol

(f) 1-cycloheptylpropyne

14-31

(a) terminal alkynes

$H-C\equiv C-(CH_2)_3CH_3$
1-hexyne

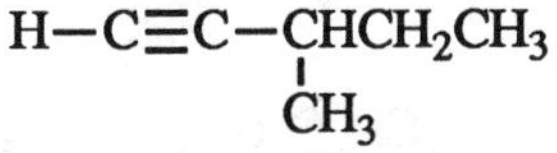
3-methyl-1-pentyne

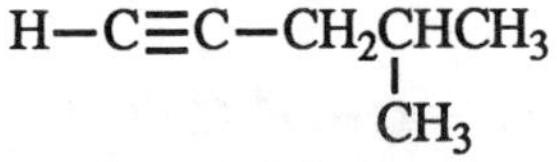
4-methyl-1-pentyne

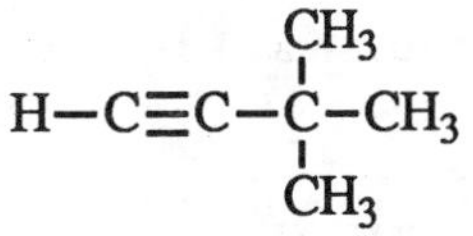
3,3-dimethyl-1-butyne

internal alkynes

$CH_3-C\equiv C-CH_2CH_2CH_3$
2-hexyne

$CH_3CH_2-C\equiv C-CH_2CH_3$
3-hexyne

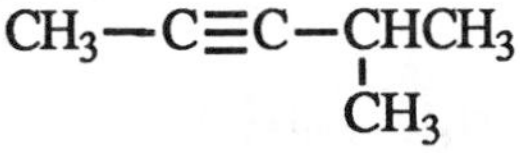
4-methyl-2-pentyne

(b) All four terminal alkynes will form precipitates with cuprous ion.

14-32 In this synthesis, bromides could be used instead of tosylates.

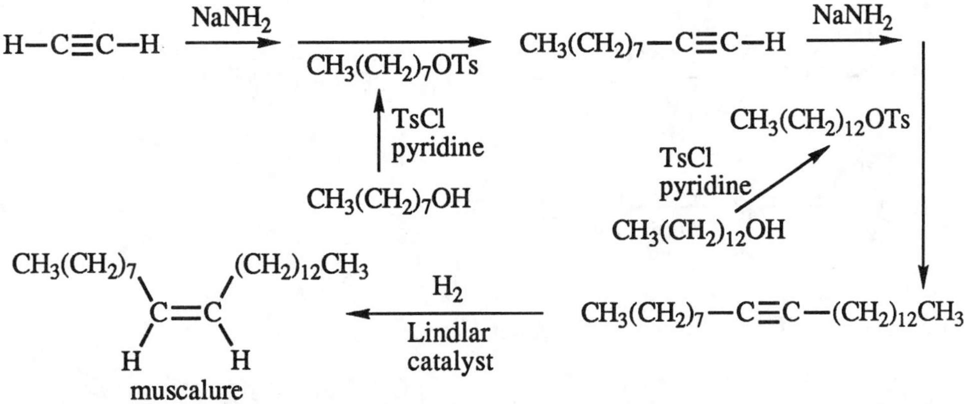

H—C≡C—H
NaNH2
CH3(CH2)7OTs
TsCl
pyridine
CH3(CH2)7OH
CH3(CH2)7—C≡C—H
NaNH2
CH3(CH2)12OTs
TsCl
pyridine
CH3(CH2)12OH
CH3(CH2)7—C≡C—(CH2)12CH3
H2
Lindlar
catalyst
CH3(CH2)7
(CH2)12CH3
C=C
H
H
muscalure

14-33 Terminal alkynes form precipitates with silver ion or cuprous ion. Filtration will remove the 1-decyne precipitate from 2-decyne.

$$H-C\equiv C-(CH_2)_7CH_3 \; + \; Cu^+ \xrightarrow{NH_3} Cu-C\equiv C-(CH_2)_7CH_3$$

precipitate

14-34

(a) 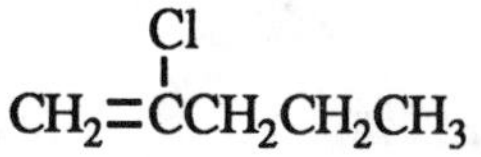

(b) 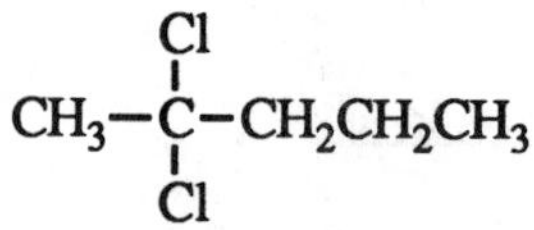

(c) $CH_3CH_2CH_2CH_2CH_3$

(d) $CH_2=CHCH_2CH_2CH_3$

(e) 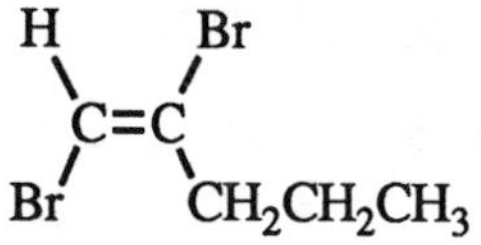

(f) 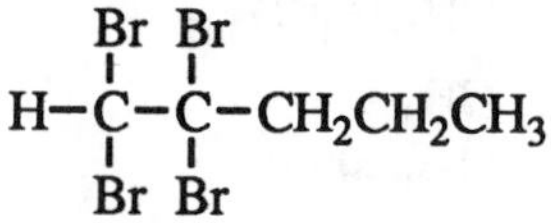

(g) 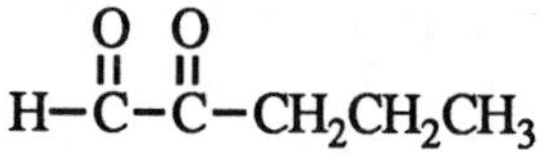

(h) 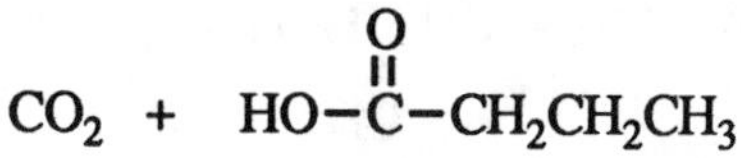

(i) $CH_2=CHCH_2CH_2CH_3$

(j) $Na^+ \ ^{-}\!:C\equiv C-CH_2CH_2CH_3$

(k) $Ag-C\equiv C-CH_2CH_2CH_3$

(l) 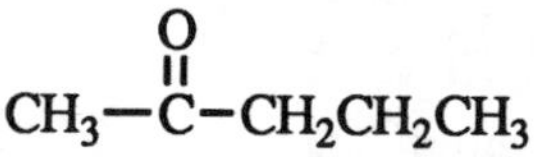

(m) 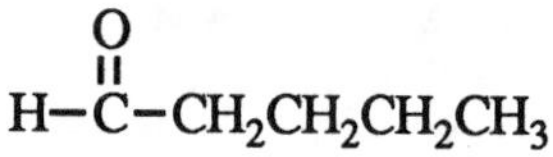

14-35

(a) 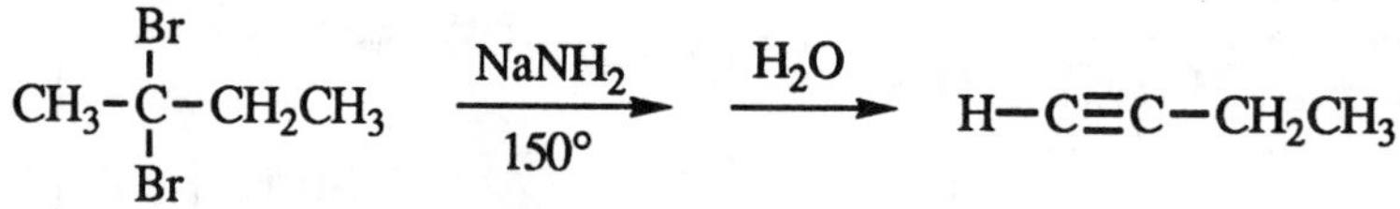

(b) 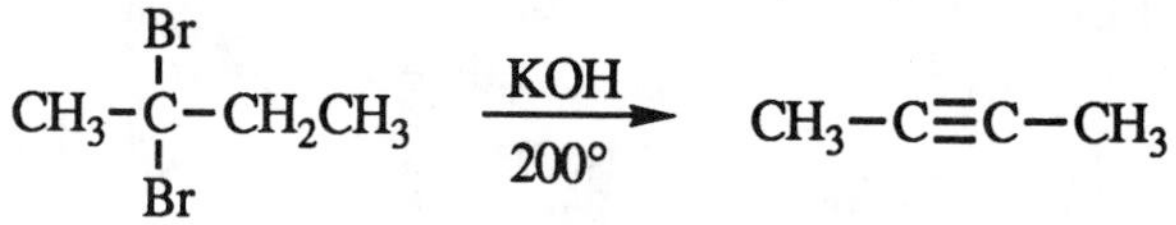

(c) 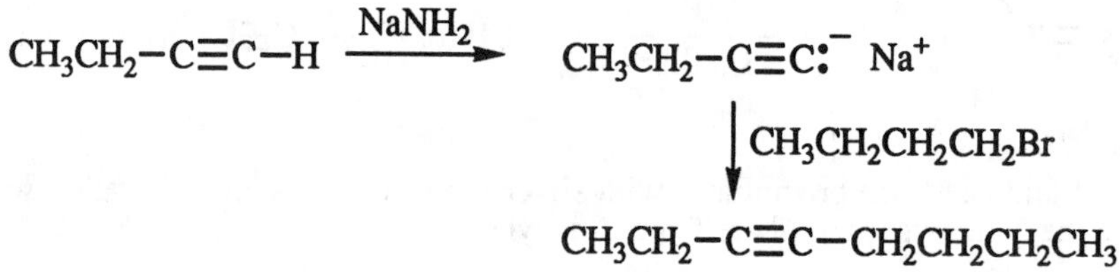

14-35 continued

(d) 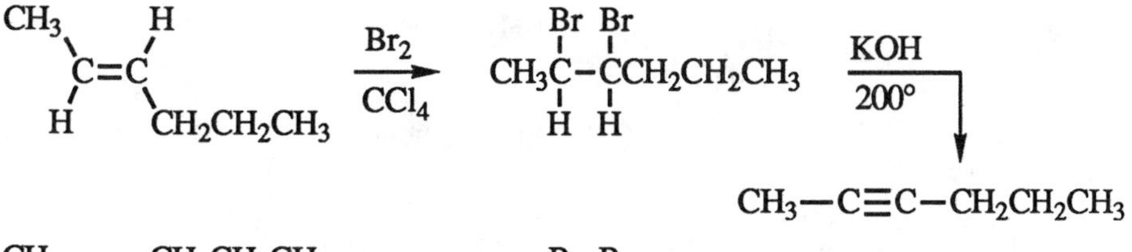

(e) 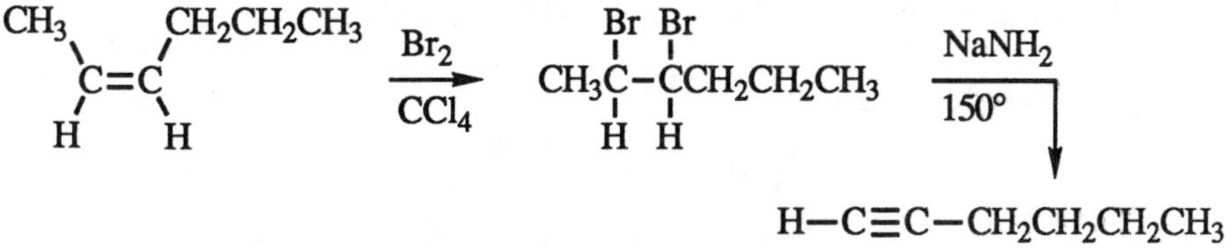

(f) 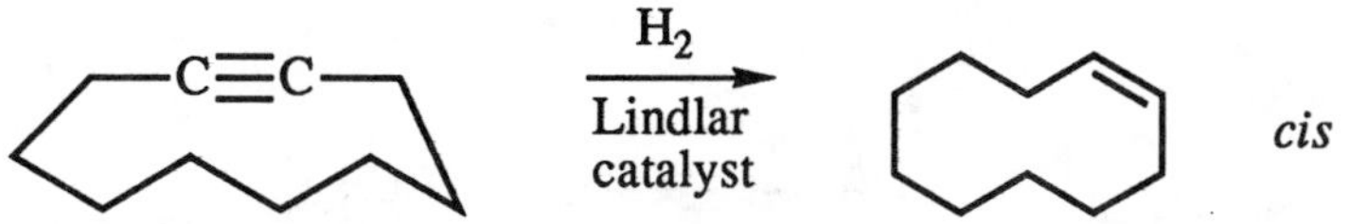

(g) 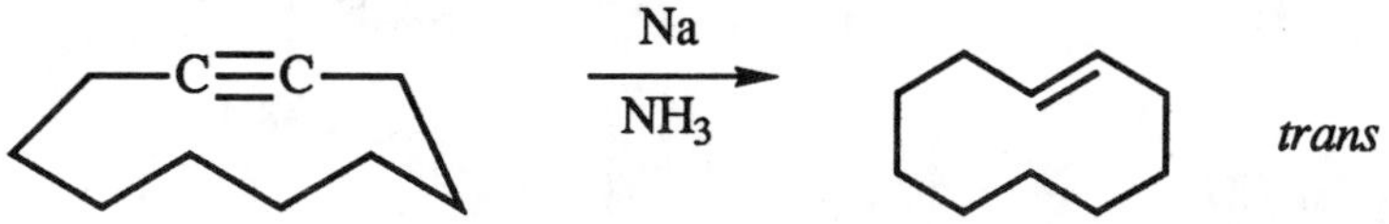

(h) 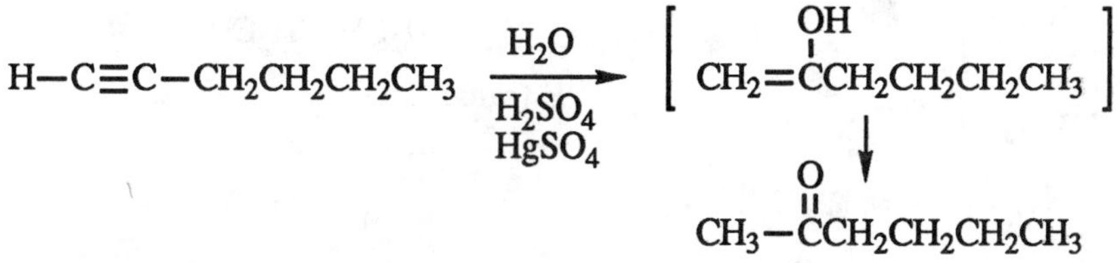

(i) 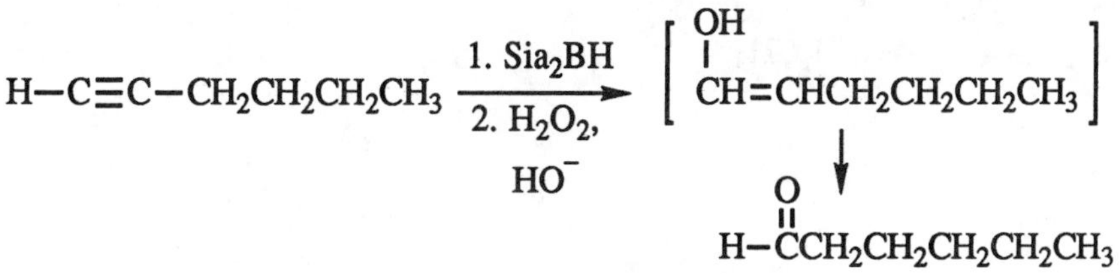

(j) 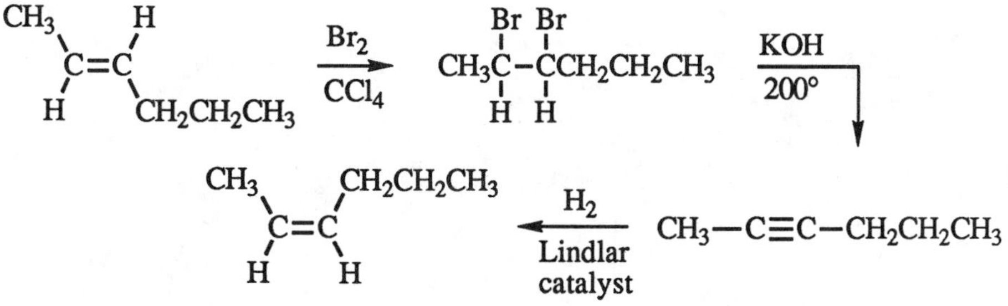

14-36

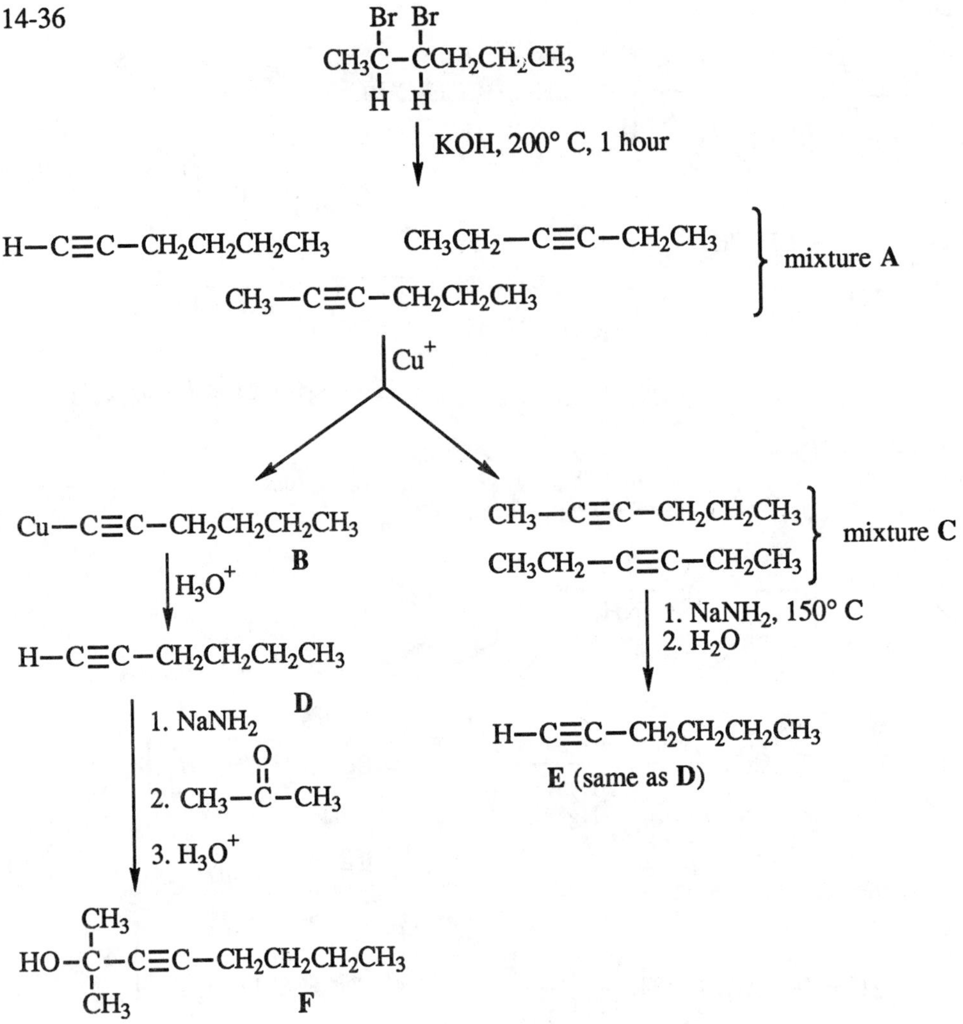
KOH, 200° C, 1 hour
mixture A
Cu+
B
H3O+
mixture C
1. NaNH2, 150° C
2. H2O
D
1. NaNH2
2. CH3–C(=O)–CH3
3. H3O+
E (same as D)
F

14-37

(a) $CH_3CH_2-C\equiv C-CH_2CH_3$

(b) 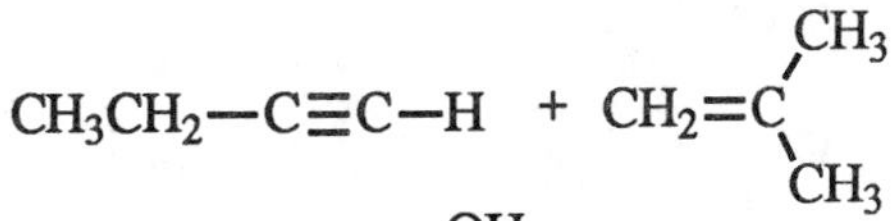

(c) $CH_3CH_2-C\equiv C-CH_2OH$
(after H_2O workup)

(d) 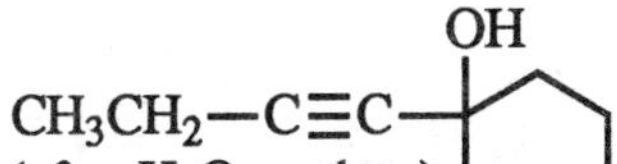
(after H_2O workup)

(e) 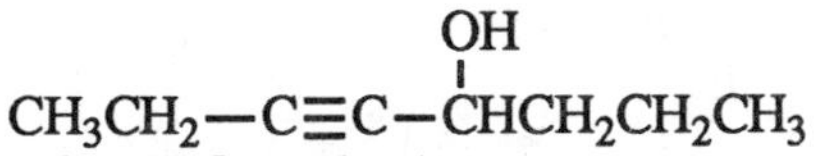
(after H_2O workup)

(f) $CH_3CH_2-C\equiv C-CH_2CH_2OH$
(after H_2O workup)

(g) 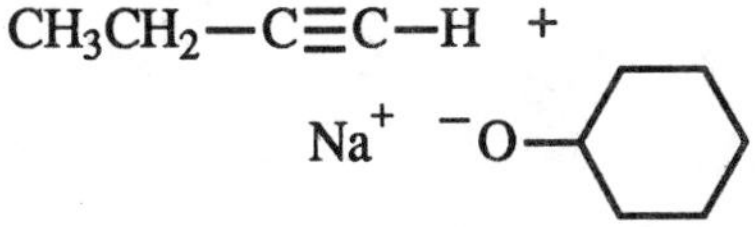

(h) 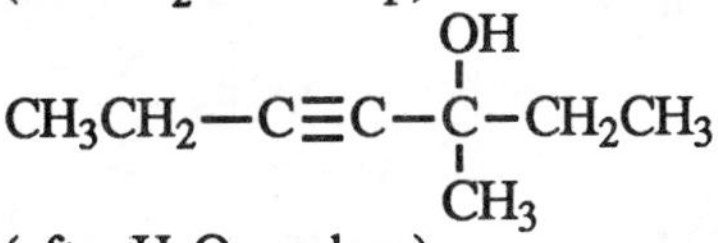
(after H_2O workup)

14-38 Transforming alcohols into bromides or tosylates will work equally well.

Sodium acetylide is prepared: $HC\equiv C-H \xrightarrow{NaNH_2} HC\equiv C{:}^- \ Na^+$

(a) $CH_3CH_2CH_2CH_2OH \xrightarrow{PBr_3} CH_3CH_2CH_2CH_2Br \xrightarrow{HC\equiv C:^-\ Na^+} HC\equiv C-CH_2CH_2CH_2CH_3$

(b) 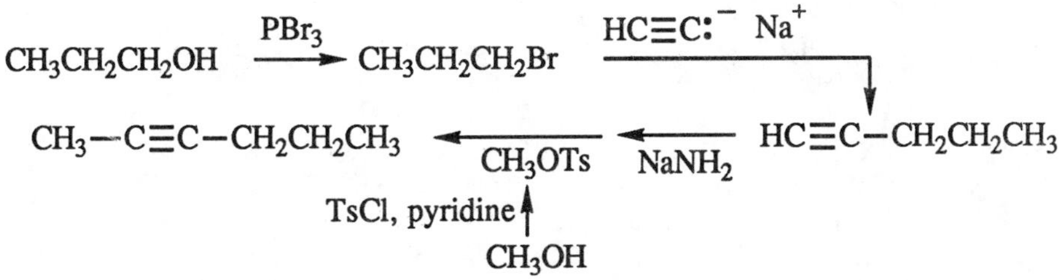

(c) 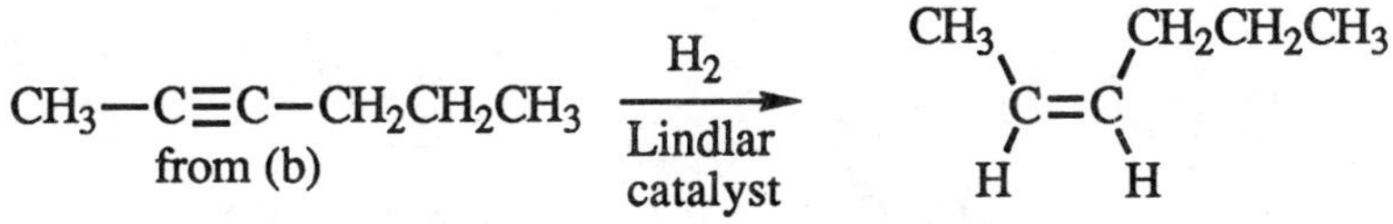

(d) 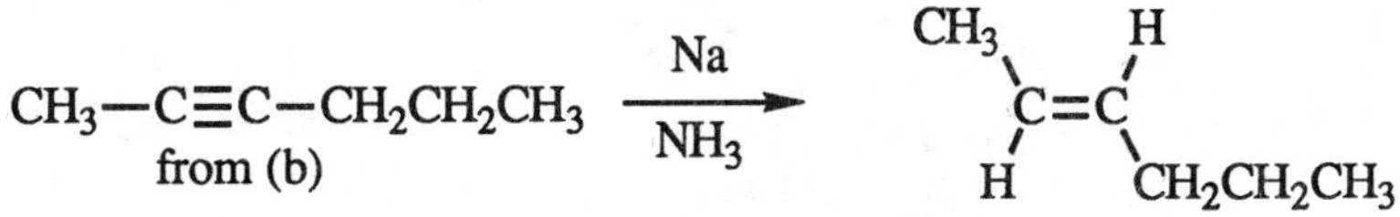

14-38 continued

(e) $$CH_3-C\equiv C-CH_2CH_2CH_3 \xrightarrow[Pt]{H_2} CH_3CH_2CH_2CH_2CH_2CH_3$$
from (b)

(f) $$HC\equiv C-CH_2CH_2CH_2CH_3 \xrightarrow{2\ HBr} CH_3-CBr_2-CH_2CH_2CH_2CH_3$$
from (a)

(g) $$H-C\equiv C-CH_2CH_2CH_3 \xrightarrow[2.\ H_2O_2,\ HO^-]{1.\ Sia_2BH} H-\overset{O}{\overset{\|}{C}}CH_2CH_2CH_2CH_3$$
from (b)

(h) $$H-C\equiv C-CH_2CH_2CH_3 \xrightarrow[H_2SO_4,\ HgSO_4]{H_2O} CH_3-\overset{O}{\overset{\|}{C}}CH_2CH_2CH_3$$
from (b)

(i) 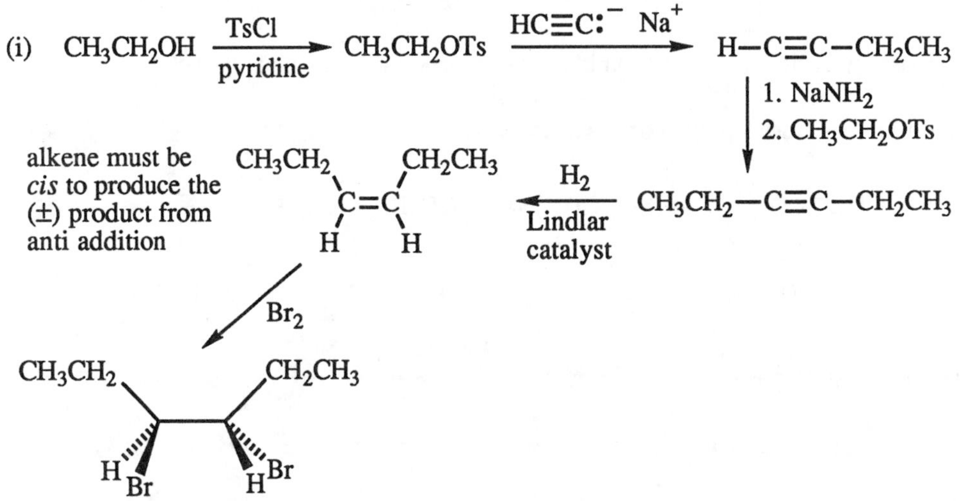
CH3CH2OH
TsCl
pyridine
CH3CH2OTs
HC≡C:⁻ Na⁺
H−C≡C−CH2CH3
1. NaNH2
2. CH3CH2OTs
CH3CH2−C≡C−CH2CH3
H2
Lindlar
catalyst
alkene must be
cis to produce the
(±) product from
anti addition
Br2

(j) 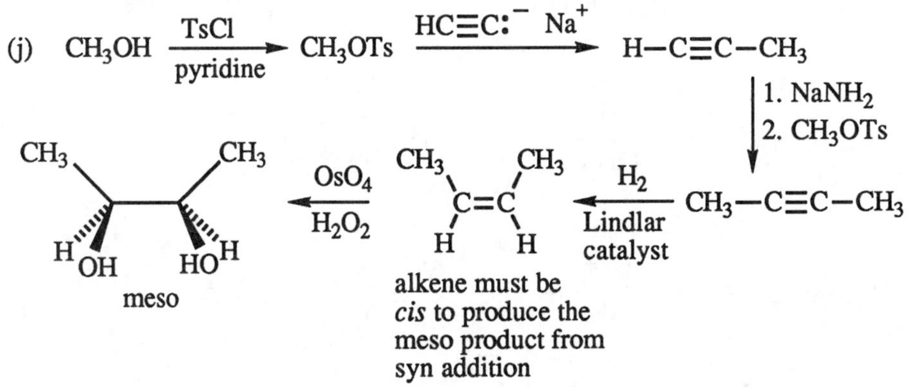
CH3OH
TsCl
pyridine
CH3OTs
HC≡C:⁻ Na⁺
H−C≡C−CH3
1. NaNH2
2. CH3OTs
CH3−C≡C−CH3
H2
Lindlar
catalyst
OsO4
H2O2
meso
alkene must be
cis to produce the
meso product from
syn addition

Alternatively, *trans*-2-butene could be anti-hydroxylated with aqueous performic acid.

14-39

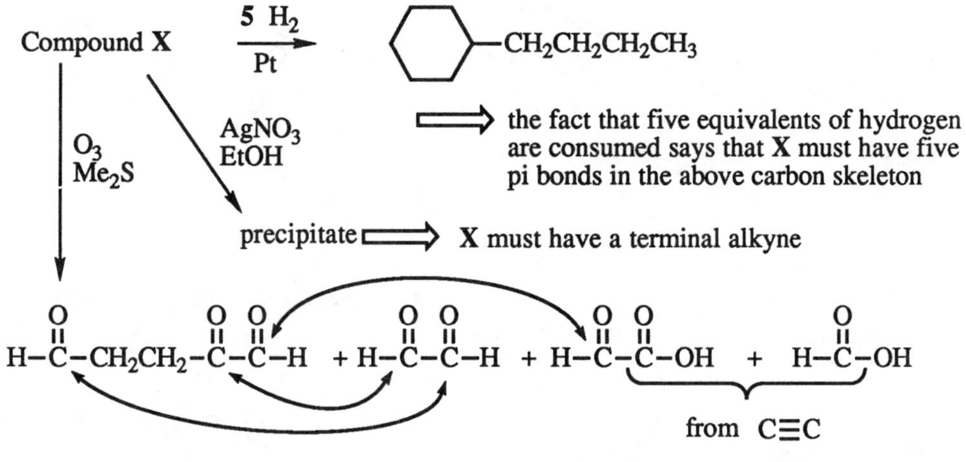

6 carbonyls ⟹ 3 alkenes   2 carboxylic acids ⟹ 1 alkyne

Compound **X**

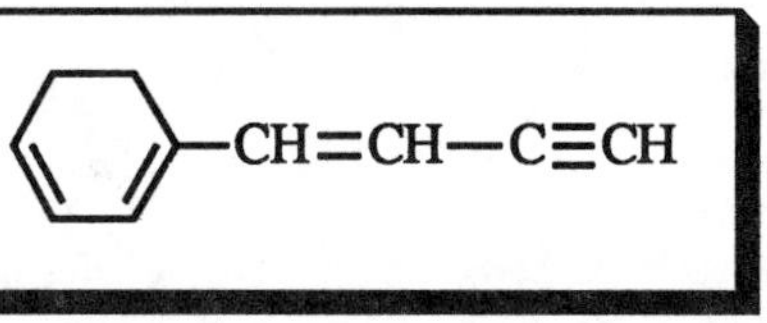

Whether the alkene is *cis* or *trans* cannot be determined from these results.

14-40

Compound **Y**:

precipitate with Ag^+ ⇒ terminal alkyne

IR: 3300 cm^{-1} ⇒ terminal alkyne
2130 cm^{-1} ⇒ alkyne
1635 cm^{-1} ⇒ alkene

NMR: 3H singlet ⇒ CH_3 with no neighbors
1H singlet ⇒ ≡C–H
2H, 2 peaks ⇒

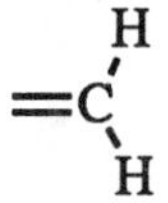

ozonolysis ⇒

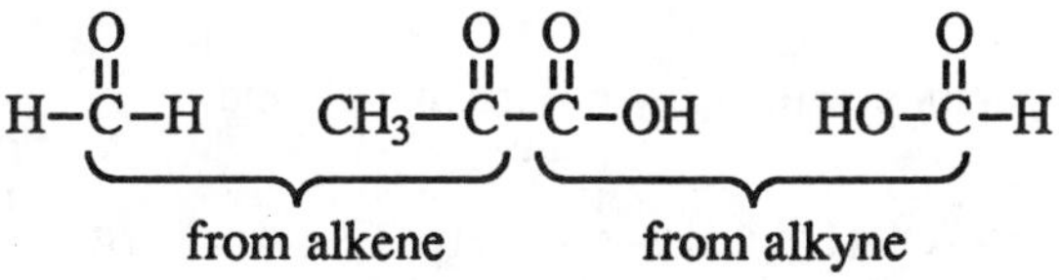

Compound **Y**

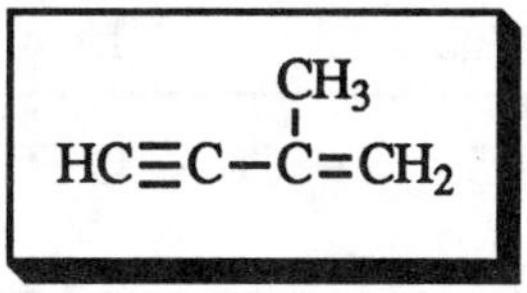

14-41

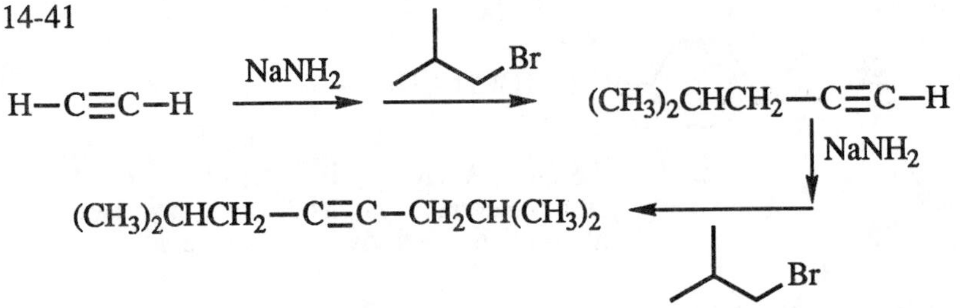

NaNH2
Br
H—C≡C—H
(CH3)2CHCH2—C≡C—H
NaNH2
(CH3)2CHCH2—C≡C—CH2CH(CH3)2
Br

(a)

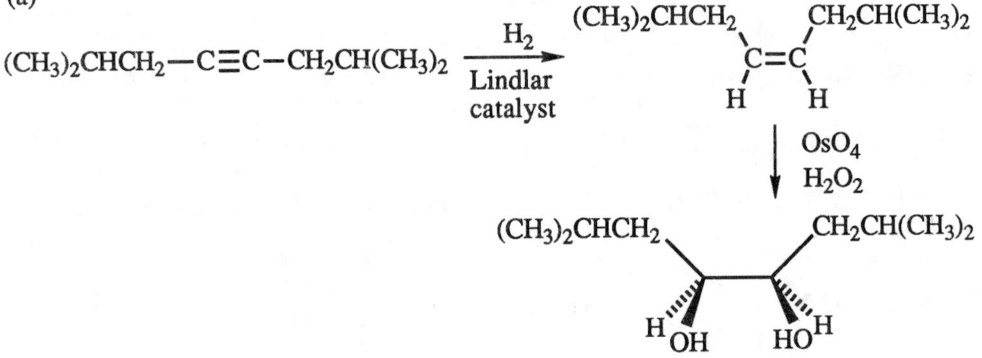

(CH3)2CHCH2—C≡C—CH2CH(CH3)2
H2
Lindlar
catalyst
(CH3)2CHCH2
CH2CH(CH3)2
C=C
H
H
OsO4
H2O2
(CH3)2CHCH2
CH2CH(CH3)2
H
OH
HO
H

meso

(b)

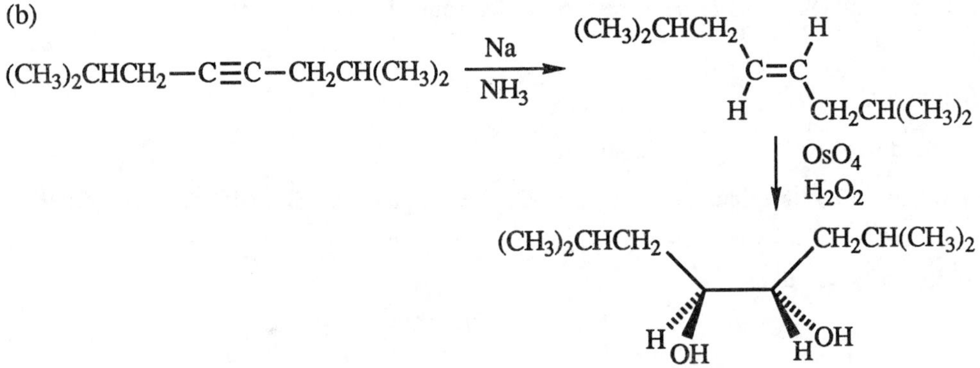

(CH3)2CHCH2—C≡C—CH2CH(CH3)2
Na
NH3
(CH3)2CHCH2
H
C=C
H
CH2CH(CH3)2
OsO4
H2O2
(CH3)2CHCH2
CH2CH(CH3)2
H
OH
H
OH

d,l

14-42 Compound **Z**: precipitate with Ag^+ ⇒ terminal alkyne

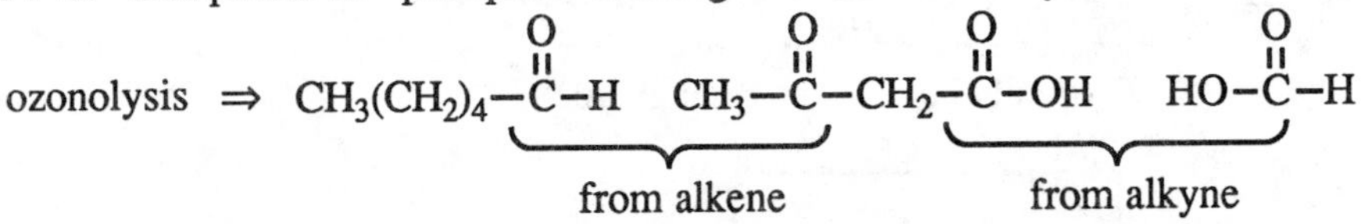

ozonolysis ⇒
CH3(CH2)4—C—H
CH3—C—CH2—C—OH
HO—C—H
O
O
O
O
from alkene
from alkyne

Compound **Z**:
Whether the alkene is *cis* or *trans* cannot be determined from this information.

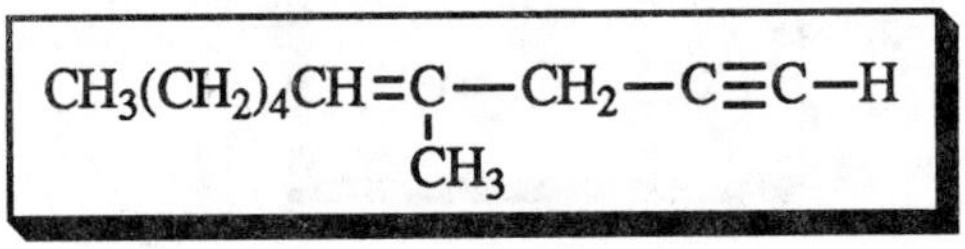

CH3(CH2)4CH=C—CH2—C≡C—H
CH3

14-43

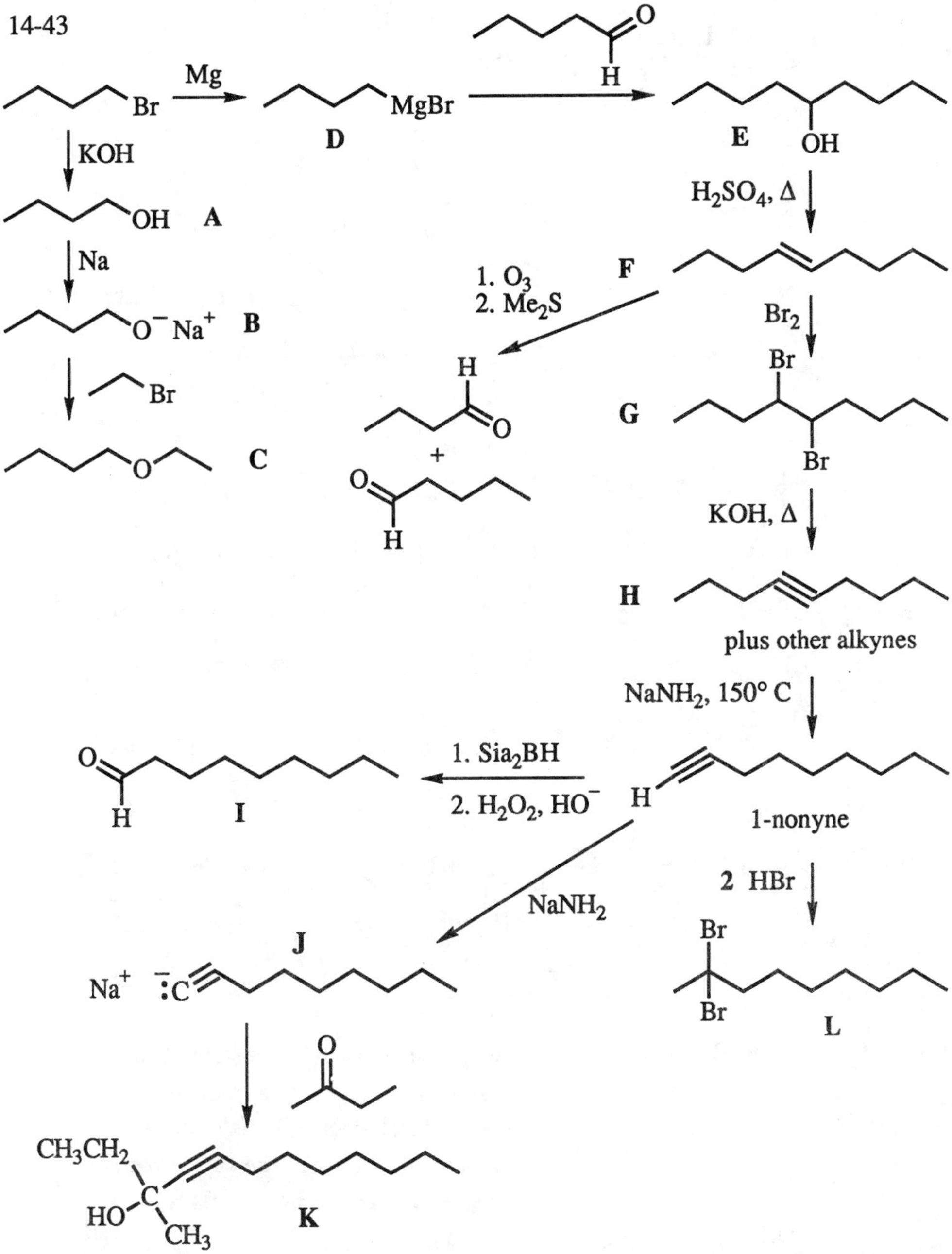

Mg
KOH
OH
A
Na
O⁻ Na⁺
B
Br
O
C
D
MgBr
H
O
E
OH
H$_2$SO$_4$, Δ
F
1. O$_3$
2. Me$_2$S
Br$_2$
Br
G
Br
KOH, Δ
H
plus other alkynes
NaNH$_2$, 150° C
1. Sia$_2$BH
2. H$_2$O$_2$, HO⁻
I
1-nonyne
2 HBr
L
NaNH$_2$
J
Na⁺
:C
CH$_3$CH$_2$
HO
C
CH$_3$
K

14-44

(a) 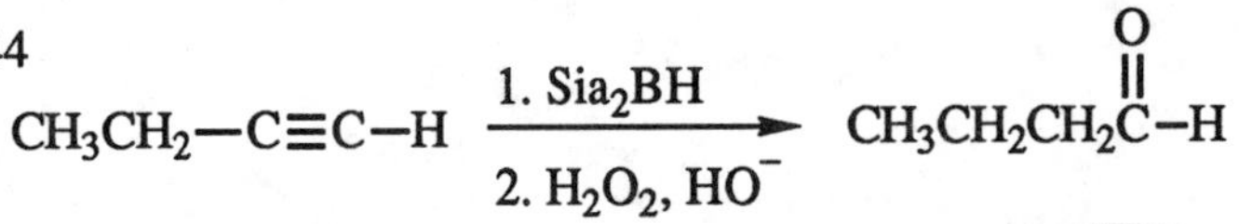

(b) 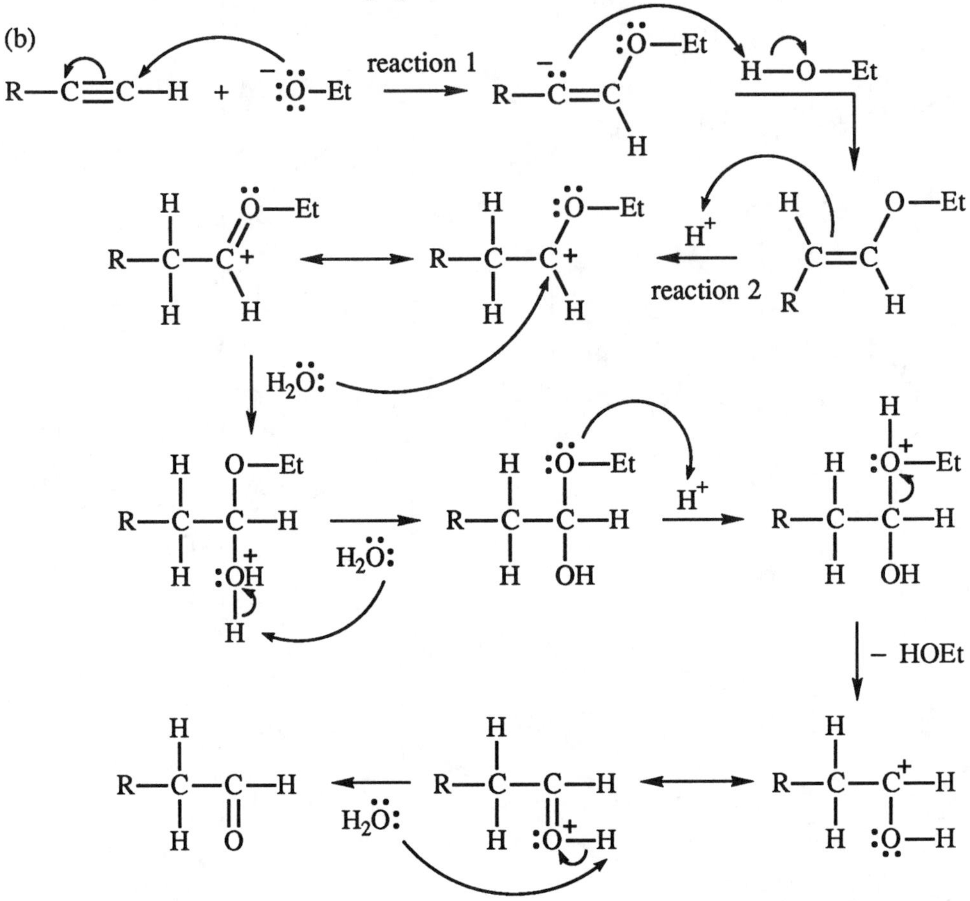

(c) alkyne

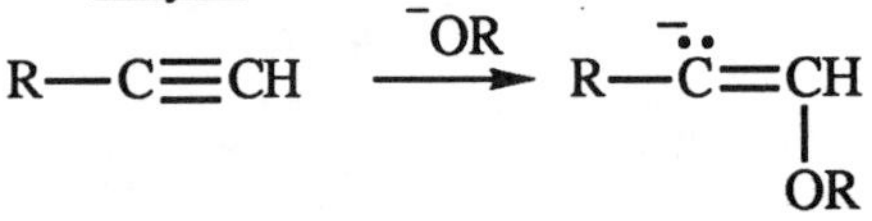

sp^2 carbanion

alkene

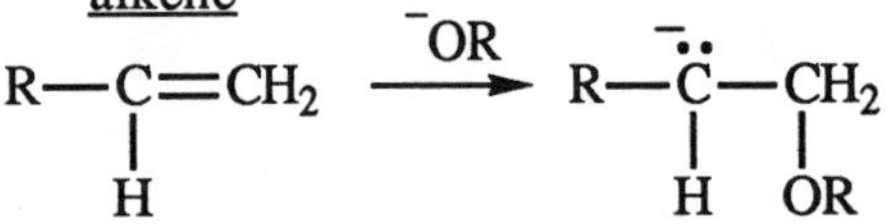

sp^3 carbanion

An sp^2 carbanion is more stable than an sp^3 carbanion: the sp^2 carbanion has more s character and is closer to the positive nucleus. The sp^2 carbanion is easier to form because of its relative stability.

14-45

IR: 3300 cm^{-1} ⇒ terminal alkyne
3080 cm^{-1} ⇒ sp^2 C—H
1600 cm^{-1} ⇒ alkene are aromatic C=C

NMR: 5H at δ 7.3 ⇒ monosubstituted benzene
1H at δ 2.9 ⇒ alkyne C—H

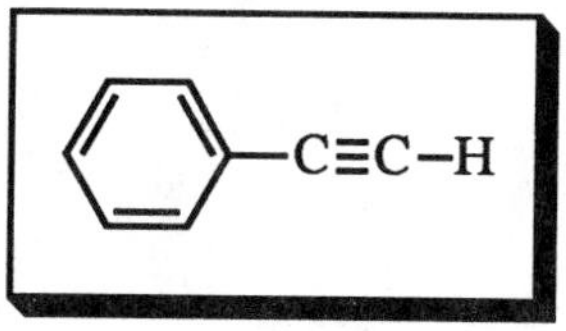

C≡C–H

mol. wt. 102

14-46 Analyze the spectra carefully—watch for overlapping peaks.

$C_8H_{14}O$ ⇒ 2 elements of unsaturation: if alkyne is present, the oxygen must be an ether or an alcohol

IR: 3500 cm^{-1} ⇒ O—H ⇒ alcohol
3300 cm^{-1} ⇒ ≡C—H (partially obscured by OH peak)
2100 cm^{-1} ⇒ C≡C (tiny blip—look carefully)

NMR: 6H doublet at δ 1.0 ⇒ isopropyl methyls $(CH_3)_2CH$—
3H singlet at δ 1.5 ⇒ CH_3 with no neighboring hydrogens
2H doublet at δ 1.6 ⇒ CH_2—CH (these peaks overlap a neighboring peak)
1H multiplet at δ 1.9 ⇒ H of isopropyl
1H singlet at δ 2.3 ⇒ ≡C—H
1H singlet at δ 3.1 ⇒ OH (exchangeable with D_2O)

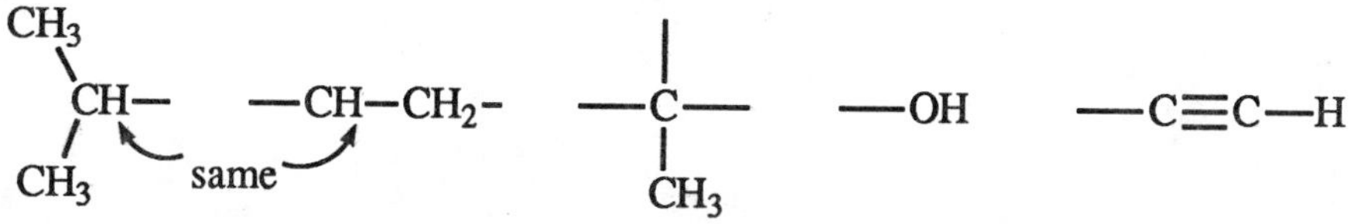

CH3
CH—
CH3
same
—CH—CH2-
—C—
CH3
—OH
—C≡C—H

Assemble the pieces

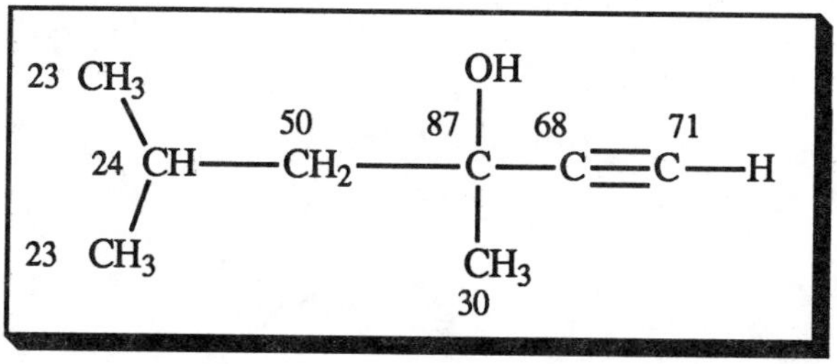

23 CH3
24 CH
23 CH3
50
CH2
OH
87
C
68
C
71
C—H
CH3
30

Numbers are δ values in ^{13}C NMR.

Synthesis

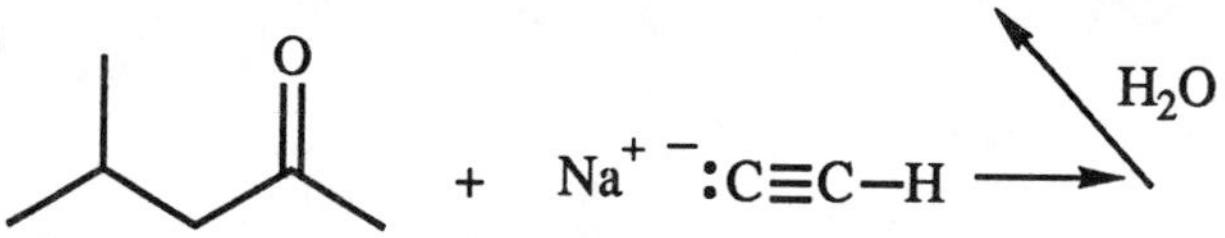

O
+
Na+ –:C≡C–H
H2O

CHAPTER 15—CONJUGATED SYSTEMS, ORBITAL SYMMETRY, AND ULTRAVIOLET SPECTROSCOPY

15-1 First, look for the number of double bonds to be hydrogenated, then for conjugation, then for degree of substitution of the alkenes.

(a)

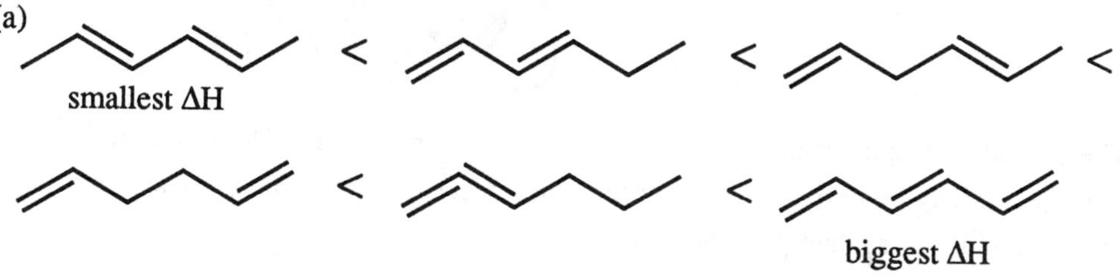

smallest ΔH
<
<
<
<
<
biggest ΔH

(b)

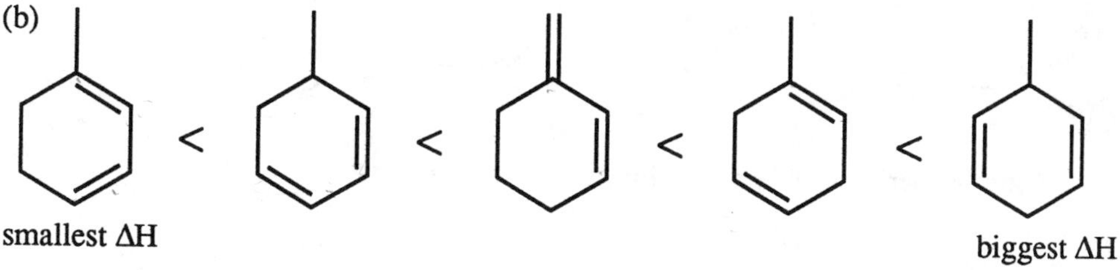

smallest ΔH
<
<
<
<
biggest ΔH

15-2

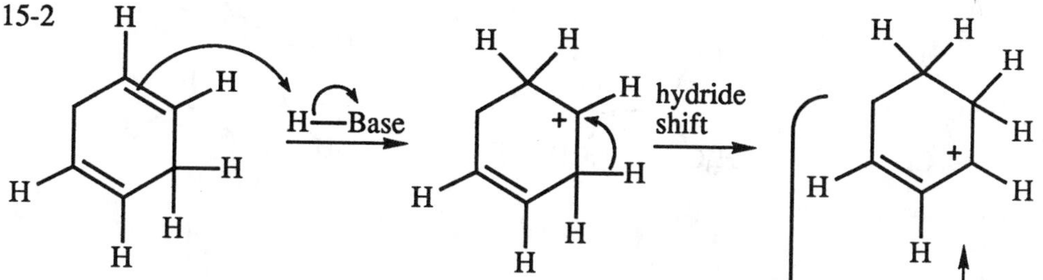

H
H
H
H
H
H
H
H—Base
H
H
H
H
+
H
H
H
hydride
shift
H
H
H
H
H
+
H
H

The key step is hydride shift from a 2° carbocation to a 2° *allylic*, resonance-stabilized carbocation, which can subsequently lose a proton to form a *conjugated* diene.

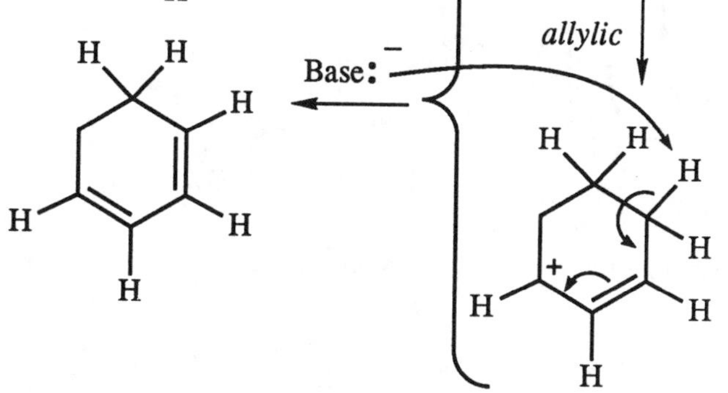

allylic
H
H
H
H
H
H
Base:⁻
H
H
H
H
H
H
+
H
H

15-3 (You may wish to refer to problem 2-6.)

(a)

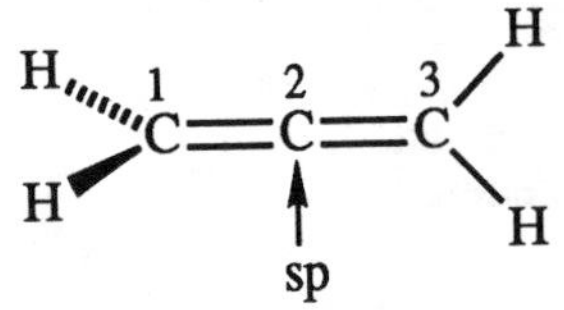
H
H
1
2
3
C
C
C
H
H
sp

orbital picture

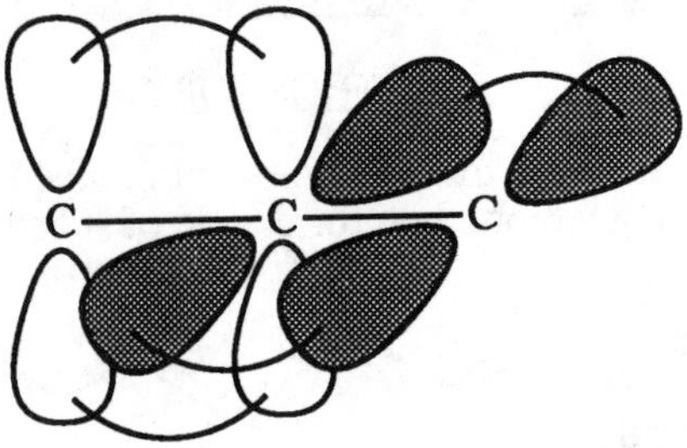
C
C
C

The central carbon atom makes two π bonds with two p orbitals. These p orbitals must necessarily be perpendicular to each other, thereby forcing the groups on the ends of the allene system perpendicular.

(b)

mirror
H
Cl
Cl
H
Cl
H
H
Cl

non-superimposable mirror images = enantiomers

15-4

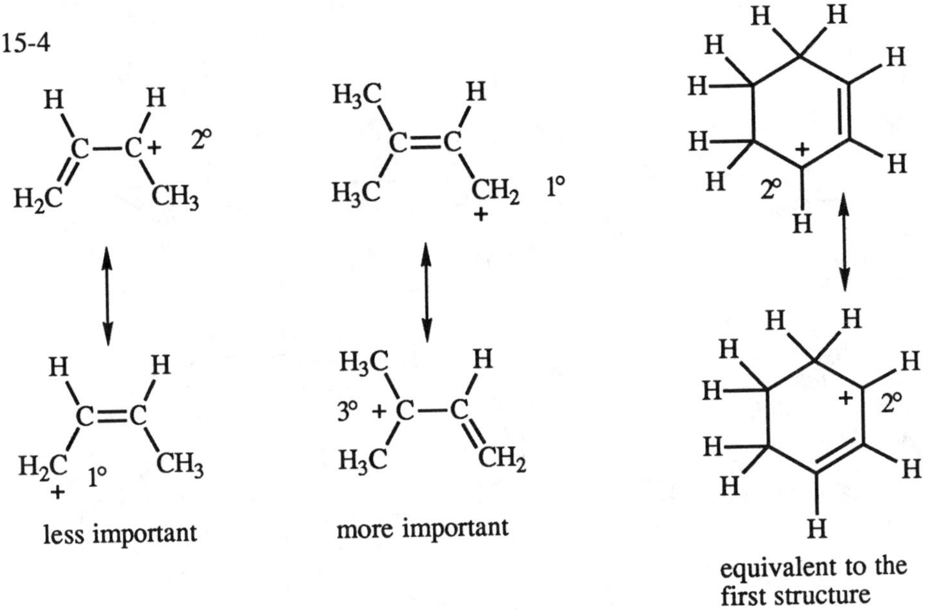
2°
1°
less important
1°
3°
more important
2°
2°
equivalent to the first structure

15-5

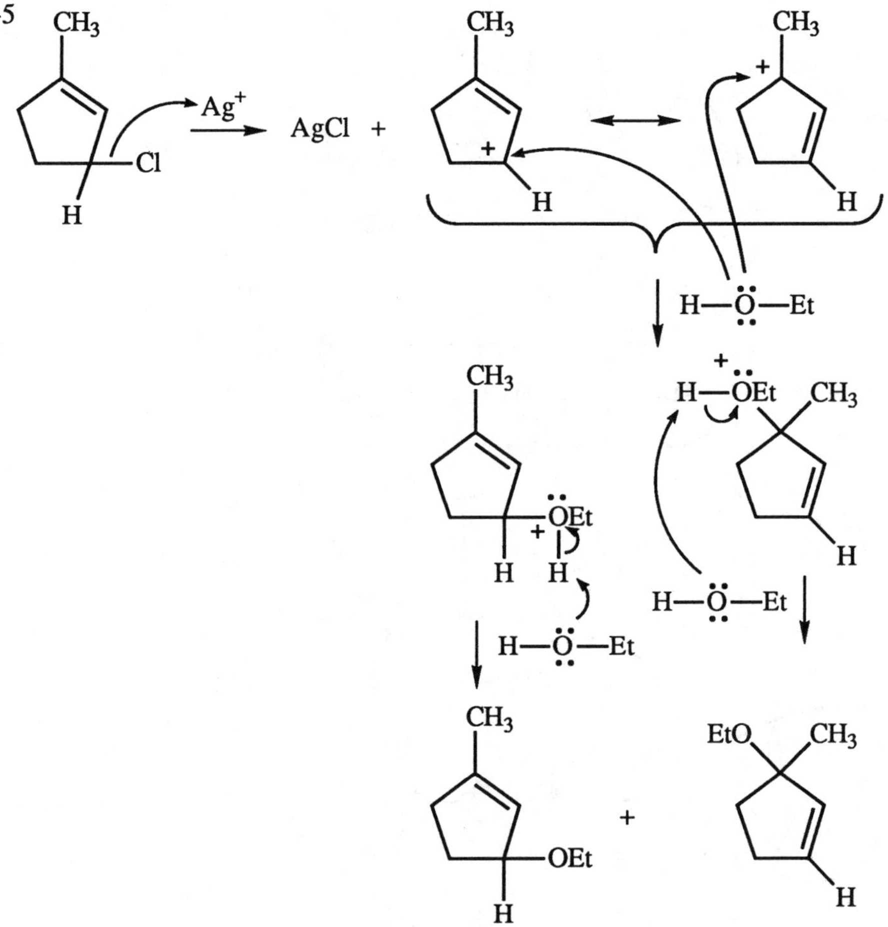

15-6

(a)

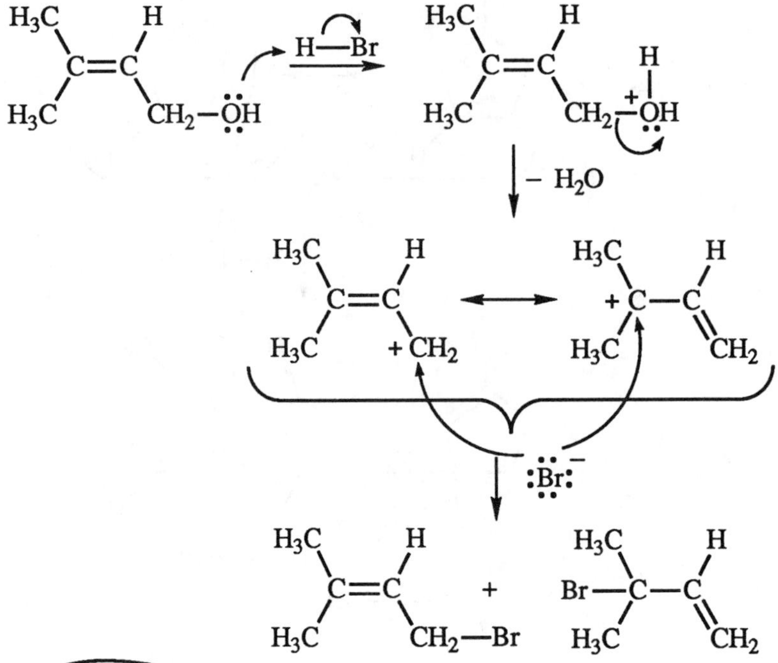
H₃C
H
C=C
H₃C
CH₂—ÖH
H—Br
H
CH₂—OH
– H₂O
+CH₂
CH₂
:Br:
CH₂—Br
Br—C—C

(b)

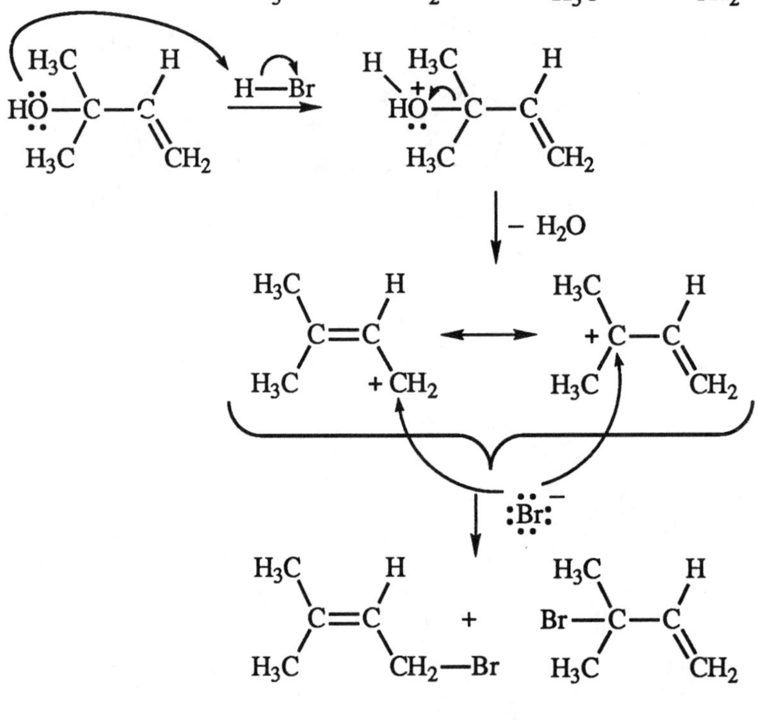
H₃C
H
HÖ—C—C
H₃C
CH₂
H—Br
– H₂O
+CH₂
:Br:
CH₂—Br
Br—C—C

15-6 continued

(c)

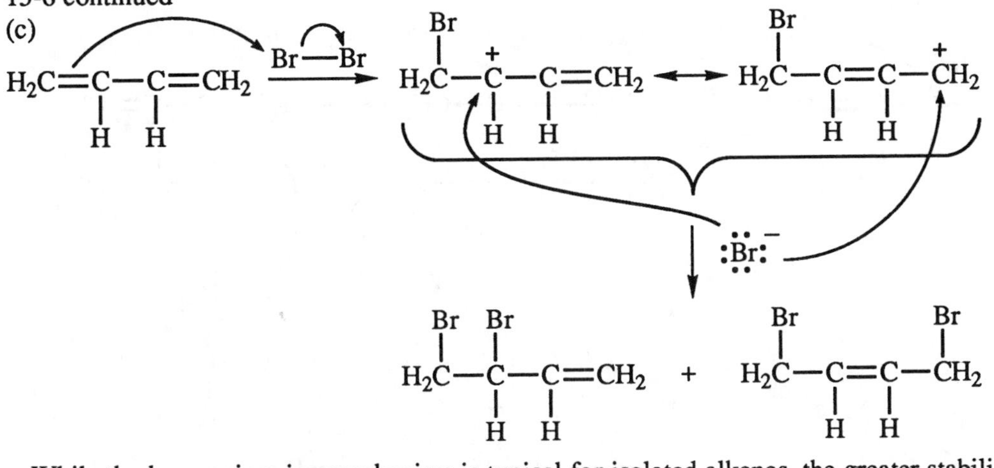

While the bromonium ion mechanism is typical for isolated alkenes, the greater stability of the resonance-stabilized carbocation will make it the lower energy intermediate for conjugated systems.

(d)

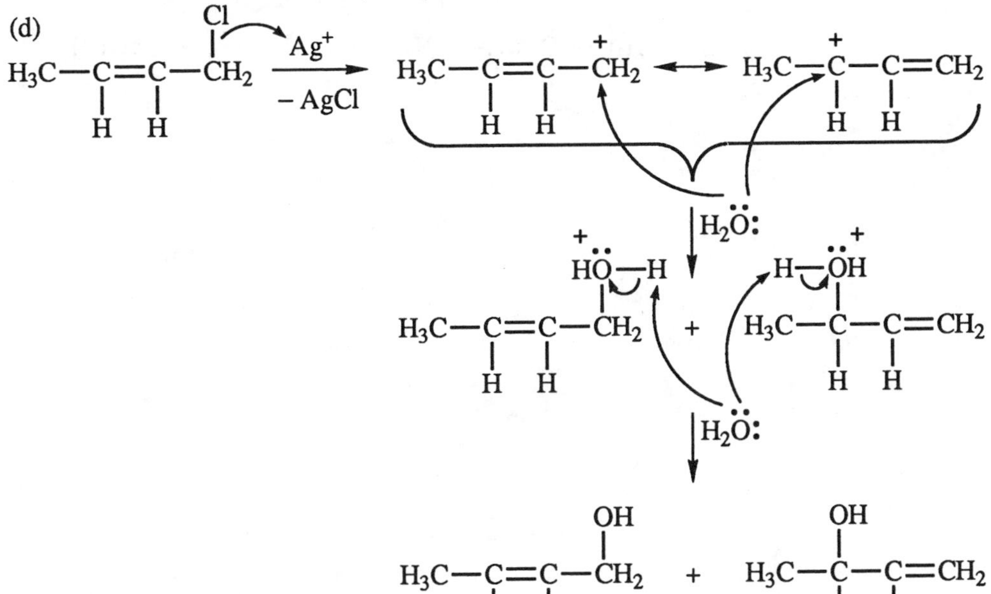

15-6 continued

(e)

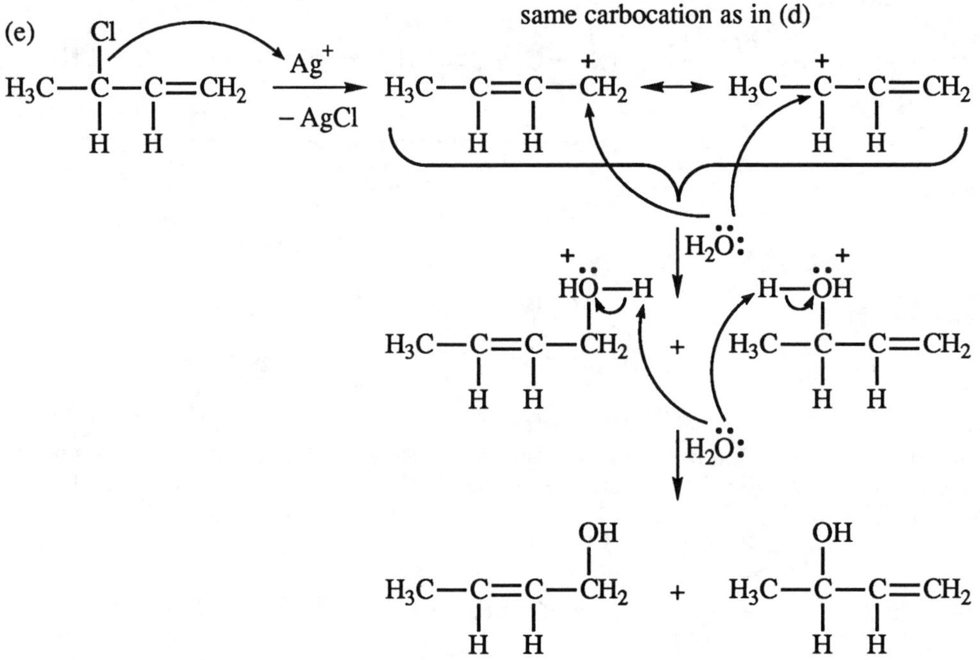

Cl
Ag+
– AgCl
same carbocation as in (d)
H₂Ö:

15-7

(a)

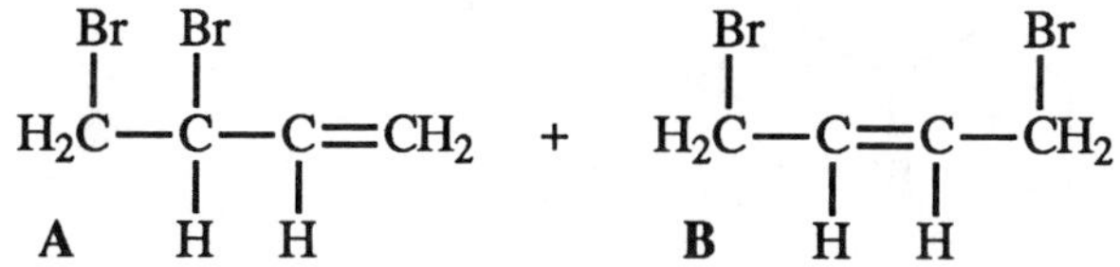

(b)

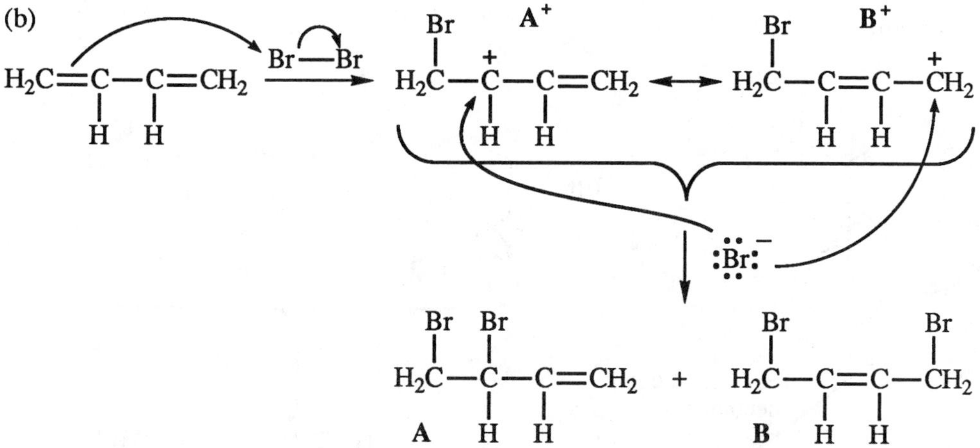

(c) The resonance form A^+, which eventually leads to product **A**, has positive charge on a 2° carbon and is a more significant resonance contributor than structure B^+. With greater positive charge on the 2° carbon than on the 1° carbon, we would expect bromide ion attack on the 2° carbon to have lower activation energy. Therefore, **A** must be the *kinetic* product. At higher temperature, however, the last step becomes reversible, and the stability of the products becomes the dominant factor in determining product ratios. As **B** has a disubstituted alkene whereas **A** is only monosubstituted, it is reasonable that **B** is the major, *thermodynamic* product at 60° C.

(d) At 60° C, ionization of **A** would lead to the same allylic carbocation as shown in (b), which would give the same product ratio as formation of **A** and **B**from butadiene.

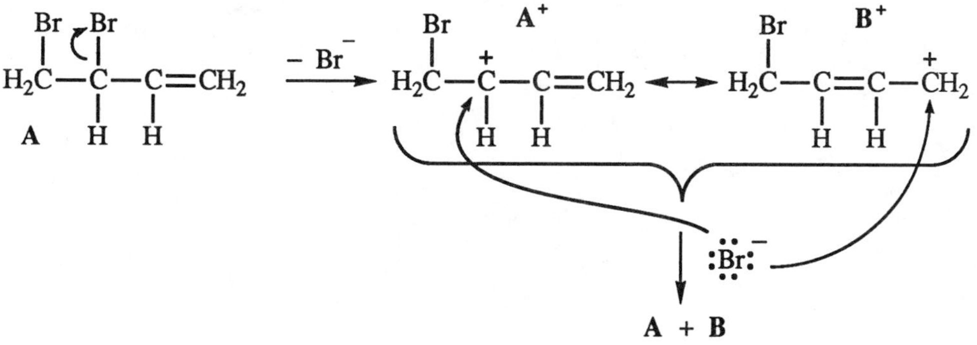

15-8

(a)

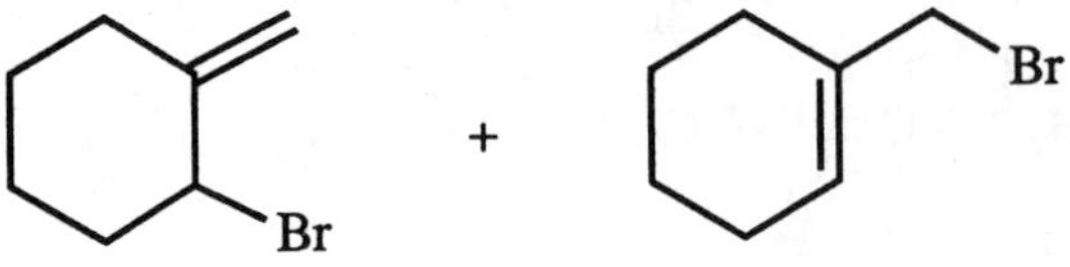
Br
+
Br

(b)

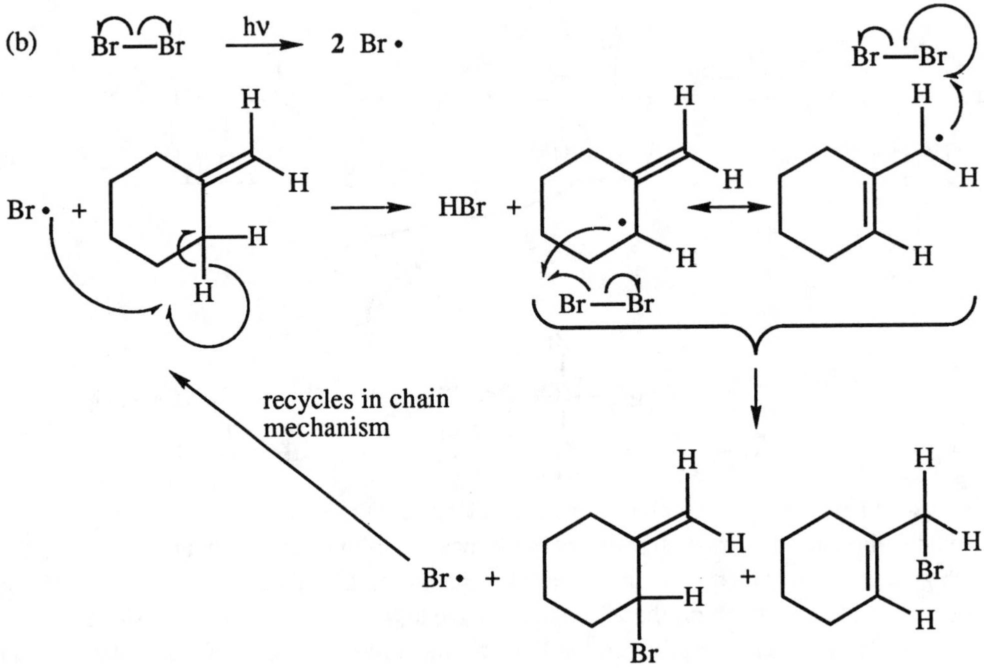
Br—Br
hν
2 Br •
Br—Br
Br • +
HBr +
Br—Br
recycles in chain mechanism
Br • +
+
H
Br

15-9 ("Pr" is the abbreviation for *n*-propyl, below.)

NBS generates a low concentration of Br_2

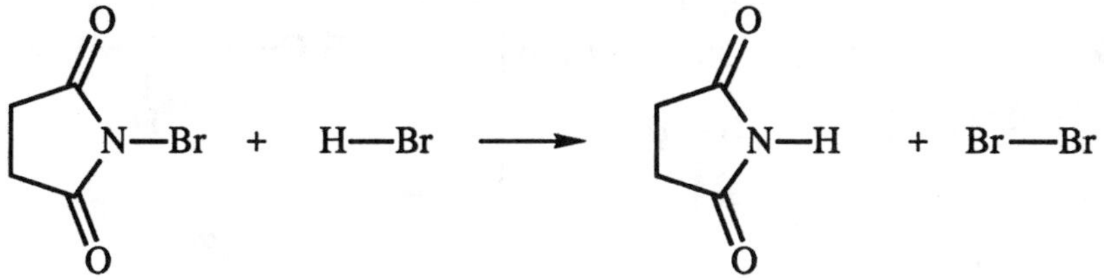

initiation

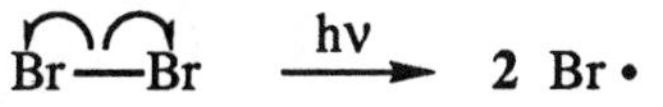

$$Br{-}Br \xrightarrow{h\nu} 2\ Br\cdot$$

propagation

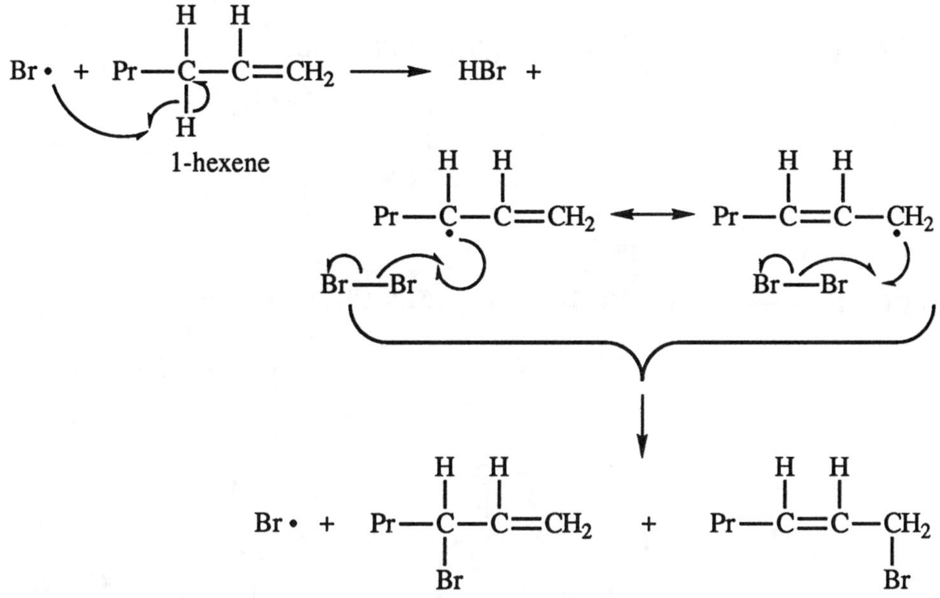

The HBr generated in the propagation step combines with NBS to produce more Br_2, continuing the chain mechanism.

15-10

(a)

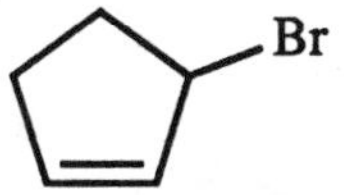

(b)

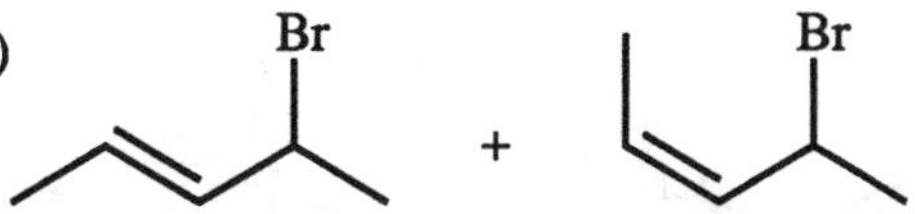

These are the major products from abstraction of a 2° allylic H.

(c)

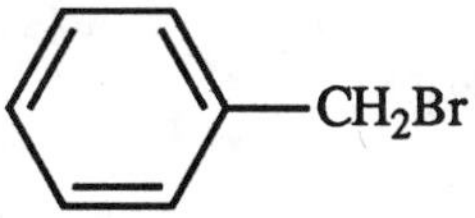

15-11 Both halides generate the same allylic carbanion.

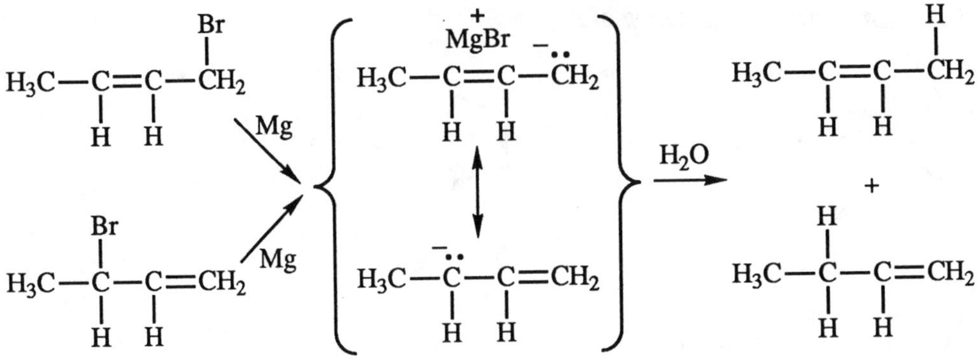

15-12

(a) $CH_3CH_2CH_2CH_2Br \xrightarrow[\text{ether}]{\text{Mg}} CH_3CH_2CH_2CH_2MgBr \xrightarrow{H_2C{=}CHCH_2Br} H_2C{=}CHCH_2(CH_2)_3CH_3$

(b) $CH_3CH(Br)CH_3 \xrightarrow[\text{ether}]{\text{Mg}} CH_3CH(MgBr)CH_3 + CH_3CH{=}CHCH_2Br \longrightarrow CH_3CH{=}CHCH_2{-}CH(CH_3)CH_3$

15-13

(a)

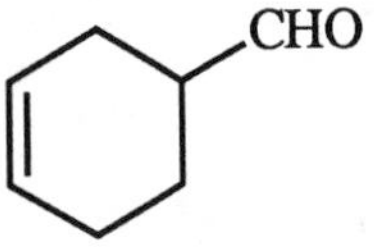

(b)

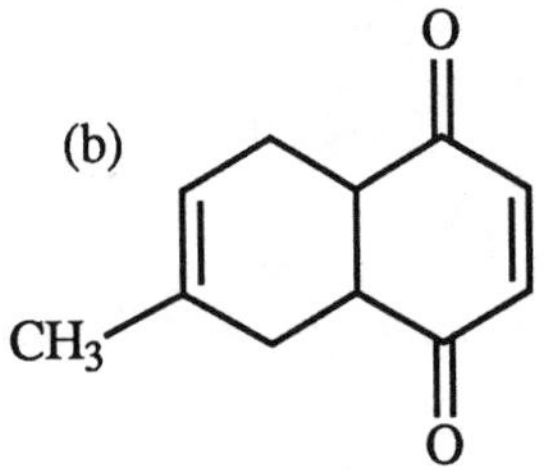

(c)

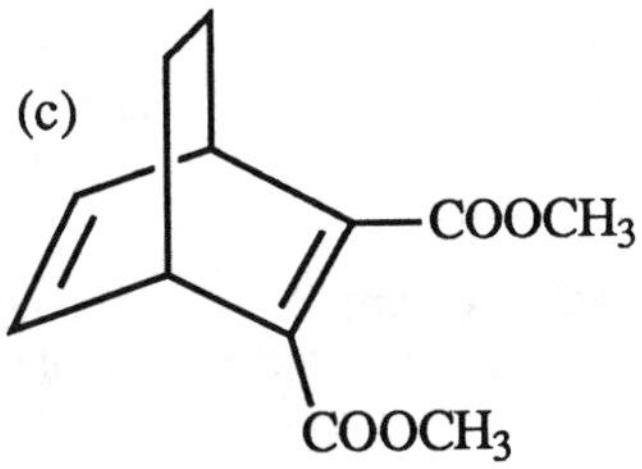

(d)

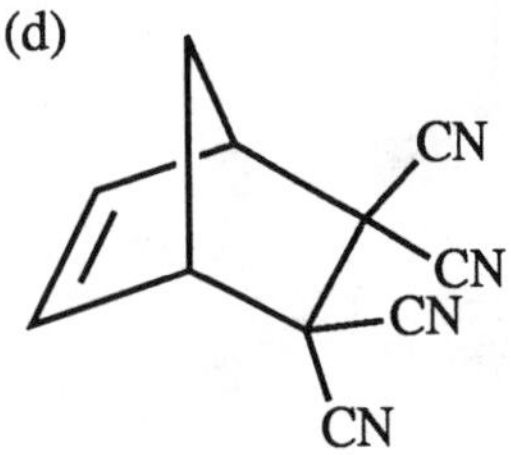

(e)

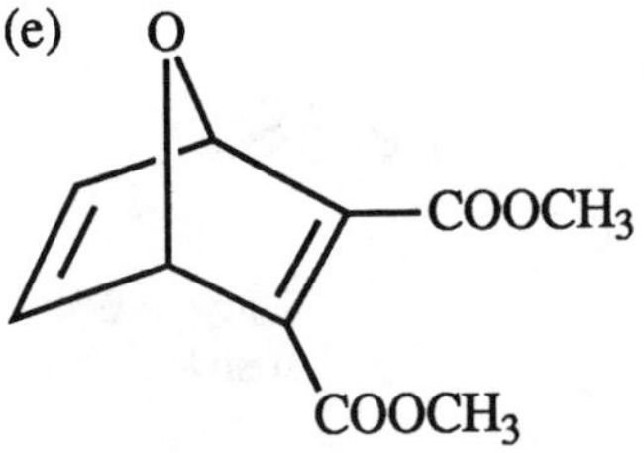

(f)

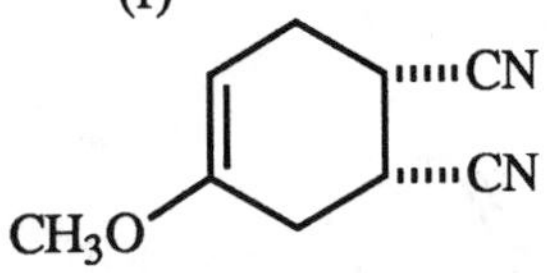

15-14

(a)

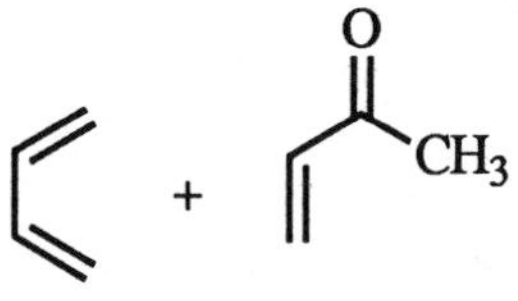

(b)

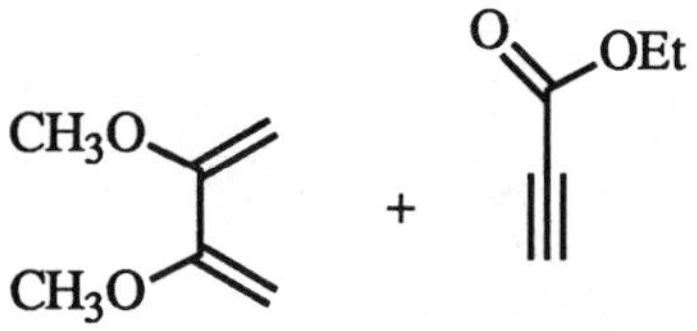

(c)

(d)

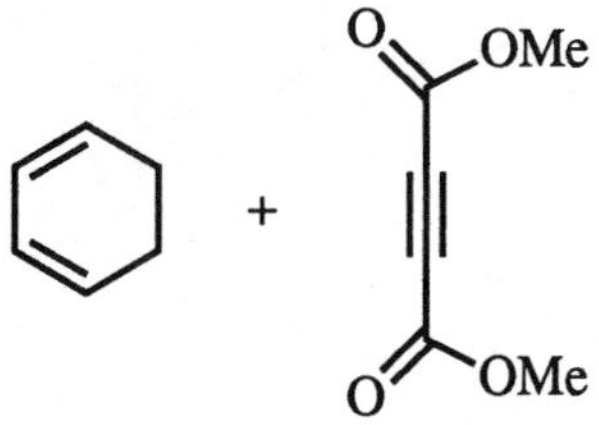

(e)

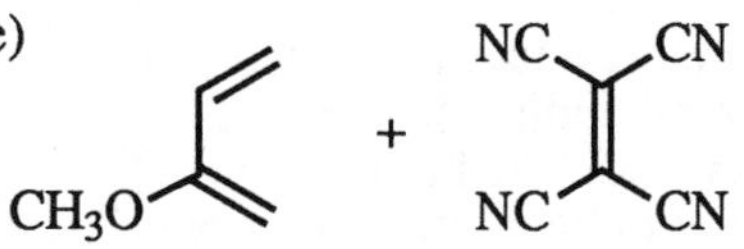

(f)

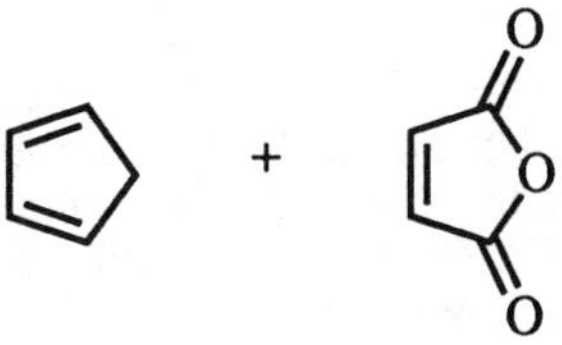

15-15

(a)

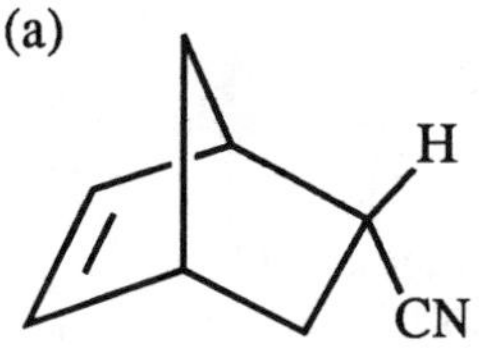

(b)

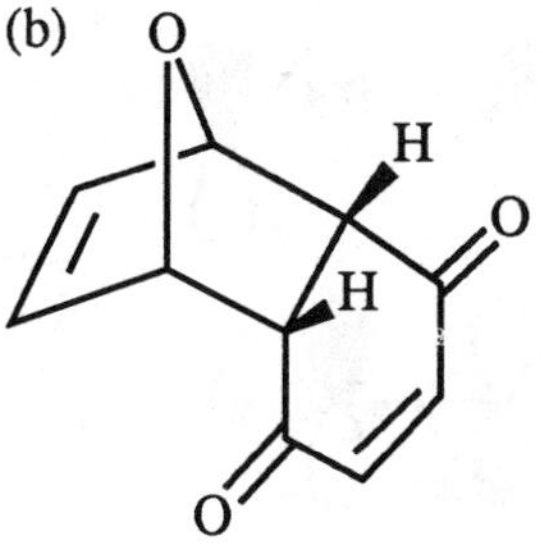

(c)

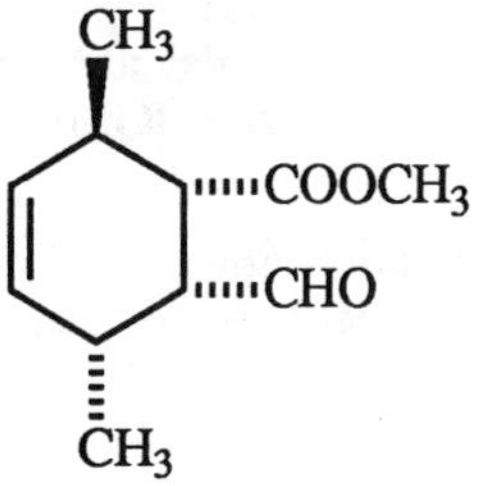

15-16

(a)

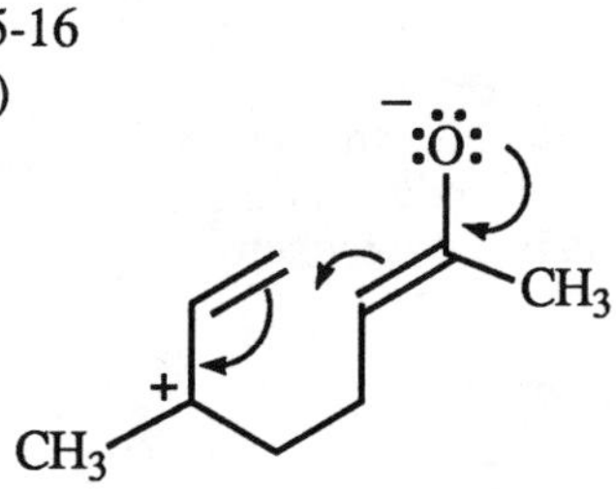

(b)

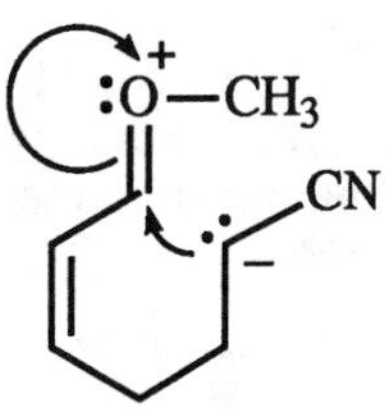

15-17

(a)

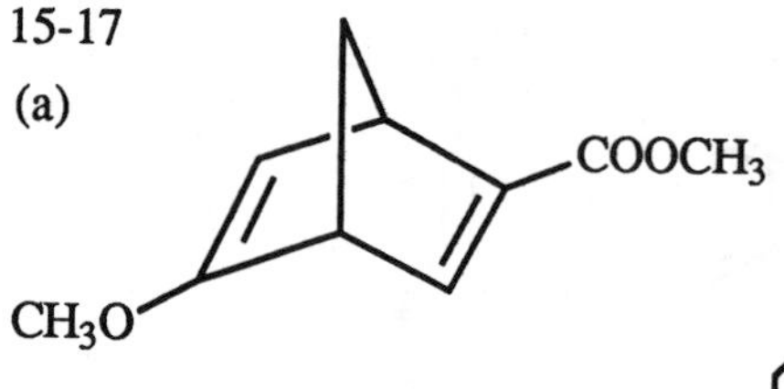

COOCH3
CH3O

(b)

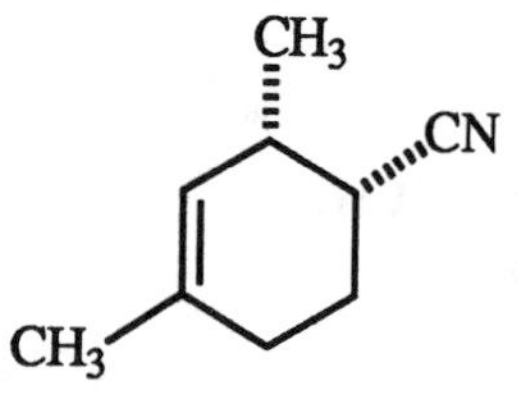

CH3
CN
CH3

(c)

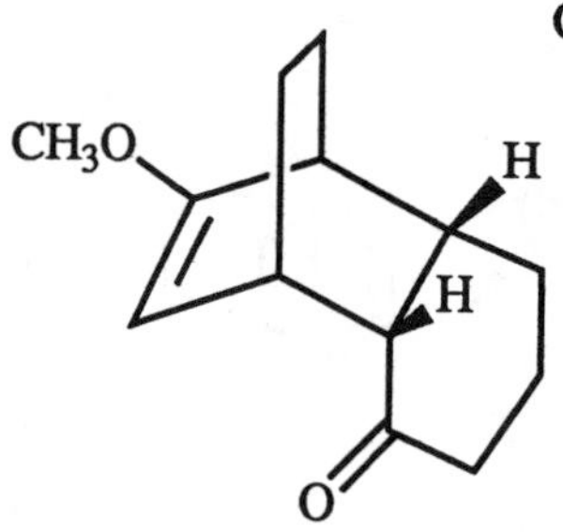

CH3O
H
H
O

15-18 For a photochemically *allowed* process, one molecule must use an excited state in which an electron has been promoted to the first antibonding orbital. All orbital interactions between the excited molecule's HOMO* and the other molecule's LUMO must be bonding for the interaction to be allowed; otherwise, it is a forbidden process.

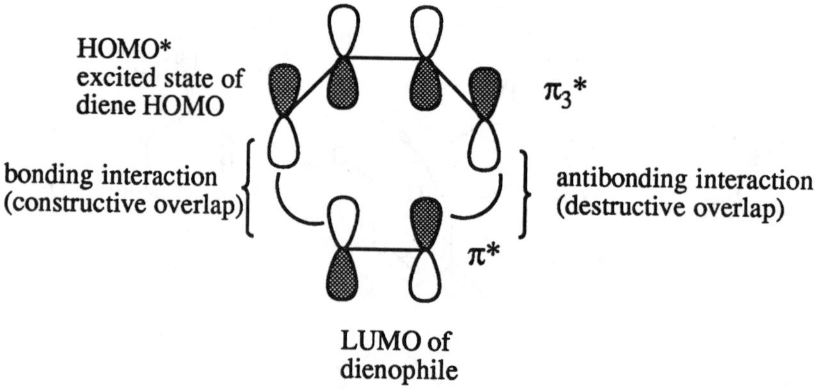

HOMO*
excited state of
diene HOMO
π3*
bonding interaction
(constructive overlap)
antibonding interaction
(destructive overlap)
π*
LUMO of
dienophile

In the Diels-Alder cycloaddition, the LUMO of the dienophile and the excited state of the HOMO of the diene (labeled HOMO*) produce one bonding interaction and one antibonding interaction. Thus, this is a photochemically forbidden process.

15-19 For a [4 + 4] cycloaddition:

(a)

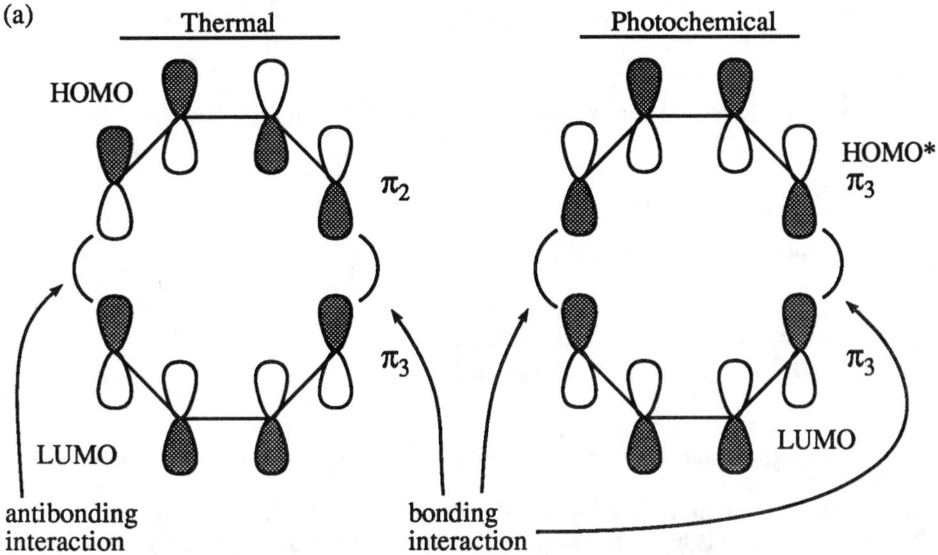
Thermal
Photochemical
HOMO
π_2
π_3
LUMO
HOMO*
π_3
π_3
LUMO
antibonding interaction
bonding interaction

one interaction is antibonding ⇒ **forbidden**

both interactions are bonding ⇒ **allowed**

(b) A [4 + 4] cycloaddition is not thermally allowed, but a [4 + 2] (Diels-Alder) is!

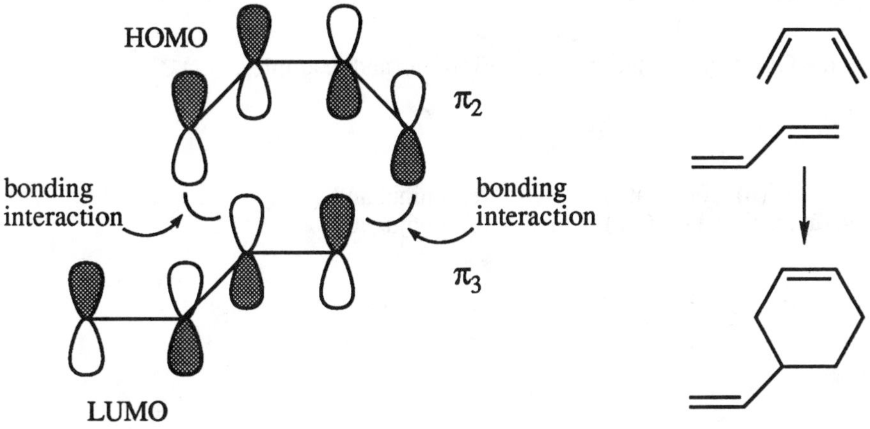
HOMO
π_2
bonding interaction
bonding interaction
π_3
LUMO

15-20

$$A = \varepsilon c l \qquad \varepsilon = \frac{A}{c l} \qquad l = 1 \text{ cm} \qquad A = 0.50$$

convert mass to moles: $1 \text{ mg} \times \frac{1 \text{ g}}{1000 \text{ mg}} \times \frac{1 \text{ mole}}{160 \text{ g}} = 6 \times 10^{-6} \text{ moles}$

$$c = \frac{6 \times 10^{-6} \text{ moles}}{10 \text{ mL}} \times \frac{1000 \text{ mL}}{1 \text{ L}} = 6 \times 10^{-4} \text{ M}$$

$$\varepsilon = \frac{0.50}{(6 \times 10^{-4})(1)} = 833 \approx \boxed{800}$$

15-21

(a) 353 nm: a conjugated tetraene—must have highest absorption maximum among these compounds;
(b) 313 nm: closest to the bicyclic conjugated triene in Table 15-2; the diene is in a more substituted ring, so it is not surprising for the maximum to be slightly higher than 304 nm;
(c) 232 nm: similar to 3-methylenecyclohexene in Table 15-2;
(d) 273 nm: 1,3 cyclohexadiene (256 nm) + 2 alkyl substituents (2 x 5 nm) = predicted value of 266 nm;
(e) 237 nm: like 3-methylenecyclohexene (232 nm) + 1 alkyl group (5 nm) = predicted value of 237 nm

15-22 Refer to the Glossary and the text for definitions and examples.

15-23

(a) isolated (b) conjugated (c) cumulated
(d) conjugated and isolated (at C-6) (e) conjugated

15-24

(a) 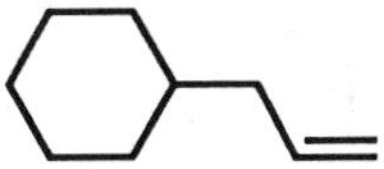

(b) 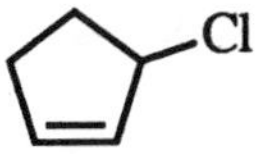

one equivalent
of HCl

(c) 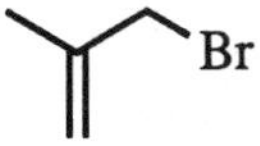

(d) 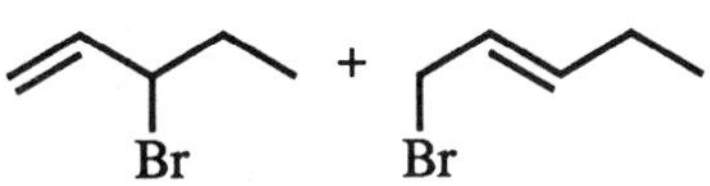

(e) 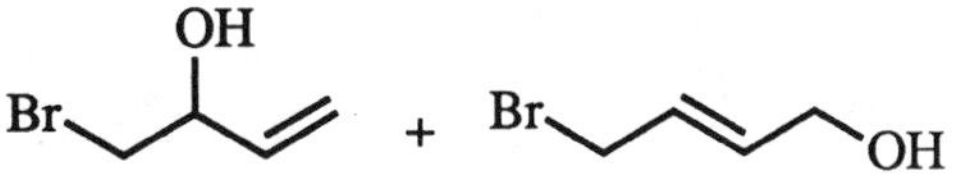

(f) 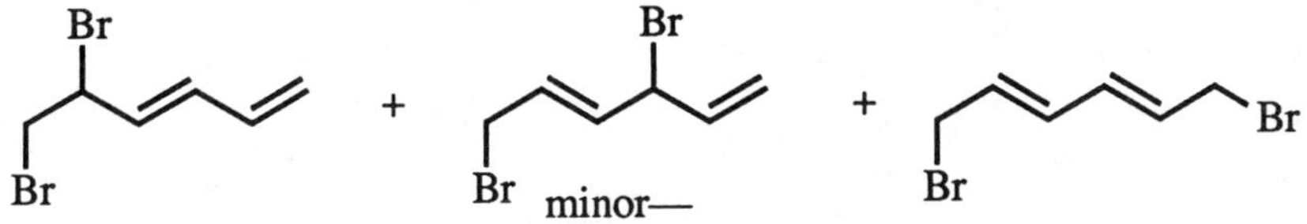

minor—
not conjugated

(g) 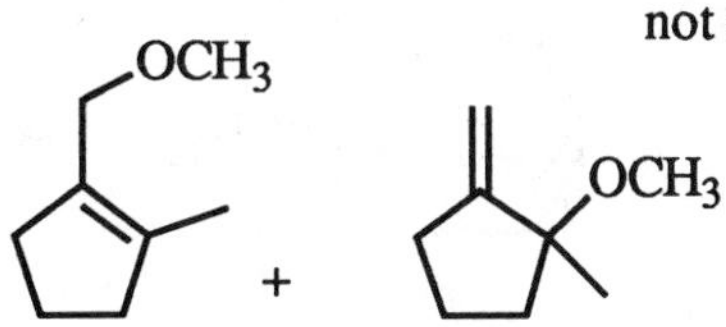

(h) 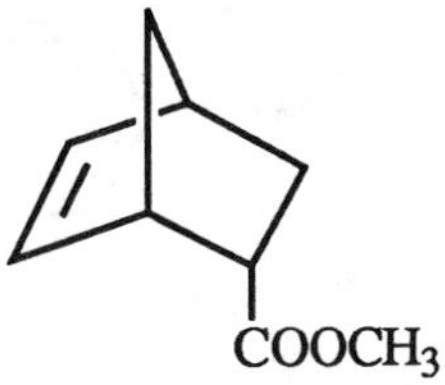

(i) 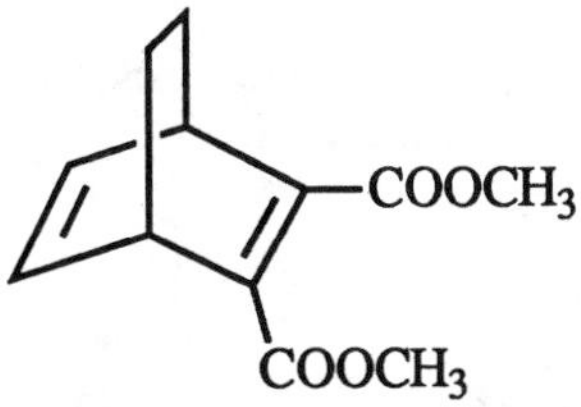

15-25

(a) 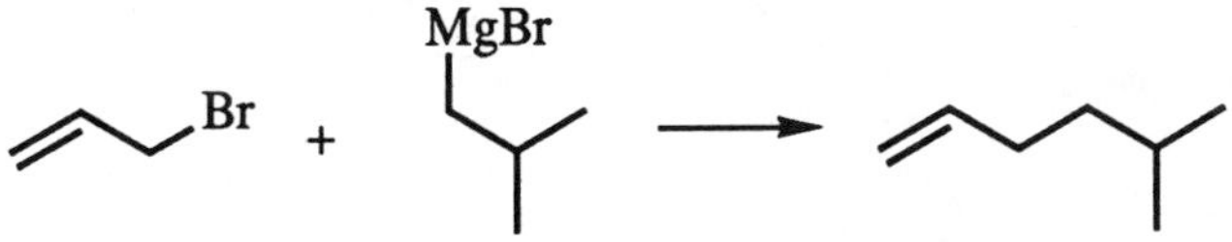

(b) 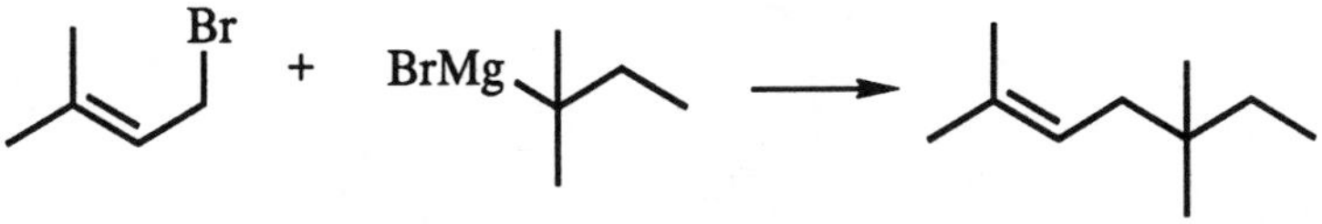

15-26

(a)

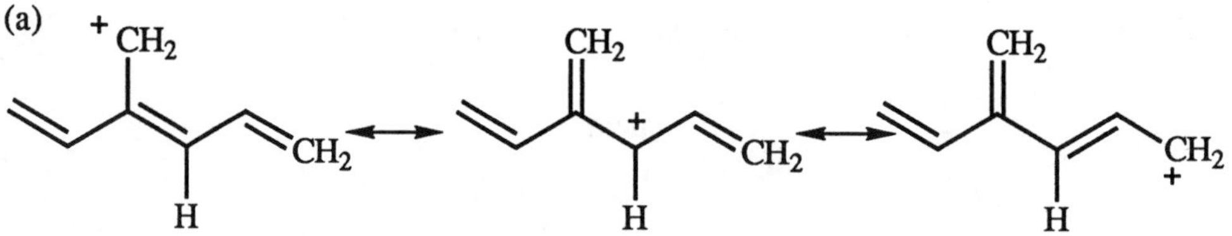

(b)

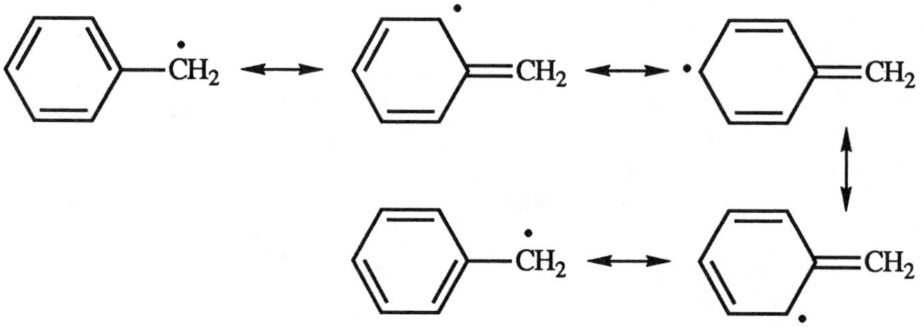

(c)

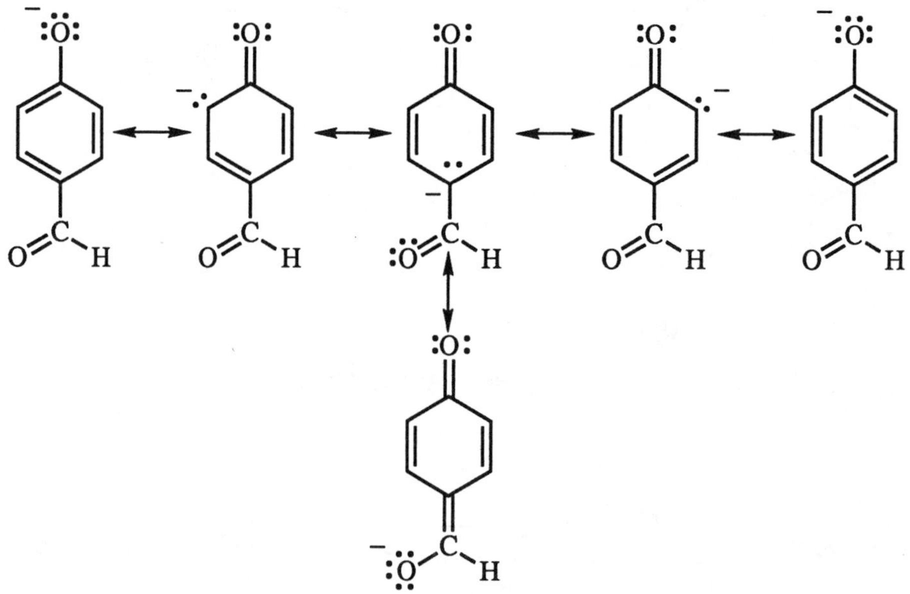

(d)

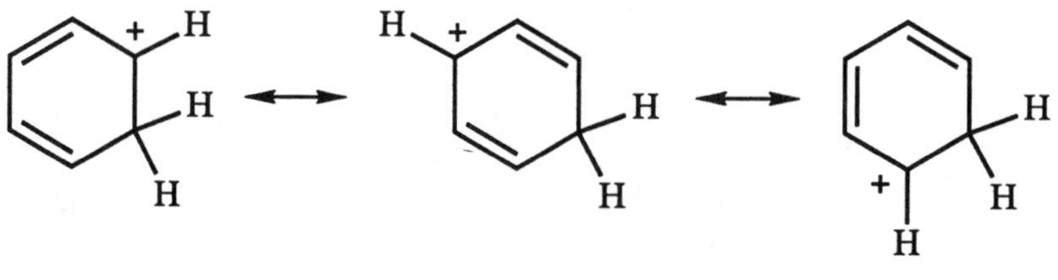

15-27

(a) $A = \varepsilon c l \qquad \varepsilon = \dfrac{A}{c l} \qquad l = 1\text{ cm} \qquad A = 0.74$

convert mass to moles: $0.0010\text{ g} \times \dfrac{1\text{ mole}}{390\text{ g}} = 2.6 \times 10^{-6}\text{ moles}$

$$c = \frac{2.6 \times 10^{-6}\text{ moles}}{100\text{ mL}} \times \frac{1000\text{ mL}}{1\text{ L}} = 2.6 \times 10^{-5}\text{ M}$$

$$\varepsilon = \frac{0.74}{(2.6 \times 10^{-5})(1)} = 28{,}860 \approx \boxed{29{,}000}$$

(b) This large value of ε could only come from a conjugated system, eliminating the first structure. The absorption maximum at 235 nm is most likely a diene rather than a triene. The most reasonable structure is:

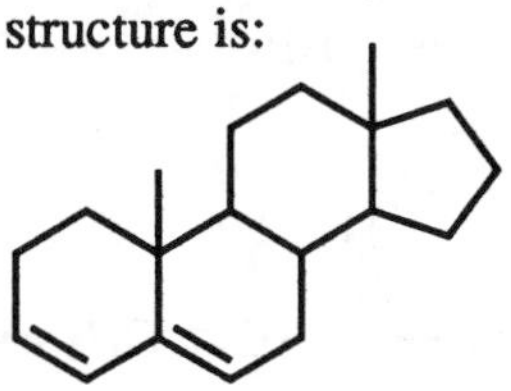

compare with:

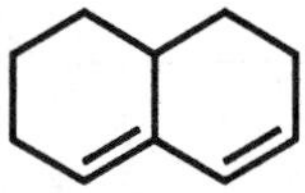

λ_{max} = 235 nm
Solved Problem 15-3

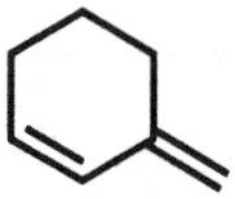

λ_{max} = 232 nm
Table 15-2

15-28

(a)

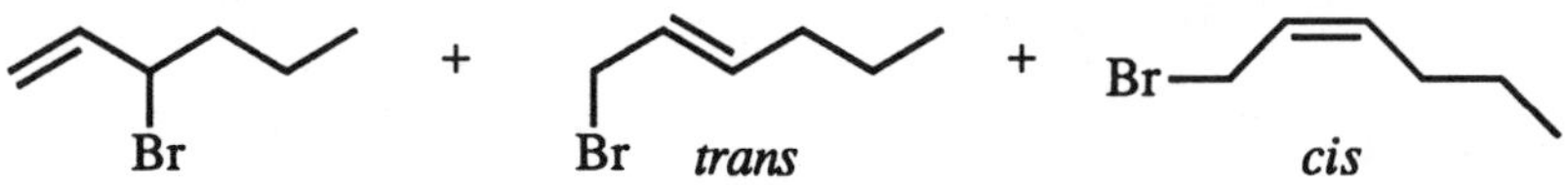
+
Br
Br
trans
+
Br
cis

15-28 continued

(b) ("Pr" is the abbreviation for *n*-propyl, below.)

NBS gnerates a low concentration of Br_2

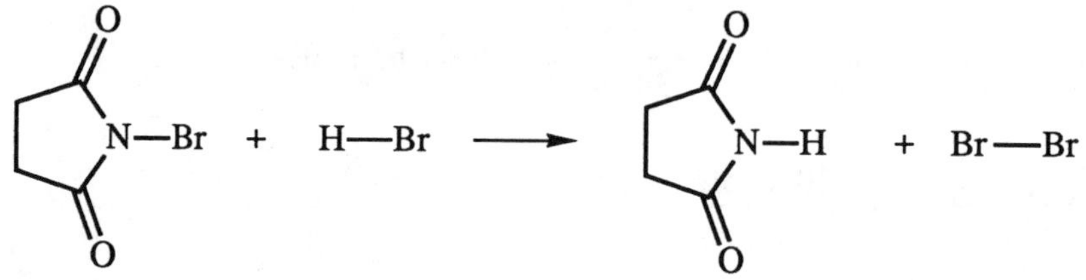

initiation

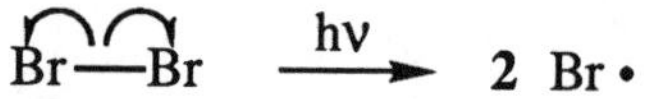

propagation

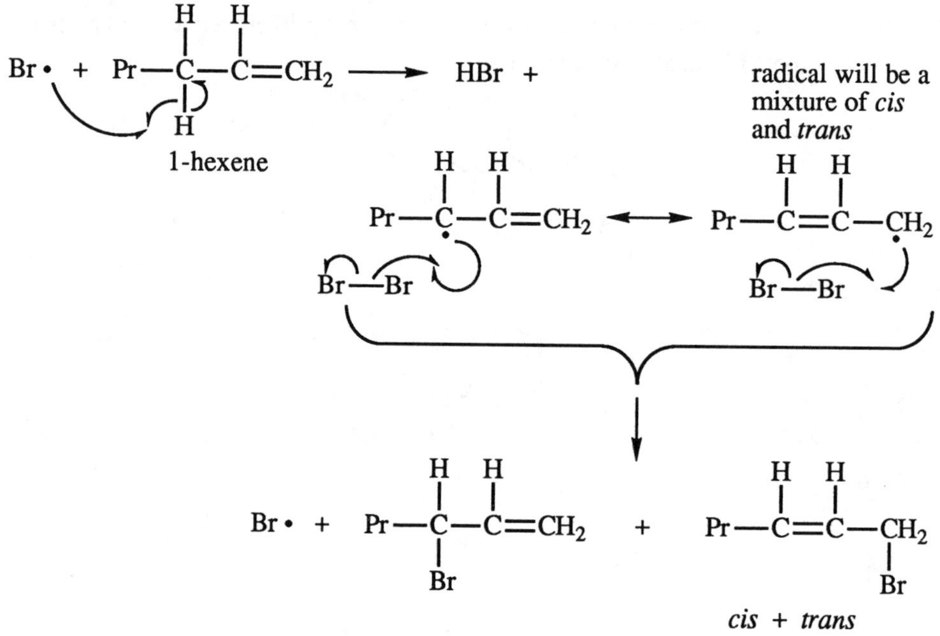
1-hexene
radical will be a mixture of *cis* and *trans*
cis + *trans*

15-29

(a)

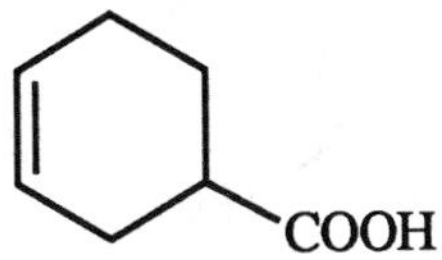

(b)

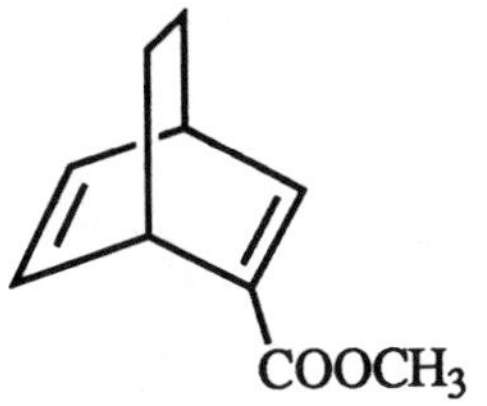

(c)

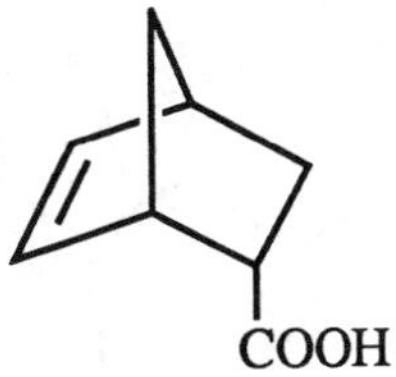

(d)

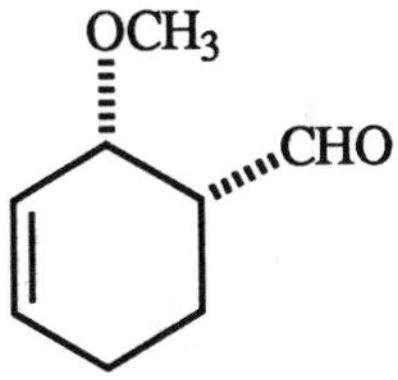

(e)

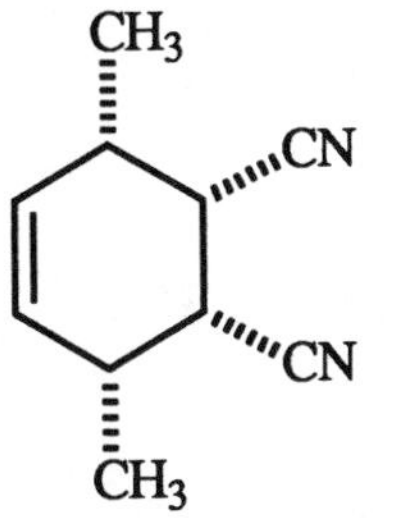

(f)

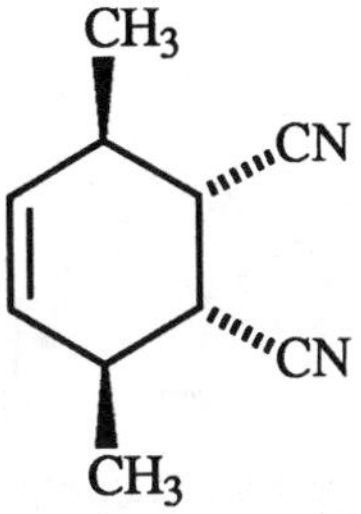

(Note: Yield of the product in (f) would be small or zero due to severe steric interaction in the *s-cis* conformation of the diene. See Figure 15-16.)

15-30

(a)

more significant contributor: 2°

(b)

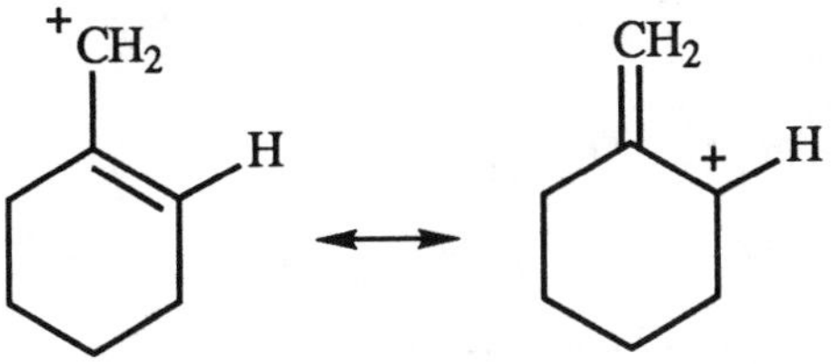

more significant contributor: 2°

(c)

more significant contributor—
negative charge on more electronegative atom

(d)

more significant contributor—
no charge separation

15-30 continued

(e)

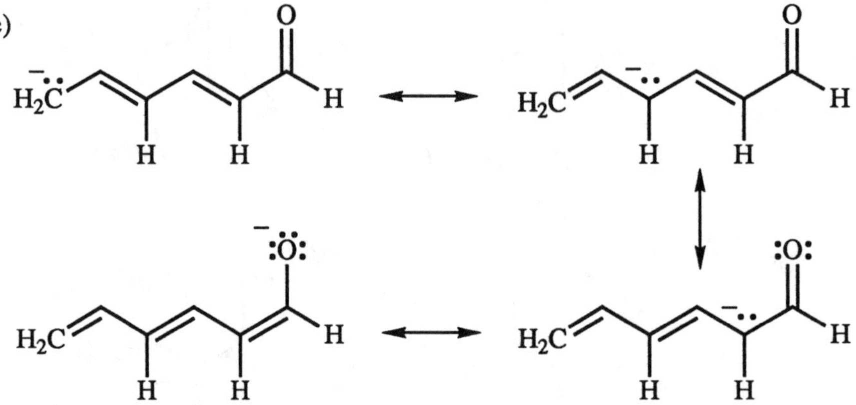

most significant contributor—
negative charge on more electronegative atom

(f)

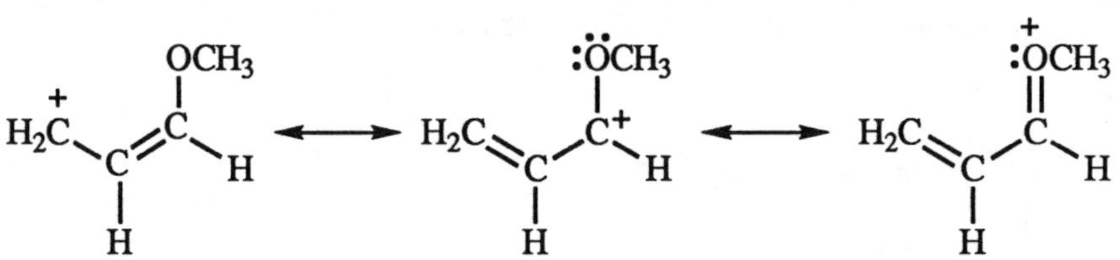

most significant contributor—
all atoms have full octets

(g)

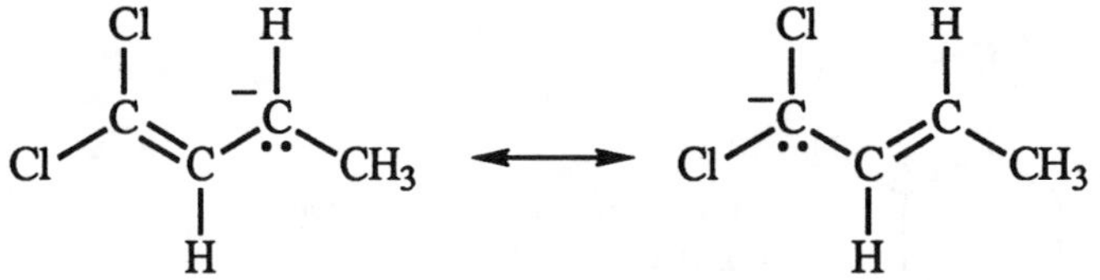

more significant contributor—
negative charge can be stabilized
by electronegative chlorines

15-31 (a) The absorption at 1630 cm^{-1} suggests a conjugated alkene. The higher temperature allowed for migration of the double bond.

(b)

desired actual

15-31 continued

(c)

expected:

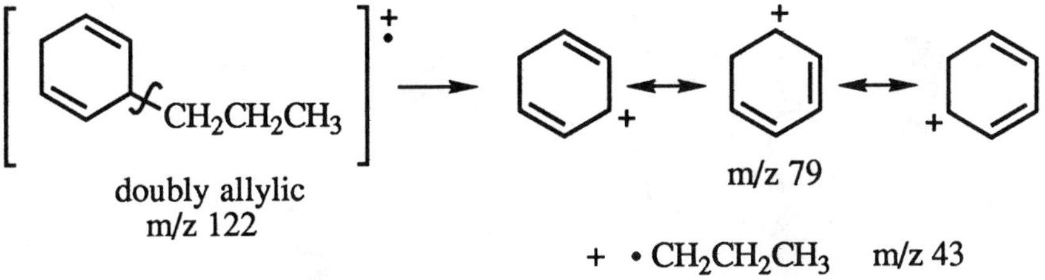
CH2CH2CH3
doubly allylic
m/z 122
m/z 79
+ • CH2CH2CH3 m/z 43

actual:

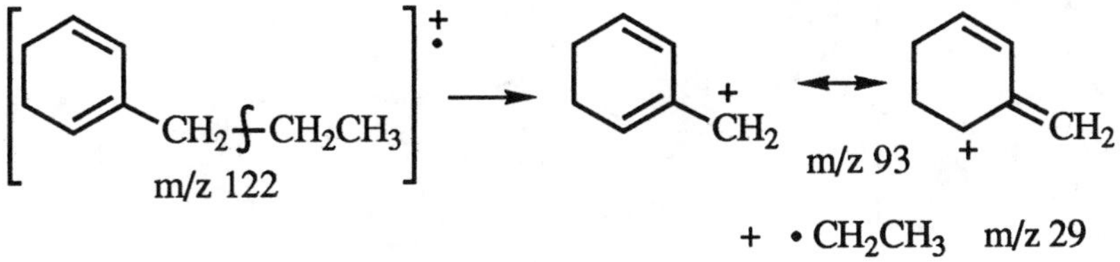
CH2 CH2CH3
m/z 122
CH2
m/z 93
CH2
+ • CH2CH3 m/z 29

15-32

(a)

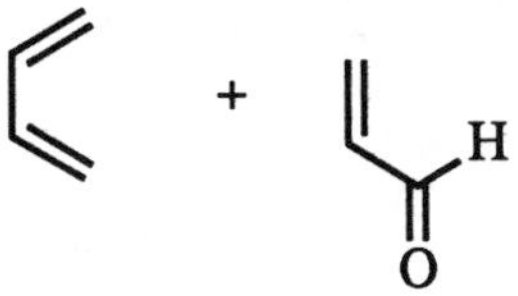
H
O

(b)

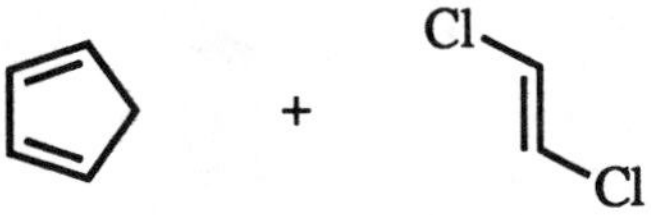
Cl
Cl

(c)

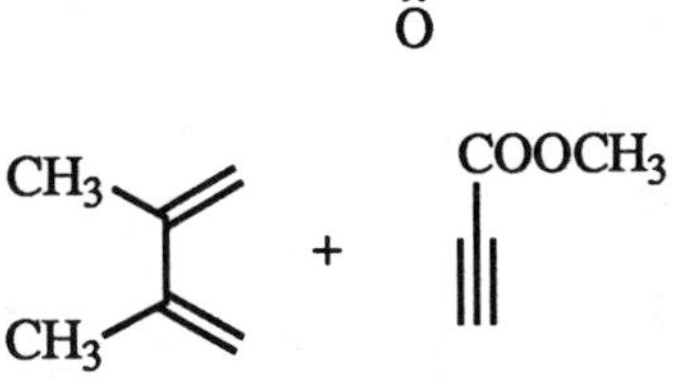
CH3
CH3
COOCH3

(d)

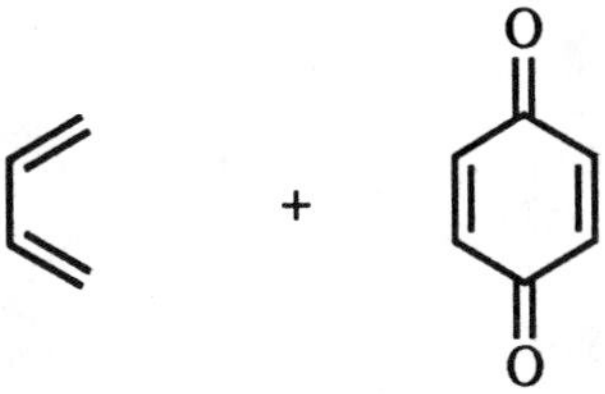
O
O

(e)

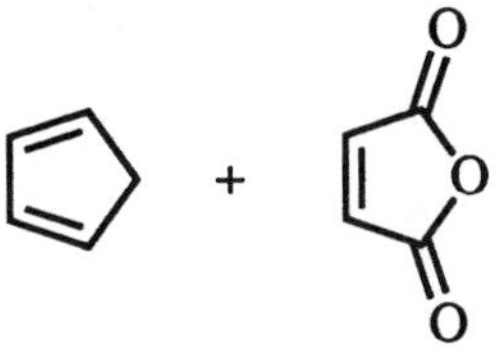
O
O
O

(f)

O
CN
CN

(g)

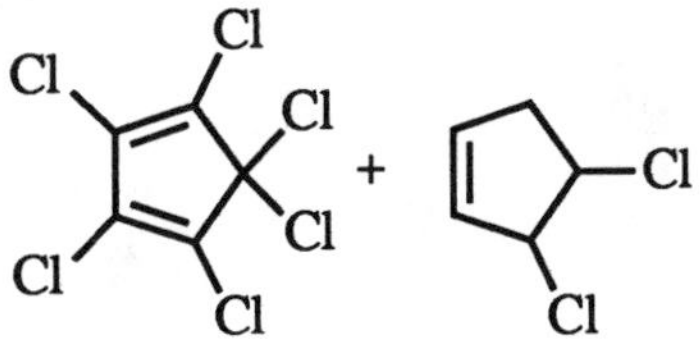
Cl
Cl
Cl
Cl
Cl
Cl
Cl
Cl

(h)

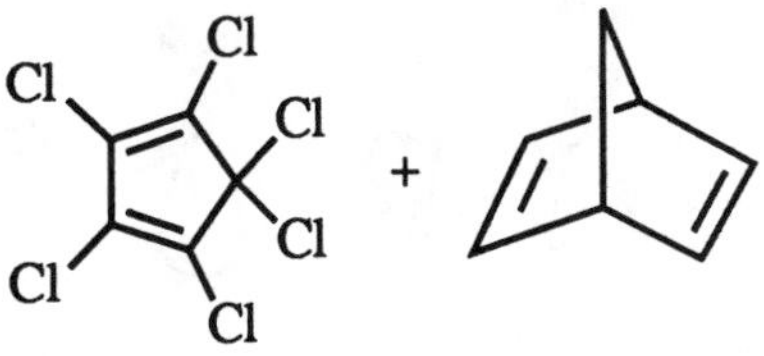
Cl
Cl
Cl
Cl
Cl
Cl

(i)

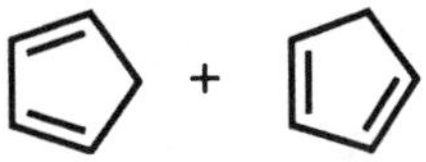

15-33 Nodes are represented by dashed lines.

(a)

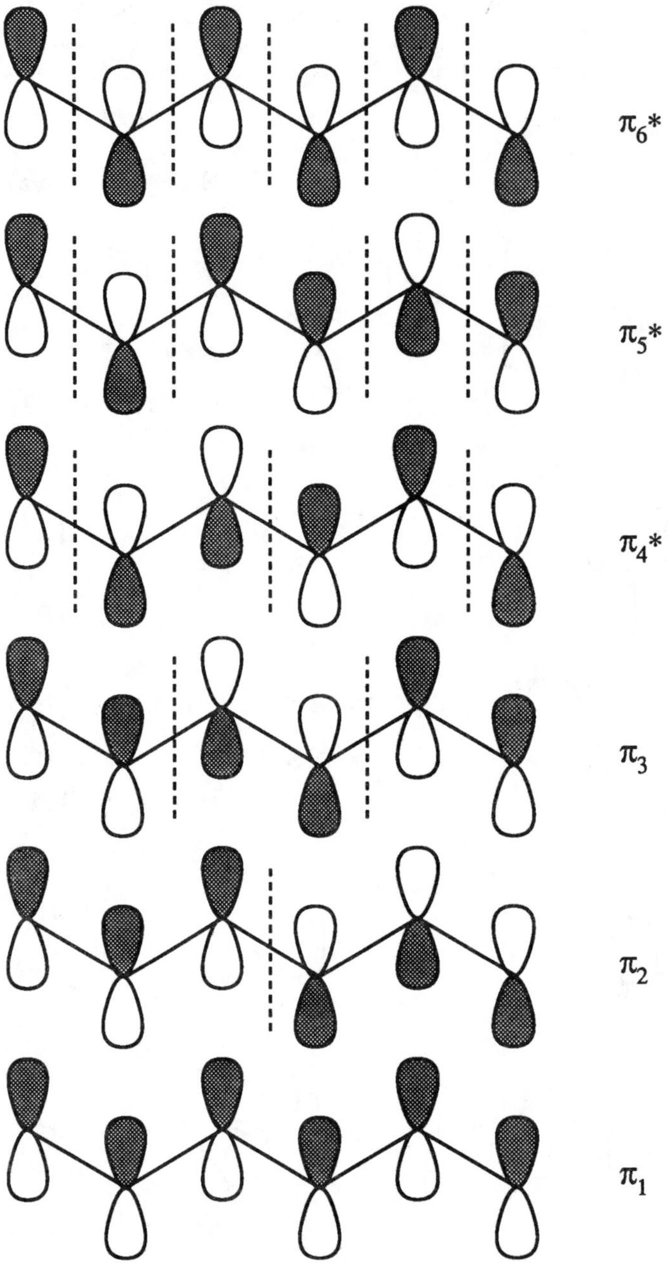

15-33 continued (b)

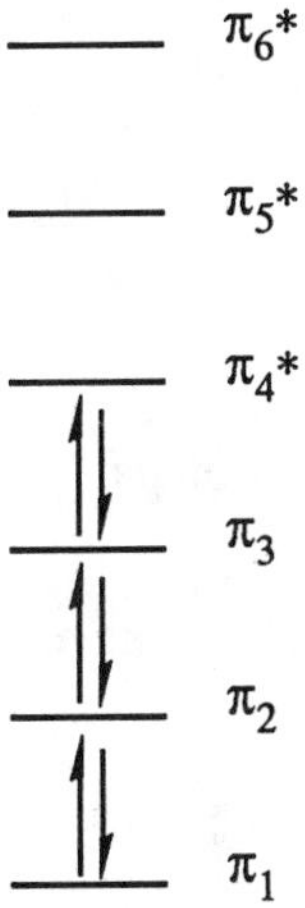

π_6*
π_5*
π_4*
π_3
π_2
π_1

(c)

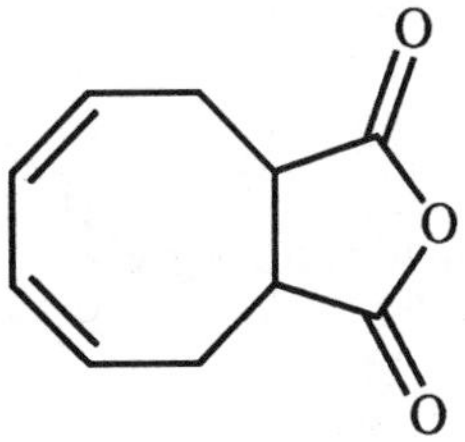

O
O
O

(d) Whether the triene is the HOMO and the alkene is the LUMO, or *vice versa*, the answer will be the same.

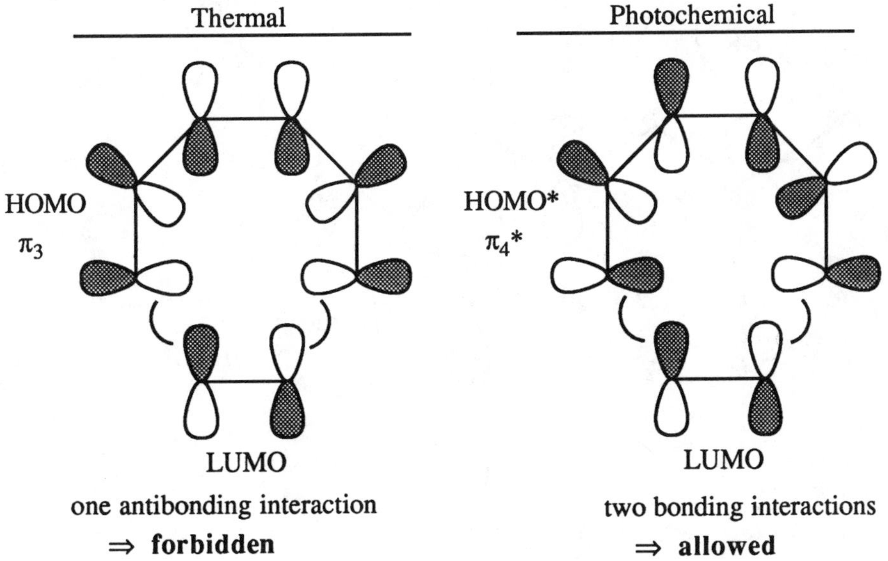

Thermal
HOMO
π_3
LUMO
one antibonding interaction
⇒ forbidden
Photochemical
HOMO*
π_4*
LUMO
two bonding interactions
⇒ allowed

(e)

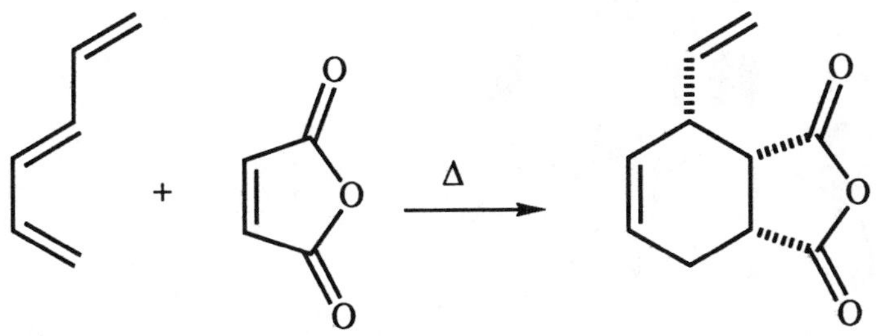

+
Δ
O
O
O
O
O
O

15-34

(a)

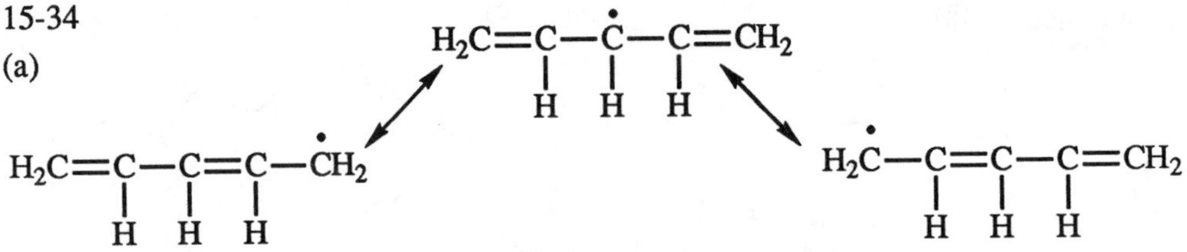

(b) Five atomic orbitals will generate five molecular orbitals.

(c) The lowest energy molecular orbital has no nodes. Each higher molecular orbital will have one more node, so the fifth molecular orbital will have four nodes.

(d) Nodes are represented by dashed lines.

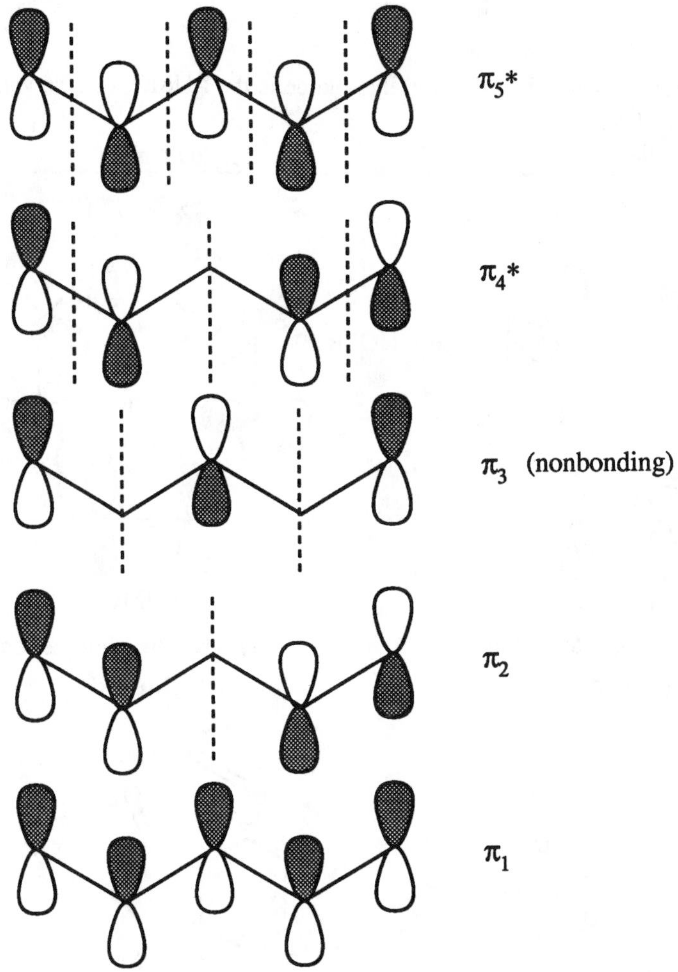
π_5*
π_4*
π_3 (nonbonding)
π_2
π_1

15-34 continued

(e)

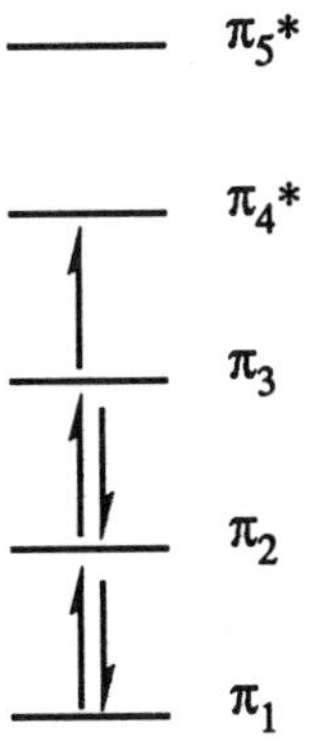

(f) The HOMO, π_3, contains an unpaired electron giving this species its radical character. The HOMO is a non-bonding orbital with lobes only on carbons 1, 3, and 5, consistent with the resonance picture.

(g)

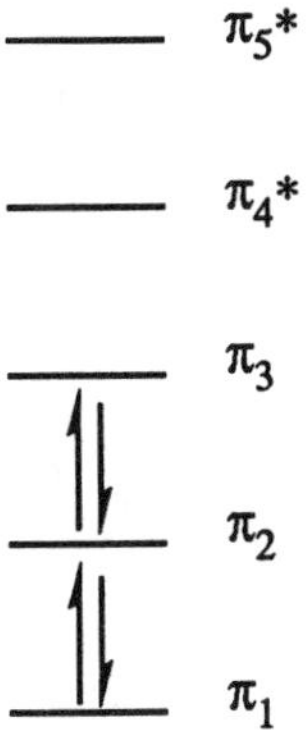

Again, it is π_3 that determines the character of this species. When the single electron in π_3 of the neutral radical is removed, positive charge appears only in the position(s) which that electron occupied. That is, the positive charge depends on the now *empty* π_3, with *empty* lobes (positive charge) on carbons 1, 3, and 5, consistent with the resonance description.

(h)

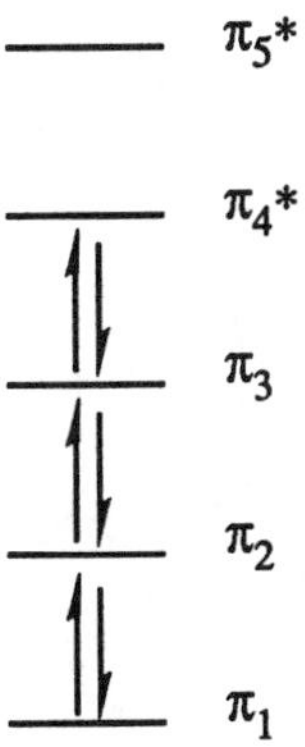

Again, it is π_3 that determines the character of this species. The negative charge depends on the *filled* π_3, with lobes (negative charge) on carbons 1, 3, and 5, consistent with the resonance description.

CHAPTER 16—AROMATIC COMPOUNDS

16-1

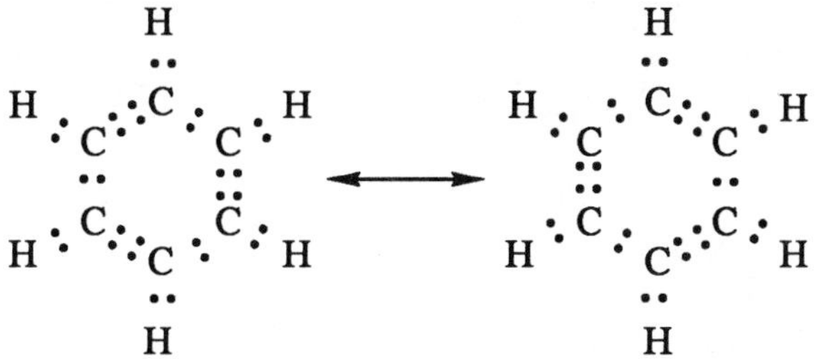

16-2 All values are per mole.

(a)

benzene	− 49.8 kcal	(− 208 kJ)
− (1,4-cyclohexadiene	− 57.4 kcal	(− 240 kJ))
ΔH =	+ 7.6 kcal	(+ 32 kJ)

(b)

benzene	− 49.8 kcal	(− 208 kJ)
− (cyclohexene	− 28.6 kcal	(− 120 kJ))
ΔH =	− 21.2 kcal	(− 88 kJ)

(c)

1,3-cyclohexadiene	− 55.4 kcal	(− 232 kJ)
− (cyclohexene	− 28.6 kcal	(− 120 kJ))
ΔH =	− 26.8 kcal	(− 112 kJ)

16-3

(a)

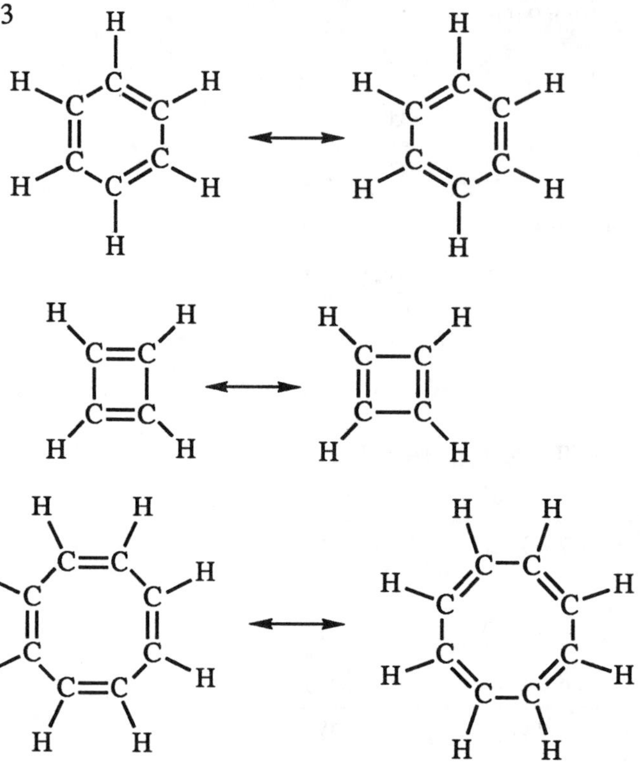

(b)

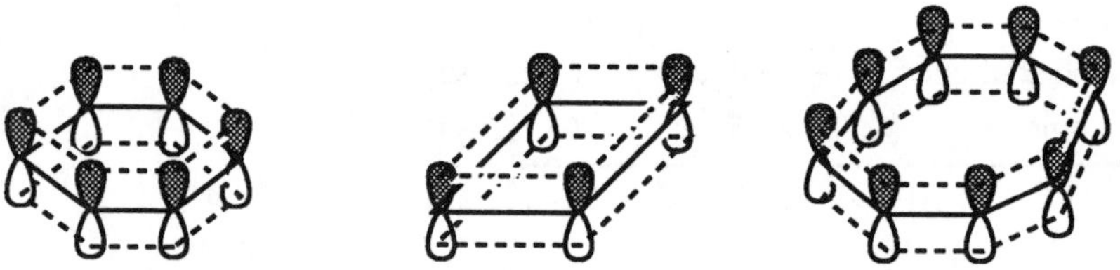

16-4

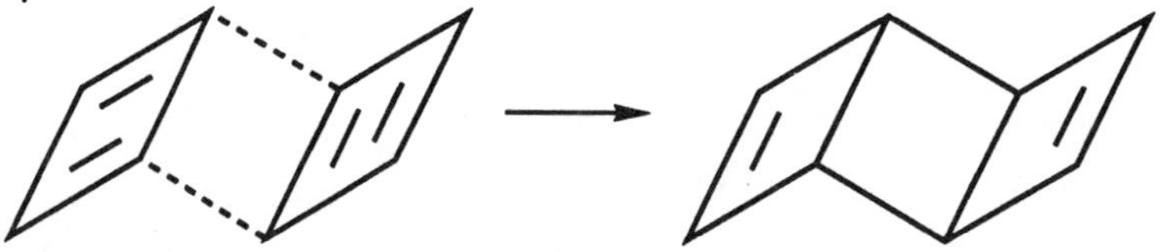

16-5 Figure 16-8 shows that the first 3 pairs of electrons are in three bonding molecular orbitals of cyclooctatetraene. Electrons 7 and 8, however, are located in two different nonbonding orbitals. As in cyclobutadiene, a planar cyclooctatetraene is predicted to be a diradical, a particularly unstable electron configuration.

16-6

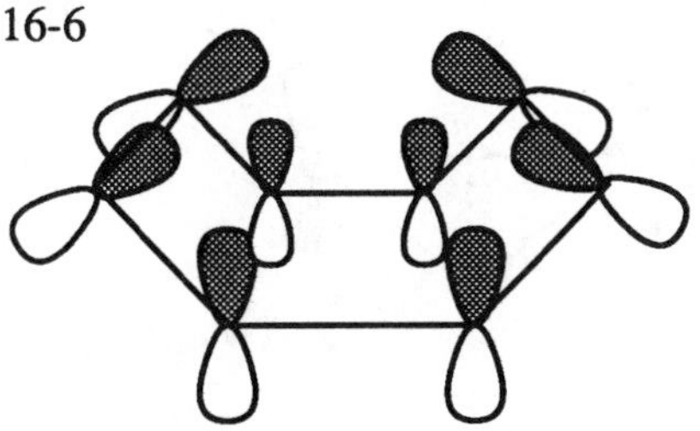

Models show that the angles between p orbitals on adjacent π bonds approach 90°.

16-7 (a) nonaromatic: internal hydrogens prevent planarity
(b) nonaromatic: not all atoms in the ring have a p orbital
(c) aromatic: [14]annulene
(d) aromatic: also a [14]annulene in the outer ring: the internal alkene is not part of the aromatic system

16-8 Azulene satisfies all the criteria for aromaticity, and it has a Huckel number of π electrons: 10. Both heptalene (12 π electrons) and pentalene (8 π electrons) are antiaromatic.

16-9

(a)

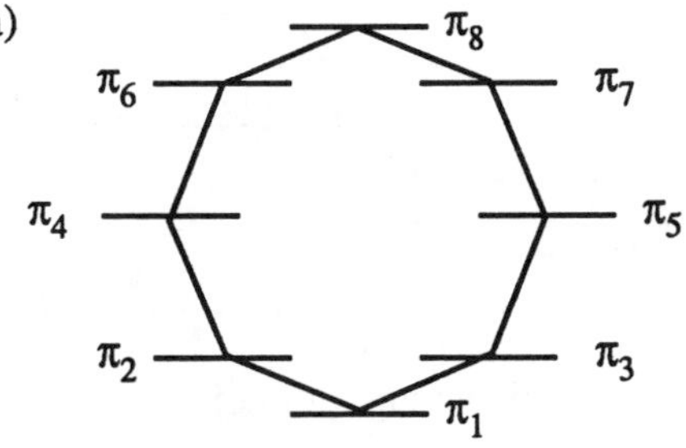
π_8
π_6
π_7
π_4
π_5
π_2
π_3
π_1

(b)

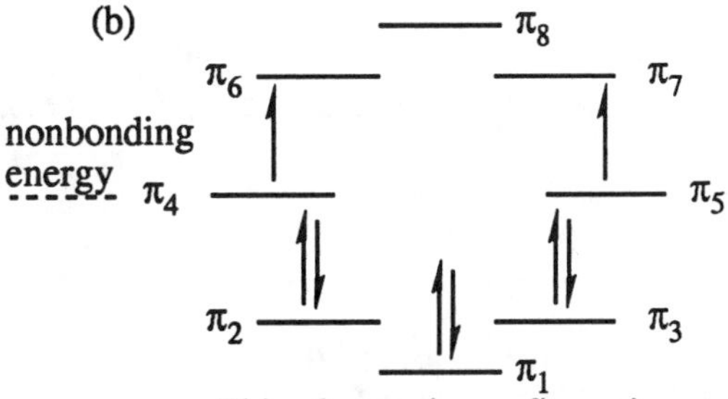
nonbonding energy
π_8
π_6
π_7
π_4
π_5
π_2
π_3
π_1

This electronic configuration is antiaromatic.

(c)

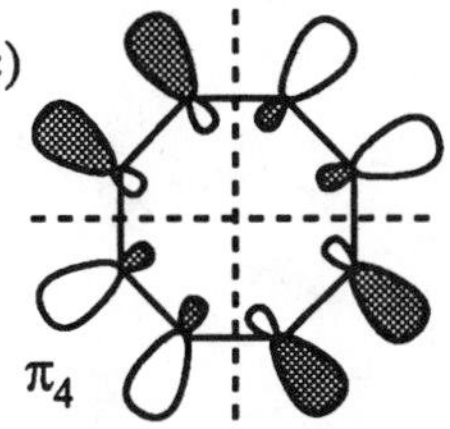
π_4

These are TOP views. Nodes are shown by dashed lines.

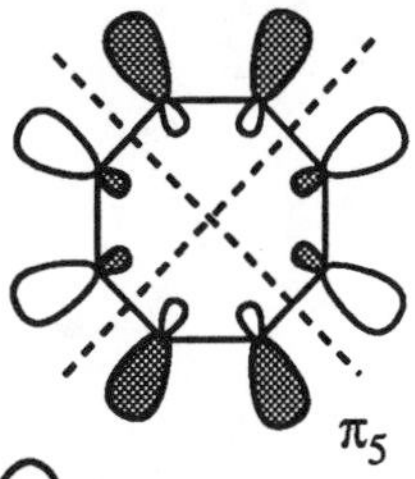
π_5

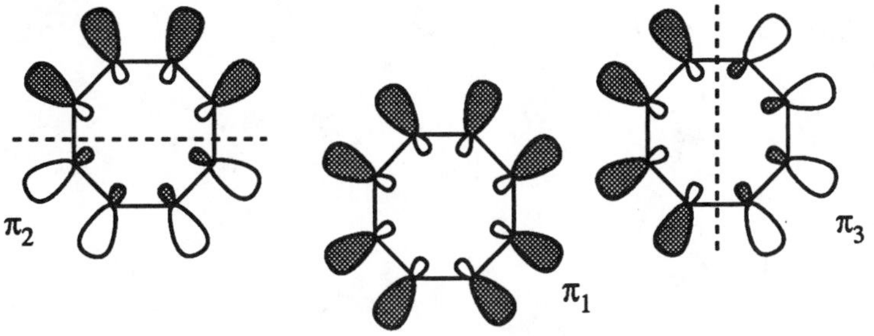
π_2
π_1
π_3

16-10

(a)

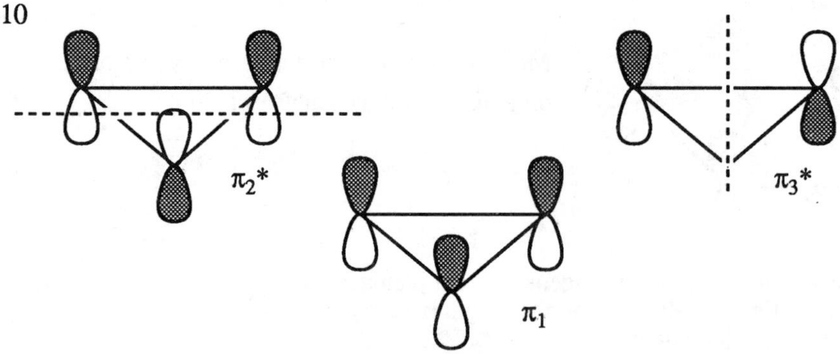

(b)

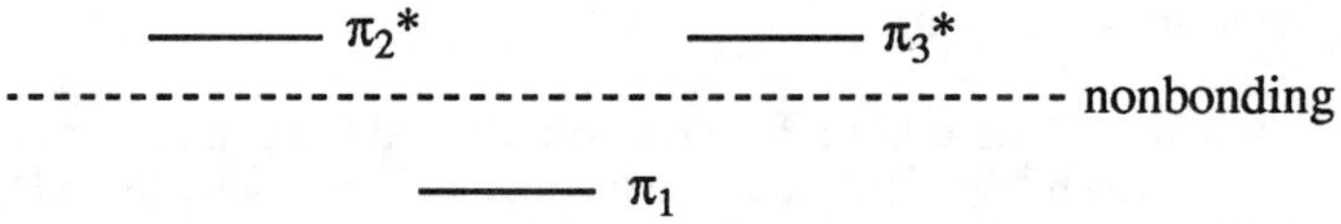

π_1 is bonding; π_2* and π_3* are antibonding.

(c)

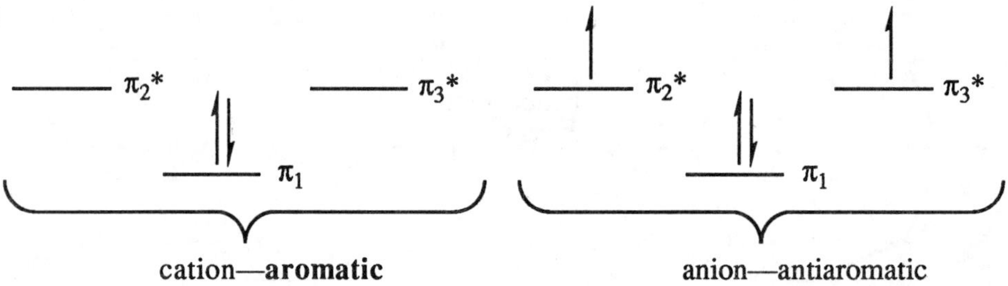

16-11

(a)

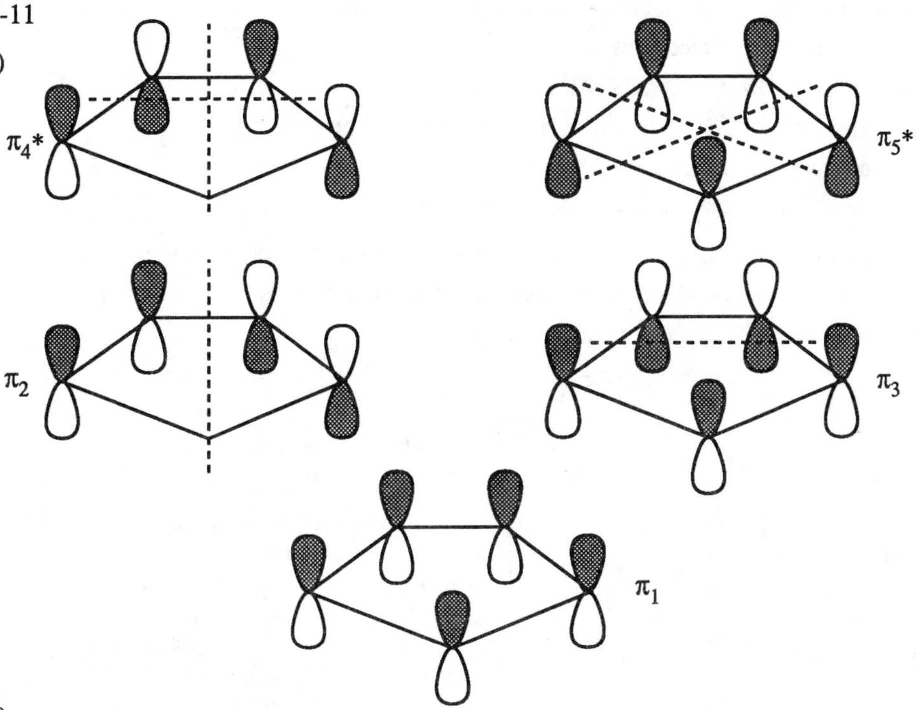

π_4*
π_5*
π_2
π_3
π_1

(b)

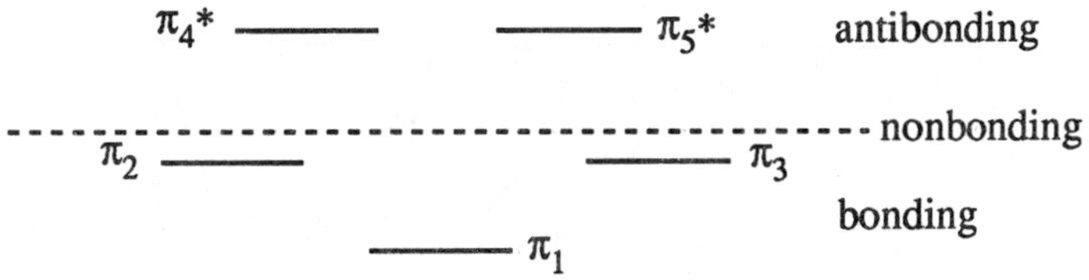

π_4*
π_5*
antibonding
nonbonding
π_2
π_3
bonding
π_1

(c)

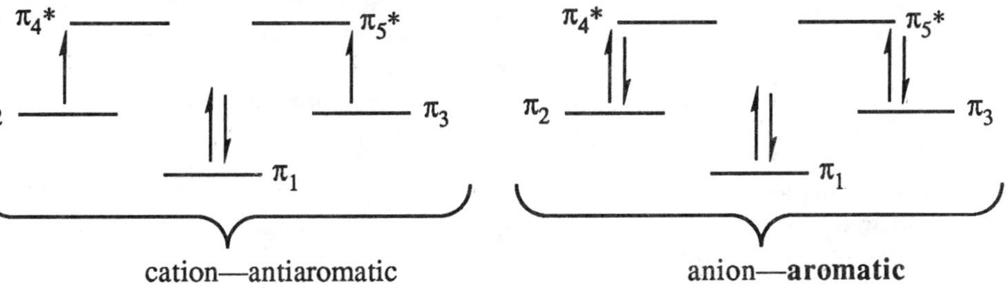

π_4*
π_5*
π_2
π_3
π_1
cation—antiaromatic
π_4*
π_5*
π_2
π_3
π_1
anion—**aromatic**

16-12 (a) antiaromatic: 8 π electrons, not a Huckel number

(b) aromatic: 10 π electrons, a Huckel number

(c) aromatic: 18 π electrons, a Huckel number

(d) antiaromatic: 20 π electrons, not a Huckel number

(e) nonaromatic: no cyclic π system

(f) aromatic: 18 π electrons, a Huckel number

16-13 The reason for the dipole can be seen in a resonance form distributing the electrons to give each ring 6 π electrons. This resonance picture gives one ring a negative charge and the other ring a positive charge.

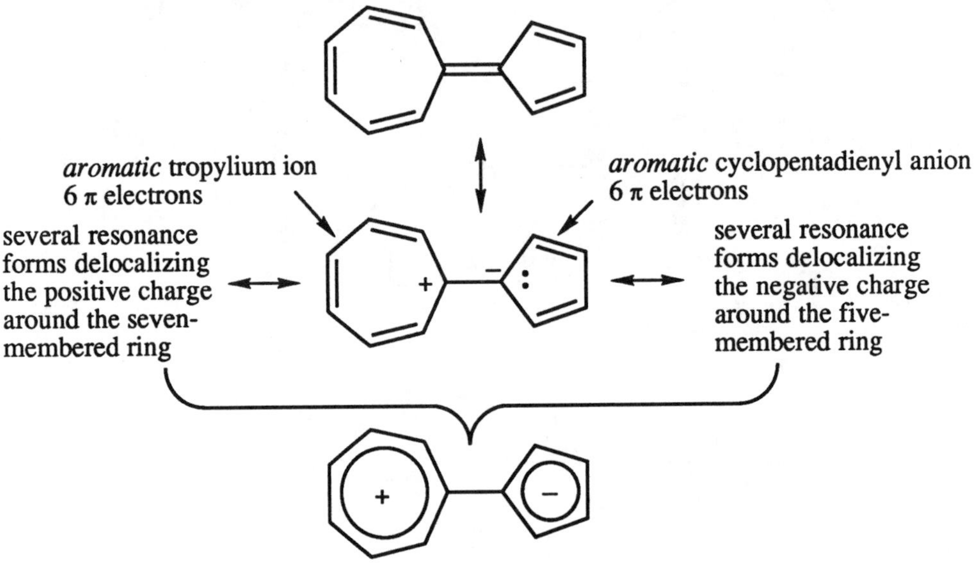

aromatic tropylium ion
6 π electrons
aromatic cyclopentadienyl anion
6 π electrons
several resonance forms delocalizing the positive charge around the seven-membered ring
several resonance forms delocalizing the negative charge around the five-membered ring

16-14

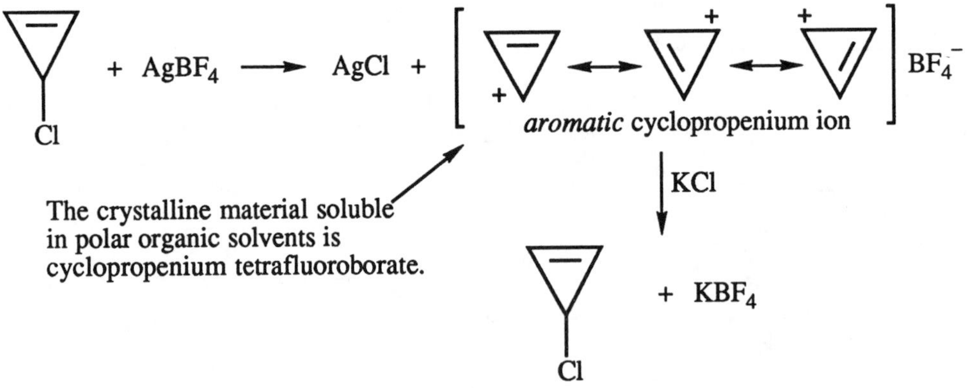

Cl
+ $AgBF_4$
AgCl +
BF_4^-
aromatic cyclopropenium ion
KCl
The crystalline material soluble in polar organic solvents is cyclopropenium tetrafluoroborate.
+ KBF_4
Cl

16-15

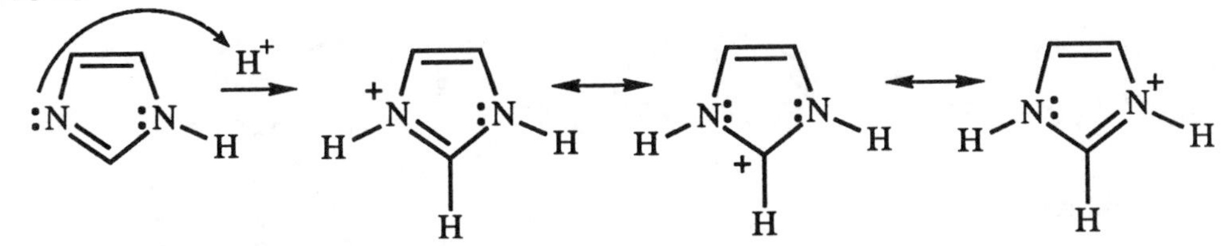

The two nitrogens of the protonated imidazole are equivalent and equally acidic.

16-16

(a) aromatic: one electron pair from oxygen joins the two π bonds to make a 6 π electron system

(b) nonaromatic: no cyclic π system

(c) aromatic: cation on carbon-4 indicates an empty p orbital; two π bonds plus a pair of electrons from oxygen makes a 6 π electron system

(d) nonaromatic: no cyclic π system

(e) aromatic: one electron pair from sulfur joins the two π bonds to make a 6 π electron system

16-17

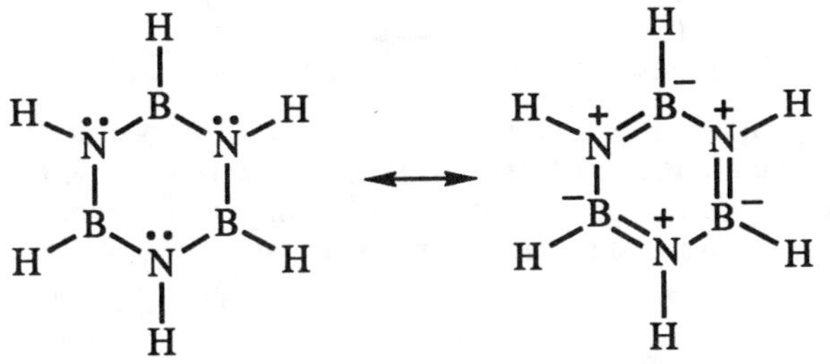

Borazole is a non-carbon equivalent of benzene. Each boron is hybridized in its normal sp^2. Each nitrogen is also sp^2 with its pair of electrons in its p orbital. The system has six π electrons in 6 p orbitals—aromatic!

16-18

(a)

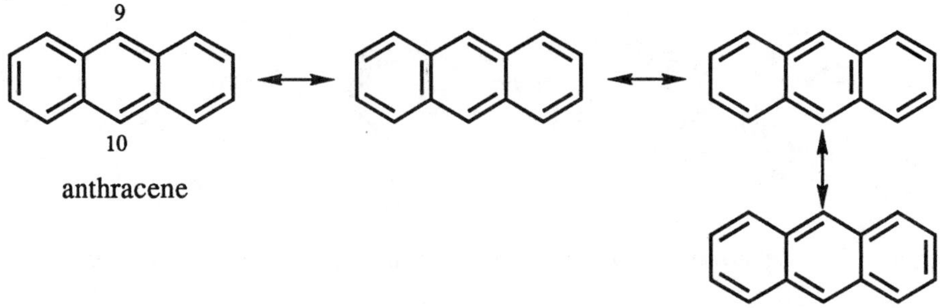

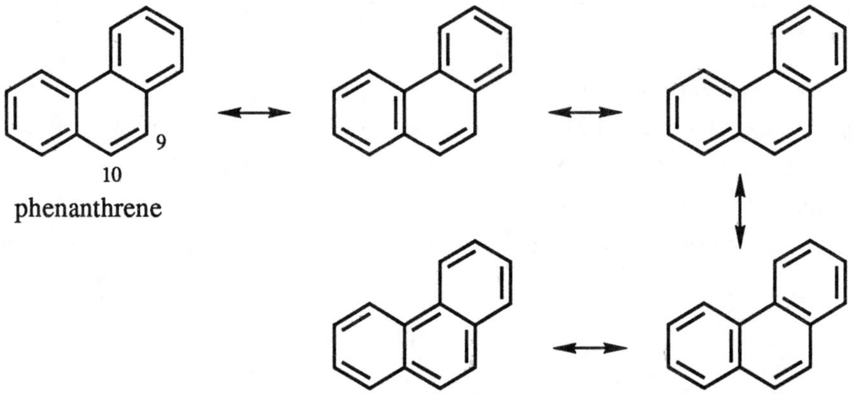

(b) Isolated double bonds undergo addition reactions. The C-9 and C-10 positions in anthracene and phenanthrene are the most isolated from the benzene-type rings on the ends of the molecules. This can be seen most readily in phenanthrene in which four of the five resonance forms show an alkene between C-9 and C-10.

(c) <u>anthracene</u>

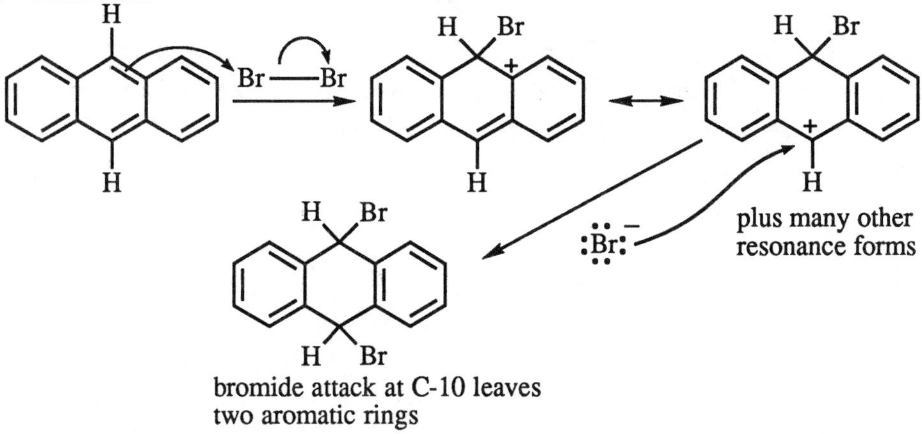

16-18 (c) continued

<u>phenanthrene</u>

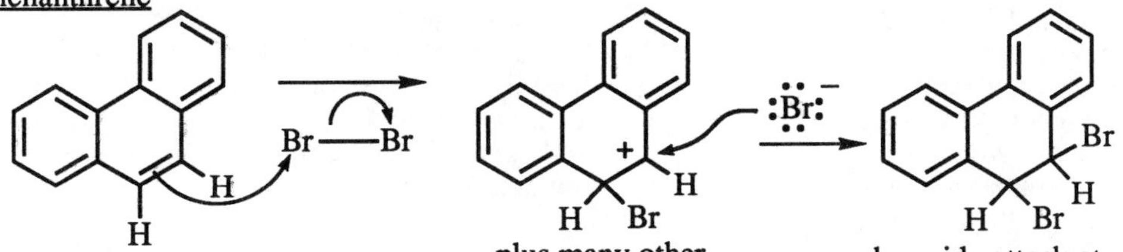

plus many other resonance forms

bromide attack at C-9 leaves two aromatic rings

16-19

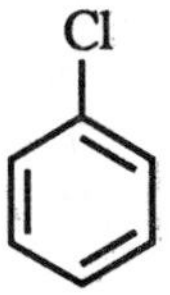

chlorobenzene

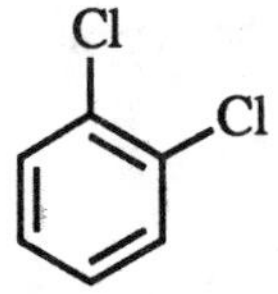

o-dichlorobenzene (1,2-dichlorobenzene)

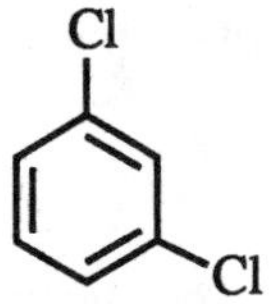

m-dichlorobenzene (1,3-dichlorobenzene)

p-dichlorobenzene (1,4-dichlorobenzene)

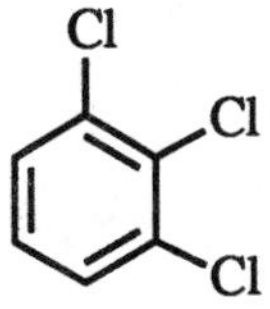

1,2,3-trichlorobenzene

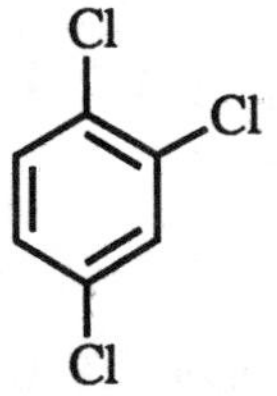

1,2,4-trichlorobenzene

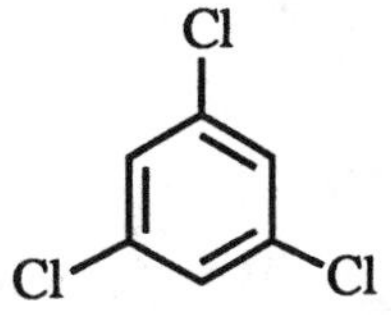

1,3,5-trichlorobenzene

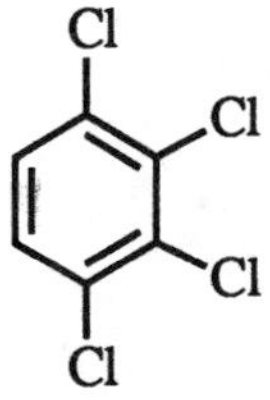

1,2,3,4-tetrachlorobenzene

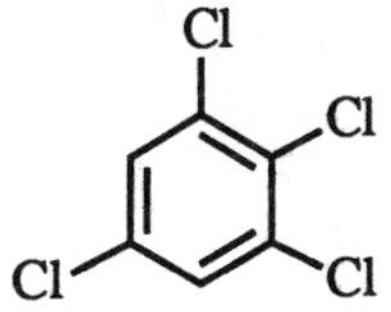

1,2,3,5-tetrachlorobenzene

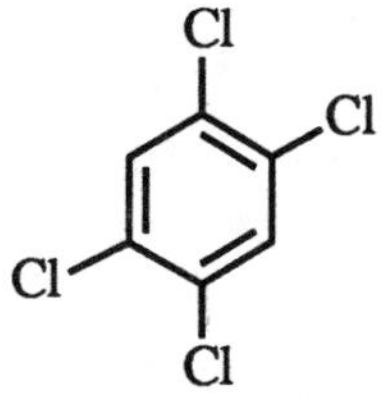

1,2,4,5-tetrachlorobenzene

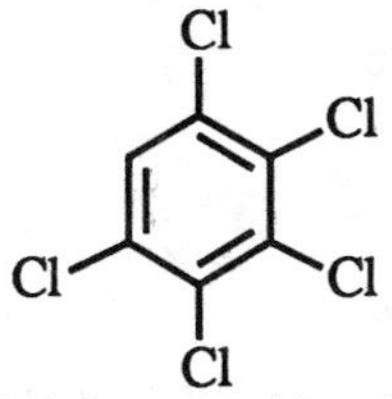

1,2,3,4,5-pentachlorobenzene

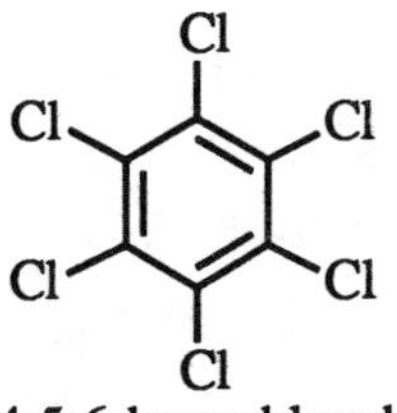

1,2,3,4,5,6-hexachlorobenzene

16-20

(a) fluorobenzene
(b) 4-phenyl-1-butyne
(c) *m*-methylphenol, or 3-methylphenol (common name: *m*-cresol)
(d) *o*-nitrostyrene
(e) *p*-bromobenzoic acid, or 4-bromobenzoic acid
(f) isopropoxybenzene, or isopropyl phenyl ether
(g) 3,4-dinitrophenol
(h) benzyl ethyl ether, or benzoxyethane, or (ethoxymethyl)benzene, or α-ethoxytoluene

16-21 These examples are representative. Your examples may be different from these and still be correct.

(a) 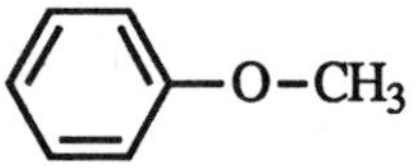
O–CH3
methyl phenyl ether, or methoxybenzene, or ansiole

(b) 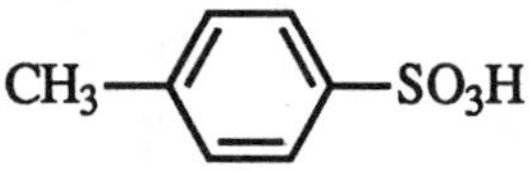
CH3
SO3H
p-toluenesulfonic acid

(c) 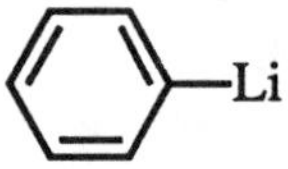
Li
phenyllithium

(d) 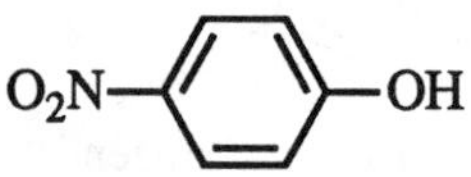
O2N
OH
These are *phenols*. This is 4-nitrophenol.

(e) 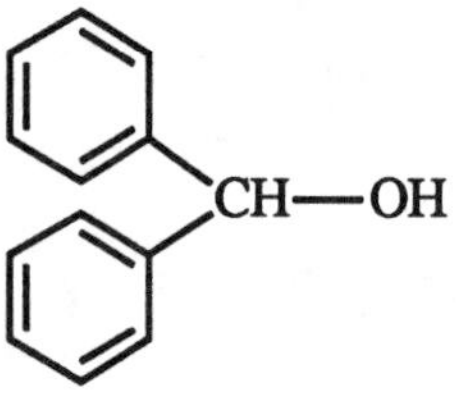
CH—OH
diphenylmethanol

(f) 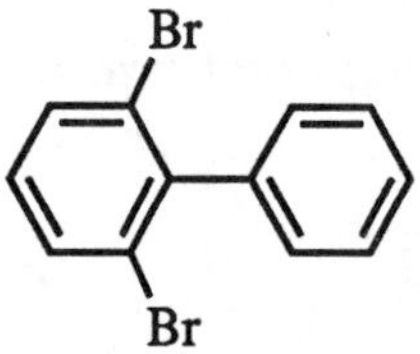
Br
Br
1,3-dibromo-2-phenylbenzene

(g) 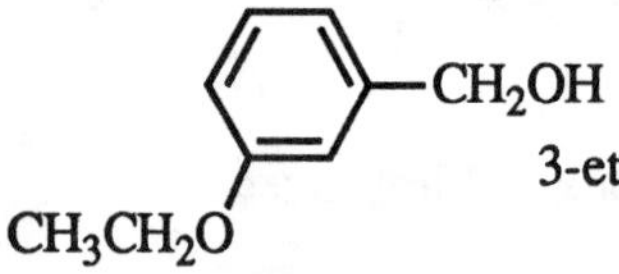
CH2OH
CH3CH2O
3-ethoxybenzyl alcohol, or 3-ethoxyphenylmethanol

16-22

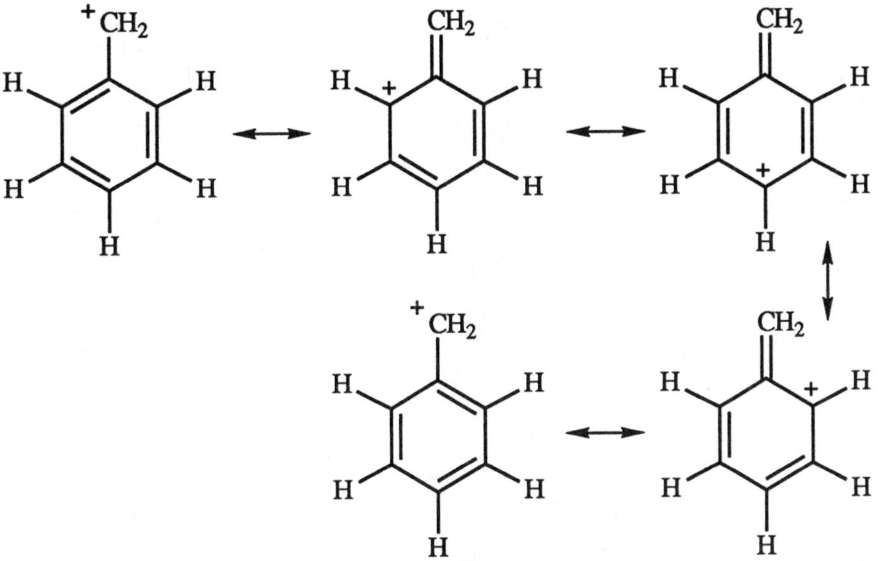

CH2
H

16-23

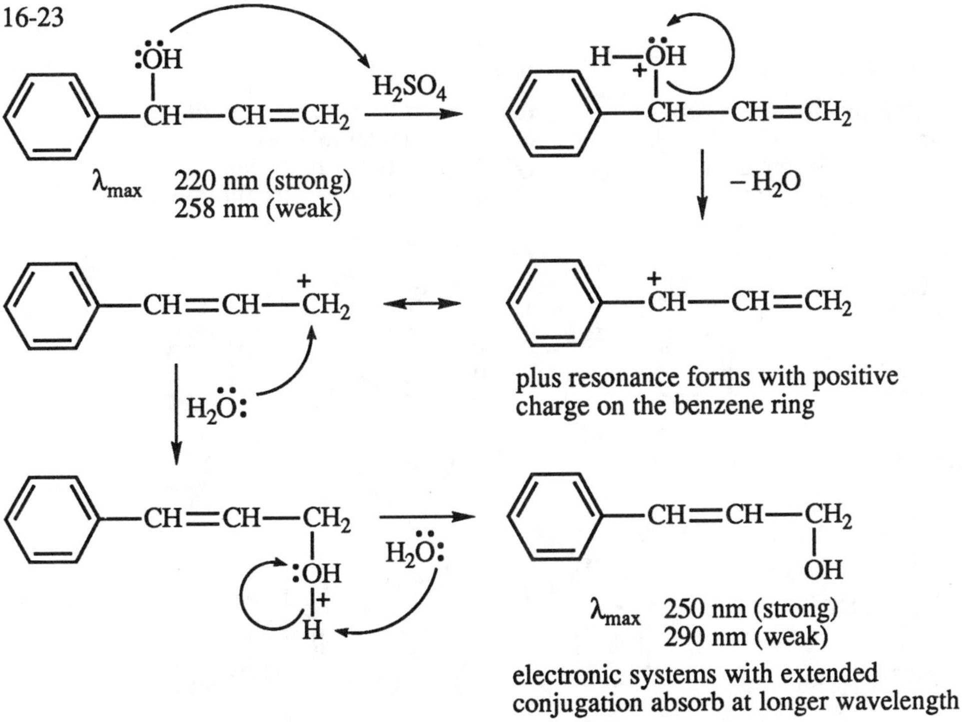

:ÖH
CH—CH=CH2
H2SO4
H—ÖH
CH—CH=CH2
– H2O
λmax 220 nm (strong)
258 nm (weak)
CH=CH—CH2
H2Ö:
plus resonance forms with positive charge on the benzene ring
:OH
H
OH
λmax 250 nm (strong)
290 nm (weak)
electronic systems with extended conjugation absorb at longer wavelength

16-24 Refer to the Glossary and the text for definitions and example.

16-25

(a) 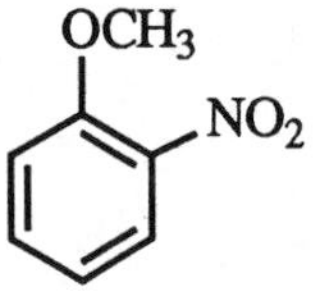

(b) 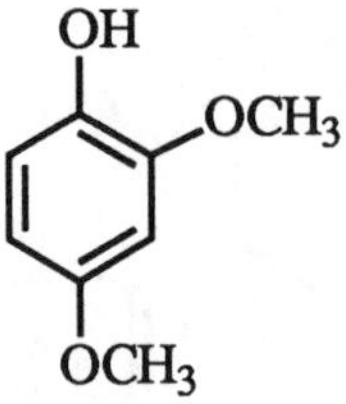

(c) 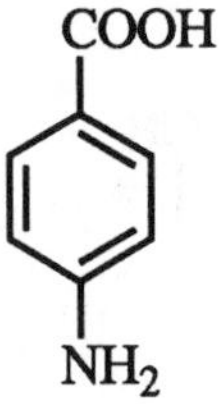

(d) 

(e) 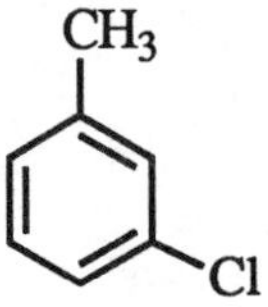

(f) 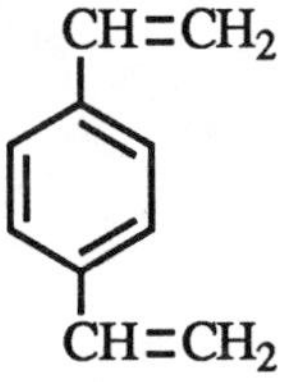

(g) 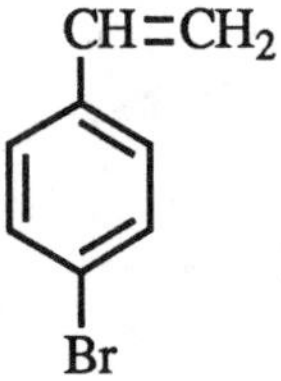

(h) 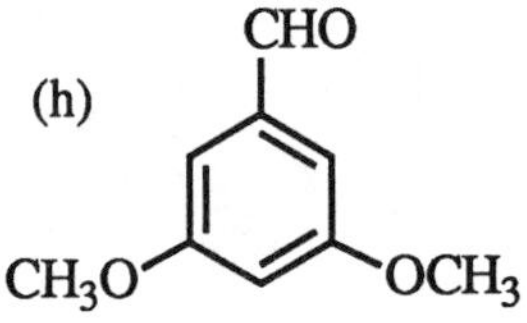

(i) 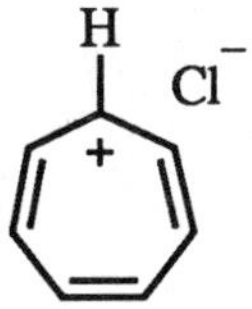

(j) 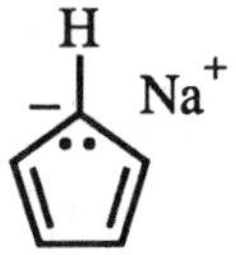

16-26

(a) 1,2-dichlorobenzene (*ortho*)
(b) 4-nitroanisole (*para*)
(c) 2,3-dibromobenzoic acid
(d) 2,7-dimethoxynaphthalene
(e) 3-chlorobenzoic acid (*meta*)
(f) 2,4,6-trichlorophenol
(g) 2-*sec*-butylbenzaldehyde (*ortho*)
(h) cyclopropenium tetrafluoroborate

16-27

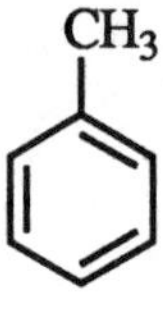

toluene

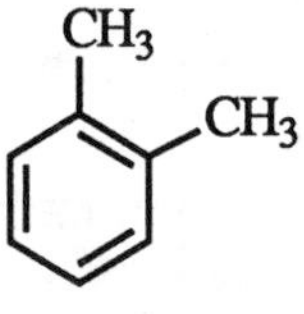

o-xylene

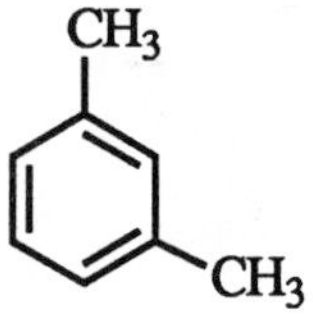

m-xylene

p-xylene

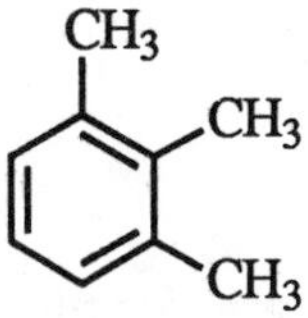

1,2,3-trimethylbenzene

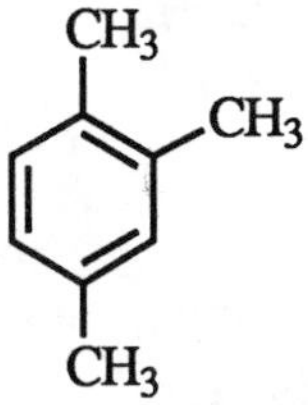

1,2,4-trimethylbenzene

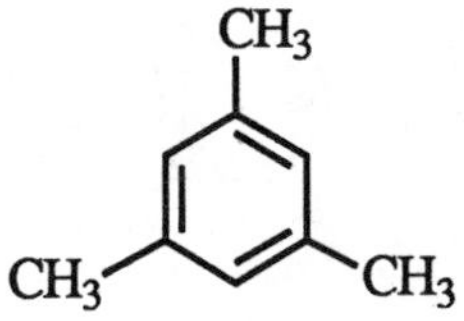

1,3,5-trimethylbenzene

16-28 Aromaticity is one of the strongest stabilizing forces in organic molecules. The cyclopentadienyl system is stabilized in the anion form where it has 6 π electrons, a Huckel number. The question then becomes: which of the four structures can lose a proton to become aromatic?

While the first, third, and fourth structures can lose protons from sp^3 carbons to give resonance-stabilized anions, only structure two can make a cyclopentadienide anion. It will lose a proton most easily of these four structures which, by definition, means it is the strongest acid.

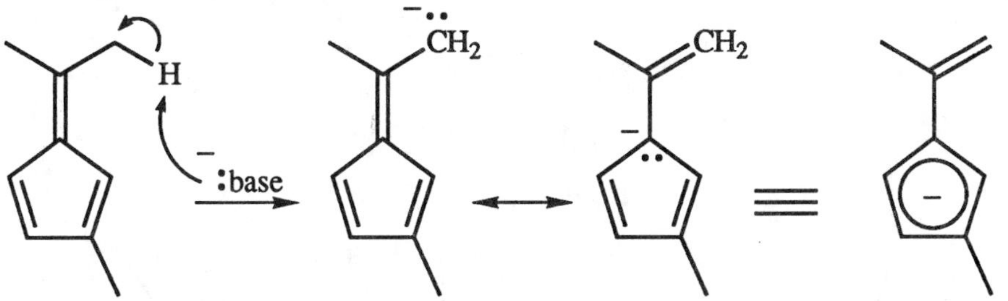

AROMATIC

16-29 Draw resonance forms showing the carbonyl polarization, leaving a positive charge on the carbonyl carbon.

2 π electrons—
AROMATIC

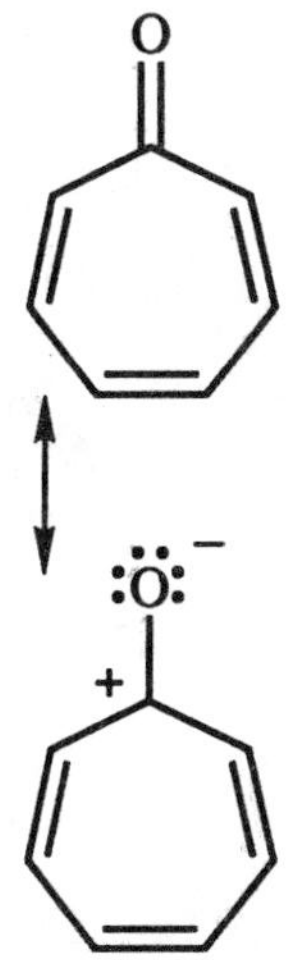

6 π electrons—
AROMATIC

4 π electrons—
ANTIAROMATIC

The three- and seven-membered rings are aromatic; the five-membered ring is antiaromatic and, not surprisingly, very reactive.

16-30

(a)

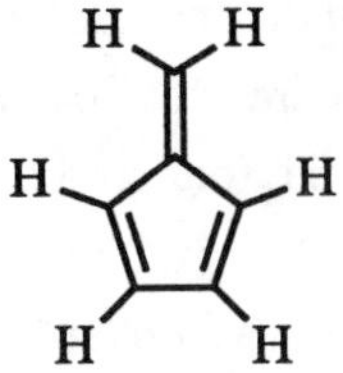

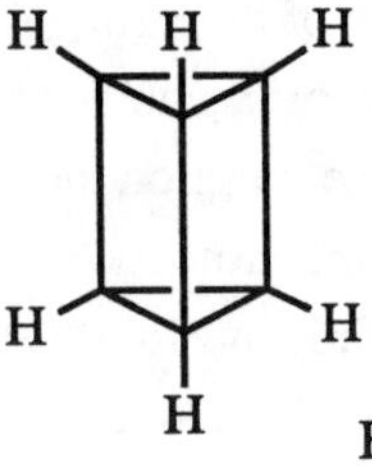

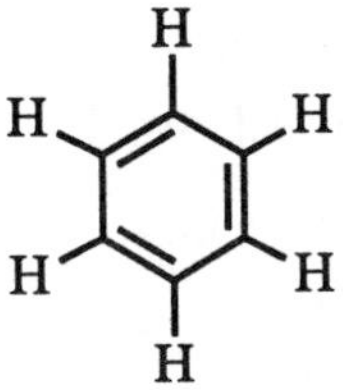

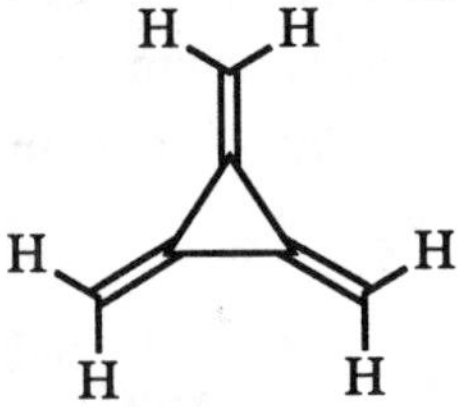

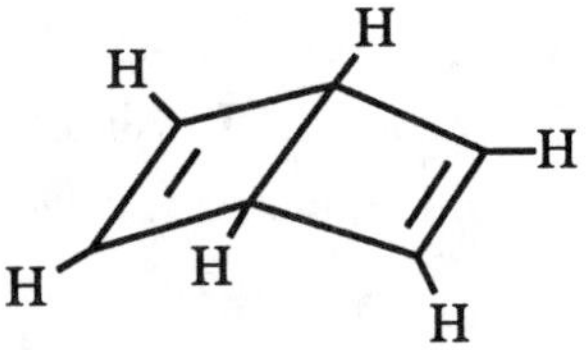

(b)

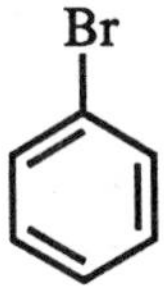

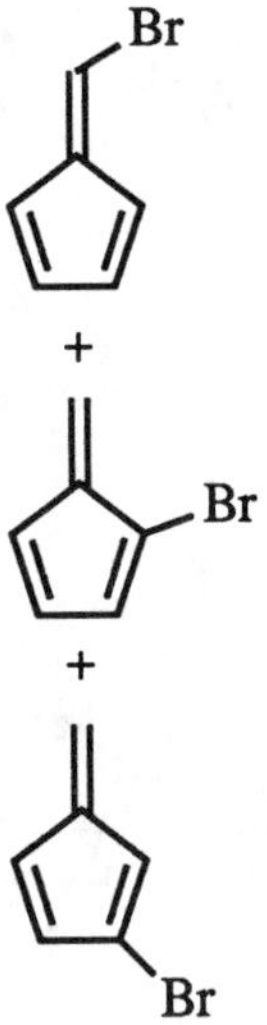

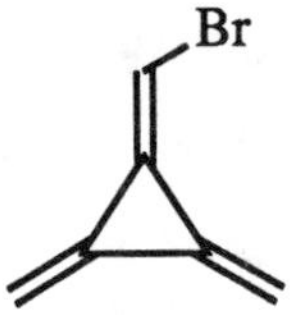

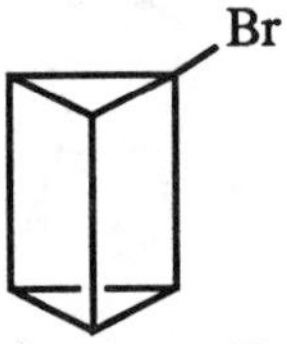

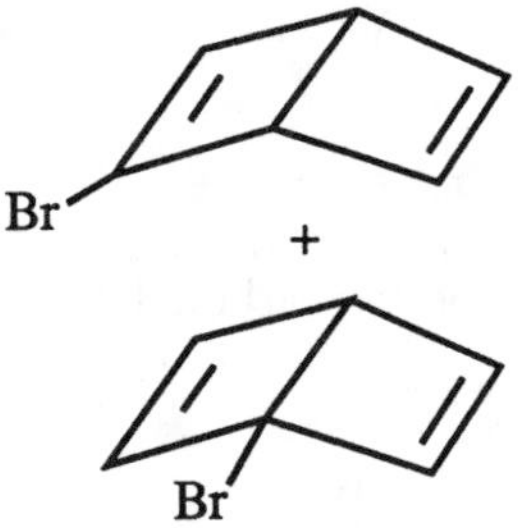

16-30 continued

(c)

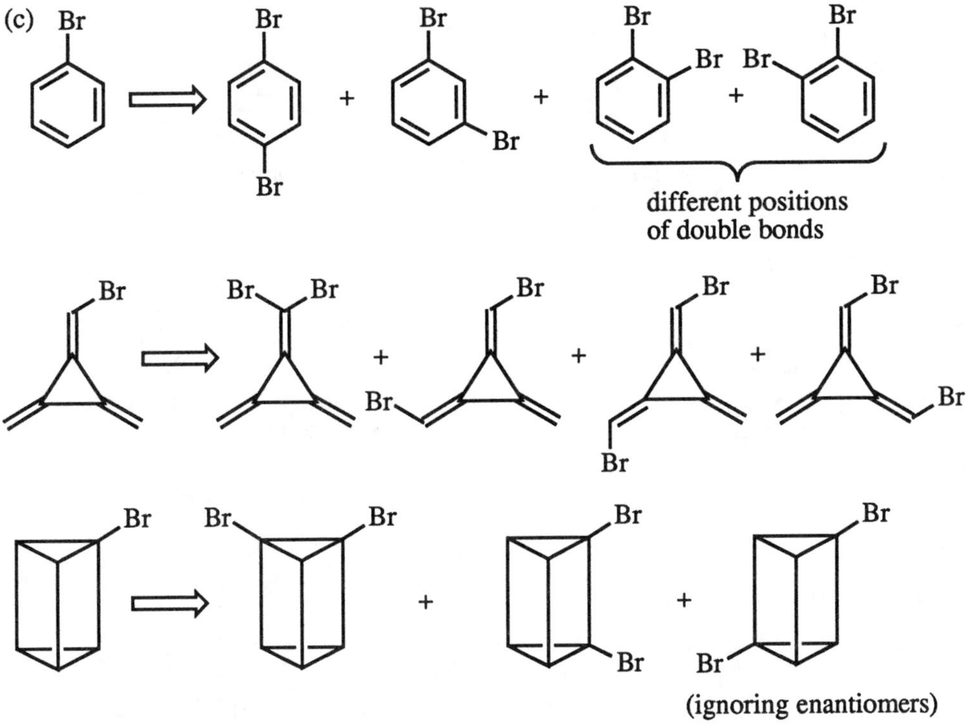

Br
Br
Br
Br
Br
Br
Br
different positions of double bonds
Br
Br
Br
Br
Br
Br
Br
Br
Br
Br
Br
Br
Br
Br
Br
Br
Br
(ignoring enantiomers)

(d) The only structure consistent with three isomers of dibromobenzene is the prism structure, called Ladenburg benzene. It also gives no test for alkenes, consistent with the behavior of benzene. (Kekulé defended his structure by claiming that the "two" structures of *ortho*-dibromobenzene were rapidly interconverted, equilibrating so quickly that they could never be separated.)

(e) We now know that three- and four-membered rings are the least stable, but this fact was unknown to chemists during the mid-1800's when the benzene controversy was raging. Ladenburg benzene has two three-membered rings and three four-membered rings (of which only four of the rings are independent), which we would predict to be unstable. (In fact, the structure has been synthesized. Called *prismane*, it is NOT aromatic, but rather, is very reactive toward addition reactions.)

16-31

(a)

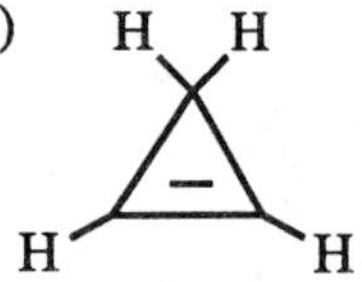

nonaromatic

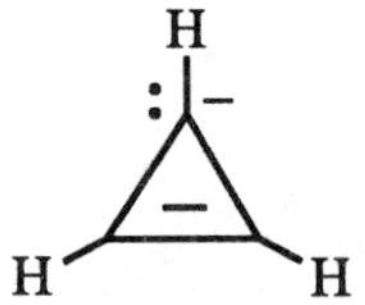

antiaromatic—
4 π electrons

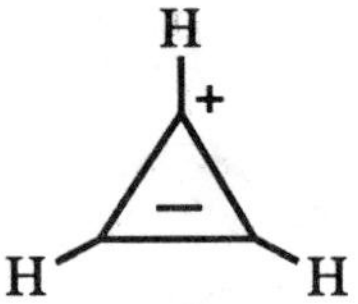

aromatic—
2 π electrons

(b)

aromatic—
6 π electrons

aromatic—
6 π electrons

nonaromatic

aromatic—
6 π electrons

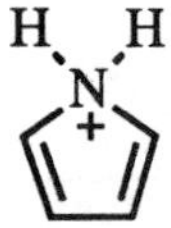

nonaromatic

(c)

nonaromatic

aromatic—
6 π electrons

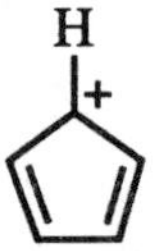

antiaromatic—
4 π electrons

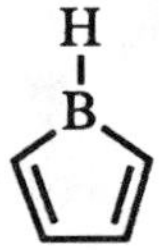

antiaromatic—
4 π electrons
(B is sp^2, but donates no electrons to the π system)

(d)

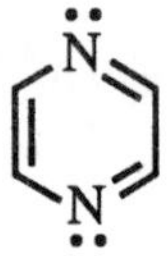

aromatic—
6 π electrons

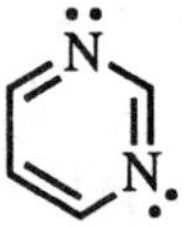

aromatic—
6 π electrons

nonaromatic

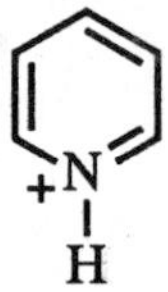

aromatic—
6 π electrons

(e)

nonaromatic

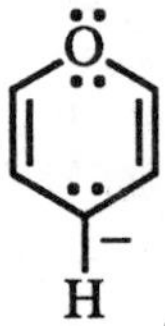

if oxygen is sp^2 ⇒ antiaromatic (8 π electrons); if oxygen is sp^3 ⇒ nonaromatic ⇒ NEVER AROMATIC

aromatic—
6 π electrons

16-31 continued

(f)

aromatic—
6 π electrons

aromatic—
6 π electrons

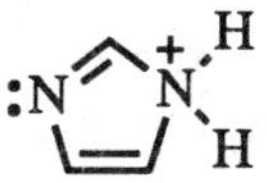

nonaromatic

(g)

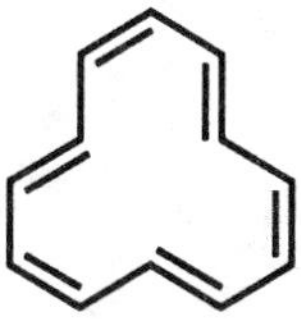

antiaromatic—
12 π electrons

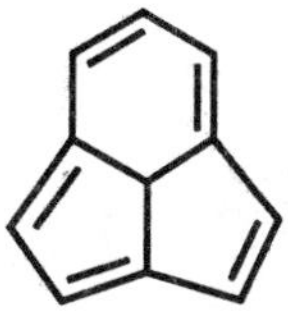

aromatic—
10 π electrons

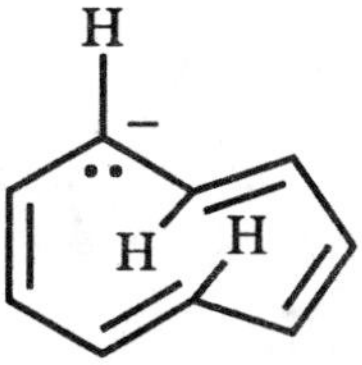

this is a tough call—it has 10 π electrons so it could be aromatic, but internal H's might force it out of planarity

(h)

nonaromatic

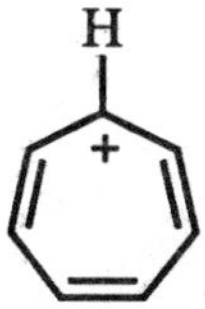

aromatic—
6 π electrons

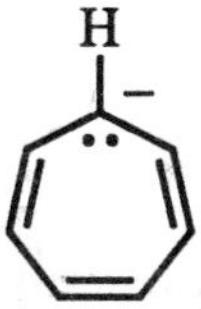

antiaromatic—
8 π electrons

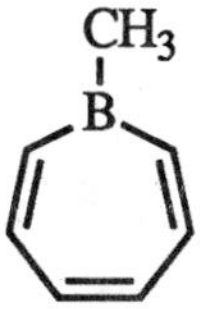

aromatic—
6 π electrons
(B is sp^2, but donates no electrons to the π system)

16-32 The clue to azulene is recognition of the five- and seven-membered rings. To attain aromaticity, a seven-membered carbon ring must have a positive charge; a five-membered carbon ring must have a negative charge. Drawing a resonance form of azulene shows this:

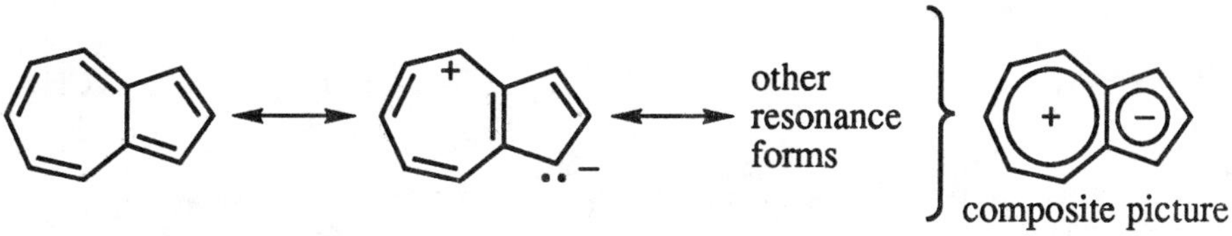
other
resonance
forms
composite picture

The composite picture shows that the negative charge is concentrated in the five-membered ring, giving rise to the dipole.

16-33 Whether a nitrogen is strongly basic or weakly basic depends on the location of its electron pair. If the electron pair is needed for an aromatic π system, the nitrogen will not be basic (shown here as "weak base"). If the electron pair is in either an sp^2 or sp^3 orbital, it is available for bonding, and the nitrogen is a "strong base".

(a) 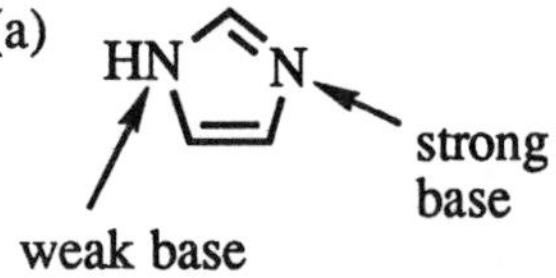
HN
N
strong base
weak base

(b) 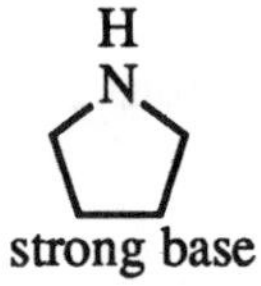
H
N
strong base

(c) 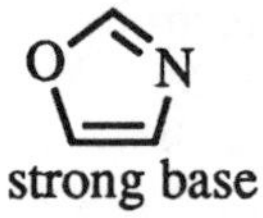
O
N
strong base

(d) 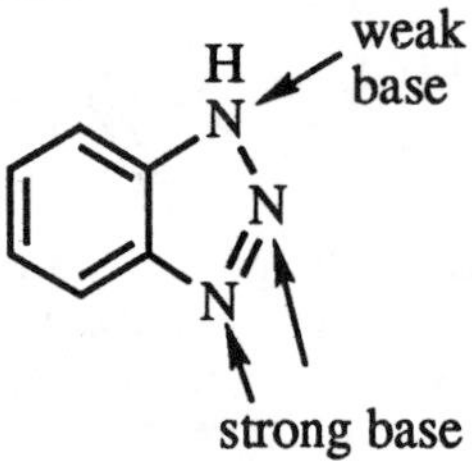
H
N
weak base
N
N
strong base

(e) 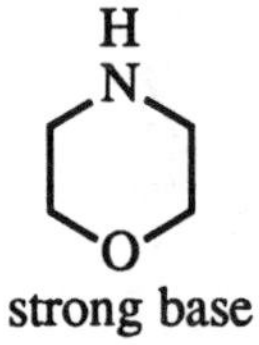
H
N
O
strong base

(f) 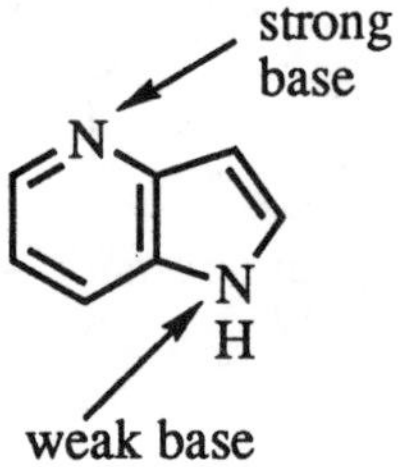
strong base
N
N
H
weak base

16-34

(a) 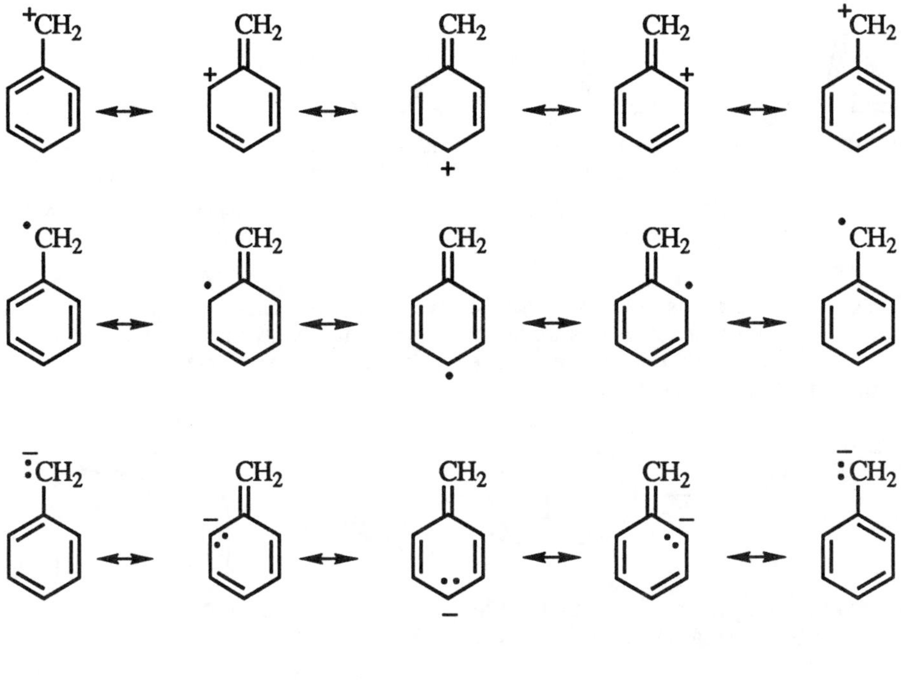

16-34 continued

(b) initiation

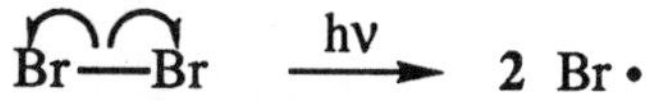

propagation

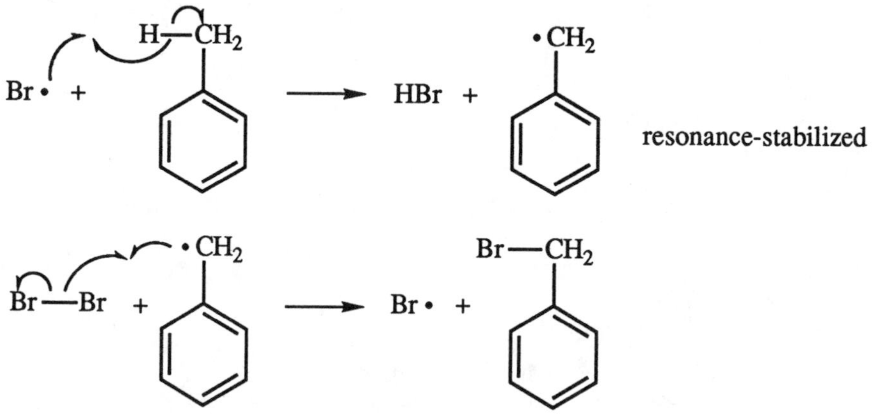

(c) Both reactions are S_N2 on primary carbons, but the one at the benzylic carbon occurs faster. In the transition state of S_N2, as the nucleophile is approaching the carbon and the leaving group is departing, the electron density resembles that of a p orbital. As such, it can be stabilized through overlap with the π system of the benzene ring.

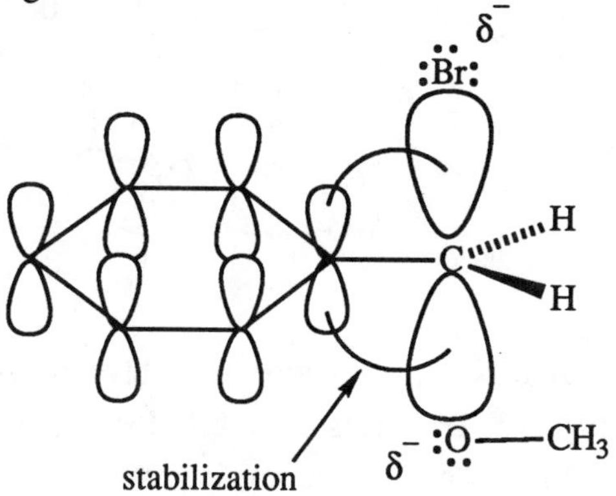

16-35

(a)

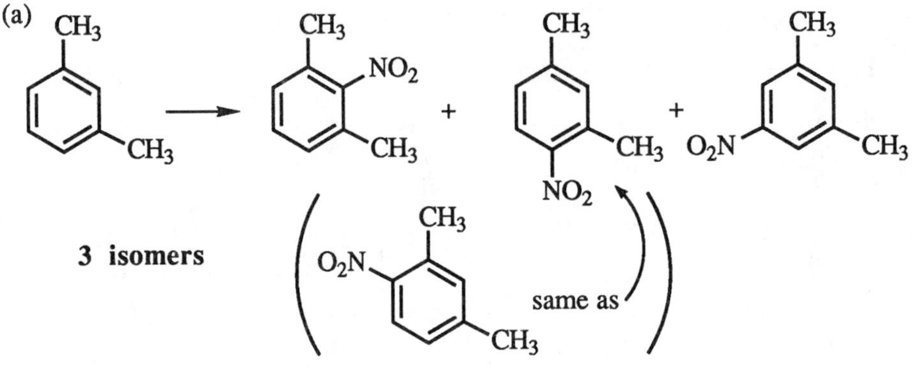

(b)

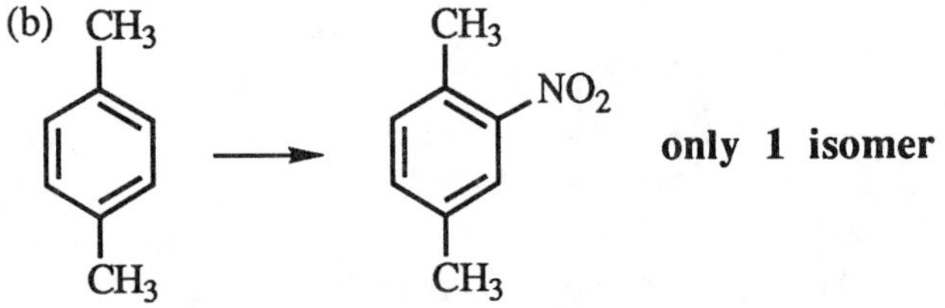

(c) The original compound had to have been *meta*-dibromobenzene as this is the only dibromo isomer that gives three mononitrated products.

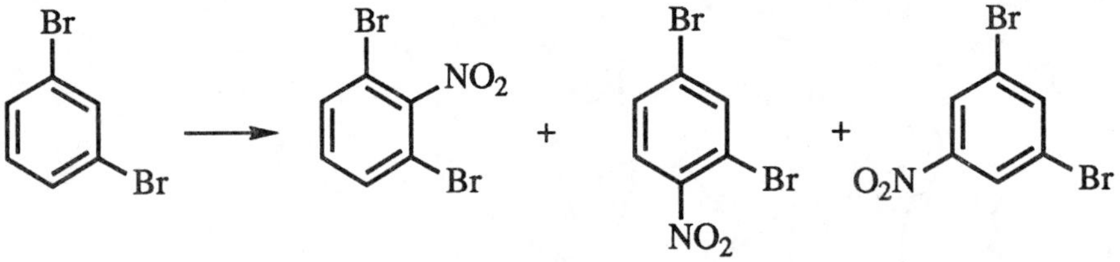

16-36

(a) The formula C_8H_7OCl has five elements of unsaturation, probably a benzene ring (4) plus either a double bond or a ring. The IR suggests a carbonyl at 1690 cm^{-1} and an aromatic ring at 1602 cm^{-1}. The NMR shows five aromatic protons, indicating a monosubstituted benzene. A 2H singlet at δ 4.8 is a deshielded methylene.

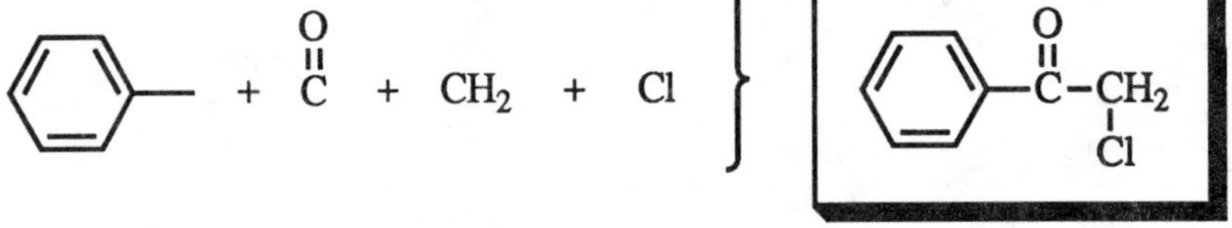

16-36 continued

(b) The mass spectral evidence of molecular ion peaks of 1 : 1 intensity at 184 and 186 shows the presence of a bromine atom. The m/z 184 minus 79 for bromine gives a mass of 105 for the rest of the molecule, which is about a benzene ring plus two carbons and a few hydrogens. The NMR shows four aromatic hydrogens in a typical *para* pattern (two doublets), indicating a *para*-disubstituted benzene. The 2H quartet and 3H triplet are characteristic of an ethyl group.

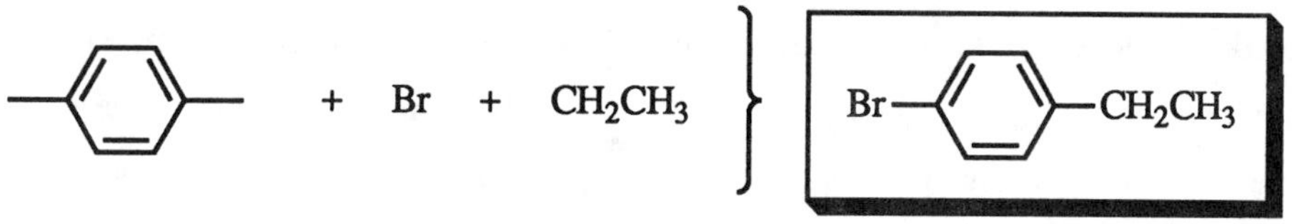

16-37

(a)

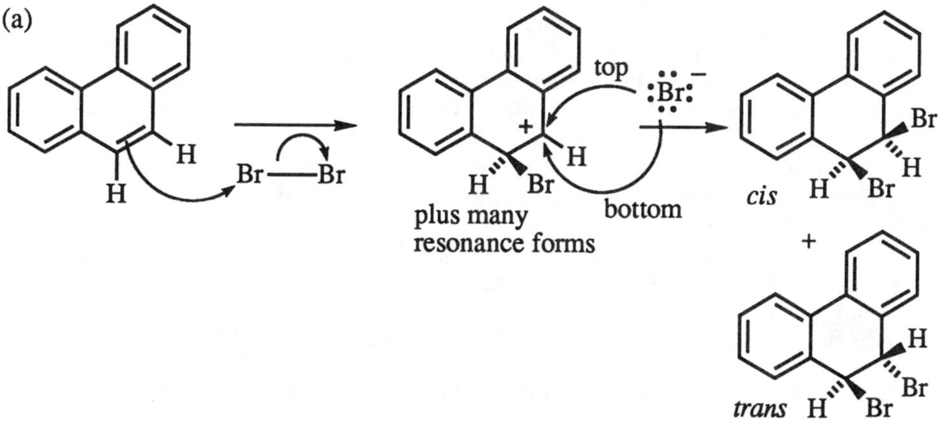

(b) A typical addition of bromine occurs with a bromonium ion intermediate which can give only anti addition. Addition of bromine to phenanthrene, however, generates a free carbocation because the carbocation is benzylic, stabilized by resonance over two rings. In the second step of the mechanism, bromide nucleophile can attack either side of the carbocation giving a mixture of *cis* and *trans* products.

(c)

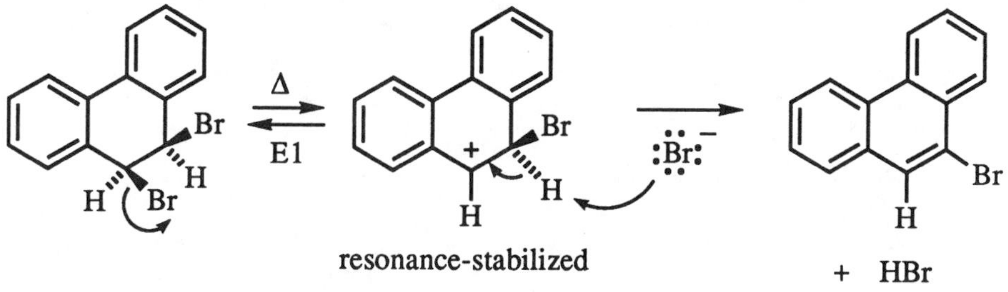

16-38

(a) No, biphenyl is not fused. The rings must share two atoms to be labeled "fused".

(b) There are 12 π electrons in biphenyl compared with 10 for naphthalene.

(c) Biphenyl has 6 "double bonds". An isolated alkene releases 28.6 kcal/mole upon hydrogenation.

predicted:	6 x 28.6 kcal/mole (120 kJ/mole) ≈	172 kcal/mole (720 kJ/mole)
observed:		100 kcal/mole (418 kJ/mole)
	resonance energy:	72 kcal/mole (302 kJ/mole)

(d) On a "per ring" basis, biphenyl is 72 ÷ 2 = 36 kcal/mole, identical to the value for benzene. Naphthalene's resonance energy is 60 kcal/mole (252 kJ/mole); on a "per ring" basis, naphthalene has only 30 kcal/mole of stabilization per ring. This is consistent with the greater reactivity of naphthalene compared with benzene. In fact, the more fused rings, the lower the resonance energy per ring, and the more reactive the compound. (Refer to Problem 16-18.)

(Here is an unconventional view of resonance stabilization energy. Are we being fair to naphthalene? Naphthalene does not have two separate rings like biphenyl, and it does not have two independent aromatic systems. What is the stabilization energy *per pair of π electrons*? For biphenyl: 72 kcal ÷ 6 pairs = 12 kcal per pair of π electrons; for benzene: 36 kcal ÷ 6 pairs = 12 kcal per pair of π electrons; for naphthalene: 60 kcal ÷ 5 pairs = 12 kcal per pair of π electrons! All these compounds have identical stabilization energies per pair of π electrons!)

16-39 Two protons are removed from sp^3 carbons to make sp^2 carbons and to generate a π system with 10 π electrons.

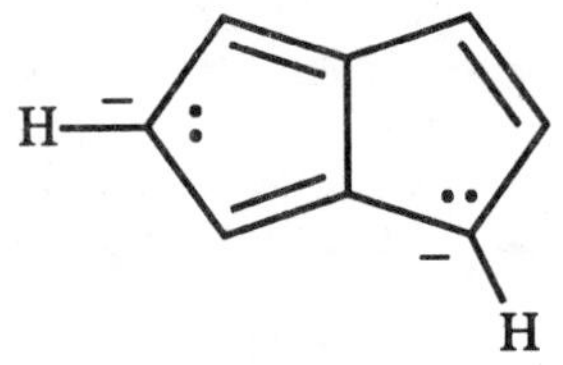

16-40

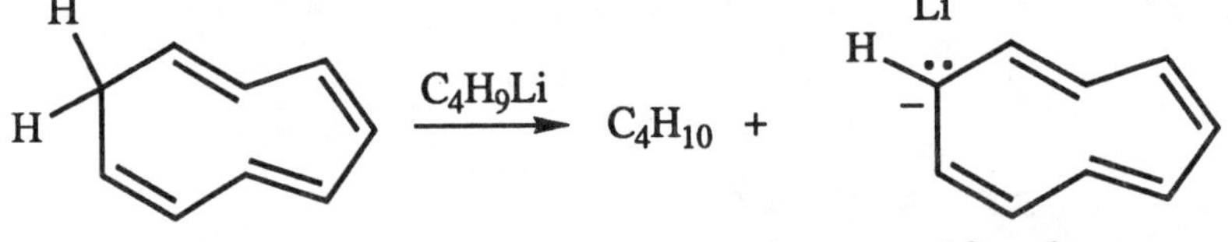

10 π electrons

aromatic (if planar)

16-41 These four bases can be aromatic, partially aromatic, or aromatic in a tautomeric form. In other words, aromaticity plays an important role in the chemistry of all four structures.

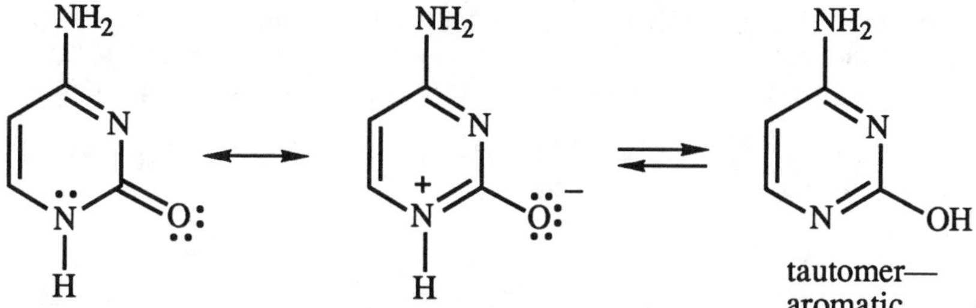

cytosine

aromatic to the extent that this resonance form contributes

tautomer—aromatic

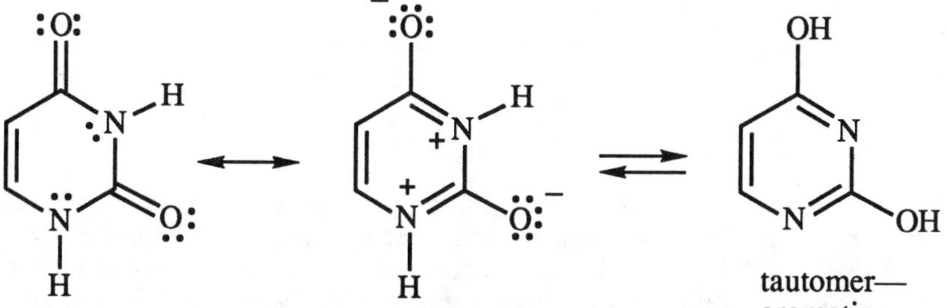

uracil

aromatic to the extent that this resonance form contributes

tautomer—aromatic

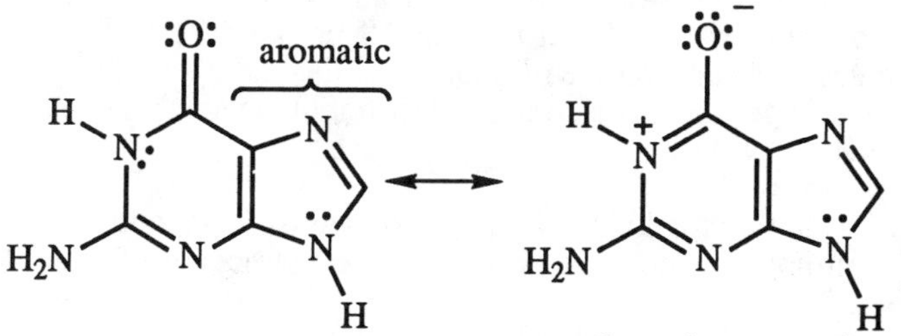

guanine

aromatic to the extent that this resonance form contributes

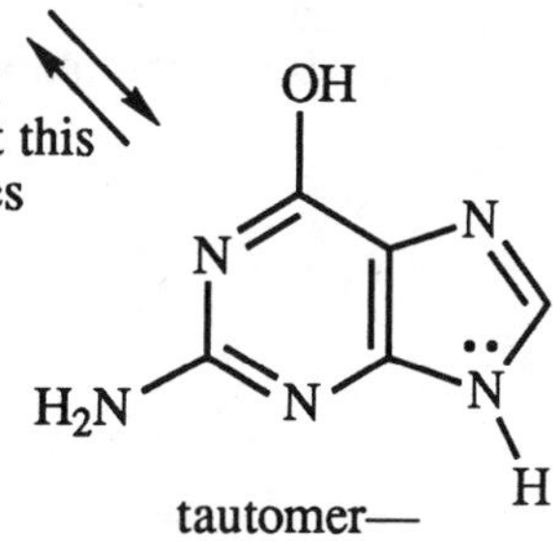

tautomer—fully aromatic

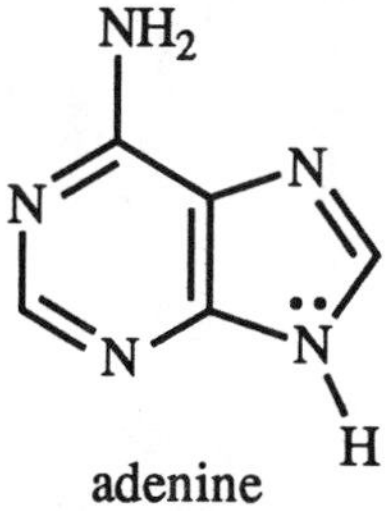

fully aromatic

adenine

16-42

(a) Antiaromatic—only 4 π electrons.

(b) This molecule is electronically equivalent to cyclobutadiene. Cyclobutadiene is unstable and undergoes a Diels-Alder reaction with another molecule of itself. The *t*-butyl groups prevent dimerization by blocking approach of any other molecule.

(c) Yes, the nitrogen should be basic. The pair of electrons on the nitrogen is in an sp^2 orbital and is not part of the π system.

(d)

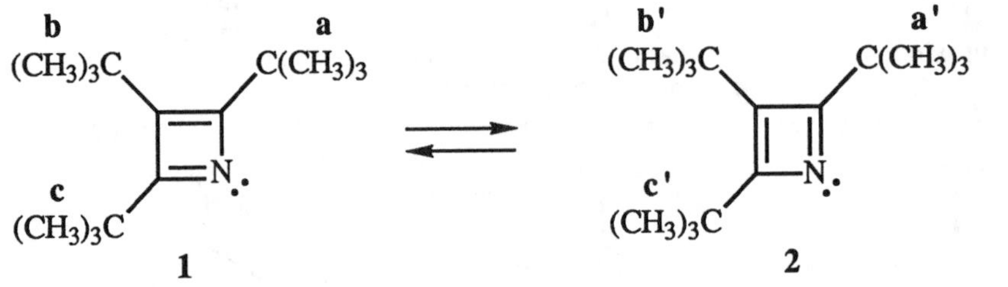

Analysis of structure **1** shows the three *t*-butyl groups in unique environments in relation to the nitrogen. We would expect three different signals in the NMR, as is observed at –110° C. Why do signals coalesce as the temperature is increased? Two of the *t*-butyl groups become equivalent—which two? Most likely, they are **a** and **c** that become equivalent as they are symmetric around the nitrogen. But they are *not* equivalent in structure **1**—what is happening here?

What must happen is an equilibration between structures **1** and **2**, very slow at –110° C, but very fast at room temperature, faster than the NMR can differentiate. So the signal which has coalesced is an average of **a** and **a'** and **c** and **c'**. (This type of low temperature NMR experiment is also used to differentiate axial and equatorial hydrogens on a cyclohexane.)

The NMR data prove that **1** is not aromatic, and that **1** and **2** are isomers, not resonance forms. If **1** were aromatic, then **a** and **c** would have identical NMR signals at all temperatures.

16-43

Mass spectrum: Molecular ion at 150; base peak at 135, M – 15, is loss of methyl.

Infrared spectrum: The broad peak at 3500 cm^{-1} is OH; thymol must be an alcohol. The peak at 1620 cm^{-1} suggests an aromatic compound.

NMR spectrum: The broad singlet at δ 4.7 is OH; it disappears upon shaking with D_2O. The 6H doublet at δ 1.2 and the 1H multiplet at δ 3.2 are an isopropyl group, apparently on the benzene ring. A 3H singlet at δ 2.2 is a methyl group, also on the benzene ring.

Analysis of the aromatic protons indicates the substitution pattern. The three aromatic hydrogens confirm that there are three substituents. The singlet at δ 6.5 is a proton between two substituents (no neighboring H's). The doublets at δ 6.7 and δ 7.1 are ortho hydrogens, splitting each other.

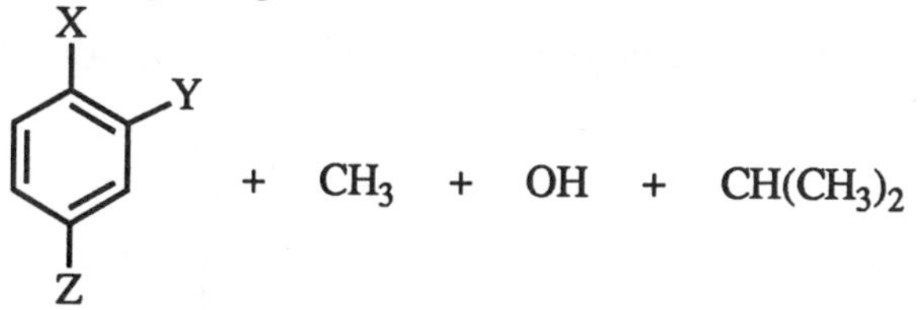

X
Y
Z
+ CH3 + OH + CH(CH3)2

Several isomeric combinations are consistent with the spectra (although the single H giving δ 6.5 suggests that either Y or Z is the OH group—an OH on a benzene ring shields hydrogens ortho to it, moving them upfield). The structure of thymol is:

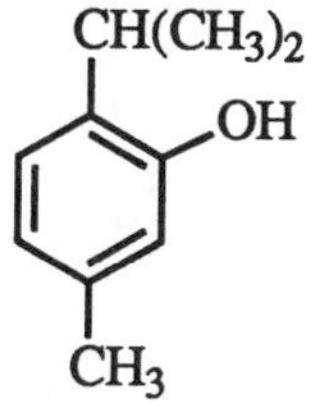

CH(CH3)2
OH
CH3

thymol

The final question is how the molecule fragments in the mass spectrometer:

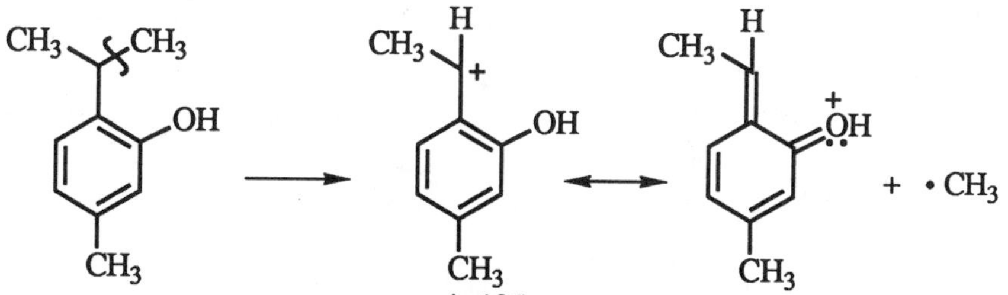

CH3
CH3
OH
CH3
H
CH3
+
OH
CH3
H
CH3
+
OH
CH3
+ • CH3

m/z 135

resonance stabilization of this benzylic cation includes forms with positive charge on three ring carbons and on oxygen (shown)

16-44

Mass spectrum: Molecular ion at 170; two prominent peaks are M – 15 (loss of methyl) and M – 43 (as we shall see, most likely the loss of acetyl, CH_3CO).

Infrared spectrum: The two most significant peaks are at 1680 cm^{-1} (conjugated carbonyl) and 1630 cm^{-1} (aromatic C=C).

NMR spectrum: A 3H singlet at δ 2.6 is methyl next to a carbonyl, shifted slightly downfield by an aromatic ring. The other signals are seven aromatic protons.

Conclusions:

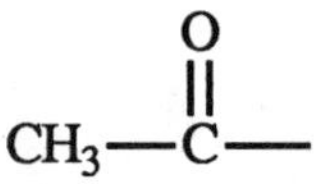

CH_3—C(=O)— + m/z 127 including 7H $\Rightarrow$ mass 120 for carbons $\Rightarrow$ 10 C

The fragment $C_{10}H_7$ is almost certainly a naphthalene. Which isomer cannot readily be determined from these spectra.

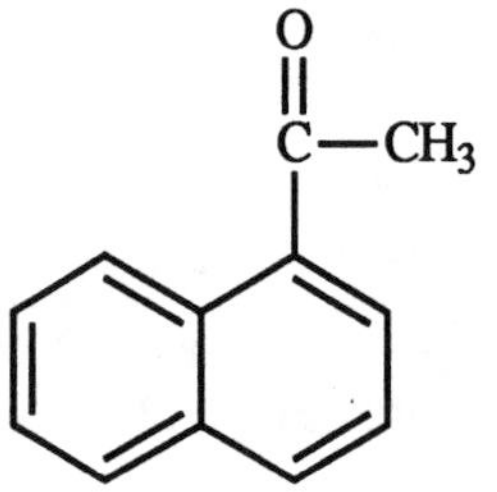

or

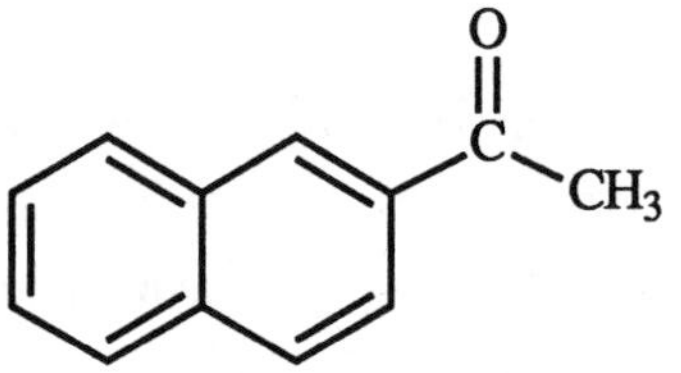

CHAPTER 17—REACTIONS OF AROMATIC COMPOUNDS

Note to the student: The representation of benzene with a circle to represent the π system is fine for questions of nomenclature, properties, isomers, and reactions. For many questions of mechanism or reactivity, however, the representation with three alternating double bonds (the Kekulé picture) is more informative. For consistency, this Solutions Manual will use the Kekulé form exclusively.

17-1

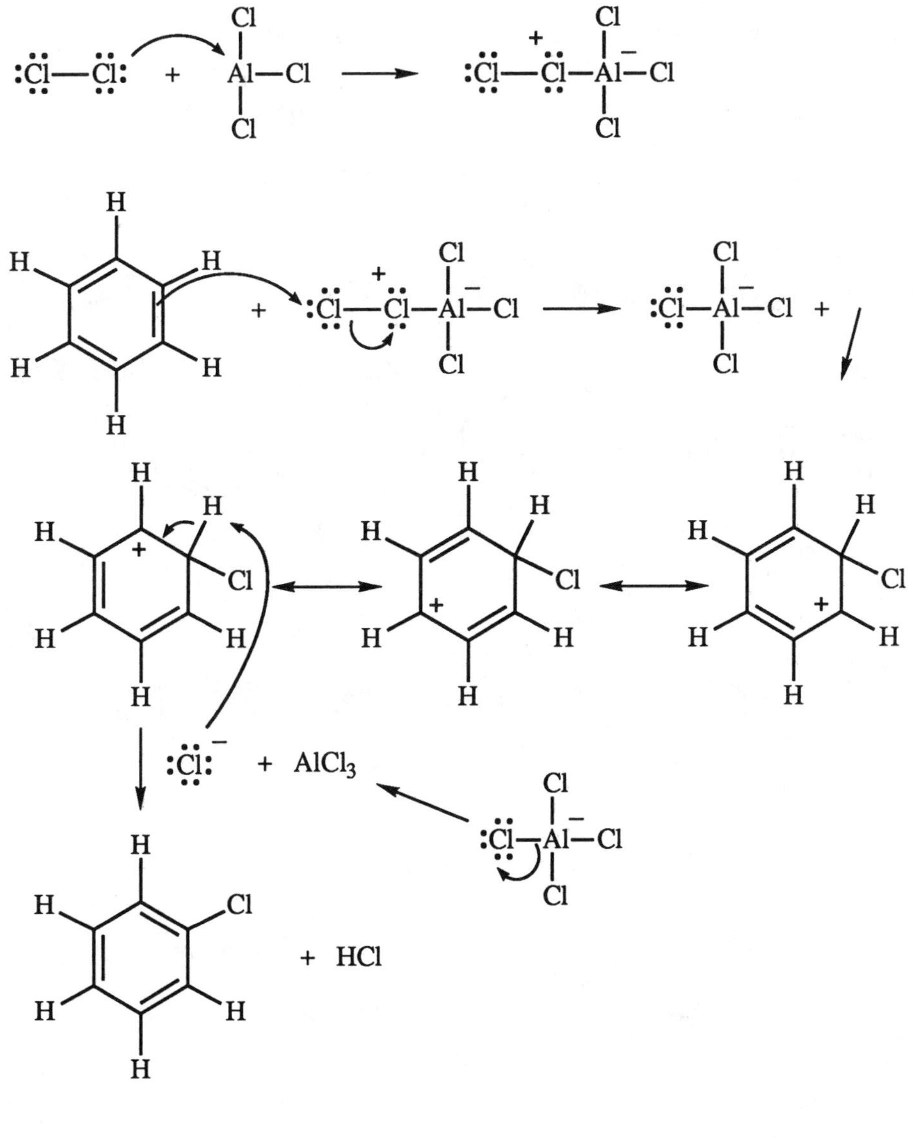

17-2

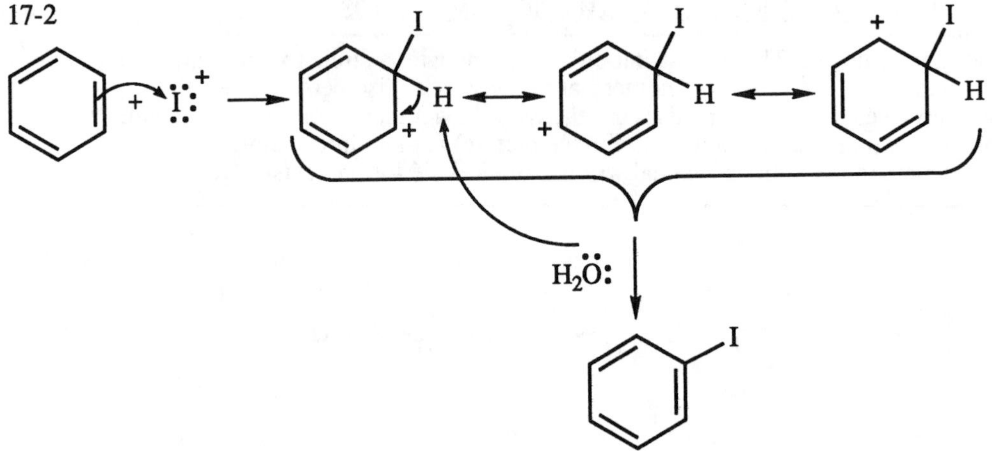

17-3

$$Tl(OCOCF_3)_3 \rightleftharpoons \overset{+}{Tl}(OCOCF_3)_2 + CF_3COO^-$$

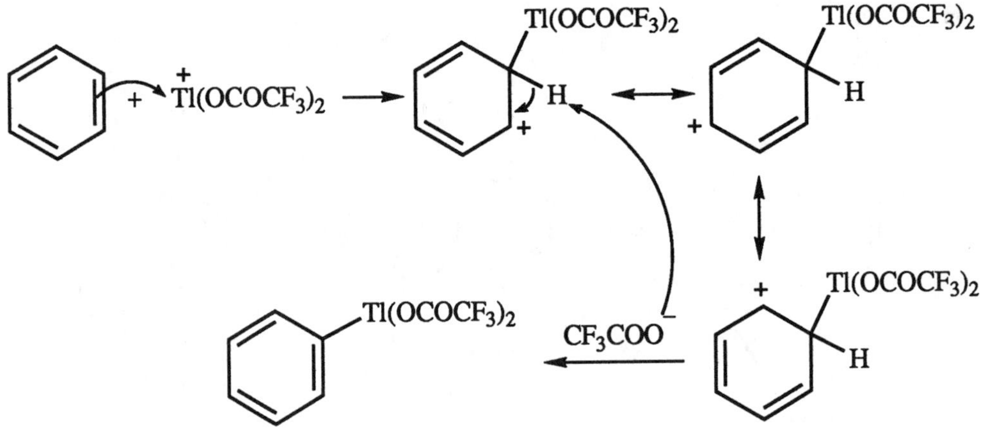

17-4

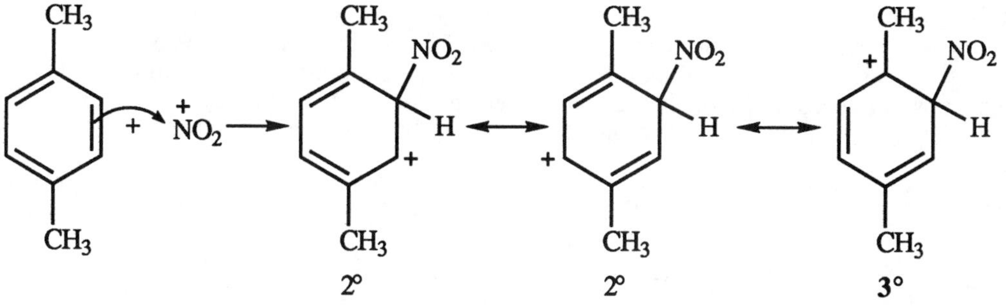

Benzene's sigma complex has positive charge on three 2° carbons. The sigma complex above shows positive charge in one resonance form on a 3° carbon, lending greater stabilization to this sigma complex. The more stable the intermediate, the lower the activation energy required to reach it, and the faster the reaction will be.

17-5

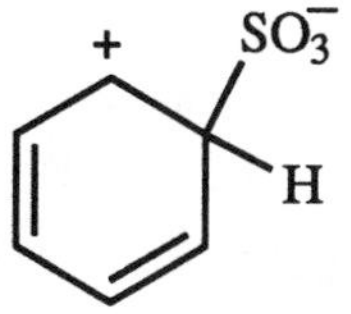

delocalization of the positive charge on the ring

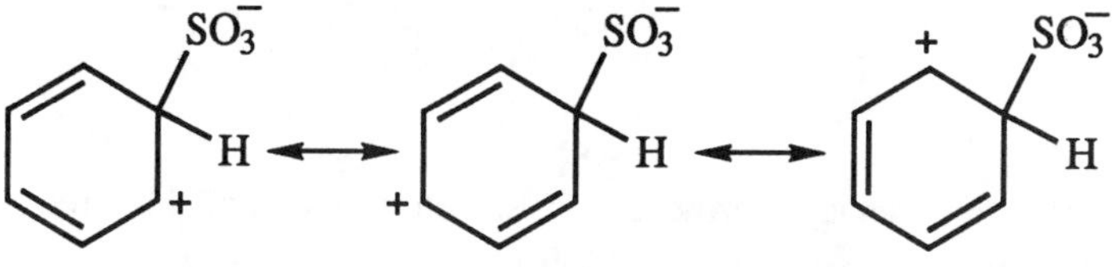

delocalization of the negative charge on the sulfonate group

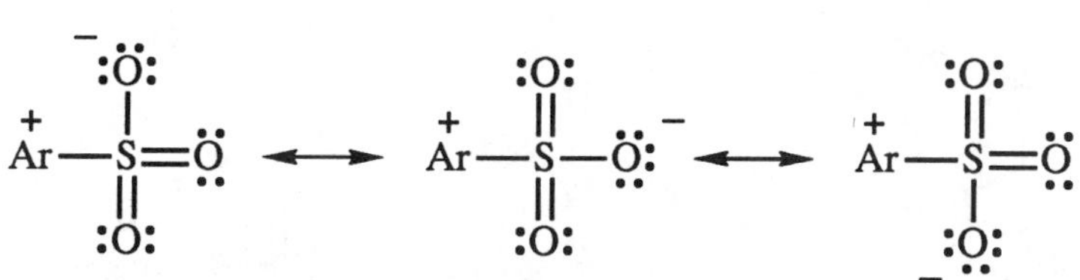

("Ar" stands for the benzene ring)

17-6

(a) The key to electrophilic aromatic substitution lies in the stability of the sigma complex. When the electrophile bonds at ortho or para positions of ethylbenzene, the positive charge is shared by the 3° carbon with the ethyl group. Bonding of the electrophile at the meta position lends no particular advantage because the positive charge in the sigma complex is never adjacent to, and therefore never stabilized by, the ethyl group.

ortho

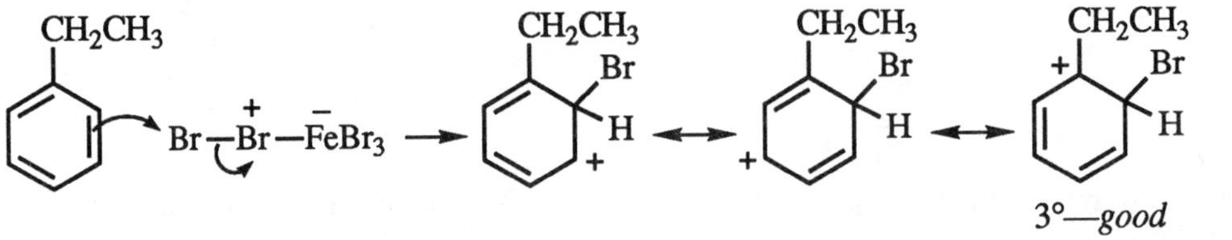

meta

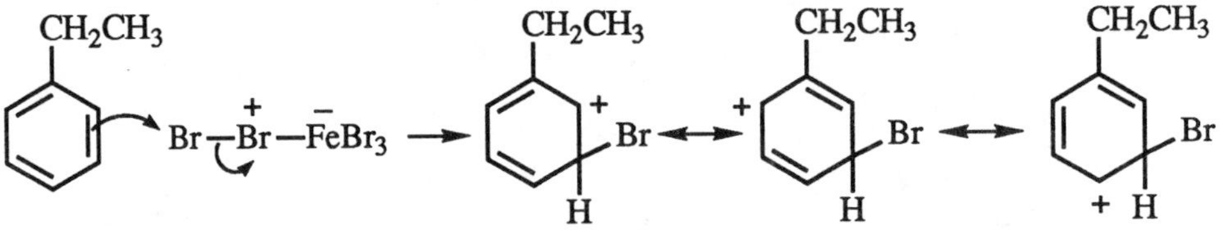

para

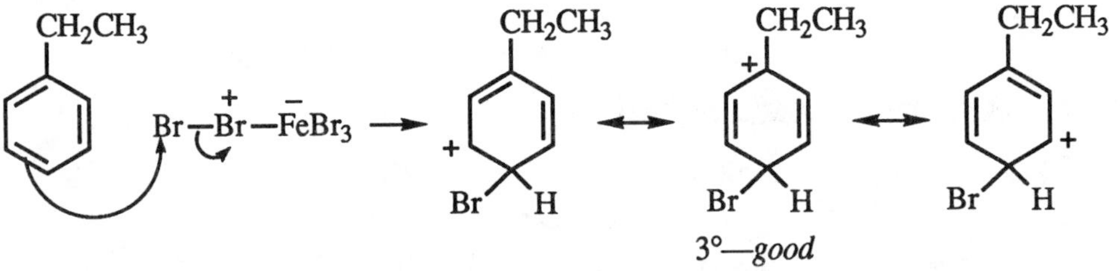

(b) Electrophilic attack on *p*-xylene gives an intermediate in which only one of the three resonance forms is stabilized by a substituent (see the solution to Problem 17-4). *m*-Xylene, however, is stabilized in two of its three resonance forms. A more stable intermediate gives a faster reaction.

m-xylene

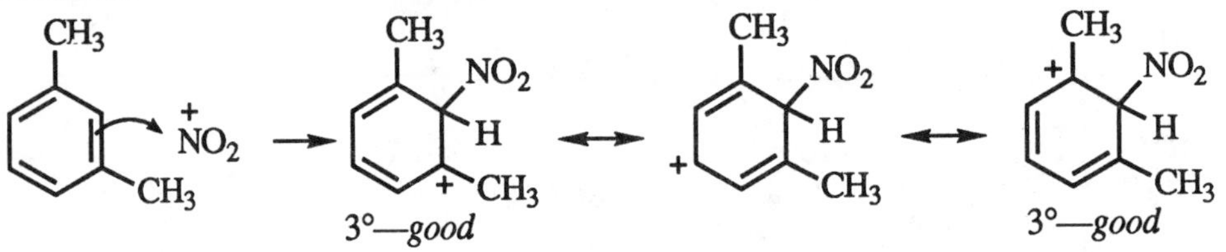

17-7 For ortho and para attack, the positive charge in the sigma complex can be shared by resonance with the vinyl group. This cannot happen with meta attack because the positive charge is never adjacent to the viny! group. (Ortho attack is shown; para attack gives an intermediate with positive charge on the same carbons.)

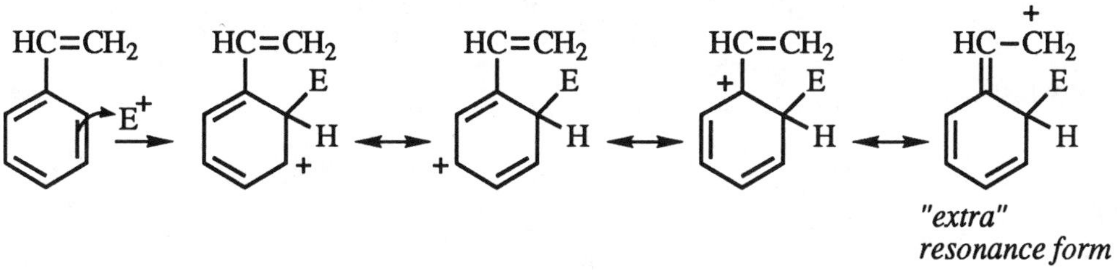

17-8 Attack at only ortho and para positions (not meta) places the positive charge on the carbon with the ethoxy group, where the ethoxy group can stabilize the positive charge by resonance donation of a lone pair of electrons. (Ortho attack is shown; para attack gives a similar intermediate.)

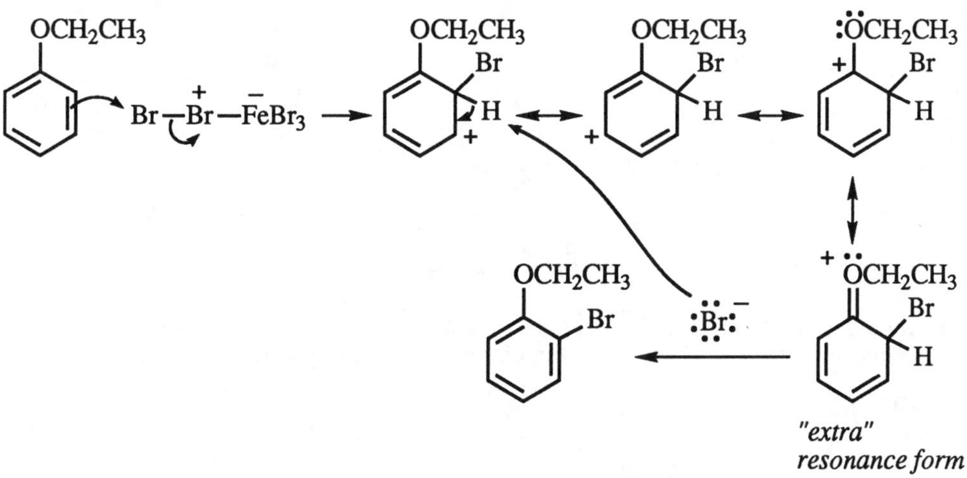

17-9

ortho

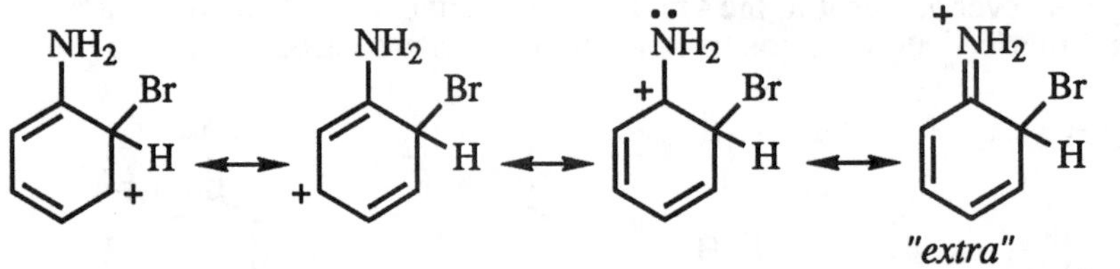

"extra" resonance form

meta

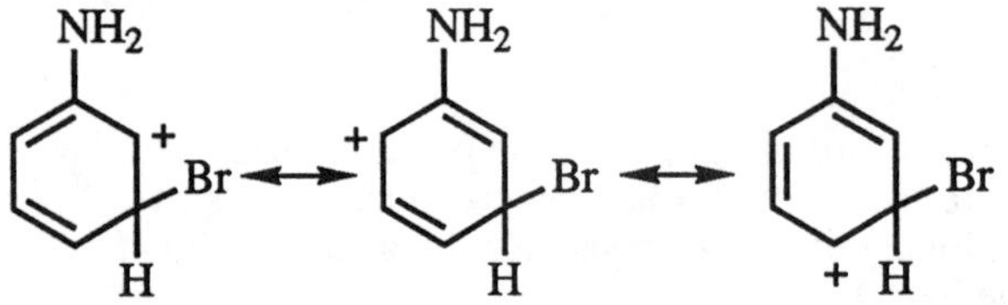

para

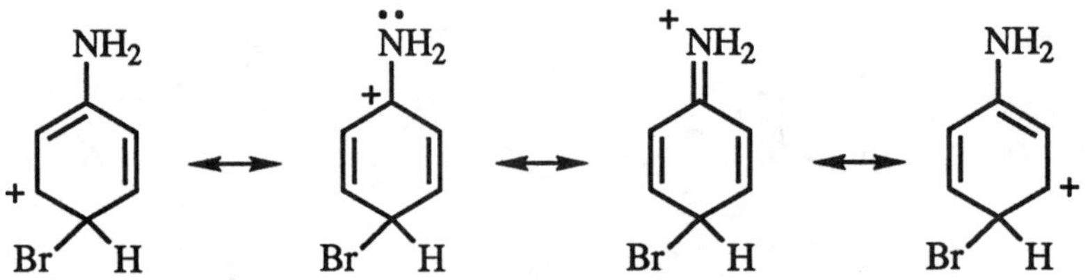

"extra" resonance form

17-10 The methoxy group is strongly activating because it can donate electrons by resonance. Its effect will dominate the effect of the weakly-activating methyl group.

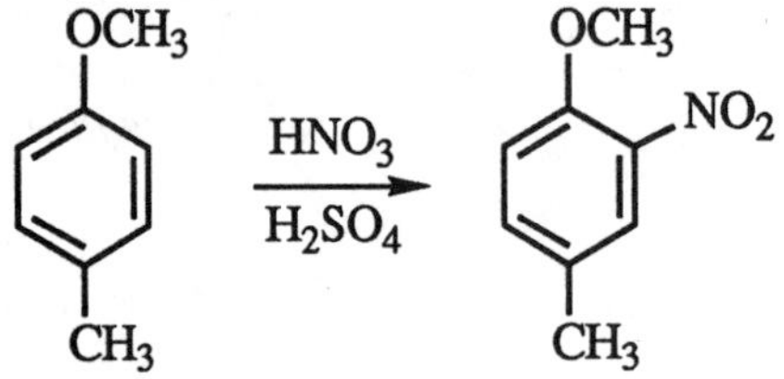

major product

17-11

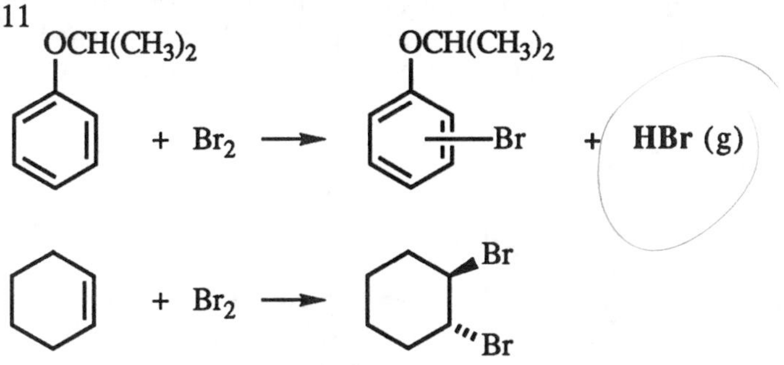

Substitution generates HBr whereas the addition does not. If the reaction is performed in an organic solvent, bubbles of HBr can be observed. If the reaction is performed in water, then adding moist litmus paper to test for acid will differentiate the results of the two compounds.

17-12

(a) Nitration is performed with nitric acid and a sulfuric acid catalyst. In strong acid, amines in general, including aniline, are protonated.

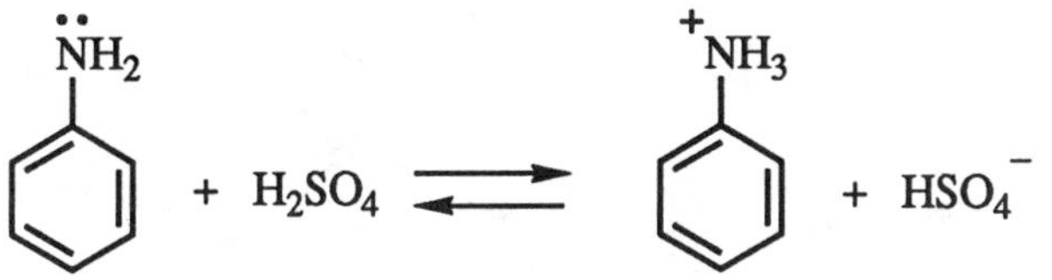

(b) The NH_2 group is a strongly activating ortho,para-director. In acid, however, it exists as the protonated ammonium ion—a strongly **deactivating meta-director**. The strongly acidic nitrating mixture itself forces the reaction to be slower.

(c) The acetyl group removes some of the electron density from the nitrogen, making it much less basic; the nitrogen of this amide is not protonated under the reaction conditions. The N retains enough electron density to share with the benzene ring, so the $NHCOCH_3$ group is still an activating ortho,para-director, albeit weaker than NH_2.

17-13 Nitronium ion attack at the ortho and para positions places positive charge on the carbon adjacent to the bromine, allowing resonance stabilization by an unshared electron pair from the bromine. Meta attack does not give a stabilized intermediate.

ortho

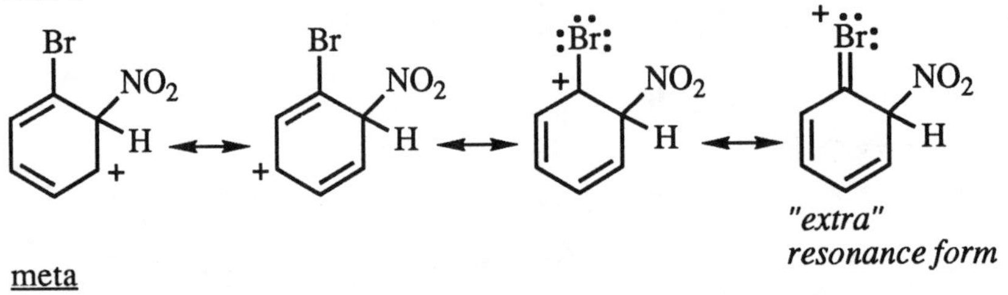

meta

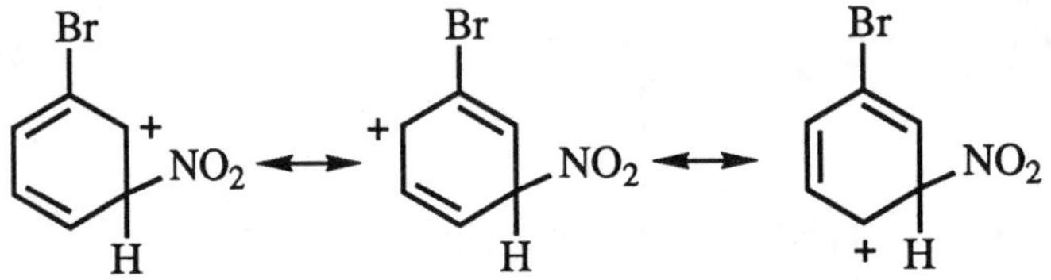

para

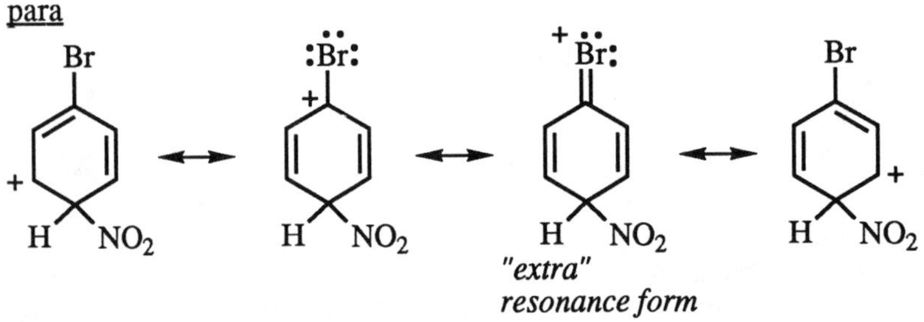

17-14

(a)

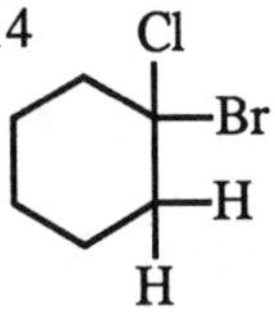

(b)

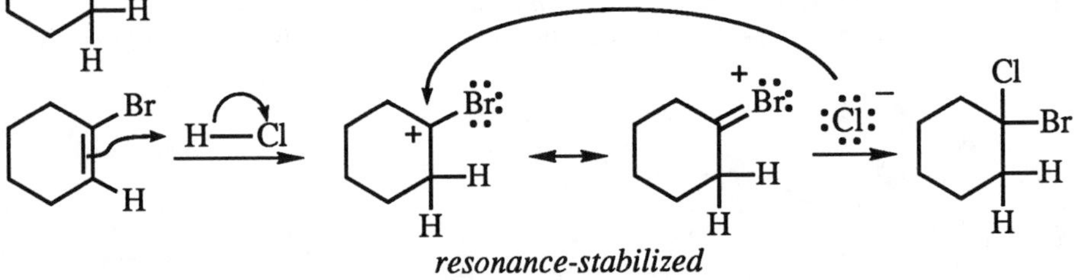

(c) A bromine atom can stabilize positive charge by sharing a pair of electrons. Bromine can do this in any cationic species, whether from electrophilic addition or electrophilic aromatic substitution.

17-15

(a)

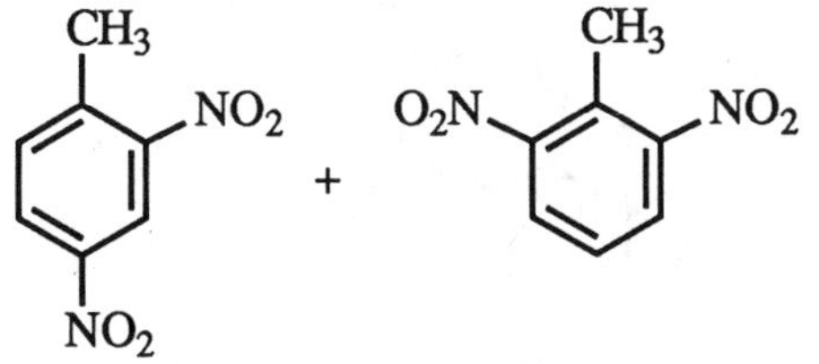

CH3
NO2
+
CH3
O2N
NO2
NO2

(b)

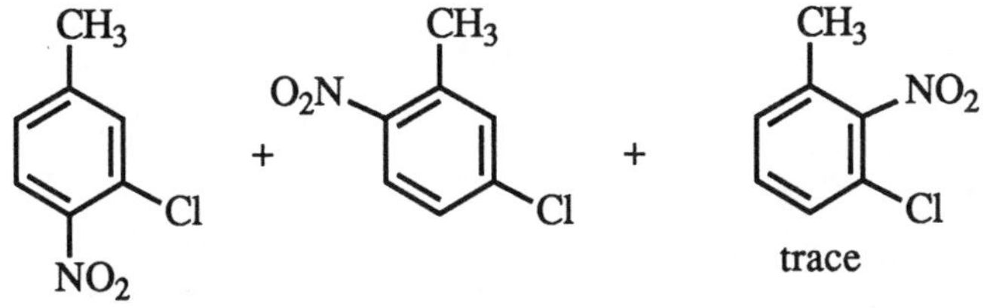

CH3
Cl
NO2
+
CH3
O2N
Cl
+
CH3
NO2
Cl
trace

(c)

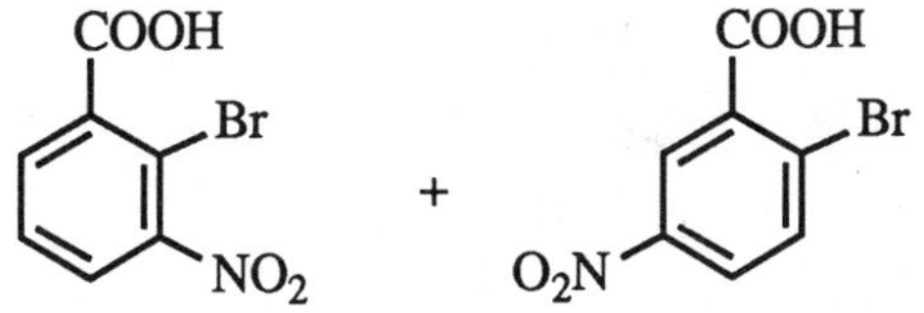

COOH
Br
NO2
+
COOH
Br
O2N

(d)

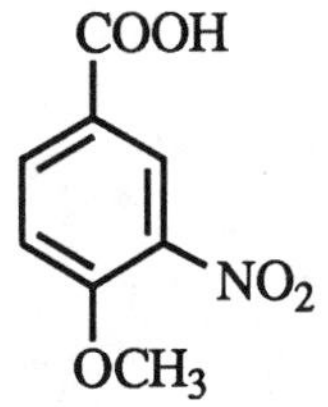

COOH
NO2
OCH3

(e)

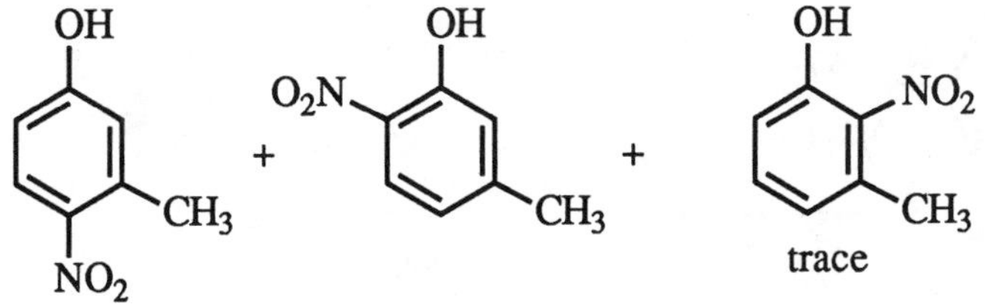

OH
CH3
NO2
+
OH
O2N
CH3
+
OH
NO2
CH3
trace

17-16

(a)

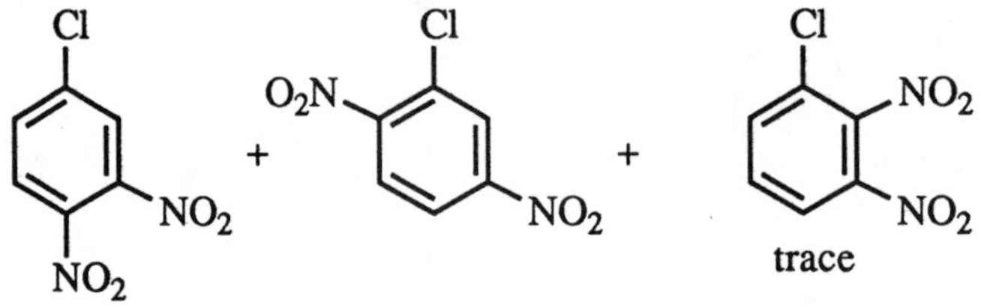

(b)

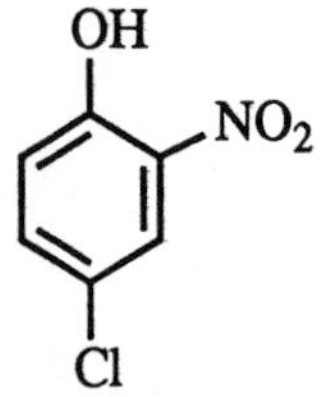

(c)

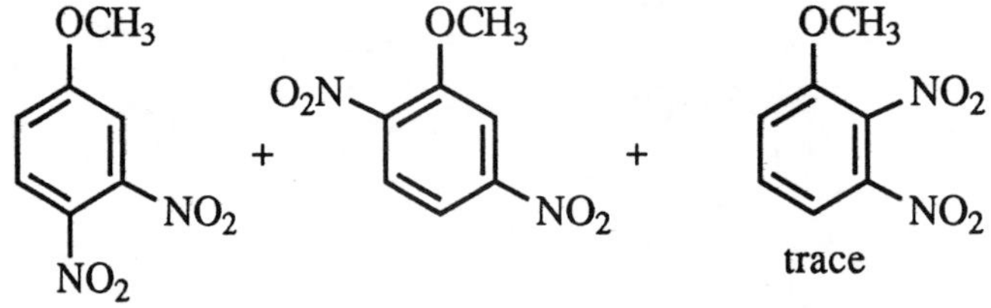

(d)

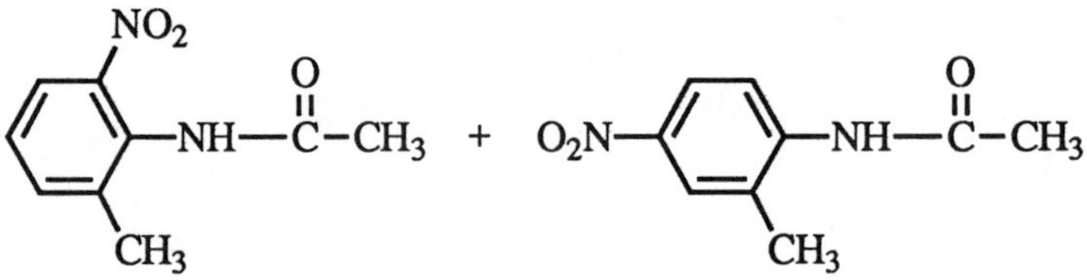

(e)

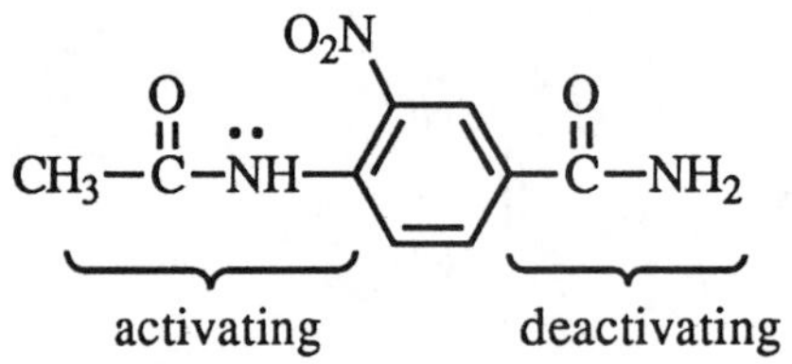

17-17

(a) Sigma complex of ortho attack—benzene stabilizes positive charge by resonance:

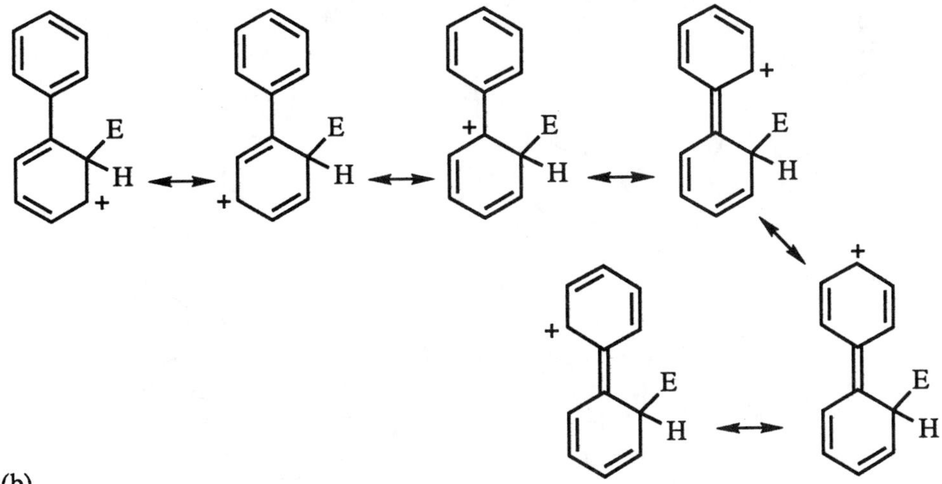

(b)

(i)

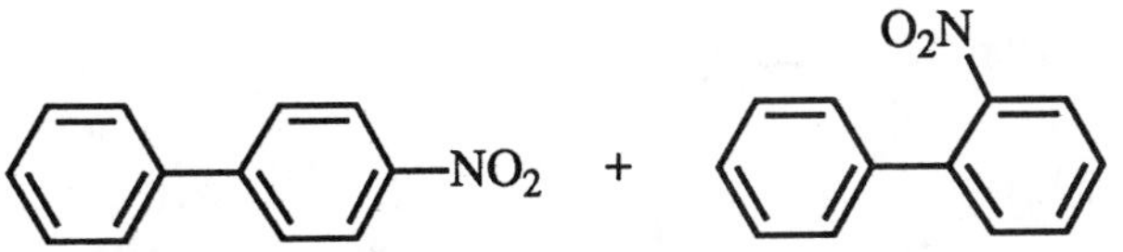

(ii)

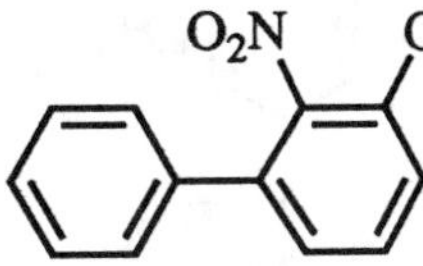

trace

+

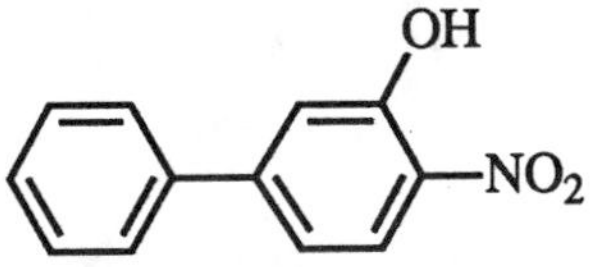

+

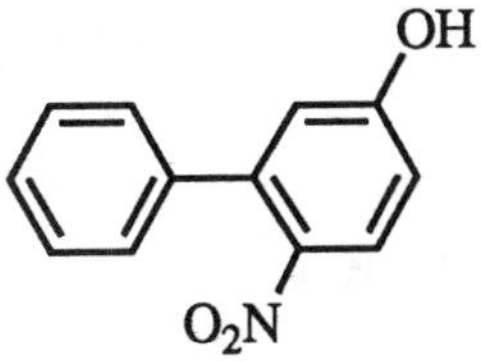

(iii)

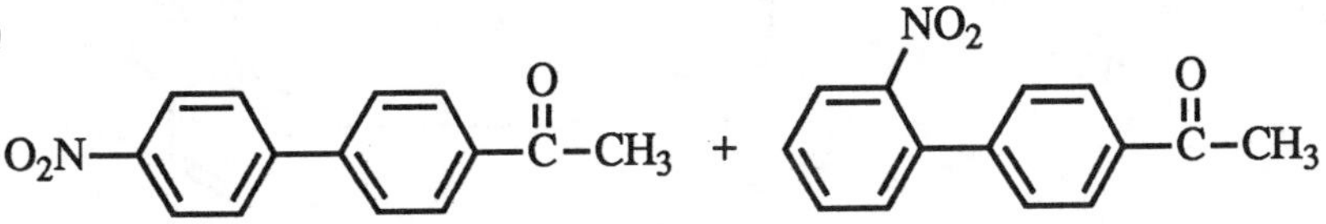

(iv)

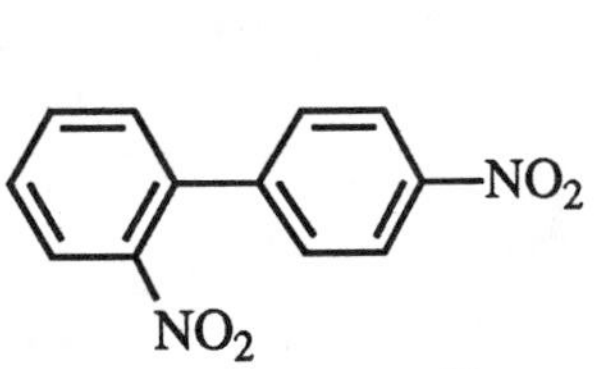

+

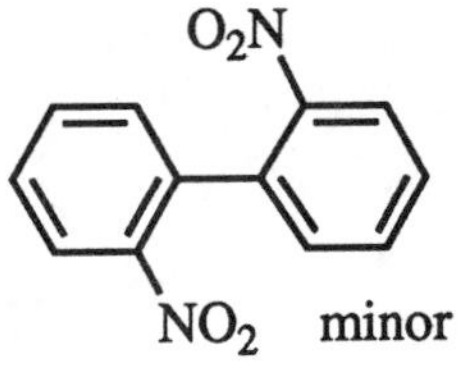

minor

(v)

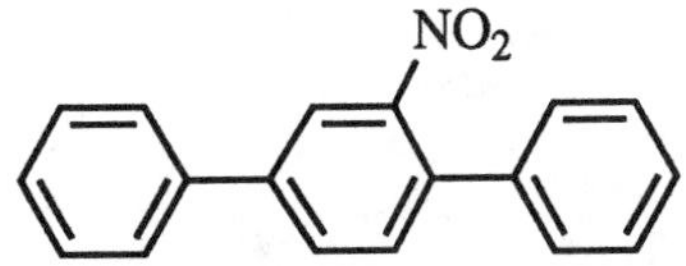

major (minor amounts of nitration on the outer rings)

17-18

(a)

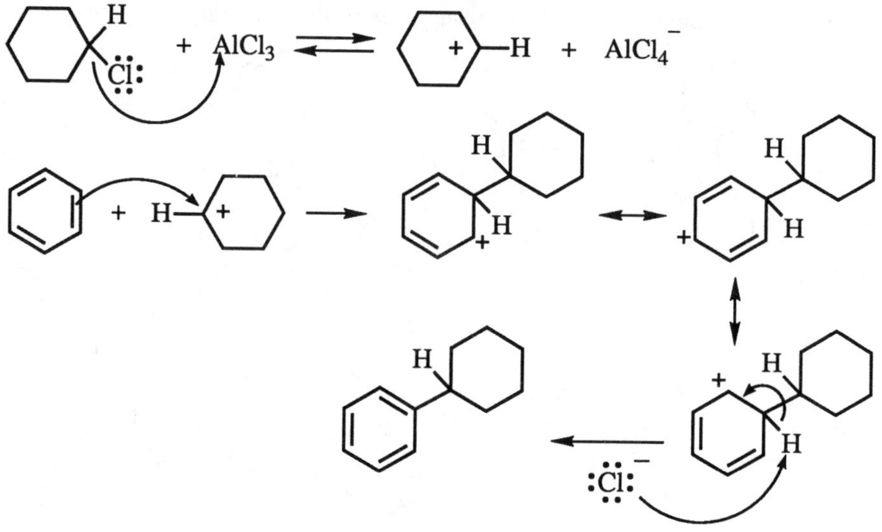

H
Cl
+ AlCl3
+
H
+ AlCl4
+ H
H
H
+
H
H
H
+
H
H
H
Cl

(b)

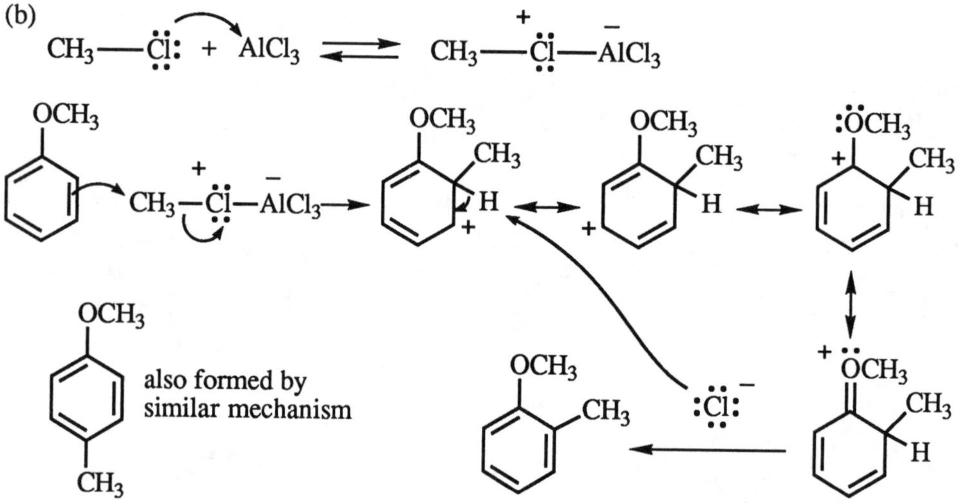

CH3
Cl
+ AlCl3
+
CH3
Cl
AlCl3
OCH3
CH3
Cl
AlCl3
OCH3
CH3
H
OCH3
CH3
H
OCH3
CH3
H
OCH3
CH3
H
OCH3
CH3
Cl
OCH3
CH3
also formed by
similar mechanism

17-18 continued

(c)

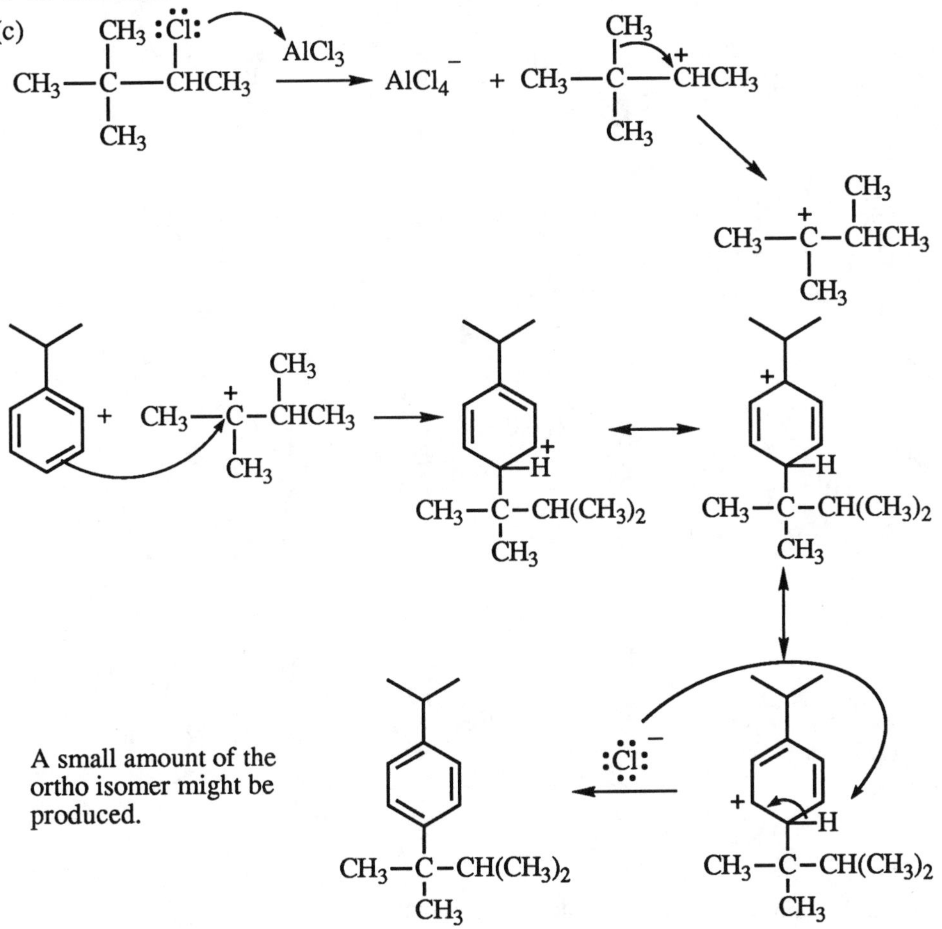

A small amount of the ortho isomer might be produced.

17-19

(a)

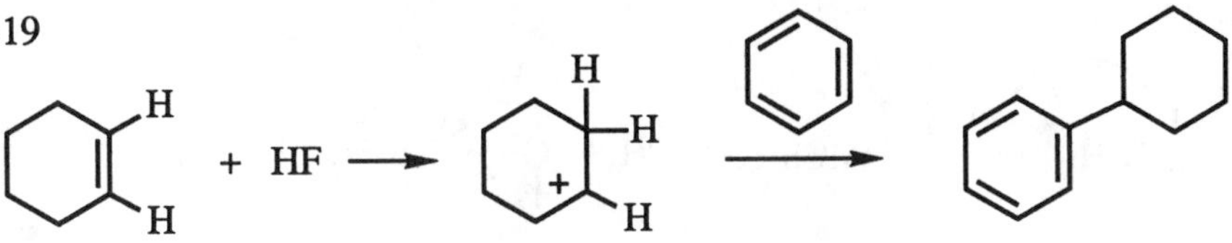

(b)

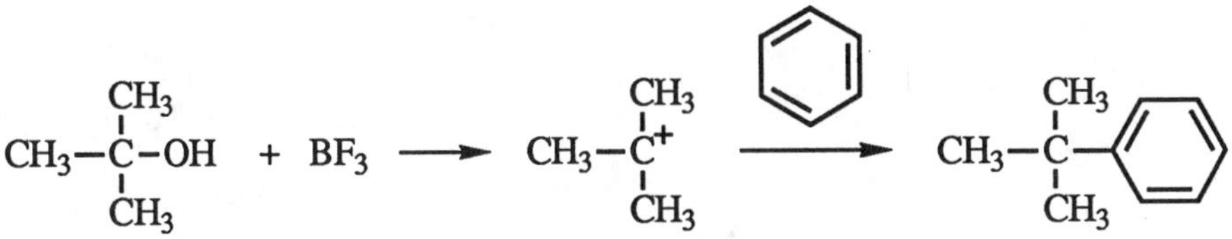

(c)

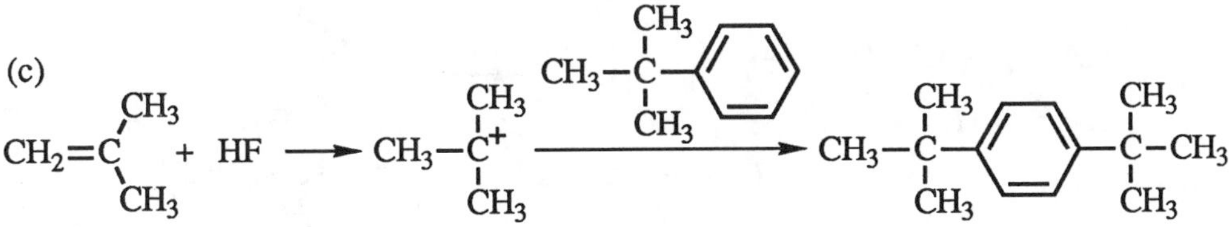

(d)

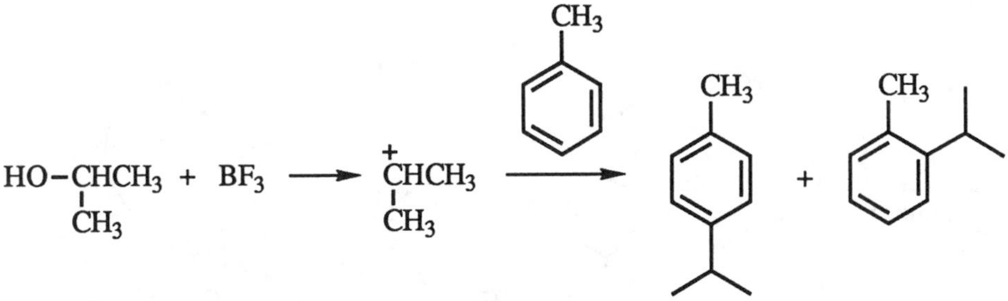

17-20 In (a), (b), and (d), the electrophile has rearranged.

(a)

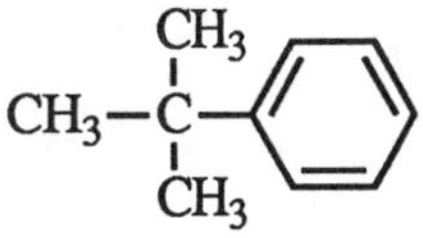

(b)

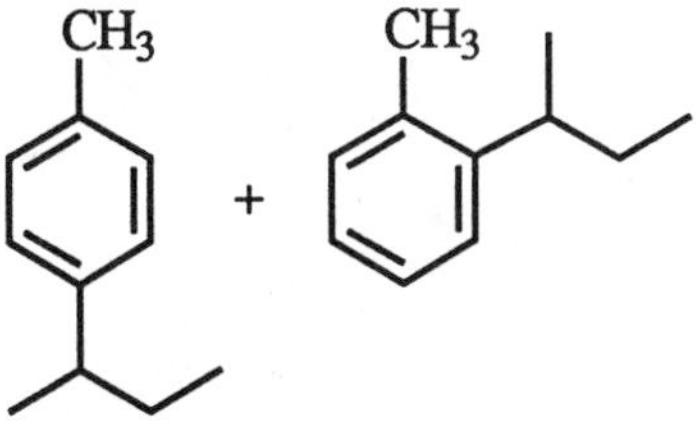

(c) No reaction: nitrobenzene is too deactivated for the Friedel-Crafts reaction to succeed.

(d)

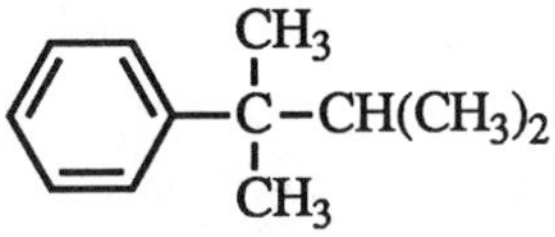

17-21

(a) 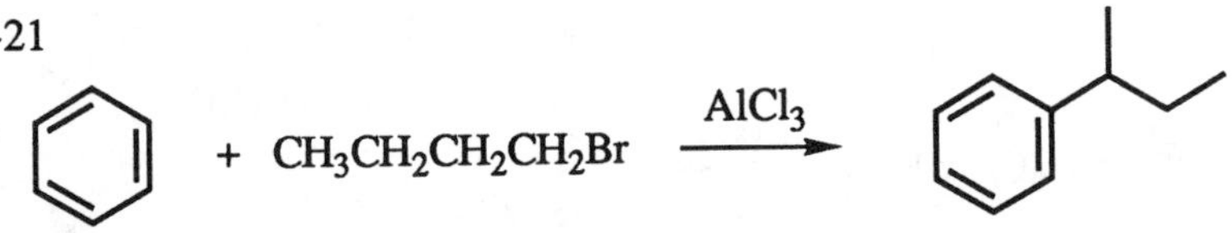
+ $CH_3CH_2CH_2CH_2Br$
$AlCl_3$

plus over-alkylation products

(b) gives desired product
(c) gives desired product, plus ortho isomer; use excess bromobenzene to avoid overalkylation
(d) gives desired product, plus ortho isomer
(e) gives desired product

17-22

(a) 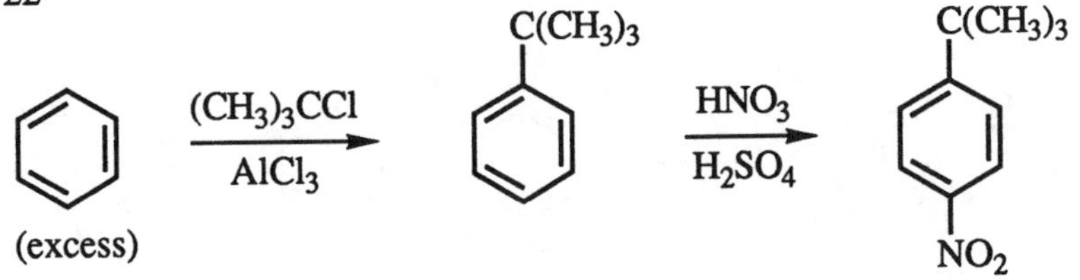
$(CH_3)_3CCl$
$AlCl_3$
(excess)
$C(CH_3)_3$
HNO_3
H_2SO_4
$C(CH_3)_3$
NO_2

(b) 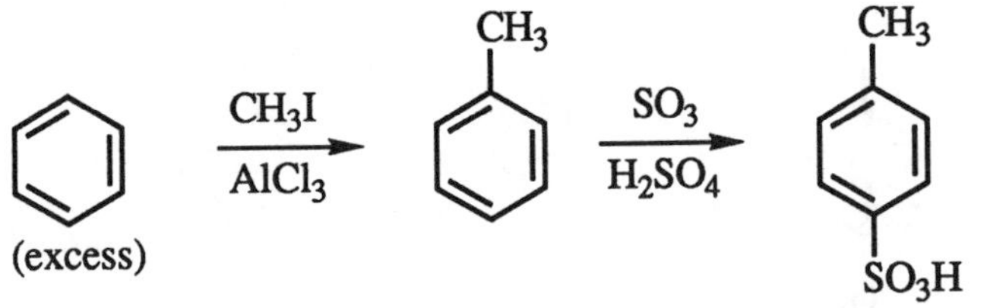
CH_3I
$AlCl_3$
(excess)
CH_3
SO_3
H_2SO_4
CH_3
SO_3H

(separate from *ortho*)

(c) 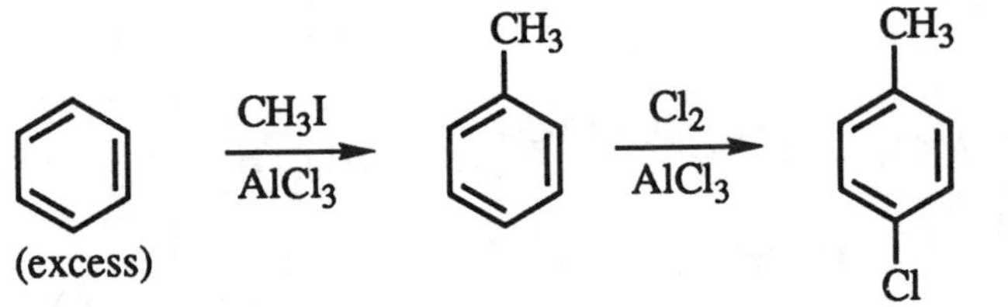
CH_3I
$AlCl_3$
(excess)
CH_3
Cl_2
$AlCl_3$
CH_3
Cl

(separate from *ortho*)

17-23

(a)

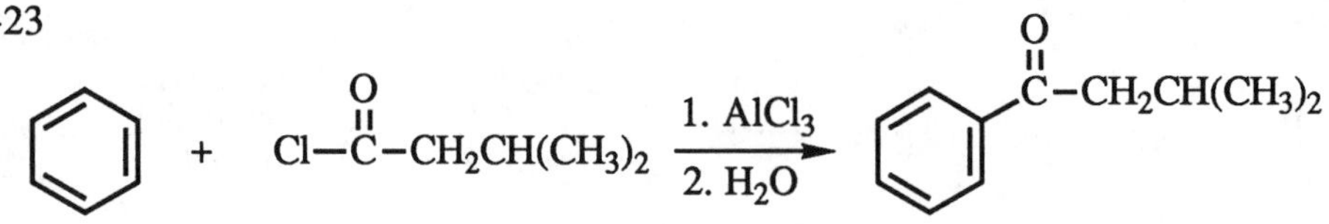

(b)

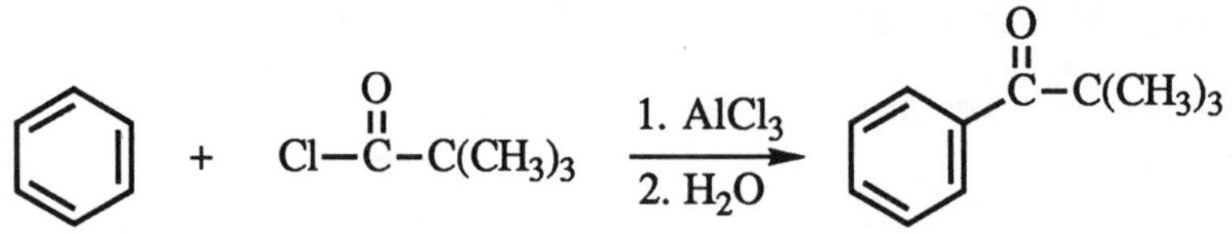

(c)

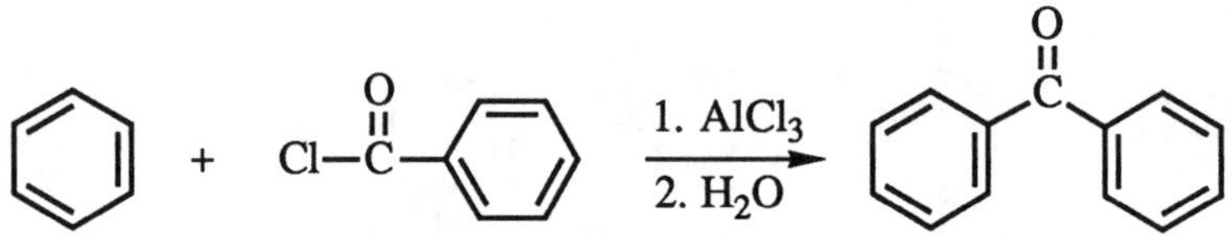

(d)

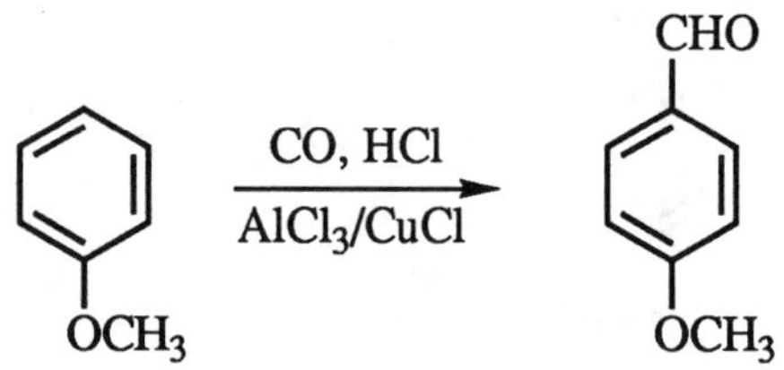

(e)

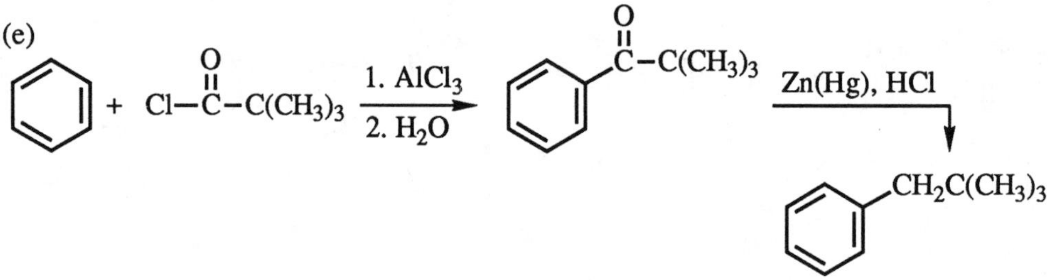

(f)

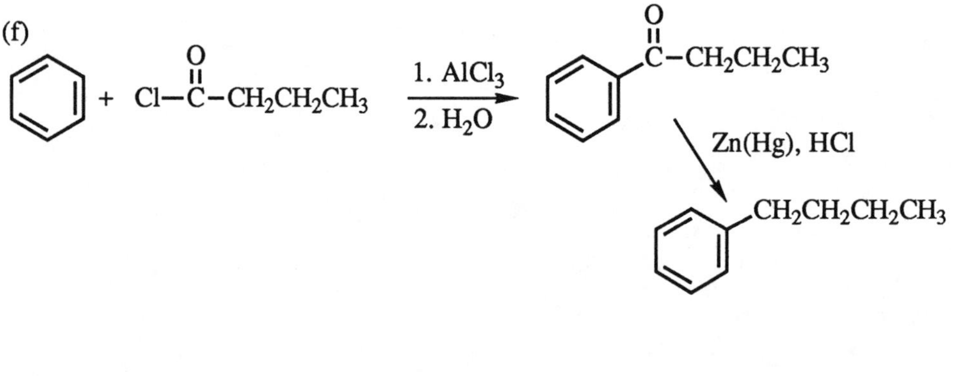

17-24 Another way of asking this question is this: Why is fluoride ion a good leaving group from **A** but not from **B**?

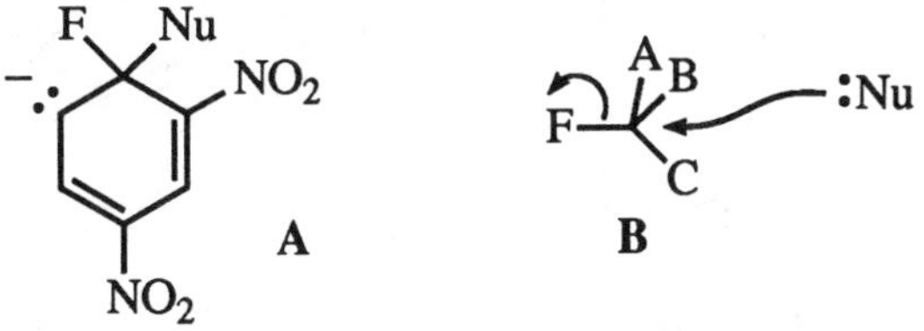

Formation of the anionic sigma complex **A** is the rate-determining (slow) step in nucleophilic aromatic substitution. The loss of fluoride ion occurs in a subsequent fast step where the nature of the leaving group does not affect the overall reaction rate. In the S_N2 mechanism, however, the carbon-fluorine bond is breaking in the rate-determining step (the only step), so the poor leaving group ability of fluoride does indeed affect the rate.

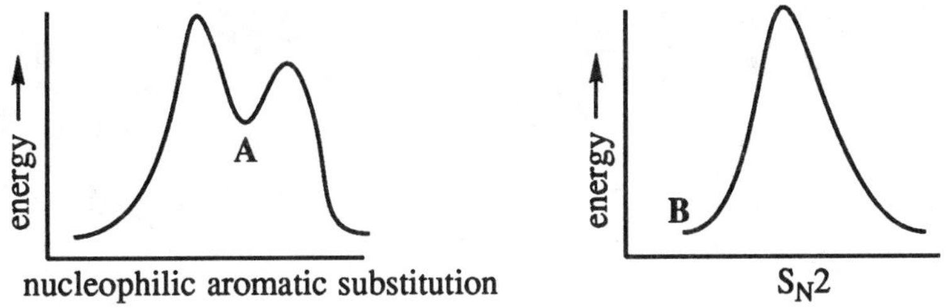

17-25

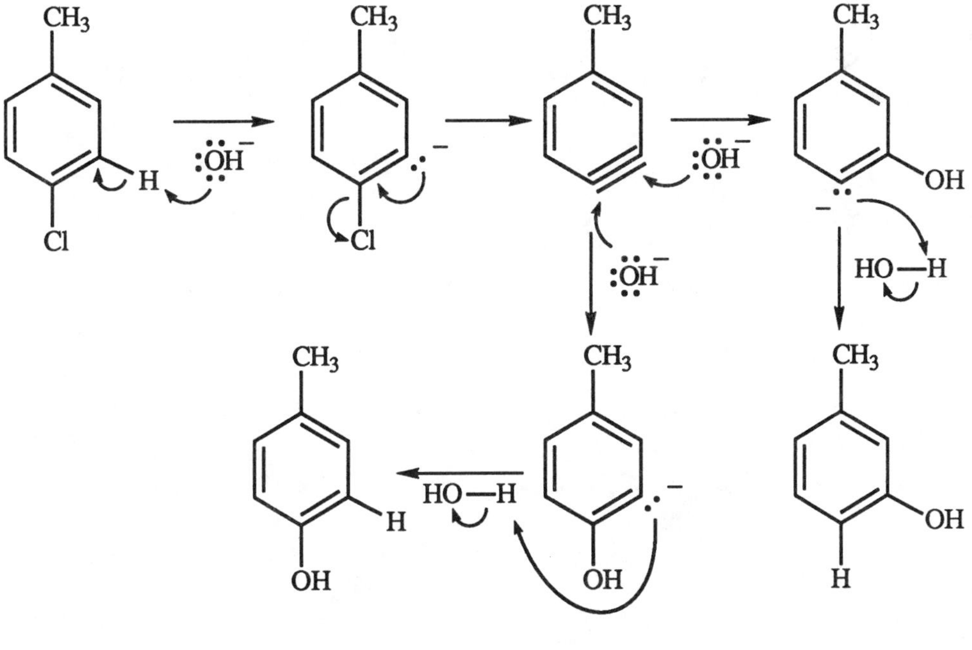

17-26

(a)

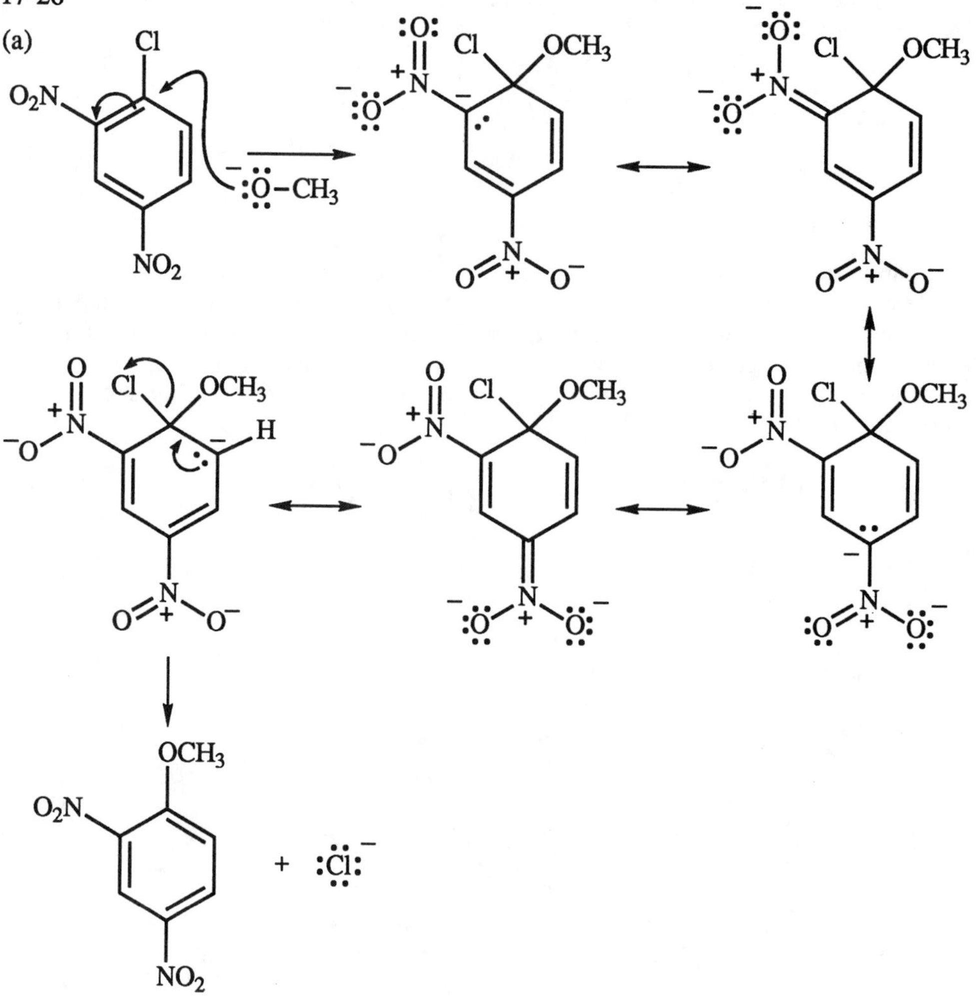

17-26 continued

(b)

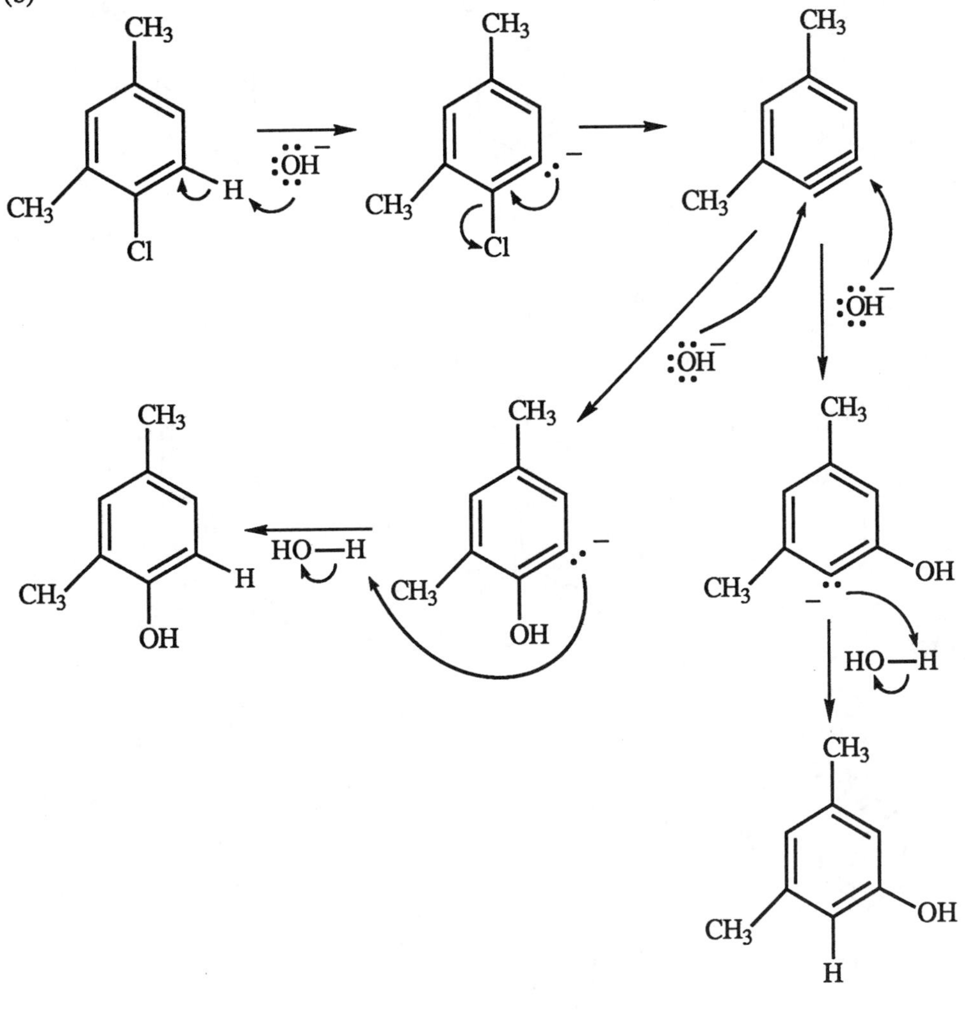

17-26 continued

(c)

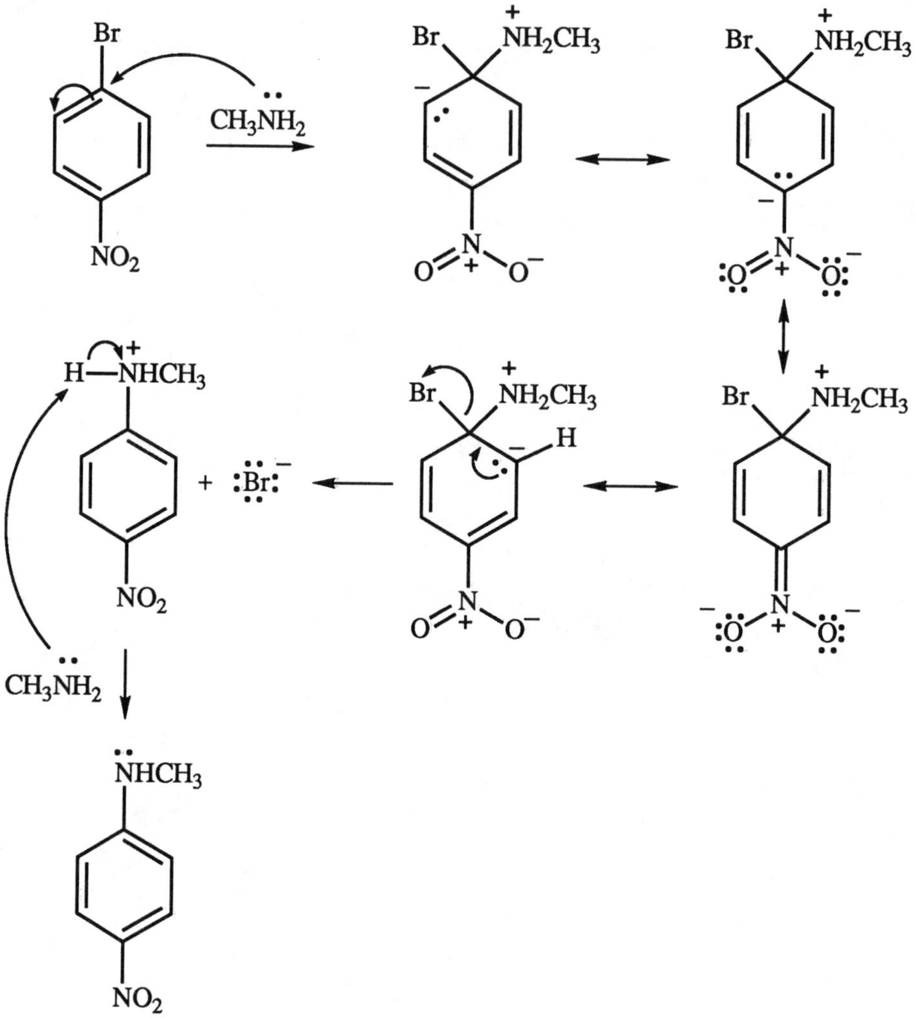

17-26 continued

(d)

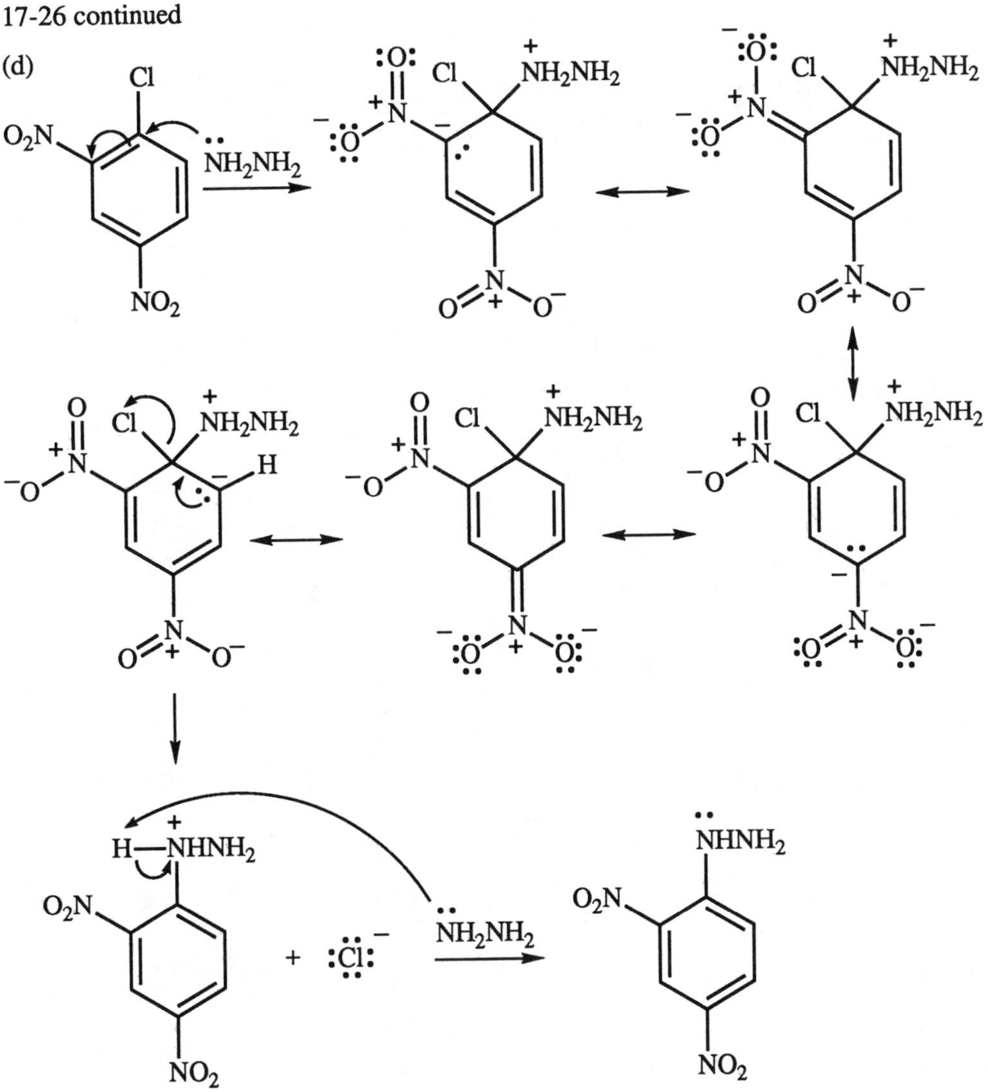

17-27

(a)

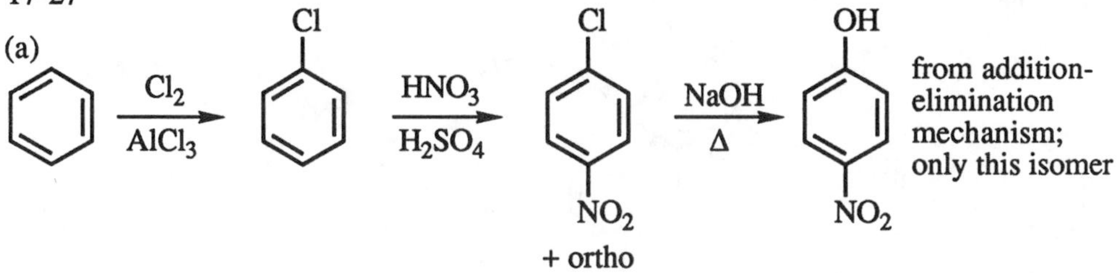
Cl
Cl_2
$AlCl_3$
HNO_3
H_2SO_4
NO_2
+ ortho
NaOH
Δ
OH
from addition-elimination mechanism; only this isomer

(b)

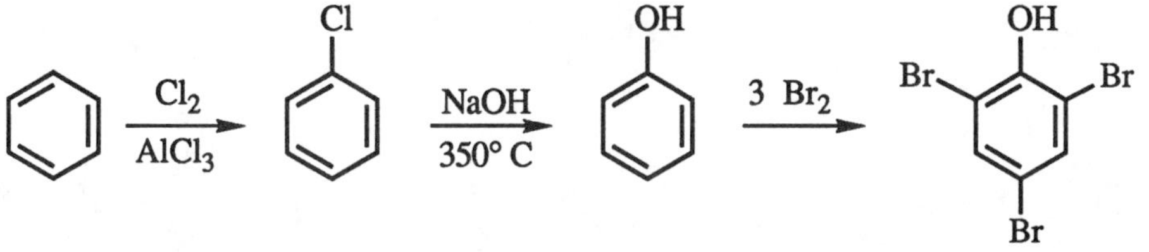
Cl
Cl_2
$AlCl_3$
NaOH
350° C
OH
3 Br_2
Br

(c)

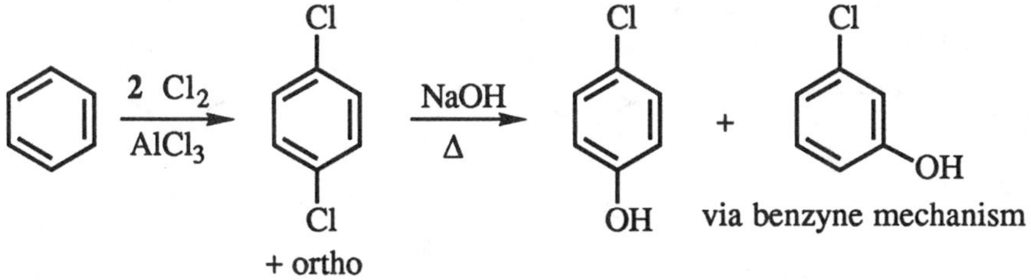
2 Cl_2
$AlCl_3$
Cl
+ ortho
NaOH
Δ
OH
+
via benzyne mechanism

(d)

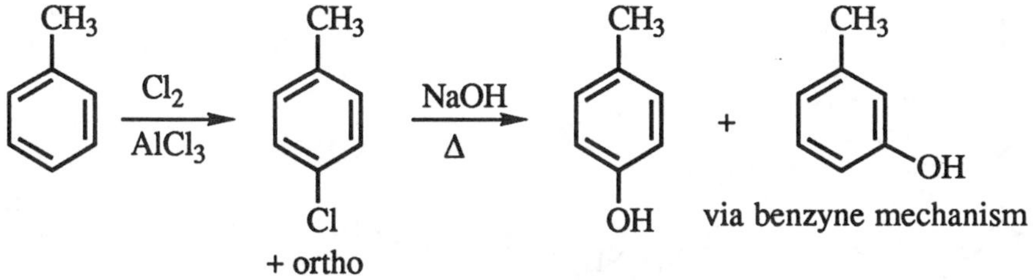
CH_3
Cl_2
$AlCl_3$
Cl
+ ortho
NaOH
Δ
OH
+
via benzyne mechanism

(e)

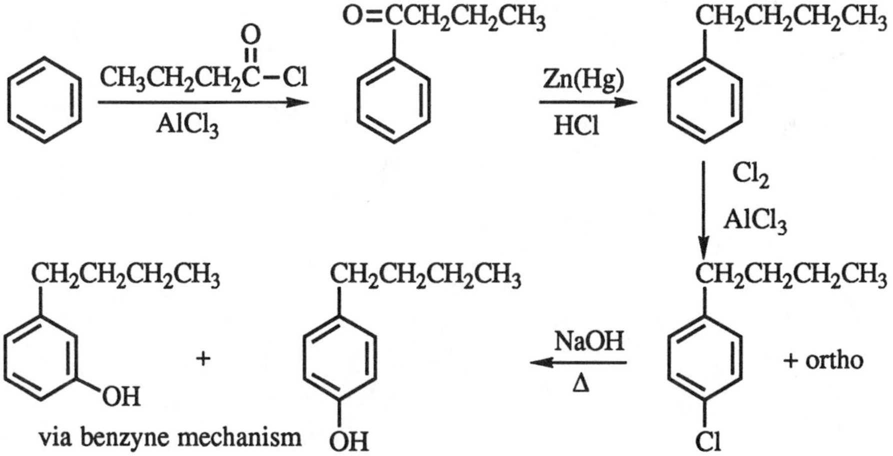
$CH_3CH_2CH_2\overset{O}{\overset{\|}{C}}-Cl$
$AlCl_3$
$O=CCH_2CH_2CH_3$
Zn(Hg)
HCl
$CH_2CH_2CH_2CH_3$
Cl_2
$AlCl_3$
Cl
+ ortho
NaOH
Δ
OH
+
via benzyne mechanism

17-28

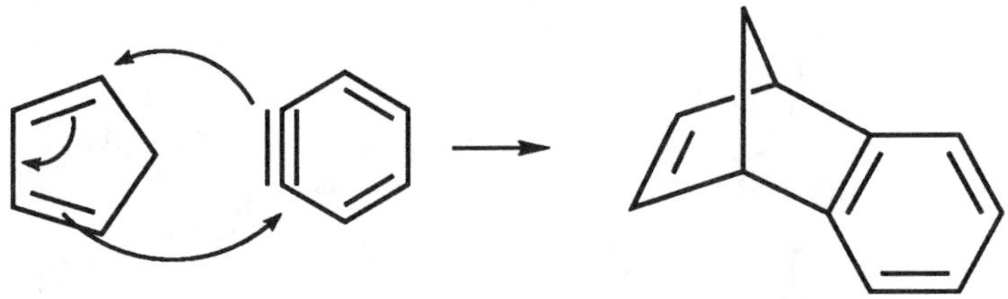

17-29

(a) First, the carboxylic acid proton is neutralized.

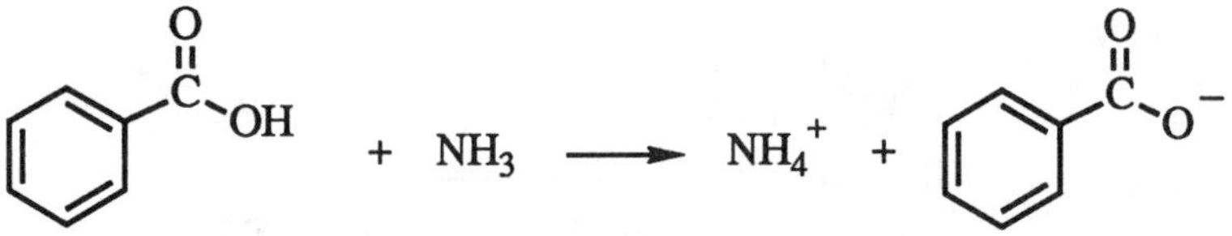

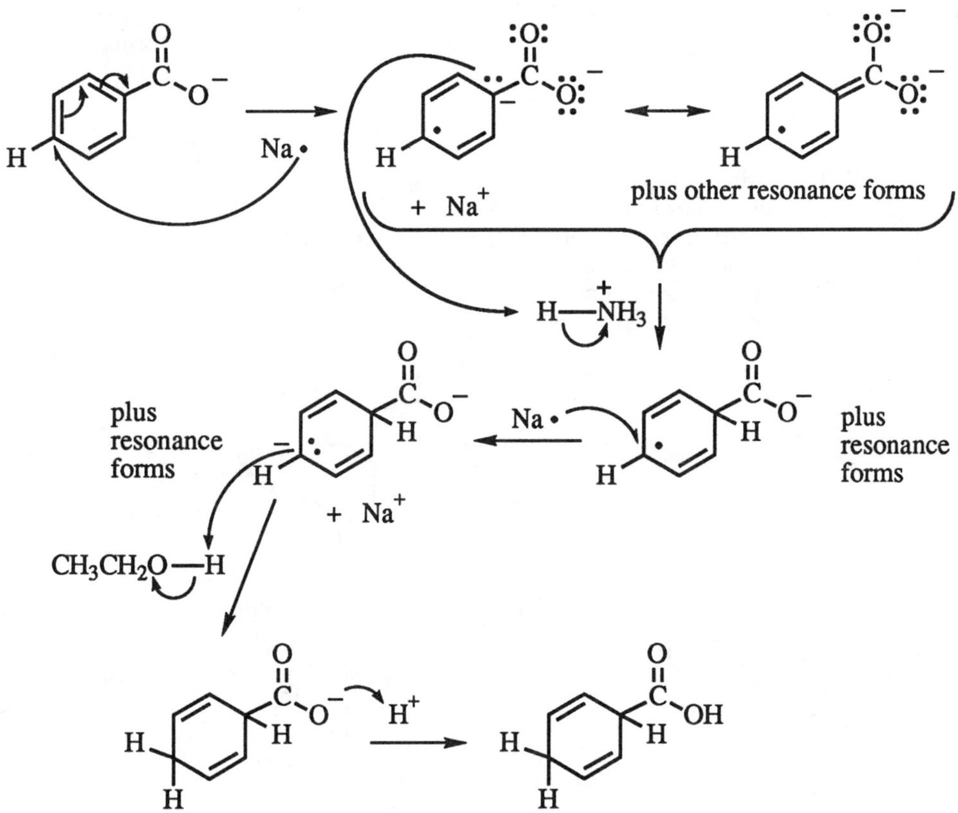
Na•
H
+ Na+
plus other resonance forms
H—NH3
plus resonance forms
+ Na+
Na•
plus resonance forms
CH3CH2O—H
H+

17-29 continued

(b)

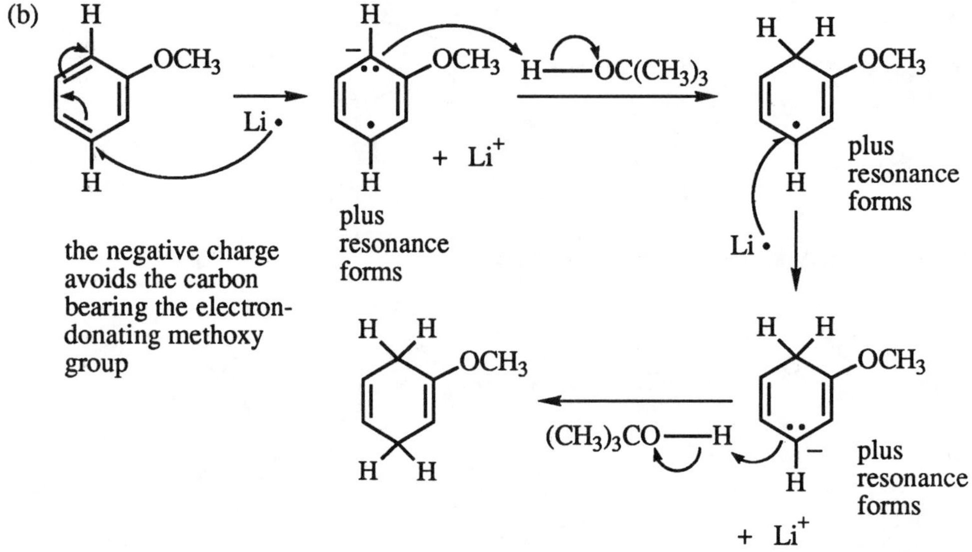

H
OCH3
Li •
H
H
OCH3
H—OC(CH3)3
+ Li+
H
plus
resonance
forms
H H
OCH3
H
plus
resonance
forms
Li •
the negative charge
avoids the carbon
bearing the electron-
donating methoxy
group
H H
OCH3
H H
(CH3)3CO—H
H H
OCH3
H
plus
resonance
forms
+ Li+

17-30

(a) 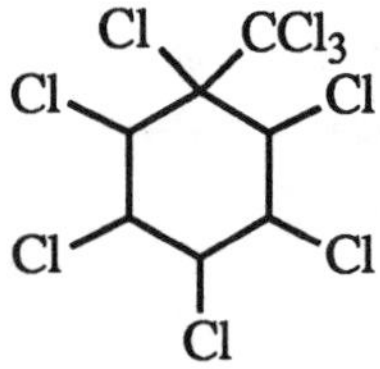

Cl
CCl3
Cl
Cl
Cl
Cl
Cl

rapid benzylic substitution, then addition of chlorine to the π system of the ring, giving a mixture of stereoisomers

(b) 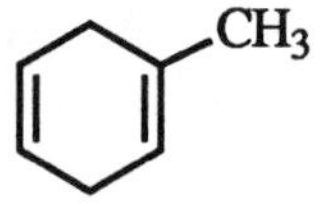

CH3

(c) 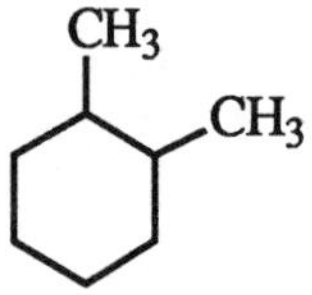

CH3
CH3

cis + trans

(d) 

CH3
CH3

(e) 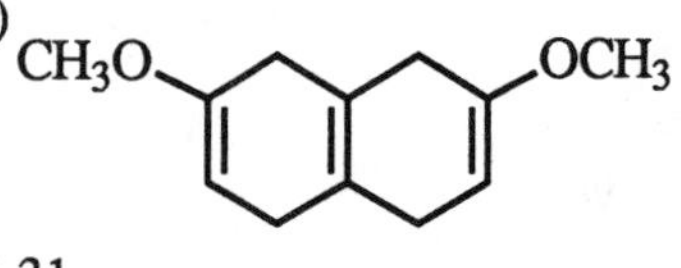

CH3O
OCH3

(f) 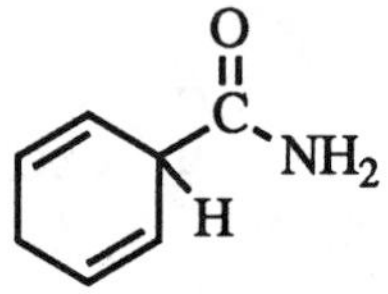

O
C
NH2
H

17-31

(a) 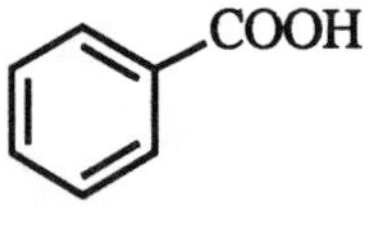

COOH

(b) 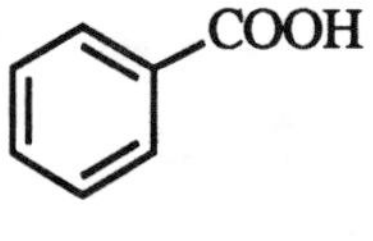

COOH

(c) 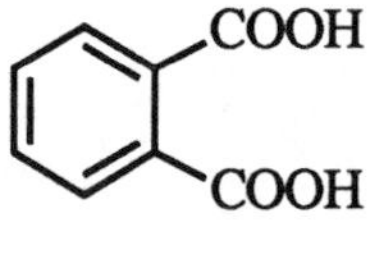

COOH
COOH

17-32

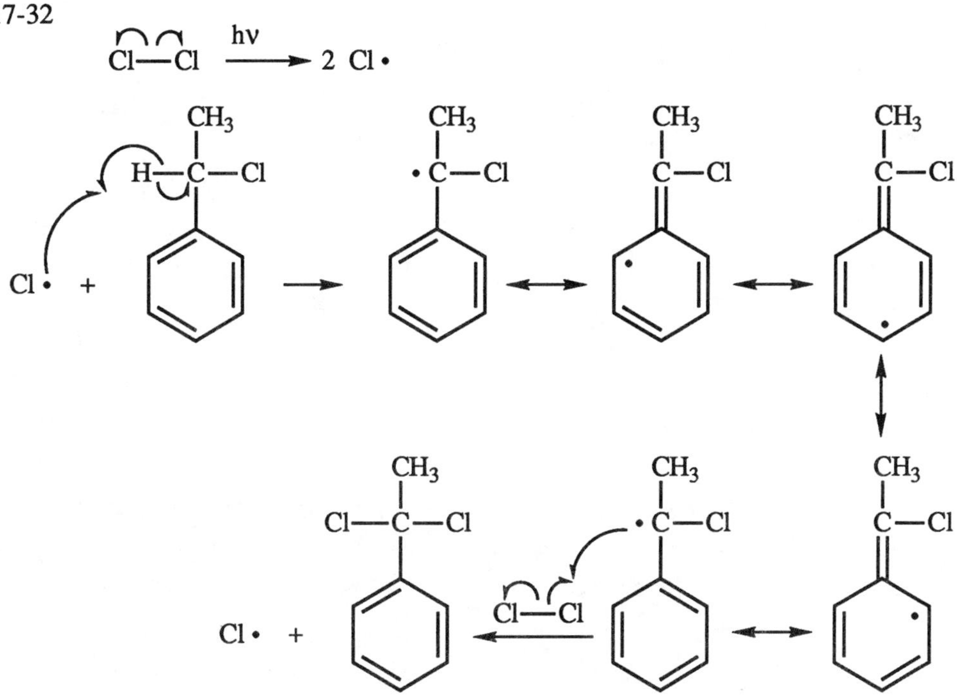

17-33 A statistical mixture would give 2 : 3 or 40% : 60% α to β. To calculate the relative reactivities, the percents must be corrected for the numbers of each type of hydrogen.

$$\alpha: \quad \frac{56\%}{2H} = 28 \text{ relative reactivity}$$

$$\beta: \quad \frac{44\%}{3H} = 14.7 \text{ relative reactivity}$$

The reactivity of α to β is $\frac{28}{14.7} = 1.9$ to 1

17-34

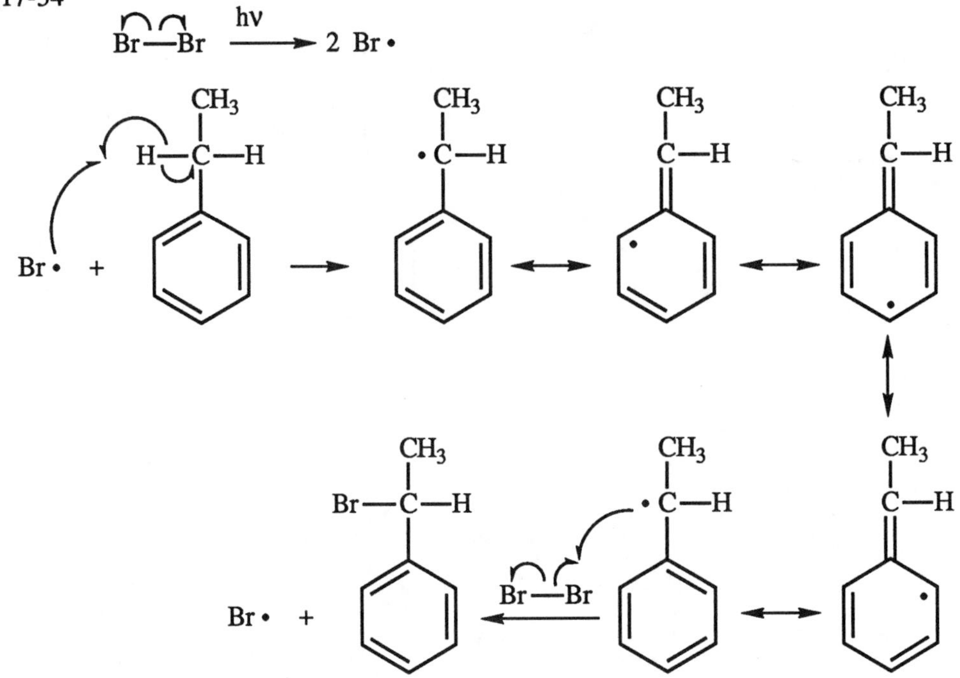

Br—Br
hν
2 Br•
CH3
H—C—H
Br•
+
•C—H
C—H
CH3
Br—C—H
Br—Br

17-35

(a) (1)

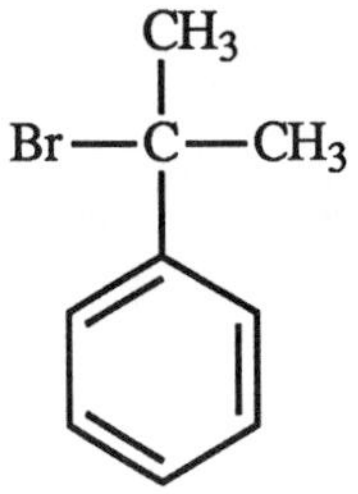

CH3
Br—C—CH3

(2)

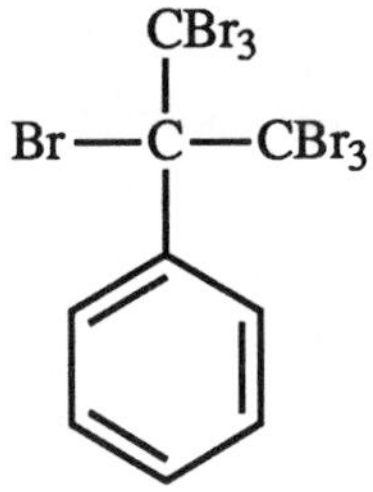

CBr3
Br—C—CBr3

(b) (1)

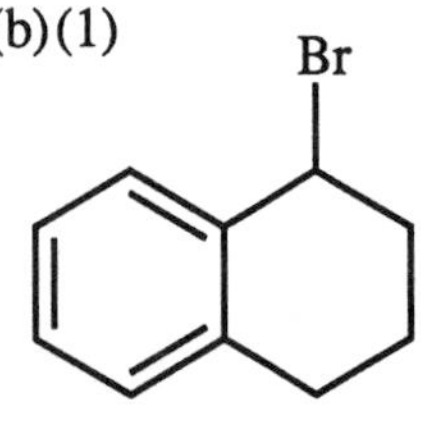

Br

(2)

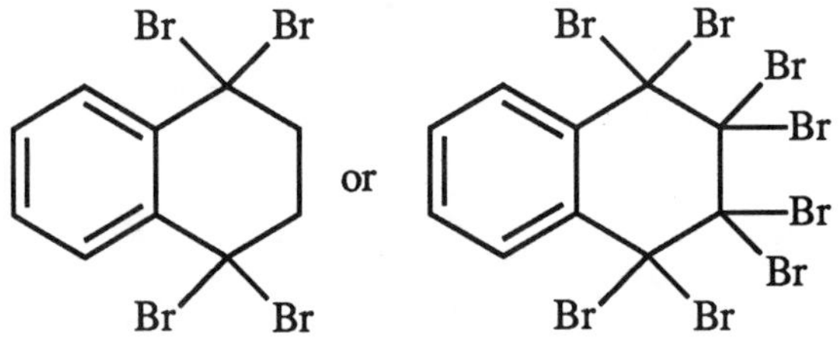

Br
or

17-36

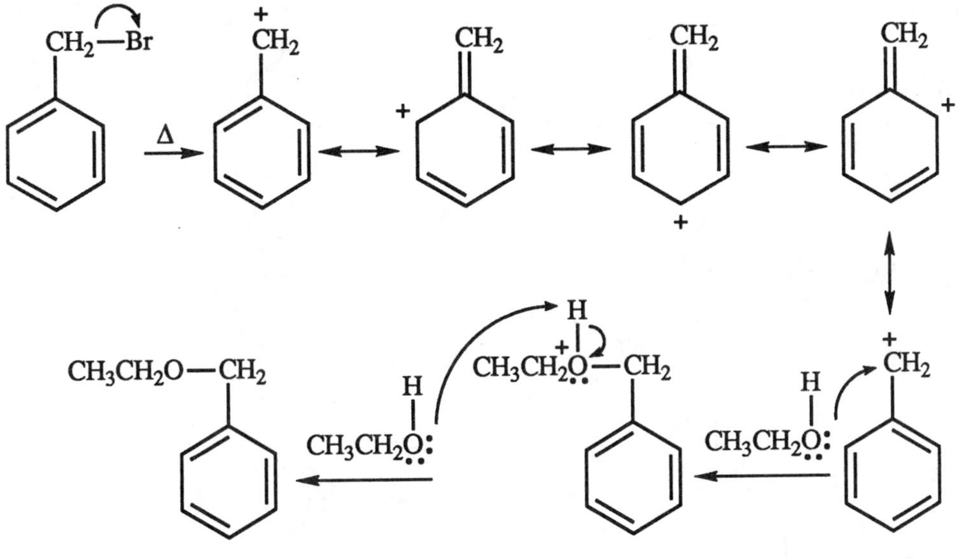

CH2—Br
Δ
CH2
CH3CH2O—CH2
CH3CH2O:
CH3CH2O—CH2
H

17-37

(a) Benzylic cations are stabilized by resonance and are much more stable than regular alkyl cations. The product is 1-bromo-1-phenylpropane.

(b)

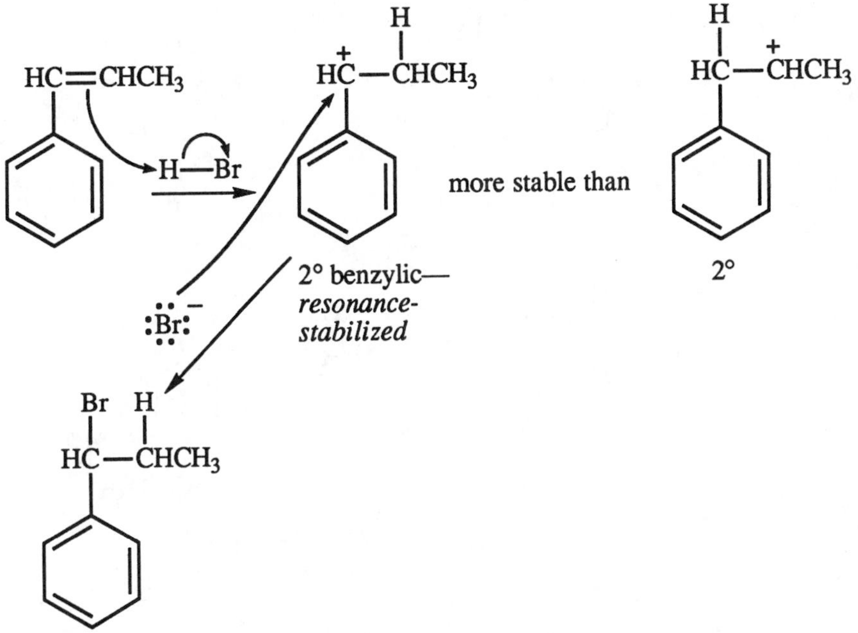

HC=CHCH3
H—Br
HC—CHCH3
more stable than
2° benzylic—
resonance-
stabilized
2°
:Br:
Br H

17-38

(a)

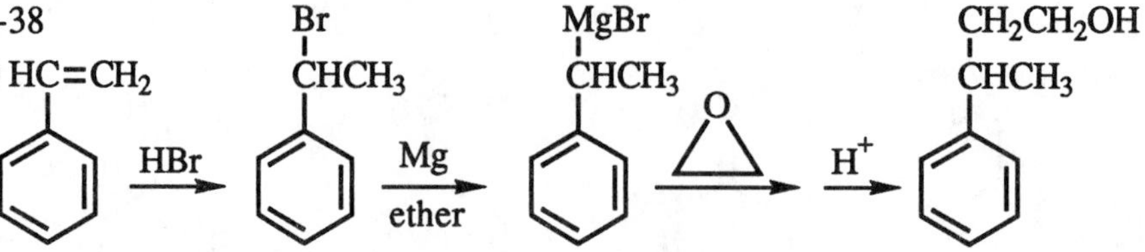

(b)

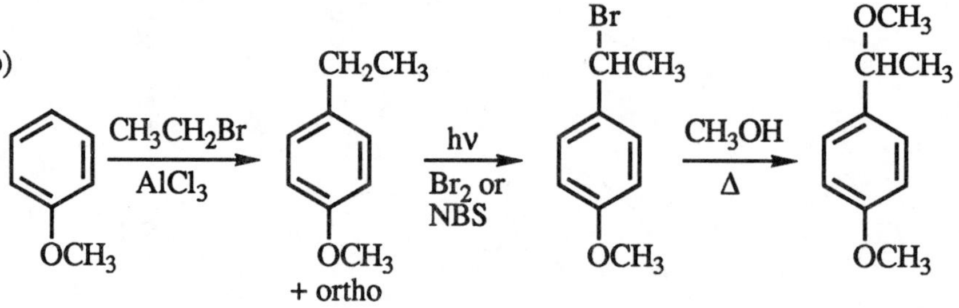

(c)

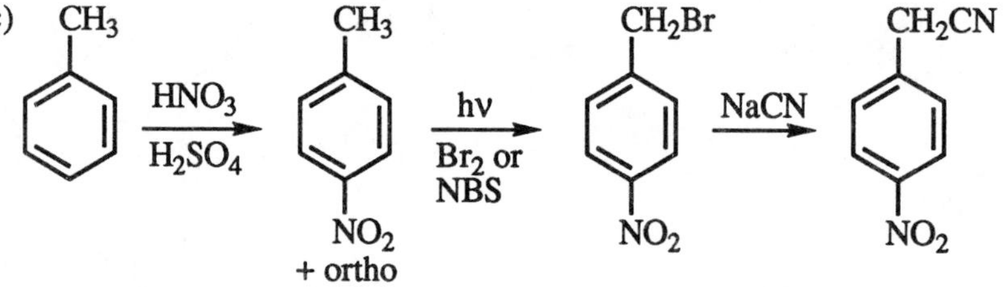

17-39

(a)

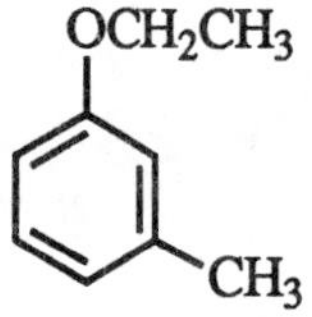

(b)

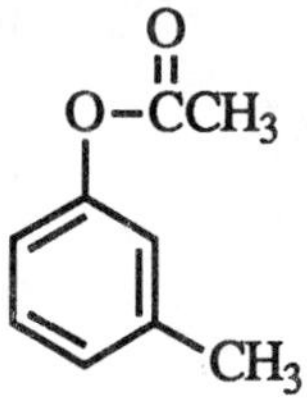

(c)

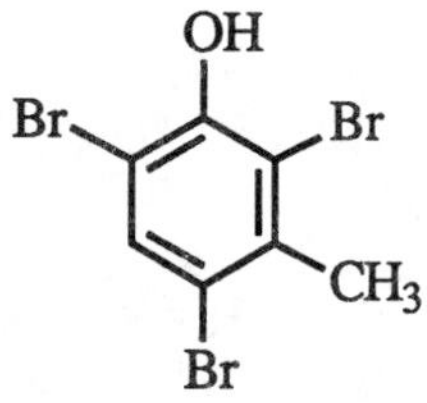

(d)

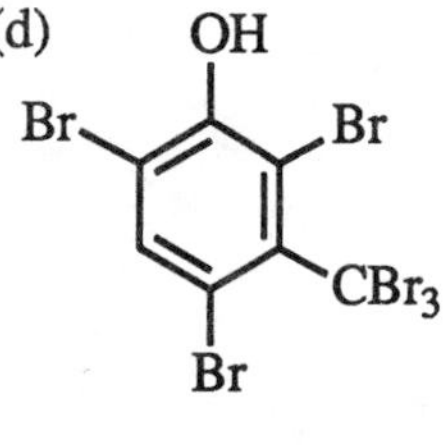

(e)

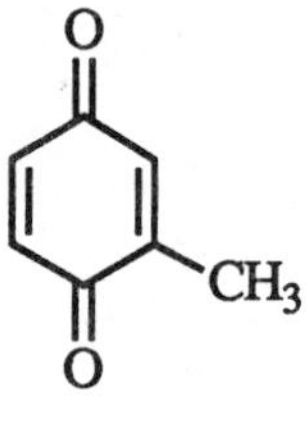

(f)

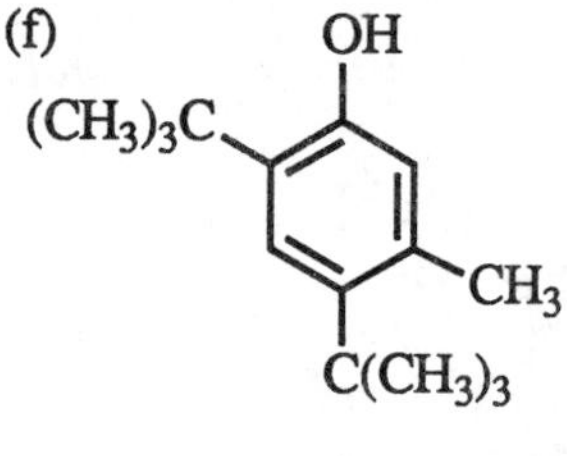

17-40

(a)

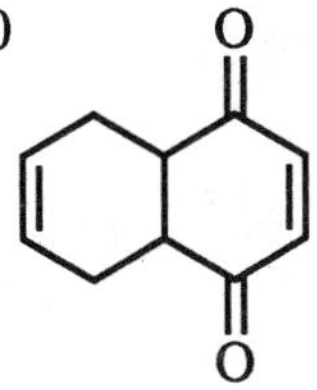
O
O

(b)

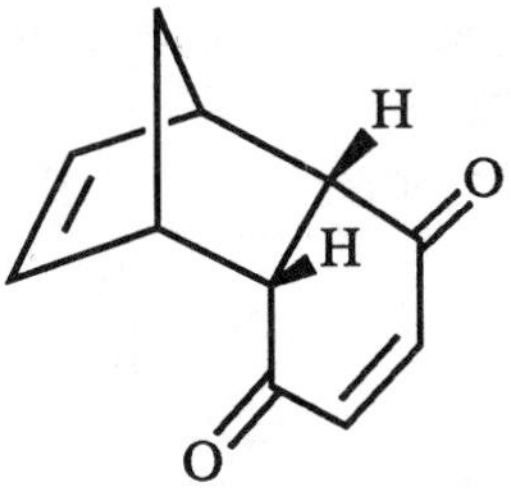
H
O
H
O

17-41

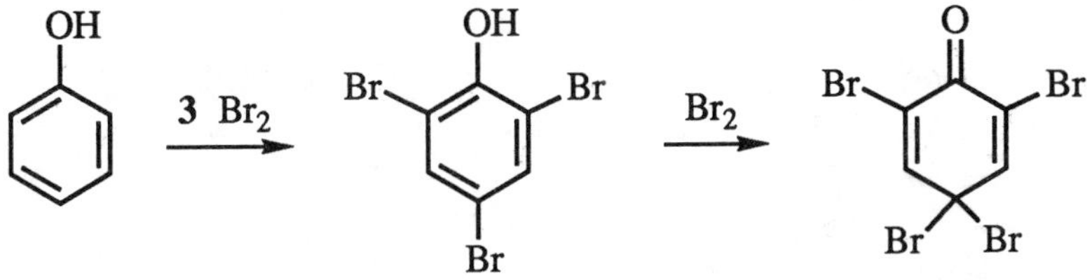
OH
3 Br$_2$
OH
Br
Br
Br
Br$_2$
O
Br
Br
Br
Br

$C_6H_3OBr_3$ $C_6H_2OBr_4$

17-42 Refer to the Glossary and the text for definitions and examples.

17-43

(a)

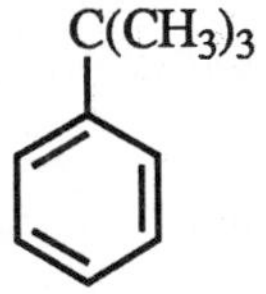
C(CH3)3

(b)

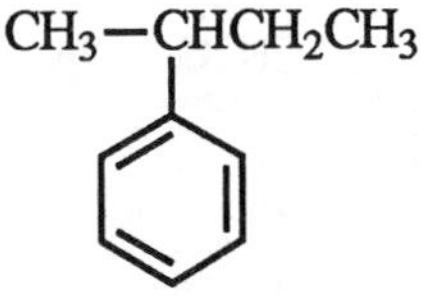
CH3−CHCH2CH3

rearrangement

(c)

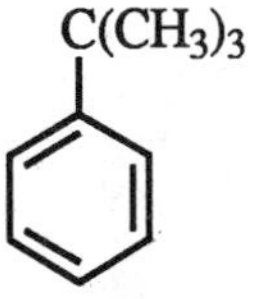
C(CH3)3

rearrangement

(d)

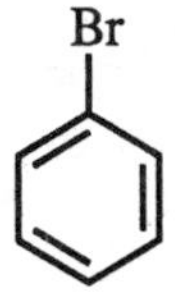
Br

(e)

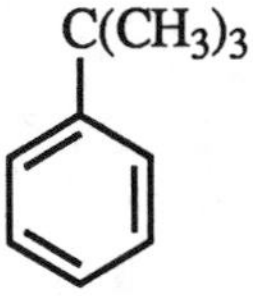
C(CH3)3

(f)

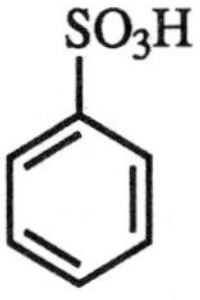
SO3H

(g)

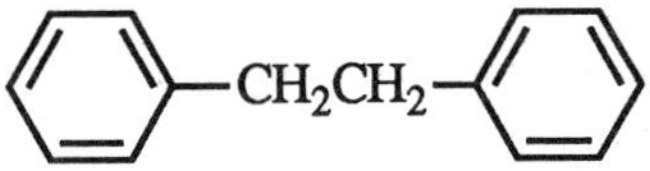
CH2CH2

(h)

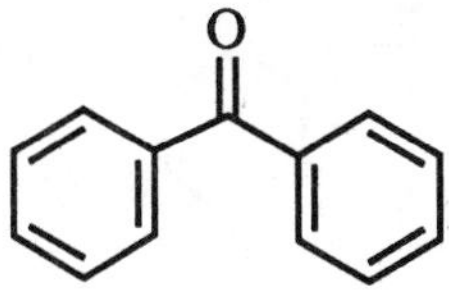
O

(i)

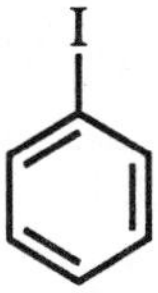
I

(j)

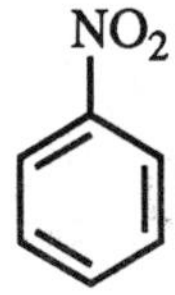
NO2

(k)

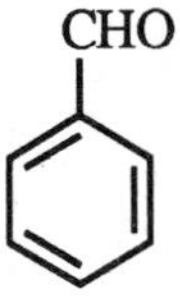
CHO

(l)

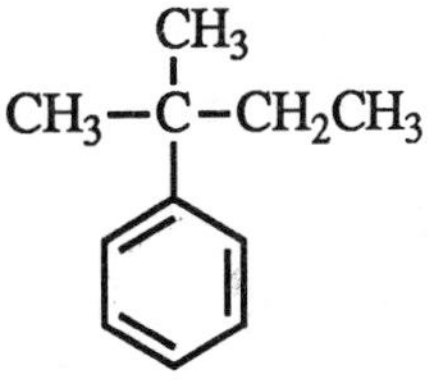
CH3
CH3−C−CH2CH3

rearrangement

17-44 Products from substitution at the ortho position will be minor because of the steric bulk of the isopropyl group.

(a)

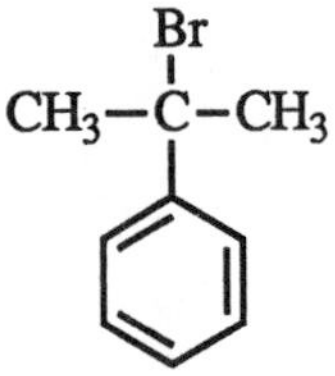

Br
CH3–C–CH3

(b)

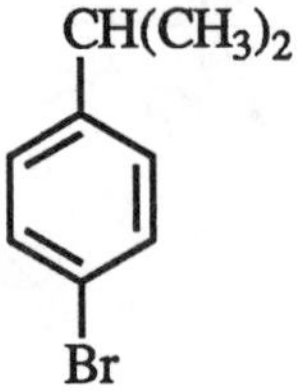

CH(CH3)2
Br

(c)

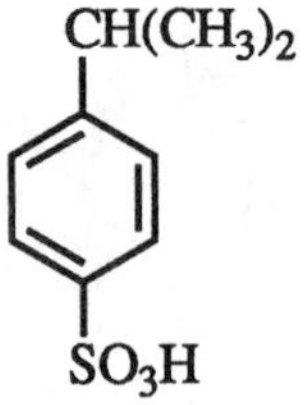

CH(CH3)2
SO3H

(d)

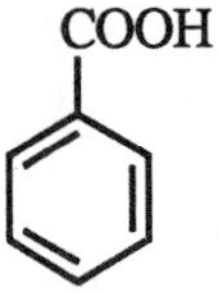

COOH

(e)

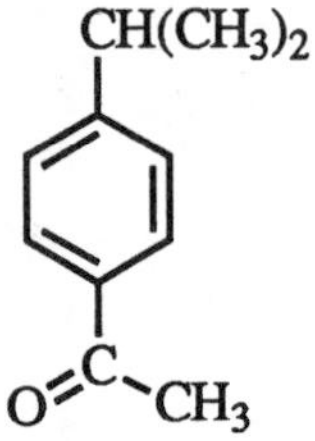

CH(CH3)2
O=C
CH3

(f)

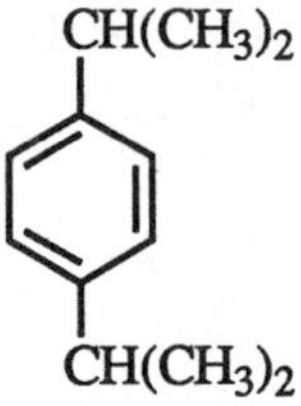

CH(CH3)2
CH(CH3)2

rearrangement

17-45

(a)

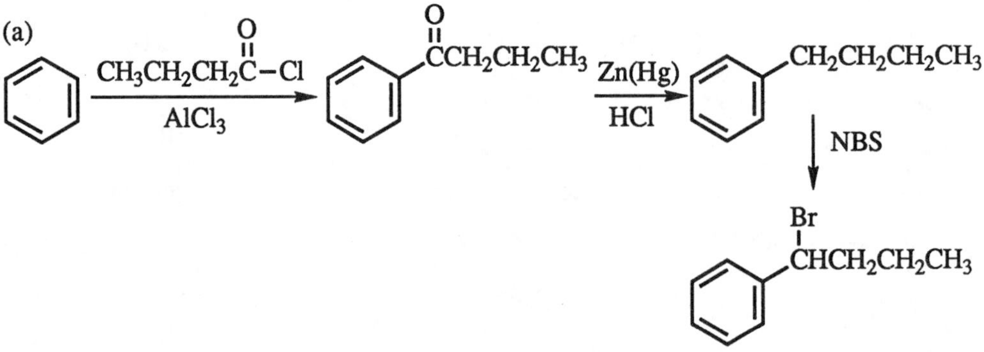

O
CH3CH2CH2C–Cl
AlCl3
O
CCH2CH2CH3
Zn(Hg)
HCl
CH2CH2CH2CH3
NBS
Br
CHCH2CH2CH3

(b)

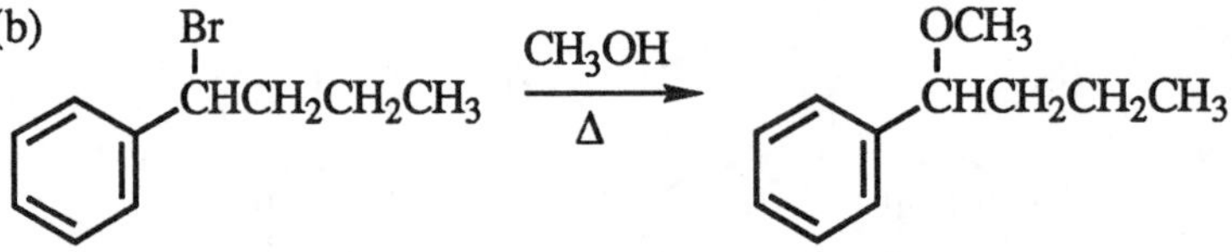

Br
CHCH2CH2CH3
CH3OH
Δ
OCH3
CHCH2CH2CH3

from (a)

17-45 continued

(c)

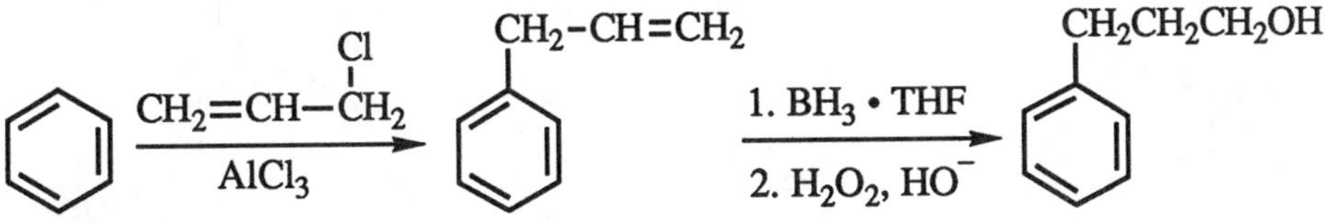

OR

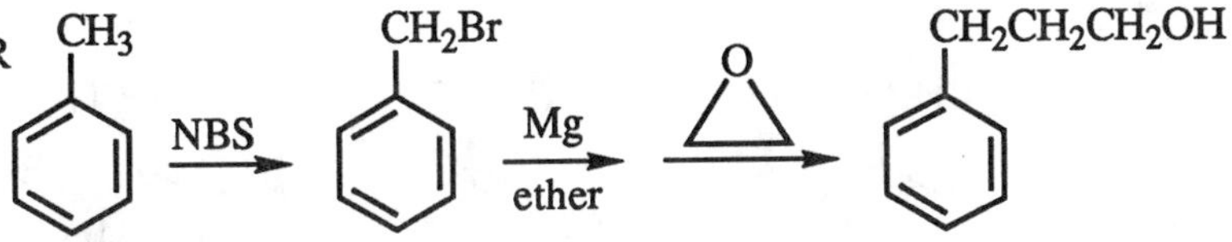

(d)

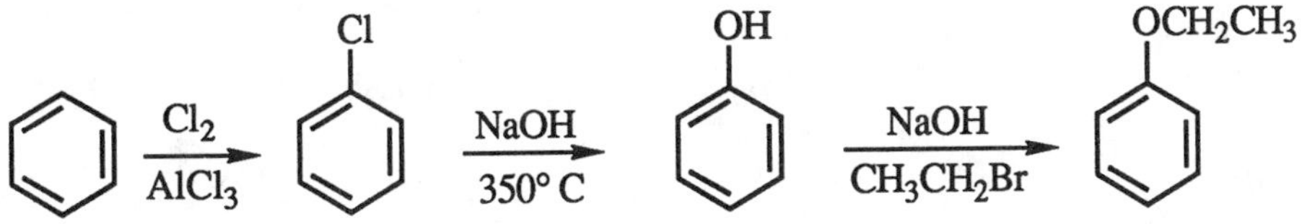

(e)

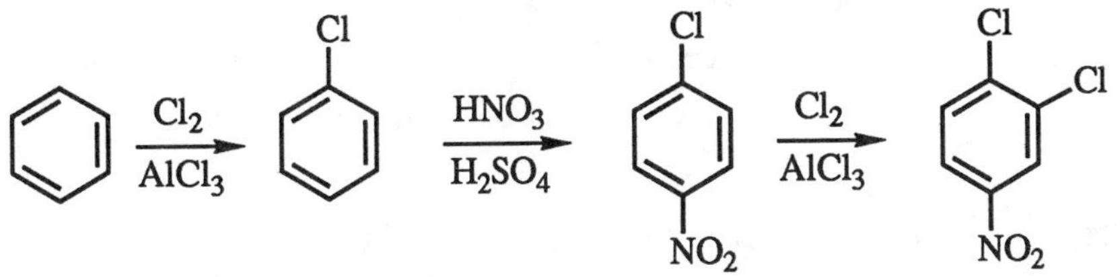

(f)

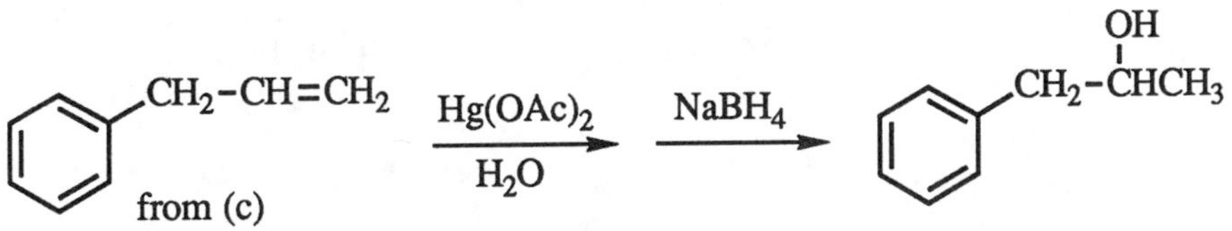

OR

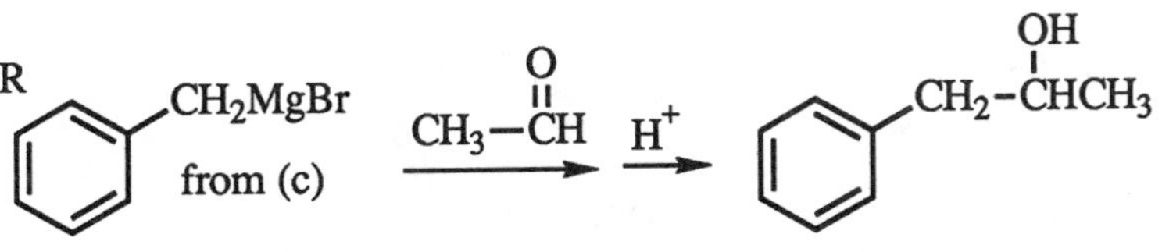

(g)

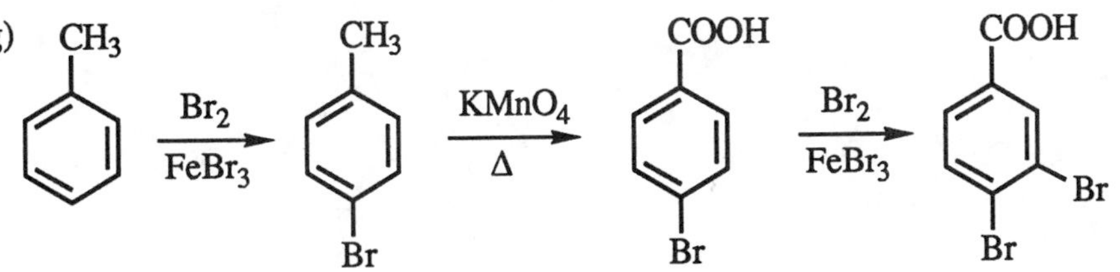

17-46

(a)

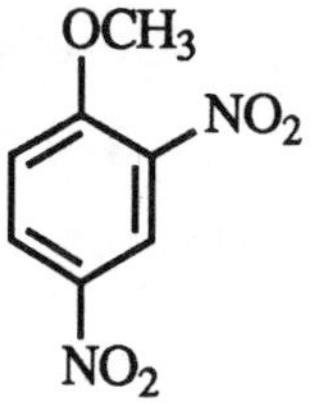

(b)

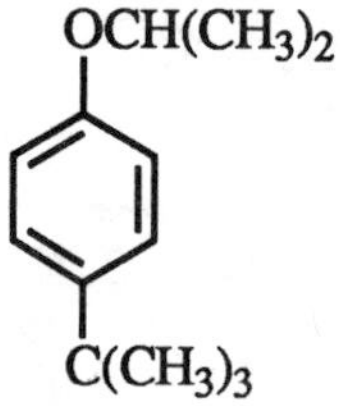

(c)

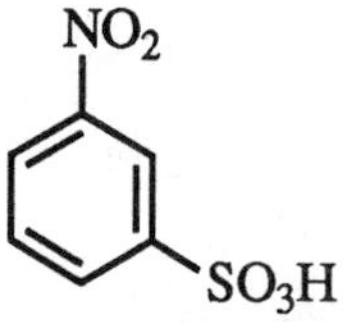

(d) no reaction

(e)

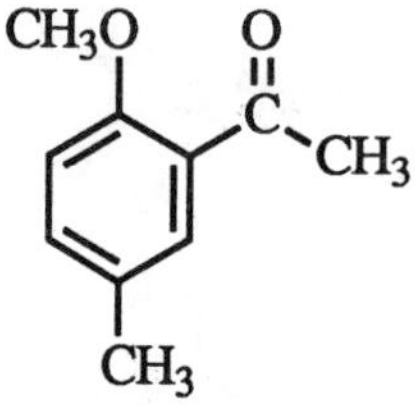

(f)

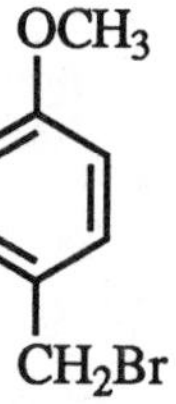

or

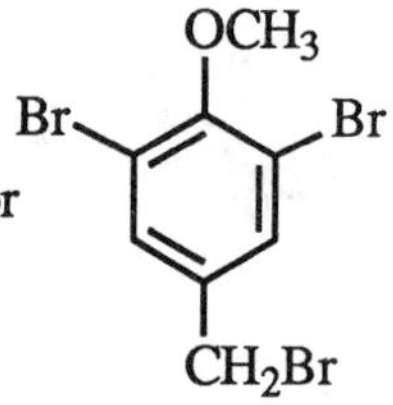

(g)

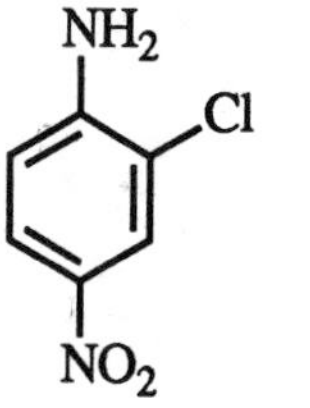

(h)

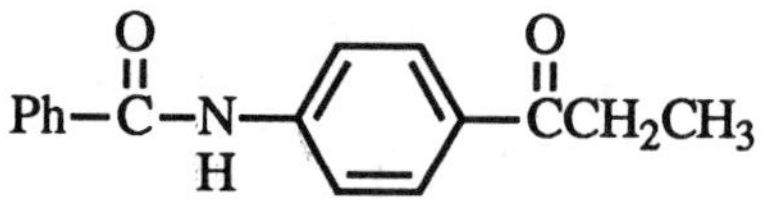

(i)

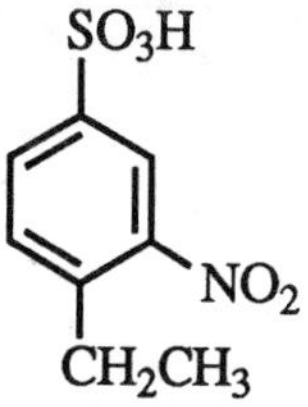

(j)

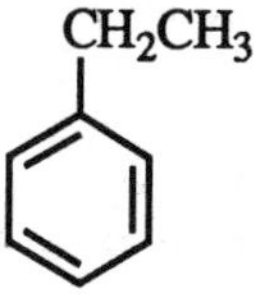

(k)

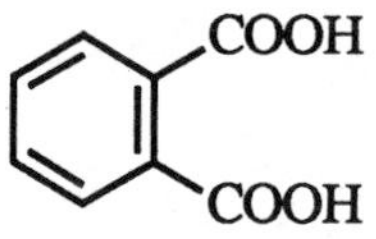

(l)

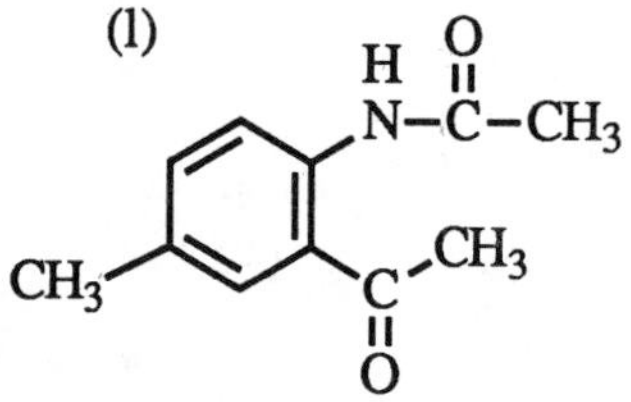

17-47 Major products are shown. Other isomers are possible.

(a)

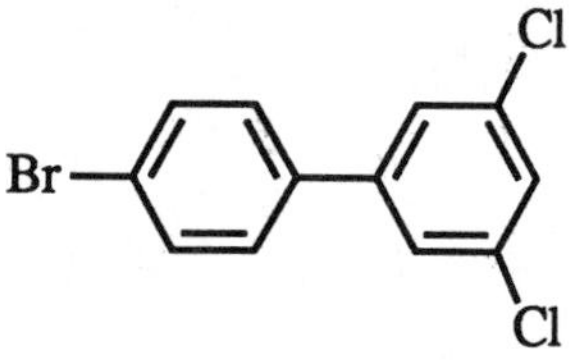

(b)

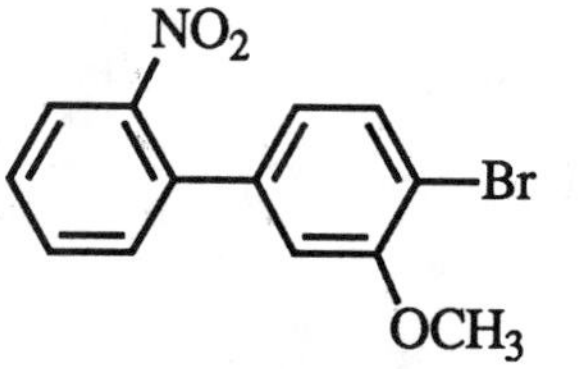

(c)

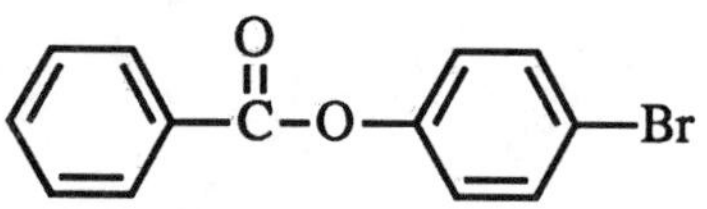

(d)

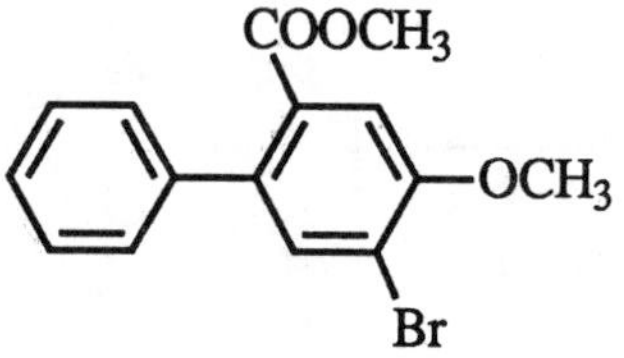

17-48

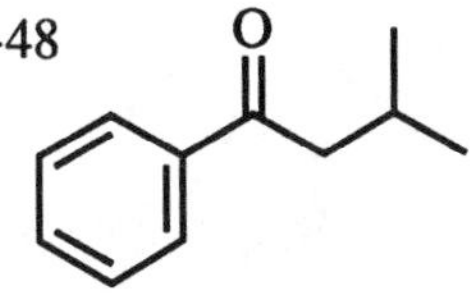

A

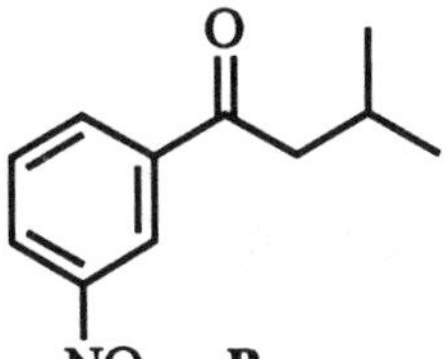

B

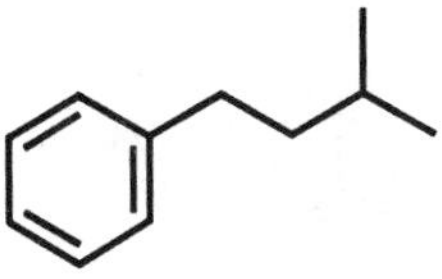

C

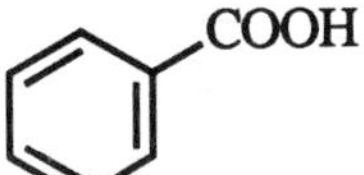

D

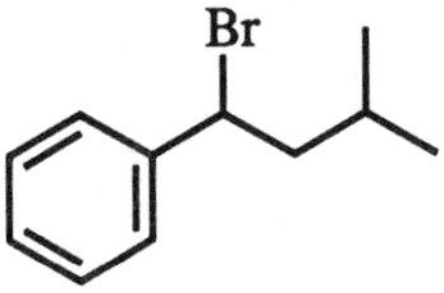

E

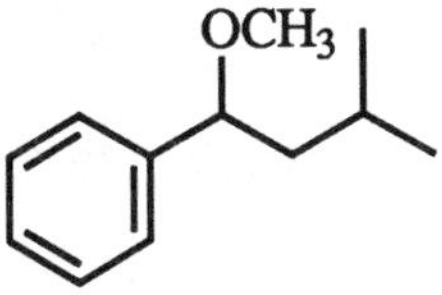

F

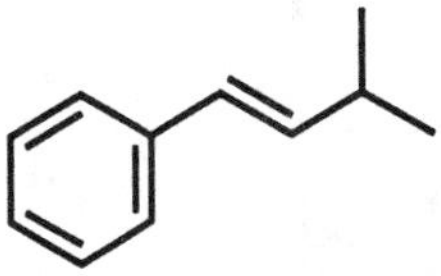

G

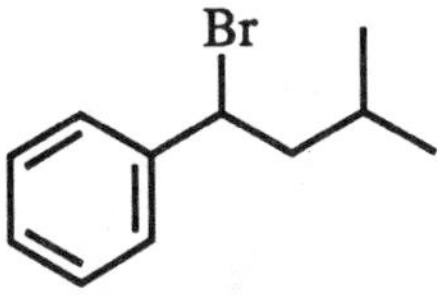

H (same as **E**)

17-49

product

starting material
molecular weight 150

molecular weight 132 =
loss of 18 = loss of H_2O

IR spectrum: The dominant peak is the carbonyl at 1700 cm^{-1}.

NMR spectrum: The splitting is complicated but the integration is helpful. In the region of δ 2.5-3.3, there are two signals with integration values of 2H; these must be the two adjacent methylenes, CH_2CH_2. The aromatic region from δ 7.2-7.9 has integration of 4H, so the ring must be disubstituted.

Carbon NMR: Only four of the aromatic signals are doublets (C—H); two are singlets, also showing a disubstituted benzene. Also indicated are carbons in a carbonyl and two methylenes.

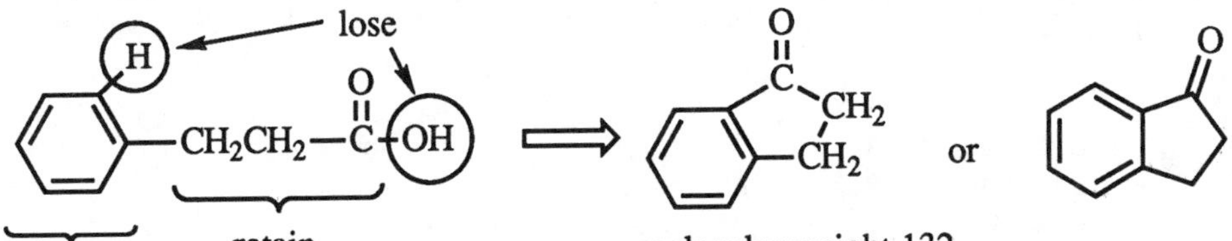

lose 1 H

molecular weight 132

The product must be the cyclized ketone, formed in an intramolecular Friedel-Crafts acylation.

17-50

(a)

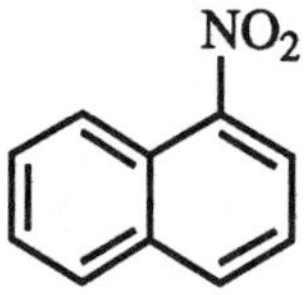

NO_2

(b)

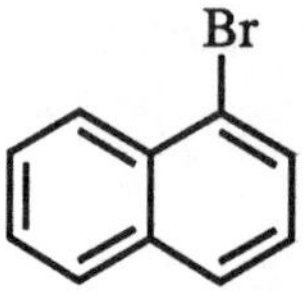

Br

(c)

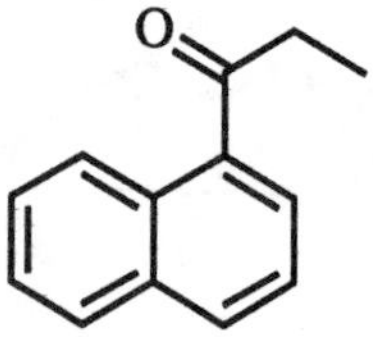

O

(d)

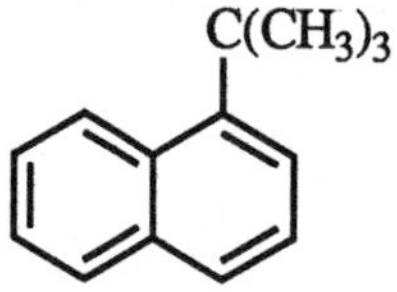

$C(CH_3)_3$

(e)

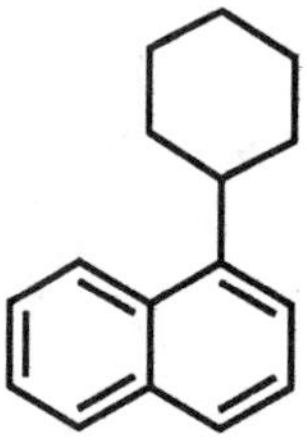

(f)

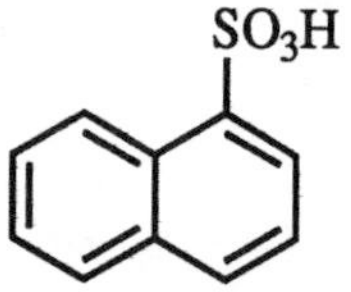

SO_3H

17-51

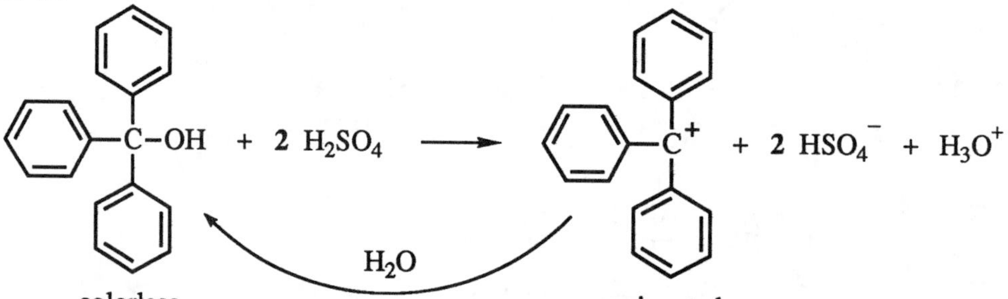

C–OH
+ 2 H_2SO_4
C^+
+ 2 HSO_4^-
+ H_3O^+
H_2O

colorless

conjugated—
yellow

Concentrated sulfuric acid "dehydrates" the alcohol, producing a highly conjugated, colored carbocation, and protonating the water to prevent the reverse reaction. Upon adding more water, however, there are too many water molecules for the acid to protonate, and triphenylmethanol is regenerated.

17-52

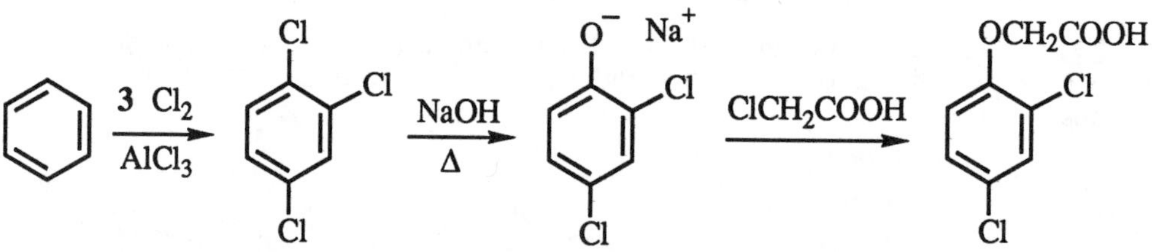

3 Cl_2
$AlCl_3$
Cl
Cl
Cl
NaOH
Δ
O^- Na^+
Cl
Cl
$ClCH_2COOH$
OCH_2COOH
Cl
Cl

(why is this the major isomer?)

(why is this the major isomer?)

17-53

(a)

<u>bromination at C-2</u>

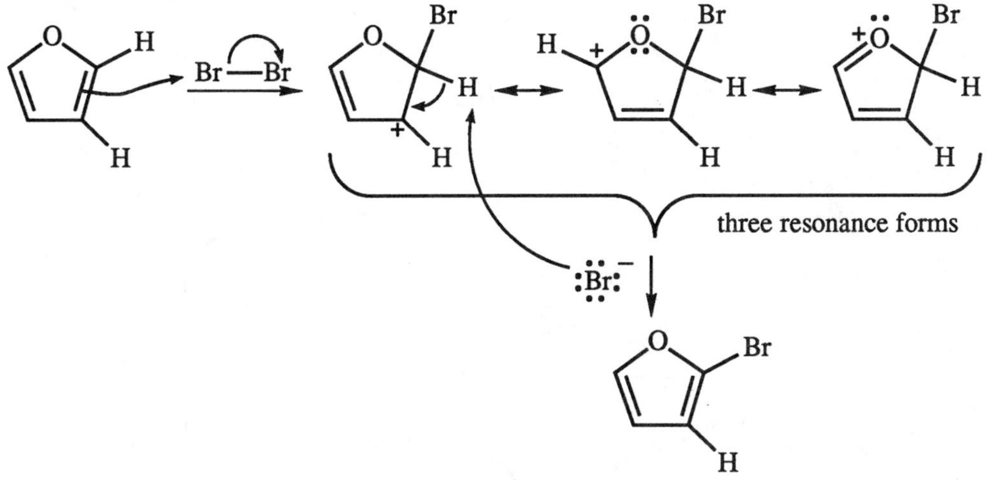

<u>bromination at C-3</u>

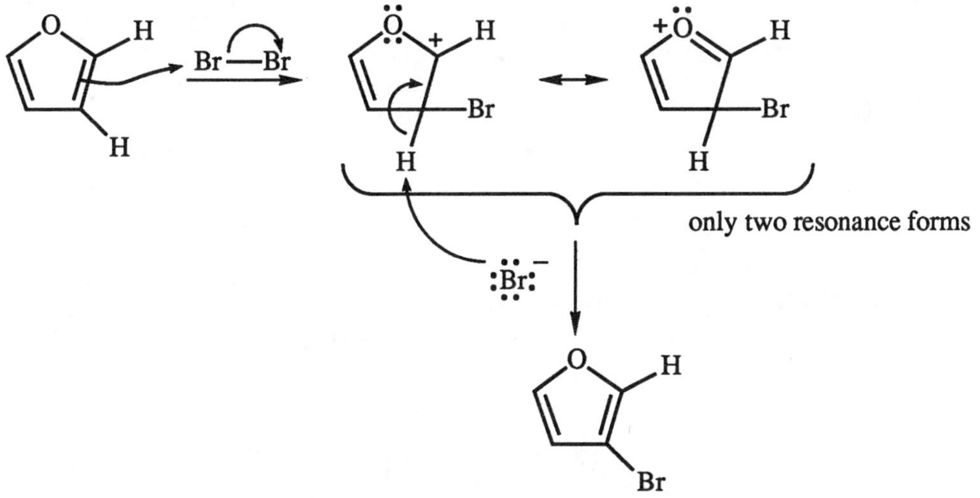

(b) Attack at C-2 gives an intermediate stabilized by three resonance forms, as opposed to only two resonance forms stabilizing attack at C-3. Bromination at C-2 will occur more readily.

17-54

abbreviate $PhCH_2-CH(NH_2)COOH$ (phenylalanine) as $R-NH_2$

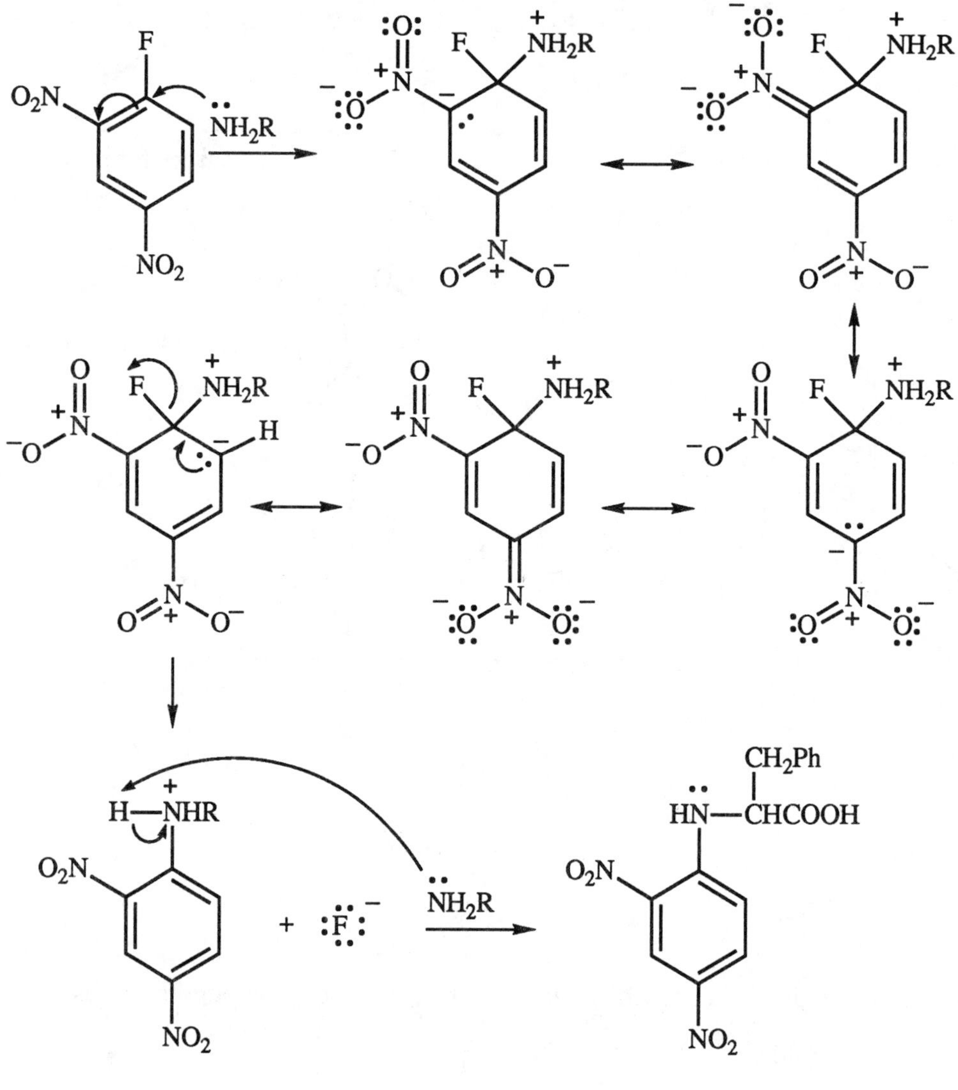

17-55

(a) This is an example of kinetic versus thermodynamic control of a reaction. At low temperature, the kinetic product predominates: in this case, almost a 1 : 1 mixture of ortho and para. These two isomers must be formed at approximately equal rates at 0° C. At 100° C, however, enough energy is provided for the *desulfonation* to occur rapidly; the large excess of the para isomer indicates the para is more stable, even though it is formed initially at the same rate as the ortho.

(b) The product from the 0° C reaction will equilibrate as it is warmed, and at 100° C will produce the same ratio of products as the reaction which was run initially at 100° C.

17-56

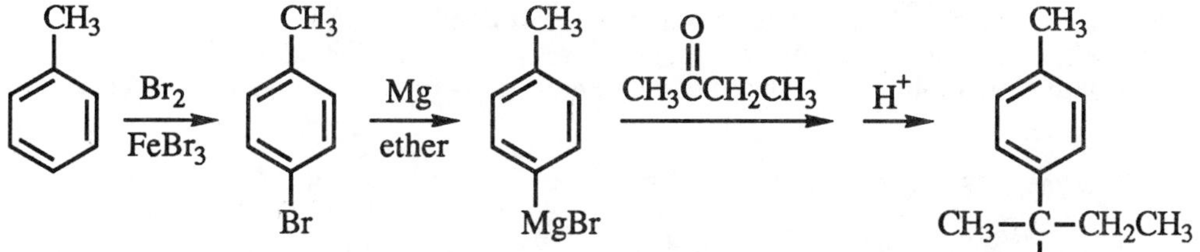

(possible alternative syntheses include acylation followed by Grignard)

17-57

(a)

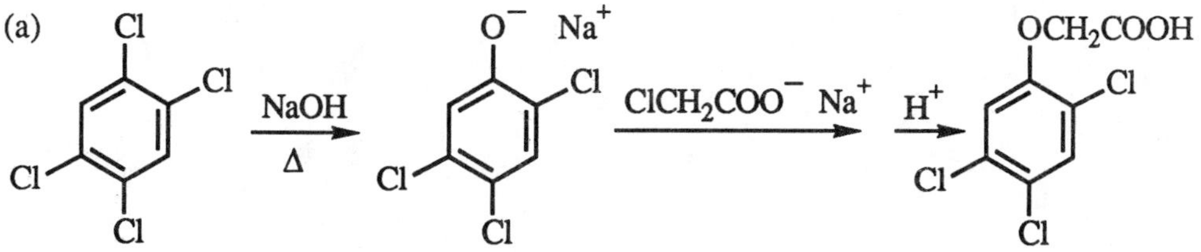

2,4,5-T

(b)

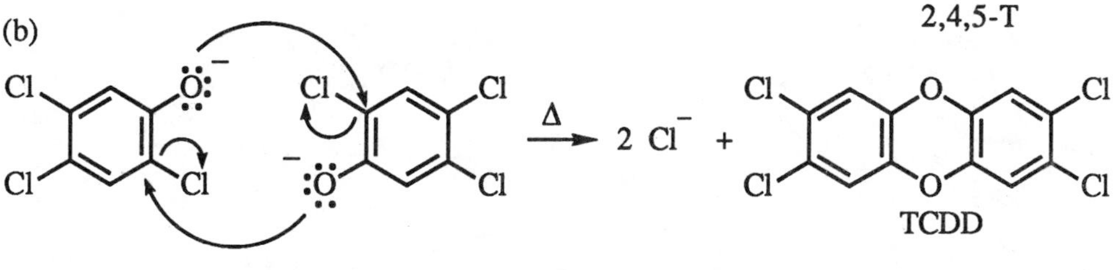

TCDD

Two nucleophilic aromatic substitutions form a new six-membered ring. (Though not shown here, this reaction would follow the standard addition-elimination mechanism.)

(c) To minimize formation of TCDD during synthesis: 1) keep the solutions dilute; 2) avoid high temperature; 3) replace chloroacetate with a more reactive molecule like bromoacetate or iodoacetate; 4) add an excess of the haloacetate.

To separate TCDD from 2,4,5-T at the end of the synthesis, take advantage of the acidic properties of 2,4,5-T. The 2,4,5-T will dissolve in an aqueous solution of a weak base like $NaHCO_3$. The TCDD will remain insoluble and can be filtered or extracted into an organic solvent like ether or dichloromethane. The 2,4,5-T can be precipitated from aqueous solution by adding acid.

17-58

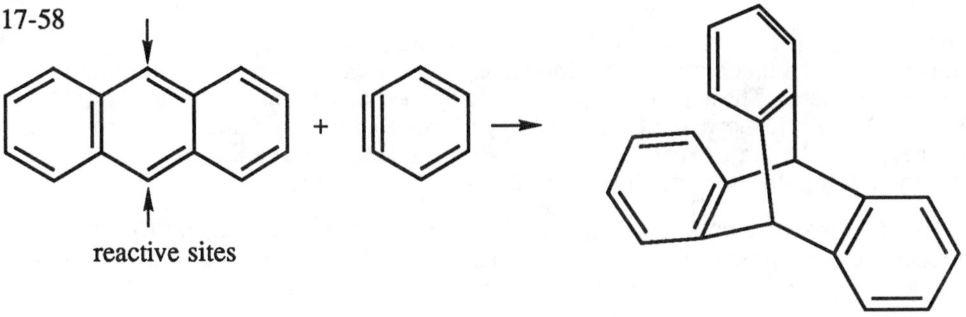

As we saw in Chapter 16, the carbons of the center ring of anthracene are susceptible to electrophilic addition, leaving two isolated benzene rings on the ends. Benzyne is such a reactive dienophile that the reluctant anthracene is forced into a Diels-Alder reaction.

17-59 A benzyne must have been generated from the Grignard reagent.

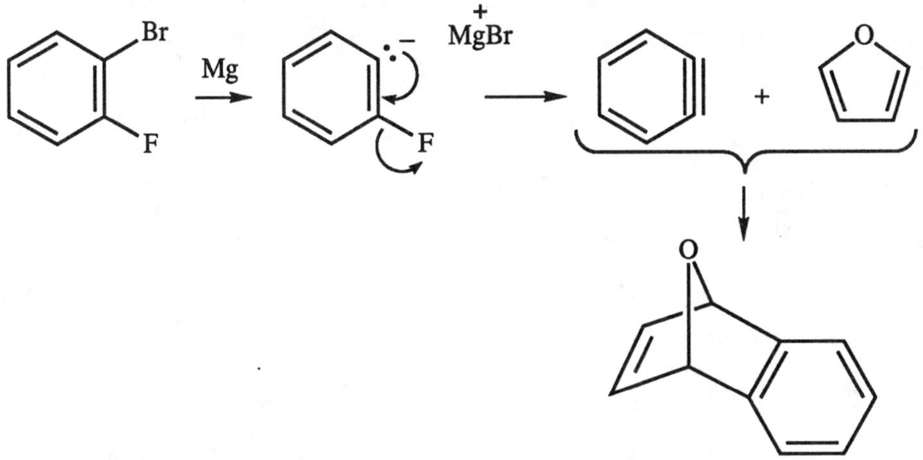

17-60

(a)

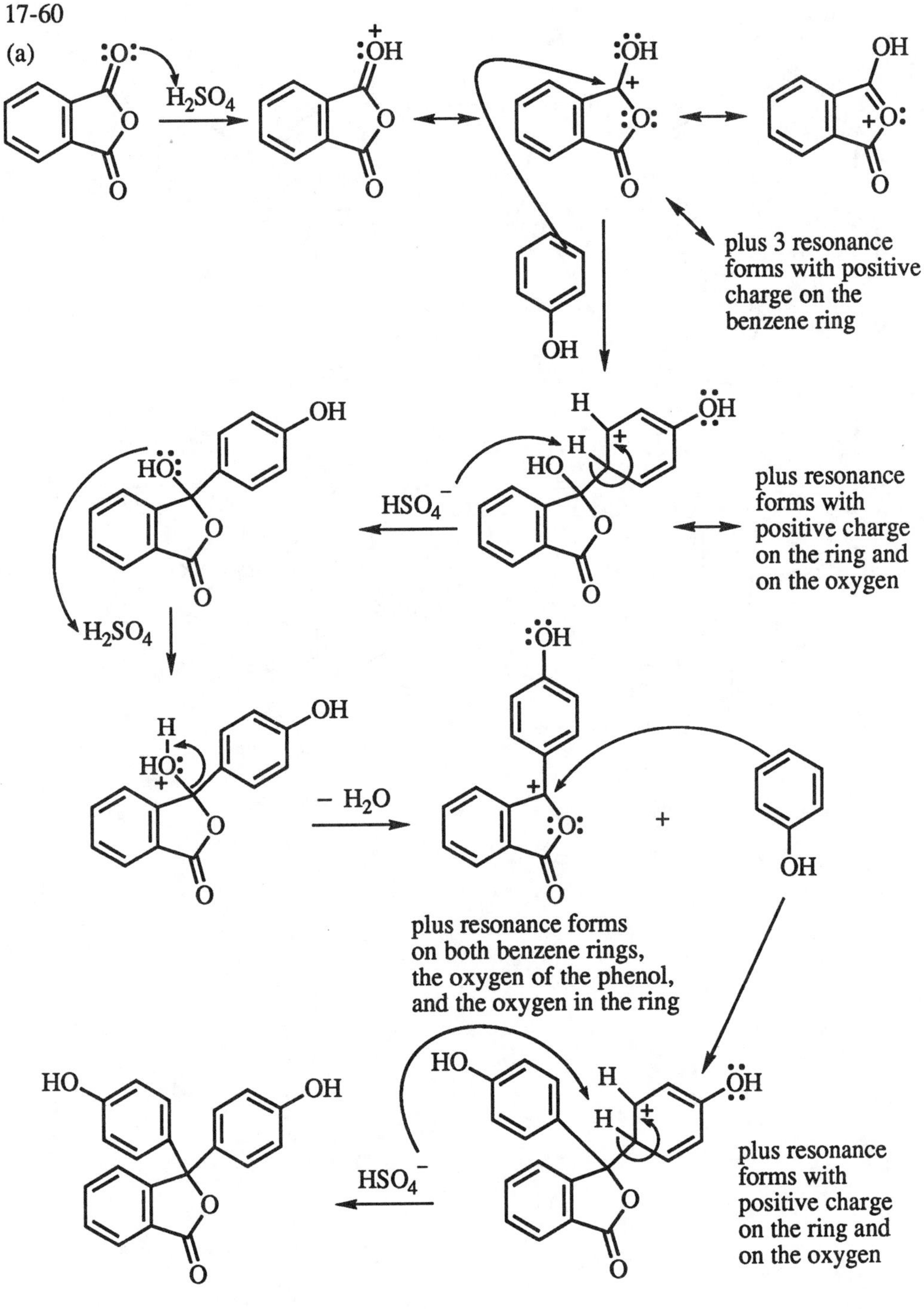

H2SO4
plus 3 resonance forms with positive charge on the benzene ring
HSO4−
plus resonance forms with positive charge on the ring and on the oxygen
H2SO4
− H2O
plus resonance forms on both benzene rings, the oxygen of the phenol, and the oxygen in the ring
HSO4−
plus resonance forms with positive charge on the ring and on the oxygen

17-60 continued

(b)

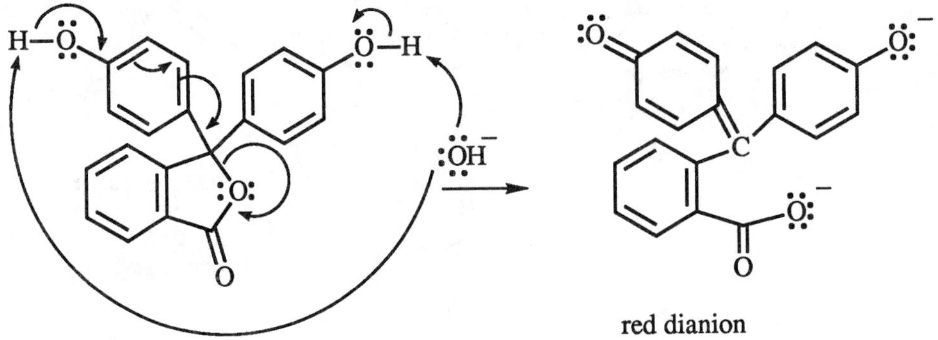

red dianion

(c)

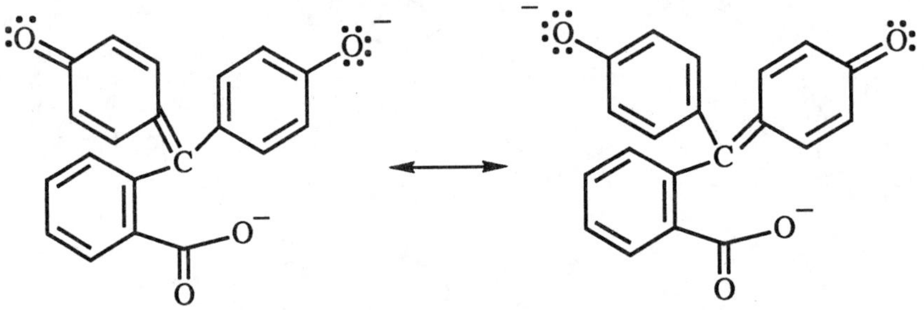

CHAPTER 18—KETONES AND ALDEHYDES

18-1

(a) 5-hydroxy-3-hexanone; ethyl β-hydroxypropyl ketone
(b) 3-phenylbutanal; β-phenylbutyraldehyde
(c) *trans*-2-methoxycyclohexanecarbaldehyde; no common name
(d) 6,6-dimethyl-2,4-cyclohexadienone; no common name

18-2

(a) $C_9H_{10}O$ ⇒ 5 elements of unsaturation
1H singlet at δ 9.7 ⇒ aldehyde hydrogen
5H singlet at δ 7.2 ⇒ monosubstituted benzene
1H quartet at δ 3.5 and 3H doublet at δ 1.4 ⇒ $CHCH_3$

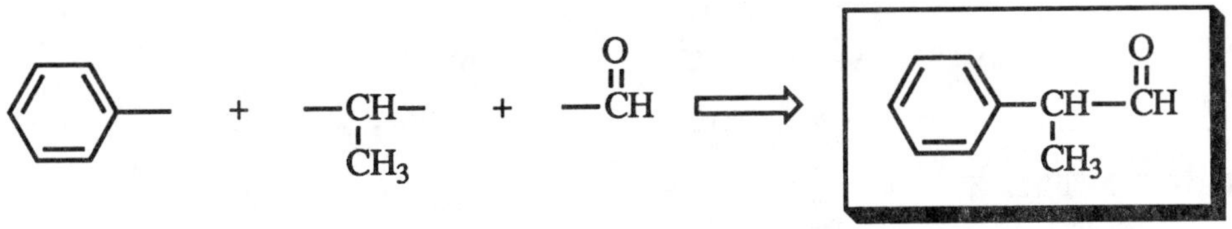

(b) C_8H_8O ⇒ 5 elements of unsaturation
cluster of 4 peaks at δ 128-137 ⇒ mono- or para-substituted benzene ring
peak at δ 197 ⇒ carbonyl carbon
peak at δ 26 ⇒ methyl next to carbonyl or benzene

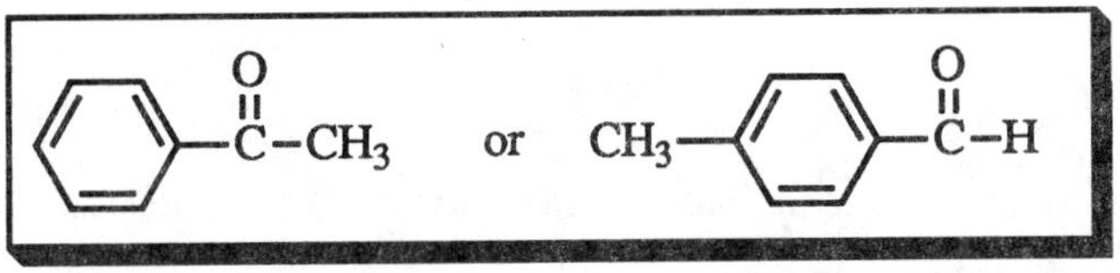

18-3 A compound has to have a hydrogen on a γ carbon (or other atom) in order for the McLafferty rearrangement to occur. 2-Butanone has no γ-hydrogen.

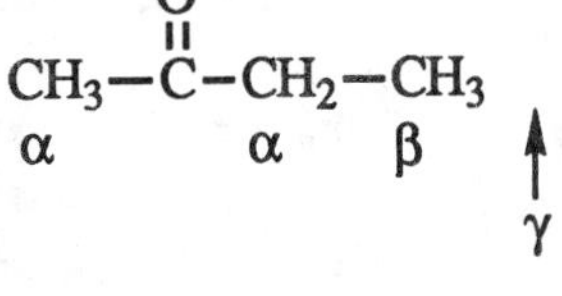

18-4

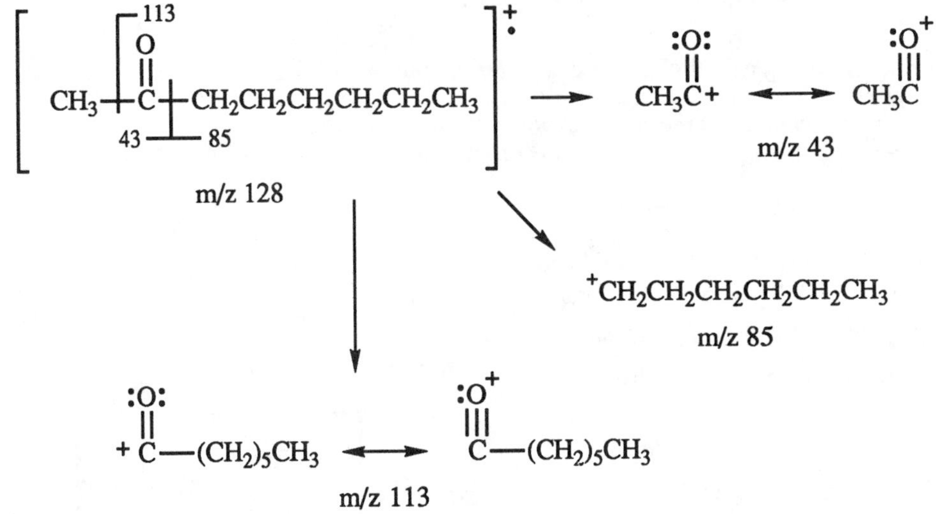

McLafferty rearrangement

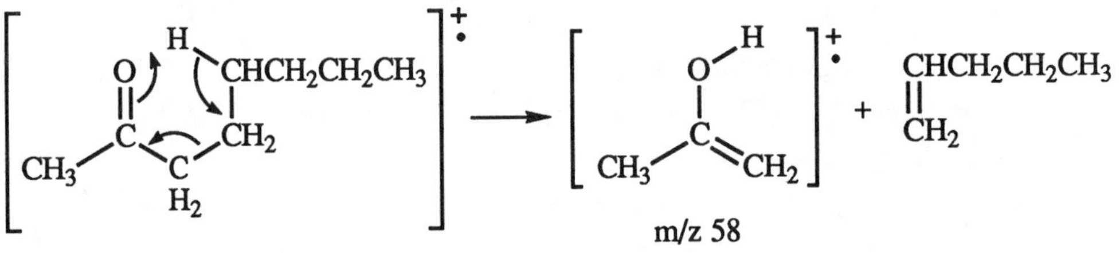

18-5

The first value is the π to π*; the second value is the n to π*. The values are approximate.

(a) < 200 nm; 280 nm; this simple ketone should have values similar to acetone

(b) 230 nm; 310 nm; conjugated system (210) plus 2 alkyl groups (20) = 230; the value of 310 nm is similar to Figure 18-7: ketone (280-300) plus two alkyl groups (20)

(c) 280 nm; 360 nm; conjugated system (210) plus 1 extra double bond (30) plus 4 alkyl groups (40) = 280; similar reasoning for the other transition, starting with an average base value of 290

(d) 270 nm; 350 nm; same as in (c) except only 3 alkyl groups instead of 4

18-6

(a) The conjugated diene has a maximum at 235 nm (see Solved Problem 15-3) and the ketone has a maximum at about 237 nm, so the π to π* transition cannot be used to differentiate the compounds.
(b) The ketone has an n to π* transition around 315 nm that the diene cannot have.

18-7

(a) 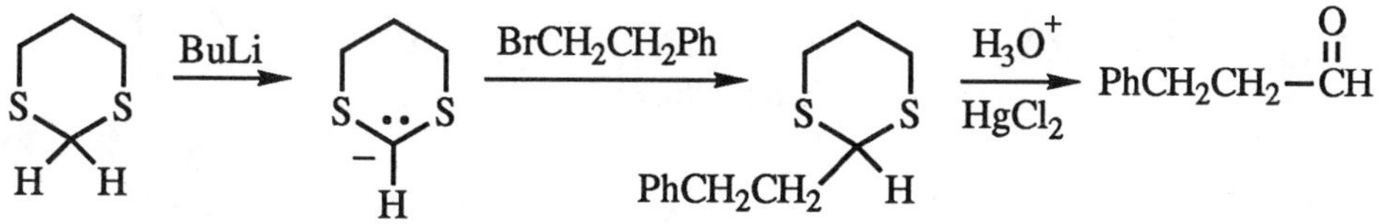

(b) 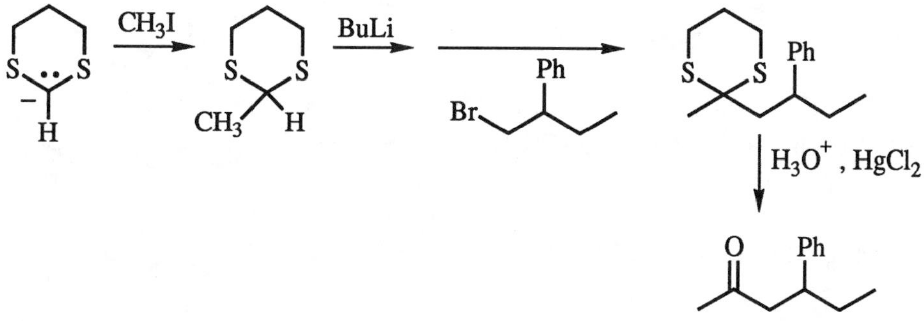

(c) 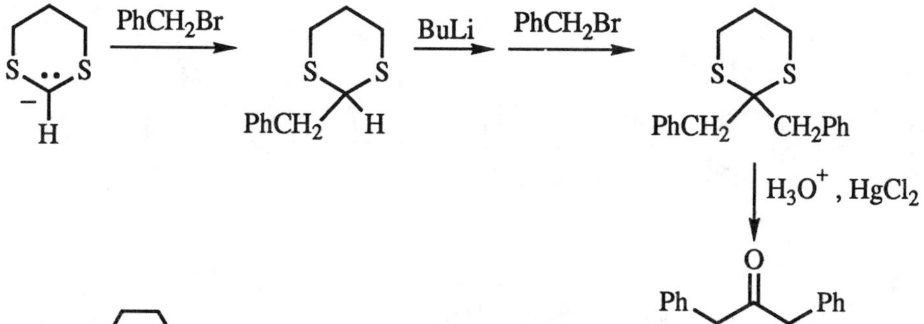

(d) 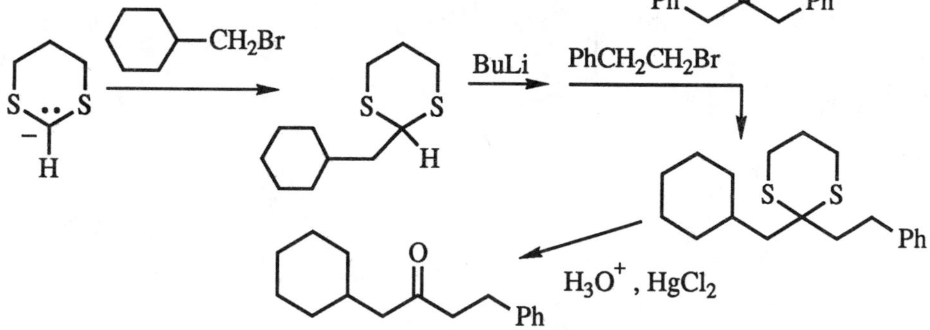

18-8

(a) 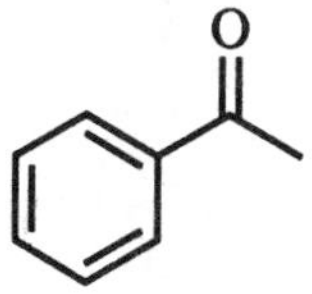

(b) 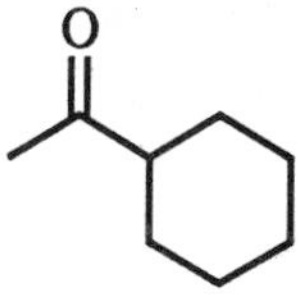

(c) $CH_3(CH_2)_3\overset{\overset{O}{\|}}{C}-CH_2CH_3$

18-9

(a) $CH_3(CH_2)_3C(=O)-CH_2CH_3$ (b) $PhCH_2CN$ (simple S_N2) (c) 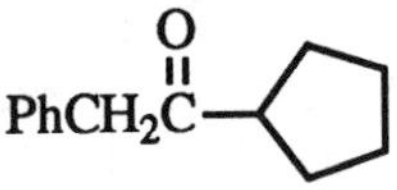
O
PhCH2C

18-10

(a) 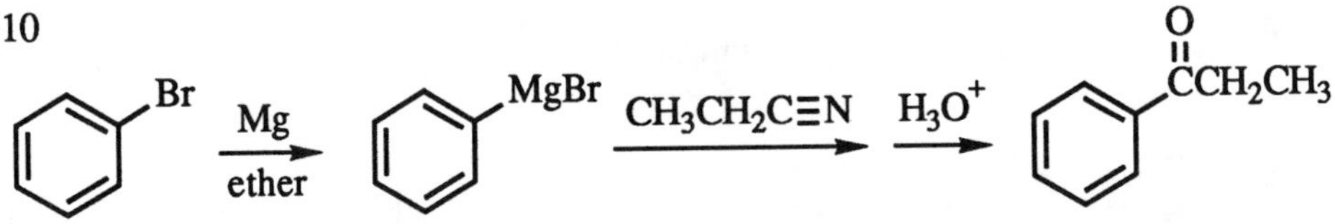
Br
Mg
ether
MgBr
CH3CH2C≡N
H3O+
O
CCH2CH3

(b) 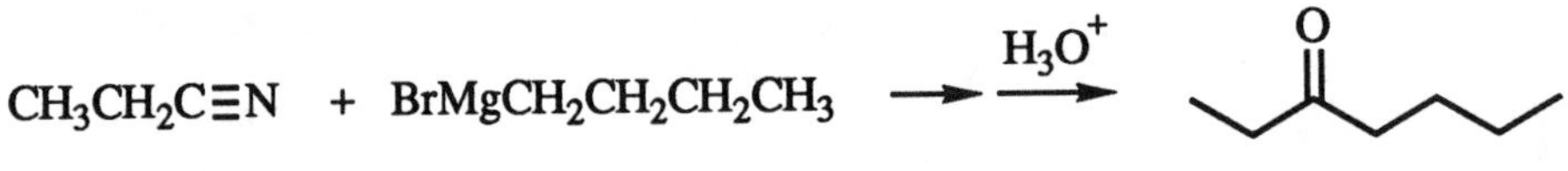
CH3CH2C≡N + BrMgCH2CH2CH2CH3
H3O+
O

(c) 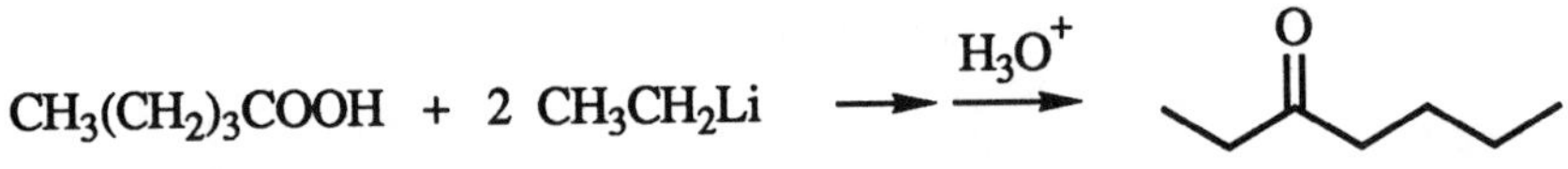
CH3(CH2)3COOH + 2 CH3CH2Li
H3O+
O

(d) 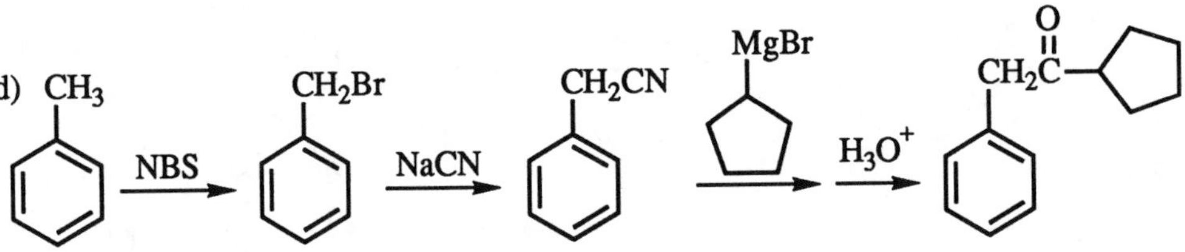
CH3
NBS
CH2Br
NaCN
CH2CN
MgBr
H3O+
O
CH2C

18-11

(a) 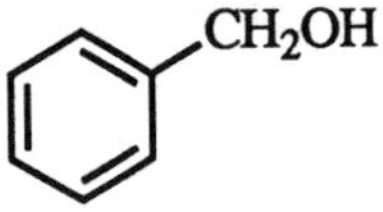
CH2OH

(b) 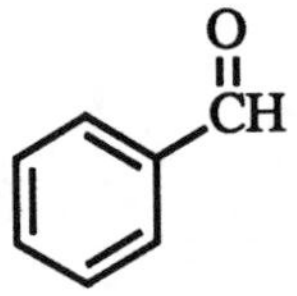
O
CH

(c) 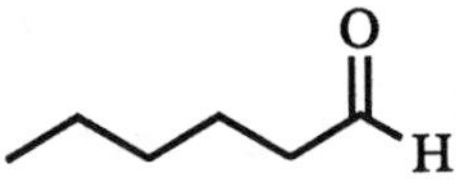
O
H

18-12

(a)

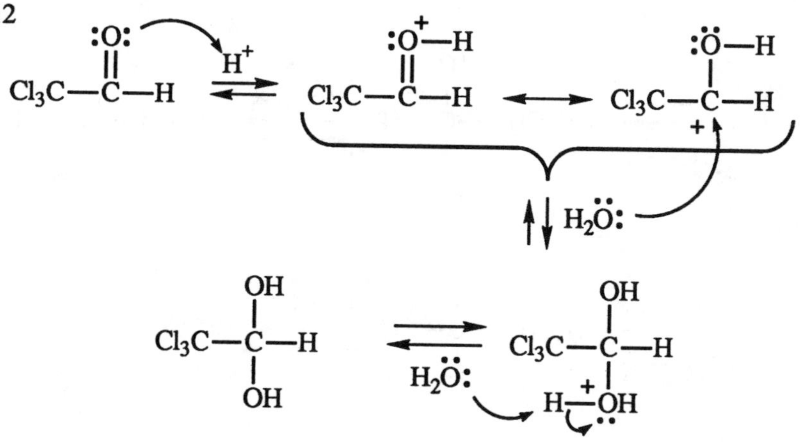

(b)

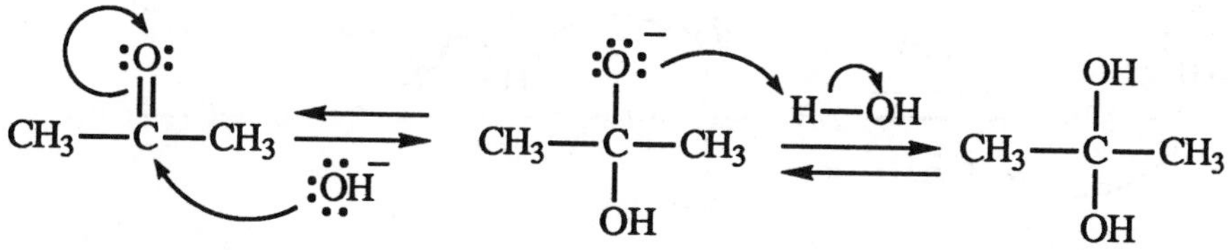

18-13

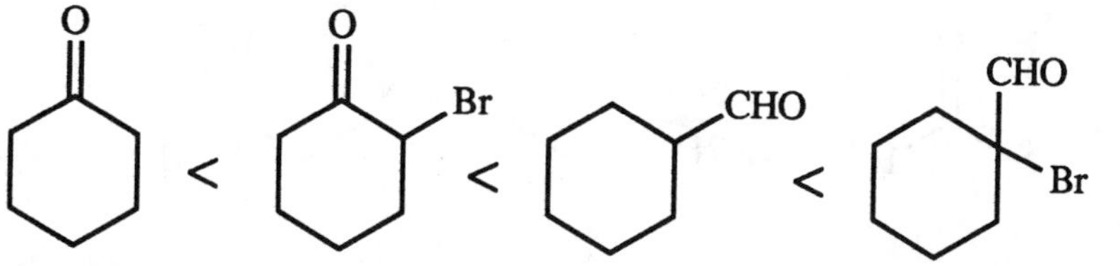

least amount of hydrate

greatest amount of hydrate

18-14

(a)

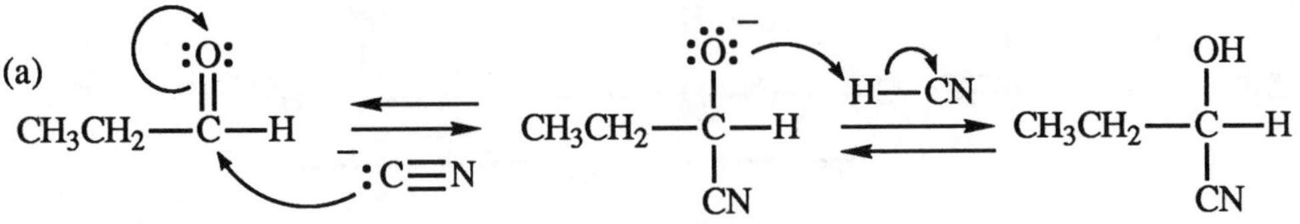

(b)

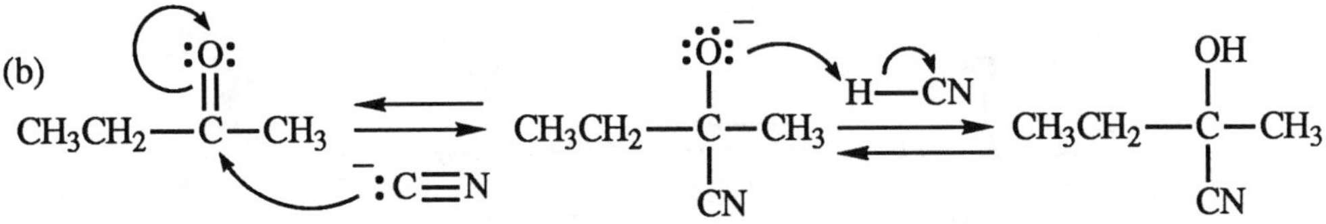

(c)

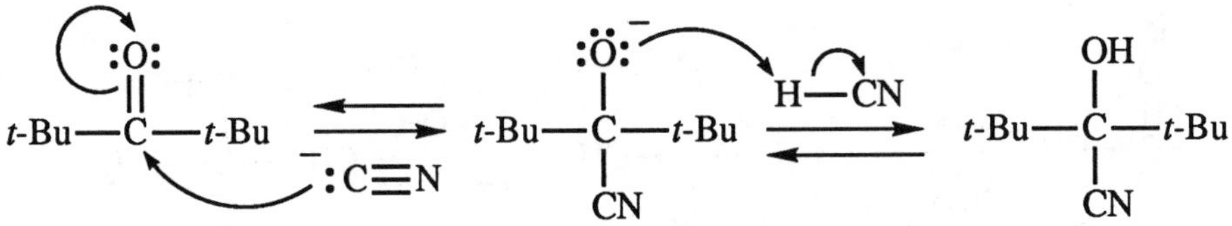

18-15

(a)

(b)

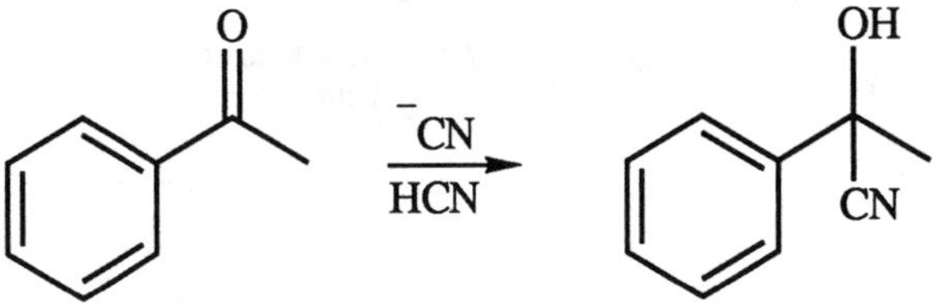

(c)

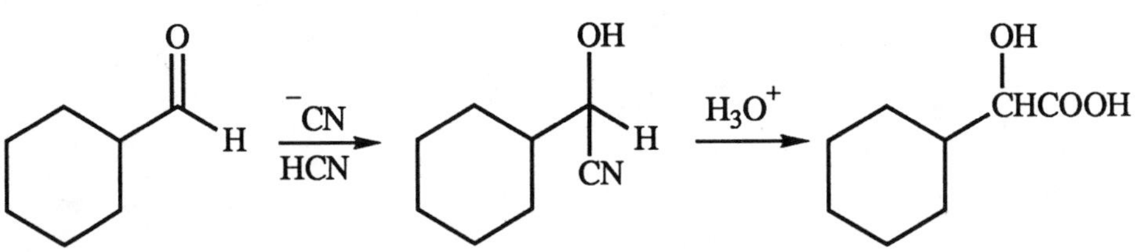

Note to the student: Mechanisms of nucleophilic attack at carbonyl carbon frequently include species with both positive and negative charges. These species are very short-lived as each charge is quickly neutralized by a rapid proton transfer; in fact, these steps are probably the fastest of the whole mechanism. In most cases, this Solutions Manual will show these *two* steps as occurring at the same time, even though you have been admonished to show *all* steps of a mechanism separately. The practice of showing these proton transfers in one step is legitimate as long as it is understood that these are *two* fast steps.

18-16

(a)

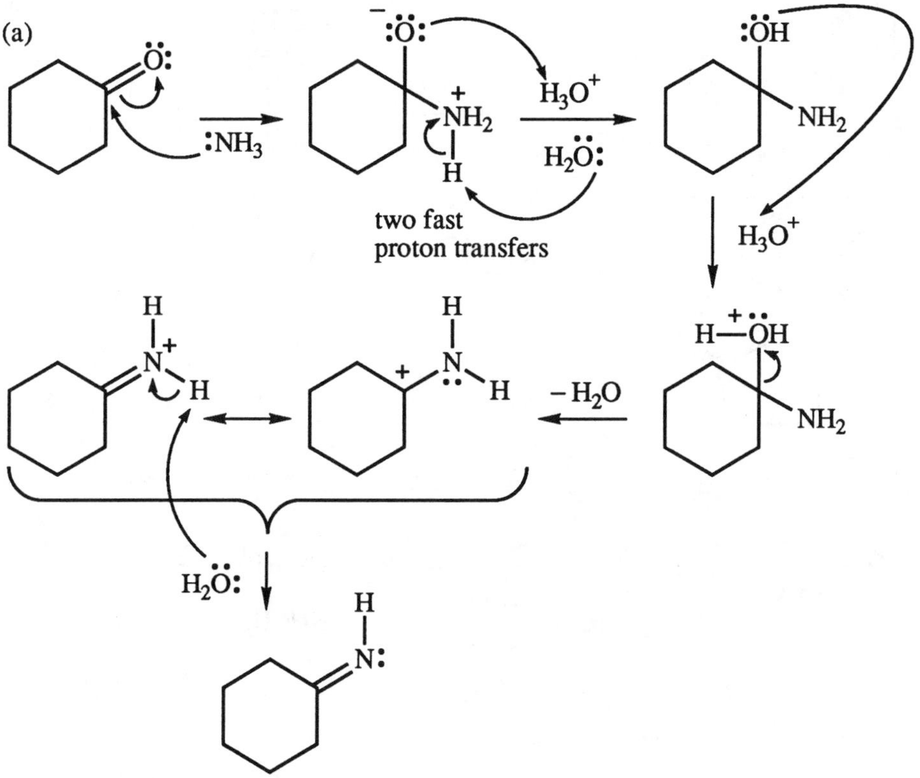

18-16 continued

(b)

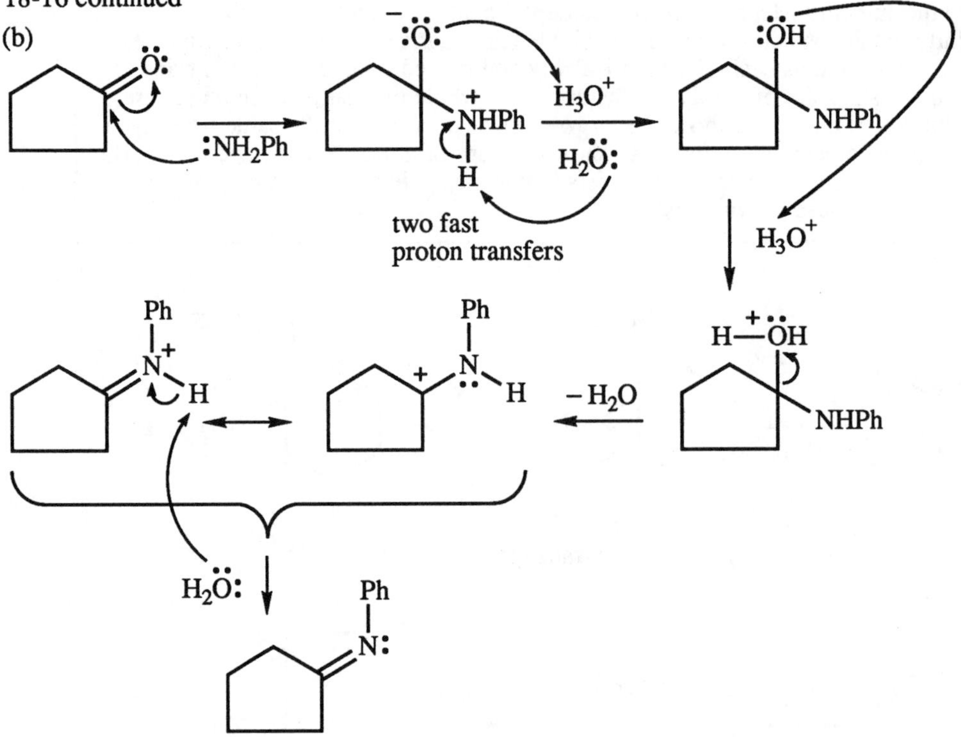

(c)

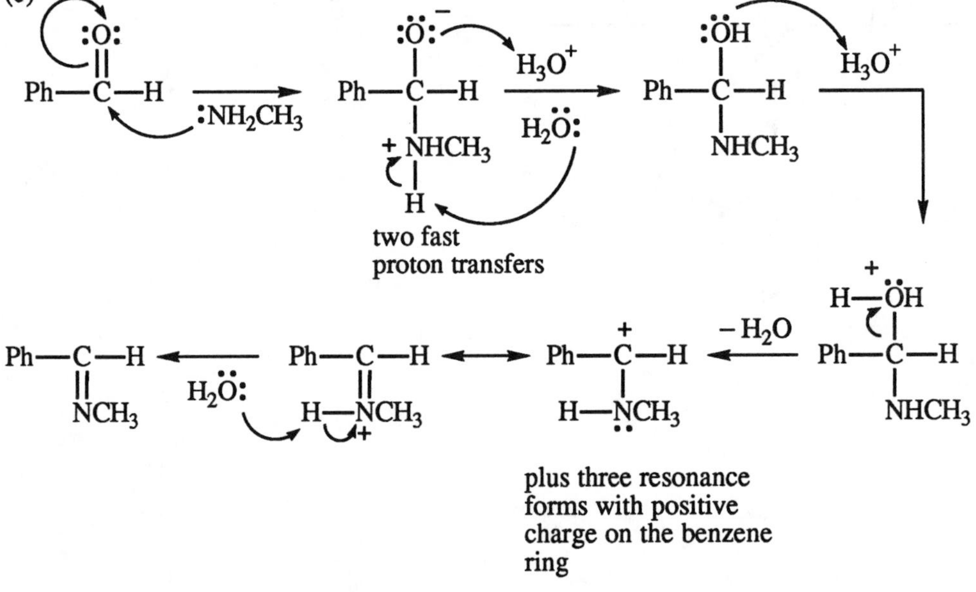

18-17 Whenever a double bond is formed, stereochemistry must be considered. The two compounds are the *Z* and *E* isomers.

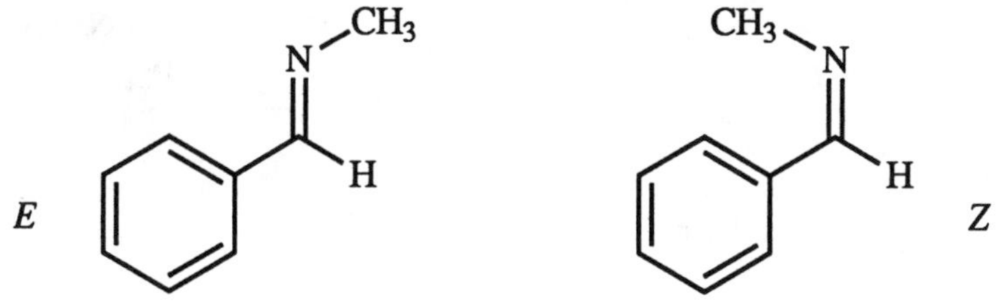

18-18

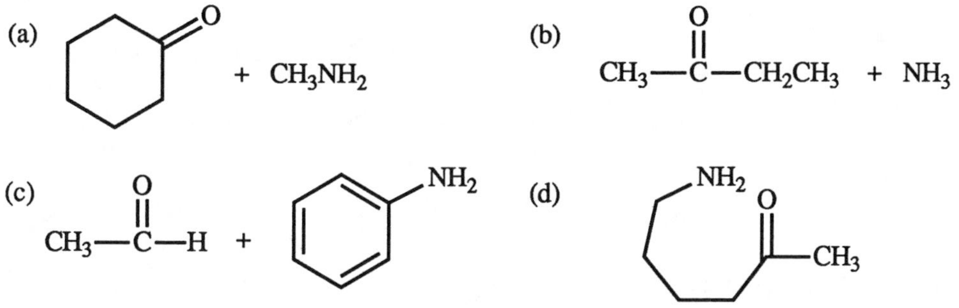

18-19

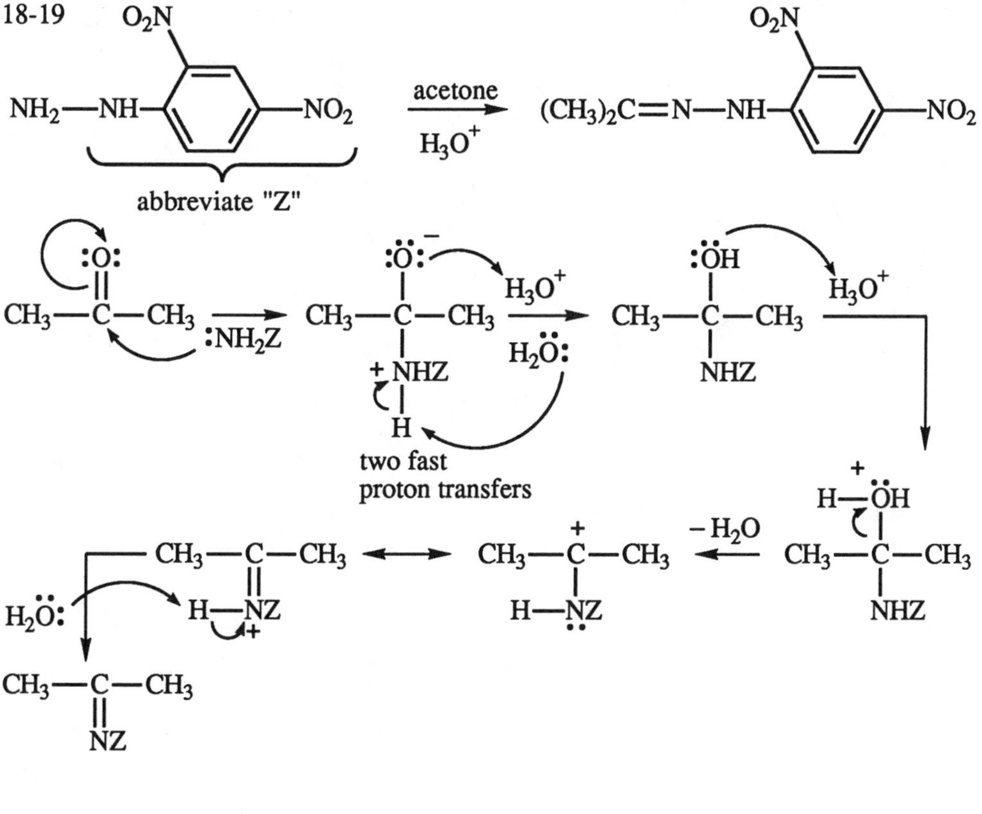

18-20

(a)

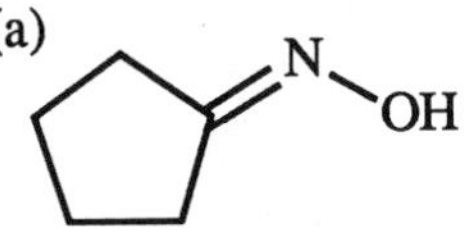

(b)

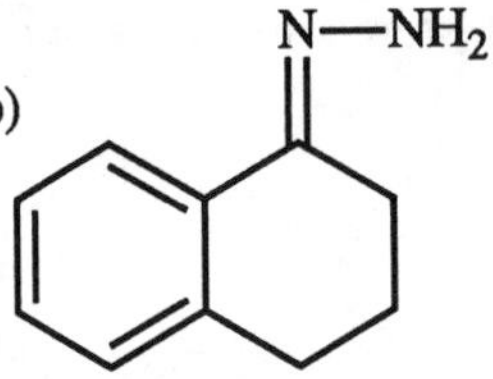

(c) $PhCH{=}CH{-}CH{=}N{-}NH{-}\overset{O}{\overset{\|}{C}}NH_2$

(d) $Ph{-}\underset{}{C}({=}N{-}NHPh){-}Ph$

18-21

(a) PhCHO + $H_2NNH{-}\overset{O}{\overset{\|}{C}}{-}NH_2$

(b)

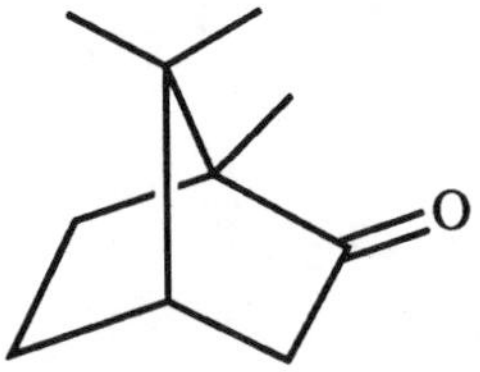

+ H_2NOH

(c)

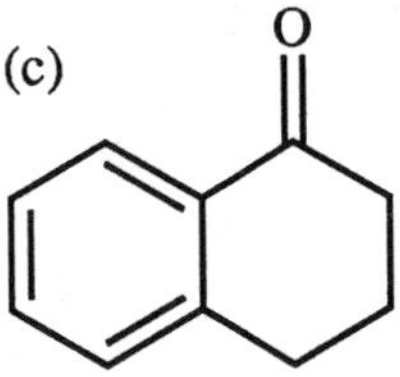

+ $H_2N{-}NHPh$

(d)

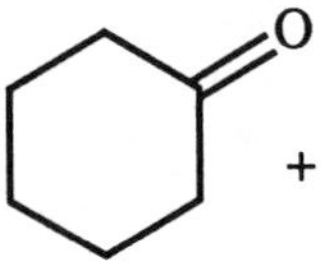

+

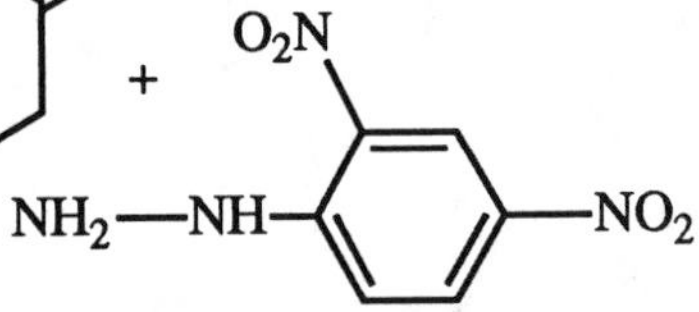

(e)

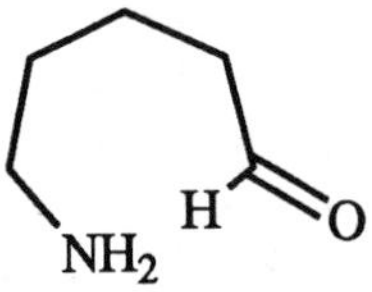

(f)

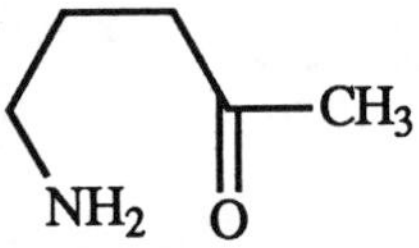

18-22

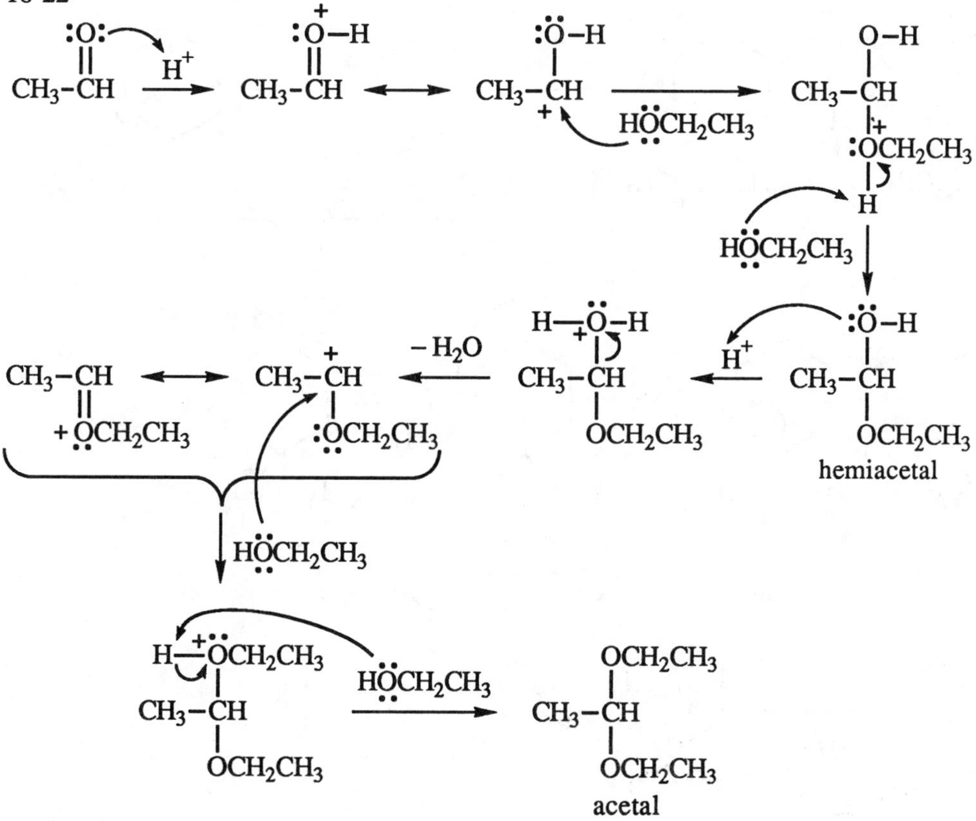
H+
HÖCH2CH3
hemiacetal
− H2O
acetal

18-23

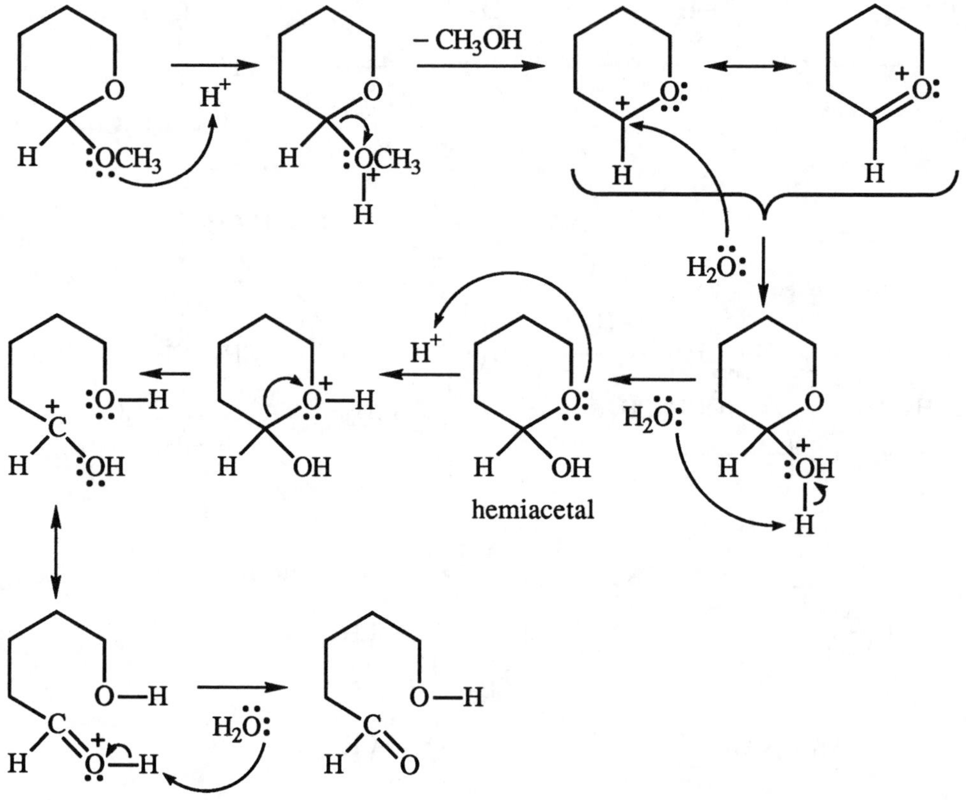
H⁺
– CH3OH
H2O:
hemiacetal
H⁺
H2O:
H2O:

18-24

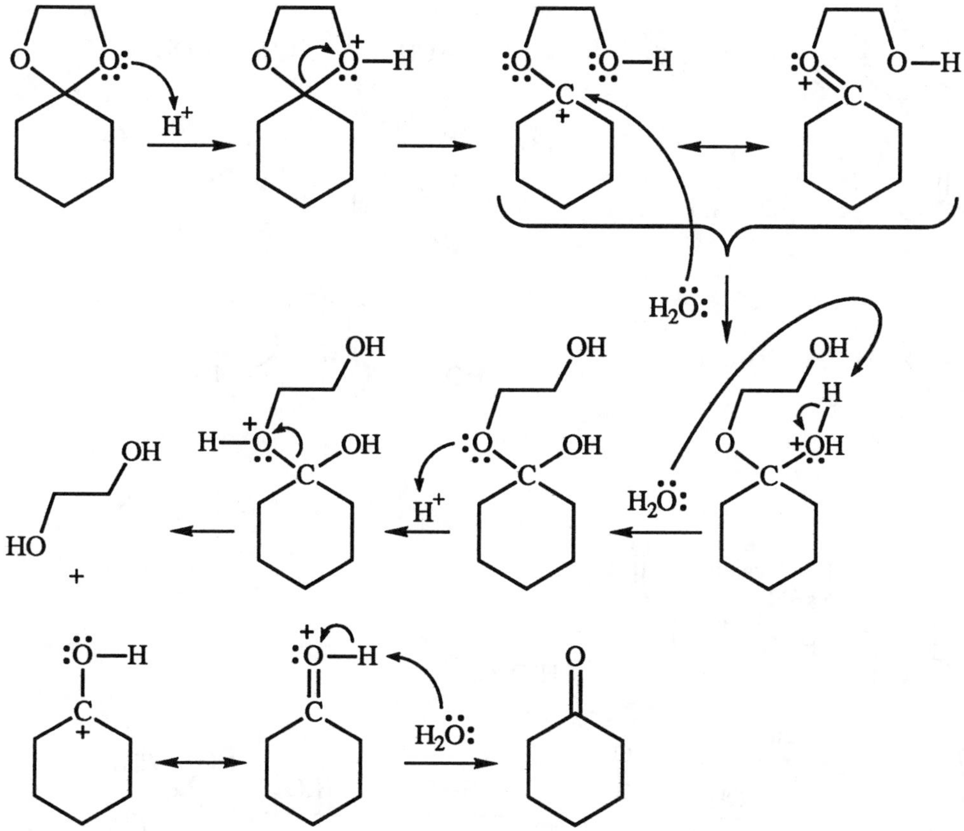

18-25

(a)

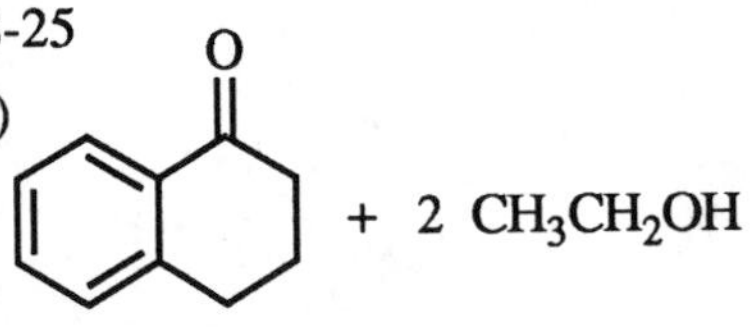
O
+ 2 CH3CH2OH

(b)

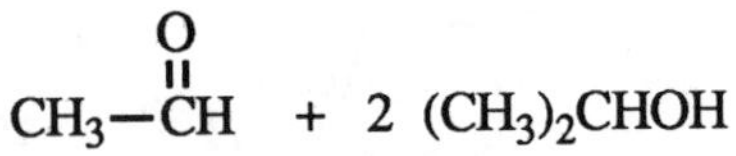
O
CH3—CH
+ 2 (CH3)2CHOH

(c)

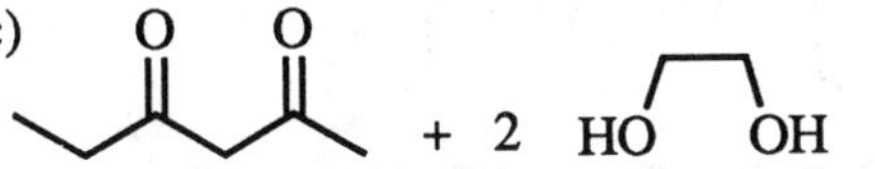
O
O
+ 2
HO
OH

(d)

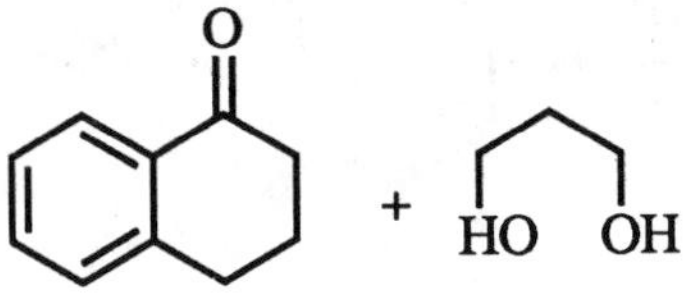
O
+
HO
OH

(e)

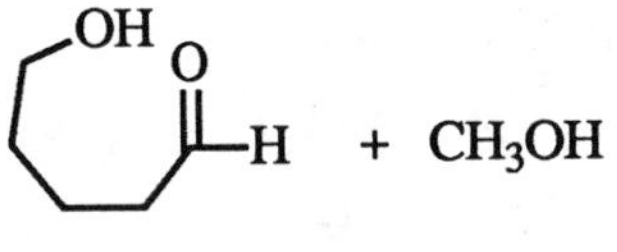
OH
O
H
+ CH3OH

(f)

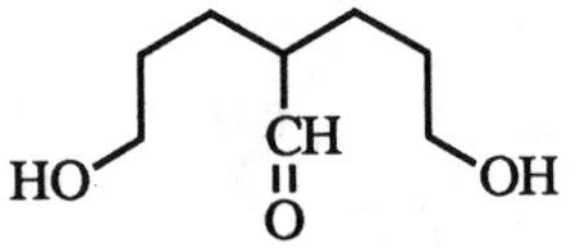
HO
CH
O
OH

18-26

(a)

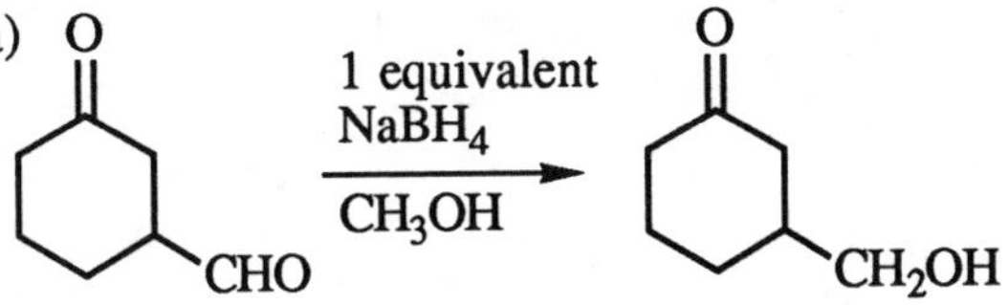
O
CHO
1 equivalent
NaBH4
CH3OH
O
CH2OH

(b)

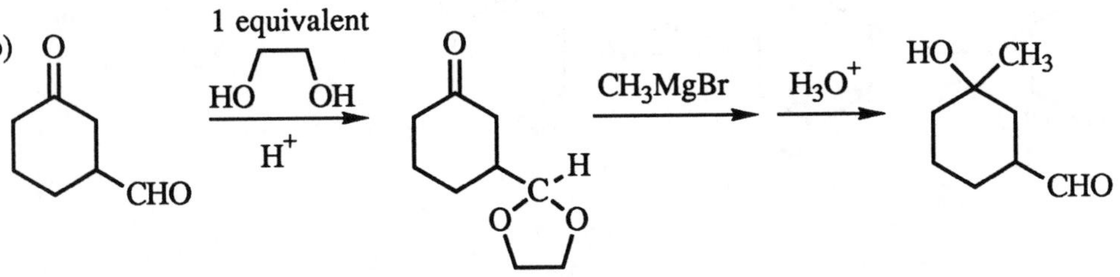
O
CHO
1 equivalent
HO
OH
H+
O
C
H
O
O
CH3MgBr
H3O+
HO
CH3
CHO

(c)

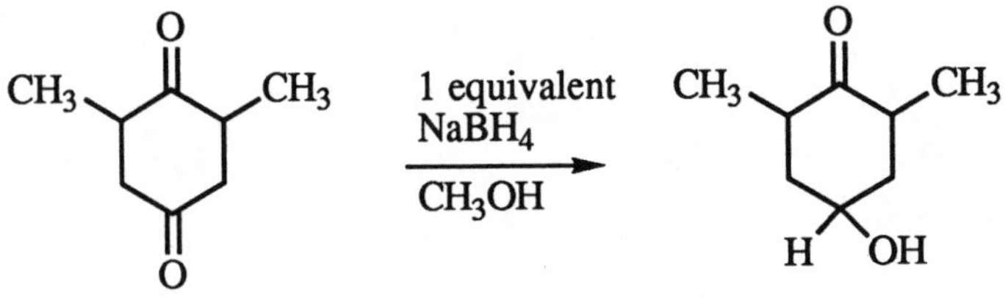
CH3
O
CH3
O
1 equivalent
NaBH4
CH3OH
CH3
O
CH3
H
OH

18-26 continued

(d)

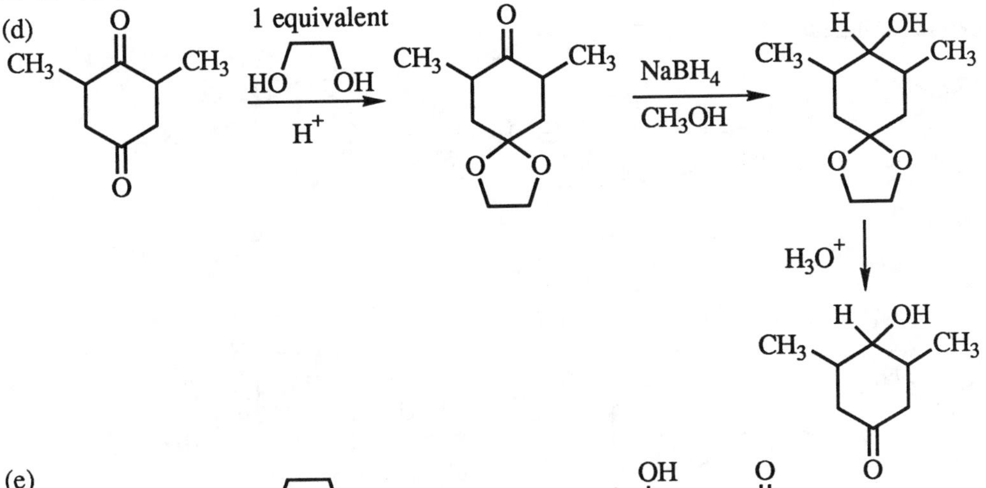

(e)

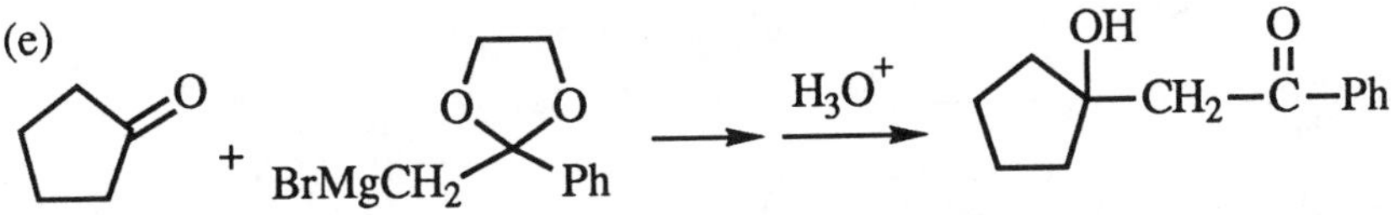

(f)

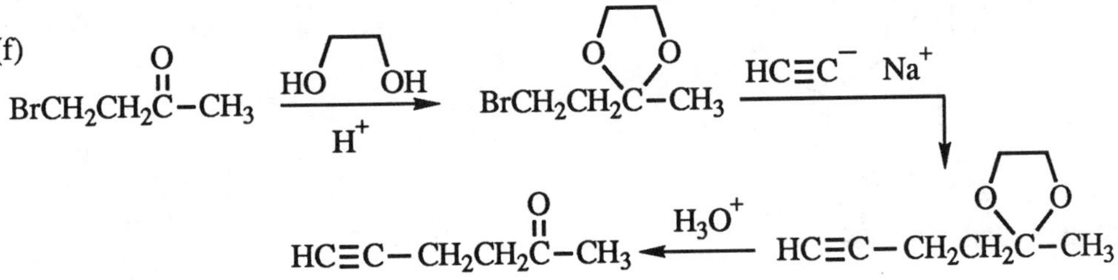

18-27

(a)

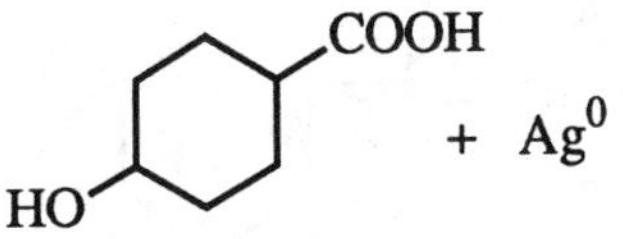

after adding H$^+$

(b)

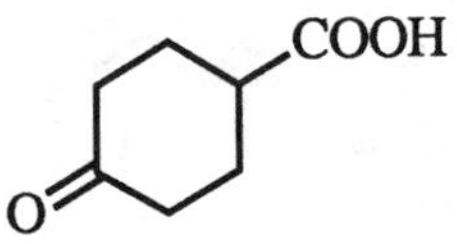

(c)

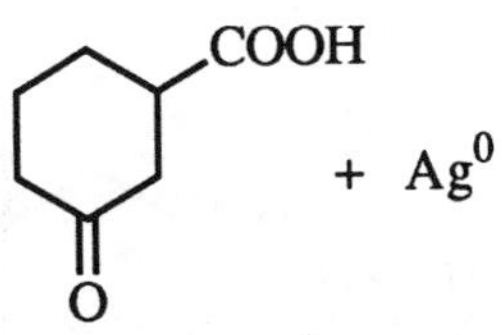

after adding H$^+$

(d)

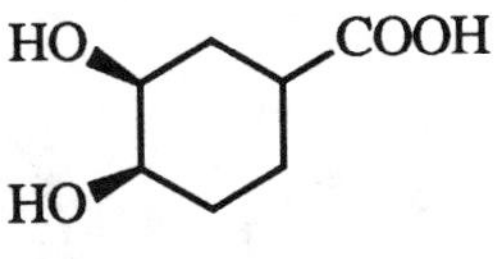

18-28

hydrazone formation

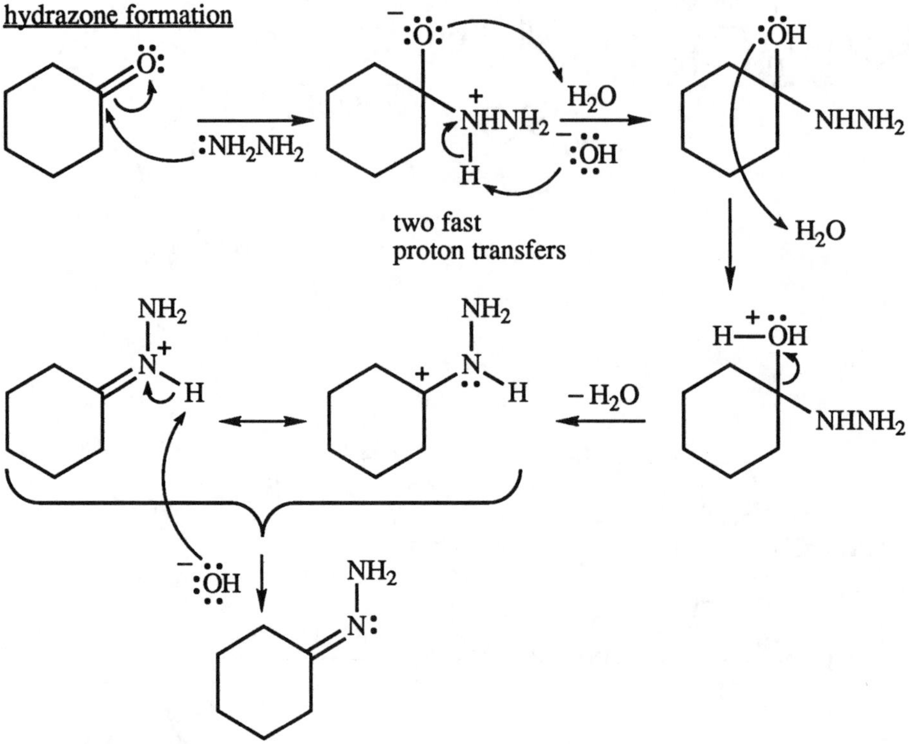

reduction of the hydrazone

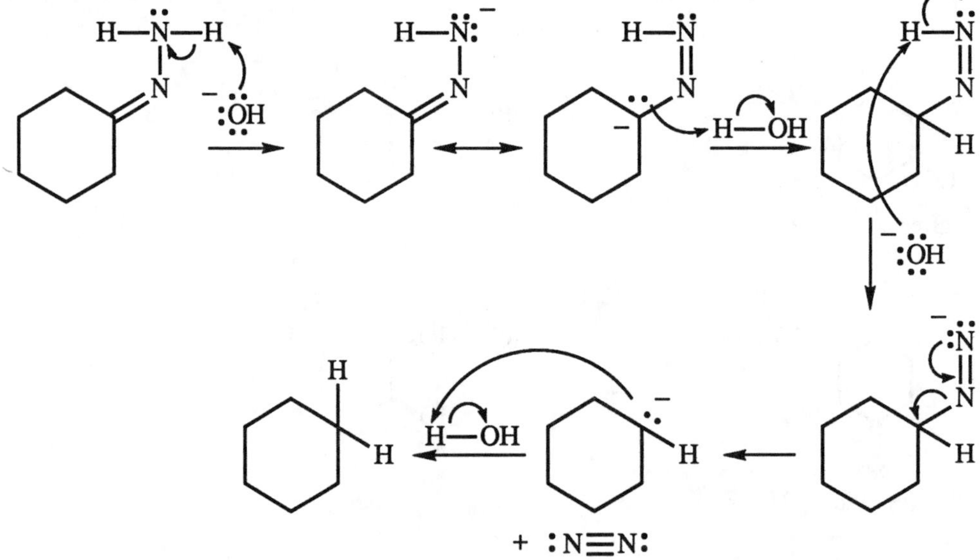

18-29

(a) 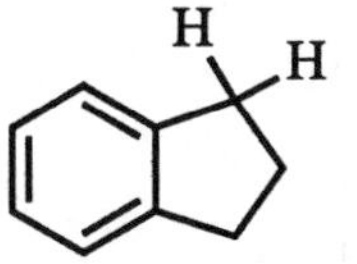

(b) 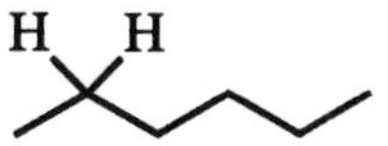

(c) 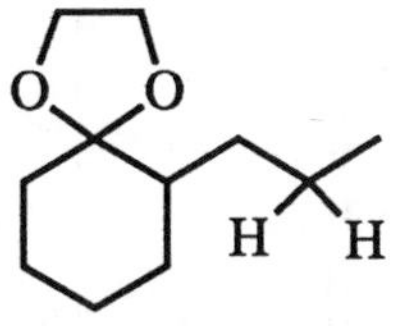

(d) 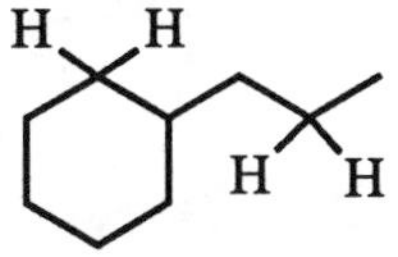

18-30 Refer to the Glossary and the text for definitions and examples.

18-31 IUPAC names first; then common names.

(a) 2-heptanone; methyl *n*-pentyl ketone
(b) 4-heptanone; di-*n*-propyl ketone
(c) heptanal; no simple common name
(d) benzophenone; diphenyl ketone
(e) butanal; butyraldehyde
(f) propanone; acetone (IUPAC accepts "acetone")
(g) 4-bromo-2-methylhexanal; no common name
(h) 3-phenyl-2-propenal; (cinnamaldehyde)
(i) 2,4-hexadienal; no common name
(j) 3-oxopentanal; no common name
(k) 3-oxocyclopentanecarbaldehyde; no common name
(l) *cis*-2,4-dimethylcyclopentanone; no common name

18-32 In order of increasing equilibrium constant for hydration:

$$CH_3-\overset{\overset{\displaystyle O}{\|}}{C}-CH_3 < CH_3-\overset{\overset{\displaystyle O}{\|}}{C}-CH_2Cl < CH_3-\overset{\overset{\displaystyle O}{\|}}{C}-H < ClCH_2-\overset{\overset{\displaystyle O}{\|}}{C}-H < H-\overset{\overset{\displaystyle O}{\|}}{C}-H$$

least amount of hydration — greatest amount of hydration

18-33

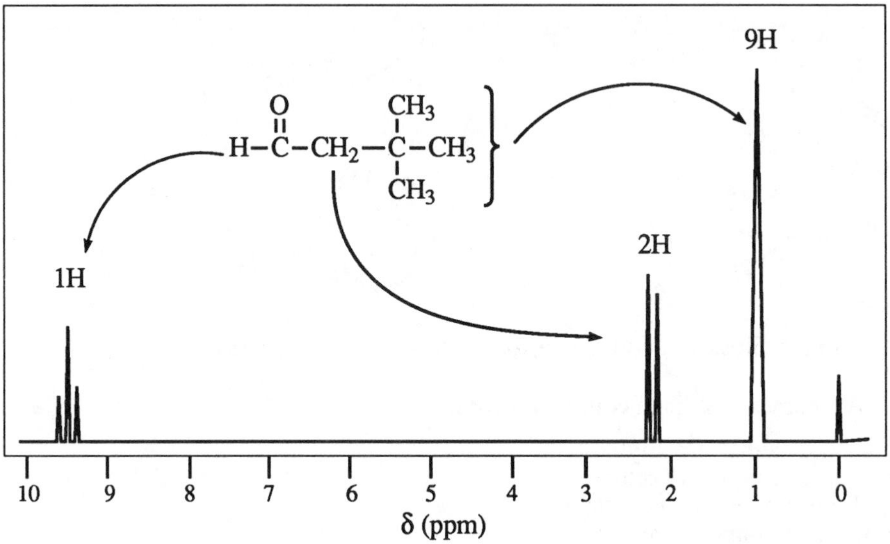

18-34

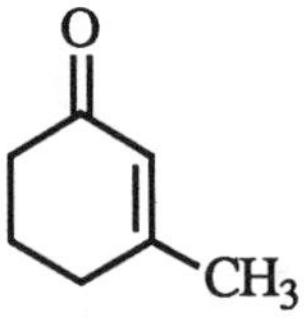

By comparison with similarly substituted molecules shown in the text:

$\pi \rightarrow \pi^*$ base value (210) plus 3 alkyl groups (30) = 240 nm

$n \rightarrow \pi^*$ 300-320 nm

18-35

$C_6H_{10}O_2$ indicates two elements of unsaturation.

The IR absorption at 1708 cm^{-1} suggests a ketone, or possibly two ketones since there are two oxygens and two elements of unsaturation. The NMR singlets in the ratio of 2 : 3 indicate a highly symmetric molecule. The singlet at δ 2.15 is probably methyl next to carbonyl, and the singlet at δ 2.67 integrating to two is likely to be CH_2 on the other side of the carbonyl.

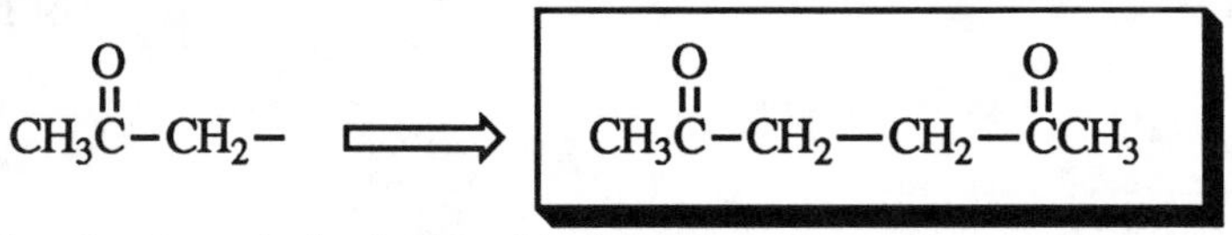

Since the molecular formula is double this fragment, the molecule must be twice the fragment.

Two questions arise. Why is the integration 2 : 3 and not 4 : 6? Integration provides a *ratio*, not absolute numbers, of hydrogens. Why don't the two methylenes show splitting? Adjacent, *identical* hydrogens, with identical chemical shifts, do not split each other; the signals for ethane or cyclohexane appear as singlets.

18-36 The formula $C_{10}H_{12}O$ indicates 5 elements of unsaturation. A solid 2,4-DNP derivative suggests an aldehyde or a ketone, but a negative Tollens test precludes the possibility of an aldehyde; therefore, the unknown must be a ketone.

The NMR shows the typical ethyl pattern at δ 1.3 (3H, triplet) and δ 2.7 (2H, quartet), and a monosubstituted benzene at δ 7.5 (5H, singlet). The singlet at δ 4.0 is a CH_2, but quite far downfield, apparently deshielded by two groups. Assemble the pieces:

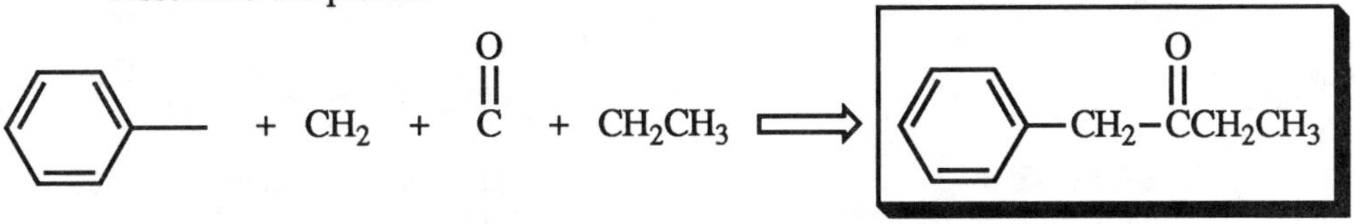

+ CH2 + C + CH2CH3
O
CH2−CCH2CH3

18-37

(a)

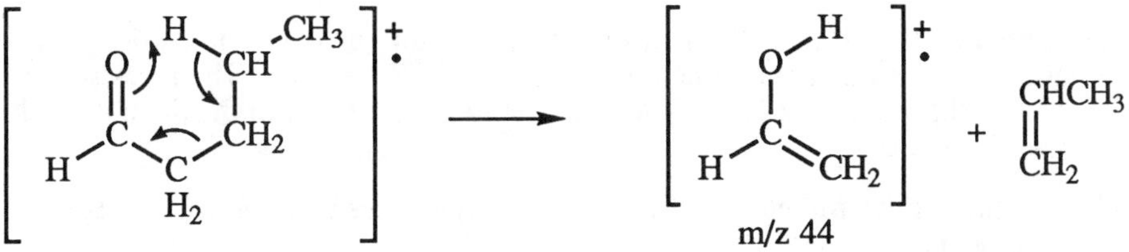

m/z 44
CHCH3
CH2

(b)

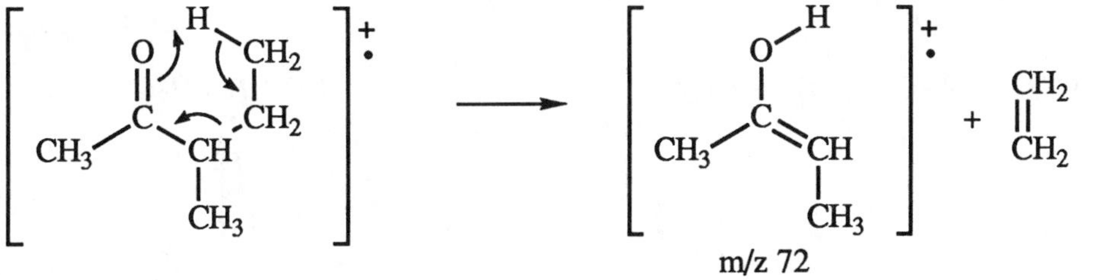

m/z 72
CH2
CH2

(c)

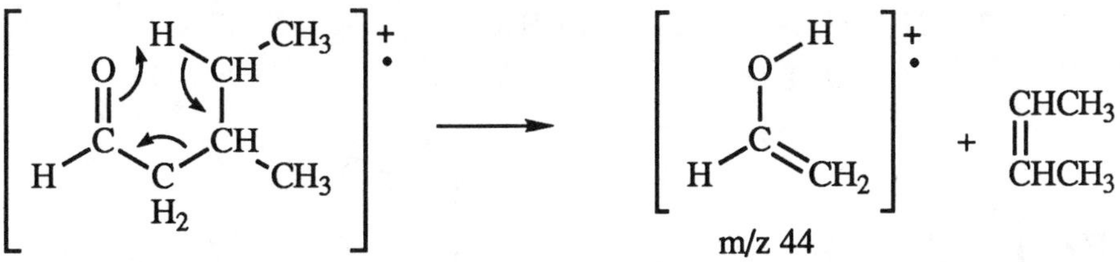

m/z 44
CHCH3
CHCH3

(d)

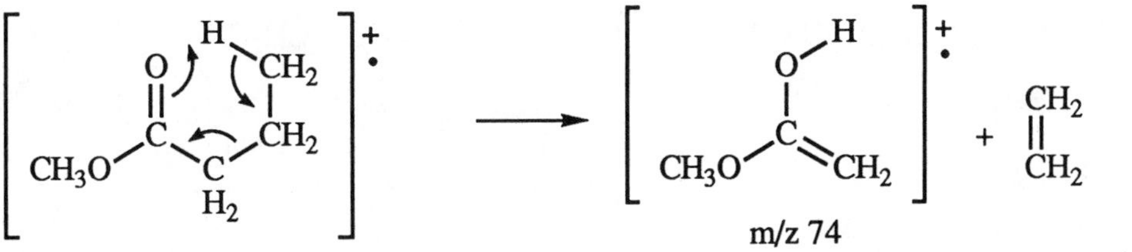

m/z 74
CH2
CH2

18-38 A solid 2,4-DNP indicates an aldehyde or a ketone. Possible structures:

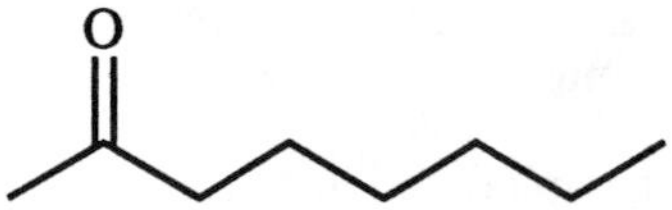

2-octanone
molecular weight 128
A

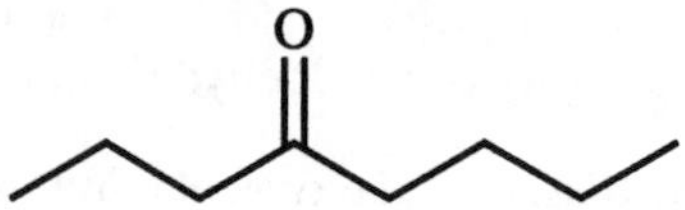

4-octanone
molecular weight 128
B

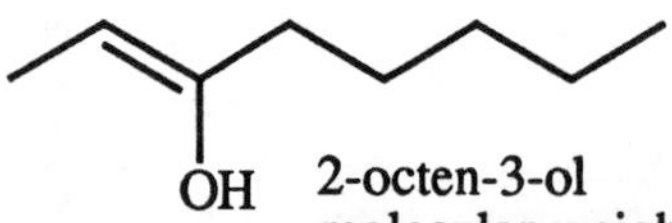

2-octen-3-ol
molecular weight 128
C

5-propoxy-1-pentanol
molecular weight 146
D

The positive 2,4-DNP test eliminates **D** as a possibility as **D** has no carbonyl. (**C** is the enol form of 3-octanone; it could give a positive 2,4-DNP.) The mass spectrum of **A** would show a strong M – 15, loss of methyl, not present in the data; thus, the unknown cannot be **A**.

What are the MS fragmentations of **B**? Two α-cleavages and *two* McLafferty rearrangements:

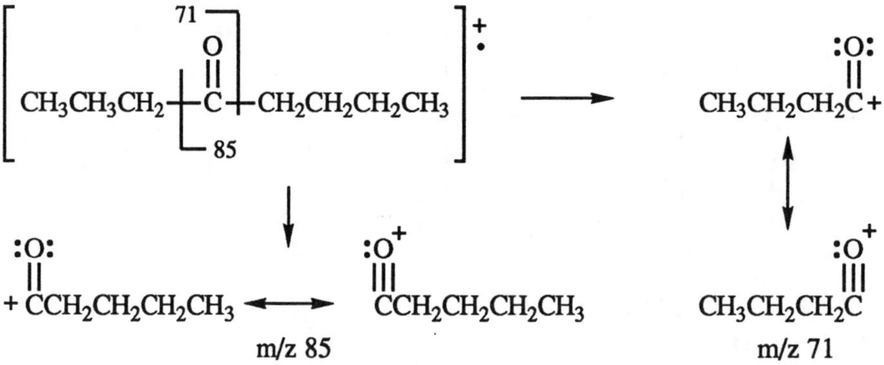

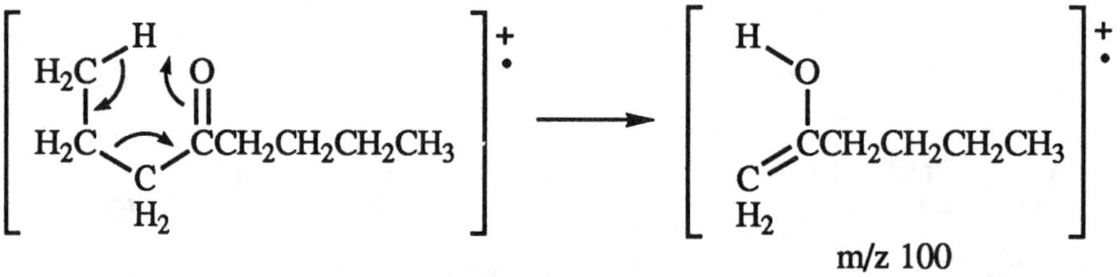

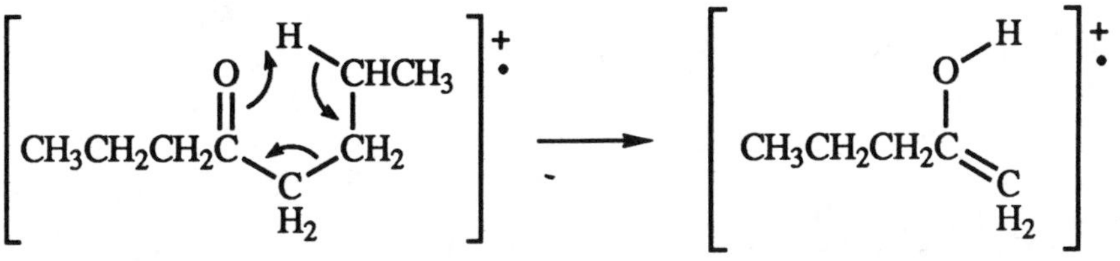

m/z 86

4-Octanone exactly fits the MS data. What about structure **C**? Because of the location of the π bond (in either the enol or keto forms), **C** could only do one McLafferty rearrangement and could not give rise to two even-mass peaks.

Therefore, the unknown must be 4-octanone, **B**.

18-39 A molecular ion of m/z 70 means a fairly small molecule. A solid semicarbazone derivative and a negative Tollens test indicate a ketone. The carbonyl (CO) has mass 28, so 70 – 28 = 42, enough mass for only 3 more carbons. The molecular formula is probably C_4H_6O (mass 70); with two elements of unsaturation, we can infer the presence of a double bond or a ring in addition to the carbonyl.

The IR shows a strong peak at 1785 cm^{-1}, indicative of a ketone in a small ring. No peak in the 1600-1650 cm^{-1} region shows the absence of an alkene. The only possibilities for a small ring ketone containing four carbons are these:

The NMR can distinguish these. No methyl doublet appears in the NMR spectrum, ruling out **B**. The NMR does show a 4H triplet at δ 3.3; this signal comes from the two methylenes (C-2 and C-4) adjacent to the carbonyl, split by the two hydrogens on C-3. The signal for the methylene at C-3 appears at δ 2.2, roughly a quintet because of splitting by four neighboring protons.

The unknown is cyclobutanone, **A**. The IR absorption of the carbonyl at 1785 is characteristic of small ring ketones; ring strain strengthens the carbon-oxygen double bond, increasing its frequency of vibration. (See Section 11-9 in the text.)

18-40

(a) hemiacetal
(old: hemiketal)

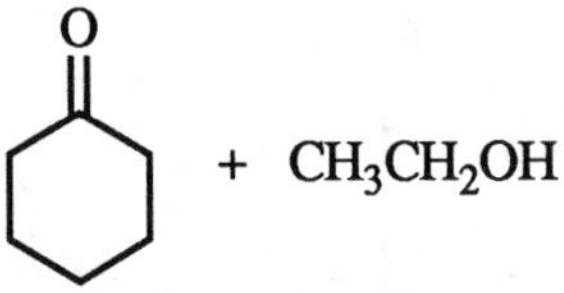

(b) acetal
(old: ketal)

$$CH_3CH_2CH_2-\overset{O}{\overset{\|}{C}}-CH_3 \;+\; \mathbf{2}\;CH_3OH$$

(c) acetal

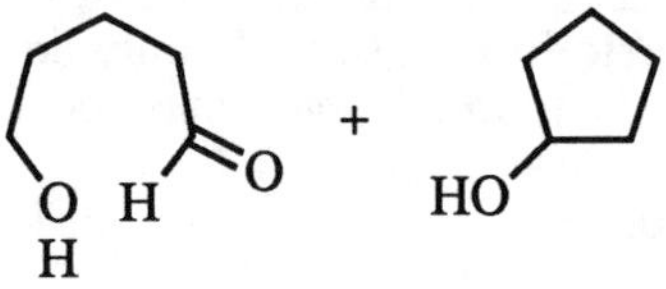

(d) acetal
(old: ketal)

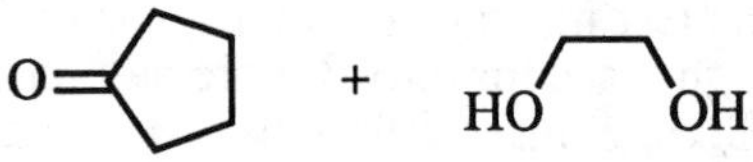

(e) acetal

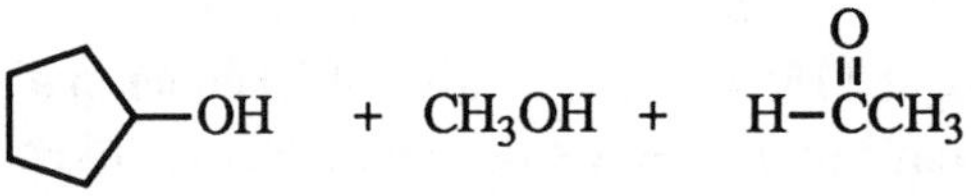

(f) dithiane
(dithioacetal)

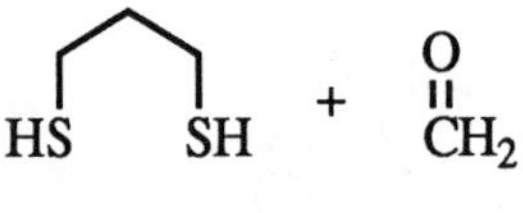

(g) imine(s)

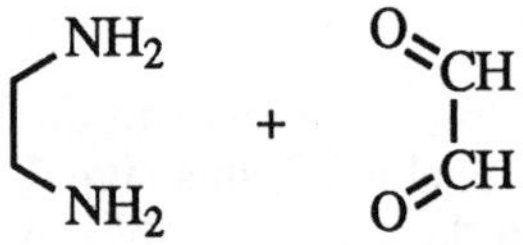

(h) imine

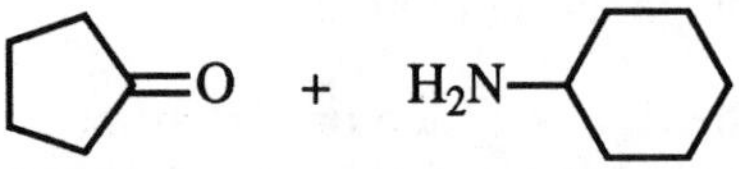

18-41

(a) $CH_3-\overset{O}{\overset{\|}{C}}H \xrightarrow[HCN]{KCN} CH_3-\overset{OH}{\overset{|}{C}}H-C\equiv N \xrightarrow{H_3O^+} CH_3-\overset{OH}{\overset{|}{C}}H-COOH$

(b)

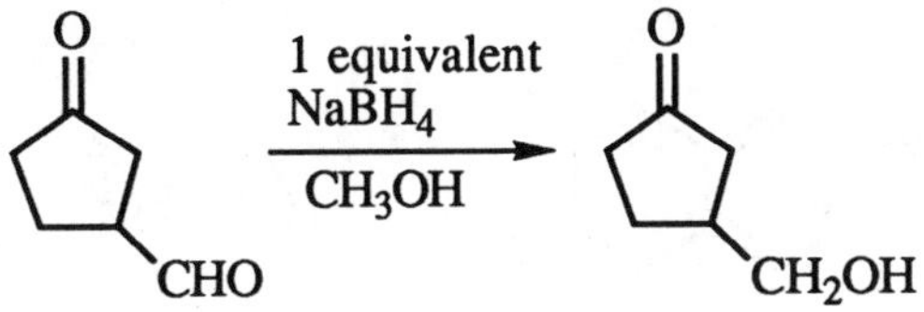

(c)

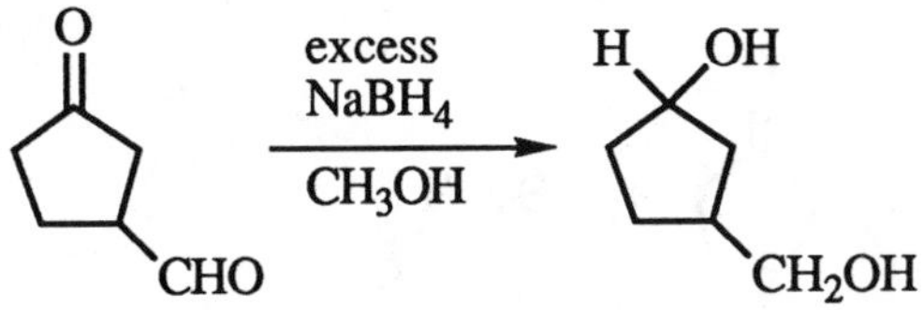

(d)

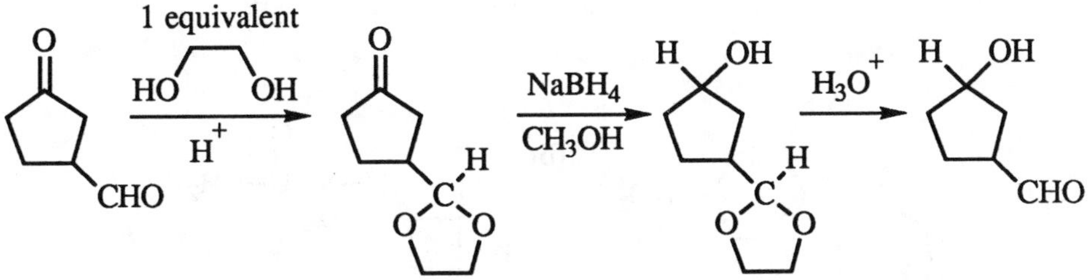

(e)

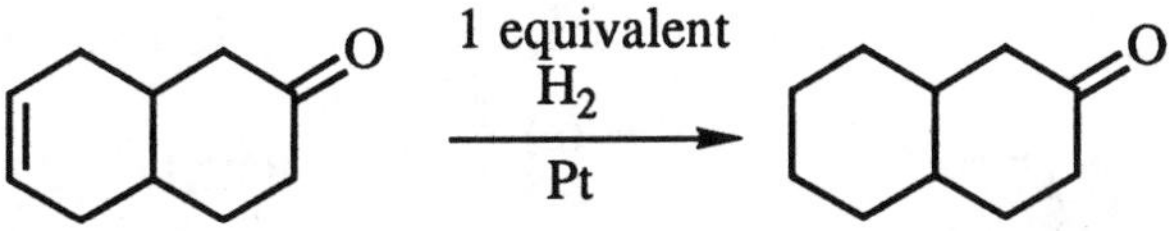

(f)

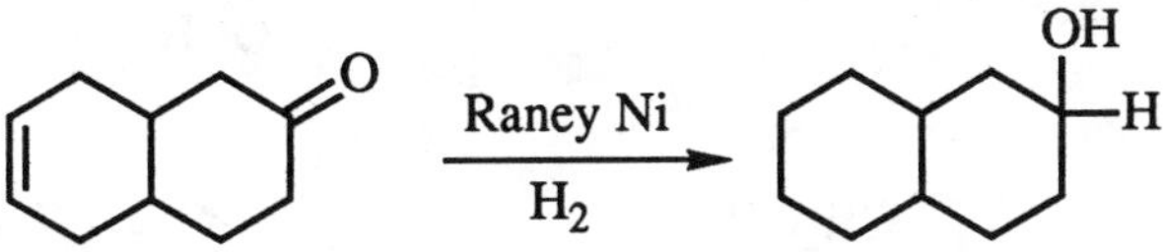

(g)

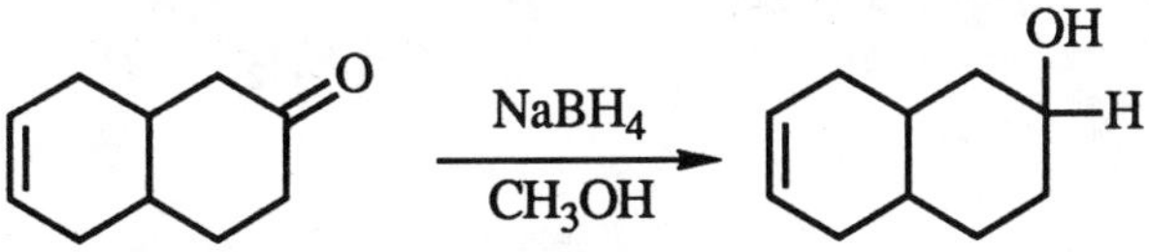

18-42 All of these reactions would be acid-catalyzed.

(a)

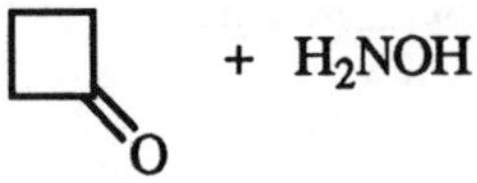

(b)

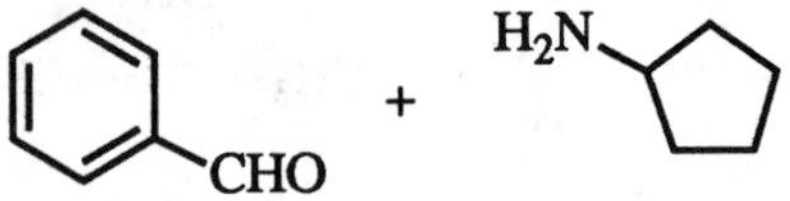

(c)

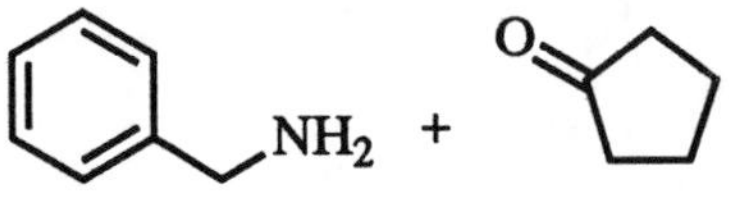

(d)

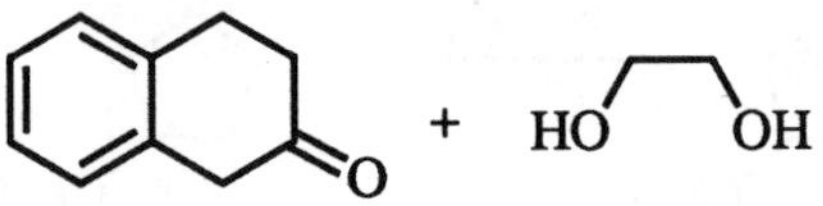

(e)

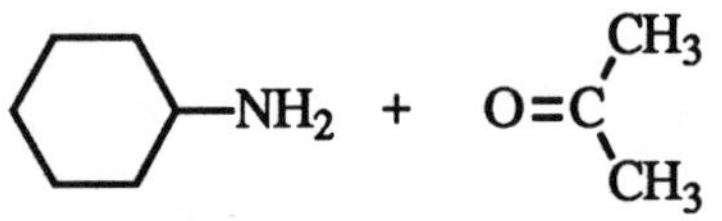

(f)

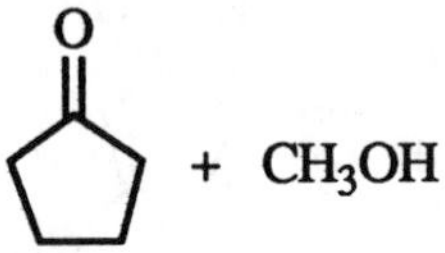

18-43

(a)

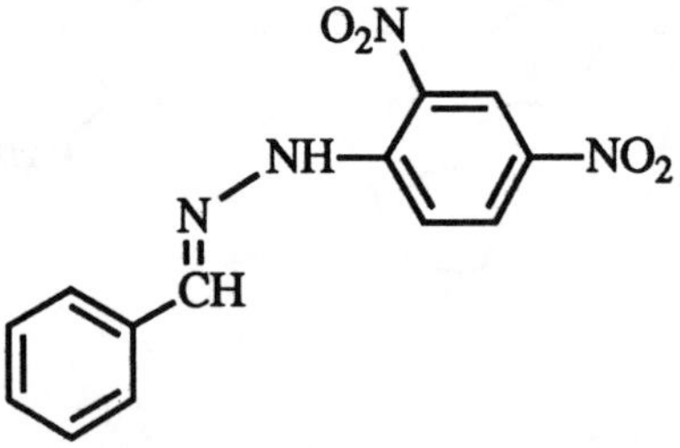

(b)

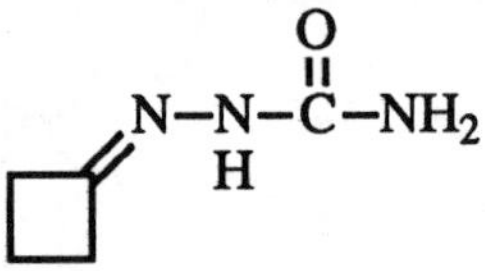

(c)

(d)

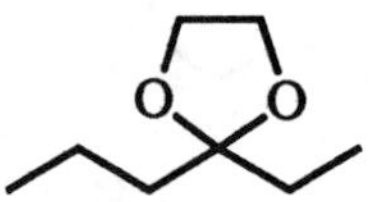

(e)

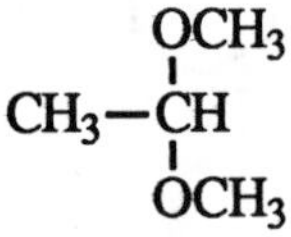

(f)

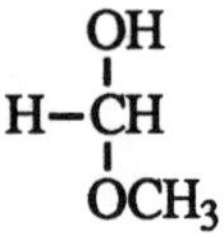

(g)

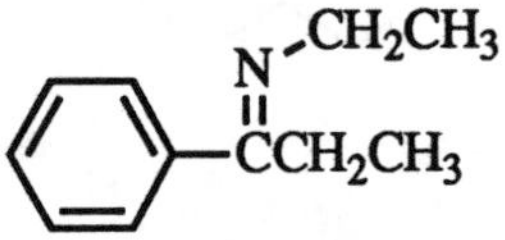

(h)

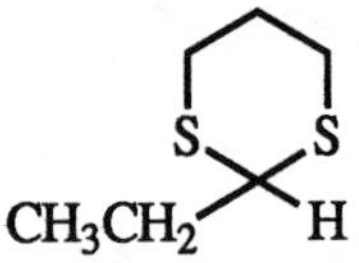

18-44

(a) 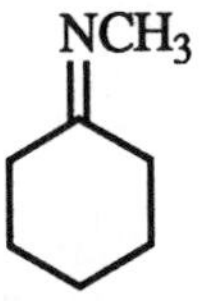
NCH3

(b) 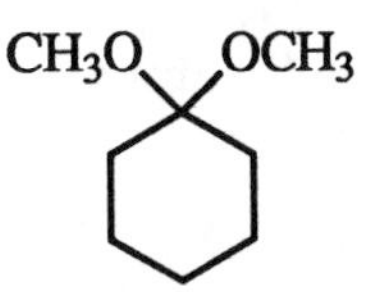
CH3O
OCH3

(c) 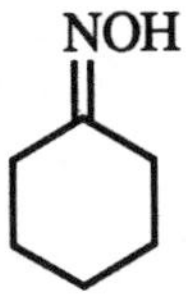
NOH

(d) 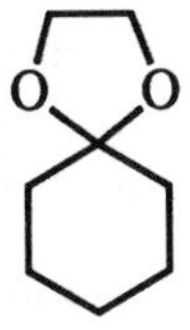
O
O

(e) 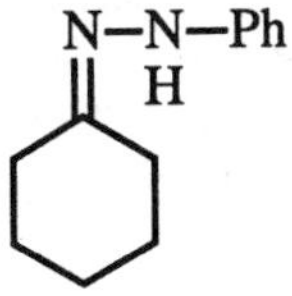
N—N—Ph
H

(f) 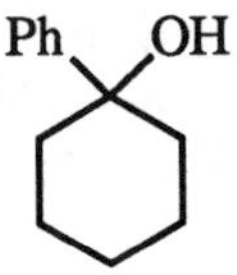
Ph
OH

(g) no reaction

(h) 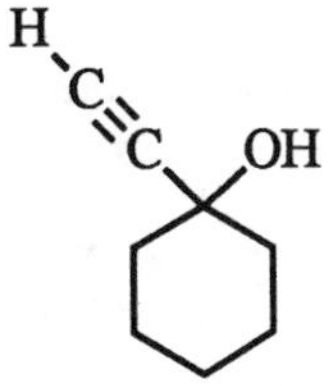
H
C
C
OH

(i) 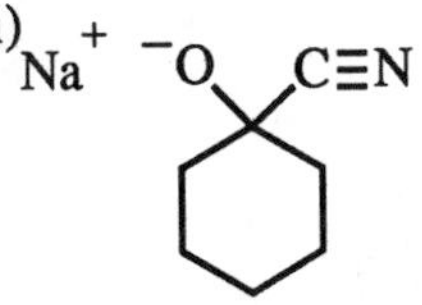
Na+ −O
C≡N

(j) 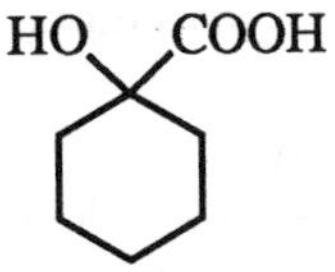
HO
COOH

(k) 

18-45 The new bond to carbon comes from the $NaBH_4$ or $NaBD_4$. The new bond to oxygen comes from the protic solvent.

(a) $CH_3-C(=O)-CH_2CH_3 \xrightarrow{NaBD_4} \xrightarrow{H_2O} CH_3-C(OH)(D)-CH_2CH_3$

(b) $CH_3-C(=O)-CH_2CH_3 \xrightarrow{NaBD_4} \xrightarrow{D_2O} CH_3-C(OD)(D)-CH_2CH_3$

(c) $CH_3-C(=O)-CH_2CH_3 \xrightarrow{NaBH_4} \xrightarrow{D_2O} CH_3-C(OD)(H)-CH_2CH_3$

18-46 While hydride is a small group, the actual chemical species supplying it, AlH_4^-, is fairly large, so it prefers to approach from the less hindered side of the molecule, that is, the side opposite the methyl. This forces the oxygen to go to the same side as the methyl, producing the *cis* isomer as the major product.

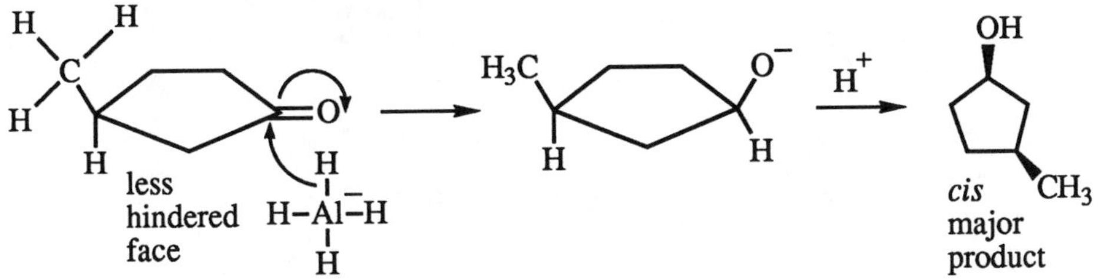
H
H
C
H
O
H
less
hindered
face
H
H−Al−H
H
H3C
O−
H
H
H+
OH
cis
CH3
major
product

18-47

A comment about the first step: this step is similar to an S_N2 reaction, but we saw earlier that RO^- is not a leaving group in nucleophilic substitution. Why does this work? What is different here?

Two factors facilitate this step. First, the Mg, a mild Lewis acid, complexes with the oxygen as it leaves, making the oxygen more neutral than ionic (similar to protonation to make a neutral leaving group). Second, the other two oxygens assist in two ways: a) they withdraw electrons by induction, making the carbon more positive and therefore more susceptible to attack by the carbanion; and b) they donate electrons by resonance, stabilizing any positive charge that develops on this carbon in the transition state.

No evidence is presented to suggest whether substitution by the Grignard reagent is one step or more; it will be shown here as one step, for simplicity.

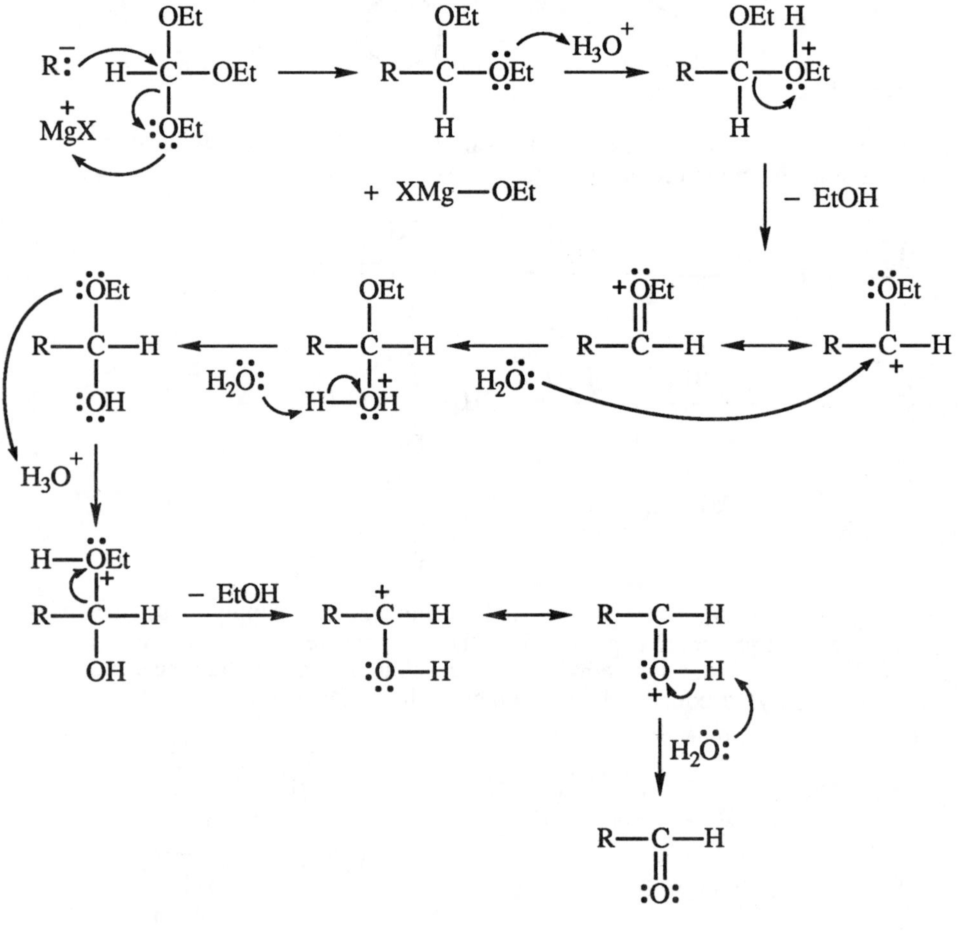

18-48

(a) 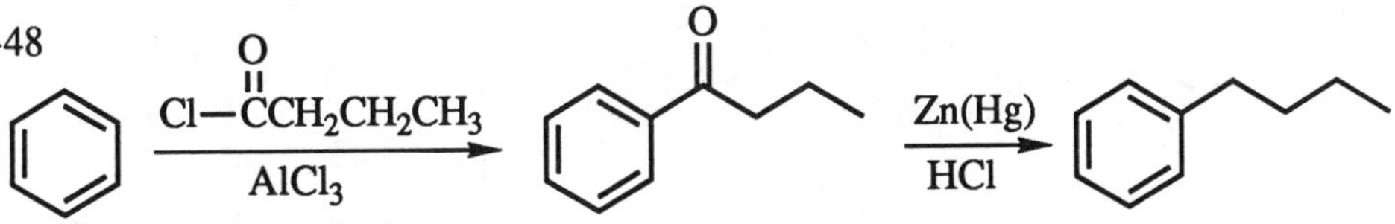

(b) 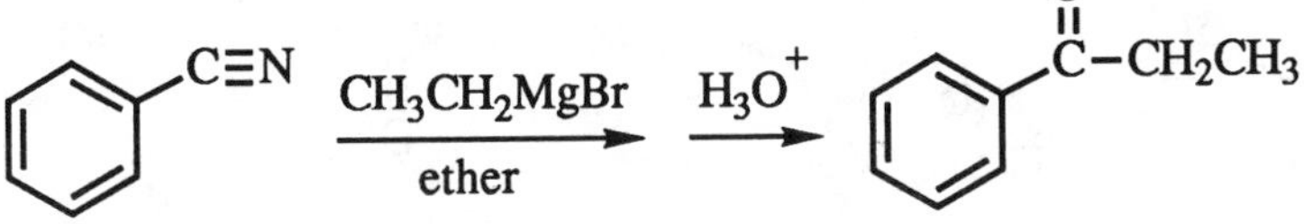

(c) 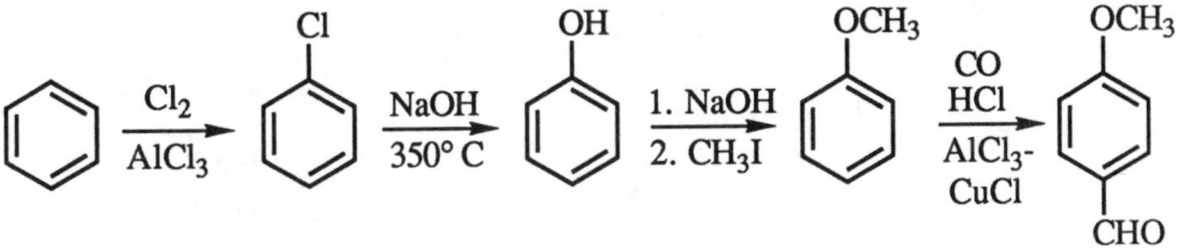

(d) 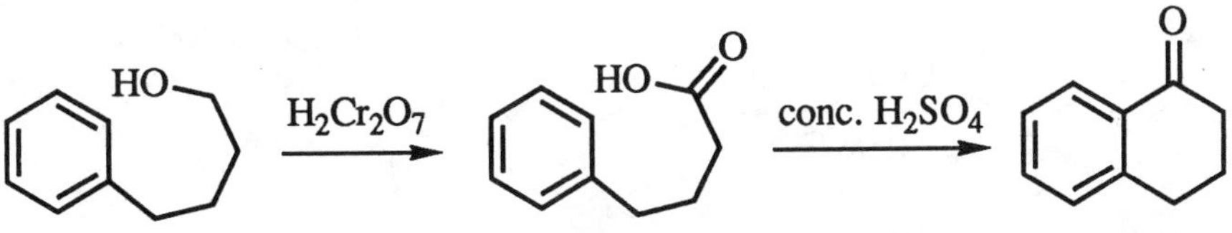

Alternatively, the acid chloride could be made with $SOCl_2$, then cyclized by Friedel-Crafts acylation with $AlCl_3$.

18-49

(a) 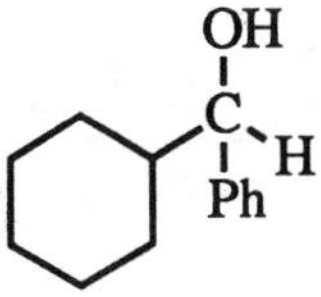

(b) 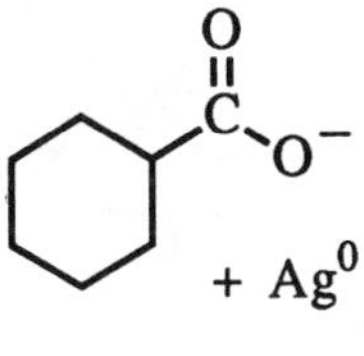

(c) 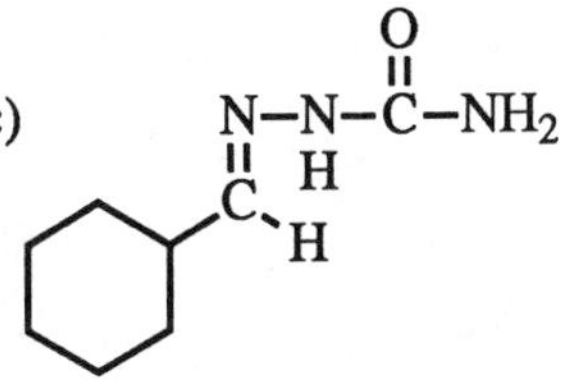

(d) 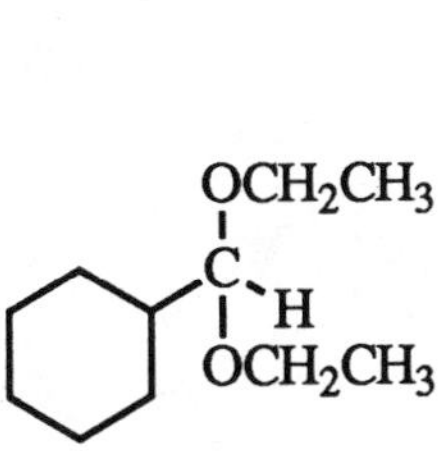

(e) 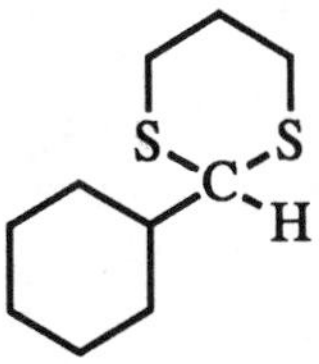

(f) 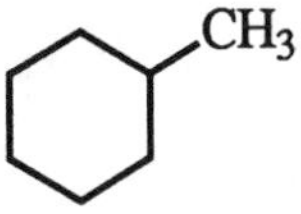

18-50

(a)

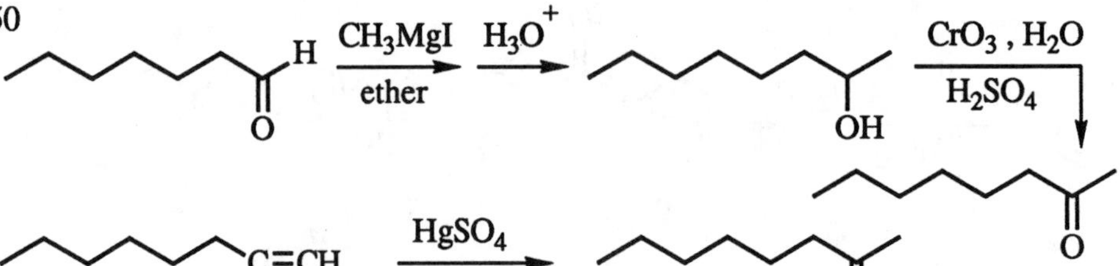

(b)

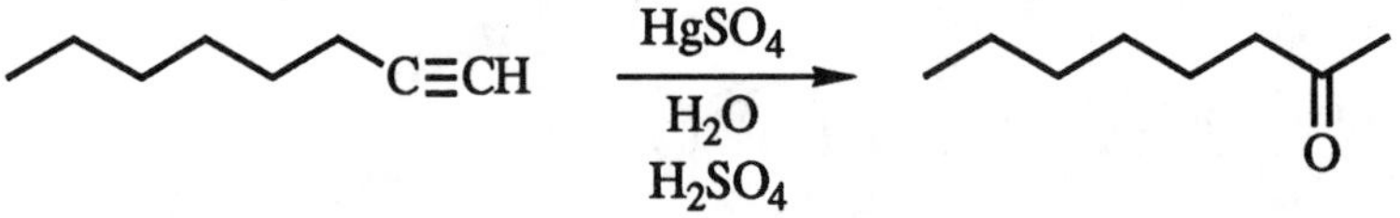

(c)

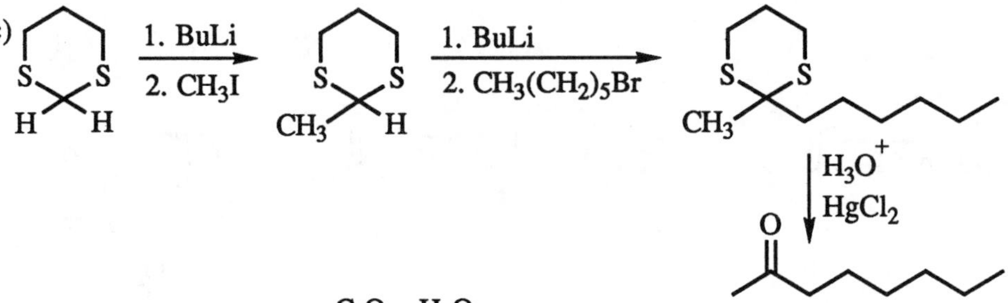

(d)

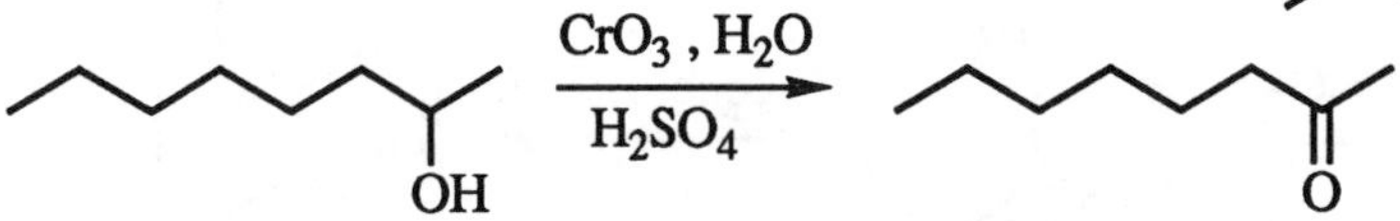

(e)

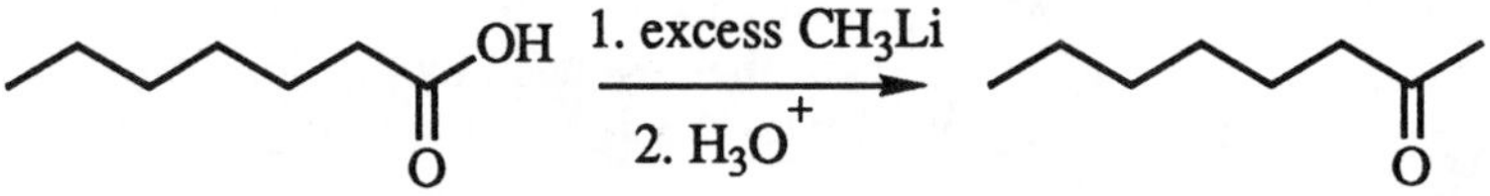

(f)

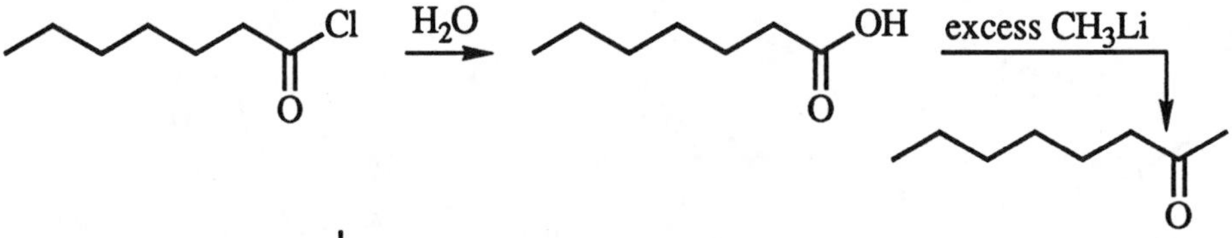

(g)

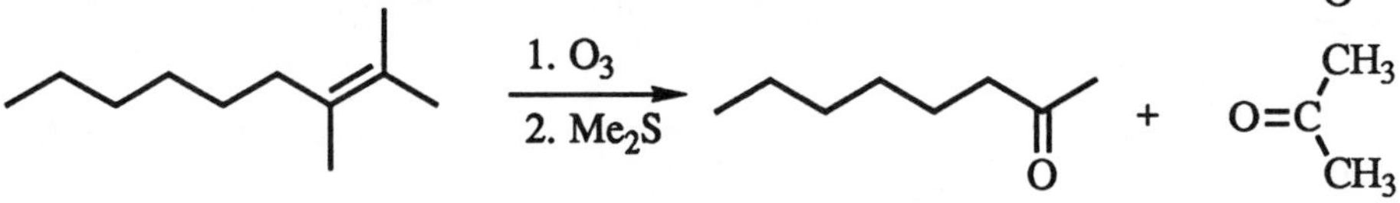

(h)

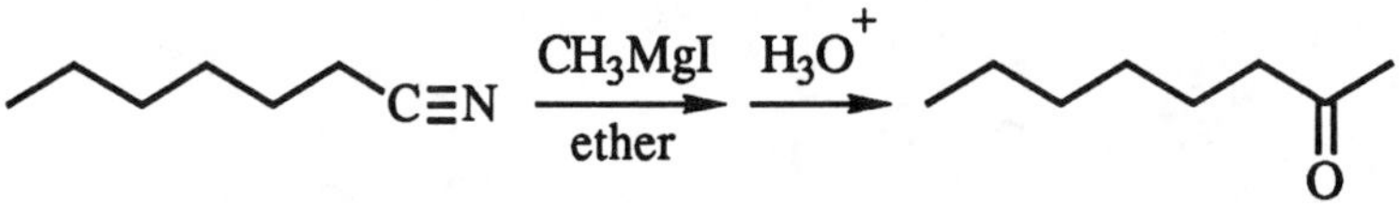

18-51

(a)

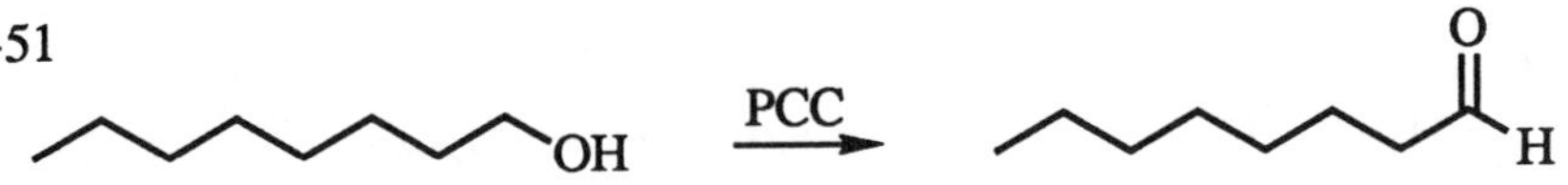

OH
PCC
O
H

(b)

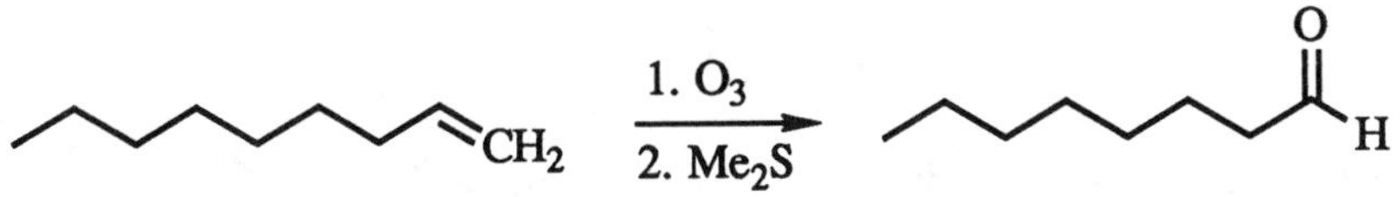

CH2
1. O3
2. Me2S
O
H

(c)

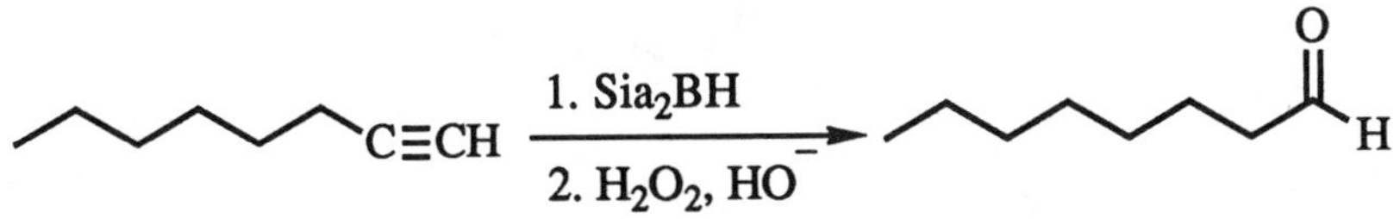

C≡CH
1. Sia2BH
2. H2O2, HO−
O
H

(d)

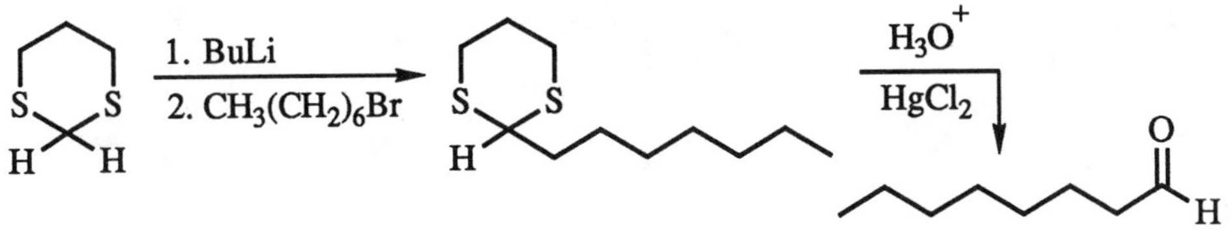

S
S
H
H
1. BuLi
2. CH3(CH2)6Br
S
S
H
H3O+
HgCl2
O
H

(e)

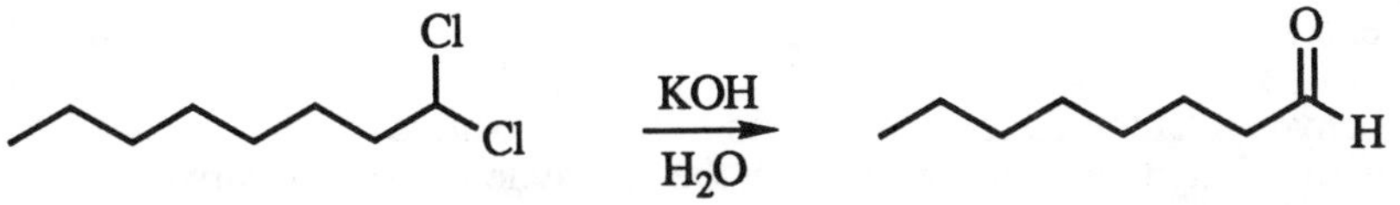

Cl
Cl
KOH
H2O
O
H

(This reaction needs a solvent like THF to keep all reactants in solution.)

(f)

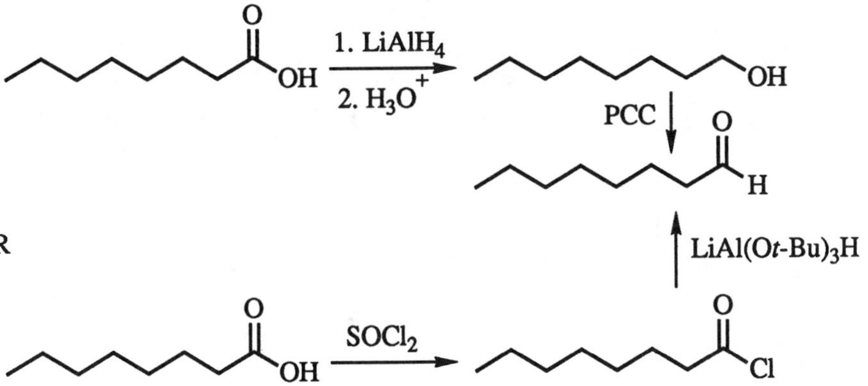

O
OH
1. LiAlH4
2. H3O+
OH
PCC
O
H
LiAl(Ot-Bu)3H
O
OH
SOCl2
O
Cl

OR

18-52

(a) $CH_3CH_2-C(=O)-CH_2CH_2CH_3$

(b) $CH_3CH_2-C(=O)-CH_2CH_2CH_3$ +

$CH_3-C(=O)-CH_2CH_2CH_2CH_3$

(c) $CH_3-C(=O)-CH_2CH_2CH_2CH_3$

(d)

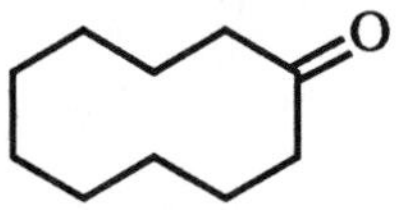

(e)

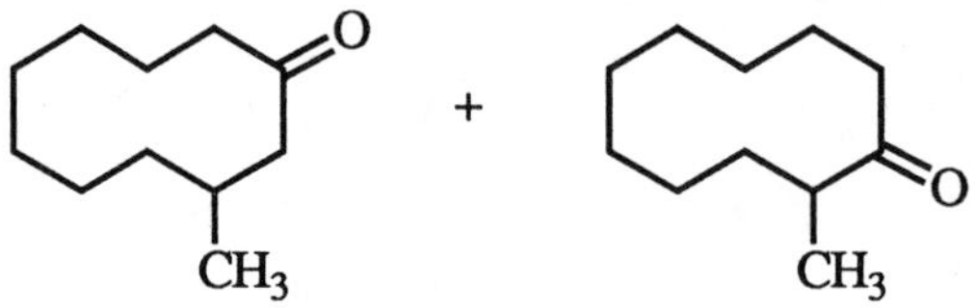

18-53

(a) ketone: no reaction
(b) aldehyde: positive
(c) enol of an aldehyde—tautomerizes to aldehyde in base: positive
(d) hemiacetal of an aldehyde in equilibrium with the aldehyde in base: positive
(e) acetal—stable in base: no reaction
(f) hemiacetal of an aldehyde in equilibrium with the aldehyde in base: positive

18-54 The structure of **A** can be deduced from its reaction with **J** and **K**. What is common to both products of these reactions is the 2-heptanol part; the reactions must be Grignard reactions with 2-heptanone, so **A** must be 2-heptanone.

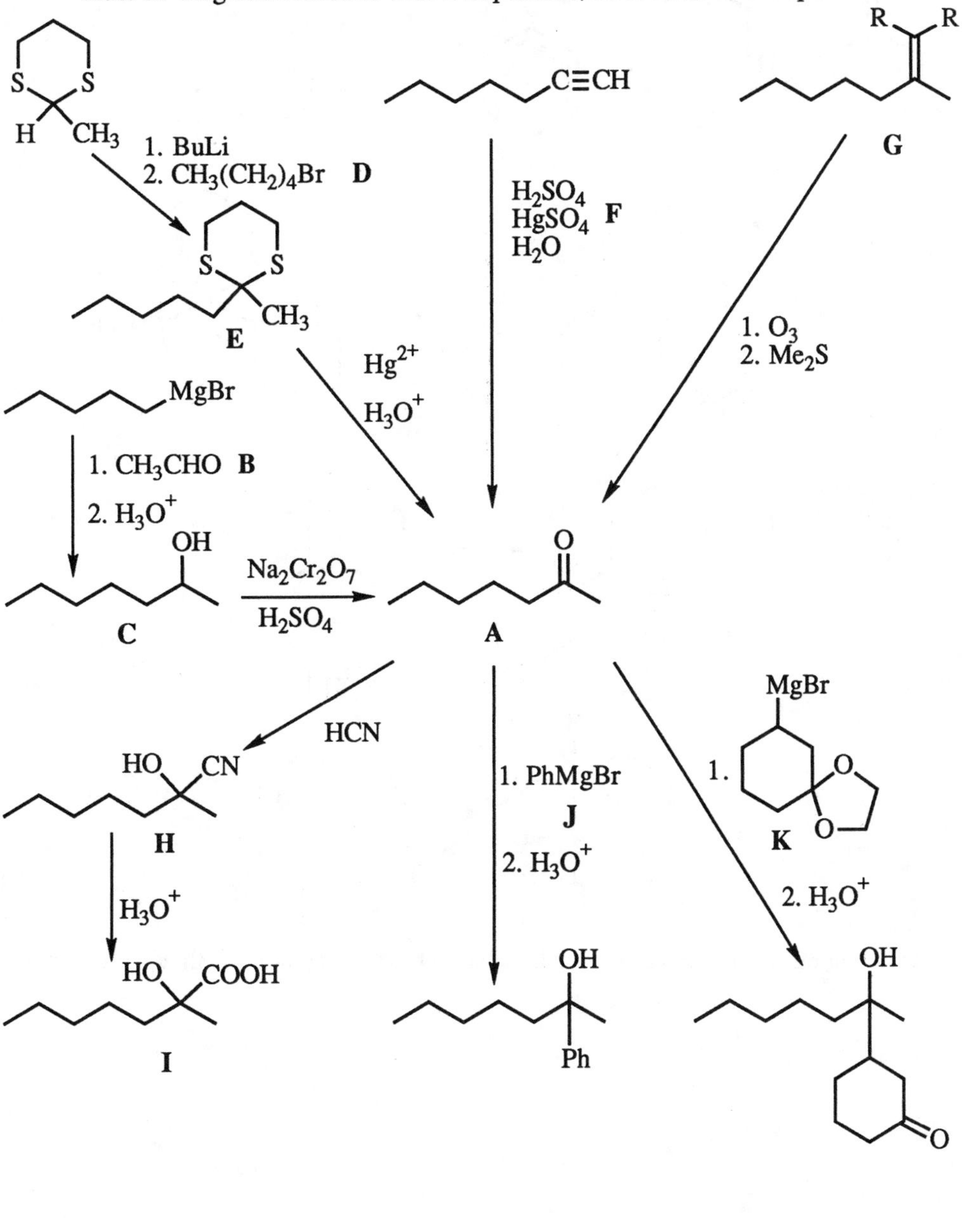

18-55

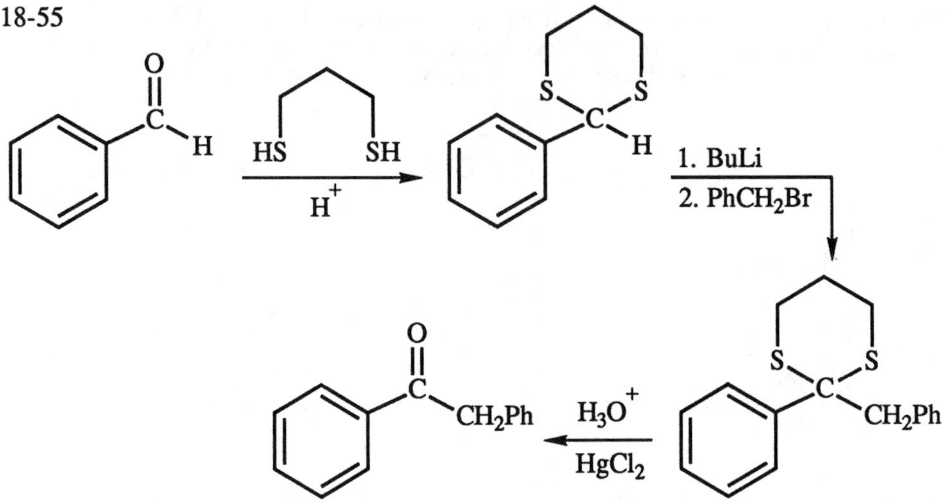

18-56

(a)

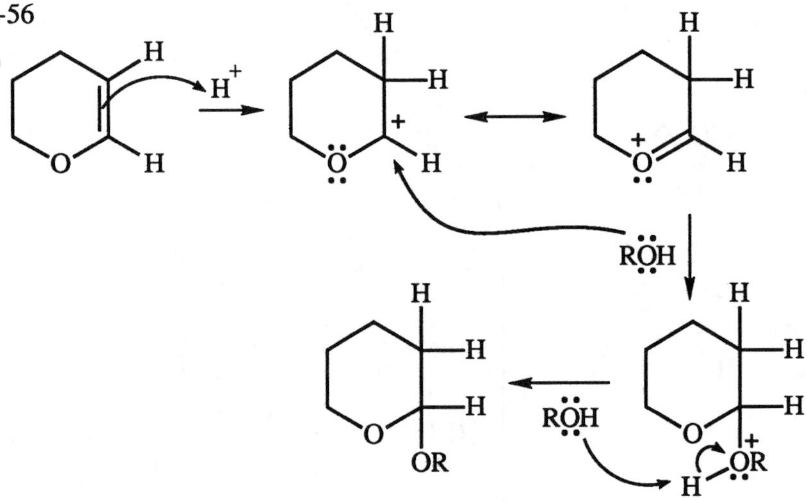

(b) This is not an ether, but rather an acetal, stable to base but reactive with aqueous acid.

18-56 continued

(c)

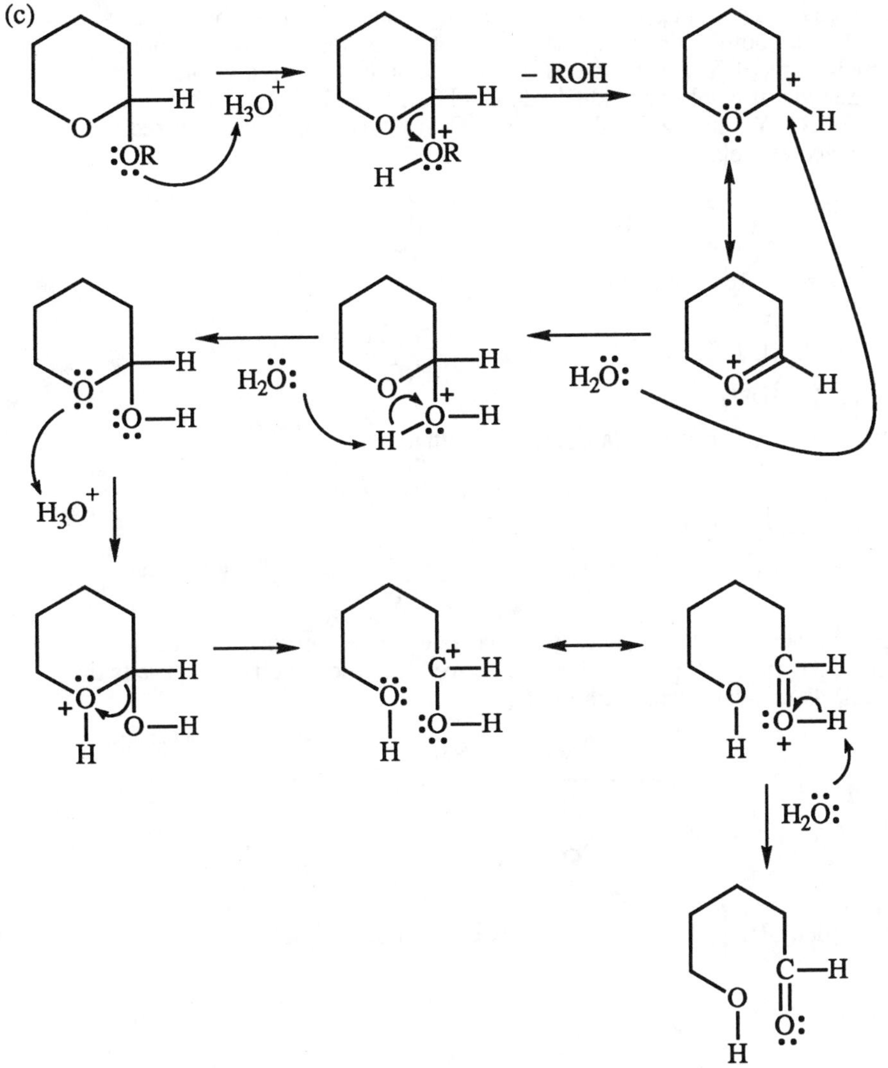

18-57 Units for peaks in the IR are cm^{-1}.

(a) First, deduce what functional groups are present in **A** and **B**. The IR of **A** shows no alkene and no carbonyl: the strongest peak is at 1060, possible a C—O bond. After acid hydrolysis of **A**, the IR of **B** shows a carbonyl at 1715: a ketone. (If it were an aldehyde, it would have aldehyde C—H around 2700-2800, absent in the spectrum of **B**.) What functional group has C—O bonds and is hydrolyzed to a ketone? An acetal (ketal)!

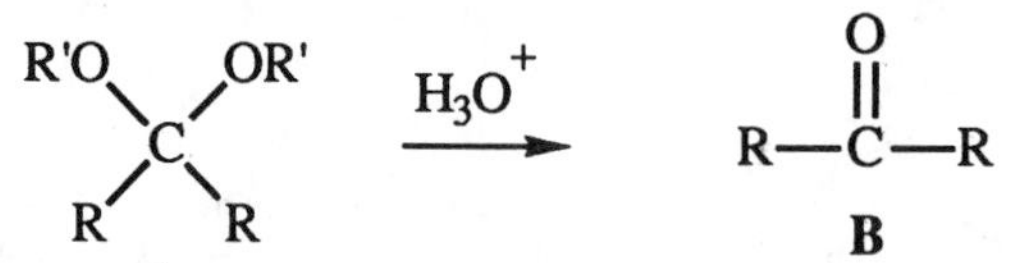

R'O
OR'
C
R
R
H_3O^+
O
R—C—R

A mol. wt. 116 ⇒ $C_6H_{12}O_2$

B mol. wt. 72 ⇒ C_4H_8O

There is only one ketone of formula C_4H_8O: 2-butanone.

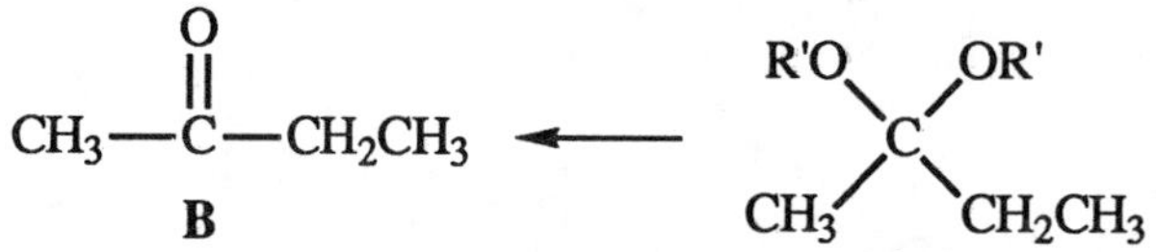

O
CH_3—C—CH_2CH_3
R'O
OR'
C
CH_3
CH_2CH_3

B **A**

A must have the same alkyl groups as **B**. **A** has one element of unsaturation and is missing only C_2H_4 from the partial structure above. The most likely structure is the ethylene ketal. Is this consistent with the NMR?

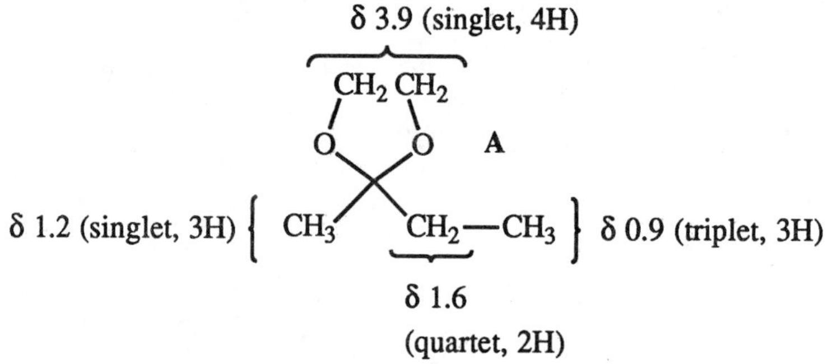

δ 3.9 (singlet, 4H)
CH_2 CH_2
O
O
A
δ 1.2 (singlet, 3H)
CH_3
CH_2—CH_3
δ 0.9 (triplet, 3H)
δ 1.6
(quartet, 2H)

What about the peak in the MS at m/z 87? This is the loss of 29 from the molecular ion at 116.

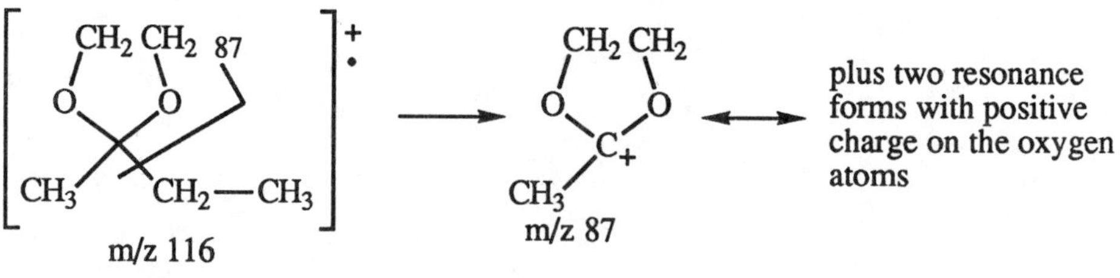

CH_2 CH_2 87
O
O
CH_3
CH_2—CH_3
m/z 116
CH_2 CH_2
O
O
C+
CH_3
m/z 87
plus two resonance forms with positive charge on the oxygen atoms

18-57 continued

(b)

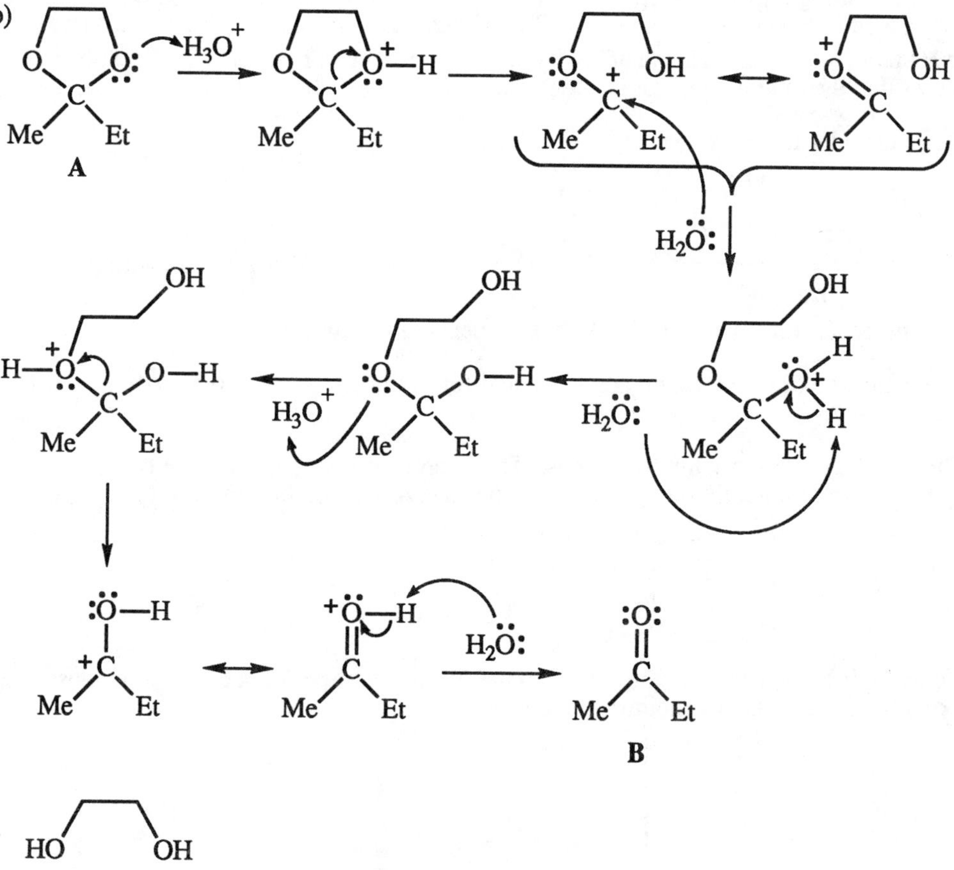

18-58

The very strong π to π* absorption at 225 nm in the UV spectrum suggests a conjugated ketone or aldehyde. The IR confirms this: strong, conjugated carbonyl at 1690 cm^{-1} and small alkene at 1620 cm^{-1}. The absence of peaks at 2700-2800 cm^{-1} shows that the unknown is not an aldehyde.

The molecular ion at 96 leads to the molecular formula:

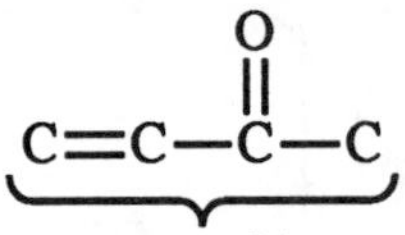

mass 64

$$\begin{array}{r} 96 \\ -\,64 \\ \hline \end{array}$$

32 mass units ⇒ add 2 carbons and 8 hydrogens

molecular formula = C_6H_8O = 3 elements of unsaturation

Two elements of unsaturation are accounted for in the enone. The other one is likely a ring.

The NMR shows two vinyl hydrogens. The doublet at δ 6.0 says that the two hydrogens are on neighboring carbons (two peaks = one neighboring H).

doublet: 1 neighboring H

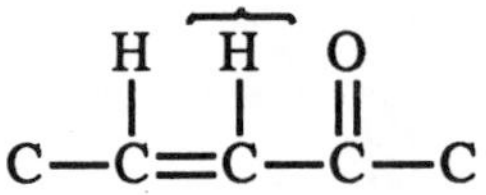

+ 1 C + 6 H + 1 ring

No methyls are apparent in the NMR, so the 6H group of peaks at δ 1.8-2.6 is most likely 3 CH_2 groups. Combining the pieces:

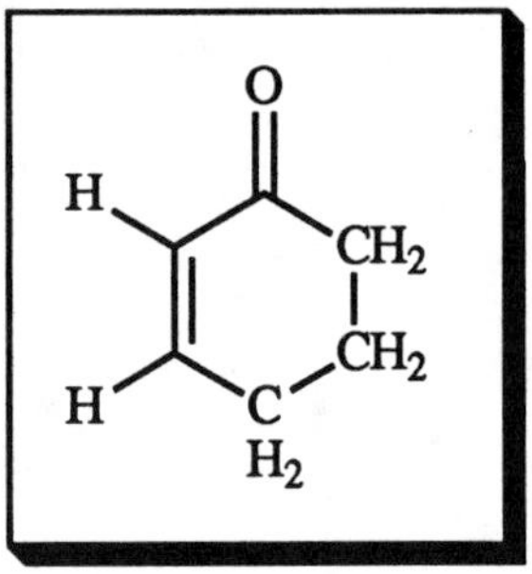

The mass spectral fragmentation can be explained by a "retro" or reverse Diels-Alder fragmentation:

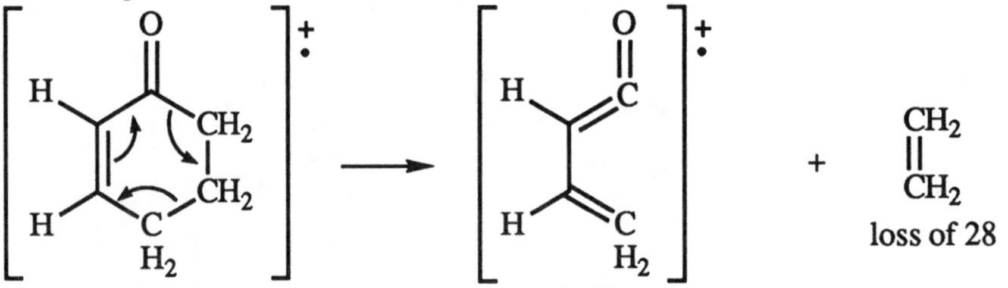

m/z 96 m/z 68 loss of 28

18-59

(a)

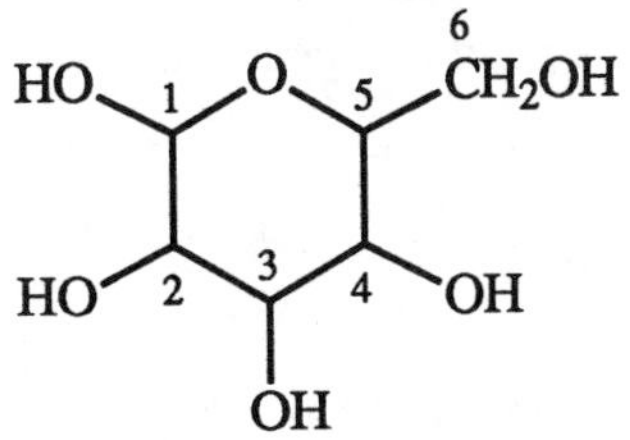

(b) The functional group at C-1 is a hemiacetal.

(c)

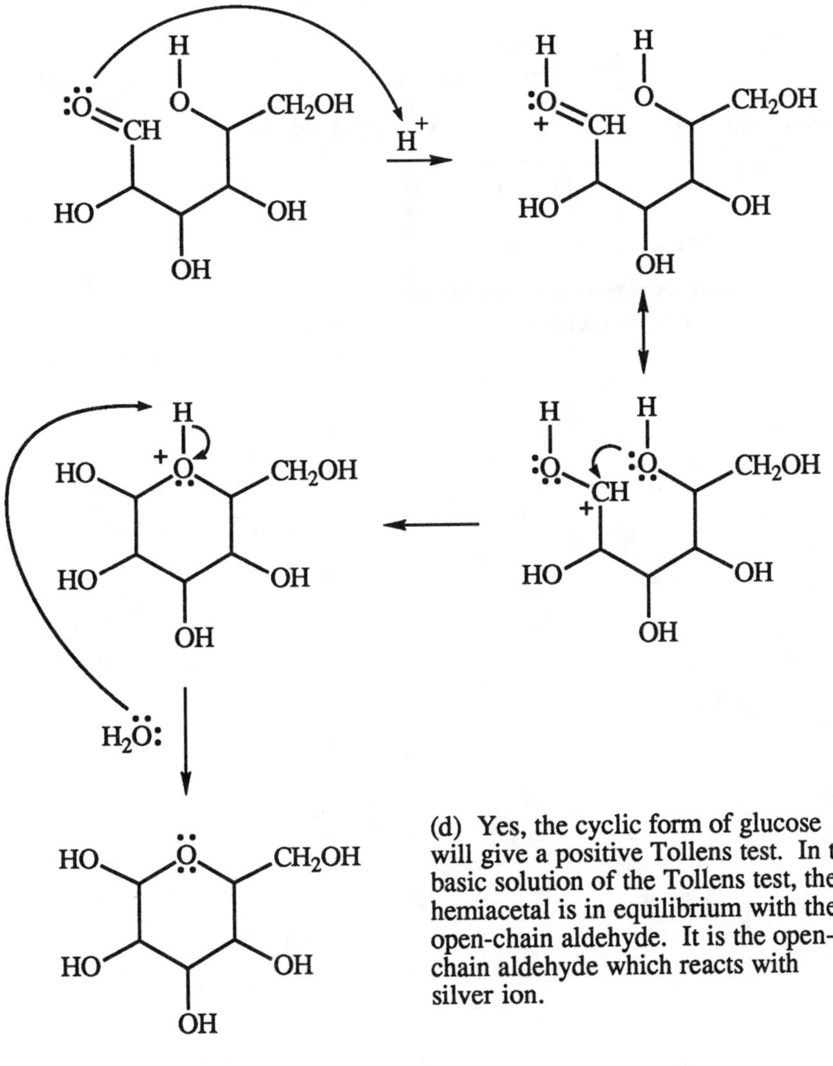

(d) Yes, the cyclic form of glucose will give a positive Tollens test. In the basic solution of the Tollens test, the hemiacetal is in equilibrium with the open-chain aldehyde. It is the open-chain aldehyde which reacts with silver ion.

18-60

The strong UV absorption at 220 nm indicates a conjugated aldehyde or ketone. The IR shows a strong carbonyl at 1690 cm^{-1}, alkene at 1645 cm^{-1}, and two peaks at 2720 cm^{-1} and 2810 cm^{-1} —aldehyde!

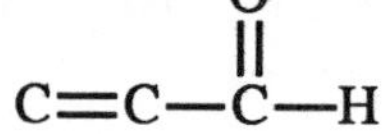

The NMR shows the aldehyde proton at δ 9.5 split into a doublet, so it has one neighboring H. There are only two vinyl protons, so there must be an alkyl group coming off the β carbon:

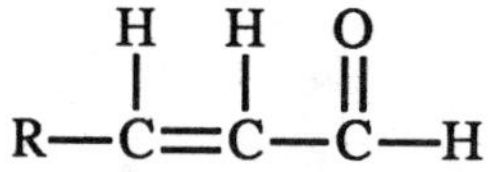

The only other NMR signal is a 3H doublet: R must be methyl.

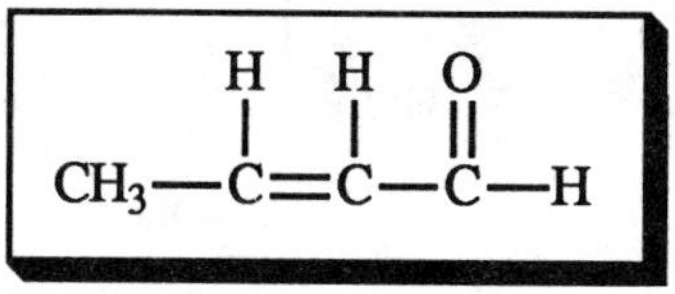

"crotonaldehyde"

CHAPTER 19—AMINES

19-1 These compounds satisfy the criteria for aromaticity (planar, cyclic π system, and the Huckel number of $4n + 2$ π electrons): pyrrole, imidazole, indole, pyridine, 2-methylpyridine, pyrimidine, and purine. The systems with 6 π electrons are: pyrrole, imidazole, pyridine, 2-methylpyridine, and pyrimidine. The systems with 10 π electrons are: indole and purine. The other nitrogen heterocycles shown are not aromatic because they do not have cyclic π systems.

19-2

(a) 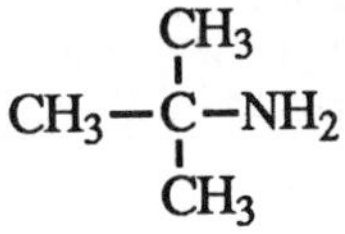

(b) 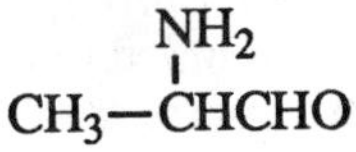

(c) 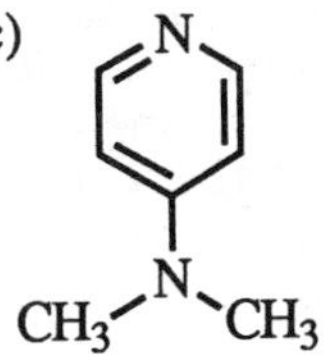

(d) 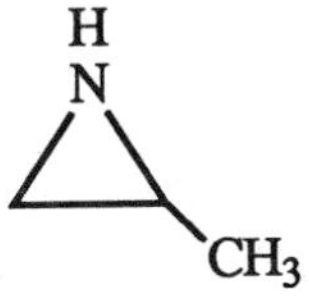

(e) 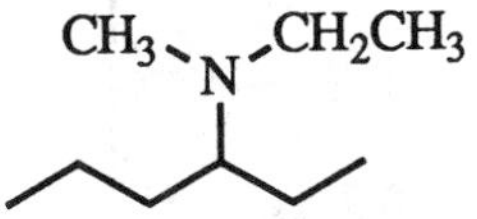

(f) 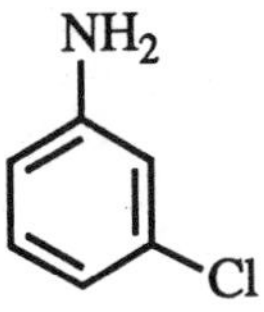

19-3

(a) 2-pentanamine
(b) *N*-methyl-2-butanamine
(c) 3-aminophenol (or *meta*-)
(d) 3-methylpyrrole
(e) *trans*-1,2-cyclopentanediamine
(f) *cis*-3-aminocyclohexanecarbaldehyde

19-4

(a) resolvable: there are two chiral carbons; carbon does not invert
(b) not resolvable: the nitrogen is free to invert
(c) not resolvable: it is symmetric
(d) not resolvable: even though the nitrogen is quaternary, one of the groups is a proton which can exchange rapidly, allowing for inversion
(e) resolvable: the nitrogen is quaternary and cannot invert when bonded to carbons

19-5 In order of increasing boiling point (increasing intermolecular hydrogen bonding):

(a) triethylamine and*n*-propyl ether have the same b.p. < di-*n*-propylamine
(b) dimethyl ether < dimethylamine < ethanol
(c) trimethylamine < diethylamine < diisopropylamine

19-6

(a) primary amine: two spikes in the 3200-3400 cm^{-1} region, indicating NH_2
(b) alcohol: strong, broad peak around 3400 cm^{-1}
(c) secondary amine: one spike in the 3200-3400 cm^{-1} region, indicating NH

19-7 $C_4H_{11}N$ has no elements of unsaturation. The NMR doublet at δ 2.5 integrates to 2H; this must be a CH_2 bonded to the nitrogen, and a doublet because of one neighboring H:

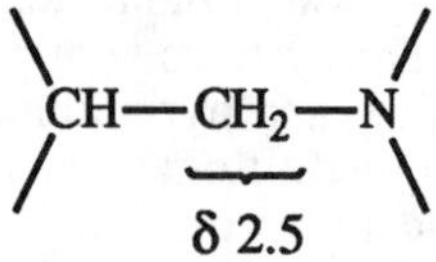
CH—CH2—N
δ 2.5

+ 2 C + 8 H

The CH appears to be the multiplet at δ 1.5, part of a typical isopropyl pattern with the methyl groups appearing as a 6H doublet at δ 0.9. If any more carbons were bonded to the N, the chemical shift of their protons would also be in the δ 2.5 region, so it is safe to infer than the nitrogen is bonded to two hydrogens (a primary amine), and the remaining two carbons and six hydrogens are two methyl groups on the methine (CH). This would account for the 6H doublet at δ 0.9.

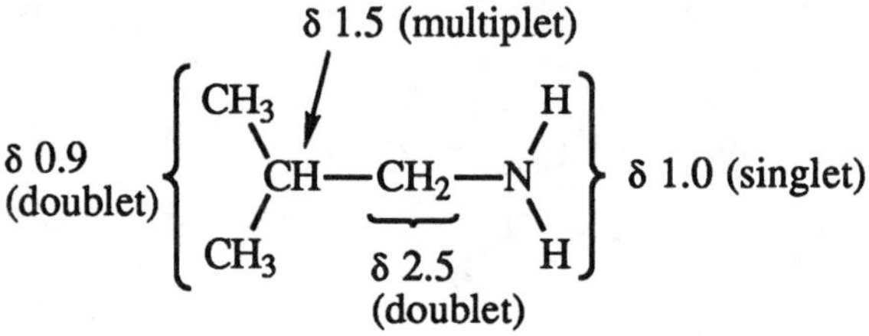
δ 1.5 (multiplet)
δ 0.9
(doublet)
δ 1.0 (singlet)
δ 2.5
(doublet)

19-8

(a)

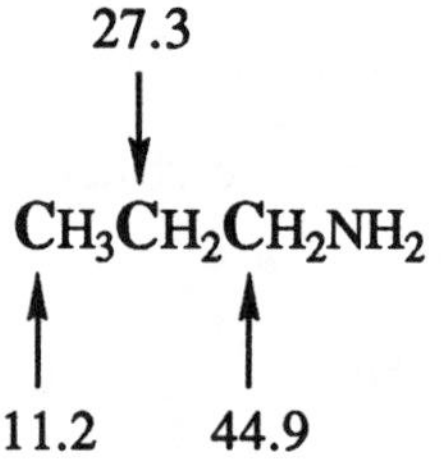
27.3
CH3CH2CH2NH2
11.2
44.9

(b)

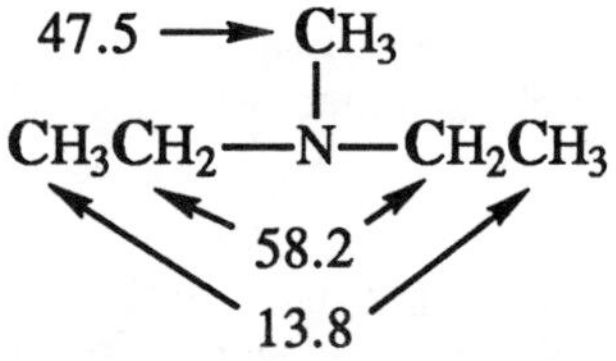
47.5
58.2
13.8

(c)

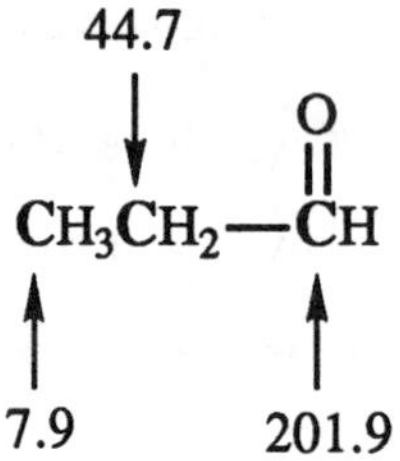
44.7
7.9
201.9

(d)

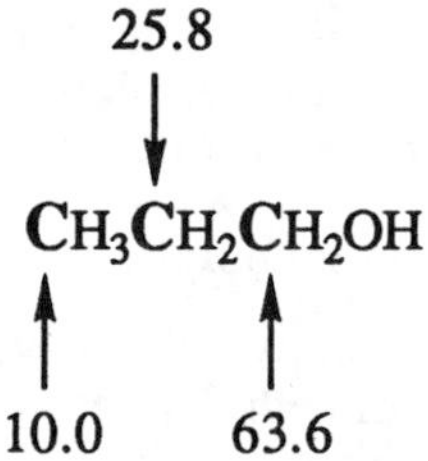
25.8
CH3CH2CH2OH
10.0
63.6

19-9

(a)

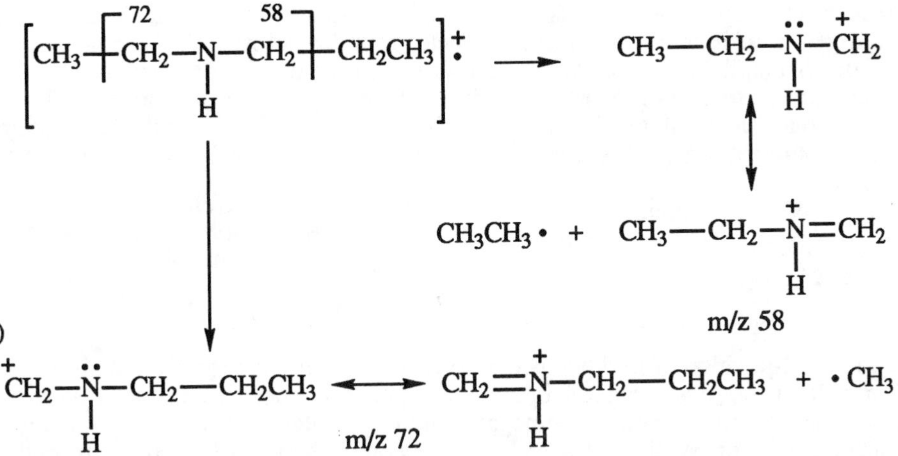
72
58
$CH_3CH_3\cdot$ +
m/z 58
m/z 72
+ $\cdot CH_3$

(b)

(c) The fragmentation in (a) occurs more often than the one in (b) because of stability of the radicals produced along with the iminium ions. Ethyl radical is much more stable than methyl radical, so pathway (a) is preferred.

19-10 In order of increasing basicity:

(a) $PhNH_2$ < NH_3 < CH_3NH_2 < NaOH
(b) *p*-nitroaniline < aniline < *p*-methylaniline (*p*-toluidine)
(c) pyrrole < aniline < pyridine
(d) 3-nitropyrrole < pyrrole < imidazole

19-11

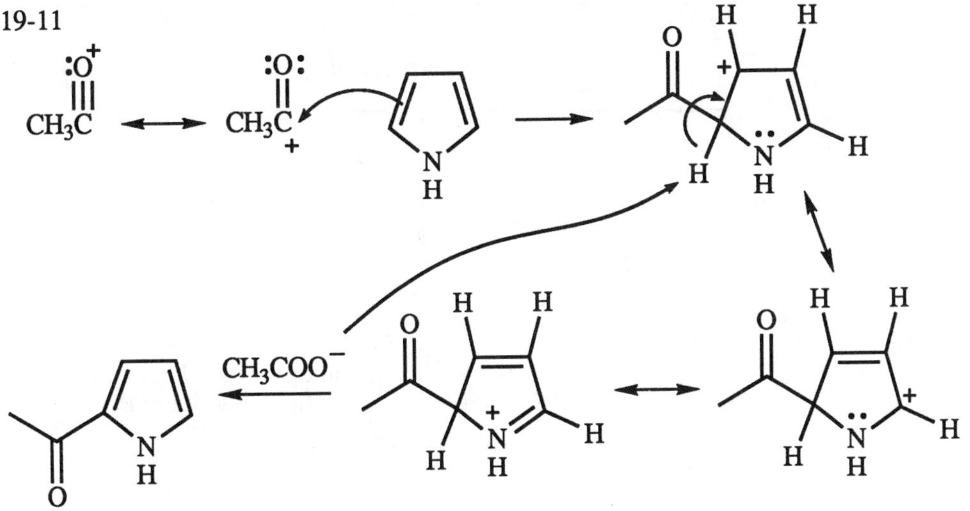
CH_3COO^-

19-12

Pyridine is deactivated in the presence of electrophiles because the pair of electrons on the nitrogen is nucleophilic and reacts with electrophiles before they can attach to the ring. In pyrrole, the pair of electrons that we draw on the nitrogen is actually delocalized in the π system of the aromatic ring. The pyrrole nitrogen is not a good nucleophile, so an electrophile will react with electrons of the ring instead. The preferred site of electrophilic attack (and of protonation by strong acid) is on C-2 of the pyrrole ring.

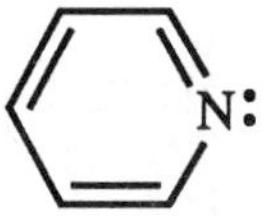

pyridine—
basic, nucleophilic

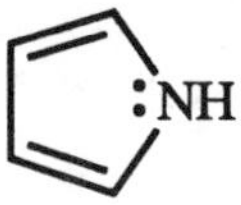

pyrrole—
not basic, not nucleophilic

19-13

Nitration at the 4-position of pyridine is not observed for the same reason that nitration at the 2-position is not observed: the intermediate puts some positive character on an electron-deficient nitrogen, and electronegative nitrogen hates that. (It is important to distinguish this type of positive nitrogen without a complete octet of electrons, from the quaternary nitrogen, also positively charged but with a full octet. It is the number of electrons around atoms that is most important; the charge itself is less important.)

GOOD: 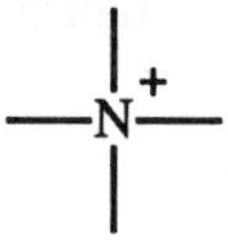

VERY BAD: 

mechanism

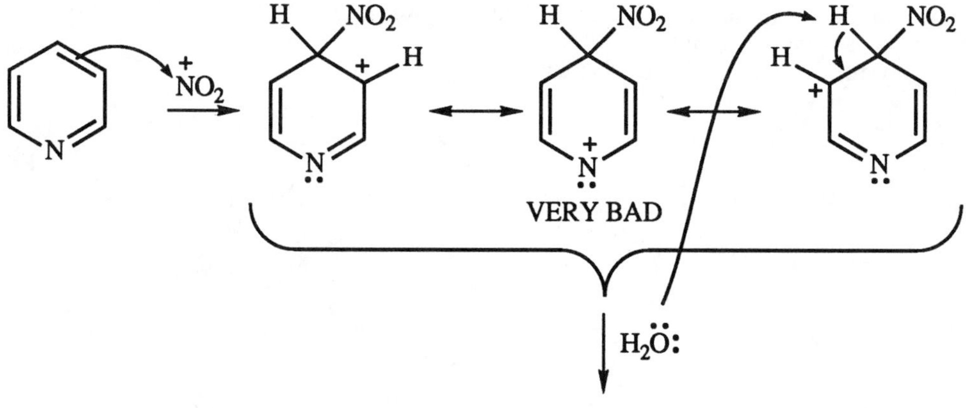

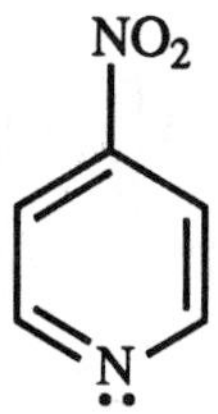

not produced because of unfavorable intermediate

19-14

Any electrophilic attack, including sulfonation, is preferred at the 3-position of pyridine because the intermediate is more stable than the intermediate from attack at either the 2-position or the 4-position. (Resonance forms of the sulfonate group are not shown, but don't forget them!)

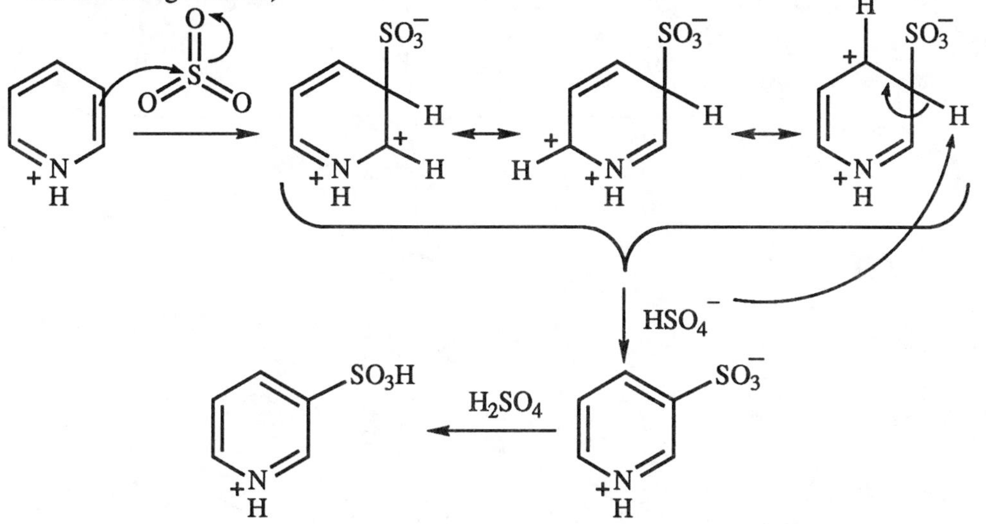

19-15

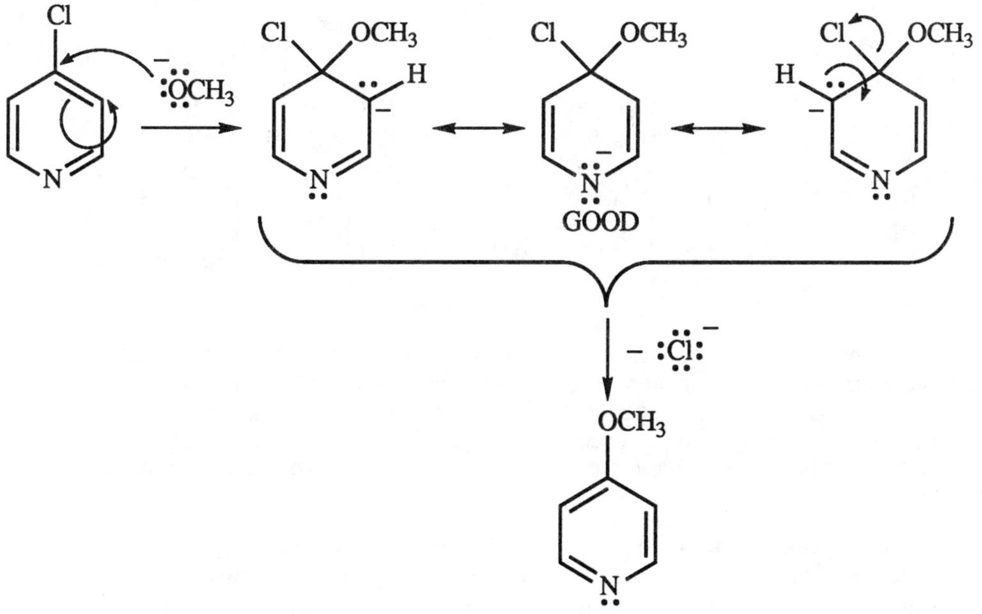

19-16

(a)

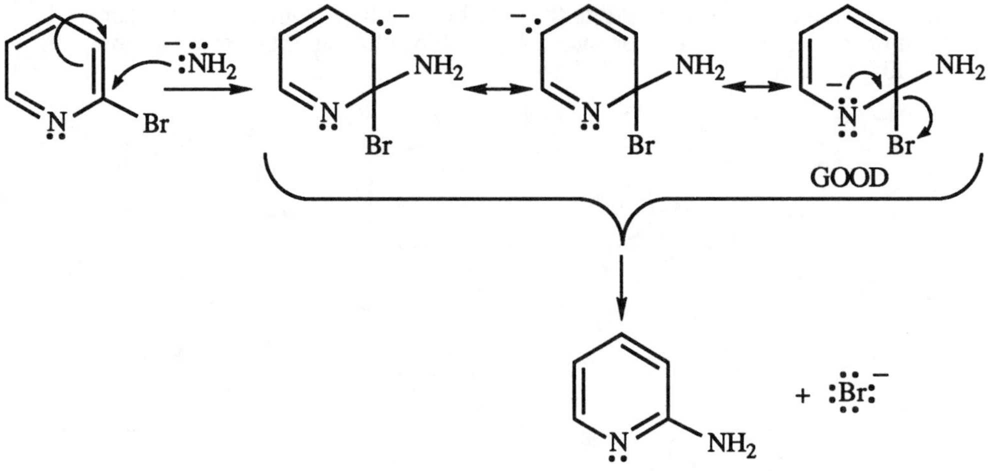
:NH2
NH2
Br
GOOD
+ :Br:

(b)

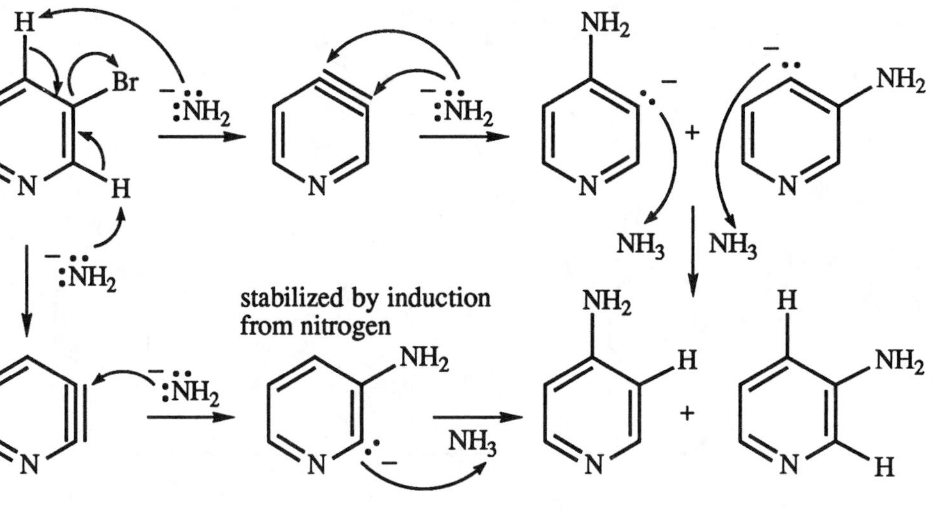
H
Br
:NH2
NH2
NH3
stabilized by induction from nitrogen
+

This is a benzyne-type mechanism. (For simplicity above, two steps of benzyne generation are shown as one step: first, a proton is abstracted by amide anion, followed by loss of bromide.) Amide ion is a strong enough base to remove a proton from 3-bromopyridine as it does from a halobenzene. Once a benzyne is generated (two possibilities), the amide ion reacts quickly, forming a mixture of products.

Why does the 3-bromo follow this extreme mechanism while the 2-bromo reacts smoothly by the addition-elimination mechanism? Stability of the intermediate! Negative charge on the electronegative nitrogen makes for a more stable intermediate in the 2-bromo substitution. No such stabilization is possible in the 3-bromo case.

19-17

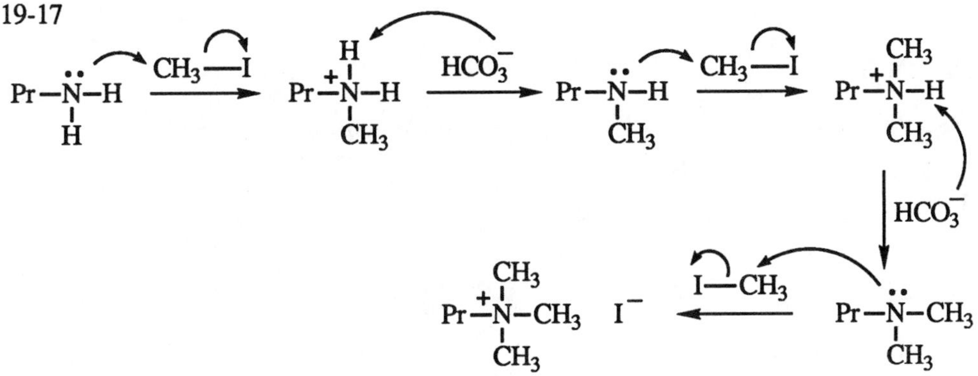

19-18

(a) $PhCH_2NH_2$ + excess CH_3I $\xrightarrow{NaHCO_3}$ $PhCH_2\overset{+}{N}(CH_3)_3\ \ I^-$

(b) 

(c) excess NH_3 + $PhCH_2Br$ $\longrightarrow$ $PhCH_2NH_2$

19-19

(a) $CH_3\overset{O}{\overset{\|}{C}}-NHCH_2CH_3$

(b) 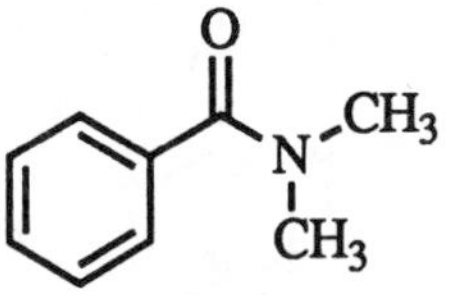

(c) 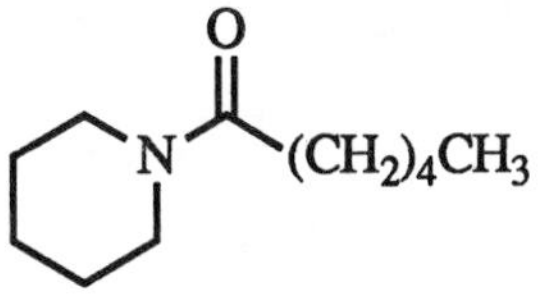

19-20 If the amino group were not protected, it would do a nucleophilic substitution on chlorosulfonic acid. Later in the sequence, this group could not be removed without cleaving the other sulfonamide group.

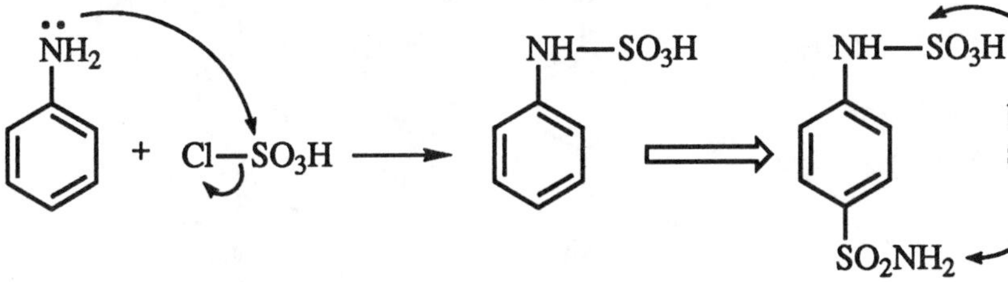

19-21

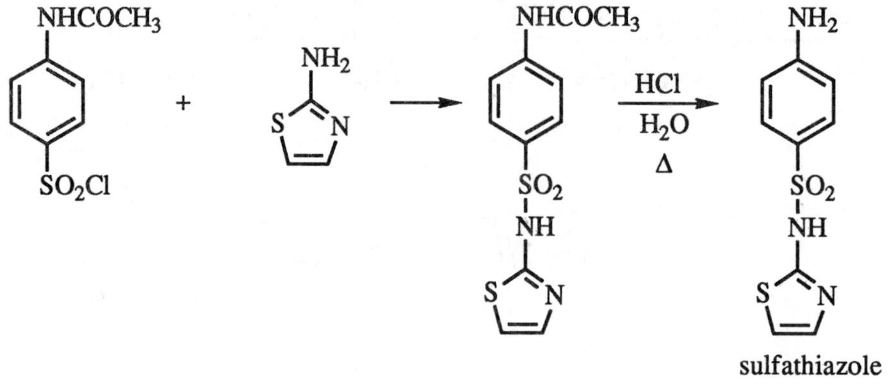
NHCOCH3
SO2Cl
NH2
S
N
NHCOCH3
HCl
H2O
Δ
SO2
NH
NH2
sulfathiazole

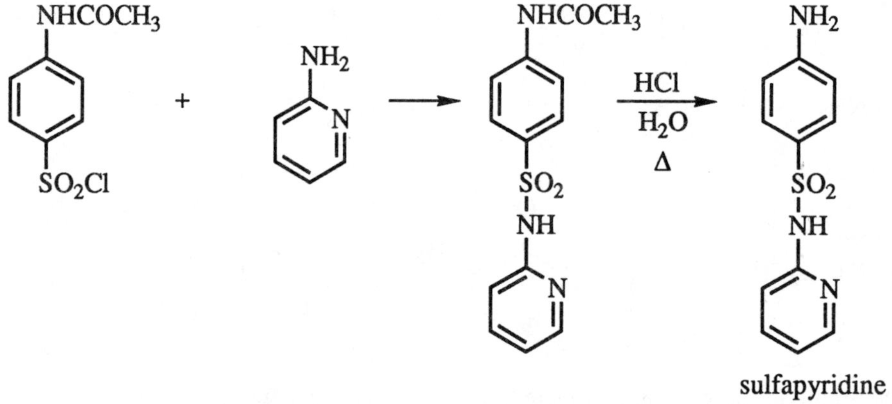
NHCOCH3
SO2Cl
NH2
N
HCl
H2O
Δ
SO2
NH
NH2
sulfapyridine

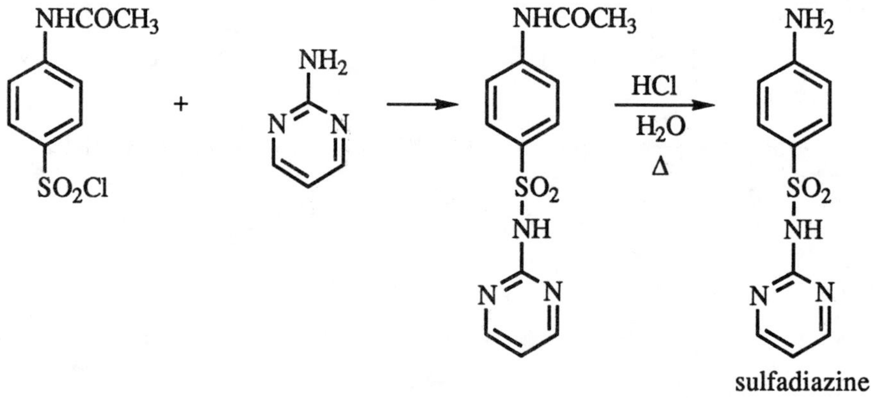
NHCOCH3
SO2Cl
NH2
N
HCl
H2O
Δ
SO2
NH
NH2
sulfadiazine

19-22

(a) 

(b) 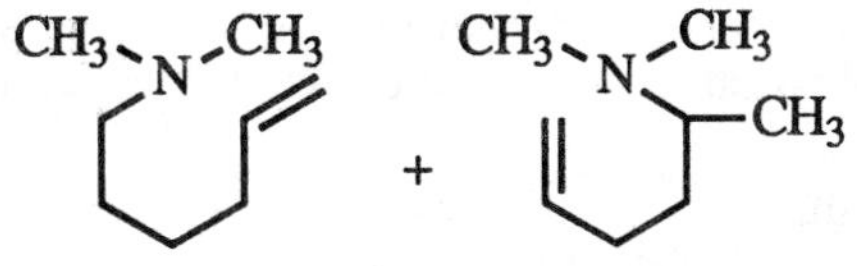

(c) 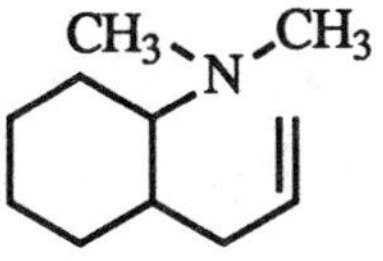

(d) 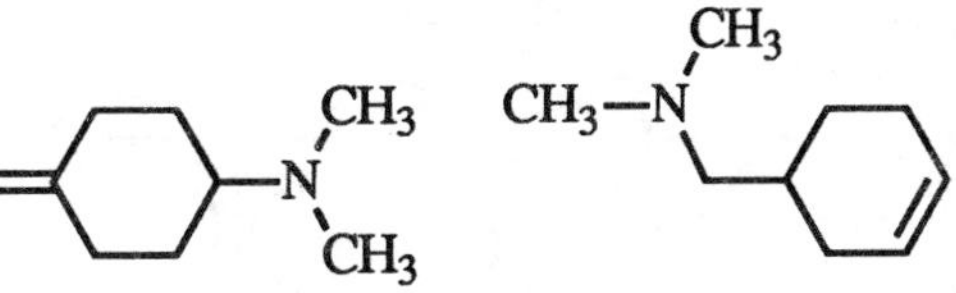

(e) 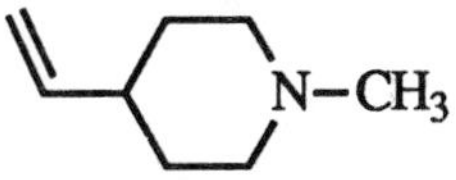

19-23 Orientation of the Cope elimination is similar to Hofmann elimination: the *less* substituted alkene is the major product.

(a) 

+ $(CH_3)_2NOH$

(b) $CH_2{=}CH_2$ + 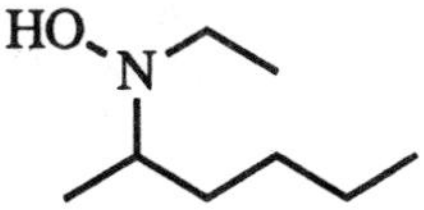

major

+  + $(CH_3CH_2)_2NOH$

minor

(c)  + $(CH_3)_2NOH$

(d) $CH_2{=}CH_2$ + 

+ 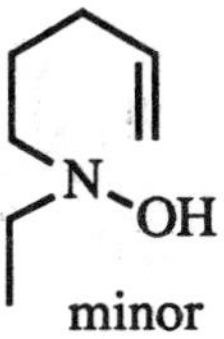

minor

19-24 The key to this problem is to understand that Hofmann elimination occurs via an E2 mechanism requiring *anti* coplanar stereochemistry, whereas Cope elimination requires *syn* coplanar stereochemistry.

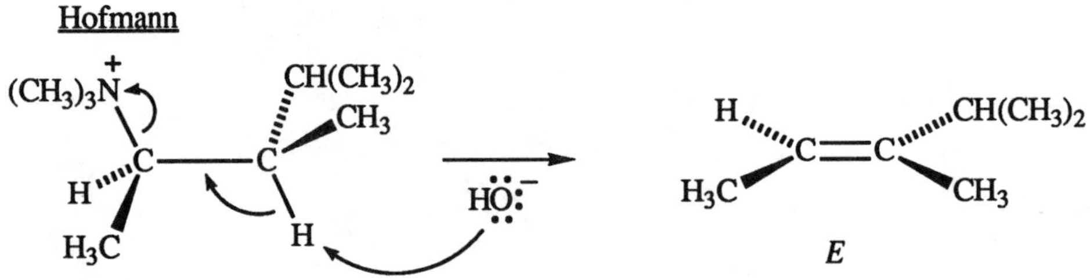

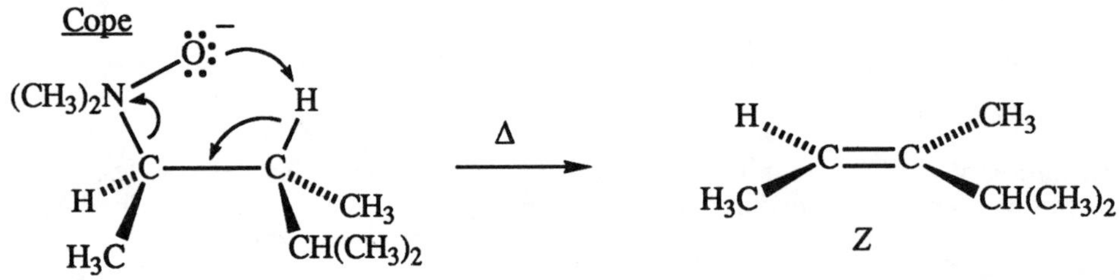

19-25

(a)

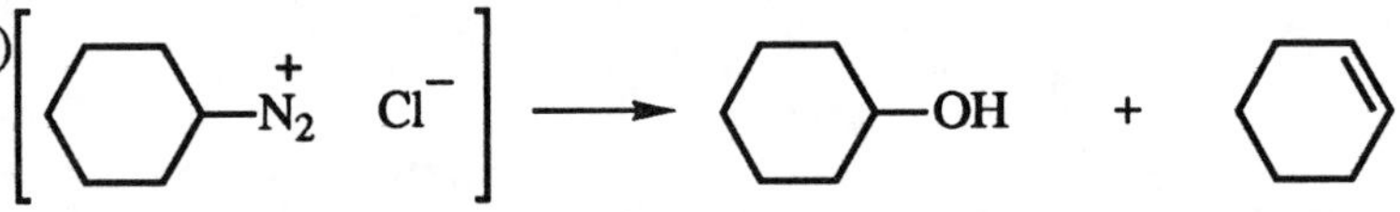

Aliphatic diazonium ions are very unstable, rapidly decomposing to carbocations.

(b)

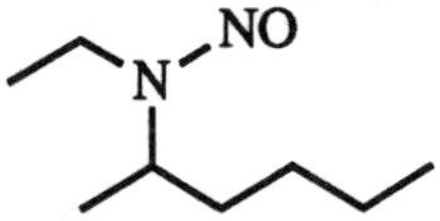

(c)

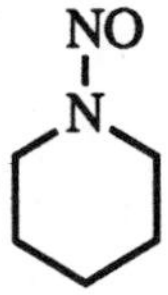

(d)

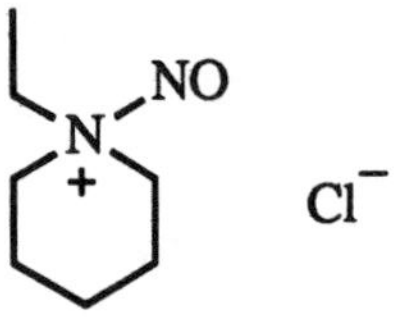

(e)

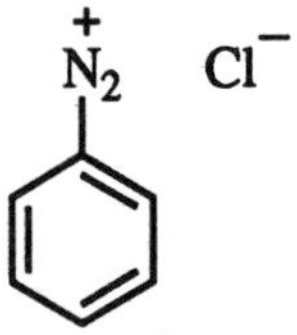

Aryl diazonium ions are relatively stable if kept cold.

19-26 The diazonium ion can do aromatic substitution like any other electrophile.

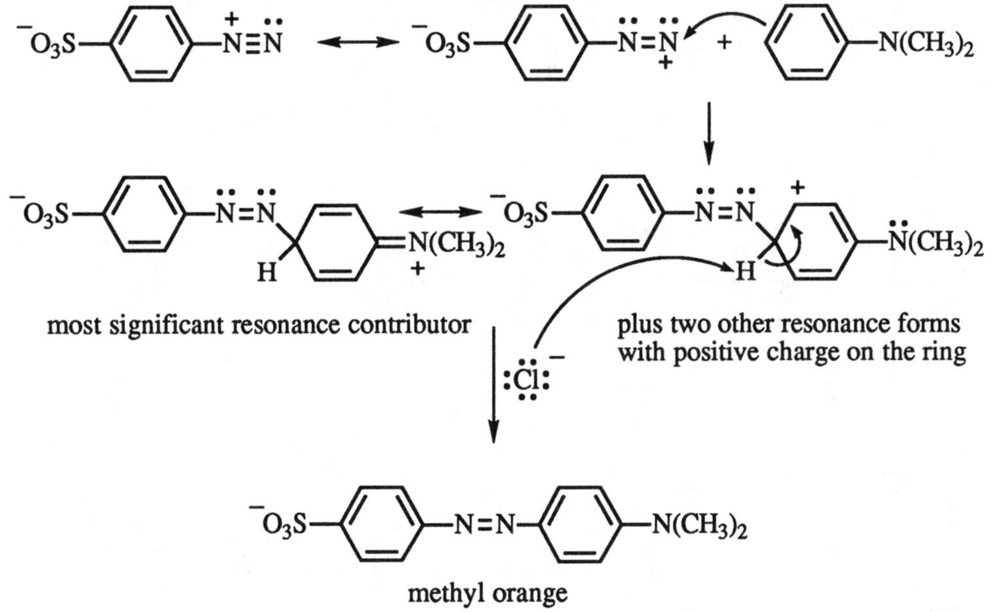

most significant resonance contributor
plus two other resonance forms with positive charge on the ring
methyl orange

19-27

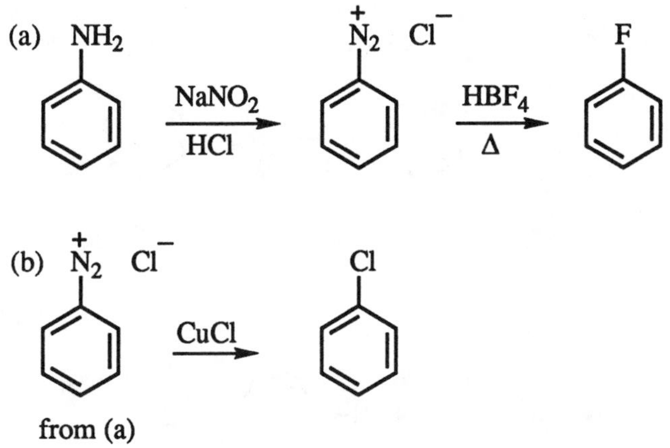

(a)
NaNO2
HCl
HBF4
Δ
(b)
CuCl
from (a)

19-27 continued

(c)

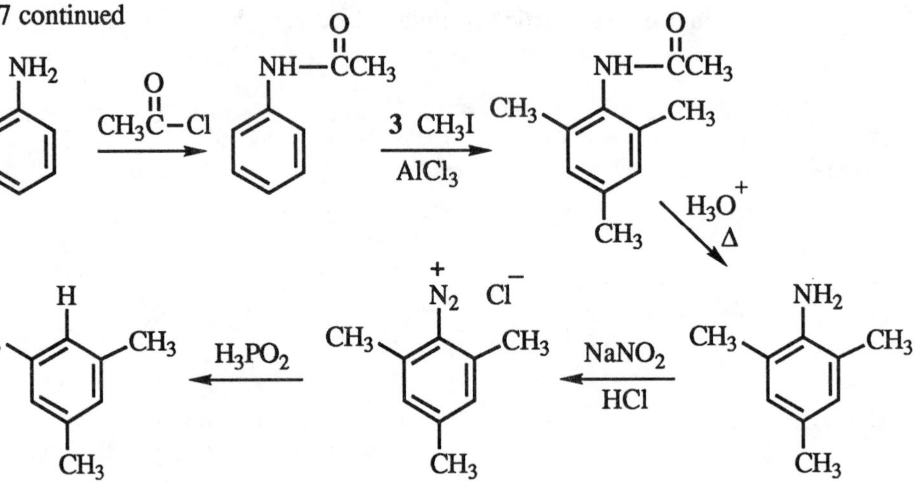

(d)

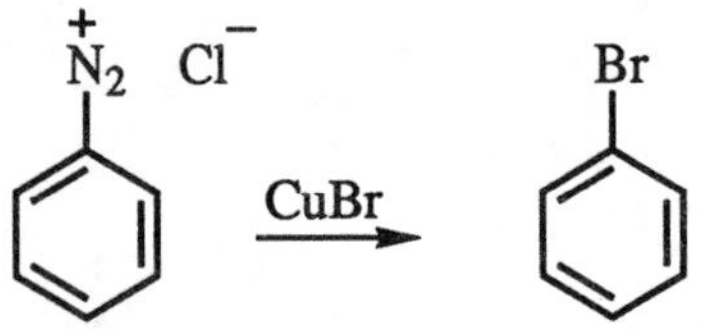

from (a)

(e)

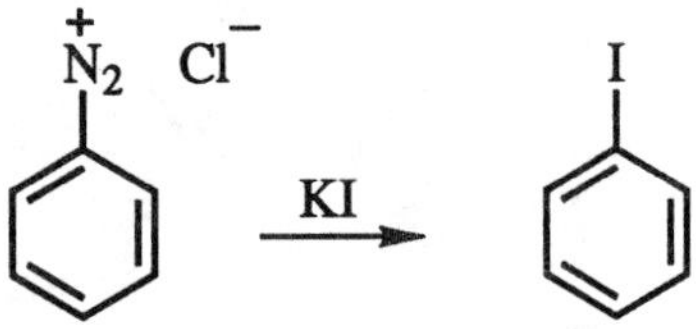

(f)

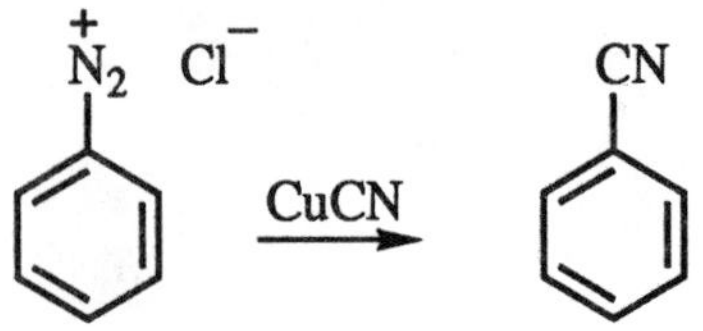

(g)

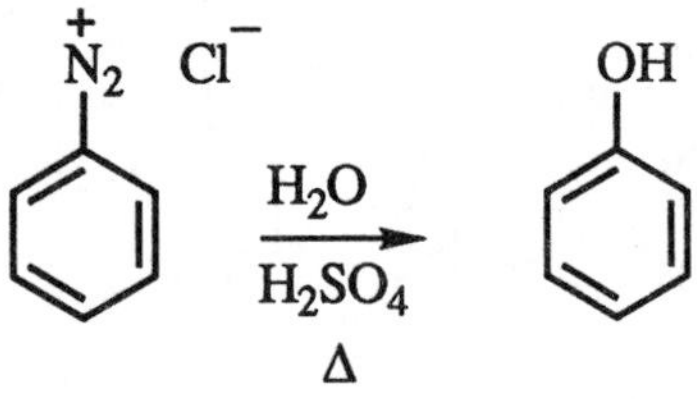

(h)

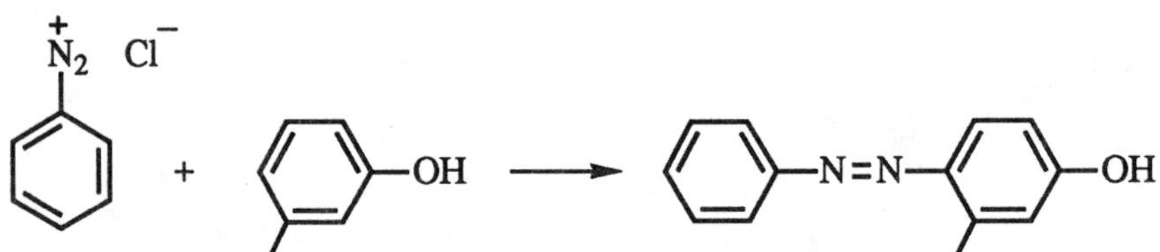

19-28 General guidelines for choice of reagent for reductive amination: use $LiAlH_4$ when the imine or oxime is isolated. Use $NaBH_3CN$ in solution when the imine or iminium ion is not isolated. Alternatively, catalytic hydrogenation works in most cases.

(a)

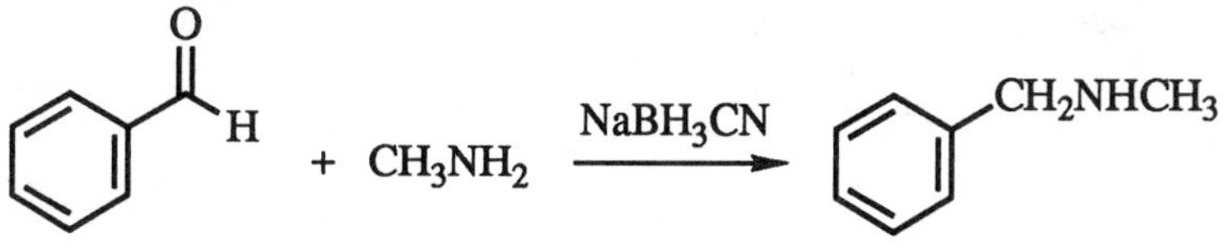
O
H
+ CH3NH2
NaBH3CN
CH2NHCH3

(b)

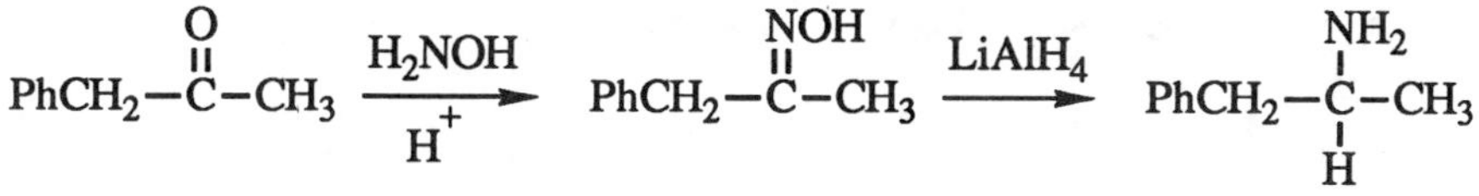
O
PhCH2—C—CH3
H2NOH
H+
NOH
PhCH2—C—CH3
LiAlH4
NH2
PhCH2—C—CH3
H

(c)

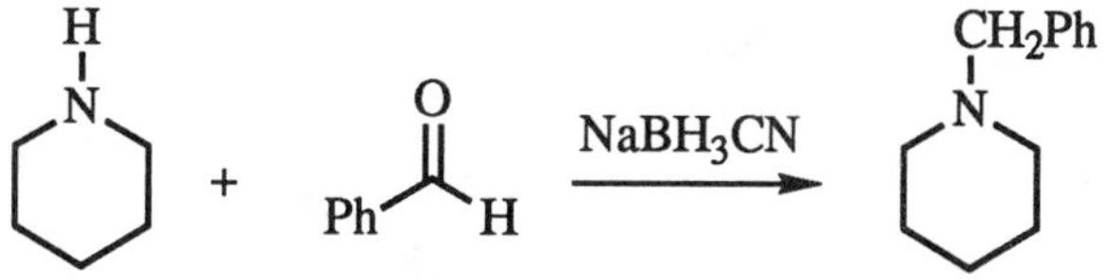
H
N
+
O
Ph
H
NaBH3CN
CH2Ph
N

(d)

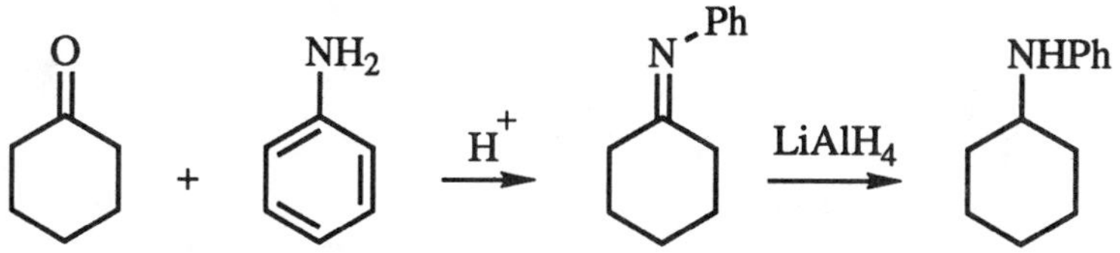
O
NH2
+
H+
N
Ph
LiAlH4
NHPh

(e)

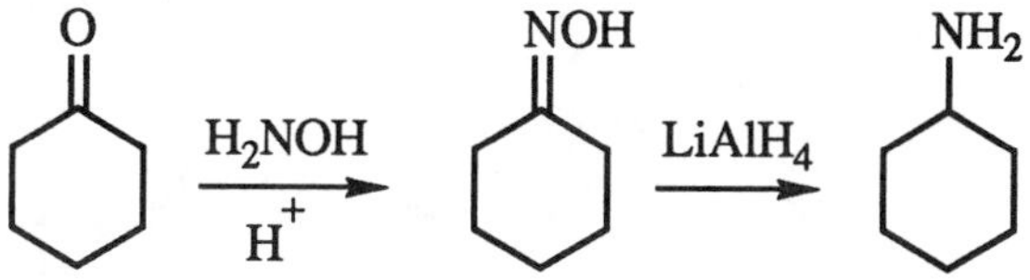
O
H2NOH
H+
NOH
LiAlH4
NH2

(f)

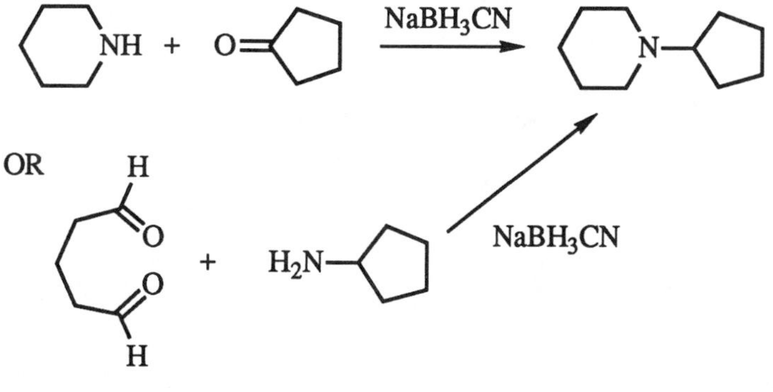
NH + O
NaBH3CN
N
OR
H
O
O
H
+
H2N
NaBH3CN

19-29 To reduce nitroaromatics, the reducing reagents (H_2 plus a metal catalyst, or a metal plus HCl) can be used virtually interchangeably.

(a)

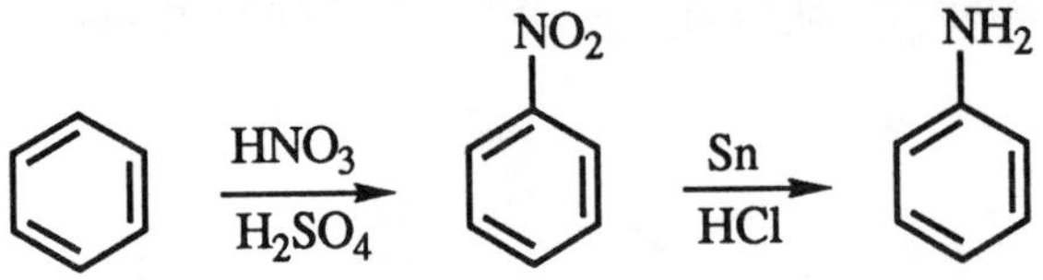

(b)

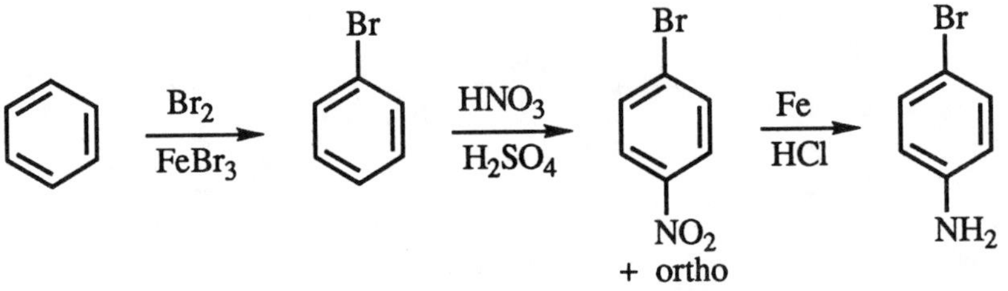

(c)

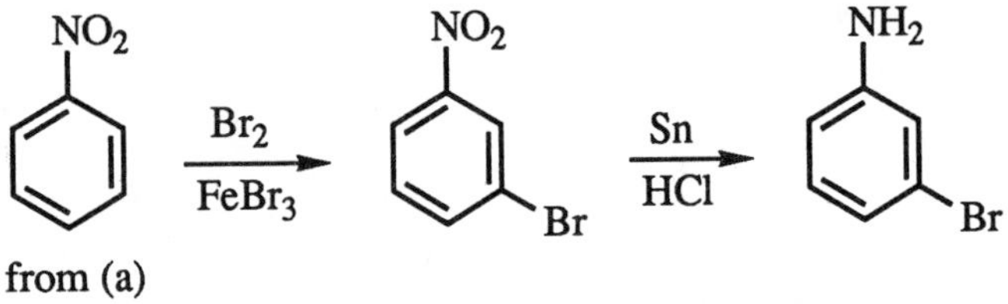

(d)

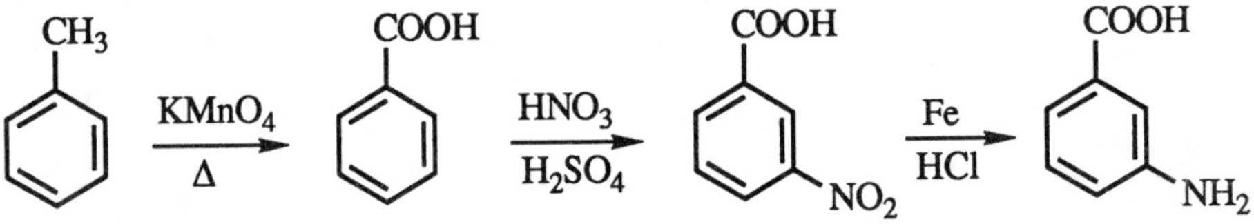

19-30

(a)

(b)

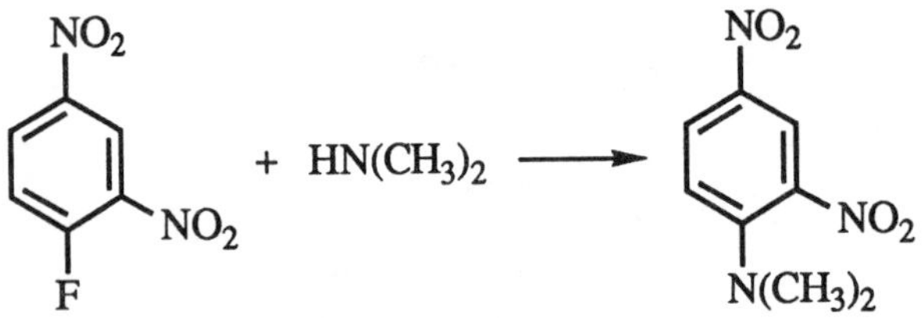

(c)

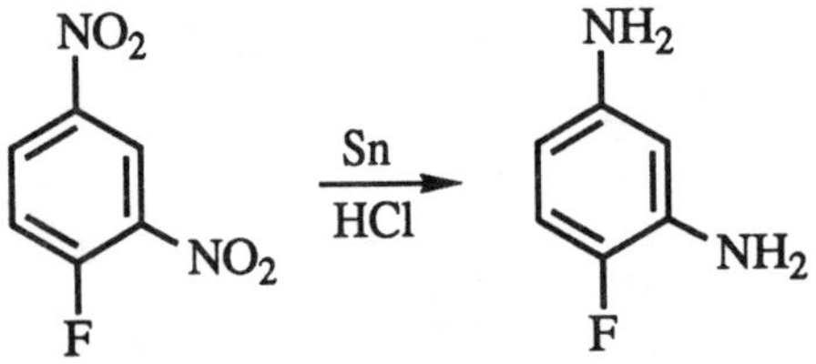

19-31

(a)

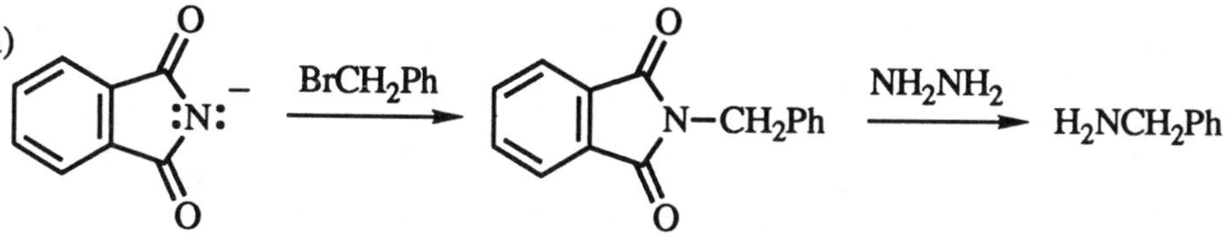

:N:⁻
BrCH2Ph
N−CH2Ph
NH2NH2
H2NCH2Ph

(b)

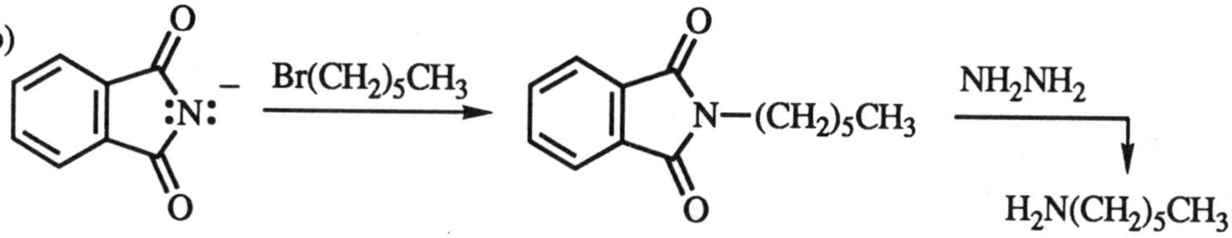

:N:⁻
Br(CH2)5CH3
N−(CH2)5CH3
NH2NH2
H2N(CH2)5CH3

(c)

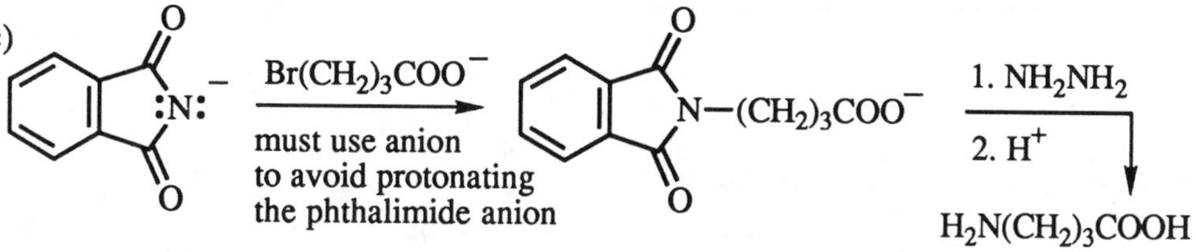

:N:⁻
Br(CH2)3COO⁻
must use anion
to avoid protonating
the phthalimide anion
N−(CH2)3COO⁻
1. NH2NH2
2. H+
H2N(CH2)3COOH

19-32 To make aniline by the Gabriel synthesis, the phthalimide anion would have to do nucleophilic substitution on benzene. Benzene is electron rich and is attacked by electrophiles; nucleophiles cannot attack a benzene without substituents.

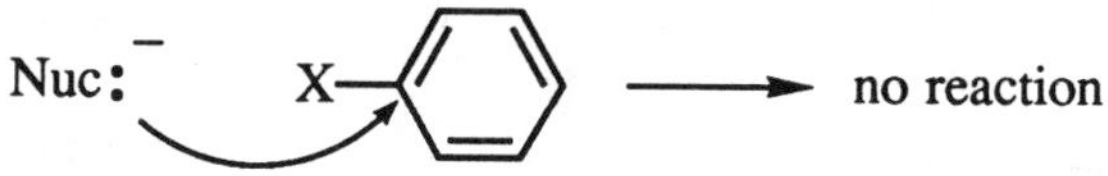

Nuc:⁻
X
no reaction

19-33

(a) $PhCH_2Br + NaCN \longrightarrow PhCH_2CN \xrightarrow[Pt]{H_2} PhCH_2CH_2NH_2$

(b)

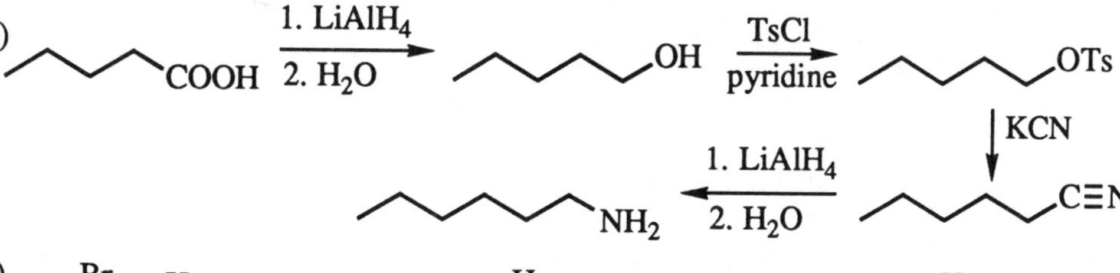

COOH
1. LiAlH4
2. H2O
OH
TsCl
pyridine
OTs
KCN
C≡N
1. LiAlH4
2. H2O
NH2

(c)

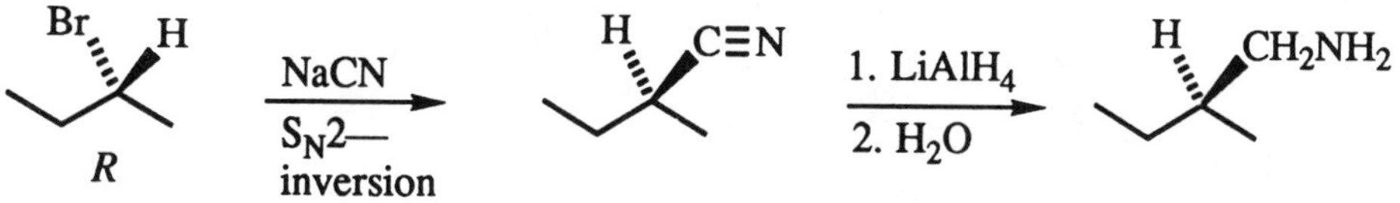

Br
H
R
NaCN
SN2—
inversion
H
C≡N
1. LiAlH4
2. H2O
H
CH2NH2

(d)

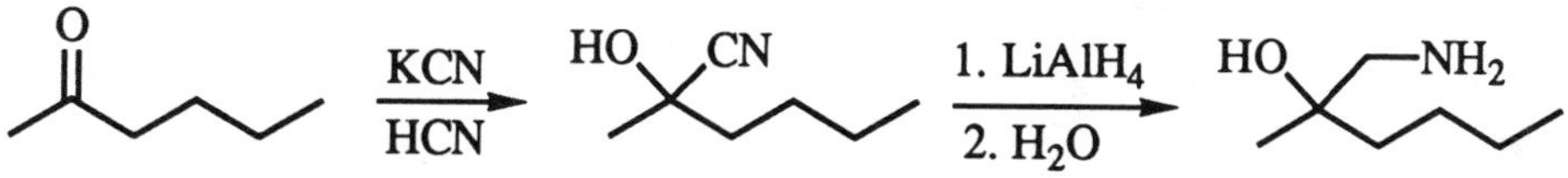

O
KCN
HCN
HO
CN
1. LiAlH4
2. H2O
HO
NH2

19-34

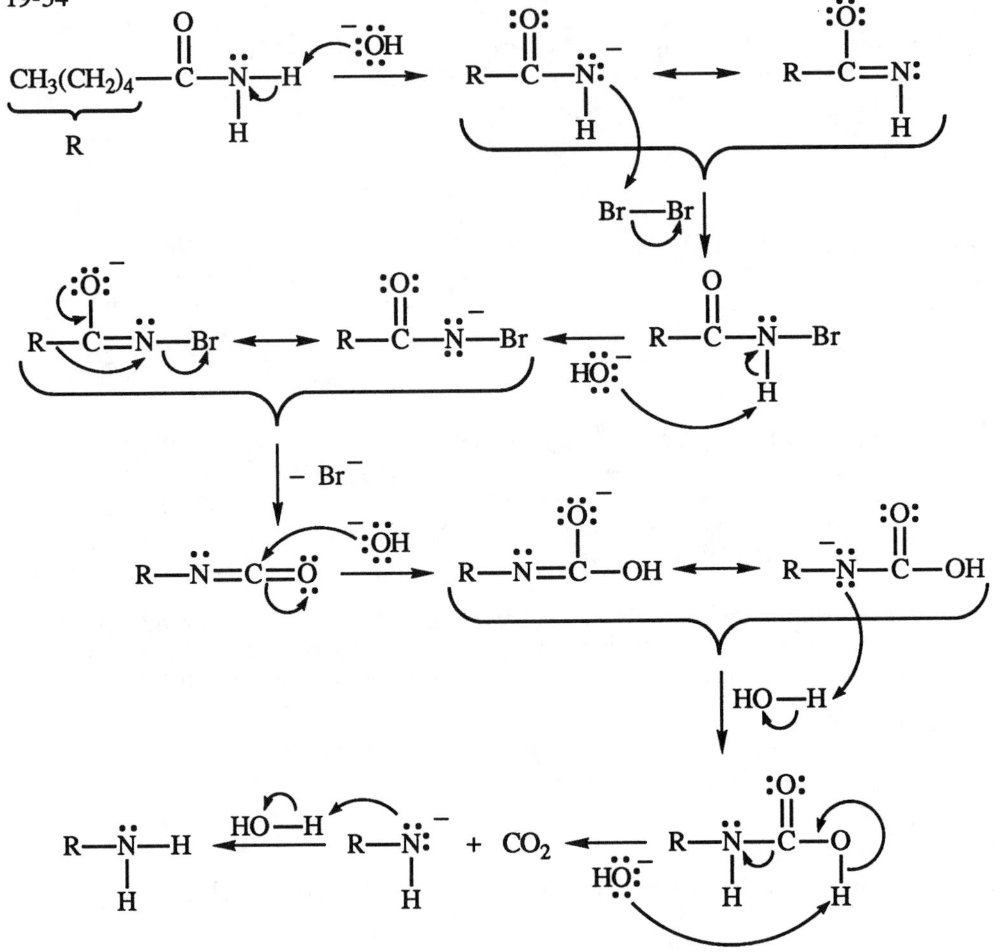

19-35

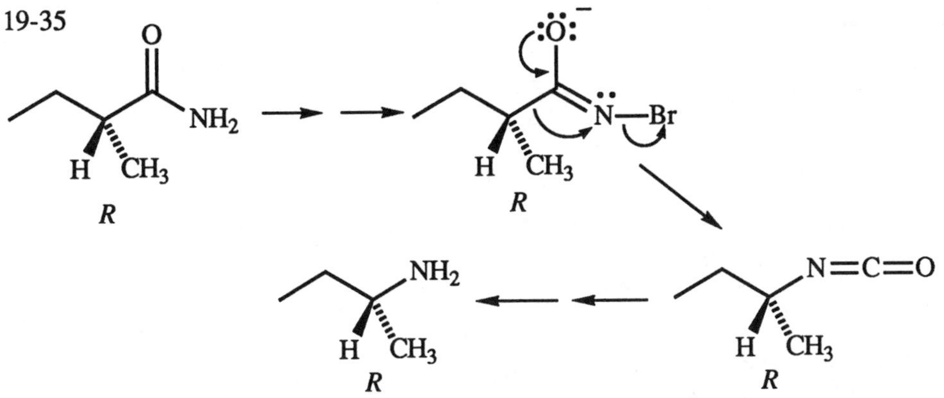

Stereochemistry at a chiral carbon is lost if the carbon goes through a planar intermediate, *e. g*., either a carbocation or a free radical. Configuration at a chiral carbon can be inverted during substitution by a nucleophile from the side opposite the leaving group. However, when the carbon retains all four pairs of electrons, as in this Hofmann rearrangement, it retains its configuration.

19-36 Refer to the Glossary and the text for definitions and examples.

19-37

(a) primary amine; 2,2-dimethyl-1-propanamine, or neopentylamine
(b) primary amine; 2-propanamine, or isopropylamine
(c) tertiary heterocyclic amine and a nitro group; 3-nitropyridine
(d) secondary amine; *N*-methylethanamine, or ethylmethylamine
(e) tertiary aromatic amine; *N*-ethyl-*N*-methylaniline
(f) tertiary heterocyclic ammonium ion; pyridinium chloride
(g) quaternary heterocyclic ammonium ion; *N*,*N*-dimethylpiperidinium iodide
(h) tertiary aromatic amine oxide; *N*-ethyl-*N*-methylaniline oxide
(i) secondary amine; *N*,4-diethyl-3-hexanamine

19-38 In order of increasing basicity:

(a)  Ph—N(H)—Ph < $PhNH_2$ < 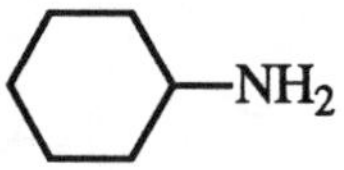

(b) 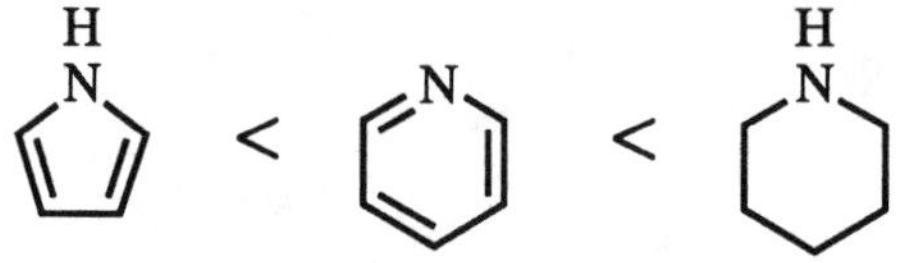

(c) 

19-39

(a) not resolvable: planar

(b) resolvable: chiral carbon

(c) not resolvable: symmetric

(d) resolvable: nitrogen inversion is very slow

(e) not resolvable: symmetric

(f) resolvable: chiral nitrogen, unable to invert

19-40

(a) 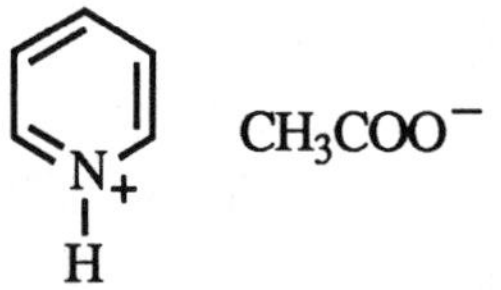

(b) no reaction

(c) 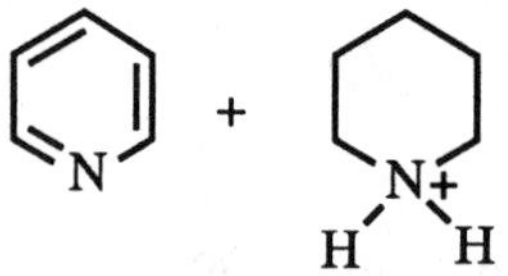

(d) 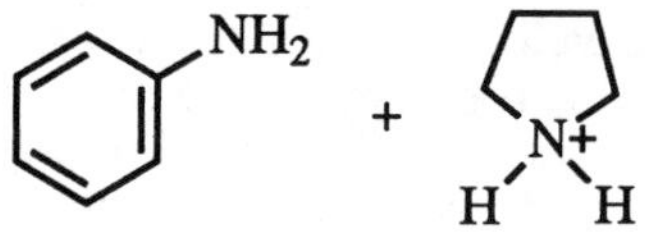

19-41

(a) $CH_3(CH_2)_6NH_2$

(b) $PhCH_2CH_2NH_2$

(c) $H_2N(CH_2)_4NH_2$

19-42

(a) $PhCH_2CH_2CH_2NH_2$

(b) 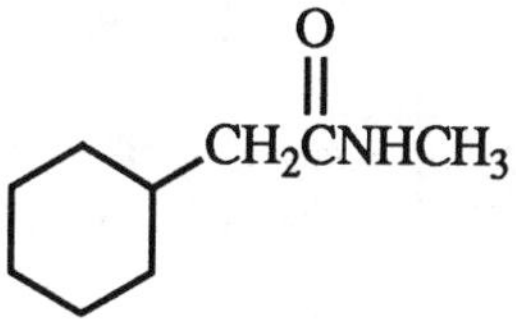

(c) 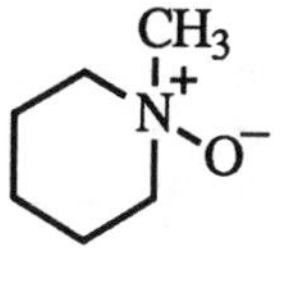

(d) 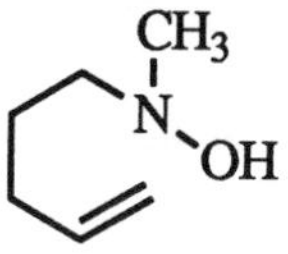

(e) 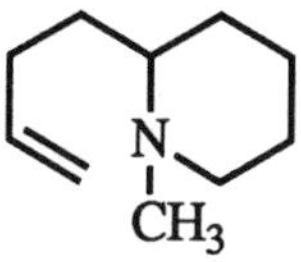

(f) 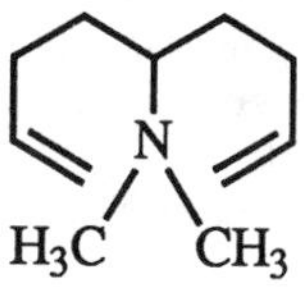

(g) 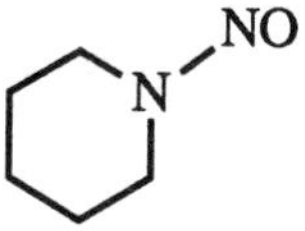

(h) 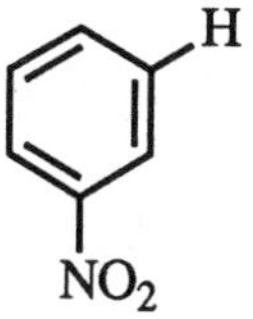

(i) 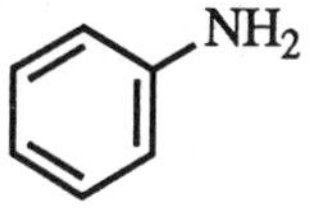

(j) 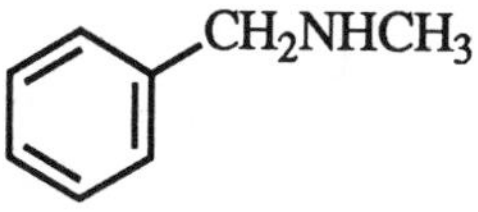

(k) 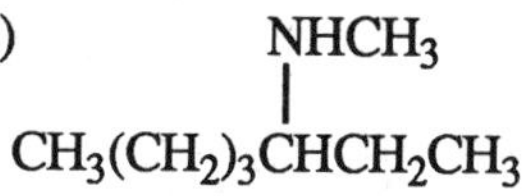

(l) 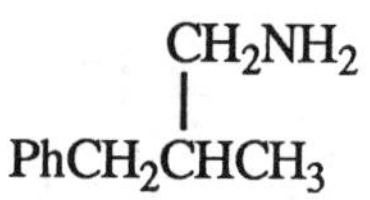

(m) 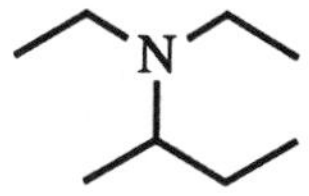

(n) 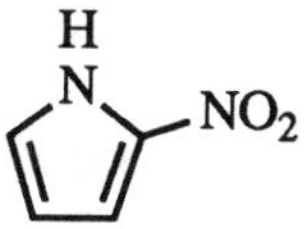

(o) 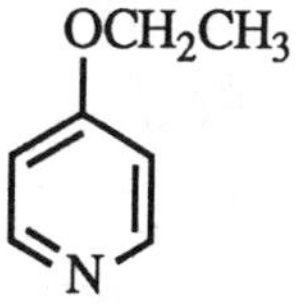

19-43

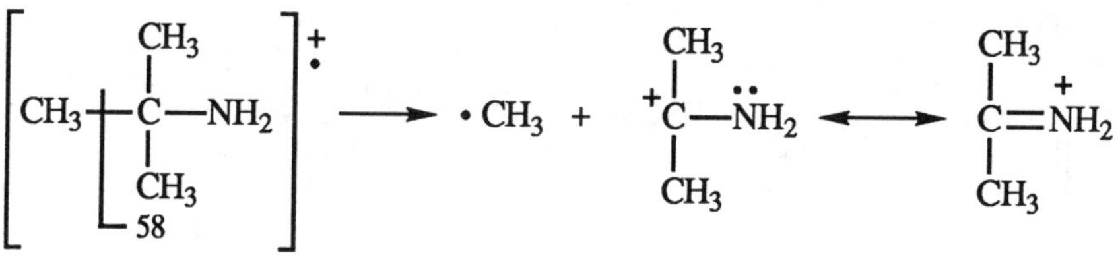

m/z 73

This fragmentation is favorable because the iminium ion produced is stabilized by resonance. Also, there are three possible cleavages that give the same ion.

19-44 1-Octanamine is basic and octanamide is not basic. Dissolve the two compounds in an organic solvent, and add dilute aqueous HCl; this protonates the amine, making it water soluble. The protonated amine is extracted into the aqueous solution, leaving the amide in the organic solvent. Separate layers.

To regenerate the amine, add base to the aqueous solution; the amine is deprotonated, making it water insoluble. Extract the amine/water mixture with organic solvent to remove the amine from the water. Evaporate the organic solvent. The pure amine will remain.

19-45

(a) 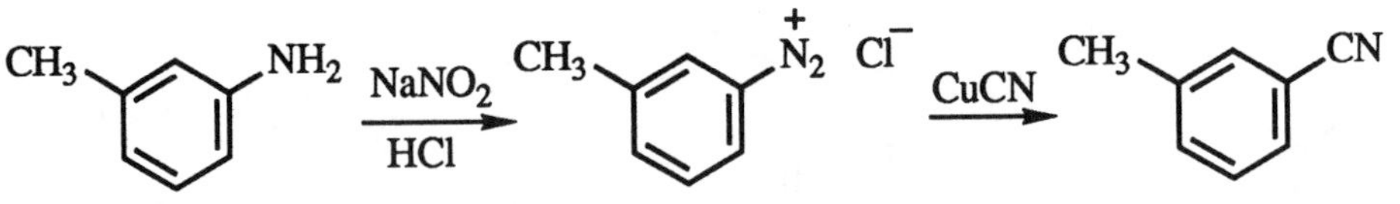
CH3
NH2
NaNO2
HCl
CH3
N2 Cl
CuCN
CH3
CN

(b) 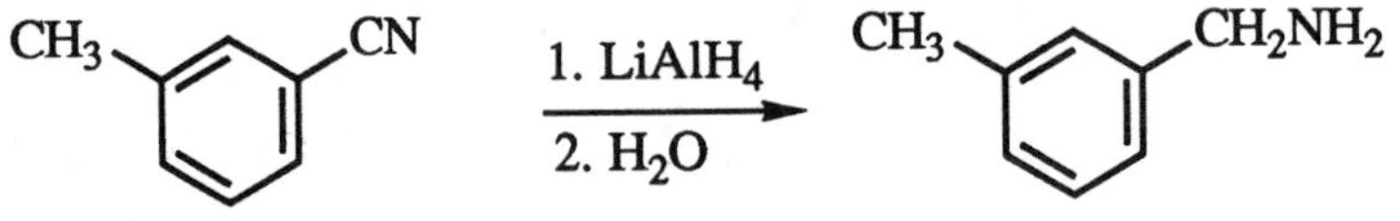
CH3
CN
1. LiAlH4
2. H2O
CH3
CH2NH2

from (a)

(c) 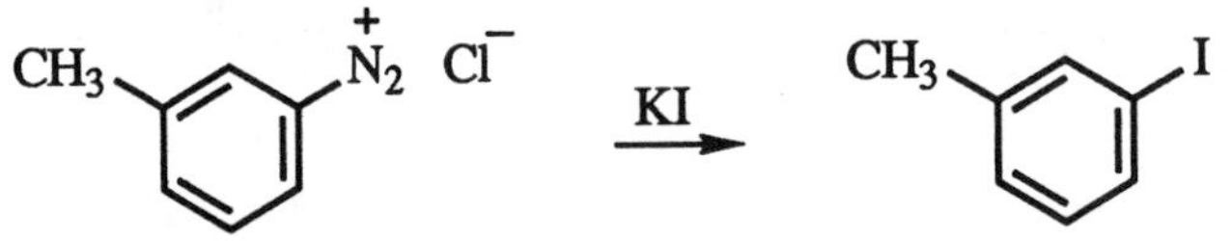
CH3
N2 Cl
KI
CH3
I

from (a)

(d) 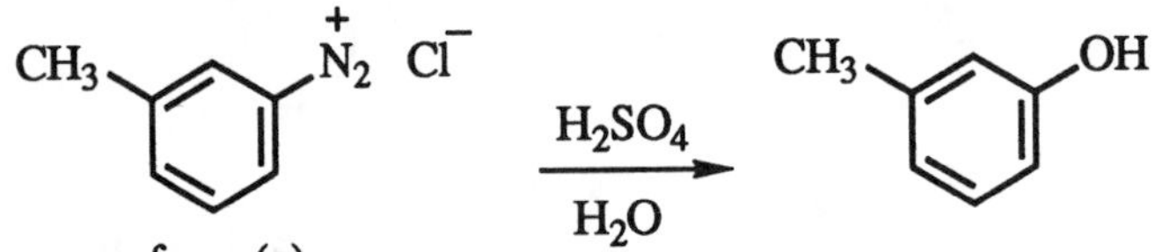
CH3
N2 Cl
H2SO4
H2O
CH3
OH

from (a)

(e) 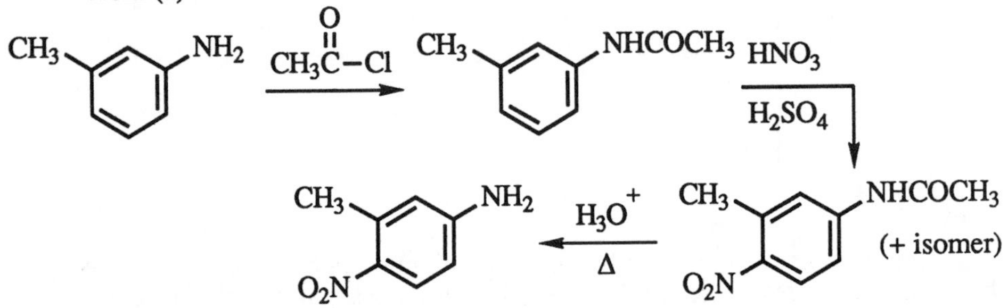
CH3
NH2
CH3C−Cl
CH3
NHCOCH3
HNO3
H2SO4
CH3
NHCOCH3
(+ isomer)
O2N
H3O+
Δ
CH3
NH2
O2N

(f) 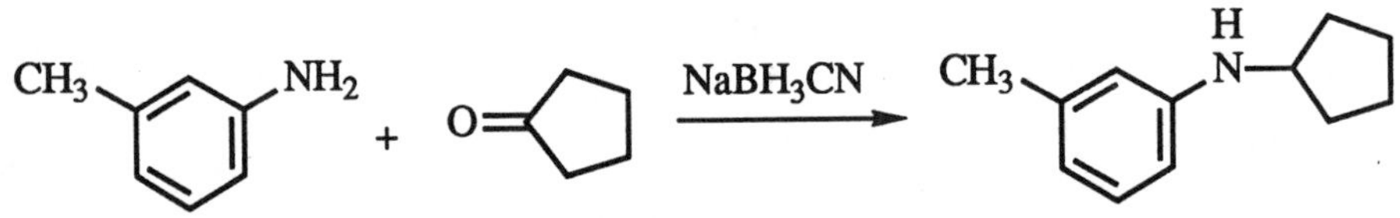
CH3
NH2
+
O
NaBH3CN
CH3
H
N

19-46

(a) 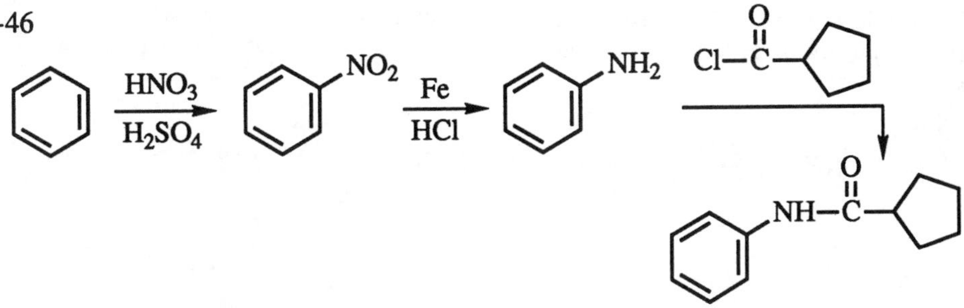
HNO3
H2SO4
NO2
Fe
HCl
NH2
Cl—C—
O
NH—C—

(b) 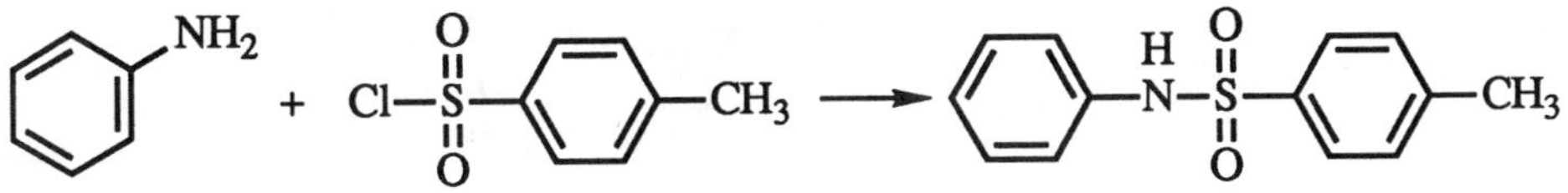
NH2
Cl—S—
O
CH3
H
N—S—

(c) 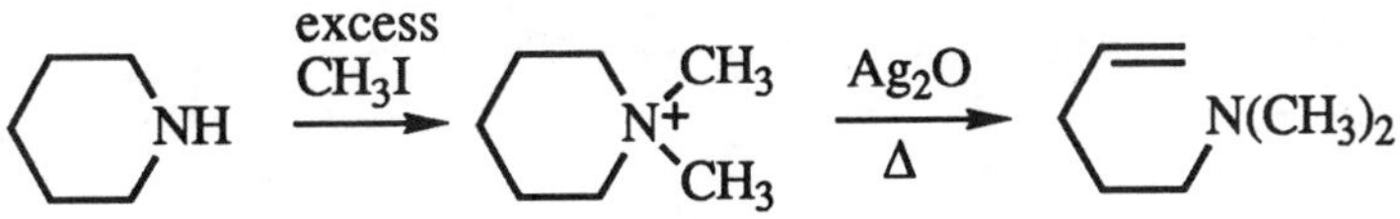
NH
excess
CH3I
N+
CH3
Ag2O
Δ
N(CH3)2

(d) 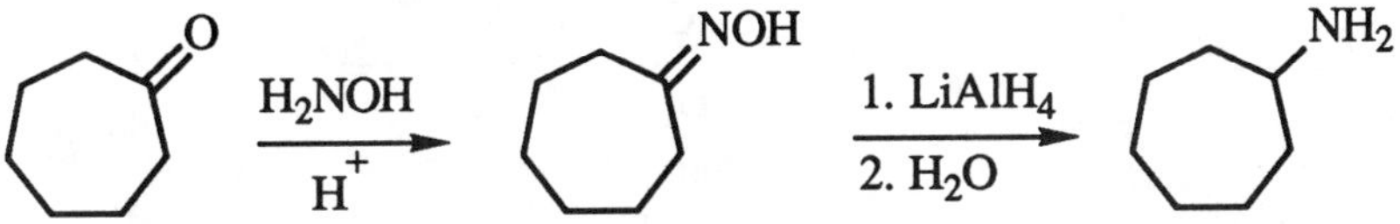
O
H2NOH
H+
NOH
1. LiAlH4
2. H2O
NH2

(e) 
N—CH3
MCPBA
O−
N+
CH3

(f) 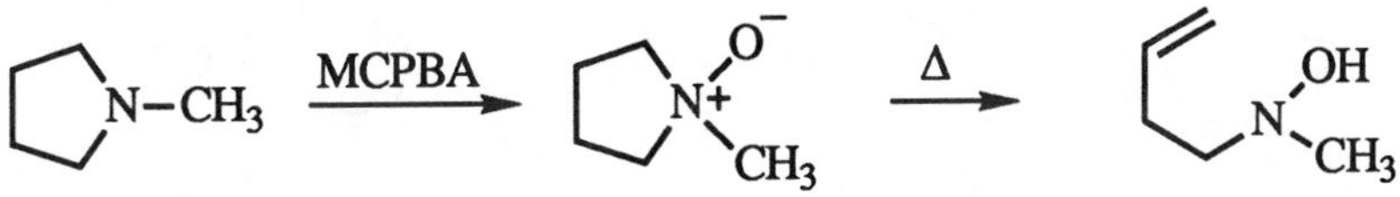
N—CH3
MCPBA
O−
N+
CH3
Δ
OH
N

(g) 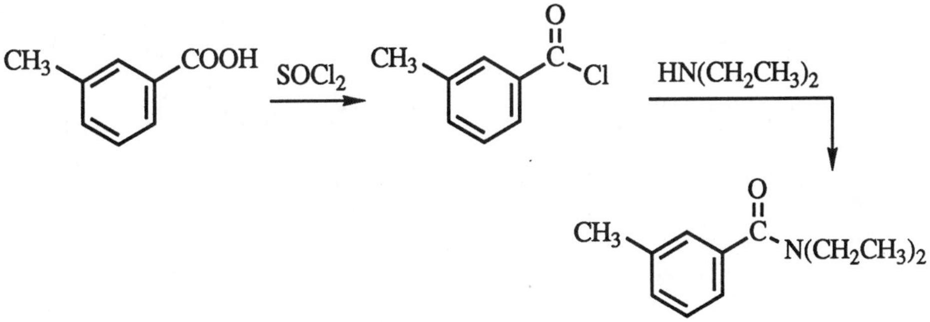
CH3
COOH
SOCl2
C
Cl
O
HN(CH2CH3)2
N(CH2CH3)2

19-47 The problem restricts the starting materials to six carbons or fewer. Always choose starting materials with as many of the necessary functional groups as possible.

(a)

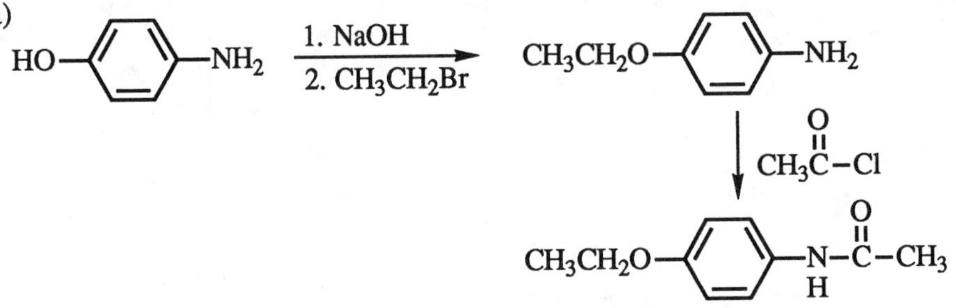

HO
NH2
1. NaOH
2. CH3CH2Br
CH3CH2O
NH2
O
CH3C–Cl
CH3CH2O
N–C–CH3
H

(b)

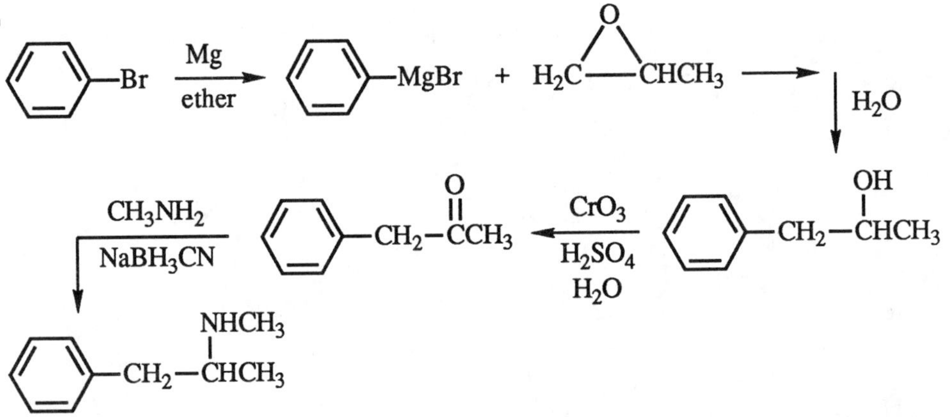

Br
Mg
ether
MgBr
+ H2C—CHCH3
O
H2O
OH
CH2–CHCH3
CrO3
H2SO4
H2O
CH2–CCH3
CH3NH2
NaBH3CN
NHCH3
CH2–CHCH3

(c)

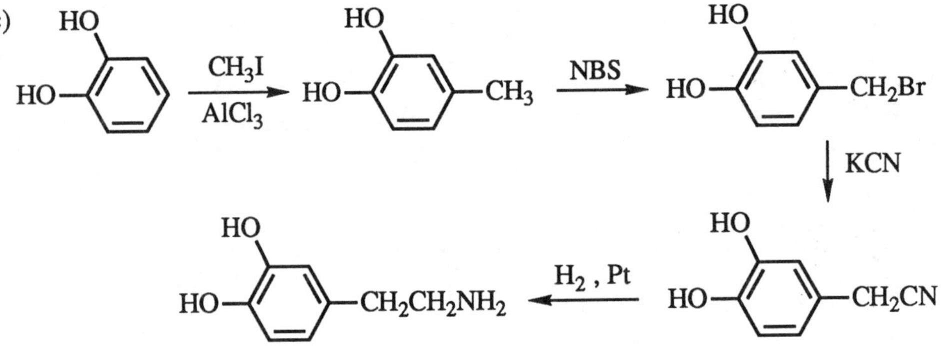

HO
HO
CH3I
AlCl3
CH3
NBS
CH2Br
KCN
CH2CN
H2 , Pt
CH2CH2NH2

19-48

(a)

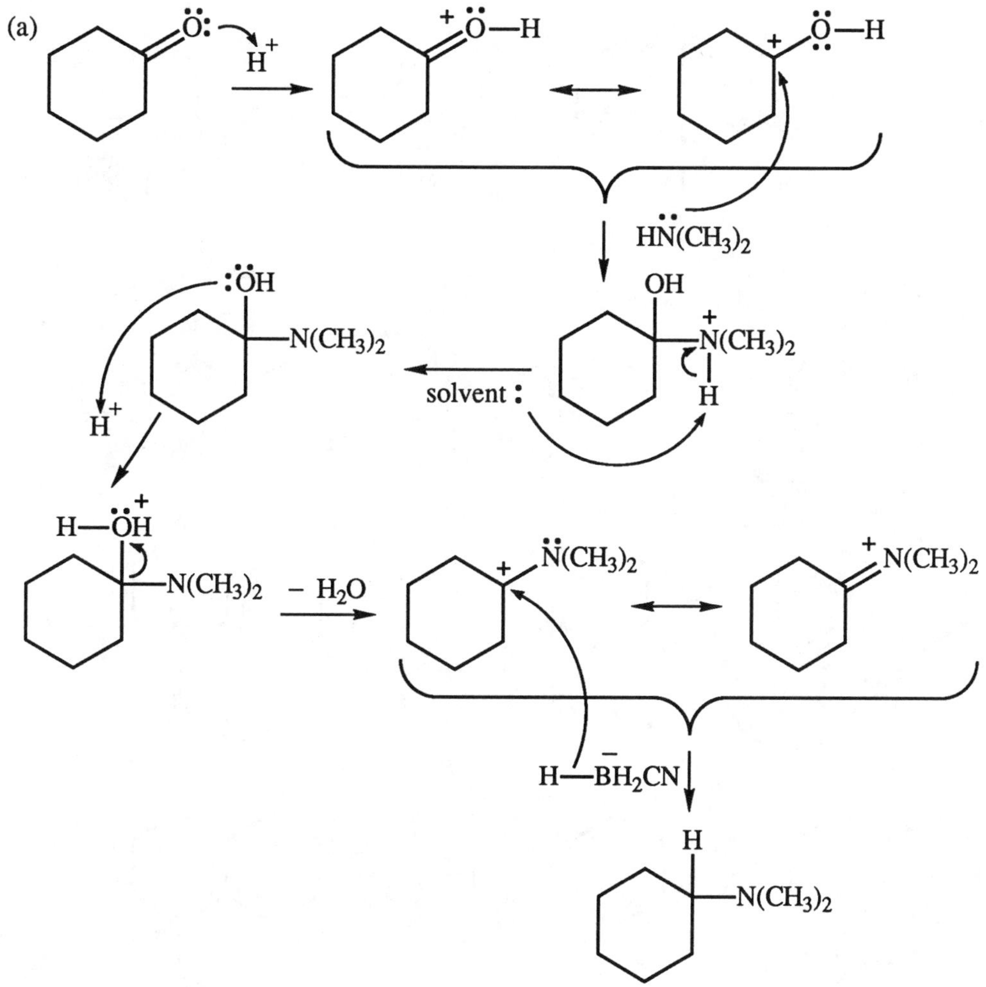

19-48 continued

(b)

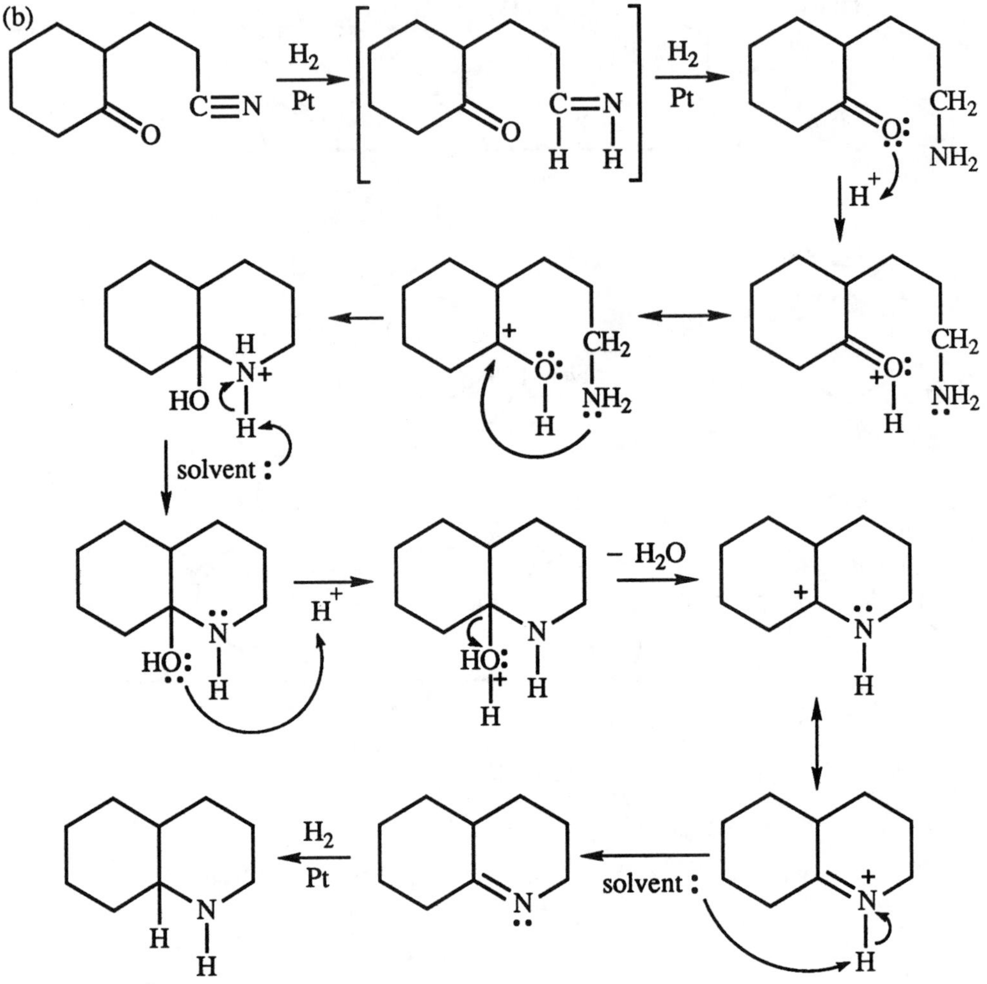

19-49

(a)

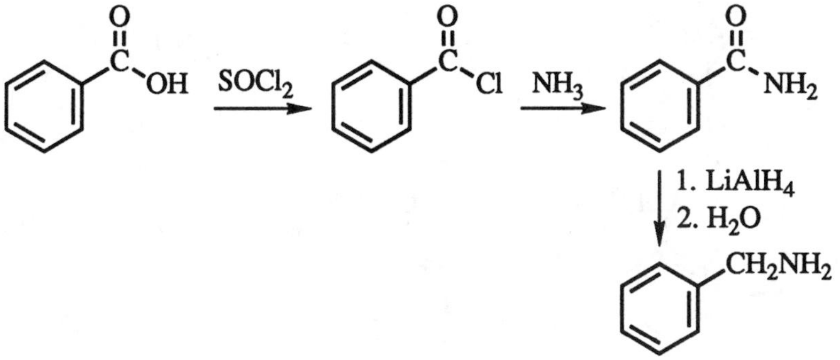

(b)

(c)

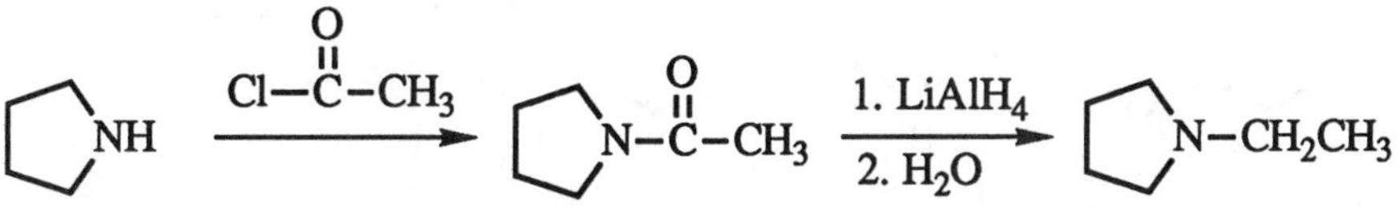

(d)

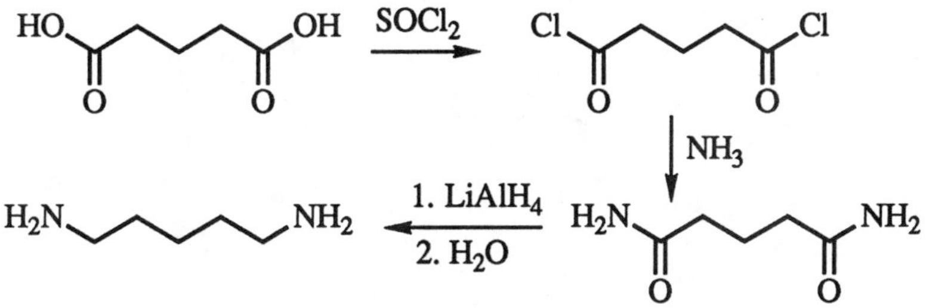

(e)

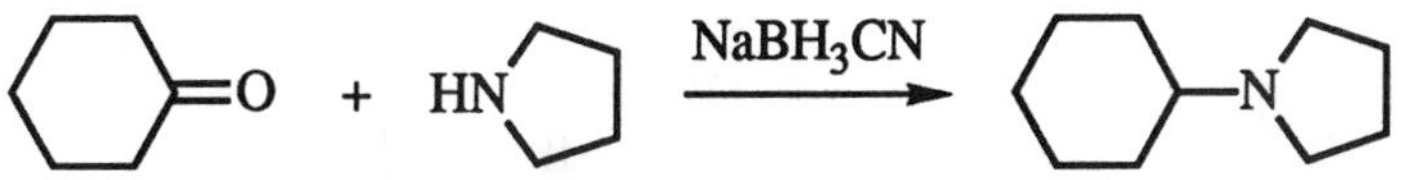

19-50

(a)

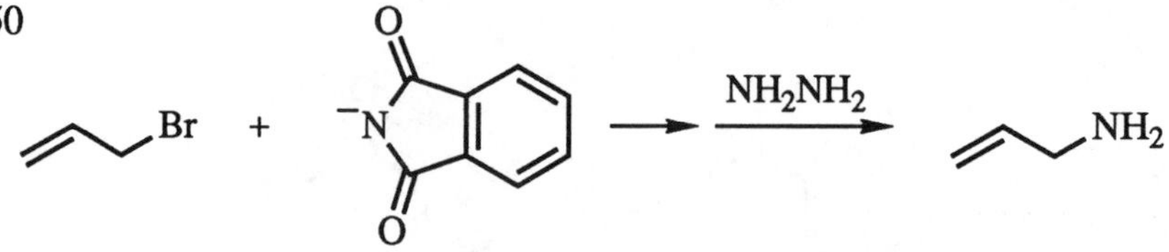
Br
+
O
N
O
NH2NH2
NH2

(b)

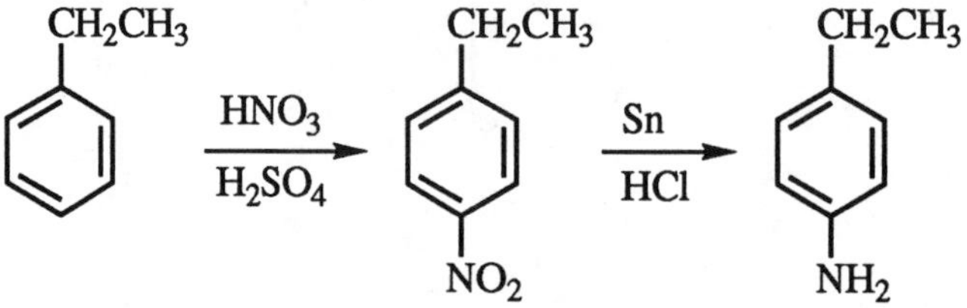
CH2CH3
HNO3
H2SO4
CH2CH3
NO2
Sn
HCl
CH2CH3
NH2

(c)

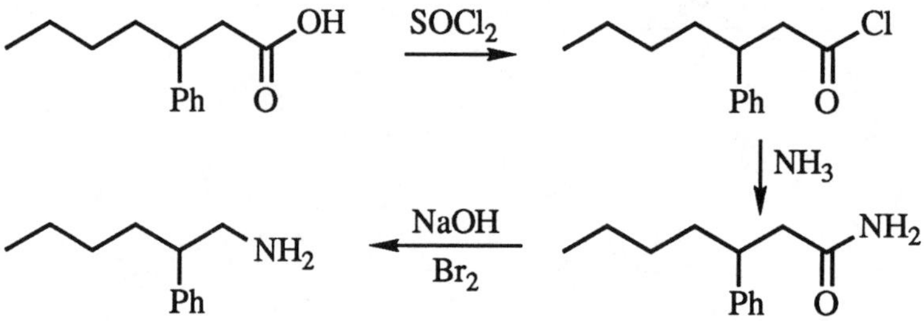
OH
Ph
O
SOCl2
Cl
Ph
O
NH3
NH2
Ph
O
NaOH
Br2
NH2
Ph

(d)

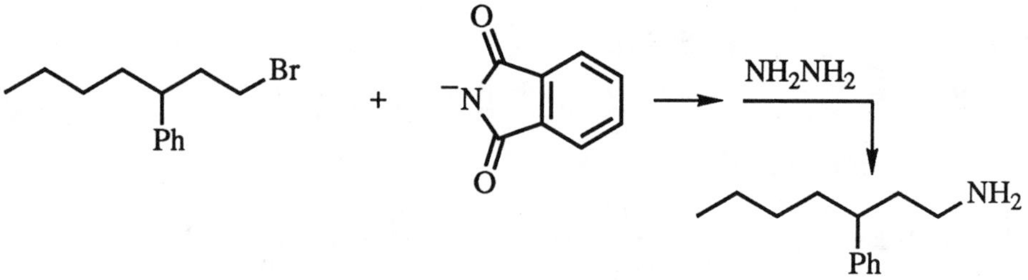
Br
Ph
+
O
N
O
NH2NH2
NH2
Ph

(e)

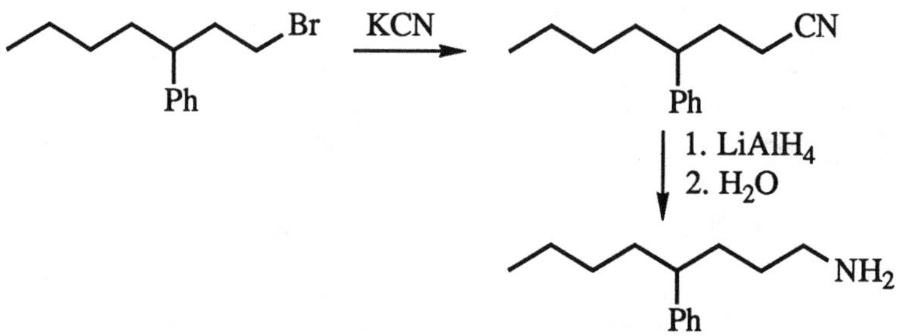
Br
Ph
KCN
CN
Ph
1. LiAlH4
2. H2O
NH2
Ph

19-51

(a) When guanidine is protonated, the cation is greatly stabilized by resonance, distributing the positive charge over all atoms (except H):

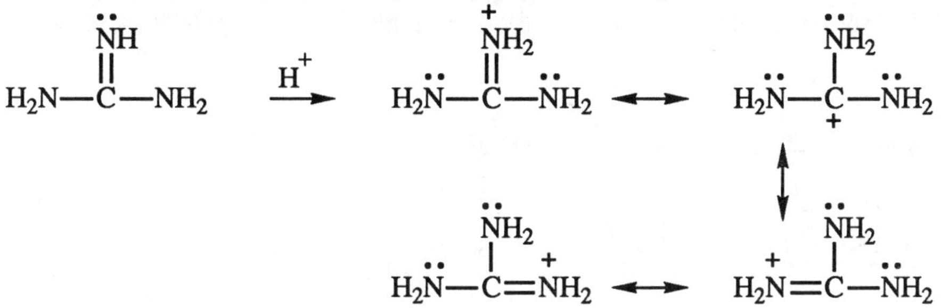

(b) The unprotonated molecule has a resonance form shown below that the protonated molecule cannot have. Therefore, the unprotonated form is stabilized relative to the protonated form. This greater stabilization of the unprotonated form is reflected in weaker basicity.

19-51 continued

(c) Anilines are weaker bases than aliphatic amines because the electron pair on the nitrogen is shared with the ring, stabilizing the system. There is a steric requirement, however: the p orbital on the N must be parallel with the p orbitals on the benzene ring in order for the electrons on N to be distributed into the π system of the ring.

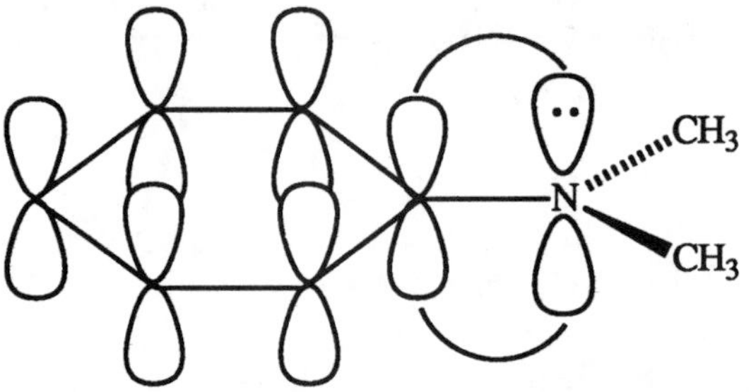

If the orbital on the nitrogen is forced out of this orientation (by substitution on C-2 and C-6, for example), the electrons are no longer shared with the ring. The nitrogen is hybridized sp^3 (no longer any reason to be sp^2), and the electron pair is readily available for bonding $\Rightarrow$ increased basicity.

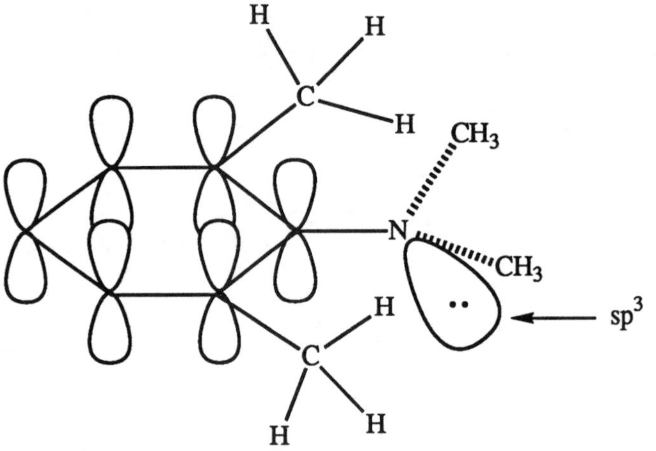

This compound is about as basic as a tertiary aliphatic amine, except that the aromatic ring substituent is electron-withdrawing *by induction*, decreasing the basicity slightly. (This phenomenon is called "steric inhibition of resonance". We will see more examples in future chapters.)

19-52

(a) 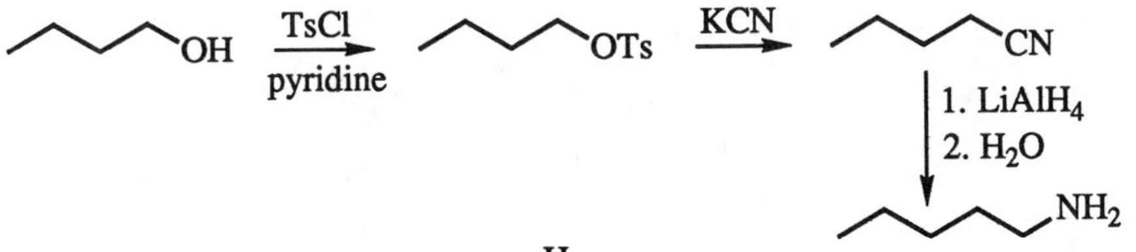

(b) 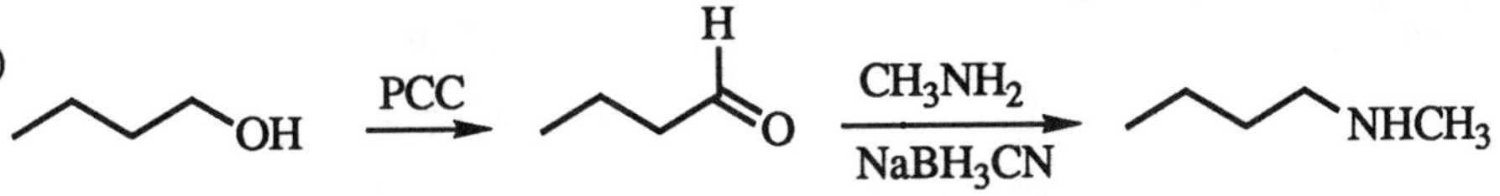

(c) 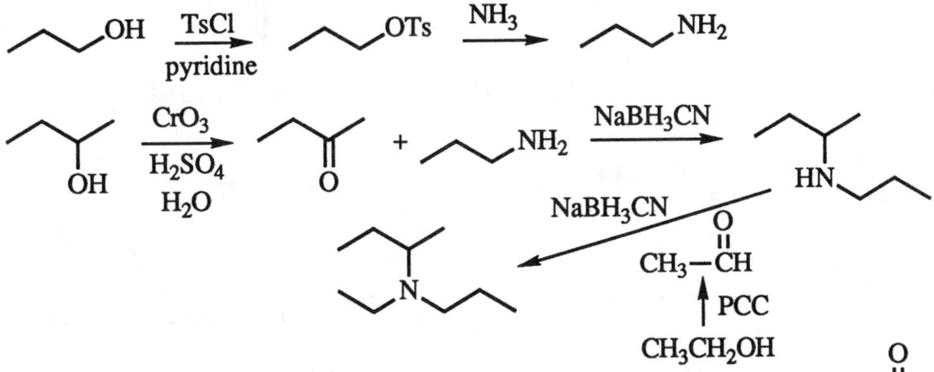

(d) 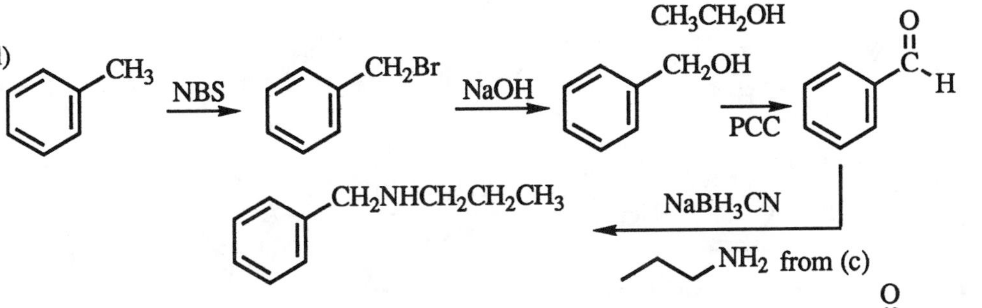

(e) 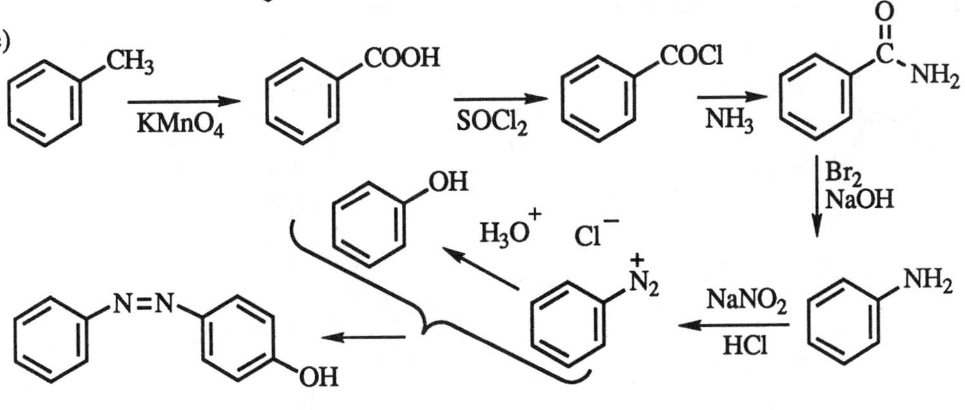

19-53

(a)

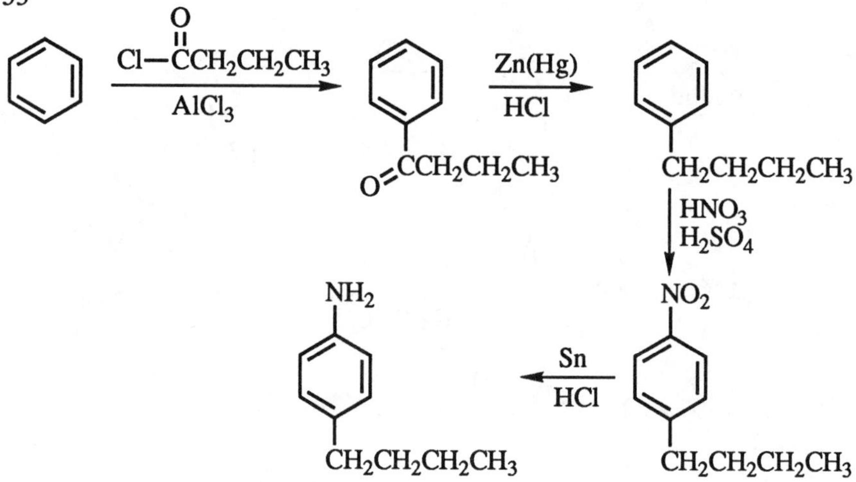

Cl—CCH2CH2CH3
AlCl3
Zn(Hg)
HCl
CCH2CH2CH3
CH2CH2CH2CH3
HNO3
H2SO4
NO2
Sn
HCl
NH2
CH2CH2CH2CH3

(b)

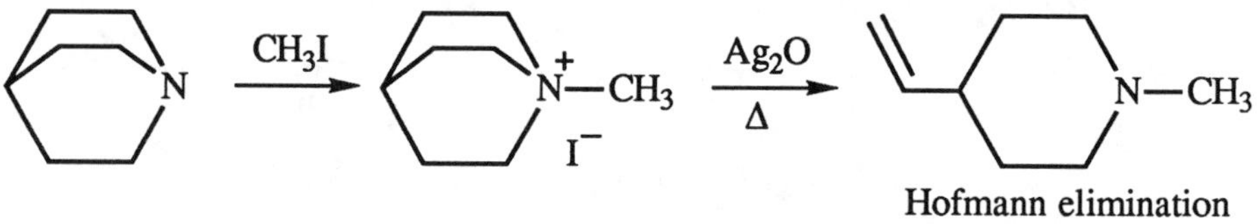

N
CH3I
N—CH3
I−
Ag2O
Δ
N—CH3
Hofmann elimination

(c)

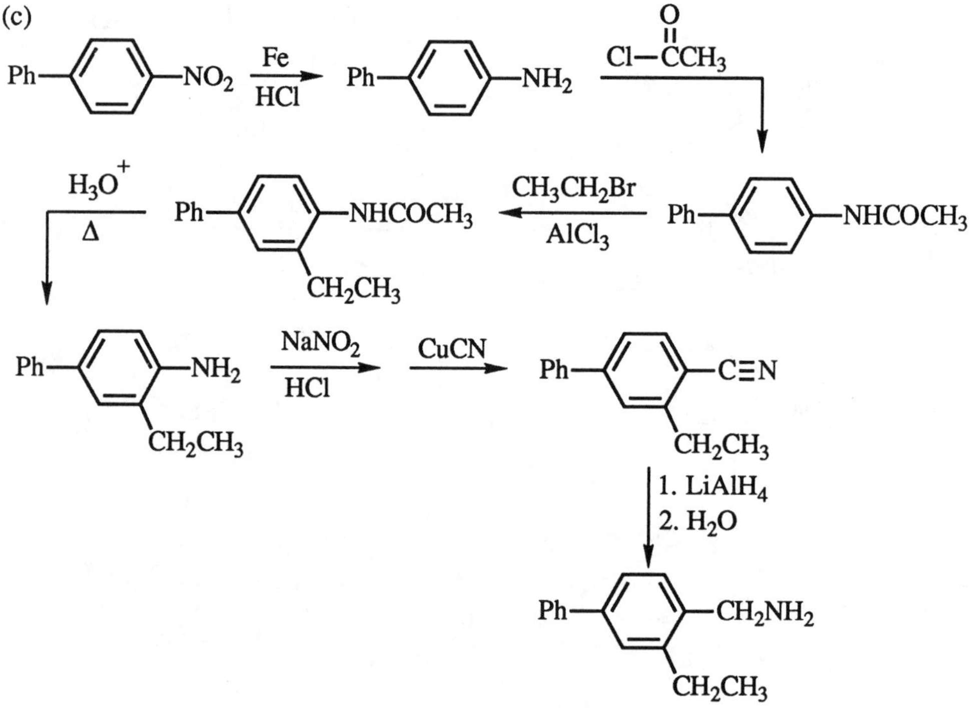

Ph
NO2
Fe
HCl
NH2
Cl—CCH3
NHCOCH3
CH3CH2Br
AlCl3
CH2CH3
H3O+
Δ
NaNO2
HCl
CuCN
C≡N
1. LiAlH4
2. H2O
CH2NH2

19-54

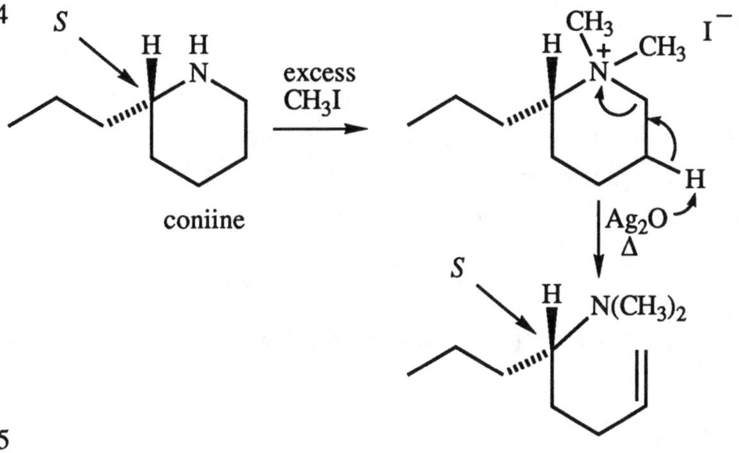

19-55

(a) <u>unknown **X**</u>

—fishy odor ⇒ amine

—molecular weight 101 ⇒ odd number of nitrogens

⇒ if one nitrogen and no oxygen, the remainder is C_6H_{15}

<u>Mass spectrum</u>:

—fragment at 86 = M − 15 = loss of methyl ⇒ the compound is likely to have this structural piece: CH_3 ┼ C—N (α-cleavage)

<u>IR spectrum</u>:

—no OH, no NH ⇒ must be a 3° amine

—no C=O or C=C or C≡N or NO_2

<u>NMR spectrum</u>:

—only a triplet and quartet, integration about 3 : 2 ⇒ ethyl group(s) only

assemble the evidence: $C_6H_{15}N$, 3° amine, only ethyl in the NMR:

$$CH_3CH_2—\underset{\displaystyle CH_2CH_3}{\underset{|}{N}}—CH_2CH_3$$

(b) React the triethylamine with HCl. The pure salt is solid and odorless.

$$(CH_3CH_2)_3N \;+\; HCl \longrightarrow \underset{\text{salt}}{(CH_3CH_2)_3\overset{+}{N}H \quad Cl^-}$$

(c) Washing her clothing in dilute acid like vinegar (dilute acetic acid) or dilute HCl would form a water-soluble salt as shown in (b). Normal washing will remove the water-soluble salt.

19-56

Compound A

Mass spectrum:

—molecular ion at 73 = odd mass = odd number of nitrogens;
if one nitrogen and no oxygen present ⇒ molecular formula $C_4H_{11}N$

—base peak at 44 is M - 29 ⇒ this fragment must be present:

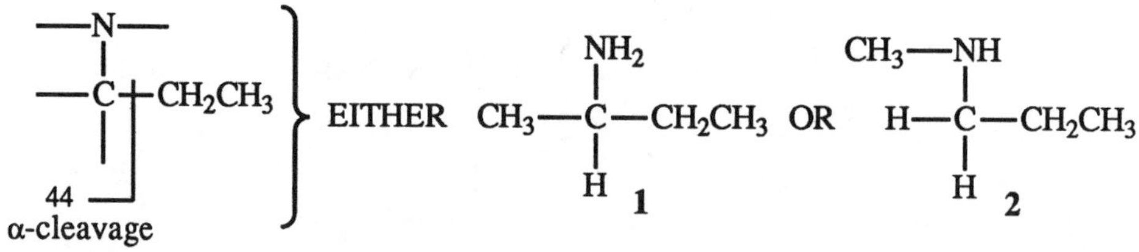

IR spectrum:

—two spikes around 3300 cm^{-1} indicate a 1° amine; no indication of oxygen

NMR spectrum:

—two exchangeable protons suggest NH_2 present

—1H multiplet at δ 2.7 means a CH—NH_2

The structure of **A** must be the same as **1** above: **A**

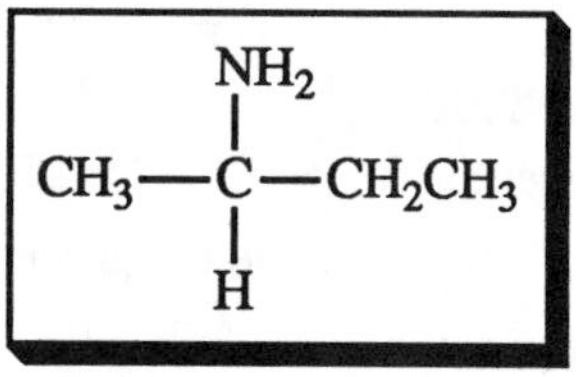

Compound B

an isomer of **A**, so its molecular formula must also be $C_4H_{11}N$

IR spectrum:

—only one spike at 3300 cm^{-1} ⇒ 2° amine

NMR spectrum:

—one exchangeable proton ⇒ NH

—two ethyls present

The structure of **B** must be: **B**

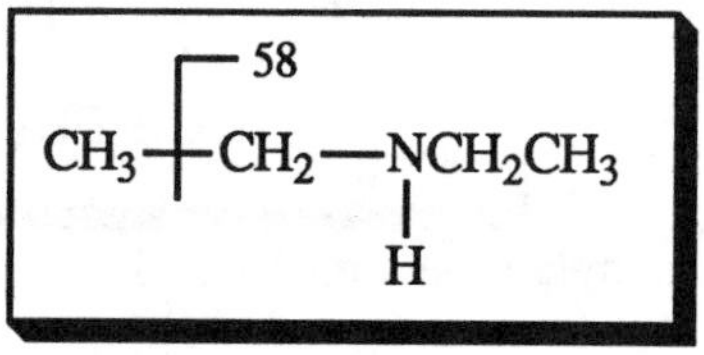

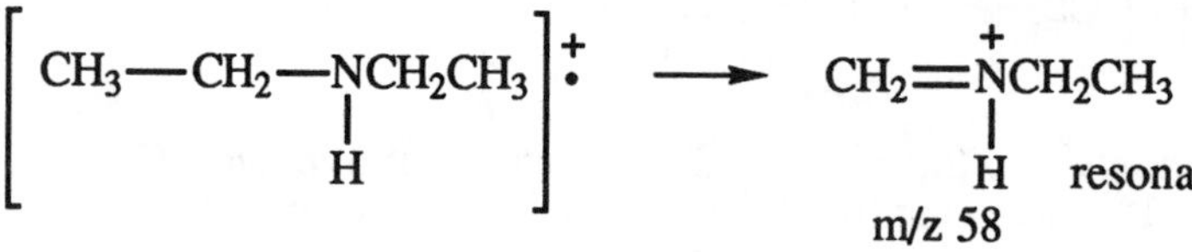

resonance-stabilized

m/z 58

19-57

Mass spectrum:

—molecular ion at 87 = odd mass = odd number of nitrogens present

—if one nitrogen and no oxygens ⇒ molecular formula $C_5H_{13}N$

—base peak at m/z 30 ⇒ structure must include this fragment

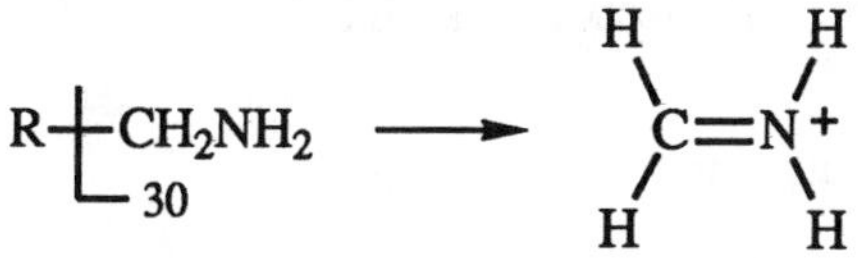

IR spectrum:

—two spikes in the 3300-3400 cm^{-1} region ⇒ 1° amine

NMR spectrum:

—singlet at δ 0.9 for 9H must be a *t*-butyl group

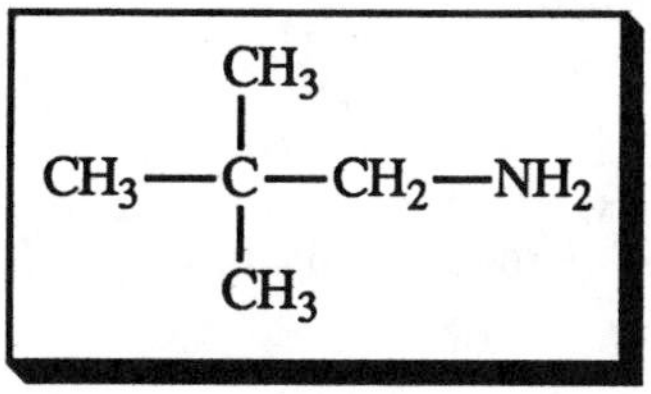

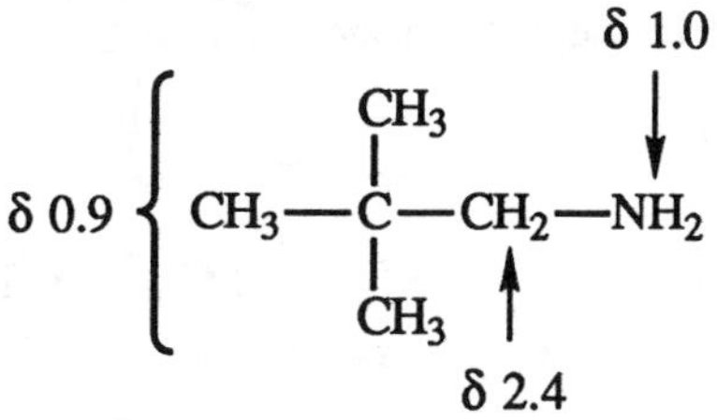

Note that the base peak in the MS arises from cleavage to give these two, relatively stable fragments:

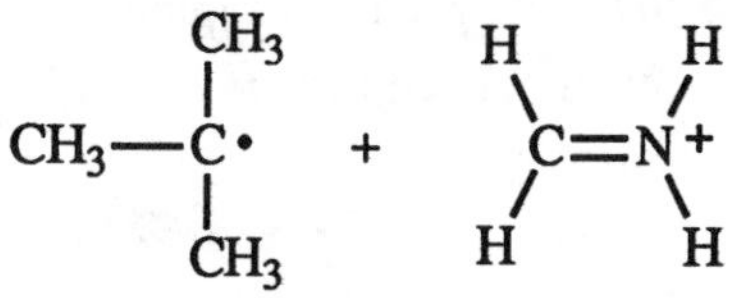

19-58 (a tough problem)

molecular formula $C_{11}H_{16}N_2$ has 5 elements of unsaturation, enough for a benzene ring; no oxygens precludes NO_2 and amide; if C≡N is present, there are not enough elements of unsaturation left for a benzene ring, so benzene and C≡N are mutually exclusive

IR spectrum:

—one spike around 3300 cm^{-1} suggests a 2° amine

—no C≡N

—CH and C=C regions suggest an aromatic ring

Proton NMR spectrum:

—5H broad singlet at δ 7.3 indicates a monosubstituted benzene ring (the fact that it is a singlet precludes N being bonded to the ring)

—1H singlet at δ 1.57 is exchangeable ⇒ NH of secondary amine

—2H singlet at δ 3.5 is CH_2; the fact that it is so strongly deshielded and unsplit suggests that it is between a nitrogen and the benzene ring

fragments so far:

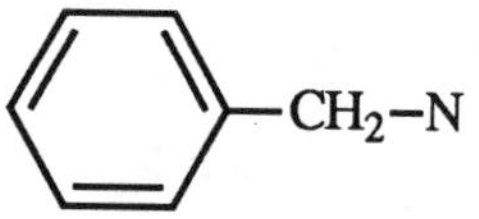

+ NH + 4 C + 8 H + 1 element of unsaturation

Carbon NMR spectrum:

—four signals around δ 140 are the aromatic carbons

—the signal at δ 65 is the CH_2 bonded to the benzene

—the other 4 carbons come as two signals at δ 46 and δ 55; each is a triplet, so there are two sets of two equivalent CH_2 groups, each bonded to N to shift it downfield

fragments so far:

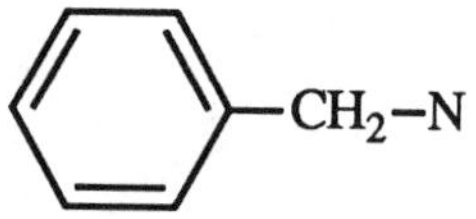

+ NH + $\begin{matrix}CH_2\\CH_2\end{matrix}$ + $\begin{matrix}CH_2\\CH_2\end{matrix}$ + 1 element of unsaturation

There is no evidence for an alkene in any of the spectra, so the remaining element of unsaturation must be a ring. The simplicity of the NMR spectra indicates a fairly symmetric compound.

Assemble the pieces:

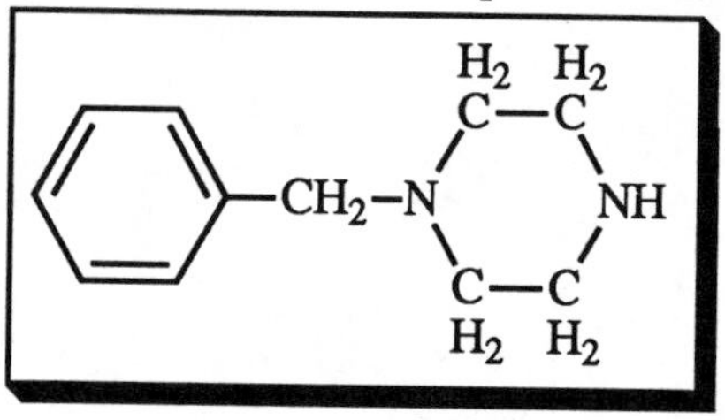

CHAPTER 20—CARBOXYLIC ACIDS

20-1

(a) 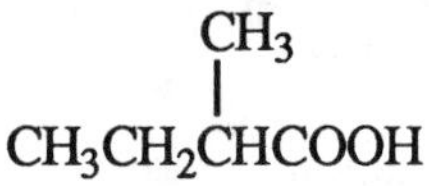

(b) 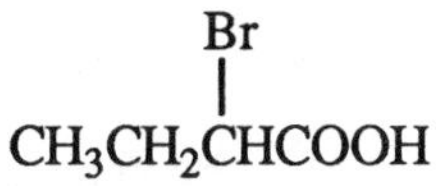

(c) 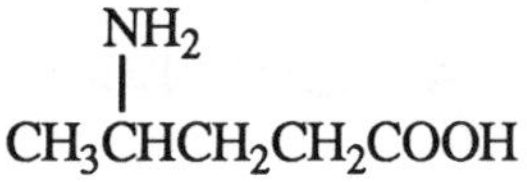

(d) 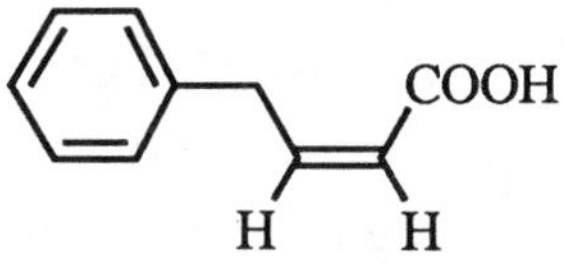

(e) 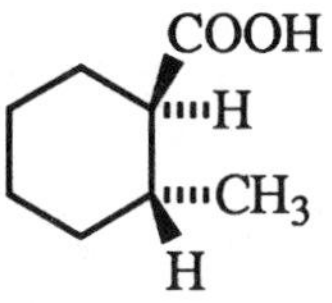

(f) 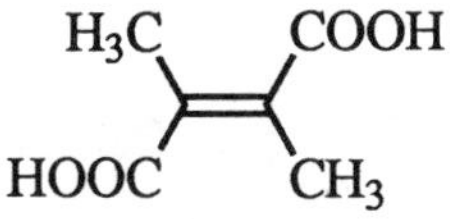

(g) 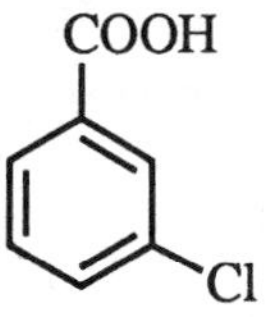

(h) 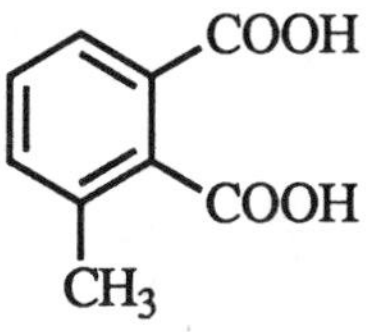

(i) 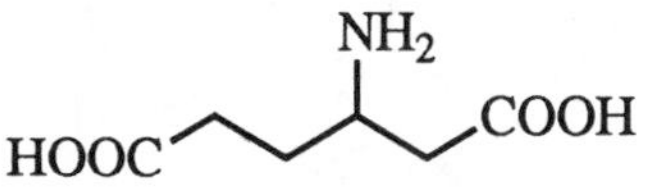

(j) 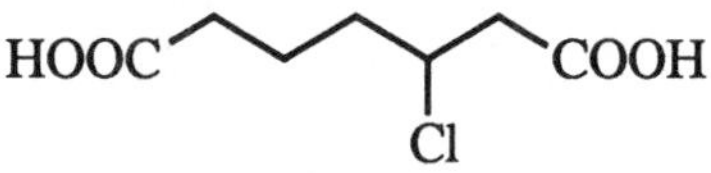

20-2 IUPAC name first; then common name.

(a) 2-iodo-3-methylpentanoic acid; α-iodo-β-methylvaleric acid
(b) (Z)-3,4-dimethyl-3-hexenoic acid
(c) 2,3-dinitrobenzoic acid; no common name
(d) *trans*-1,2-cyclohexanedicarboxylic acid; (*trans*-hexahydrophthalic acid)
(e) 2-chlorobenzene-1,4-dicarboxylic acid; 2-chloroterephthalic acid
(f) 3-methylhexanedioic acid; β-methyladipic acid

20-3 In order of increasing acid strength (weakest acid first):

(a) CH_3CH_2COOH < $CH_3-CH(Br)COOH$ < $CH_3-C(Br)_2COOH$

The greater the number of electron-withdrawing substituents, the greater the stabilization of the carboxylate ion.

CONTINUED ON NEXT PAGE

20-3 continued

(b) $CH_3-\underset{\displaystyle Br}{\underset{|}{C}H}CH_2CH_2COOH < CH_3CH_2-\underset{\displaystyle Br}{\underset{|}{C}H}CH_2COOH < CH_3CH_2CH_2-\underset{\displaystyle Br}{\underset{|}{C}H}COOH$

The closer the electron-withdrawing group, the greater the stabilization of the carboxylate ion.

(c) $CH_3CH_2COOH < CH_3-\underset{\displaystyle Cl}{\underset{|}{C}H}COOH < CH_3-\underset{\displaystyle C\equiv N}{\underset{|}{C}H}COOH < CH_3-\underset{\displaystyle NO_2}{\underset{|}{C}H}COOH$

The stronger the electron-withdrawing effect of the substituent, the greater the stabilization of the carboxylate ion.

20-4

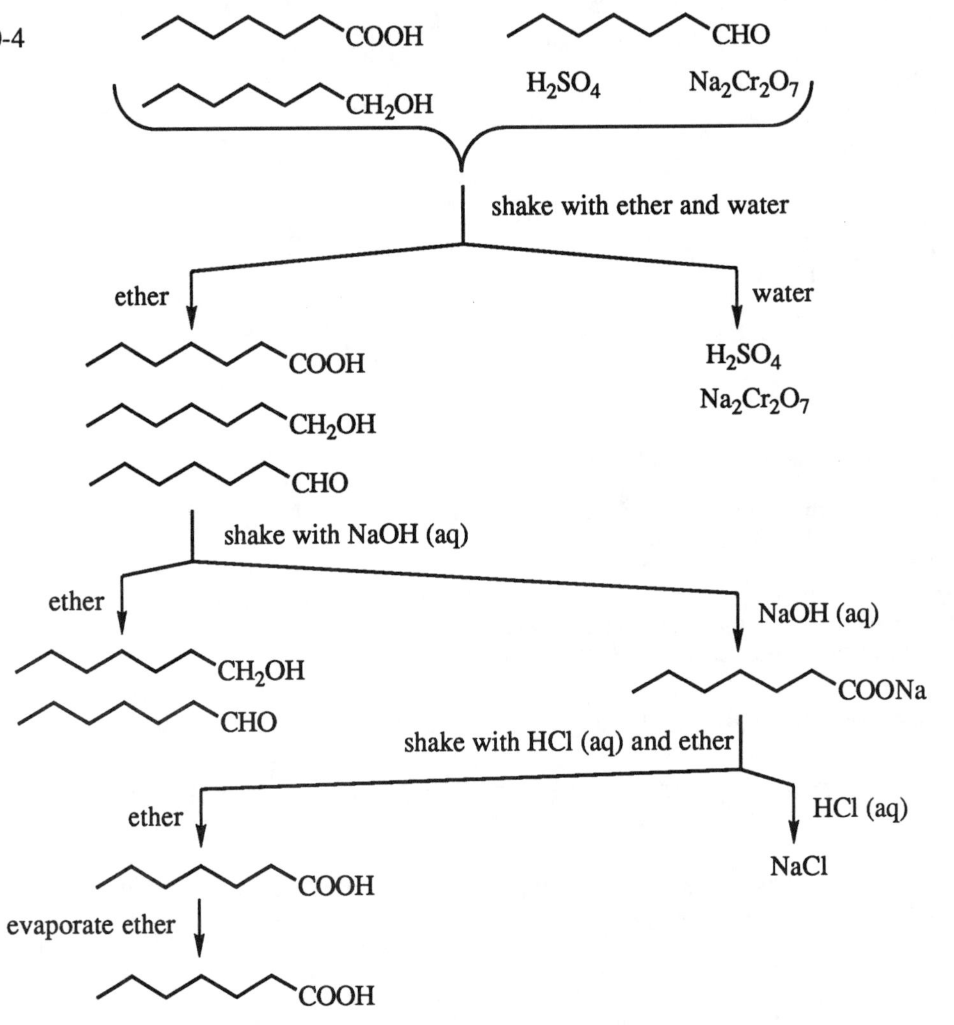

20-5 The principle used to separate a carboxylic acid (a stronger acid) from a phenol (a weaker acid) is to neutralize with a weak base ($NaHCO_3$)— a base strong enough to ionize the stronger acid but not strong enough to ionize the weaker acid.

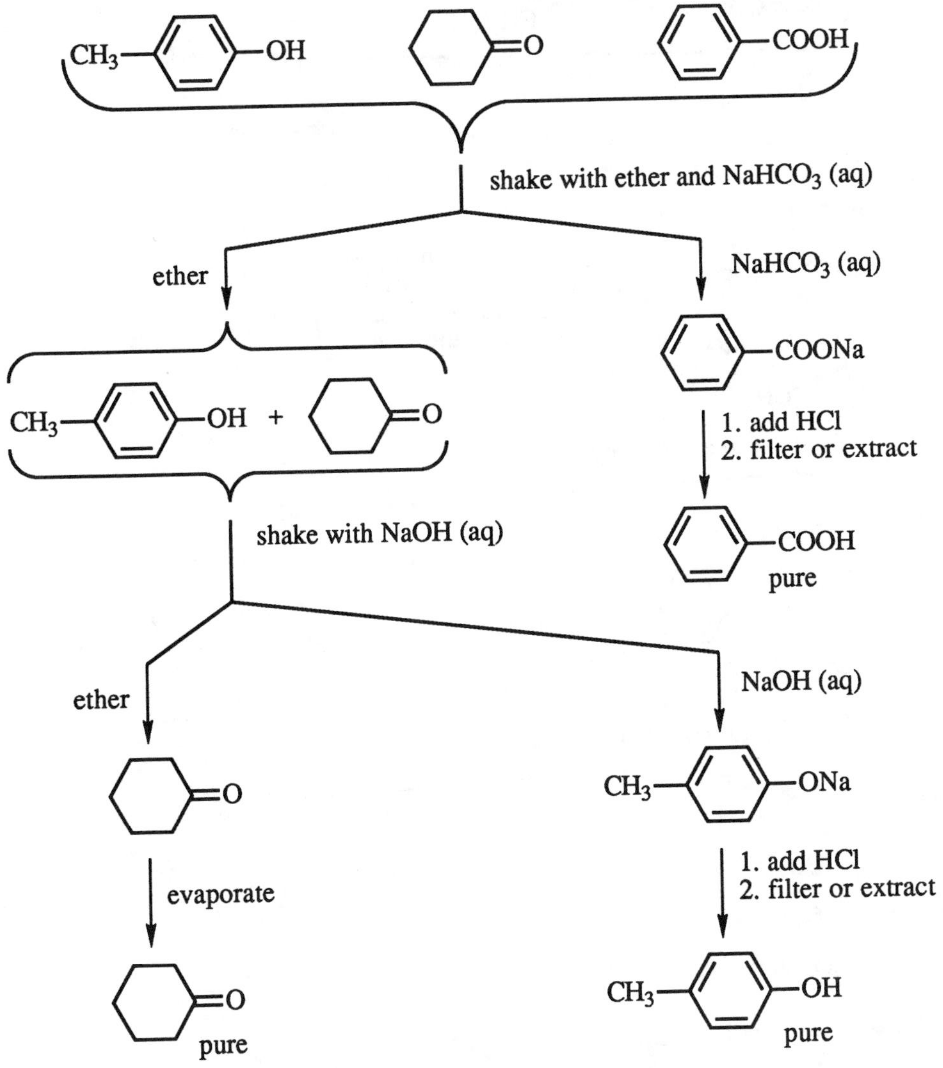

20-6

(a)

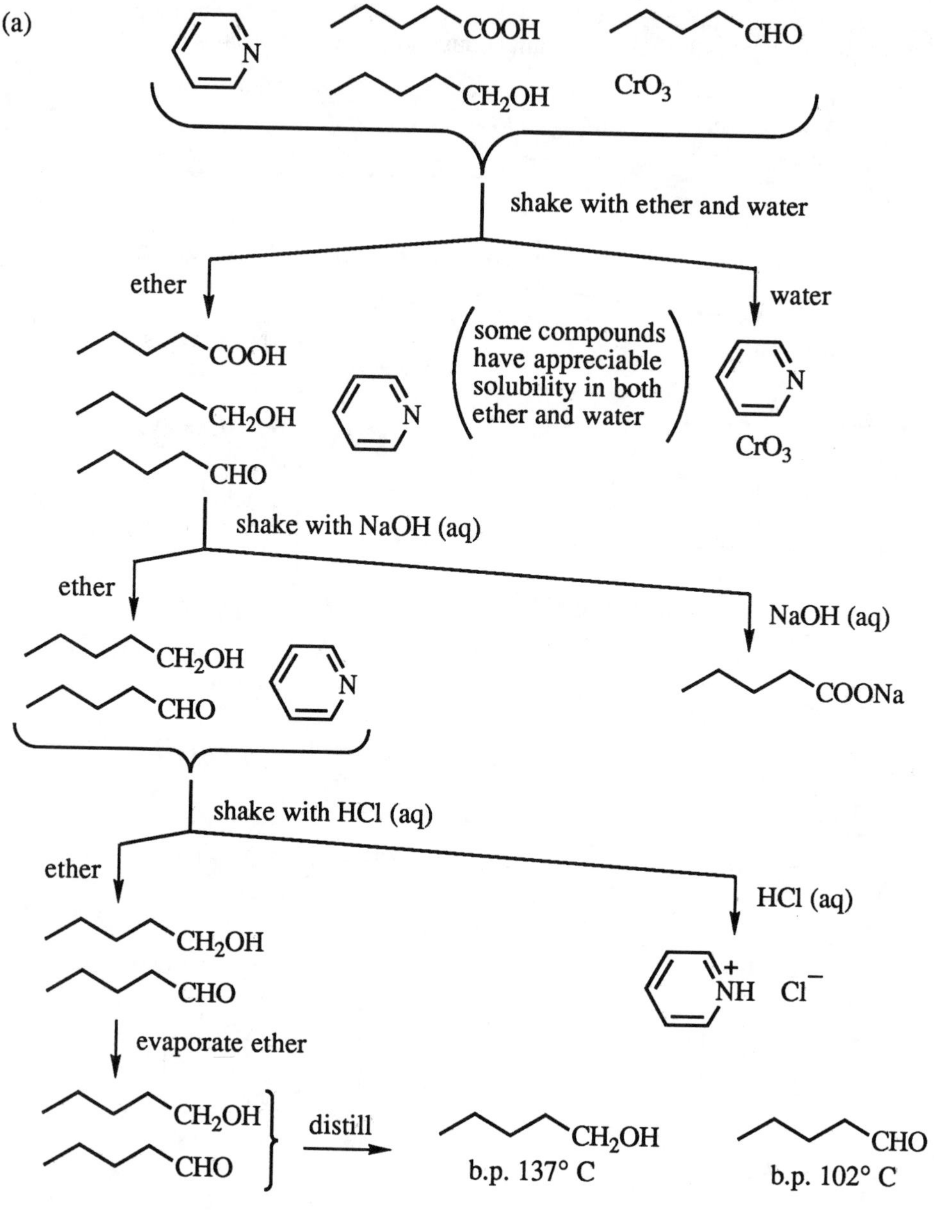

(b) 1-Pentanol cannot be removed from pentanal by acid-base extraction. These two remaining products can be separated by distillation, the alcohol having the higher boiling point because of hydrogen-bonding.

20-7 The COOH has a characteristic IR absorption: a broad peak from 3500-2500 cm^{-1}, with a "shoulder" around 2700 cm^{-1}. The carbonyl stretch at 1690 cm^{-1} is a little lower than the standard 1710 cm^{-1}, suggesting conjugation. The strong alkene absorption at 1650 cm^{-1} also suggests it is conjugated.

20-8 (a) The ethyl pattern is obvious: a 3H triplet at δ 1.0 and a 2H quartet at δ 2.4. The only other peak is the COOH at δ 11.3 (a 300 Hz offset in a 60 MHz spectrum is 5 δ units).

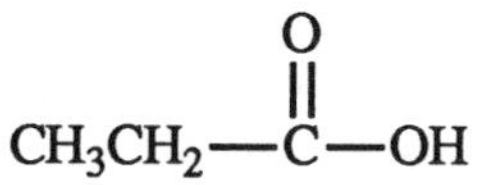

(b)

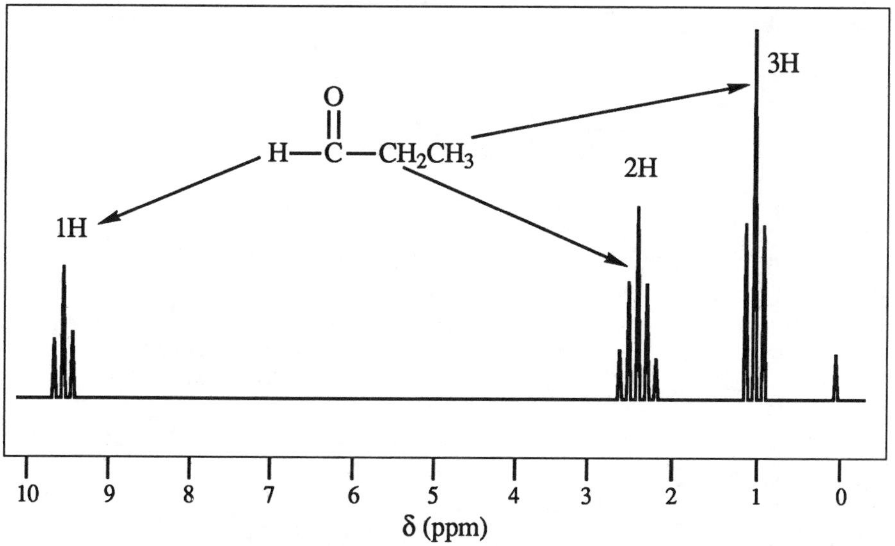

The multiplet between δ 2 and δ 3 is drawn as a pentet as though it were split equally by the aldehyde proton and the CH_2 group. These coupling constants are probably unequal, in which case the actual splitting pattern will be a complex multiplet.

(c) The chemical shift of the aldehyde proton is between δ 9-10, not as far downfield as the carboxylic acid proton. Also, the aldehyde proton is split into a triplet by the CH_2, unlike the COOH proton which always appears as a singlet. Finally, the CH_2 is split by an extra proton, so it will give a multiplet with complex splitting, instead of the quartet shown in the acid.

20-9

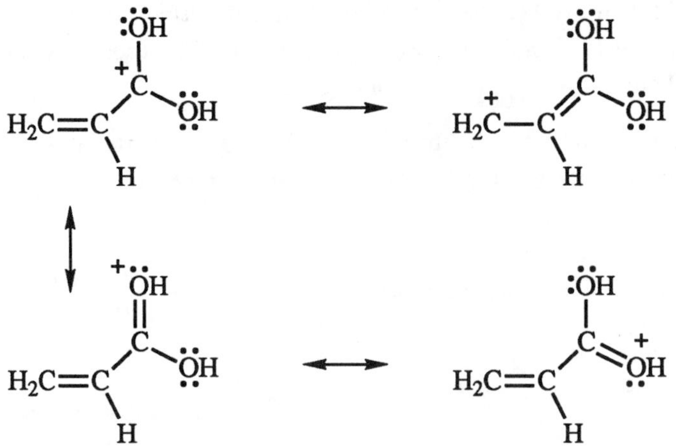

20-10

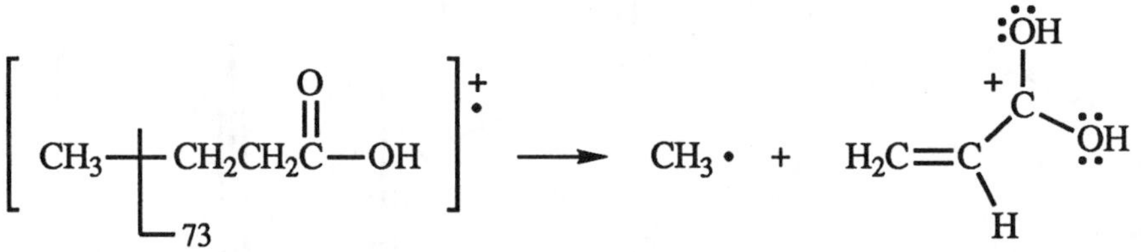

plus resonance forms
as shown in 20-9

m/z 73

McLafferty rearrangement

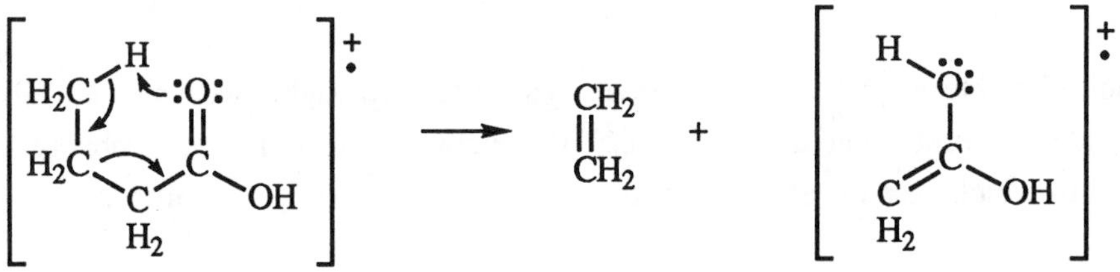

m/z 60

20-11

(a) 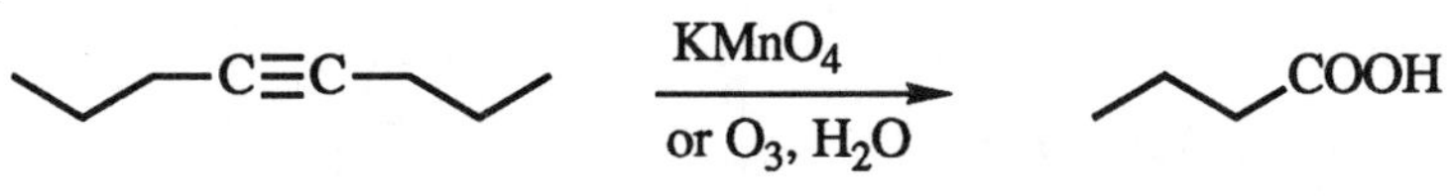

(b) 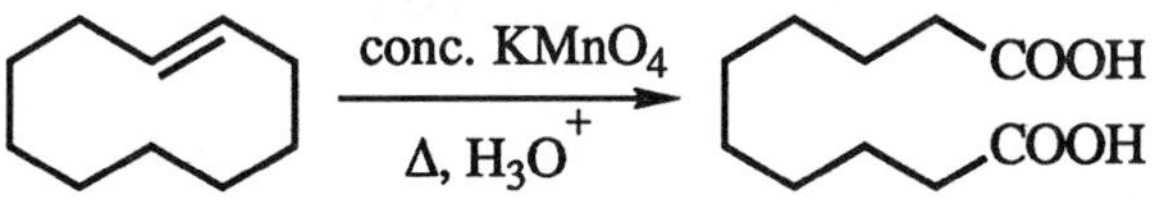

(c) 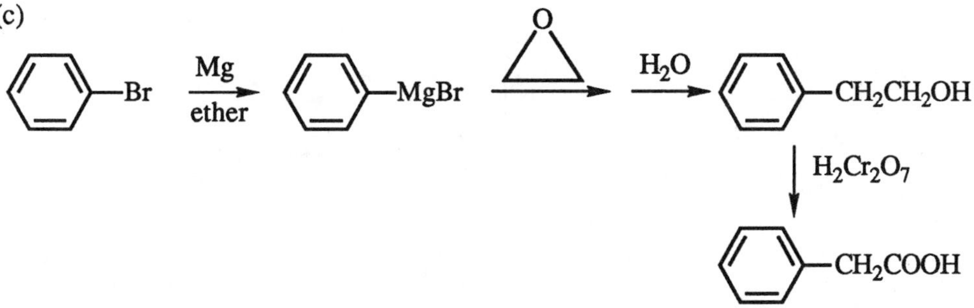

(d) 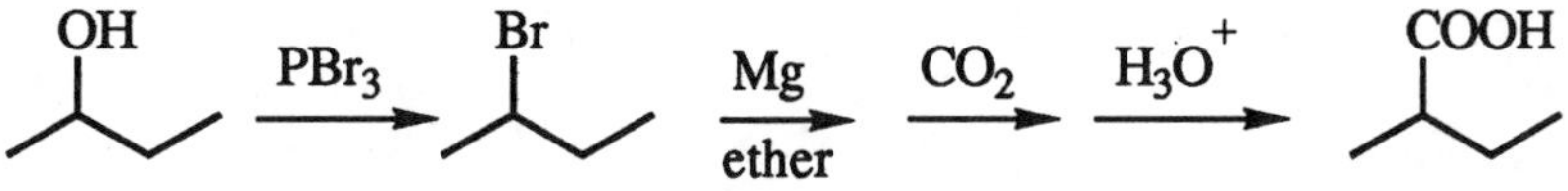

(e) 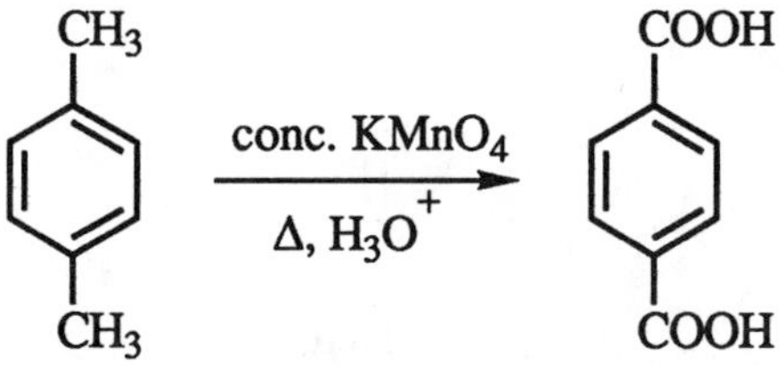

(f) 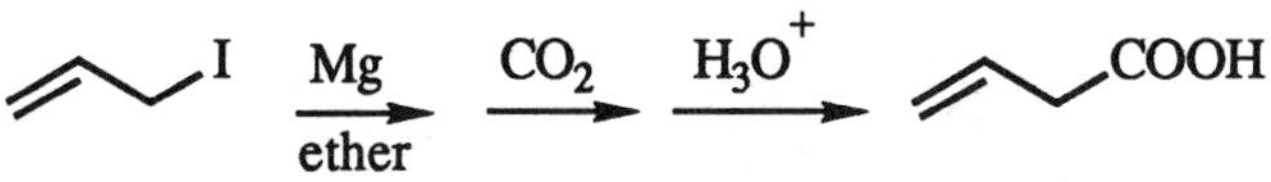

OR 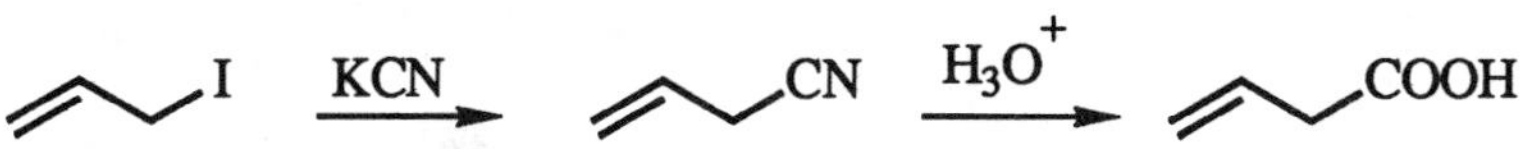

20-12

(a)

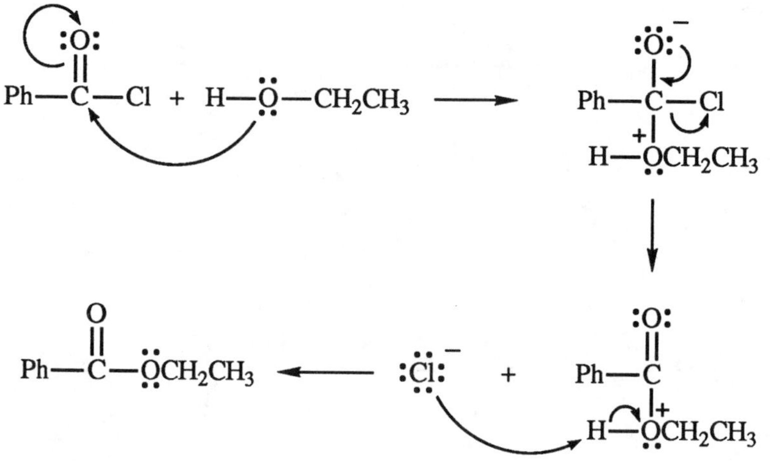

(b)

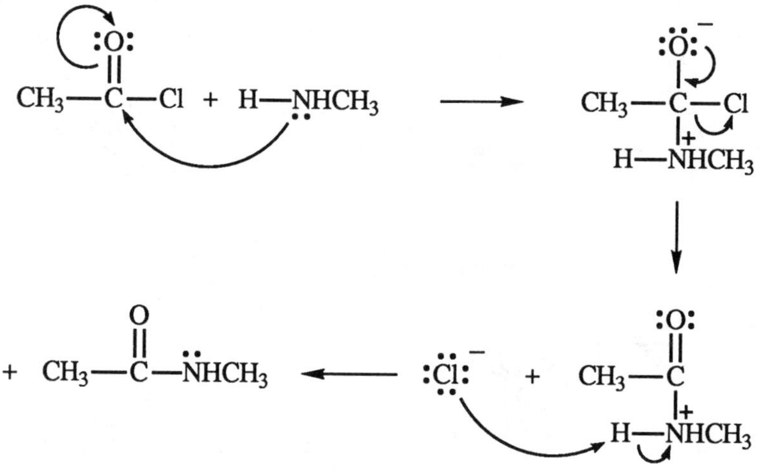

20-13

(a)

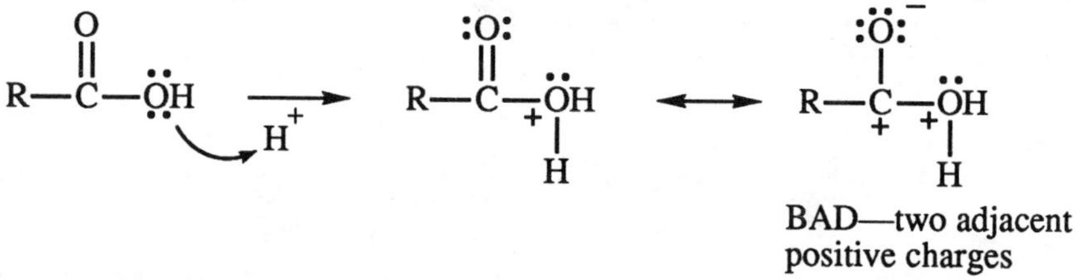

(b) Protonation on the OH gives only two resonance forms, one of which is bad because of adjacent positive charges. Protonation on the C=O is good because of three resonance forms distributing the positive charge over three atoms, with no additional charge separation.

20-13 continued

(c) The carbonyl oxygen is more "basic" because, by definition, it reacts with a proton more readily. It does so because the intermediate it produces is more stable than the intermediate from protonation of the OH.

20-14

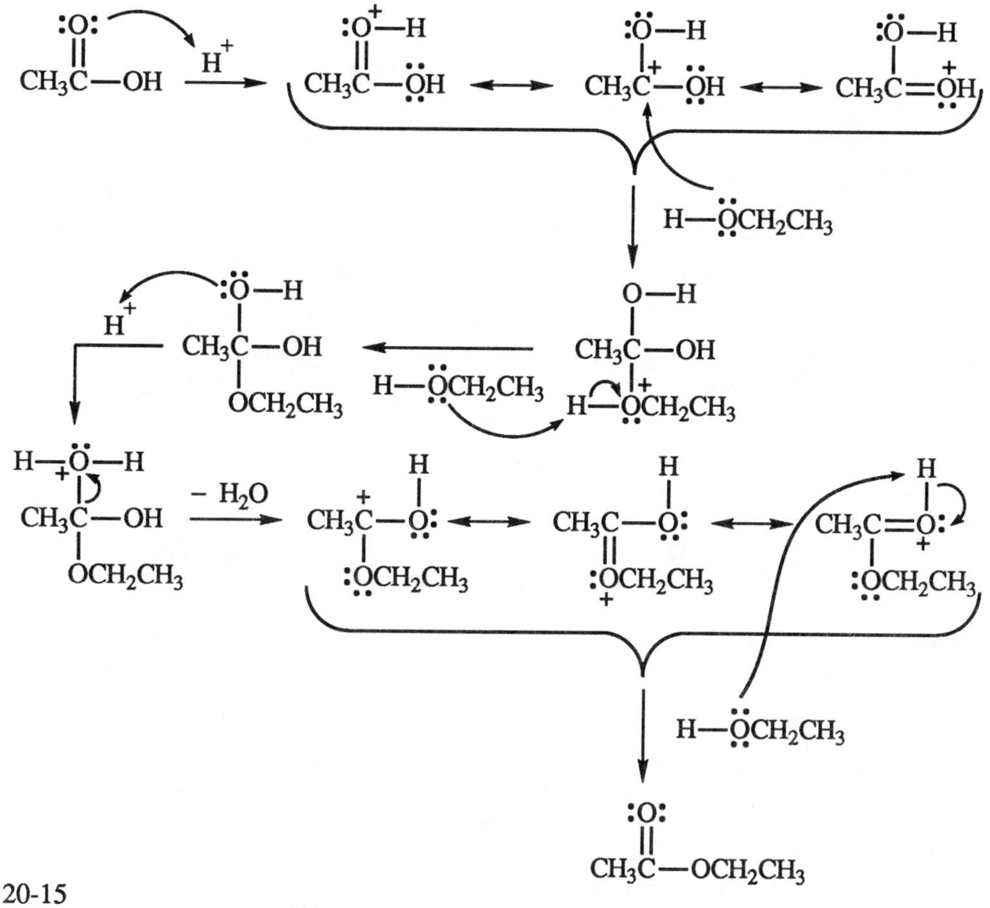

H^+
$H—\ddot{O}CH_2CH_3$
$- H_2O$
$H—\ddot{O}CH_2CH_3$
$CH_3C—OCH_2CH_3$

20-15

(a)

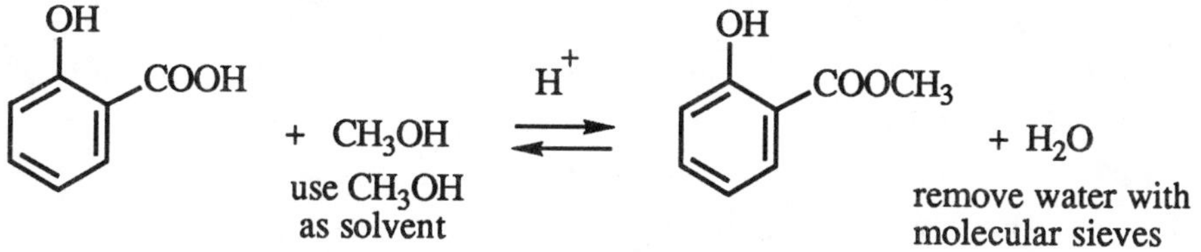

OH
COOH
+ CH_3OH
use CH_3OH as solvent
H^+
OH
$COOCH_3$
+ H_2O
remove water with molecular sieves

(b)

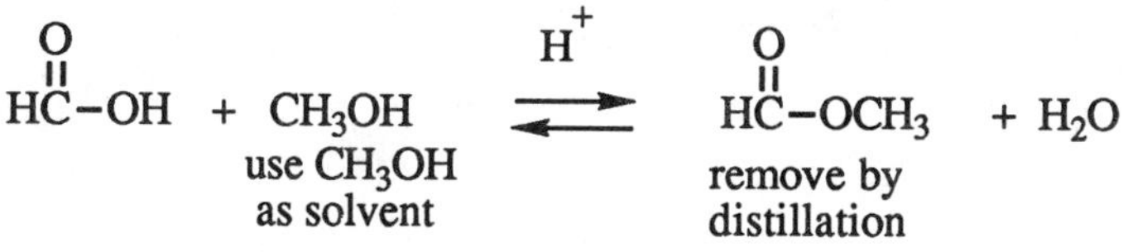

HC−OH
+ CH_3OH
use CH_3OH as solvent
H^+
$HC−OCH_3$
remove by distillation
+ H_2O

20-15 continued

(c)

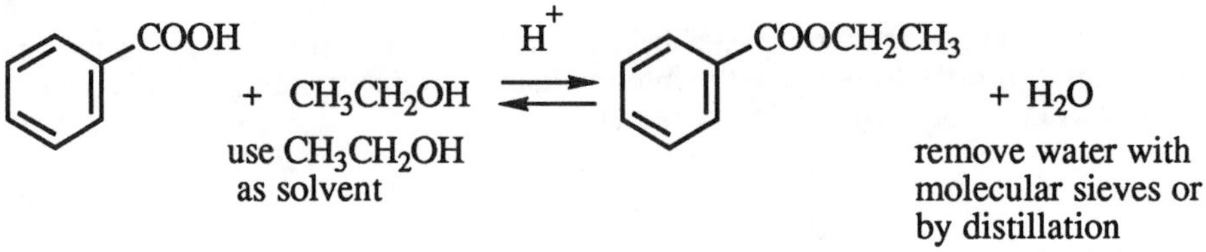
COOH
+ CH3CH2OH
use CH3CH2OH as solvent
H+
COOCH2CH3
+ H2O
remove water with molecular sieves or by distillation

20-16

(a) (1) (2)

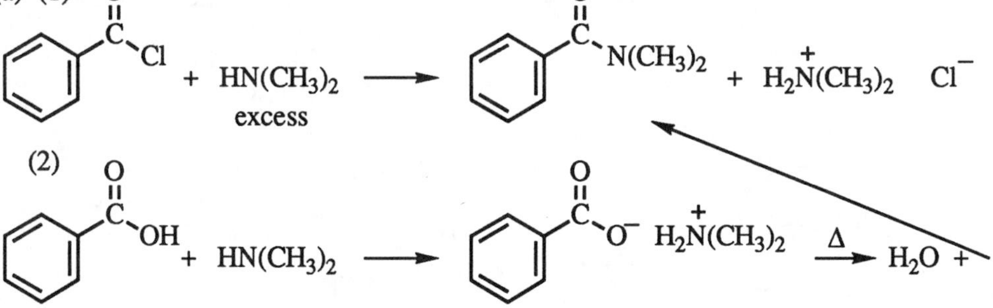
(1)
C
Cl
+ HN(CH3)2
excess
N(CH3)2
+ H2N(CH3)2 Cl−
(2)
OH
+ HN(CH3)2
O− H2N(CH3)2
Δ
H2O +

(b) (1) (2)

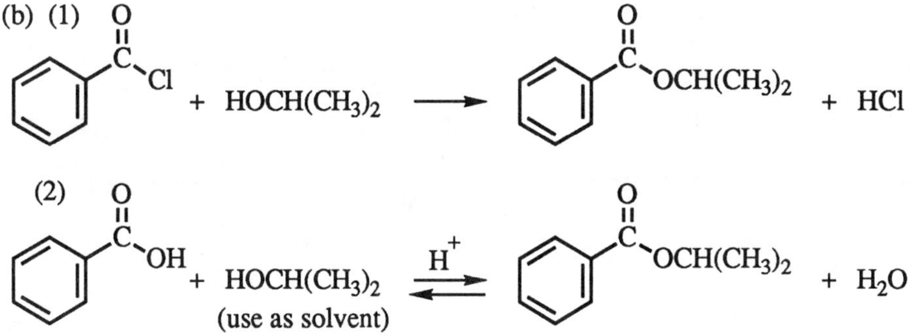
(2)
Cl
+ HOCH(CH3)2
OCH(CH3)2
+ HCl
OH
+ HOCH(CH3)2
(use as solvent)
H+
OCH(CH3)2
+ H2O

(c) (1) (2)

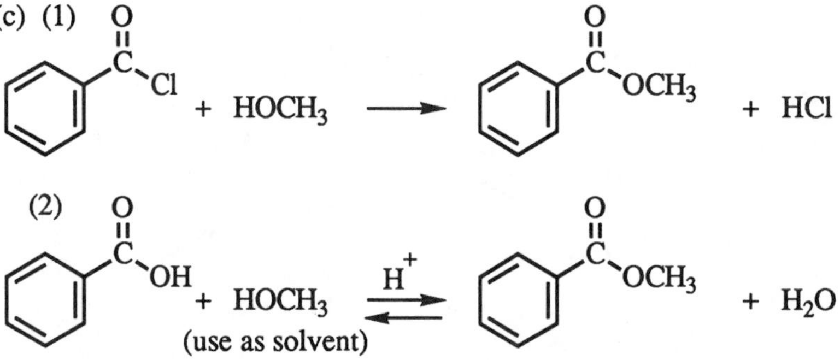
(2)
Cl
+ HOCH3
OCH3
+ HCl
OH
+ HOCH3
(use as solvent)
H+
OCH3
+ H2O

20-17

(a)

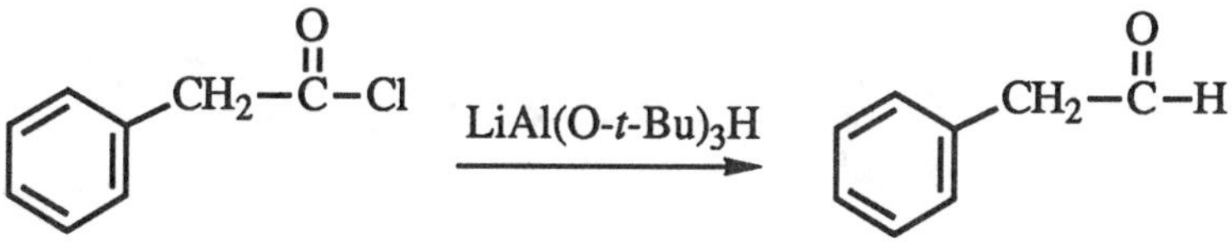

$LiAl(O\text{-}t\text{-}Bu)_3H$

(b)

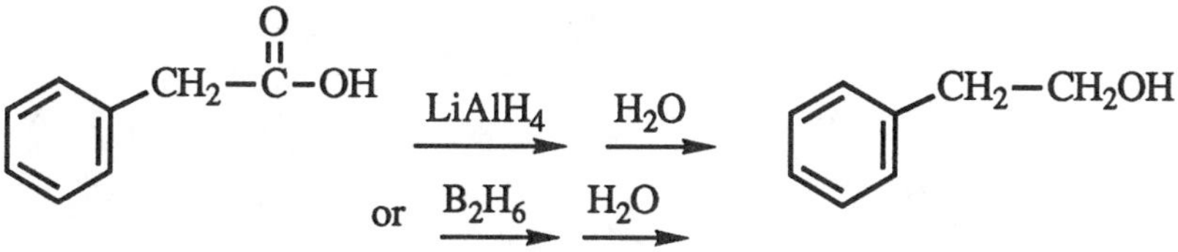

$LiAlH_4$
H_2O
or
B_2H_6
H_2O

(c)

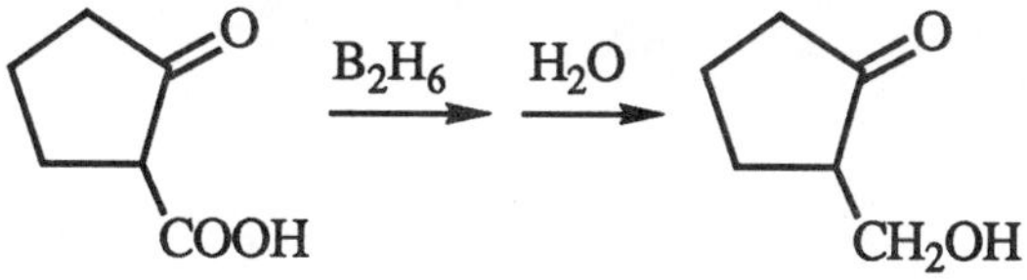

B_2H_6
H_2O

B_2H_6 selectively reduces a carboxylic acid in the presence of a ketone

(d)

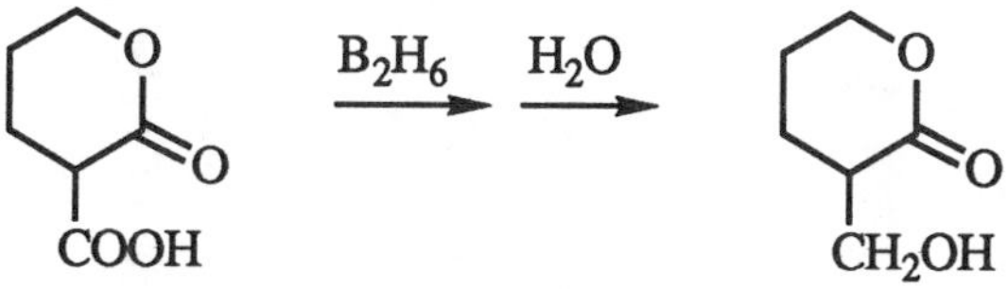

B_2H_6
H_2O

B_2H_6 selectively reduces a carboxylic acid in the presence of an ester

20-18

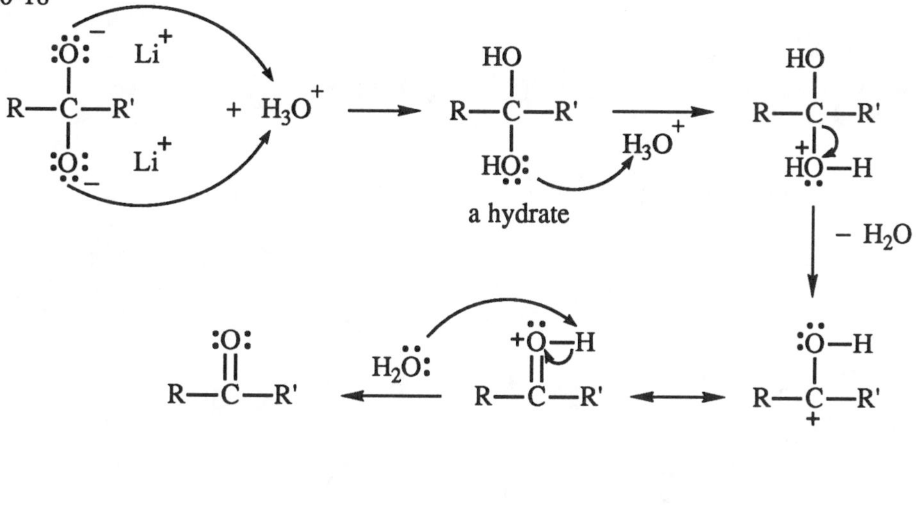

a hydrate
$- H_2O$

20-19

(a)

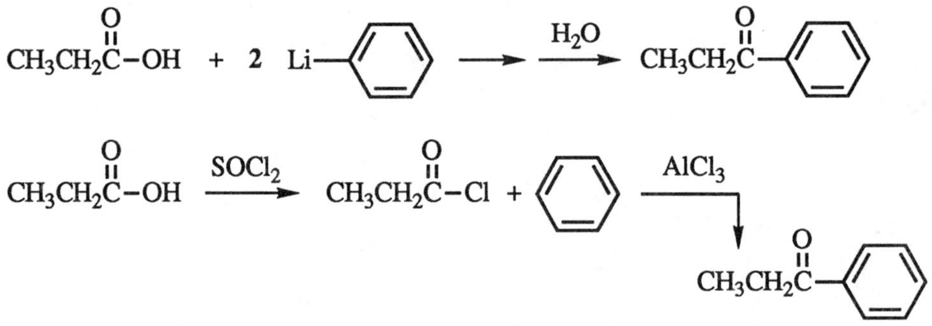

(b)

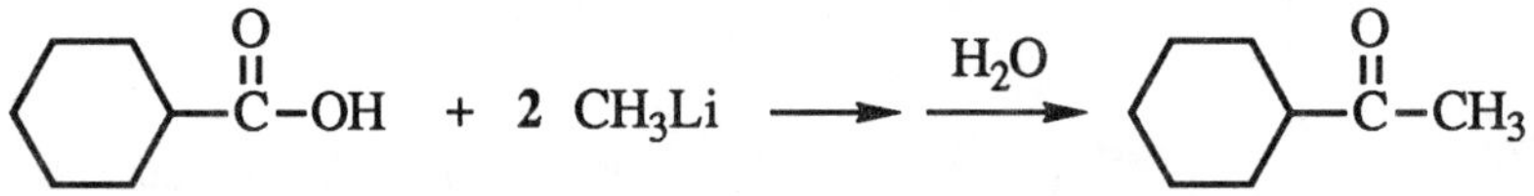

20-20

Initiation

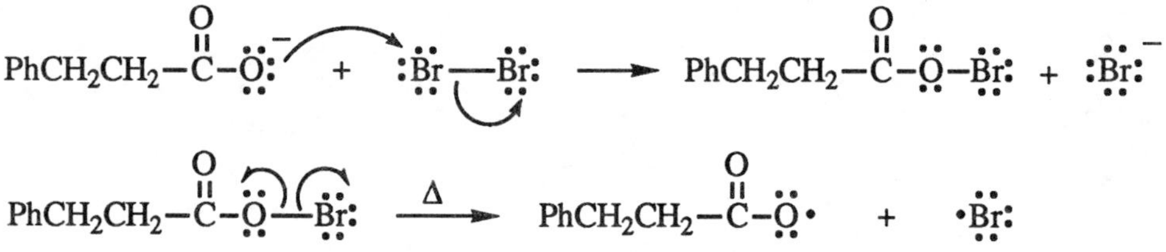

Propagation

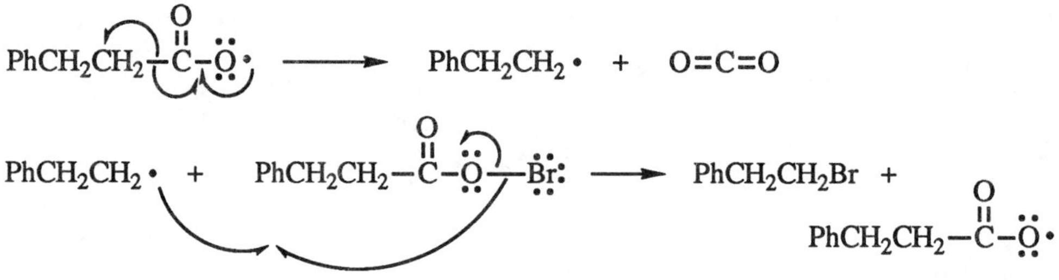

20-21

(a)

1-bromononane

+ AgBr + CO_2

(b)

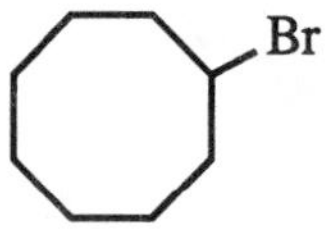

+ $HgBr_2$

+ CO_2

20-22 Refer to the Glossary and the text for definitions and examples.

20-23

(a) 3-phenylpropanoic acid
(b) 2-methylbutanoic acid
(c) 2-bromo-3-methylbutanoic acid
(d) 3-methylpentanedioic acid
(e) sodium 2-methylbutanoate
(f) 3-methyl-2-butenoic acid
(g) *trans*-2-methylcyclopentanecarboxylic acid
(g) 2,4,6-trinitrobenzoic acid

20-24

(a) β-phenylpropionic acid
(b) α-methylbutyric acid
(c) α-bromo-β-methylbutyric acid
(d) β-methylglutaric acid
(e) sodium β-methylbutyrate
(f) β-aminobutyric acid
(g) *o*-bromobenzoic acid
(h) magnesium oxalate

20-25

(a)

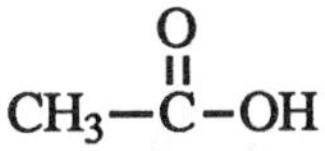

(b)

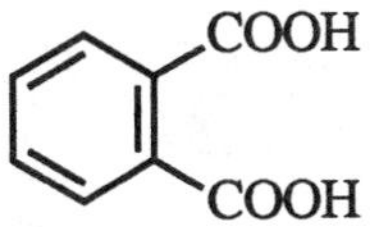

(c)

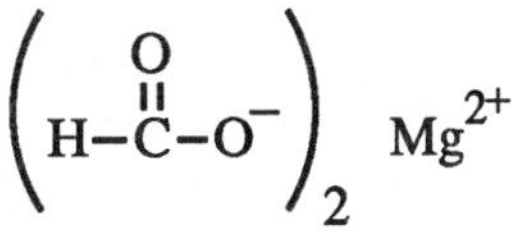

(d)

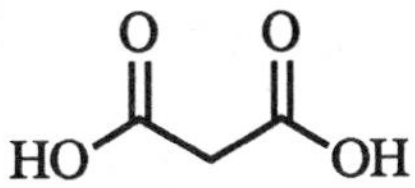

(e)

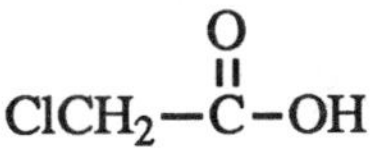

(f)

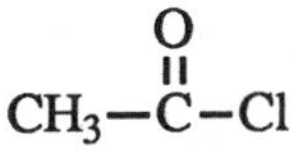

(g)

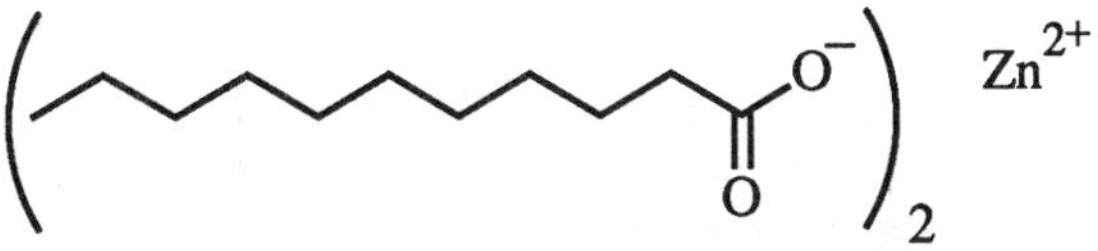

(h)

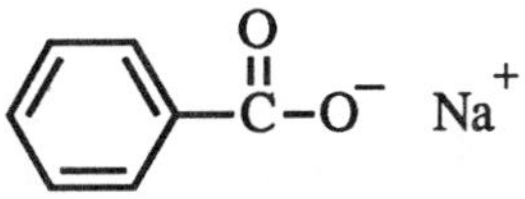

(i)

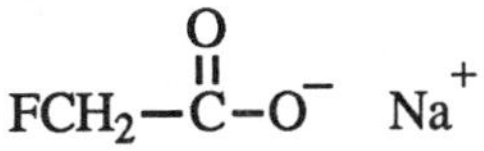

20-26 Stronger base listed first. (Stronger bases come from *weaker* conjugate acids.)

(a) CH_3COO^- > $ClCH_2COO^-$

(b) $HC{\equiv}C^-\ Na^+$ > $CH_3COO^-\ Na^+$

(c) $CH_3CH_2O^-\ Na^+$ > $CH_3COO^-\ Na^+$

20-27

(a) $CH_3COOH + NH_3 \longrightarrow CH_3COO^- \overset{+}{N}H_4$

(b)

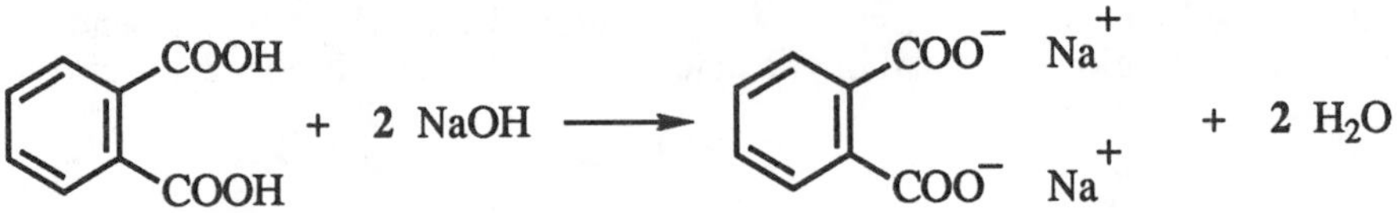
COOH
COOH
+ 2 NaOH
COO⁻ Na⁺
COO⁻ Na⁺
+ 2 H₂O

(c)

CH₃
COOH
+ CF₃COO⁻ K⁺
no reaction

(d) $CH_3-\underset{Br}{CH}COOH + CH_3CH_2COO^-\ Na^+ \longrightarrow CH_3-\underset{Br}{CH}COO^-\ Na^+$

$+ CH_3CH_2COOH$

(e)

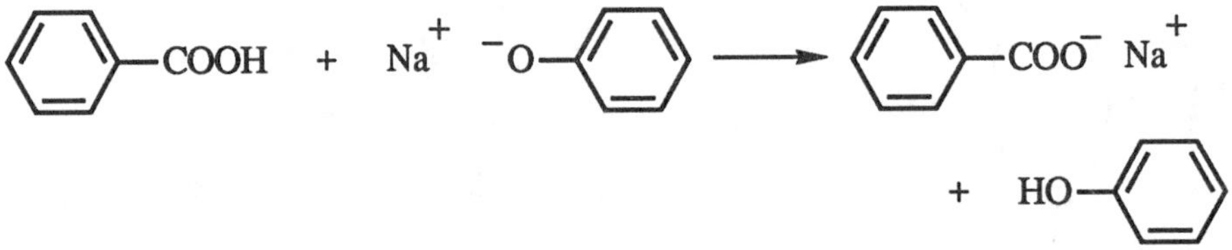
COOH
+ Na⁺ ⁻O
COO⁻ Na⁺
+ HO

20-28

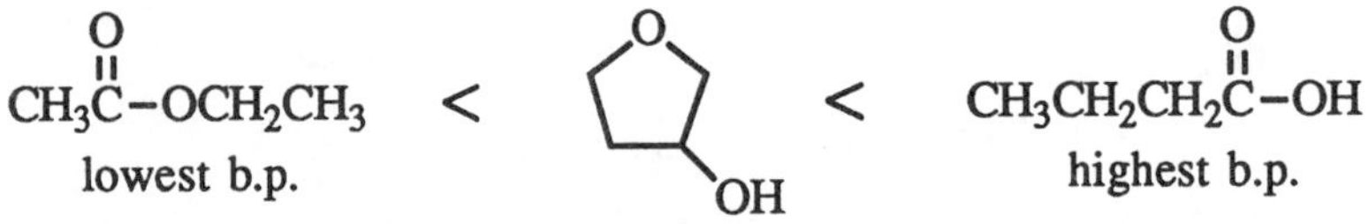
CH₃C−OCH₂CH₃
<
O
OH
<
CH₃CH₂CH₂C−OH
lowest b.p.
highest b.p.

The ester cannot hydrogen bond and will be the lowest boiling. The alcohol can hydrogen bond. The carboxylic acid forms two hydrogen bonds and boils as the dimer, the highest boiling among these three compounds.

20-29 In order of increasing acidity (weakest acid first):

(a) ethanol < phenol < acetic acid

(b) acetic acid < chloroacetic acid < *p*-toluenesulfonic acid

(c) benzoic acid < *m*-nitrobenzoic acid < *o*-nitrobenzoic acid

(d) butyric acid < β-bromobutyric acid < α-bromobutyric acid

20-30

(a)

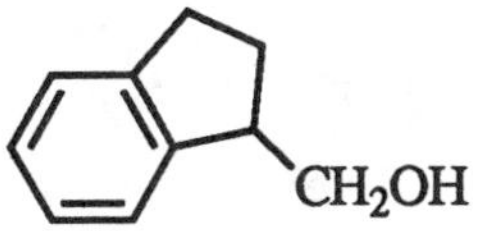
CH2OH

(b)

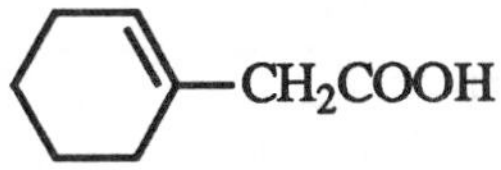
CH2COOH

(c)

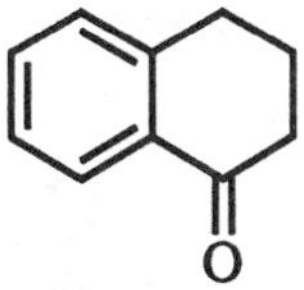
O

(d)

2 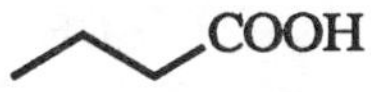
COOH

(e)

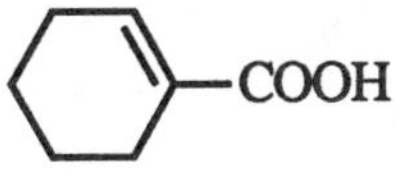
COOH

(f)

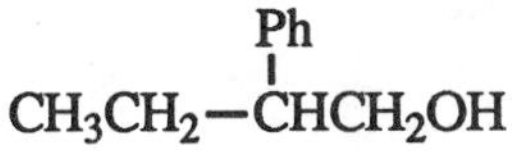
Ph
CH3CH2—CHCH2OH

(g)

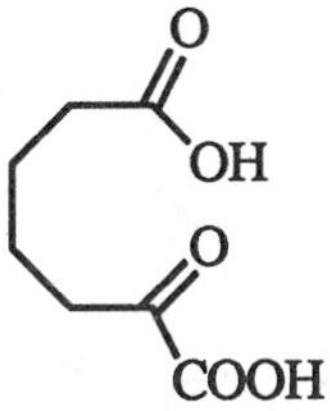
O
OH
O
COOH

(h)

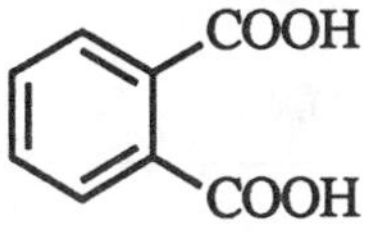
COOH
COOH

(i)

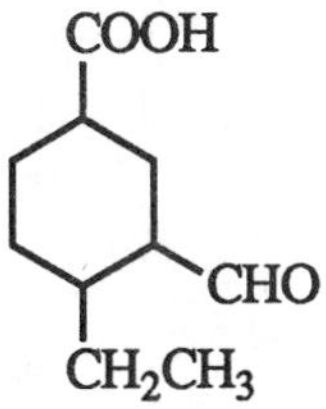
COOH
CHO
CH2CH3

(j)

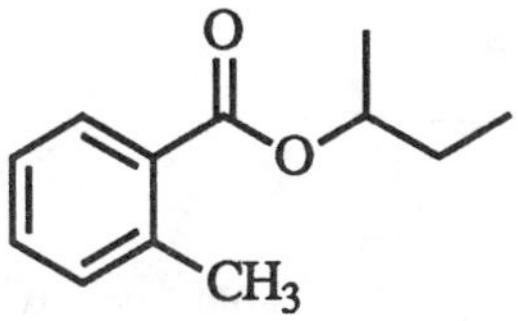
O
O
CH3

(k)

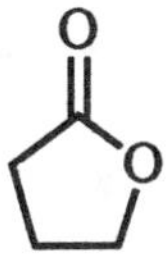
O
O

20-31

(a)

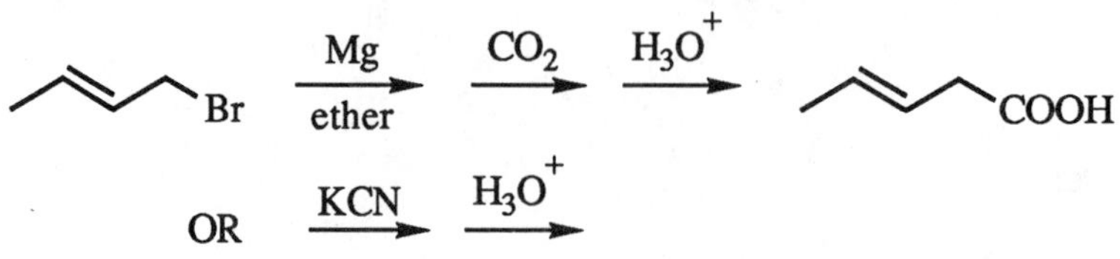
Br
Mg
ether
CO2
H3O+
COOH
OR
KCN
H3O+

(b)

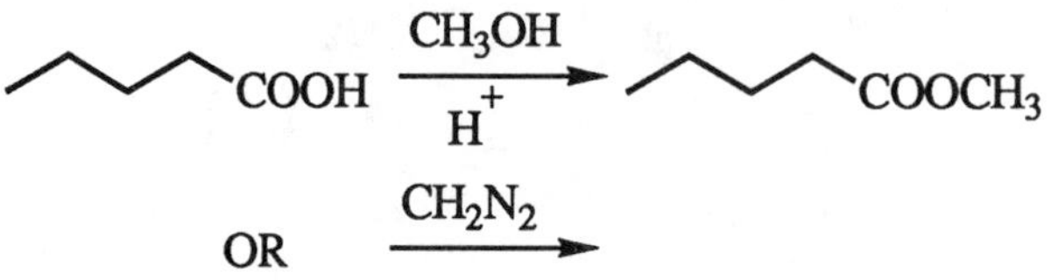
COOH
CH3OH
H+
COOCH3
OR
CH2N2

(c)

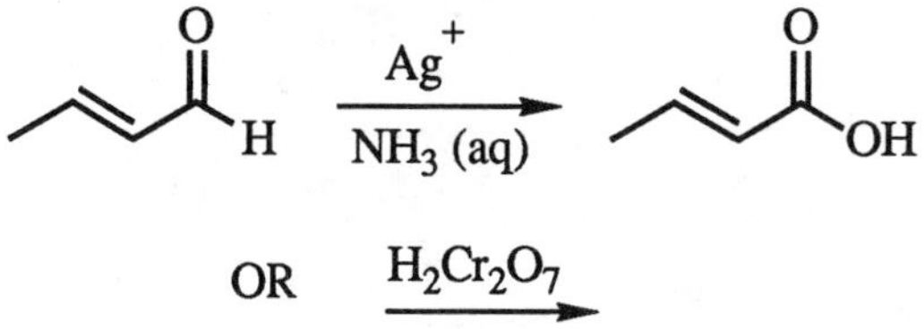
O
H
Ag+
NH3 (aq)
O
OH
OR
H2Cr2O7

(d)

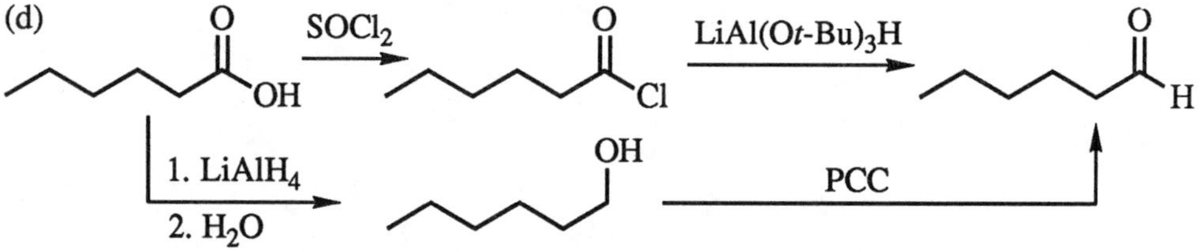
O
OH
SOCl2
O
Cl
LiAl(Ot-Bu)3H
O
H
1. LiAlH4
2. H2O
OH
PCC

(e)

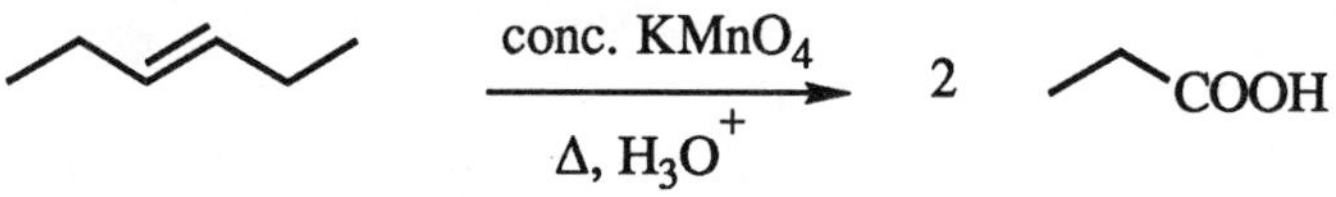
conc. KMnO4
Δ, H3O+
2
COOH

(f)

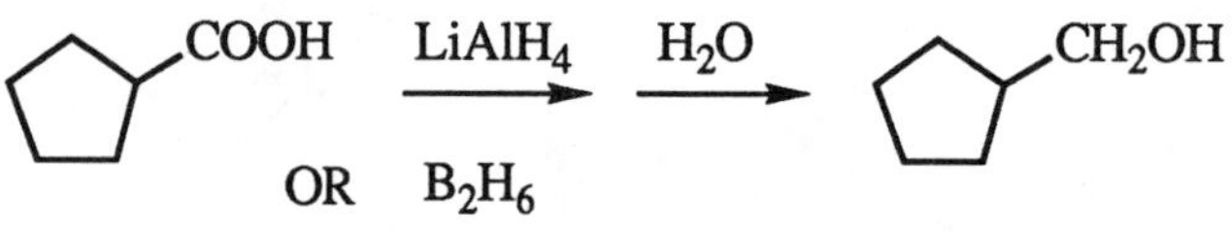
COOH
LiAlH4
H2O
CH2OH
OR
B2H6

(g)

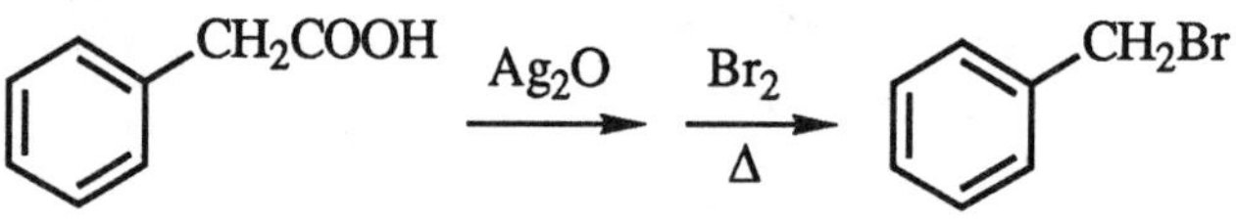
CH2COOH
Ag2O
Br2
Δ
CH2Br

20-32

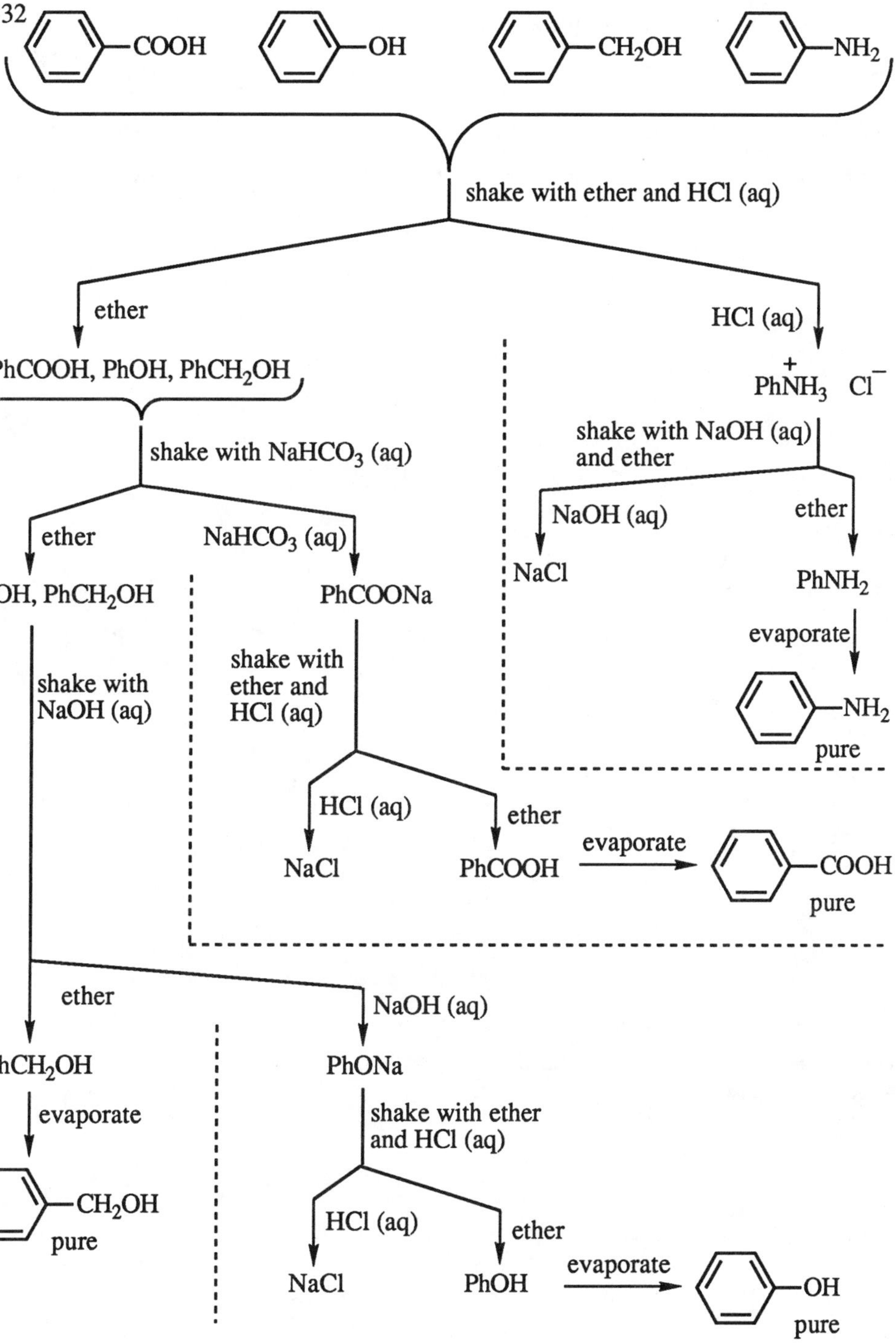
COOH
OH
CH2OH
NH2
shake with ether and HCl (aq)
ether
HCl (aq)
PhCOOH, PhOH, PhCH2OH
PhNH3+ Cl−
shake with NaHCO3 (aq)
shake with NaOH (aq) and ether
ether
NaHCO3 (aq)
NaOH (aq)
ether
NaCl
PhOH, PhCH2OH
PhCOONa
PhNH2
evaporate
shake with NaOH (aq)
shake with ether and HCl (aq)
NH2
pure
HCl (aq)
ether
NaCl
PhCOOH
evaporate
COOH
pure
ether
NaOH (aq)
PhCH2OH
PhONa
evaporate
shake with ether and HCl (aq)
CH2OH
pure
HCl (aq)
ether
NaCl
PhOH
evaporate
OH
pure

20-33 The asterisk ("*") denotes the ^{18}O isotope.

(a) and (b)

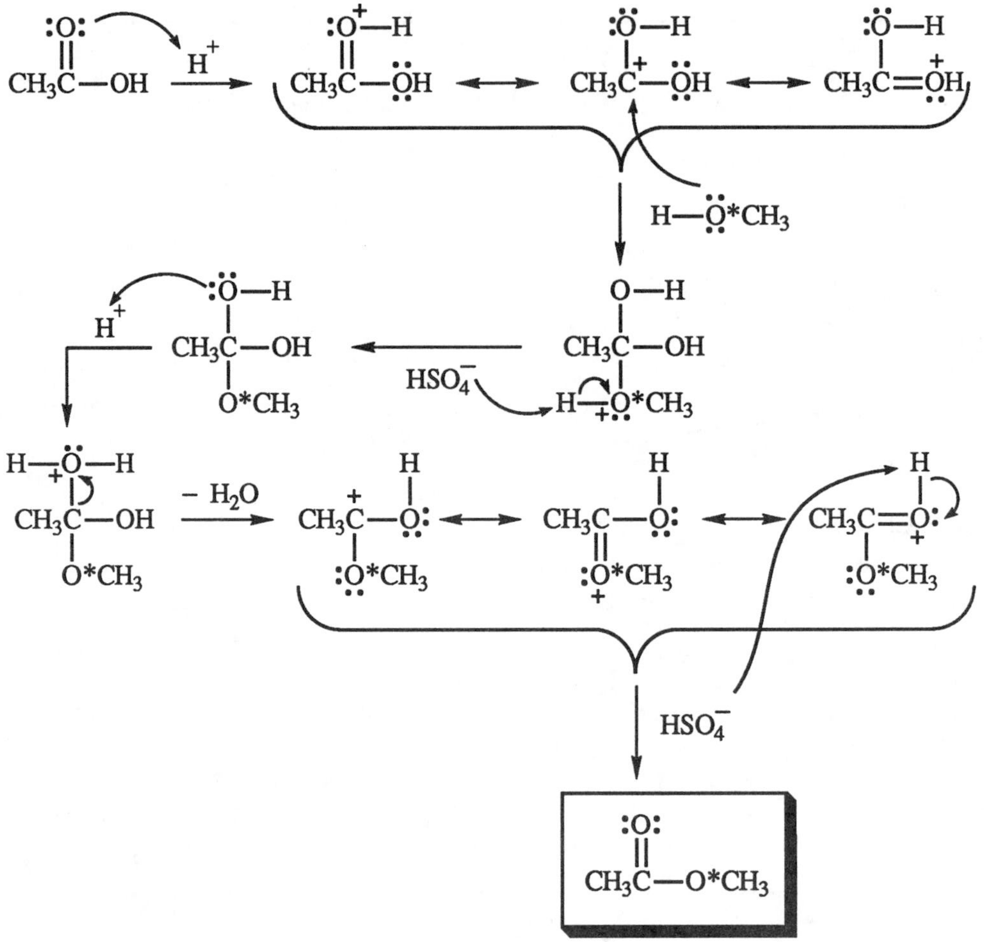

(c) The ^{18}O has two more neutrons, and therefore two more mass units, than ^{16}O. The instrument ideally suited to analyze compounds of different mass is the mass spectrometer.

20-34

(a) Mass spectrum:

—m/z 152 ⇒ molecular ion ⇒ molecular weight 152

—m/z 107 ⇒ M − 45 ⇒ loss of COOH

—m/z 77 ⇒ monosubstituted benzene ring,

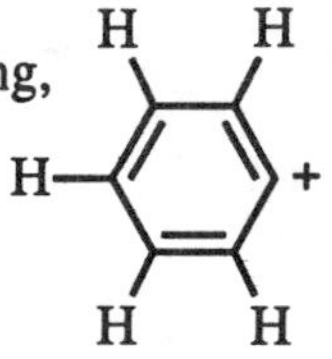
H
H
H
+
H
H

IR spectrum:

—3300-2700 cm^{-1}, broad ⇒ COOH

—1710 cm^{-1} ⇒ C=O

—1220 cm^{-1} ⇒ C−O

—1600 cm^{-1} ⇒ aromatic C=C

NMR spectrum:

—integration ⇒ 8H

—δ 11.0, 1H ⇒ COOH

—δ 6.7-7.6, 5H ⇒ monosubstituted benzene ring

—δ 4.5, 2H singlet ⇒ CH_2, deshielded

(b) Fragments indicated in the spectra:

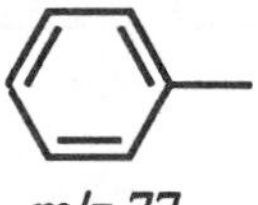

m/z 77

CH_2
m/z 14

COOH
m/z 45

This appears deceptively simple. The problem is that the mass of these fragments adds to 136, not 152—we are missing 16 mass units ⇒ oxygen! Where can the oxygen be? There are only two possibilities:

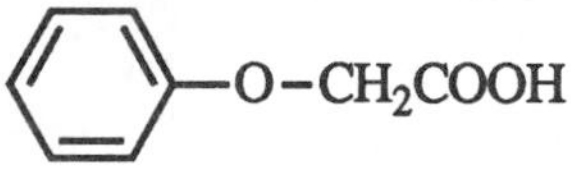
O−CH2COOH

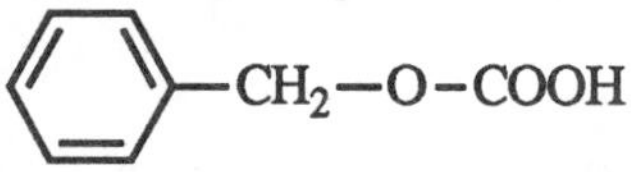
CH2−O−COOH

How can we differentiate? Mass spectrometry!

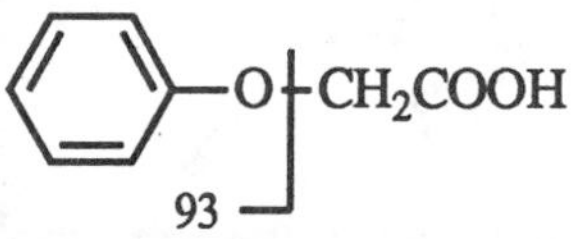
O
CH2COOH
93

phenoxyacetic acid

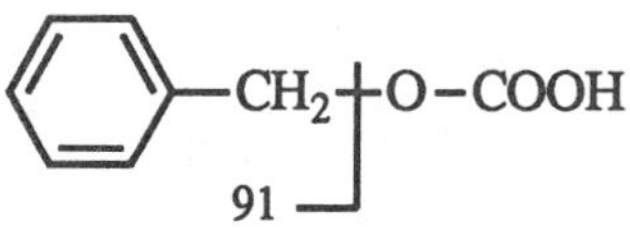
CH2
O−COOH
91

The m/z 93 peak in the MS confirms the structure is phenoxyacetic acid. The CH_2 is so far downfield in the NMR because it is between two electron-withdrawing groups, the O and the COOH.

20-35

(a)

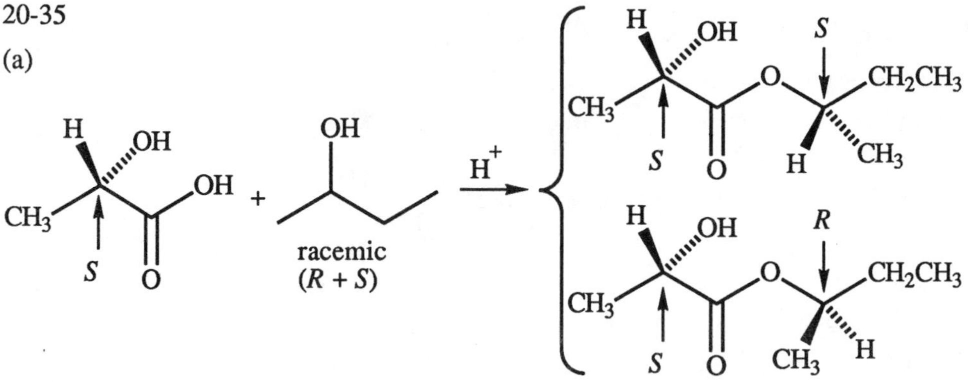

(b) Isomers which are *R,S* and *S,S* are diastereomers.

20-36

<u>Spectrum A</u>: $C_9H_{10}O_2$ ⇒ 5 elements of unsaturation

δ 11.8, 1H ⇒ COOH

δ 7.3, 5H ⇒ monosubstituted benzene ring

δ 3.8, 1H quartet ⇒ $CHCH_3$

δ 1.5, 3H doublet ⇒ $CHCH_3$

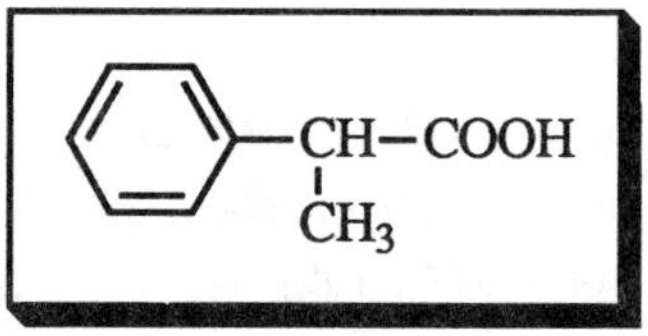

<u>Spectrum B</u>: $C_4H_6O_2$ ⇒ 2 elements of unsaturation

δ 11.5, 1H ⇒ COOH

δ 6.2, 1H singlet ⇒ H—C=C

δ 5.8, 1H singlet ⇒ H—C=C

δ 2.0, 3H singlet ⇒ vinyl CH_3 with no H neighbors CH_3—C=C

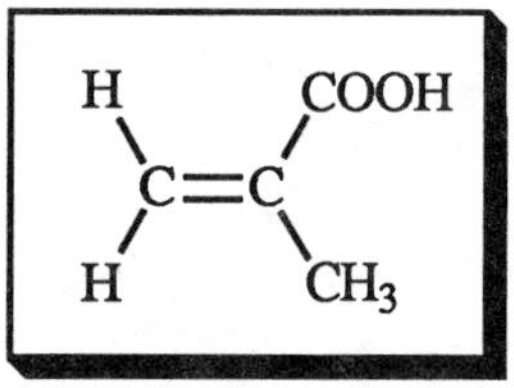

<u>Spectrum C</u>: $C_6H_{10}O_2$ ⇒ 2 elements of unsaturation

δ 12.0, 1H ⇒ COOH

δ 7.0, 1H multiplet ⇒ H—C=C—COOH

δ 5.7, 1H doublet ⇒ C=C(H)—COOH

δ 2.2-0.8 ⇒ $CH_2CH_2CH_3$

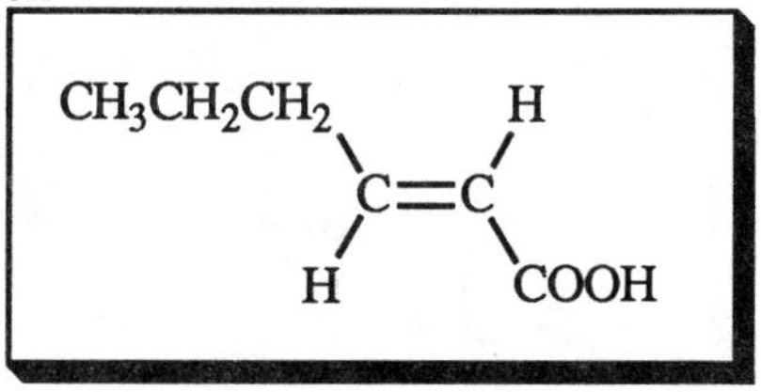

must be *trans* due to large coupling constant in doublet at δ 5.7

20-37

(a) $PhCH_2CH_2OH \xrightarrow[\text{pyridine}]{\text{TsCl}} \xrightarrow{\text{KCN}} PhCH_2CH_2CN \xrightarrow[\Delta]{H_3O^+} PhCH_2CH_2COOH$

$PhCH_2CH_2OH \xrightarrow{PBr_3} PhCH_2CH_2Br \xrightarrow[\text{ether}]{\text{Mg}} \xrightarrow{CO_2} \xrightarrow{H_3O^+} PhCH_2CH_2COOH$

(b)

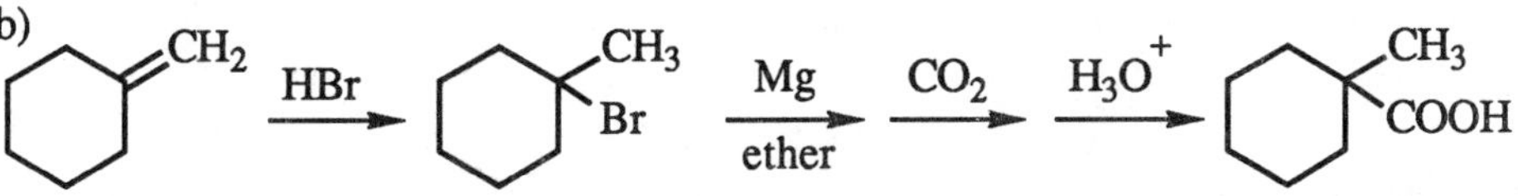

(c)

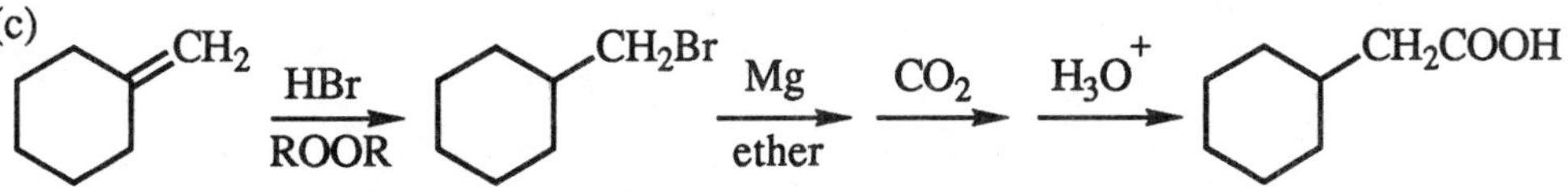

(d)

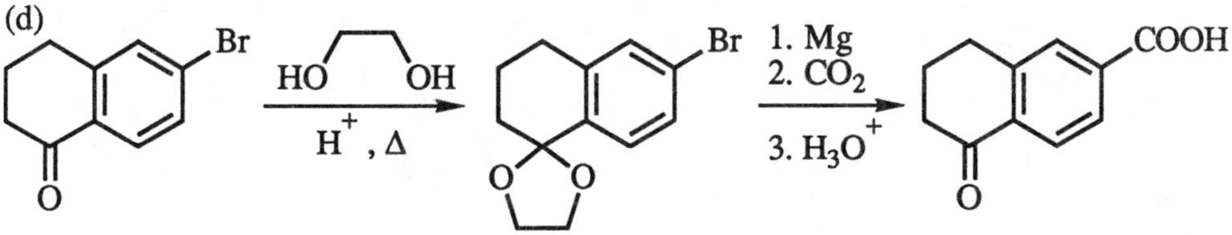

(e)

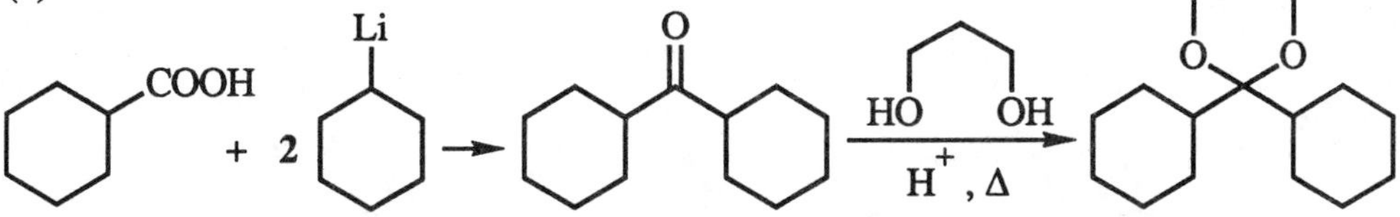

20-38

(a)

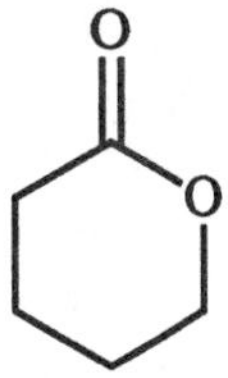
δ-valerolactone

(b)

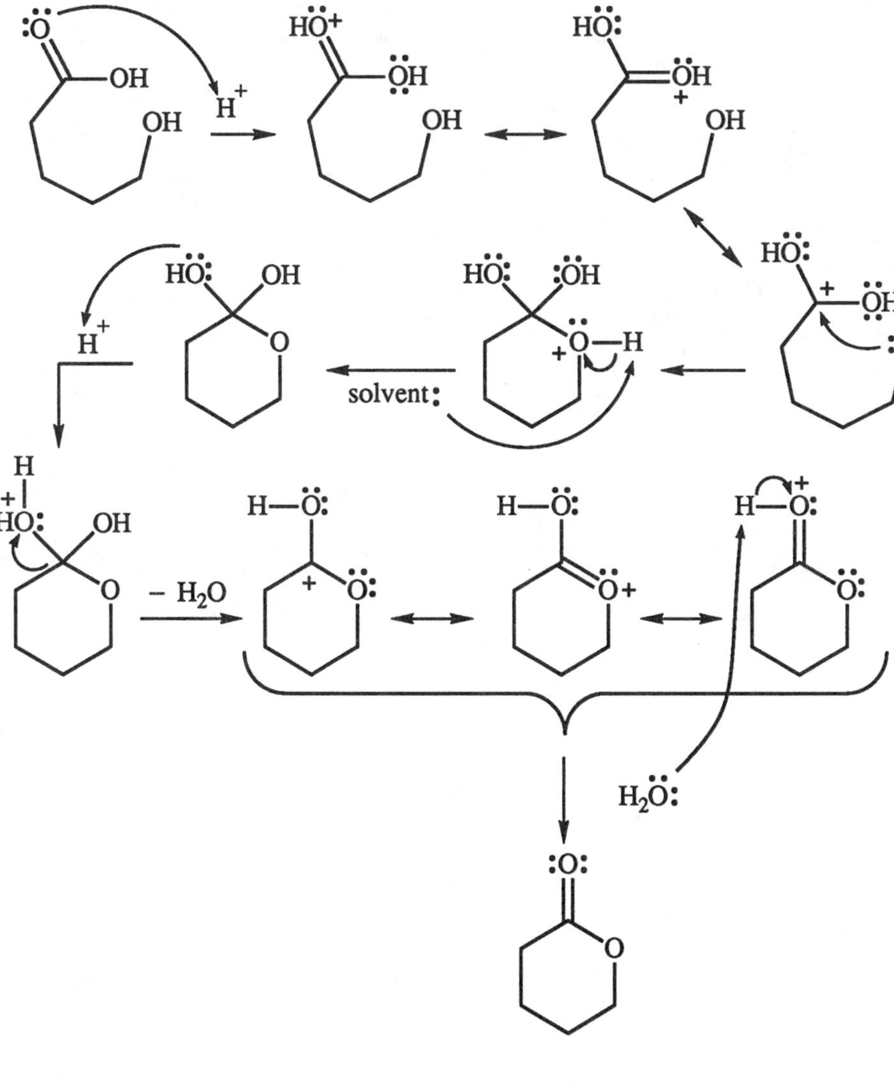
solvent
H$_2$O

20-39

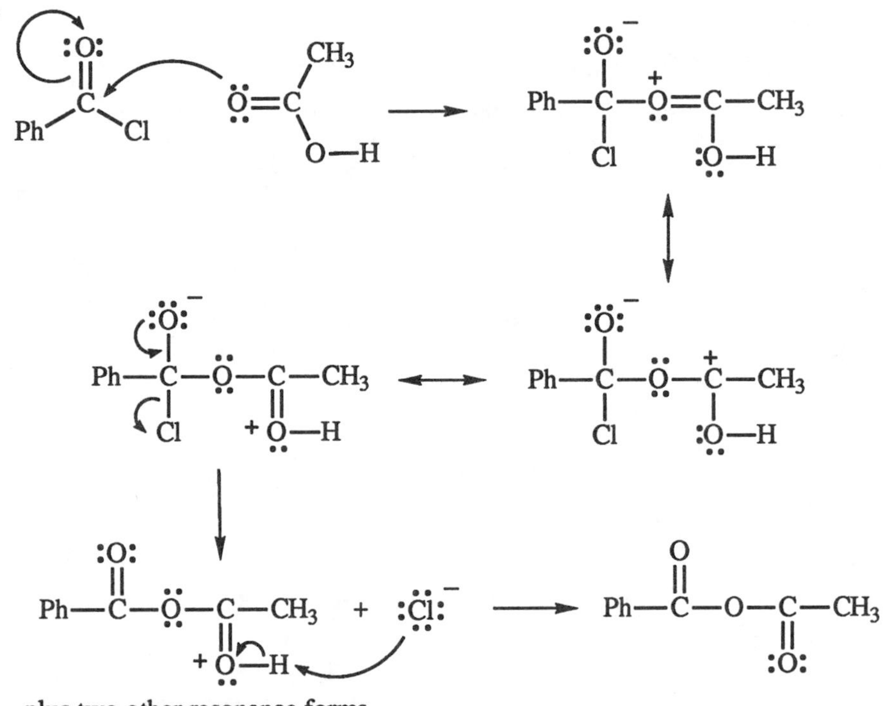

plus two other resonance forms

20-40

A few words about the two types of electronic effects: induction and resonance. Inductive effects are a result of polarized σ bonds, usually because of electronegative atom substituents. Resonance effects work through π systems, requiring overlap of p orbitals to delocalize electrons.

All substituents have an inductive effect compared to hydrogen (the reference). Many groups also have a resonance effect; all that is required to have a resonance effect is that the atom or group have at least one p orbital for overlap.

The most interesting groups have both inductive and resonance effects. In such groups, how can we tell the direction of electron movement, that is, whether a group is electron-donating or electron-withdrawing? And do the resonance and inductive effects reinforce or conflict with each other? We can never "turn off" an inductive effect from a resonance effect; that is, any time a substituent is expressing its resonance effect, it is also expressing its inductive effect. We can minimize a group's inductive effect by moving it farther away; inductive effects decrease with distance. The other side of the coin is more accessible to the experimenter: we can "turn off" a resonance effect in order to isolate an inductive effect. We can do this by interrupting a conjugated π system by inserting an sp^3-hybridized atom, or by making resonance overlap impossible for steric reasons (steric inhibition of resonance).

20-40 continued

These three problems are examples of separating inductive effects from resonance effects.

(a) In electrophilic aromatic substitution, the phenyl substituent is an ortho,para-director because it can stabilize the intermediate from electrophilic attack at the ortho and para positions. The phenyl substituent is electron-donating *by resonance.*

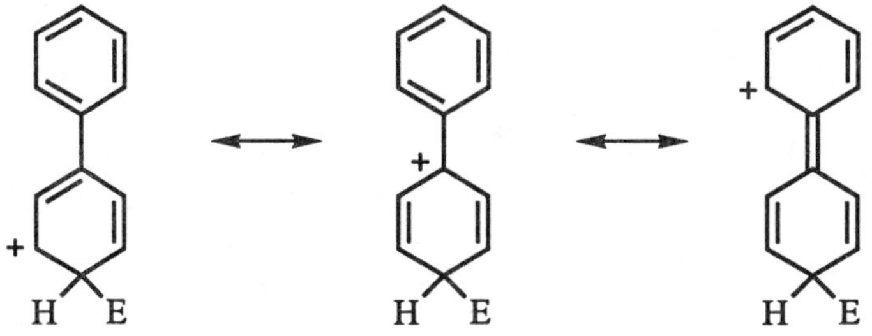

plus other resonance forms

BUT:

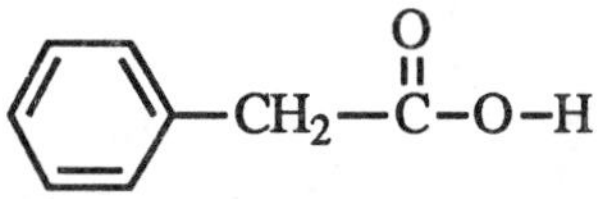

is a stronger acid than $H-CH_2-C(=O)-O-H$

The greater acidity of phenylacetic acid shows that the phenyl substituent is electron-withdrawing, thereby stabilizing the product carboxylate's negative charge. Does this contradict what was said above? Yes and no. What is different is that, since there is no p-orbital overlap between the phenyl group and the carboxyl group because of the CH_2 group in between, the increased acidity must be from a pure *inductive effect.* This structure isolates the inductive effect (which can't be "turned off") from the resonance effect of the phenyl group.

We can conclude three things: (1) phenyl is electron-withdrawing by induction; (2) phenyl is (in this case) electron-donating by resonance; (3) for phenyl, the resonance effect is stronger than the inductive effect (since it is an ortho,para-director).

(b) The simpler case first—induction only:

$$CH_3O-CH_2-C(=O)-O-H \quad \text{is a stronger acid than} \quad H-CH_2-C(=O)-O-H$$

There is no resonance overlap between the methoxy group and the carboxyl group, so this is a pure inductive effect. The methoxy substituent increases the acidity, so methoxy must be electron-withdrawing by induction. This should come as no surprise as oxygen is the second most electronegative element.

The anomaly comes in the decreased acidity of 4-methoxybenzoic acid:

20-40 (b) continued

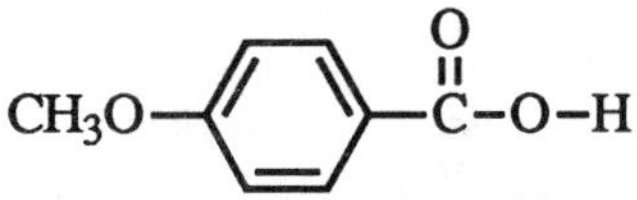 is a weaker acid than 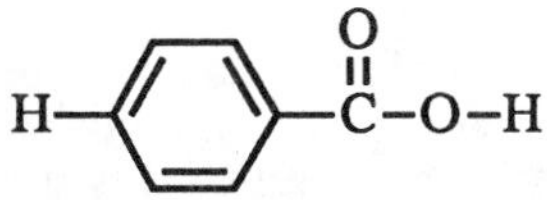

Through resonance, a pair of electrons from the methoxy oxygen can be donated through the benzene ring to the carboxyl group—a stabilizing effect. However, this electron donation *destabilizes* the carboxylate anion as there is already a negative charge on the carboxyl group; the resonance donation intensifies the negative charge. Since the product of the equilibrium would be destabilized relative to the starting material, the proton donation would be less favorable, which we define as a weaker acid.

Methoxy is another example of a group which is electron-withdrawing by induction but electron-donating by resonance.

(c) This problem gives three pieces of data to interpret:

(1) $CH_3-CH_2-C(=O)-O-H$ is a weaker acid than $H-CH_2-C(=O)-O-H$

Interpretation: the methyl group is electron-donating by induction.

(2) 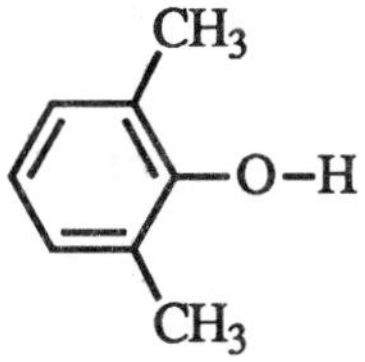 is a weaker acid than 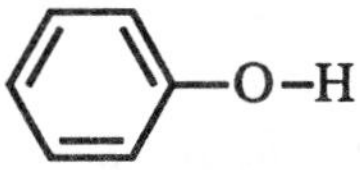

Interpretation: the methyl group is electron-donating by induction. This interpretation is consistent with (1), as expected, since methyl cannot have any resonance effect.

(3) 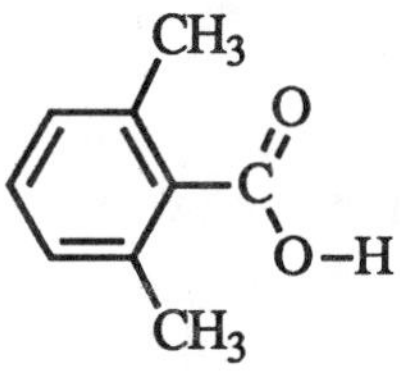 is a **stronger** acid than 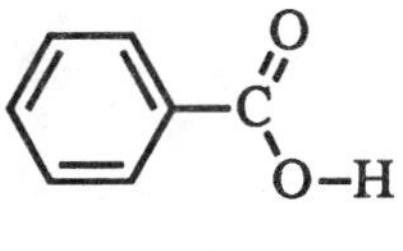

Interpretation: this is the anomaly. Contradictory to the data in (1) and (2), by putting on two methyl groups, the substituent seems to have become electron-withdrawing instead of electron-donating. How?

20-40 (c) continued

Steric inhibition of resonance! In benzoic acid, the phenyl ring and the carboxyl group are all in the same plane, and benzene is able to donate electrons by resonance overlap through parallel p orbitals. This stabilizes the starting acid (and destabilizes the carboxylate anion) and makes the acid weaker than it would be without resonance.

plus other resonance forms

Putting substituents at the 2- and 6-positions prevents the carboxyl or carboxylate from coplanarity with the ring. Resonance is interrupted, and now the carboxyl group sees a phenyl substituent which cannot stabilize the acid through resonance; the stabilization of the acid is lost. At the same time, the *electron-withdrawing inductive effect* of the benzene ring stabilizes the carboxylate anion. These two effects work together to make this acid unusually strong. (Apparently, the slight electron-donating inductive effect of the methyls is overpowered by the stronger electron-withdrawing inductive effect of the benzene ring.)

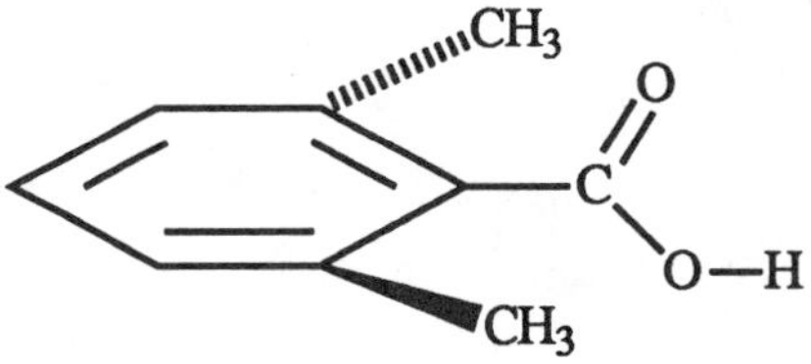

COOH group is perpendicular to the plane of the benzene ring—no resonance interaction.

20-41

(a) 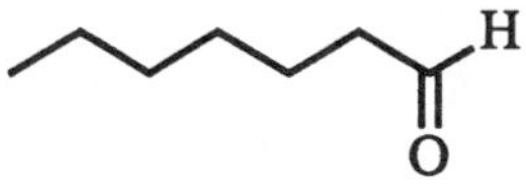

stock bottle

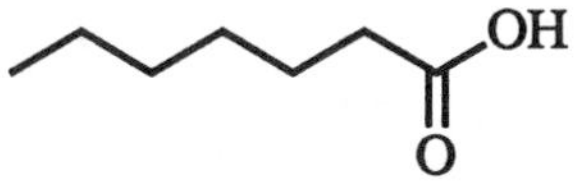

students' samples

(b) The spectrum of the students' samples shows the carboxylic acid present. Apparently, contact with oxygen from the air oxidized the sensitive aldehyde group to the acid.

(c) Storing the aldehyde in an inert atmosphere like nitrogen or argon will prevent oxidation. Freshly prepared unknowns will avoid the problem.

CHAPTER 21—CARBOXYLIC ACID DERIVATIVES

21-1 IUPAC name first; then common name

(a) isobutyl benzoate (both IUPAC and common)
(b) phenyl methanoate; phenyl formate
(c) methyl 2-phenylpropanoate; methyl α-phenylpropionate
(d) *N*-phenyl-3-methylbutanamide; β-methylbutyranilide
(e) *N*-benzylethanamide; *N*-benzylacetamide
(f) 3-hydroxybutanenitrile; β-hydroxybutyronitrile
(g) 3-methylbutanoyl bromide; isovaleryl bromide
(h) dichloroethanoyl chloride; dichloroacetyl chloride
(i) 2-methylpropanoic methanoic anhydride; isobutyric formic anhydride
(j) cyclopentyl cyclobutanecarboxylate (both IUPAC and common)
(k) 5-hydroxyhexanoic acid lactone; δ-caprolactone
(l) *N*-cyclopentylbenzamide (both IUPAC and common)
(m) propanedioic anhydride; malonic anhydride
(n) 1-hydroxycyclopentanecarbonitrile; cyclopentanone cyanohydrin
(o) *cis*-4-cyanocyclohexanecarboxylic acid; no common name
(p) 3-bromobenzoyl chloride; *m*-bromobenzoyl chloride
(q) *N*-methyl-5-aminoheptanoic acid lactam; no common name
(r) *N*-ethanoylpiperidine; *N*-acetylpiperidine

21-2 An aldehyde has a C—H absorption (usually 2 peaks) at 2700-2800 cm^{-1}. A carboxylic acid has a strong, broad absorption between 2500-3500 cm^{-1}. The spectrum of methyl benzoate has no peaks in this region.

21-3 The C—O single bond stretch in ethyl octanoate appears at 1150 cm^{-1}, while methyl benzoate shows this absorption at 1280 cm^{-1}.

21-4

(a) acid chloride: single C=O peak at 1820 cm^{-1}; no other carbonyl comes so high
(b) primary amide: C=O at 1650 cm^{-1} and two N—H peaks between 3200-3400 cm^{-1}
(c) carboxylic acid: C=O at 1710 cm^{-1} and broad O—H at 2500-3500 cm^{-1}
(d) anhydride: two C=O absorptions at 1730 and 1800 cm^{-1}

21-5

(a) The formula C_3H_5NO has two elements of unsaturation. The IR spectrum shows two peaks between 3200-3400 cm^{-1}, an NH_2 group. The strong peak at 1690 cm^{-1} is a C=O, and the peak at 1600 cm^{-1} is a C=C. This accounts for all of the atoms.

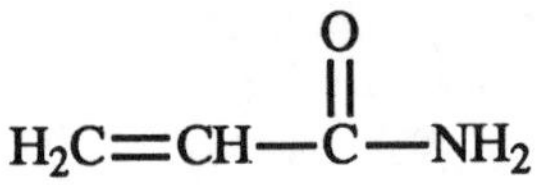

The NMR corroborates the assignment. The 1H multiplet at δ 5.6 is the vinyl H next to the carbonyl. The 2H multiplet at δ 6.3 is the vinyl hydrogen pair on carbon-3. The broad 2H hump at δ 7.3 is the amide hydrogens.

(b) The formula $C_5H_8O_2$ has two elements of unsaturation. The IR spectrum shows no OH, so this compound is neither an alcohol nor a carboxylic acid. The strong peak at 1730 cm^{-1} is likely an ester carbonyl. The C—O appears at 1200 cm^{-1}. The IR shows no C=C absorption, so the other element of unsaturation is likely a ring. The carbon NMR spectrum shows the carbonyl carbon at δ 173, the C—O carbon at δ 67, and three more carbons in the aliphatic region.

Several structures are consistent with an ester and a ring. Some possibilities:

21-6 Reactions which go from a more reactive functional group to a less reactive functional group ("downhill reactions") will occur readily. All of these reactions except (a) fit this pattern and will take place as written. In part (a), an amide will not readily form an acid chloride.

21-7

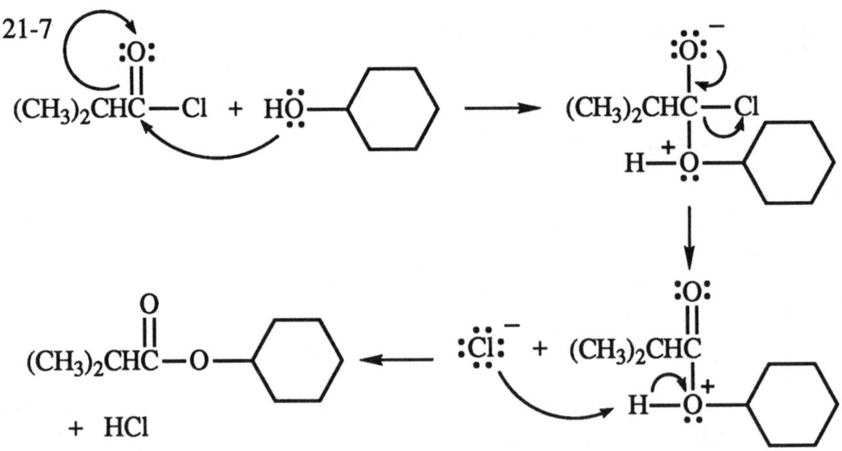

21-8

(a)

$$CH_3CH_2\overset{O}{\overset{\|}{C}}-Cl + HOCH_2CH_3 \longrightarrow CH_3CH_2\overset{O}{\overset{\|}{C}}-OCH_2CH_3 + HCl$$

(b)

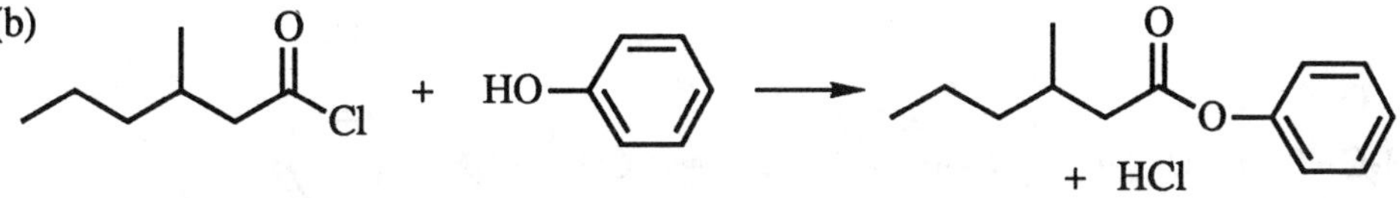

(c)

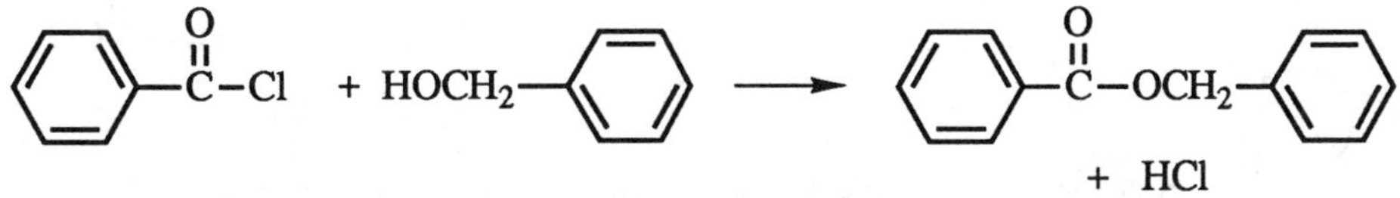

(d)

21-9

(a)

$$CH_3\overset{O}{\overset{\|}{C}}-Cl + HN(CH_3)_2 \longrightarrow CH_3\overset{O}{\overset{\|}{C}}-N(CH_3)_2 + HCl$$

(b)

(c)

(d)

21-10

(a) (i)

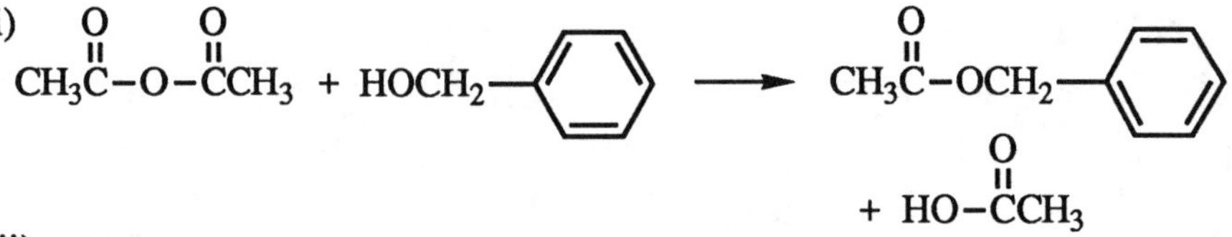

(ii)

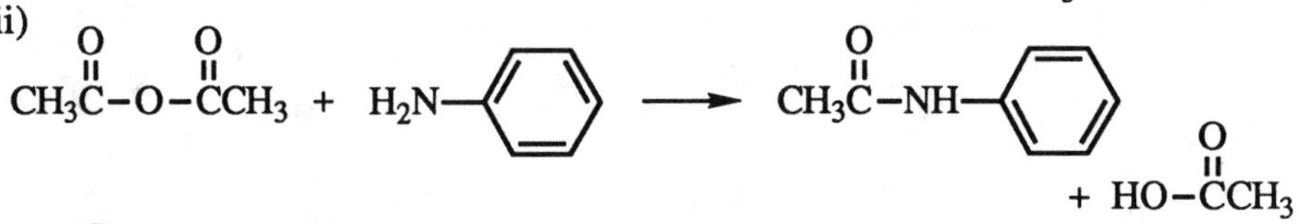

(b) (i)

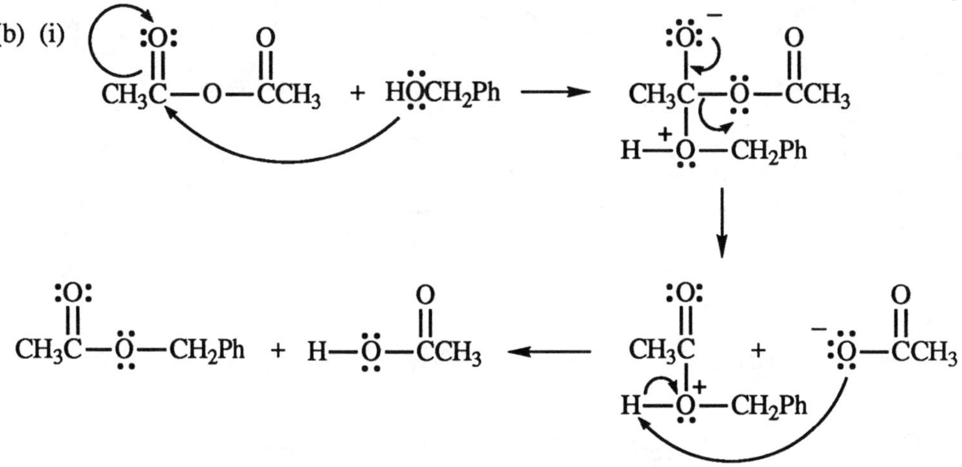

(ii)

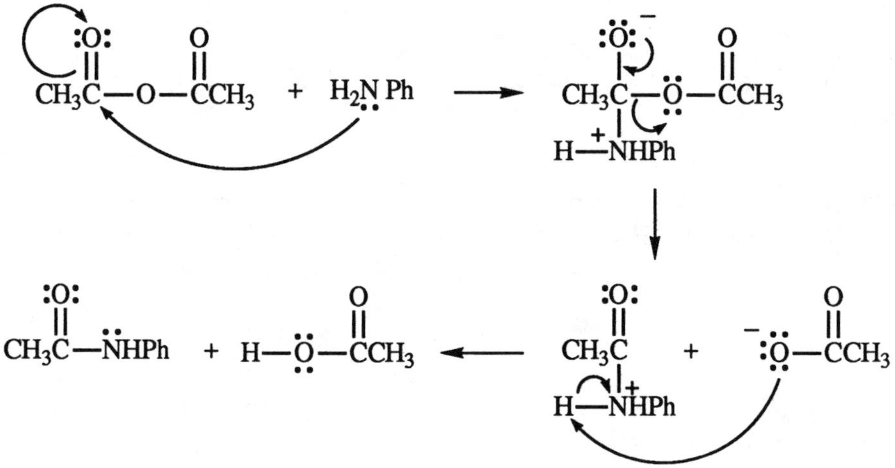

21-11

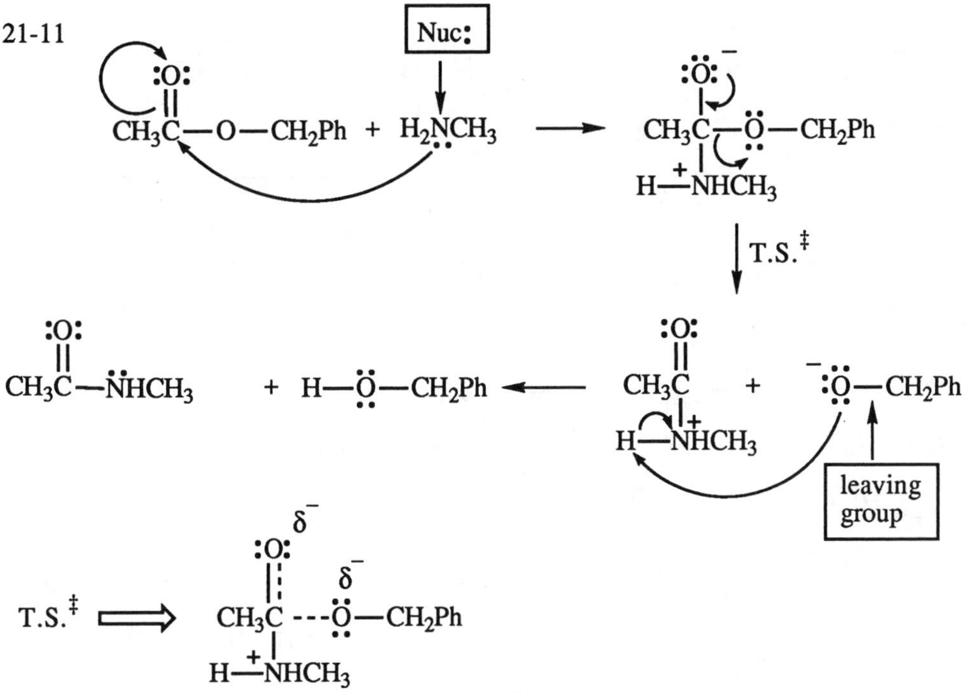

21-12

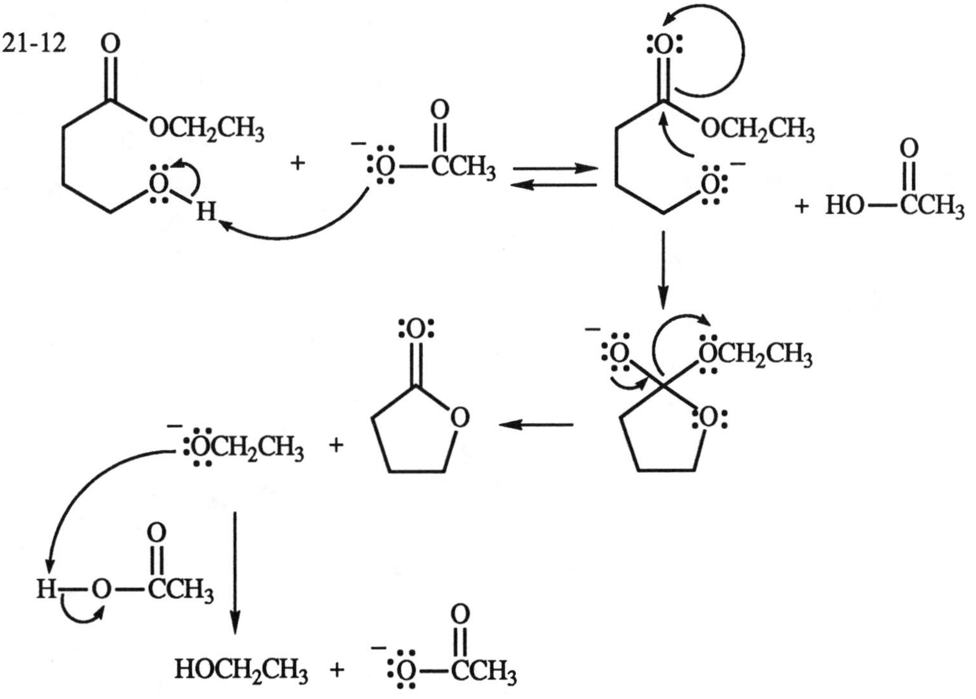

21-13

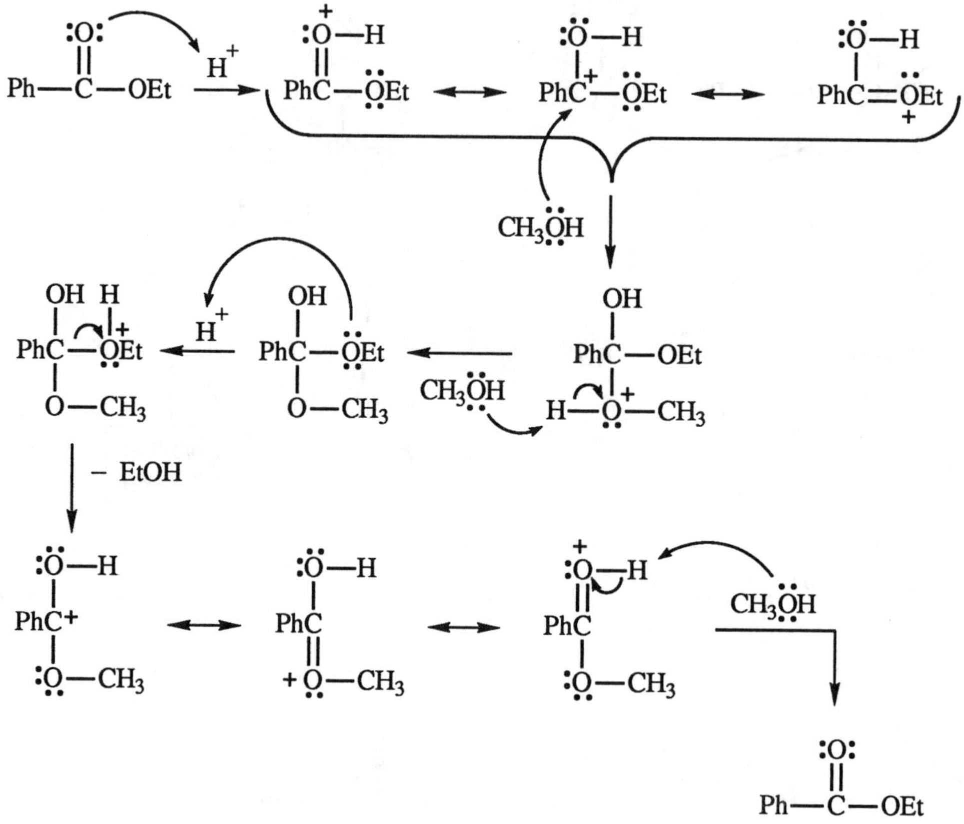

21-14

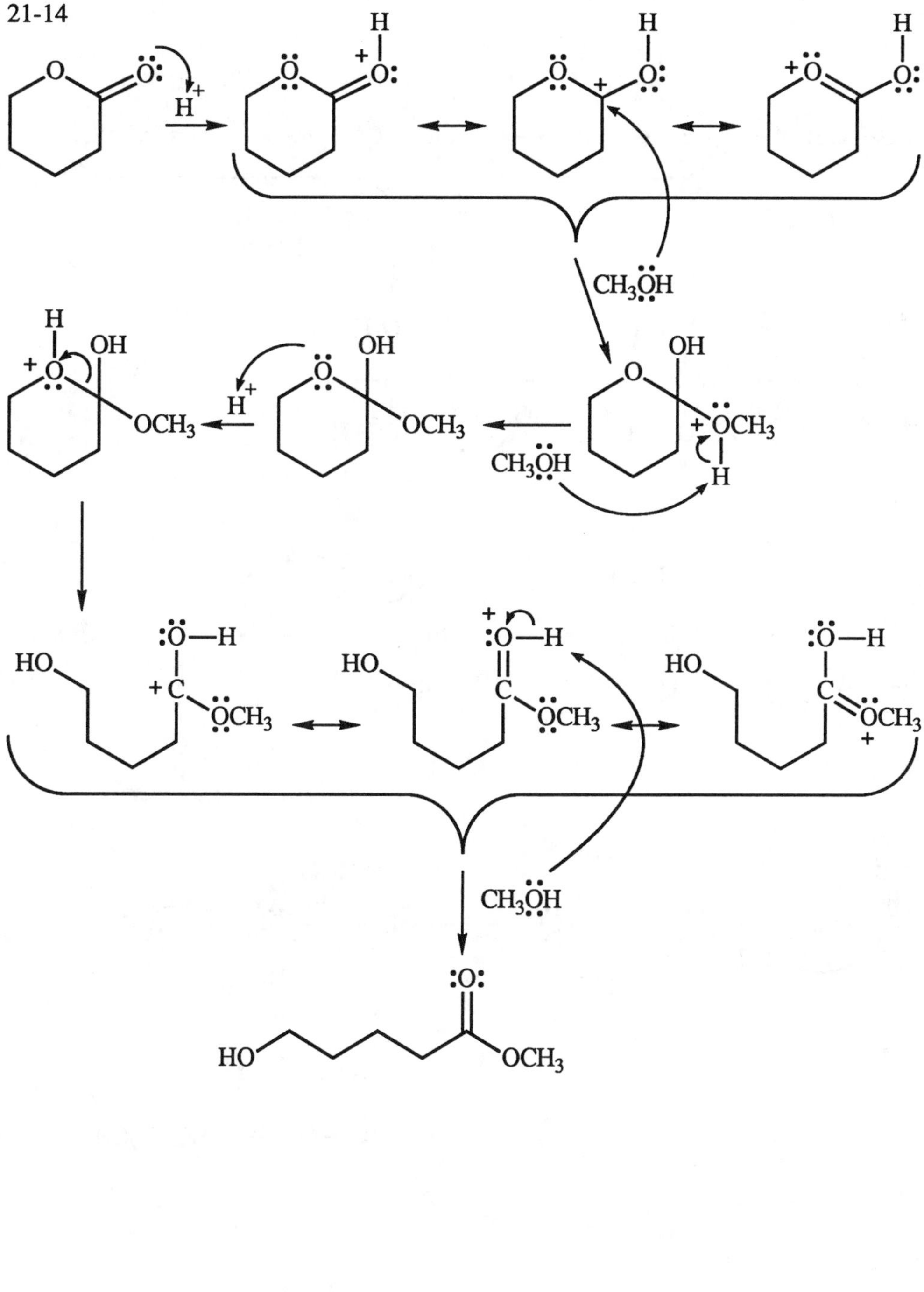

21-15

(a)

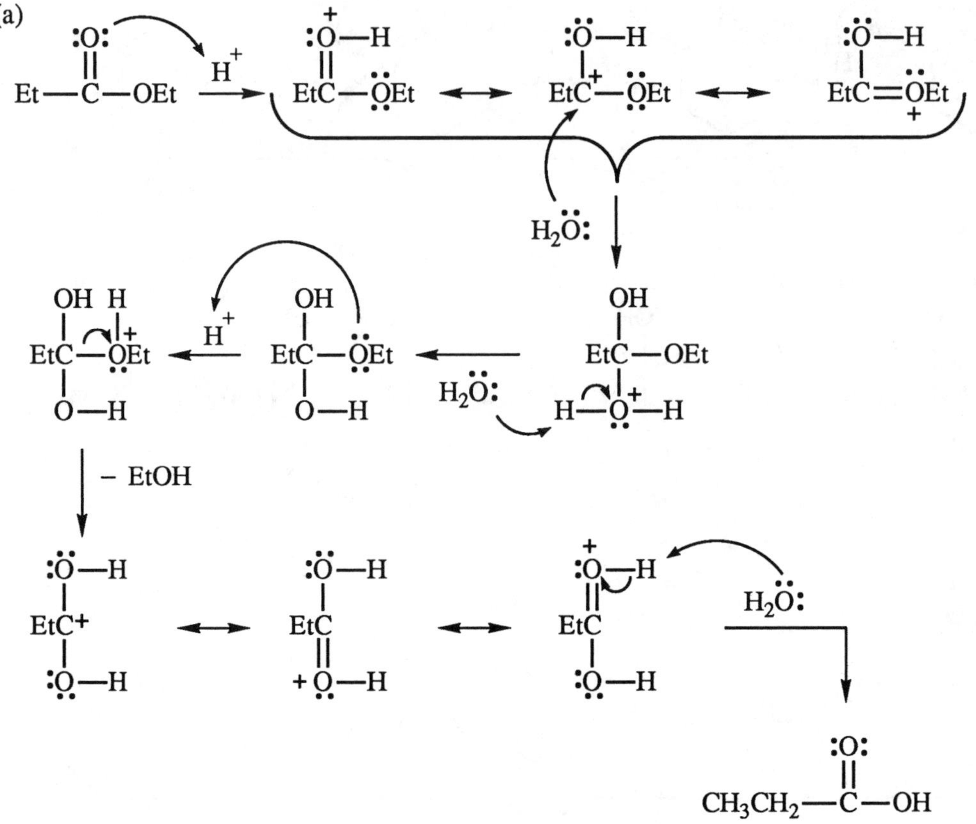

(b)

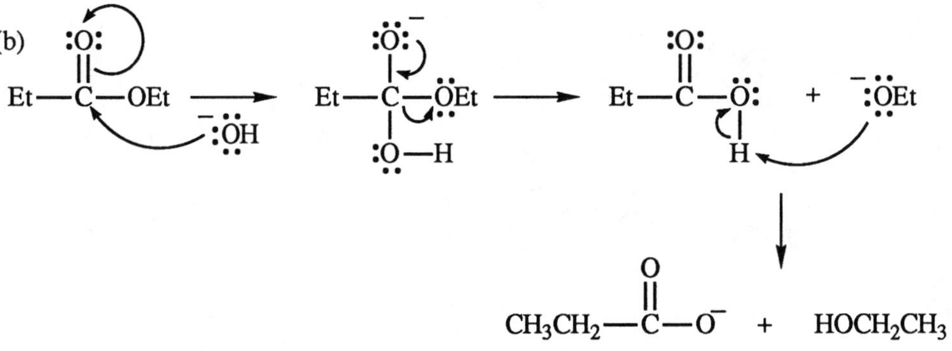

21-16

(a) A catalyst is defined as a chemical species that speeds a reaction but is not consumed in the reaction. In the acidic hydrolysis, acid is used in the first and fourth steps of the mechanism but is regenerated in the third and last steps. Acid is not consumed; the final concentration of acid is the same as the initial concentration. In the basic hydrolysis, however, the hydroxide that initially attacks the carbonyl is never regenerated. An alkoxide leaves from the carbonyl, but it quickly neutralizes the carboxylic acid. For every molecule of ester, one molecule of hydroxide is consumed; the base *promotes* the reaction but does not *catalyze* the reaction.

(b) Basic hydrolysis is not reversible. Once an ester molecule is hydrolyzed in base, the carboxylate cannot form ester. Acid catalysis, however, is an equilibrium: the mixture will always contain some ester, and the yield will never be as high as in basic hydrolysis. Second, long chain fatty acids are not soluble in water until they are neutralized; they are soluble as their sodium salts (soap). Basic hydrolysis is preferred for higher yield and greater solubility of the product.

21-17

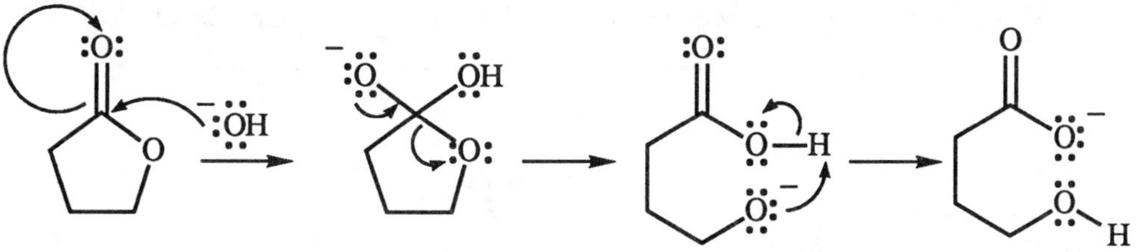

21-18

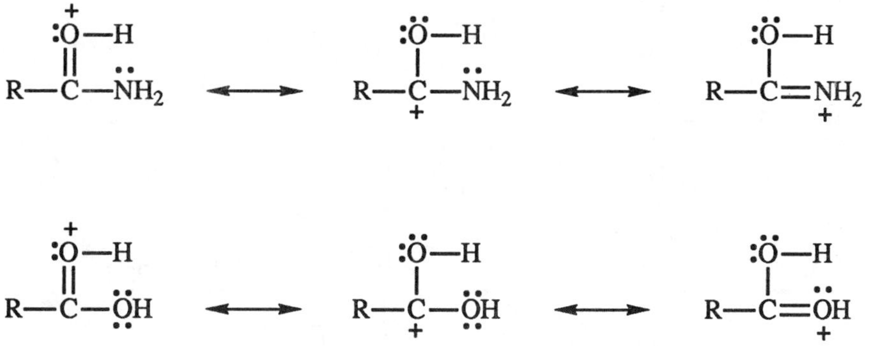

21-19

(a)

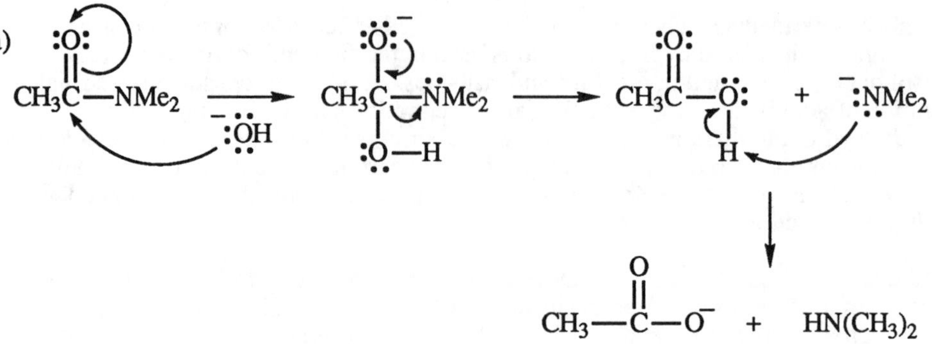

(b)

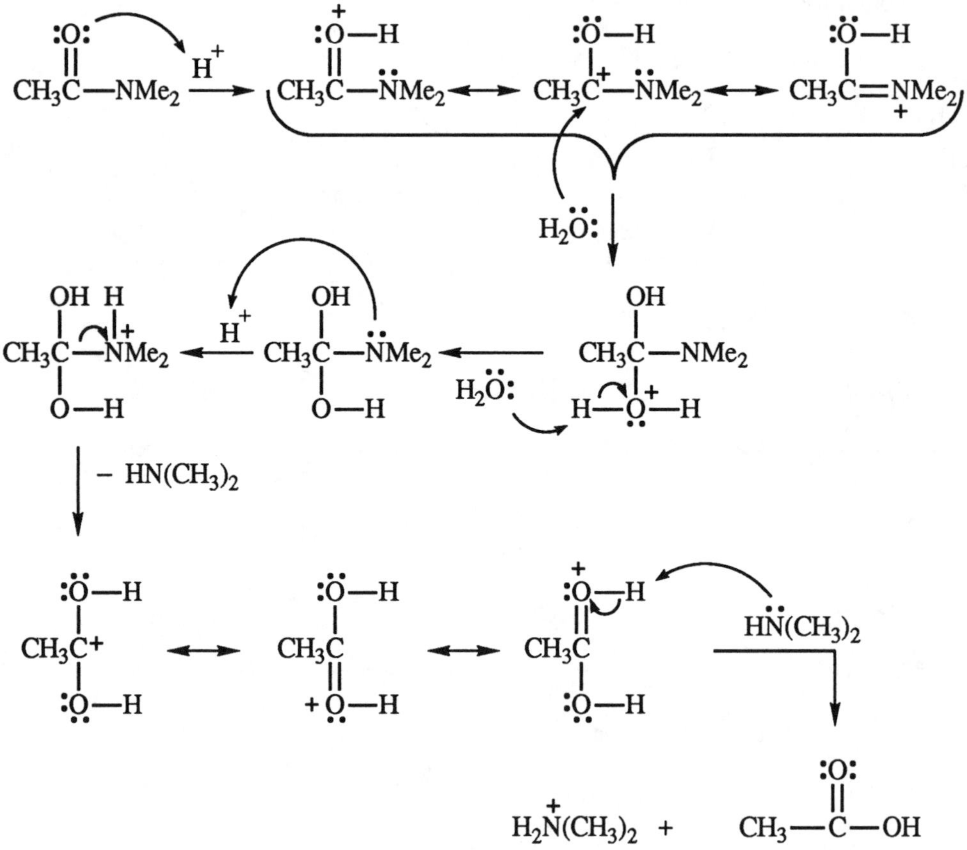

21-20

In the basic hydrolysis (21-19(a)), the step which drives the reaction to completion is the final step, the deprotonation of the carboxylic acid by the amide anion. In the acidic hydrolysis (21-19(b)), protonation of the amine by acid is exothermic and it prevents the reverse reaction by tying up the pair of electrons on the nitrogen so that the amine is no longer nucleophilic.

21-21

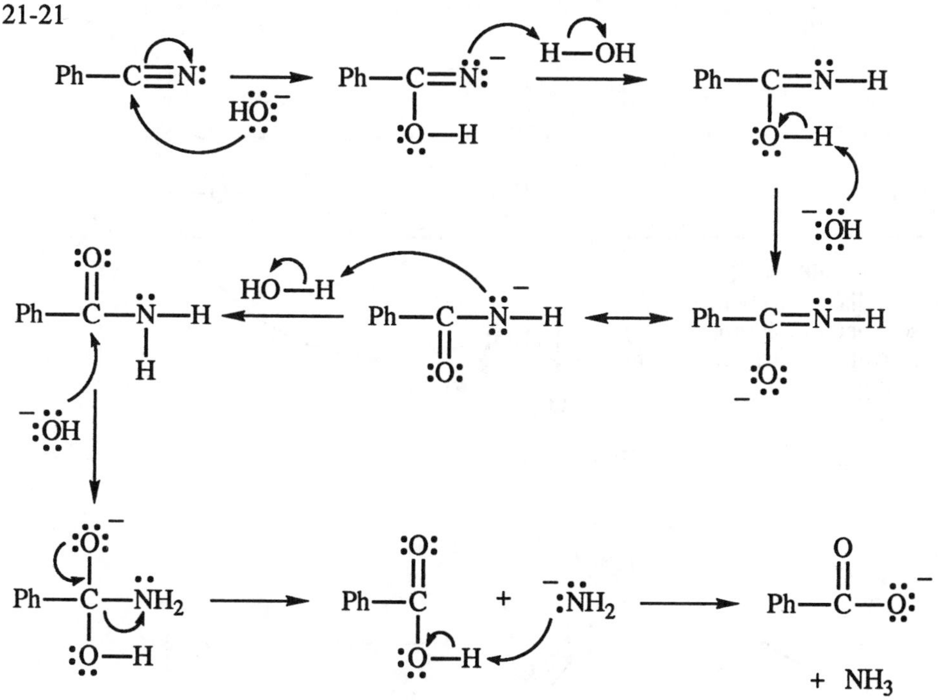

21-22

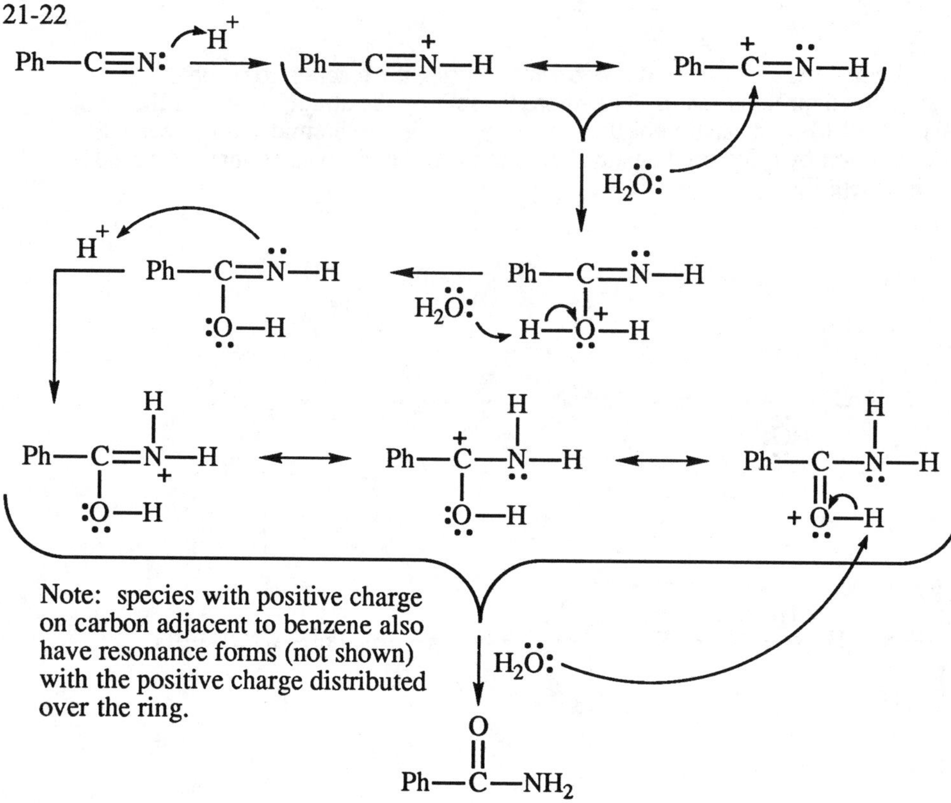

Note: species with positive charge on carbon adjacent to benzene also have resonance forms (not shown) with the positive charge distributed over the ring.

21-23

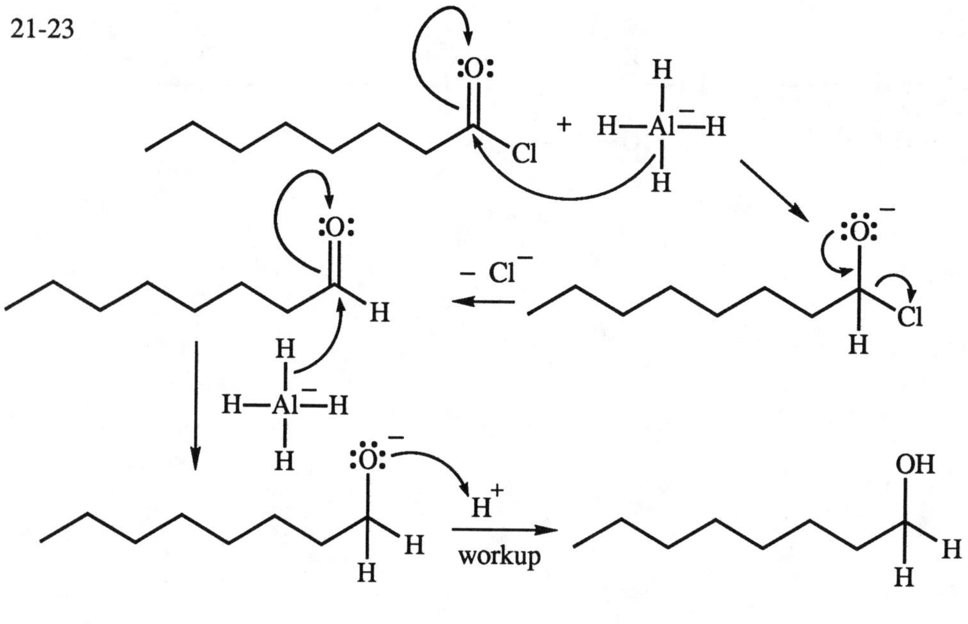

21-24

(a)

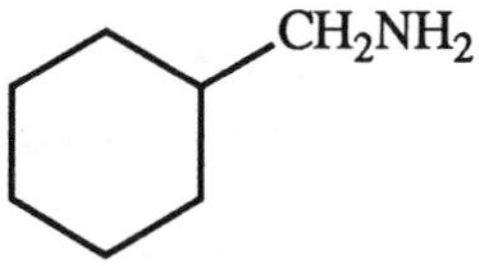

(b)

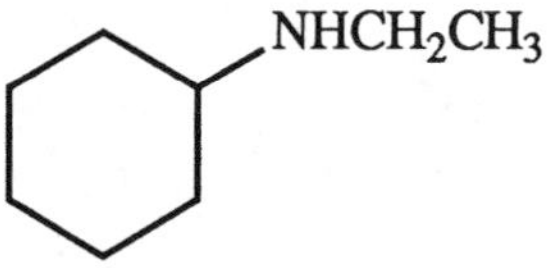

(c)

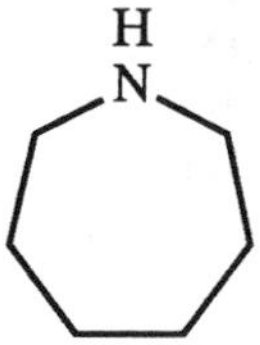

(d)

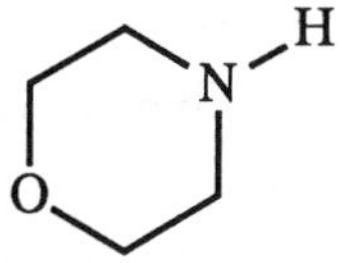

(e)

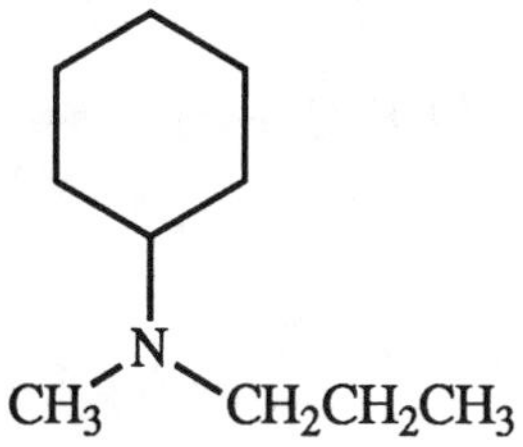

(f)

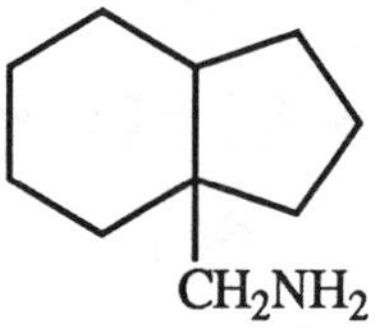

21-25

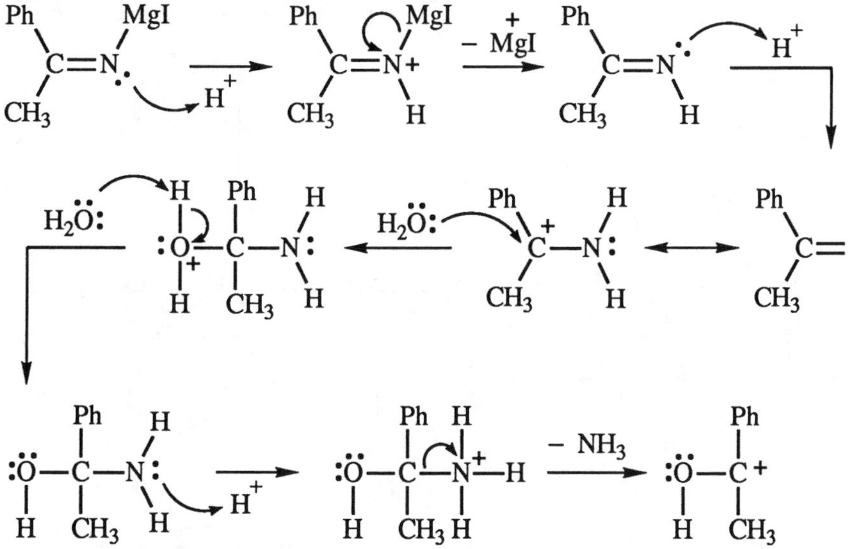

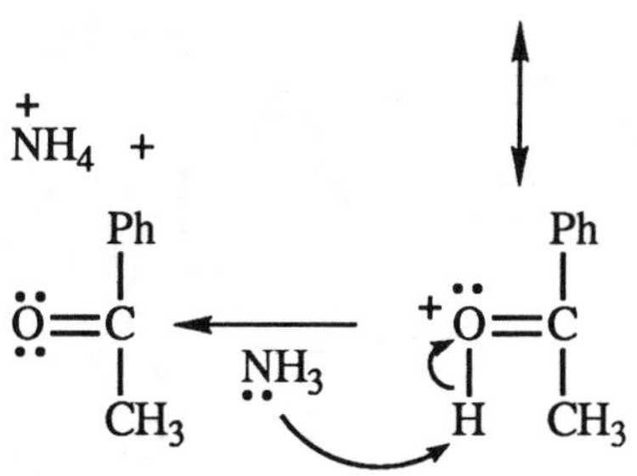

Note: species with positive charge on carbon adjacent to benzene also have resonance forms (not shown) with the positive charge distributed over the ring.

21-26

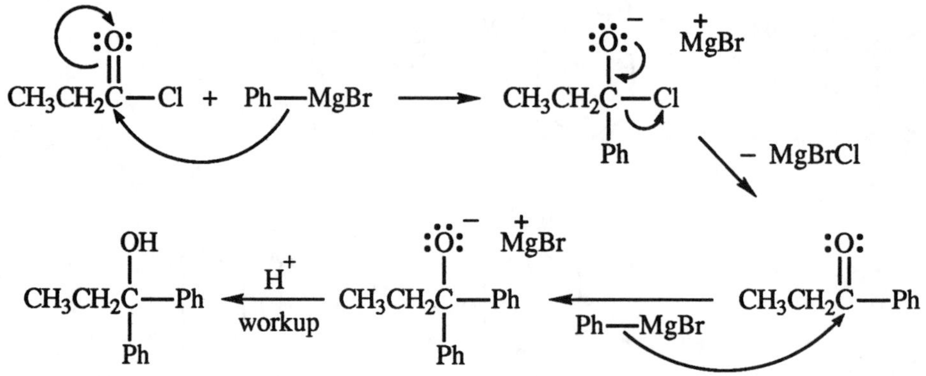

CH3CH2C—Cl
Ph—MgBr
MgBr
CH3CH2C—Cl
Ph
– MgBrCl
OH
CH3CH2C—Ph
Ph
H+
workup
MgBr
CH3CH2C—Ph
Ph
Ph—MgBr
CH3CH2C—Ph

21-27

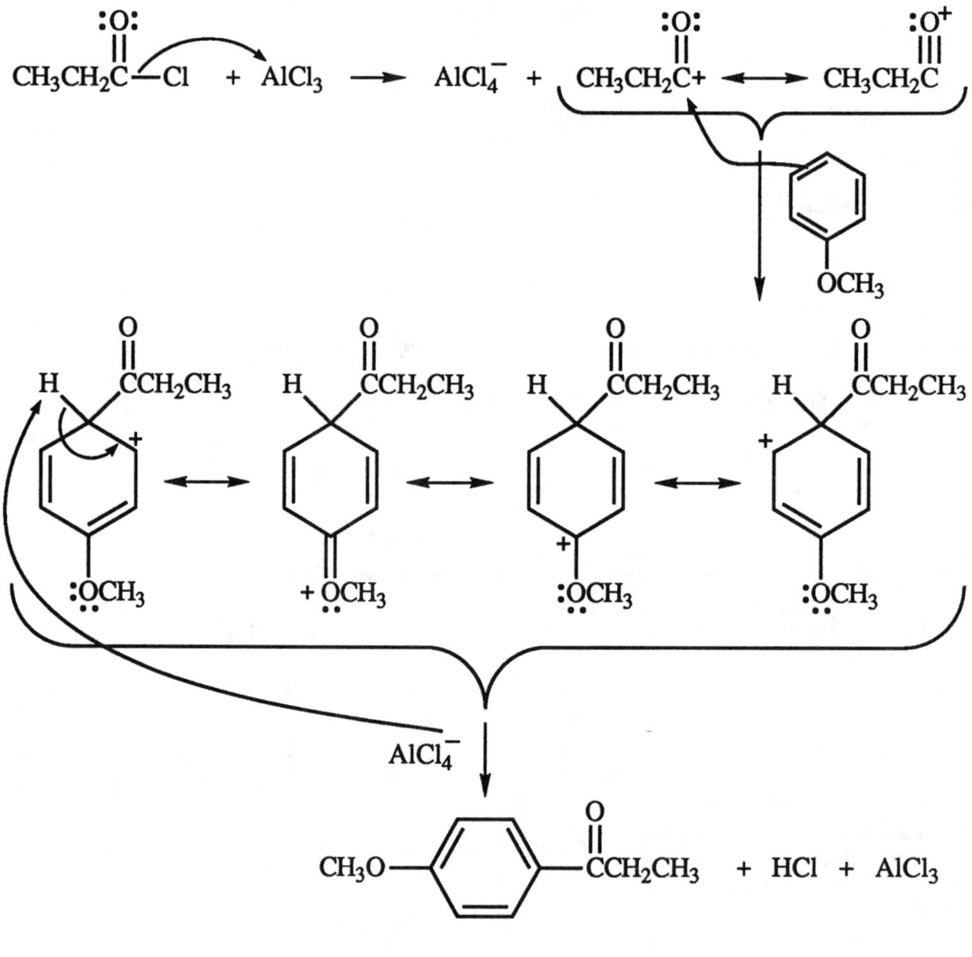

CH3CH2C—Cl
AlCl3
AlCl4–
CH3CH2C+
CH3CH2C
OCH3
O
CCH2CH3
H
:OCH3
+ OCH3
AlCl4–
CH3O
CCH2CH3
+ HCl + AlCl3

21-28

(a)

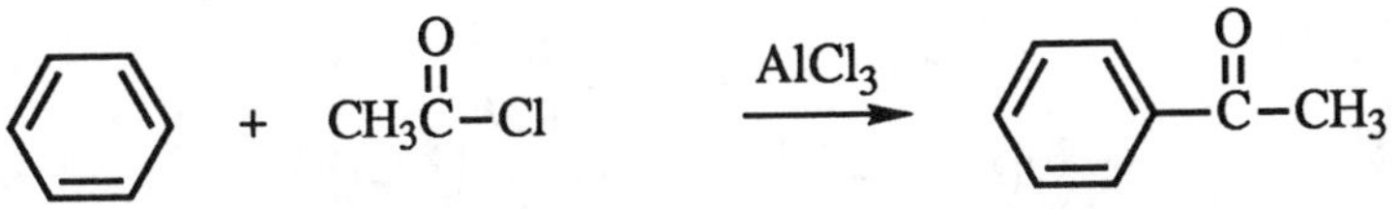

(b)

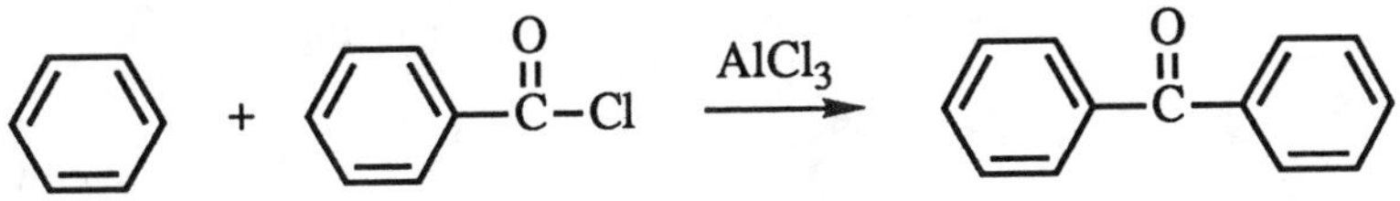

(c)

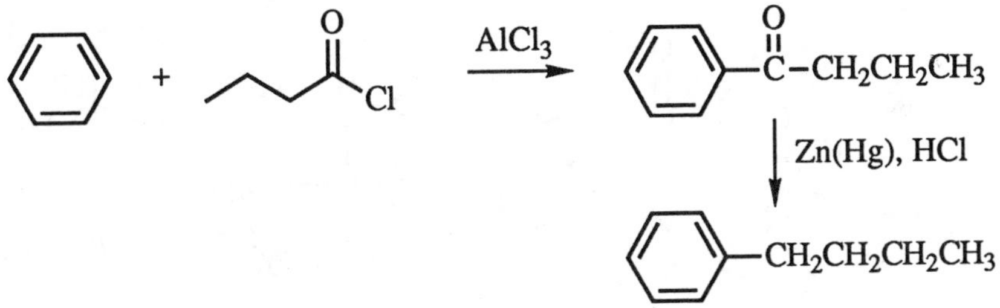

21-29

(a) (i)

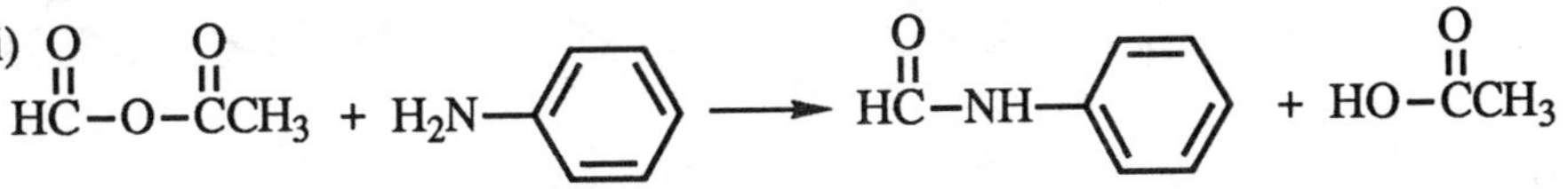

(ii)

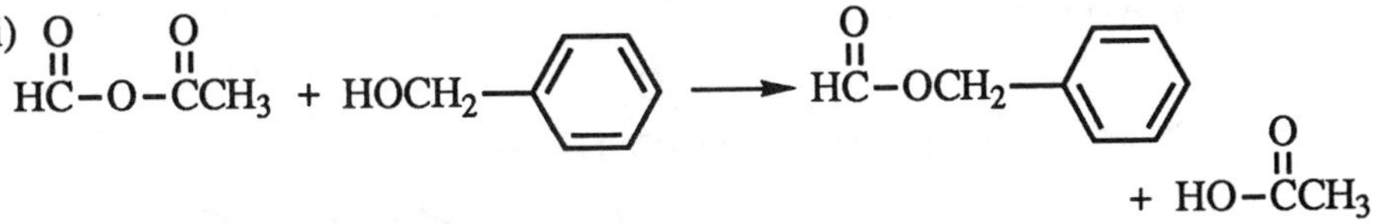

(b) (i)

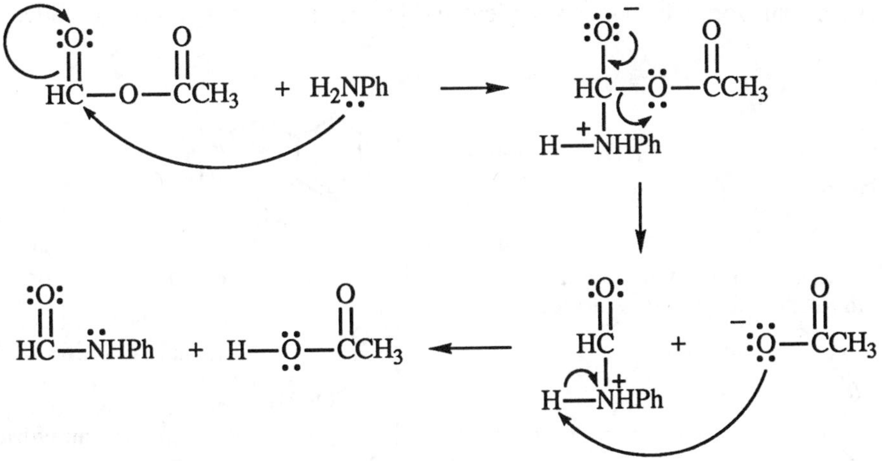

21-29 (b) continued

(ii)

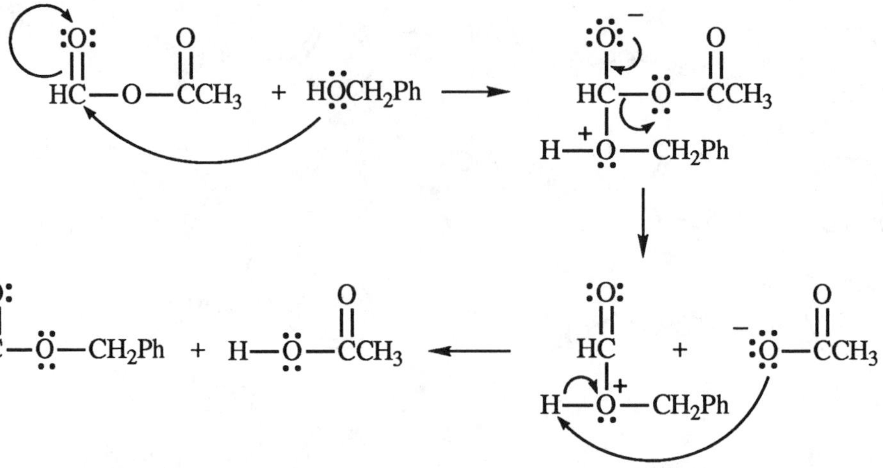

21-30

(a)

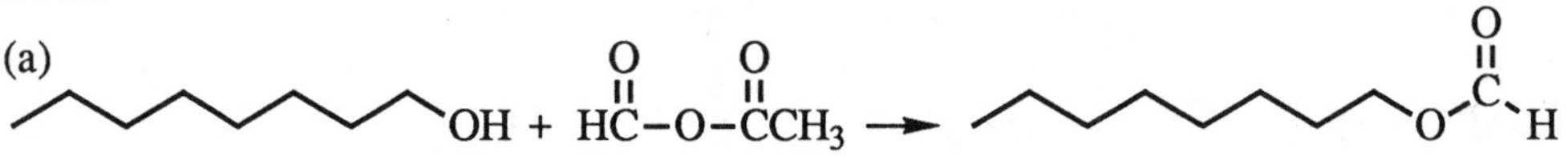

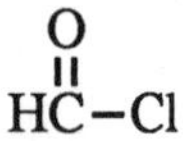 does not exist, so acetic formic anhydride is the most practical way to formylate the alcohol.

(b)

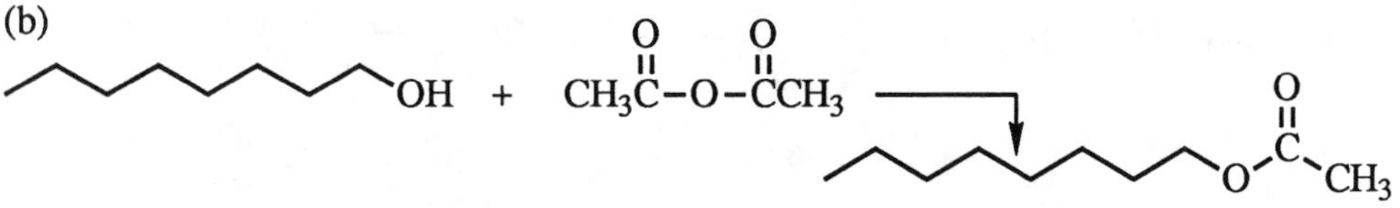

Acetic anhydride is more convenient and less expensive than acetyl chloride.

(c)

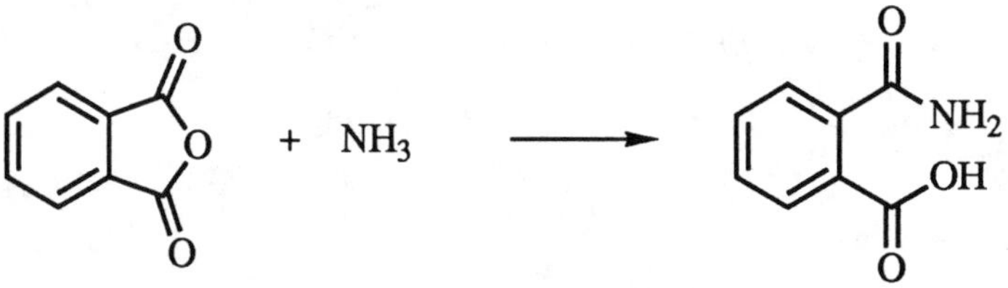

The acid chloride would tend to react at both carbonyls instead of just one; only the anhydride will give this product.

(d)

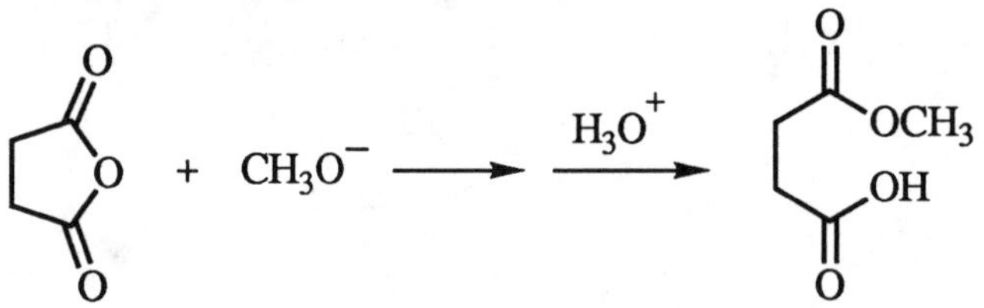

The acid chloride would tend to react at both carbonyls instead of just one; only the anhydride will give this product.

21-31

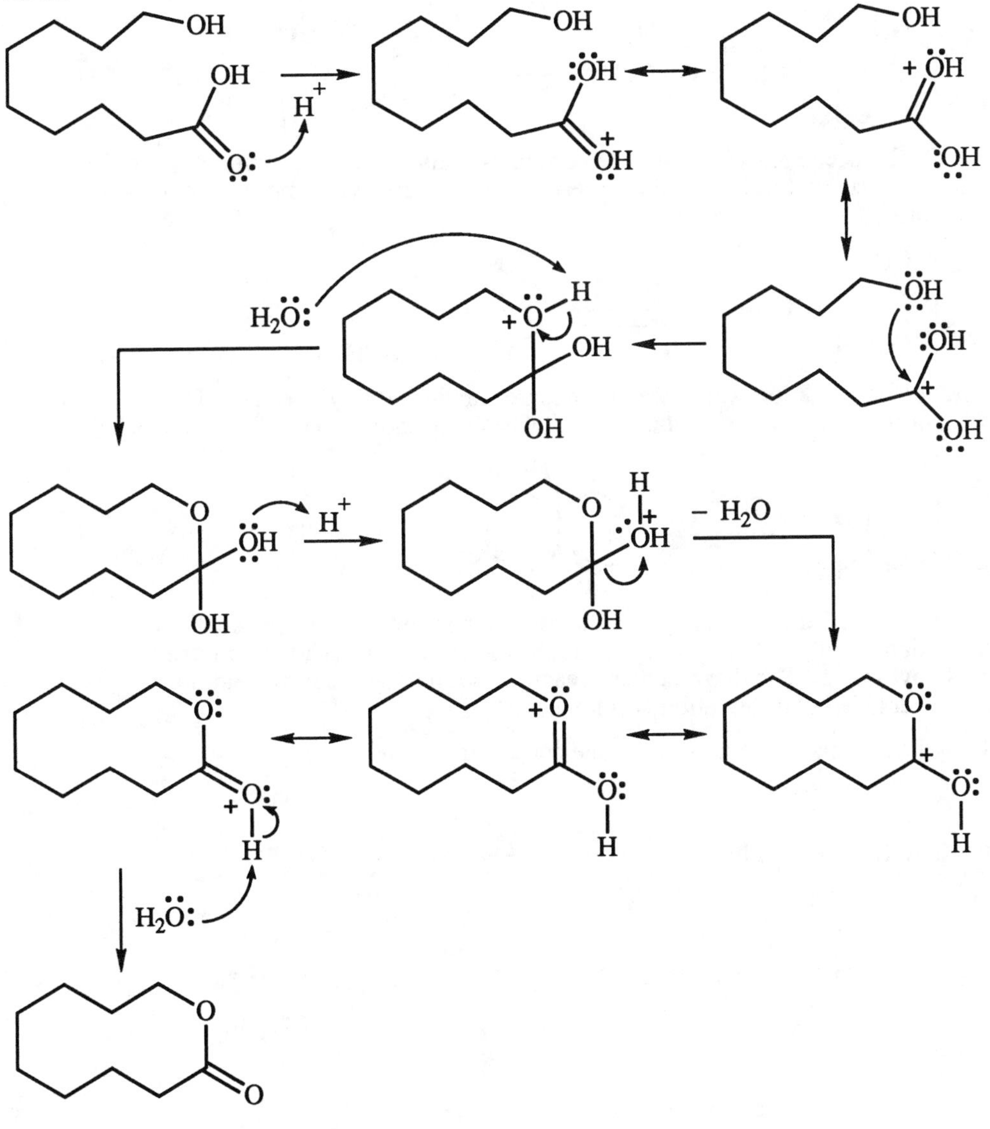

21-32

(a) 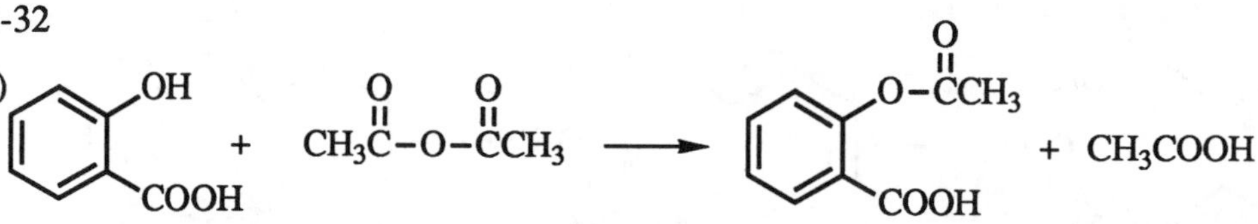

Generally, acetic anhydride is the optimum reagent for the preparation of acetate esters. Acetyl chloride would also react with the carboxylic acid to form a mixed anhydride.

(b) 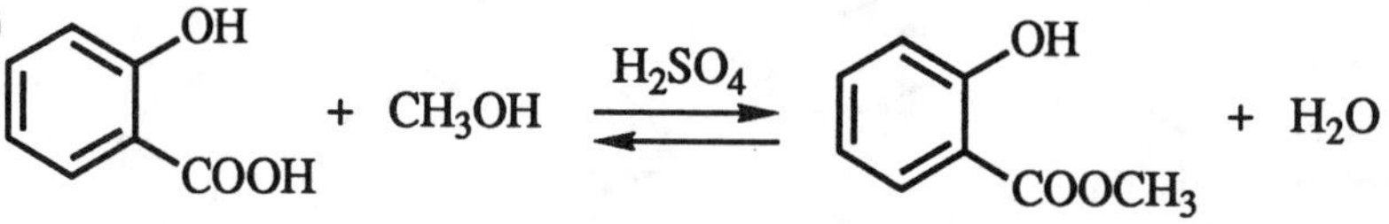

Fischer esterification works well to prepare simple carboxylic esters. The diazomethane method would also react with the phenol, making the phenyl ether.

(c) 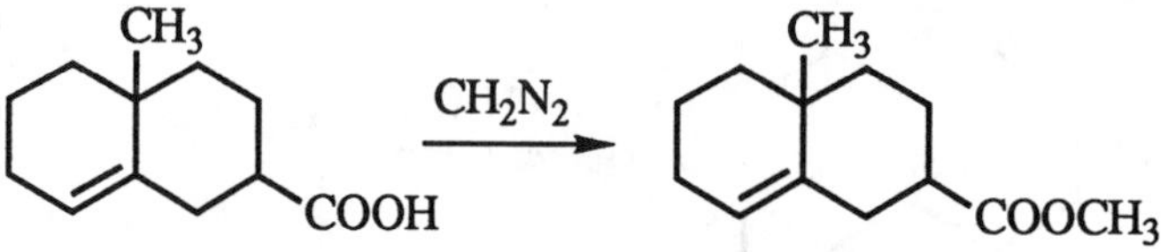

Fischer esterification would make the ester, but in the process, the acidic conditions would risk migrating the double bond into conjugation with the carbonyl group. The diazomethane reaction is run under neutral conditions where double bond migration will not occur.

21-33 Syntheses may have more than one correct approach.

(a)
$$\mathrm{Ph{-}\overset{O}{\overset{\|}{C}}{-}OCH_3} \;+\; 2\ \mathrm{PhMgBr} \xrightarrow{\text{ether}} \xrightarrow{\mathrm{H_2O}} \mathrm{Ph{-}\underset{Ph}{\overset{OH}{C}}{-}Ph}$$

(b)
$$\mathrm{H{-}\overset{O}{\overset{\|}{C}}{-}OCH_2CH_3} \;+\; 2\ \mathrm{PhCH_2MgBr} \xrightarrow{\text{ether}} \xrightarrow{\mathrm{H_2O}} \mathrm{H{-}\underset{CH_2Ph}{\overset{OH}{C}}{-}CH_2Ph}$$

(c)
$$\mathrm{Ph{-}\overset{O}{\overset{\|}{C}}{-}OCH_3} \;+\; \mathrm{H_2NCH_2CH_3} \longrightarrow \mathrm{Ph{-}\overset{O}{\overset{\|}{C}}{-}NHCH_2CH_3}$$

(d)
$$\mathrm{H{-}\overset{O}{\overset{\|}{C}}{-}OCH_2CH_3} \;+\; 2\ \mathrm{PhMgBr} \xrightarrow{\text{ether}} \xrightarrow{\mathrm{H_2O}} \mathrm{H{-}\underset{Ph}{\overset{OH}{C}}{-}Ph}$$

21-33 continued

(e)

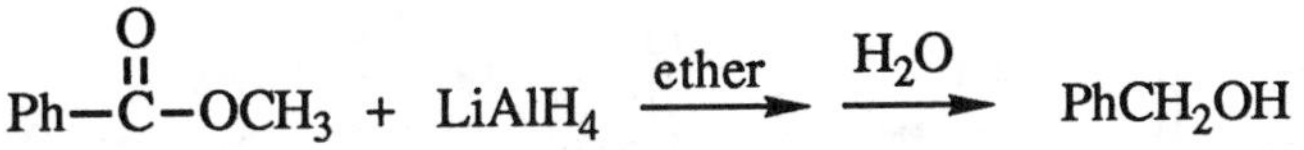

(f)

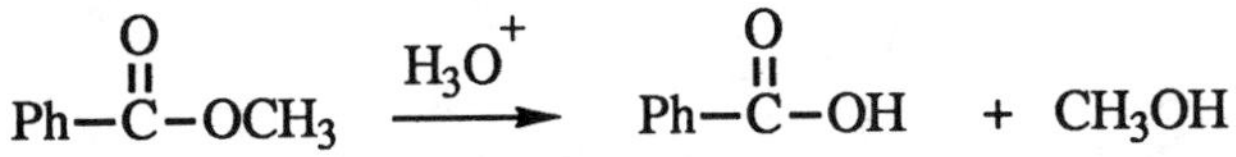

(g)

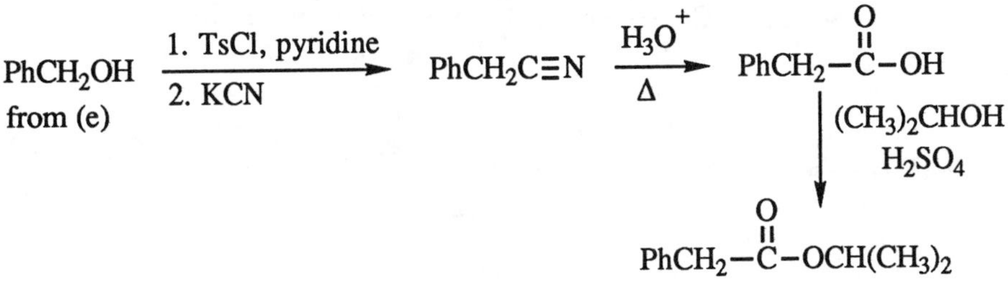

(h)

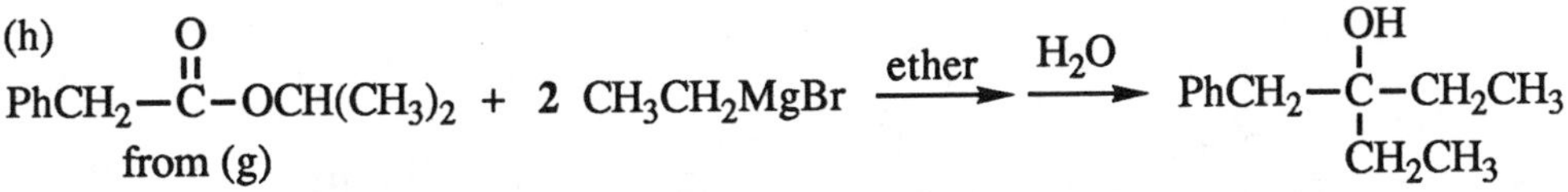

21-34 There may be additional correct approaches to these problems.

(a)

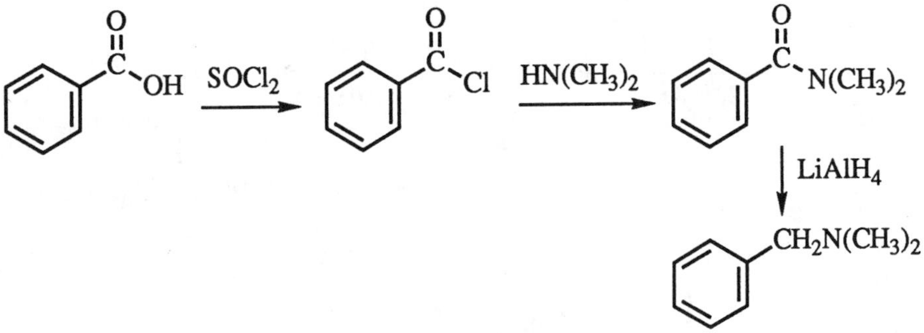

(b)

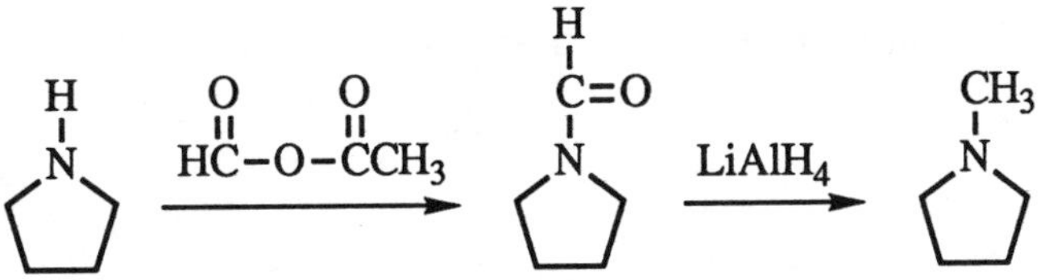

21-35 Prepare the amide, then dehydrate.

(a)

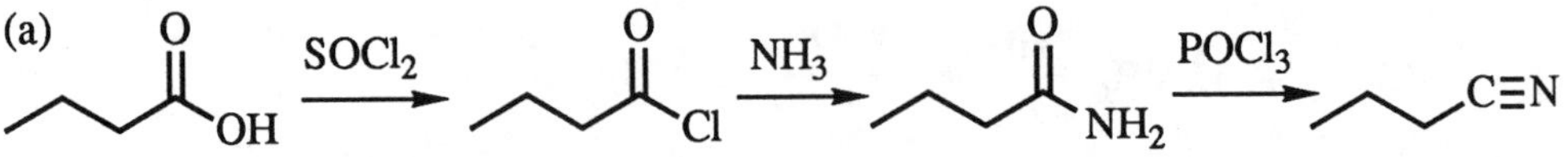

(b)

$$\text{PhC(=O)}-\text{OH} \xrightarrow{SOCl_2} \text{PhC(=O)}-\text{Cl} \xrightarrow{NH_3} \text{PhC(=O)}-\text{NH}_2 \xrightarrow{POCl_3} \text{PhC}\equiv\text{N}$$

(c)

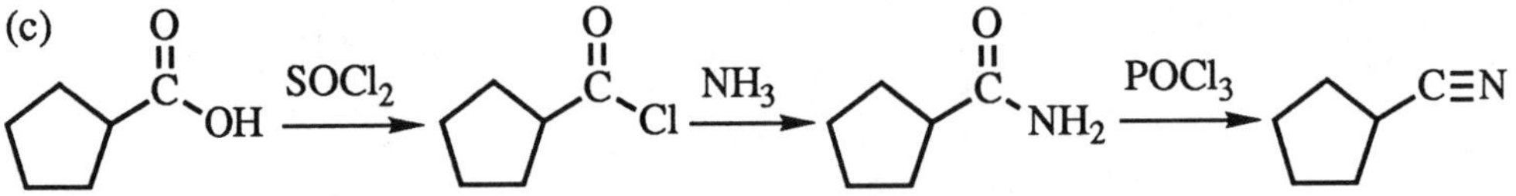

21-36

(a)

$$\text{PhCH}_2-\text{C(=O)}-\text{OH} \xrightarrow{SOCl_2} \text{PhCH}_2-\text{C(=O)}-\text{Cl} \xrightarrow{NH_3} \text{PhCH}_2-\text{C(=O)}-\text{NH}_2 \xrightarrow{POCl_3} \text{PhCH}_2-\text{C}\equiv\text{N}$$

(b)

$$\text{PhCH}_2-\text{C(=O)}-\text{OH} \xrightarrow[2.\ H_2O]{1.\ LiAlH_4} \text{PhCH}_2\text{CH}_2\text{OH} \xrightarrow[\text{pyridine}]{TsCl} \text{PhCH}_2\text{CH}_2\text{OTs} \xrightarrow{NaCN} \text{PhCH}_2\text{CH}_2\text{CN}$$

(c)

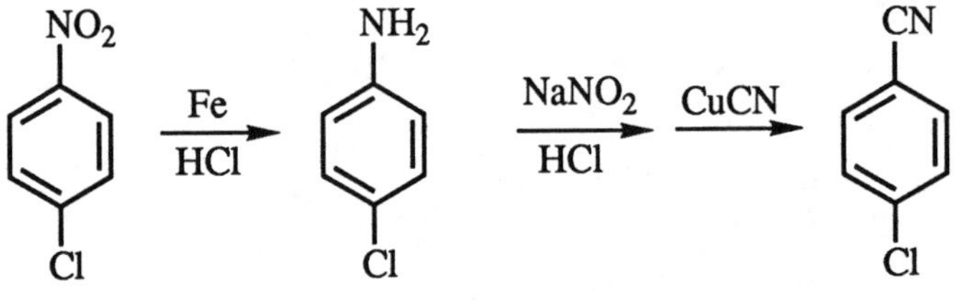

21-37

(a)
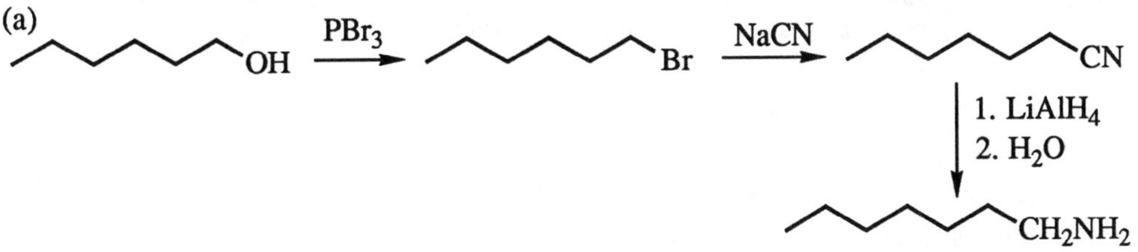

(b)
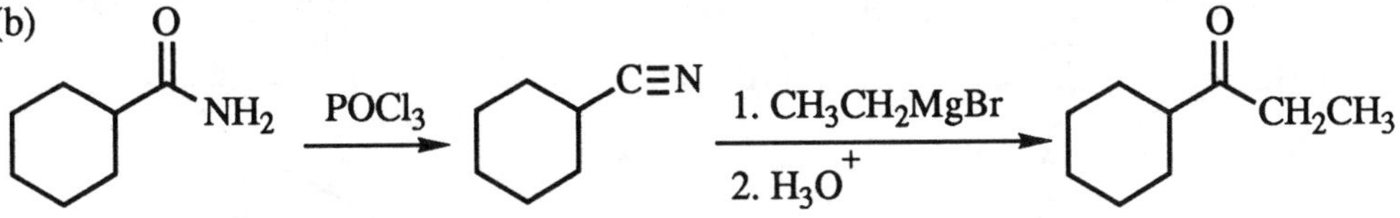

(c)
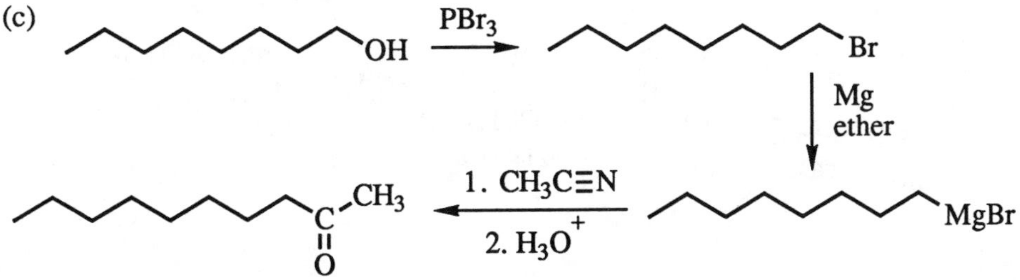

21-38 Refer to the Glossary and the text for definitions and examples.

21-39

(a) 3-methylpentanoyl chloride
(b) benzoic formic anhydride
(c) acetanilide; *N*-phenylethanamide
(d) *N*-methylbenzamide
(e) phenyl acetate; phenyl ethanoate
(f) methyl benzoate
(g) benzonitrile
(h) 4-phenylbutane nitrile; γ-phenylbutyronitrile
(i) dimethyl phthalate
(j) *N,N*-diethyl-3-methylbenzamide
(k) 4-hydroxypentanoic acid lactone; γ-valerolactone
(l) 3-aminopentanoic acid lactam; β-valerolactam

21-40

(a) $PhC(=O)-OCH_2CH_3$

(b) $PhC(=O)-O-C(=O)CH_3$

(c)
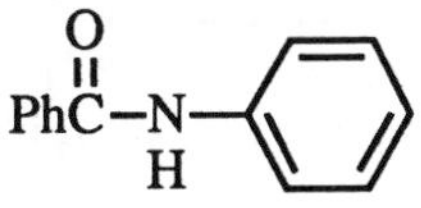

(d)
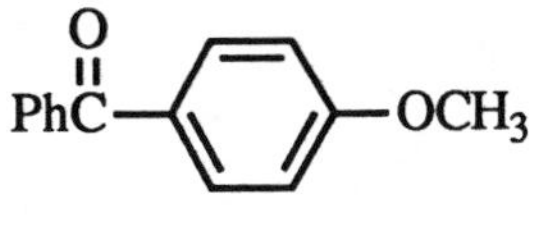

(e) $Ph-C(OH)(Ph)-Ph$

21-41 In nucleophilic acyl substitution on an anhydride, the acid catalyst protonates a carbonyl oxygen, increasing the positive charge on the carbonyl carbon, making the carbonyl more susceptible to nucleophilic attack.

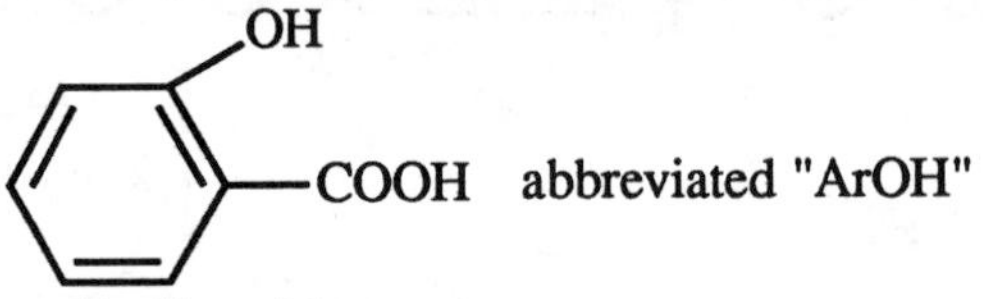
OH
COOH
abbreviated "ArOH"

salicylic acid

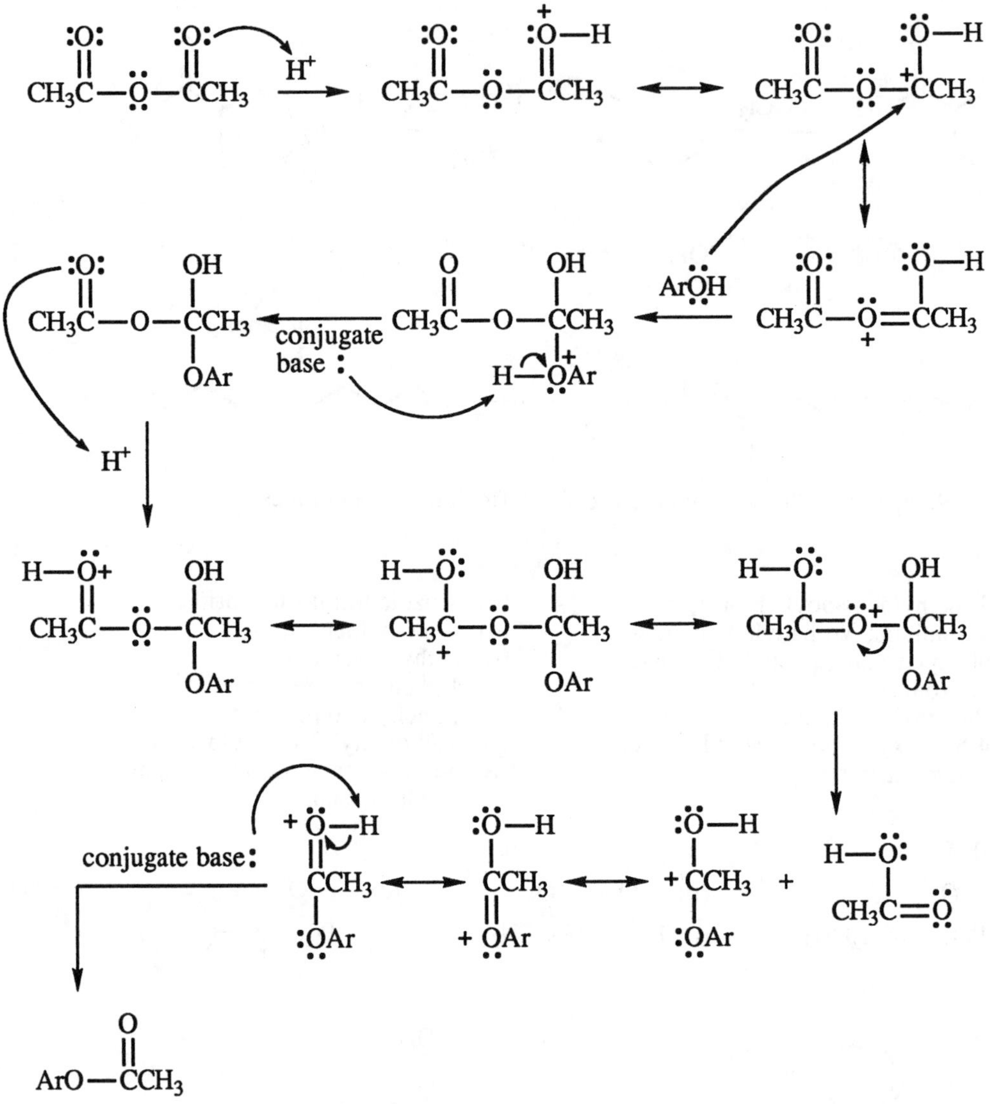
H+
ArOH
conjugate base
H+
conjugate base

21-42

When a carboxylic acid is treated with a basic reagent, the base removes the acidic proton rather than attacking at the carbonyl (proton transfers are much faster than formation or cleavage of other types of bonds). Once the carboxylate anion is formed, the carbonyl is no longer susceptible to nucleophilic attack: nucleophiles do not attack sites of negative charge. By contrast, in acidic conditions, the protonated carbonyl has a positive charge and is activated to nucleophilic attack.

basic conditions

anion—not susceptible to nucleophilic attack

acidic conditions

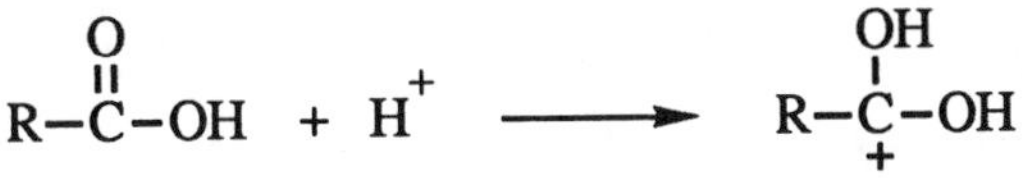

rapidly attacked by R'OH nucleophile

21-43

(a)

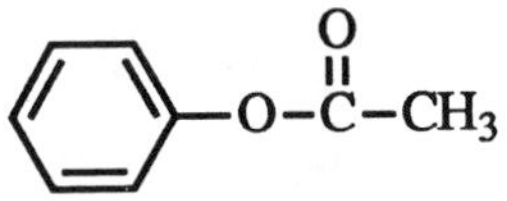

(b)

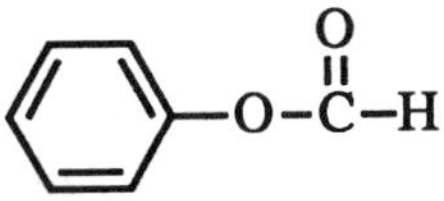

(c)

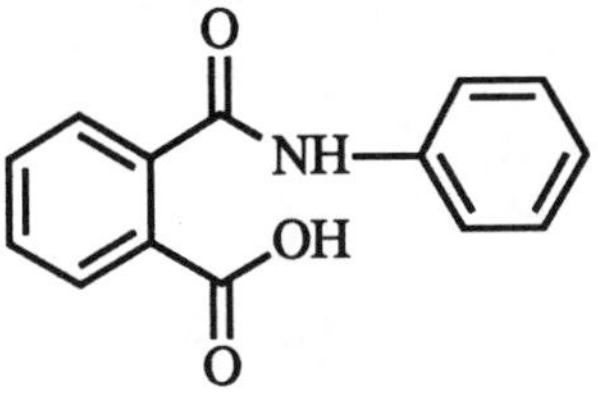

anhydrides react only once

(d)

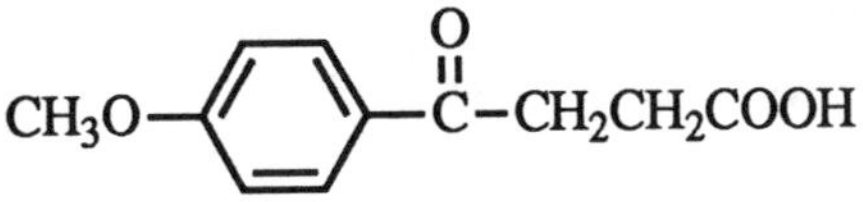

(e)

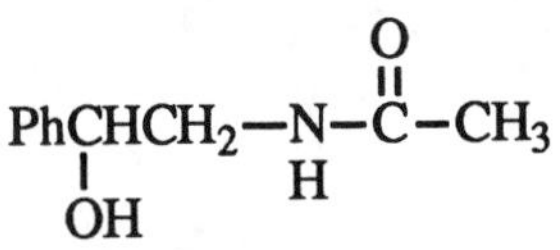

amines are more nucleophilic than alcohols

(f)

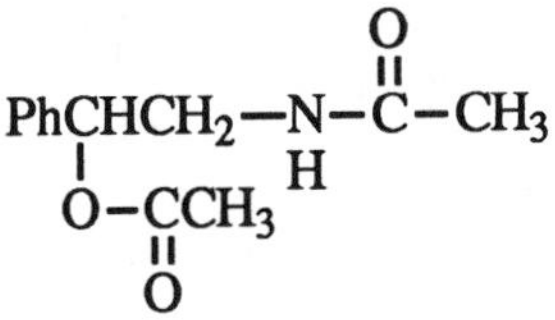

21-44

(a)

HO–C(=O)–C6H4–C(=O)–OH
SOCl2
Cl–C(=O)–C6H4–C(=O)–Cl

(b)

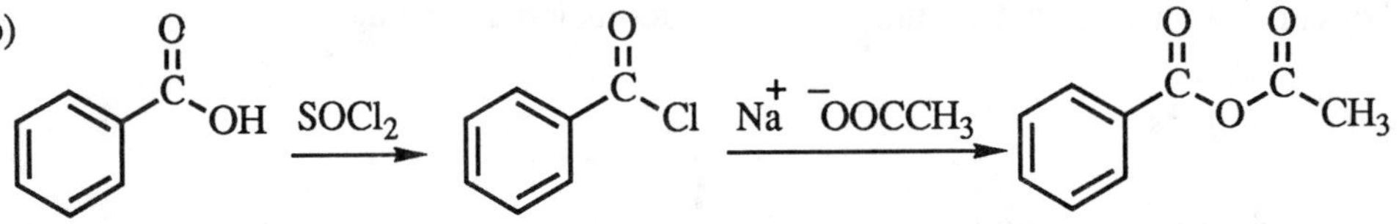
OH
SOCl2
Cl
Na+ –OOCCH3
O
CH3

(c)

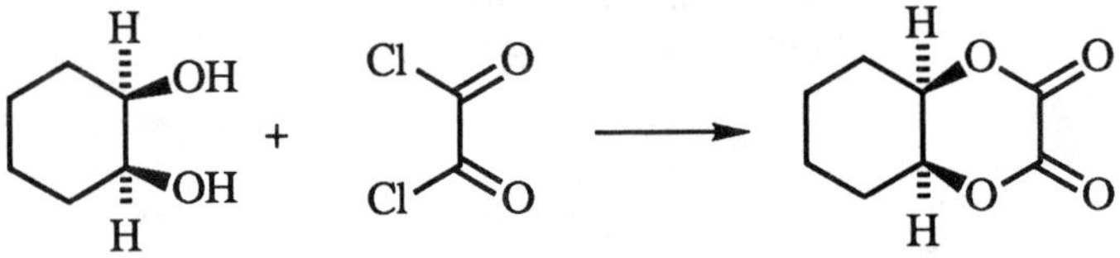
H
OH
OH
H
+
Cl
O
Cl
O
H
O
O
O
O
H

(d)

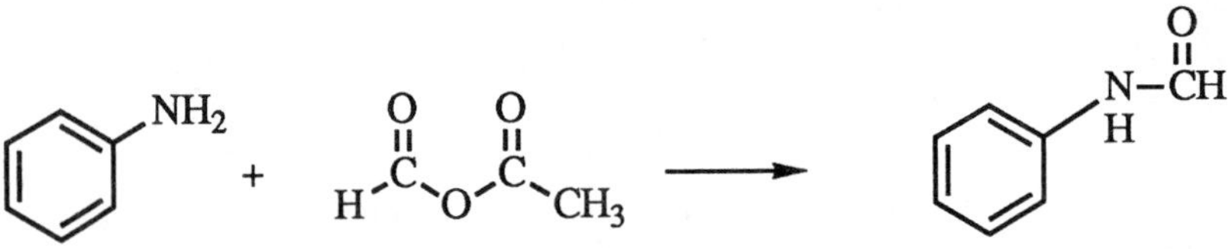
NH2
+
H
O
CH3
N–CH
H

(e)

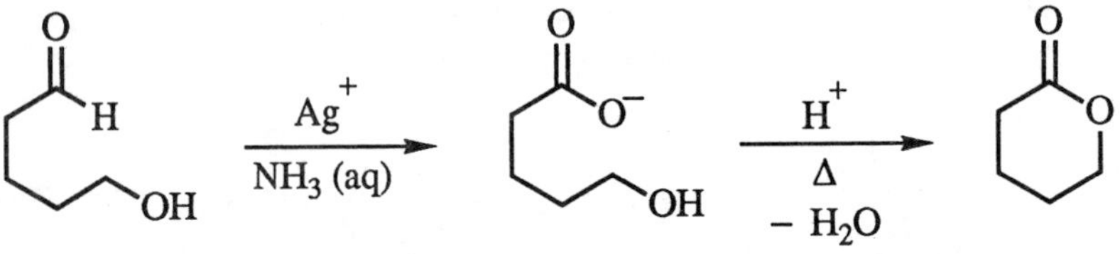
H
OH
Ag+
NH3 (aq)
O–
OH
H+
Δ
– H2O
O

(f)

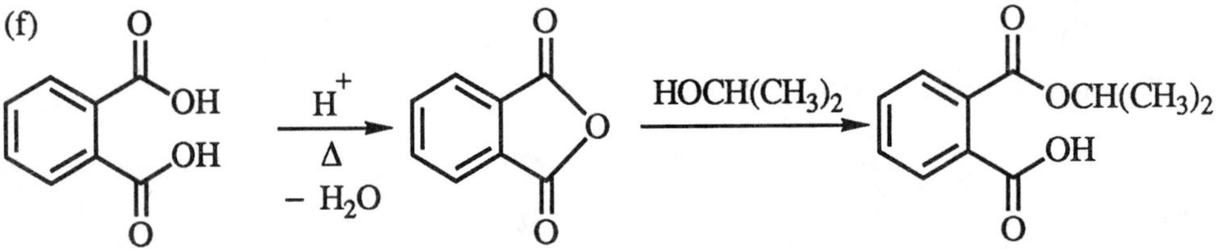
OH
OH
H+
Δ
– H2O
O
HOCH(CH3)2
OCH(CH3)2
OH

21-45

(a)

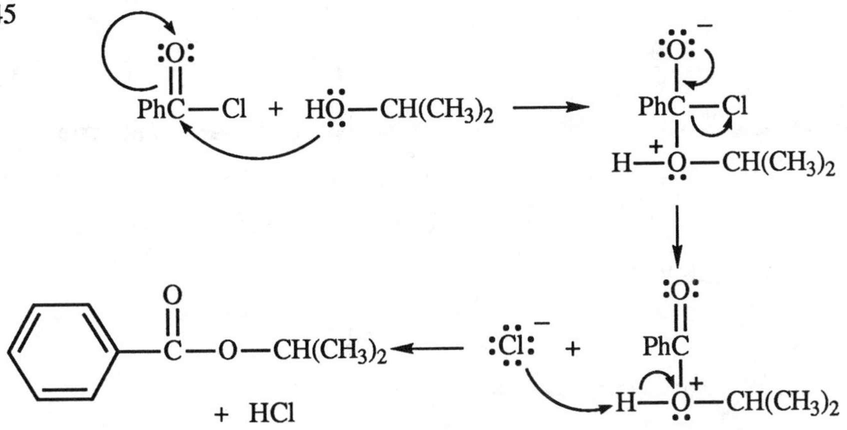

:O:
PhC—Cl + HO—CH(CH3)2
PhC—Cl
H—O—CH(CH3)2
:O:
PhC
H—O—CH(CH3)2
:Cl:
O
C—O—CH(CH3)2
+ HCl

(b)

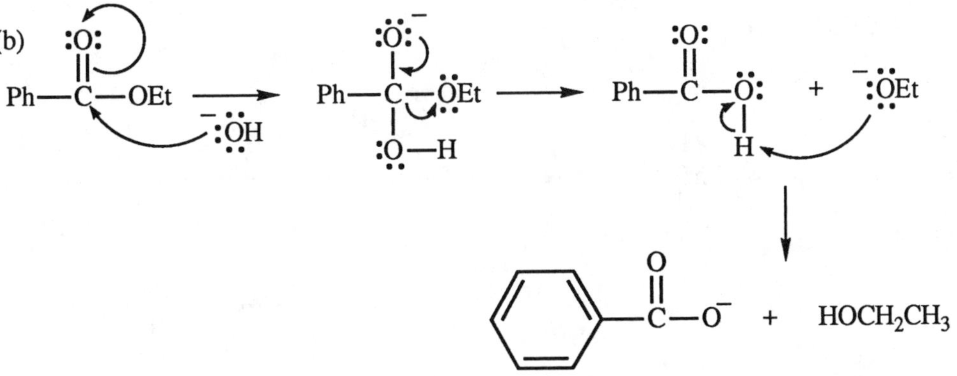

:O:
Ph—C—OEt
:OH
Ph—C—OEt
:O—H
:O:
Ph—C—O:
H
:OEt
O
C—O
HOCH2CH3

21-45 continued

(c)

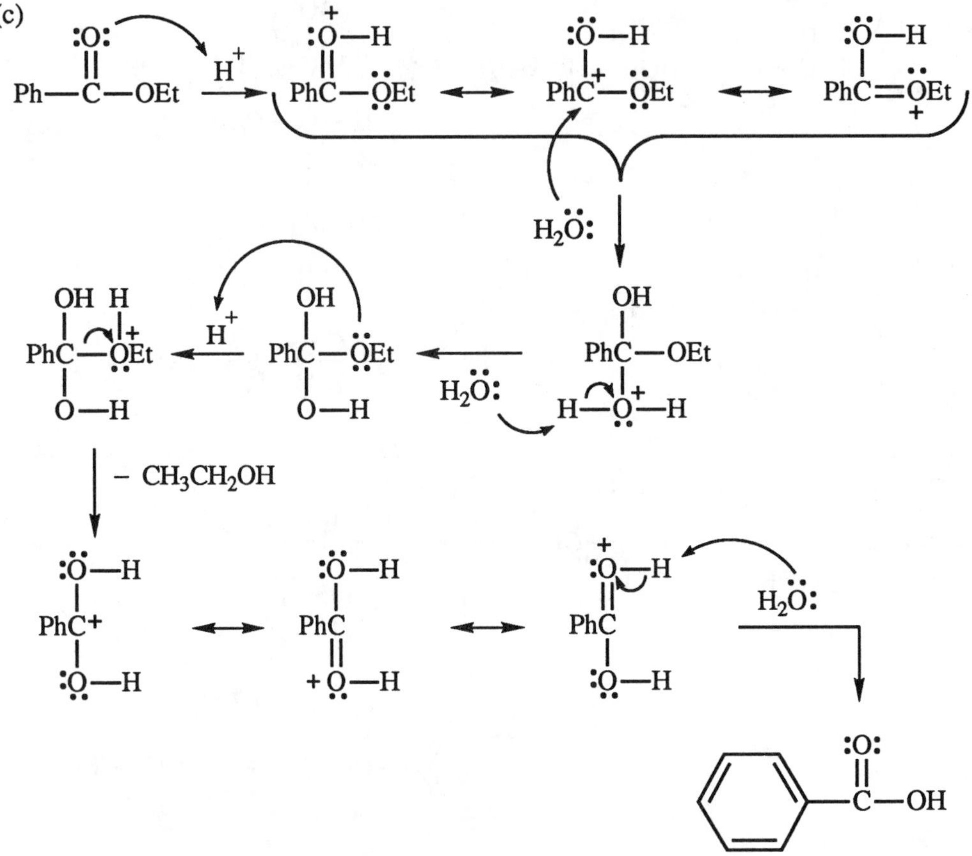

(d)

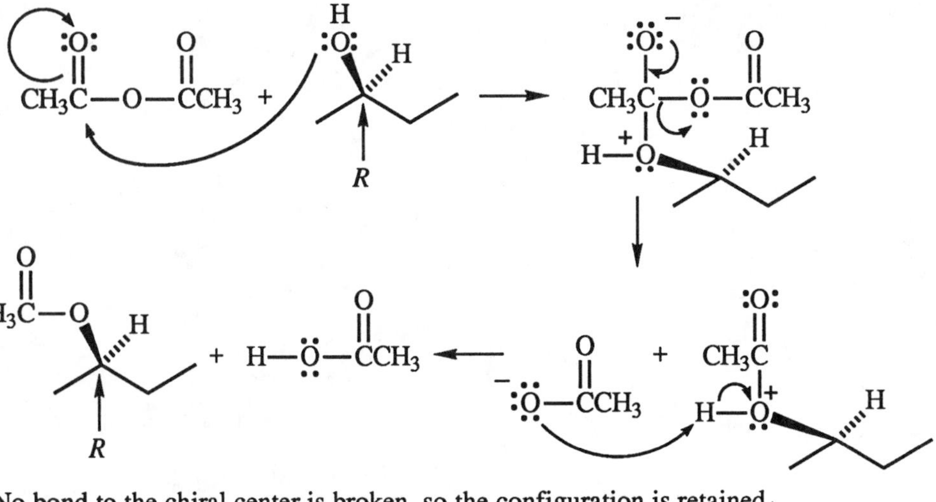

No bond to the chiral center is broken, so the configuration is retained.

21-46

(a)

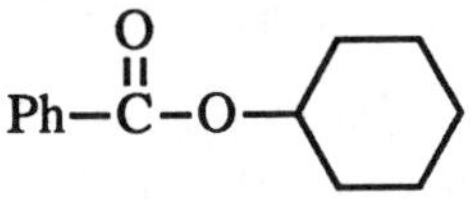

(b)

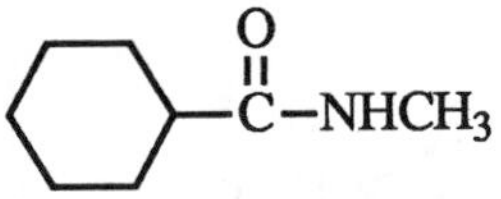

(c)

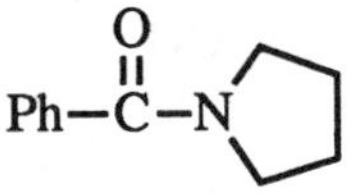

(d)

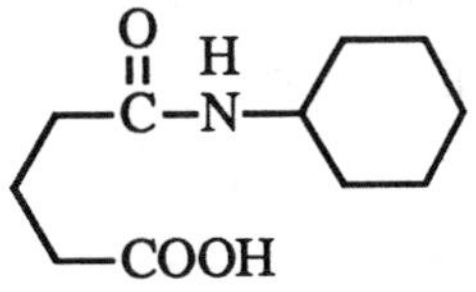

(e) $PhCH_2OH$

(f)

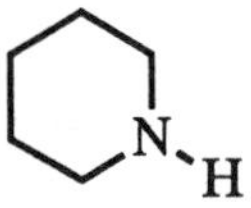

(g)

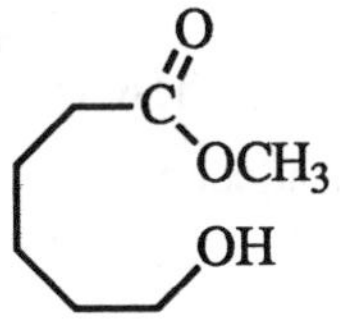

(h)

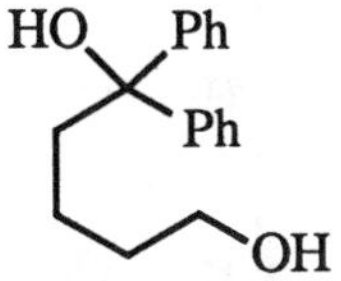

(i)

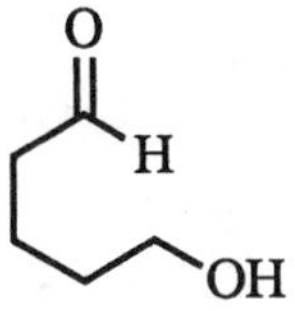

(j)

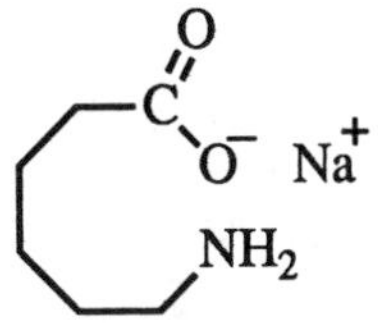

(k) $PhCH_2CH(CH_3)CH_2NH_2$

(l)

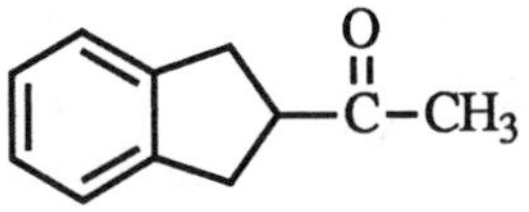

(m)

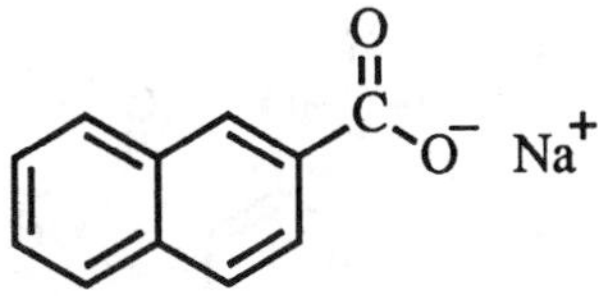

+ NH_3

21-47 Products after adding dilute acid in the workup:

(a) $HC(=O)-OH$ + HO–Ph

(b) 2 CH_3CH_2OH + CO_2

(c)

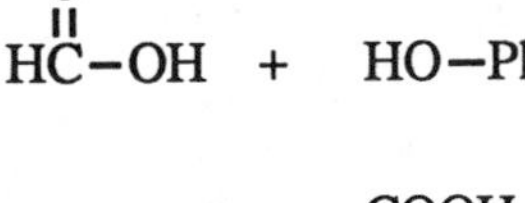

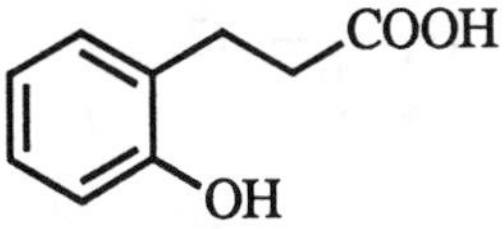

(d)

+

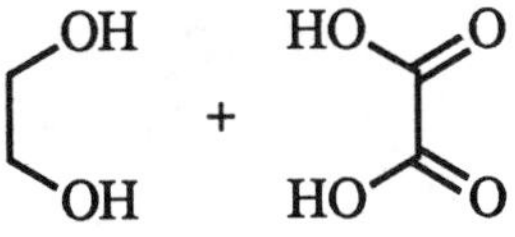

21-48

(a)

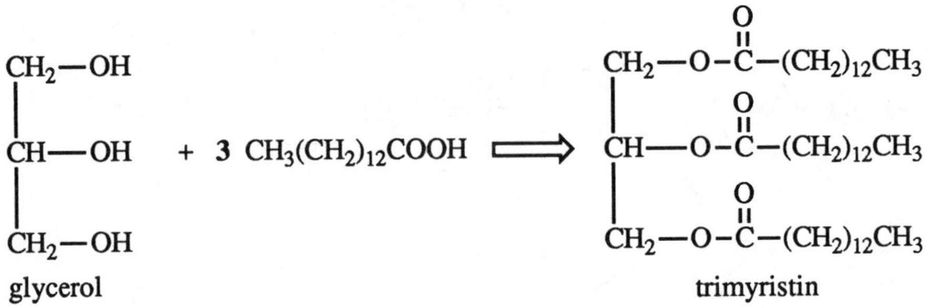

CH_2-OH
$CH-OH$
CH_2-OH
glycerol
+ 3 $CH_3(CH_2)_{12}COOH$
$CH_2-O-C(=O)-(CH_2)_{12}CH_3$
$CH-O-C(=O)-(CH_2)_{12}CH_3$
$CH_2-O-C(=O)-(CH_2)_{12}CH_3$
trimyristin

(b)

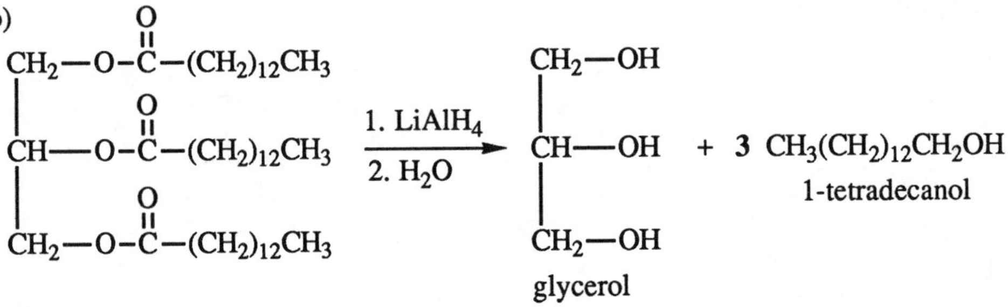

$CH_2-O-C(=O)-(CH_2)_{12}CH_3$
$CH-O-C(=O)-(CH_2)_{12}CH_3$
$CH_2-O-C(=O)-(CH_2)_{12}CH_3$
1. $LiAlH_4$
2. H_2O
CH_2-OH
$CH-OH$
CH_2-OH
glycerol
+ 3 $CH_3(CH_2)_{12}CH_2OH$
1-tetradecanol

21-49

(a)

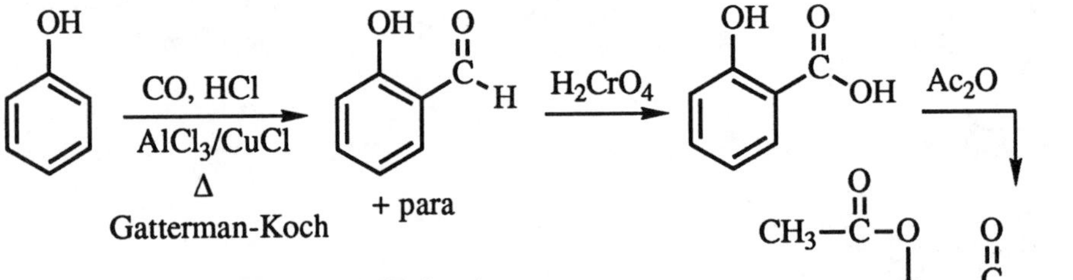

OH
CO, HCl
$AlCl_3/CuCl$
Δ
Gatterman-Koch
OH
H
+ para
H_2CrO_4
OH
OH
Ac_2O
$CH_3-C(=O)-O$
OH

(See text Section 17-15B for the industrial preparation of salicylic acid.)

(b)

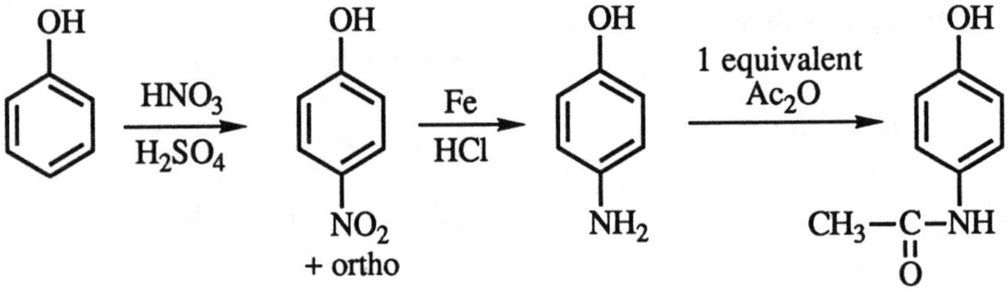

OH
HNO_3
H_2SO_4
OH
NO_2
+ ortho
Fe
HCl
OH
NH_2
1 equivalent
Ac_2O
OH
$CH_3-C(=O)-NH$

21-50

(a)

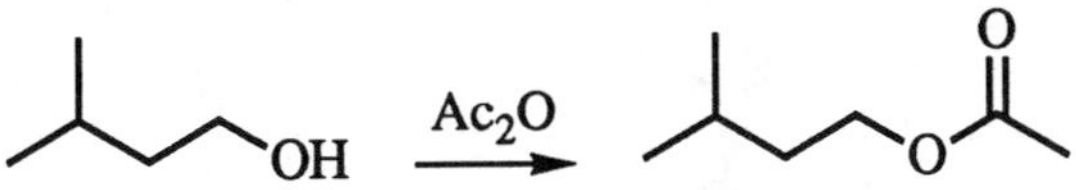

(b)

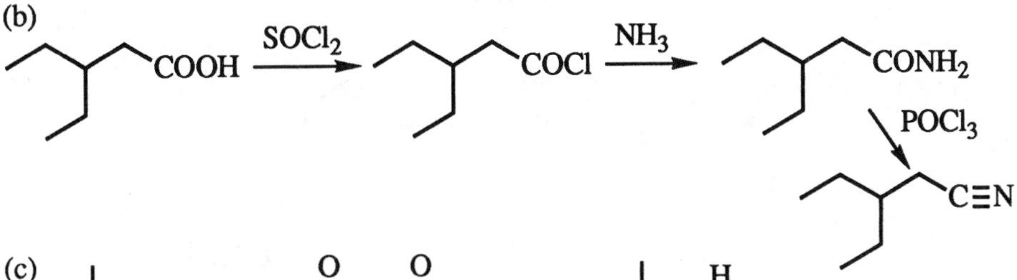

(c)

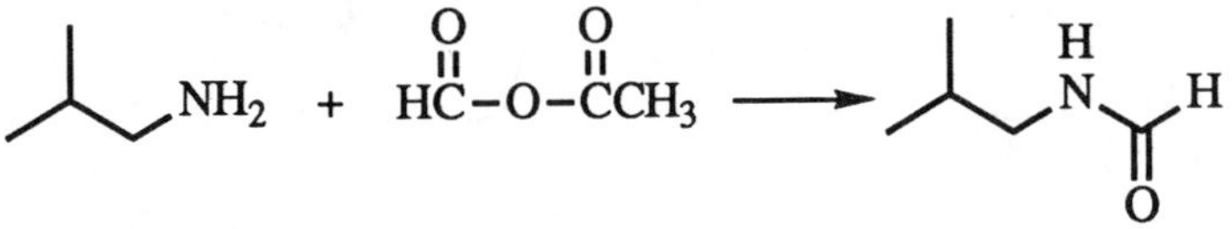

(d)

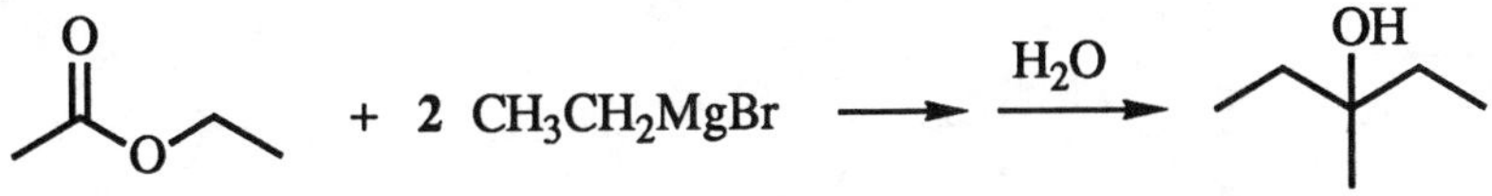

(e) See the answer to 21.34(b) for one method. Here is another: reductive alkylation.

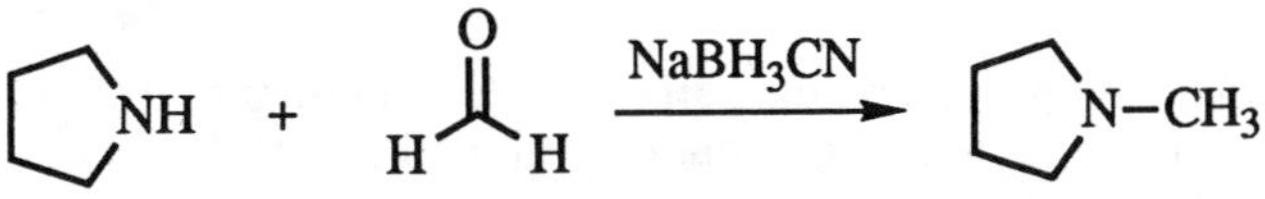

(f)

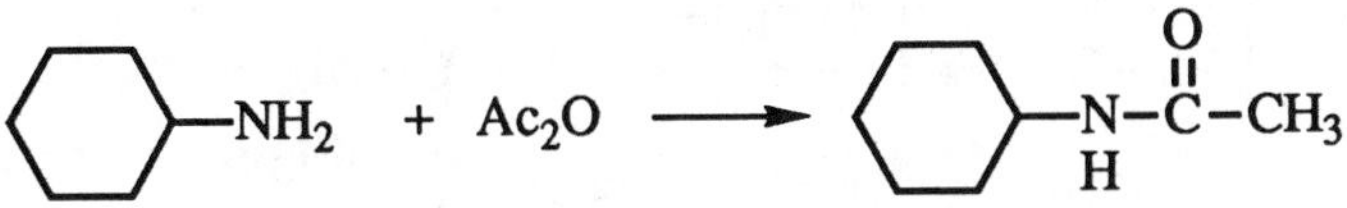

(g)

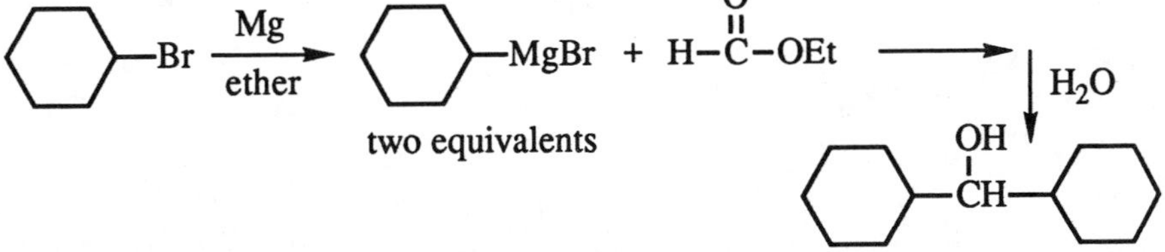

21-50 continued

(h)

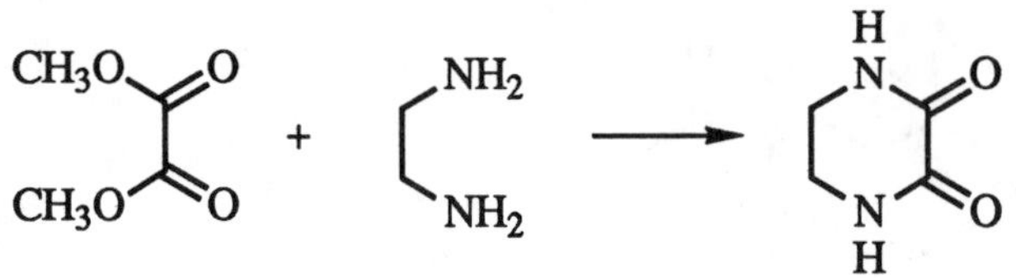

(i)

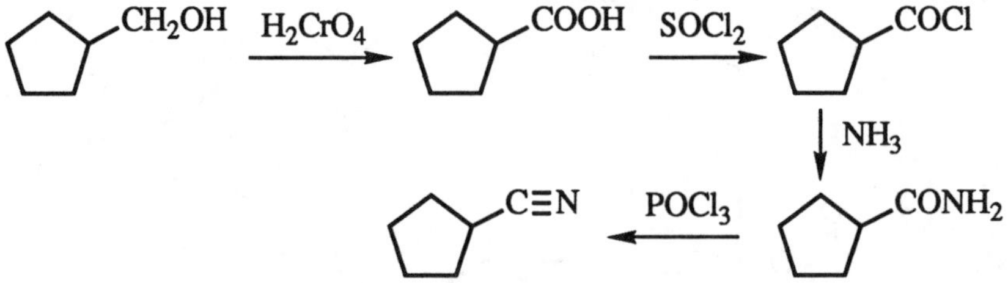

(j)

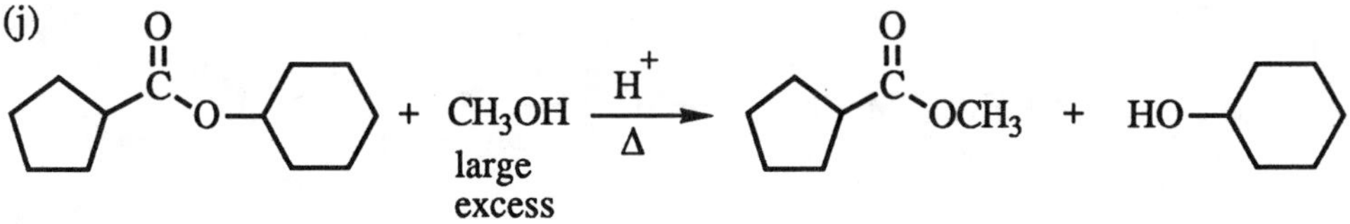

21-51 Diethyl carbonate has *two* leaving groups on the carbonyl. It can undergo *two* nucleophilic acyl substitutions, followed by one nucleophilic addition.

(a)

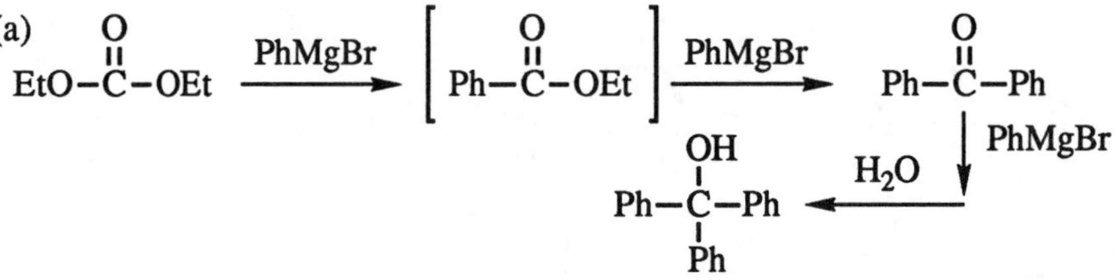

(b)

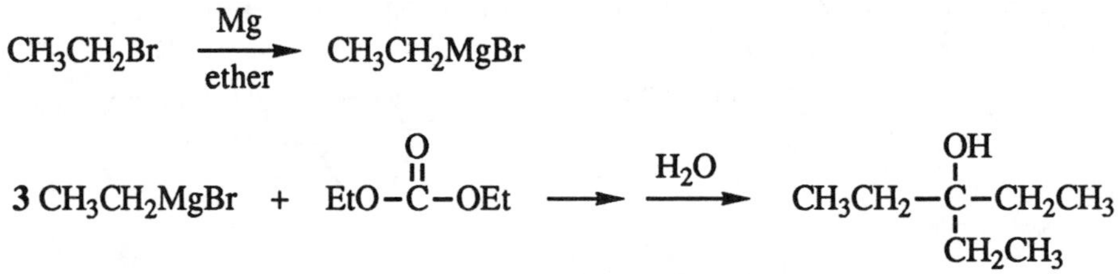

21-52

Triethylamine is nucleophilic, but it has no H on nitrogen to lose, so it forms a salt instead of a stable amide.

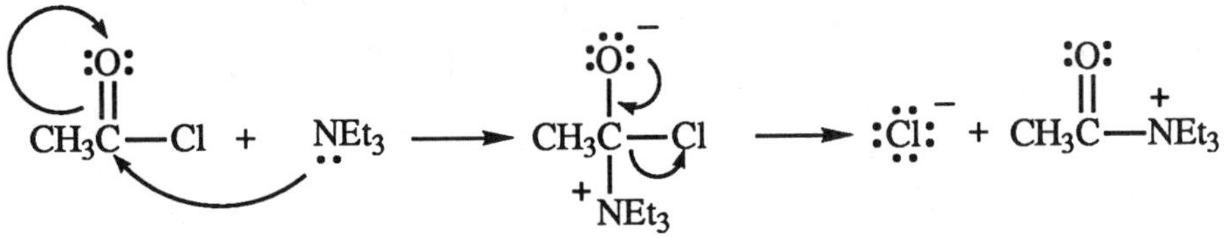

When ethanol is added, it attacks the carbonyl of the salt, with triethylamine as the leaving group.

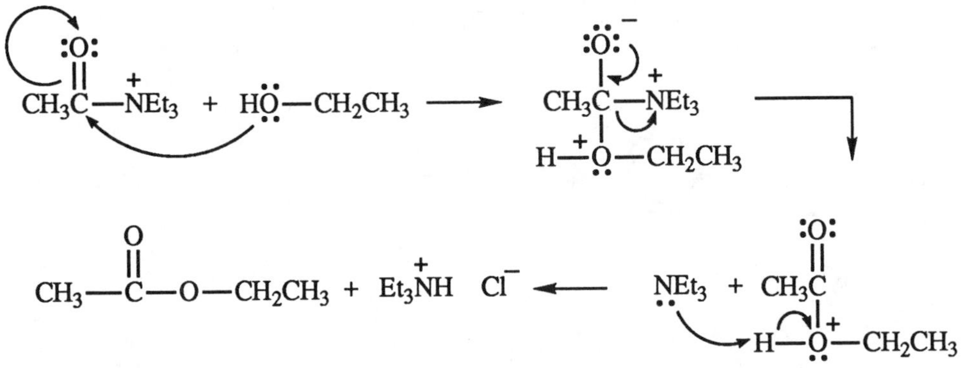

An alternate mechanism explains the same products, and is more likely with hindered bases:

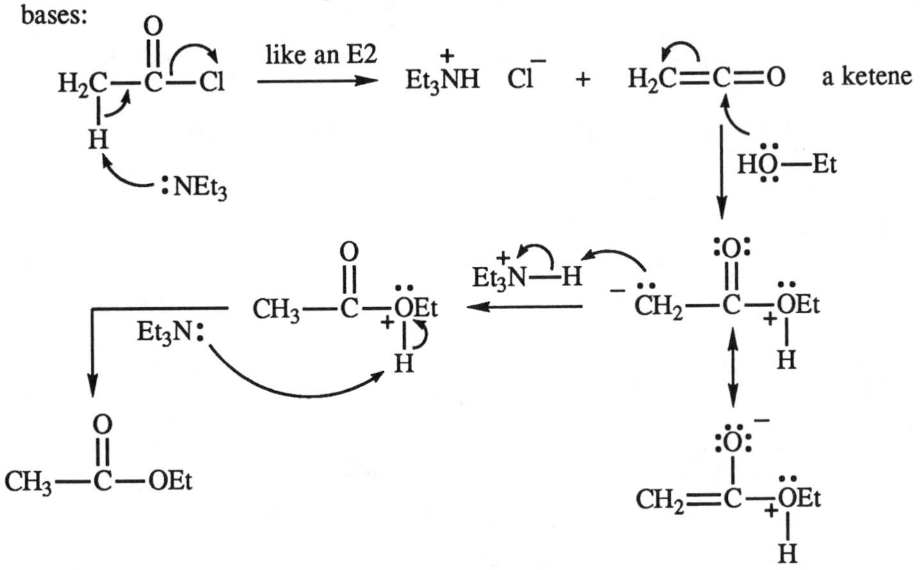
like an E2
a ketene

21-53

(a)

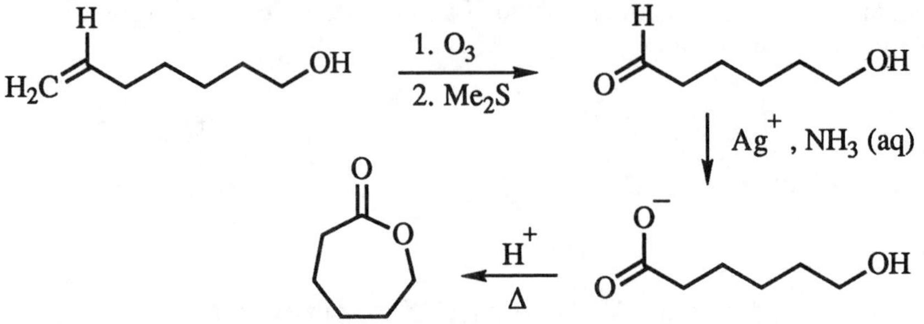

(b)

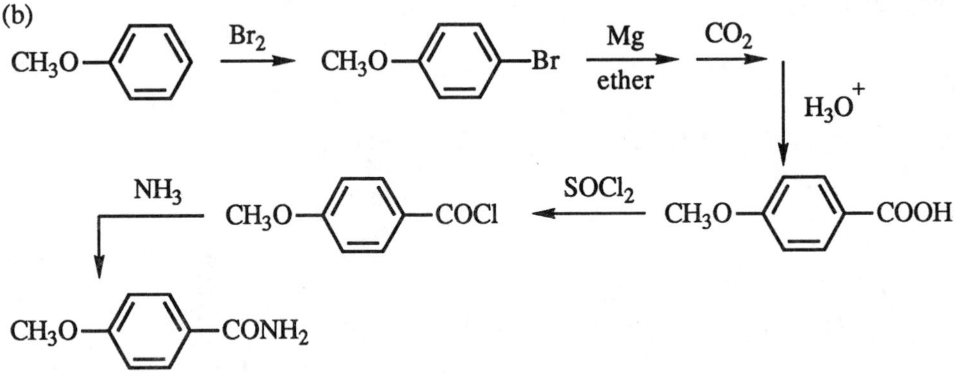

(c)

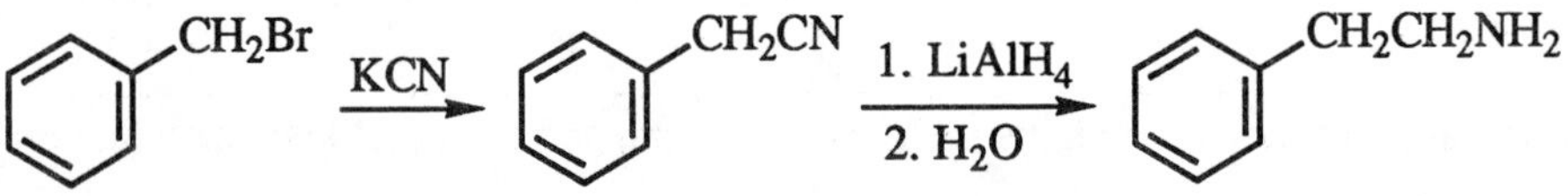

(d)

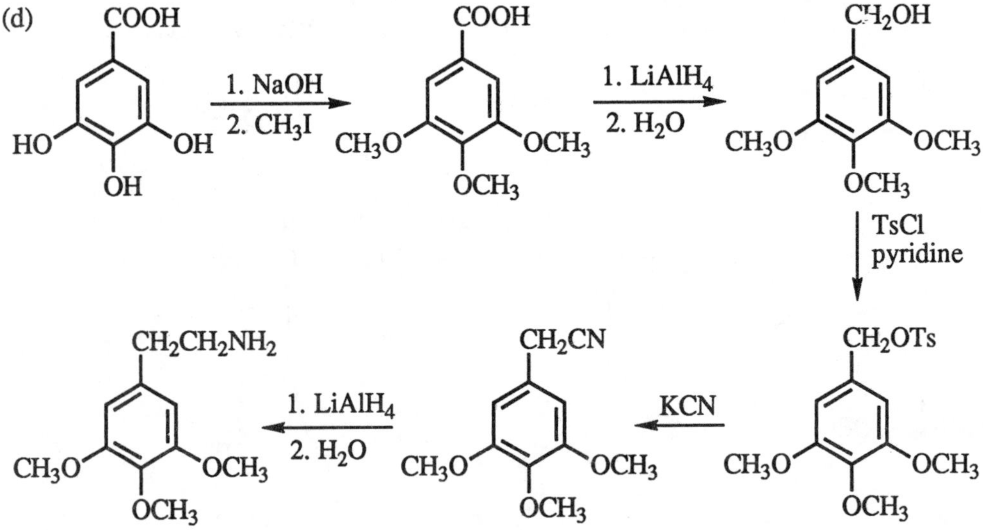

21-54 The asterisk (*) will denote ^{18}O.

(a)

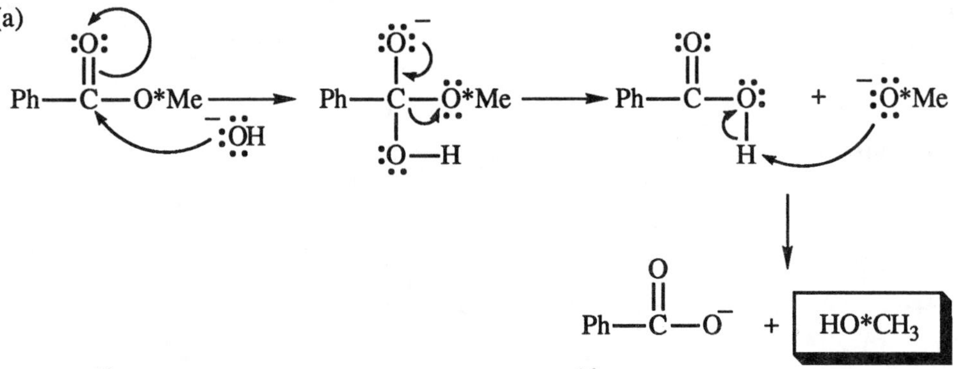

(b) The ^{18}O has 2 more neutrons in its nucleus than ^{16}O. Mass spectra of these products would show the molecular ion of benzoic acid at its standard value of m/z 122, whereas the molecular ion of methanol would appear at m/z 34 instead of m/z 32, proving that the heavy isotope of oxygen went with the methyl. This demonstrates that the bond between oxygen and the carbonyl carbon is broken, not the bond between the oxygen and the alkyl carbon.

(c) The products do not change, even though the mechanism is different.

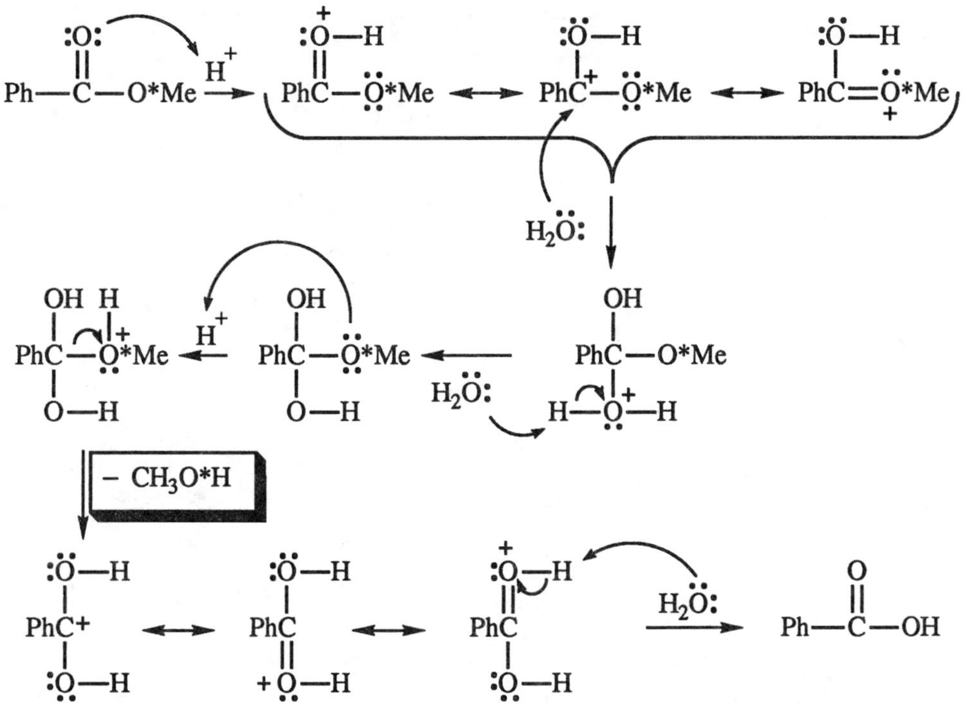

21-55

(a) 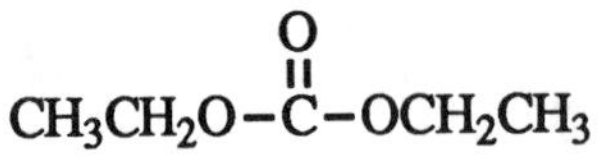

(b) 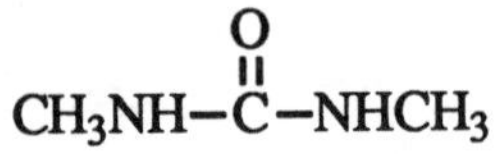

(c) 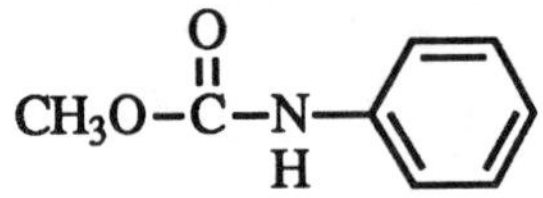

(d) 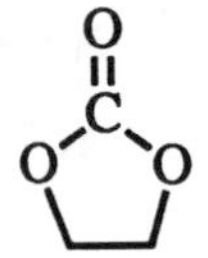

(e) 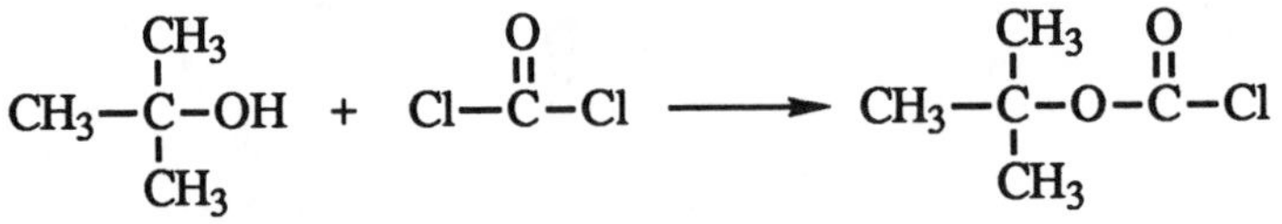

21-56

(a) 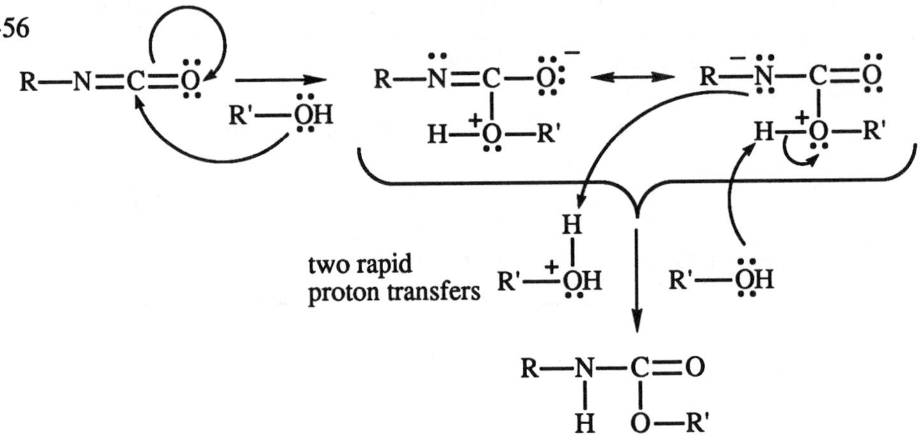
two rapid
proton transfers

(b) 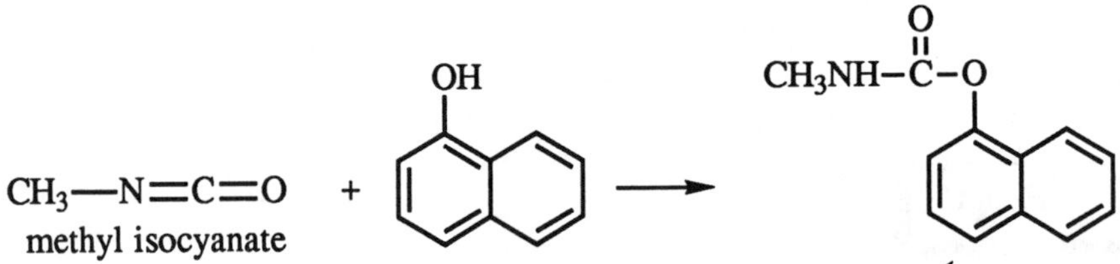
methyl isocyanate

(c) 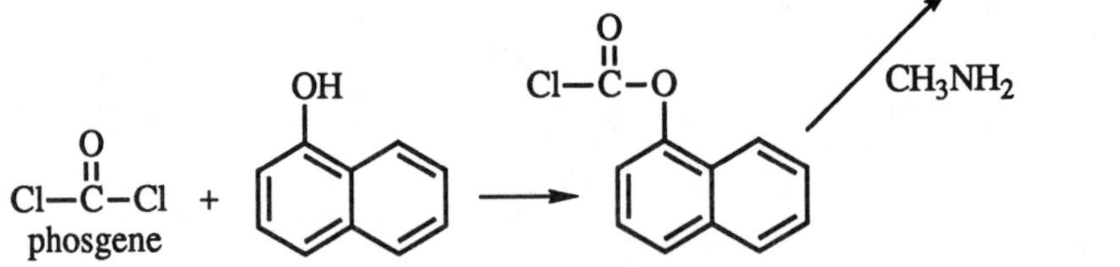
phosgene
CH_3NH_2

21-57

(a) Both structures are β-lactam antibiotics, a penicillin and a cephalosporin.

(b) "Cephalosporin N" has a 5-membered, sulfur-containing ring. This belongs in the penicillin class of antibiotics.

21-58

<u>IR spectrum</u>:
—sharp spike at 2250 cm^{-1} ⇒ C≡N
—1740 cm^{-1} ⇒ C=O } maybe an ester
—1200 cm^{-1} ⇒ C—O }

<u>NMR spectrum</u>:
—triplet and quartet ⇒ CH_3CH_2
—this quartet at δ 4.2 ⇒ CH_3CH_2—O
—2H singlet at δ 3.6 ⇒ isolated CH_2

$$CH_3CH_2O \;+\; \overset{O}{\overset{\|}{C}} \;+\; CH_2 \;+\; C{\equiv}N$$

sum of the masses is 113, consistent with the MS

The fragments can be combined in only two possible ways:

A B

The NMR proves the structure to be **A**. If the structure were **B**, the CH_2 between oxygen and the carbonyl would come farther downfield that the CH_2 of the ethyl (deshielded by oxygen and carbonyl instead of by oxygen alone). As this is not the case, the structure cannot be **B**.

The peak in the mass spectrum at m/z 68 is due to α-cleavage of the ester:

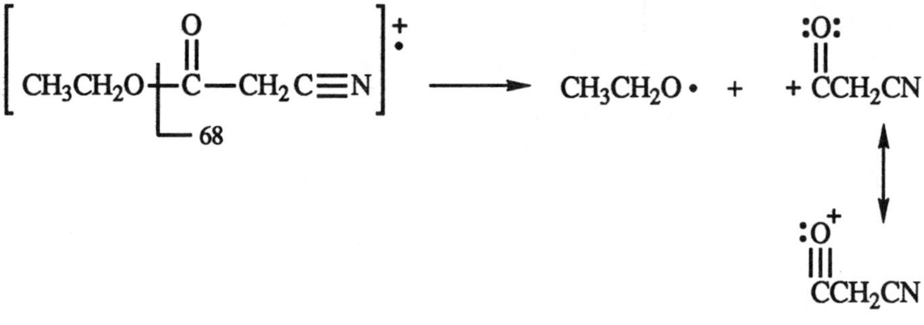

m/z 68

21-59

The formula C_5H_9NO has 2 elements of unsaturation.

<u>IR spectrum</u>: The strongest peak at 1670 cm^{-1} comes low in the carbonyl region; in the absence of conjugation (no alkene peak observed), a carbonyl this low is almost certainly an amide. There is one broad peak in the NH/OH region, hinting at the likelihood of a secondary amide.

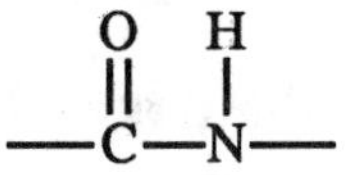

<u>NMR spectrum</u>: The broad peak at δ 8.4 is exchangeable with D_2O; this is an amide proton. A broad, 2H peak at δ 3.3 is a CH_2 next to nitrogen. A broad, 2H peak at δ 2.3 is a CH_2 next to carbonyl. The 4H peak at δ 1.8 is probably two more CH_2 groups. There appears to be coupling among these protons but it is not resolved enough to be useful for interpretation. This is often the case when the compound is cyclic, with restricted rotation around carbon-carbon bonds, giving *non-equivalent* (axial and equatorial) hydrogens on the same carbon.

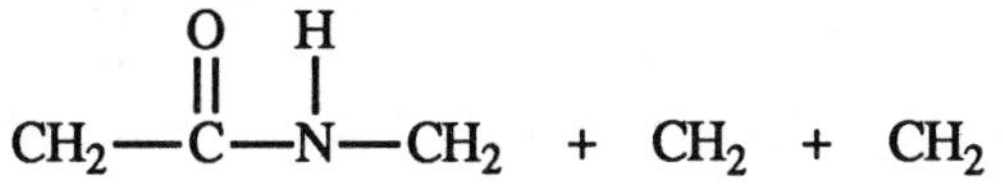

One element of unsaturation is the π bond in the carbonyl. There is no alkene so the other element of unsaturation must be a ring. The most consistent structure:

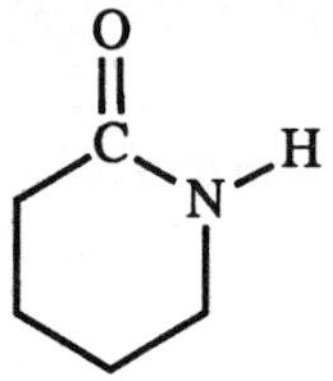

δ-valerolactam

21-60

<u>IR spectrum</u>: A strong carbonyl peak at 1720 cm^{-1}, in conjunction with the C—O peak at 1200 cm^{-1}, suggests the presence of an ester. An alkene peak appears at 1650 cm^{-1}.

<u>NMR spectrum</u>: The typical ethyl pattern stands out: 3H triplet at δ 1.3 and 2H quartet at δ 4. The chemical shift of the CH_2 suggests it is bonded to an oxygen. The other groups are: a 3H doublet at δ 1.8, likely to be a CH_3 next to one H; a 1H doublet at δ 5.7, a vinyl hydrogen with one neighboring H; and a 1H multiplet at δ 6.9, another vinyl H with many neighbors. The large coupling constant in the doublet at δ 5.7 shows that the two vinyl hydrogens are *trans*.

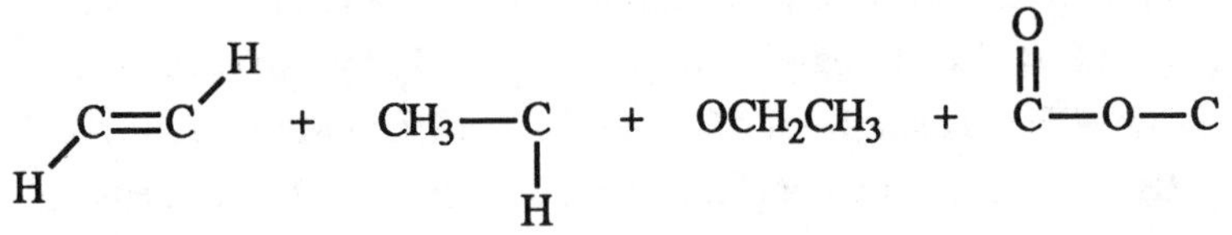

There is only one possible way to assemble these pieces:

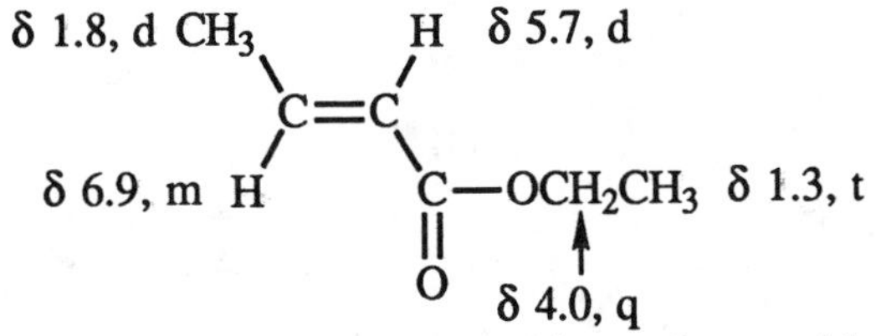
δ 1.8, d CH3
H δ 5.7, d
δ 6.9, m H
C—OCH2CH3 δ 1.3, t
δ 4.0, q

<u>Mass spectrum</u>: This structure has mass 114, consistent with the molecular ion.
Major fragmentations:

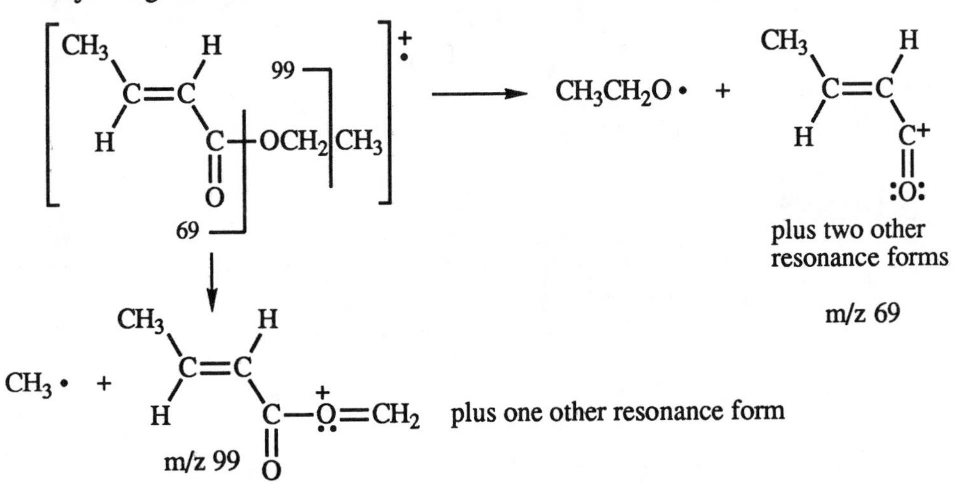
99
69
CH3CH2O • +
plus two other resonance forms
m/z 69
CH3 • +
plus one other resonance form
m/z 99

21-61

The formula $C_6H_8O_3$ indicates 3 elements of unsaturation.

IR spectrum: The absence of OH peaks shows that the compound is neither an alcohol nor a carboxylic acid. There are two carbonyl absorptions: the one about 1760 cm^{-1} is likely a strained cyclic ester (reinforced with the C—O peak around 1150 cm^{-1}), while the one at 1710 cm^{-1} is probably a ketone. (An anhydride also has two peaks, but they are of higher frequency than the ones in this spectrum.)

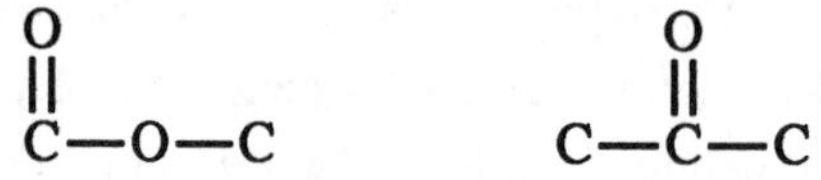

Proton NMR spctrum: The NMR shows four types of protons. The 2H triplet at δ 4.3 is a CH_2 group next to an oxygen on one side, with a CH_2 on the other. The 1H triplet at δ 3.8 is also strongly deshielded (probably by two carbonyls), a CH next to a CH_2. The 3H singlet at δ 2.4 is a CH_3 on one of the carbonyls. The remaining 2H signal is highly coupled, a CH_2 between other hydrogens. There are no vinyl hydrogens (and no alkene carbon in the carbon NMR), so the remaining element of unsaturation must be a ring.

Assemble the pieces:

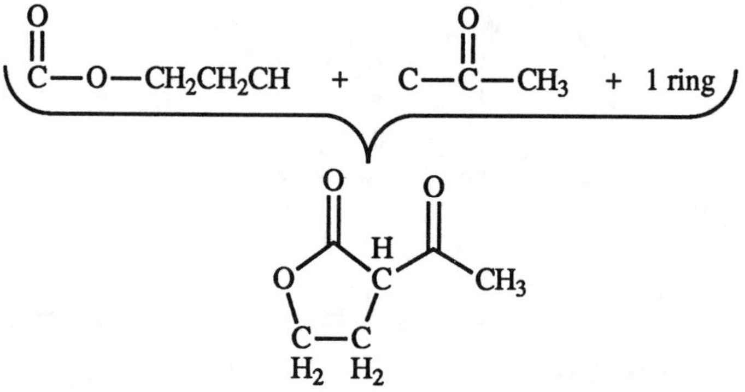

Carbon NMR:

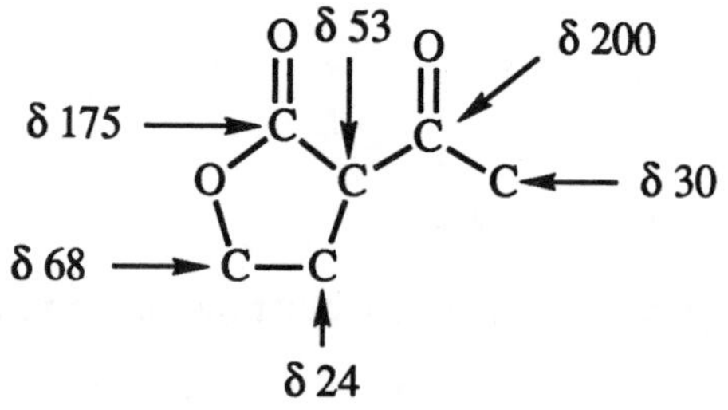

21-62

The rate of a reaction depends on its activation energy, that is, the difference in energy between starting material and the transition state. The transition state in saponification is similar in structure, and therefore in energy, to the tetrahedral intermediate:

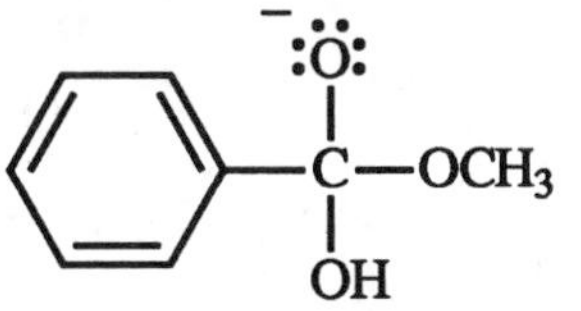

tetrahedral intermediate

The tetrahedral carbon has no resonance overlap with the benzene ring, so any resonance effect of a substituent on the ring will have very little influence on the energy of the transition state.

What will have a big influence on the activation energy is whether a substituent stabilizes or destabilizes the starting material. Anything that stabilizes the starting material will therefore increase the activation energy, slowing the reaction; anything that destabilizes the starting material will decrease the activation energy, speeding the reaction.

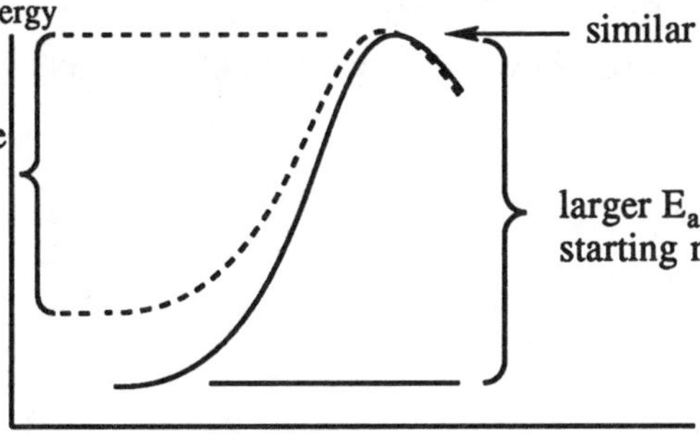
energy
similar transition state energy
smaller E_a because of destabilized starting material
larger E_a because of stabilized starting material
reaction

(a) One of the resonance forms of methyl *p*-nitrobenzoate has a positive charge on the benzene carbon adjacent to the positive carbonyl carbon. This resonance form destabilizes the starting material, lowering the activation energy, speeding the reaction.

poor resonance contributor, destabilizing the starting material; no effect in the transition state

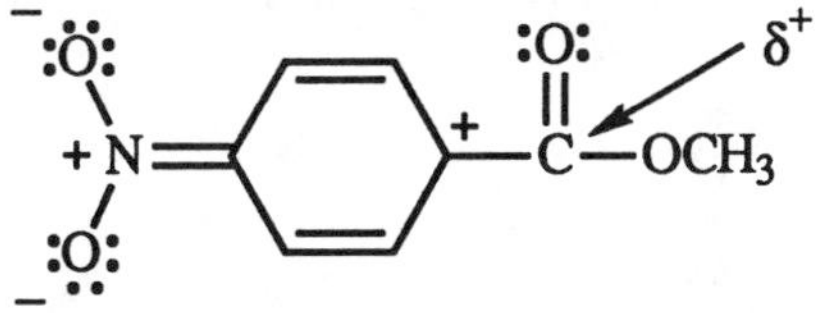

(b) One of the resonance forms of methyl *p*-methoxybenzoate has all atoms with full octets, and negative charge on the most electronegative atom. This resonance form stabilizes the starting material, increasing the activation energy, slowing the reaction.

good resonance contributor, stabilizing the starting material; no effect in the transition state

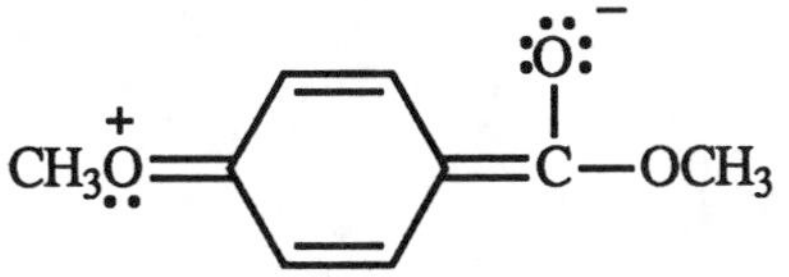

21-63

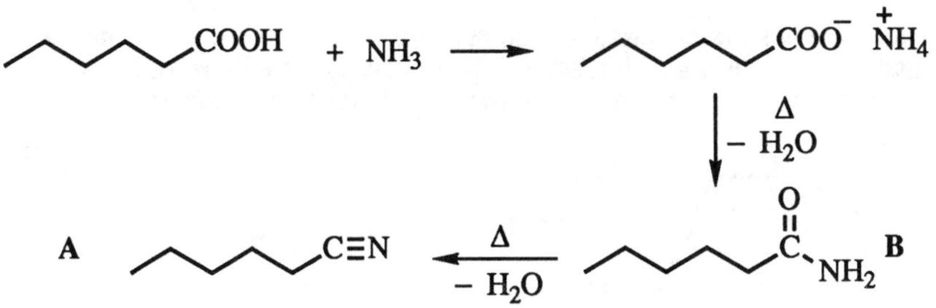

21-64

A singlet at δ 2.15 is H on carbon next to carbonyl, the only type of proton in the compound. The IR spectrum shows no OH, and shows two carbonyl absorptions at high frequency, characteristic of an anhydride. Mass of the molecular ion at 102 proves that the anhydride must be acetic anhydride, a reagent commonly used in aspirin synthesis.

$$CH_3\overset{\overset{\displaystyle O}{\|}}{C}-O-\overset{\overset{\displaystyle O}{\|}}{C}CH_3$$

Acetic anhydride can be disposed of by hydrolyzing (carefully! exothermic!) and neutralizing in aqueous base.

22-1

(a)

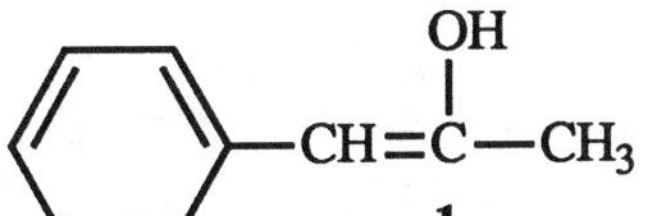

1

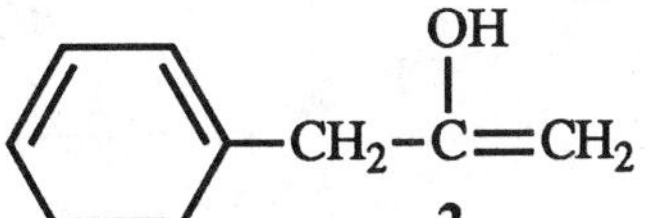

2

(b) Enol **1** will predominate at equilibrium as its double bond is conjugated with the benzene ring, making it more stable than **2**.

(c)

basic conditions

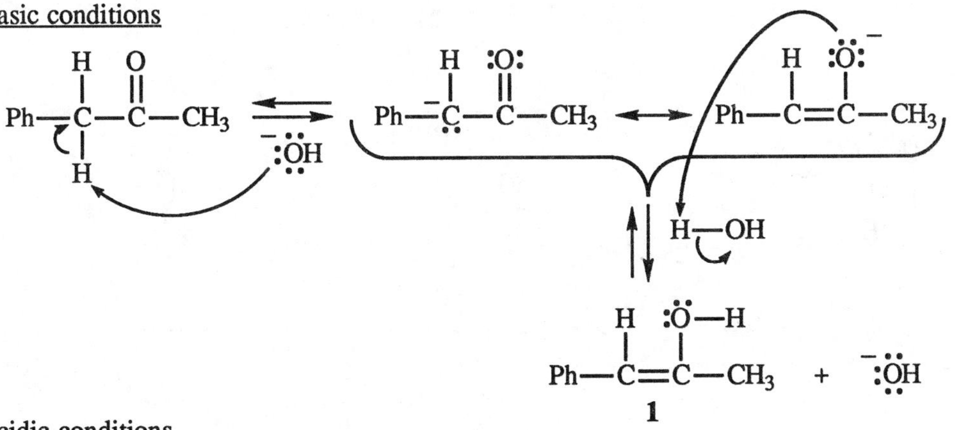

acidic conditions

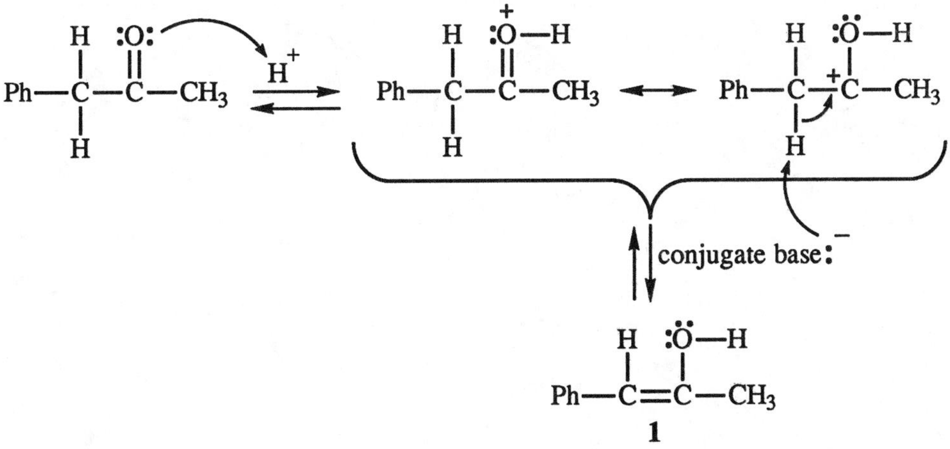

22-1 (c) continued

basic conditions

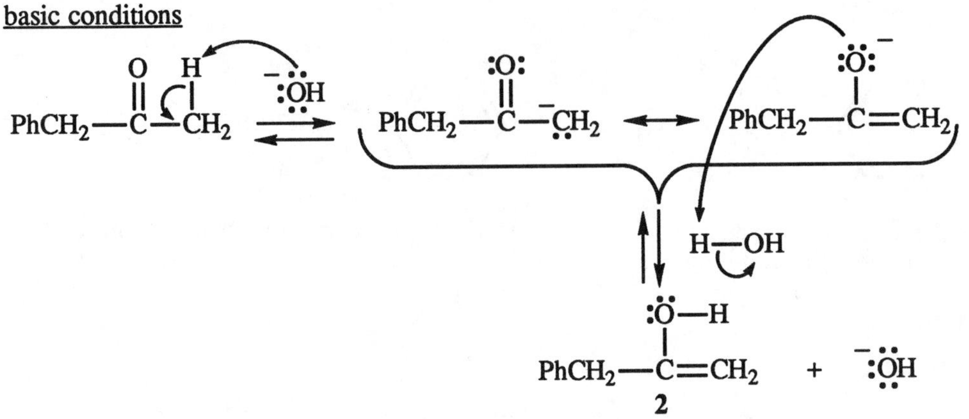

acidic conditions

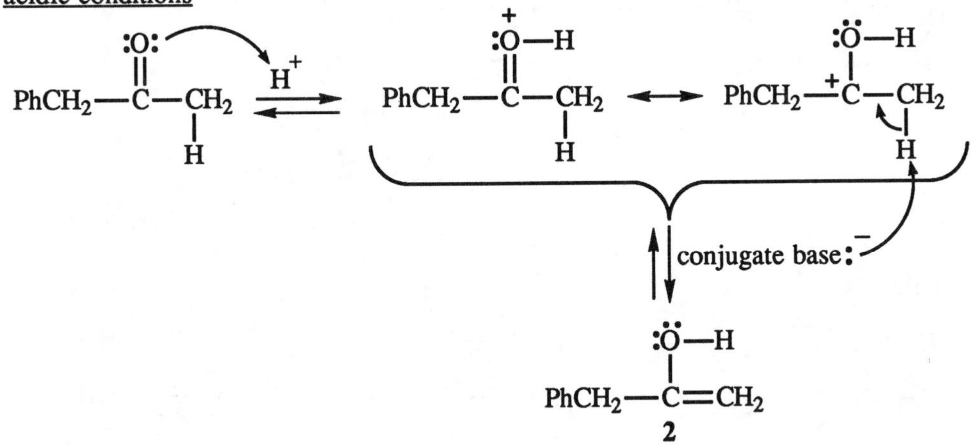

22-2

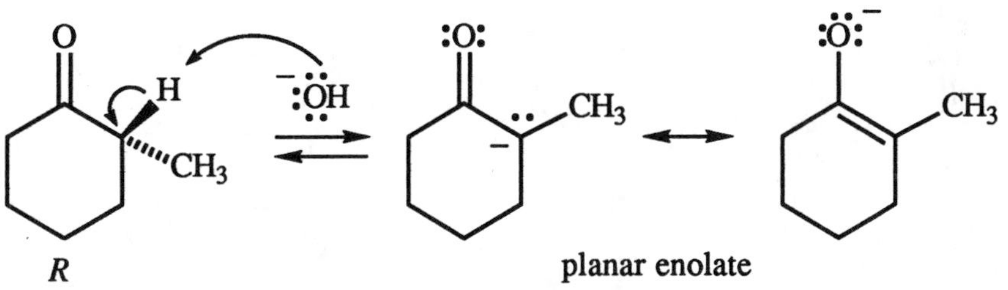

This planar enolate entermediate has lost all chirality. Protonation can occur at either the oxygen or the carbon. Protonation on oxygen (not shown here) gives the enol which can then be deprotonated to return to the enolate. Protonation on carbon leads to racemic product.

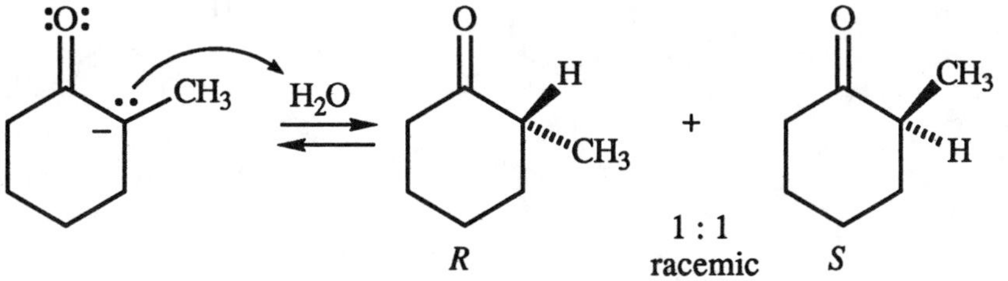

22-3

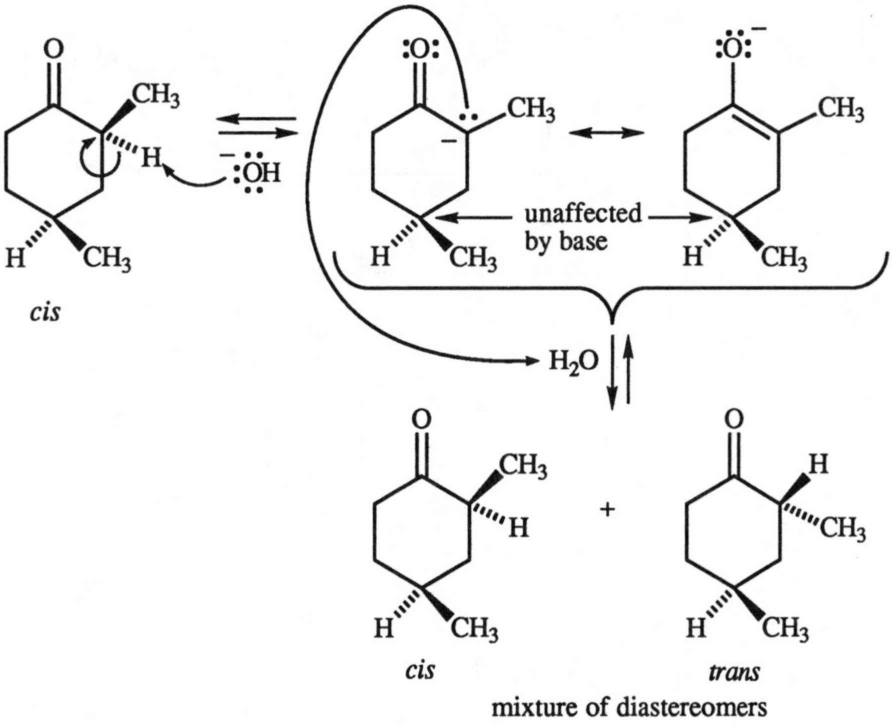

22-4

(a)

(b)

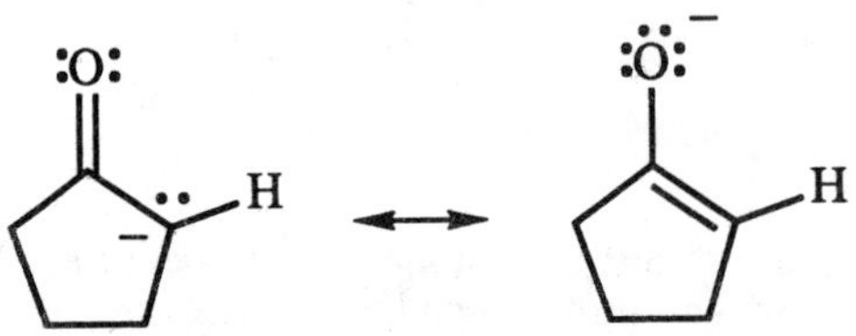

(c)

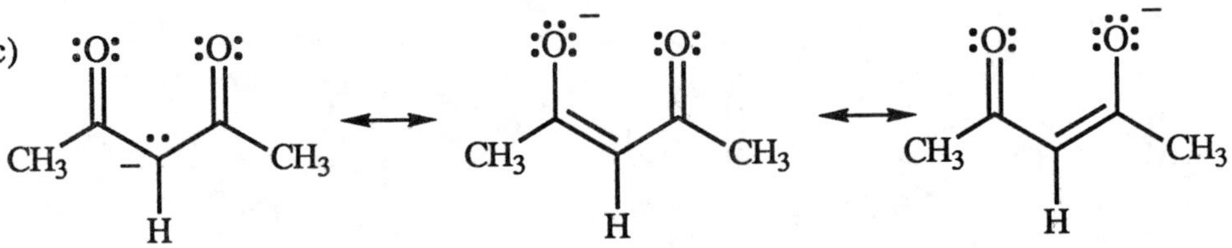

22-5

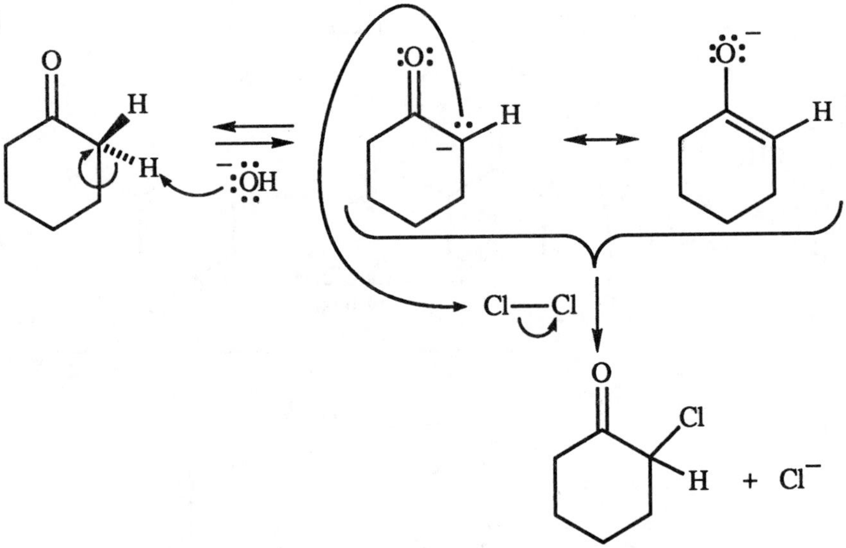

22-6

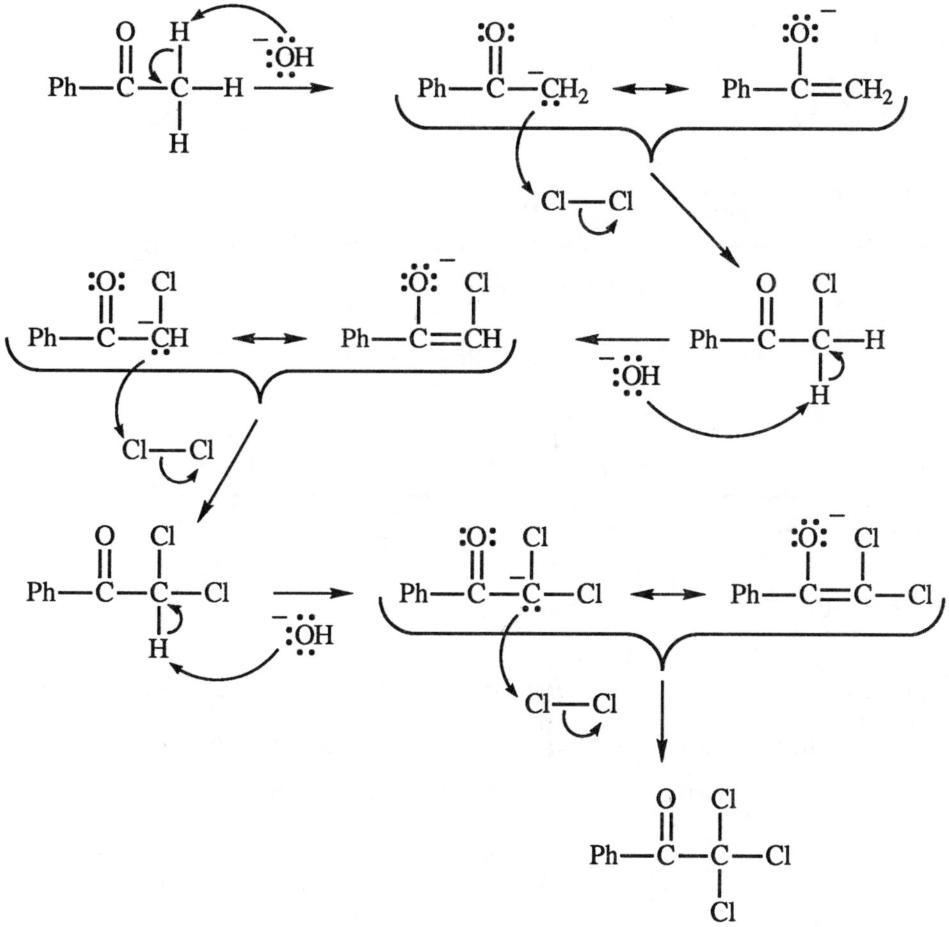

22-7 For this problem, the cyclohexyl group is abbreviated "Cy".

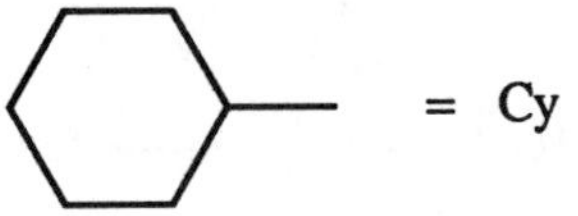

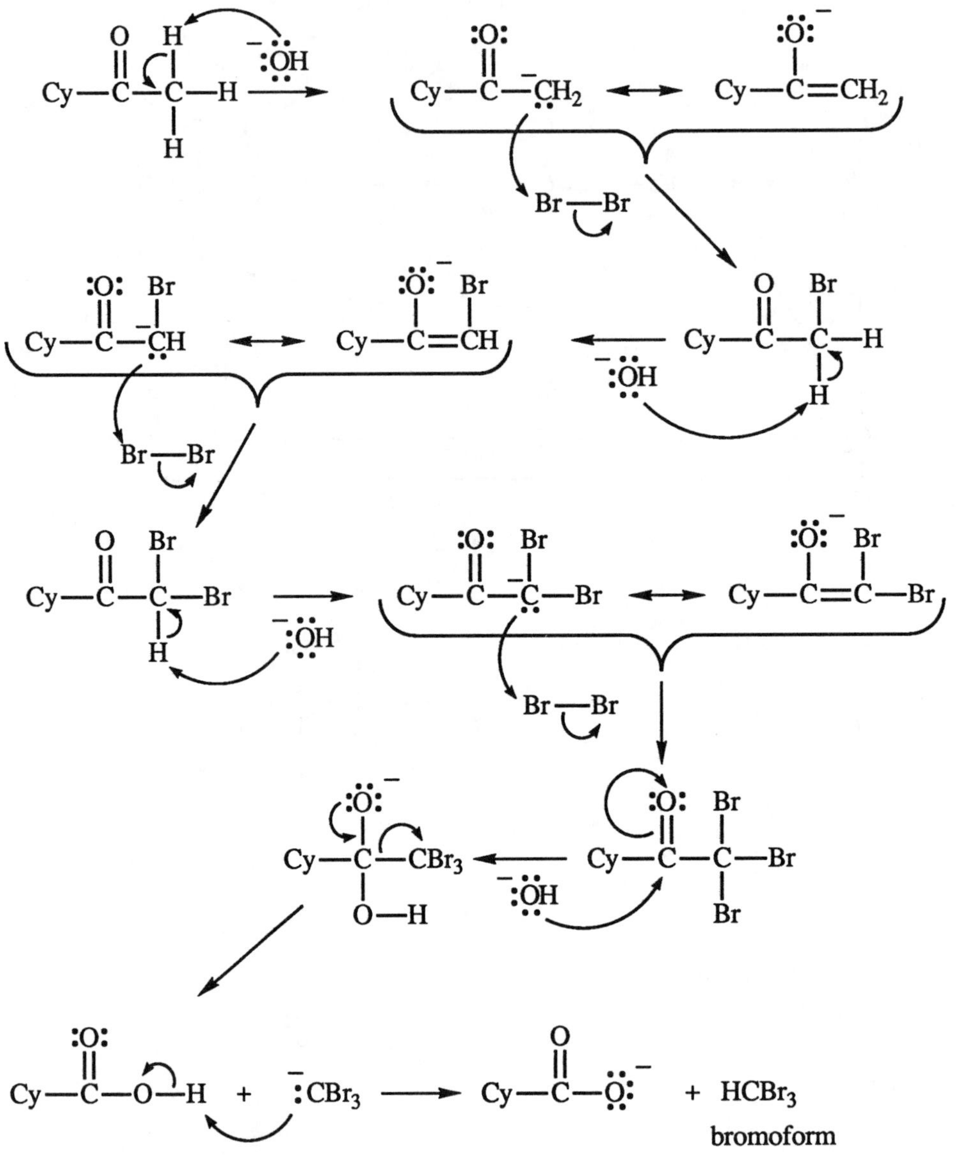

22-8

(a) 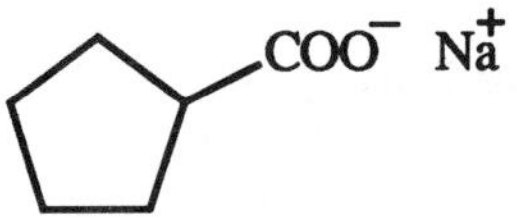 + $CHCl_3$

(b) 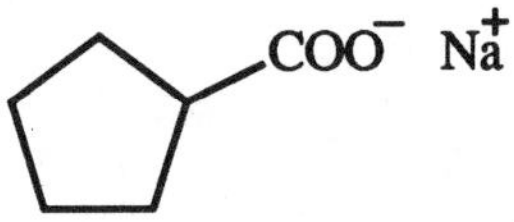 + CHI_3 (precipitate)

(c) 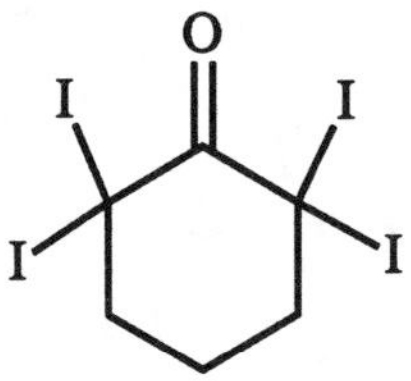

(d) 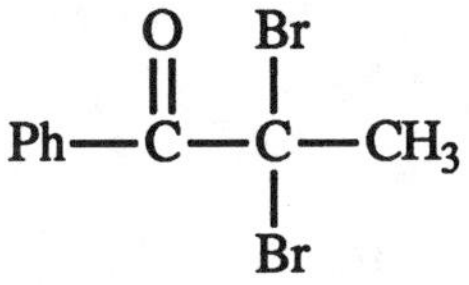

22-9 Methyl ketones, and alcohols which are oxidized to methyl ketones, will give a positive iodoform test. All of the compounds in this problem except 3-pentanone (part (d)) will give a positive iodoform test.

22-10

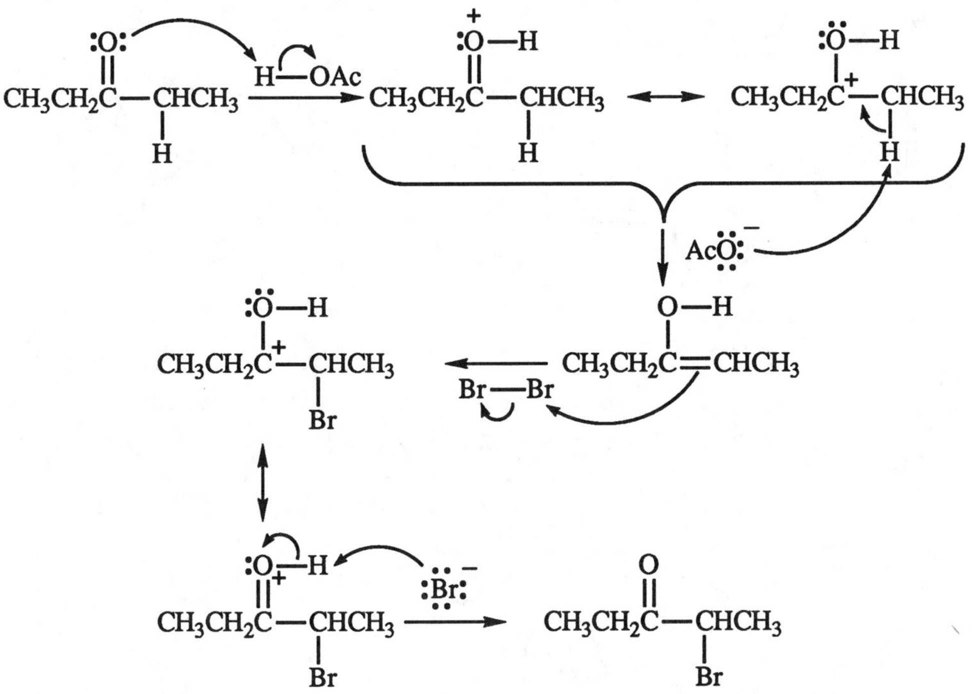

22-11

(a)

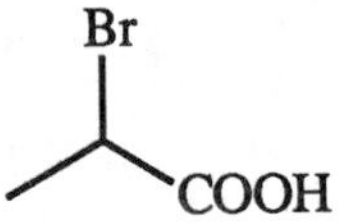

(b) no reaction: no α-hydrogen

(c)

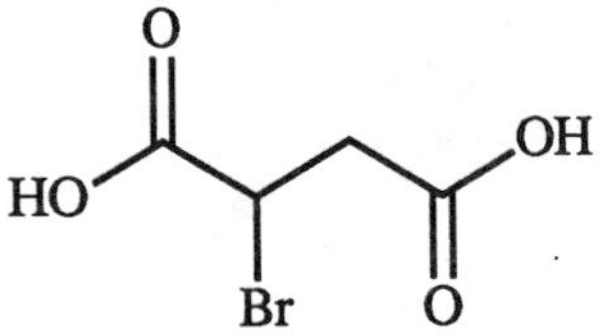

(d) no reaction: no α-hydrogen

22-12

(a)

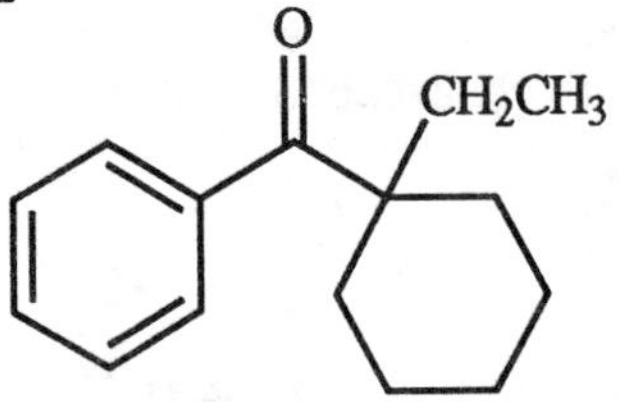

(b)

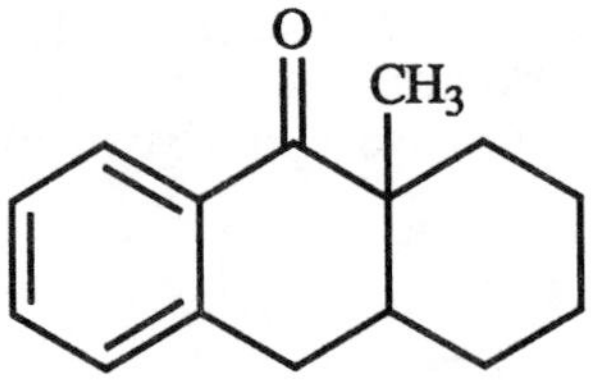

22-13

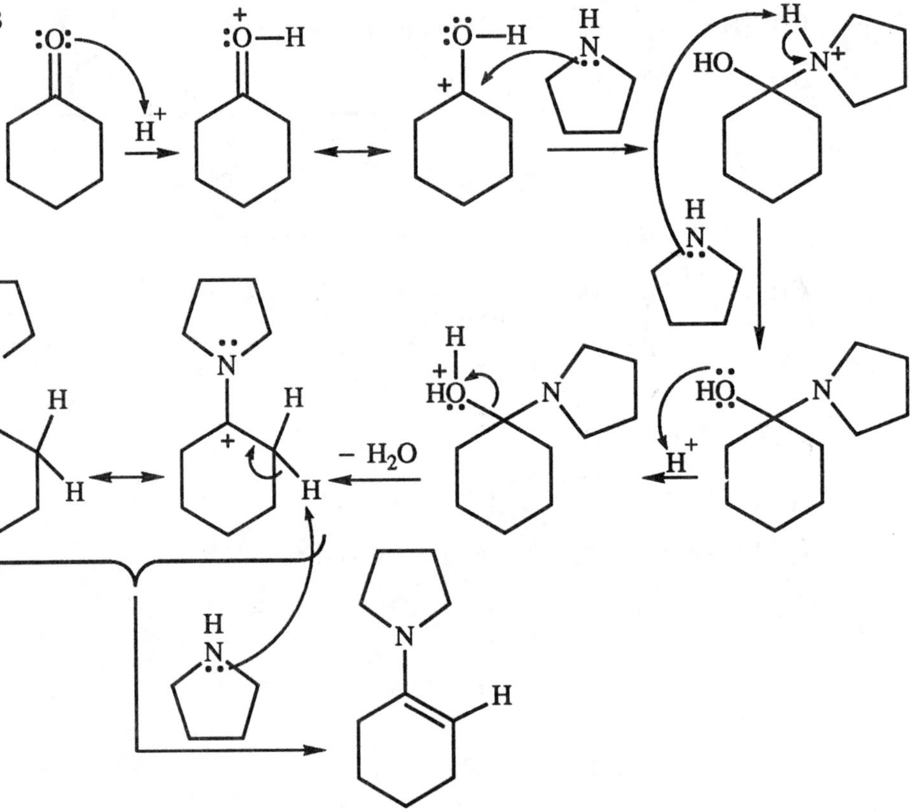

22-14

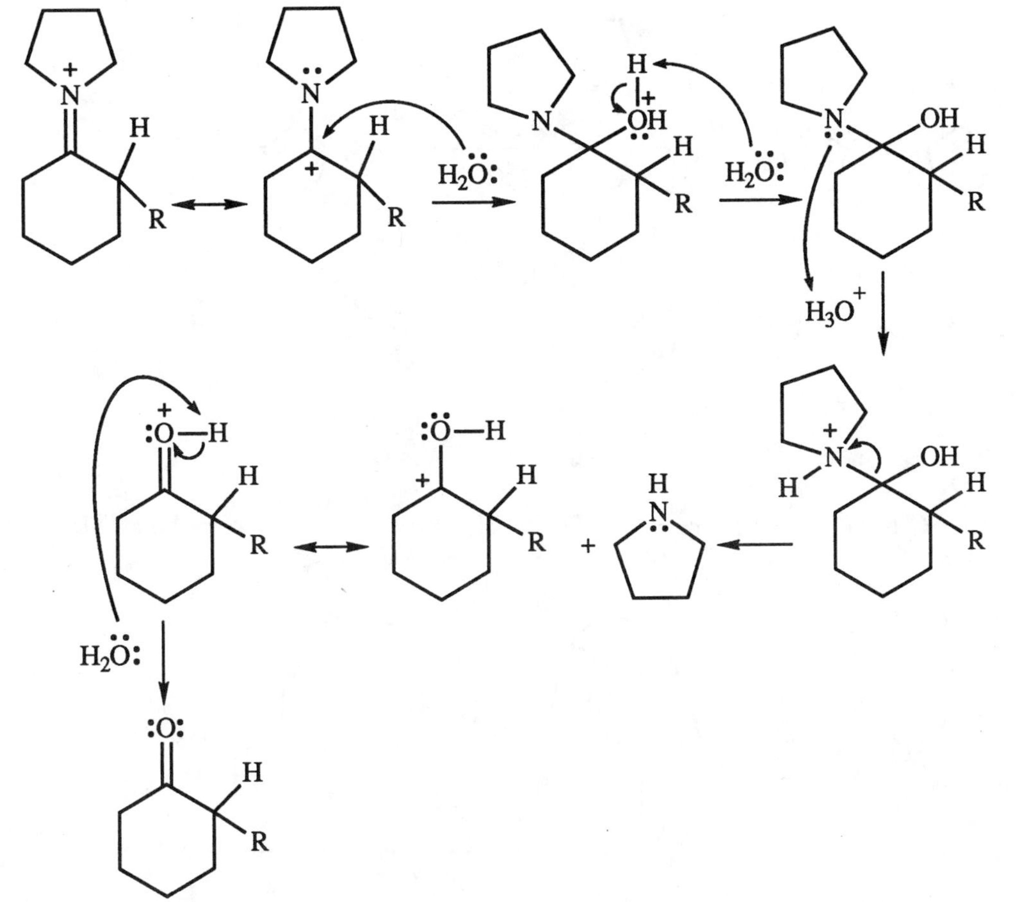

22-15

(a)

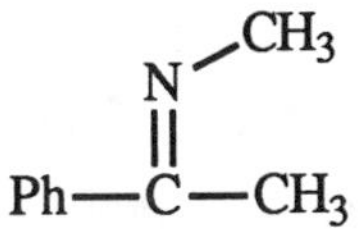

(b)

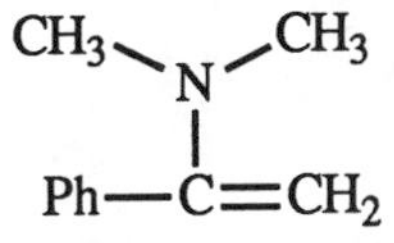

(c)

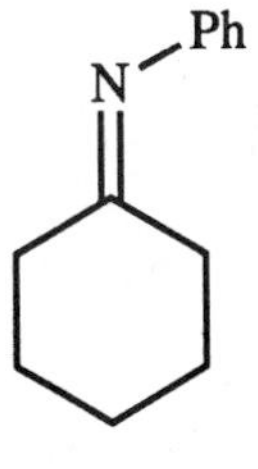

(d)

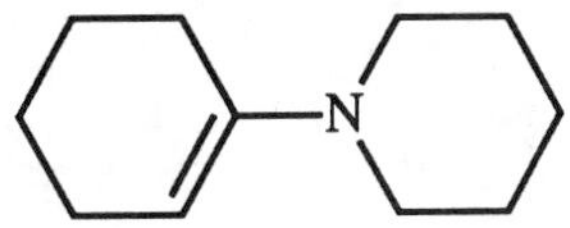

22-16 The particular 2° amines you choose for this problem do not matter.

(a)

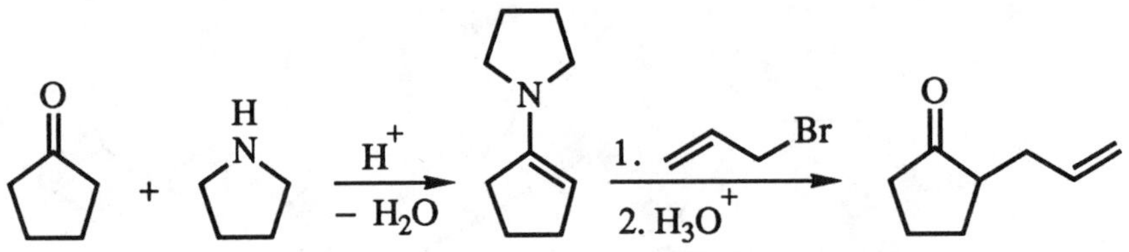

(b)

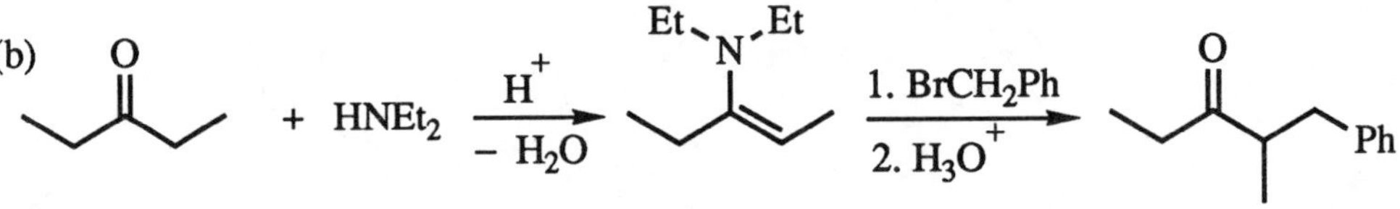

(c)

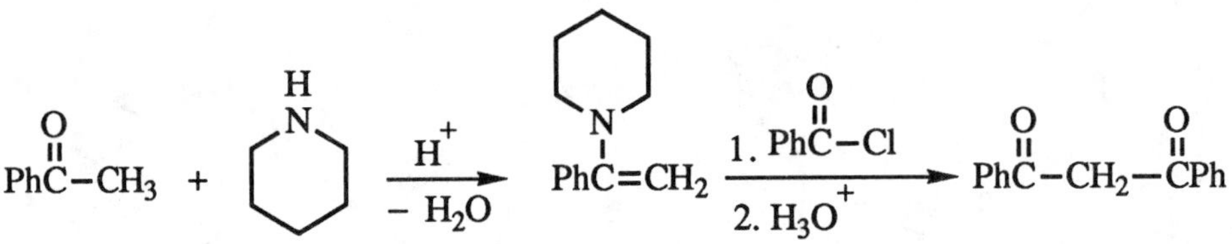

22-17

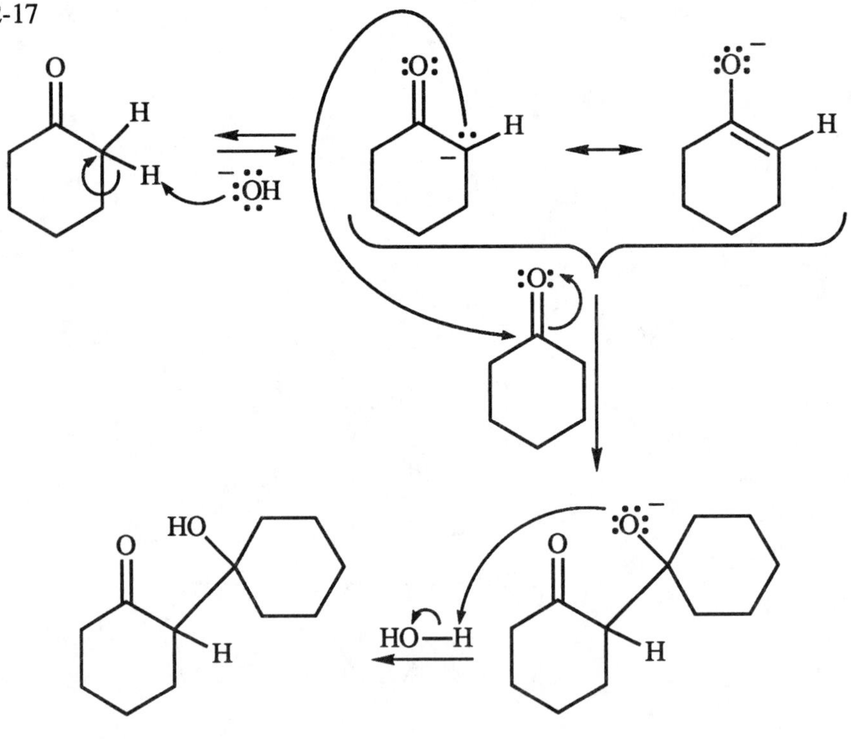

22-18

(a) 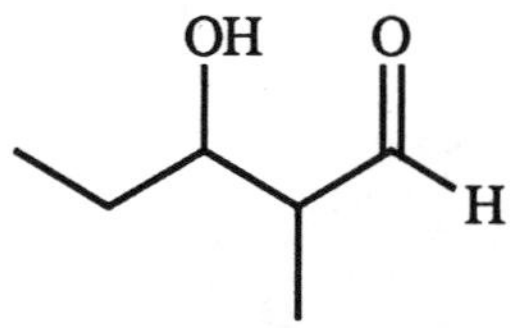

(b) 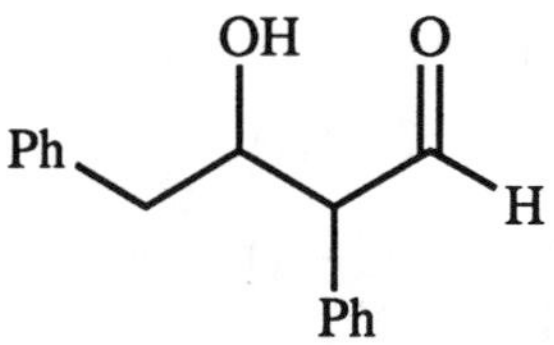

22-19 All the steps in the aldol condensation are reversible. Adding base to diacetone alcohol promoted the reverse aldol reaction. The equilibrium greatly favors acetone.

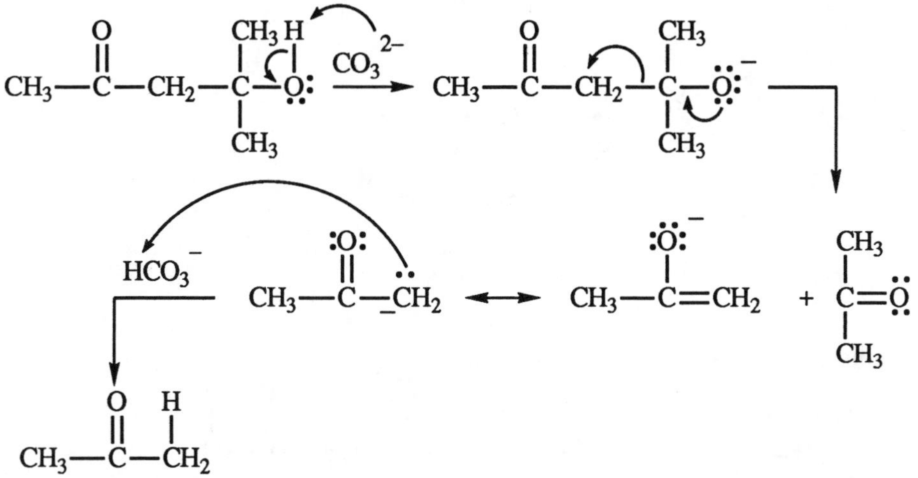

22-20

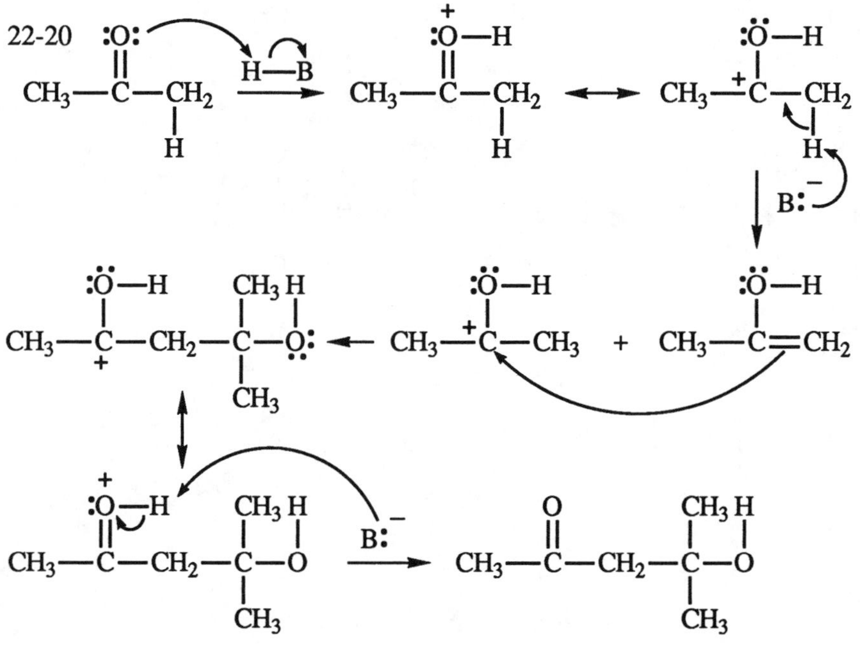

22-21

(a) acidic conditions

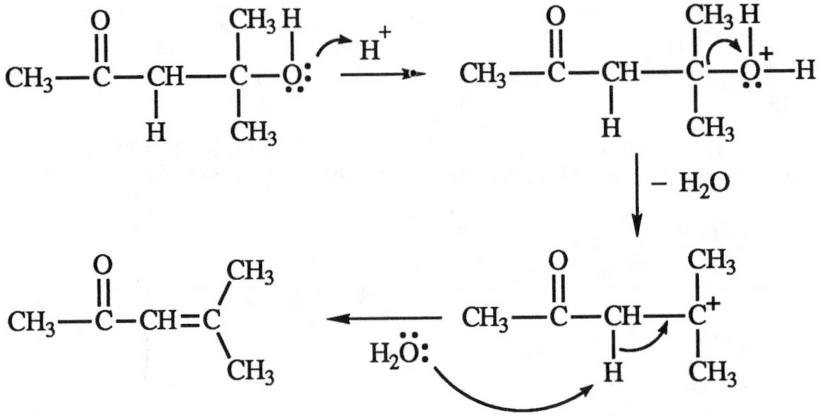
O
CH3 H
CH3—C—CH—C—O:
H
CH3
H+
CH3—C—CH—C—O—H
– H2O
CH3—C—CH=C
CH3
H2O:
C+

(b) basic conditions

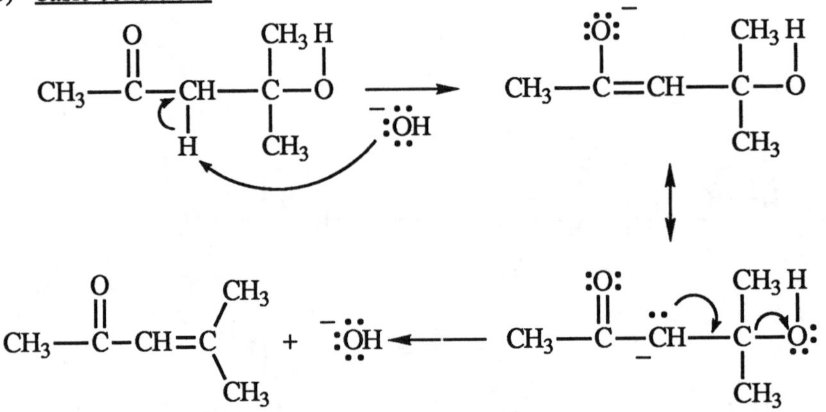
O
CH3 H
CH3—C—CH—C—O
:OH
:O:
CH3—C=CH—C—O
CH3—C—CH=C
+ :OH
CH3—C—CH—C—O:

22-22

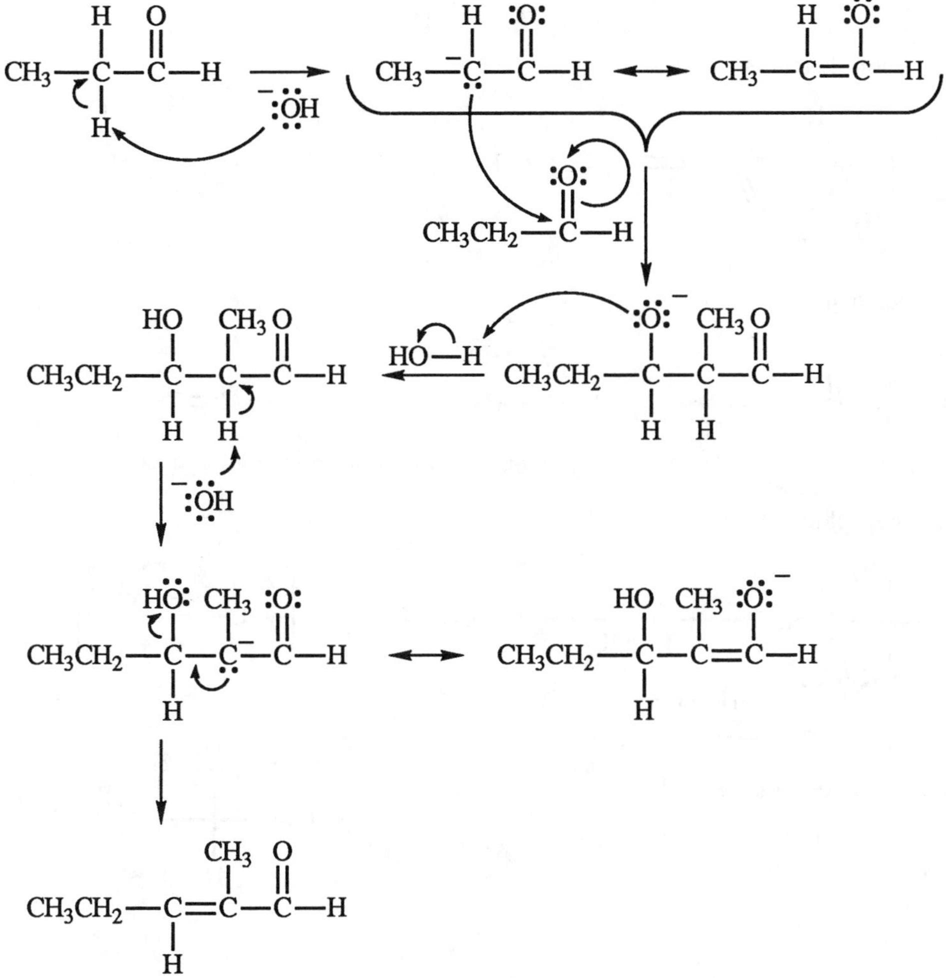

22-23

(a)

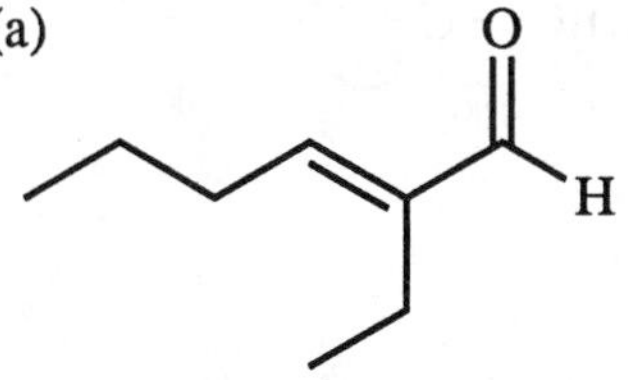

(b)

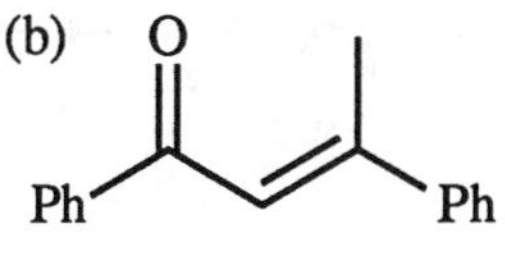

(c)

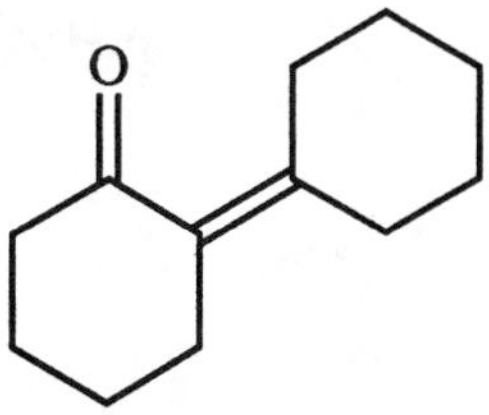

22-24

(a)

Step 1: carbon skeletons

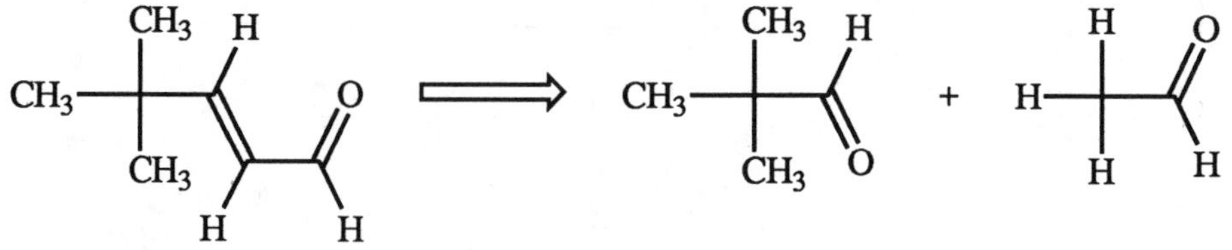

Step 2: nucleophile generation

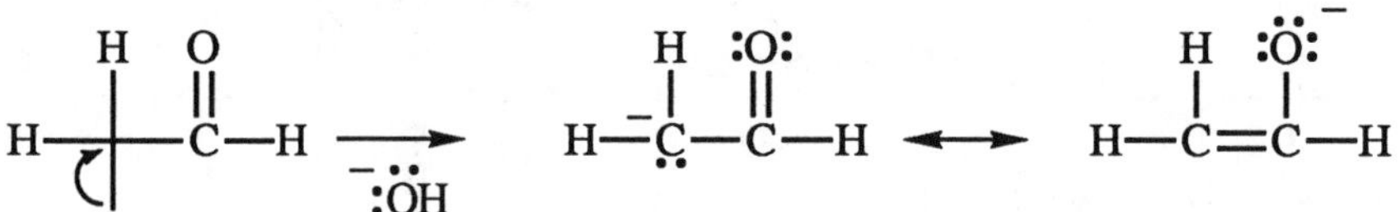

(2,2-Dimethylpropanal has no α-hydrogen.)

Step 3: nucleophilic attack

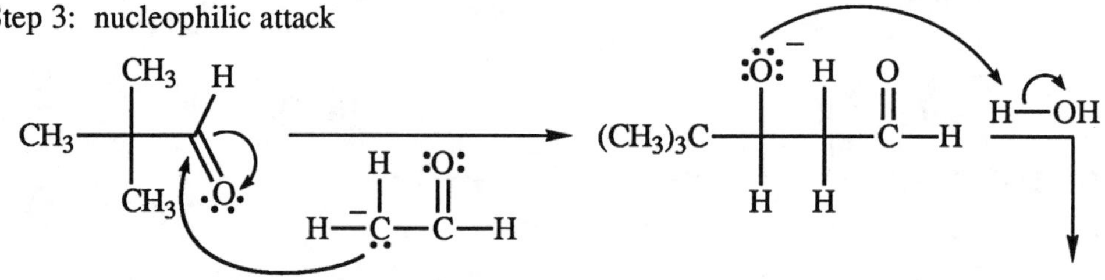

Step 4: conversion to final product

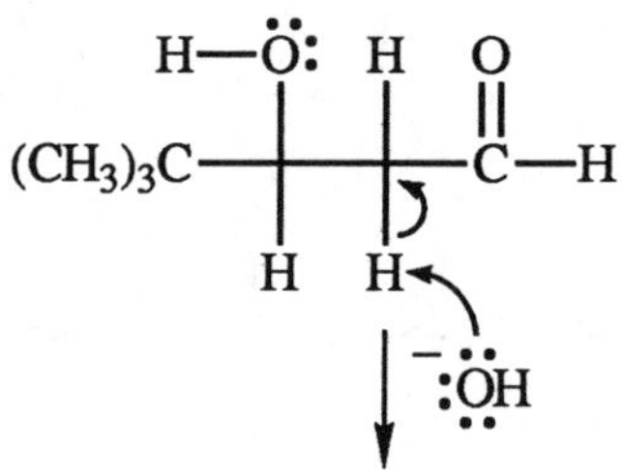

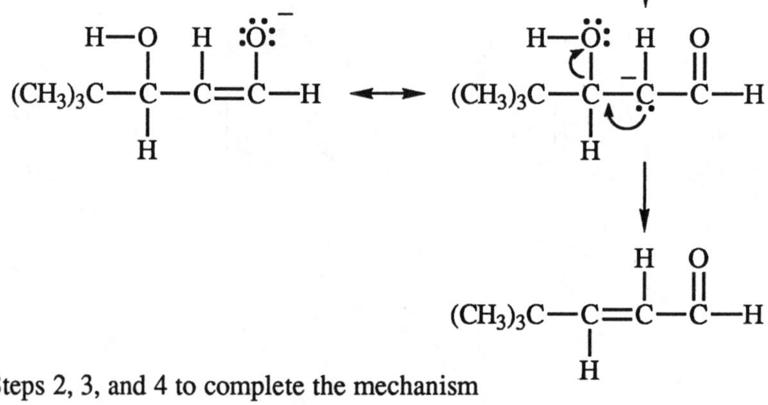

Step 5: combine Steps 2, 3, and 4 to complete the mechanism

22-24 continued

(b)

Step 1: carbon skeletons

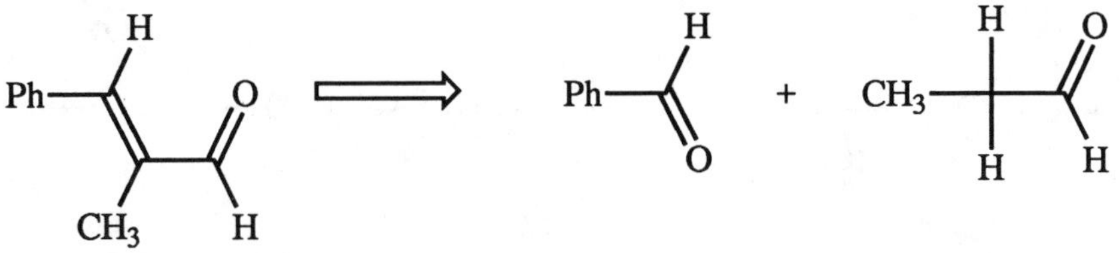

Step 2: nucleophile generation

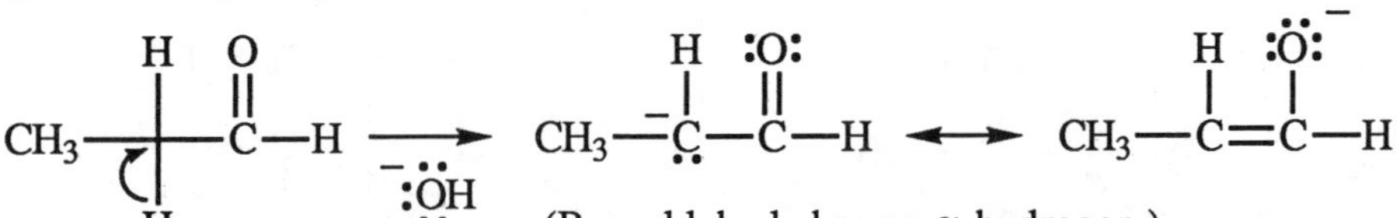

(Benzaldehyde has no α-hydrogen.)

Step 3: nucleophilic attack

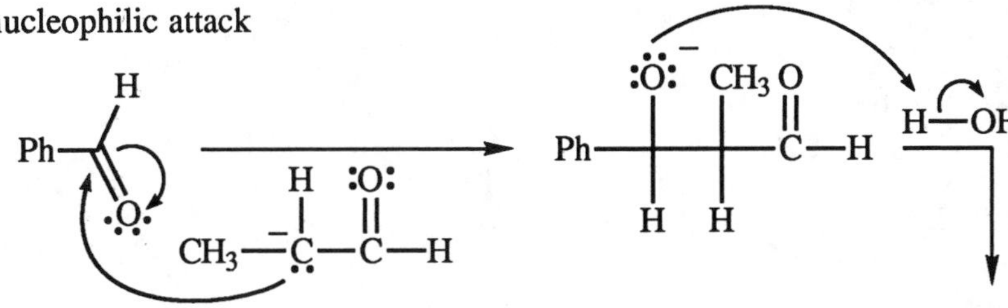

Step 4: conversion to final product

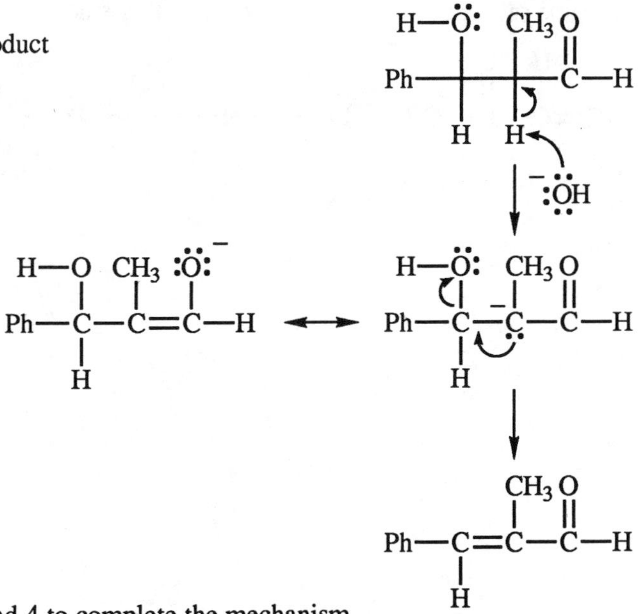

Step 5: combine Steps 2, 3, and 4 to complete the mechanism

22-25 This solution presents the sequence of reactions leading to the product, following the format of the Problem-Solving feature. This is not a complete mechanism.

Step 2: generation of the nucleophile

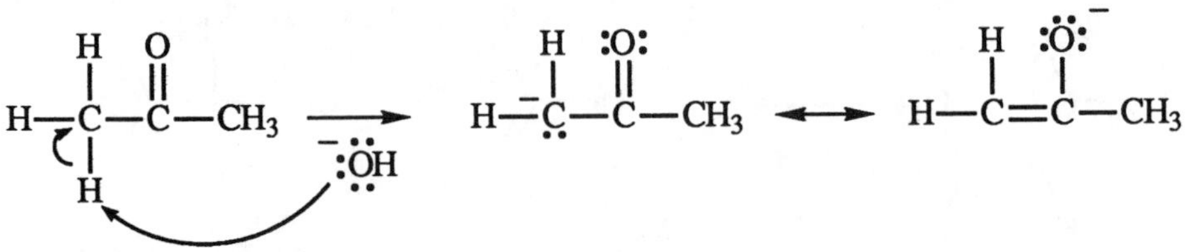

Step 3: nucleophilic attack

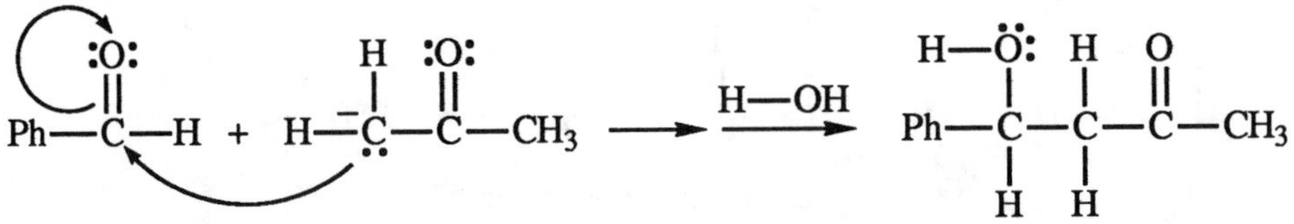

Step 4: dehydration

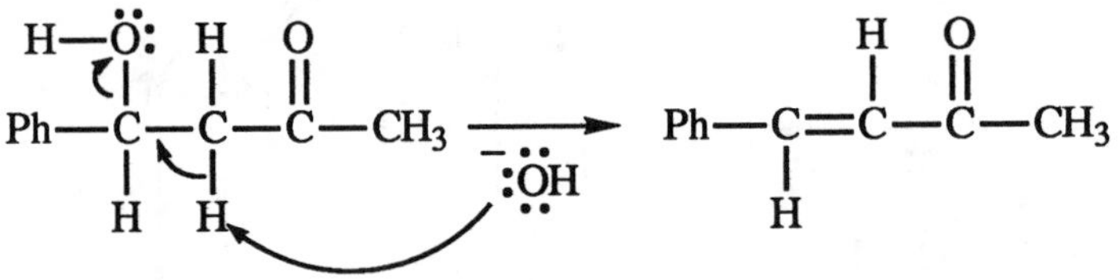

The same sequence of steps occurs on the other side.

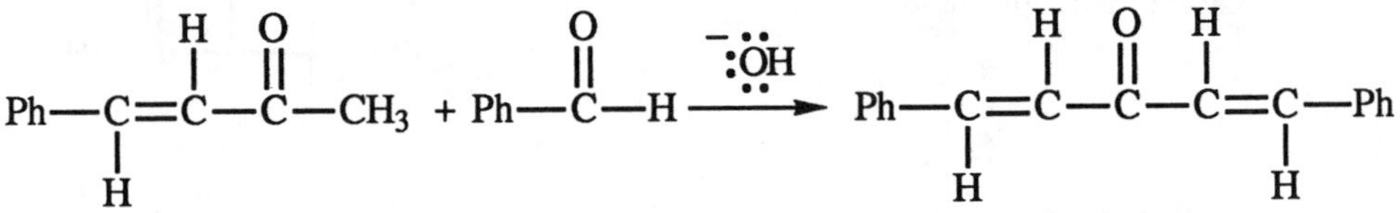

final product

22-26

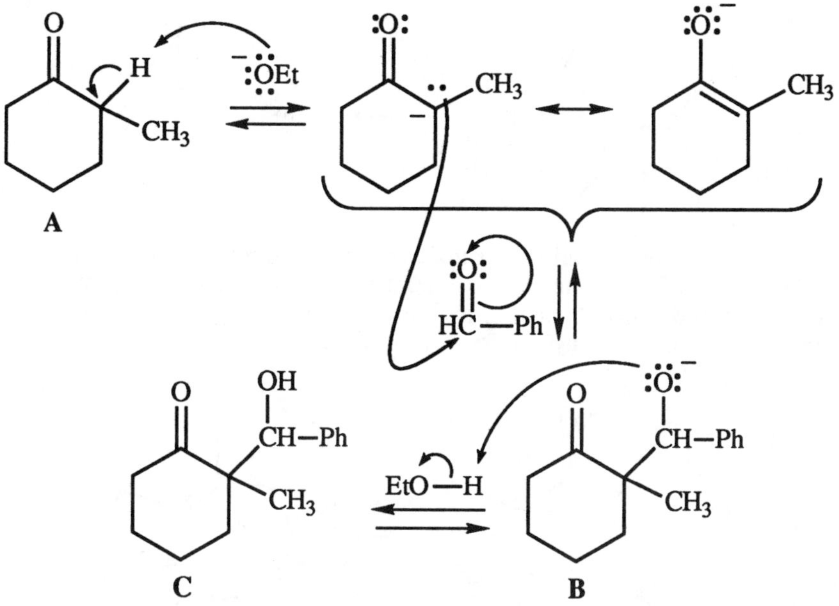

There are three problems with the reaction as shown:

1. Hydrogen on a 3° carbon (structure **A**) is less acidic than hydrogen on a 2° carbon. The 3° hydrogen will be removed at a slower rate than the 2° hydrogen.
2. Nucleophilic attack by the 3° carbon will be more hindered, and therefore slower, than attack by the 2° carbon. Structure **B** is quite hindered.
3. Once a normal aldol product is formed, dehydration gives a conjugated system which has great stability. The aldol product **C** cannot dehydrate because no α-hydrogen remains. Some **C** will form, but eventually the reverse-aldol process will return **C** to starting materials which, in turn, will react at the other α-carbon to produce the conjugated system. (This reason is the Kiss of Death for **C**.)

22-27

(a)

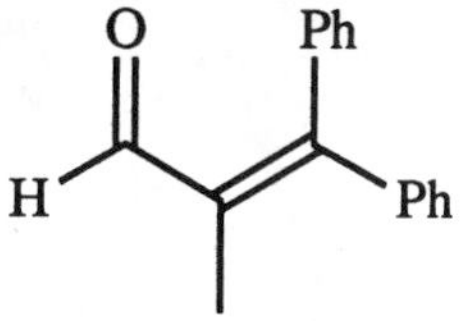

(b)

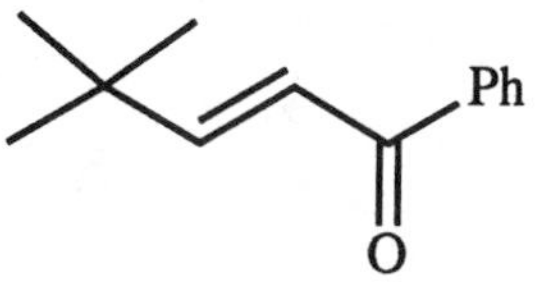

22-28

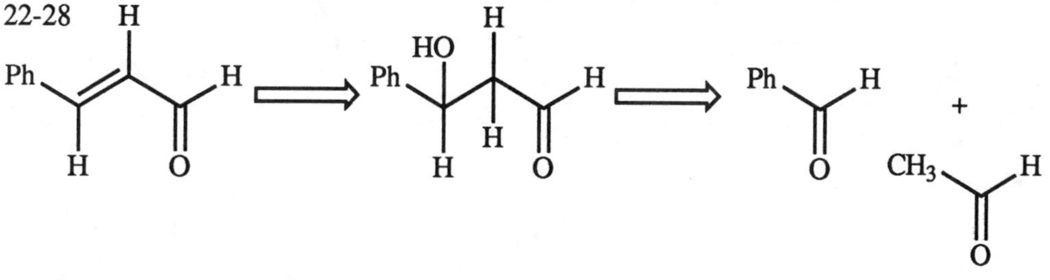

22-29

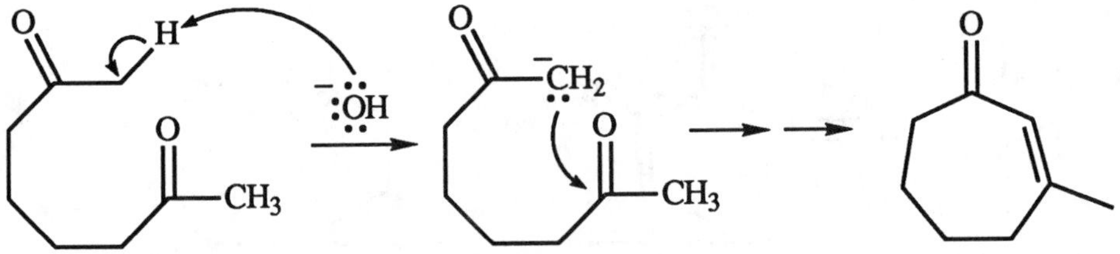

The formation of a seven-membered ring is unfavorable for entropy reasons: the farther apart the nucleophile and the electrophile, the harder time they will have finding each other. If the molecule has a possibility of forming a 5- or a 7-membered ring, it will almost always prefer to form the 5-membered ring.

22-30

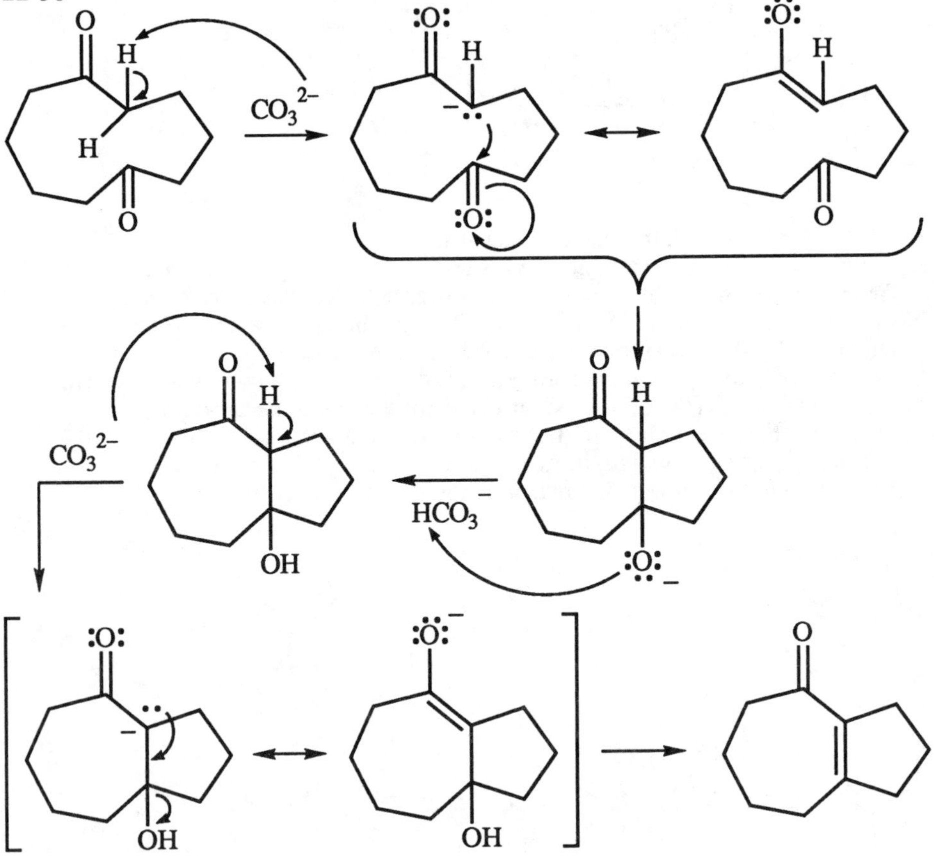

22-31

(a)

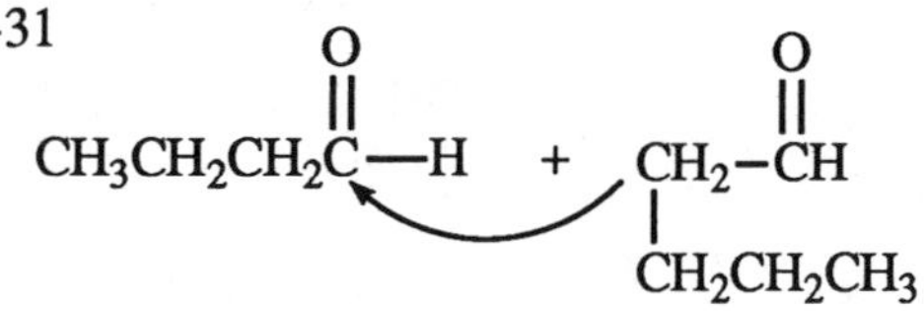

not feasible: requires condensation of two different aldehydes, each with α-hydrogens

(b)

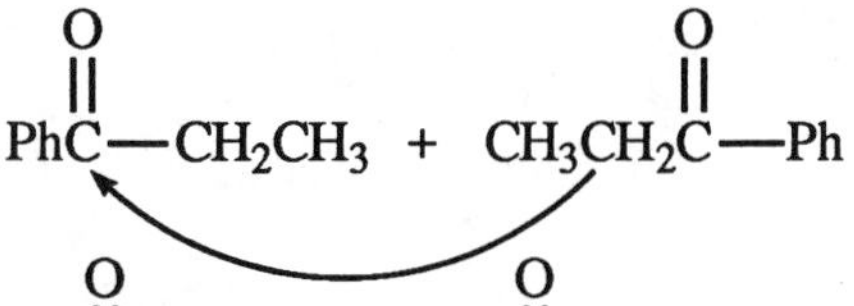

feasible: self-condensation

(c)

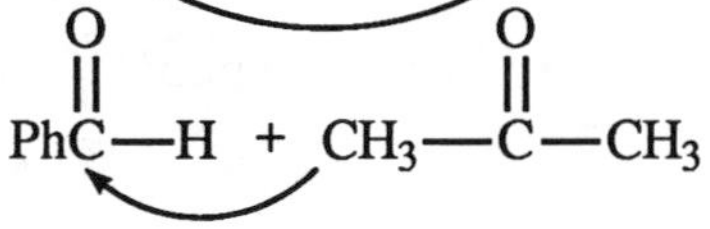

feasible: only one reactant has α-hydrogen

(d)

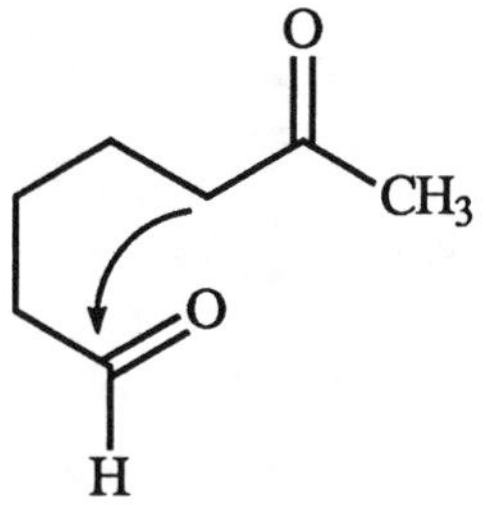

feasible; however, the cyclization from the carbon α to the aldehyde to the ketone carbonyl is also possible

(e)

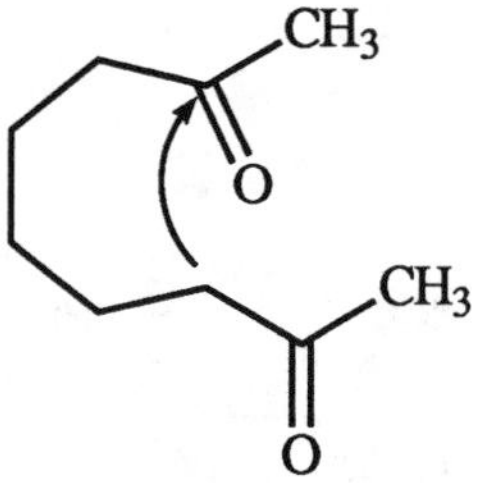

feasible: symmetric reagent will give the same product in either direction of cyclization

22-32

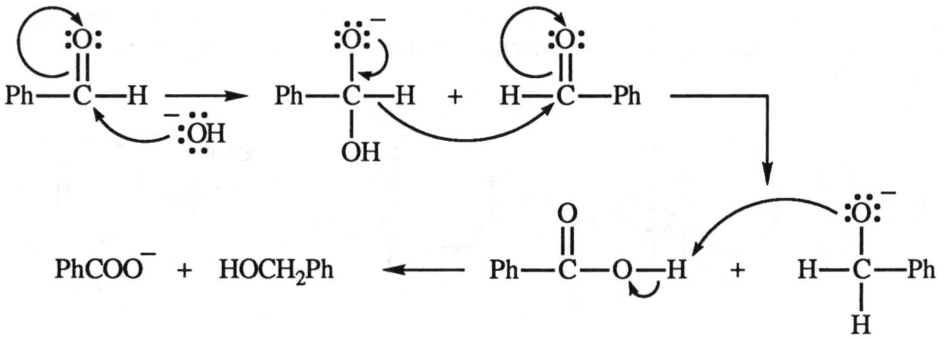

22-33

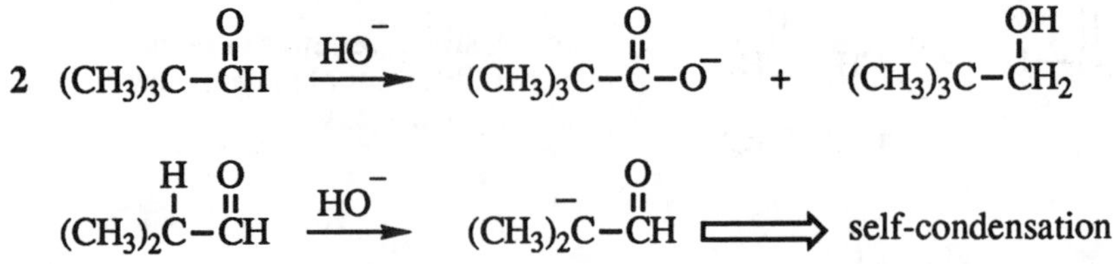

2-Methylpropanal has an α-hydrogen. Self-condensation will occur instead of the Cannizzaro reaction.

22-34 Trimethylphosphine has α-hydrogens that could be removed by butyllithium, generating undesired ylides.

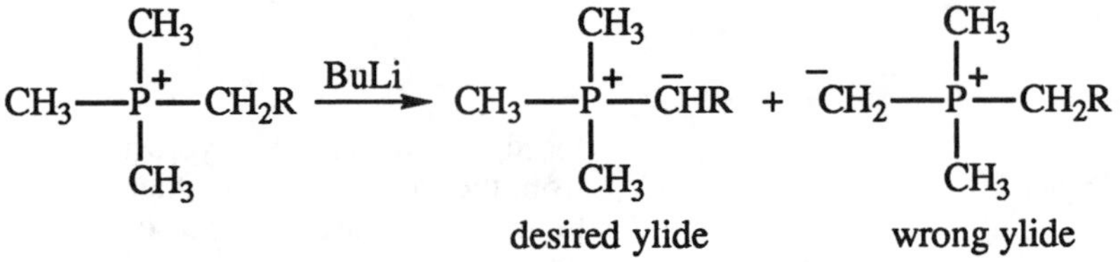

22-35

(a)

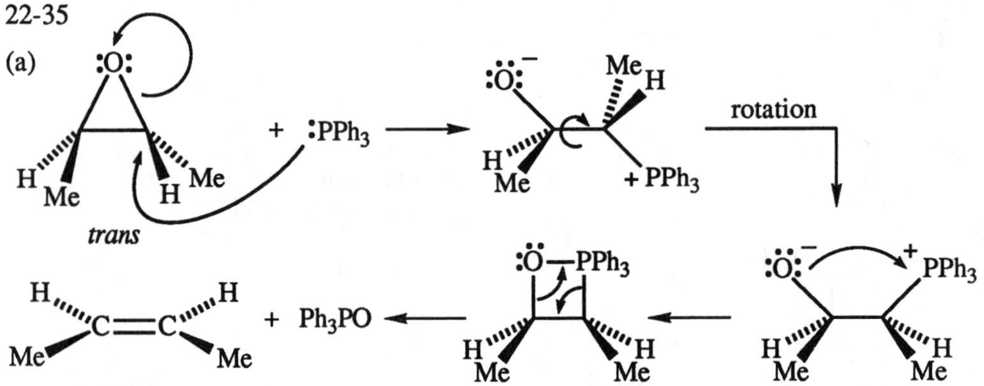

cis-2-butene

The stereochemistry is inverted. The nucleophile triphenylphosphine must attack the epoxide in an anti fashion, yet the triphenylphosphine oxide must eliminate with syn geometry.

(b)

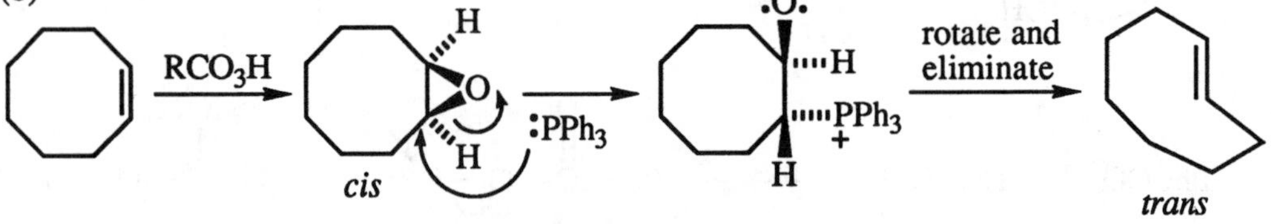

22-36

(a) $CH_2{=}CHCH_2Br \xrightarrow[\text{2. BuLi}]{\text{1. } Ph_3P} CH_2{=}CH\overset{-}{C}H{-}\overset{+}{P}Ph_3 \xrightarrow{PhCHO} PhCH{=}CH{-}CH{=}CH_2$

$PhCH_2Br \xrightarrow[\text{2. BuLi}]{\text{1. } Ph_3P} Ph\overset{-}{C}H{-}\overset{+}{P}Ph_3 \xrightarrow{O{=}CH{-}CH{=}CH_2} PhCH{=}CH{-}CH{=}CH_2$

(b) $PhCH{=}CH{-}CH_2Br \xrightarrow[\text{2. BuLi}]{\text{1. } Ph_3P} PhCH{=}CH{-}\overset{-}{C}H{-}\overset{+}{P}Ph_3 \xrightarrow{CH_2O} PhCH{=}CH{-}CH{=}CH_2$

OR $CH_3I \xrightarrow[\text{2. BuLi}]{\text{1. } Ph_3P} \overset{-}{C}H_2{-}\overset{+}{P}Ph_3 \xrightarrow{PhCH{=}CH{-}CH{=}O} PhCH{=}CH{-}CH{=}CH_2$

22-37 Many alkenes can be synthesized by two different Wittig reactions (as in the previous problem). The ones shown here form the phosphonium salt from the less hindered alkyl halide.

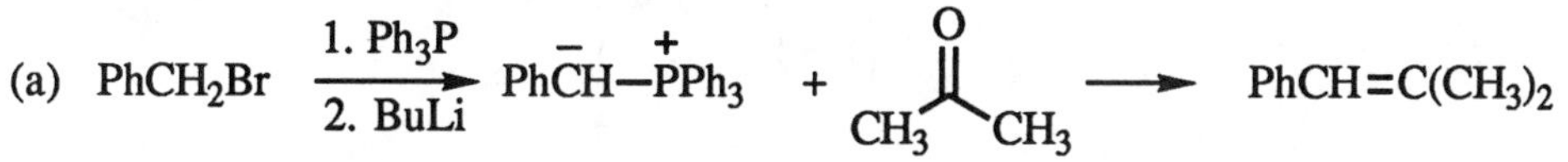

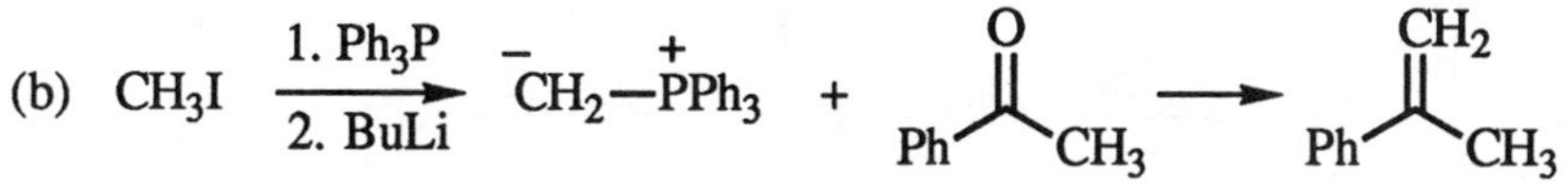

(c)
$PhCH_2Br \xrightarrow[\text{2. BuLi}]{\text{1. } Ph_3P} Ph\overset{-}{C}H{-}\overset{+}{P}Ph_3 \xrightarrow{PhCH{=}CH{-}CH{=}O} PhCH{=}CH{-}CH{=}CHPh$

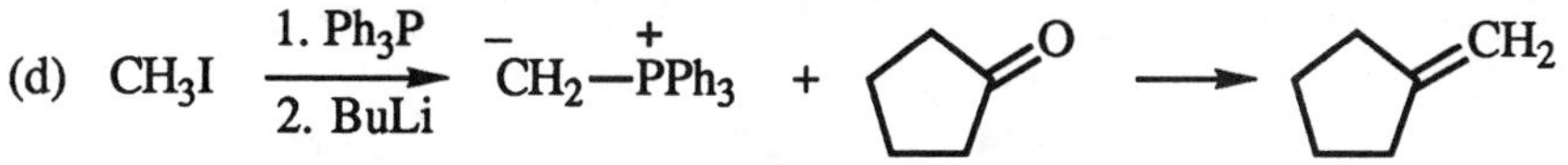

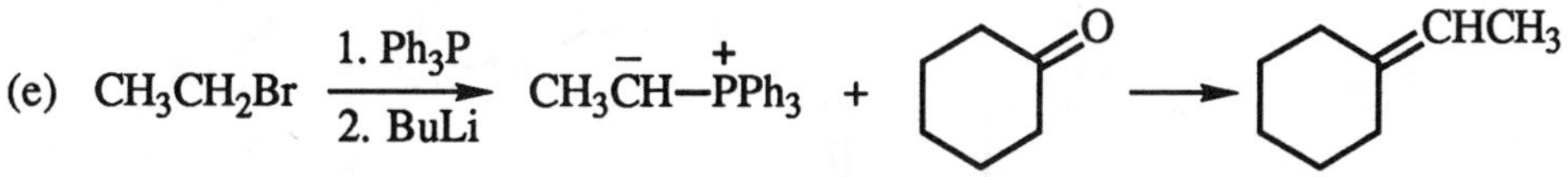

22-38

(a) The side reaction with sodium methoxide is transesterification. The starting material, and therefore the product, would be a mixture of methyl and ethyl esters.

CH3C—OCH2CH3 + NaOCH3 ⇌ CH3C—OCH3 + NaOCH2CH3

(b) Sodium hydroxide would irreversibly saponify the ester, completely stopping the Claisen condensation as the carbonyl no longer has a leaving group attached to it.

CH3C—OCH2CH3 + NaOH → CH3C—O⁻ Na⁺ + HOCH2CH3

22-39

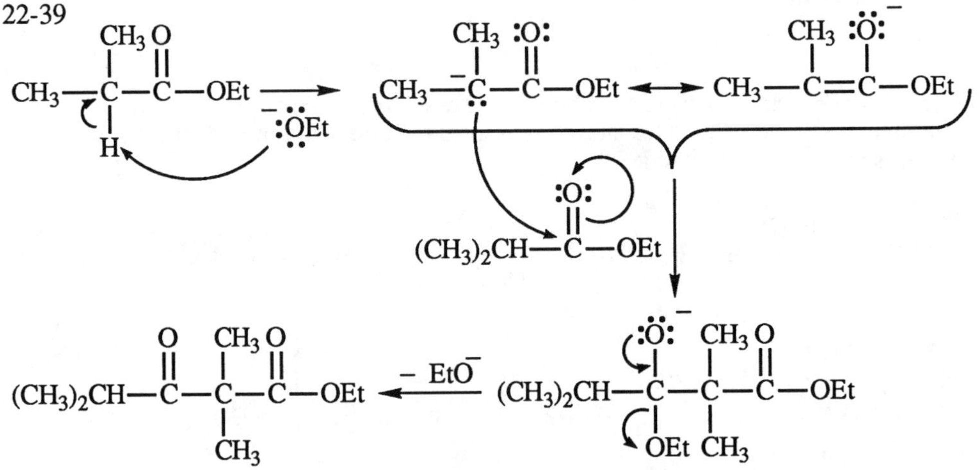

There are two reasons why this reaction gives a poor yield. The nucleophilic carbon in the enolate is 3° and attack is hindered. More important, the final product has no hydrogen on the α-carbon, so the deprotonation by base which is the driving force in other Claisen condensations cannot occur here. What is produced is an *equilibrium mixture* of product and starting materials; the conversion to product is low.

22-40

(a)

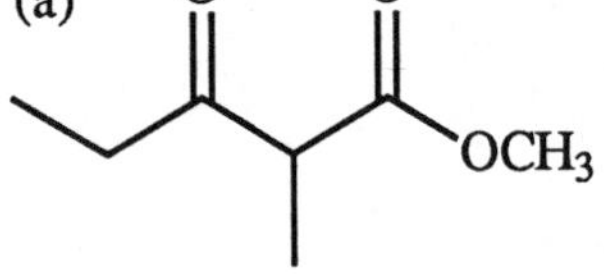

(b)

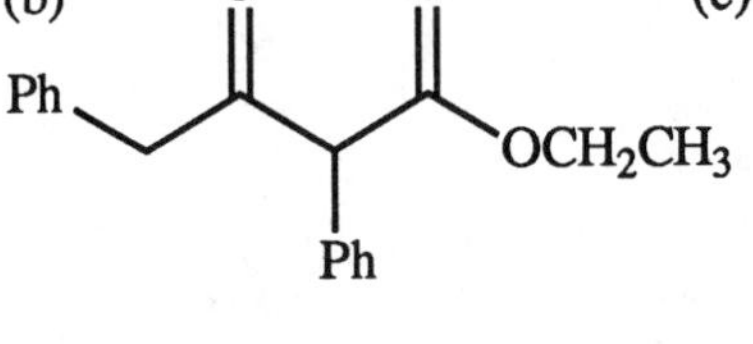

(c)

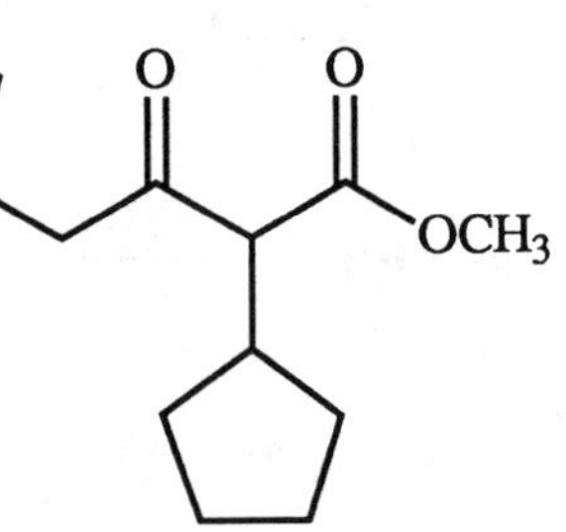

22-41

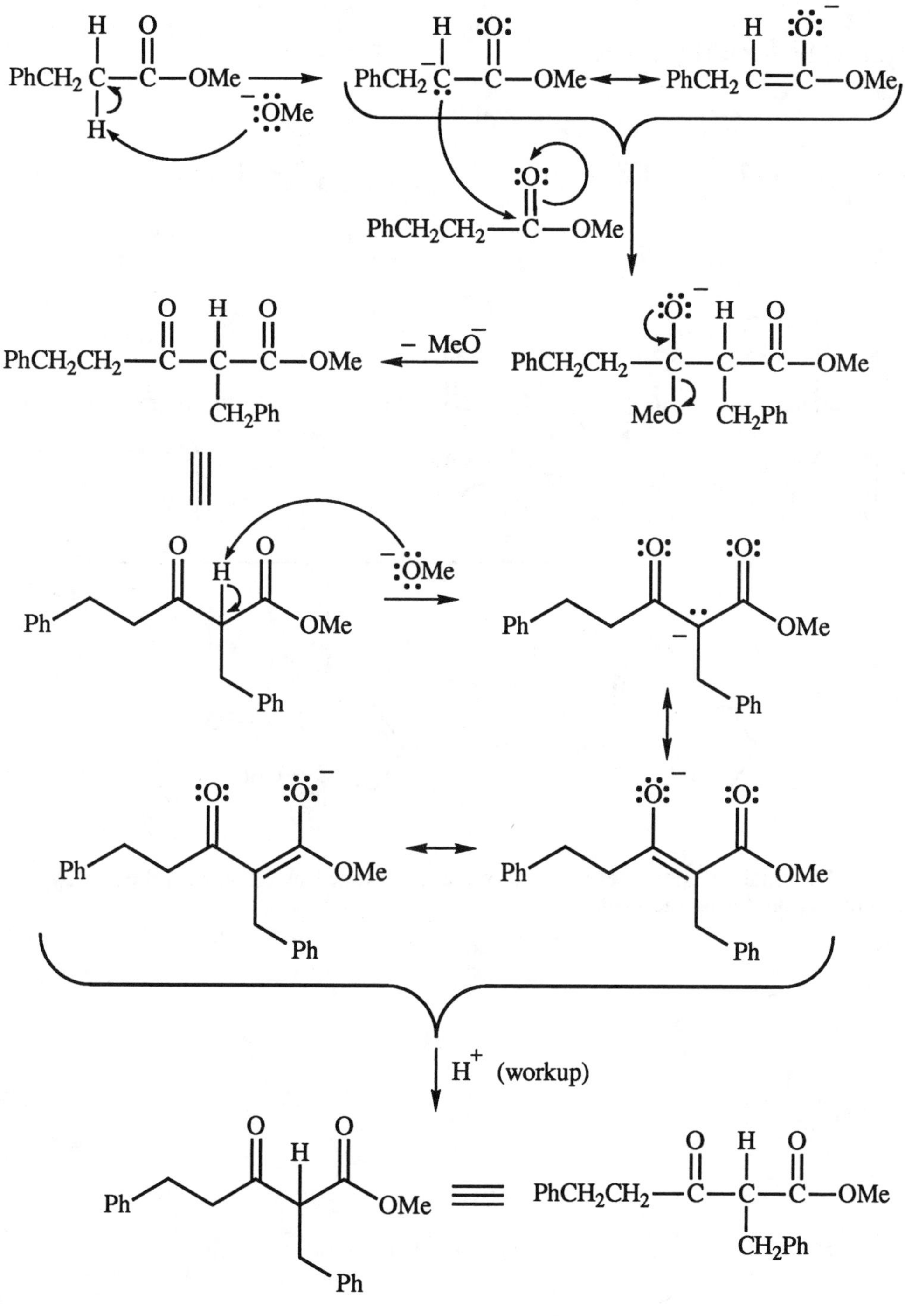

22-42

(a)

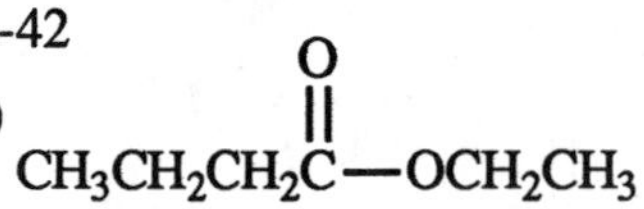

(b)

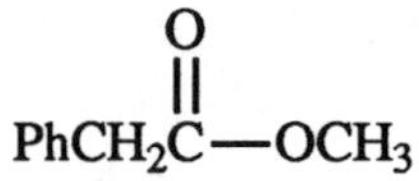

(c)

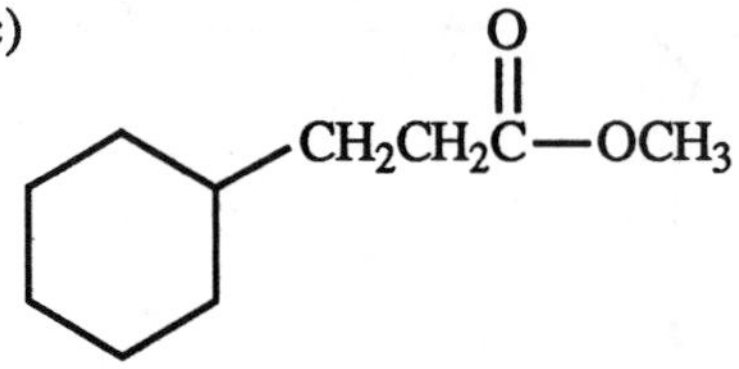

(d)

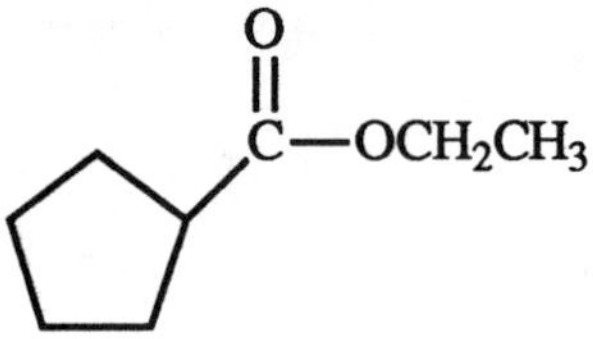

22-43

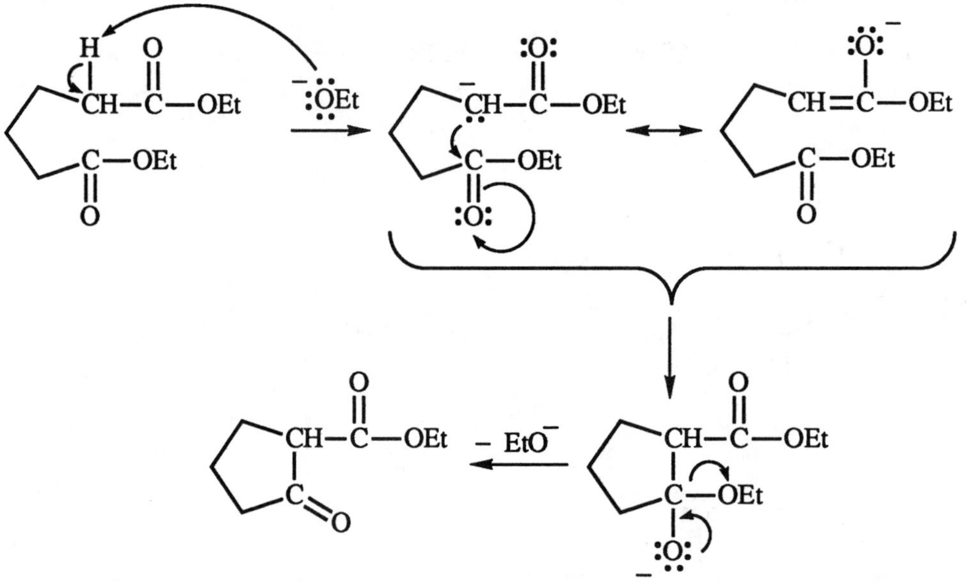

This is the final product, after removal of the α-hydrogen by ethoxide, followed by reprotonation during the workup.

22-43 continued

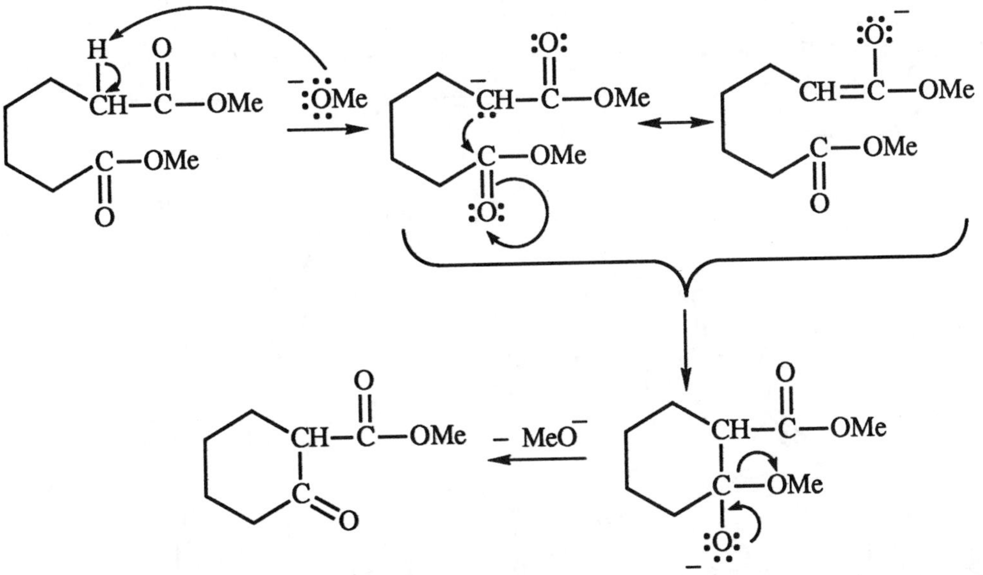

This is the final product, after removal of the α-hydrogen by methoxide, followed by reprotonation during the workup.

22-44

(a) not possible by Dieckmann—not a β-keto ester

(b)

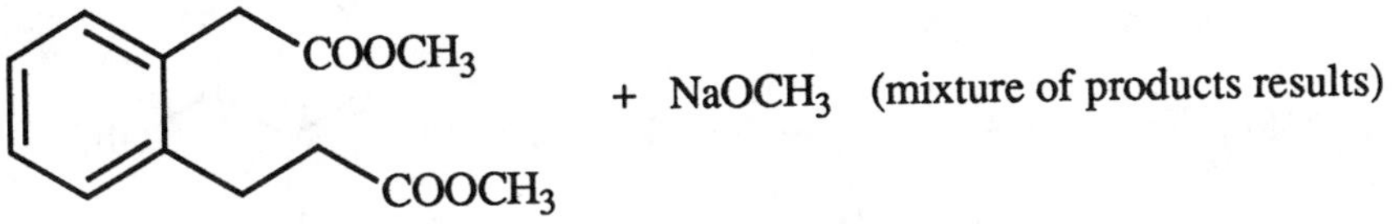

(c)

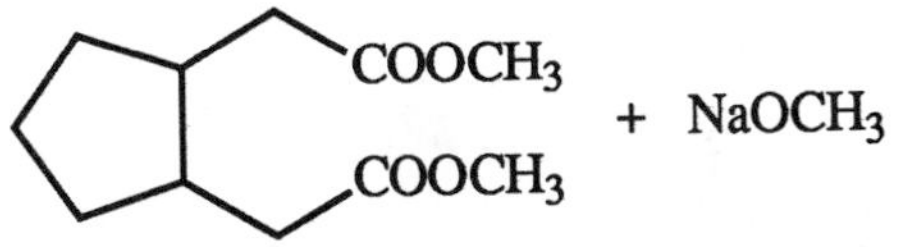

(d)

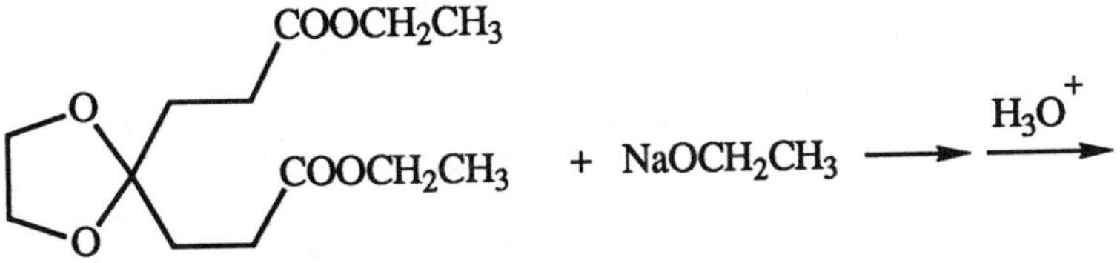

The protecting group is necessary to prevent aldol condensation. Aqueous acid workup removes the protecting group.

22-45

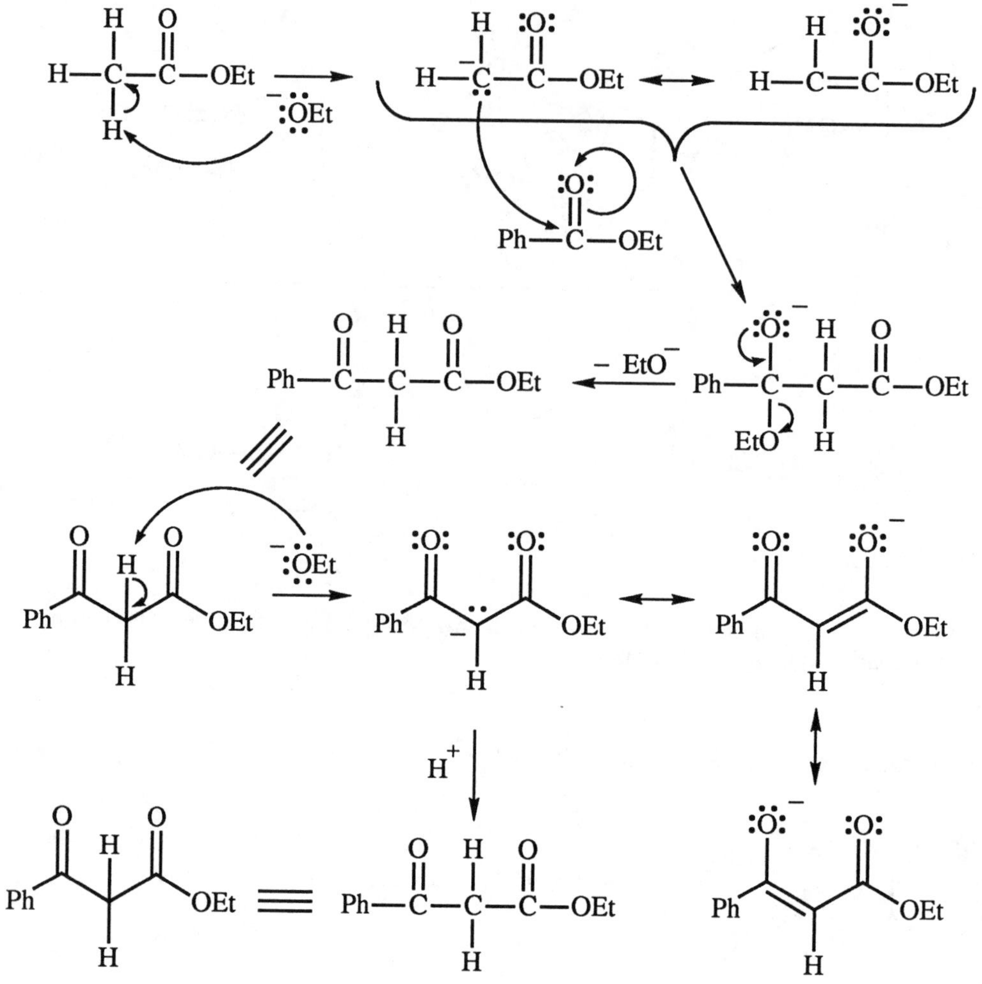

22-46

(a)

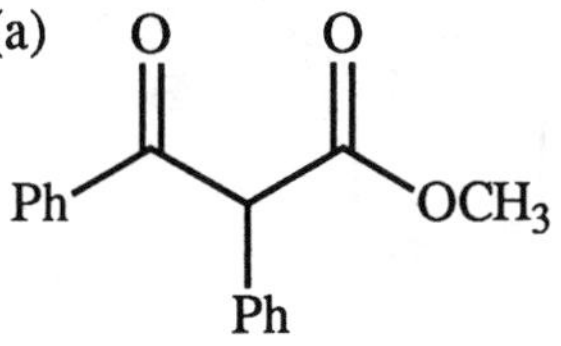

O
O
Ph
OCH3
Ph

(b)

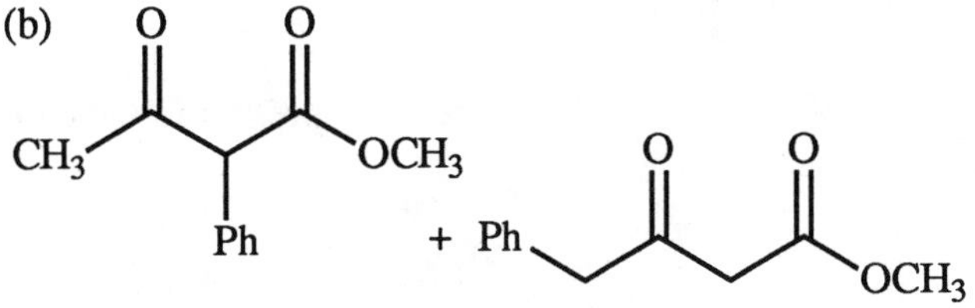

O
O
CH3
OCH3
Ph
O
O
+ Ph
OCH3

plus 2 self-condensation products—a poor choice because both esters have α-hydrogens

(c)

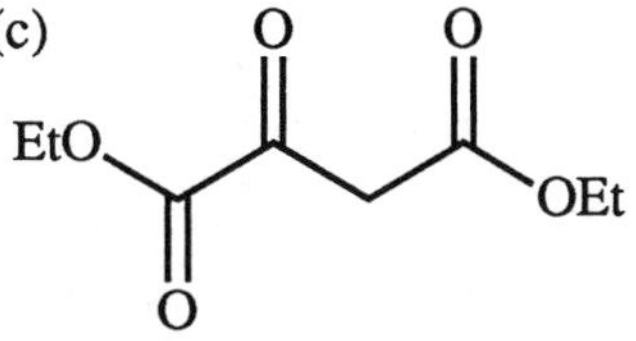

O
O
EtO
OEt
O

(d)

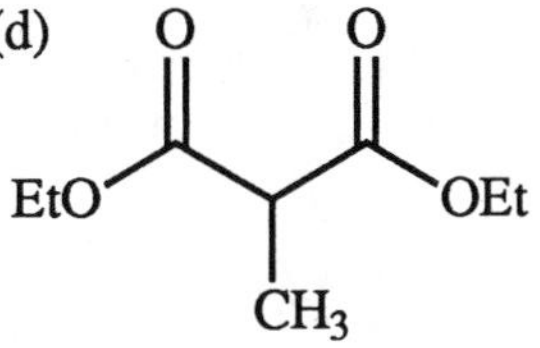

O
O
EtO
OEt
CH3

22-47

(a)

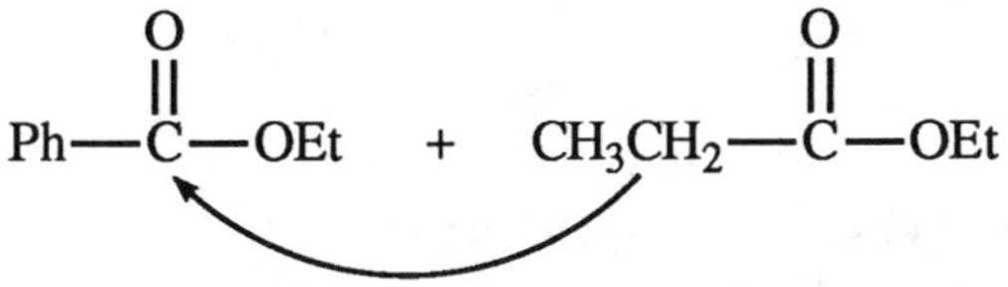

Ph—C—OEt + CH3CH2—C—OEt

(b)

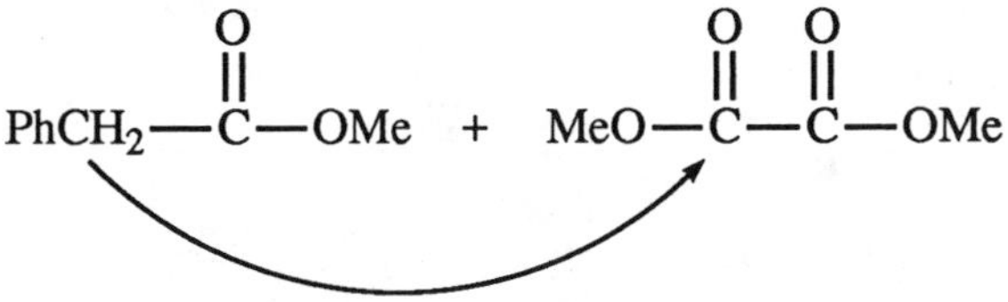

PhCH2—C—OMe + MeO—C—C—OMe

(c)

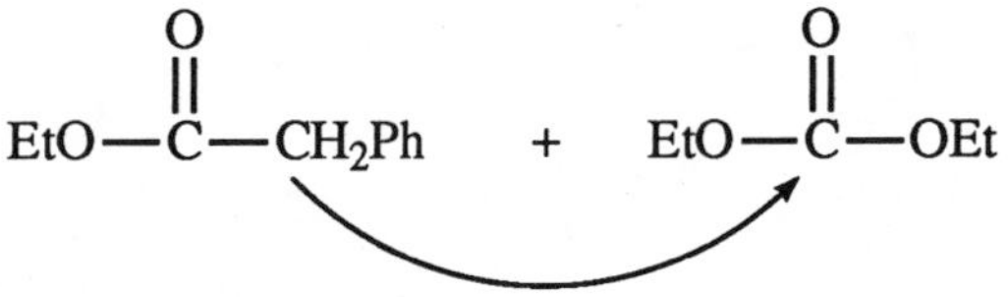

EtO—C—CH2Ph + EtO—C—OEt

(d)

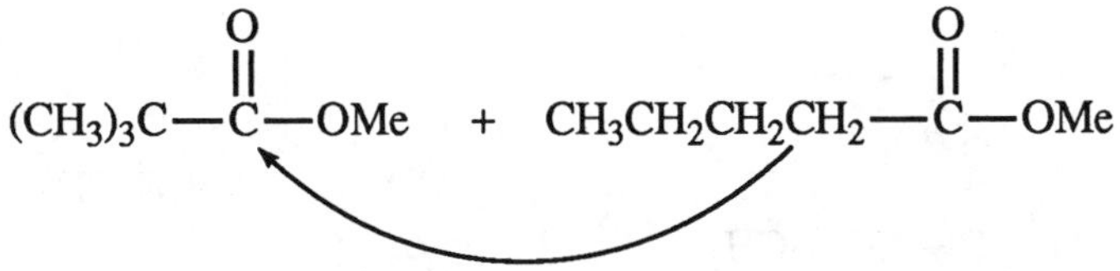

(CH3)3C—C—OMe + CH3CH2CH2CH2—C—OMe

22-48

(a)

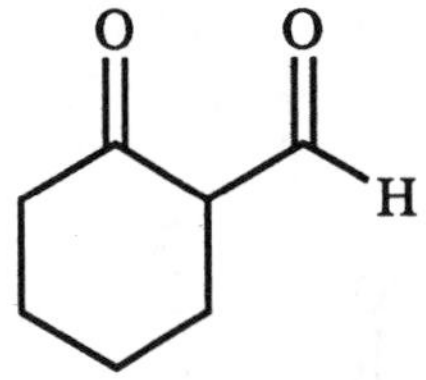

actually present in the enol form:

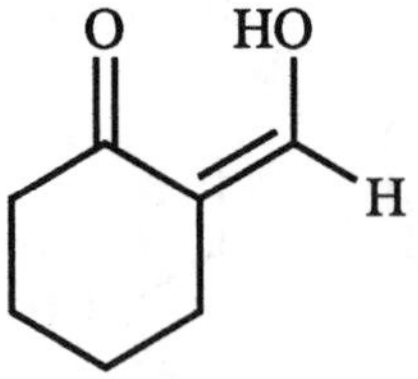

(b)

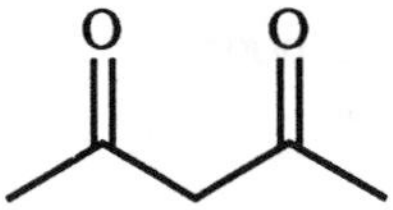

(c)

22-49

(a) two ways:

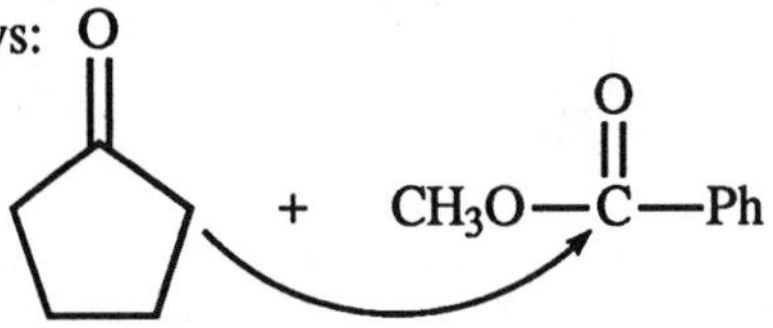

OR

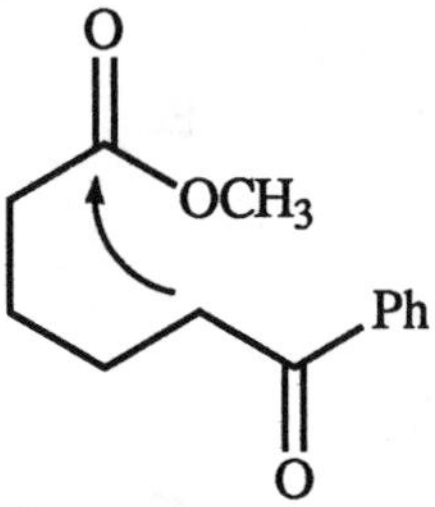

(b)

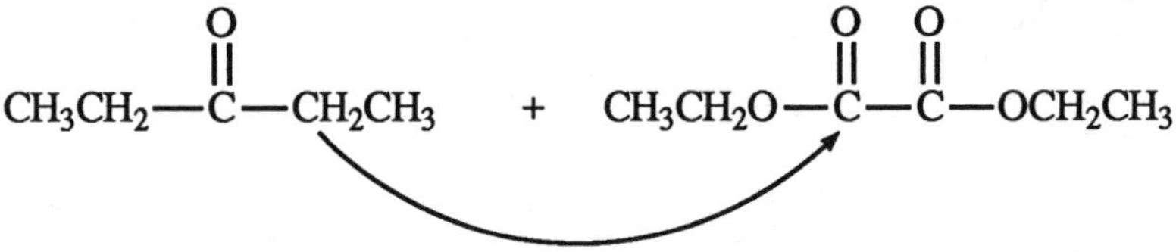

(c)

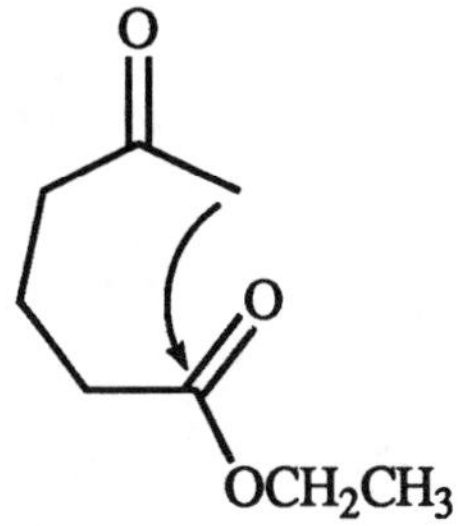

(d) two ways:

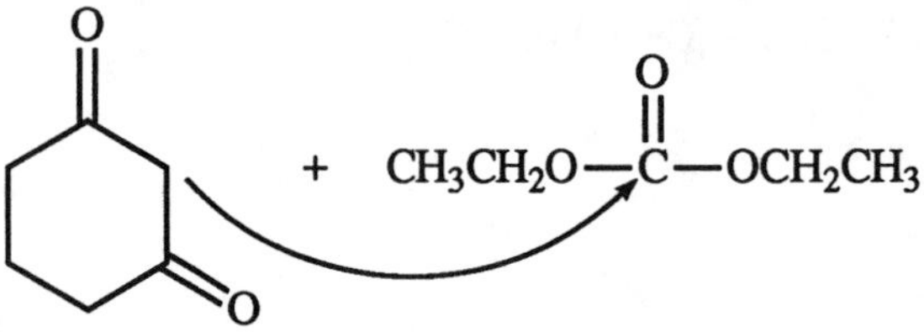

OR

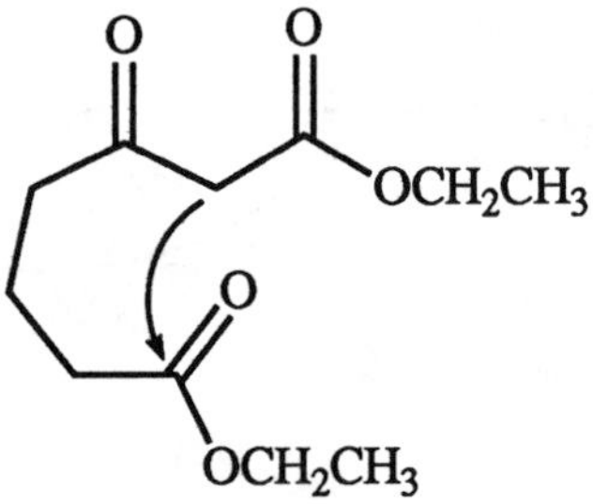

22-50

(a)

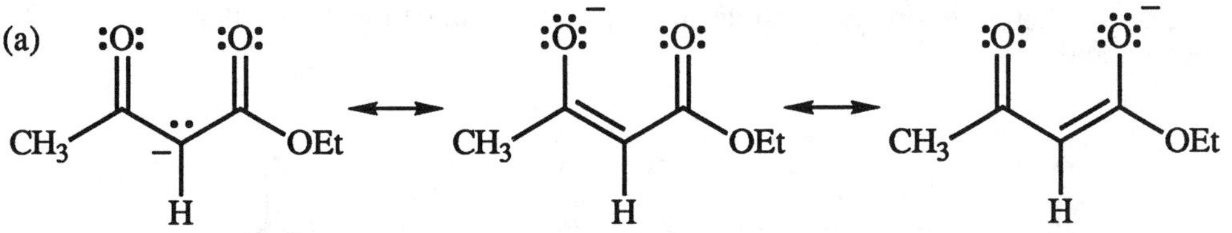

:O:
CH3
OEt
H

(b)

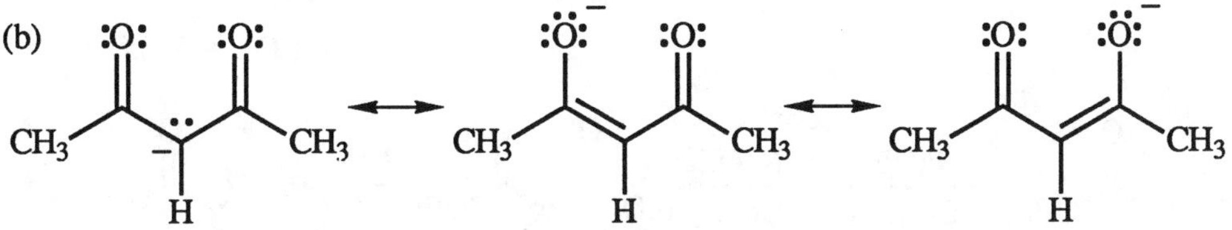

:O:
CH3
H

(c)

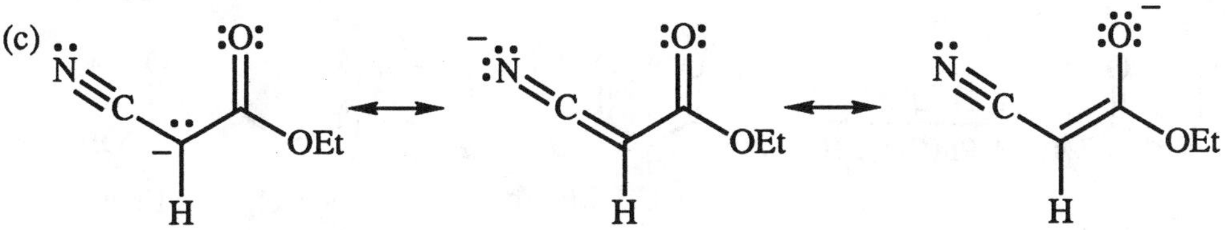

N
C
:O:
OEt
H

(d)

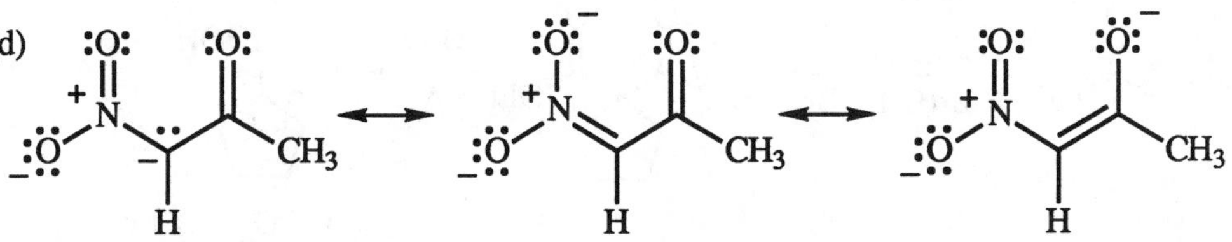

:O:
N
CH3
H

(other resonance forms of the nitro group are not shown)

22-51 In the products, the wavy lines indicate the bonds that must be made by alkylation, before hydrolysis and decarboxylation produce the substituted acetic acid.

(a) 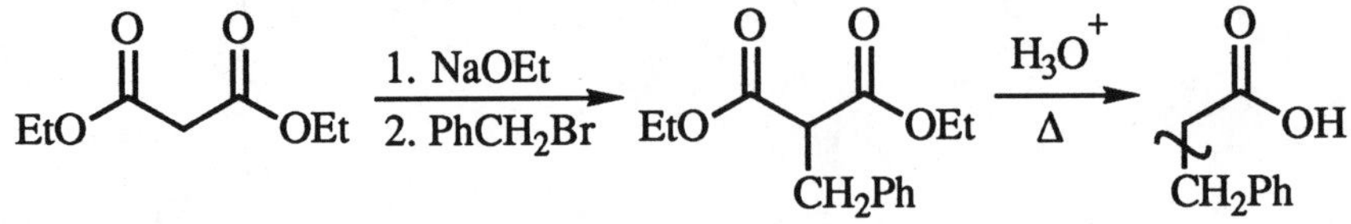

1. NaOEt
2. PhCH2Br
H3O+
Δ
EtO
OEt
OH
CH2Ph

+ CO_2 + 2 EtOH

(b) 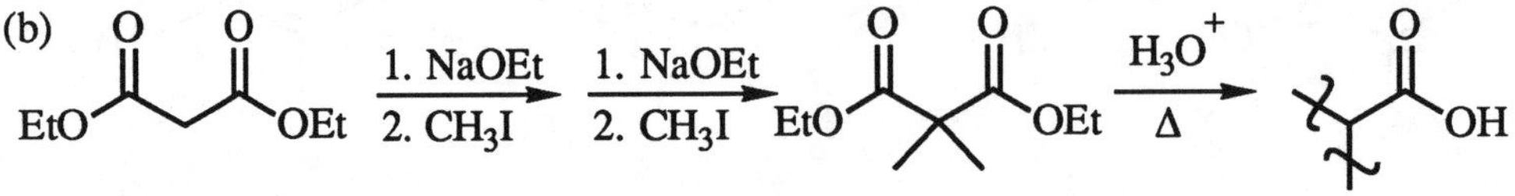

1. NaOEt
2. CH3I
1. NaOEt
2. CH3I
H3O+
Δ
EtO
OEt
OH

+ CO_2 + 2 EtOH

(c) 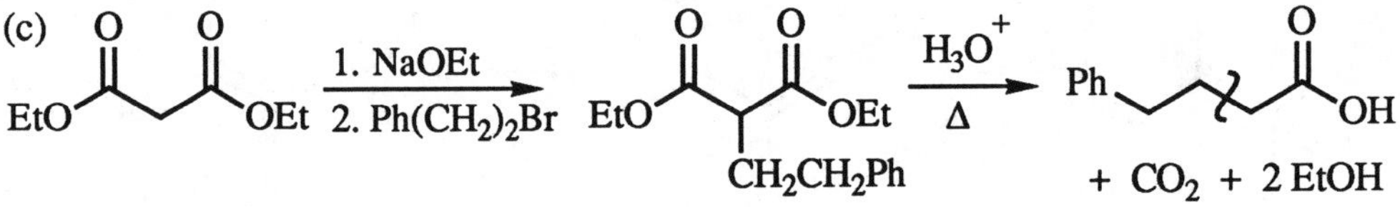

1. NaOEt
2. Ph(CH2)2Br
H3O+
Δ
EtO
OEt
CH2CH2Ph
Ph
OH

+ CO_2 + 2 EtOH

(d) 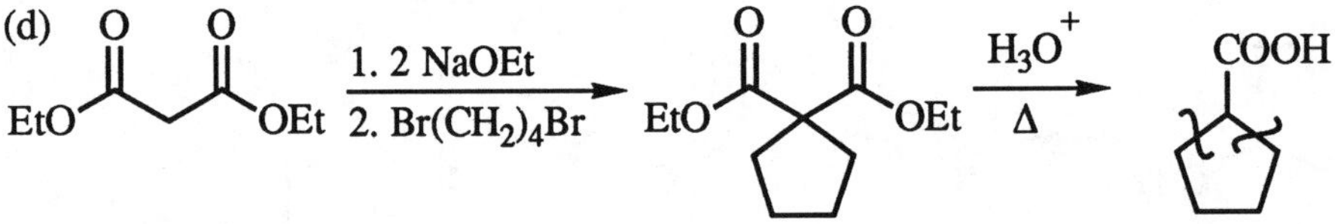

1. 2 NaOEt
2. Br(CH2)4Br
H3O+
Δ
EtO
OEt
COOH

+ CO_2 + 2 EtOH

22-52

(a) $PhCH_2CH_2-\overset{O}{\overset{\|}{C}}-CH_3$

+ CO_2 + EtOH

(b) 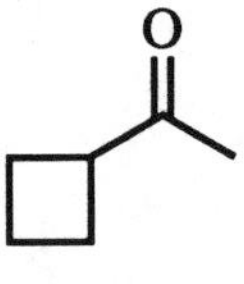

+ CO_2 + EtOH

(c) 

+ CO_2 + EtOH

22-53 In the products, the wavy lines indicate the bonds that must be made by alkylation, before hydrolysis and decarboxylation produce the substituted acetone.

(a)

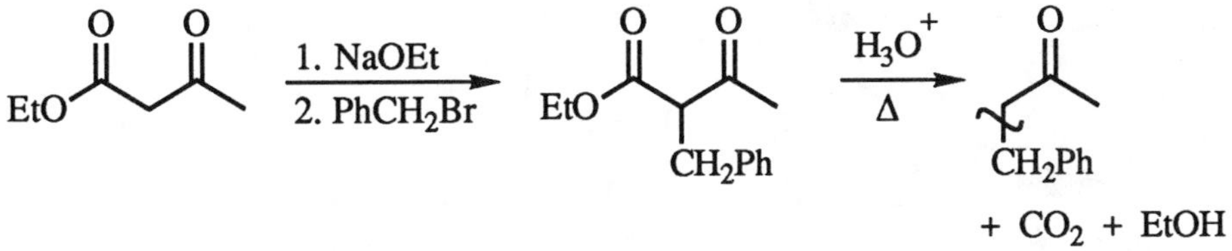

(b)

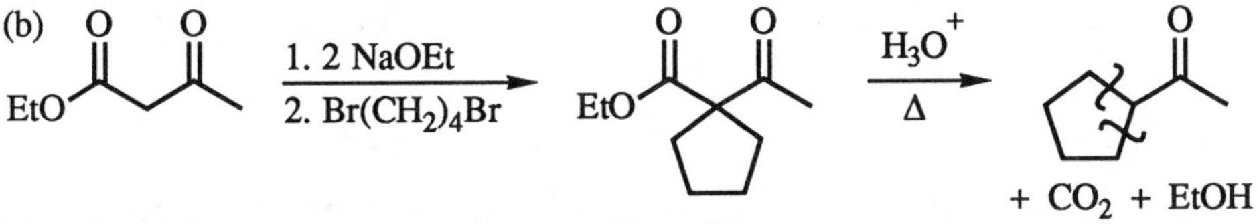

(c)

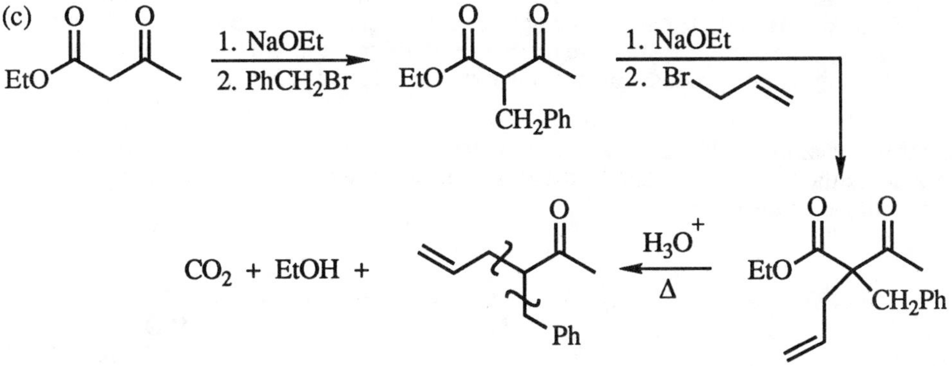

22-54

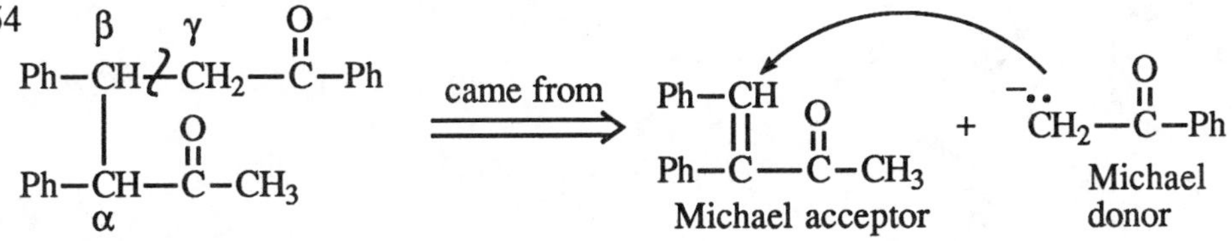

forward direction

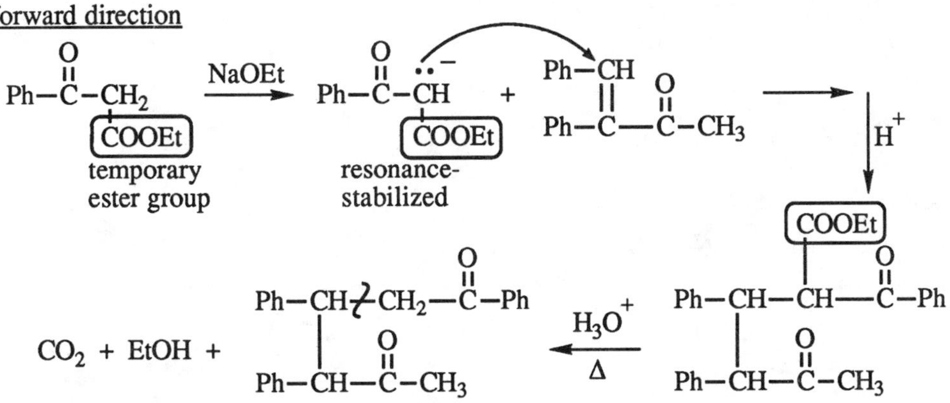

22-55

First, you might wonder why this sequence does not make the desired product:

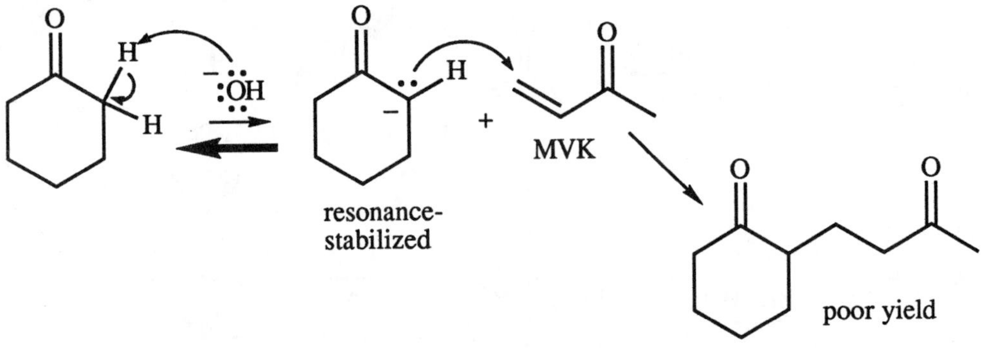

H
:OH
H
H
MVK
resonance-
stabilized
poor yield

The poor yield in this conjugate addition is due primarily to the numerous competing reactions: the ketone enolate can self-condense (aldol), can condense with the ketone of MVK (aldol), or can deprotonate the methyl of MVK to generate a new nucleophile. The complex mixture of products makes this route practically useless.

What permits enamines (or other stabilized enolates) to work are: a) the certainty of which atom is the nucleophile, and b) the lack of self-condensation. Enamines can also do conjugate addition:

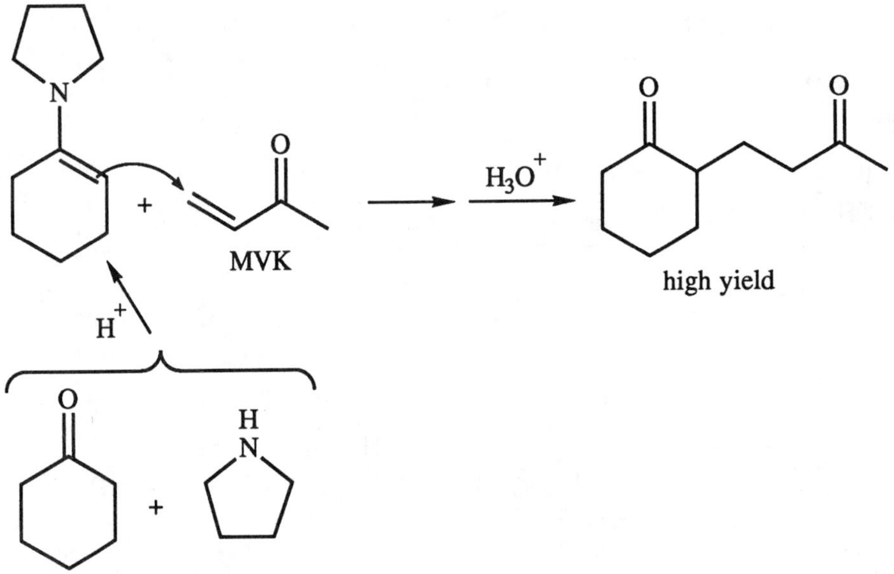

N
MVK
H3O+
high yield
H+
H
N

22-56 The enolate of acetoacetic ester can be used in a Michael addition to an α,β-unsaturated ketone like MVK.

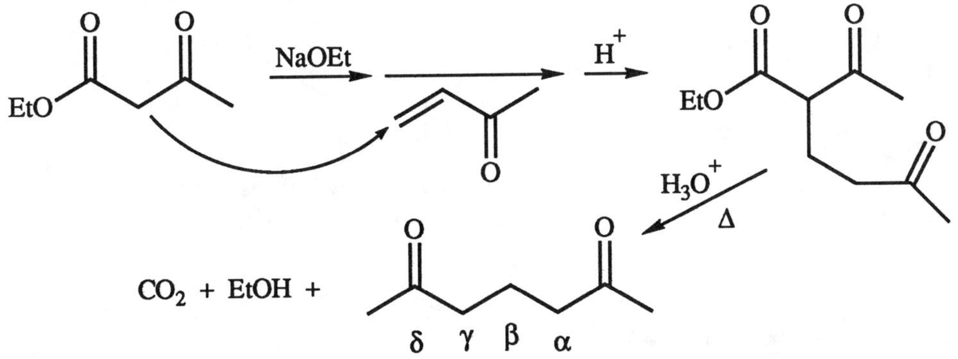
EtO
NaOEt
H+
EtO
H3O+
Δ
CO2 + EtOH +
δ γ β α

22-57

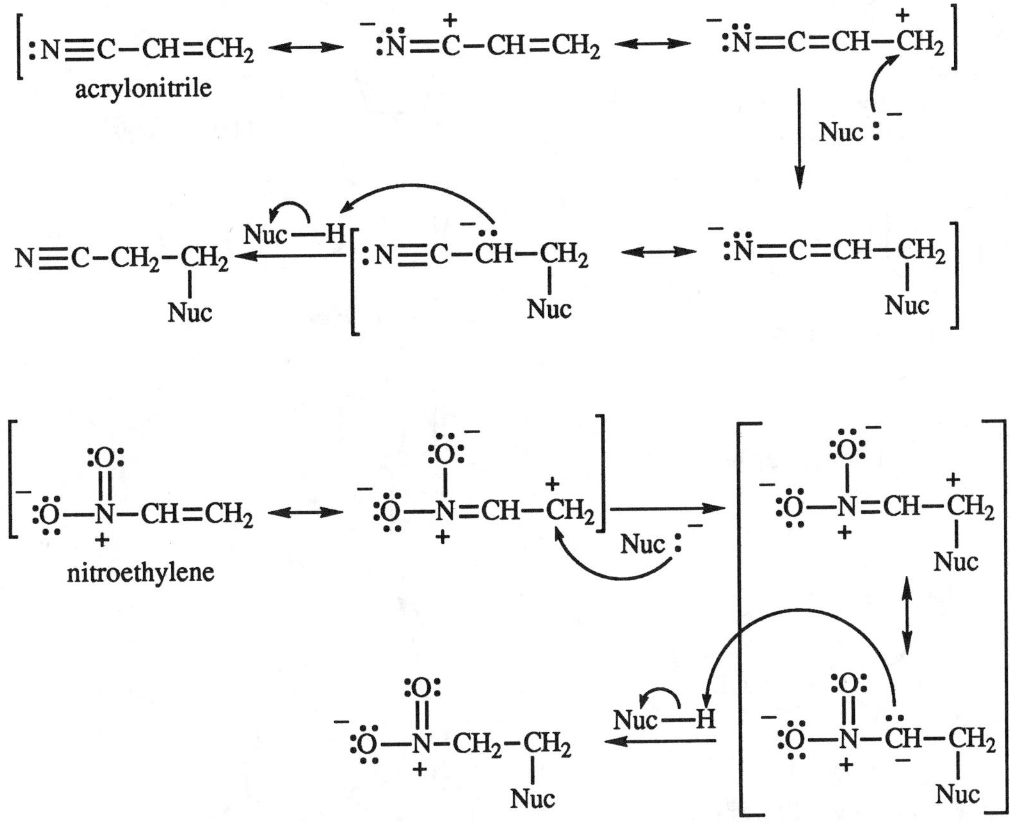
acrylonitrile
Nuc
Nuc—H
Nuc
nitroethylene

22-58

(a)

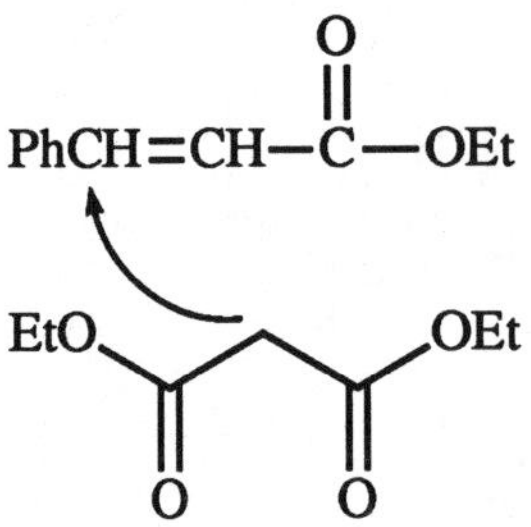

PhCH=CH—C—OEt
O
EtO
OEt
O
O

(b)

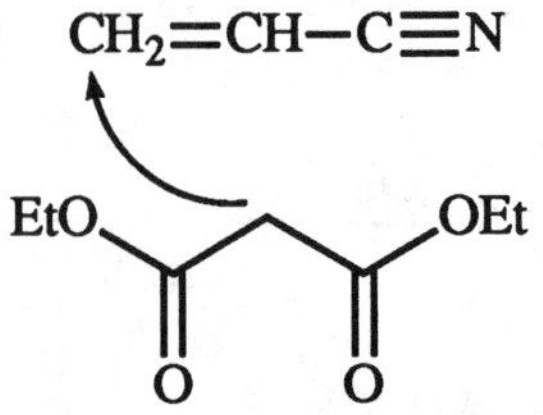

CH2=CH—C≡N
EtO
OEt
O
O

followed by hydrolysis
and decarboxylation

(c) two ways

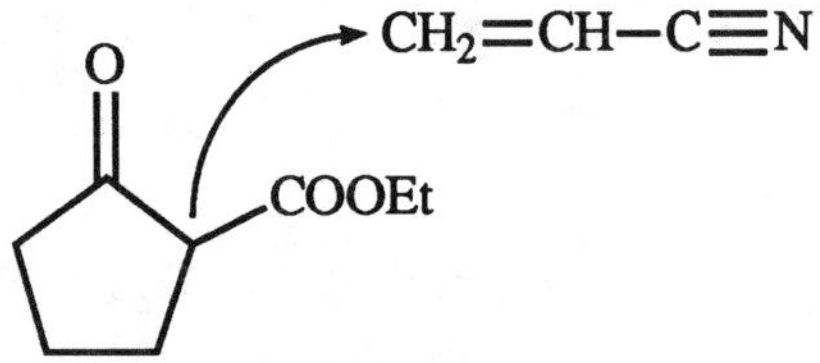

O
CH2=CH—C≡N
COOEt

followed by hydrolysis
and decarboxylation

OR

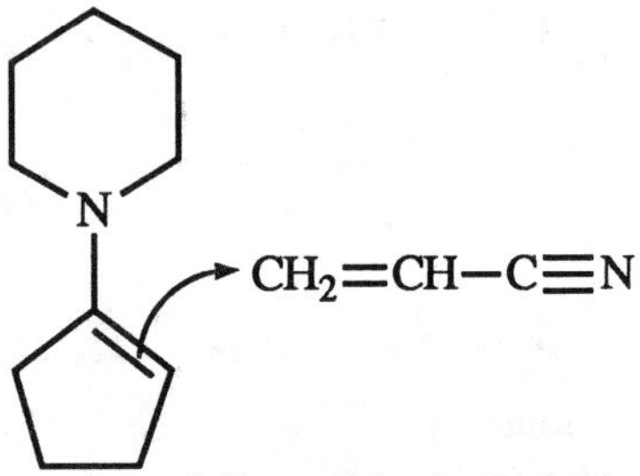

N
CH2=CH—C≡N

followed by hydrolysis

(d)

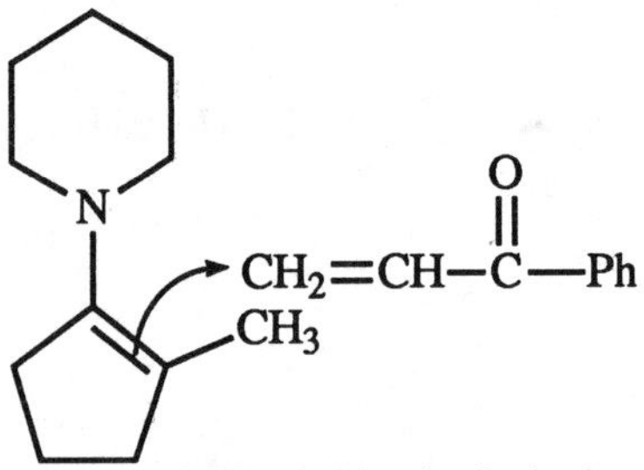

N
O
CH2=CH—C—Ph
CH3

followed by hydrolysis

(e)

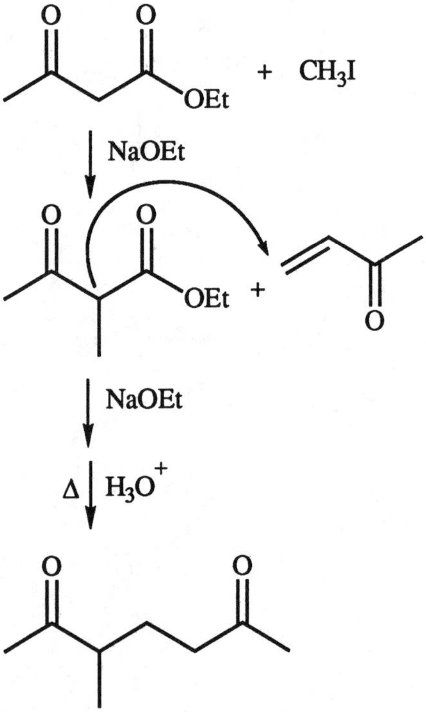

O
O
OEt
+ CH3I
NaOEt
O
O
OEt
+
O
NaOEt
Δ
H3O+
O
O

(could also be synthesized by the
Stork enamine reactions)

(f)

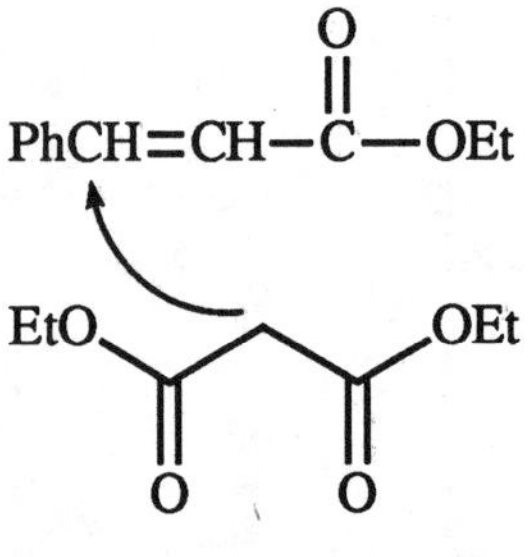

O
PhCH=CH—C—OEt
EtO
OEt
O
O

followed by hydrolysis
and decarboxylation

22-59

Step 1: carbon skeleton

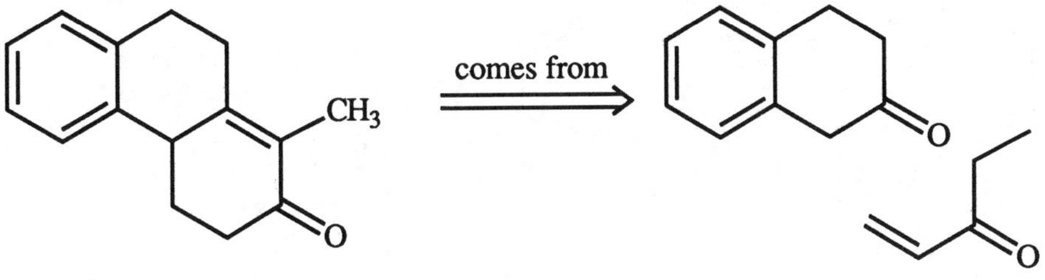
CH3
comes from
O
O
O

Step 2: nucleophile generation

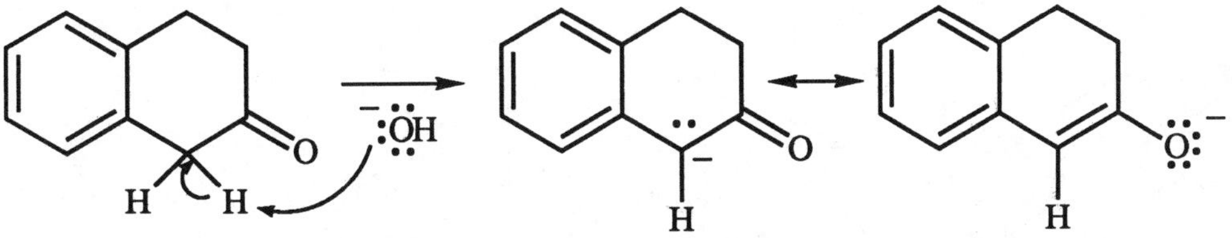
OH
H
H
O
O
H
O
H

plus resonance forms with negative charge on the benzene ring

Step 3: nucleophilic attack (Michael addition)

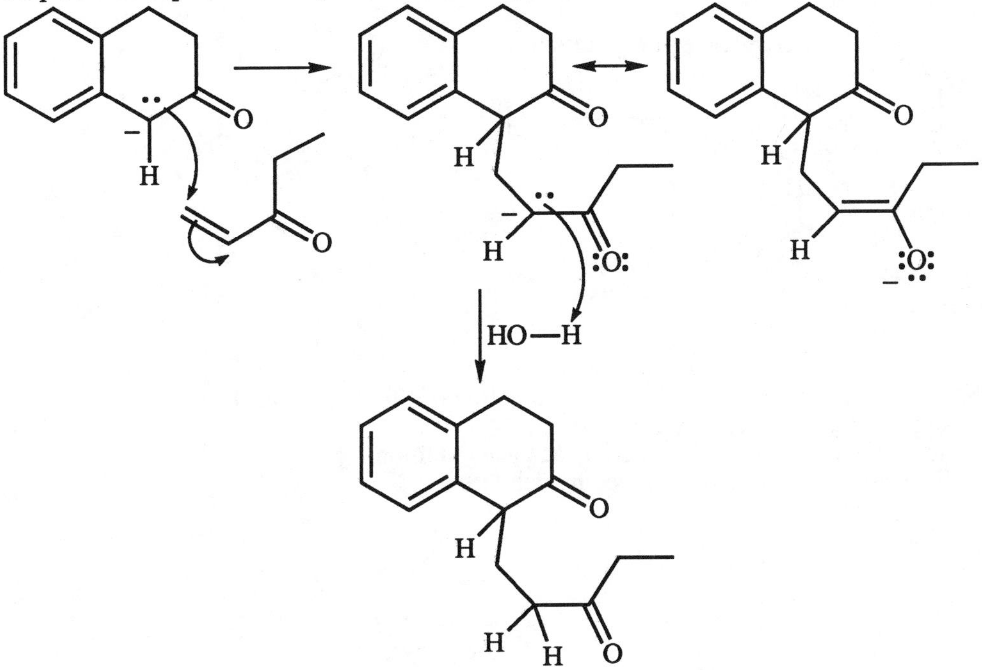
HO—H

22-59 continued

Step 4: conversion to final product

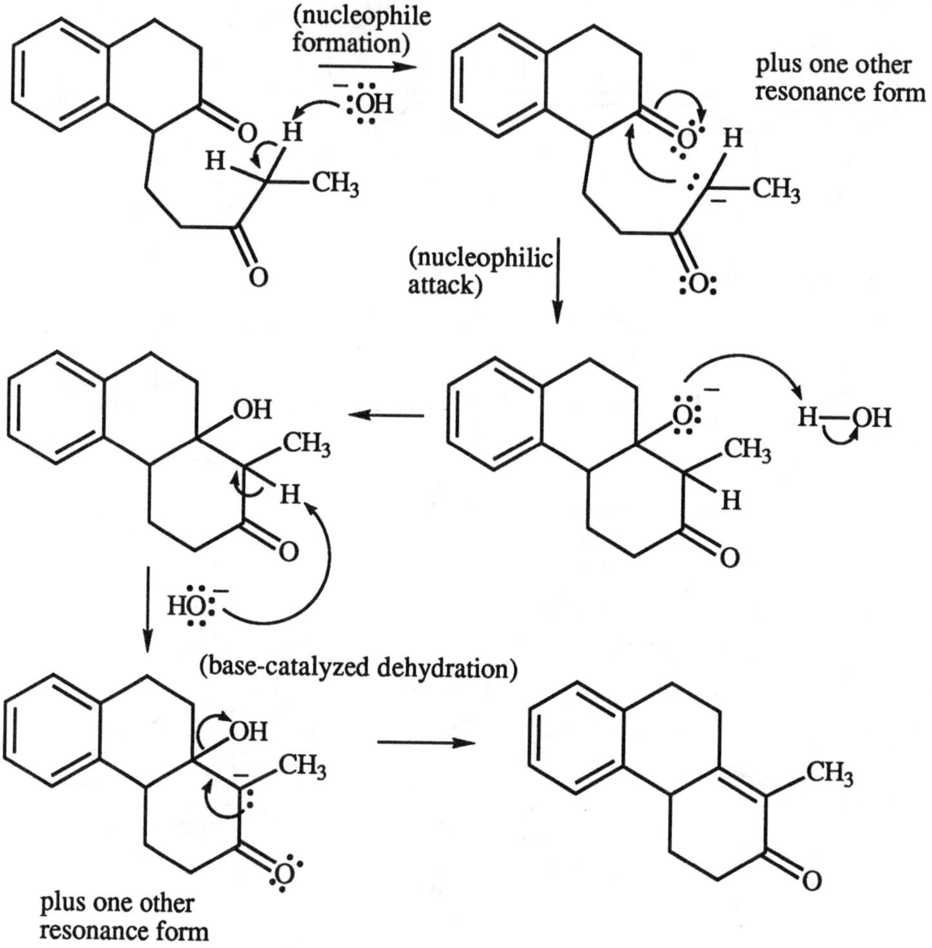

Step 5: The complete mechanism is the combination of Steps 2, 3, and 4. Notice that this mechanism is simply described by:

1) Enolate formation, followed by Michael addition;
2) Aldol condensation, followed by dehydration.

22-60

Step 1: carbon skeleton

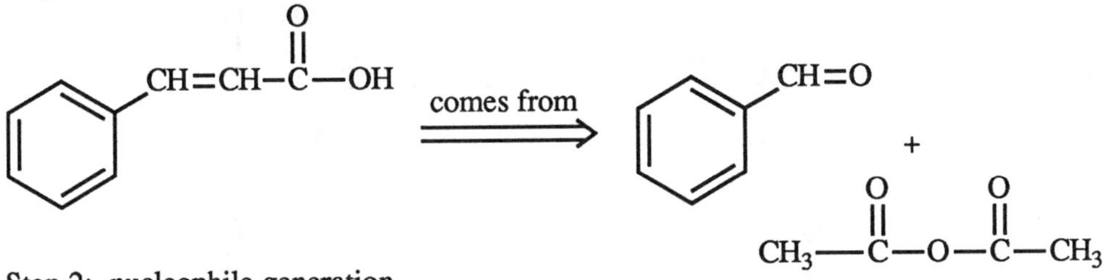

Step 2: nucleophile generation

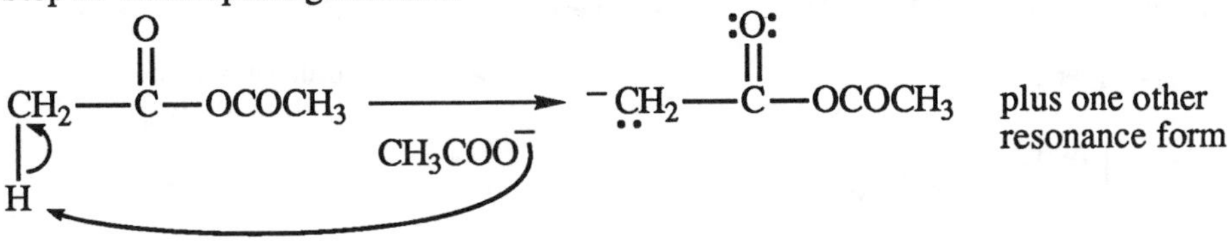

Step 3: nucleophilic attack

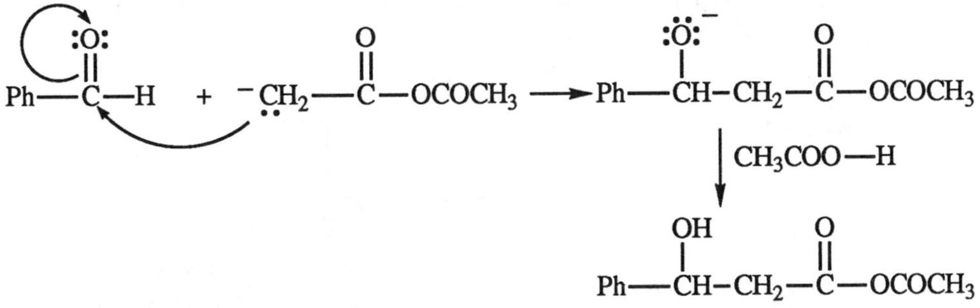

Step 4: conversion to final product

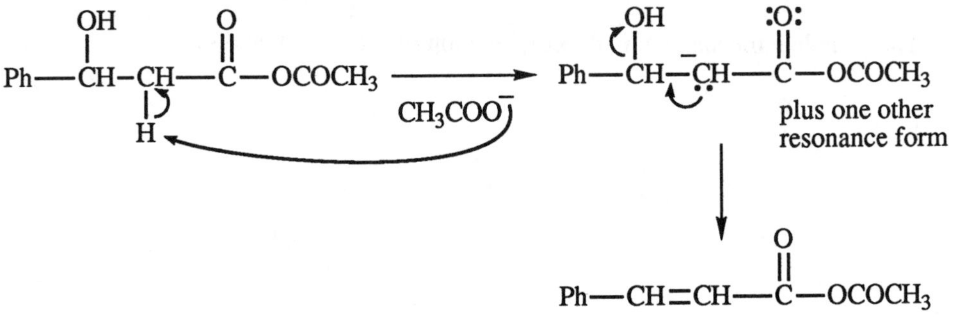

22-60 continued

hydrolysis

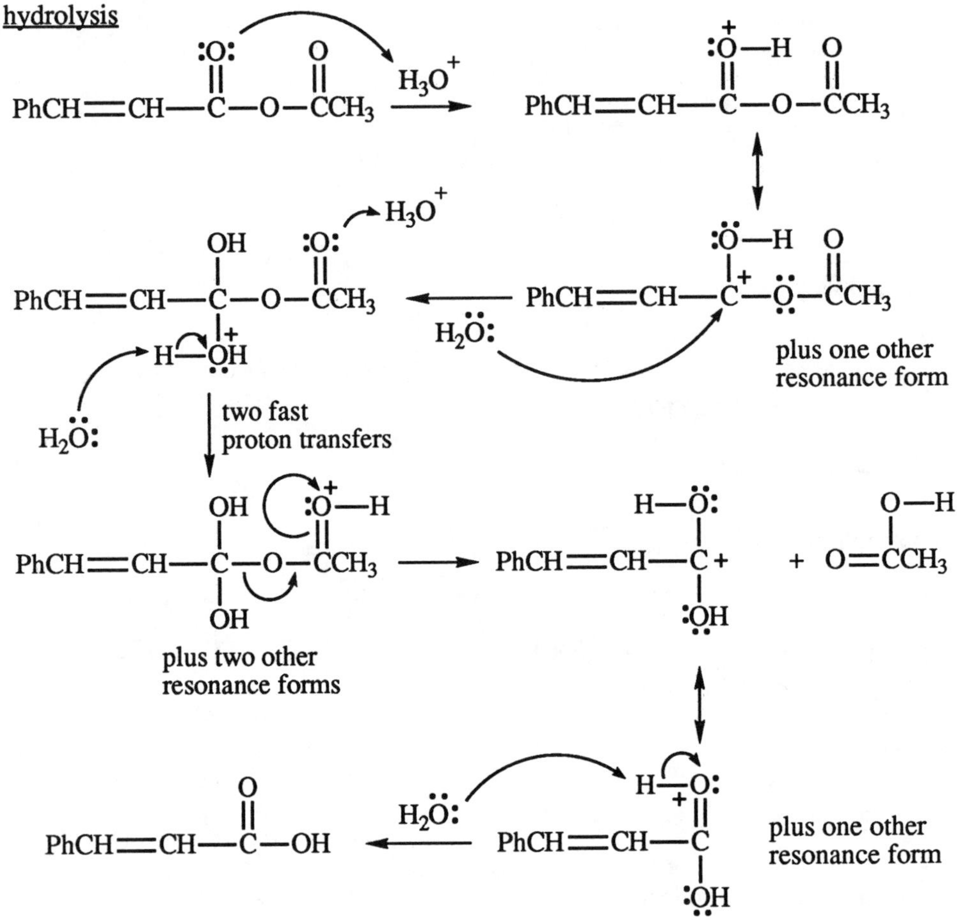

:O:
O
H_3O^+
PhCH=CH—C—O—CCH3
:O—H
O
PhCH=CH—C—O—CCH3
OH
:O:
H_3O^+
PhCH=CH—C—O—CCH3
H—OH
H2O:
:O—H
O
PhCH=CH—C—O—CCH3
H2O:
plus one other resonance form
two fast proton transfers
OH
:O—H
PhCH=CH—C—O—CCH3
OH
plus two other resonance forms
H—O:
PhCH=CH—C+
:OH
O—H
+ O=CCH3
O
PhCH=CH—C—OH
H2O:
H—O:
PhCH=CH—C
:OH
plus one other resonance form

Step 5: The complete mechanism is the combination of Steps 2, 3, and 4.

22-61 The Robinson annulation consists of a Michael addition followed by aldol cyclization with dehydration. In the retrosynthetic direction, disconnect the alkene formed in the aldol/dehydration, then disconnect the Michael addition to discover the reactants.

(a) aldol and dehydration forms the α,β double bond:

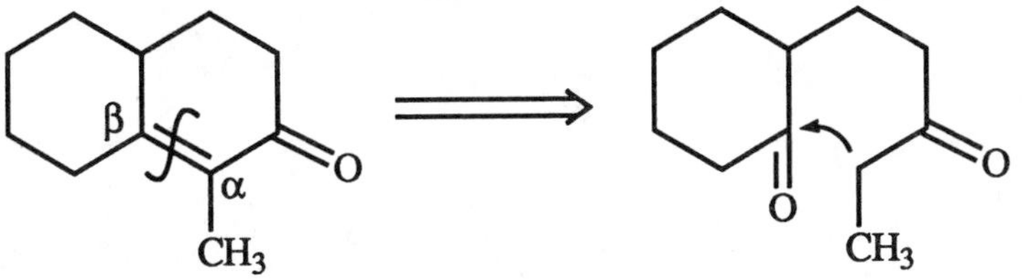

Michael addition forms a bond to the β' carbon:

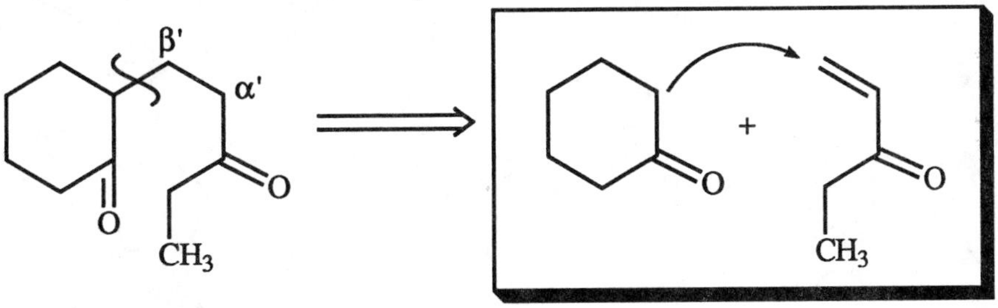

(b) aldol and dehydration forms the α,β double bond:

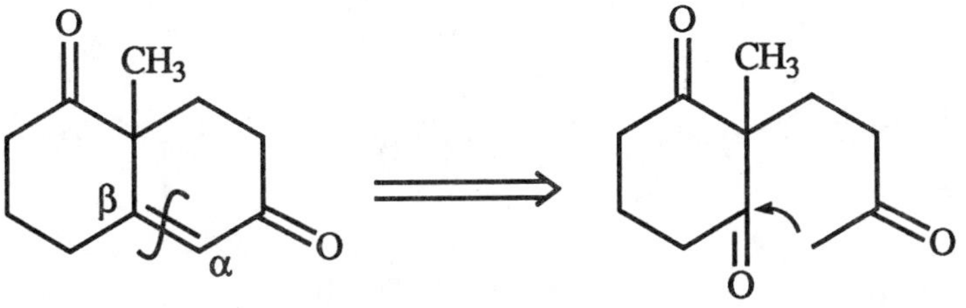

Michael addition forms a bond to the β' carbon:

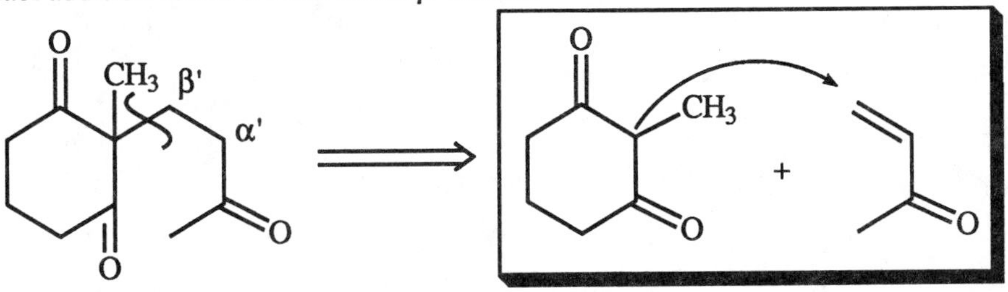

22-62 Refer to the Glossary and the text for definitions and examples.

22-63 The most acidic hydrogens are shown in boldface.

(a)

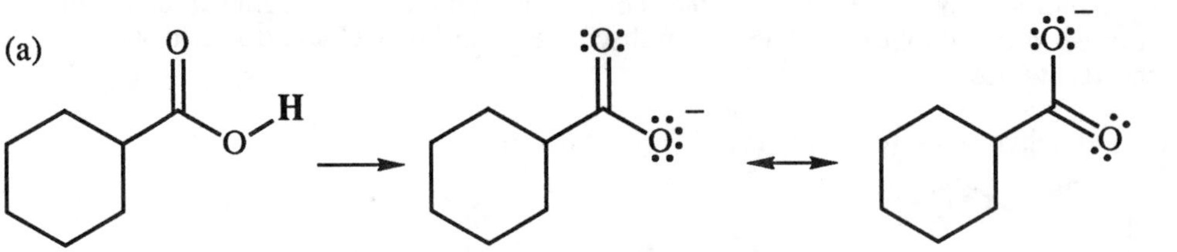

(b)

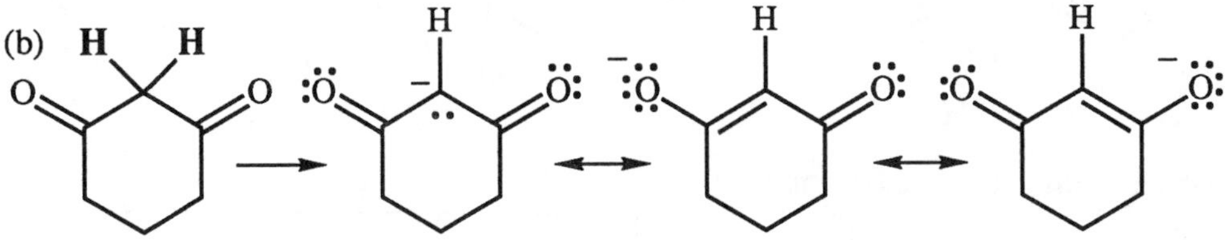

(c)

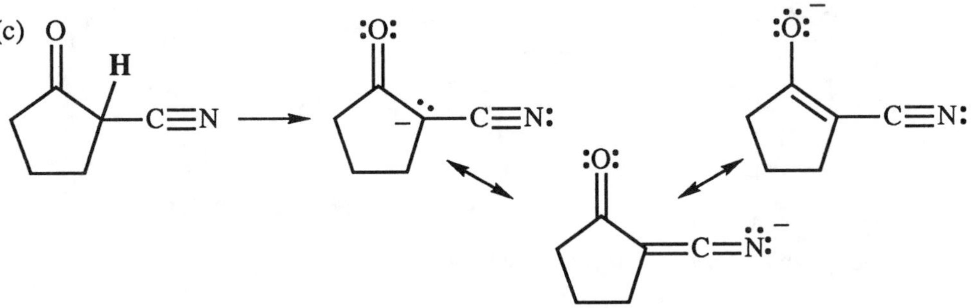

(d)

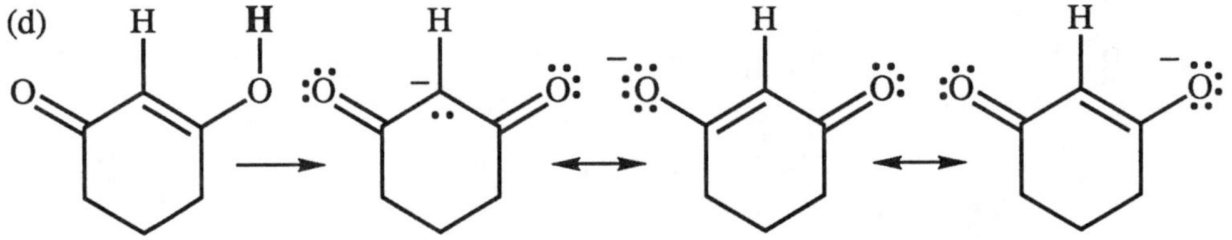

same enolate as in (b)

(e)

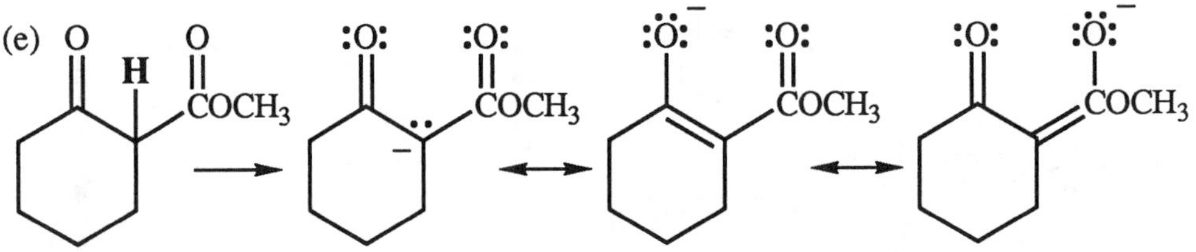

22-63 continued

(f)

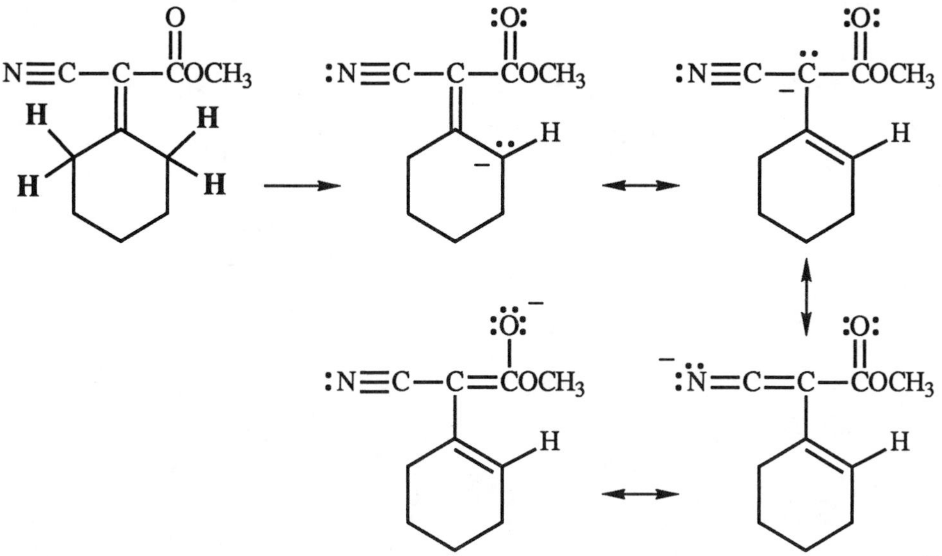

(g)

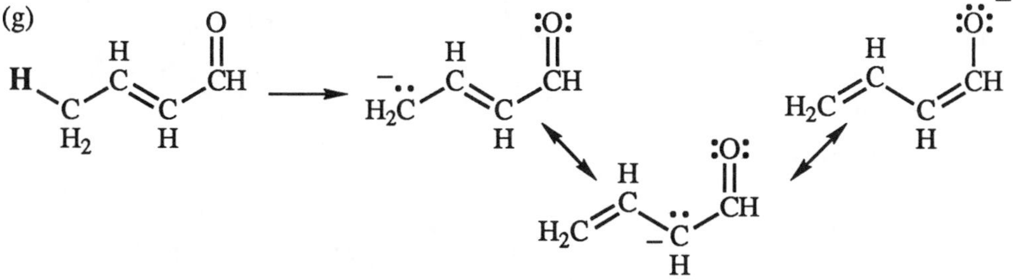

(h)

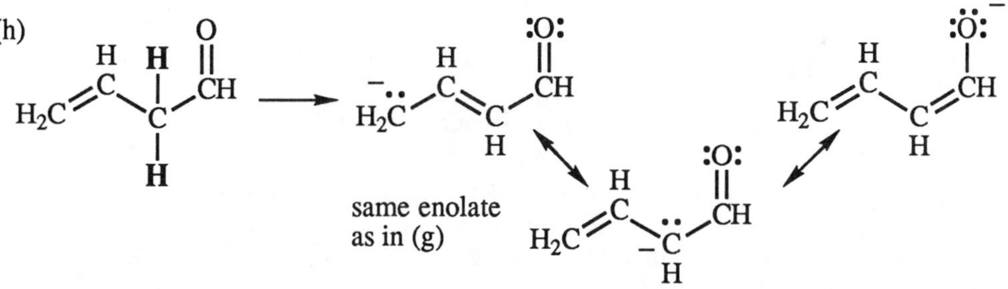

22-64 In order of increasing acidity. The most acidic protons are shown in boldface. (The approximate pK_a values are shown for comparison.)

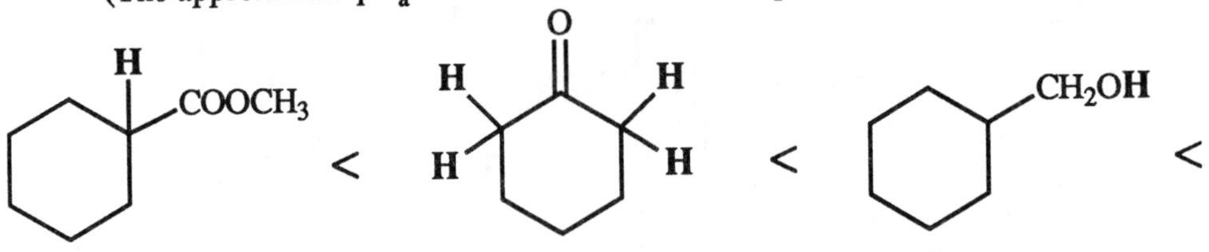

(g) pK_a 25 least acidic | (b) pK_a 20 | (f) pK_a 18

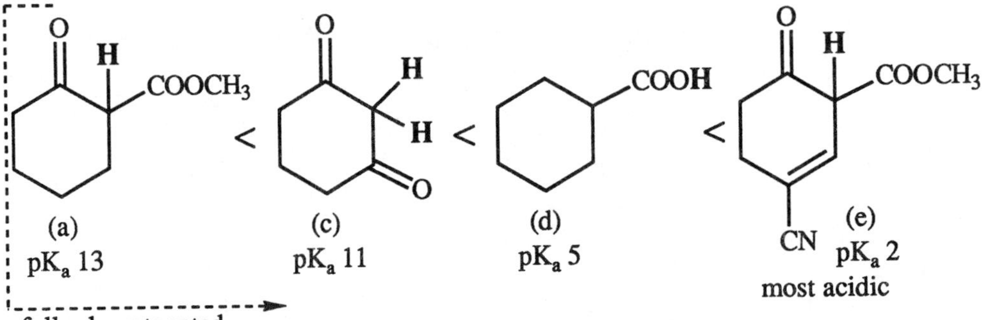

(a) pK_a 13 | (c) pK_a 11 | (d) pK_a 5 | (e) pK_a 2 most acidic

fully deprotonated by ethoxide ion

22-65

keto enol

The enol form is stable because of the conjugation and because of intramolecular hydrogen-bonding in a six-membered ring.

In dicarbonyl compounds in general, the weaker the electron-donating ability of the group G, the more it will exist in the enol form: aldehydes (G = H) are almost completely enolized, then ketones, esters, and finally amides which have virtually no enol content.

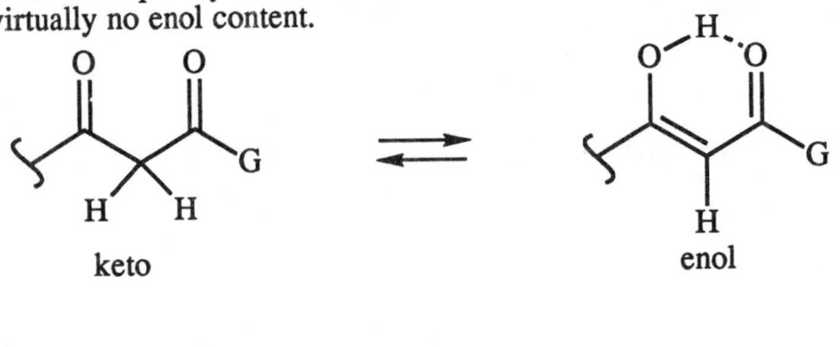

keto enol

22-66

(a)

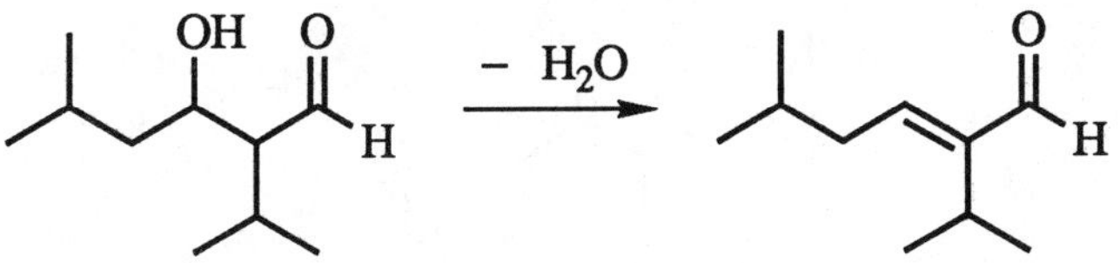

(b)

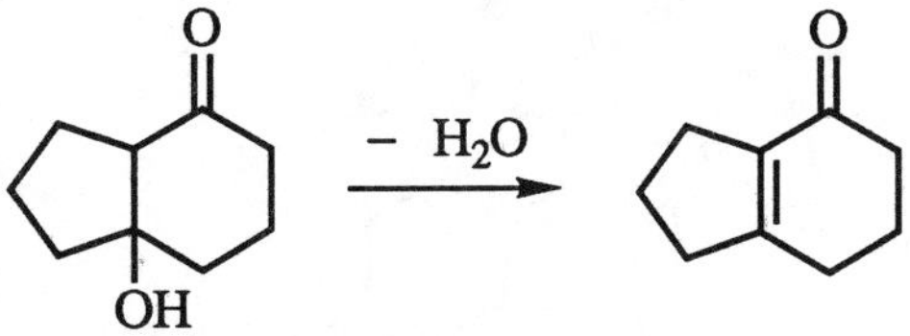

(c)

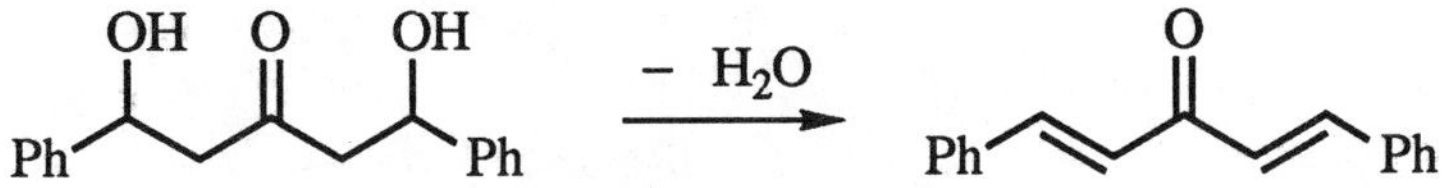

(d)

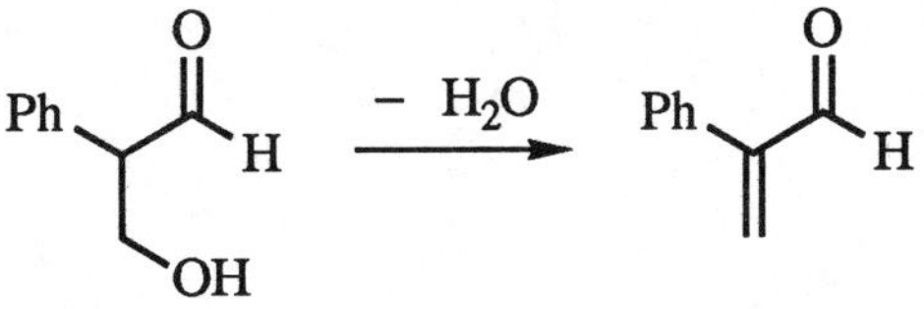

22-67

(a)

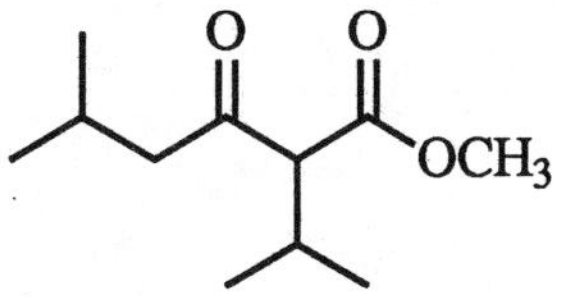

(b)

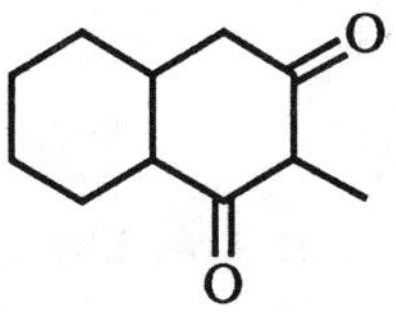

(c)

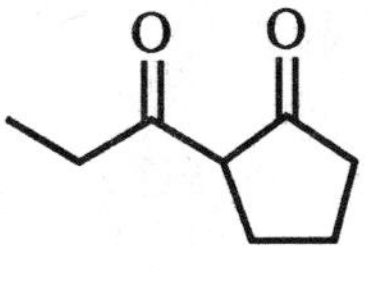

(d)

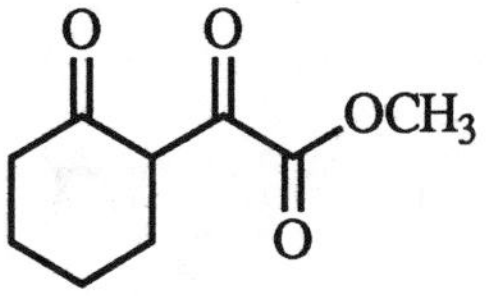

(e)

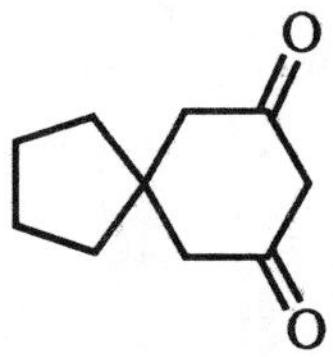

22-68

(a) $CH_3CH_2-C(=O)-CH_2CH_3$

$+ CO_2 + CH_3OH$

(b) 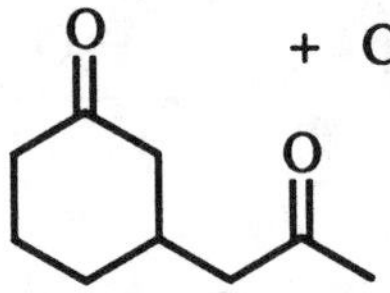 $+ CO_2 + CH_3OH$

(c) 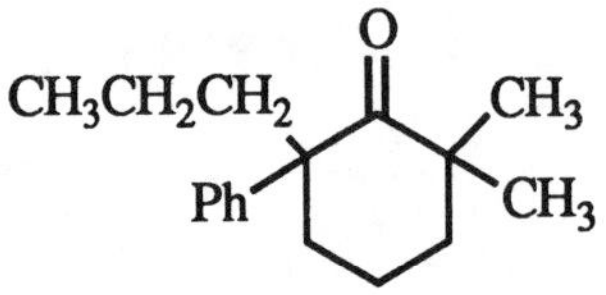

(d) 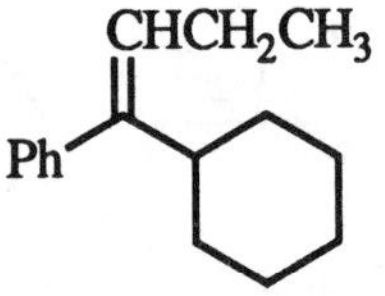

(e) 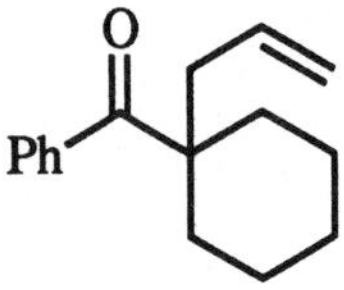

(f) 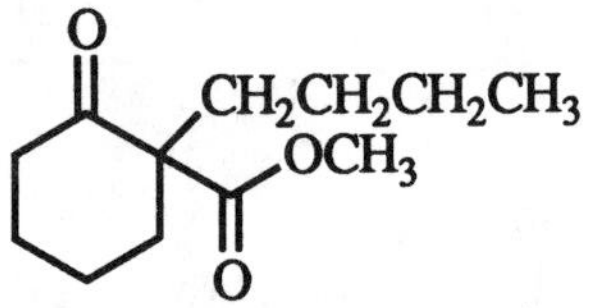

(g) 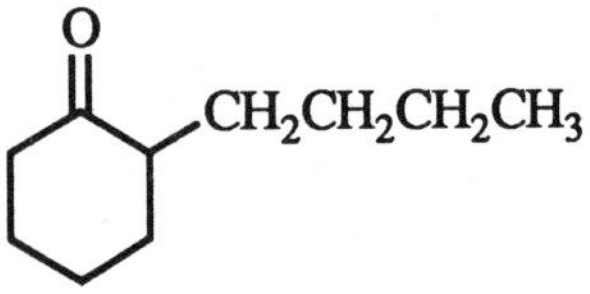

22-69

(a) reagents: Br_2, H^+

(b) reagents: Br_2, PBr_3, followed by H_2O

(c) reagents: I_2 (or Br_2 or Cl_2), NaOH

(d) $Ph-C(=O)-H \; + \; Ph_3\overset{+}{P}-\overset{-}{C}HCH_3 \longrightarrow PhCH{=}CHCH_3 \; + \; Ph_3PO$

(e) 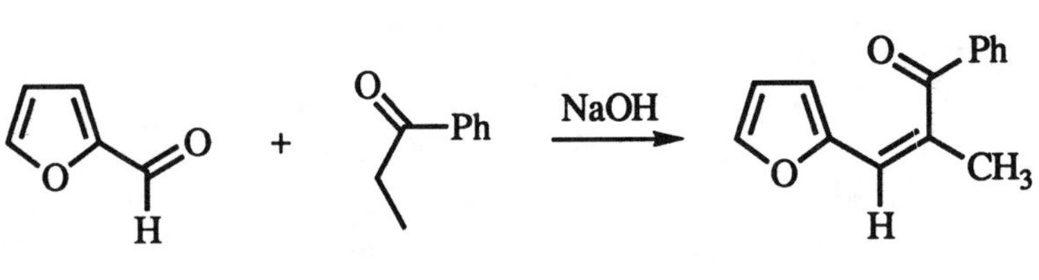

22-70 In the products, the wavy lines indicate the bonds that must be made by alkylation, before hydrolysis and decarboxylation produce the substituted acetic acid.

(a)

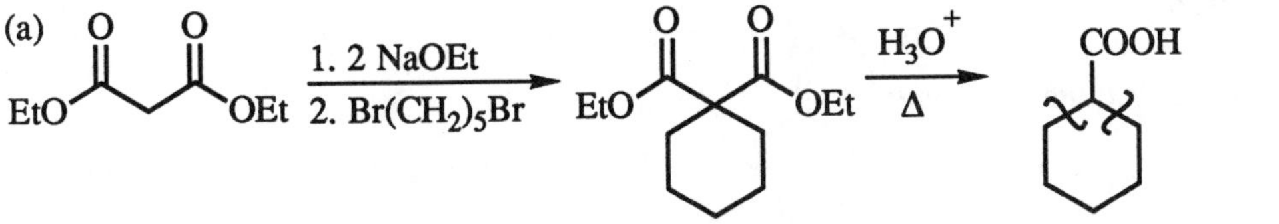
EtO
OEt
1. 2 NaOEt
2. Br(CH$_2$)$_5$Br
H_3O^+
Δ
COOH

+ CO_2 + 2 EtOH

(b)

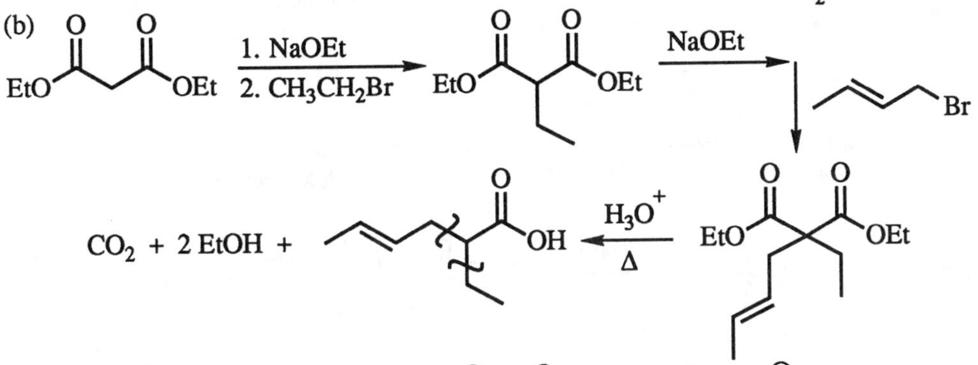
EtO
OEt
1. NaOEt
2. CH_3CH_2Br
NaOEt
Br
H_3O^+
Δ
OH
CO_2 + 2 EtOH +

(c)

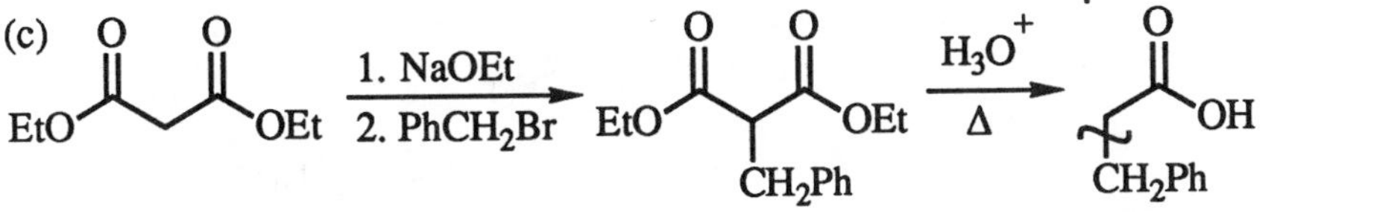
EtO
OEt
1. NaOEt
2. $PhCH_2Br$
CH_2Ph
H_3O^+
Δ
OH

+ CO_2 + 2 EtOH

22-71 In the products, the wavy lines indicate the bonds that must be made by alkylation, before hydrolysis and decarboxylation produce the substituted acetone.

(a)

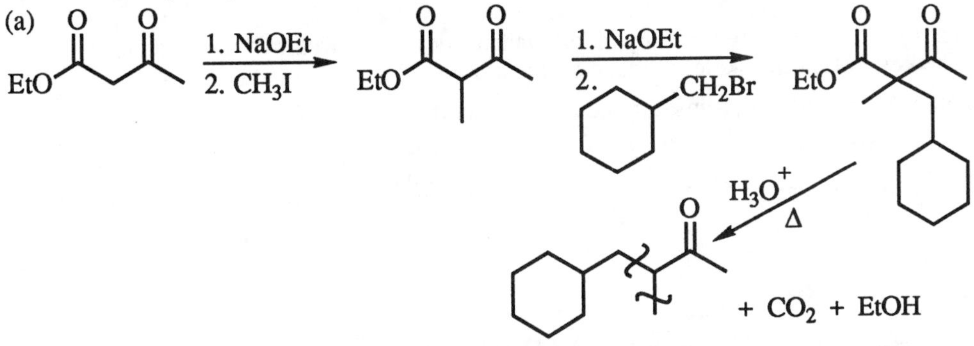
EtO
1. NaOEt
2. CH_3I
1. NaOEt
2.
CH_2Br
H_3O^+
Δ
+ CO_2 + EtOH

(b)

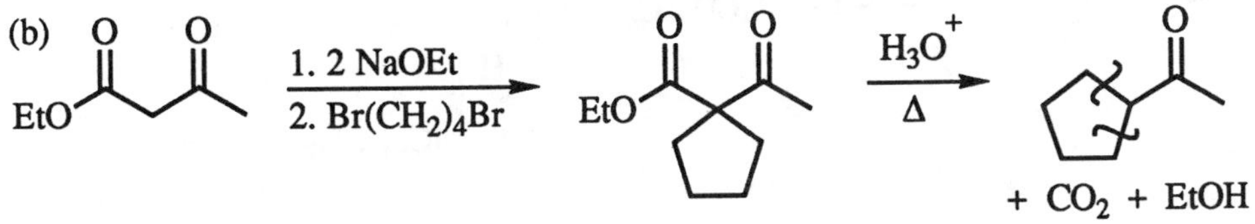
EtO
1. 2 NaOEt
2. Br(CH$_2$)$_4$Br
H_3O^+
Δ

+ CO_2 + EtOH

22-71 continued

(c) The acetoacetic ester synthesis makes substituted acetone, so where is the acetone in this product?

substituted acetone

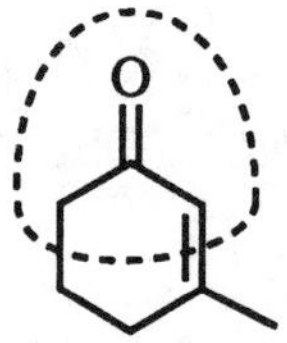

The single bond to this substituted acetone can be made by the acetoacetic ester synthesis. How can we make the α,β double bond? Aldol condensation!

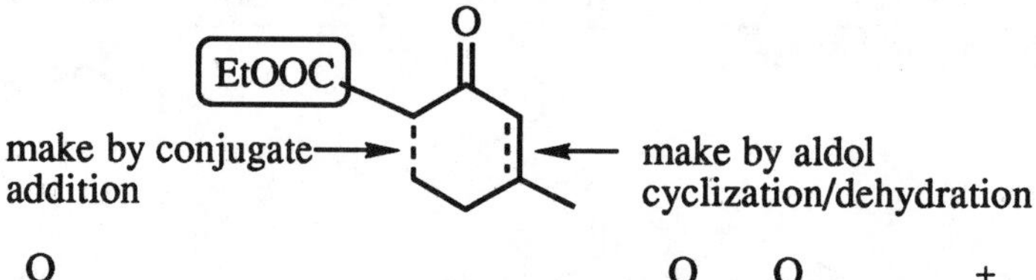
EtOOC
make by conjugate addition
make by aldol cyclization/dehydration

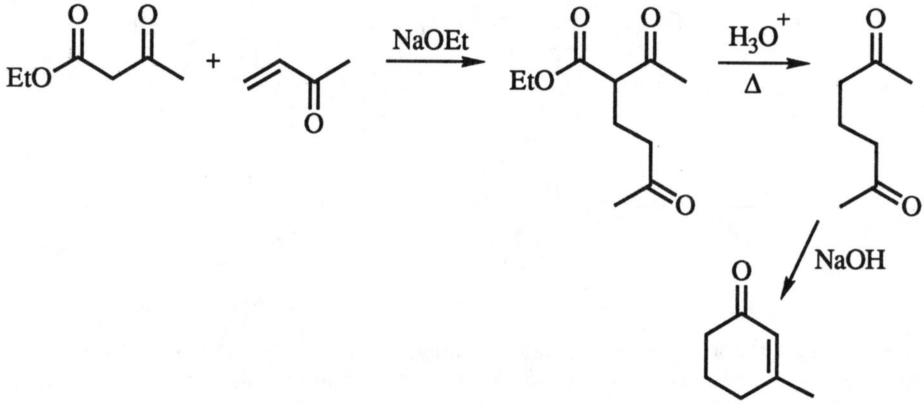
EtO
NaOEt
H3O+
Δ
NaOH

22-72 These compounds are made by aldol condensations followed by other reactions. The key to find the skeleton make by the aldol.

(a) Where is the possible α,β-unsaturated carbonyl in this skeleton?

$$\mathrm{PhCH_2CH_2{-}\overset{\displaystyle OH}{\overset{|}{C}}HPh} \Longrightarrow \mathrm{PhCH{\nmid}CH{-}\overset{\displaystyle O}{\overset{\|}{C}}{-}Ph} \xrightarrow[\text{aldol}]{\text{reverse}} \mathrm{Ph{-}\overset{\displaystyle O}{\overset{\|}{C}}{-}H + CH_3{-}\overset{\displaystyle O}{\overset{\|}{C}}{-}Ph}$$

forward synthesis

$$\mathrm{Ph{-}\overset{\displaystyle O}{\overset{\|}{C}}{-}H + CH_3{-}\overset{\displaystyle O}{\overset{\|}{C}}{-}Ph} \xrightarrow{\text{NaOH}} \mathrm{PhCH{=}CH{-}\overset{\displaystyle O}{\overset{\|}{C}}{-}Ph} \xrightarrow{\mathrm{H_2, Pt}} \mathrm{PhCH_2CH_2{-}\overset{\displaystyle OH}{\overset{|}{C}}HPh}$$

22-72 continued

(b) The aldol skeleton is not immediately apparent in this formidable product. What can we see from it? Most obvious is the β-dicarbonyl (β-ketoester) which we know to be a good nucleophile, capable of substitution or Michael addition. In this case, Michael addition is most likely as the site of attack is β to another carbonyl.

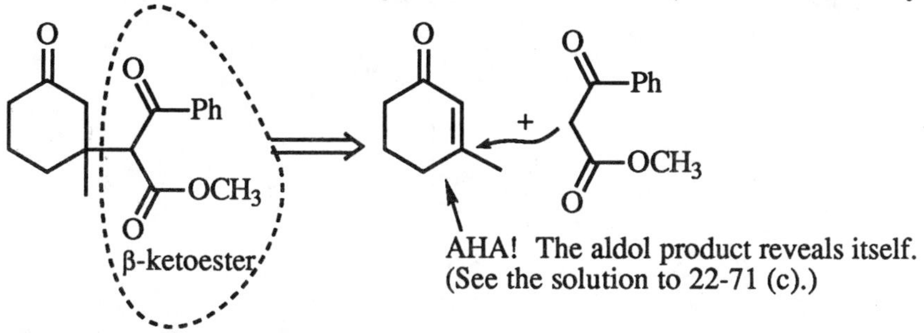

Ph
OCH3
β-ketoester
AHA! The aldol product reveals itself.
(See the solution to 22-71 (c).)

forward synthesis

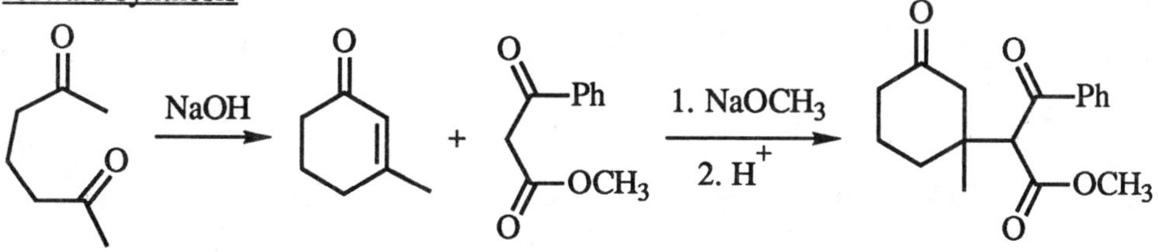

NaOH
Ph
OCH3
1. NaOCH3
2. H+

(c) The key in this product is the α-nitroketone, the equivalent of a β-dicarbonyl system, capable of doing Michael addition to the β-carbon of the other carbonyl.

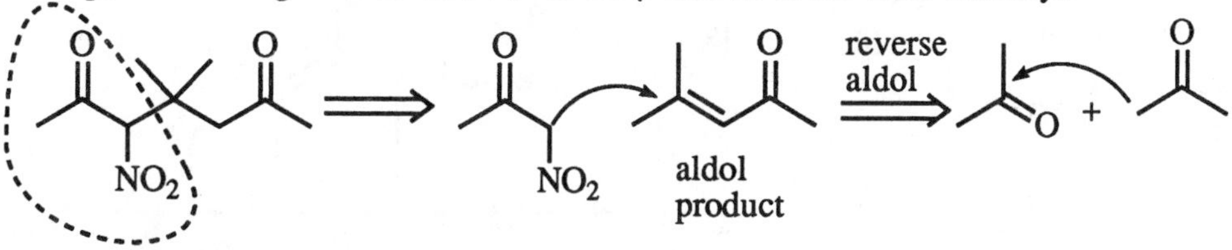

NO2
aldol
product
reverse
aldol

forward synthesis

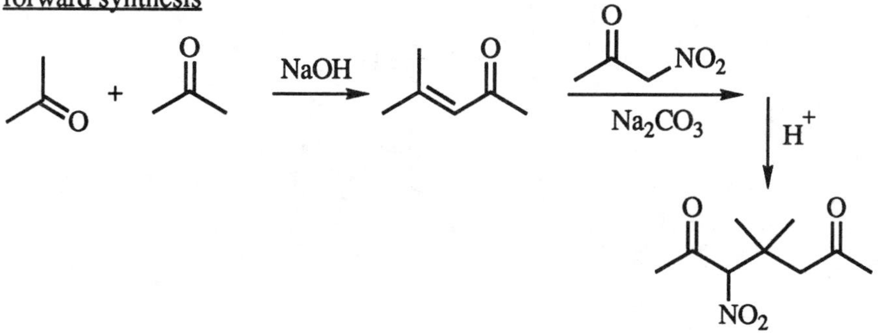

NaOH
NO2
Na2CO3
H+

22-73

(a)

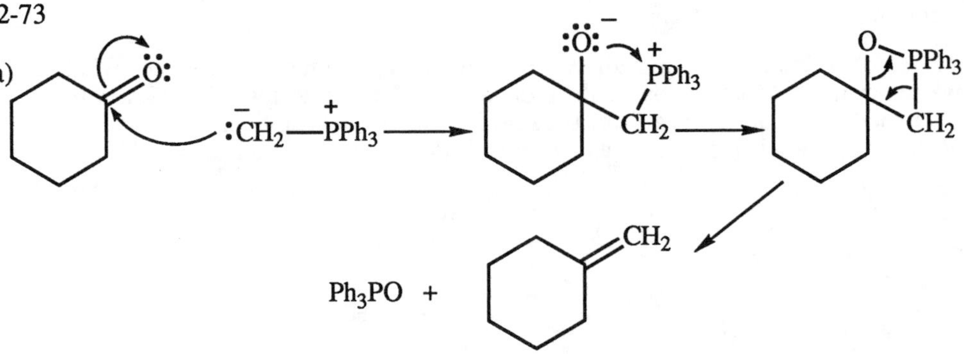

(b)

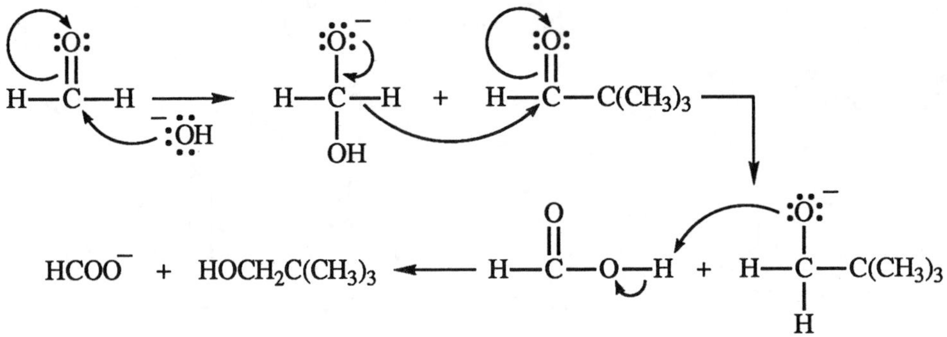

(c)

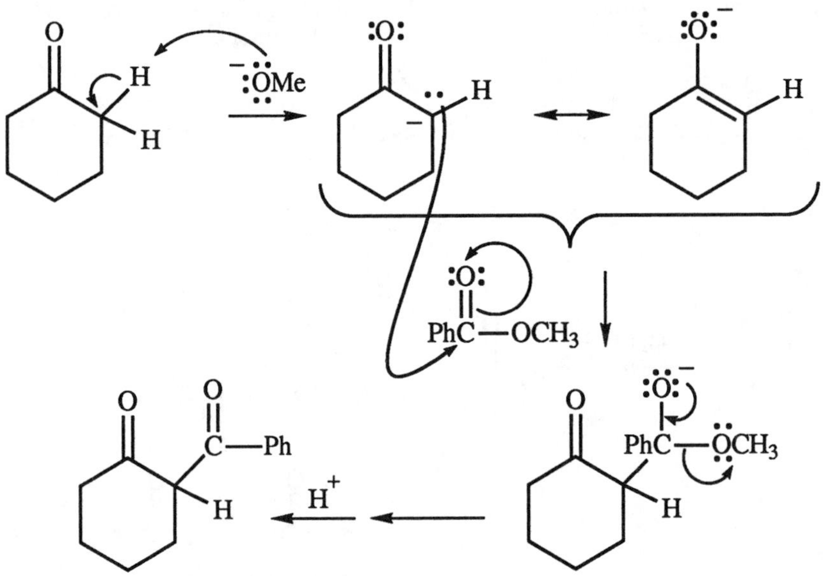

22-73 continued

(d) Robinson annulations are explained most easily by remembering that the first step is a Michael addition, followed by aldol cyclization with dehydration.

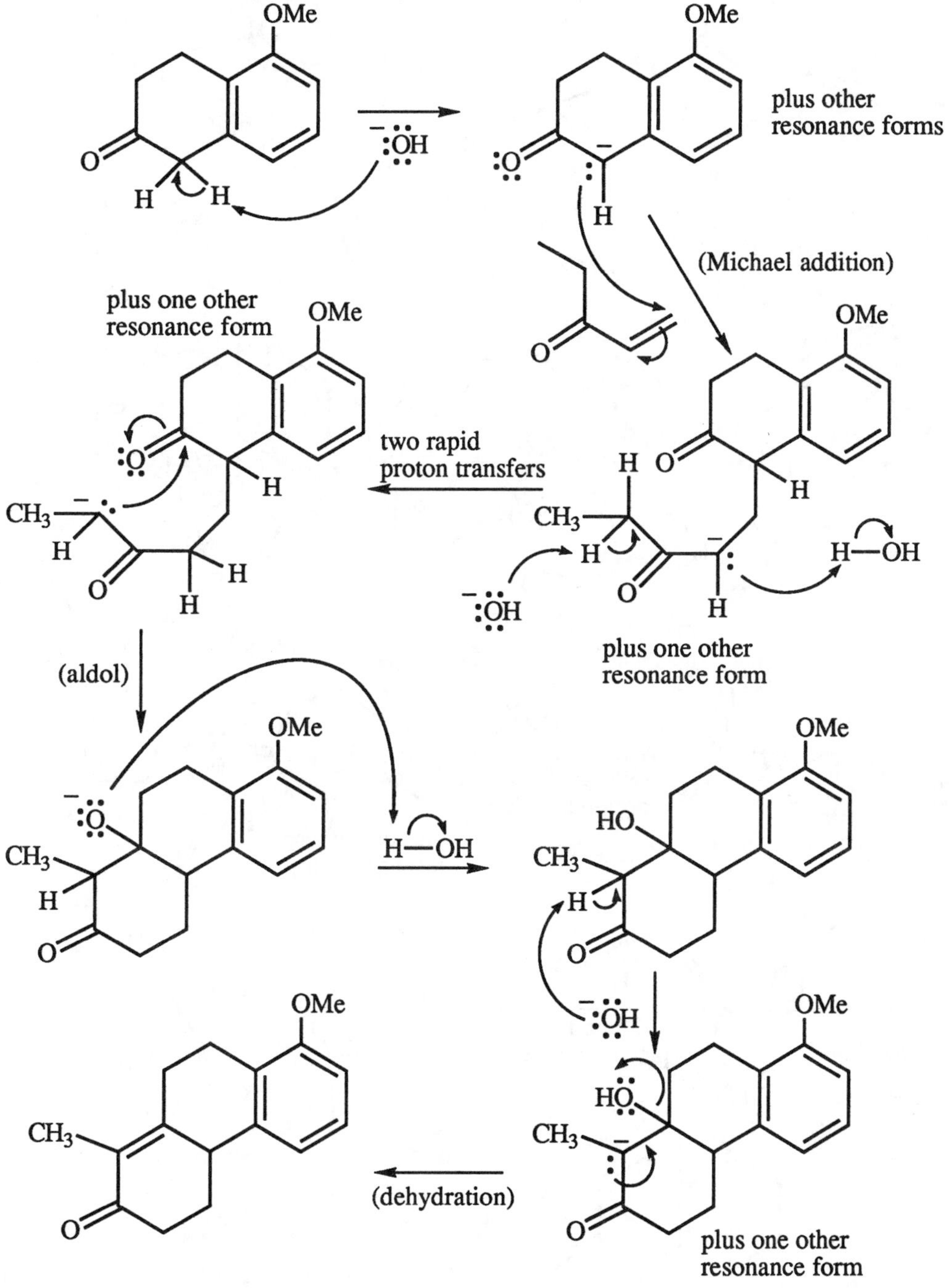

OMe
OH
O
H
H
OMe
plus other resonance forms
O
H
(Michael addition)
O
plus one other resonance form
OMe
O
CH3
H
O
H
H
two rapid proton transfers
OMe
H
O
CH3
H
O
H
H—OH
OH
H
plus one other resonance form
(aldol)
OMe
O
CH3
H
O
H—OH
OMe
HO
CH3
H
O
OH
OMe
CH3
O
(dehydration)
OMe
HO
CH3
O
plus one other resonance form

22-73 continued

(e)

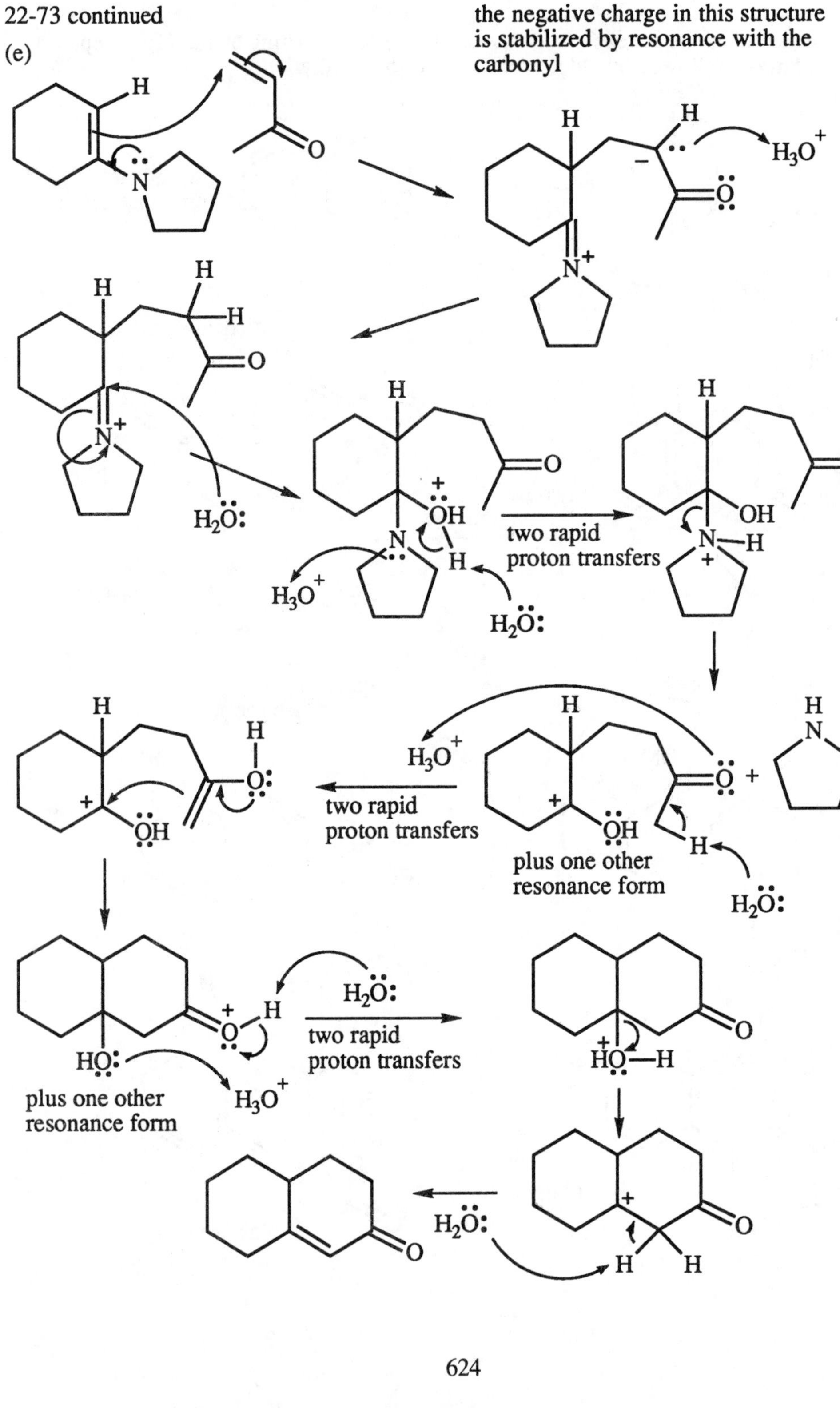

the negative charge in this structure is stabilized by resonance with the carbonyl
two rapid proton transfers
plus one other resonance form
two rapid proton transfers
plus one other resonance form
two rapid proton transfers

22-74

(a)

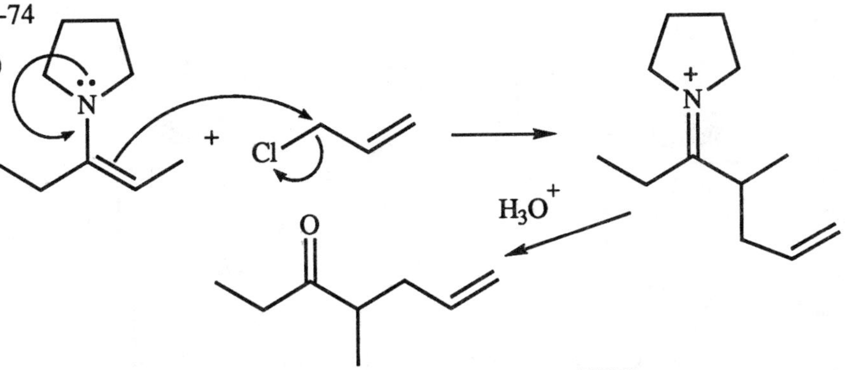

(b)

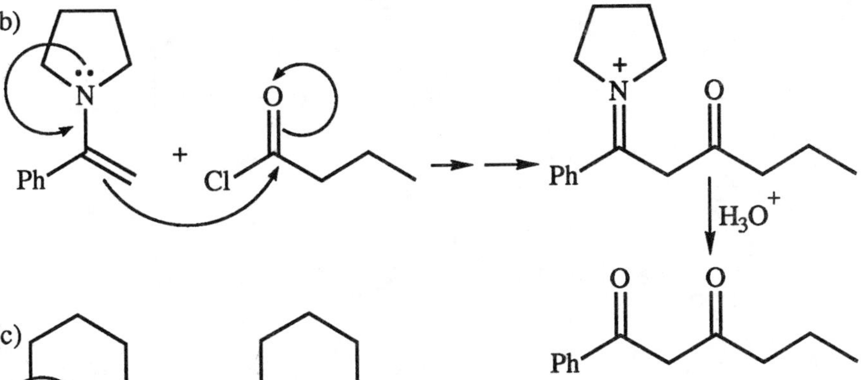

(c)

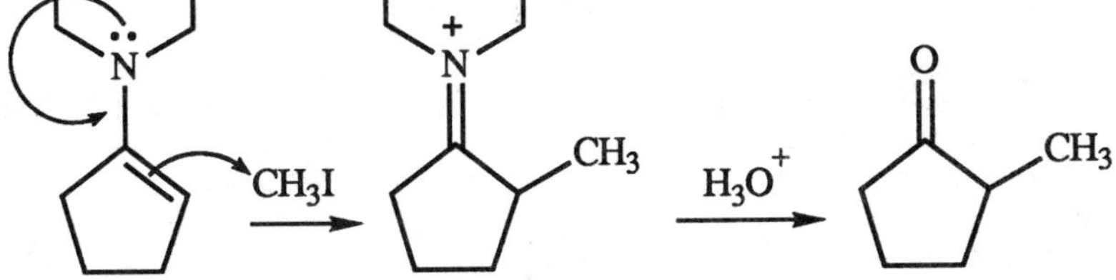

(d)

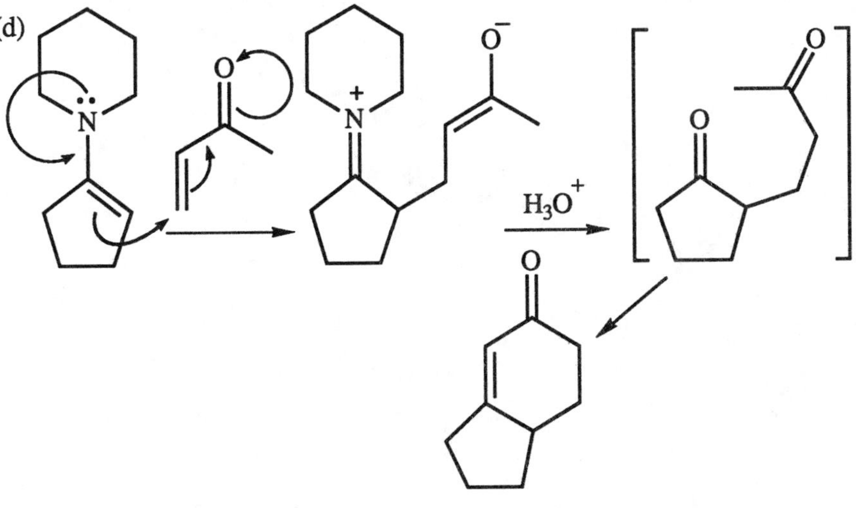

22-75

(a)

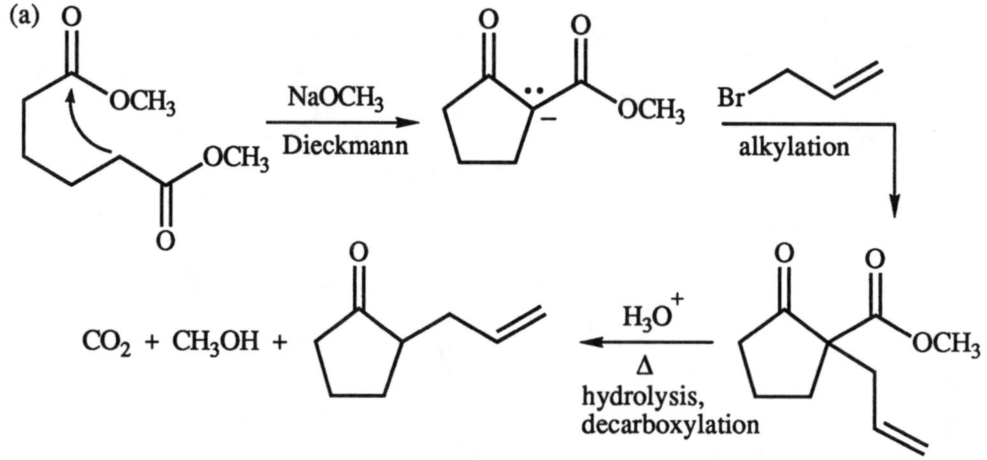

O
OCH3
OCH3
O
NaOCH3
Dieckmann
O
O
OCH3
Br
alkylation
O
O
OCH3
H3O+
Δ
hydrolysis,
decarboxylation
CO2 + CH3OH +
O

(b)

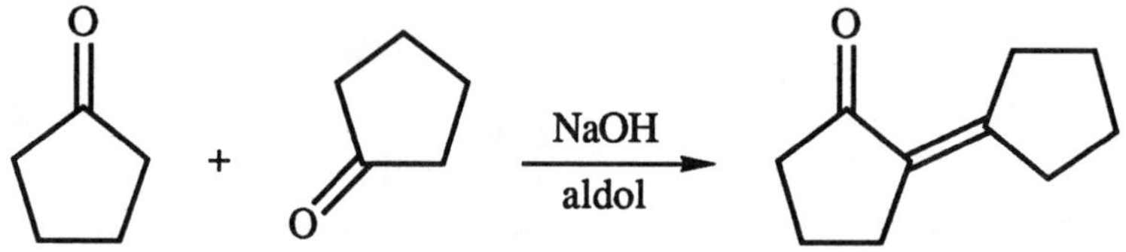

O
+
O
NaOH
aldol
O

(c)

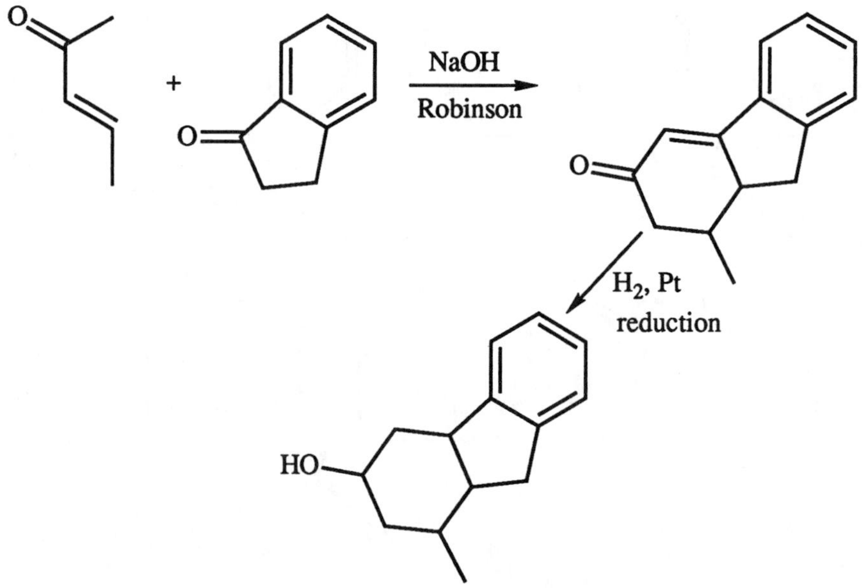

O
+
O
NaOH
Robinson
O
H2, Pt
reduction
HO

22-76

(a)

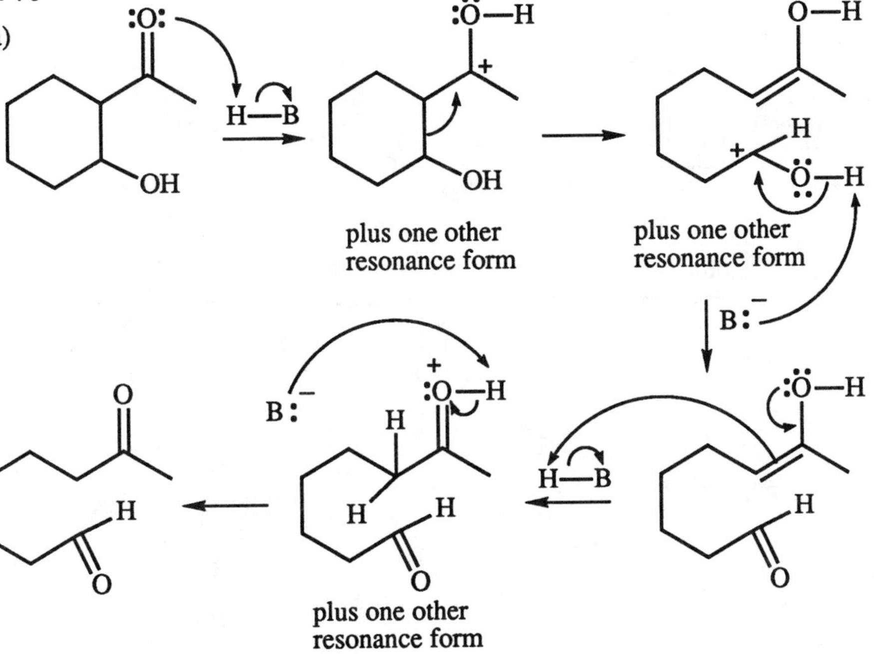

plus one other resonance form
plus one other resonance form
plus one other resonance form

(b)

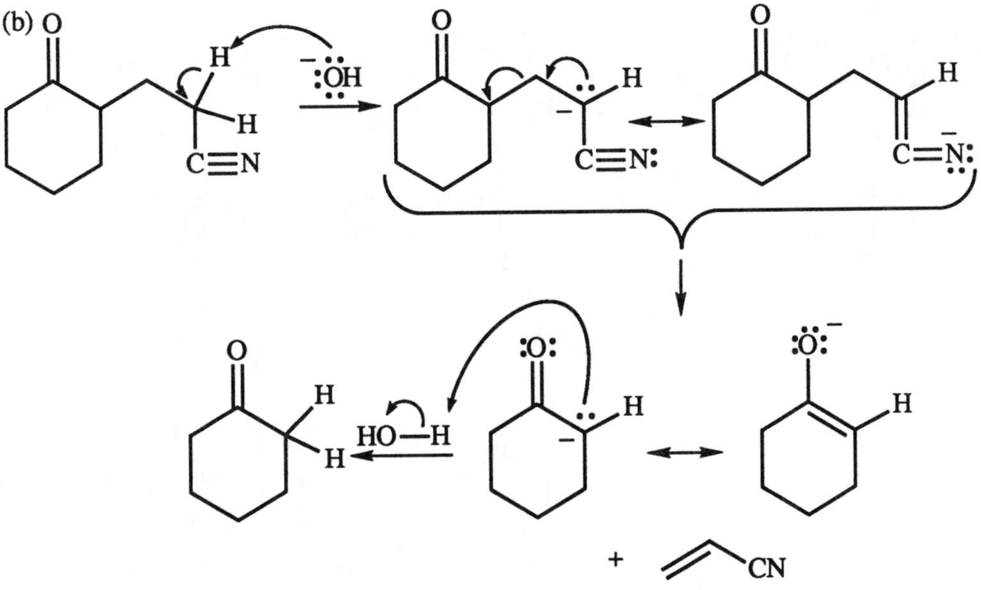

22-76 continued

(c)

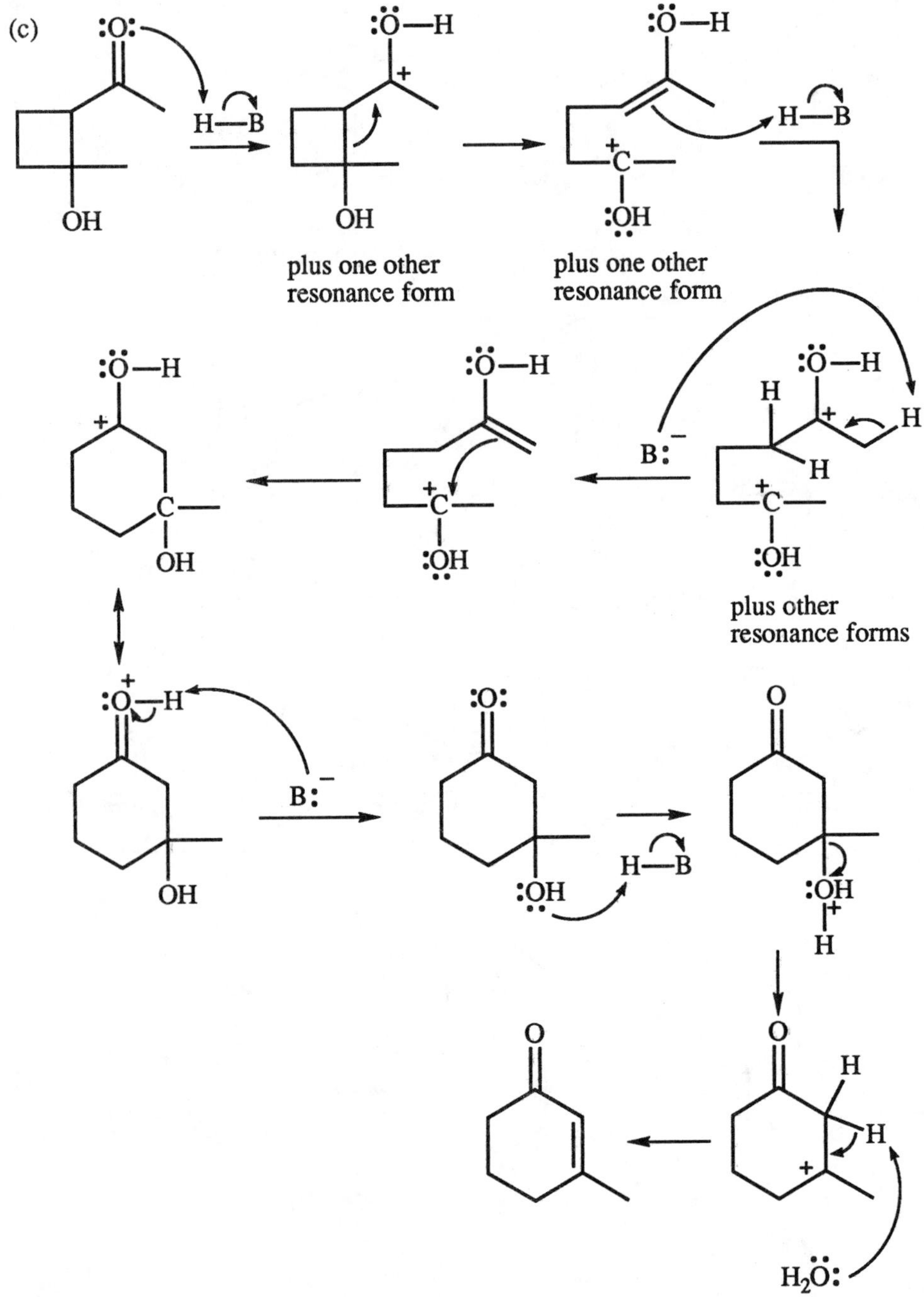

22-77 All of these Robinson annulations are catalyzed by NaOH.

(a)

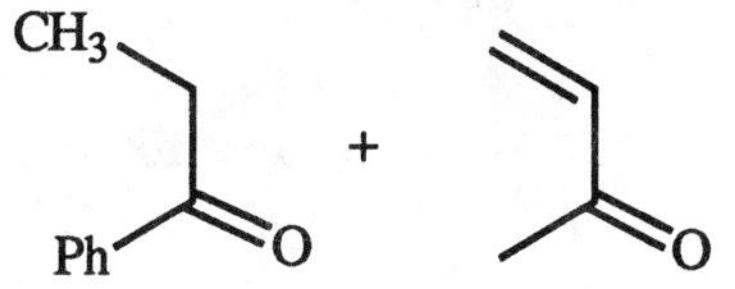

(b)

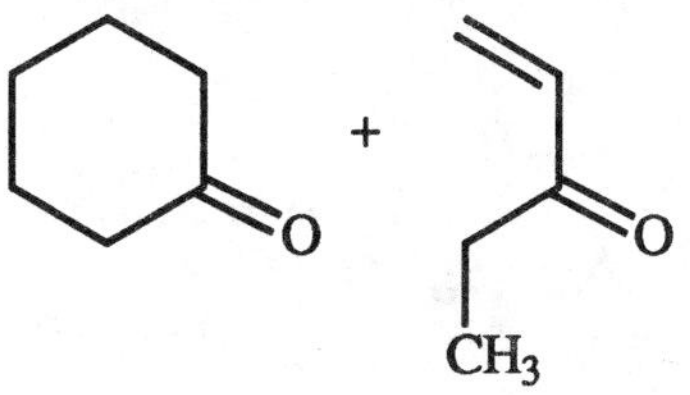

(c)

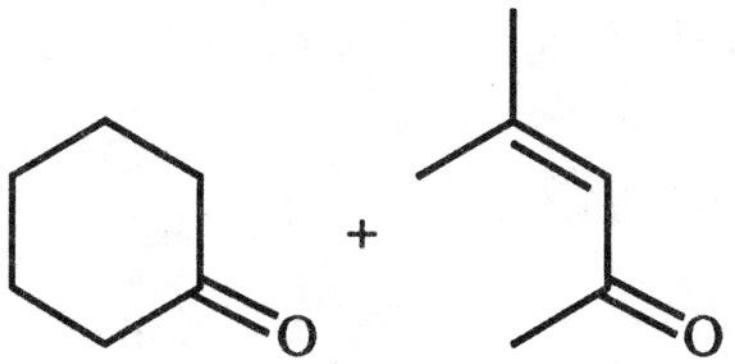

22-78

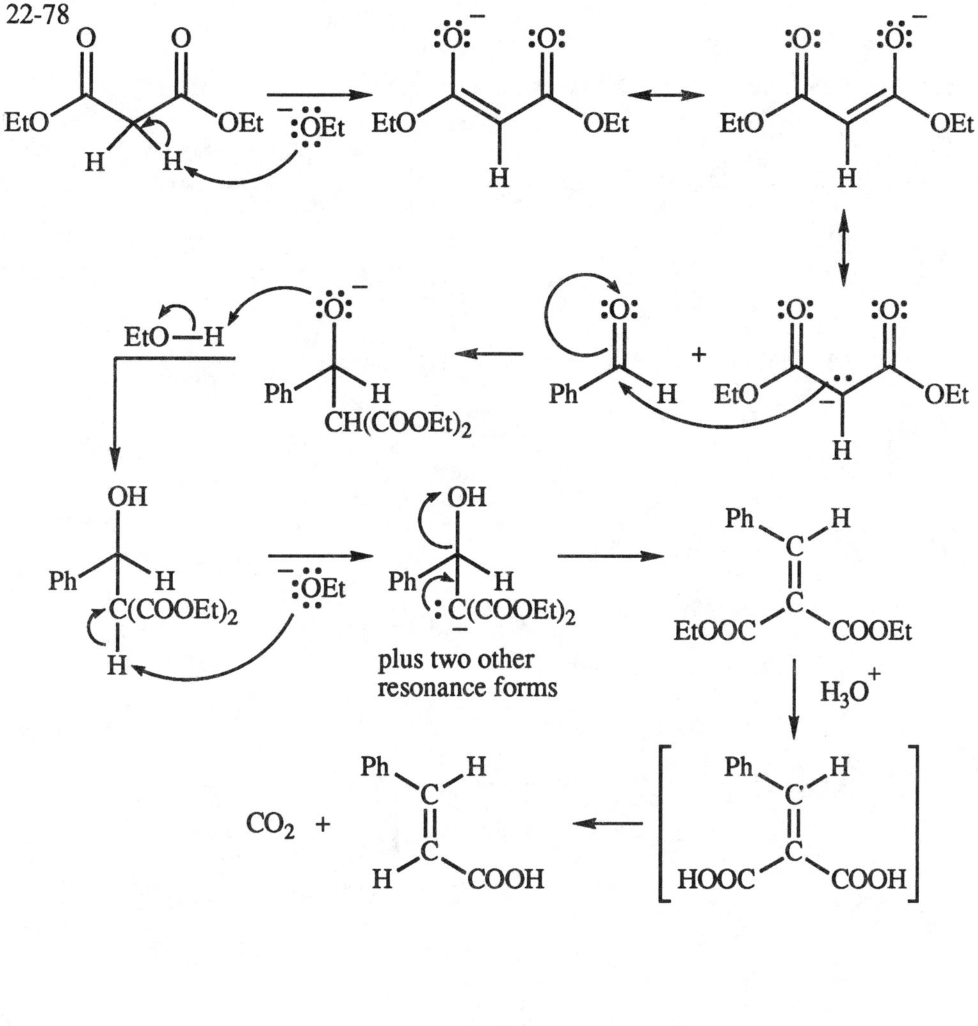

22-79

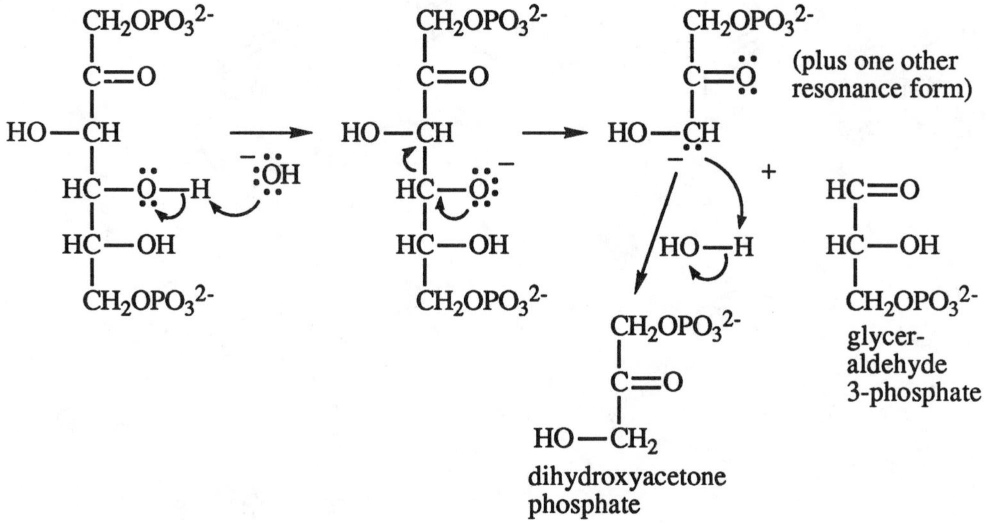
(plus one other resonance form)
glyceraldehyde 3-phosphate
dihydroxyacetone phosphate

22-80 This is an aldol condensation. **P** stands for a protein chain in this problem.

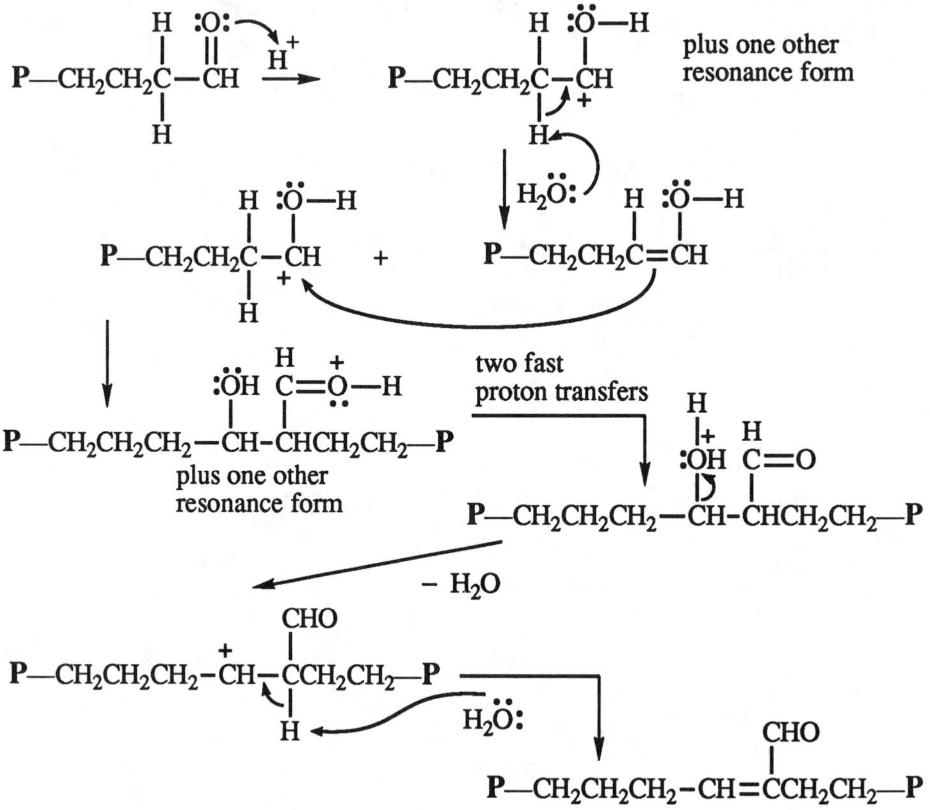
plus one other resonance form
plus one other resonance form
two fast proton transfers
− H₂O

22-81

(a) aldol followed by Michael

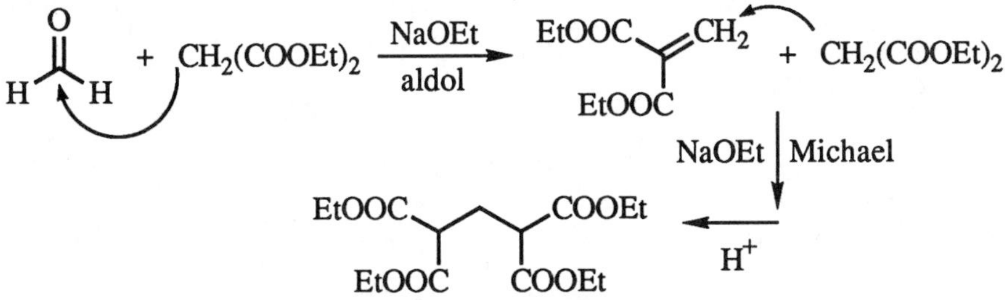

(b) Michael followed by Claisen condensation; hydrolysis and decarboxylation

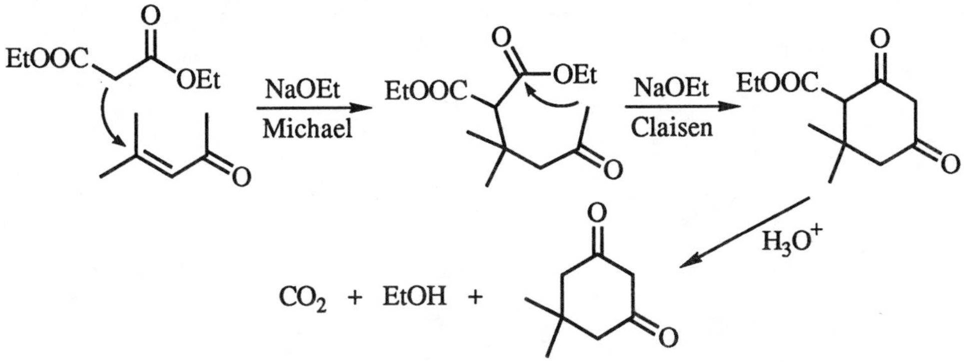

(c) aldol, Michael, aldol cyclization, decarboxylation

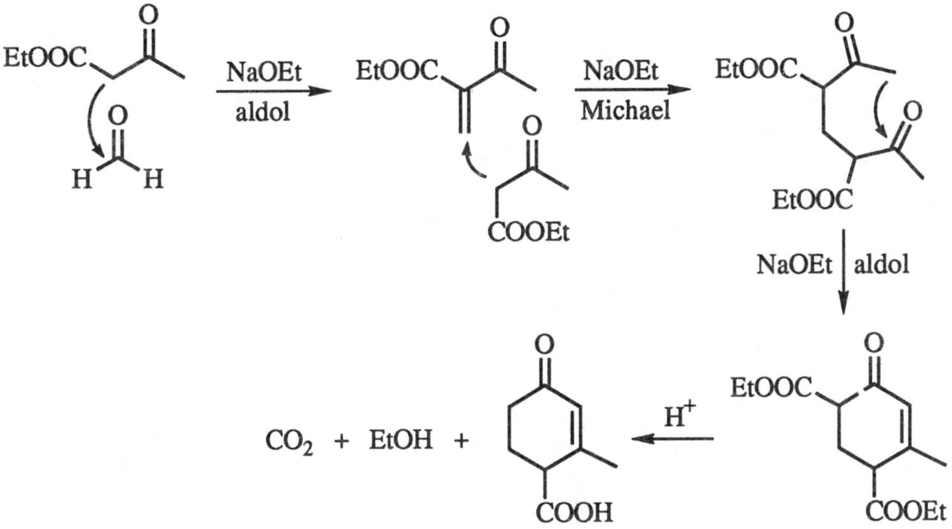

CHAPTER 23—CARBOHYDRATES AND NUCLEIC ACIDS

Note to the student: CAUTION! This chapter of the Solutions Manual contains a significant deviation from standard organic notation. Historically, chemists working with carbohydrates have used bonds with no atoms on the end to indicate a bond to *hydrogen*, not to methyl as we have seen until now. For consistency with tradition in structures of carbohydrates, and for simplicity of these sometimes cluttered pictures, the structures in this chapter and the structures *of carbohydrates* in subsequent chapters will follow this symbolism. Examples:

Fischer projection:

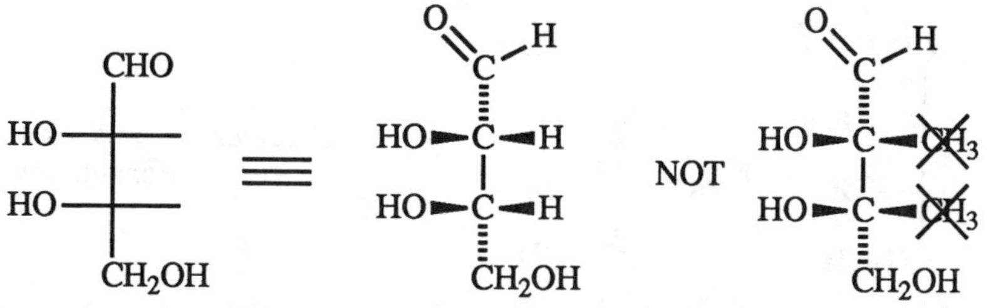

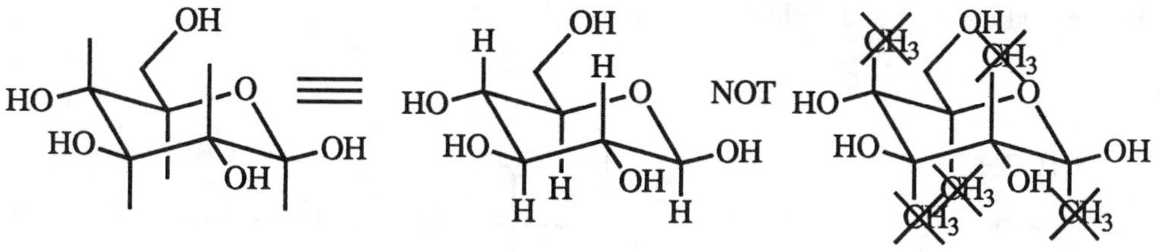

23-1

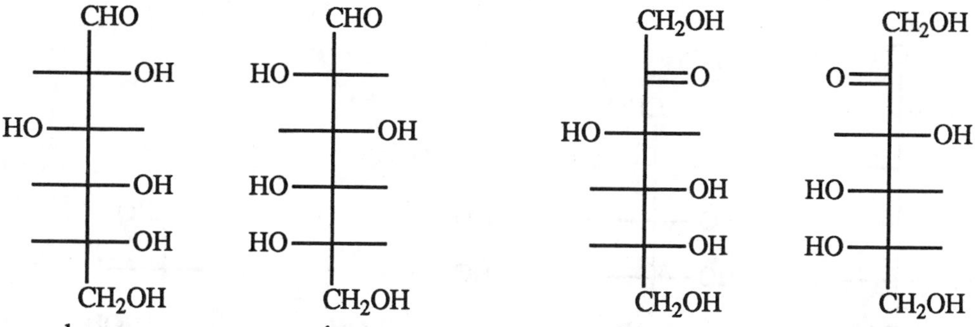

glucose — mirror image — fructose — mirror image

All four of these compounds are chiral and optically active.

23-2

(a)

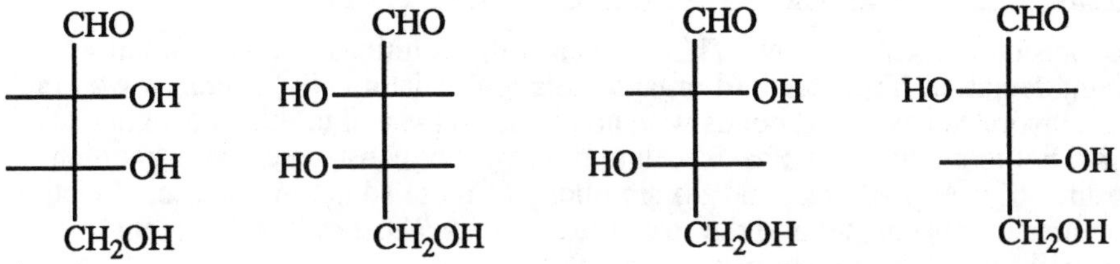

two chiral centers ⇒ four stereoisomers (two pairs of enantiomers) if none are meso

(b)

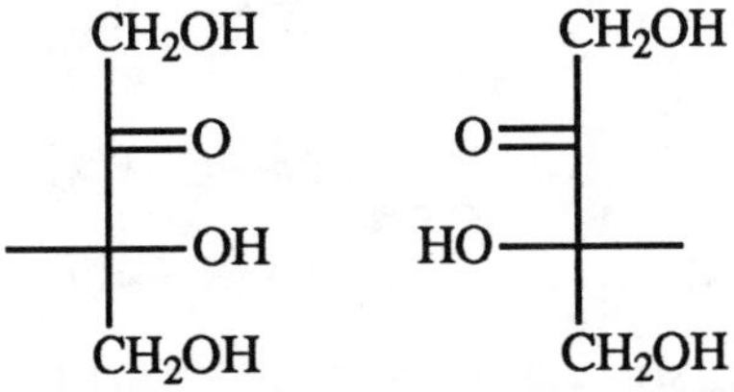

one chiral center ⇒ two stereoisomers (enantiomers)

(c) An aldohexose has four chiral carbons and sixteen stereoisomers. A ketohexose has three chiral carbons and eight stereoisomers.

23-3

(a)

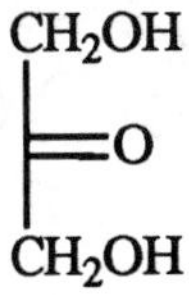

(b)

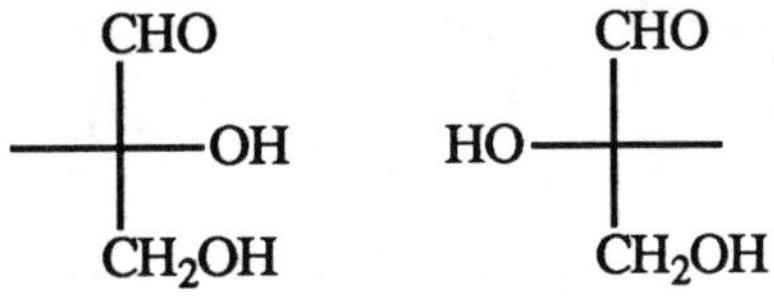

23-4

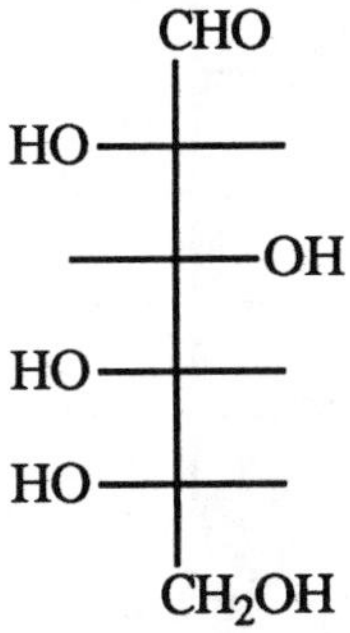

L-(−)-glucose

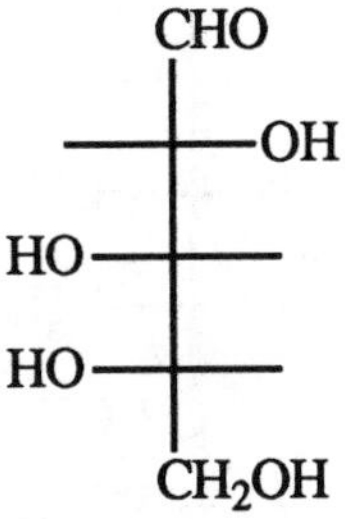

L-(+)-arabinose

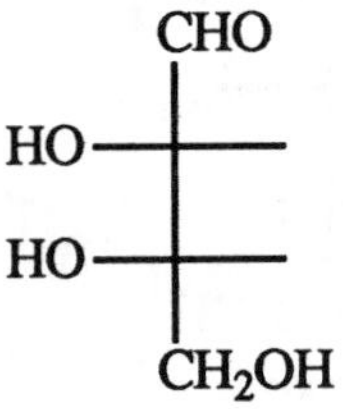

L-(+)-erythrose

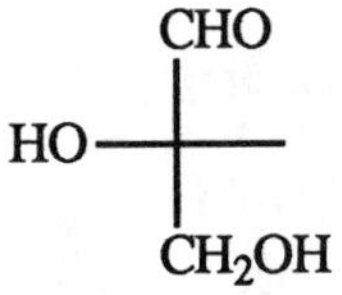

L-(−)-glyceraldehyde

23-5

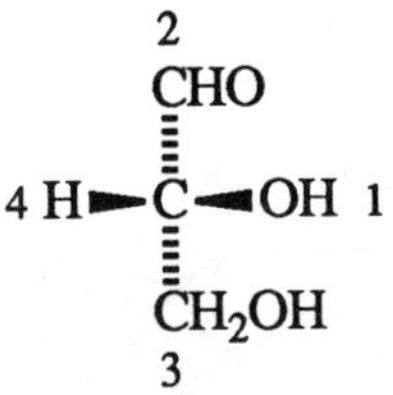

D = *R*

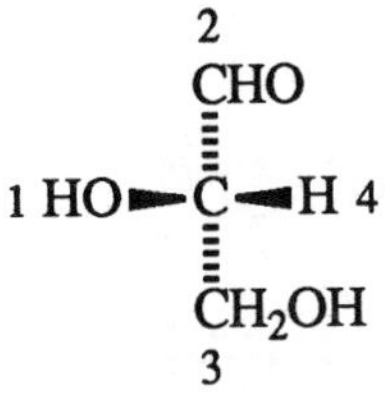

L = *S*

23-6

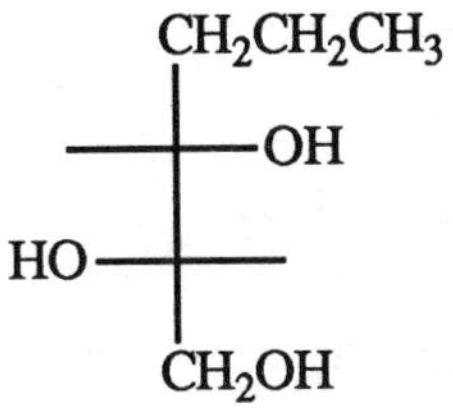

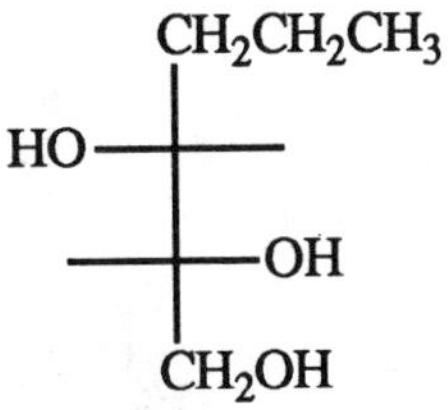

23-7

(a)

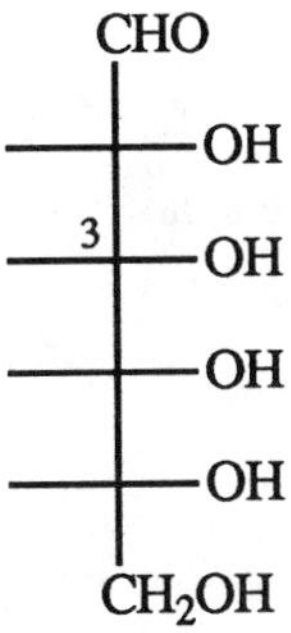

D-allose

(b)

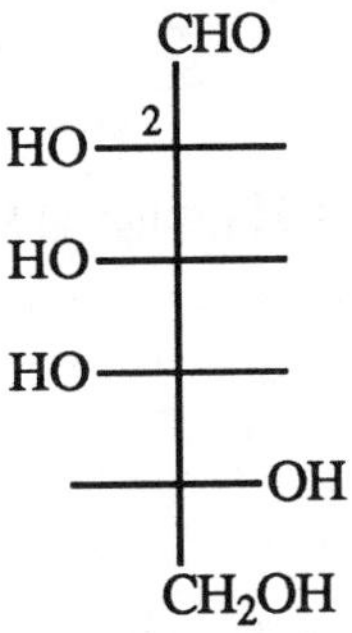

D-talose

(c)

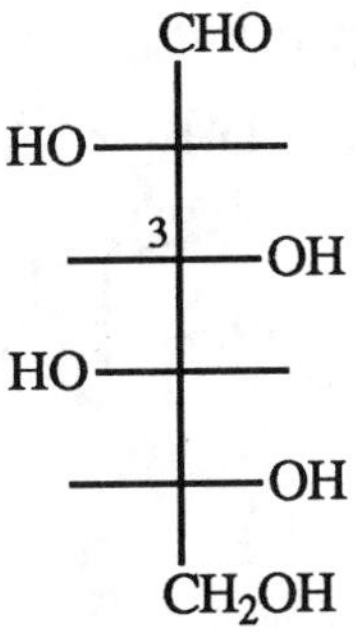

D-idose

(d)

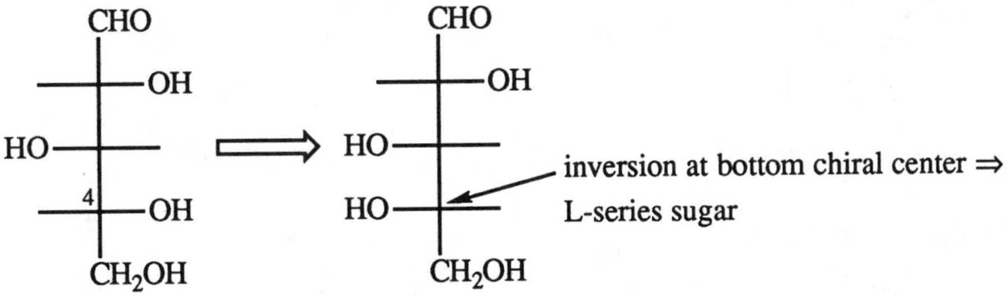
inversion at bottom chiral center ⇒
L-series sugar

D-xylose L-arabinose

23-8 D-mannose

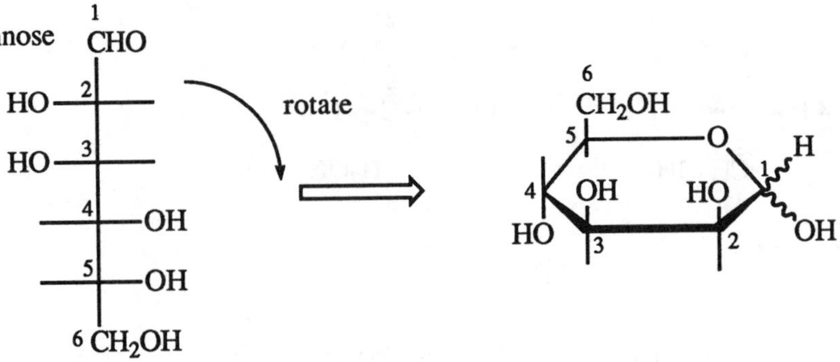

rotate

23-9 D-allose

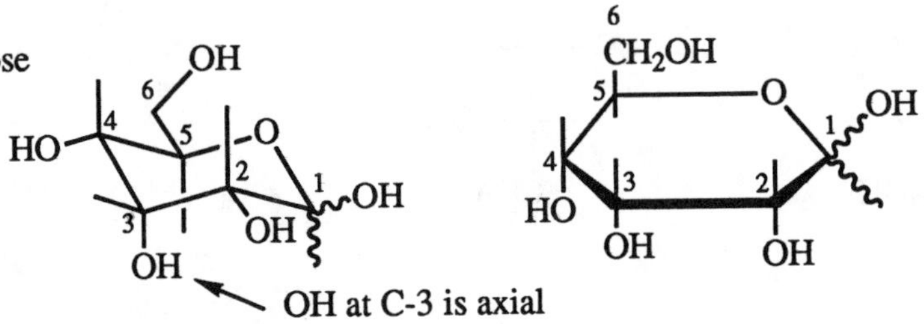

OH at C-3 is axial

23-10 D-talopyranose

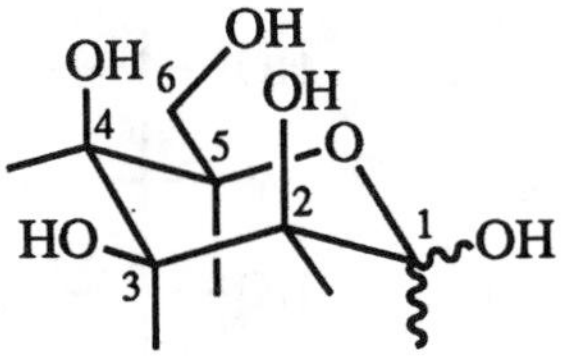

OH groups at C-2 and C-4 are axial

23-11

(a)

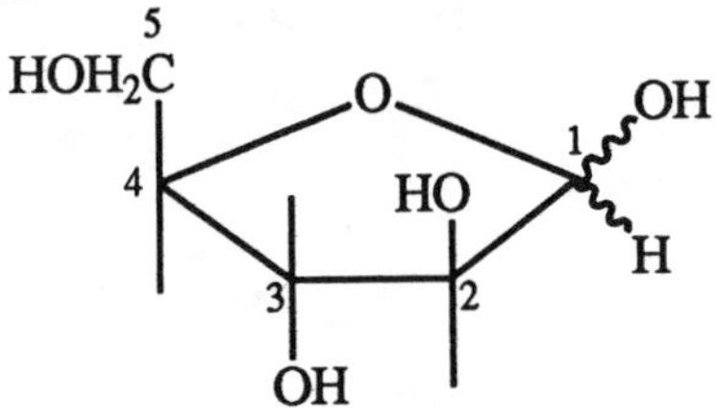

D-arabinofuranose

(b)

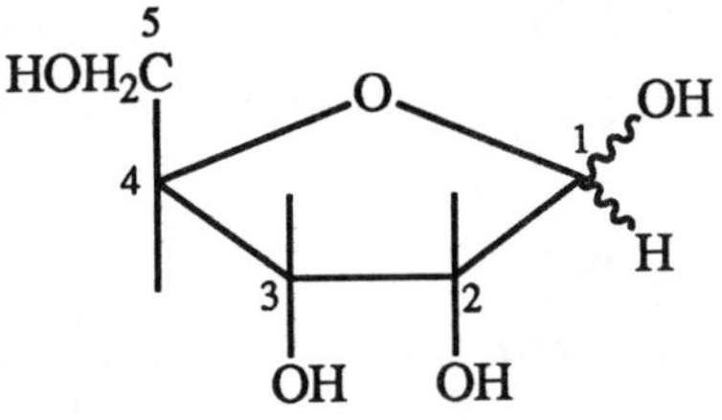

D-ribofuranose

23-12

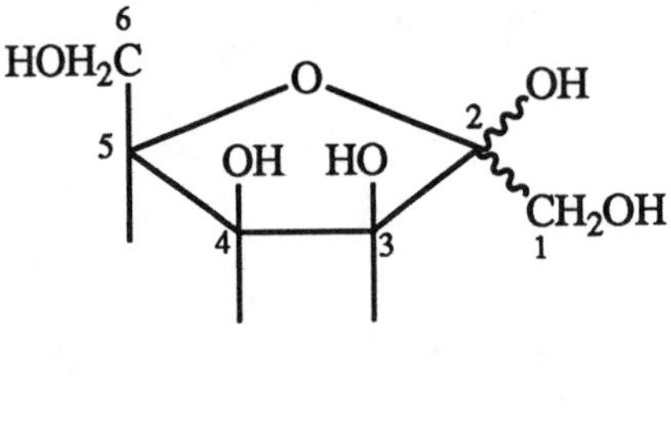

23-13

(a) α-D-mannopyranose

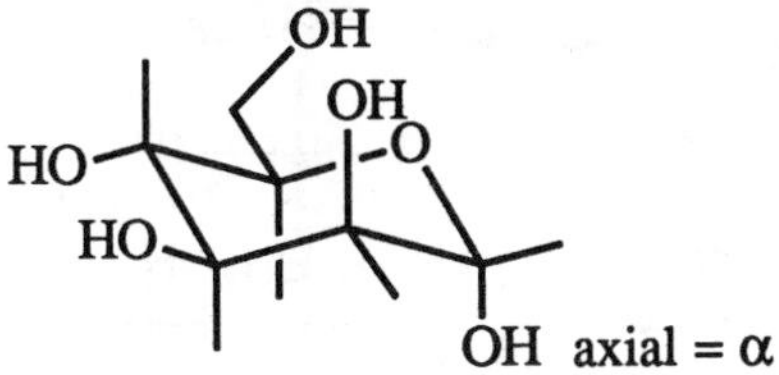
OH
OH
HO
HO
O
OH axial = α

(b) β-D-galactopyranose

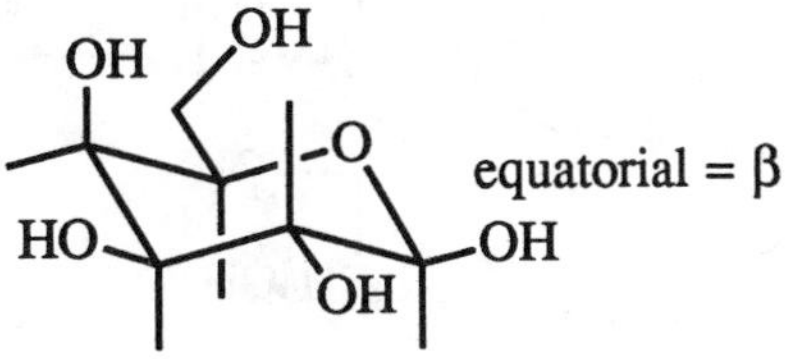
OH
OH
O
equatorial = β
HO
OH
OH

(c) β-D-allopyranose

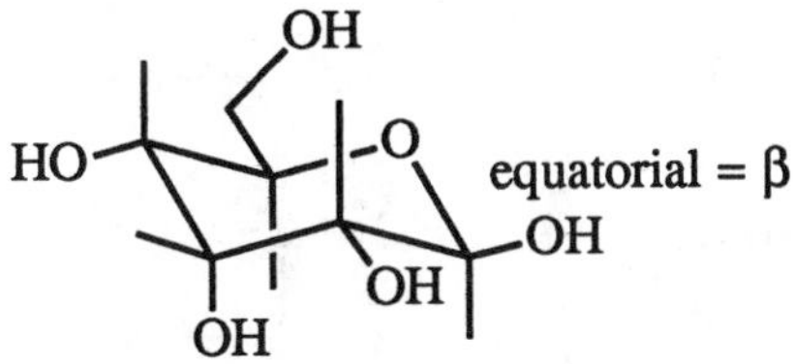
OH
HO
O
equatorial = β
OH
OH
OH

(d) α-D-arabinofuranose

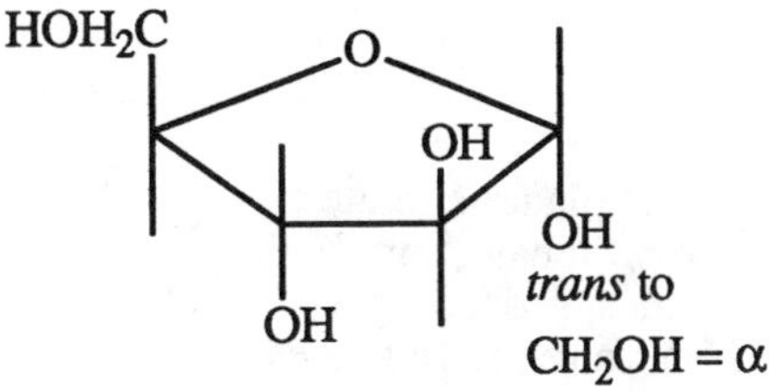
HOH2C
O
OH
OH
trans to
CH2OH = α
OH

(e) β-D-ribofuranose

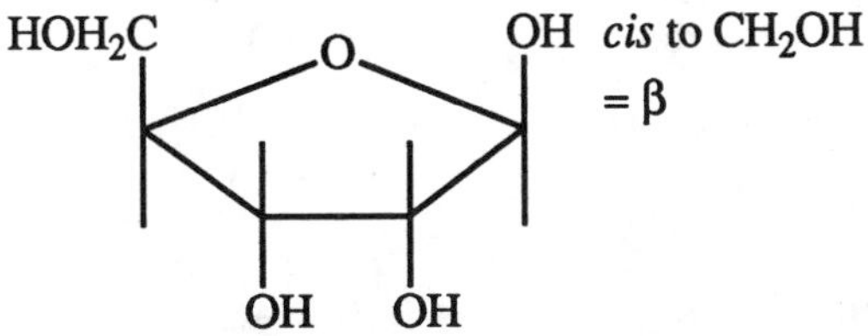
HOH2C
O
OH *cis* to CH2OH
= β
OH
OH

23-14

a = fraction of galactose as the α anomer; b = fraction of galactose as the β anomer

$$a\,(+150.7°) + b\,(+52.8°) = +80.2°$$

$$a + b = 1; \quad b = 1 - a$$

$$a\,(+150.7°) + (1 - a)\,(+52.8°) = +80.2° \xrightarrow{\text{solve for "a"}} a = 0.28; \quad b = 0.72$$

The equilibrium mixture contains 28% of the α anomer and 72% of the β anomer.

23-15

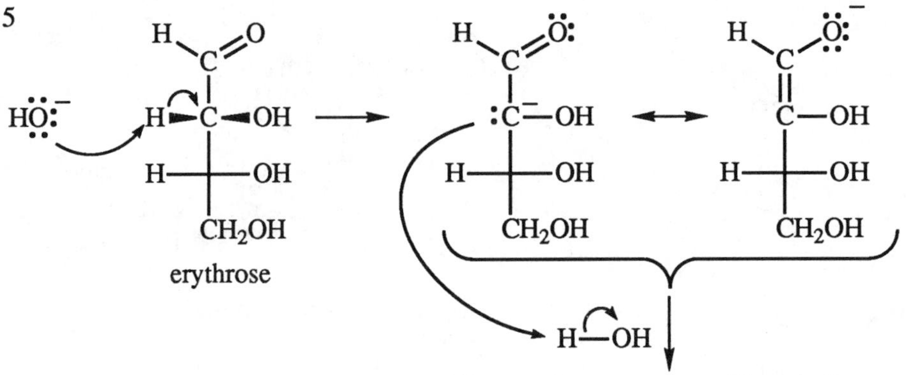
erythrose

The planar enolate can reprotonate from both sides, producing a mixture of erythrose and threose.

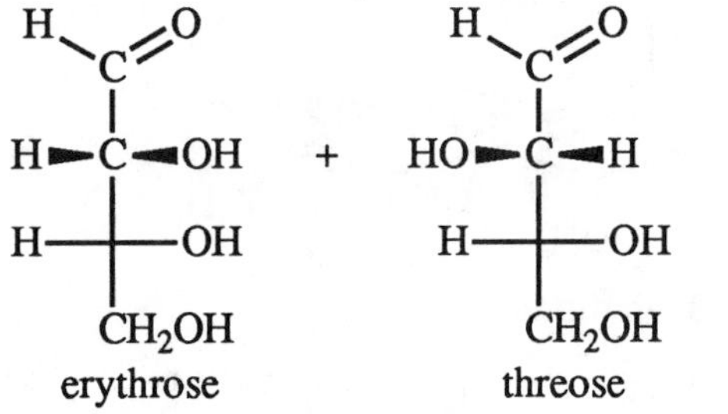
erythrose
threose

23-16

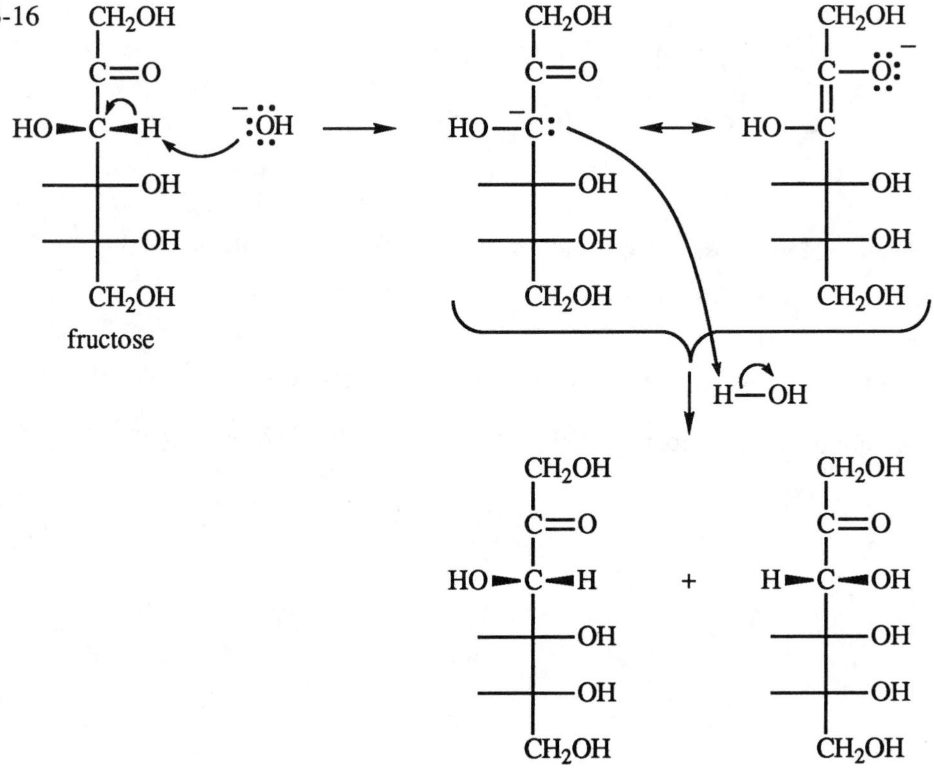
fructose

23-17

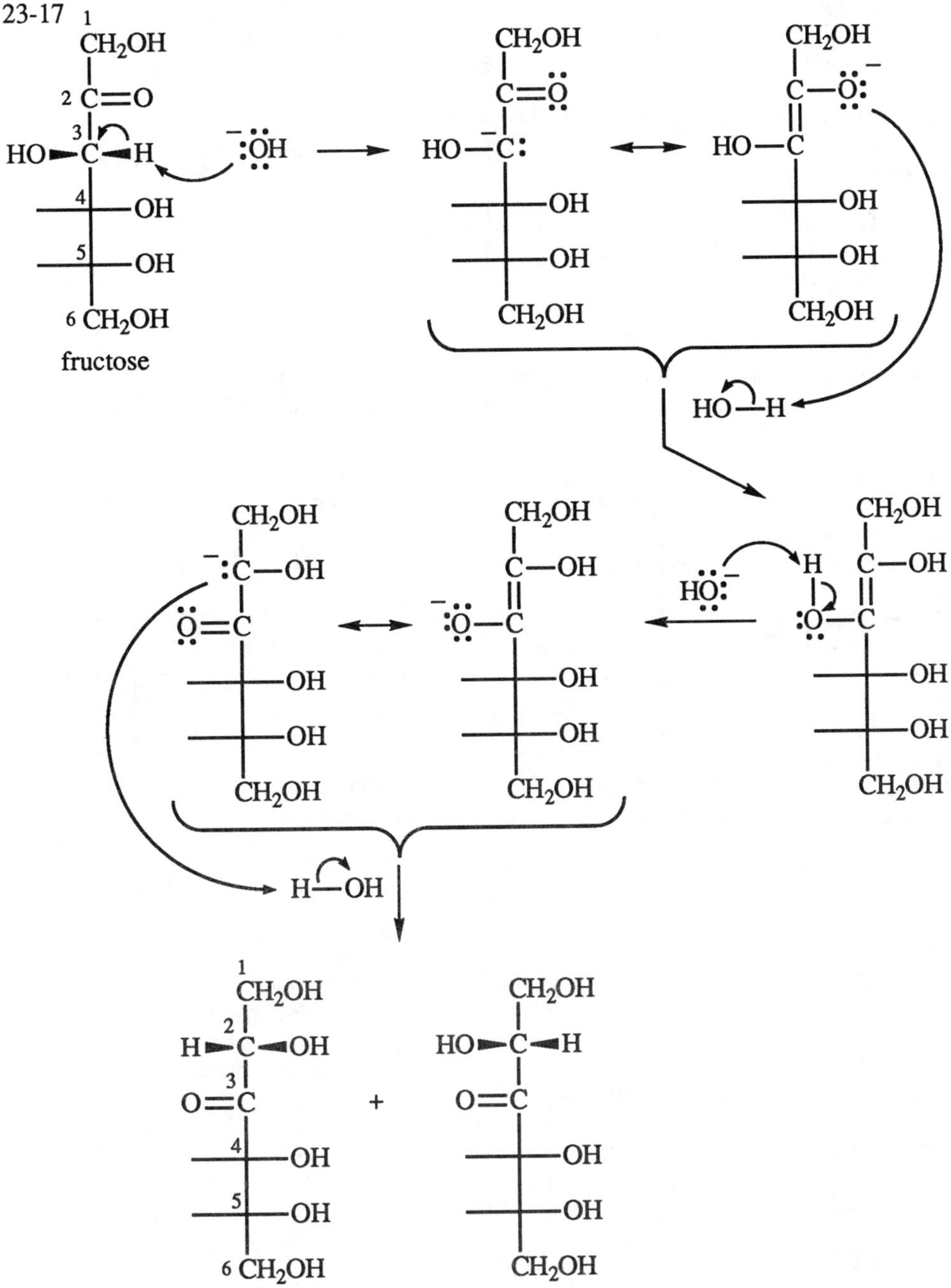

23-18

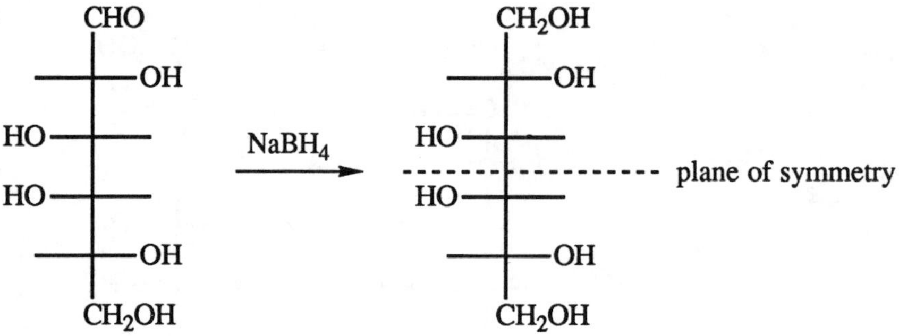
CHO
OH
HO
HO
OH
CH2OH
NaBH4
CH2OH
OH
HO
HO
OH
CH2OH
plane of symmetry

D-galactose

The reduction product of D-galactose still has four chiral centers but also has a plane of symmetry; it is a meso compound and is therefore optically inactive.

23-19

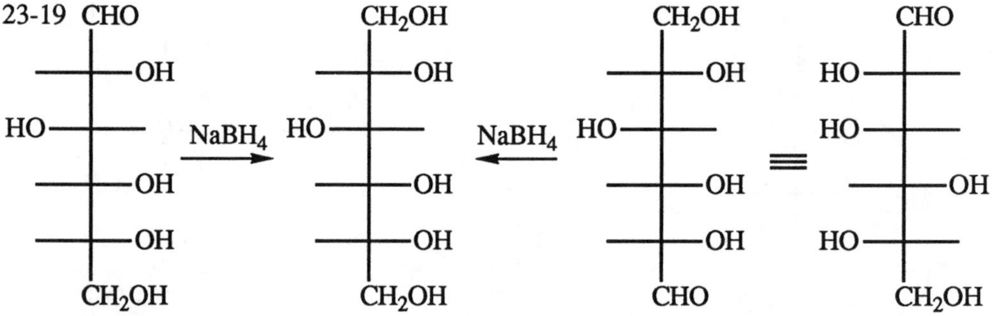
CHO
OH
HO
OH
OH
CH2OH
NaBH4
CH2OH
OH
HO
OH
OH
CH2OH
NaBH4
CH2OH
OH
HO
OH
OH
CHO
≡
CHO
HO
HO
OH
HO
CH2OH

D-glucose D-glucitol L-gulose L-gulose

23-20

(a)

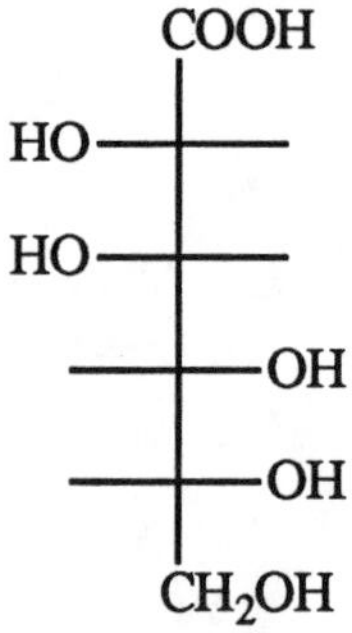
COOH
HO
HO
OH
OH
CH2OH

D-mannonic acid

(b)

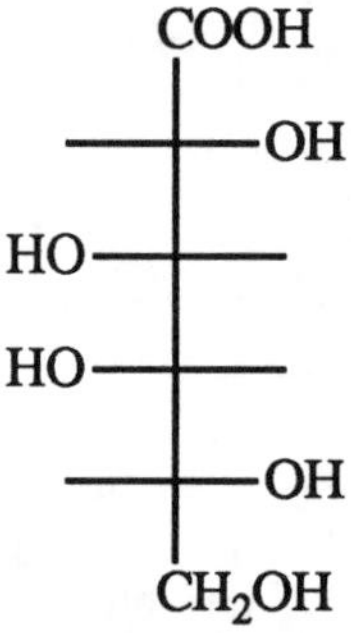
COOH
OH
HO
HO
OH
CH2OH

D-galactonic acid

(c) no reaction—
D-fructose is a
ketohexose; only
aldoses react

23-21

(a)

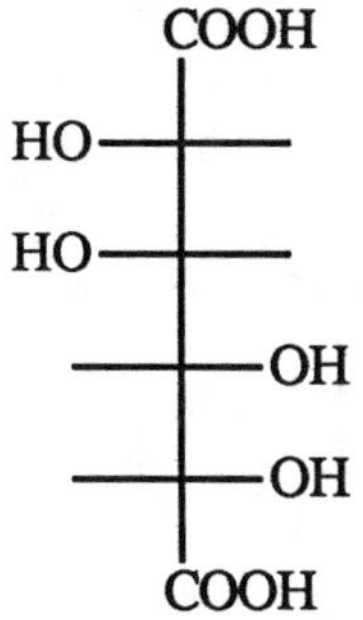
COOH
HO
HO
OH
OH
COOH

mannaric acid

(b)

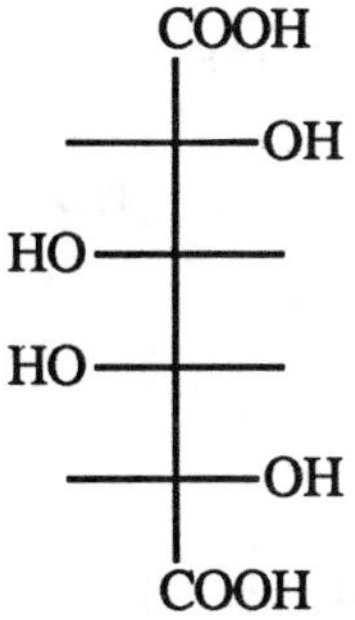
COOH
OH
HO
HO
OH
COOH

galactaric acid (meso)

23-22

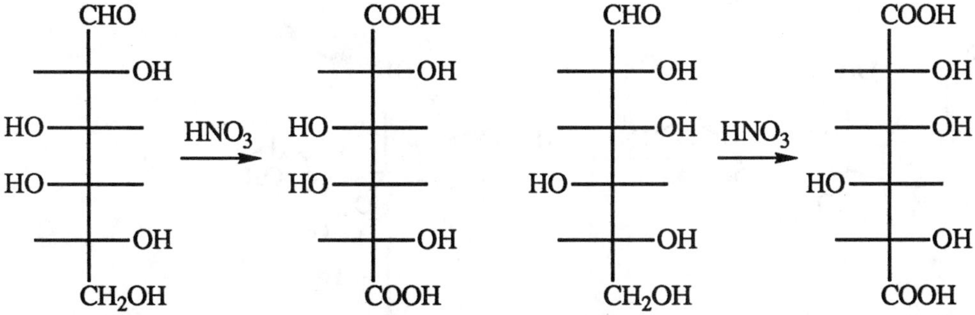
CHO
OH
HO
HO
OH
CH_2OH
HNO_3
COOH
OH
HO
HO
OH
COOH
CHO
OH
OH
HO
OH
CH_2OH
HNO_3
COOH
OH
OH
HO
OH
COOH

galactose **A** — meso— optically inactive — glucose **B** — optically active

23-23

(a) not reducing: an acetal ending in "oside"
(b) reducing: a hemiacetal ending in "ose"
(c) reducing: a hemiacetal ending in "ose"
(d) not reducing: an acetal ending in "oside"
(e) reducing: one of the rings has a hemiacetal
(f) not reducing: all anomeric carbons are in acetal form; sucrose, a common name, is misleading as it suggests a hemiacetal form

23-24

(a)

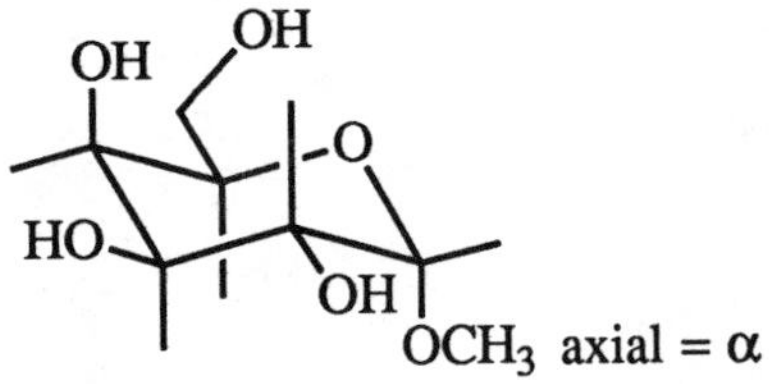

OH
OH
O
HO
OH
OCH3 axial = α

(c)

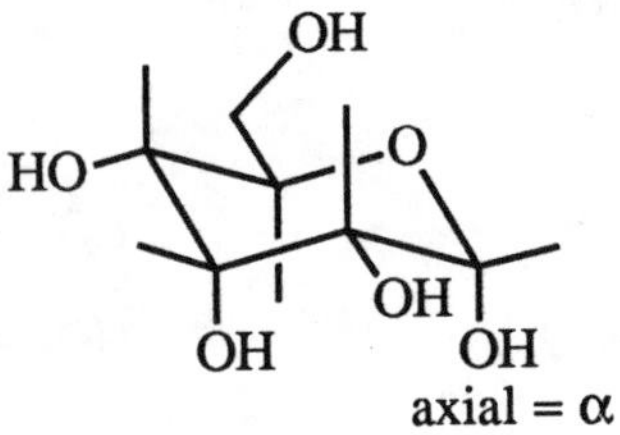

OH
HO
O
OH
OH
OH
axial = α

(d)

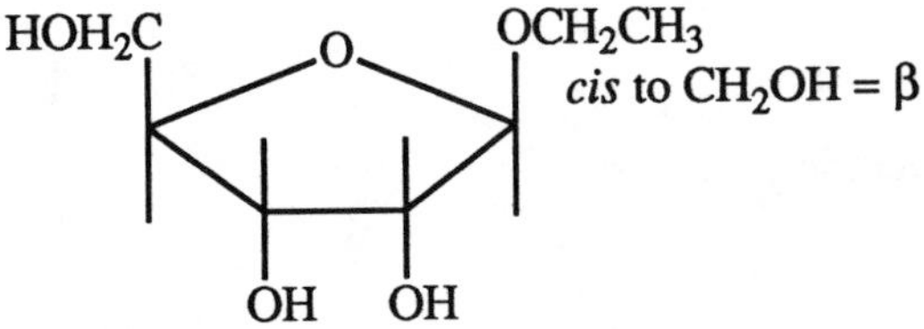

HOH2C
O
OCH2CH3
cis to CH2OH = β
OH
OH

23-25

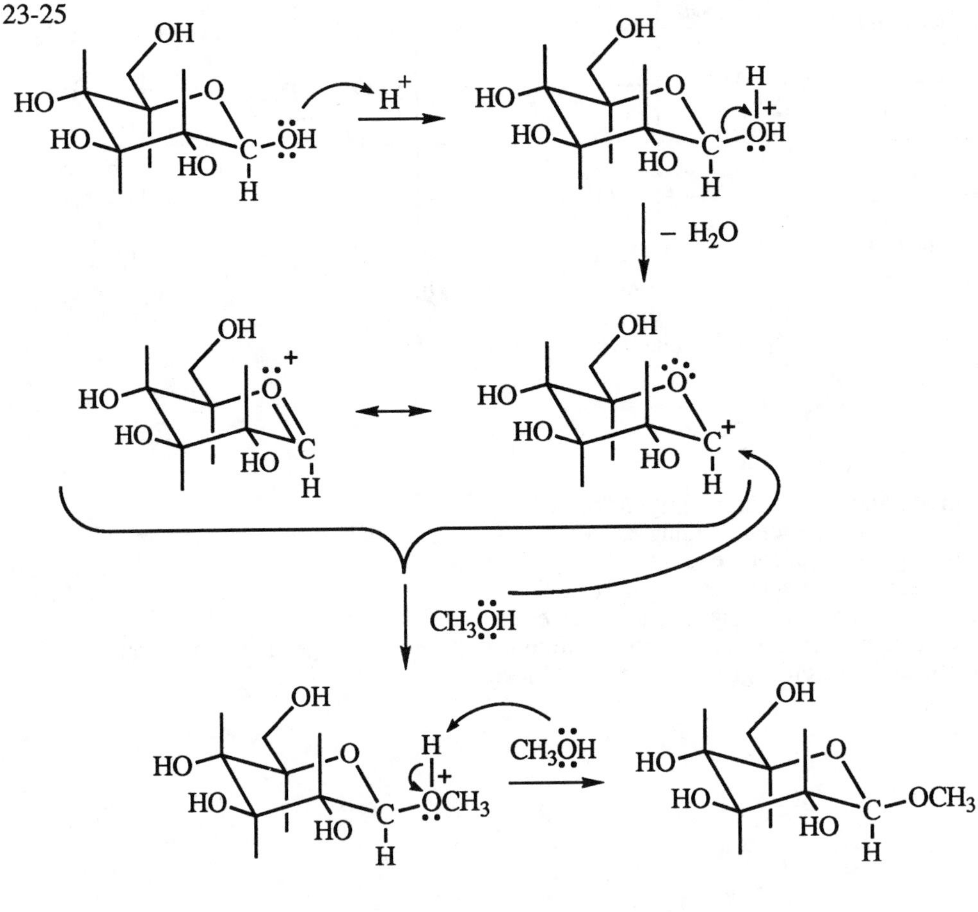

H+
− H2O
CH3OH
CH3OH
OCH3

23-26

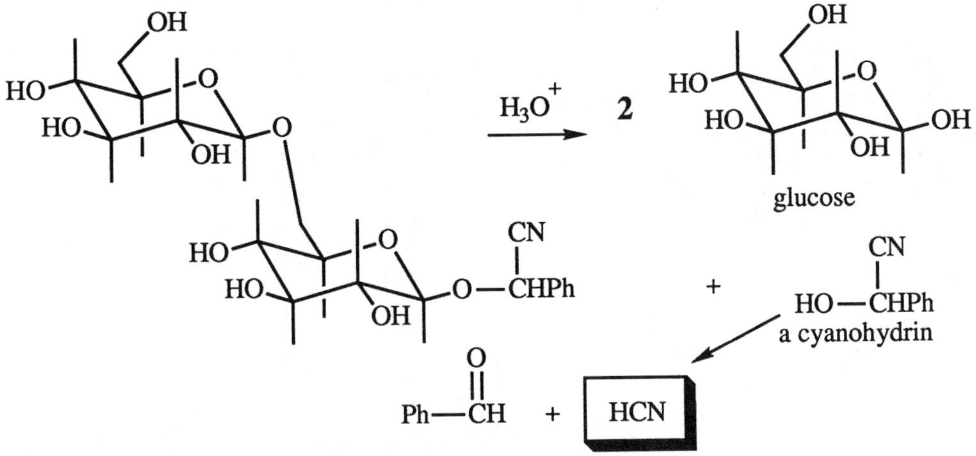
H_3O^+
2
glucose
a cyanohydrin
HCN

HCN is released from amygdalin. HCN is a potent cytotoxic (cell-killing) agent, particularly toxic to nerve cells.

23-27

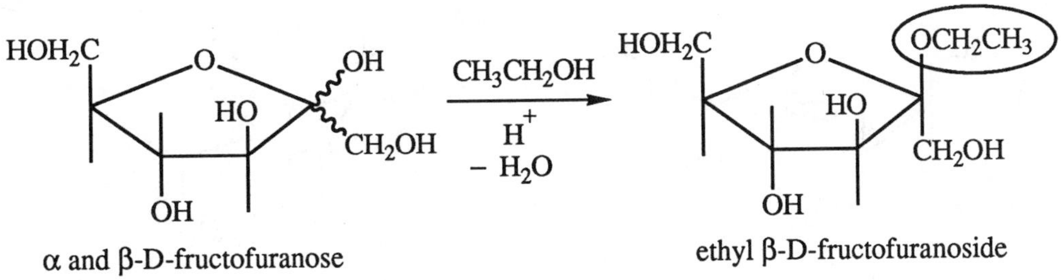
CH_3CH_2OH
H^+
$- H_2O$
α and β-D-fructofuranose
ethyl β-D-fructofuranoside
+

The aglycone in each product is circled.

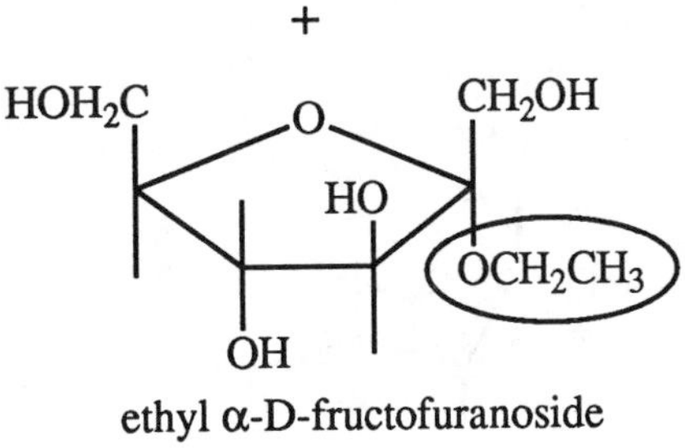
ethyl α-D-fructofuranoside

23-28

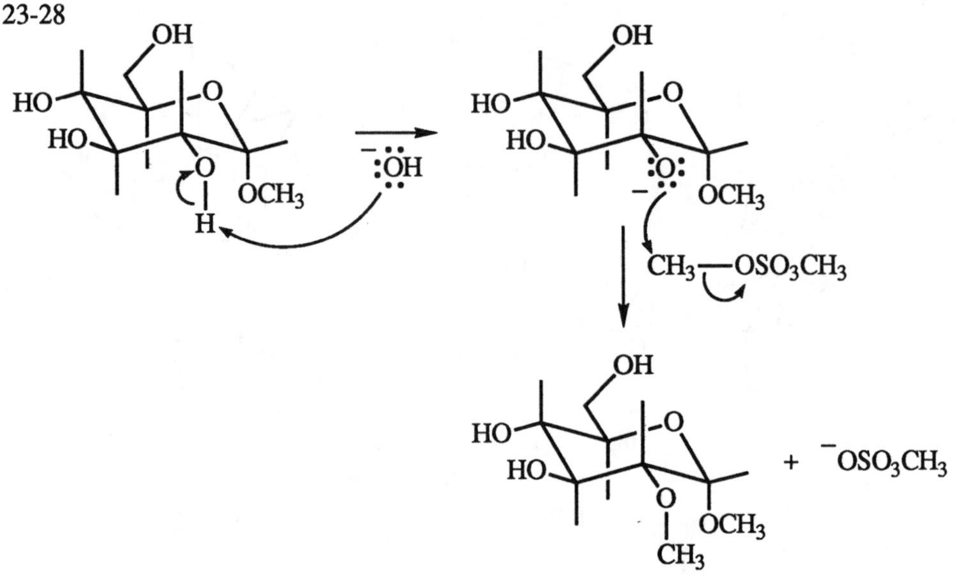

23-29

(a)

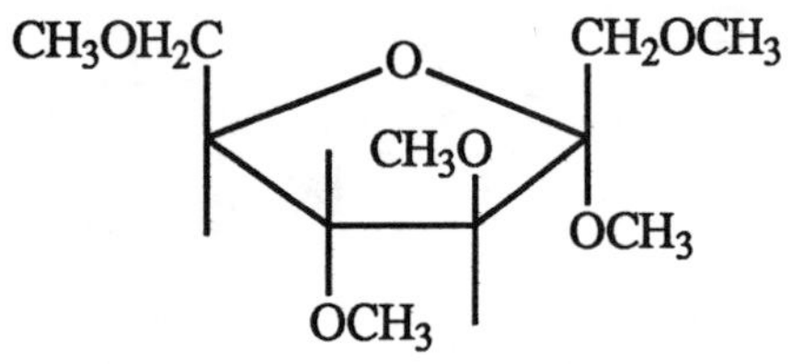

(b)

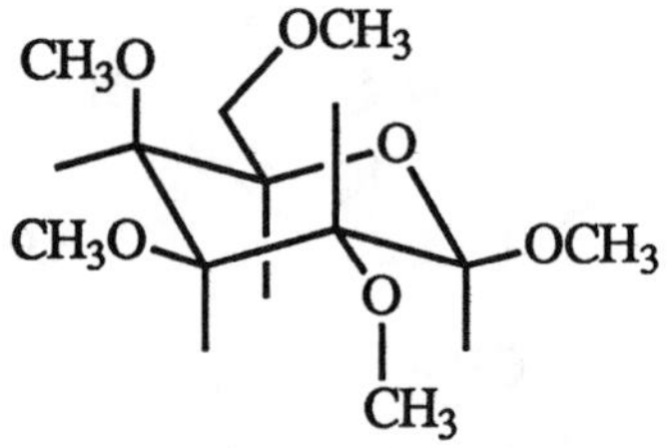

23-30

(a)

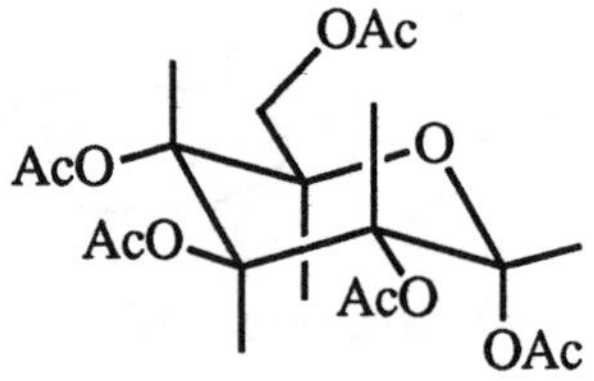

(b)

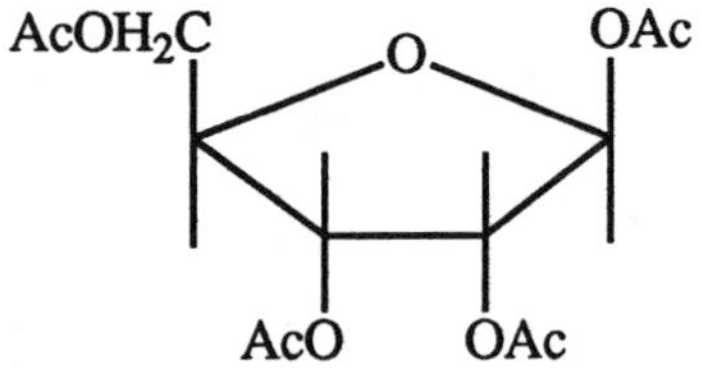

23-31

(a)

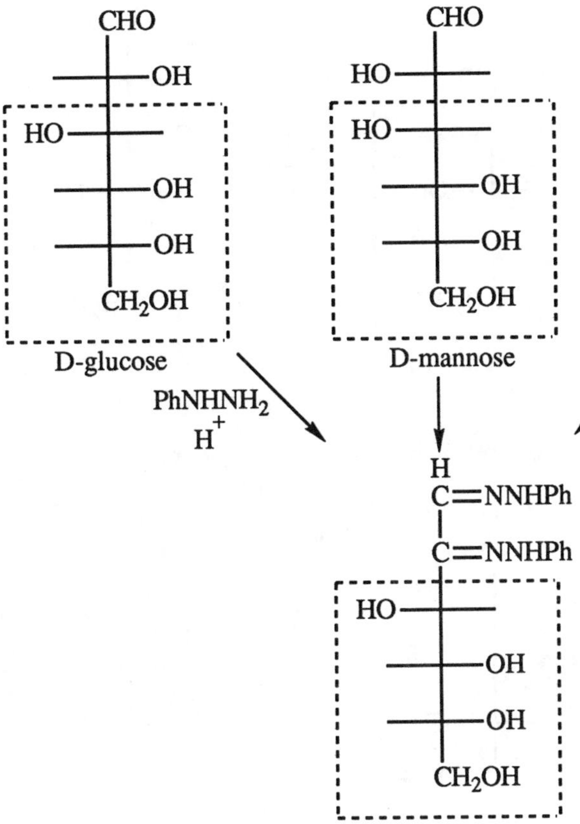

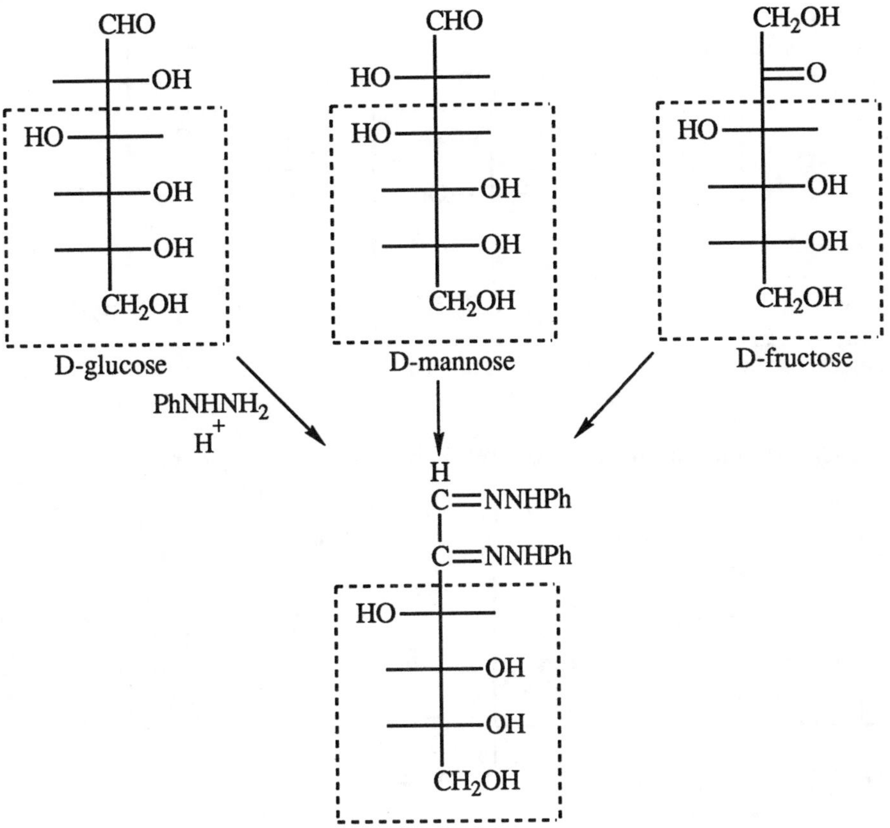

(b)

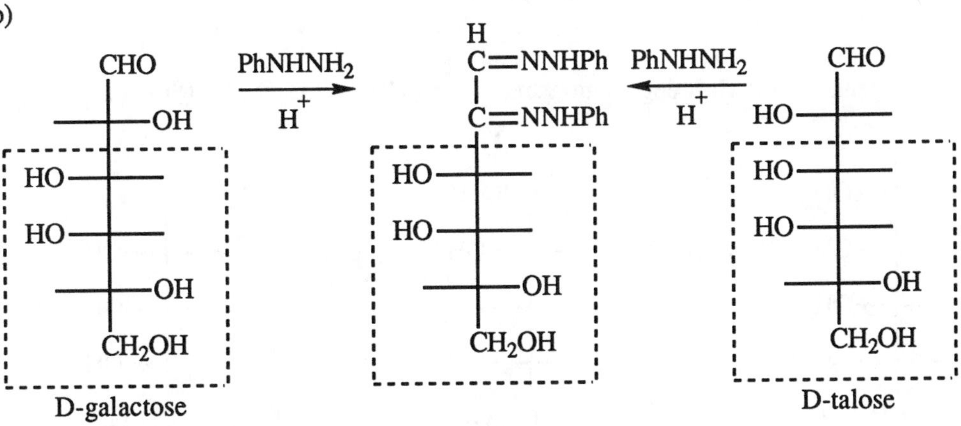

D-Talose must be the C-2 epimer of D-galactose.

23-32 Reagents for the Ruff degradation are: 1. Br_2, H_2O; 2. H_2O_2, $Fe_2(SO_4)_3$.

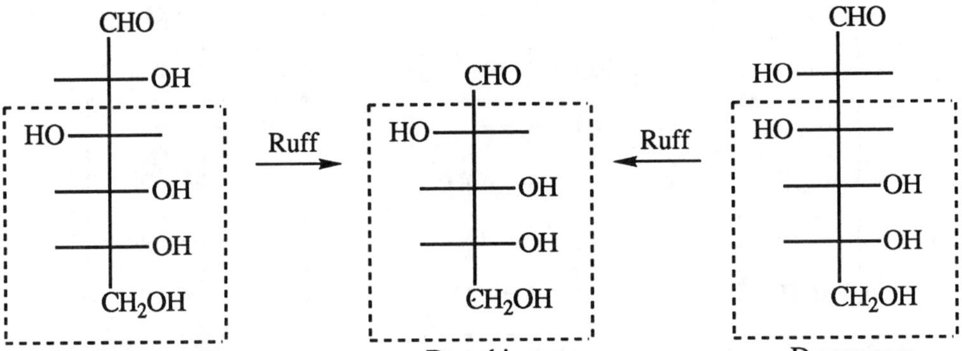

D-glucose D-arabinose D-mannose

23-33 Reagents for the Ruff degradation are: 1. Br_2, H_2O; 2. H_2O_2, $Fe_2(SO_4)_3$.

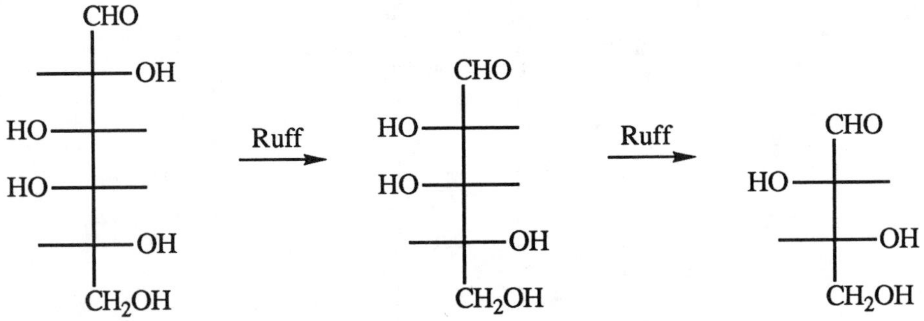

D-galactose D-lyxose D-threose

23-34 Reagents for the Ruff degradation are: 1. Br_2, H_2O; 2. H_2O_2, $Fe_2(SO_4)_3$.

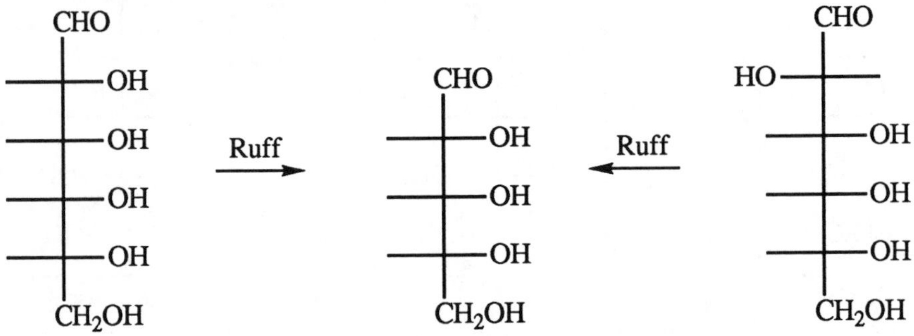

D-allose D-altrose

23-35

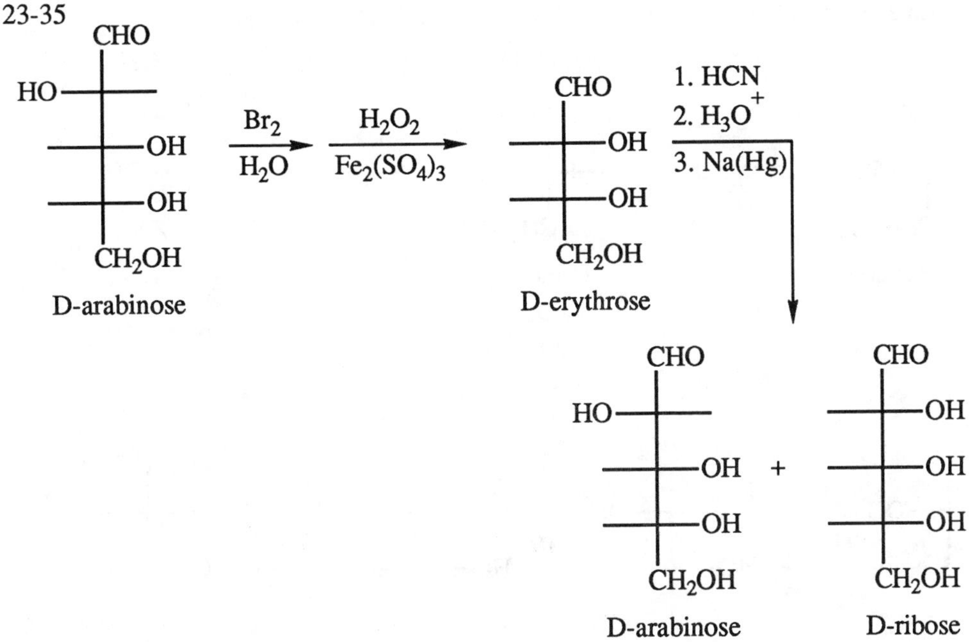
CHO
HO
OH
OH
CH₂OH
D-arabinose
Br₂
H₂O
H₂O₂
Fe₂(SO₄)₃
CHO
OH
OH
CH₂OH
D-erythrose
1. HCN
2. H₃O⁺
3. Na(Hg)
CHO
HO
OH
OH
CH₂OH
D-arabinose
+
CHO
OH
OH
OH
CH₂OH
D-ribose

23-36

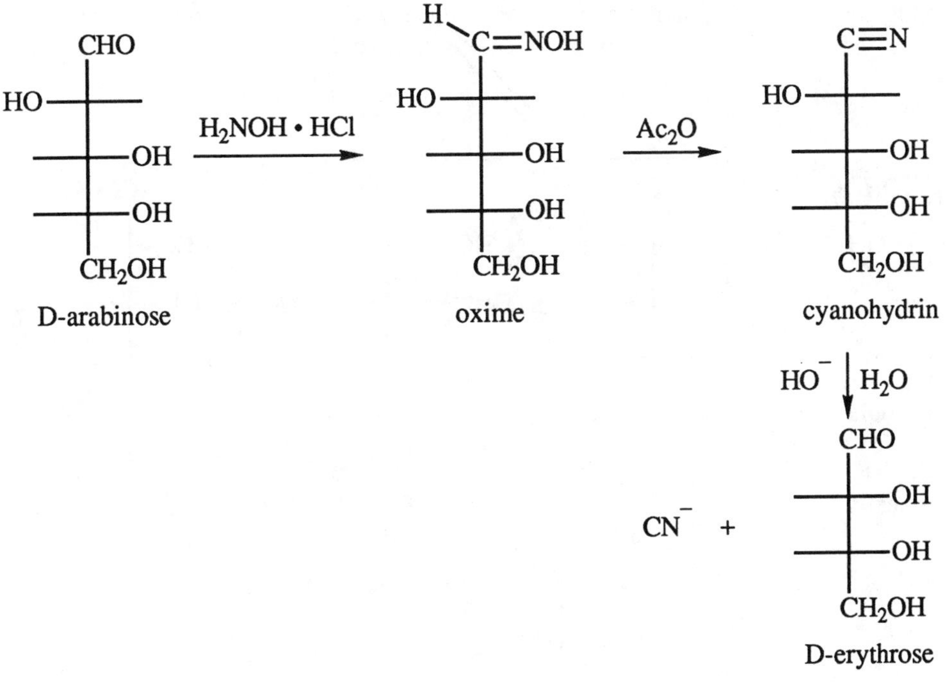
CHO
HO
OH
OH
CH₂OH
D-arabinose
H₂NOH • HCl
H
C=NOH
HO
OH
OH
CH₂OH
oxime
Ac₂O
C≡N
HO
OH
OH
CH₂OH
cyanohydrin
HO⁻
H₂O
CN⁻
+
CHO
OH
OH
CH₂OH
D-erythrose

23-37 Solve this problem by working backward from (+)-glyceraldehyde.

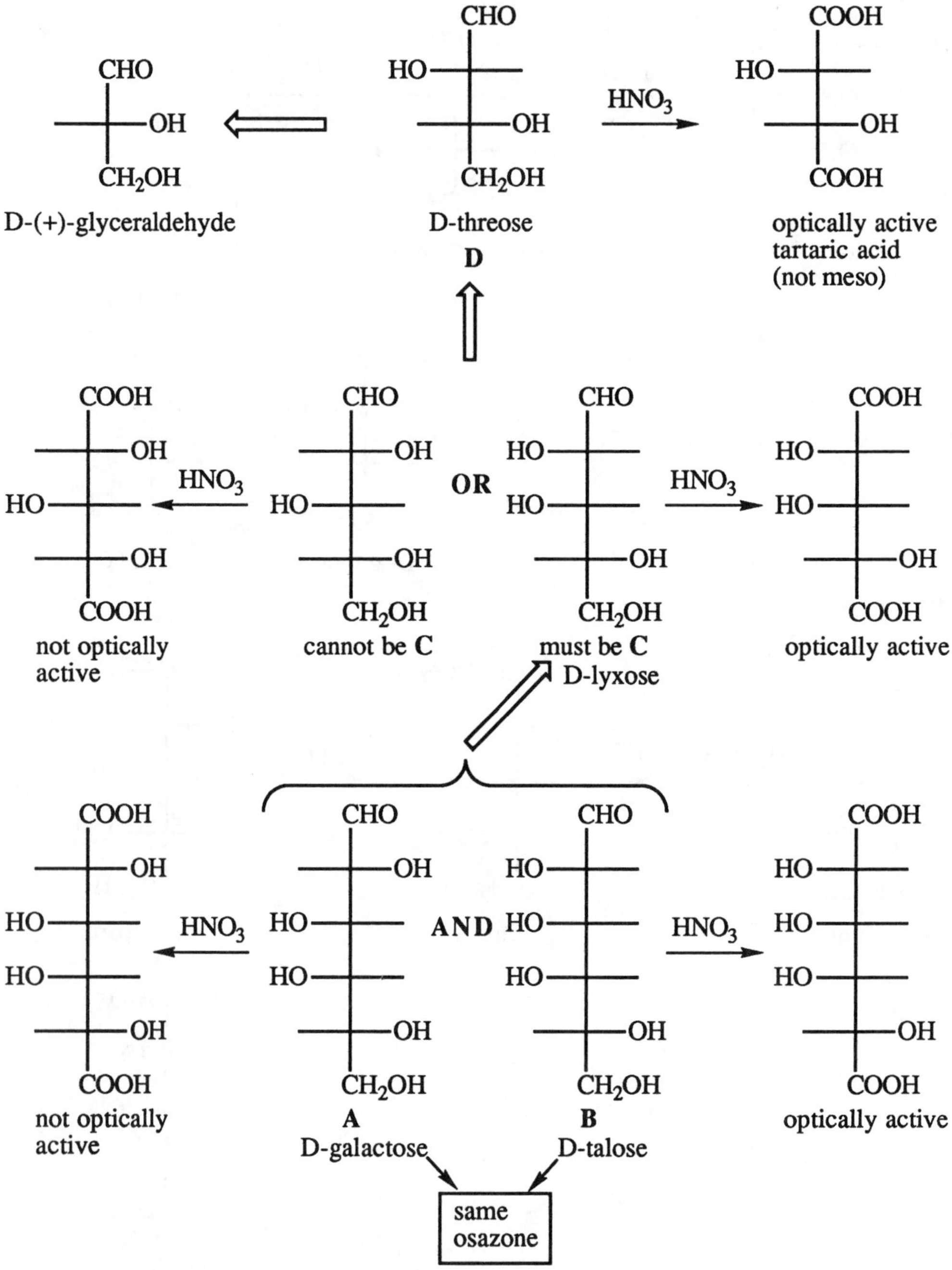

23-38 Solve this problem by working backward from (+)-glyceraldehyde.

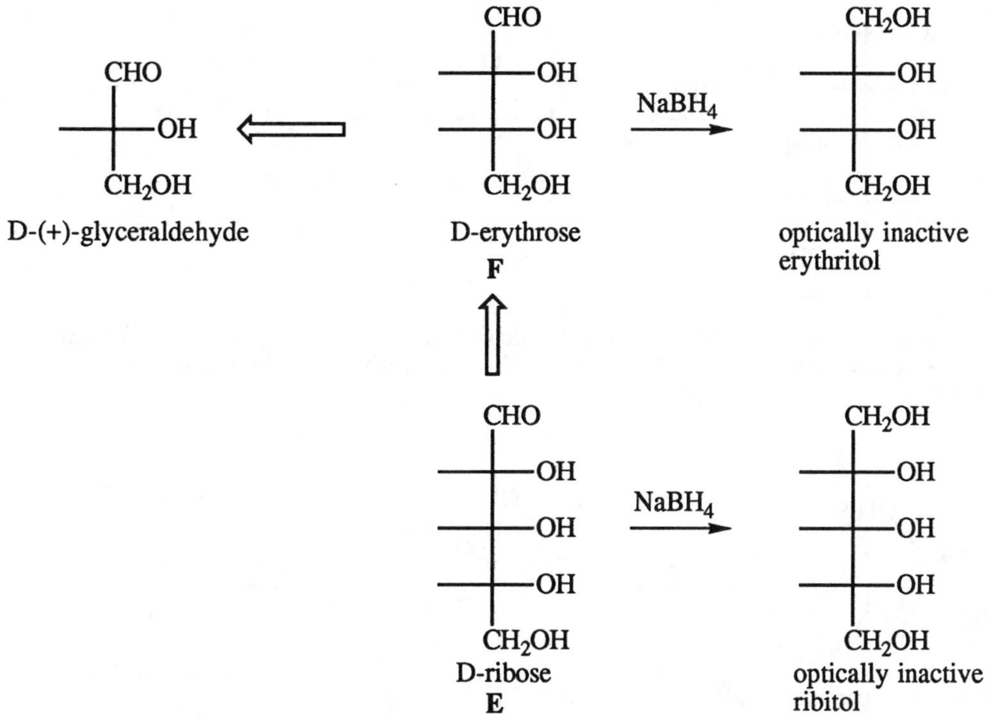

23-39

(a)

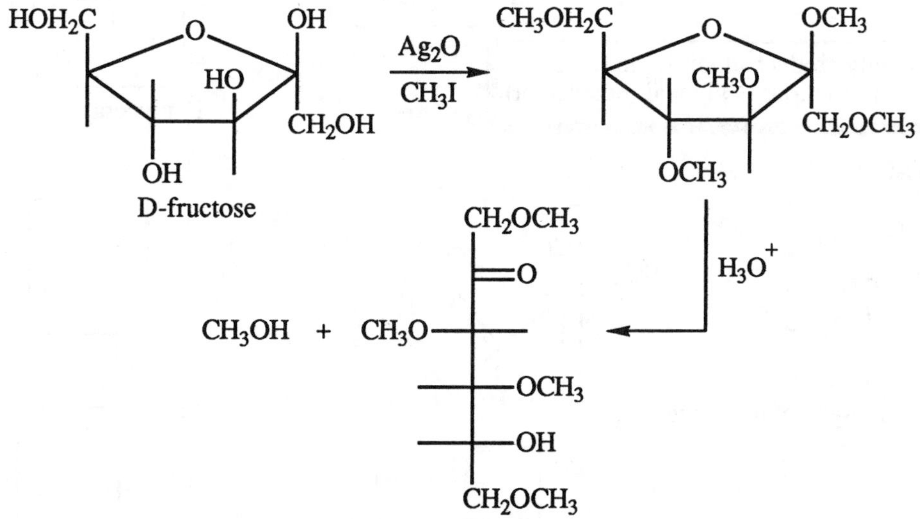

(b)

(c) Determining that the open chain form of fructose is a ketone at C-2 with a free OH at C-5 shows that fructose exists as a furanose hemiacetal.

23-40

(a)

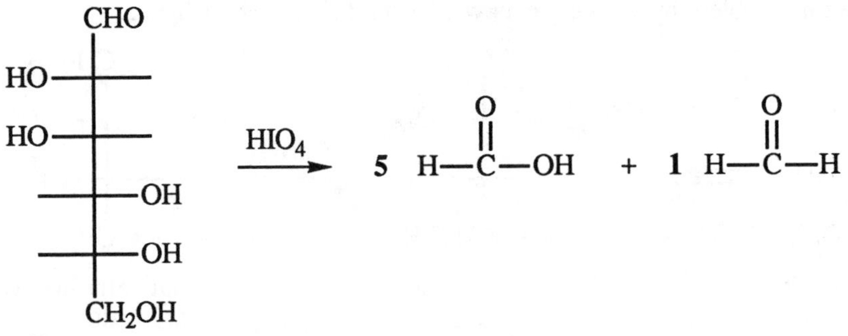

D-mannose

(b) The aldehyde carbonyl will give a molecule of formic acid upon periodate oxidation. The ketone carbonyl will produce a molecule of CO_2 which will bubble out of solution.

23-41

(a)

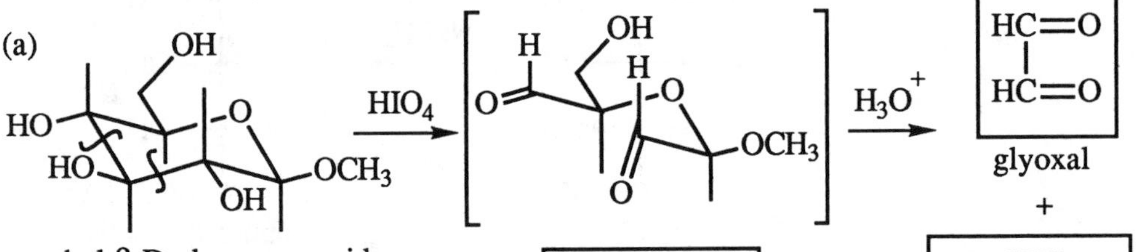

methyl β-D-glucopyranoside

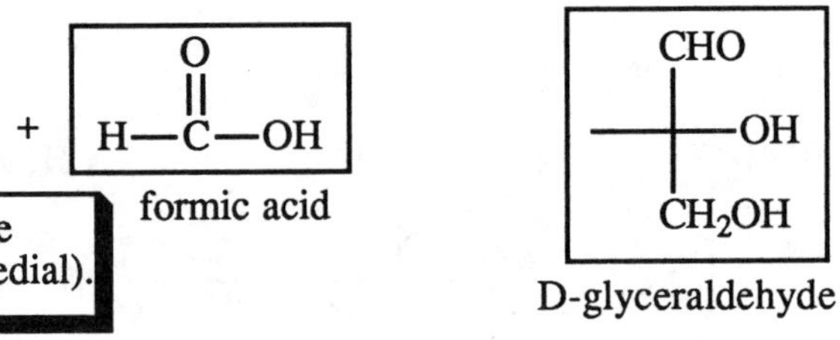

> The only product which is the same in both reactions is glyoxal (ethanedial).

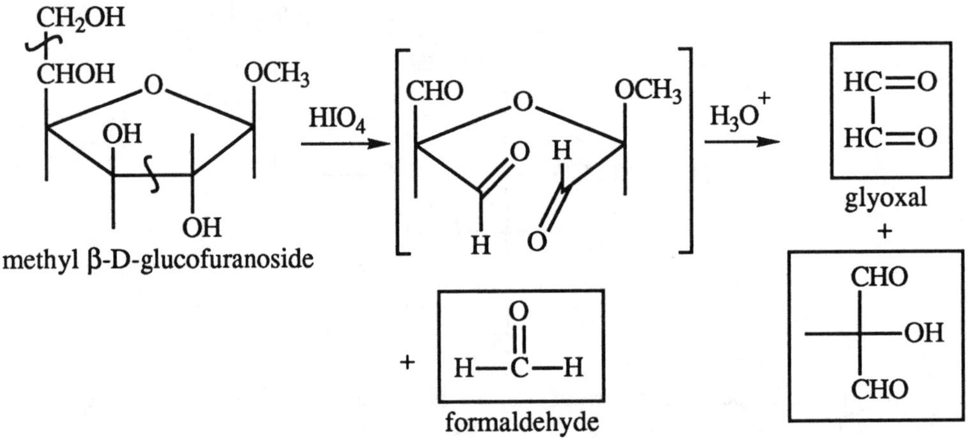

23-41 continued

(b)

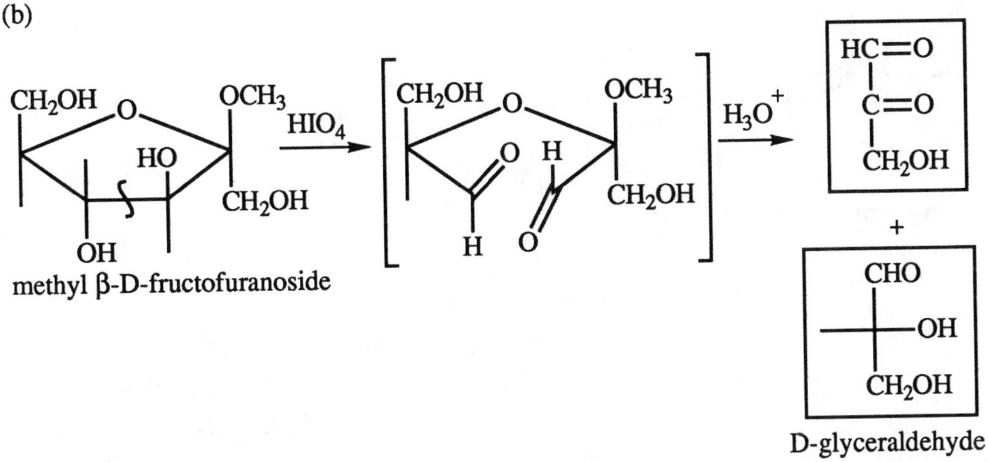

Had the fructose acetal been in a six-membered ring, one molecule of formic acid would have been produced, as in part (a). Isolation of two three-carbon molecules proves the presence of a furanoside.

23-42

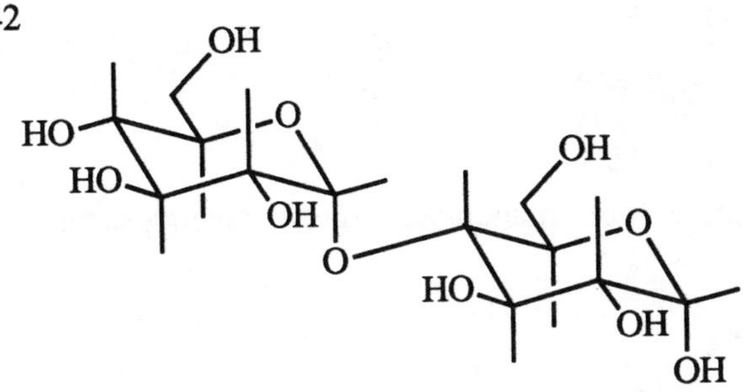

α anomer of maltose

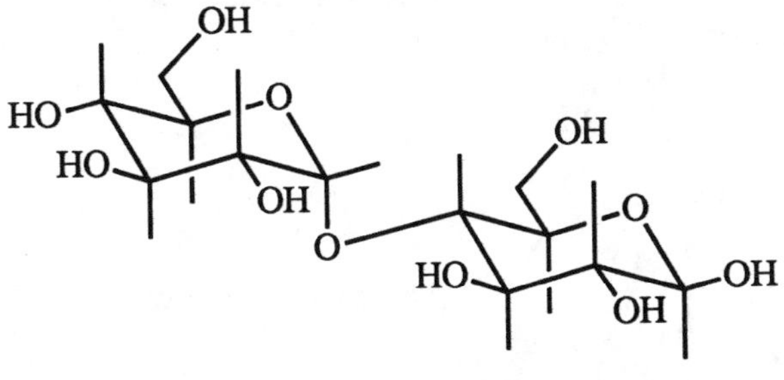

β anomer of maltose

23-43

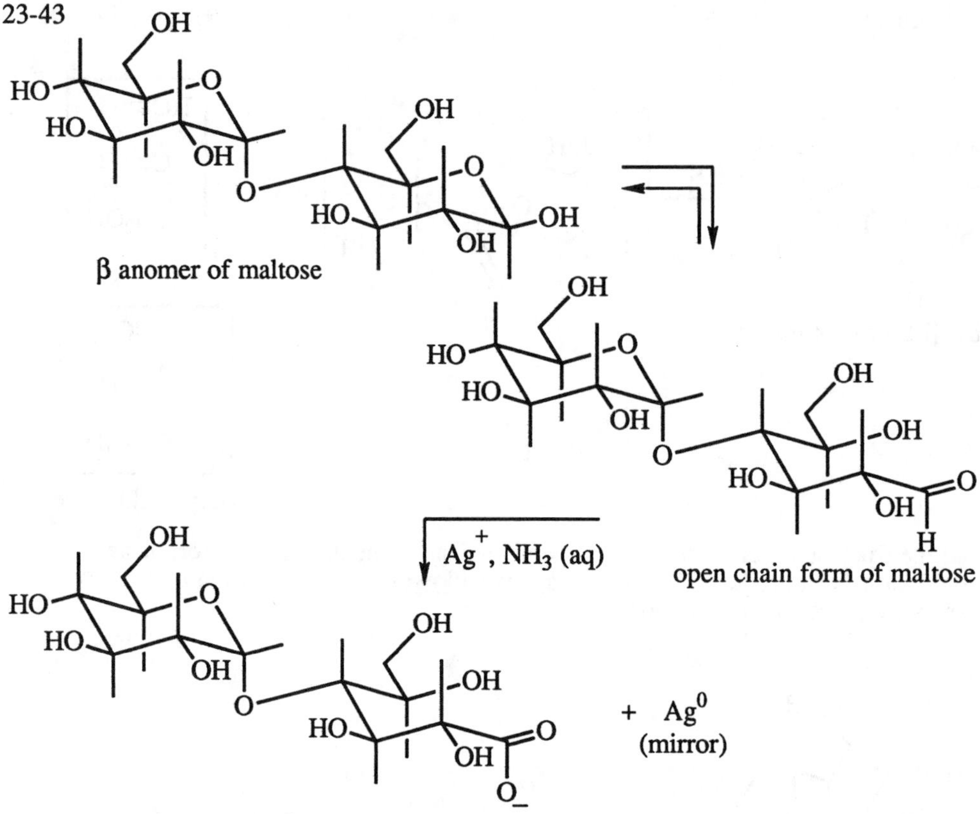

β anomer of maltose
open chain form of maltose
Ag^+, NH_3 (aq)
+ Ag^0 (mirror)

23-44 Lactose is a hemiacetal. Therefore, it can mutarotate and is a reducing sugar.

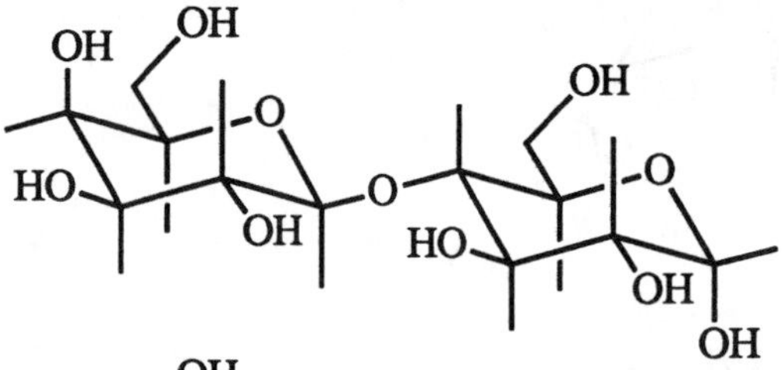

α anomer of lactose

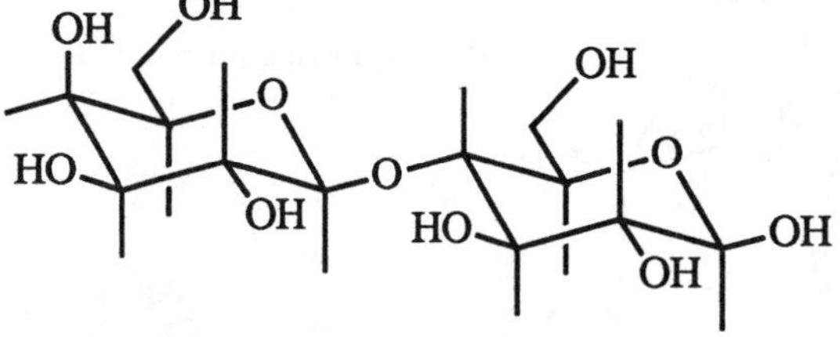

β anomer of lactose

23-45 Gentiobiose is a hemiacetal; in water, the hemiacetal is in equilibrium with the open-chain form and can react as an aldehyde. Gentiobiose can mutarotate and is a reducing sugar.

23-46 Trehalose must be two glucose molecules connected by an α-1,1'-glycoside.

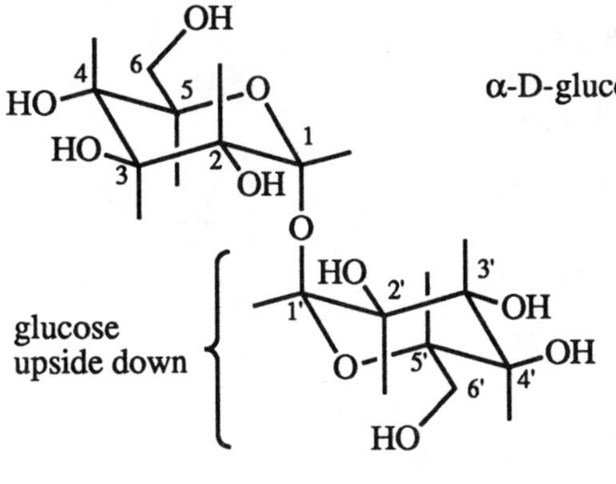

OH
HO
HO
OH
O
HO
OH
OH
HO
glucose upside down

α-D-glucopyranosyl-α-D-glucopyranoside

23-47

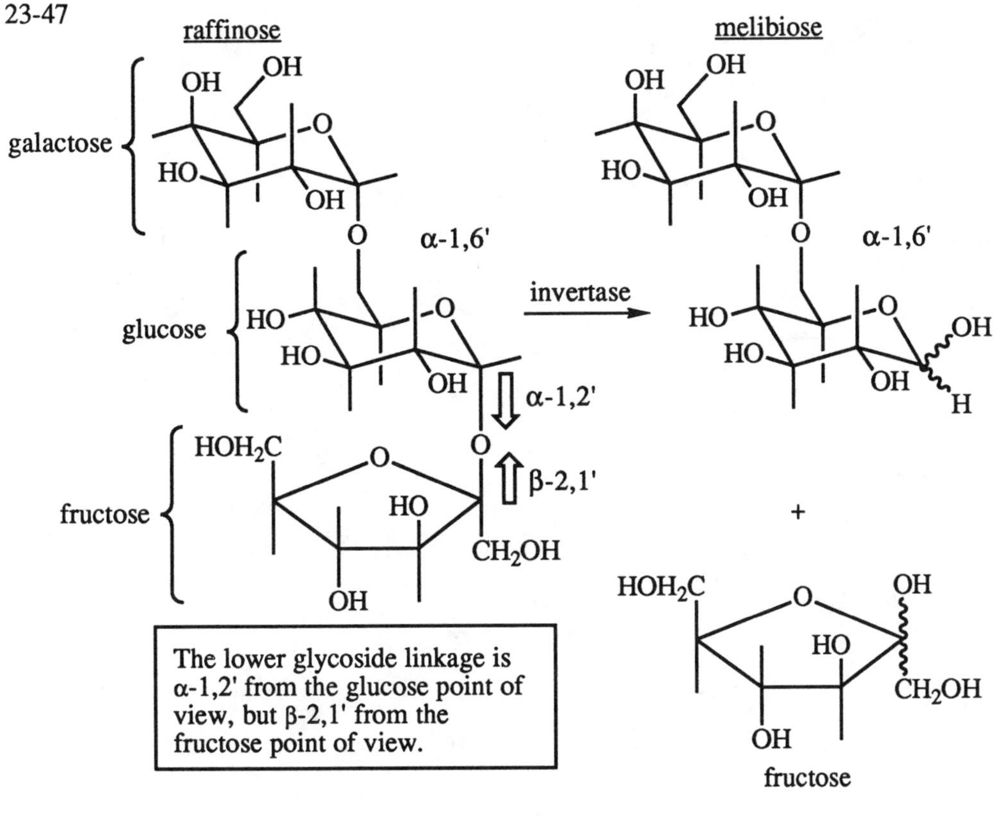

raffinose
galactose
glucose
fructose
α-1,6'
α-1,2'
β-2,1'
HOH2C
CH2OH
invertase
melibiose
α-1,6'
+
fructose

The lower glycoside linkage is α-1,2' from the glucose point of view, but β-2,1' from the fructose point of view.

melibiose = 6-O-(α-galactopyranosyl)-D-glucopyranose

23-48

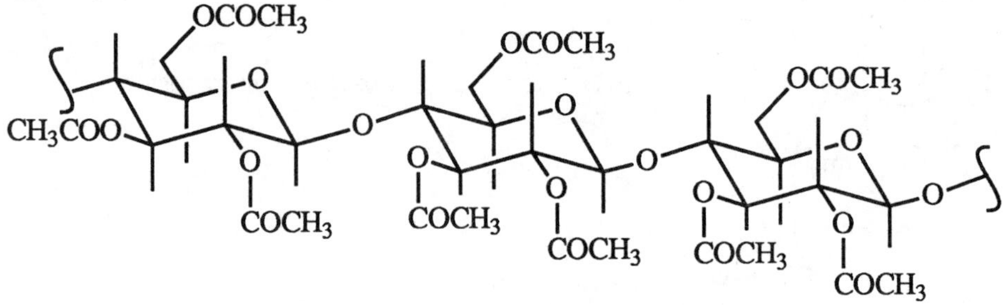

23-49

cytosine:

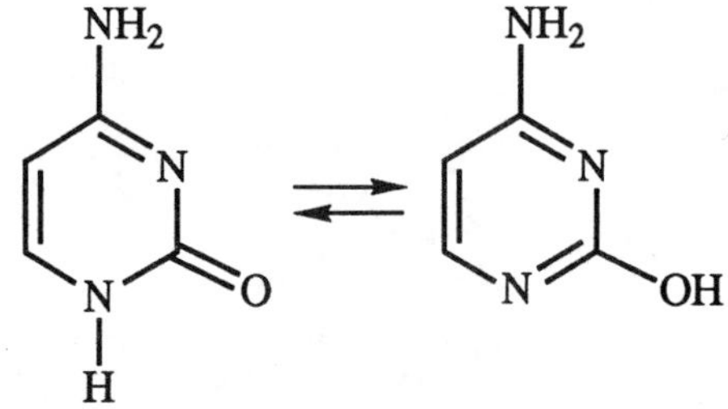

uracil:

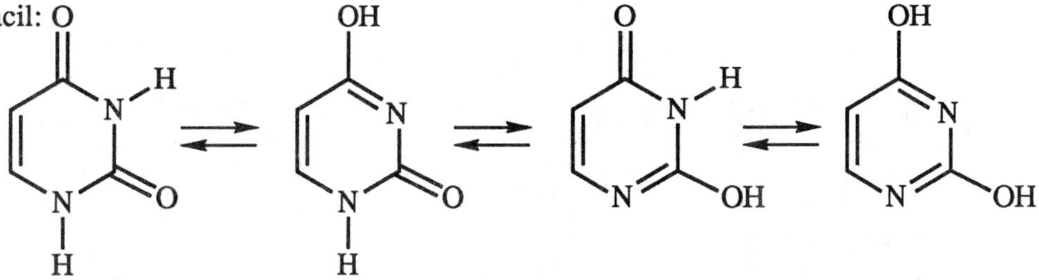

guanine:

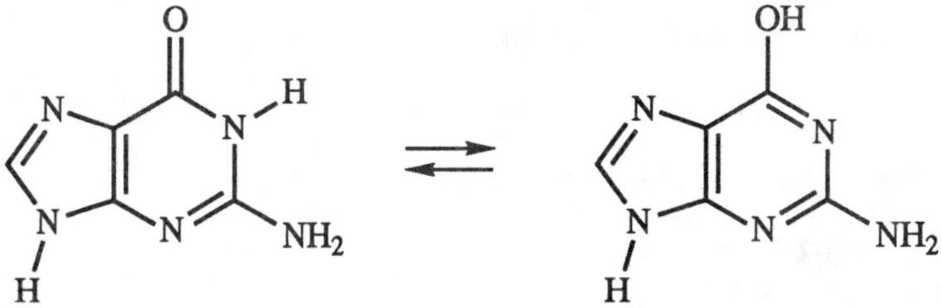

23-50 Cytidine will be used as an example in this problem, but the principles apply to all of the nucleosides.

In order for hydroxide to cleave the riboside, presumably by an S_N2 mechanism, either the oxygen or the nitrogen would have to be the anionic leaving group. These strong bases are *never* leaving groups in S_N2.

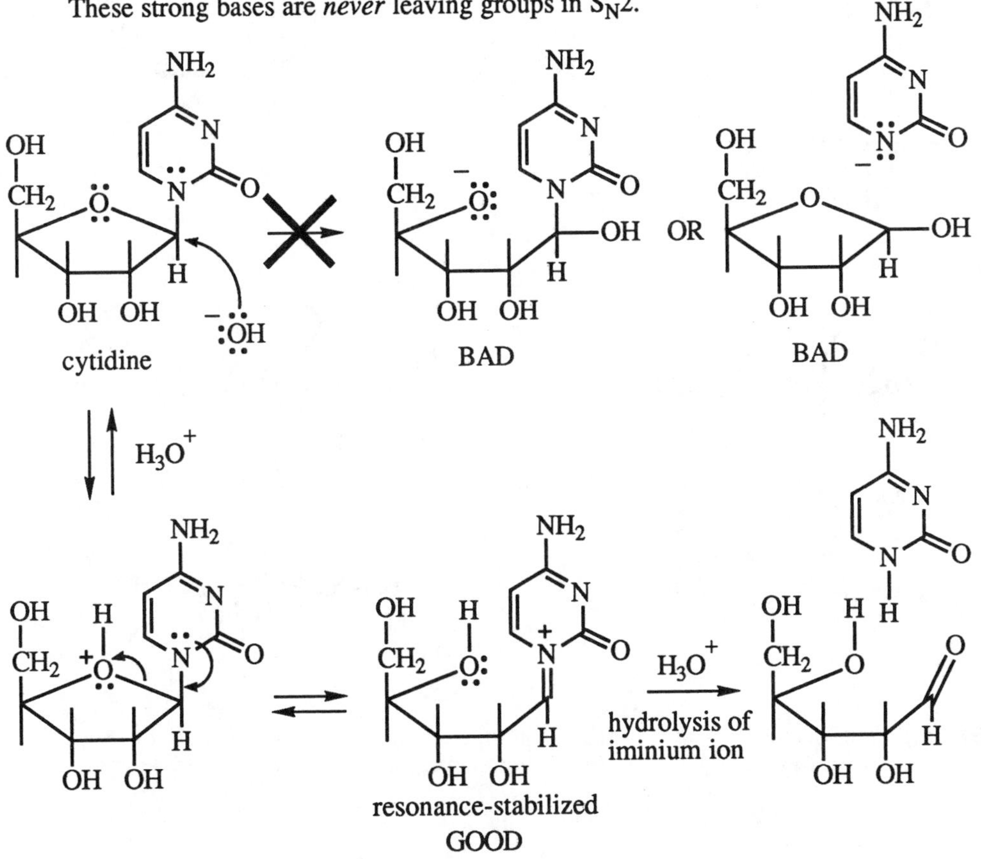

Protonation of the riboside oxygen allows the leaving group to be neutral and for the intermediate to be resonance-stabilized. Aqueous acid quickly cleaves ribosides to ribose plus the purine or pyrimidine base.

23-51 Refer to the Glossary and the text for definitions and examples.

23-52

(a)

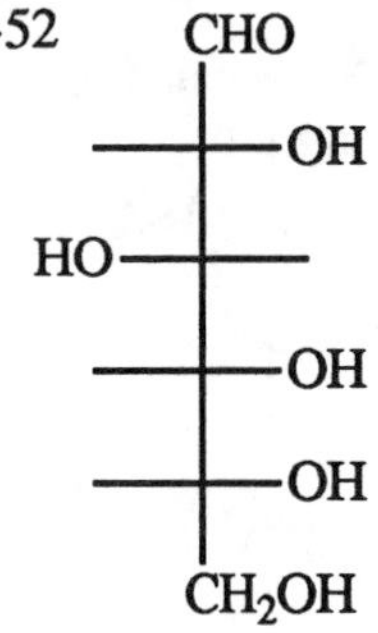

CHO
OH
HO
OH
OH
CH2OH

(b)

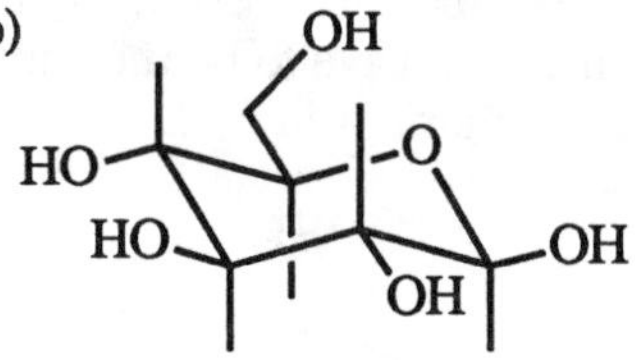

OH
HO
O
HO
OH
OH

(c)

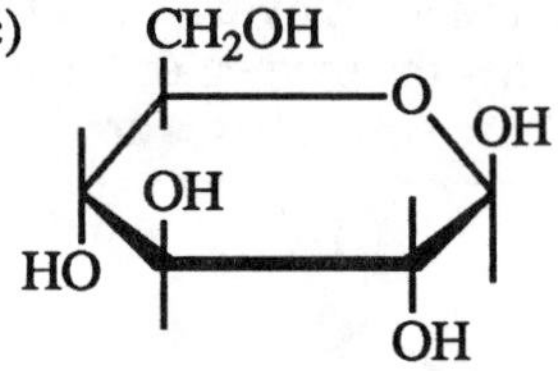

CH2OH
O
OH
OH
HO
OH

23-53

(a)

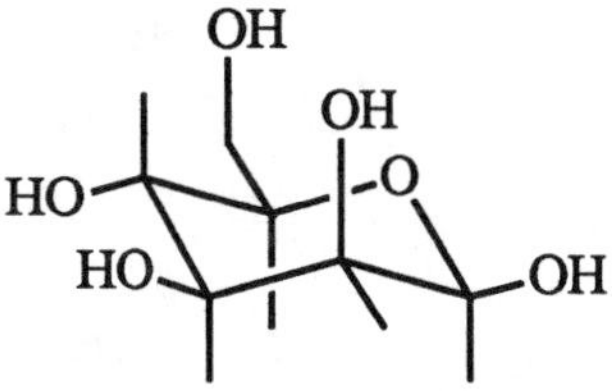

OH
OH
HO
O
HO
OH

(b)

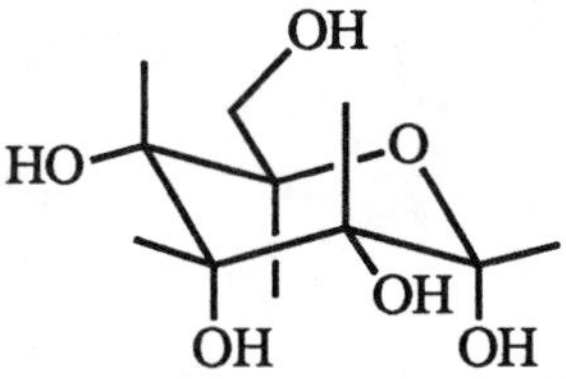

OH
HO
O
OH
OH
OH

(c)

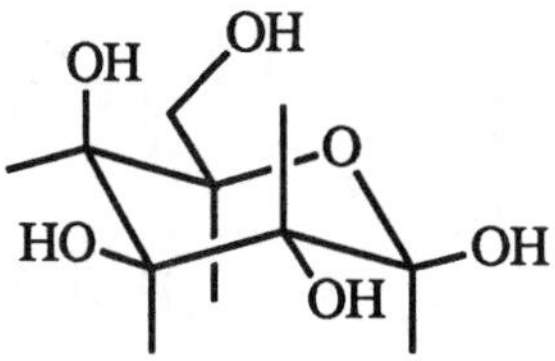

OH
OH
O
HO
OH
OH

(d)

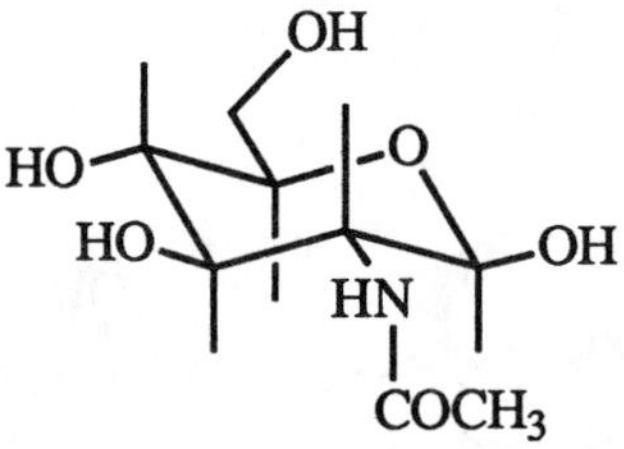

OH
HO
O
HO
OH
HN
COCH3

23-54 (a) D-aldohexose (b) D-aldopentose (c) L-ketohexose
(d) L-aldohexose (e) D-ketopentose (f) L-aldotetrose

23-55

(a)

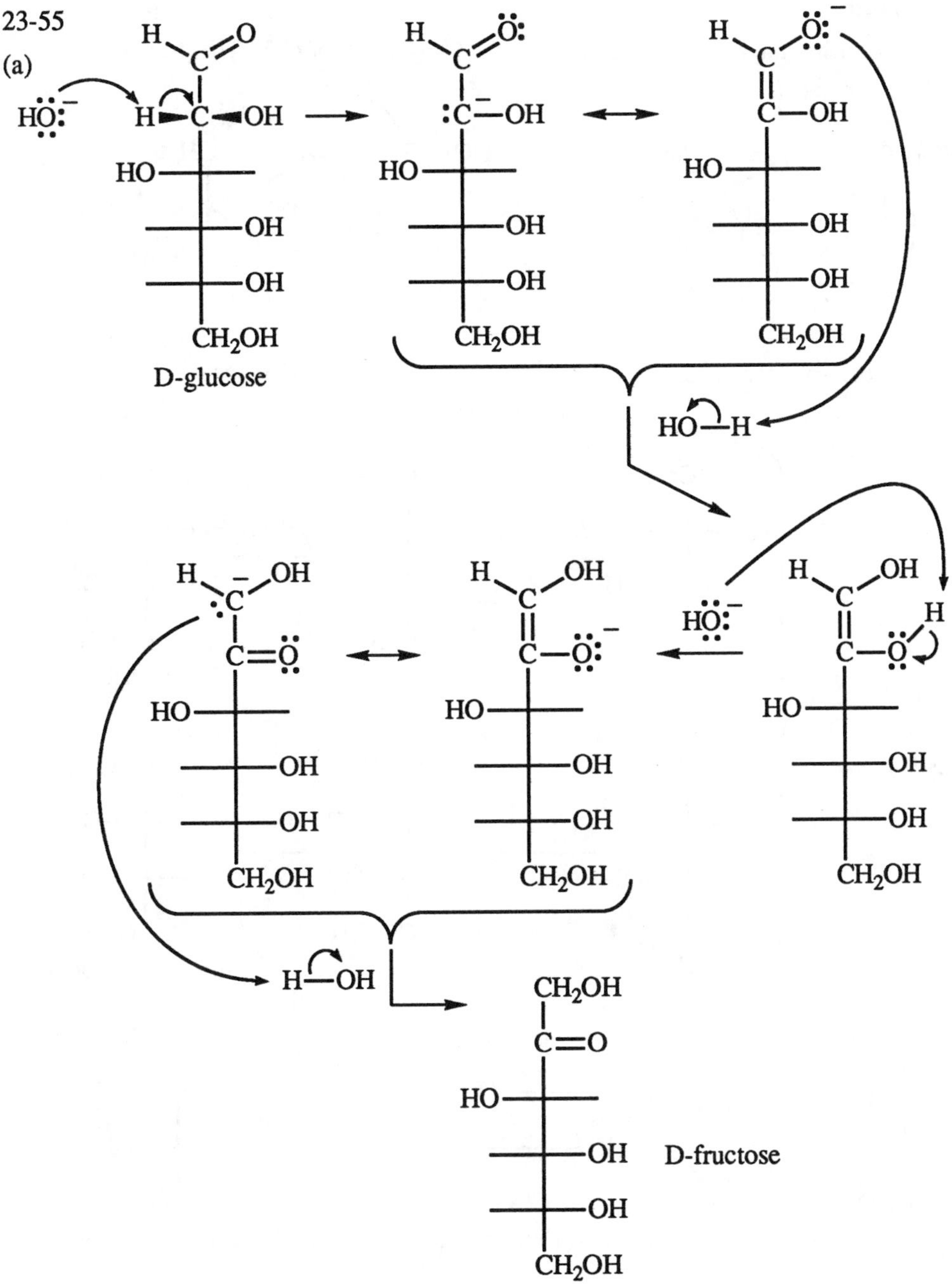

23-55 continued

(b) The α isomer has the anomeric OH *trans* to the CH_2OH off of C-5. The β-anomer has these groups *cis*.

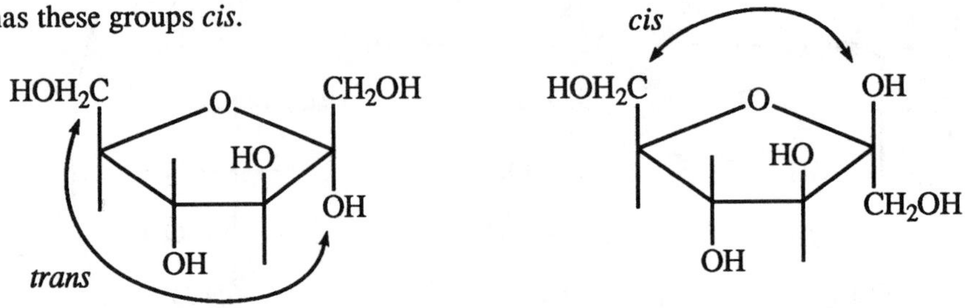

cis
HOH2C
O
CH2OH
HO
OH
OH
trans
OH
HOH2C
O
HO
CH2OH
OH

α-D-fructofuranose β-D-fructofuranose

23-56

(a)

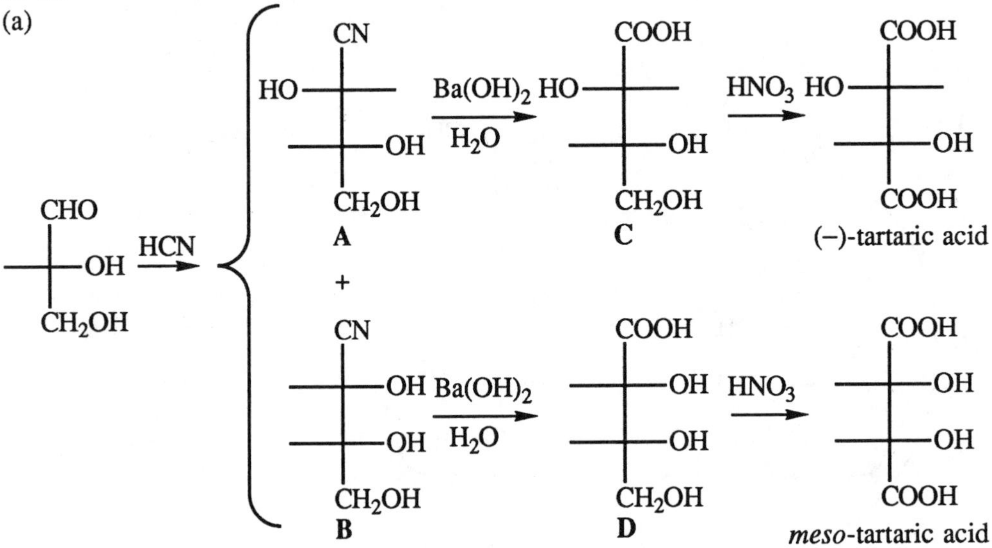

CHO
OH
CH2OH
HCN
CN
HO
OH
CH2OH
A
Ba(OH)2
H2O
COOH
HO
OH
CH2OH
C
HNO3
COOH
HO
OH
COOH
(−)-tartaric acid
+
CN
OH
OH
CH2OH
B
Ba(OH)2
H2O
COOH
OH
OH
CH2OH
D
HNO3
COOH
OH
OH
COOH
meso-tartaric acid

(b)

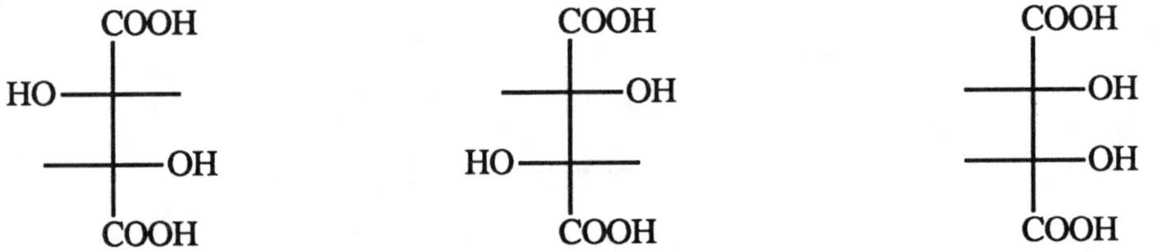

COOH
HO
OH
COOH
COOH
OH
HO
COOH
COOH
OH
OH
COOH

(*S*,*S*)-(−)-tartaric acid (*R*,*R*)-(+)-tartaric acid (*R*,*S*)-*meso*-tartaric acid

23-57 (a) D-(−)-ribose
(b) D-(+)-altrose
(c) L-(+)-erythrose
(d) L-(−)-galactose
(e) L-(+)-idose

23-58

(a)

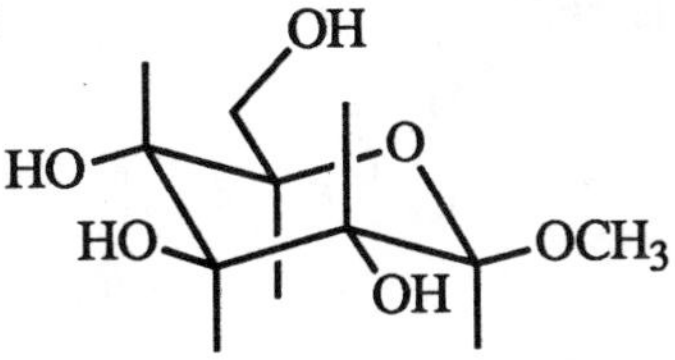

(b)

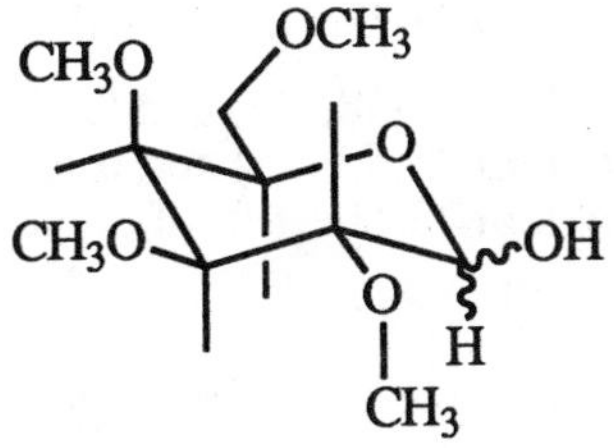

(c)

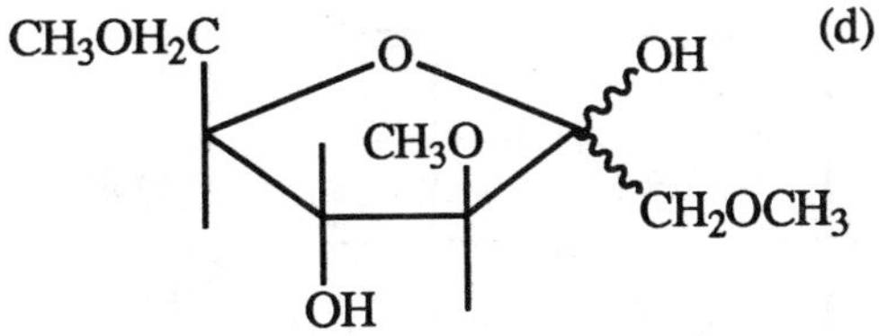

(d)

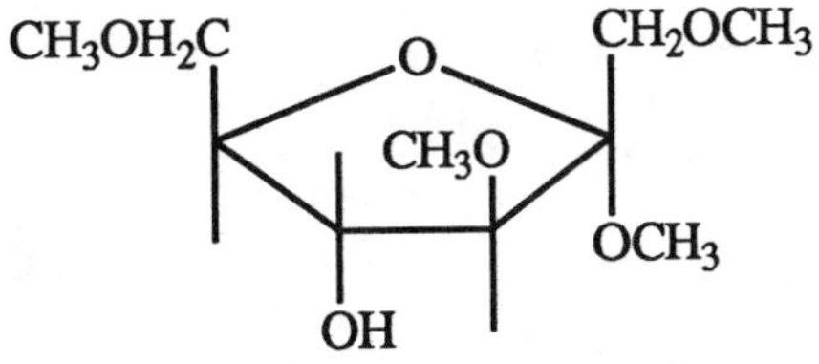

23-59

(a)

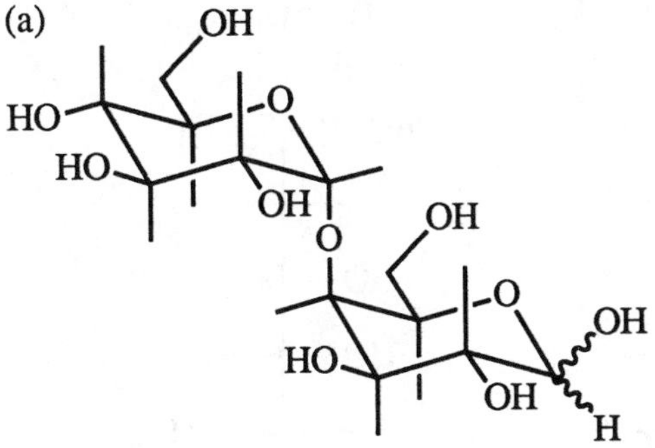

(b)

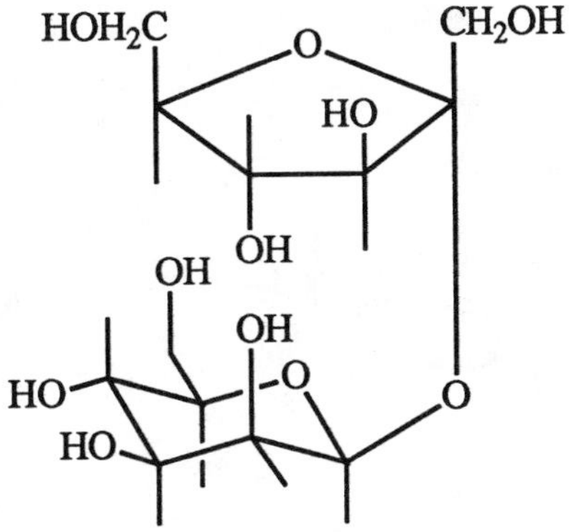

(c)

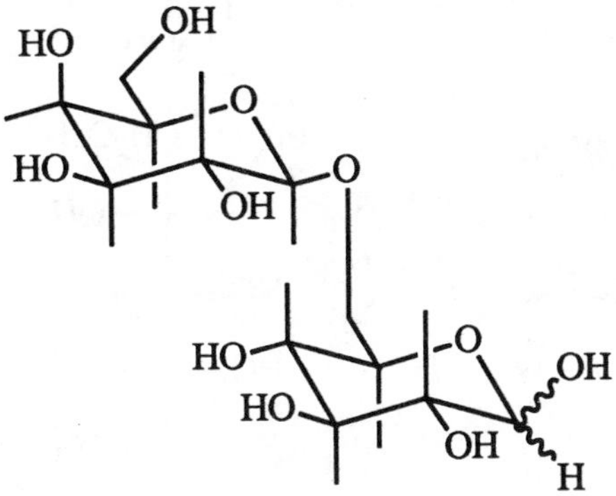

23-60

(a) methyl β-D-fructofuranoside
(b) 3,6-di-O-methyl-β-D-mannopyranose
(c) 4-O-(α-D-fructofuranosyl)-β-D-galactopyranose
(d) β-D-*N*-acetylgalactopyranosamine, or 2-acetamido-2-deoxy-β-D-galactopyranose

23-61 These are reducing sugars and would undergo mutarotation:

—in problem 23-58: (b) and (c);
—in problem 23-59: (a) and (c);
—in problem 23-60: (b), (c), and (d)

23-62

(a)

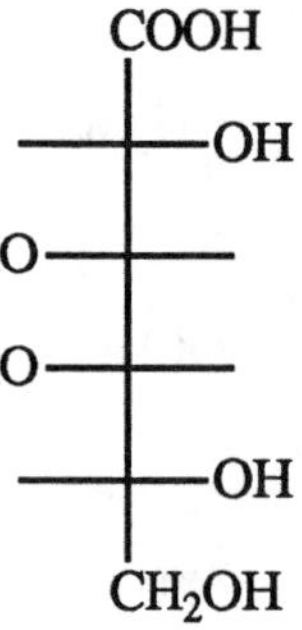

(b)

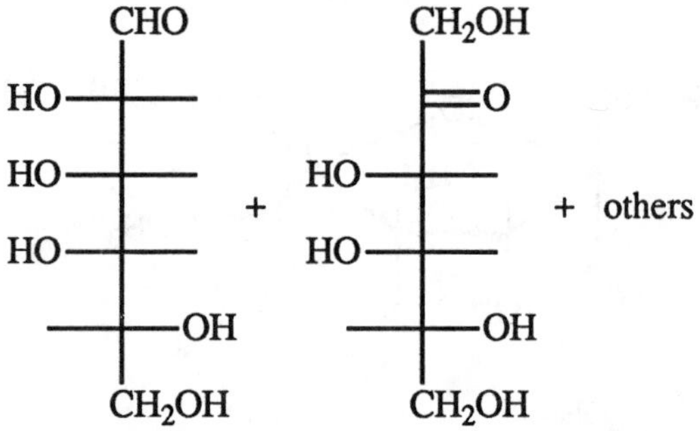

(c)

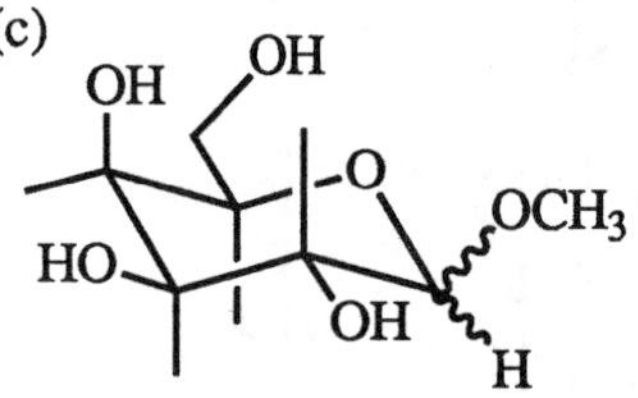

(d)

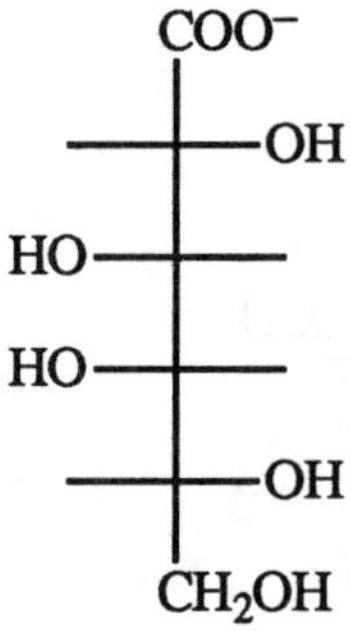

(e)

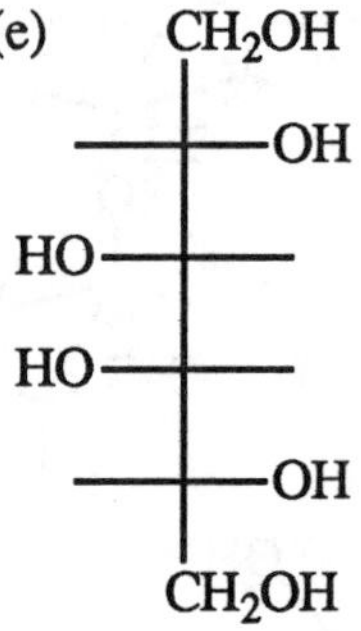

(f)

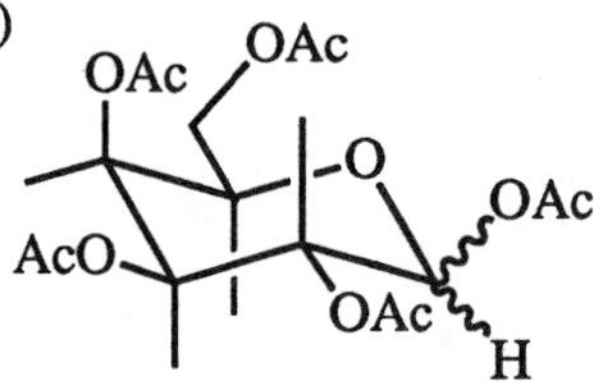

(g)

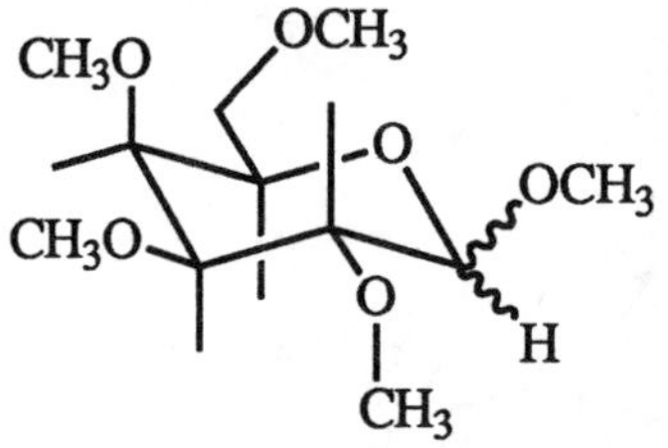

(h)

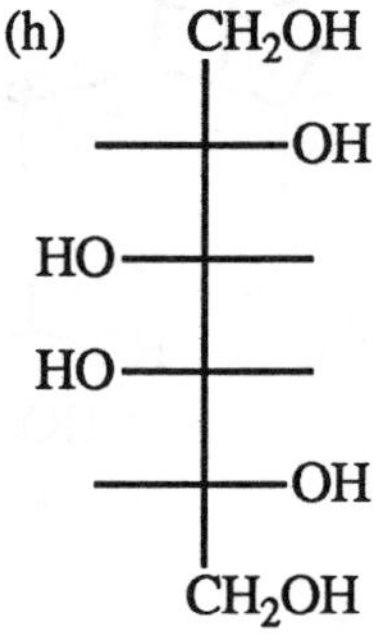

23-62 continued

(i)

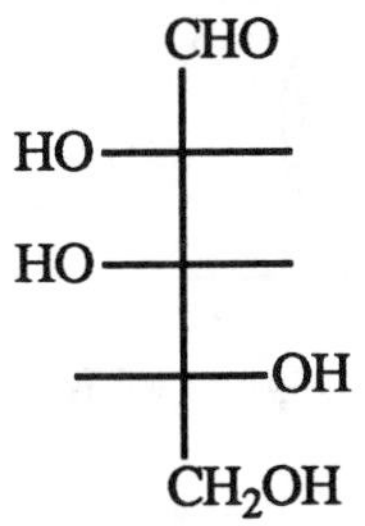

CHO
HO
HO
OH
CH2OH

(j)

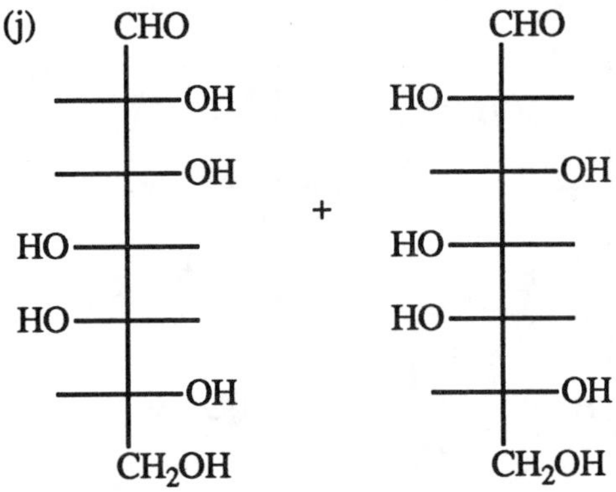

CHO
OH
OH
HO
HO
OH
CH2OH
+
CHO
HO
OH
HO
HO
OH
CH2OH

(k)

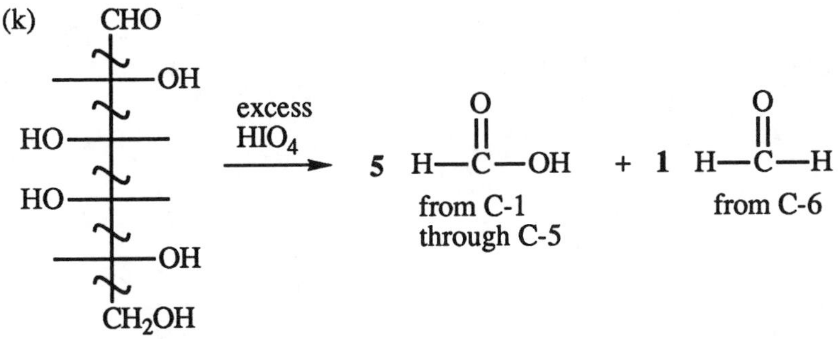

CHO
OH
HO
HO
OH
CH2OH
excess
HIO4
5 H—C—OH
O
+ 1 H—C—H
O
from C-1
through C-5
from C-6

23-63 Use the milder reagent, CH_3I/Ag_2O, when the sugar is in the hemiacetal form; the mild conditions prevent isomerization. When the carbohydrate is present as an acetal (a glycoside), use the more basic reagent, $NaOH/(CH_3)_2SO_4$; an acetal is stable to basic conditions.

(a)

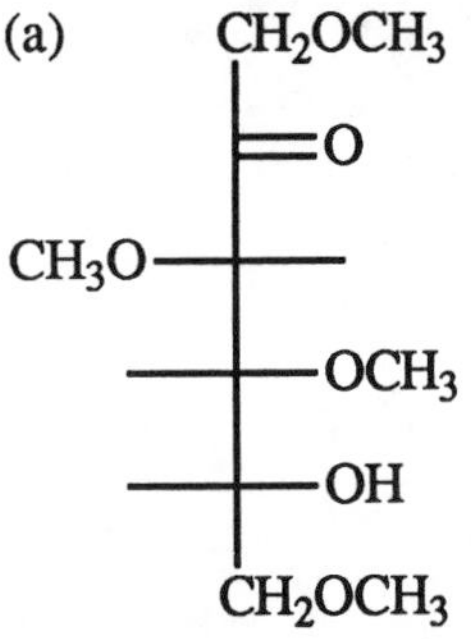

CH2OCH3
=O
CH3O
OCH3
OH
CH2OCH3

using CH_3I, Ag_2O

(b)

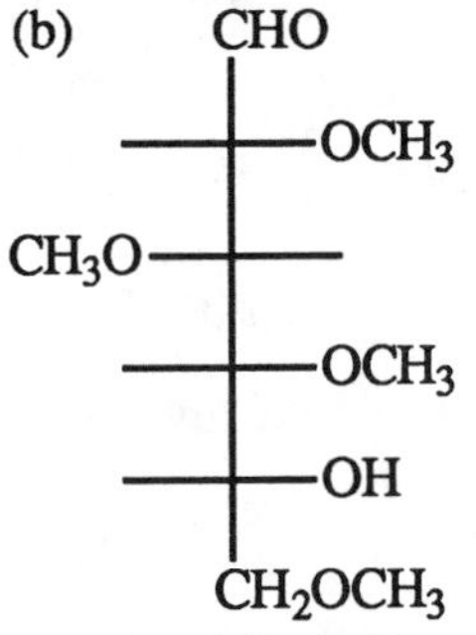

CHO
OCH3
CH3O
OCH3
OH
CH2OCH3

using $(CH_3)_2SO_4$, NaOH

(c)

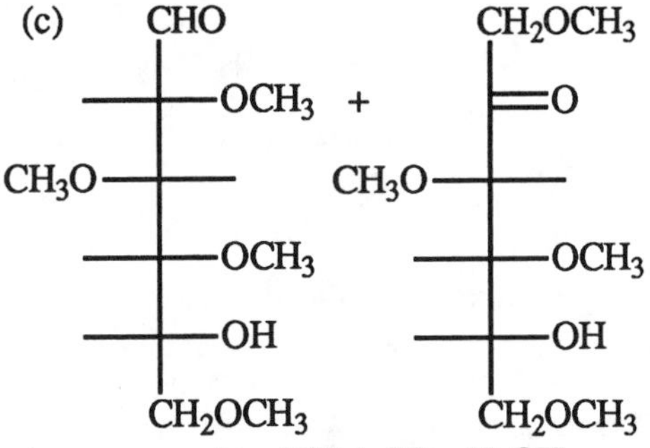

CHO
OCH3
CH3O
OCH3
OH
CH2OCH3
+
CH2OCH3
=O
CH3O
OCH3
OH
CH2OCH3

using $(CH_3)_2SO_4$, NaOH

23-63 continued

(d)

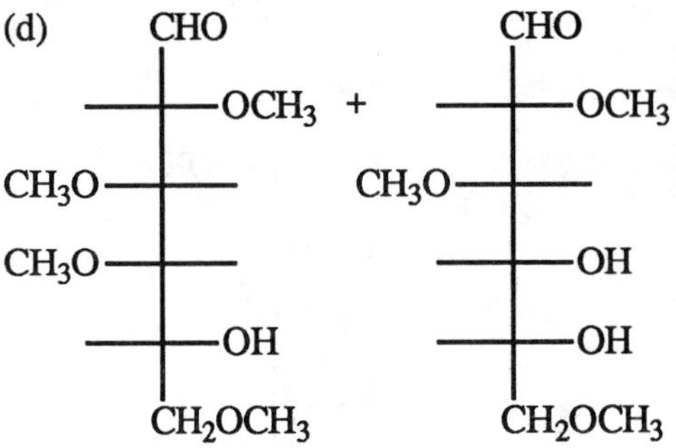

using CH_3I, Ag_2O

(e)

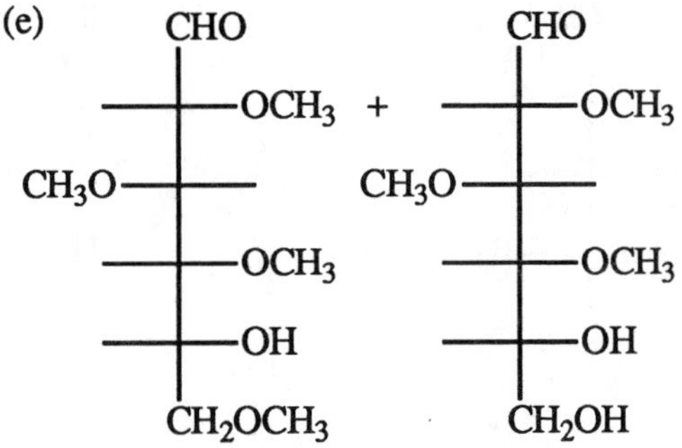

using CH_3I, Ag_2O

(f)

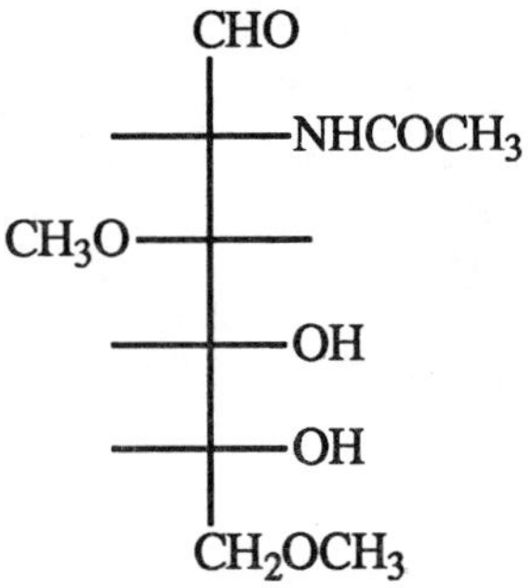

using $(CH_3)_2SO_4$, NaOH

23-64

(a) These D-aldopentoses will give optically active aldaric acids.

D-arabinose

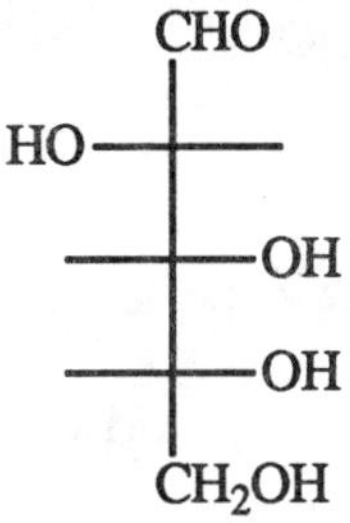

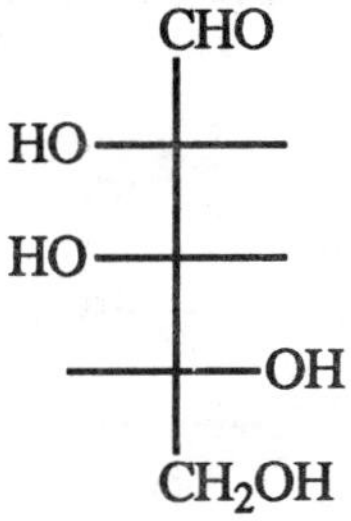

D-lyxose

(b) Only D-threose (of the aldotetroses) will give an optically active aldaric acid.

D-threose

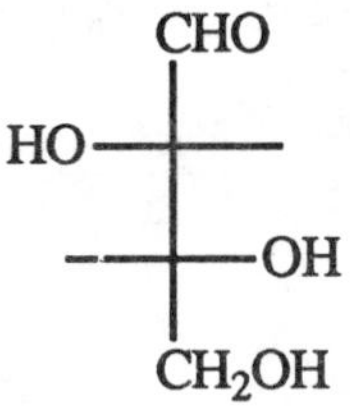

23-64 continued

(c) **X** is D-galactose.

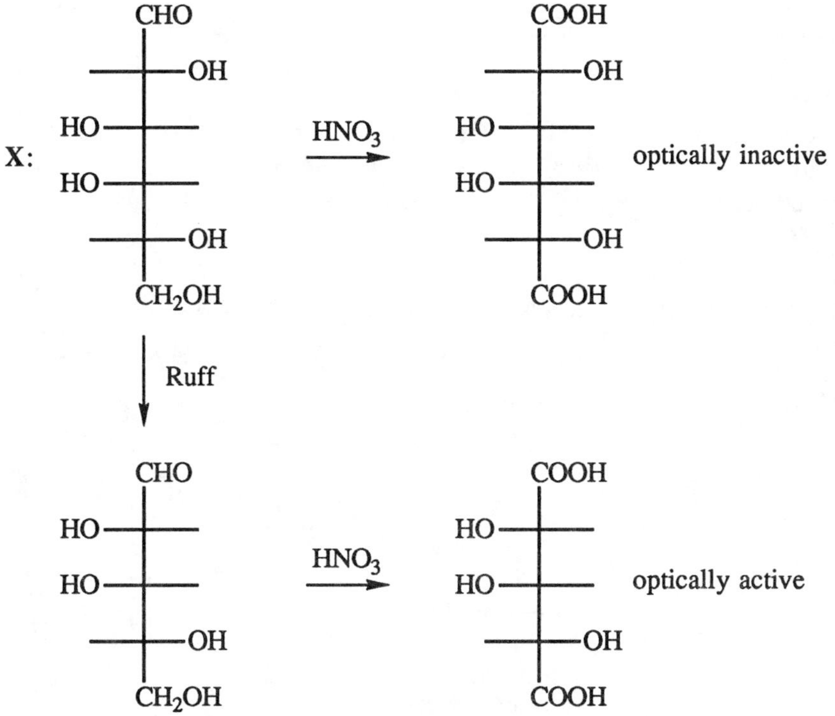

The other aldohexose that gives an optically inactive aldaric acid is D-allose, with all OH groups on the right side of the Fischer projection. Ruff degradation followed by nitric acid gives an optically *inactive* aldaric acid, however, so **X** cannot be D-allose.

(d) The optically active, five-carbon aldaric acid comes from the optically active pentose, not from the optically inactive, six-carbon aldaric acid. The principle is not violated.

(e)

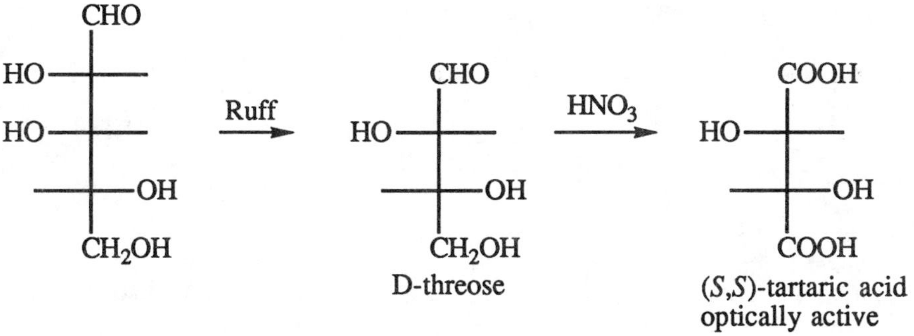

23-65

(a)

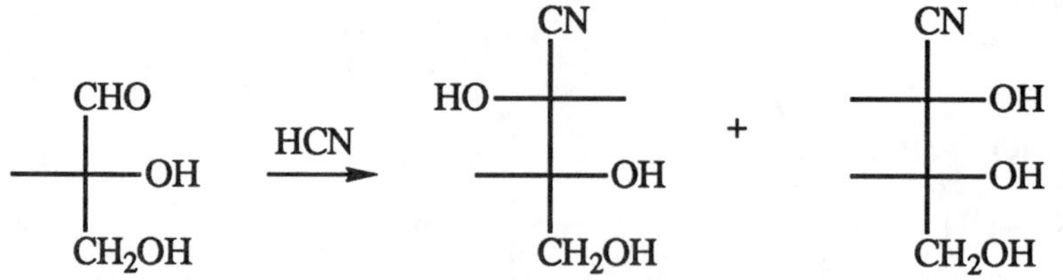

CHO
OH
CH2OH
HCN
CN
HO
OH
CH2OH
+
CN
OH
OH
CH2OH

(b) The products are diastereomers with different physical properties. They could be separated by crystallization, distillation, or chromatography.

(c) Both products are optically active. Each has two chiral centers and no plane of symmetry.

23-66

(a) The Tollens reaction is run in aqueous base which isomerizes carbohydrates.

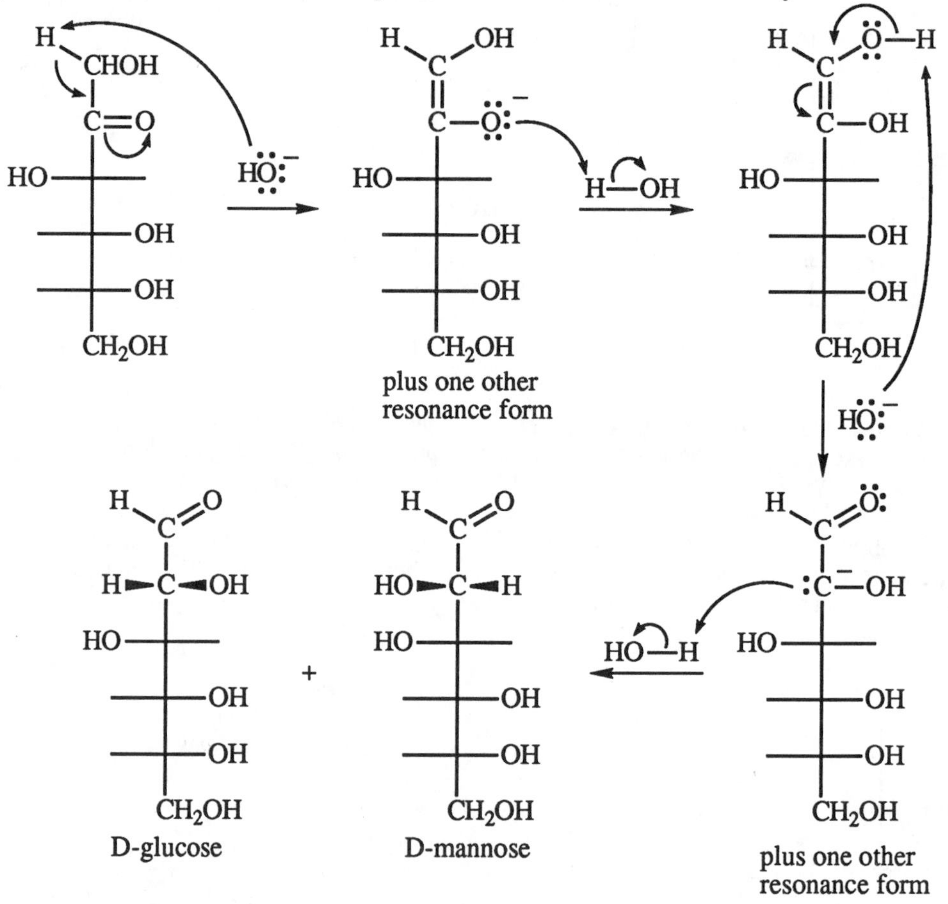

H
CHOH
C=O
HO
OH
OH
CH2OH
HO:⁻
plus one other
resonance form
H—OH
HO:⁻
HO—H
D-glucose
D-mannose
plus one other
resonance form

(b) Bromine water is acidic, not basic like the Tollens reagent. Carbohydrates isomerize quickly in base, but only very slowly in acid, so bromine water can oxidize without isomerization.

23-67 Tagatose is a monosaccharide, a ketohexose, that is found in the pyranose form.

(a)

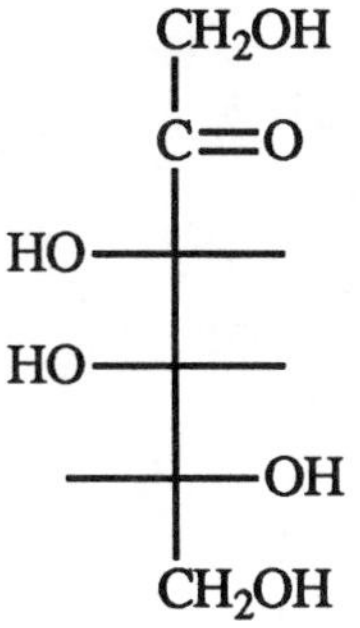

(b)

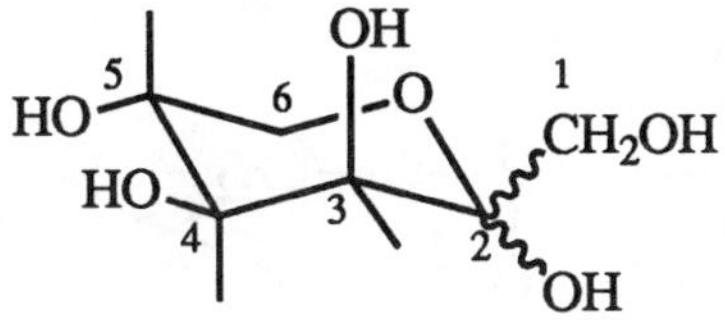

23-68

(a)

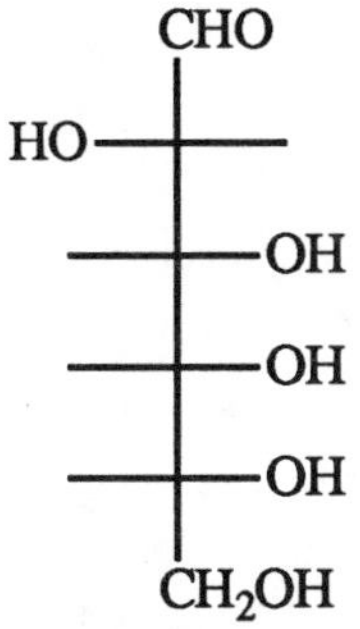

D-altrose

(b)

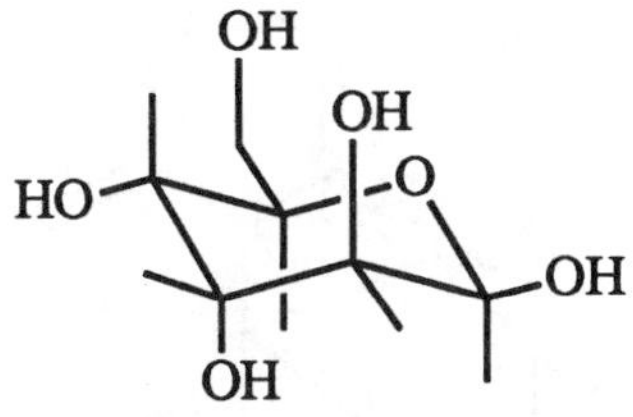

β-D-altropyranose

23-69

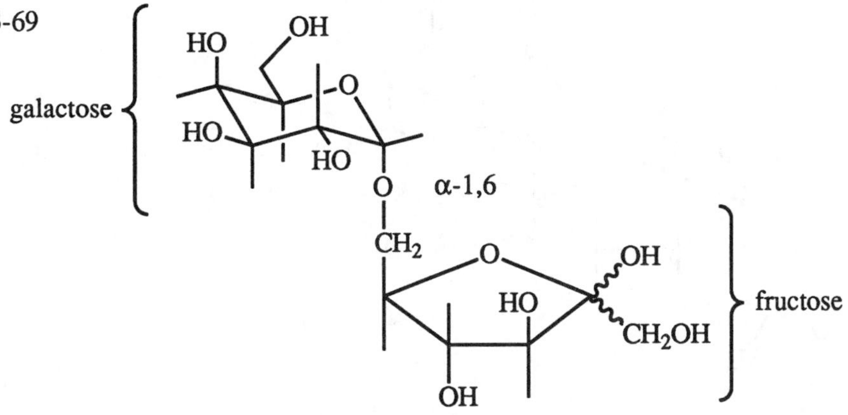
galactose
α-1,6
fructose

6-O-(α-D-galactopyranosyl)-D-fructofuranose

23-70

(a)

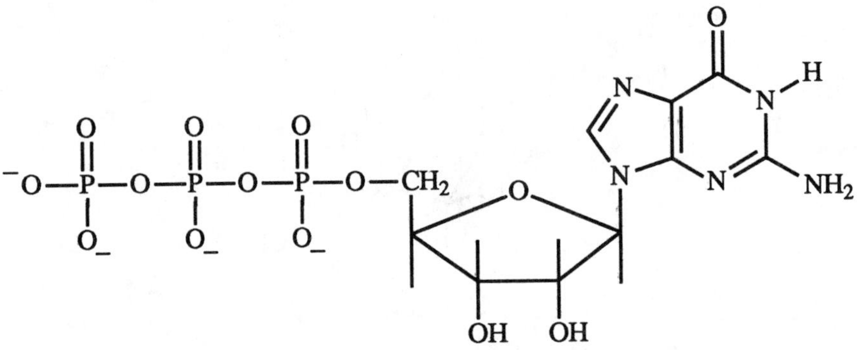
O
N
H
N
O
O
O
−O—P—O—P—O—P—O—CH2
N
N
NH2
O
O−
O−
O−
OH
OH

(b)

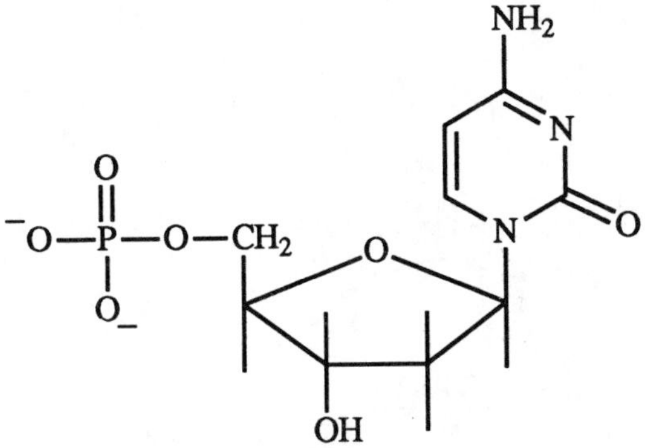
NH2
N
O
−O—P—O—CH2
N
O
O
O−
OH

(c)

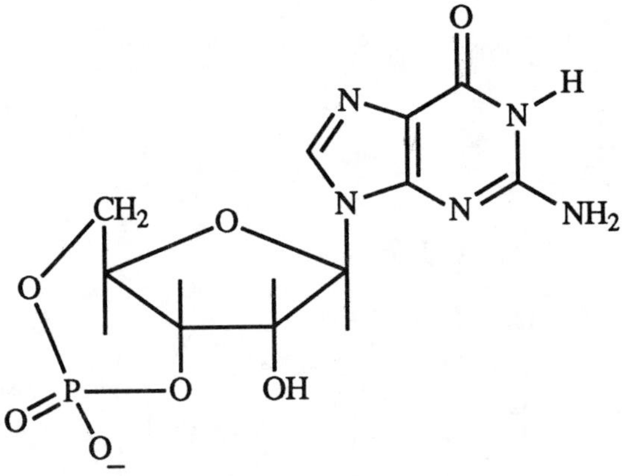
O
H
N
N
N
N
NH2
CH2
O
O
O=P
O
OH
O−

23-71 Bonds to H are omitted for simplicity.

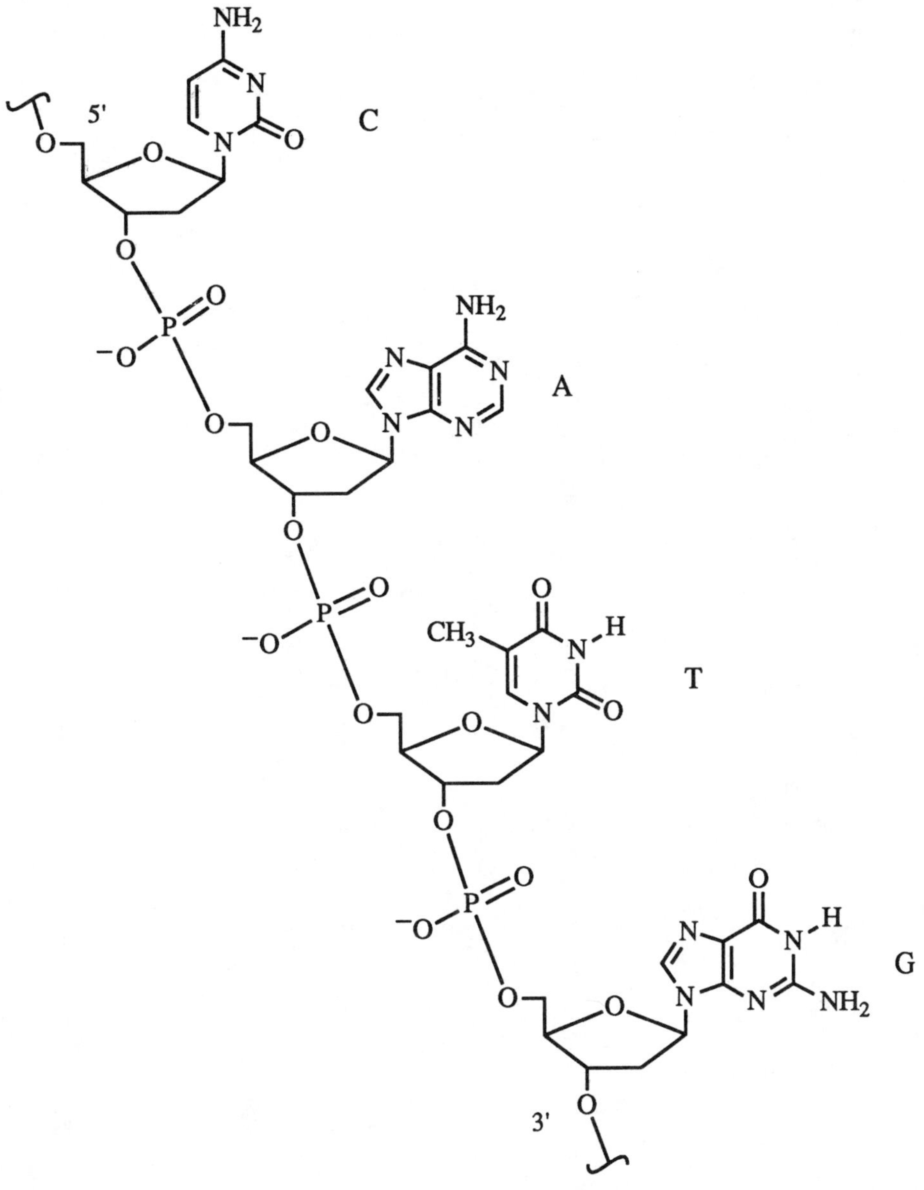

23-72

(a) No, there is no relation between the amount of G and A.
(b) Yes, this must be true mathematically.
(c) Chargaff's rule must apply only to double-stranded DNA. For each G in one strand, there is a complementary C in the opposing strand, but there is no correlation between G and C *in the same strand.*

24-1

(a)

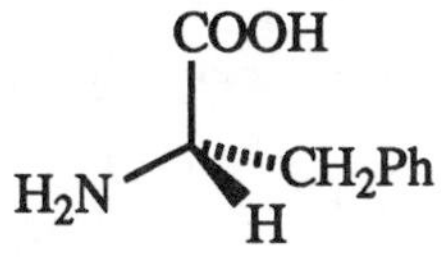

(b)

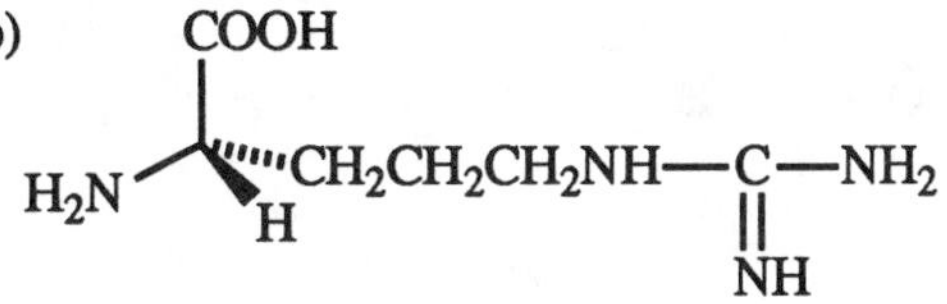

(c)

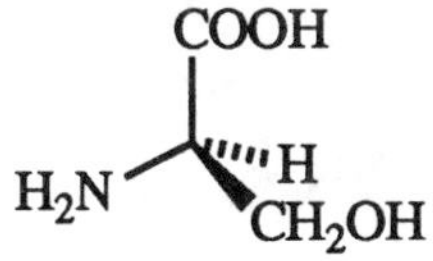

(d)

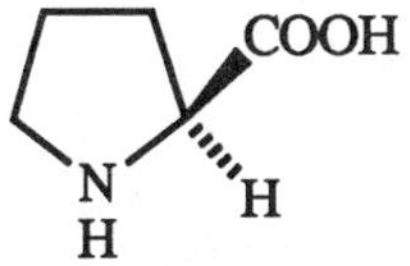

24-2

(a)

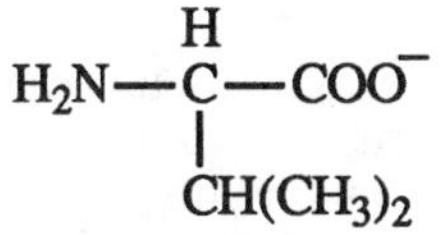

(b)

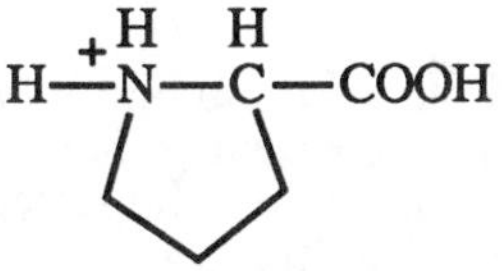

(c)

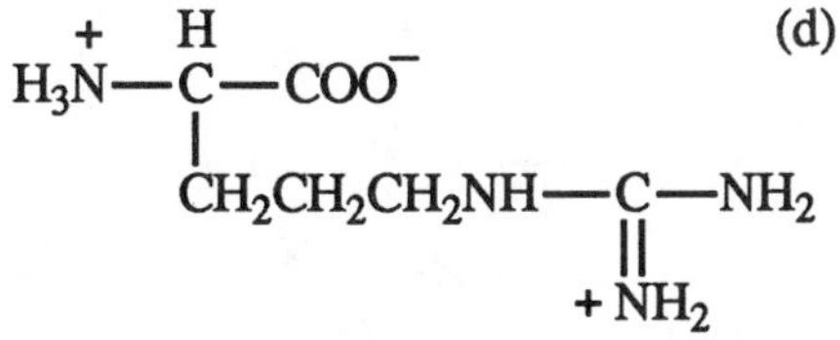

(d)

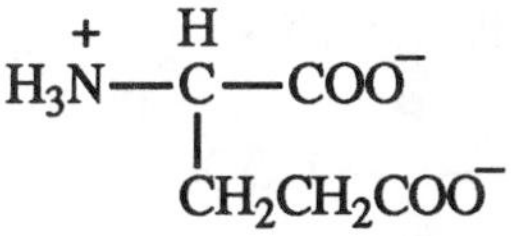

(e)

	alanine	lysine	aspartic acid
(i) pH 6	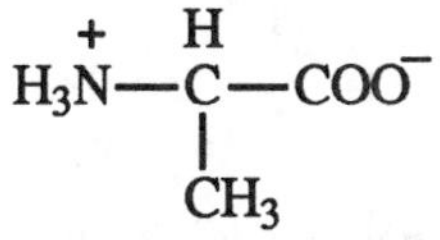	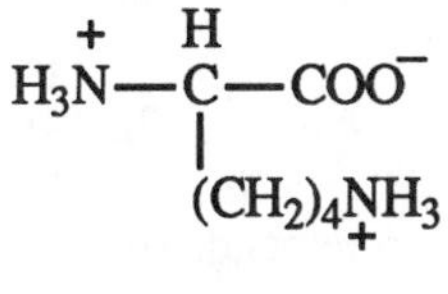	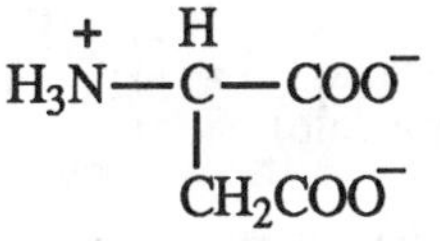
(ii) pH 11	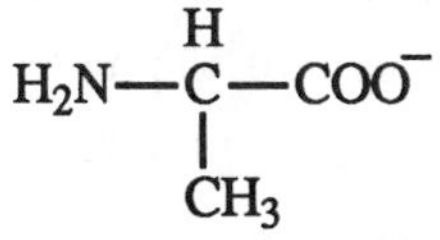	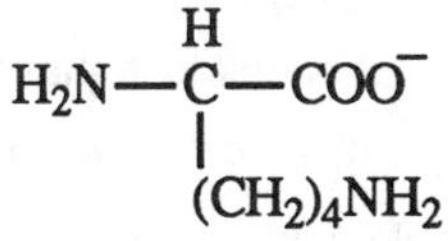	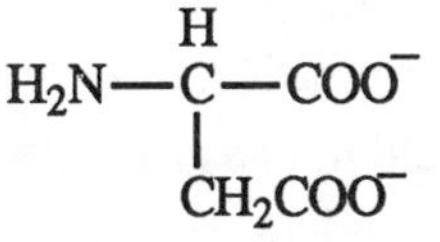
(iii) pH 2	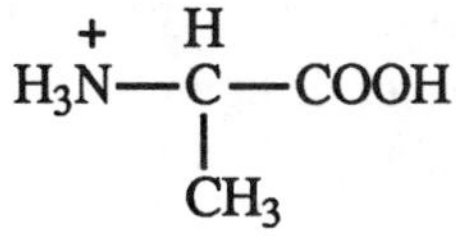	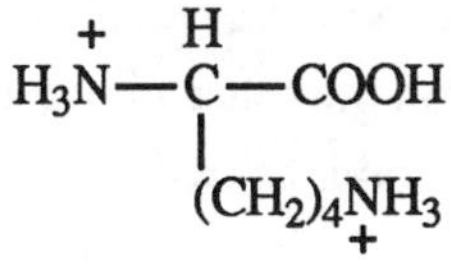	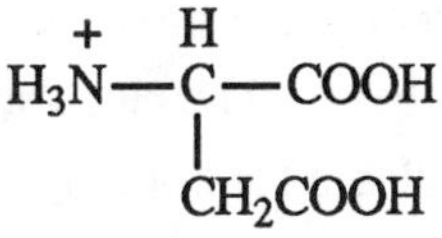

24-3

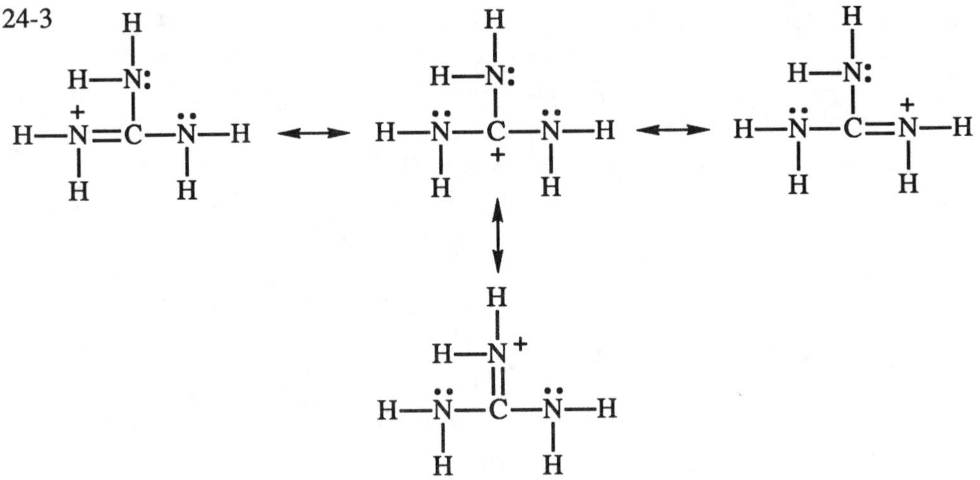

Protonation of the guanidino group gives a resonance-stabilized cation with all octets filled and the positive charge delocalized over three nitrogen atoms. Arginine's strongly basic isoelectric point reflects the unusual basicity of the guanidino group due to this resonance stabilization in the protonated form. (See Problems 1-37 and 19-51.)

24-4

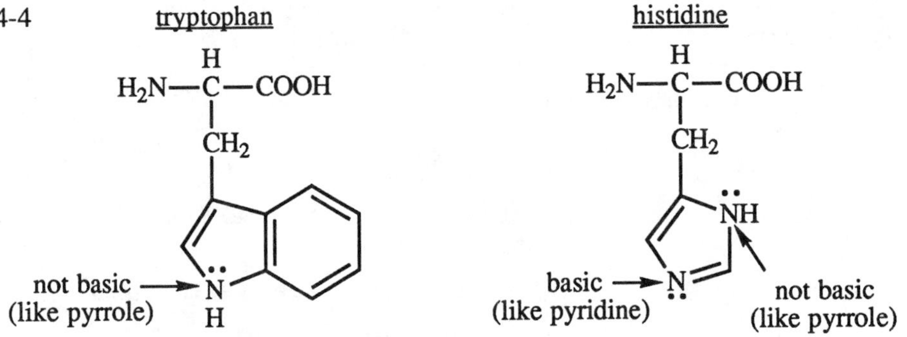

The basicity of any nitrogen depends on its electron pair's availability for bonding with a proton. In tryptophan, the nitrogen's electron pair is part of the aromatic π system; without this electron pair in the π system, the molecule would not be aromatic. Using this electron pair for bonding to a proton would therefore destroy the aromaticity—not a favorable process.

In the imidazole ring of histidine, the electron pair of one nitrogen is also part of the aromatic π system and is unavailable for bonding; this nitrogen is not basic. The electron pair on the other nitrogen, however, is in an sp^2 orbital available for bonding, and is about as basic as pyridine.

24-5 At pH 9.7, alanine (isoelectric point (IEP) 6.0) has a charge of −1 and will migrate to the anode. Lysine (IEP 9.7) is at its isoelectric point and will not move. Aspartic acid (IEP 2.8) has a charge of −2 and will also migrate to the anode, faster than alanine.

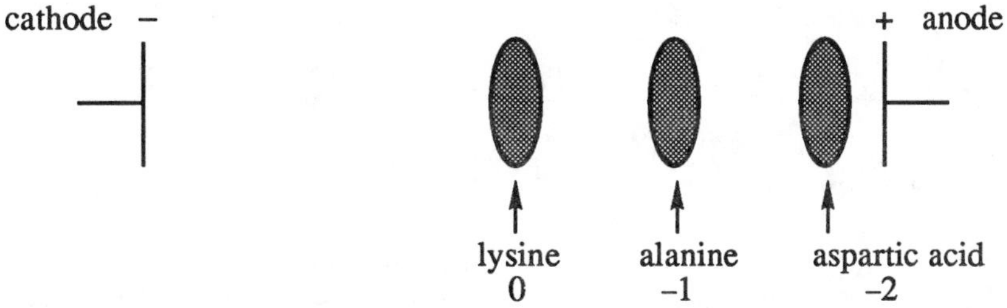

24-6 At pH 6.0, tryptophan (IEP 5.9) has a charge of zero and will not migrate. Cysteine (IEP 5.0) has a partial negative charge and will move toward the anode. Histidine (IEP 7.6) has a partial positive charge and will move toward the cathode.

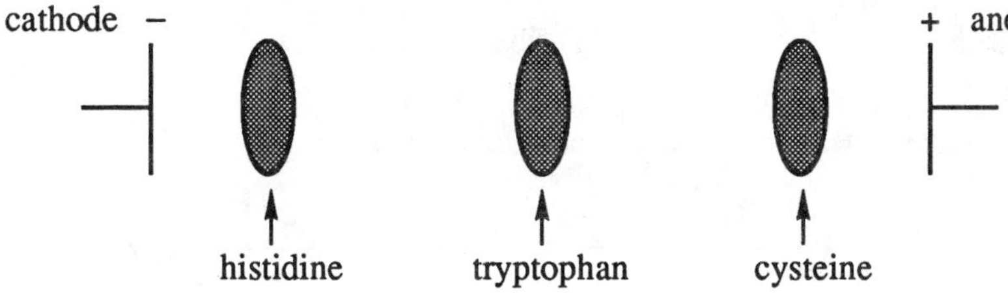

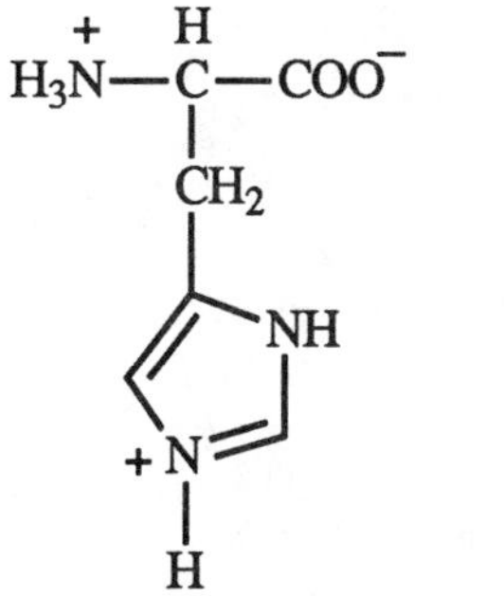

histidine
(partially protonated
imidazole ring)

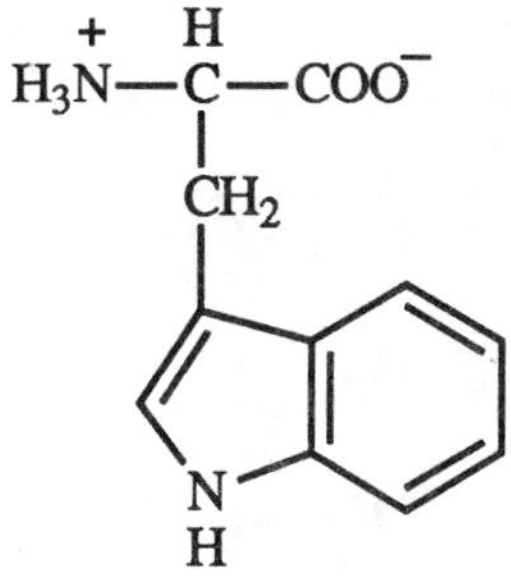

tryptophan

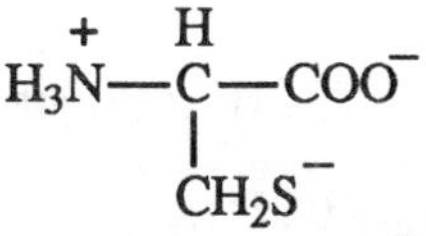

cysteine
(partially
deprotonated
sulfur)

(The SH is more acidic than the NH_3^+ group.)

24-7

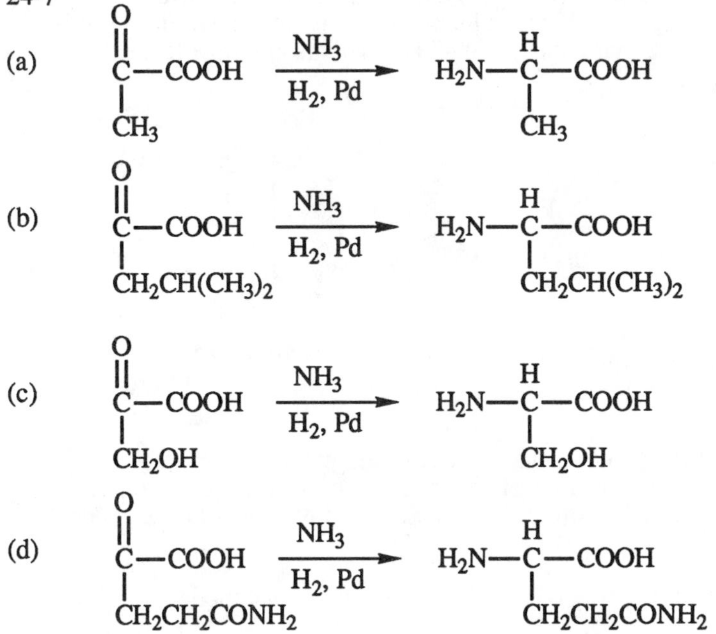

24-8 All of these reactions use: first arrow: Br_2/PBr_3, followed by H_2O workup; second arrow: excess NH_3, followed by neutralizing workup.

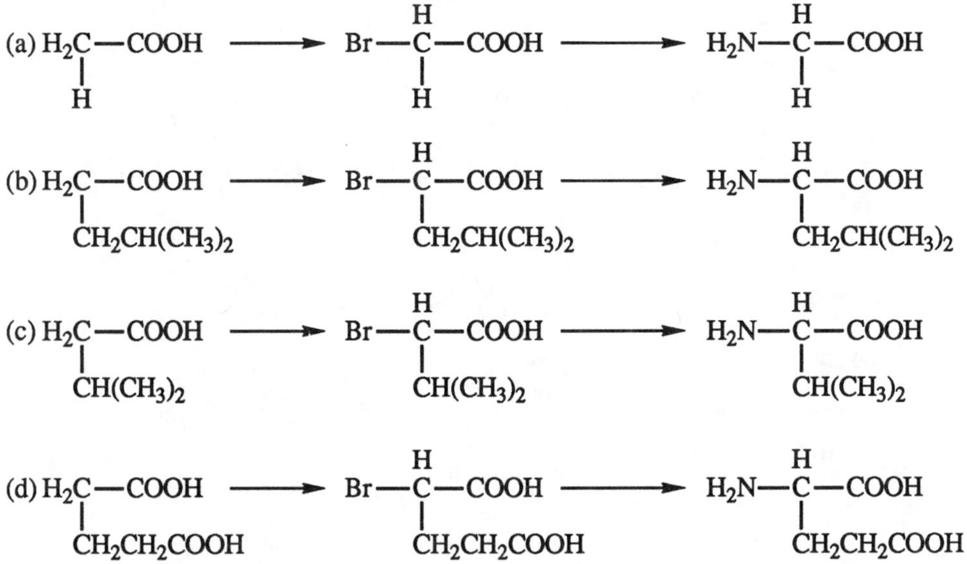

In part (d), care must be taken to avoid reaction α to the other COOH. In practice, this would be accomplished by using less than one-half mole of bromine per mole of the diacid.

24-9

(a)

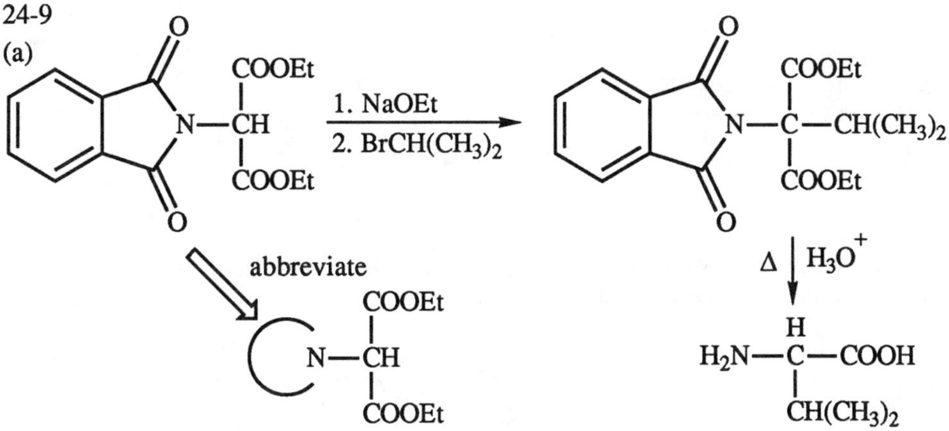
COOEt
N—CH
COOEt
1. NaOEt
2. BrCH(CH3)2
N—C—CH(CH3)2
abbreviate
Δ
H3O+
H2N—C—COOH
CH(CH3)2

(b)

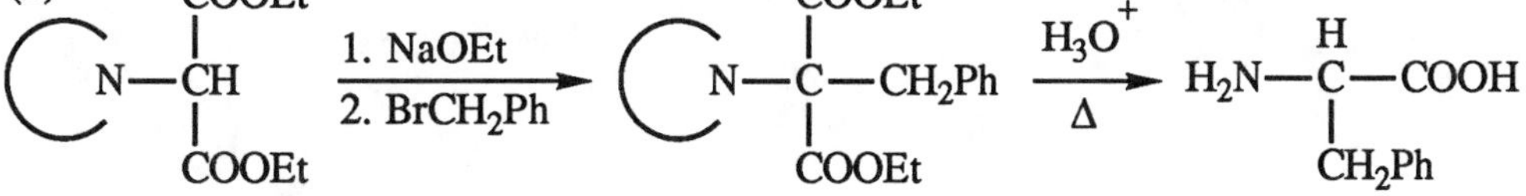
COOEt
N—CH
COOEt
1. NaOEt
2. BrCH2Ph
N—C—CH2Ph
H3O+
Δ
H2N—C—COOH
CH2Ph

(c)

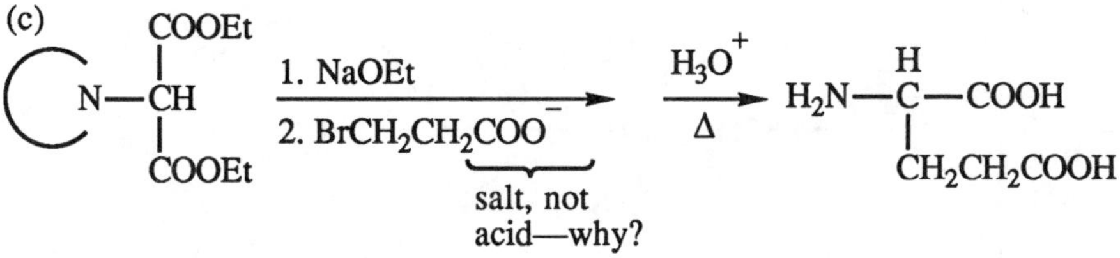
COOEt
N—CH
COOEt
1. NaOEt
2. BrCH2CH2COO−
salt, not acid—why?
H3O+
Δ
H2N—C—COOH
CH2CH2COOH

(d)

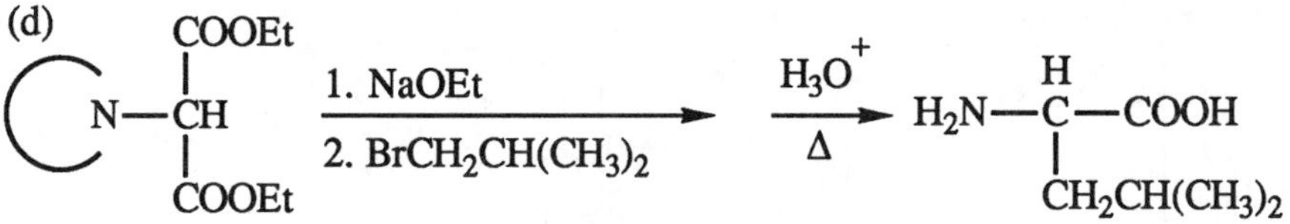
COOEt
N—CH
COOEt
1. NaOEt
2. BrCH2CH(CH3)2
H3O+
Δ
H2N—C—COOH
CH2CH(CH3)2

24-10

(a)

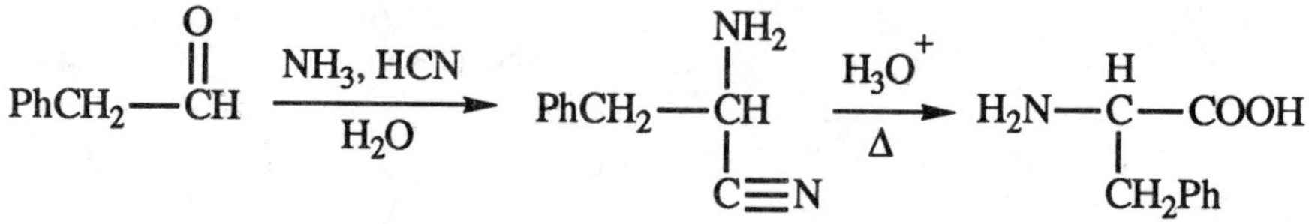

PhCH2—CH(=O)
NH3, HCN
H2O
PhCH2—CH(NH2)—C≡N
H3O+
Δ
H2N—C(H)(CH2Ph)—COOH

(b) While the solvent for the Strecker synthesis is water, the proton acceptor is ammonia and the proton donor is ammonium ion.

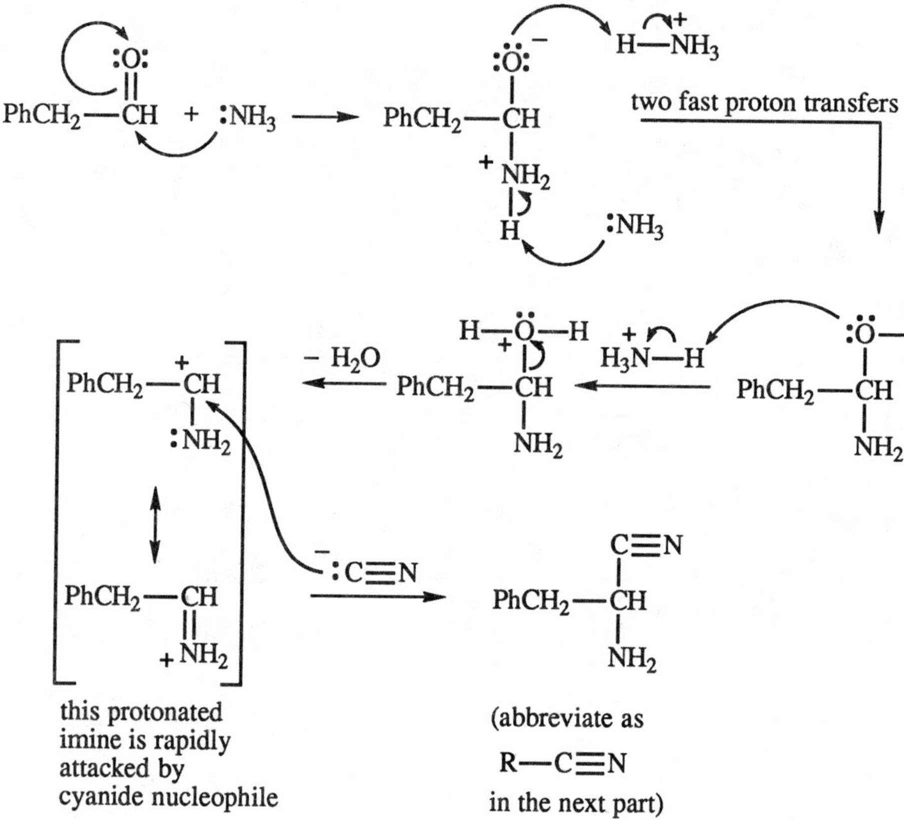

two fast proton transfers
− H2O
this protonated imine is rapidly attacked by cyanide nucleophile
(abbreviate as
R—C≡N
in the next part)

24-10 (b) continued

mechanism of acid hydrolysis of the nitrile

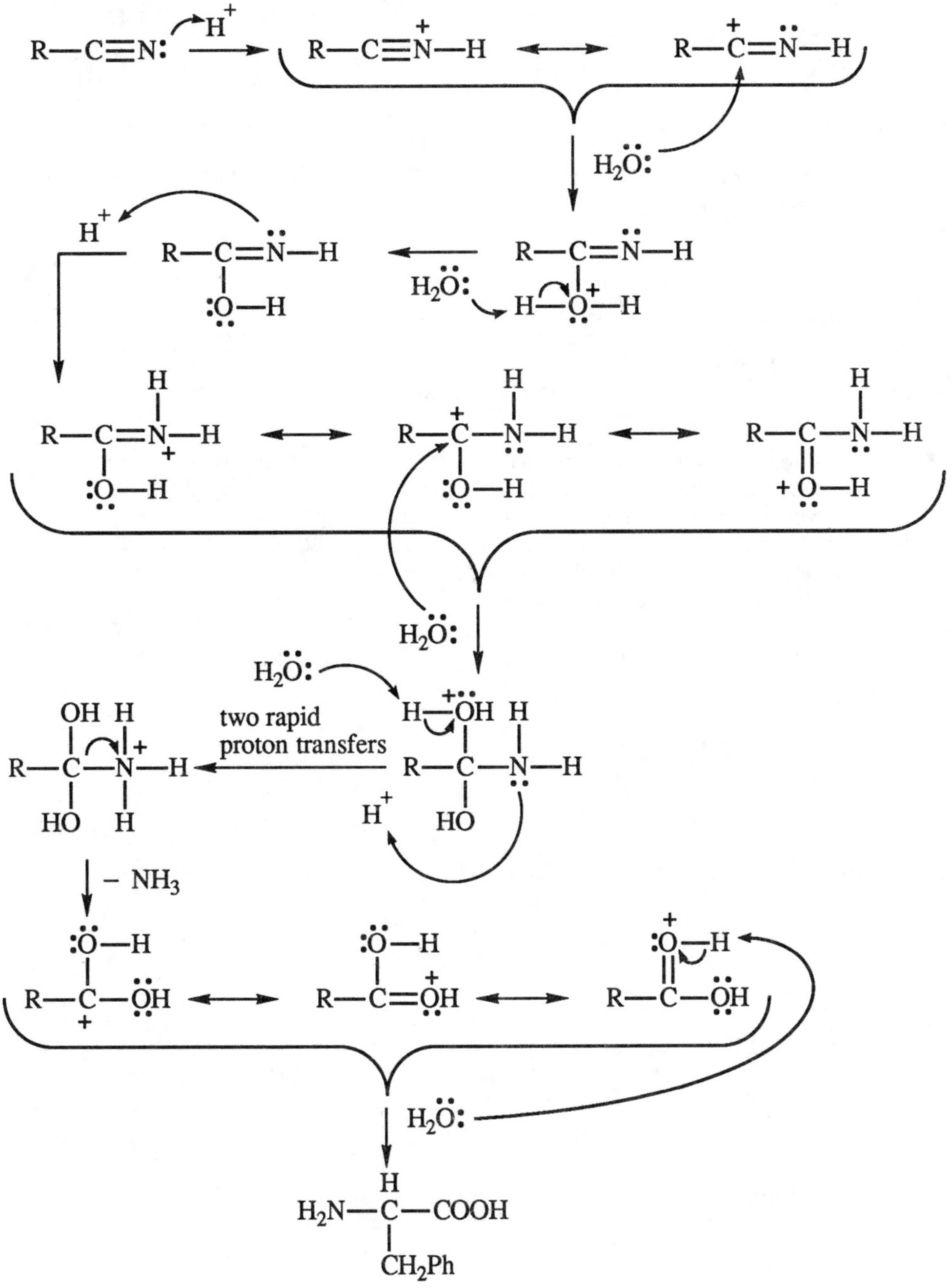

24-11

(a) 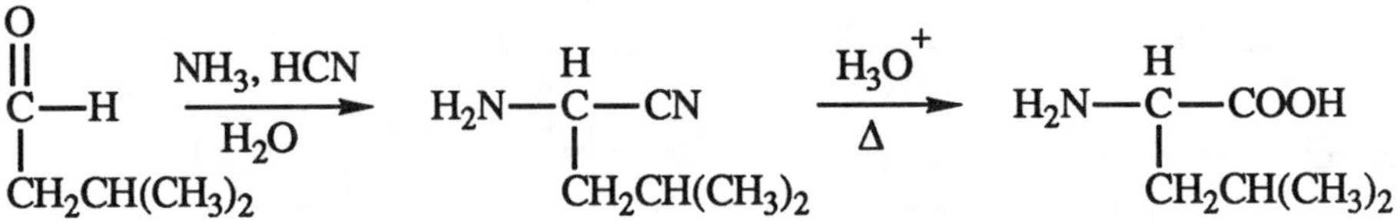

(b) 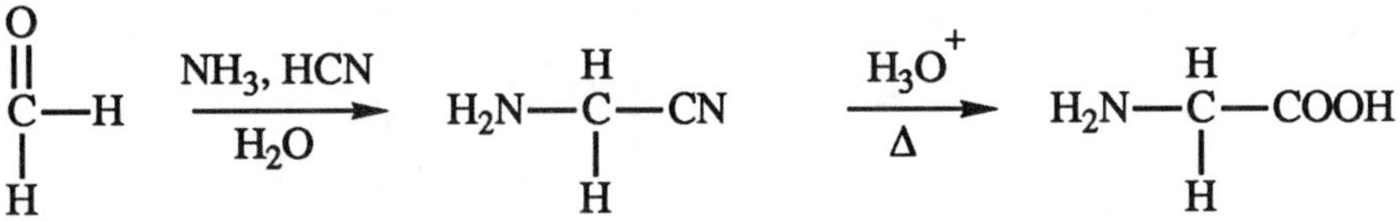

(c) 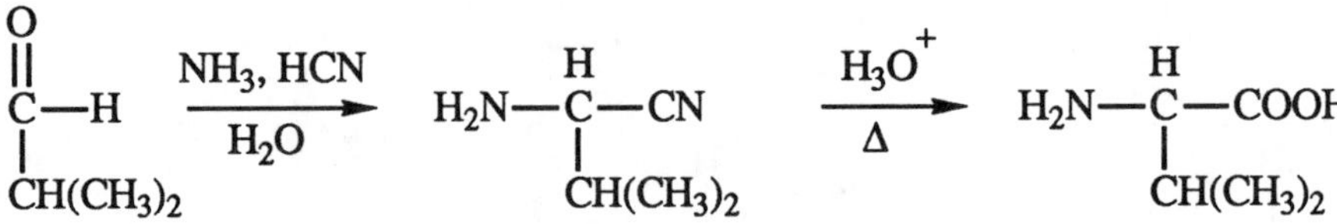

24-12 In acid solution, the free amino acid will be protonated, with a positive charge, and probably soluble in water as are other organic ions. The acylated amino acid, however, is not basic since the nitrogen is present as an amide. In acid solution, the acylated amino acid is neutral and not soluble in water. Water extraction or ion-exchange chromatography (Figure 24-10) would be practical techniques to separate these compounds.

24-13

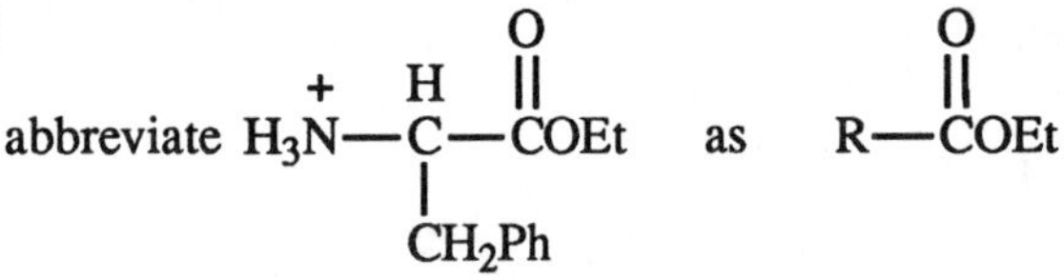

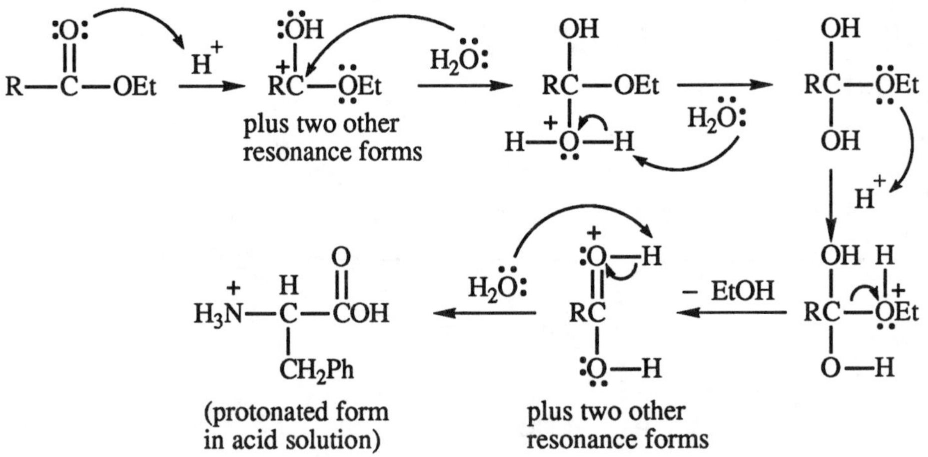

24-14

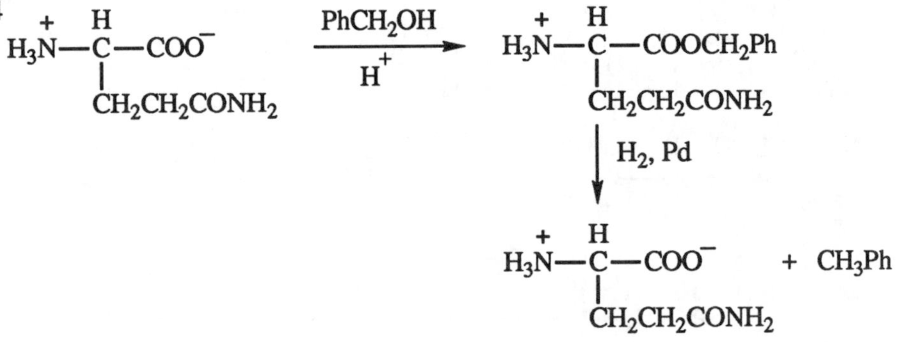

24-15

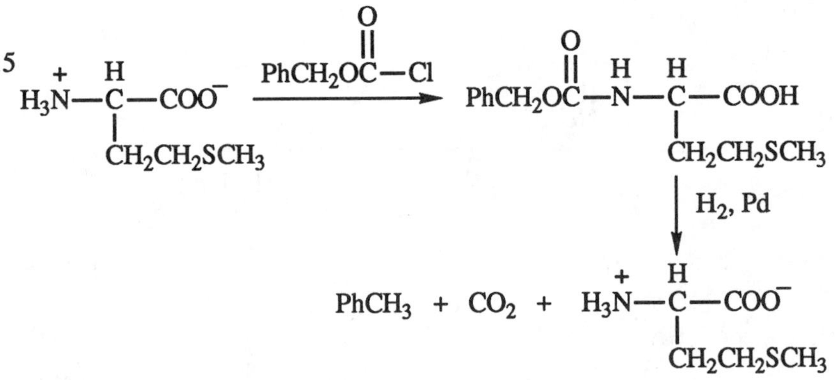

24-16

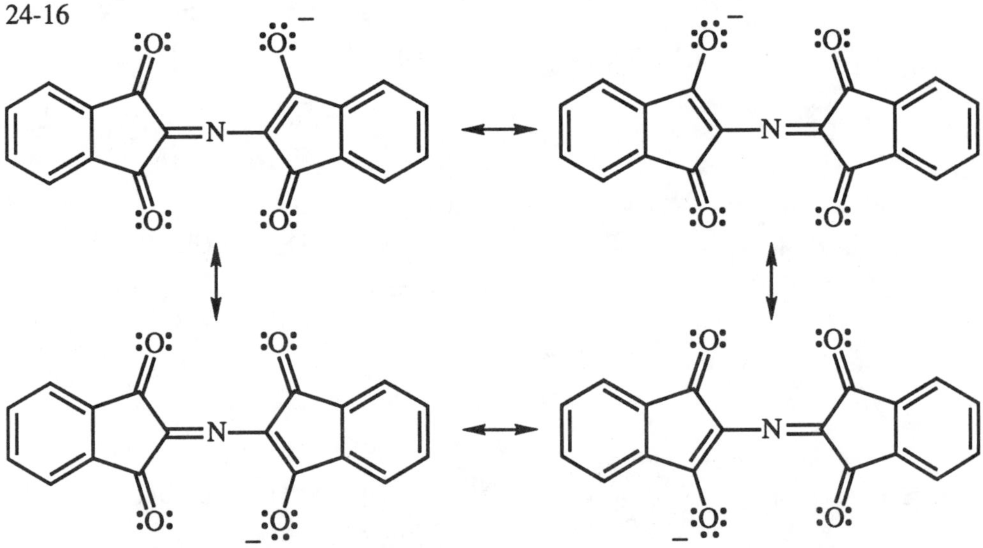

These are the most significant resonance contributors in which the electronegative oxygens carry the negative charge. There are also two other forms in which the negative charge is on the carbons bonded to the nitrogen, plus the usual resonance forms involving the alternate Kekulé structures of the benzene rings.

24-17

(a)

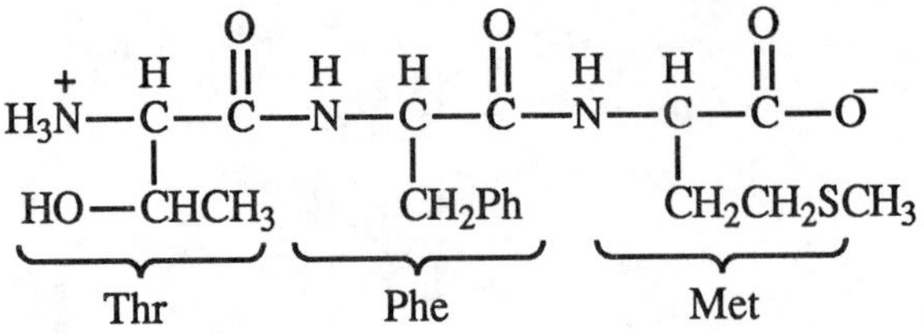
Thr
Phe
Met

(b)

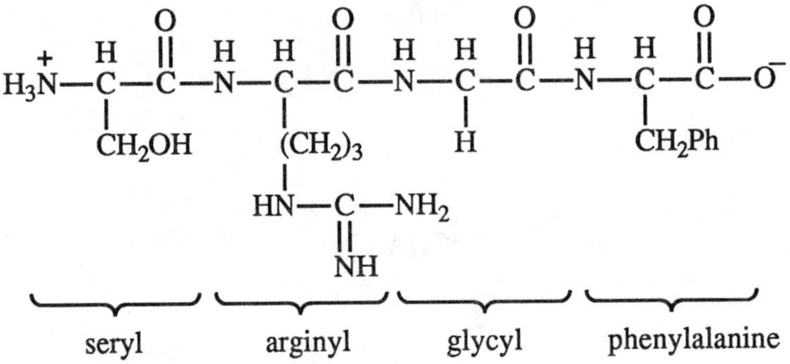
seryl
arginyl
glycyl
phenylalanine

24-18

(a)

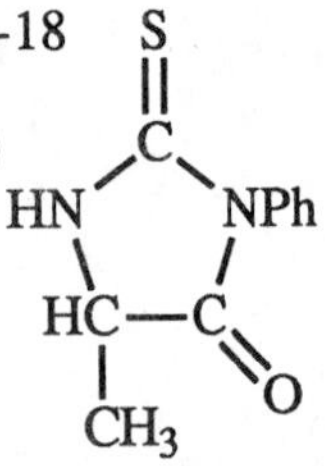

(b)

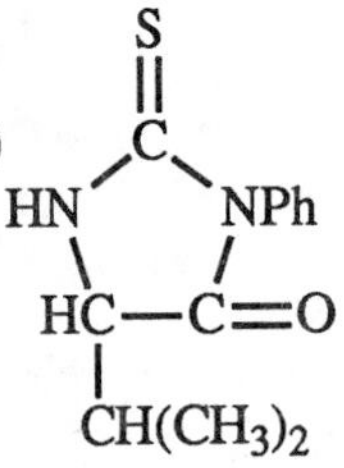

(c)

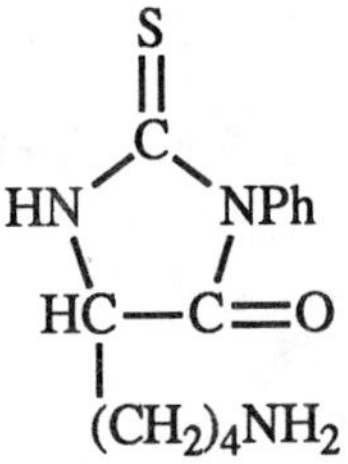

(d)

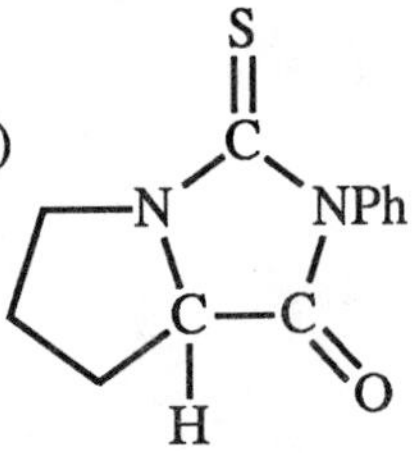

24-19

Step 3

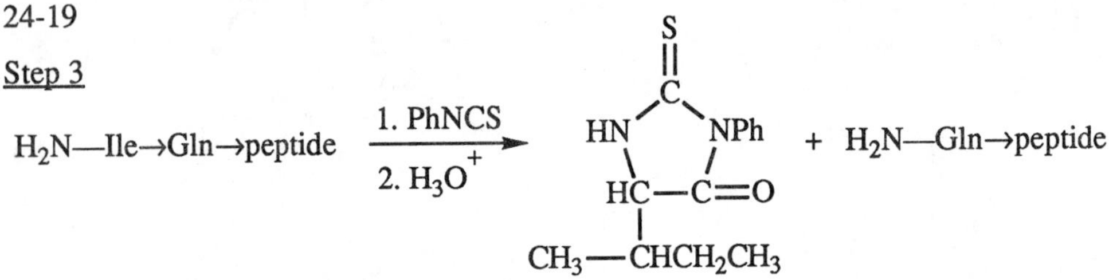

Step 4

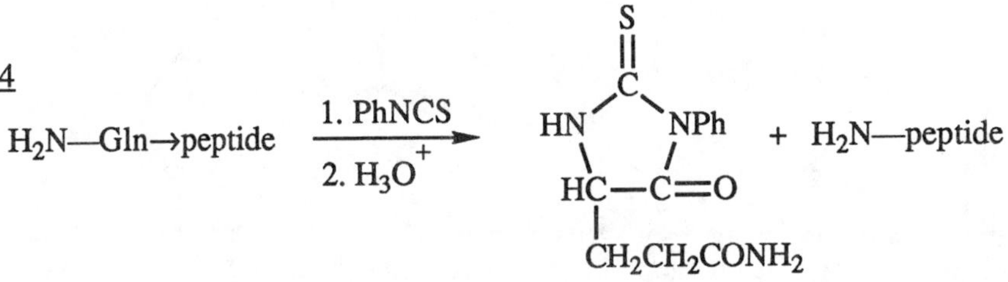

24-20

(a) This is a nucleophilic aromatic substitution by the addition-elimination mechanism. The presence of two nitro groups makes this reaction feasible under mild conditions.

abbreviate the N-terminus of the peptide chain as NH_2R

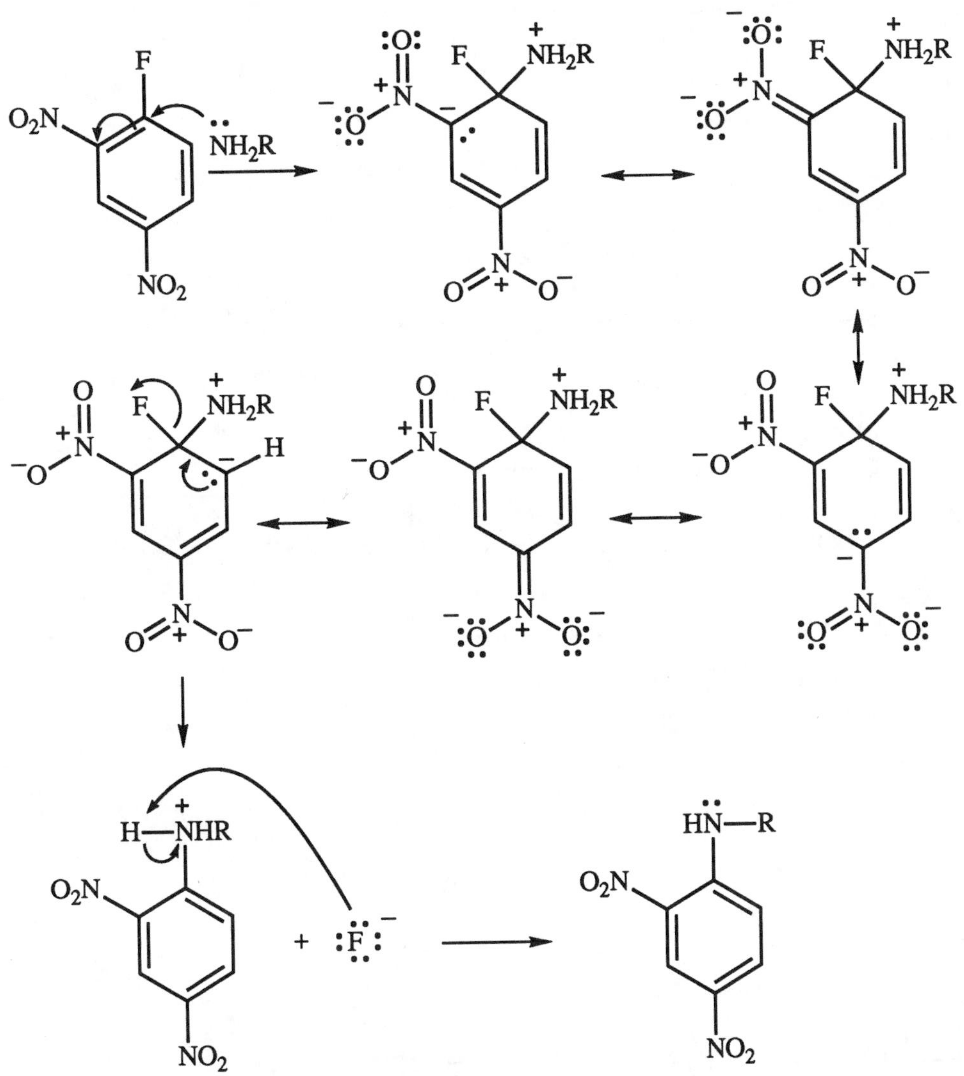

(b) The main drawback of the Sanger method is that only one amino acid is analyzed per sample of protein. The Edman degradation can usually analyze more than 20 amino acids per sample of protein.

24-21

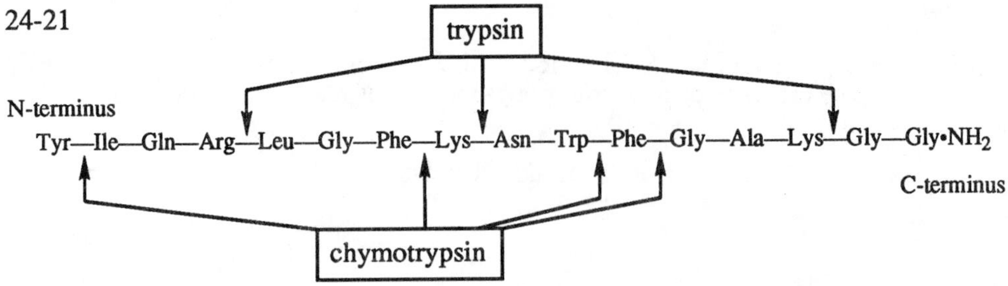

24-22

Phe—Gln—Asn

Pro—Arg—Gly•NH$_2$

Cys—Tyr—Phe

Asn—Cys—Pro—Arg

Tyr—Phe—Gln—Asn

Cys—Tyr—Phe—Gln—Asn—Cys—Pro—Arg—Gly•NH$_2$

24-23 abbreviations used in this problem:

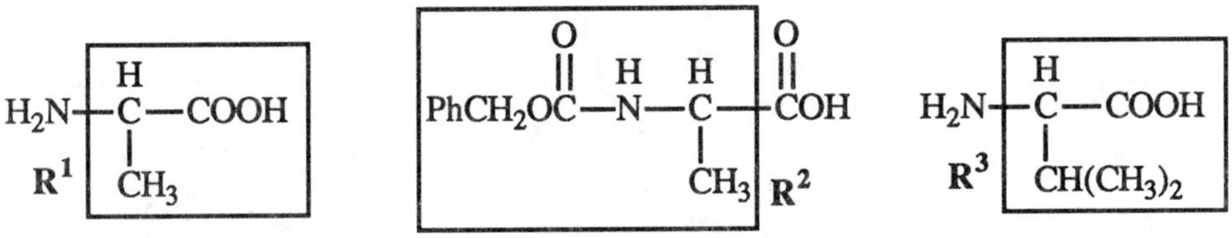

mechanism of formation of Z-Ala

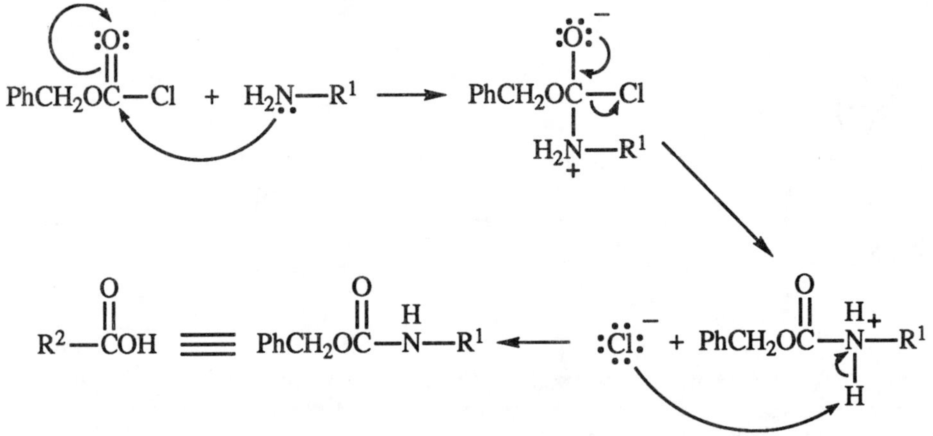

24-23 continued

two possible mechanisms of ethyl chloroformate activation

mechanism 1

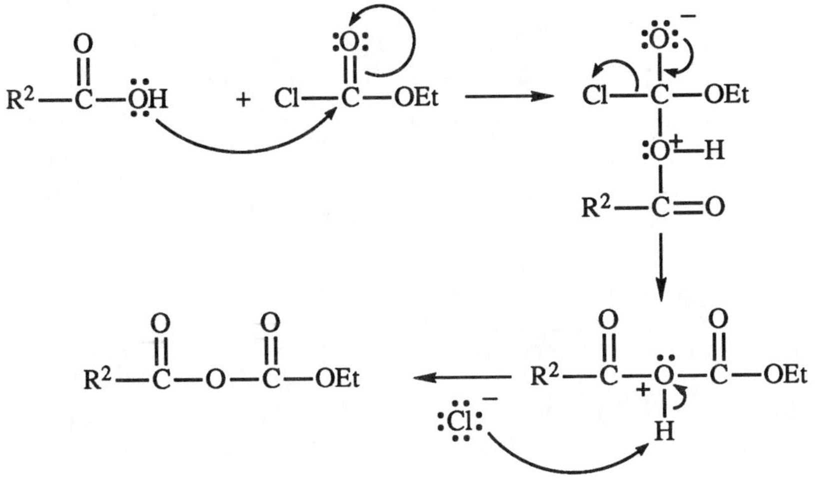

mechanism 2

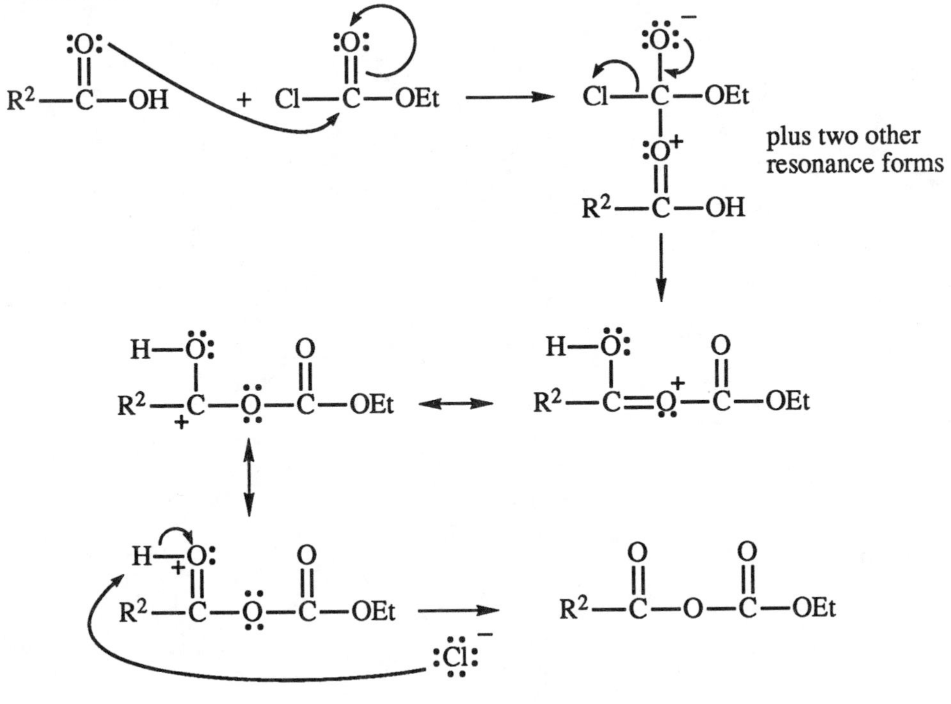

24-23 continued

mechanism of the coupling with valine

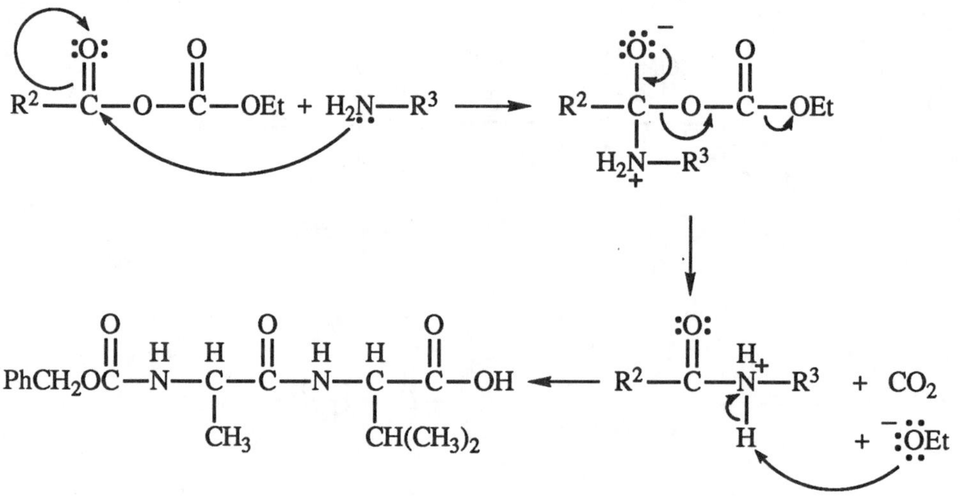

R2—C—O—C—OEt + H2N—R3
R2—C—O—C—OEt
H2N—R3
PhCH2OC—N—C—C—N—C—C—OH
CH3
CH(CH3)2
R2—C—N—R3 + CO2
H
+ :OEt

24-24

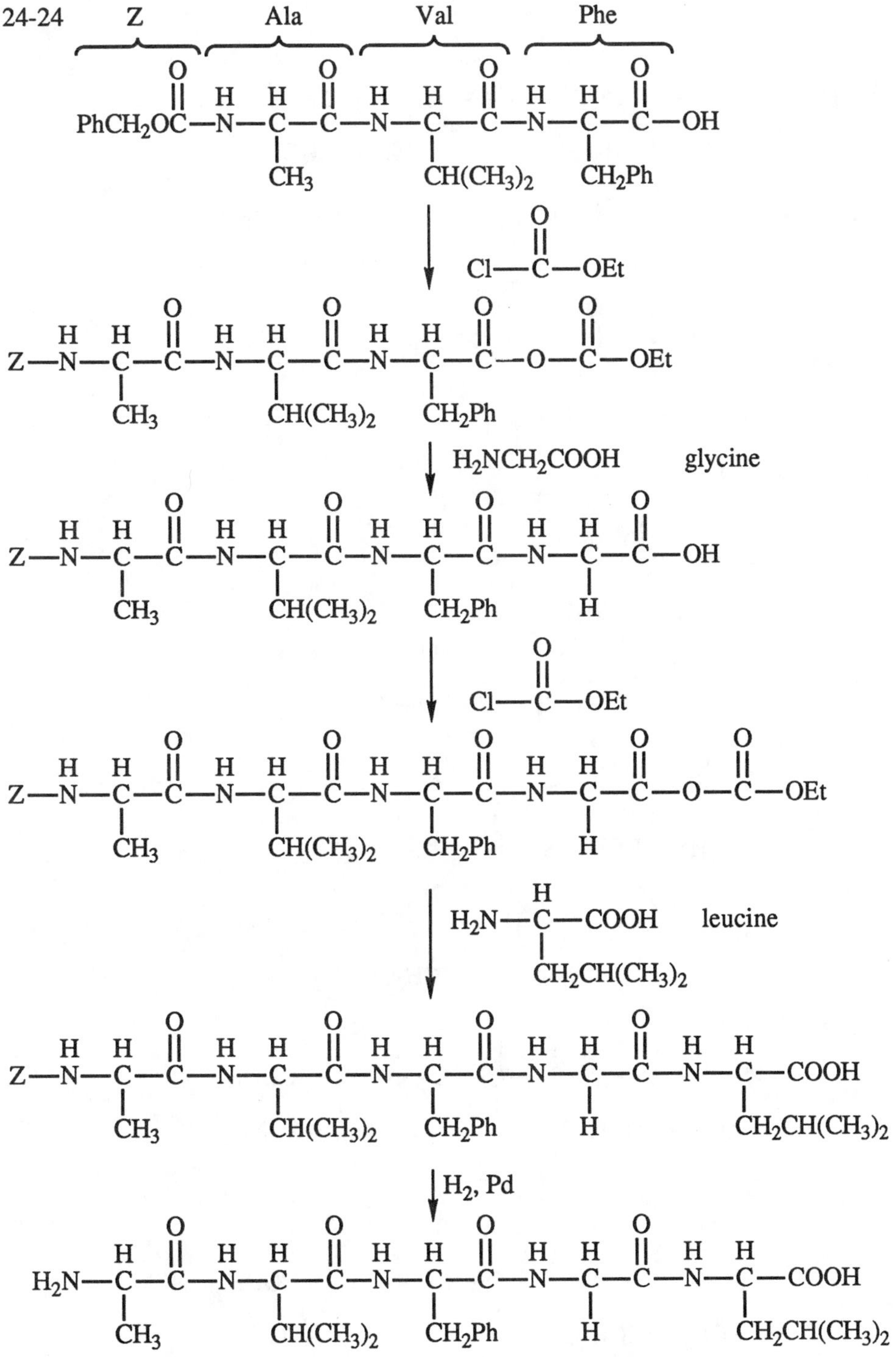

24-25

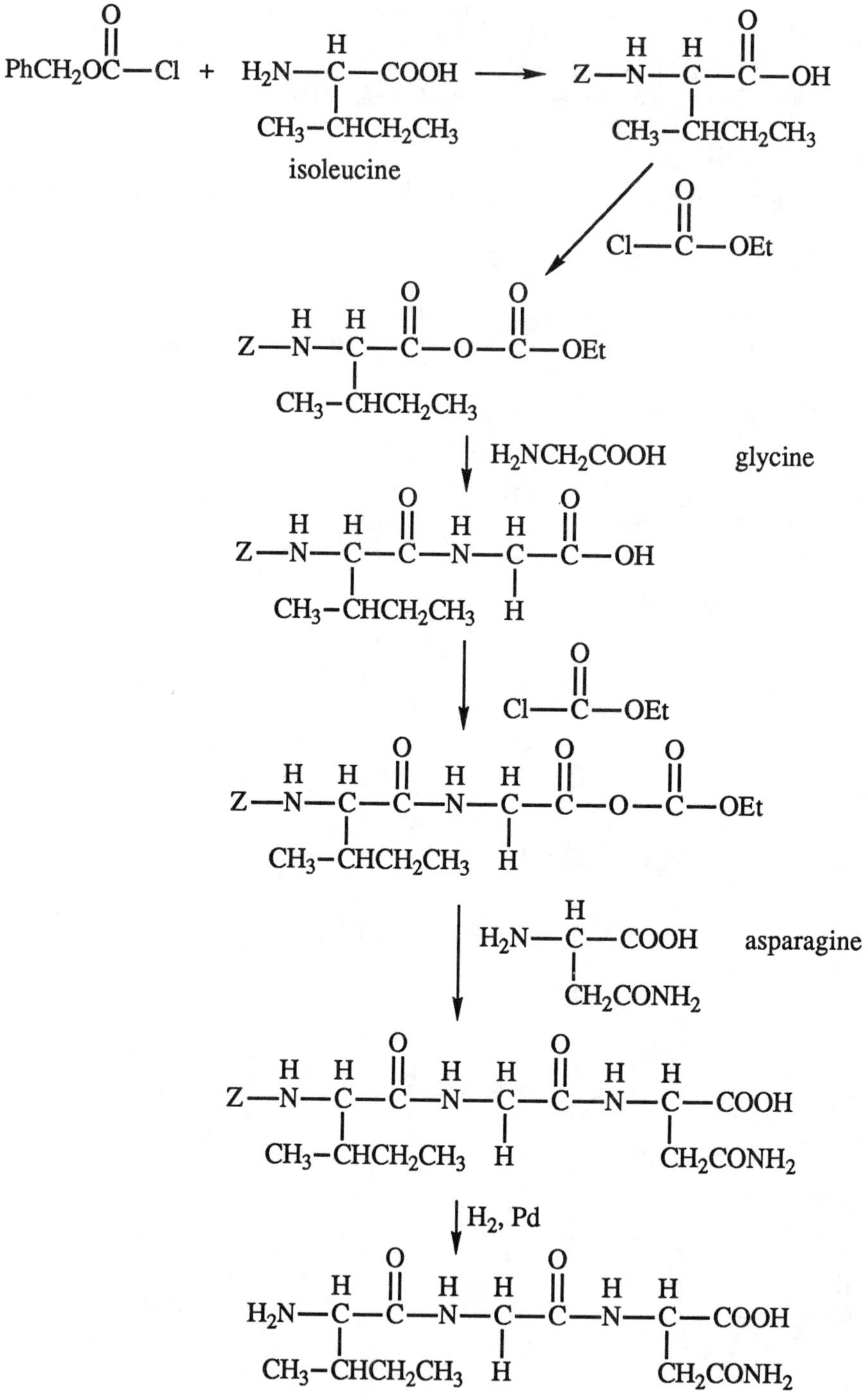

PhCH2OC—Cl
isoleucine
Z—N—C—C—OH
Cl—C—OEt
H2NCH2COOH
glycine
asparagine
H2, Pd

24-26 In this problem, "Cy" stands for "cyclohexyl".

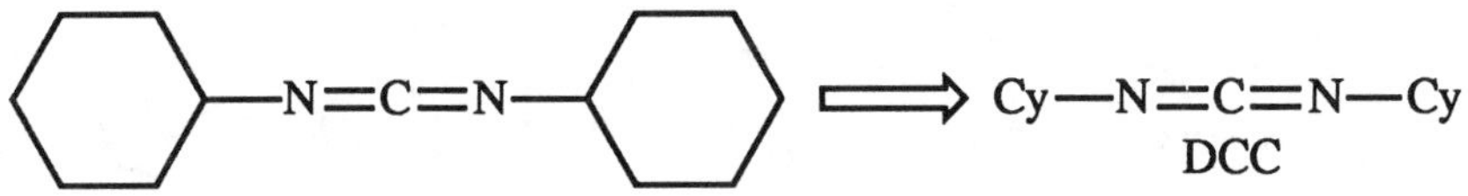

mechanism

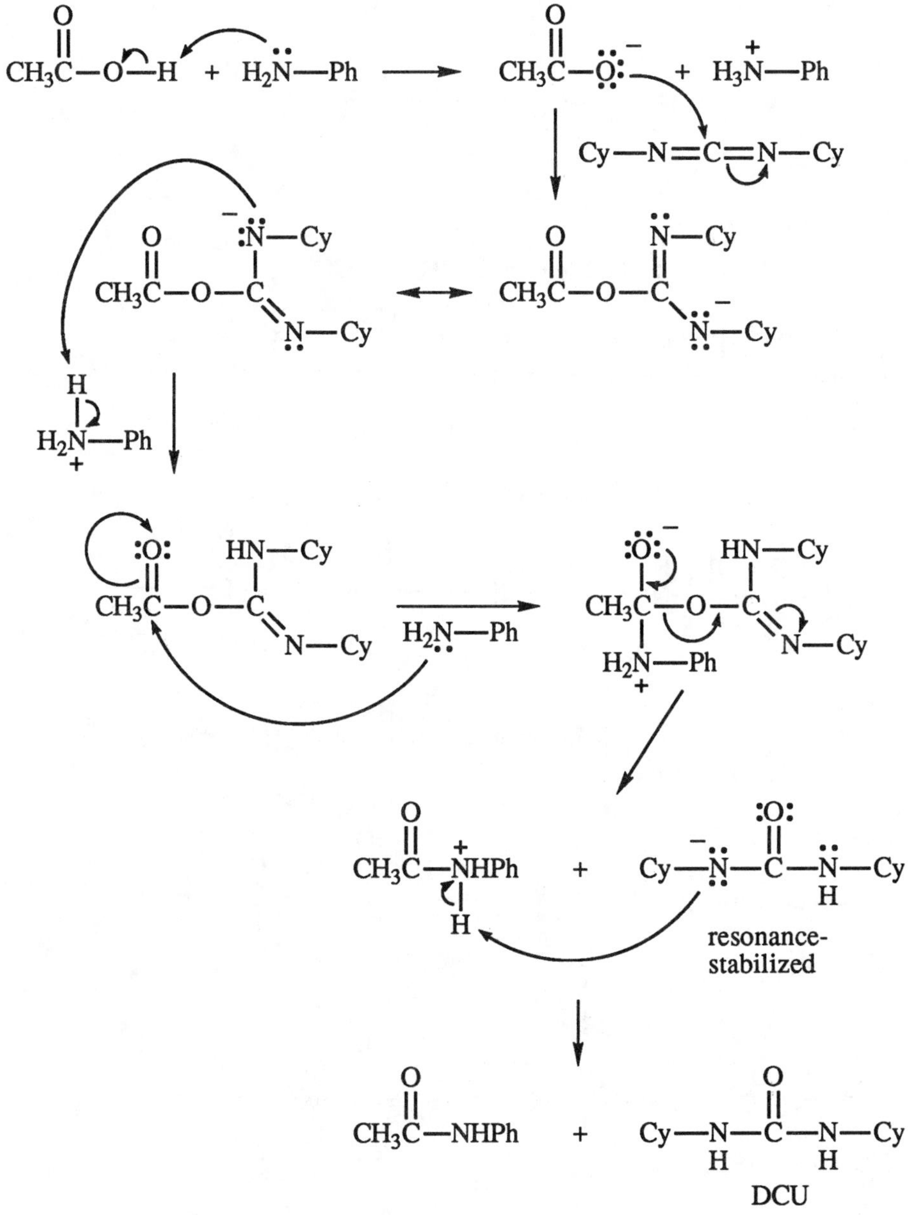

24-27

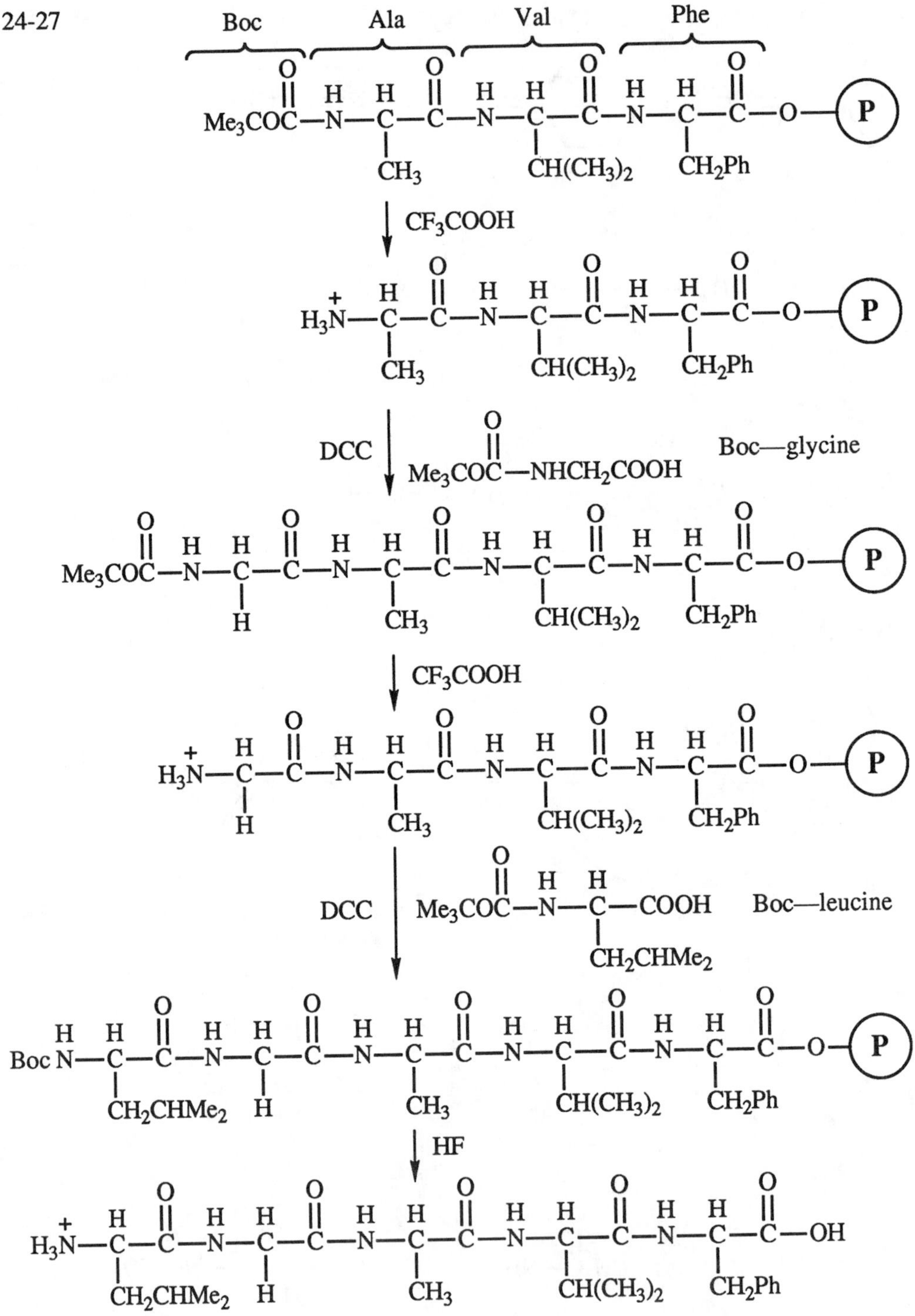
Boc
Ala
Val
Phe
CF3COOH
DCC
Boc—glycine
CF3COOH
DCC
Boc—leucine
HF

24-28

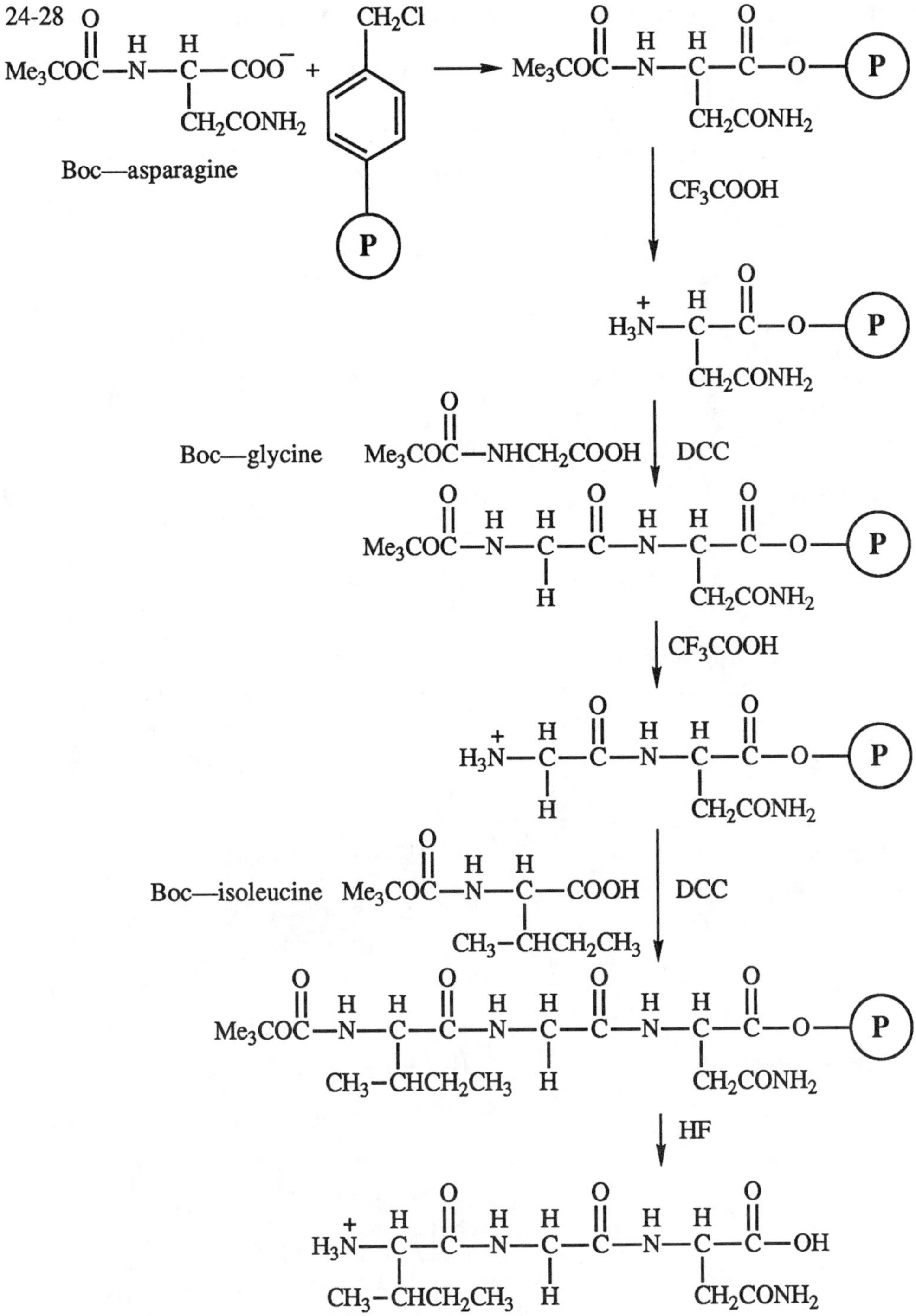
Boc—asparagine
CF₃COOH
Boc—glycine
DCC
CF₃COOH
Boc—isoleucine
DCC
HF

24-29 Refer to the Glossary and the text for definitions and examples.

24-30

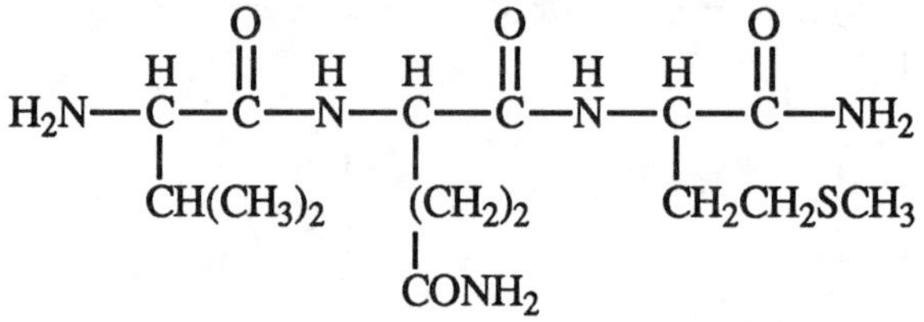

24-31

(a)

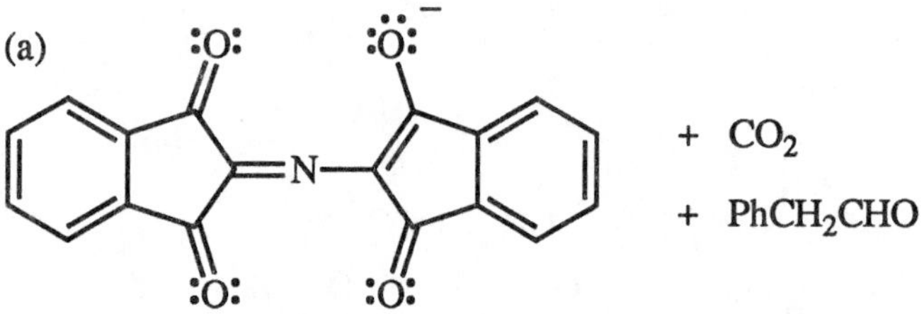

(b)

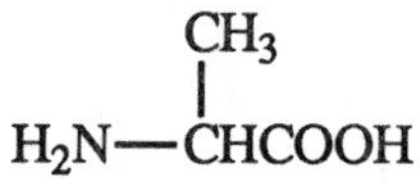

(c)

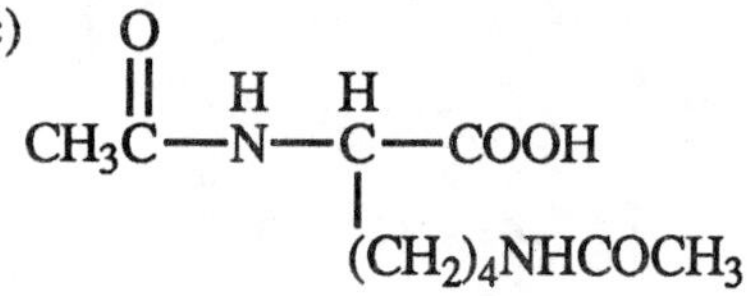

(d)

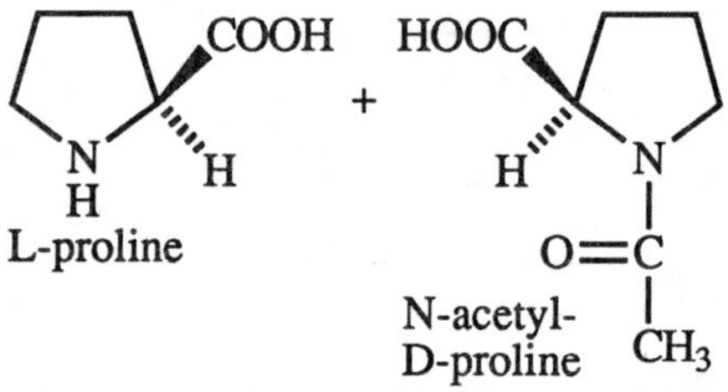
L-proline
N-acetyl-
D-proline

(e)

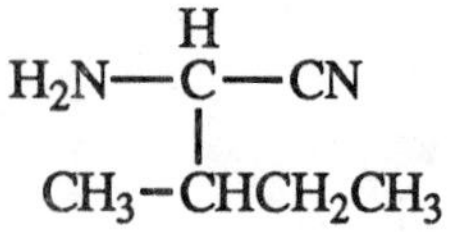

(f)

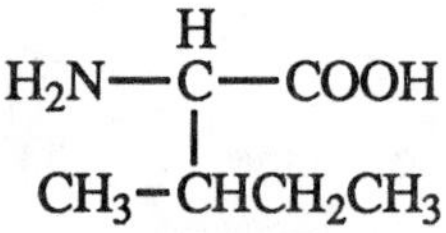

isoleucine

(g)

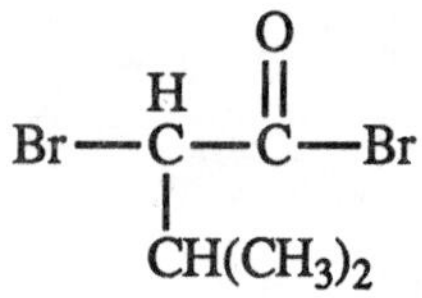

(h)

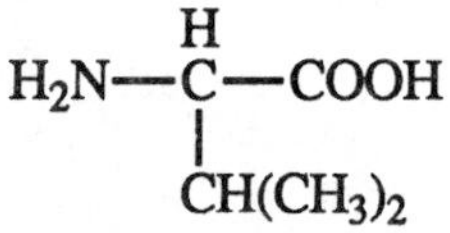

valine

another possible answer, depending on whether the acid bromide is hydrolyzed before adding ammonia

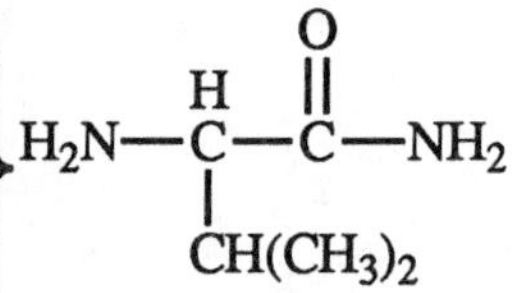

24-32

(a)

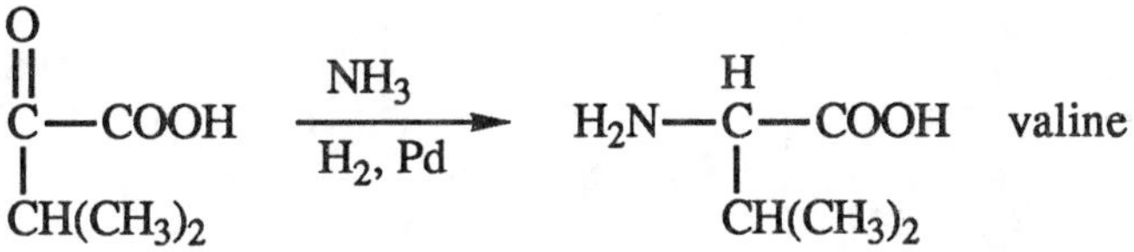
C—COOH
CH(CH3)2
NH3
H2, Pd
H2N—C—COOH
CH(CH3)2
valine

(b)

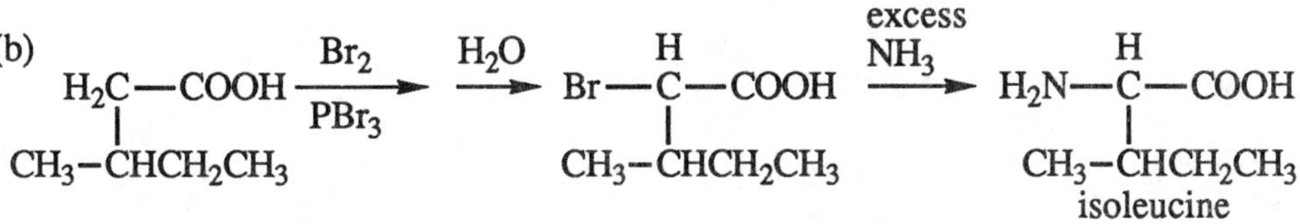
H2C—COOH
CH3–CHCH2CH3
Br2
PBr3
H2O
Br—C—COOH
excess
NH3
H2N—C—COOH
isoleucine

(c)

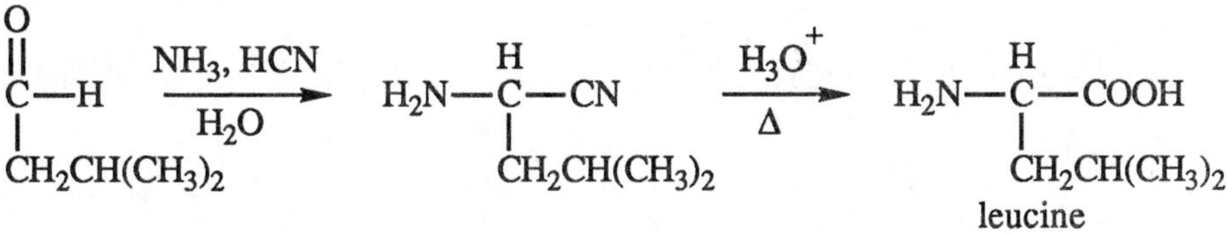
C—H
CH2CH(CH3)2
NH3, HCN
H2O
H2N—C—CN
H3O+
Δ
H2N—C—COOH
leucine

(d)

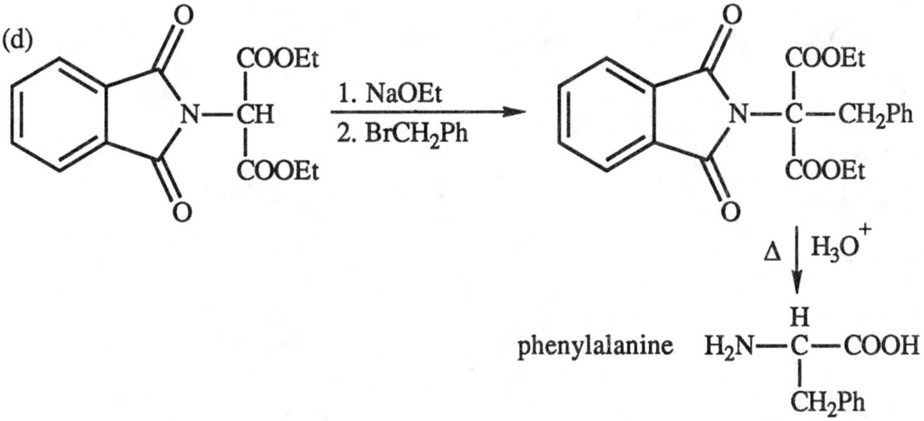
COOEt
N—CH
1. NaOEt
2. BrCH2Ph
N—C—CH2Ph
Δ
H3O+
phenylalanine
H2N—C—COOH
CH2Ph

24-33

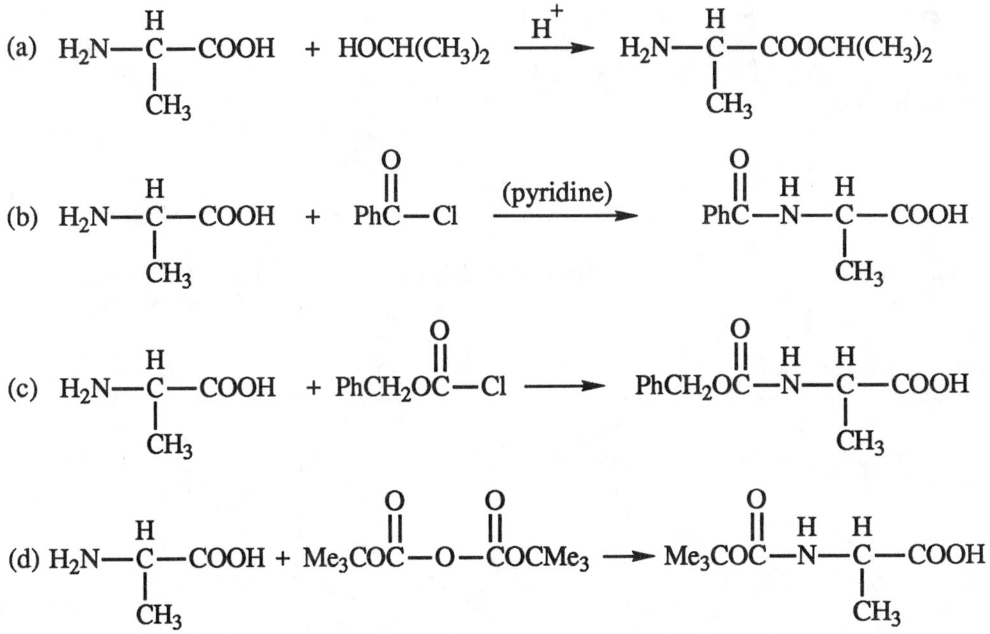

24-34

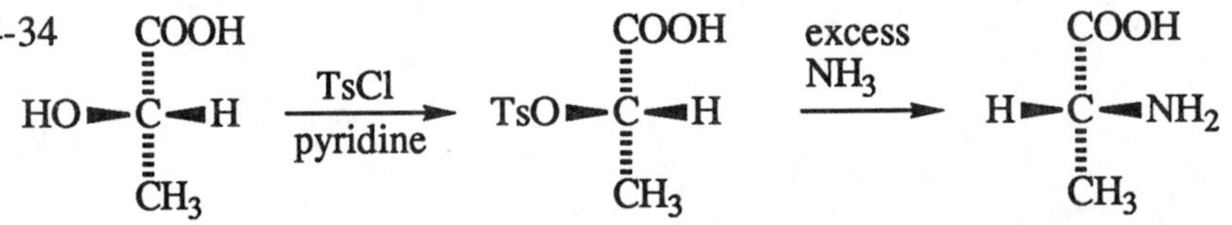

D-alanine

24-35

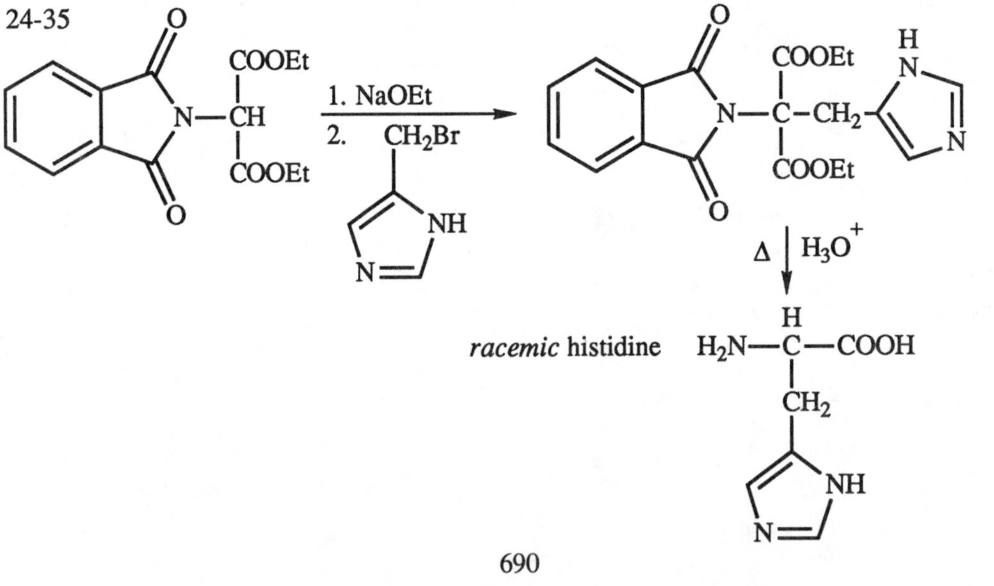

1. NaOEt
2. CH_2Br
Δ
H_3O^+
racemic histidine

24-36

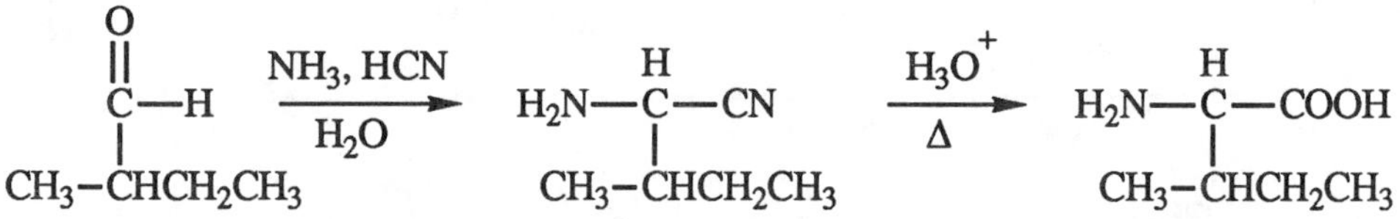

racemic isoleucine

24-37

(a)

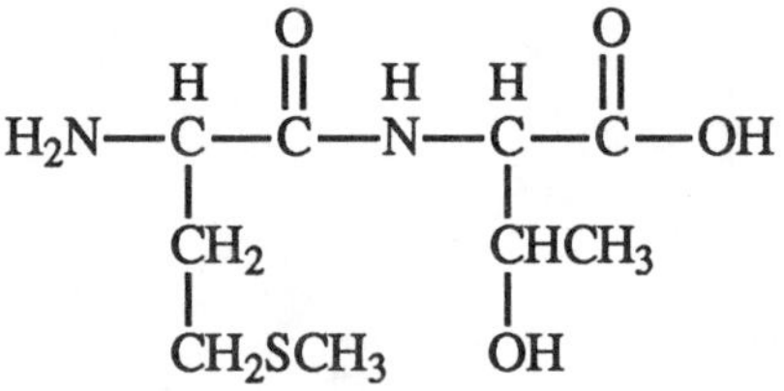

neutral

(b)

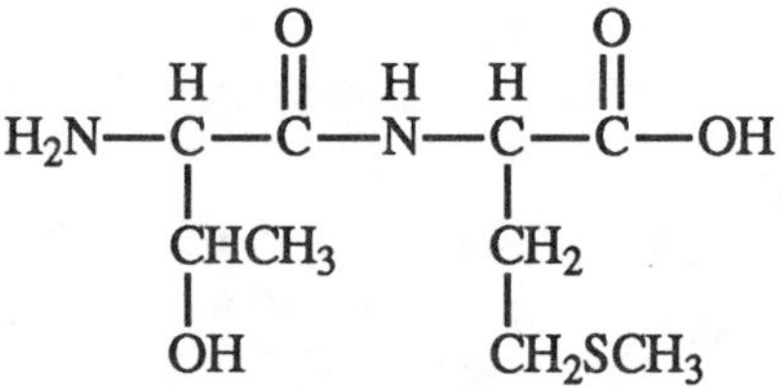

neutral

(c)

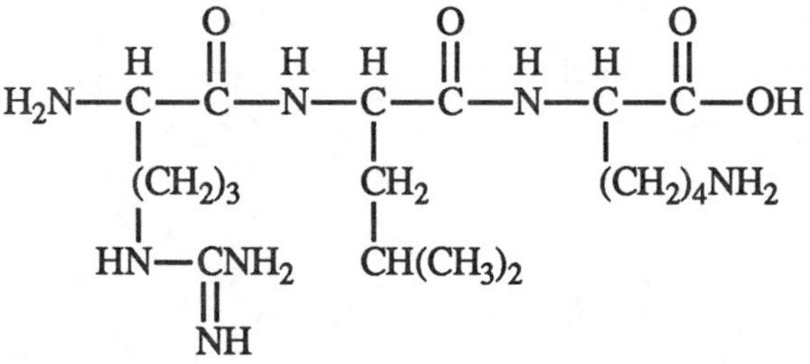

basic: two basic side chains

(d)

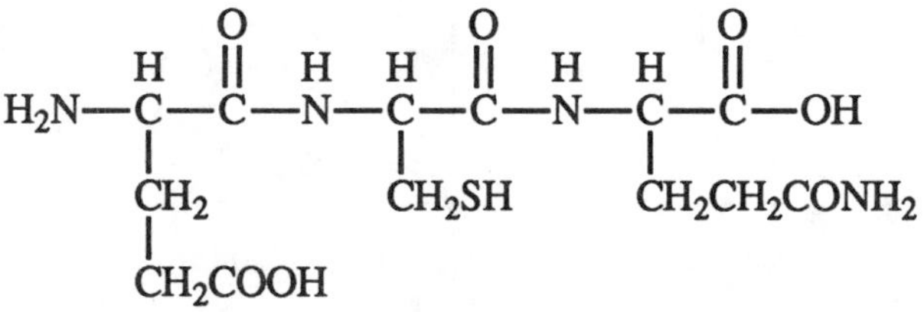

acidic: carboxylic acid side chain, and the SH is weakly acidic

24-38

(a), (b), (c)

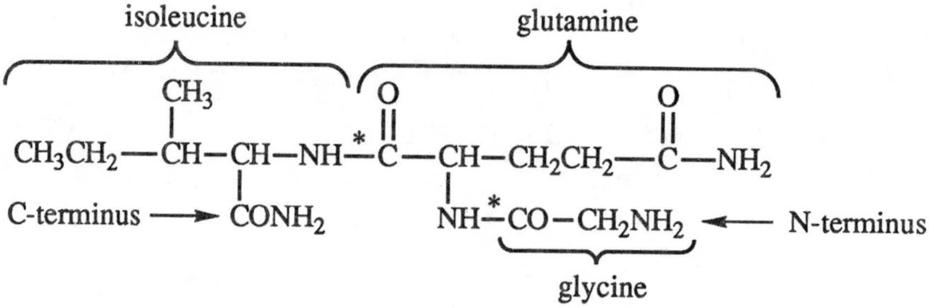
isoleucine
glutamine
C-terminus
N-terminus
glycine

Peptide bonds are denoted with asterisks (*).

(d) glycylglutamylisoleucinamide; Gly—Glu—Ile • NH_2

24-39

Aspartame:

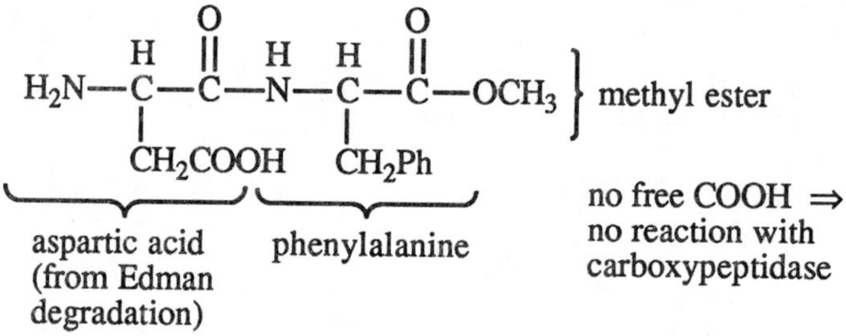
methyl ester
aspartic acid (from Edman degradation)
phenylalanine
no free COOH ⇒ no reaction with carboxypeptidase

Aspartame is aspartylphenylalanine methyl ester.

24-40

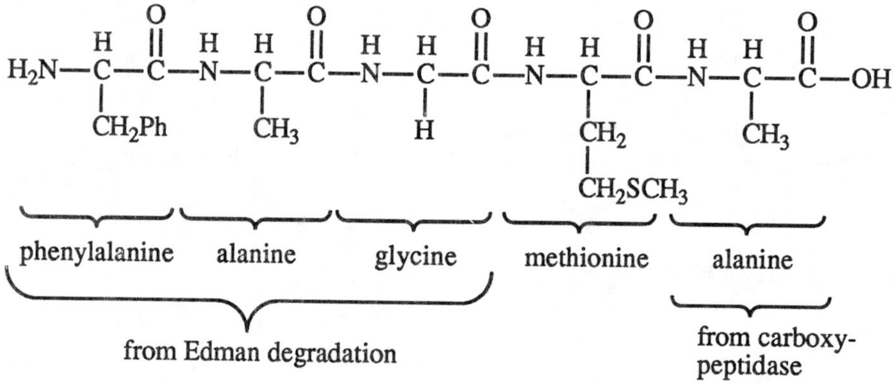
phenylalanine
alanine
glycine
methionine
alanine
from Edman degradation
from carboxy-peptidase

24-41

(a)

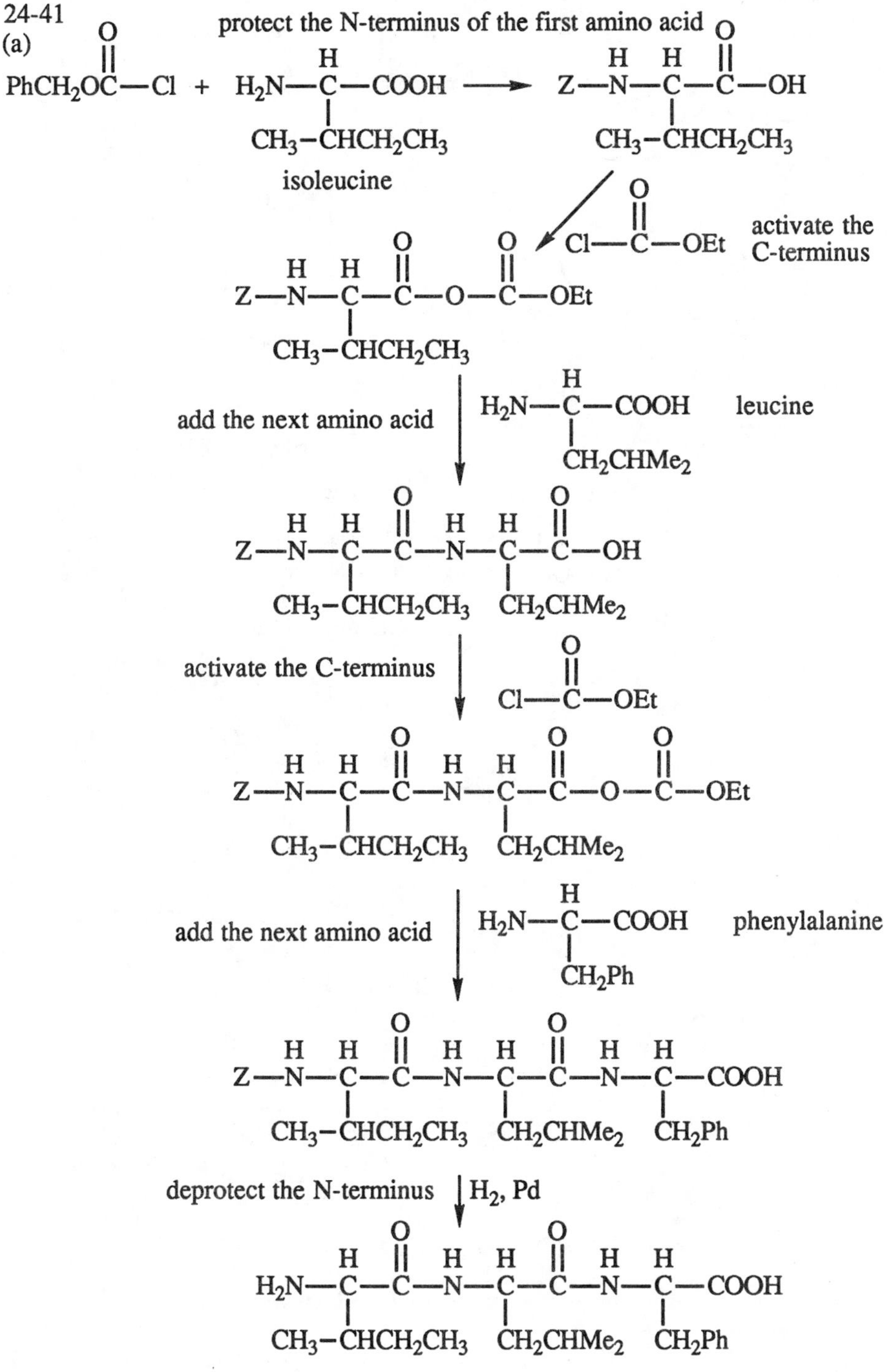
protect the N-terminus of the first amino acid
isoleucine
activate the C-terminus
add the next amino acid
leucine
activate the C-terminus
add the next amino acid
phenylalanine
deprotect the N-terminus
H2, Pd

24-41 continued
(b)

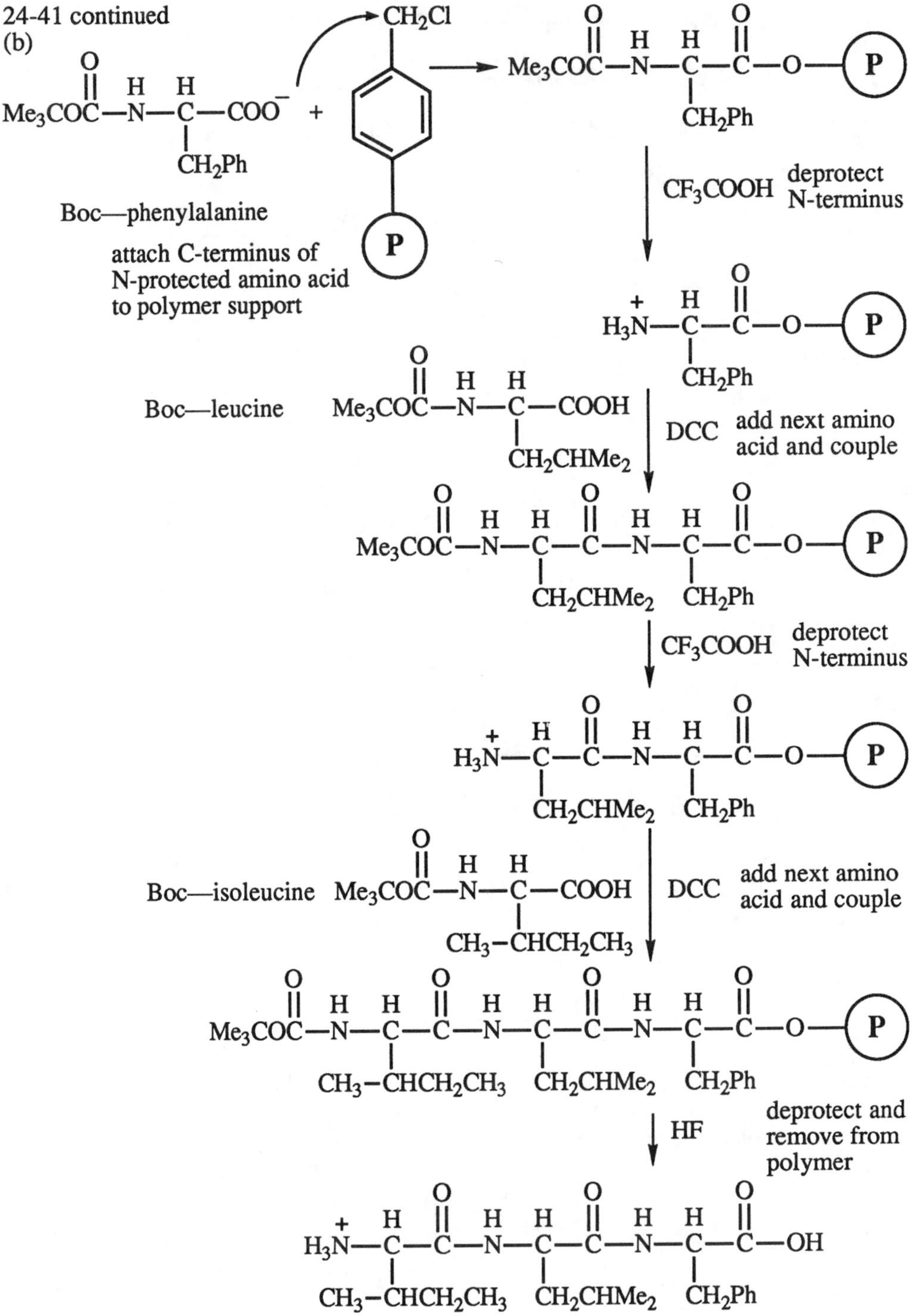
Boc—phenylalanine
attach C-terminus of N-protected amino acid to polymer support
CF3COOH
deprotect N-terminus
Boc—leucine
DCC
add next amino acid and couple
CF3COOH
deprotect N-terminus
Boc—isoleucine
DCC
add next amino acid and couple
HF
deprotect and remove from polymer

24-42

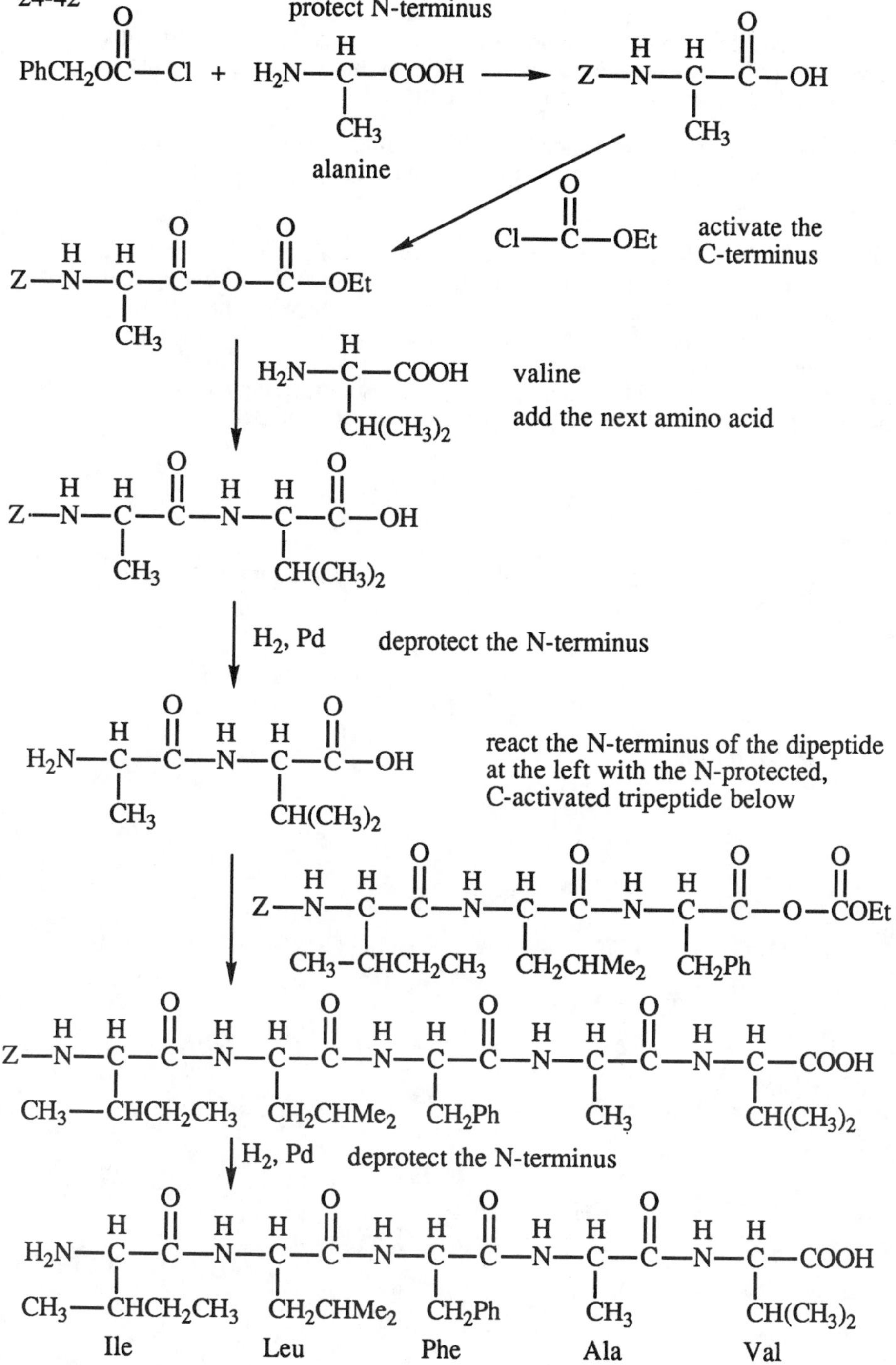
protect N-terminus
alanine
activate the C-terminus
valine
add the next amino acid
H2, Pd
deprotect the N-terminus
react the N-terminus of the dipeptide at the left with the N-protected, C-activated tripeptide below
H2, Pd
deprotect the N-terminus
Ile
Leu
Phe
Ala
Val

24-43

(a) There are two possible sources of ammonia in the hydrolysate. The C-terminus could have been present as the amide instead of the carboxyl, or the glutamic acid could have been present as its amide, glutamine.

(b) The C-terminus is present as the amide. The N-terminus is present as the lactam (cyclic amide) combining the amino group with the carboxyl group of the glutamic acid side chain.

(c) The fact that hydrolysis does not release ammonia implies that the C-terminus is not an amide. Yet, carboxypeptidase treatment gives no reaction, showing that the C-terminus is not a free carboxyl group. Also, treatment with phenyl isothiocyanate gives no reaction, suggesting no free amine at the N-terminus. The most plausible explanation is that the N-terminus has reacted with the C-terminus to produce a cyclic amide, a lactam. (These large rings, called macrocycles, are often found in nature as hormones or antibiotics.)

24-44

(a) Lipoic acid is a mild oxidizing agent. In the process of oxidizing another reactant, lipoic acid is reduced.

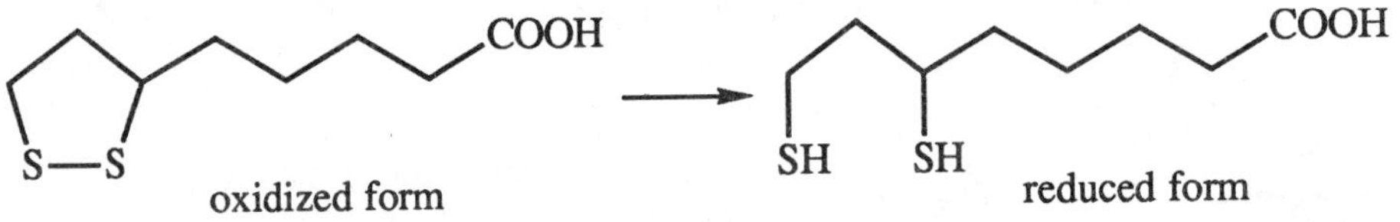

(b)

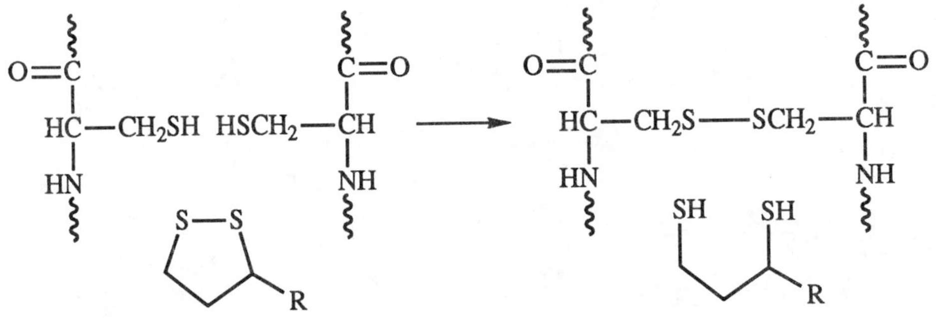

(c)

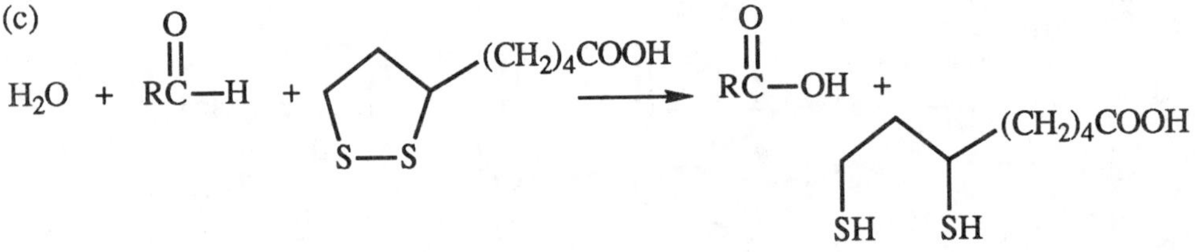

24-45

(a) histidine:

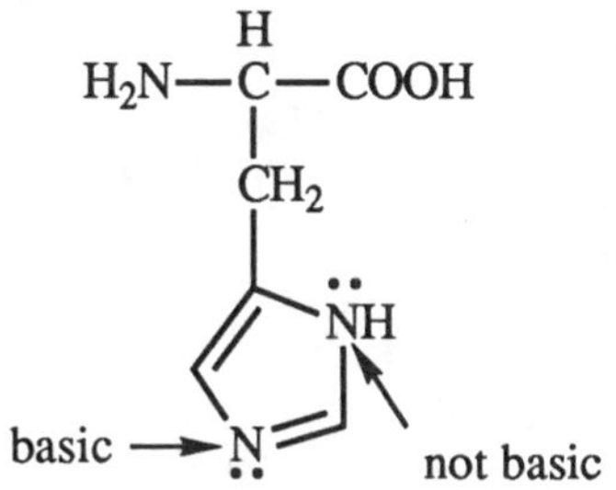

(b)

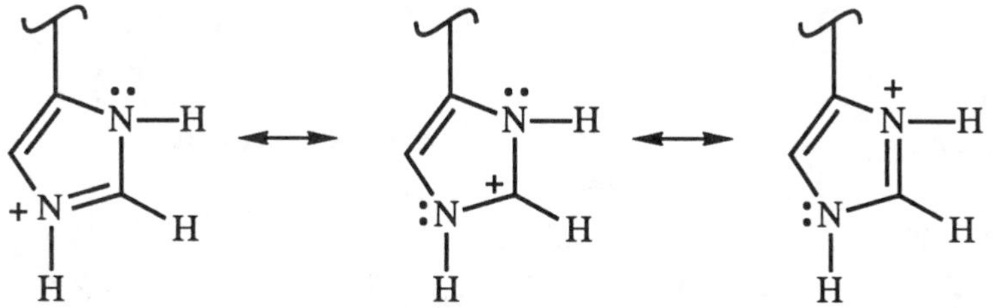

In the protonated imidazole, the two N's are similar in structure, and both NH groups are acidic.

(c)

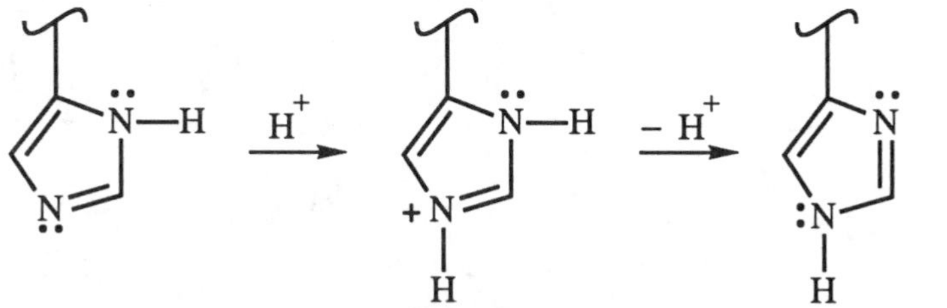

resonance-
stabilized

We usually think of protonation-deprotonation reactions occurring in solution where protons can move with solvent molecules. In an enzyme active-site, there is no "solvent", so there must be another mechanism for movement of protons. Often, conformational changes in the protein will move atoms closer or farther. Histidine serves the function of moving a proton toward or away from a particular site by using its different nitrogens in concert as a proton acceptor and a proton donor.

24-46 The high isoelectric point suggests a strongly basic side chain as in lysine. The N—CH_2 bond in the side chain of arginine is likely to have remained intact during the metabolism. (Can you propose a likely mechanism for this reaction?)

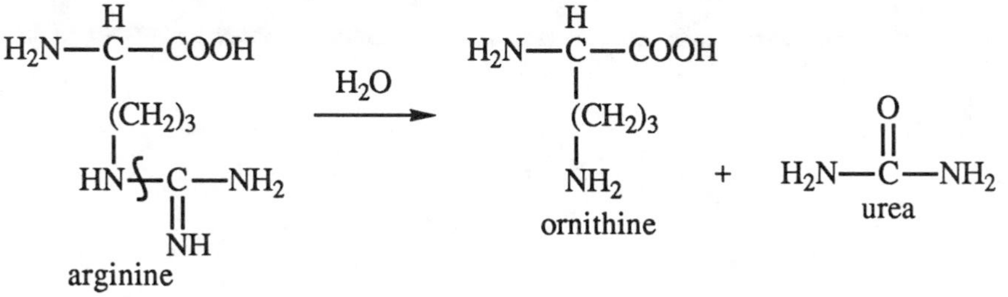

24-47

glutathione:

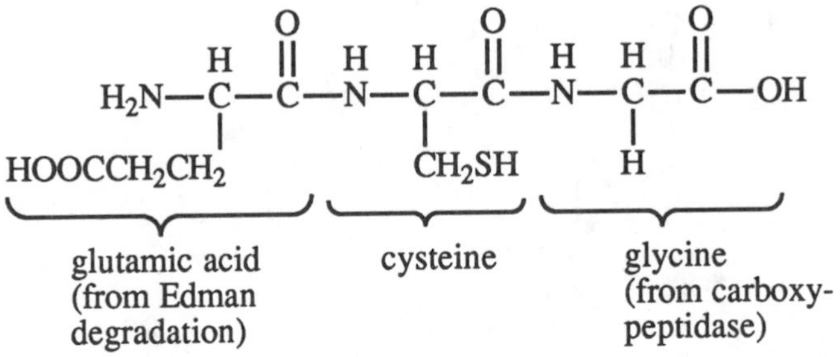

HOOCCH2CH2
CH2SH
glutamic acid (from Edman degradation)
cysteine
glycine (from carboxy-peptidase)

24-48

end groups: N-terminus Ala————————Ile C-terminus

<u>chymotrypsin fragments</u>

A Glu—Gly—Tyr (middle)

B Ala—Lys—Phe
N-terminus

C Arg—(Ser? Leu?)—Ile
C-terminus

Ala—Lys—Phe—Glu—Gly—Tyr—Arg—(Ser? Leu?)—Ile

<u>trypsin fragments</u>

D Ala—Lys

E Phe—Glu—Gly—Tyr—Arg

F Ser—Leu—Ile

Ala—Lys—Phe—Glu—Gly—Tyr—Arg—Ser—Leu—Ile

CHAPTER 25—LIPIDS

25-1

trimyristin

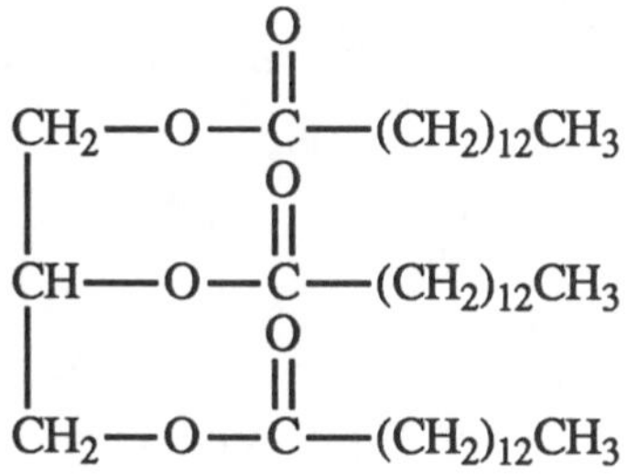

25-2

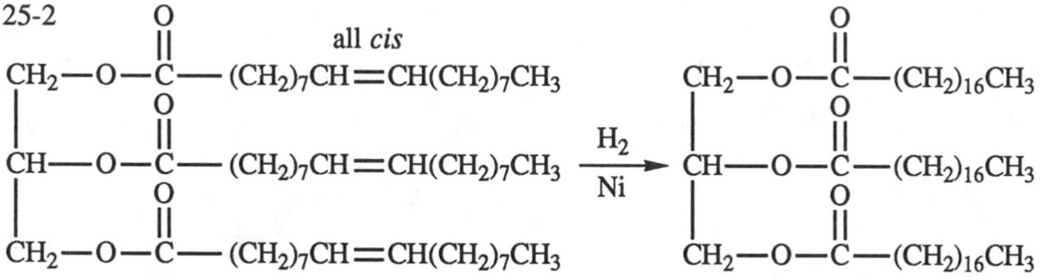

triolein, m.p. −4° C
(liquid at room temperature)

tristearin, m.p. 72° C
(solid at room temperature)

25-3

(a)

$$2\ CH_3(CH_2)_{16}{-}\overset{\displaystyle O}{\overset{\|}{C}}{-}O^-\ Na^+ + Ca^{2+} \longrightarrow \left[CH_3(CH_2)_{16}{-}\overset{\displaystyle O}{\overset{\|}{C}}{-}O\right]_2 Ca\ +\ 2\ Na^+$$

(b)

$$2\ CH_3(CH_2)_{16}{-}\overset{\displaystyle O}{\overset{\|}{C}}{-}O^-\ Na^+ + Mg^{2+} \longrightarrow \left[CH_3(CH_2)_{16}{-}\overset{\displaystyle O}{\overset{\|}{C}}{-}O\right]_2 Mg\ +\ 2\ Na^+$$

(c)

$$3\ CH_3(CH_2)_{16}{-}\overset{\displaystyle O}{\overset{\|}{C}}{-}O^-\ Na^+ + Fe^{3+} \longrightarrow \left[CH_3(CH_2)_{16}{-}\overset{\displaystyle O}{\overset{\|}{C}}{-}O\right]_3 Fe\ +\ 3\ Na^+$$

25-4

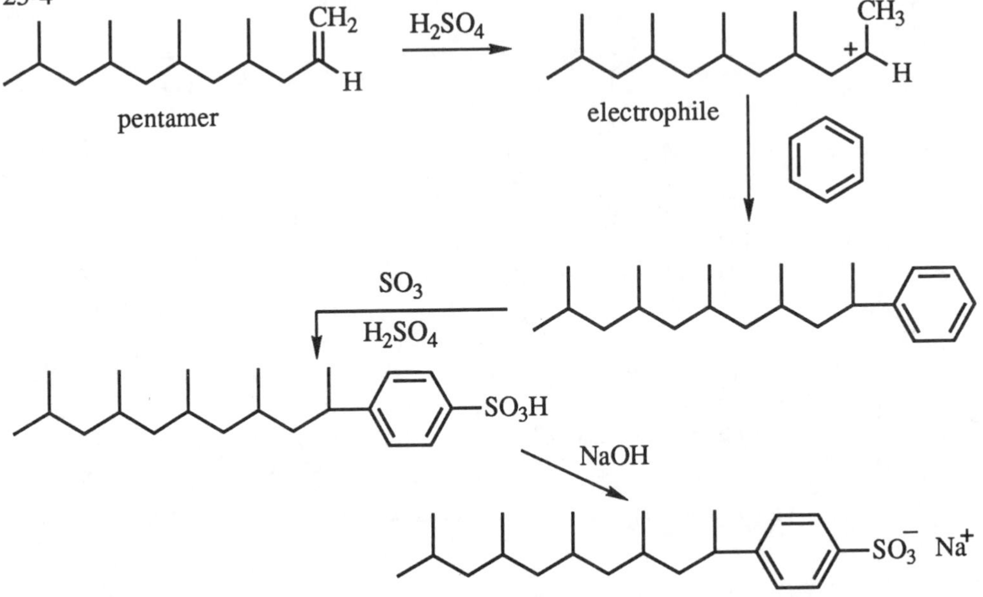
CH2
H
pentamer
H2SO4
CH3
+
H
electrophile
SO3
H2SO4
SO3H
NaOH
SO3− Na+

25-5

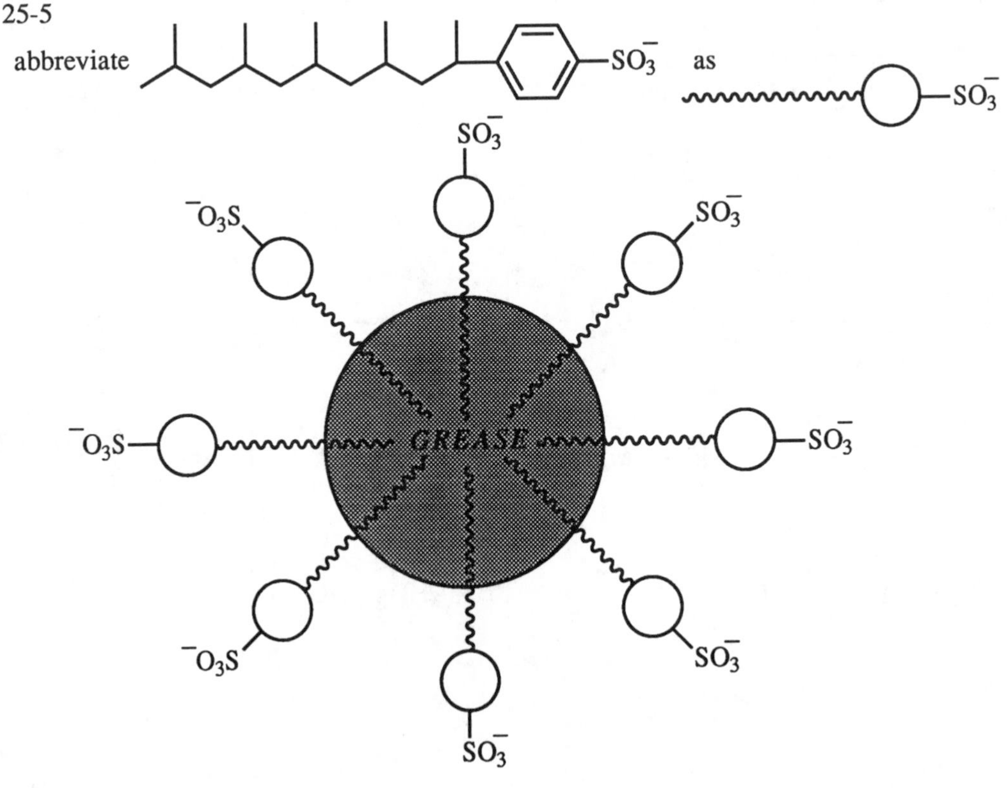
abbreviate
SO3−
as
SO3−
GREASE
−O3S

25-6

(a)

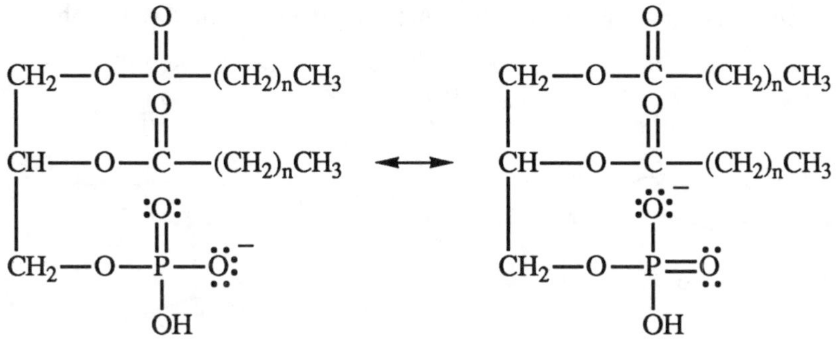

(b)

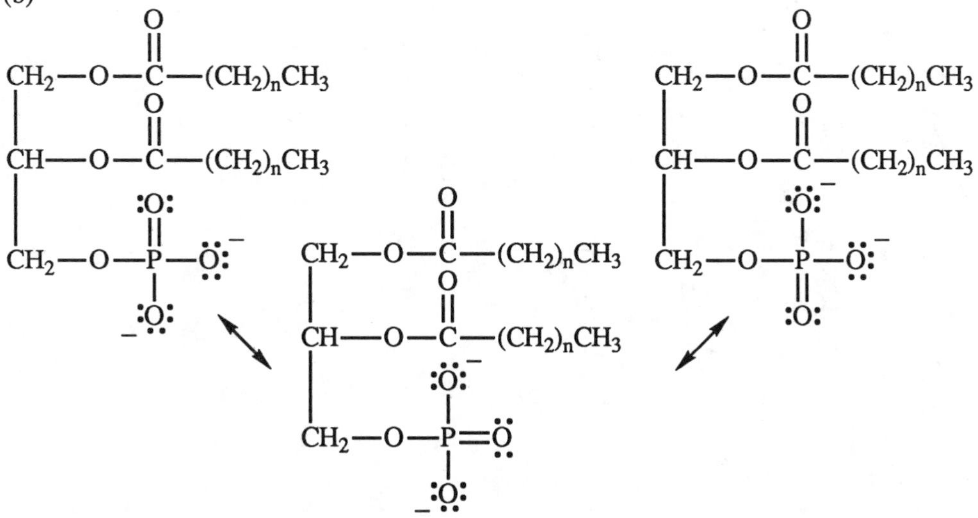

25-7 Estradiol is a phenol and can be ionized with aqueous NaOH. Testosterone does not have any hydrogens acidic enough to react with NaOH. Treatment of a solution of estradiol and testosterone in organic solvent with aqueous base will extract the phenoxide form of estradiol into the aqueous layer, leaving testosterone in the organic layer. Acidification of the aqueous base will precipitate estradiol which can be filtered. Evaporation of the organic solvent will leave testosterone.

25-8 Models may help. Abbreviations: "ax" = axial; "eq" = equatorial. Note that substituents at *cis*-fused ring junctures are axial to one ring and equatorial to another.

(a)

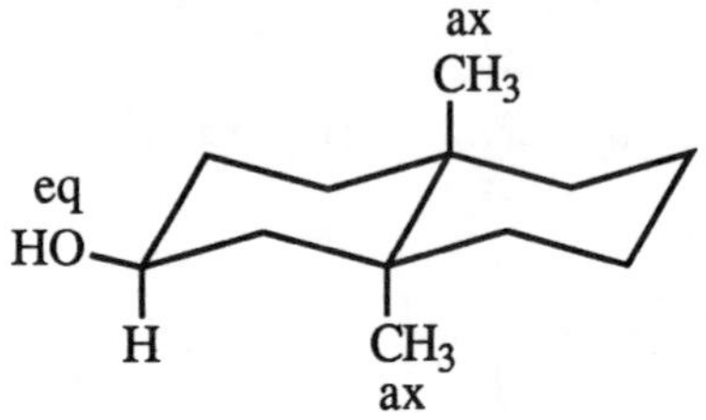

(b)

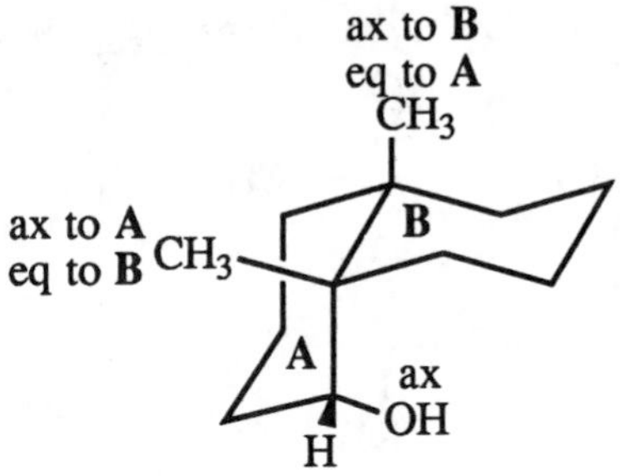

(c)

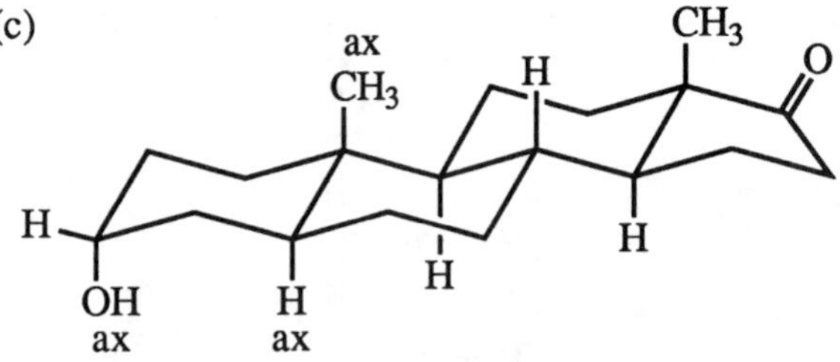

(d)

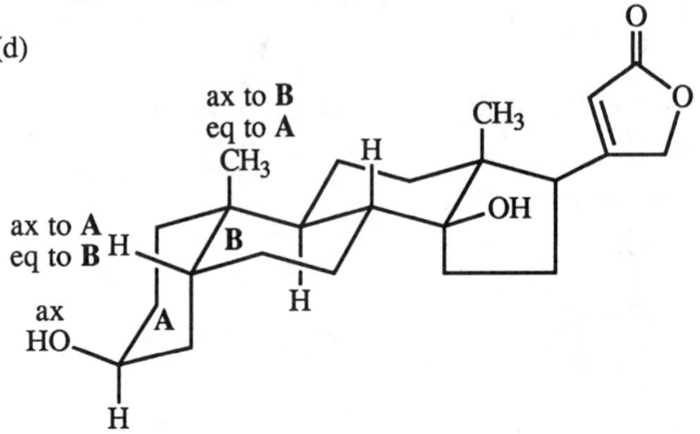

25-9

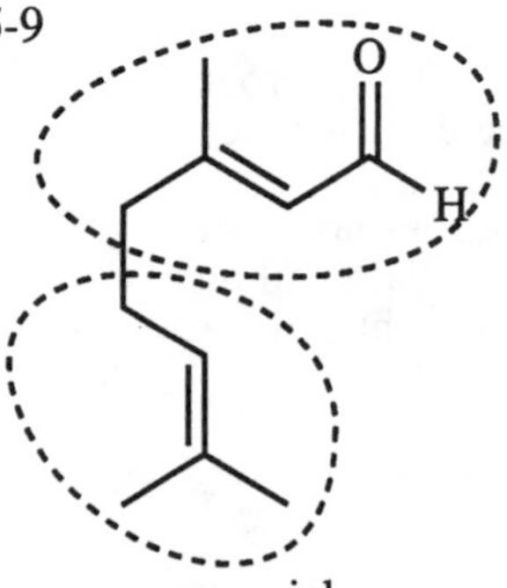

geranial

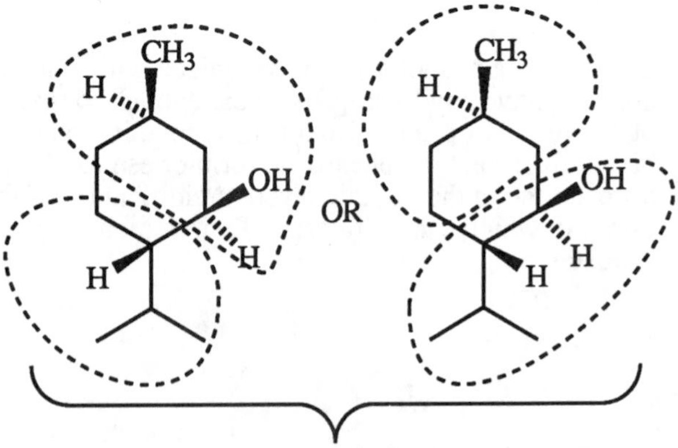

menthol

25-9 continued

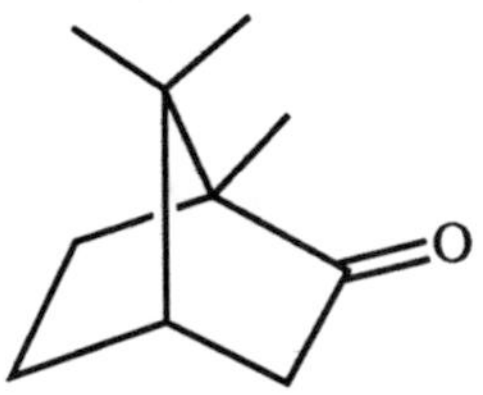

camphor

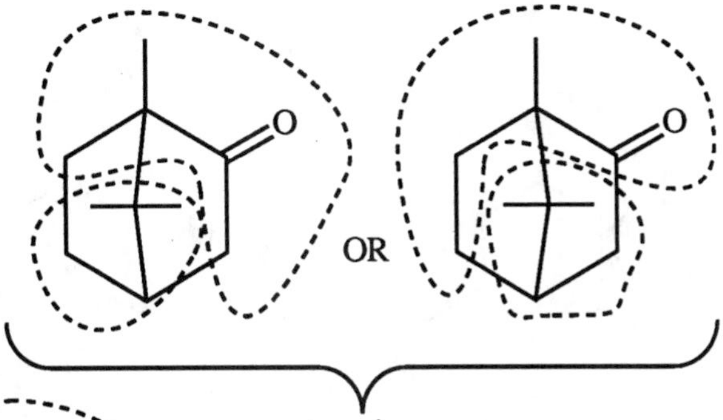

top view

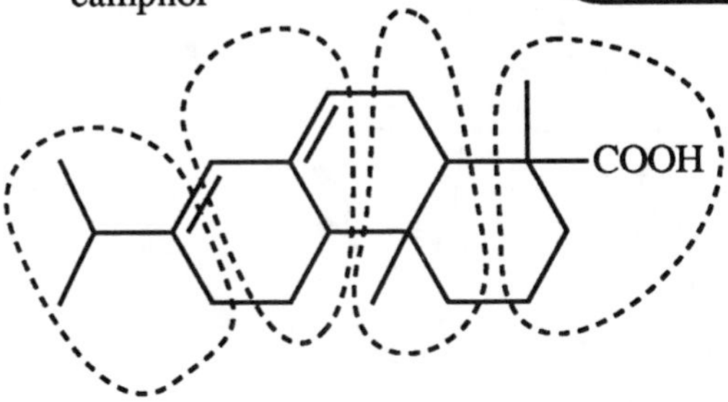

abietic acid

25-10 β-carotene

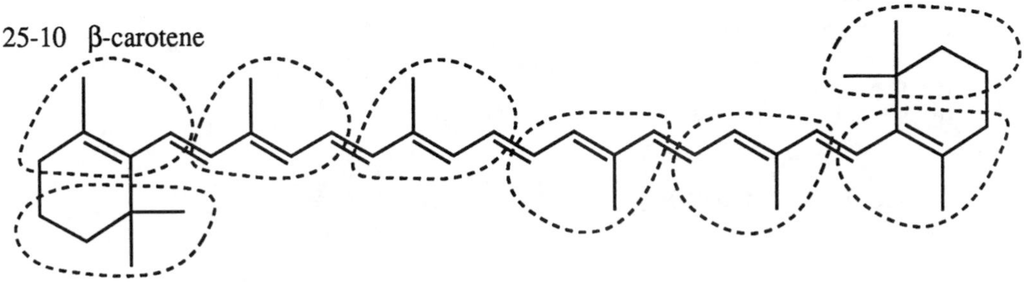

25-11

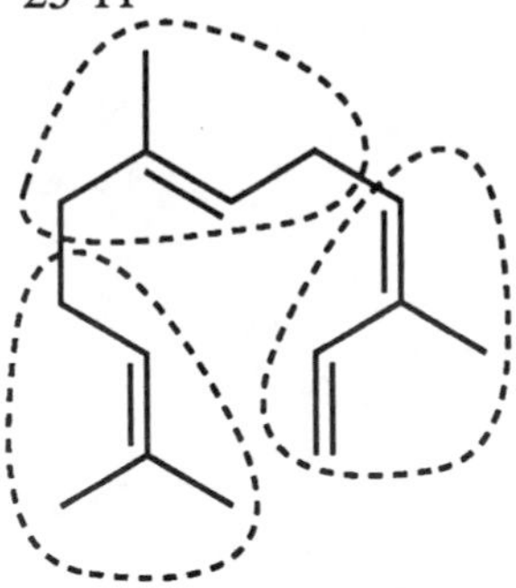

α-farnesene
sesquiterpene

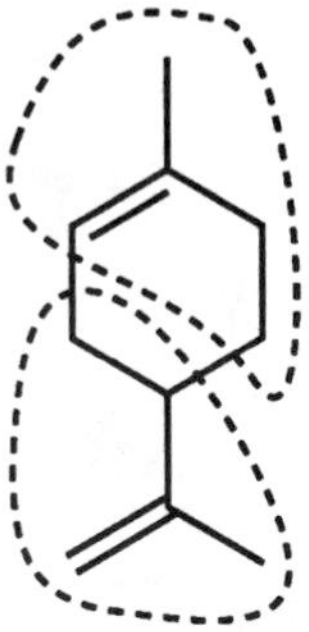

limonene
monoterpene
(Limonene can also be circled in the other direction around the ring—see menthol in 25-9.)

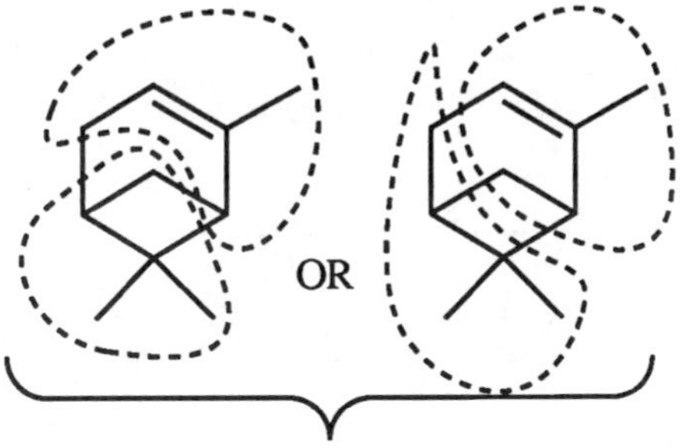

α-pinene
monoterpene

25-11 continued

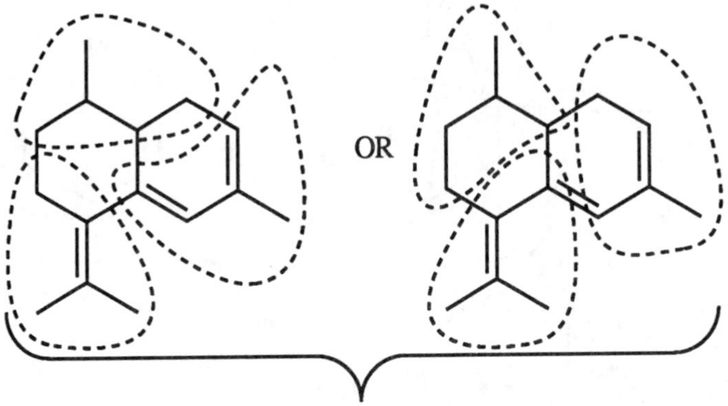

zingiberene
sesquiterpene

25-12 Refer to the Glossary and the text for definitions and examples.

25-13 (a) triglyceride
(b) synthetic detergent
(c) wax
(d) sesquiterpene
(e) steroid

25-14

(a) **3** $CH_3(CH_2)_7CH{=}CH(CH_2)_7COO^- \; Na^+$ + $HOCH(CH_2OH)_2$
soap — glycerol

(b)

$$\begin{array}{l} CH_2{-}O{-}\overset{\displaystyle O}{\overset{\|}{C}}{-}(CH_2)_{16}CH_3 \\ | \\ CH{-}O{-}\overset{\displaystyle O}{\overset{\|}{C}}{-}(CH_2)_{16}CH_3 \\ | \\ CH_2{-}O{-}\overset{\displaystyle O}{\overset{\|}{C}}{-}(CH_2)_{16}CH_3 \end{array}$$

tristearin

25-14 continued

(c)

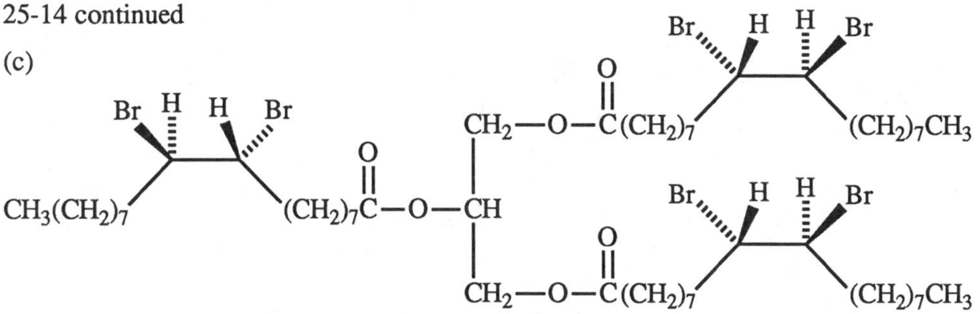

(mixture of diastereomers)

(d)

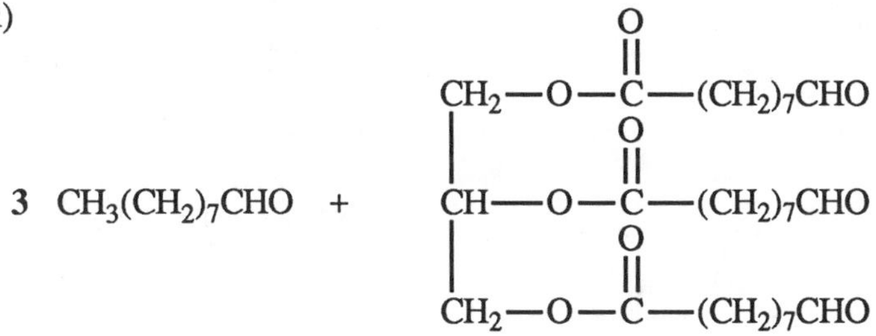

(e)

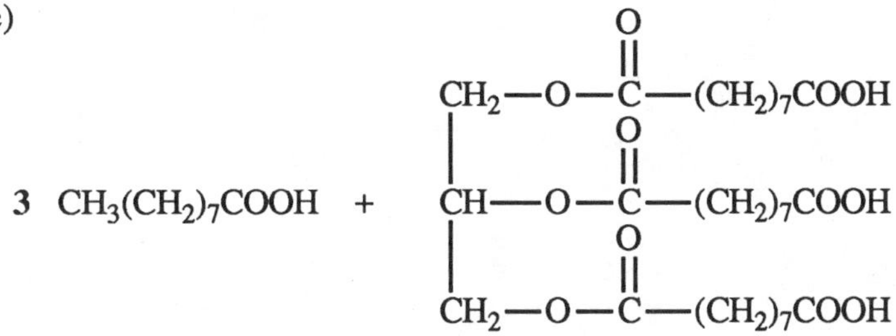

(f)

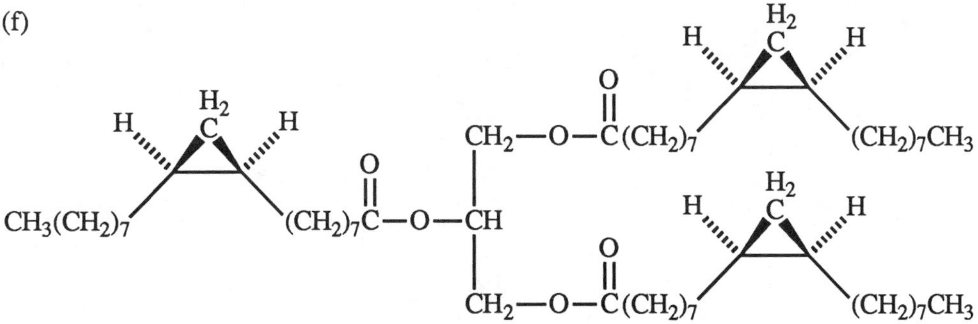

(mixture of diastereomers)

(g) **3** $CH_3(CH_2)_7CH{=}CH(CH_2)_7CH_2OH$ + $HOCH(CH_2OH)_2$
glycerol

25-15

(a) $CH_3(CH_2)_7CH{=}CH(CH_2)_7COOH \xrightarrow[Ni]{H_2} \xrightarrow[2.\ H_3O^+]{1.\ LiAlH_4} CH_3(CH_2)_{16}CH_2OH$

(b) $CH_3(CH_2)_7CH{=}CH(CH_2)_7COOH \xrightarrow[Ni]{H_2} CH_3(CH_2)_{16}COOH$

(c) $CH_3(CH_2)_{16}COOH$ (from (b)) + $HOCH_2(CH_2)_{16}CH_3$ (from (a)) $\xrightarrow[\Delta]{H^+}$ $CH_3(CH_2)_{16}COOCH_2(CH_2)_{16}CH_3$

(d) $CH_3(CH_2)_7CH{=}CH(CH_2)_7COOH \xrightarrow[2.\ Me_2S]{1.\ O_3} CH_3(CH_2)_7CH{=}O$ (nonanal) + $O{=}CH(CH_2)_7COOH$

(e) $CH_3(CH_2)_7CH{=}CH(CH_2)_7COOH \xrightarrow[H_2O,\ \Delta]{KMnO_4} CH_3(CH_2)_7COOH$ + $HOOC(CH_2)_7COOH$ (nonanedioic acid)

(f) $CH_3(CH_2)_7CH{=}CH(CH_2)_7COOH \xrightarrow[PBr_3]{\text{excess } Br_2}$ $\xrightarrow{H_2O}$

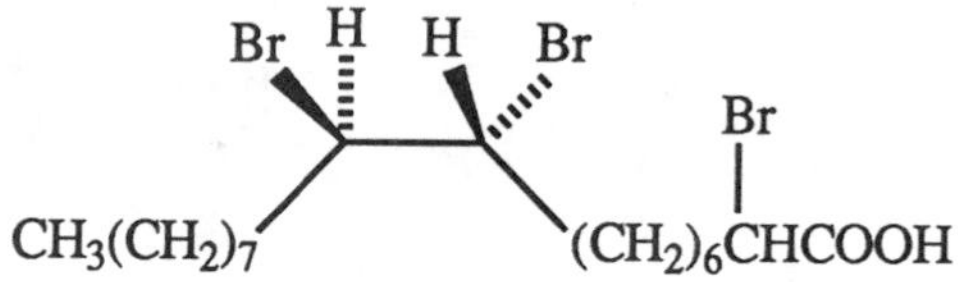

25-16

(a)

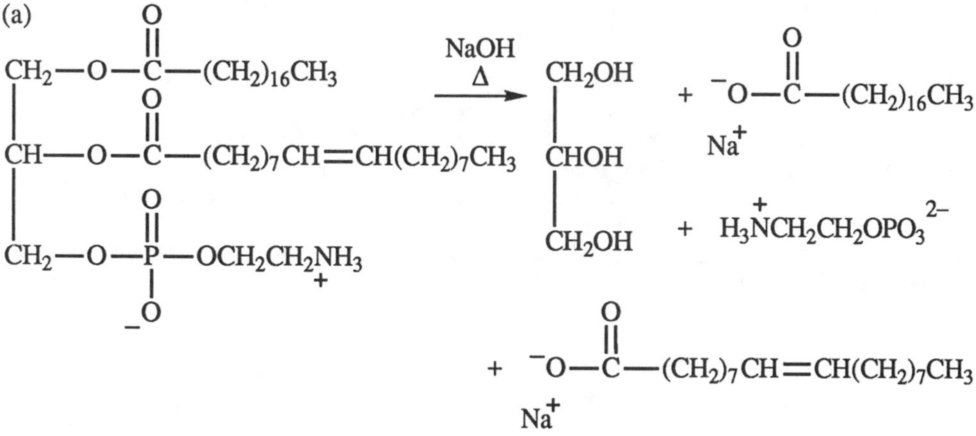

(b)

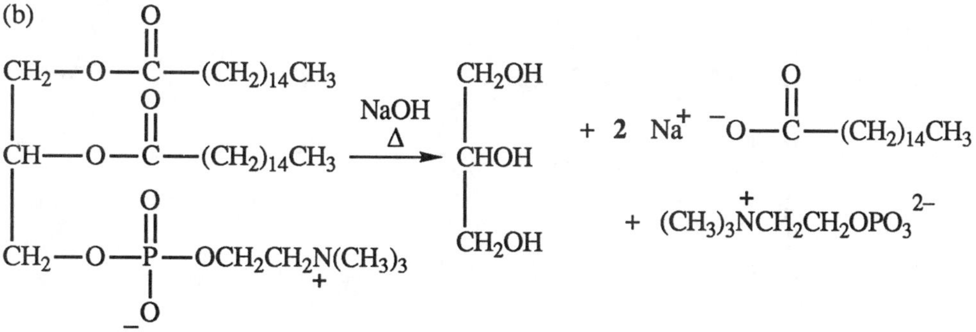

25-17

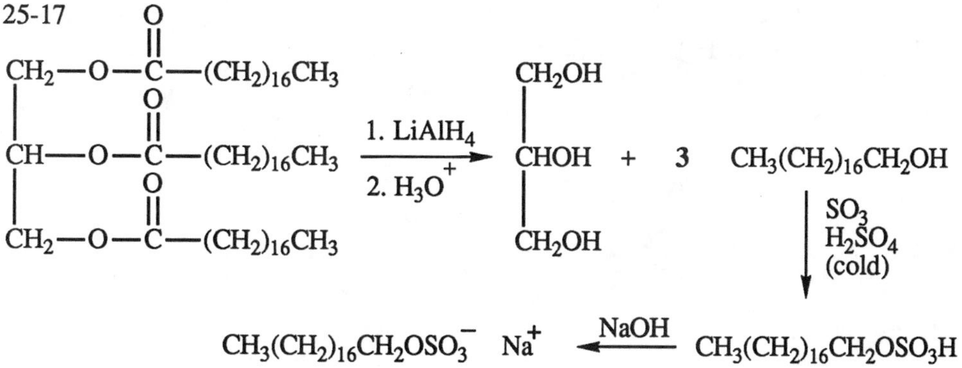

25-18 Reagents in parts (a), (b), and (d) would react with alkenes. If both samples contained alkenes, these reagents could not distinguish the samples. Saponification (part (c)), however, is a reaction of an ester, so only the vegetable oil would react, not the hydrocarbon oil mixture.

25-19 (a) Add an aqueous solution of calcium ion or magnesium ion. Sodium stearate will produce a precipitate, while the sulfonate will not precipitate.
(b) Beeswax, an ester, can be saponified with NaOH. Paraffin wax is a solid mixture of alkanes and will not react.
(c) Myristic acid will dissolve (or be emulsified) in dilute aqueous base. Trimyristin will remain unaffected.
(d) Triolein (an unsaturated oil) will decolorize bromine in CCl_4, but trimyristin (a saturated fat) will not.

25-20

(a)

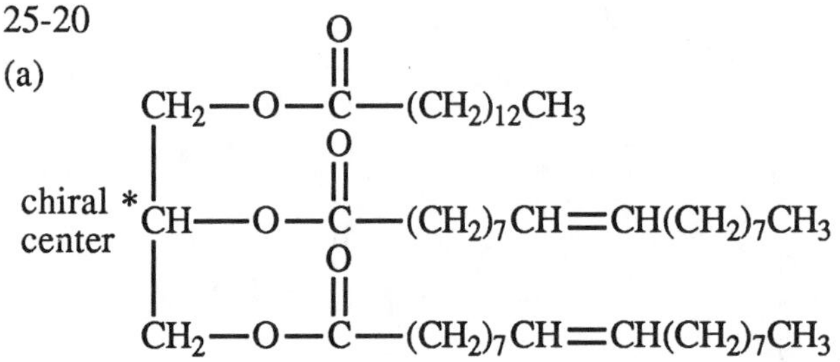
chiral *
center

(b)

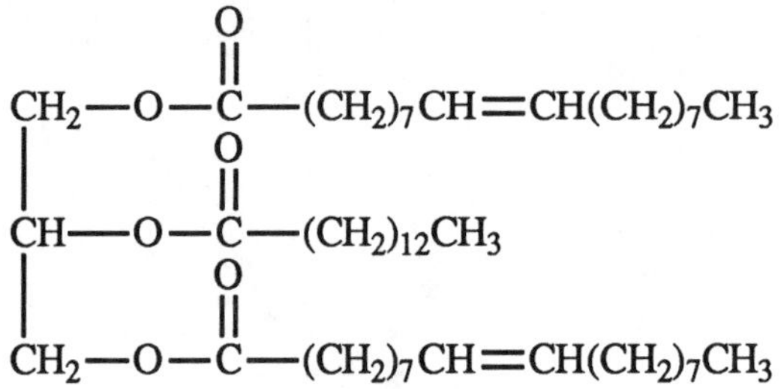

25-21

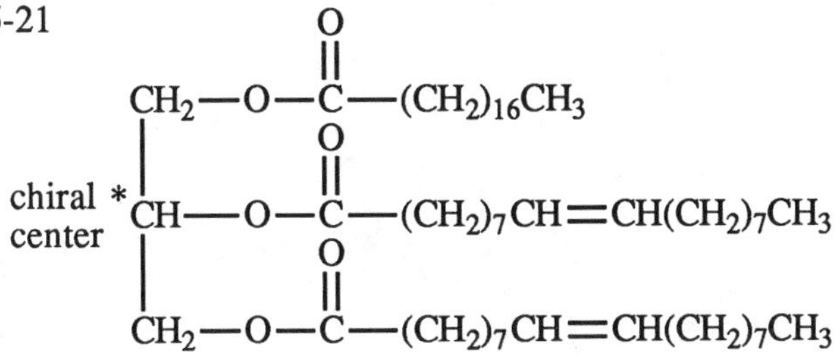
chiral center

optically active

(a)

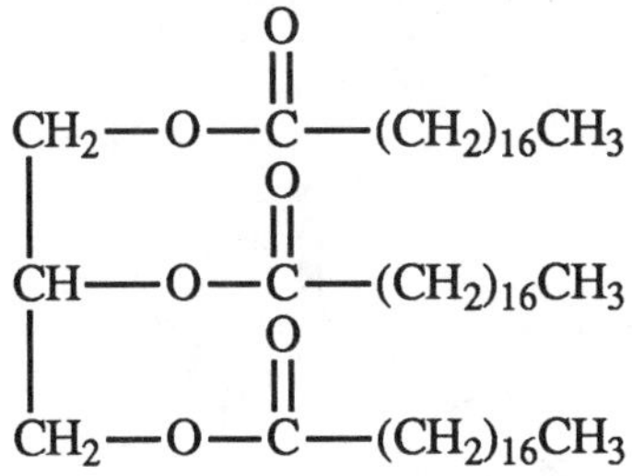

not optically active

(b)

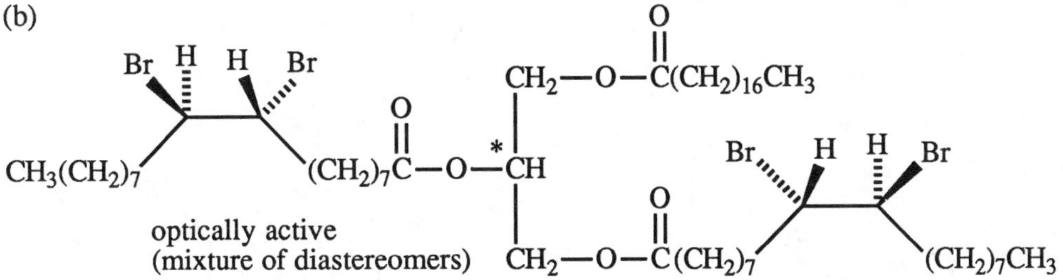

optically active
(mixture of diastereomers)

(c)

$$HOCH(CH_2OH)_2 + HOOC(CH_2)_{16}CH_3 + 2\ HOOC(CH_2)_7CH{=}CH(CH_2)_7CH_3$$

products are not optically active

(d)

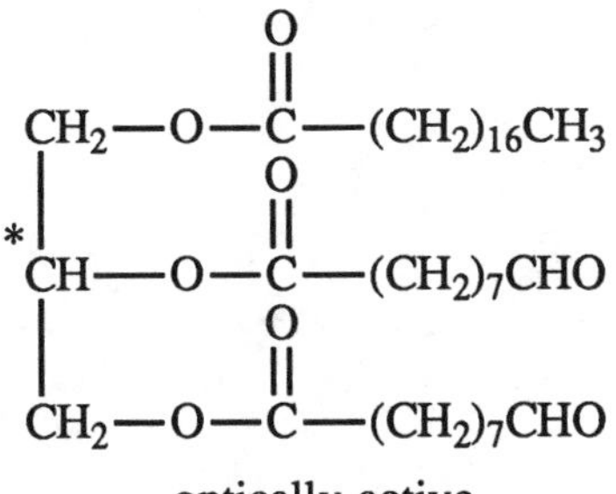

optically active

$+\ 2\ O{=}CH(CH_2)_7CH_3$

not optically active

25-22

(a) sesquiterpene

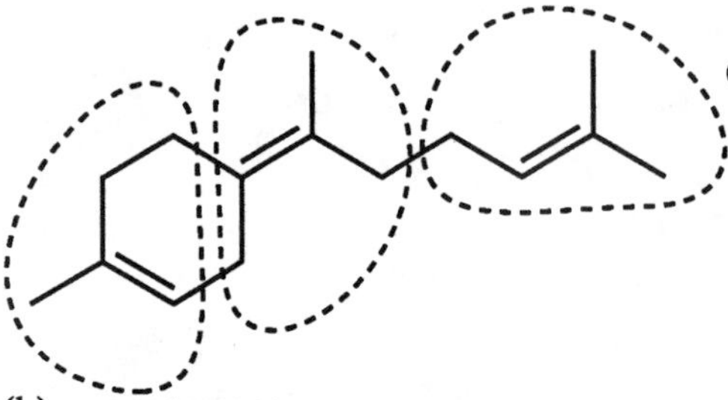

OR

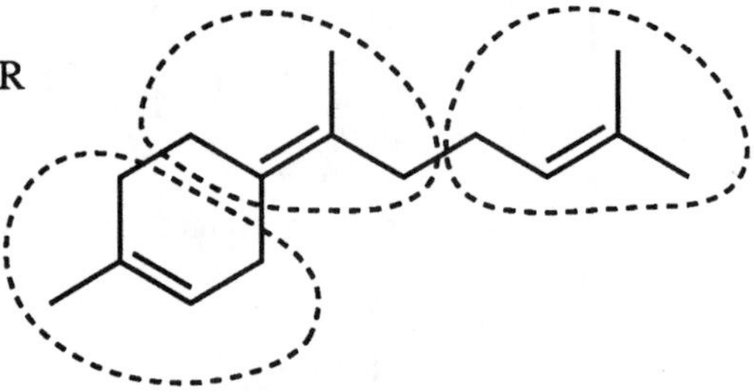

(b) monoterpene

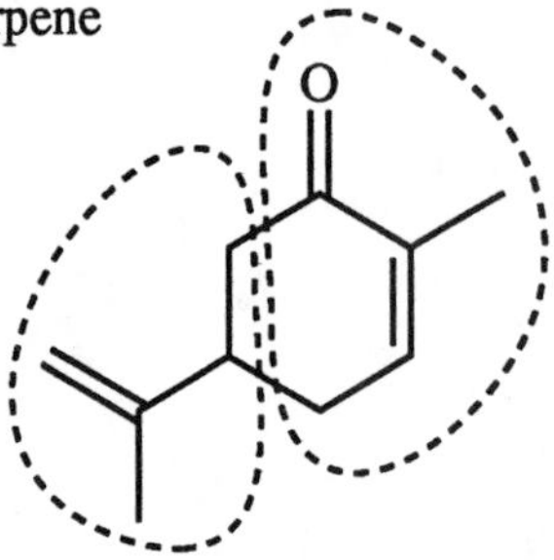

OR

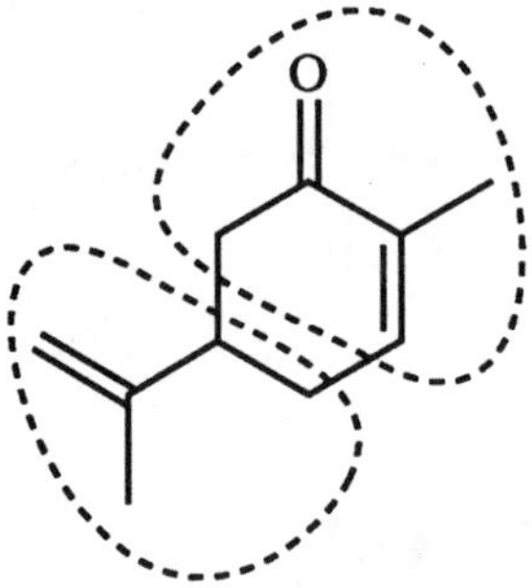

(c) sesquiterpene

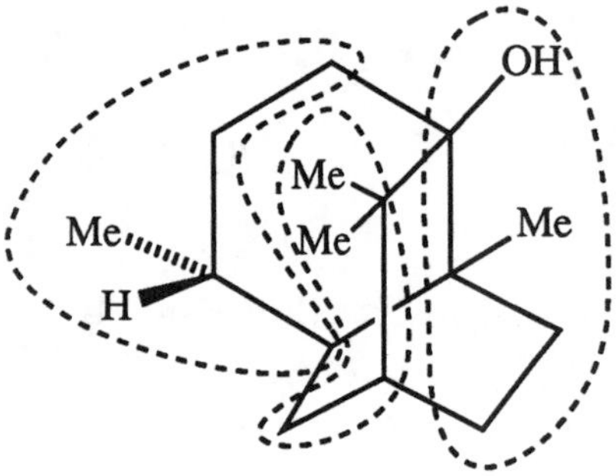

OR

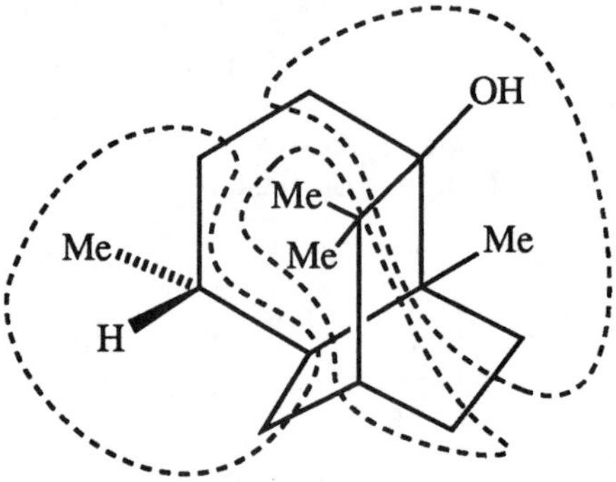

(d) sesquiterpene

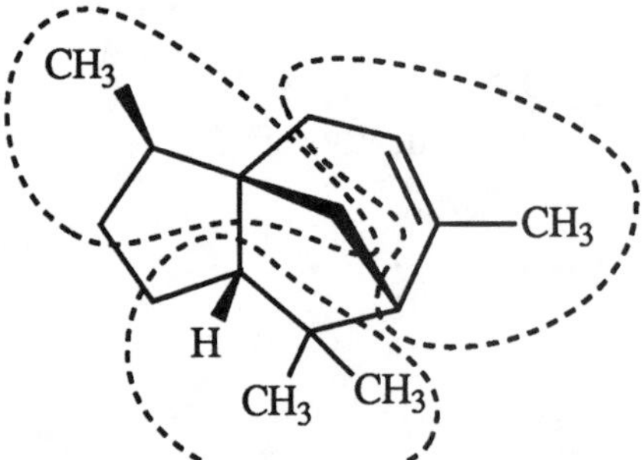

OR

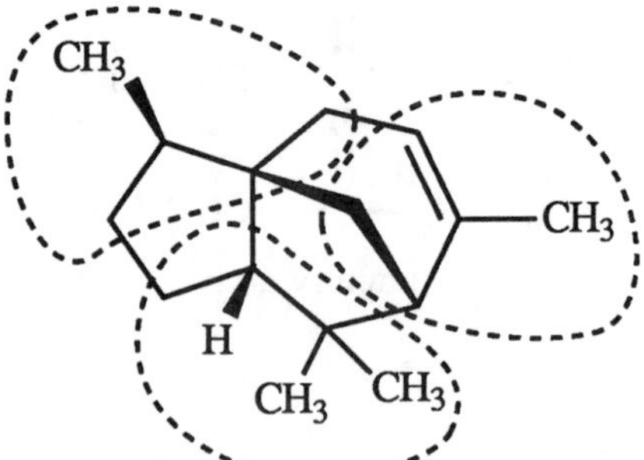

25-23 The products in (b), (c) and (f) are mixtures of stereoisomers.

(a)

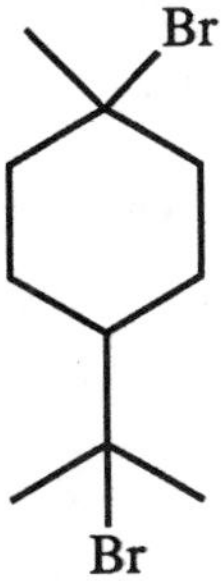

(b)

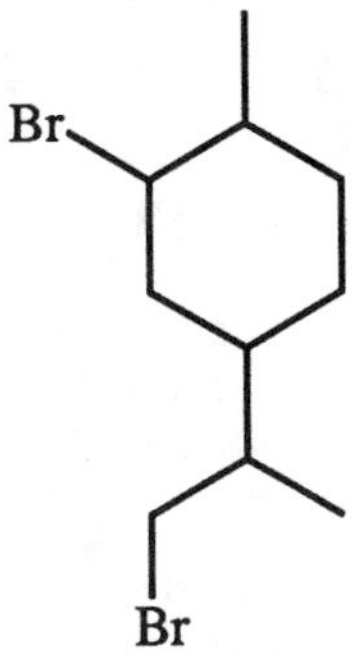

(c)

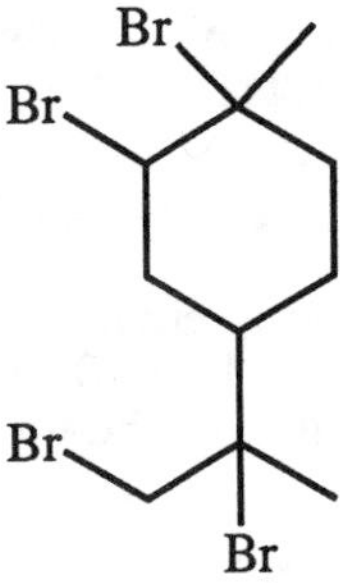

(d)

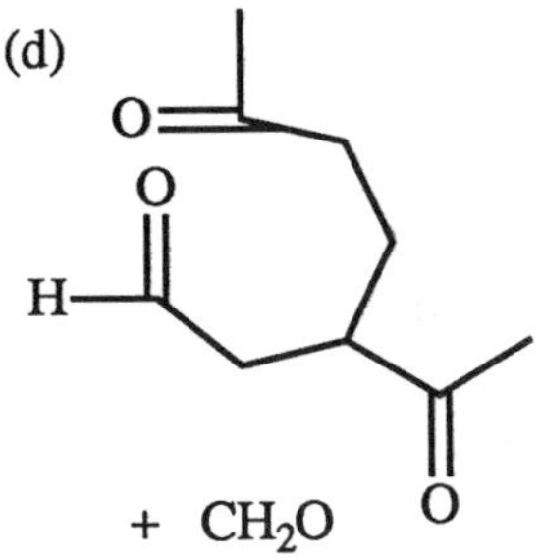

+ CH_2O

(e)

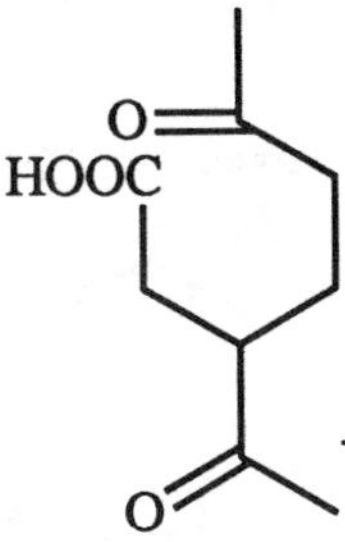

+ CO_2

(f)

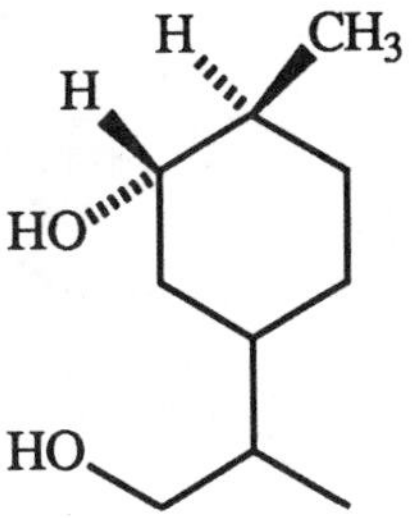

25-24

(a)

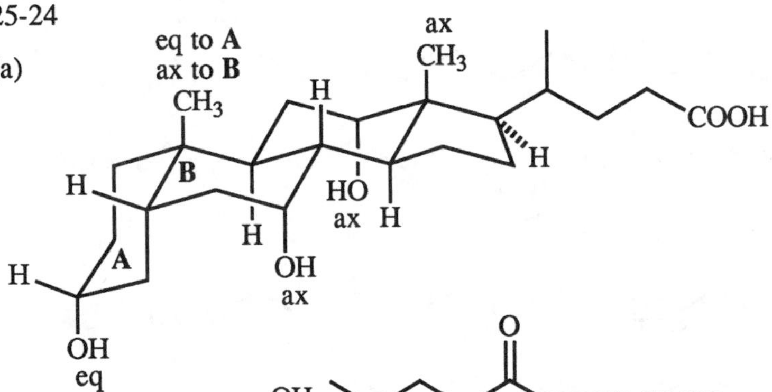

(b)

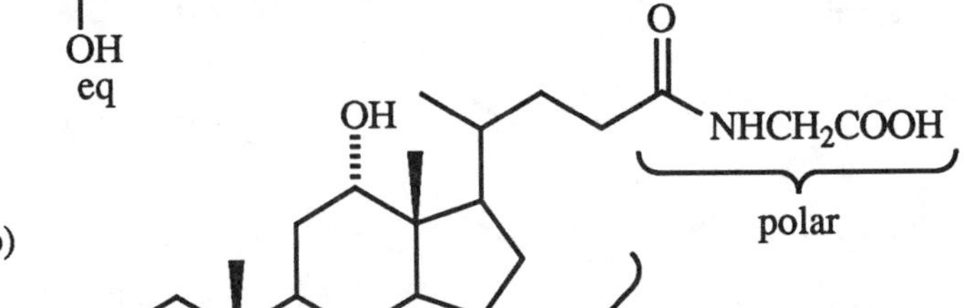

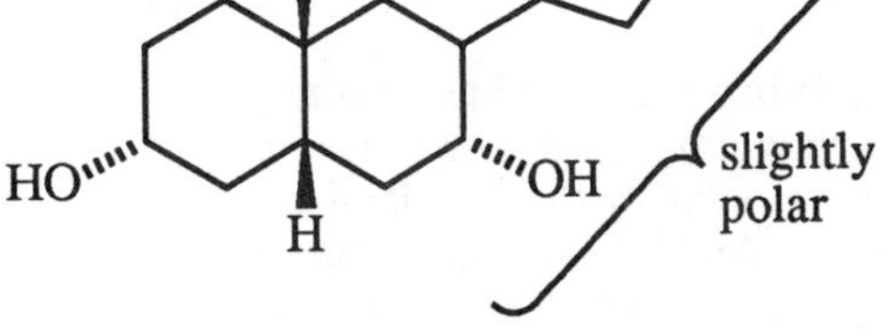

Like a soap or detergent, this cholic acid—glycine combination has a very polar "head" and a slightly polar "tail". The "tail" can dissolve non-polar molecules and the polar "head" can carry the complex into polar media.

25-25 The formula $C_{18}H_{34}O_2$ has two elements of unsaturation; one is the carbonyl, so the other must be an alkene or a ring. Catalytic hydrogenation gives stearic acid, so the carbon cannot include a ring; it must contain an alkene. The products from KMnO4 oxidation determine the location of the alkene:

$$HOOC(CH_2)_4COOH \ + \ HOOC(CH_2)_{10}CH_3 \ \Longrightarrow HOOC(CH_2)_4CH{=}CH(CH_2)_{10}CH_3$$

If the alkene were *trans*, the coupling constant for the vinyl protons would be about 15 Hz; a 10 Hz coupling constant indicates a *cis* alkene. The 7 Hz coupling is from the vinyl H's to the neighboring CH_2's.

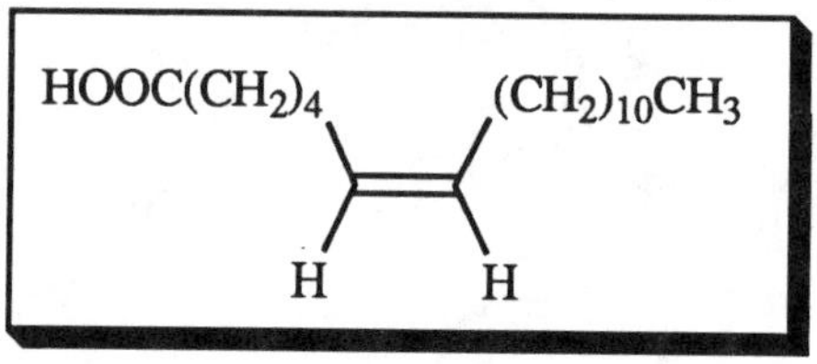
HOOC(CH2)4
(CH2)10CH3
H
H

25-26

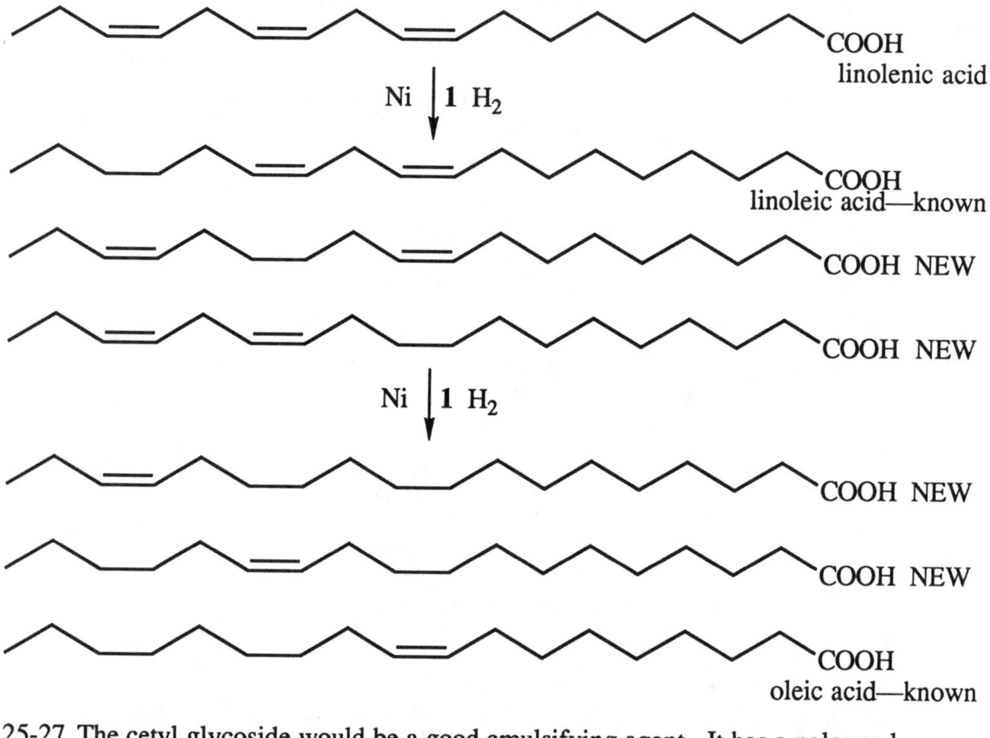
COOH
linolenic acid
Ni
1 H2
COOH
linoleic acid—known
COOH NEW
COOH NEW
Ni
1 H2
COOH NEW
COOH NEW
COOH
oleic acid—known

25-27 The cetyl glycoside would be a good emulsifying agent. It has a polar end (glucose) and a non-polar end (the hydrocarbon chain), so it can dissolve non-polar molecules, then carry them through polar media in micelles.

This is found in Nature where non-polar molecules such as steroid hormones, antibiotics, and other physiologically-active compounds are carried through the bloodstream (aqueous) by attaching saccharides (usually mono, di, or tri), making the non-polar group water soluble.

CHAPTER 26—SYNTHETIC POLYMERS

> Note to the student: In this chapter, the "wavy bond" symbol means the continuation of a polymer chain. ~~~~~~~~

26-1

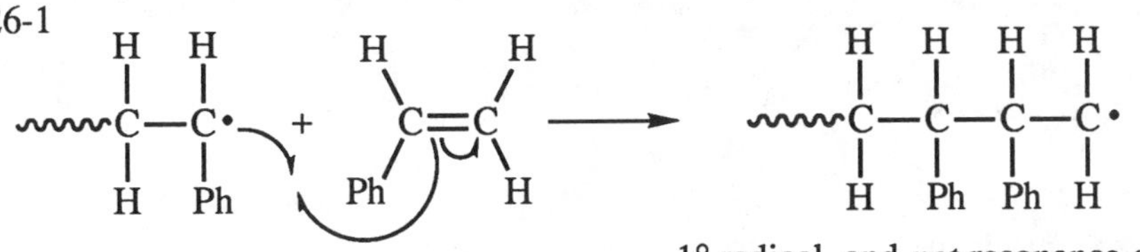

1° radical, and *not* resonance-stabilized—
this orientation is not observed

Orientation of addition always generates the more stable intermediate; the energy difference between a 1° radical (shown above) and a benzylic radical is huge. The phenyl substituents must necessarily be on alternating carbons because the orientation of attack is always the same—not a random process.

26-2

$$PhCOO{-}OOCPh \longrightarrow 2\ Ph\bullet + 2\ CO_2$$

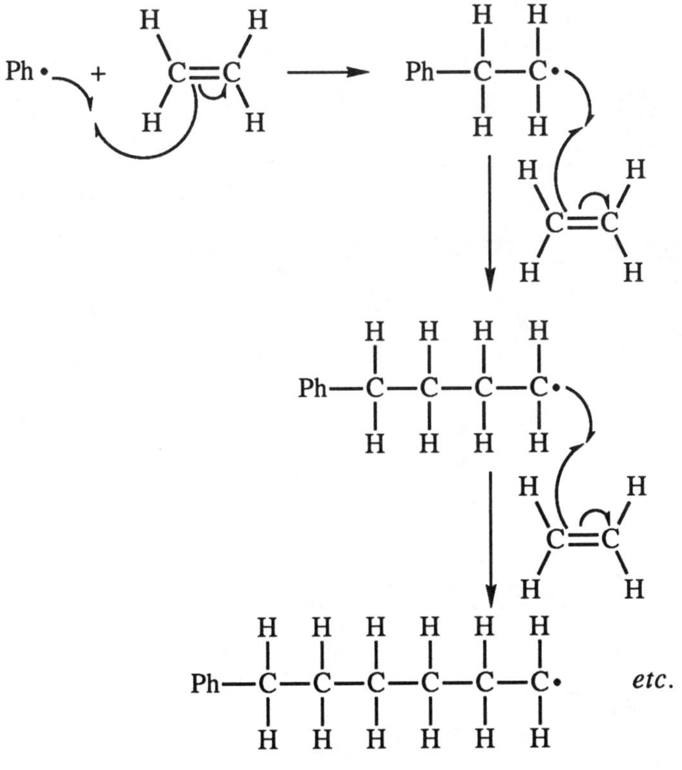

26-3 The benzylic hydrogen will be abstracted in preference to a 2° hydrogen because the benzylic radical is both 3° and resonance-stabilized, and the 2° radical is neither.

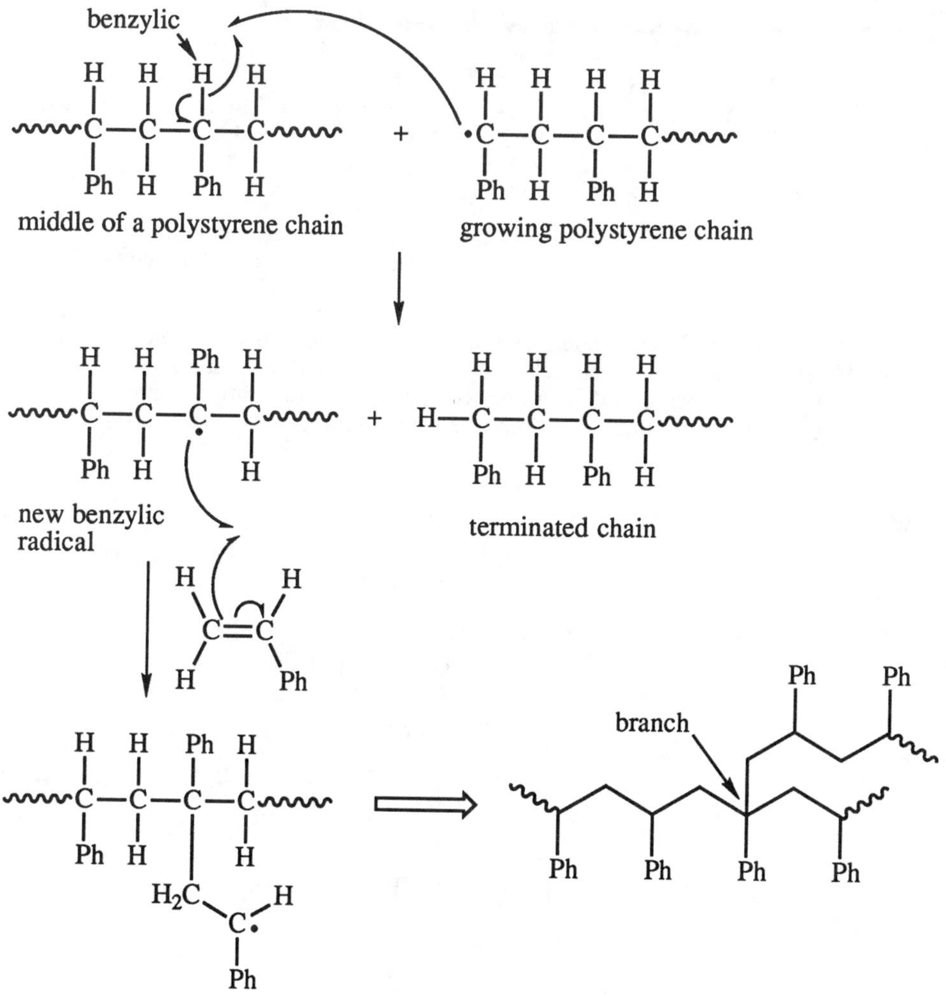
benzylic
middle of a polystyrene chain
growing polystyrene chain
new benzylic radical
terminated chain
branch

26-4 Addition occurs with the orientation giving the more stable intermediate. In the case of isobutylene, the growing chain will bond at the less substituted carbon to generate the more highly substitited carbocation.

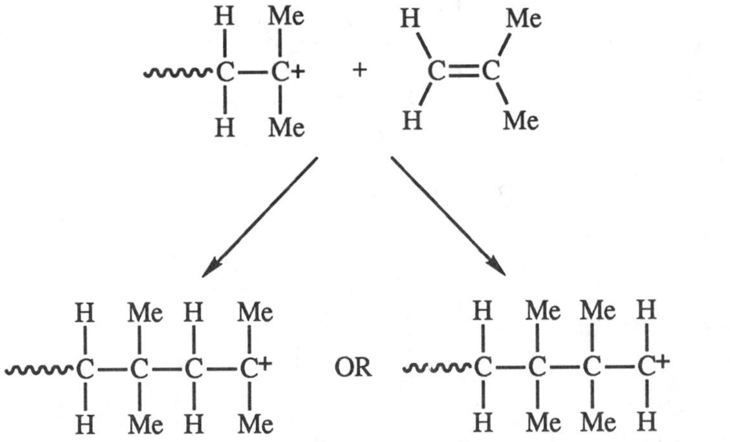

3° carbocation—favored

1° carbocation—disfavored (also more steric hindrance)

26-5

(a) chlorine can stabilize a carbocation intermediate by resonance

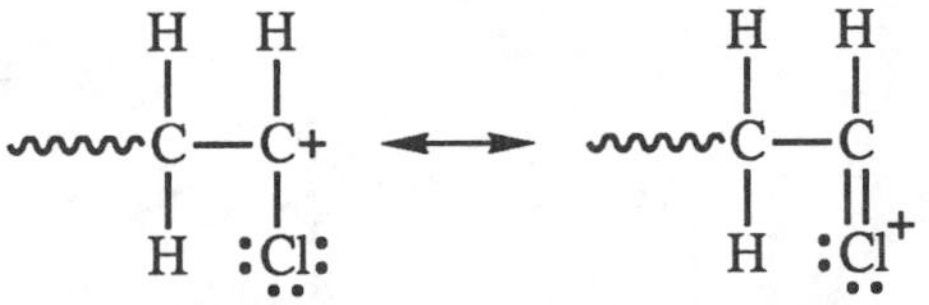

(b) CH_3 can stabilize the carbocation intermediate by induction

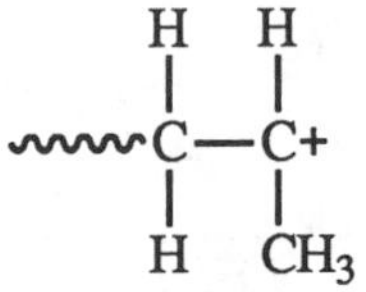

2° (not the best case imaginable, but still possible)

(c) terrible for cationic polymerization: both substituents are electron-withdrawing and would *destabilize* the carbocation intermediate

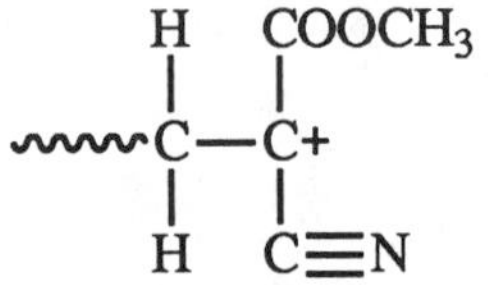

destabilized carbocation

26-6

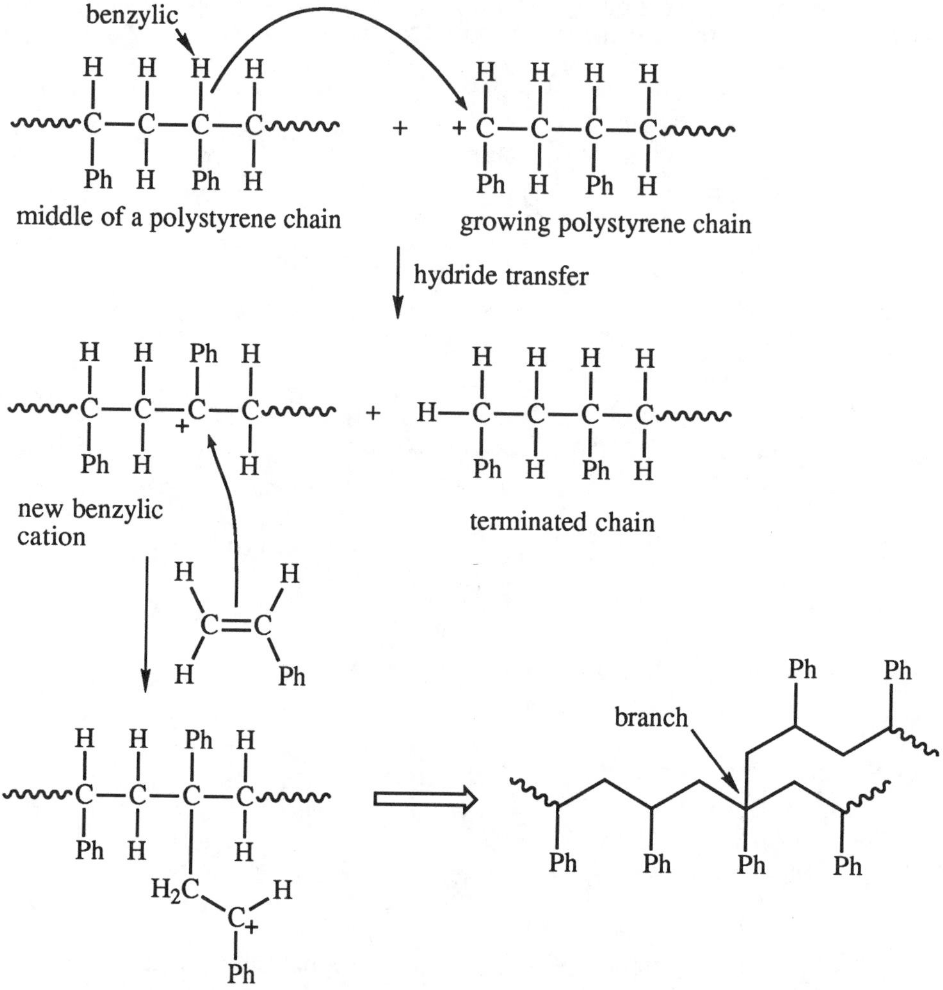
benzylic
middle of a polystyrene chain
growing polystyrene chain
hydride transfer
new benzylic cation
terminated chain
branch

Polystyrene is particularly susceptible to branching because the 3° benzylic cation produced by a hydride transfer is so stable. In poly(isobutylene), there is no hydrogen on the carbon with the stabilizing substituents; any hydride transfer would generate a 2° carbocation at the expense of a 3° carbocation at the end of a growing chain—this is an increase in energy and therefore unfavorable.

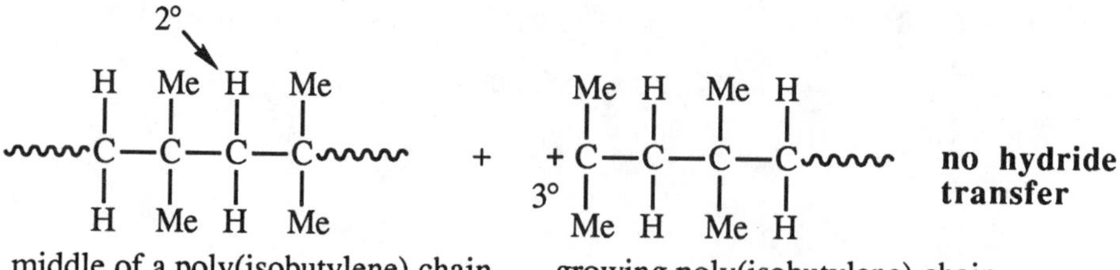
2°
middle of a poly(isobutylene) chain
growing poly(isobutylene) chain
3°
no hydride transfer

26-7

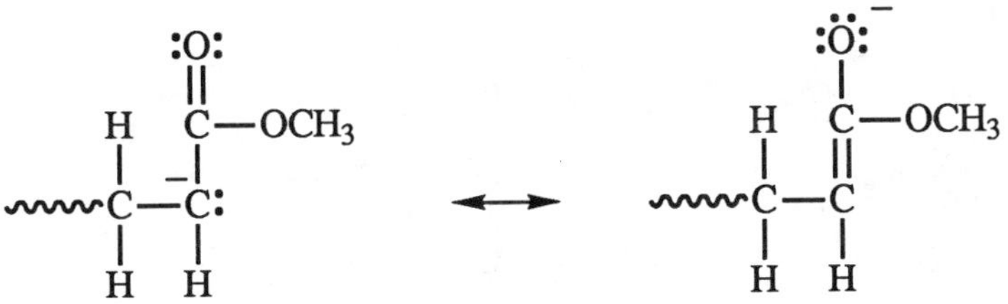

26-8

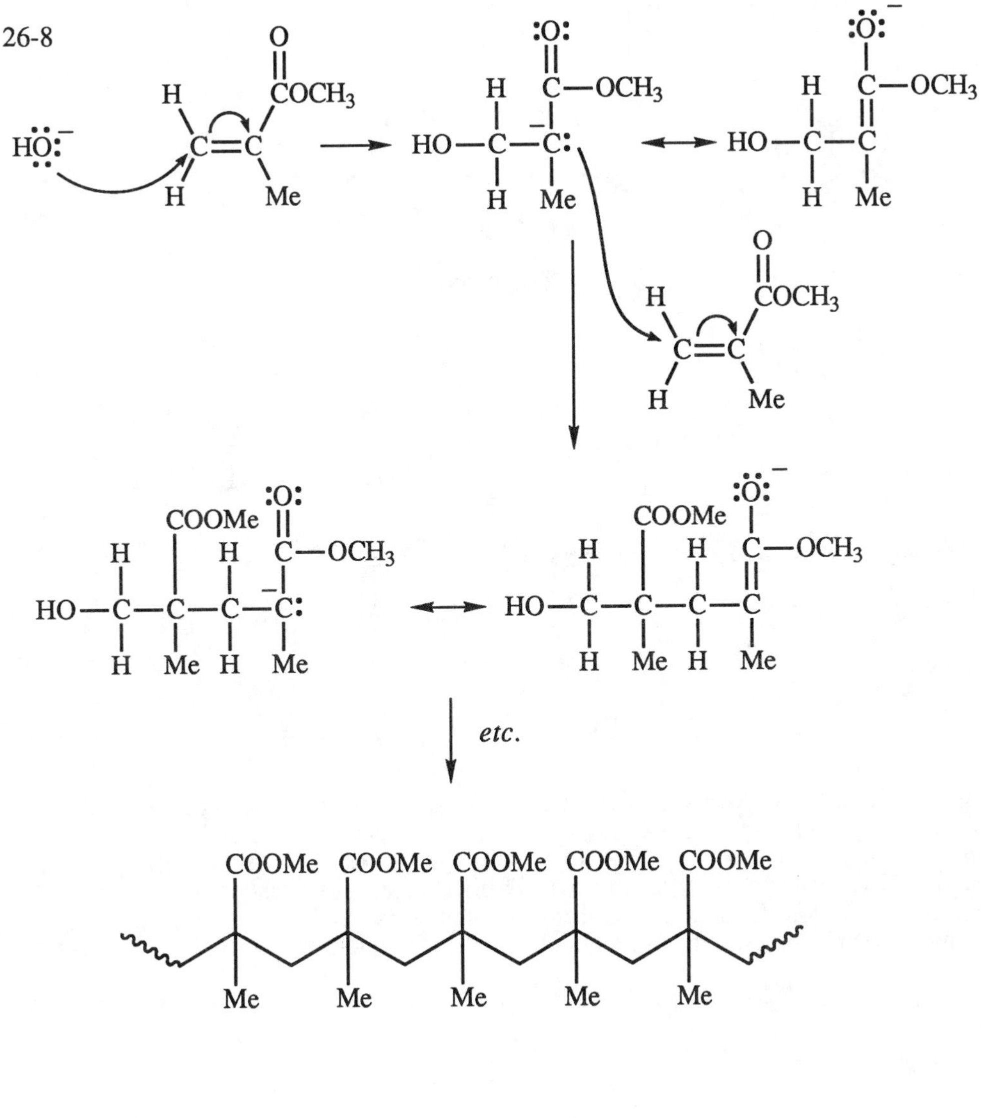

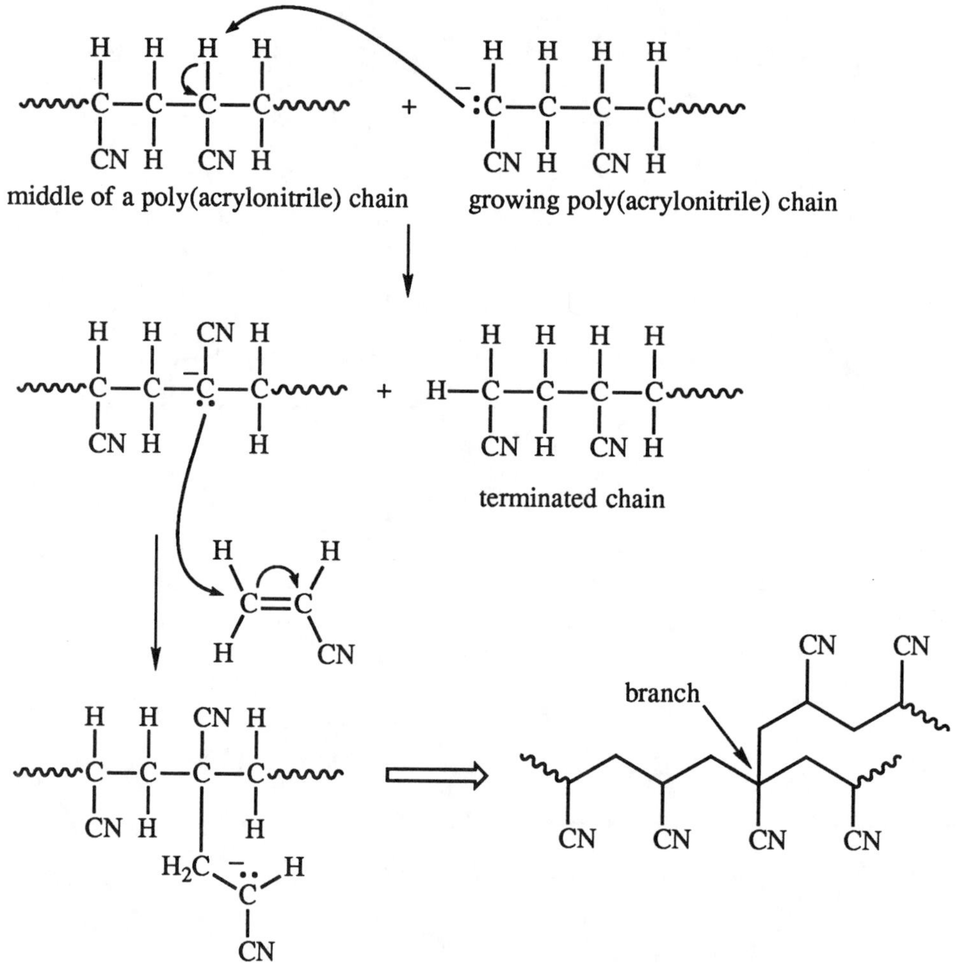
middle of a poly(acrylonitrile) chain
growing poly(acrylonitrile) chain
terminated chain
branch

The chain-branching hydride transfer (from a cationic mechanism) or proton transfer (from an anionic mechanism) ends a less-highly-substituted end of a chain and generates an intermediate on a more-highly-substituted middle of a chain (a 3° carbon in these mechanisms). This stabilizes a carbocation, but greater substitution *destabilizes* a carbanion. Branching can and does happen in anionic mechanisms, but it is less likely than in cationic mechanisms.

26-10

<u>isotactic poly(acrylonitrile)</u>

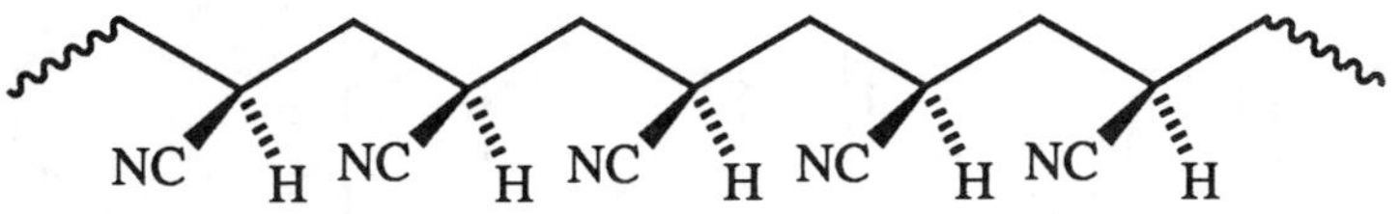

NC
H
NC
H
NC
H
NC
H
NC
H

<u>syndiotactic polystyrene</u>

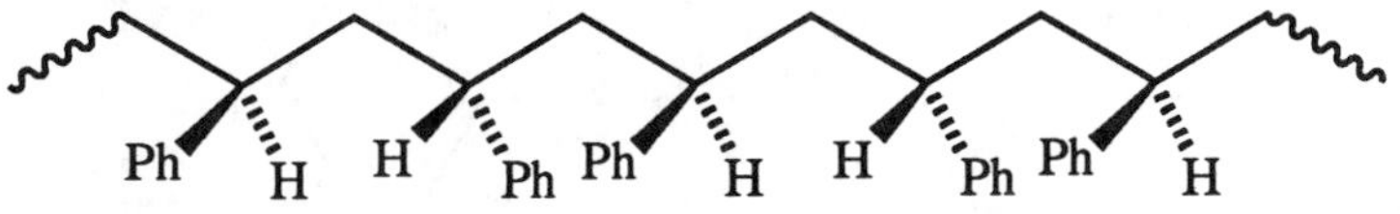

Ph
H
H
Ph
Ph
H
H
Ph
Ph
H

26-11

(a)

all *trans*

(b) The *trans* double bonds in gutta-percha allow for more ordered packing of the chains, that is, a higher degree of crystallinity. (Recall how *cis* double bonds in fats and oils lower the melting points because the *cis* orientation disrupts the ordering of the packing of the chains.) The more crystalline a polymer is, the less elastic it is.

26-12 Whether the alkene is *cis* or *trans* is not specified.

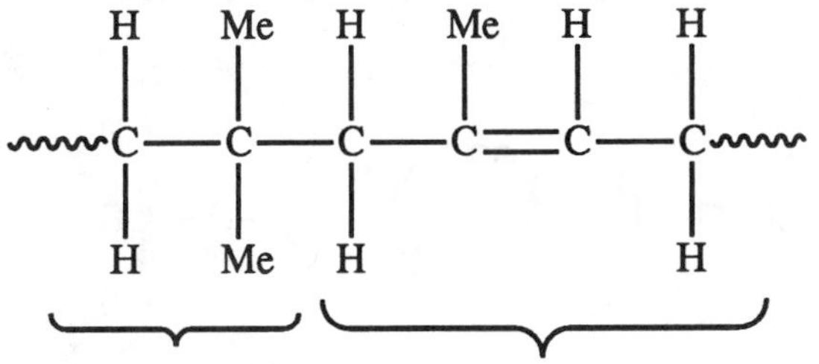

H
Me
H
Me
H
H
C
C
C
C
C
C
H
Me
H
H

isobutylene isoprene

26-13 The repeating unit in each polymer is boxed.

(a) Nomex

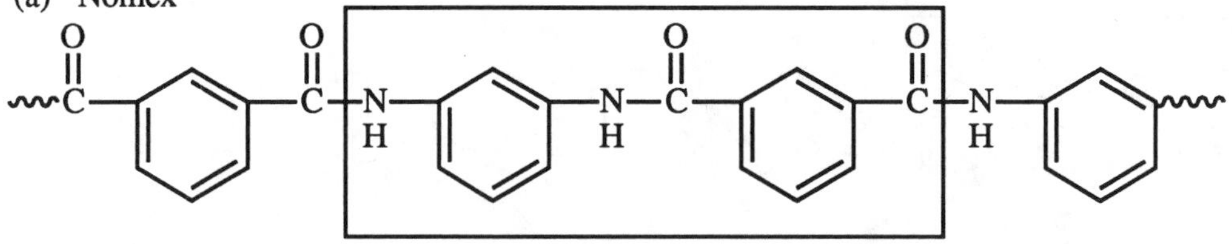

(b) Kevlar

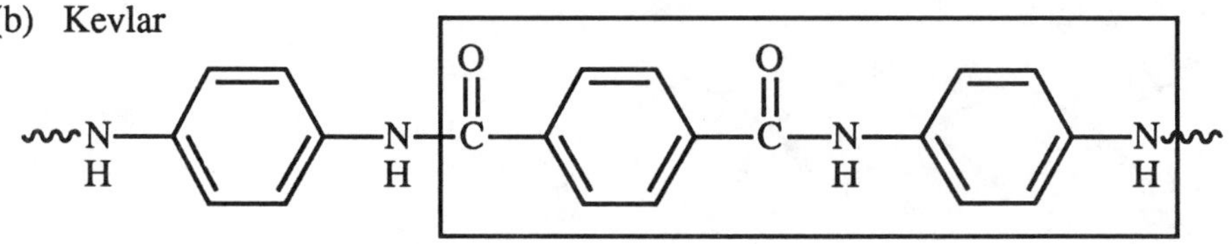

26-14 Kodel polyester (only one repeating unit shown)

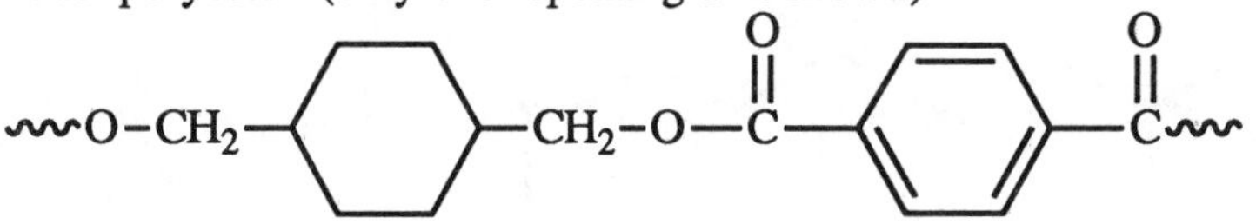

26-15

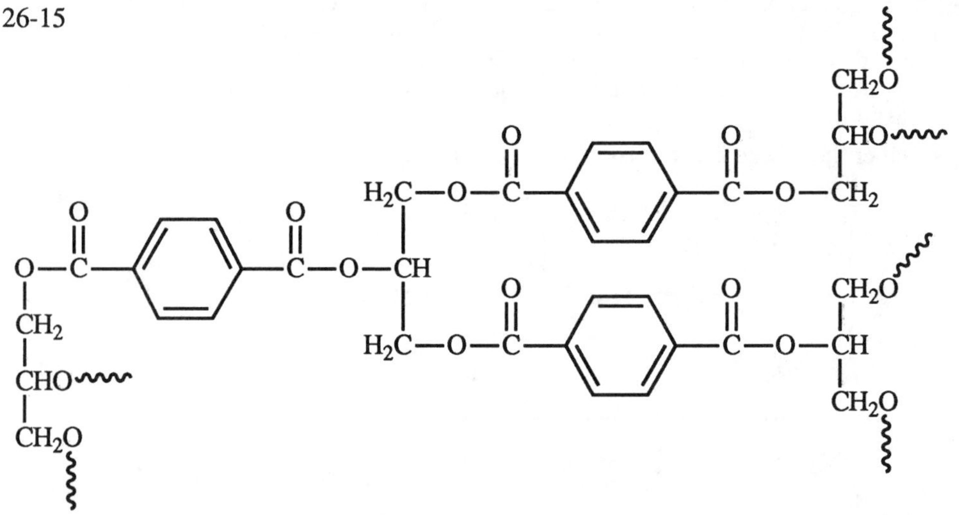

Glycerol is a trifunctional molecule, so not only does it grow in two directions to make a chain, it grows in three directions. All of its chains are cross-linked, forming a three-dimensional lattice with very little motion possible. The more cross-linked the polymer is, the more rigid it is.

26-16

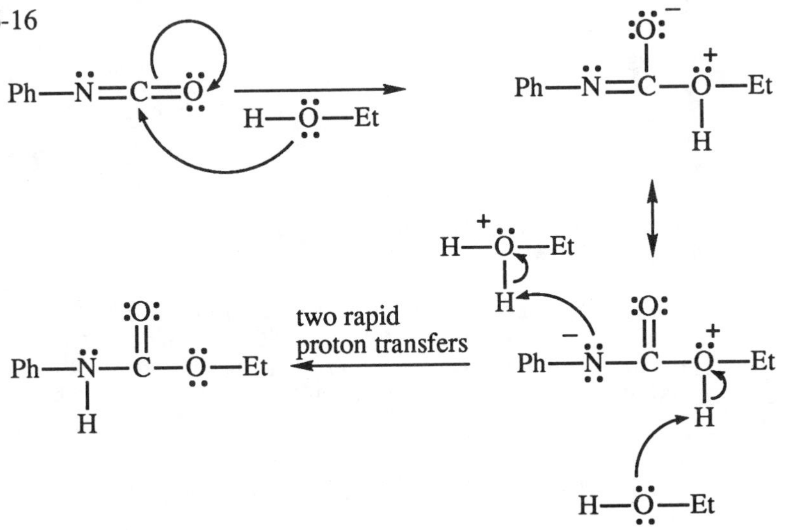

Ph—N=C=O
H—O—Et
Ph—N=C—O—Et
H
H—O—Et
H
two rapid
proton transfers
Ph—N—C—O—Et
H
Ph—N—C—O—Et
H
H—O—Et

26-17 Glycerol is a trifunctional alcohol. It uses two of its OH groups in a growing chain. The third OH group cross-links with another chain. The more cross-linked a polymer, the more rigid it is.

26-18

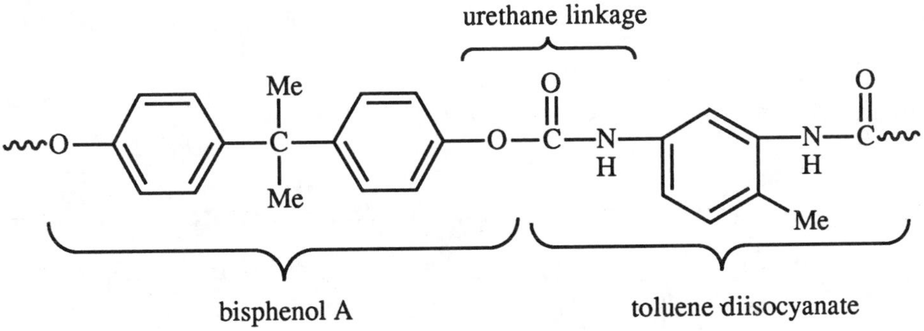

urethane linkage
Me
Me
O
O
O—C—N
H
N—C
H
Me
bisphenol A
toluene diisocyanate

26-19 Refer to the Glossary and the text for definitions and examples.

26-20

(a)

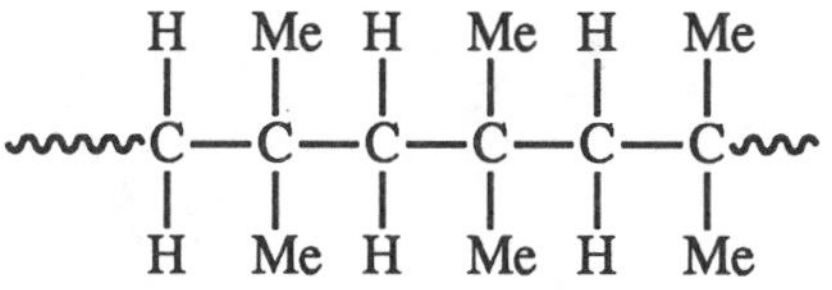

(b) Polyisobutylene is an addition polymer. No small molecule is lost, so this cannot be a condensation polymer.

(c) Either cationic polymerization or free-radical polymerization would be appropriate. The carbocation or free-radical intermediate would be 3° and therefore relatively stable. Anionic polymerization would be inappropriate as there is no electron-withdrawing group to stabilize the anion.

26-21

(a) It is a polyurethane.

(b) As with all polyurethanes, it is a condensation polymer.

(c)

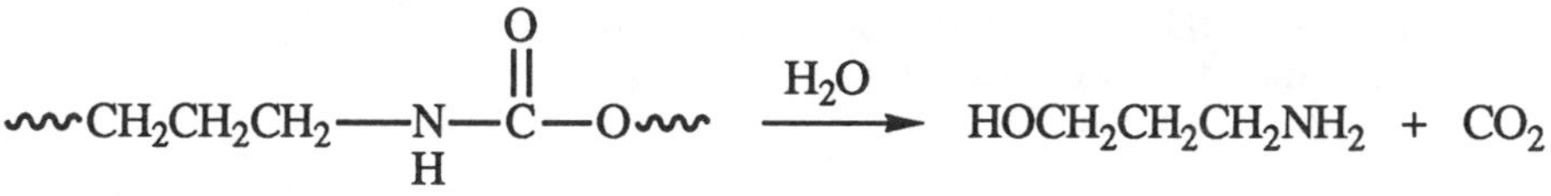

26-22

(a) It is a polyester.

(b) As with all polyesters, it is a condensation polymer.

(c)

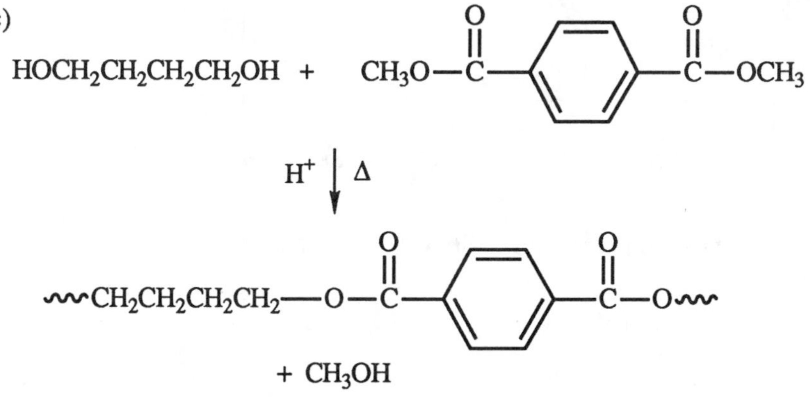

Using the dicarboxylic acid instead of the ester would produce water as the small neutral molecule lost in this condensation.

26-23

(a) Urylon is a polyurea.

(b) A polyurea is a condensation polymer.

(c)

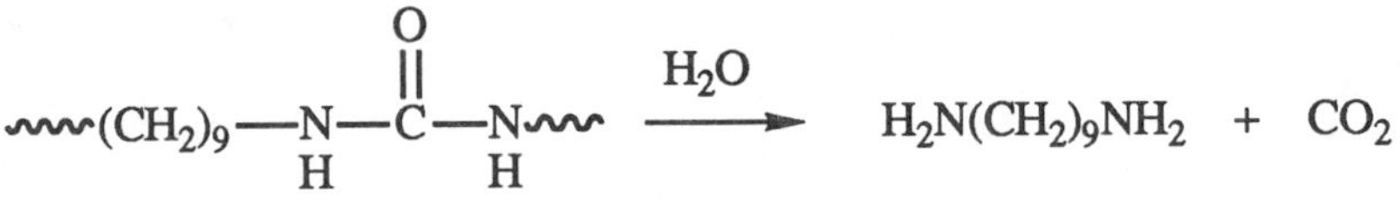

(CH2)9—N—C—N
O
H H
H2O
H2N(CH2)9NH2 + CO2

26-24

(a) Polyethylene glycol, abbreviated PEG, is a polyether.

(b) PEG is usually made from ethylene oxide (first reaction shown). In theory, PEG could also be made by intermolecular dehydration of ethylene glycol (second reaction shown), but the yields are low and the chains are short.

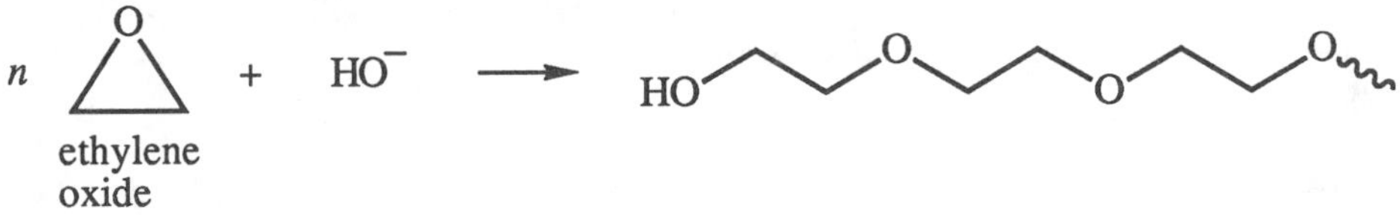

n
+ HO−
HO
ethylene
oxide

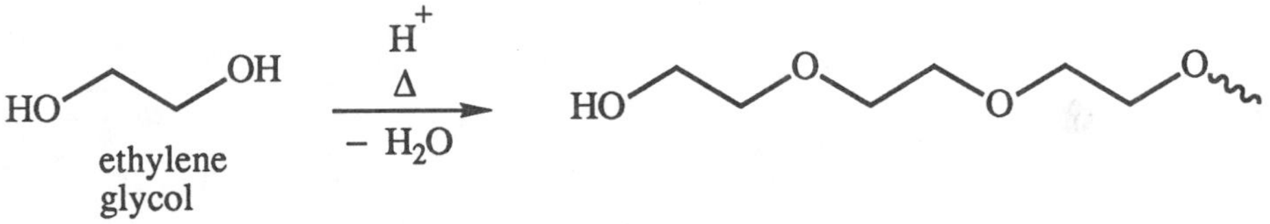

HO
OH
H+
Δ
− H2O
ethylene
glycol

(c) Basic catalysts are most likely as they open the epoxide to generate a new nucleophile. Acid catalysts are possible but they risk dehydration and ether cleavage.

(d) Mechanism of ethylene oxide polymerization (showing hydroxide as the base):

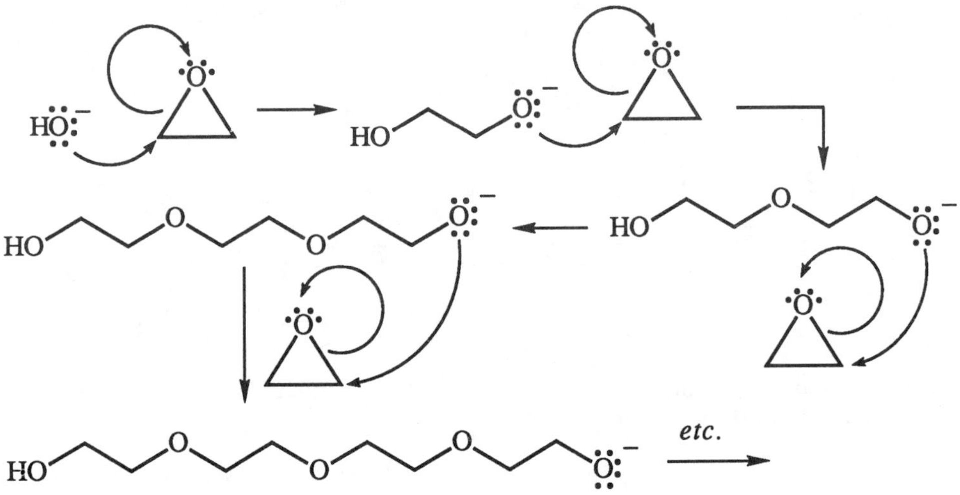

HO
etc.

26-25

(a) Polychloroprene (Neoprene) is an addition polymer.

(b) Polychloroprene comes from the diene, chloroprene, just as natural rubber comes from isoprene:

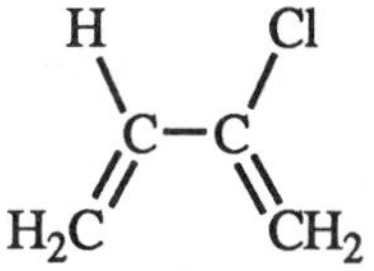

chloroprene

26-26

(a)

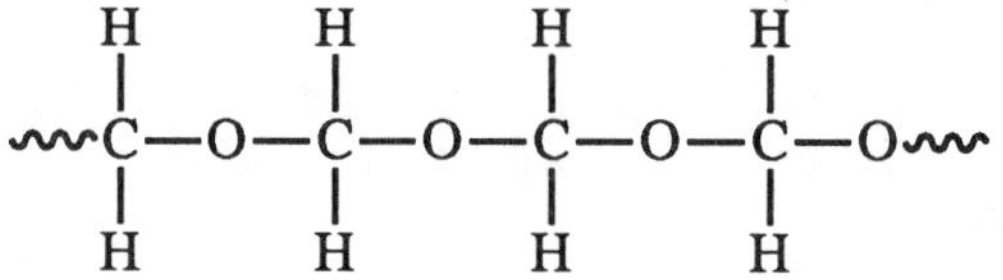

Delrin (polyformaldehyde)

(b) All of these intermediates are resonance-stabilized.

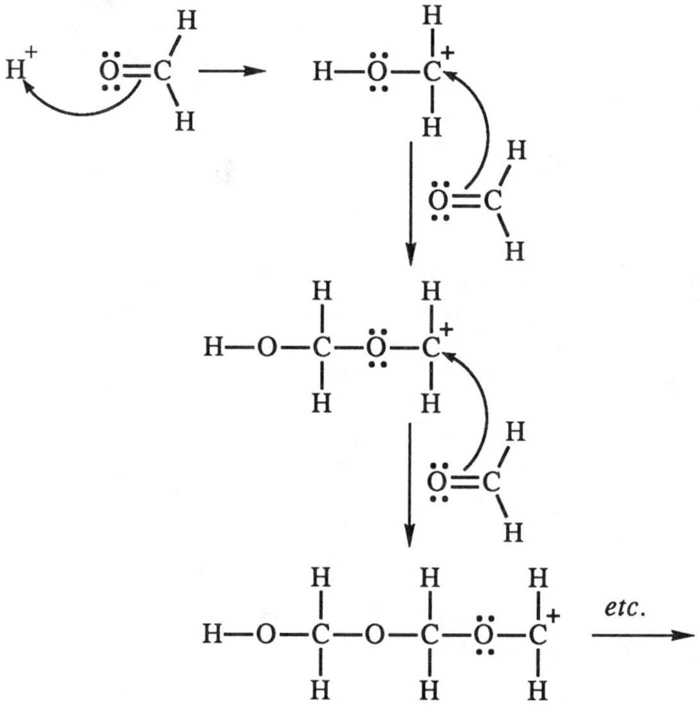

trimer

(c) Delrin is an addition polymer; instead of adding across the double bond of an alkene, addition occurs across the double bond of a carbonyl group.

26-27

(a) *cis*

trans

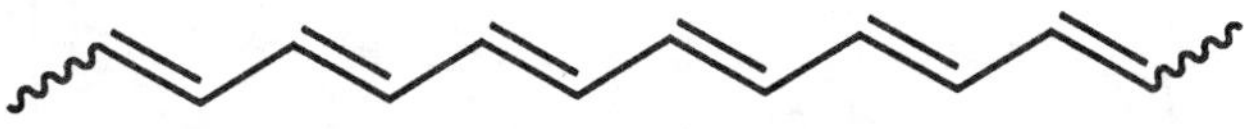

(b) Each structure has a fully conjugated chain. It is reasonable to expect electrons to be able to be transferred through the π system, just as resonance effects can work over long distances through conjugated systems.

(c) It is not surprising that the conductivity is directional. Electrons must flow along the π system of the chain, so it the chains were aligned, conductivity would be greater in the direction parallel to the polymer chains. (It is possible, though less likely, that electrons could pass from the π system of one chain to the π system of another, that is, perpendicular to the direction of the chain; we would expect reduced conductivity in that direction.)

26-28

(a) A Nylon is a polyamide. Amides can be hydrolyzed in aqueous acid, cleaving the polymer chain in the process.

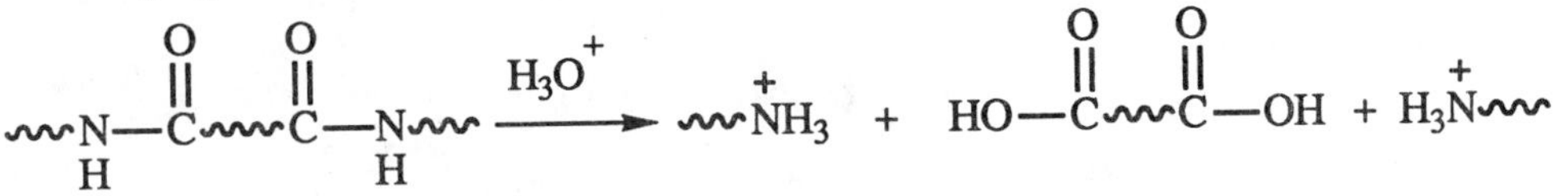

(b) A polyester can be saponified in aqueous base, cleaving the polymer chain in the process.

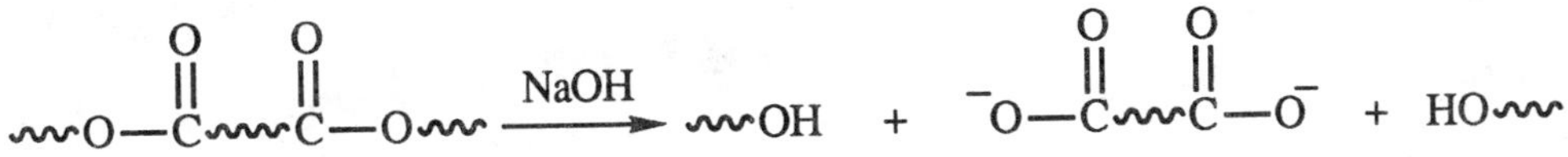

26-29

(a)

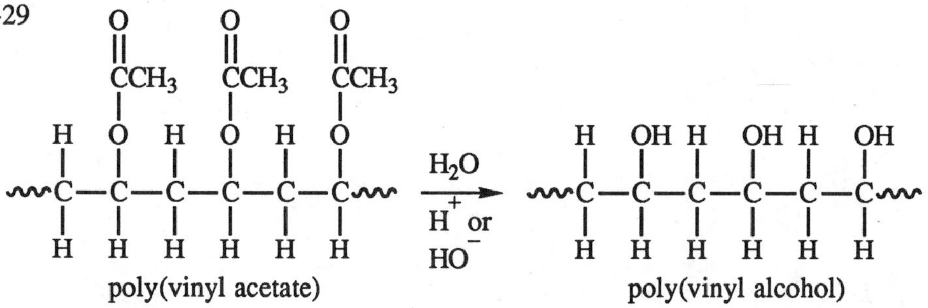

(b) A polyester is a condensation polymer in which monomer units are linked through ester groups as part of the polymer chain. Poly(vinyl acetate) is really a substituted polyethylene, an **addition** polymer, with only carbons in the chain; the ester groups are in the side chains, not in the polymer backbone.

(c) Hydrolysis of the esters in poly(vinyl acetate) does not affect the chain because the ester groups do not occur in the chain as they do in Dacron.

(d) Vinyl alcohol cannot be polymerized because it is unstable, tautomerizing to acetaldehyde.

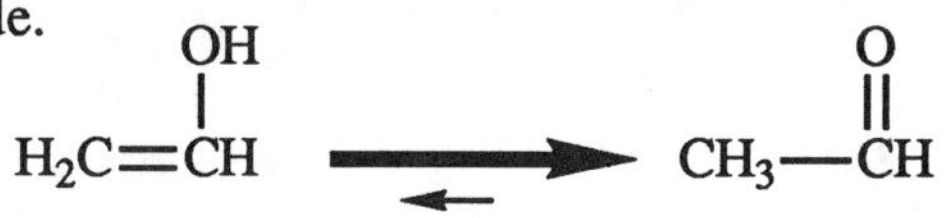

26-30

(a)

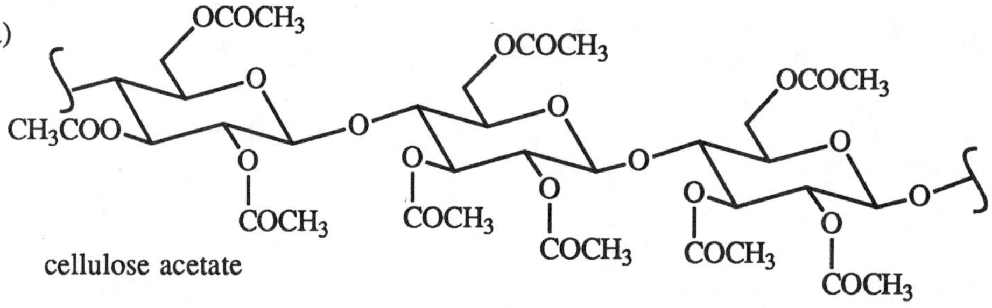

cellulose acetate

(b) Cellulose has three OH groups per glucose monomer, which form hydrogen bonds with other polar groups. Transforming these OH groups into acetates makes the polymer much less polar and therefore more soluble in organic solvents.

(c) The acetone dissolved the cellulose acetate in the fibers. As the acetone evaporated, the cellulose acetate remained but no longer had the fibrous, woven structure of cloth. It recrystallized as white fluff.

(d) Any article of clothing made from synthetic fibers is susceptible to the ravages of organic solvents. Solvent splashes leave dimples or blotches on Corfam shoes. (Yet, Corfam shoes could still provide protection for the toenail polish!)